Chemistry

An Introduction to General, Organic, and Biological Chemistry

Seventh Edition
Media Update

D0002743

Chemistry

An Introduction to General, Organic, and Biological Chemistry

Seventh Edition
Media Update

Karen C. Timberlake
LOS ANGELES VALLEY COLLEGE

Benjamin Cummings

An imprint of Addison Wesley Longman, Inc.
Originally published by HarperCollins College Publishers

San Francisco • Reading, Massachusetts • New York • Harlow, England
Don Mills, Ontario • Sydney • Mexico City • Madrid • Amsterdam

Publisher	Robin J. Heyden
Executive Editor	Ben Roberts
Sponsoring Editor	Maureen Kennedy
Project Editor/Production Editor	Lisa Weber
Project Editor of Media Update	Claudia Herman
Market Development Manager	Chalon Bridges
Director of Marketing	Stacy Treco
Marketing Manager	Christy Lawrence
Art Coordinators	Anita Wagner, Vivian McDougal, Lisa Weber
Composition and Film Buyer	Vivian McDougal
Permissions Editor	Ariane DePree-Kajifez
Manufacturing Supervisor	Vivian McDougal
Publishing Associates	Anthony J. Asaro, Paige Farmer
Copyeditor	Anita Wagner
Proofreaders	Cathy Cobb, Martha Ghent
Text and Cover Designer	Mark Ong, Side by Side Studios
Illustrator	Rolin Graphics
Photo Editor	Kathleen Cameron
Cover Photograph	© David Stewart/Tony Stone Images

Copyright © 2001 by Benjamin Cummings, an imprint of Addison Wesley.

For permission to use copyrighted material, grateful acknowledgment is made to the copyright holders on page 722, which is hereby made part of this copyright page.

All rights reserved. No part of this publication may be reproduced, stored in a retrieval system, or transmitted in any form or by any means, electronic, mechanical, photocopying, recording, or any other media embodiments now known or hereafter to become known, without the prior written permission of the publisher. Manufactured in the United States of America. Published simultaneously in Canada.

Many of the designations used by manufacturers and sellers to distinguish their products are claimed as trademarks. Where those designations appear in this book, and the publisher was aware of a trademark claim, the designations have been printed in initial caps or all caps.

Library of Congress Cataloging-in-Publication Data
Timberlake, Karen.
Chemistry : an introduction to general, organic, and biological chemistry / Karen C. Timberlake.— Media ed.
p. cm.
Includes index.
ISBN 0-8053-3568-4
1. Chemistry. I. Title.

QD31 .T615 2001
540—dc21

00-047486

3 4 5 6 7 8 9 10—QWV—03 02 01 00

The Benjamin/Cummings Publishing Company, Inc.
1301 Sansome Street
San Francisco, CA 94111

TRADEMARKS

Accutane and Invirase are registered trademarks of Hoffmann-LaRoche, Inc.

Advil is a registered trademark of Boots Company, PLC.

Alka-Seltzer is a registered trademark of Miles, Inc. (Assigned by Bayer Corporation).

Amphojel is a registered trademark of American Home Products Corporation.

Antabuse, Fluothane, and Inderal are registered trademarks of American Home Products Corporation/Delaware Corporation.

Bisodol is a registered trademark of American Home Products Corporation and Bisodol Company, Inc.

Benzedrine is an expired trademark of Smith, Kline, and French laboratories.

Breathalyzer is a registered trademark of Robert F. Borkenstein (individual).

Dacron (synthetic fibers), Freon, Mylar, Orlon, and Teflon are registered trademarks of E. I. Du Pont De Nemours and Company/Delaware Corporation.

Demerol is a registered trademark of Alba Pharmaceutical Company, Inc.

Di-Gel is a registered trademark of Plough, Inc./Delaware Corporation.

Drano is a registered trademark of the Drackett Company.

Ethrane is a registered trademark of Anaquest, Inc./Delaware Corporation.

Forane is a registered trademark of Ohmeda Pharmaceutical Products Division, Inc./Delaware Corporation.

Gelusil is a registered trademark of William R. Warner and Co., Inc., New York.

Kaopectate and Motrin are registered trademarks of The Upjohn Company/Delaware Corporation.

Lactaid is a registered trademark of Lactaid, Inc./New Jersey Corporation.

Lifesavers and Life Savers are registered trademarks of Nabisco Brands, Inc./Delaware Corporation.

Lucite is a registered trademark of Ici Acrylics, Inc./Delaware Corporation.

Lysol is a registered trademark of Sterling Drug, Inc./Delaware Corporation.

Maalox is a registered trademark of Rhone-Poulene Rorer Pharmaceuticals, Inc.

Mylanta is a registered trademark of ICI Americas, Inc./Delaware Corporation.

Necco is a registered trademark of New England Confectionery Company/Massachusetts Corporation.

Neo-Synephrine is a registered trademark of Frederick Stearns & Company.

Nuprin is a registered trademark of The Upjohn Company.

Norlutin is a registered trademark of Parke, Davis & Company.

Penthrane is a registered trademark of Abbot Laboratories/Illinois Corporation.

Pepto-Bismol is a registered trademark of Norwich Eaton Pharmaceuticals, Inc.

Percodan is a registered trademark of Du Pont Pharmaceuticals Caribe, Inc./Delaware Corporation.

Ping-Pong is a registered trademark of Indian Industries, Inc.

Plexiglas is a registered trademark of Rohm and Haas Company/Delaware Corporation.

Pyrex is a registered trademark of Corning Incorporated.

Retin-A and Tylenol are registered trademarks of Johnson & Johnson/New Jersey Corporation.

Riopan is a registered trademark of BYK-Gulden Lumberg, Chemische Fabrik GMBH.

Slinky is a registered trademark of James Industries, Inc./Pennsylvania Corporation.

Styrofoam and Saran Wrap are registered trademarks of The Dow Chemical Corporation.

Thorazine and Tums are registered trademarks of SmithKline Beecham Corporation/Pennsylvania Corporation.

Titralac is a registered trademark of Riker Laboratories, Inc./Delaware Corporation.

Toradol is a registered trademark of Sythex (USA), Inc./Delaware Corporation.

Vaseline is a registered trademark of Chesebrough-Pond's Inc./New York Corporation.

Voltaren is a registered trademark of CIBA-Geigy Corporation/New York Corporation.

Xylocaine is a registered trademark of Astra Pharmaceutical Products, Inc.

About the Cover

Running and Chemistry

You may wonder why a runner is pictured on the cover of a chemistry book. During running or any other exercise, many chemical changes occur in the body. In this course, you will discover how each of these changes involves chemistry and contributes to good health. For instance, during running or any moderate-to-strenuous exercise, you have probably experienced increases in your breathing and heart rates—this is chemistry in action. Our runner's muscles consume oxygen at a rapid rate, producing carbon dioxide that passes into the bloodstream. If her body's oxygen supply becomes too low, lactic acid forms in her muscles. An increase in lactic acid spurs several chemical changes: the blood becomes more acidic and the rate and depth of breathing increase. In fact, the breathing rate during exercise can increase by as much as 30 times more than the resting rate. As lactic acid accumulates, our runner may experience so much fatigue and pain that she eventually slows down or stops. She may also begin gasping for air, which supplies more oxygen to the blood and reduces the lactic acid in the cells.

The energy she needs to run comes from the food she has eaten. Carbohydrates provide glucose that is stored as a chemical compound called glycogen in the muscles. During exercise, glycogen is broken down to maintain a supply of glucose in the blood. In the muscle cells, glucose is oxidized to provide energy to form another chemical compound called ATP (adenosine triphosphate), the fuel required for muscle contraction. As long as ATP energy is available, the runner can continue to move the muscles that propel her during her run. A high-carbohydrate diet after running will help to restore her muscle glycogen more rapidly than a fat and protein diet.

Running produces heat, which increases body temperature. In a process called homeostasis, the body reduces dangerously high temperatures by expanding the capillaries to let more blood reach the skin surface. You know that during intense exercise, a large amount of water can be lost. Sweat glands increase the production of perspiration, which then evaporates from the skin and cools the body. Even the breath exhaled by the runner contains warm water vapor. If water is not replaced, production of saliva will decrease, the water level in the blood will drop, and the body will become dehydrated. All these changes will cause the runner to feel thirsty. When she quenches her thirst, the water in the body will return to normal levels.

By now, you may begin to see how chemistry is involved in nearly everything we do—and how our runner serves to illustrate this point. This book will teach you what happens when you eat or run, when acid rain forms, when food is irradiated, or when you give a patient a saline solution. Learning about all these processes as you study chemistry will give you a better understanding of your body and the world around you.

About the Author

This book is dedicated to my family, our son John, and my husband for his loving support and late meals, and my students whose hard work and wonderful comments continue to influence my writing.

The whole art of teaching is only the art of awakening the natural curiosity of young minds.

—Anatole France

One must learn by doing the thing; though you think you know it, you have no certainty until you try.

—Sophocles

Discovery consists of seeing what everybody has seen and thinking what nobody has thought.

—Albert Szent-Gyorgi

Karen Timberlake is a professor of chemistry and has been teaching the foundations of chemistry for 34 years at Los Angeles Valley College in Van Nuys, California. She received her bachelor's degree in chemistry from the University of Washington in 1962 and her master's degree in biochemistry from University of California at Los Angeles in 1965. She has also taken graduate courses in science education during a sabbatical at the University of Northern Colorado.

Professor Timberlake belongs to numerous science and educational organizations including the American Chemical Society (ACS), National Science Teachers Association (NSTA), and International Society for Exploring Teaching Alternatives (ISETA). In 1987 she was the western regional winner of the Excellence in College Chemistry Teaching award given by the Chemical Manufacturers Association.

Professor Timberlake has been writing chemistry textbooks for 25 years with an emphasis on applying the concepts of chemistry to real-life situations concerning health and environment. In the past 5 years, she has participated in education grants for science teaching with the Los Angeles Collaborative for Teaching Excellence (LACTE) and Title III grant at her college. She often speaks at conferences and educational meetings on using student-centered teaching methods in chemistry to promote learning success. Her Lecture PLUS learning system utilizes PowerPoint slides in mini-lectures, formative assessments, and activity worksheets for group work.

Her husband, Bill, is also a chemistry instructor and contributed to the writing of *CHEMISTRY.* When Professor Timberlake is not teaching or writing, she relaxes by hiking, traveling, playing tennis, and playing the piano. The Timberlakes' son, John, is a business development manager in San Jose, California.

To the Student

I hope that this textbook helps you discover exciting new ideas and gives you a rewarding experience as you develop an understanding and appreciation of the role of chemistry in your life. If you would like to share your experience with chemistry or comments about this text, I would appreciate hearing from you.

Karen C. Timberlake
Los Angeles Valley College
Department of Chemistry
5800 Fulton Avenue
Valley Glen, CA 91401
Email: khemist@aol.com
karen@karentimberlake.com

Brief Contents

1	Measurements	1
2	Atoms and Elements	46
3	Nuclear Radiation	80
4	Compounds and Their Bonds	117
5	Chemical Reactions and Quantities	160
6	Energy and States of Matter	200
7	Gases	244
8	Solutions	281
9	Acids and Bases	326
10	Alkanes and Aromatic Hydrocarbons	362
11	Haloalkanes, Alkenes, and Alkynes	401
12	Alcohols, Phenols, Ethers, Aldehydes, and Ketones	436
13	Carbohydrates	478
14	Carboxylic Acids, Esters, Amines, and Amides	515
15	Lipids	557
16	Amino Acids, Proteins, and Enzymes	594
17	Metabolic Pathways and Energy Production	636
18	Nucleic Acids	681

Contents

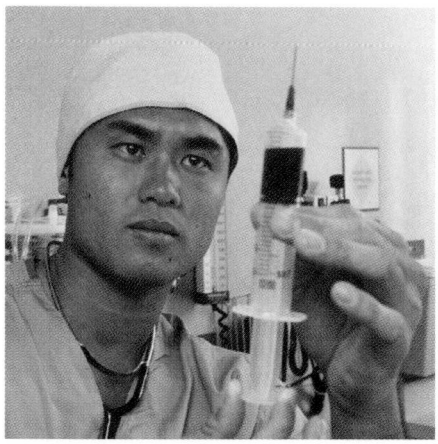

1 Measurements 1

1.1 Units of Measurement 2

EXPLORE YOUR WORLD
Units Listed on Labels 4

1.2 Scientific Notation 5

1.3 Measured and Exact Numbers 7

EXPLORE YOUR WORLD
Temperature Scales on Thermometers 8

1.4 Significant Figures 9

1.5 Significant Figures in Calculations 12

1.6 SI and Metric Prefixes 16

1.7 Conversion Factors 20

EXPLORE YOUR WORLD
SI and Metric Equalities on Product Labels 23

1.8 Using Conversion Factors 25

1.9 Density 33

EXPLORE YOUR WORLD
Sink or Float? 35

HEALTH NOTE
Determination of Percentage of Body Fat 37

HEALTH NOTE
Specific Gravity of Urine 39

Concept Guide *40*
Chapter Summary *40*
Key Terms *41*
Chapter Practice *42*
Answers *43*

2 Atoms and Elements 46

2.1 Elements and Symbols 47

HEALTH NOTE
Elements Essential to Health 49

HEALTH NOTE
Latin Names for Elements in Clinical Usage 49

EXPLORE YOUR WORLD
Origins of Element Names 49

EXPLORE YOUR WORLD
Physical Properties of Aluminum 50

2.2 The Periodic Table 51

HEALTH NOTE
Calcium and Strontium 54

HEALTH NOTE
Some Important Trace Elements in the Body 55

2.3 The Atom 57

EXPLORE YOUR WORLD
Repulsion and Attraction 58

2.4 Atomic Number and Mass Number 60

2.5 Isotopes and Atomic Mass 64

2.6 Electron Arrangement and the Periodic Law 67

EXPLORE YOUR WORLD
Light Released from Candy 71

ENVIRONMENTAL NOTE
Biological Reactions to UV Light 71

Concept Guide *75*
Chapter Summary *75*
Key Terms *76*
Chapter Practice *77*
Answers *78*

3 Nuclear Radiation 80

3.1 Natural Radioactivity 81

HEALTH NOTE
Biological Effects of Radiation 83

EXPLORE YOUR WORLD
Dental X Rays 85

3.2 Nuclear Equations 87

ENVIRONMENTAL NOTE
Radon in Our Homes 88

EXPLORE YOUR WORLD
Smoke Detectors 89

HEALTH NOTE
Beta Emitters in Medicine 91

3.3 Producing Radioactive Isotopes 92

3.4 Radiation Detection and Measurement 95

HEALTH NOTE
Radiation and Food 96

3.5 Medical Applications Using Radioactivity 99

HEALTH NOTE
Radiation Doses in Diagnostic and Therapeutic Procedures 101

HEALTH NOTE
Other Imaging Methods 102

3.6 Half-Life of a Radioisotope 103

EXPLORE YOUR WORLD
Modeling Half-Lives 105

ENVIRONMENTAL NOTE
Dating Ancient Objects 106

3.7 Nuclear Fission and Fusion 108

Concept Guide *112*
Chapter Summary *112*
Key Terms *113*
Chapter Practice *113*
Answers *115*

4 Compounds and Their Bonds 117

4.1 Valence Electrons 118

4.2 Ions and the Octet Rule 121

HEALTH NOTE
Some Uses for Noble Gases 122

HEALTH NOTE
Some Important Ions in the Body 125

4.3 Ionic Compounds 127

4.4 Naming and Writing Ionic Formulas 130

4.5 Covalent Bonds 136

4.6 Covalent Compounds 141

4.7 Bond Polarity 143

4.8 Polyatomic Ions 147

HEALTH NOTE
Polyatomic Ions in Bone and Teeth 149

EXPLORE YOUR WORLD
Names of Compounds Listed on Products 150

Concept Guide *155*
Chapter Summary *155*
Key Terms *156*
Chapter Practice *156*
Answers *157*

5 Chemical Reactions and Quantities 160

5.1 Chemical Changes 161

EXPLORE YOUR WORLD
Evidence of a Chemical Change 162

5.2 Chemical Equations 163

5.3 Balancing a Chemical Equation 166

5.4 Types of Reactions 170

HEALTH NOTE
Smog and Health Concerns 172

5.5 A Closer Look: Oxidation-Reduction Reactions 174

EXPLORE YOUR WORLD
Oxidation of Fruits and Vegetables 176

5.6 The Mole 178

5.7 Calculations Using Molar Mass 183

5.8 Mole Relationships in Chemical Equations 186

5.9 A Closer Look: Mass Calculations for Reactions 191

Concept Guide *195*
Chapter Summary *195*
Key Terms *196*
Chapter Practice *196*
Answers *197*

6 Energy and States of Matter 200

6.1 Energy 201

ENVIRONMENTAL NOTE
Global Warming 203

6.2 Measuring Temperature 204

HEALTH NOTE
Variation in Body Temperature 205

6.3 Measuring Heat Energy 209

EXPLORE YOUR WORLD
Loss and Gain of Heat 209

EXPLORE YOUR WORLD
Specific Heat 210
6.4 A Closer Look: Energy and Nutrition 212
EXPLORE YOUR WORLD
Counting Calories 215
HEALTH NOTE
Losing and Gaining Weight 215
EXPLORE YOUR WORLD
Calories in One Day 215
6.5 States of Matter 216
EXPLORE YOUR WORLD
States of Matter 218
6.6 Melting and Freezing 219
6.7 Boiling and Condensation 221
EXPLORE YOUR WORLD
Evaporation, Boiling, and Condensation 221
HEALTH NOTE
Steam Burns 223
6.8 Heating and Cooling Curves 224
6.9 Energy in Chemical Reactions 228
HEALTH NOTE
Hot Packs and Cold Packs 229
EXPLORE YOUR WORLD
Hotter or Colder? 230
EXPLORE YOUR WORLD
How Fast Can It Go? 231
6.10 A Closer Look: Chemical Equilibrium 233
EXPLORE YOUR WORLD
Modeling an Equilibrium System 234
HEALTH NOTE
Oxygen, Hemoglobin, and Carbon Monoxide Poisoning 235
HEALTH NOTE
Homeostasis: Regulation of Body Temperature 236
Concept Guide *238*
Chapter Summary *238*
Key Terms *239*
Chapter Practice *240*
Answers *241*

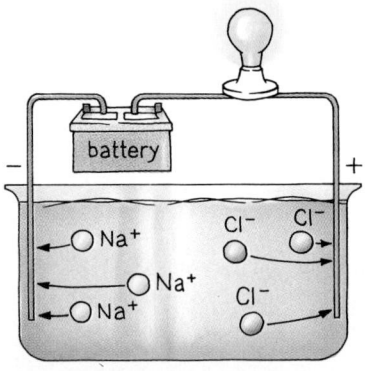

strong electrolyte

7 Gases 244
7.1 Properties of Gases 245
EXPLORE YOUR WORLD
Forming a Gas 247
7.2 Gas Pressure 248
HEALTH NOTE
Measuring Blood Pressure 249
EXPLORE YOUR WORLD
Vapor Pressure 251
7.3 Pressure and Volume (Boyle's Law) 253
HEALTH NOTE
Pressure–Volume Relationship in Breathing 254
EXPLORE YOUR WORLD
Medicine Dropper "Diver" 256
7.4 Temperature and Volume (Charles' Law) 259
EXPLORE YOUR WORLD
Temperature and Volume 260
7.5 Temperature and Pressure (Gay-Lussac's Law) 262
7.6 The Combined Gas Law 265
7.7 Volume and Moles (Avogadro's Law) 266
7.8 Partial Pressures (Dalton's Law) 269
EXPLORE YOUR WORLD
Oxygen in the Air 272
HEALTH NOTE
Hyperbaric Chambers 274
Concept Guide *276*
Chapter Summary *276*
Key Terms *277*
Chapter Practice *277*
Answers *278*

8 Solutions 281
8.1 Properties of Water 282
HEALTH NOTE
Water in the Body 283
EXPLORE YOUR WORLD
Surface Tension I 284
EXPLORE YOUR WORLD
Surface Tension II 284
8.2 Solutions 285
EXPLORE YOUR WORLD
Like Dissolves Like 287
8.3 Electrolytes and Nonelectrolytes 290
HEALTH NOTE
Electrolytes in Body Fluids 294
8.4 Solubility 296

HEALTH NOTE
Gout: A Problem of Saturation in Body Fluids 297
EXPLORE YOUR WORLD
Preparing Solutions 297
EXPLORE YOUR WORLD
Rock Candy: A Saturated Sugar Solution 298
HEALTH NOTE
Kidney Stones Are Insoluble Salts 300
8.5 Percent Concentration 302
8.6 Colloids and Suspensions 306
EXPLORE YOUR WORLD
Tyndall Effect 307
HEALTH NOTE
Colloids and Solutions in the Body 309
8.7 A Closer Look: Osmosis and Dialysis 310
EXPLORE YOUR WORLD
Everyday Osmosis 313
HEALTH NOTE
Dialysis by the Kidneys and the Artificial Kidney 314
8.8 Molarity 316
HEALTH NOTE
Osmolarity of Physiological Solutions and Body Fluids 319
Concept Guide *321*
Chapter Summary *321*
Key Terms *322*
Chapter Practice *323*
Answers *323*

9 Acids and Bases 326
9.1 Acids and Bases 327
EXPLORE YOUR WORLD
Acids and Bases in Products at Home 330
9.2 Ionization of Water 331
9.3 The pH Scale 334
EXPLORE YOUR WORLD
Using Vegetables and Flowers as pH Indicators 336
HEALTH NOTE
Stomach Acid, HCl 336
ENVIRONMENTAL NOTE
Acid Rain 339
9.4 Strengths of Acids and Bases 341
9.5 Making Dilutions 343
9.6 Acid–Base Neutralization 347
EXPLORE YOUR WORLD
Everyday Neutralization Reactions 348
HEALTH NOTE
Antacids 349
9.7 Buffers 350
HEALTH NOTE
CO_2 and the Bicarbonate Buffer 352
9.8 A Closer Look: Acid–Base Titration 354
Concept Guide *357*
Chapter Summary *357*
Key Terms *358*
Chapter Practice *358*
Answers *359*

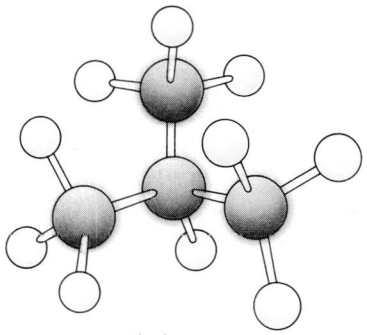

methylpropane

10 Alkanes and Aromatic Hydrocarbons 362
10.1 Organic Compounds 363
10.2 Bonding in Hydrocarbons 365
10.3 Naming Alkanes 370
ENVIRONMENTAL NOTE
Petroleum Products 370
10.4 Properties of Alkanes 373
EXPLORE YOUR WORLD
Solubility of Alkanes 374
EXPLORE YOUR WORLD
Combustion 374
HEALTH NOTE
Incomplete Combustion 375
10.5 Branched Alkanes 376
10.6 Structural Formulas 382
10.7 Cycloalkanes 387
10.8 Aromatic Hydrocarbons 391
HEALTH NOTE
Aromatic Compounds in Health and Medicine 391
HEALTH NOTE
Polycyclic Aromatic Hydrocarbons 392

Concept Guide 394
Chapter Summary 394
Summary of Naming 395
Summary of Reactions 395
Key Terms 396
Chapter Practice 396
Answers 397

11 Haloalkanes, Alkenes, and Alkynes 401

11.1 Functional Groups 402
11.2 Haloalkanes 404
HEALTH NOTE
Haloalkanes as Anesthetics 405
ENVIRONMENTAL NOTE
Chlorinated Cyclic Compounds as Pesticides 405
ENVIRONMENTAL NOTE
CFCs and Ozone Depletion 407
11.3 Alkenes and Alkynes 409
EXPLORE YOUR WORLD
Ripening Fruit 410
ENVIRONMENTAL NOTE
Fragrant Alkenes 412
11.4 Geometric Isomers of Alkenes 413
EXPLORE YOUR WORLD
Modeling Cis–Trans Isomers 415
ENVIRONMENTAL NOTE
Pheromones in Insect Communication 416
HEALTH NOTE
Cis–Trans Isomers for Night Vision 416
11.5 Addition Reactions 418
HEALTH NOTE
Hydrogenation of Unsaturated Fats 419
EXPLORE YOUR WORLD
Unsaturation in Fats and Oils 420
11.6 Polymers of Alkenes 425
EXPLORE YOUR WORLD
Polymers and Recycling Plastics 428

Concept Guide 429
Chapter Summary 429
Summary of Naming 430
Summary of Addition Reactions 430
Key Terms 430
Chapter Practice 431
Answers 432

12 Alcohols, Phenols, Ethers, Aldehydes, and Ketones 436

12.1 Alcohols, Phenols, and Thiols 437
EXPLORE YOUR WORLD
Alcohols in Household Products 440
HEALTH NOTE
Alcohols in Health and Medicine 441
HEALTH NOTE
Derivatives of Phenol Are Found in Antiseptics, Essential Oils, and Antioxidants 442
12.2 Ethers 445
HEALTH NOTE
Ethers as Anesthetics 447
12.3 Reactions of Alcohols 448
EXPLORE YOUR WORLD
Combustion of Alcohol 449
12.4 Aldehydes and Ketones 452
HEALTH NOTE
Aldehydes and Ketones in Health and Medicine 455
ENVIRONMENTAL NOTE
Vanilla 456
12.5 Oxidation and Reduction 458
12.6 Aldehydes and Ketones React with Alcohols 466

Concept Guide 469
Chapter Summary 470
Summary of Naming 470
Summary of Reactions 471
Key Terms 472
Chapter Practice 473
Answers 475

13 Carbohydrates 478

13.1 Classification of Carbohydrates 479
EXPLORE YOUR WORLD
Photosynthesis 480
13.2 Monosaccharides 481
13.3 Chiral Carbon Atoms 483
HEALTH NOTE
Chiral Compounds in Nature 487
13.4 Structures of Some Important Monosaccharides 490
HEALTH NOTE
Hyperglycemia and Hypoglycemia 490
13.5 Cyclic Structures of Monosaccharides 493
HEALTH NOTE
Testing for Glucose in Urine 497
13.6 Disaccharides 499
EXPLORE YOUR WORLD
Sugar and Sweeteners 501
HEALTH NOTE
How Sweet Is My Sweetener? 502
HEALTH NOTE
Blood Types and Carbohydrates 503
13.7 Polysaccharides 505
HEALTH NOTE
Fiber in the Diet 507
EXPLORE YOUR WORLD
Iodine Test for Starch 508
EXPLORE YOUR WORLD
Polysaccharides 508
ENVIRONMENTAL NOTE
Some Other Polysaccharides 509

Concept Guide *510*
Chapter Summary *511*
Summary of Carbohydrates *511*
Summary of Reactions *512*
Key Terms *512*
Chapter Practice *513*
Answers *513*

14 Carboxylic Acids, Esters, Amines, and Amides 515

14.1 Carboxylic Acids 516
HEALTH NOTE
Alpha Hydroxy Acids 518
14.2 Properties of Carboxylic Acids 521
14.3 Esters 524
EXPLORE YOUR WORLD
Carboxylic Acids and Esters in Foods and Cosmetics 524
14.4 Esterification and Hydrolysis 527
HEALTH NOTE
Salicylic Acid and Aspirin 528
ENVIRONMENTAL NOTE
Plastics 529
14.5 Amines 533
HEALTH NOTE
Amines in Health and Medicine 536
14.6 Reactions of Amines 537
14.7 Amides 543
HEALTH NOTE
Amides in Health and Medicine 544
14.8 Reactions of Amides 546
HEALTH NOTE
Amide Hydrolysis in Nature 547

Concept Guide *548*
Chapter Summary *549*
Summary of Naming *549*
Summary of Reactions *550*
Key Terms *552*
Chapter Practice *552*
Answers *553*

15 Lipids 557

15.1 Types of Lipids 558
EXPLORE YOUR WORLD
Solubility of Fats and Oils 558
15.2 Fatty Acids 560
HEALTH NOTE
Omega-3 Fatty Acids in Fish Oils 562

HEALTH NOTE
Prostaglandins 563
15.3 Waxes, Fats, and Oils 564
EXPLORE YOUR WORLD
Fats and Oils 567
HEALTH NOTE
Olestra: A Fat Substitute 568
15.4 Chemical Properties of Triglycerides 569
EXPLORE YOUR WORLD
Types of Fats 570
HEALTH NOTE
Trans Fatty Acids and Hydrogenation 571
ENVIRONMENTAL NOTE
Cleaning Action of Soaps 573
15.5 Phospholipids and Glycolipids 574
HEALTH NOTE
Lipid Diseases 578
15.6 Steroids and Cholesterol 580
HEALTH NOTE
Cholesterol in the Body 581
HEALTH NOTE
Lipoproteins: Transporting Lipids in Solution 582
HEALTH NOTE
Anabolic Steroids 586
15.7 Plasma (Cell) Membranes 587
Concept Guide *589*
Chapter Summary *589*
Key Terms *590*
Chapter Practice *590*
Answers *591*

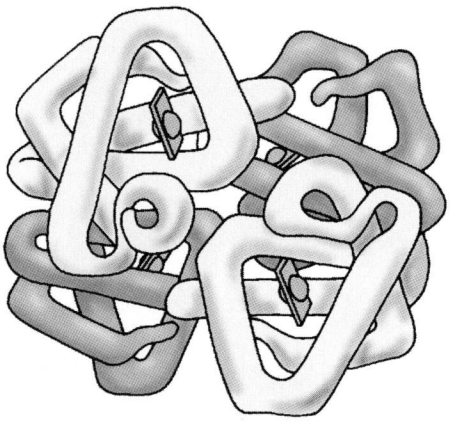

16 Amino Acids, Proteins, and Enzymes 594
16.1 Types of Proteins 595
16.2 Amino Acids 596
16.3 The Peptide Bond 601
16.4 Primary and Secondary Structure 603
HEALTH NOTE
Natural Opiates in the Body 605
16.5 Tertiary and Quaternary Structure 606
HEALTH NOTE
Hair Permanents 608
HEALTH NOTE
Sickle-Cell Anemia 610
16.6 Protein Hydrolysis and Denaturation 612
EXPLORE YOUR WORLD
Denaturation of Milk Protein 614
16.7 Enzymes 615
16.8 Enzyme Action 617
HEALTH NOTE
Enzymes as Diagnostic Tools 619
16.9 Factors Affecting Enzyme Action 621
EXPLORE YOUR WORLD
Enzyme Activity 624
16.10 Enzyme Inhibition 626
HEALTH NOTE
Competitive Inhibitors in Medicine 628
Concept Guide *630*
Chapter Summary *631*
Key Terms *631*
Chapter Practice *632*
Answers *633*

17 Metabolic Pathways and Energy Production 636

17.1 Metabolism and ATP Energy 637

HEALTH NOTE
ATP Energy and Ca^{2+} Needed to Contract Muscles 638

17.2 Digestion of Foods 640

EXPLORE YOUR WORLD
Digestion of Foods 642

17.3 Important Coenzymes in Metabolic Pathways 643

17.4 Glycolysis: Oxidation of Glucose 646

17.5 Pathways for Pyruvate 649

17.6 Citric Acid Cycle 652

17.7 Electron Transport Chain 656

17.8 ATP Energy from Glucose 661

17.9 Oxidation of Fatty Acids 664

HEALTH NOTE
Ketone Bodies 668

17.10 Metabolic Pathways for Amino Acids 670

HEALTH NOTE
Why Do I Get Hungry? 672

Concept Guide 674
Chapter Summary 675
Summary of Key Reactions 676
Key Terms 676
Chapter Practice 677
Answers 677

18 Nucleic Acids 681

18.1 Nucleic Acids 682

18.2 Structures of Nucleic Acids 685

18.3 DNA Replication 690

18.4 RNA and Transcription 692

18.5 The Genetic Code 695

18.6 Protein Synthesis 697

HEALTH NOTE
Many Antibiotics Inhibit Protein Synthesis 700

18.7 Recombinant DNA 701

18.8 Genetic Mutations 704

EXPLORE YOUR WORLD
A Model for DNA Replication and Mutation 705

HEALTH NOTE
Genetic Diseases 706

18.9 Regulation of Protein Synthesis 708

HEALTH NOTE
Cancer 711

18.10 Viruses and Aids 712

Concept Guide 716
Chapter Summary 717
Key Terms 718
Chapter Practice 719
Answers 719

Credits 722

Glossary/Index 723

Applications and Activities

HEALTH NOTES

Determination of Percentage of Body Fat 37
Specific Gravity of Urine 39
Elements Essential to Health 49
Latin Names for Elements in Clinical Usage 49
Calcium and Strontium 54
Some Important Trace Elements in the Body 55
Biological Effects of Radiation 83
Beta Emitters in Medicine 91
Radiation and Food 96
Radiation Doses in Diagnostic and Therapeutic Procedures 101
Other Imaging Methods 102
Some Uses for Noble Gases 122
Some Important Ions in the Body 125
Polyatomic Ions in Bone and Teeth 149
Smog and Health Concerns 172
Variation in Body Temperature 205
Losing and Gaining Weight 215
Steam Burns 223
Hot Packs and Cold Packs 229
Oxygen, Hemoglobin, and Carbon Monoxide Poisoning 235
Homeostasis: Regulation of Body Temperature 236
Measuring Blood Pressure 249
Pressure–Volume Relationship in Breathing 254
Hyperbaric Chambers 274
Water in the Body 283
Electrolytes in Body Fluids 294
Gout: A Problem of Saturation in Body Fluids 297
Kidney Stones Are Insoluble Salts 300
Colloids and Solutions in the Body 309
Dialysis by the Kidneys and the Artificial Kidney 314
Osmolarity of Physiological Solutions and Body Fluids 319
Stomach Acid, HCl 336
Antacids 349
CO_2 and the Bicarbonate Buffer 352
Incomplete Combustion 375
Aromatic Compounds in Health and Medicine 391
Polycyclic Aromatic Hydrocarbons 392
Haloalkanes as Anesthetics 405
Cis–Trans Isomers for Night Vision 416
Hydrogenation of Unsaturated Fats 419
Alcohols in Health and Medicine 441
Derivatives of Phenol Are Found in Antiseptics, Essential Oils, and Antioxidants 442
Ethers as Anesthetics 447
Aldehydes and Ketones in Health and Medicine 455
Chiral Compounds in Nature 487
Hyperglycemia and Hypoglycemia 490
Testing for Glucose in Urine 497
How Sweet Is My Sweetener? 502
Blood Types and Carbohydrates 503
Fiber in the Diet 507
Alpha Hydroxy Acids 518
Salicylic Acid and Aspirin 528
Amines in Health and Medicine 536
Amides in Health and Medicine 544
Amide Hydrolysis in Nature 547
Omega-3 Fatty Acids in Fish Oils 562
Prostaglandins 563
Olestra: A Fat Substitute 568
Trans Fatty Acids and Hydrogenation 571
Lipid Diseases 578
Cholesterol in the Body 581
Lipoproteins: Transporting Lipids in Solution 582
Anabolic Steroids 586
Natural Opiates in the Body 605
Hair Permanents 608
Sickle-Cell Anemia 610
Enzymes as Diagnostic Tools 619
Competitive Inhibitors in Medicine 628
ATP Energy and Ca^{2+} Needed to Contract Muscles 639
Ketone Bodies 668

page 172

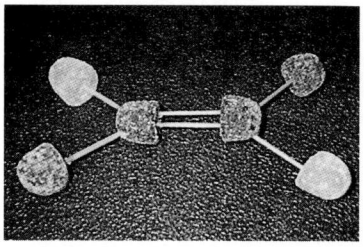

page 415

page 456

Why Do I Get Hungry? 672
Many Antibiotics Inhibit Protein Synthesis 700
Genetic Diseases 706
Cancer 711

ENVIRONMENTAL NOTES

Biological Reactions to UV Light 71
Radon in Our Homes 88
Dating Ancient Objects 106
Global Warming 203
Acid Rain 339
Petroleum Products 370
Chlorinated Cyclic Compounds as Pesticides 405
CFCs and Ozone Depletion 407
Fragrant Alkenes 412
Pheromones in Insect Communication 416
Vanilla 456
Some Other Polysaccharides 509
Plastics 529
Cleaning Action of Soaps 573

EXPLORE YOUR WORLD

Units Listed on Labels 4
Temperature Scales on Thermometers 8
SI and Metric Equalities on Product Labels 23
Sink or Float? 35
Origins of Element Names 49
Physical Properties of Aluminum 50
Repulsion and Attraction 58
Light Released from Candy 71
Dental X Rays 85
Smoke Detectors 89
Modeling Half-Lives 105
Names of Compounds Listed on Products 150
Evidence of a Chemical Change 162
Oxidation of Fruits and Vegetables 176
Loss and Gain of Heat 209
Specific Heat 210
Counting Calories 215
Calories in One Day 215
States of Matter 218
Evaporation, Boiling, and Condensation 221
Hotter or Colder? 230
How Fast Can It Go? 231
Modeling an Equilibrium System 234
Forming a Gas 247
Vapor Pressure 251
Medicine Dropper "Diver" 256
Temperature and Volume 260
Oxygen in the Air 272
Surface Tension I 284
Surface Tension II 284
Like Dissolves Like 287
Preparing Solutions 297
Rock Candy: A Saturated Sugar Solution 298
Tyndall Effect 307
Everyday Osmosis 313
Acids and Bases in Products at Home 330
Using Vegetables and Flowers as pH Indicators 336
Everyday Neutralization Reactions 348
Solubility of Alkanes 374
Combustion 374
Ripening Fruit 410
Modeling Cis–Trans Isomers 415
Unsaturation in Fats and Oils 420
Polymers and Recycling Plastics 428
Alcohols in Household Products 440
Combustion of Alcohol 449
Photosynthesis 480
Sugar and Sweeteners 501
Iodine Test for Starch 508
Polysaccharides 508
Carboxylic Acids and Esters in Foods and Cosmetics 526
Solubility of Fats and Oils 558
Fats and Oils 567
Types of Fats 570
Denaturation of Milk Protein 614
Enzyme Activity 624
Digestion of Foods 642
A Model for DNA Replication and Mutation 705

Preface

Welcome to the Media Update of the seventh edition of *CHEMISTRY: An Introduction to General, Organic, and Biological Chemistry.* This book is designed for an introductory course in chemistry and assumes no prior knowledge of chemistry by the student. It is geared to students preparing for careers in health-related professions, such as nursing and respiratory therapy. It can also be used as a laboratory science course for liberal arts students and future teachers.

For me, this Media Update is a special edition. Its publication combines my 25th year of writing textbooks for allied health students with my desire to use technology to make new learning resources available to students and instructors.

The key features of the Media Update are shown on the following pages in *A Visual Guide to the Media Update.* Each chapter functions as a complete package, so that an instructor can easily skip some chapters or change the order of chapters. Throughout each chapter, Health Notes, Environmental Notes, and Explore Your World activities present real-life applications of chemical concepts at the point where each concept is discussed. Every section has a set of Sample Problems, Study Checks, and Questions & Problems. Students are encouraged to solve problems and apply ideas as they progress through the chapter instead of waiting until they finish. A set of Practice Problems at the end of the chapter integrates material from the entire chapter to promote critical thinking. Answers to the odd-numbered problems now appear at the end of each chapter, rather than at the end of the book.

Most important, a rich array of media resources is available. In each chapter, icons refer students to dynamic learning tools on two CD-ROMs and *The Chemistry Place, Special Edition for Timberlake* web site. *The Study Pack 2.0* CD-ROM and *The Chemistry of Life for Introductory Chemistry* CD-ROM feature tutorials, reviews, practice exams, and interactive exercises to help students visualize new concepts. On *The Chemistry Place, Special Edition for Timberlake*, students have access to tutorials, review slides, a glossary, quizzes for every chapter, a periodic table, and a calculator. The web site also offers instructors a variety of teaching tools, including an online instructor's manual, lab exercises, PowerPoint slide modules for classroom presentation, and a syllabus manager.

It is my goal that every student understands the scientific concepts that will help him or her to make informed decisions concerning health and the environment. In producing this Media Update, my hope is that the many pedagogical features of the text and all the media resources will make teaching and learning chemistry an enthusiastic and positive experience.

A Visual Guide to the Media Update

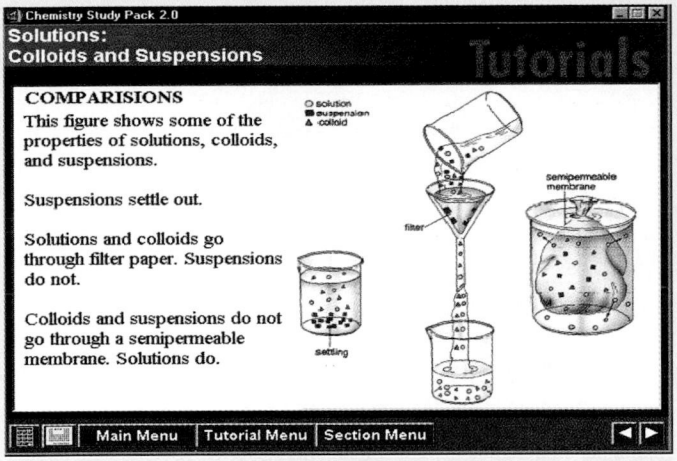

The *Chemistry Study Pack, Version 2.0* CD-ROM features **interactive tutorials** for all 18 chapters of the book, including randomized **practice tests** for each tutorial. A **quick quiz** is available for each of the 168 sections of the text and **chapter quizzes** test students' knowledge at the end of each tutorial.

Look for the Study Pack icon for cross-references to this CD.

Chemistry Study Pack CD-ROM

Through more than 600 high-quality **animations** and **interactive simulations**, the *Chemistry of Life for Introductory Chemistry* CD-ROM brings the molecular world to life. Correlated to the book, this CD-ROM reinforces concepts introduced in the text. In addition, students can prepare for tests with the 250 **quiz questions** and the illustrated **glossary**.

Look for the Chem of Life icon for cross-references to this CD.

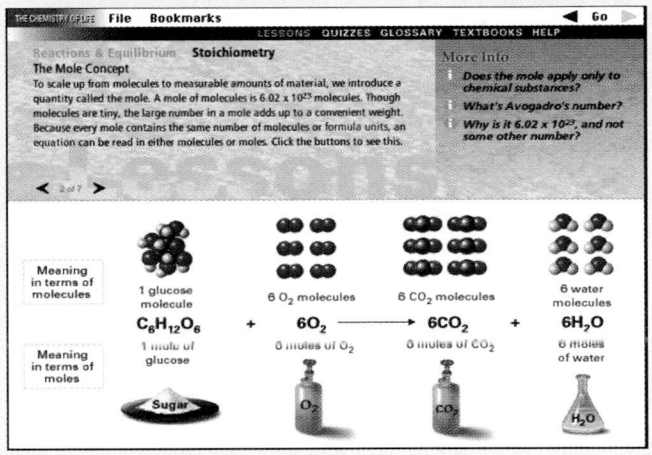

Chemistry of Life for Introductory Chemistry CD-ROM

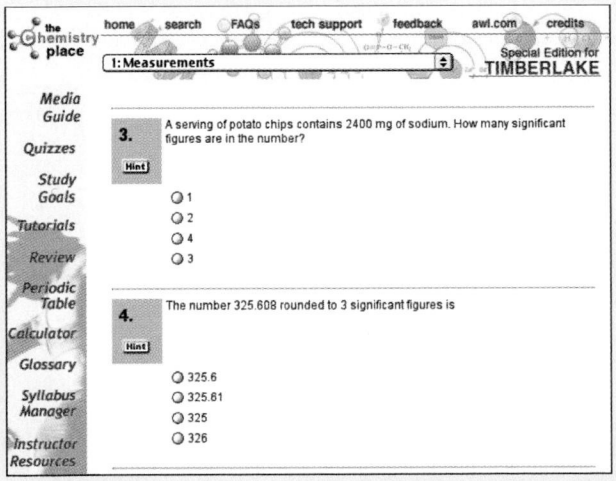

The Chemistry Place, Special Edition for Timberlake contains many tools to help students prepare for exams. This web site includes **quiz questions** for each chapter, **review slides**, web links, **interactive tutorials**, a glossary and an interactive periodic table.

The Chemistry Place, Special Edition for Timberlake
www.chemplace.com/college/timberlake

A Visual Guide to this Book

Each chapter opens with **Looking Ahead,** an outline of the topic covered in each section of the text.

Looking Ahead

7

Gases

Looking Ahead

7.1 Properties of Gases
7.2 Gas Pressure
7.3 Pressure and Volume (Boyle's Law)
7.4 Temperature and Volume (Charles' Law)
7.5 Temperature and Pressure (Gay-Lussac's Law)
7.6 The Combined Gas Law
7.7 Volume and Moles (Avogadro's Law)
7.8 Partial Pressures (Dalton's Law)

www.chemplace.com/college/timberlake

Visit the web site above to test yourself, run simulations, and review key concepts from this chapter. For even more interactive review and quizzing, use the two CD-ROMs included with the book.

Figure 7.6 As the air inside a hot-air balloon is heated, its volume expands and its density decreases. The hot-air balloon rises and floats in the air.

7.4

7.4 Temperature and Volume (Charles' Law)

Study Pack

Learning Goal

Use the temperature–volume relationship (Charles' law) to determine the new temperature or volume of a certain amount of gas at a constant pressure.

When preparing a hot-air balloon for a flight, the air in the balloon is heated with a small propane heater. As the air warms, its volume increases. The resulting decrease in density allows the balloon to rise. (See Figure 7.6.)

To study the effect of changing temperature on the volume of a gas, we must not change the pressure or the amount of the gas. Suppose we increase the Kelvin temperature of a gas sample. The kinetic theory shows that the activity (kinetic energy) of the gas will also increase. To keep pressure constant, the volume of the container must increase. (See Figure 7.7.) By contrast, if the temperature of the gas is lowered, the volume of the container must be reduced to maintain the same pressure.

Suppose that you are going to take a ride in a hot air balloon. The captain turns on a propane burner to heat the air inside the balloon. As the temperature rises, the air particles move faster and spread out, causing the volume of the balloon to increase. Eventually, the air in the balloon becomes less dense than the air outside, and the balloon and its passengers rise. In fact, it was in 1787 that Jacques Charles, a balloonist as well as a physicist, proposed that the volume of a gas is related to the temperature. This became **Charles' law,** which states that the volume (V) of a gas is directly related to the temperature (K) when there is no change in the pressure (P) or amount (n) of gas. A

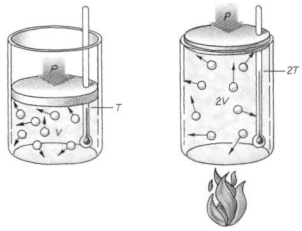

Figure 7.7 The Kelvin temperature of a gas is directly related to the volume of the gas when there is no change in pressure or amount.

Learning Goals at the beginning of each new section in a chapter give the outcome that can be expected after reading the material in the section and working the problems.

Learning Goals

A rich array of **Health Notes** in each chapter apply chemical concepts to real-life issues and relevant topics of health and medicine such as weight loss and weight gain, artificial fats, anabolic steroids, alcohol, genetic diseases, viruses, and cancer.

Health Note

To determine the equivalent or rem dose, the absorbed dose (rads) is multiplied by a factor that adjusts for biological damage caused by a particular form of radiation. For beta and gamma radiation, and X rays, the factor is 1, so the biological damage in rems is the same as the absorbed radiation (rads). For high-energy protons and neutrons, the factor is about 10, and for alpha particles it is 20.

Biological damage (rem) = absorbed dose (rad) × factor

Often the measurement for an equivalent dose will be in units of **millirems (mrem)**. One rem is equal to 1000 mrem. A newer unit is the **sievert (Sv).** One sievert is equal to 100 rems. Table 3.4 summarizes the units used to measure radiation.

3.4

HEALTH NOTE

Radiation and Food

Foodborne illnesses caused by pathogenic bacteria such as *Salmonella*, *Listeria*, and *Escherichia coli* have become a major health concern in the United States. The Centers for Disease Control and Prevention estimates that each year *E. coli* in contaminated foods infects 20,000 people in the United States, and that 500 people die. A recent outbreak of *E. coli* O157:H7 resulted in more than 700 cases of illness and four deaths. This same *E. coli* has been responsible for at least nine outbreaks of illness from contaminated fruit juices, lettuce, and alfalfa sprouts.

The Food and Drug Administration (FDA) has approved the use of 0.3 kilogray (0.3 kGy) to 1 kGy of ionizing radiation produced by cobalt-60 or cesium-137 for the treatment of foods. The irradiation technology is much like that used to sterilize medical supplies. Cobalt pellets are placed in stainless steel tubes, which are arranged in racks. When food moves through the series of racks, the gamma rays pass through the food and kill the bacteria.

It is important for consumers to understand that when food is irradiated, it never comes in contact with the radioactive source. The gamma rays pass through the food to kill bacteria, but that does not make the food radioactive. The radiation kills bacteria because it stops their ability to divide and grow. We cook or heat food thoroughly for the same purpose. Radiation, as well as heat, has little effect on the food itself because its cells are no longer dividing or growing. Thus irradiated food is not harmed although a small amount of vitamin B_1 and C may be lost.

Currently, tomatoes, blueberries, and mushrooms are being irradiated to allow them to be harvested when completely ripe and extend their shelf life (see Figure 3.10). The FDA has also approved the irradiation of pork, poultry, and beef in order to decrease potential infections and to extend shelf life. Currently, irradiated vegetable and meat products are available in retail markets in South Africa. Apollo 17 astronauts ate irradiated foods on the moon, and some U.S. hospitals and nursing homes now use irradiated poultry to reduce the possibility of infections among patients. The extended shelf life of irradiated food also makes it useful for campers and military personnel. Soon consumers concerned about food safety will have a choice of irradiated meats, fruits, and vegetables at the market.

Figure 3.10 **(a)** The FDA requires this symbol to appear on irradiated retail foods. **(b)** After 2 weeks, the irradiated blueberries on the right show no spoilage. Mold is starting to grow on the nonirradiated ones on the left.

b

ENVIRONMENTAL NOTE

Acid Rain

Rain typically has a pH of 6.2. It is slightly acidic because carbon dioxide in the air combines with water to form carbonic acid. However, in many parts of the world, rain has become considerably more acidic, with pH values as low as 3 being reported. One cause of acid rain is the sulfur dioxide (SO_2) gas produced when coal that contains sulfur is burned.

In the air, the SO_2 gas reacts with oxygen to produce SO_3, which then combines with water to form sulfuric acid, H_2SO_4, a strong acid.

$$S + O_2 \longrightarrow SO_2$$

$$2SO_2 + O_2 \longrightarrow 2SO_3$$

$$SO_3 + H_2O \longrightarrow H_2SO_4$$

In parts of the United States, acid rain has made lakes so acidic they are no longer able to support fish and plant life. Limestone ($CaCO_3$) is sometimes added to these lakes to neutralize the acid. In Eastern Europe, acid rain has brought about an environmental disaster. Nearly 40% of the forests in Poland have been severely damaged, and some parts of the land are so acidic that crops will not grow. Throughout Europe and the United States, monuments made of marble (a form of $CaCO_3$) are deteriorating as acid rain dissolves the marble (see Figure 9.8).

$$2H^+ + CaCO_3 \longrightarrow Ca^{2+} + H_2O + CO_2$$

Efforts to slow or stop the damaging effects of acid rain include the reduction of sulfur emissions. This will require installation of expensive equipment in coal-burning plants to absorb more of the SO_2 gases before they are emitted. In some outdated plants, this may be impossible, and they will need to be closed. It is a difficult problem for engineers and scientists, but one that must be solved.

9.3

a

b

Figure 9.8 **(a)** Detail of the marble that is part of the Washington Square Arch in Washington Square Park, New York City, completed on July 17, 1935. **(b)** The destructive effect of acid rain on the same marble, June 1994.

QUESTIONS & PROBLEMS

The pH Scale

9.19 Why does a neutral solution have a pH of 7?

9.20 If you know the $[OH^-]$, how would you determine the pH of a solution?

9.21 State whether each of the following solutions is acidic, basic, or neutral:
a. blood, pH 7.4 **b.** vinegar, pH 2.8 **c.** drain cleaner, pH 11.2
d. coffee, pH 5.5 **e.** milk, pH 7.0

9.22 State whether each of the following solutions is acidic, basic, or neutral:
a. soda, pH 3.2 **b.** shampoo, pH 6.0
c. hot tub water, pH 7.8 **d.** acid rain, pH 6.2
e. laundry detergent, pH 9.5

Environmental Notes delve into issues such as global warming, radon, ozone depletion, acid rain, food irradiation, pheromones, and recycling of plastics.

Environmental Note

Explore Your World are hands-on activities that use everyday materials to encourage students to actively explore selected chemistry topics, either individually or in group learning environments. Each activity is followed by questions to encourage critical thinking.

Explore Your World

Figure 9.7 The dye in red cabbage juice acts as a pH indicator by turning different colors that correspond to different pH values.

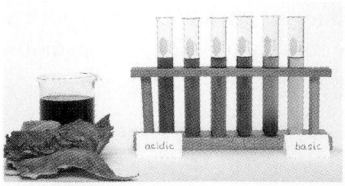

9.3

EXPLORE YOUR WORLD

Using Vegetables and Flowers as pH Indicators

Many flowers and vegetables with strong color, especially reds and purples, contain compounds that change color with changes in pH. Some examples are red cabbage, cranberry juice, and cranberry drinks.

Materials Needed

Red cabbage, water, and a saucepan; or cranberry juice or drinks

Several glasses or small glass containers and some tape and a pen or pencil to mark the containers

Several colorless household solutions such as vinegar, lemon juice, other fruit juices, baking soda, antacids, aspirin, window cleaners, soaps, shampoos, and detergents

Procedure

1. Obtain a bottle of cranberry juice or cranberry drink, or use a red cabbage to prepare the red cabbage pH indicator, as follows: Tear up several red cabbage leaves and place them in a saucepan and cover with water. Boil for about 5 minutes. Cool, strain, and collect the purple indicator solution.
2. Place small amounts of each household solution into separate, clear glass containers and mark what each one is. If the sample is a solid or a thick liquid, add a small amount of water. Add some cranberry juice or some red cabbage indicator until you obtain a color.
3. Observe the colors of the various samples. The colors that indicate acidic solutions are the pink and orange colors (pH 1–4) and the pink to lavender colors (5–6). A neutral solution has about the same purple color as the indicator. Bases will give blue to green color (pH 8–11) or a yellow color (pH 12–13).
4. Arrange your samples by color and pH. Classify each of the solutions as acidic (1–6), neutral (7), or basic (8–13).
5. Try to make an indicator using other colorful fruits or flowers.

Questions

1. Which products that tested acidic listed acids on their labels?
2. Which products that tested basic listed a base on their labels?
3. How many products were neutral?
4. Which flowers or vegetables behaved as indicators?

HEALTH NOTE

Stomach Acid, HCl

When a person sees, smells, thinks about, and/or tastes food, the gastric glands in the stomach begin to secrete a HCl solution that is strongly acidic. In a single day, a person may secrete as much as 2000 mL of gastric juice.

The HCl in the gastric juice activates a digestive [illegible]led pepsin that breaks down proteins in food entering the stomach. The secretion of HCl continues until the stomach has a pH of about 2, which is the optimum pH for activating the digestive enzymes without ulcerating the stomach lining. Normally, large quantities of viscous mucus are secreted within the stomach to protect its lining from acid and enzyme damage.

SAMPLE 8.14 PROBLEM

Calculating Percent Concentration

A student prepared a solution by dissolving 5 g of KI with enough water to give a final volume of 250 mL. (See Figure 8.16.) What is the mass/volume percent of the KI solution?

8.5

Figure 8.16 A 2% (m/v) KI solution is prepared by placing 5 g of KI in a 250-mL flask and filling the flask to the 250-mL mark.

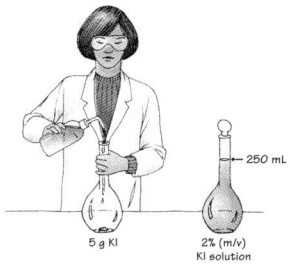

Solution

Placing grams of solute and milliliters of solution in the expression for percent (m/v) concentration gives

$$\text{mass/volume \%} = \frac{\overset{\text{Mass of solute}}{5\text{ g KI}}}{\underset{\text{Volume of solution}}{250\text{ mL solution}}} \times 100 = 2\%\text{ (m/v) KI}$$

Study Check

What is the mass/volume percent, % (m/v), of Br_2 in a solution prepared by dissolving 12 g of bromine (Br_2) in enough carbon tetrachloride to make 250 mL of solution?

Percent Concentrations as Conversion Factors

In the preparation of solutions, we often need to calculate the amount of solute or solution. Then the percent concentration is useful as a conversion factor. Some examples of percent concentrations, their meanings, and possible conversion factors are given in Table 8.8.

Table 8.8 Conversion Factors from Percent Concentrations

Percent Concentration	Meaning	Conversion Factors		
10% (m/m) KCl	There are 10 g of KCl in exactly 100 g of solution.	$\frac{10\text{ g KCl}}{100\text{ g solution}}$	and	$\frac{100\text{ g solution}}{10\text{ g KCl}}$
5% (m/v) glucose	There are 5 g of glucose in exactly 100 mL of solution.	$\frac{5\text{ g glucose}}{100\text{ mL solution}}$	and	$\frac{100\text{ mL solution}}{5\text{ g glucose}}$

Sample Problems appear throughout the text to immediately demonstrate the application of each new concept. The worked-out solution gives step-by-step explanations, provides a problem-solving model, and illustrates required calculations.

Study Checks at the end of each Sample Problem ask the student to work a similar problem.

Sample Problems

Study Checks

Questions & Problems at the end of each section in the chapters encourage students to immediately apply concepts and begin problem solving after learning a workable chunk of information, instead of waiting until the end of the chapters to work problems.

Questions & Problems

QUESTIONS & PROBLEMS

Pressure and Volume (Boyle's Law)

7.13 Why do scuba divers need to exhale air when they ascend to the surface of the water?

7.14 Why does a sealed bag of chips expand when you take it to a higher altitude?

7.15 What happens to the volume of your lungs during expiration when you get rid of some CO_2?

7.16 How do respirators (or CPR) help a person obtain oxygen and expel carbon dioxide?

7.17 The air in a cylinder with a piston has a volume of 220 mL and a pressure of 650 mm Hg.

a. If a change results in a higher pressure inside the cylinder, does cylinder A or B represent the final volume? Explain your choice.

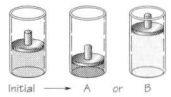

b. If the pressure inside the cylinder increases to 1.2 atm, what is the final volume of the cylinder? Complete the following data table:

Property	Initial	Final
Pressure (P)		
Volume (V)		

7.18 A balloon is filled with helium gas. When the following changes are made at constant temperature, which of these diagrams (A, B, or C) shows the new volume of the balloon?

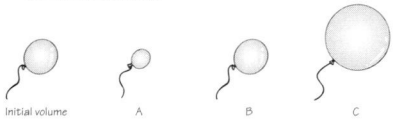

a. The balloon floats to a higher altitude where the outside pressure is lower.
b. The balloon is taken inside the house, but the atmospheric pressure remains the same.
c. The balloon is put in a hyperbaric chamber in which the pressure is increased.

7.19 Complete the following using the pressure–volume relationship.

	Pressure	Volume
a. decreases		_____
b. increases		_____

Chapter Review

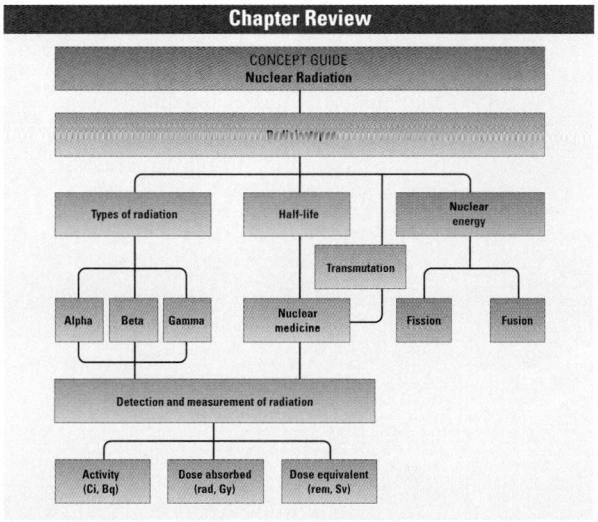

New! **Concept Guide** artwork at the end of each chapter visually links concepts and shows key relationships among the topics within a chapter.

Chapter Summary

3.1 Natural Radioactivity
Radioactive isotopes have unstable nuclei that break down (decay), spontaneously emitting alpha (α), beta (β), and gamma (γ) radiation. Because radiation can damage the cells in the body, proper protection must be used: shielding, limiting the time of exposure, and distance.

3.2 Nuclear Equations
A balanced equation is used to represent the changes that take place in the nuclei of the reactants and products. The new isotopes and the type of radiation emitted can be determined from the symbols that show the mass numbers and atomic numbers of the isotopes in the nuclear reaction.

3.3 Producing Radioactive Isotopes
A radioisotope is produced artificially when a nonradioactive isotope is bombarded by a small particle such as a proton, or an alpha or beta particle. Many radioactive isotopes used in nuclear medicine are produced in this way.

3.4 Radiation Detection and Measurement
In a Geiger counter, radiation ionizes gas in a metal tube, which produces an electrical current. The curie (Ci) measures the number of nuclear transformation of a radioactive sample. Activity is also measured in becquerel (Bq) units. The amount of radiation absorbed by a substance is measured in rads or the gray (Gy). The rem and the sievert (Sv) are units used to determine the biological damage from the different types of radiation.

3.5 Medical Applications Using Radioactivity
In nuclear medicine, radioisotopes are given that go to specific sites in the body. By detecting the radiation they emit, an evaluation can be made about the location and extent of

Chapter Summaries at the end of each chapter briefly review the main ideas in each section.

Concept Guides

Chapter Summaries

Organic and biological chemistry chapters contain additional summaries such as this **Summary of Reactions.**

Key Terms appear in boldfaced type in a chapter and are listed as a glossary at the end of the chapter to encourage vocabulary review.

Summary of Reactions

Key Terms

Summary of Reactions

Glycoside (Acetal) Formation

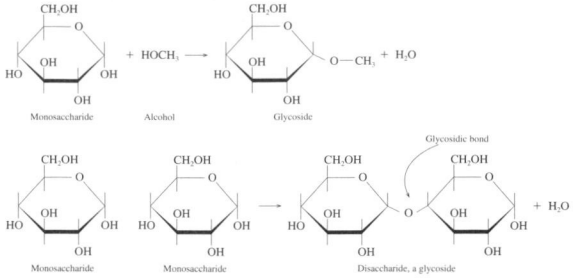

Hydrolysis of Disaccharides

Sucrose + H_2O ⟶ glucose + fructose
Lactose + H_2O ⟶ glucose + galactose
Maltose + H_2O ⟶ glucose + glucose

Key Terms

aldose Monosaccharides that contain an aldehyde group.
amylopectin A branched-chain polymer of starch composed of glucose units joined by α-1,4- and α-1,6-glycosidic bonds.
amylose An unbranched polymer of starch composed of glucose units joined by α-1,4-glycosidic bonds.
anomers The isomers that occur when the formation of a hemiacetal or hemiketal of monosaccharides produces a new hydroxyl group on carbon 1 (or carbon 2). In the α anomer, the OH is drawn downward; in the β isomer the OH is up.
carbohydrate A simple or complex sugar composed of carbon, hydrogen, and oxygen.
cellulose An unbranched polysaccharide composed of glucose units linked by β-1,4-glycosidic bonds that cannot be hydrolyzed by the human digestive system.
chiral A carbon atom that is attached to four different atoms or groups of atoms; the mirror images of a chiral molecule are not superimposable.
disaccharides Carbohydrates composed of two monosaccharides joined by a glycosidic bond.
enantiomers Stereoisomers that are mirror images of each other.
fermentation A reaction of glucose, fructose, maltose, or sucrose, in which the sugar reacts with enzymes in yeast to give ethanol and carbon dioxide gas.
Fischer projection A two-dimensional straight-chain representation of a molecule.
fructose A monosaccharide found in honey and fruit juices; it is combined with glucose in sucrose. Also called levulose and fruit sugar.
galactose A monosaccharide that occurs combined with glucose in lactose.
glucose The most prevalent monosaccharide in the diet. An aldohexose that is found in fruits, vegetables, corn syrup, and honey. Also known as blood sugar and dextrose. Combines in glycosidic bonds to form most of the polysaccharides.
glycogen A polysaccharide formed in the liver and muscles for the storage of glucose as an energy reserve. It is composed of glucose in a highly branched polymer joined by α-1,4- and α-1,6-glycosidic bonds.
glycosides Acetal products of a monosaccharide reacting with an alcohol or another sugar.
glycosidic bond The acetal bond that forms when an alcohol or a hydroxyl group of a monosaccharide adds to

Chapter Practice

2.57 Why is Co the symbol for cobalt, but not CO?

2.58 Which of the following is correct? Write the correct symbol if needed.
a. copper, Co **b.** silicon, SI **c.** iron, Fe
d. fluorine, Fl **e.** potassium, P **f.** sodium, Na
g. gold, Au **h.** lead, PB

2.59 Give the symbol and name of the element found in the following group and period on the periodic table.
a. Group 2, Period 3 **b.** Group 7, Period 4
c. Group 3, Period 3 **d.** Group 6, Period 2

2.60 Give the group and period number for the following elements:
a. potassium **b.** phosphorus
c. carbon **d.** neon

2.61 The following statements are *false*. Reword the underlined word or phrase to make a *true* statement.
a. The proton is a neutral particle.
b. The electrons are found in the nucleus.
c. The nucleus is the largest part of the atom.
d. The neutron has a negative charge.
e. [illegible] of the mass of an atom is du[illegible]

c. Which atoms have the same mass number?
d. What atoms have the same number of neutrons?

2.68 Five isotopes of zinc are zinc-64, zinc-66, zinc-67, zinc-68, and zinc-70.
a. Write the atomic symbols, including atomic number and mass number for each of these atoms.
b. Give the number of protons, electrons, and neutrons for each of the zinc isotopes.

2.69 A sample of copper has two naturally occurring isotopes. The isotope Cu-63 (mass 62.9) makes up 69.2% of all the copper atoms, and Cu-65 (mass 64.9) makes up 30.8% of the sample. What average atomic mass would you calculate for copper?

2.70 In a sample of magnesium, Mg-24 (mass 24.0) is 79.0% of the sample, Mg-25 (mass 25.0) is 10.0%, and Mg-26 (mass 26.0[illegible] mass would you c[illegible]

2.71 I[illegible] sodium atom is 3.14 × 10[illegible] sodium atoms would fit along a line ex[illegible] 1 inch long?

Answers

Answers to Study Checks

16.1 hormones

16.2 **a.** nonpolar **b.** polar

16.3

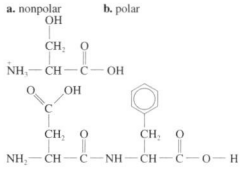

16.4

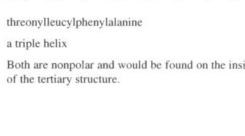

16.5 threonylleucylphenylalanine

16.6 a triple helix

16.7 Both are nonpolar and would be found on the inside of the tertiary structure.

16.8 quaternary

16.9 The heavy metal Ag^+ denatures the protein in bacteria that cause gonorrhea.

16.10 hydrolase

16.11 The active site has a rigid shape in the lock-and-key model but is flexible in the induced-fit model and adapts to the shape of the substrate.

16.12 The vitamin B_6 in part c.

16.13 Urease will lose enzymatic activity as denaturation of its protein occurs.

16.14 noncompetitive inhibitor

Answers to Selected Questions and Problems

16.1 **a.** transport **b.** structural
c. structural **d.** enzyme

16.3 All amino acids contain a carboxylic acid group and an amino group on the α carbon.

Chapter Practice problems at the end of each chapter combine concepts from that chapter and previous chapters. These problems are more complex mathematically and require more critical thinking; several involve real-life relationships.

Answers to all Study Checks and odd-numbered problems are given at the end of each chapter.

Chapter Practice Answers

Key Content Changes

Chapter 1: Measurements contains features from the old appendices on scientific notation and working with a calculator. A new discussion on accuracy and precision is now included with the topic of significant figures. Temperature equations have been moved to Chapter 6, where temperature is needed for heat calculations.

Chapter 2: Atoms and Elements appears earlier in the sequence with heat and energy moved to a later chapter. The new IUPAC names for elements 104–109 are included in *CHEMISTRY* along with spaces for elements 110–118. The section on the structure of the atom has been rewritten and now includes Rutherford's gold foil experiment. The term "atomic mass" replaces "atomic weight." In agreement with reviewer comments, the section in the sixth edition on subshells and orbitals has been deleted.

Chapter 3: Nuclear Radiation has been moved to follow the chapter on atoms because a discussion of radioactivity relates to subatomic particles, particularly the nucleus. The definitions of terms such as "mass numbers" and "isotopes" in Chapter 2 are applied immediately to radioisotopes. It is a continuation of a study of the behavior of atoms, in this case radioactive ones, and their subatomic particles. A new health note discusses the controversy of food irradiation, and the SI units for measurement of radiation have been added.

Chapter 4: Compounds and Their Bonds has a revised order. The discussion of bonding and naming of binary ionic and covalent compounds now precedes the topics of polarity and polyatomic ions in ternary compounds.

Chapter 5: Chemical Reactions and Quantities now begins with chemical change, equations, and balancing equations. A new section on types of reactions has been added along with a section on oxidation–reduction and combustion reactions. Some problems now include biochemistry examples. The mole concept is then introduced with a final, optional section on using the mole to calculate the mass of reactants and products in a reaction. The discussion of molar mass has been rewritten and the term "formula mass" deleted.

Chapter 6: Energy and States of Matter presents energy, temperature equations, heat and nutrition calculations, and states of matter later in the sequence so they can be combined with energy in reactions and chemical equilibrium. A section on melting and freezing with the heat of fusion is followed by a separate section on boiling and condensation with the heat of vaporization. Calculations for energy in changes of states have been simplified and several steps omitted. The new section on chemical equilibrium ties together rates of reactions and reversible reactions and discusses the effects of adding or removing a reactant or product, and changing temperature on equilibrium. The concept of equilibrium is applied to biological systems with the examples of hemoglobin and oxygen and carbon monoxide.

Chapter 7: Gases retains the sequence from the sixth edition except that the equation and calculations using the ideal gas law have been deleted as suggested by several reviewers. More conceptual problems have been added and some of the mathematical discussion omitted.

Chapter 8: Solutions now combines the section on solutes and solvents with polar and nonpolar solutes and solvents. A discussion on electrolytes and nonelectrolytes now follows the discussion on forming solutions with ionic substances. The topics of colloids, suspensions, osmosis, and dialysis are combined with percent concentration and precede the introduction of molarity. The discussion of molarity at the end of the chapter is in preparation for the use of molarity in acids and bases and pH calculation in the next chapter.

Chapter 9: Acids and Bases now uses the properties and theories of acids and bases to introduce the chapter. The dissociation of water and the topic of pH and the pH scale precede reactions of acids and bases. Students are encouraged to make pH indicators from fruits, vegetables, and flowers with intense colors.

Chapter 10: Alkanes and Aromatic Hydrocarbons introducing organic chemistry, now includes aromatic hydrocarbons. The haloalkane discussion has been moved to the next chapter so that Chapter 10 now deals only with compounds of carbon and hydrogen. Naming of compounds has been de-emphasized and the discussion of branched-chain alkanes has been rewritten. The reactions of alkanes is now limited to combustion. The writing of structural formulas of isomers was placed later in the chapter due to its complexity. The section on aromatic hydrocarbons has been shortened and complex-naming systems deleted.

Chapter 11: Haloalkanes, Alkenes, and Alkynes opens by describing functional groups and giving a list of the groups that will be discussed in *CHEMISTRY.* This addition is in response to reviewers who want to move faster through the organic material, concentrate more on functional groups, and limit the amount of time spent on naming compounds. Information about alkenes and alkynes and their naming is now combined into one section. A new environmental note describes the effect of ethene on the ripening process of fruits. The section on geometric isomers has been shortened and simplified. A new Explore Your World activity asks students to make models of cis–trans isomers using their hands or gumdrops and toothpicks. All the addition reactions are now in one section. A new section on addition polymers has been written in preparation for the idea of polymers in biomolecules. Recycling information is included here.

Chapter 12: Alcohols, Phenols, Ethers, Aldehydes, and Ketones combines the classification of alcohols with the oxidation of alcohols. IUPAC rules have been simplified. The oxidation of alcohol in the body is now part of the section on oxidation of alcohols. A discussion of aldehydes and ketones, including a new note on vanilla, now precedes the oxidation of alcohols. A new section on hemiacetals and hemiketals along with acetals and ketals completes the chapter. This is in preparation for the cyclic structures of monosaccharides in the following chapter on carbohydrates.

Chapter 13: Carbohydrates now immediately follows the organic chapter that describes the same functional groups found in carbohydrates. The discussion on optical activity has been deleted. Mutarotation is now discussed as an equilibrium system between the open chain and its anomers. Explore Your World activities include photosynthesis, sugars, sweetness of sugars, and the iodine test for polysaccharides.

Chapter 14: Carboxylic Acids, Esters, Amines, and Amides places carboxylic acids and their solubility into one section and simplifies naming. An Explore Your World activity asks students to list names and write structures of some carboxylic acids and esters that they identify in foods and cosmetics.

Chapter 15: Lipids presents a rewritten section on the physical properties of fatty acids. New Health Notes on omega-3 fatty acids, fat substitutes, and trans-fatty acids have been added. The discussion of plasma membranes expands on the role of phospholipids in the lipid bilayer of cell membranes. The structures of steroid compounds have been improved. Several Explore Your World activities on fats and oils and their solubility have been added.

Chapter 16: Amino Acids, Proteins, and Enzymes includes a rewritten section of the ionization of amino acids. The terms N- and C-terminus were deleted from the peptide discussion. Secondary structures now include alpha helix, beta-pleated sheet, and triple helix. Tertiary structures now include fibrous and globular proteins. A new discussion of specificity of enzymes has been added. The section on enzyme action is completely rewritten to clarify active site, lock-and-key model, and induced fit model, and uses redesigned illustrations.

Chapter 17: Metabolic Pathways and Energy (formerly Chapter 18) describes the metabolic relationship among the biomolecules discussed in the preceding chapters and their role in producing energy. A new section on coenzymes precedes the discussion of metabolic pathways. The sections on glycolysis and citric acid cycle have been rewritten and new figures drawn to give a better set of structures and pathway steps. Complex equations have been simplified. The section on the electron transport chain has been completely rewritten with electron carriers described first, followed by the electron transfer sequence using enzyme clusters and the concept of proton pumps. The detailed discussion of fatty acid synthesis has been deleted. The amino acids in transamination and oxidation deamination have been written as their anions to make their structures consistent with the anions in the citric acid cycle. Several drawings have been reworked to promote better understanding.

Chapter 18: Nucleic Acids (formerly Chapter 17) describes how DNA controls the metabolic processes in the cells. The discussion on recombinant DNA has been rewritten and expanded as a complete section with a new photograph of DNA fingerprinting.

Instructional Package

The Media Update of the seventh edition of *CHEMISTRY: An Introduction to General, Organic, and Biological Chemistry* is the nucleus of an integrated teaching and learning package of support materials for both professors and students.

For Students

Laboratory Manual, seventh edition by Karen Timberlake. Coordinates the topics and the sequence in *CHEMISTRY* to use new terms during the lab and to explore chemical concepts. Early experiments introduce students to basic laboratory skills, safety, and use of standard equipment. Laboratory investigations develop skills of manipulating laboratory equipment, gathering and reporting data, solving problems, making calculations, and drawing conclusions. Some questions at the end of labs require essay-type answers to promote writing skills in science. (0-321-03603-X). To inquire about adopting a customized version of the ***Laboratory Manual,*** please contact Addison Wesley customer service at (800) 282-0683 or visit us on the World Wide Web at www.aw.com/bc.

Essential Laboratory Manual by Karen Timberlake. Offering 21 experiments from the successful main edition of the ***Laboratory Manual,*** this slimmer version maintains the same quality and emphasis on critical thinking. Features safety procedures in the lab, pre-lab questions and report pages. (0-8053-2982-X)

Study Guide with Selected Solutions, seventh edition, by Karen Timberlake. Uses step-by-step learning exercises to review all the basic concepts for each chapter. Each *Study Guide* section is keyed to the learning goals in *CHEMISTRY* to provide a cross-reference with the working tools. The *Study Guide* is designed to promote active learning with a variety of exercises, all with answers, and a mastery test. After checking answers, students can determine if they have mastered the material or if they need further study. In this way, areas of difficulty can be identified and the material reviewed again. For the first time, complete solutions to the odd-numbered problems from *CHEMISTRY* are included for additional understanding of the solution process in problem solving. (0-321-03691-3)

Chemistry Study Pack, Version 2.0 **CD-ROM** by Tom Hall (Windows only). This CD-ROM contains 18 interactive, multimedia tutorials and Practice Tests, which are based on the 18 chapters in this textbook. The tutorials guide the student through the main concepts of the book, making use of video clips from *The World of Chemistry* (by The Annenberg/CPB Project) and color graphics from the book. The students practice what they learn by taking a 5-question Quick Quiz for each of the 145 sections of the book, and a 10-question Chapter Quiz at the end of each tutorial. To check their knowledge, students can take a Practice Test of up to 40 questions. The tests are scored, so the student knows how he or she is progressing. The CD can be used alone or with the *Study Guide.* The CD is included with the Media Update, but you can also order it as a stand-alone. (0-321-02977-1)

The Chemistry of Life for Introductory Chemistry CD-ROM by Robert M. Thornton of University of California, Davis. (For Mac and Windows) This lively tutorial teaches the essential concepts of chemistry through simulated lessons and interactive quizzes. With correlations to *CHEMISTRY,* this CD will help students picture atomic structure, reactions and equilibrium, properties of water, acids and bases, and the structure and function of macromolecules in a whole new light. The CD is included with the Media Update, but you can also order it as a stand-alone. (0-8053-3109-3)

For Instructors

The following items are available only to qualified adopters. To access online instructor resources, contact your Benjamin/Cummings sales representative. Visit www.aw.com and click "Find your rep." For further information or to order any of our products, please visit us at www.aw.com/bc on the World Wide Web, visit your campus bookstore, or contact our customer service department at (800) 282-0683.

Online Instructor Resources by Karen Timberlake. The Chemistry Place, Special Edition for Timberlake includes many online resources for instructors. The Instructor's Manual highlights chapter topics, provides suggestions for laboratory and lecture demonstration, and includes answers and solutions to problems in *CHEMISTRY.* Lab Exercises for both versions of the Laboratory Manual provide solutions to lab exercises. PowerPoint slides can be downloaded for classroom presentations. Visit www.chemplace.com/college/timberlake.

Benjamin Cummings Science Digital Library. This easy-to-search CD-ROM includes every figure and table from the book. The images can be exported into PowerPoint presentations, other presentation tools, and web sites. (0-8053-2981-1)

Test Bank by Karen Timberlake and Karen Wiechelman (Southeastern Louisiana University). Enhanced to include 35% more multiple-choice, true-false, and matching questions and their answers for a total of 850 test items. Available in a printed version (0-321-03694-8) as well as on TestGen-EQ software for Windows (0-321-03755-3) and Macintosh (0-321-03756-1) that lets instructors scramble questions, add new questions, and select questions based on the level of difficulty.

Transparency Acetates A set of 140 full-color transparency acetates contains key figures, tables, and illustrations from *CHEMISTRY.* (0-321-03692-1)

Videotape

Promoting Student Success in Chemistry by Karen Timberlake, developed with the University of California, Los Angeles. Shows actual student-centered classroom activities, including problem solving using collaborative learning in a large lecture setting, peer presentations, and study teams in discussion hours. (0-321-03759-6)

Acknowledgments

I wish to thank my husband, Bill, for his enthusiasm, invaluable assistance and suggestions, loving support, and late dinners. Thanks and appreciation to our son, John, for his understanding and support of my many hours spent writing. You are both dear to me and a precious part of my life.

This book is the result of the dedication of many people who put in hours of tireless effort to produce a high-quality book that provides an outstanding learning package. I am most appreciative of the work of the editorial team at AWL/BC. Robin Heyden, publisher, continually encouraged me on the progress of this new edition. Ben Roberts, sponsoring editor, supported my vision of a new format and learning package. Lisa Weber, production editor, brilliantly coordinated all phases of the manuscript to final pages. Anita Wagner, copy editor and art supervisor, precisely edited the manuscript and supervised a delightful art program. Kathleen Cameron, photo editor, and Vivian McDougal, prepress supervisor and emergency art coordinator, and Rolin Graphics were absolutely invaluable. In addition, I applaud the work of Robine Andrau and the composition team at Pre-Press Company who made sense of all my changes and created a beautiful text. I thank all of you for your contributions to the production of this edition.

CHEMISTRY also reflects the contributions of many professors who took the time to review and edit the manuscript, providing outstanding comments, help, and suggestions. In particular, I am indebted to accuracy checkers Dr. Elizabeth Friedman (Los Angeles Valley College) and Karen Wiechelman (University of Southwestern Louisiana) for their eagle eyes and sound suggestions. Finally, I am grateful to an incredible group of peers (see page xxviii) for carefully assessing all the new ideas for *CHEMISTRY,* suggesting additions, corrections, changes, and deletions, and for providing invaluable feedback about the best direction for the seventh edition. I appreciate every one of you.

Karen C. Timberlake
Los Angeles Valley College
Department of Chemistry
5800 Fulton Avenue
Valley Glen, CA 91401
Email: khemist@aol.com
karen@karentimberlake.com

Reviewers

SEVENTH EDITION REVIEWERS

Jim Armstrong
City College of San Francisco

Jack Bissey
Rio Hondo College

Florence (Nini) Cardoza
Sierra College

Ana Ciereszko
Miami-Dade Community College North

Vanessa Dykstra
Tarrant County Junior College Southeast

Elizabeth Friedman
Los Angeles Valley College

Harold Gladstone
Middlesex County College

Kevin A. Gratton
Johnson County Community College

Jingrong Huang
Borough of Manhattan Community College

Marvin Jaffe
Borough of Manhattan Community College

Mushtaq Kahn
Union County College

Sharon Kapica
County College of Morris

Catherine MacGowan
Armstrong Atlantic State University

Susan May
East New Mexico University

Laura Precedo
Broward Community College

Mona Rampy
Columbia Basin College

Joel Shelton
Bakersfield College

Rex Smith
Skyline College

Harry Ungar
Cabrillo College

Phil Wenzel
Monterey Peninsula College

Karen Wiechelman
University of Southwestern Louisiana

Donald Williams
Hope College

Linda Wilson
Middle Tennessee State University

Pamela Zelmer
Miami Dade Community College

SIXTH EDITION REVIEWERS

Mary Lee Abkemeier
La Guardia Community College

Edith Bartley
Tarrant County Junior College South

Thomas P. Carey
Berkshire Community College

Raymond Chamberlain
City College of San Francisco

John Ferrara
Cuyahoga Community College East

Kevin A. Gratton
Johnson County Community College

Ann A. Hicks
Thomas More College

Terri Jakuboski
Lewis and Clark Community College

Mani Jayaswal
Florida Community College

Craig R. Johnson
Carlow College

Sharon Kapica
County College of Morris

Edward King
South Plains College

Victor N. Kingery
Garland County Community College

Joanne Kirvaitis
Moraine Valley College

Bernard Koser
Westchester Community College

William Lehman
Johnson County Community College

Irving Lillien
Miami Dade Community College South

David Macaulay
William Rainey Harper College

Voya Moon
Western Iowa Tech College

Deborah Nycz
Broward Community College Central

Thomas Nycz
Broward Community College North

John A. Paparelli
San Antonio College

Phil Reedy
Metropolitan Community College

Diane Riebeth
Riverside Community College

David Saltzman
Santa Fe Community College

Dennis Sardella
Boston College

Somnath Sengupta
Quincy College

S. B. Sharma
Columbus College

Ursula Simonis
San Francisco State University

Darold Skerritt
Hartnell College

Robert Smith
Skyline College

Roy Stein
University of Toledo Community and Technical College

Michael J. Uricheck
Western Connecticut State University

Mary Vennos
Essex Community College

Dave Waggoner
Monroe County Community College

David Ziegler
Hannibal-LaGrange College

W.C. Zipperer
Armstrong College

CHEMISTRY STUDY PACK CD-ROM REVIEWERS

Cheryl Baxa
Pine Manor College

Ana Ciereszko
Miami-Dade Community College North

Kevin A. Gratton
Johnson County Community College

Gary O. Gray
Southwest Baptist University

Sharon Kapica
County College of Morris

Susan May
East New Mexico University

Bill Peacy
Olney Central College

Rex Smith
Skyline College

Linda A. Wilson
Middle Tennessee State University

Bill Younger
Red River College

1

Measurements

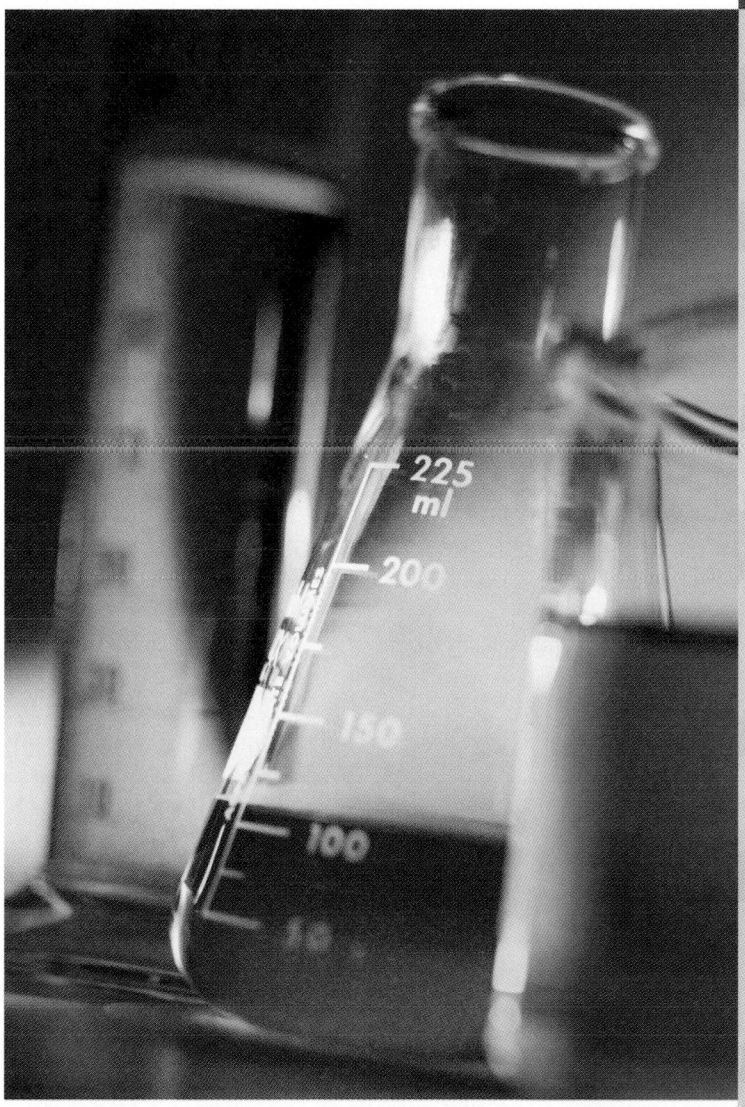

The metric system is used worldwide for measurement by scientists and health professionals. On this Erlenmeyer flask, volume is marked in milliliters (mL), a metric unit.

Looking Ahead

1.1	Units of Measurement
1.2	Scientific Notation
1.3	Measured and Exact Numbers
1.4	Significant Figures
1.5	Significant Figures in Calculations
1.6	SI and Metric Prefixes
1.7	Conversion Factors
1.8	Using Conversion Factors
1.9	Density

www.chemplace.com/college/timberlake

Visit the web site above to test yourself, run simulations, and review key concepts from this chapter. For even more interactive review and quizzing, use the two CD-ROMs included with the book.

What kinds of measurements did you make today? Perhaps you checked your weight by stepping on a scale this morning. Perhaps you measured out 2 cups of flour and a cup of milk to make your pancake batter. You may have filled the gas tank of your car with 10 gallons of gasoline and driven 12 miles to school. In each case, you were measuring mass (weight), volume, or distance. Try to recall some of the measurements you made today.

Men and women in the health sciences use measurement every day to evaluate the health of a patient. Temperatures are taken, weights and heights are recorded, and samples of blood and urine are collected for laboratory testing. Medications are given in dosages that must be measured accurately. In the dental office, hygienists measure solutions used in fluoride treatments and in the preparation of dental materials.

1.1

Study Pack

1.1 Units of Measurement

Learning Goal

Write the names and abbreviations for the metric or SI units used in measurements of length, volume, and mass.

Suppose that during a recent medical examination the nurse recorded your mass (weight) as 70.0 kilograms (70.0 kg), your height as 1.78 meters (1.78 m), and your temperature as 37.0 degrees Celsius (37.0°C). The system of measurement used in these clinical evaluations is the *metric system.* Perhaps you would be more familiar with these measurements if they were stated in the U.S. system. Then your weight would be 154 pounds (154 lb), your height 5 feet 10 inches (5 ft 10 in.), and your temperature 98.6 degrees Fahrenheit (98.6°F).

The **metric system** is used by scientists and health professionals throughout the world. It is also the common measuring system in all but a few countries in the world. In 1960, a modification of the metric system called the *International System of Units,* Système International **(SI),** was adopted to provide additional uniformity. In this text, we will use metric units and introduce some of the SI units that are in use today.

Length

The **meter (m)** is used to measure length in both the metric system and SI. It is 39.4 inches (in.), which makes a meter slightly longer than a yard (yd). A smaller unit of length, the **centimeter (cm),** is about as wide as your little finger. For comparison, there are 2.54 cm in 1 in.

1 m = 39.4 in.

2.54 cm = 1 in.

Figure 1.1 compares metric and U.S. units for length.

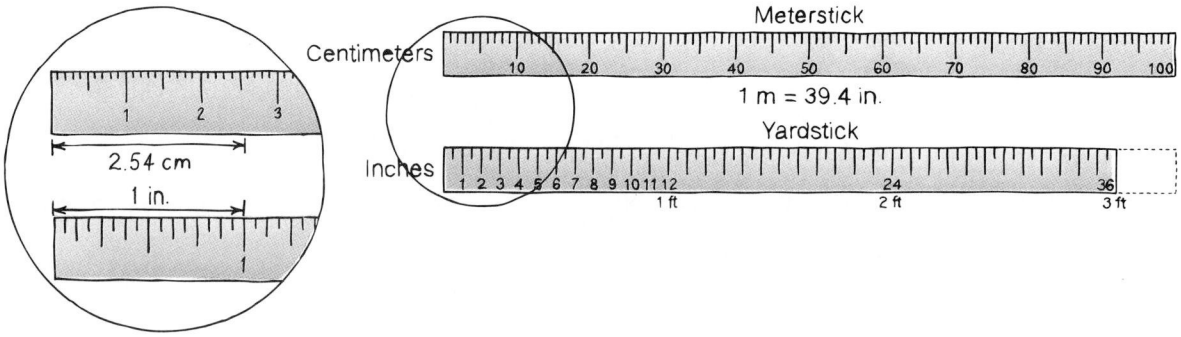

Figure 1.1 Comparison of the metric and U.S. units for measuring length.

Volume

Volume is the amount of space occupied by a substance. The metric unit *liter* is commonly used to measure volume. A **liter (L)** is slightly larger than the quart (qt), which is used in the U.S. system. The **milliliter (mL)** is more convenient for measuring smaller volumes of fluids in hospitals and laboratories.

1 L = 1.06 qt

946 mL = 1 qt

A comparison of metric and U.S. units for volume appears in Figure 1.2.

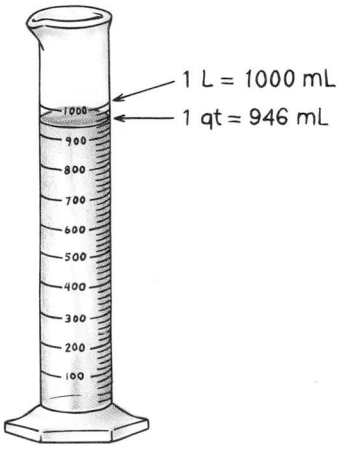

Figure 1.2 Comparison of the metric and U.S. units for measuring volume.

Mass

The **mass** of an object is a measure of the quantity of material it contains. Everything has mass, including rocks, water, people, and dogs. In the metric system, the unit for mass is the **gram (g).** The SI unit of mass, the **kilogram (kg),** is used for larger masses such as body weight. It takes 2.20 lb to make 1 kg, and 454 g is needed to equal 1 pound.

1 kg = 2.20 lb

454 g = 1 lb

Figure 1.3 illustrates the relationship between kilogram and pounds. A summary of some metric and SI units is given in Table 1.1.

You may be more familiar with the term *weight* than with mass. Weight is a measure of the gravitational pull on an object. On Earth, an astronaut with a mass of 75.0 kg has a weight of 165 lb. On the moon where the gravitational pull is one-sixth that of Earth, the astronaut has a weight of 27.5 lb. However, the mass of the astronaut is the same as on Earth, 75.0 kg. Scientists measure mass rather than weight because mass does not vary.

Table 1.1 Units of Measurement

Measurement	Metric	SI
Length	Meter (m)	Meter (m)
Volume	Liter (L)	Cubic meter (m^3)
Mass	Gram (g)	Kilogram (kg)
Time	Second (s)	Second (s)
Temperature	Celsius (°C)	Kelvin (K)

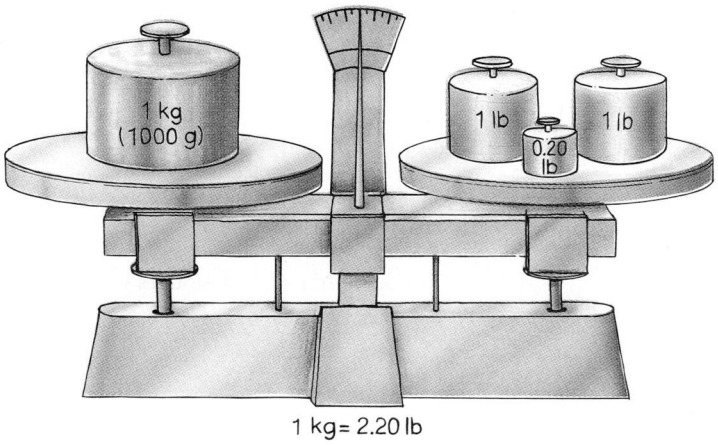

Figure 1.3 Comparison of the metric and U.S. units for measuring mass.

Figure 1.4 The mass of a sample is given as a digital readout on an electronic balance.

1.1

In a chemistry laboratory, a balance is used to measure the mass of a substance as shown in Figure 1.4.

Temperature

You probably use a thermometer to see how hot something is, or how cold it is outside, or perhaps to determine if you have a fever. (See Figure 1.5.) The **temperature** of an object tells us how hot or cold that object is. In the metric system, temperature is measured on the Celsius temperature scale. On a **Celsius (°C)** thermometer, water has a freezing point of 0°C and a boiling point of 100°C. On a Fahrenheit (°F) thermometer, water has a freezing point of 32°F and a boiling point of 212°F. In the SI system, temperature is based on the **kelvin (K)** temperature scale, where the lowest temperature possible is assigned as 0 K. We will discuss the relationships between these temperature scales in Chapter 6.

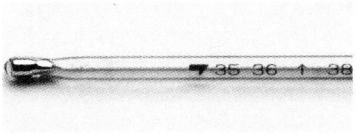

Figure 1.5 A Celsius thermometer measures a person's body temperature.

S·A·M·P·L·E 1.1 ·P·R·O·B·L·E·M

Units of Measurement

Complete the following table:

Type of Measurement	Metric Unit	Abbreviation
length	_____	_____
_____	liter	_____
_____	_____	g

Solution

length	meter	m
volume	liter	L
mass	gram	g

Study Check

What type of measurement uses the unit centimeter?

Answers to the Study Checks can be found at the end of this chapter in the Answer section. Checking your answers will let you know if you understand the material in this section.

EXPLORE YOUR WORLD

Units Listed on Labels

Read the labels on a variety of products such as sugar, salt, soft drinks, vitamins, and dental floss. List the metric or SI amounts of the products. Example: A box of sugar contains 454 g.

Questions

1. What metric or SI units of measurement are listed on the labels?
2. What type of measurement (mass, volume, etc.) do they indicate?
3. Write the metric or SI amounts in terms of a number plus a unit.

Q·U·E·S·T·I·O·N·S & P·R·O·B·L·E·M·S

Units of Measurement

Answers for odd-numbered Questions and Problems are given at the end of Chapter 1. Complete solutions to odd-numbered Questions and Problems are in the *Study Guide.*

1.1 At a grocery store in Mexico, tomatoes are priced by the kilogram, not by the pound. Why?

1.2 What are some typical units of measurement you might use in the kitchen?

1.3 Compare the units you would use and the units that a student in France would use to measure the following:

a. your weight **b.** your height
c. the amount of gasoline to fill the gas tank **d.** your temperature

1.4 Suppose that a friend tells you the following information. Why are the statements confusing and how would you make them clear?

a. I rode my bicycle for 15 today. **b.** My dog weighs 45.
c. It is hot today. It is 40. **d.** I lost 3.5 last week.

1.5 What are some common units used for measuring length in the metric system and the U.S. system?

1.6 What are some common units used for measuring mass in the metric system and the U.S. system?

1.7 What is the name of the unit and the type of measurement (mass, volume, length, or temperature) indicated for each of the following quantities?

a. 4.8 m **b.** 325 g **c.** 1.5 mL **d.** 480 m **e.** 28°C

1.8 What is the name of the unit and the type of measurement (mass, volume, length, or temperature) indicated for each of the following quantities?

a. 0.8 L **b.** 3.6 m **c.** 4 kg **d.** 35 g **e.** 373 K

1.2 Scientific Notation

Learning Goal

Write a number in scientific notation.

In chemistry and science, we use numbers that are very, very small to measure things as tiny as the width of a human hair, which is 0.000 008 m, or things that are even smaller. Or perhaps we want to know the number of hairs on the average scalp, which is about 100 000 hairs. In both measurements, it is convenient to write the numbers in scientific notation. (In this section we have added spaces to help make the places easier to count.)

Width of a human hair	0.000 008 m	8×10^{-6} m
Hairs on a human scalp	100 000 hairs	1×10^{5} hairs

Powers of 10

The number 1000 is the same as multiplying 10 three times. The three is written as a power (exponent) of 10.

Power of 10

$$1000 = 10 \times 10 \times 10 = 1 \times 10^{3}$$

1.2

Writing a Number in Scientific Notation

When a number is written in **scientific notation,** there are two parts: a coefficient and a power of 10. For example, the number 2400 is written as 2.4×10^3. The value 2.4 is the coefficient and 10^3 is the power of 10. The coefficient was determined by moving the decimal point to the left to give a number between 1 and 10. Counting the number of places (three) we moved the decimal gives the power of 10. For a large number, the power of 10 is always positive.

$$\underbrace{2400}_{\leftarrow\ 3\text{ places}} = 2.4 \times 1000 = \underbrace{2.4}_{\text{Number between 1 and 10}} \times \underbrace{10^3}_{\text{Power of 10}}$$

When a small number is written in scientific notation, the power of 10 will be negative. For example, 0.01 is the same as dividing by 10 two times, which is written as 10 with a power of −2.

$$0.01 = \frac{1}{10 \times 10} = \frac{1}{10^2} = 1 \times 10^{-2}$$

To write 0.0086 in scientific notation, we write two parts, 8.6×10^{-3}. The coefficient is 8.6 and 10^{-3} is the power of 10. For this small number, the decimal point was moved to the right three places, which gives 10^{-3} as the power of 10.

$$\underbrace{0.0086}_{3\text{ places}\ \rightarrow} = 8.6 \times \frac{1}{10 \times 10 \times 10} = \underbrace{8.6}_{\text{Number between 1 and 10}} \times \underbrace{10^{-3}}_{\text{Power of 10}}$$

S·A·M·P·L·E 1.2 P·R·O·B·L·E·M

Scientific Notation

Write the following measurements in scientific notation.

a. 350 g **b.** 0.000 16 L **c.** 5 220 000 m

Solution

a. 3.5×10^2 g **b.** 1.6×10^{-4} L **c.** 5.22×10^6 m

Study Check

Write the following measurements in scientific notation.

a. 425 000 m **b.** 0.000 000 8 g

Q·U·E·S·T·I·O·N·S & P·R·O·B·L·E·M·S

Scientific Notation

1.9 Write the following measurements in scientific notation:

a. 55 000 m **b.** 480 g
c. 0.000 005 cm **d.** 0.000 14 s

1.10 Write the following measurements in scientific notation:

a. 180 000 000 g **b.** 0.000 06 m
c. 750 000 g **d.** 0.15 m

1.11 Which number in each pair is larger?
a. 7.2×10^3 or 8.2×10^2 **b.** 4.5×10^{-4} or 3.2×10^{-2}
c. 1×10^4 or 1×10^{-4} **d.** 0.000 52 or 6.8×10^{-2}

1.12 Which number in each pair is smaller?
a. 4.9×10^{-3} or 5.5×10^{-9} **b.** 1250 or 3.4×10^2
c. 0.000 000 4 or 5×10^{-8} **d.** 4×10^8 or 4×10^{-10}

1.13 Place the following numbers in order from smallest to largest:
a. 1×10^{-3}, 1×10^{-6}, 1×10^3
b. 0.035, 3.5×10^4, 3.5×10^{-5}
c. 24 000, 3.5×10^3, 7.8×10^{-3}

1.14 Place the following numbers in order from smallest to largest:
a. 1×10^3, 1×10^{10}, 1×10^5
b. 2.5×10^4, 1.4×10^{-2}, 8.5×10^{-4}
c. 0.000 000 4, 4×10^{-3}, 4×10^1

1.3 Measured and Exact Numbers

Learning Goal
Identify a number as measured or exact.

Whenever you make a measurement, you use some type of measuring device. For example, you may use a meterstick to measure your height, a scale to check your weight, and a thermometer to take your temperature. **Measured numbers** are the numbers you obtain when you measure a quantity such as your height, weight, or temperature.

Measured Numbers

Suppose you are going to measure the lengths of the dark lines in Figure 1.6. You would select a ruler with a scale marked on it. By observing the lines on the scale, you determine the measurement for each line. Perhaps the divisions on the scale are marked as 1 cm. Another ruler might be marked in divisions of 0.1 cm. To report the length of the line, you would first read the numerical value of the marked line. Finally, you would make an *estimation* of the distance between the smallest marked lines. This estimated value is the final digit in a measured number.

An estimation is made by visually breaking up the space between the smallest marked lines, usually into tenths. For example, in Figure 1.6a, the end of the line falls between the lines marked 4 cm and 5 cm. That means that the length of the line is 4 cm

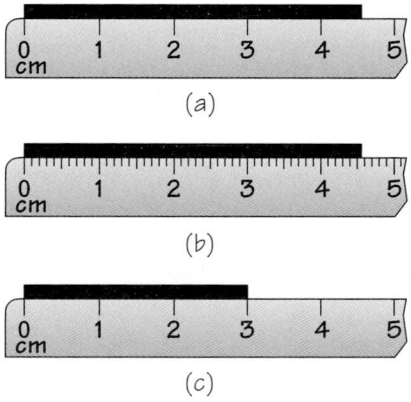

Figure 1.6 The lengths of the dark lines were measured as **(a)** 4.5 cm, **(b)** 4.55 cm, and **(c)** 3.0 cm. What is the estimated figure in each measurement? Why?

plus an estimated digit. If you estimate that the line is halfway between 4 cm and 5 cm, you would report its length as 4.5 cm. However, someone else might report the length of the line as 4.4 cm. The last digit in measured numbers will differ because people do not estimate in the same way. The ruler shown in Figure 1.6b is marked with lines at 0.1 cm. With this ruler, you can estimate to the 0.01-cm place. Perhaps you would report the length of the line now as 4.55 cm, while someone else may report its length as 4.56 cm. Both results are acceptable.

Because there are variations in estimation, there is always some *uncertainty* in every measurement. Sometimes a measurement ends right on a marked line. In order to indicate the uncertainty of the measurement, a zero is written as the estimated figure. The zero in the uncertainty place indicates that the measurement was not above or below the value of that line. For example, in Figure 1.6c, we would write the measurement as 3.0 cm to indicate that the length is not 2.9 cm or 3.1 cm.

EXPLORE YOUR WORLD

Temperature Scales on Thermometers

Obtain some different thermometers such as an oral thermometer, a meat thermometer, a candy thermometer, an outdoor thermometer, or a car temperature gauge. Determine the temperature unit for each and the temperature range from low to high for each.

Questions

1. What is the value of the marked lines on the temperature scale for each thermometer?
2. What is the estimated place that would be reported in a measurement from the scale on each type of thermometer?

Exact Numbers

Exact numbers are numbers we obtain by counting items or from a definition that compares two units in the same measuring system. Suppose a friend asks you to tell her the number of coats in your closet or the number of bicycles in your garage or the number of classes you are taking in school. To do this you must have counted the items. It was not necessary for you to use any type of measuring tool. Suppose an instructor asks you to state the number of seconds in one minute or the number of hours in one day. Again without using any measuring device, you know the answers. The relationships for time are defined as 60 seconds in one minute, and 24 hours in one day. For more examples of exact numbers, see Table 1.2.

S·A·M·P·L·E 1.3 P·R·O·B·L·E·M

Types of Numbers

Identify each of the following as an exact or a measured number:

a. number of eggs in a 3-egg omelet
b. amount of coffee needed to make 10-cup pot of coffee
c. number of tea bags needed to make a pot of tea
d. amount of sugar needed to sweeten the coffee or tea

Solution

a. exact **b.** measured **c.** exact **d.** measured

Study Check

Are the 2 pieces of toast or the 6 pancakes you ate for breakfast exact or measured numbers?

Table 1.2 Examples of Some Exact Numbers

	Defined Equalities	
Counted Numbers	**U.S. System**	**Metric System**
Eight doughnuts	1 ft = 12 in.	1 L = 1000 mL
Two baseballs	1 qt = 4 cups	1 m = 100 cm
Five capsules	1 lb = 16 ounces	1 kg = 1000 g

QUESTIONS & PROBLEMS

1.4

Measured and Exact Numbers

1.15 What is the estimated digit in each of the following measured numbers?
a. 8.6 m **b.** 45.25 g **c.** 25.0°C

1.16 What is the estimated digit in each of the following measured numbers?
a. 125.04 g **b.** 5.057 m **c.** 525.8°C

1.17 There are four chairs at the dining table. Why is the number 4 here called an exact number?

1.18 There are 7 days in one week. Why is the number 7 here called an exact number?

1.19 Are the numbers in each of the following statements measured or exact?
a. A patient weighs 155 lb.
b. The basket holds 8 apples.
c. In the metric system, 1 kg is equal to 1000 g.
d. The distance from Denver, Colorado, to Houston, Texas, is 1720 km.

1.20 Are the numbers in each of the following statements measured or exact?
a. There are 31 students in the laboratory.
b. The oldest known flower lived 120 000 000 years ago.
c. The largest gem ever found, an aquamarine, has a mass of 104 kg.
d. A laboratory test shows a blood cholesterol level of 184 mg/dL.

1.4 Significant Figures

Learning Goal

Determine the number of significant digits in measured numbers.

When we work in a medical environment or in a laboratory, it is important to make measurements that are accurate and that can be duplicated. A measurement is **accurate** when it is close to the real size or actual value of what we are measuring. That means we need to be careful when we determine a patient's temperature or take a blood pressure. It is also important to use a measuring instrument that is properly adjusted. Using a measuring device that is not in good working order will not give an accurate result even with careful measuring.

A measurement is **precise** if it can be repeated many times with results that are very close in value to each other. For example, suppose that four students measure your height as 152.3 cm, 152.4 cm, 152.1 cm, and 152.3 cm. We would say that the measurements were precise. If your actual height is 152.3 cm, each of the measurements was also accurate.

S·A·M·P·L·E· 1.4 ·P·R·O·B·L·E·M

Accuracy and Precision

The standard diameter for a large pizza at the Pizza Shack is 14 inches. However, the new pizza maker has prepared four pizzas that are each 10 inches in diameter.

a. Is the size of the pizzas accurate? Why?

b. Is the size of the pizzas precise? Why?

Solution

1.4

a. No. They are not accurate because the diameter of each pizza is not close to the standard diameter required for a large pizza.

b. Yes. They are precise because all of the pizzas are close in size to each other.

Study Check

What size do the large pizzas need to be to be both accurate and precise?

Counting Significant Figures

In measured numbers, the **significant figures** are all the reported numbers including the estimated digit. When we do calculations, we will need to count the significant figures in each of the measured numbers used in the calculations. All *nonzero* numbers are counted as significant figures. Zeros may or may not be significant depending on their position in a number. Table 1.3 gives the rules and examples of counting significant figures.

When one or more zeros in a large number are significant digits, writing the number in scientific notation can show them more clearly. For example, if the first zero in the measurement 500 m is significant, then it is shown by writing 5.0×10^2 m. In this text, we will also place a decimal point after a zero if that zero is part of the measurement. For example, writing the measurement 250. g would indicate that the zero is included

Table 1.3 Significant Figures in Measured Numbers

Rule	Measured Number	Number of Significant Figures
1. A number is a *significant figure* if it is:		
a. a nonzero digit	4.5 g	2
	122.35 m	5
b. a zero between nonzero digits	205 m	3
	5.082 kg	4
c. a zero at the end of a decimal number	50. L	2
	25.0°C	3
	16.00 g	4
d. any digit in the coefficient of a number written in scientific notation	4.0×10^5 m	2
	5.70×10^{-3} g	3
2. A number is *not significant* if it is:		
a. a zero at the beginning of a decimal number	0.0004 lb	1
	0.075 m	2
b. a zero used as a placeholder in a large number without a decimal point	850 000 m	2
	1 250 000 people	3

in the three significant figures. It could also be written as 2.50×10^2 g. Unless noted otherwise, we will assume that zeros at the end of large numbers are not significant.

Number	Number of Significant Digits	Scientific Notation
400 000 g	1	4×10^5 g
400 000 g	2	4.0×10^5 g
400 000 g	3	4.00×10^5 g

S·A·M·P·L·E **1.5** ·P·R·O·B·L·E·M

Significant Figures

Identify each of the following numbers as measured or exact; give the number of significant figures in each of the measured numbers.

a. 42.2 g **b.** 3 eggs **c.** 0.0005 cm
d. 450,000 km **e.** 9 planets

Solution

a. measured; three **b.** exact **c.** measured; one
d. measured; two **e.** exact

Study Check

State the number of significant figures in each of the following measured numbers:

a. 0.00035 g **b.** 2000 m **c.** 2.0045 L

Q·U·E·S·T·I·O·N·S & P·R·O·B·L·E·M·S

Significant Figures

1.21 Why do measured numbers have a limited number of significant figures?

1.22 When are zeros in a measured number counted as significant figures? When are they not significant?

1.23 How many significant figures are in each of the following measured quantities?

a. 11.005 g **b.** 0.000 32 m **c.** 36 000 000 km
d. 1.80×10^1 g **e.** 0.8250 L **f.** 30.0°C

1.24 How many significant figures are in each of the following measured quantities?

a. 20.60 mL **b.** 1036.48 g **c.** 4.00 m
d. 20.8°C **e.** 60 800 000 g **f.** 5.0×10^{-3} L

1.25 Write each of the following numbers in scientific notation with two significant figures:

a. 5000 L **b.** 30 000 g **c.** 100 000 m

1.26 Write each of the following numbers in scientific notation with three significant figures:

a. 5 100 000 **b.** 2600 **c.** 40 000

Study Pack

Learning Goal

Adjust calculated answers to the correct number of significant figures.

1.5

1.5 Significant Figures in Calculations

In science and medicine, we measure many things: the length of a bacterium, the volume of a medication, and the mass of cholesterol in a blood sample. The numbers from these measurements are often used in calculations. However, the calculation cannot change the precision of a measured number. Therefore, we report a calculated answer that reflects the precision of the original measurements.

Using a calculator will usually help you do calculations faster than you can without one. However, calculators cannot think for you. It is up to you to enter the numbers correctly, press the right function keys, and adjust the calculator display to give the appropriate answer.

Rounding Off

If we calculate the area of a carpet that measures 5.5 m by 3.5 m, we obtain the number 19.25 as the square meters in the area. However, all four digits cannot be significant because an answer cannot be more precise than the measured numbers used. Because each of the measurement numbers has just two significant figures, the calculated result must be *rounded off* to give an answer that also has two significant figures: 19 m^2. A final answer *cannot be more precise than any measurement used in the calculation.*

For exact numbers we do not have to determine the number of significant figures in the final answer. There is no uncertainty in an exact number to affect the outcome of any calculation.

When you obtain a calculator result, determine the number of significant figures you should keep for the answer and round off the calculator results using the following rules:

Rules for Rounding Off

1. If the first digit to be dropped is *4 or less,* it and all following digits are simply dropped from the number.
2. If the first digit to be dropped is *5 or greater,* the last retained digit of the number is increased by 1.

		Three Significant Figures	Two Significant Figures
Example 1:	8.4234 rounds off to	8.42	8.4
Example 2:	14.780 rounds off to	14.8	15

S·A·M·P·L·E· 1.6 ·P·R·O·B·L·E·M

Rounding Off

Round off each of the following numbers to three significant figures:

a. 35.7823 m **b.** 0.002627 L
c. 3826.8 g **d.** 1.2836 kg

Solution

a. 35.8 m **b.** 0.00263 L
c. 3830 g **d.** 1.28 kg

Study Check

Round off each of the numbers in Sample Problem 1.6 to two significant figures.

Multiplication and Division

In multiplication and division, the final answer is written so it has the same number of digits as the measurement with the *fewest* significant figures (SFs).

Example 1

Multiply the following measured numbers: 24.65 × 0.67.

24.65	⊠	0.67	=	16.5155	⟶	17
Four SFs		Two SFs		Calculator display		Final answer, rounded to two SFs

The measurement 0.67 has the least number of significant figures, two. Therefore, the calculator answer is rounded off to two significant figures.

Example 2

Solve the following:

$$\frac{2.85 \times 67.4}{4.39}$$

To do this problem on a calculator, enter the number and then press the operation key. In this case, we might press the keys in the following order:

2.85	⊠	67.4	÷	4.39	=	43.756264	⟶	43.8
Three SFs		Three SFs		Three SFs		Calculator display		Final answer, rounded to three SFs

All of the measurements in this problem have three significant figures. Therefore, the calculator result is rounded off to give an answer, 43.8, that has three significant figures.

Adding Significant Zeros

Sometimes, a calculator displays a small whole number. To give an answer with the correct number of significant figures, significant zeros are written after the calculator result. For example, suppose the calculator display is 4, but you used measurements that have three significant numbers. Placing two significant zeros after the 4 gives the correct answer, 4.00.

$\frac{8.00}{2.00}$ =	4	⟶	4.00
Three SFs	Calculator display		Final answer, two zeros added to give three SFs

S·A·M·P·L·E· 1.7 ·P·R·O·B·L·E·M

Significant Figures in Measured Numbers

Perform the following calculations of measured numbers. Give the answers with the correct number of significant figures.

a. 56.8 × 0.37 **b.** $\frac{71.4}{11}$ **c.** $\frac{(2.075)(0.585)}{(8.42)(0.0045)}$ **d.** $\frac{25.0}{5.00}$

Solution

a. Calculator answer: 21.016 ⟶ 21
b. Calculator answer: 6.4909091 ⟶ 6.5
c. Calculator answer: 32.036817 ⟶ 32
d. Calculator answer: 5 ⟶ 5.00 (Must add significant zeros)

Study Check

Solve:

a. 45.26×0.01088 **b.** $2.6 \div 324$ **c.** $\dfrac{4.0 \times 8.00}{16}$

1.5 Addition and Subtraction

In addition or subtraction, the answer is written so it has the same number of decimal places as the measurement having the fewest decimal places.

Example 3

Add:

	2.045	Three decimal places
+	34.1	One decimal place
	36.145	Calculator display
	36.1	Answer, rounded to one decimal place

Example 4

Subtract:

	255	Ones place
−	175.65	Two decimal places
	79.35	Calculator display
	79	Answer, rounded to ones place

When numbers are added or subtracted to give answers ending in zero, the zero does not appear after the decimal point in the calculator display. For example, 14.5 g − 2.5 g = 12.0 g. However, if you do the subtraction on your calculator, the display shows 12. To give the correct answer, a significant zero is written after the decimal point.

Example 5

	14.5 g	One decimal place
−	2.5 g	One decimal place
	12.	Calculator display
	12.0 g	Answer, zero written after the decimal point

S·A·M·P·L·E 1.8 P·R·O·B·L·E·M

Significant Figures in Addition and Subtraction

Perform the following calculations and give the answer with the correct number of significant figures:

a. 27.8 cm + 0.235 cm
b. 104.45 mL + 0.838 mL + 46 mL
c. 153.247 g − 14.82 g

Solution

a. 28.0 cm **b.** 151 mL **c.** 138.43 g

Study Check

Solve:

a. 82.45 mg + 1.245 mg + 0.00056 mg **b.** 4.259 L − 3.8 L

QUESTIONS & PROBLEMS

Significant Figures in Calculations

1.27 Why do we usually need to round off calculations that use measured numbers?

1.28 Why do we sometimes add a zero to a number in a calculator display?

1.29 How is the number of significant figures determined for multiplication and/or division?

1.30 How is the number of significant figures determined for addition and/or subtraction?

1.31 Round off each of the following numbers to three significant figures.

a. 1.854 **b.** 184.2038 **c.** 0.004738265
d. 8807 **e.** 1.832149

1.32 Round off each of the numbers in problem 1.31 to two significant figures.

1.33 For the following problems, give answers with the correct number of significant figures:

a. 45.7×0.034 **b.** 0.00278×5

c. $\dfrac{34.56}{1.25}$ **d.** $\dfrac{(0.2465)(25)}{1.78}$

1.34 For the following problems, give answers with the correct number of significant figures:

a. 400×185 **b.** $\dfrac{2.40}{(4)(125)}$

c. $0.825 \times 3.6 \times 5.1$ **d.** $\dfrac{3.5 \times 0.261}{8.24 \times 20.0}$

1.35 For the following problems, give answers with the correct number of significant figures:

a. 45.48 cm + 8.057 cm **b.** 23.45 g + 104.1 g + 0.025 g
c. 145.675 mL − 24.2 mL **d.** 1.08 L − 0.585 L

1.36 For the following problems, give answers with the correct number of significant figures:

a. 5.08 g + 25.1 g **b.** 85.66 cm + 104.10 cm + 0.025 cm
c. 24.568 mL − 14.25 mL **d.** 0.2654 L − 0.2585 L

Study Pack

Learning Goal

Use the numerical values of prefixes to write a metric equality.

Table 1.4 Daily Values for Selected Nutrients

Nutrient	Amount Recommended
Protein	44 grams
Vitamin C	60 milligrams
Vitamin B_{12}	6 micrograms
Calcium	1 gram
Iron	18 milligrams
Iodine	150 micrograms
Sodium	2400 milligrams
Zinc	15 milligrams

1.6

1.6 SI and Metric Prefixes

The U.S. Food and Drug Administration has determined the daily values (DV) of nutrients for adults and children age 4 or older. Some of these recommended daily values are listed in Table 1.4.

The special feature of the metric system of units is that a **prefix** can be attached to any unit to increase or decrease its size by some factor of 10. For example, in the daily values, the prefixes *milli* and *micro* are used to make the smaller units, milligram (mg) and microgram (μg). Table 1.5 lists some of the metric prefixes, their symbols, and their decimal values.

The prefix *centi* is like cents in a dollar. One cent would be a centidollar or $\frac{1}{100}$ of a dollar. That also means that one dollar is the same as 100 cents. The prefix *deci-* is like dimes in a dollar. One dime would be a decidollar or $\frac{1}{10}$ of a dollar. That also means that one dollar is the same as 10 dimes.

The relationship of a unit to its base unit can be expressed by replacing the prefix with its numerical value. For example, when the prefix *kilo* in *kilometer* is replaced with its value of 1000, we find that a kilometer is equal to 1000 meters. Other examples follow.

1 **kilo**meter (1 km) = **1000** meters (1000 m)

1 **kilo**liter (1 kL) = **1000** liters (1000 L)

1 **kilo**gram (1 kg) = **1000** grams (1000 g)

S·A·M·P·L·E **1.9** ·P·R·O·B·L·E·M

Prefixes

Fill in the blanks with the correct numerical value:

a. kilogram = _____ grams
b. millisecond = _____ second
c. deciliter = _____ liter

Solution

a. The numerical value of *kilo* is 1000; 1000 grams.
b. The numerical value of *milli* is 0.001; 0.001 second.
c. The numerical value of *deci* is 0.1; 0.1 liter.

Study Check

Write the correct prefix in the blanks:

a. 1 000 000 seconds = _____ second
b. 0.01 meter = _____ meter

Measuring Length

An ophthalmologist may measure the diameter of the retina of the eye in centimeters (cm), whereas a surgeon may need to know the length of a nerve in millimeters (mm). When the prefix *centi* is used with the unit *meter,* it indicates the unit *centimeter,* a length that is one-hundredth of a meter (0.01 m). A *millimeter* measures a length of 0.001 m. There are 1000 mm in a meter.

If we compare the lengths of a millimeter and a centimeter, we find that 1 mm is 0.1 cm; there are 10 mm in 1 cm. These comparisons are examples of **equalities,** which

Table 1.5 Metric and SI Prefixes

Prefix	Symbol	Meaning	Numerical Value	Scientific Notation
Prefixes That Increase the Size of the Unit				
mega	M	one million	1 000 000	10^6
kilo	k	one thousand	1 000	10^3
Prefixes That Decrease the Size of the Unit				
deci	d	one-tenth	0.1	10^{-1}
centi	c	one-hundredth	0.01	10^{-2}
milli	m	one-thousandth	0.001	10^{-3}
micro	μ	one-millionth	0.000 001	10^{-6}
nano	n	one-billionth	0.000 000 001	10^{-9}

show the relationship between two units that measure the same quantity. For example, in the equality 1 m = 100 cm, each quantity describes the same length but in a different unit. Note that each quantity in the equality expression has both a number and a unit.

An Equality

First Quantity		Second Quantity	
1	m =	100	cm
↑	↑	↑	↑
Number +	unit	Number +	unit

Some Length Equalities

1 m = 100 cm

1 m = 1000 mm

1 cm = 10 mm

Some metric units for length are compared in Figure 1.7, and examples are shown in Figure 1.8.

Table 1.6 Some Typical Laboratory Test Values

Substance in Blood	Typical Range
Albumin	3.5–5.0 g/dL
Ammonia	20–150 μg/dL
Calcium	8.5–10.5 mg/dL
Cholesterol	105–250 mg/dL
Iron (male)	80–160 μg/dL
Protein (total)	6.0–8.0 g/dL

Measuring Volume

Volumes of 1 L or smaller are common in the health sciences. When a liter is divided into 10 equal portions, each portion is a deciliter (dL). There are 10 dL in 1 L. Laboratory results for blood work are often reported in deciliters. Notice the values listed in Table 1.6 for some substances in the blood.

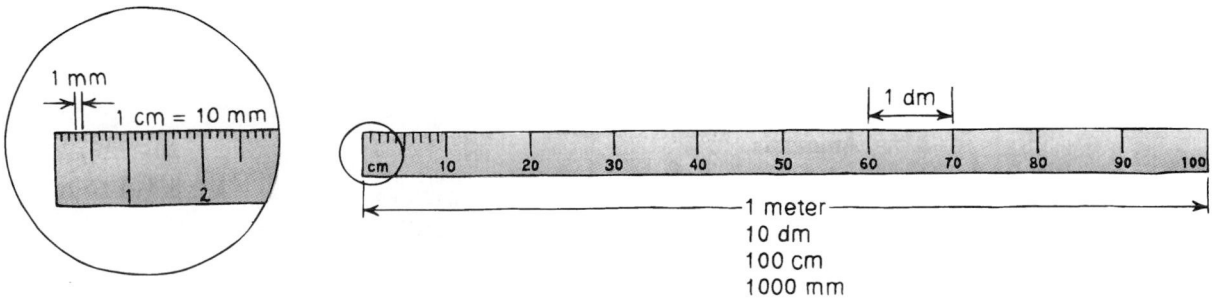

Figure 1.7 Comparison of some metric units used to measure length.

Figure 1.8 Examples of some metric measurements of length.

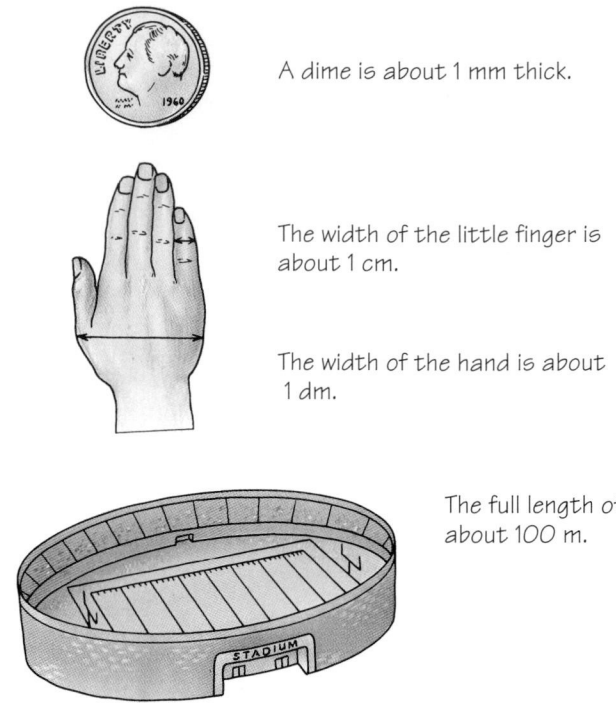

1.6

When a liter is divided into a thousand parts, each of the smaller volumes is called a milliliter. In a 1-L bottle of physiological saline, there are 1000 mL of solution. Bottles of intravenous (IV) liquids typically contain solution volumes of 500 mL or 250 mL. Small amounts of liquids measured in milliliters or microliters (μL) may be added to the IV solution or given by injection (see Figure 1.9).

Some Volume Equalities

1 L = 10 dL

1 L = 1000 mL

1 dL = 100 mL

The **cubic centimeter** (**cm^3** or **cc**) is the volume of a cube whose dimensions are 1 cm on each side. A cubic centimeter has the same volume as a milliliter, and the units are often used interchangeably.

$1\ cm^3 = 1\ cc = 1\ mL$

Figure 1.9 A technician measures a specific volume of blood.

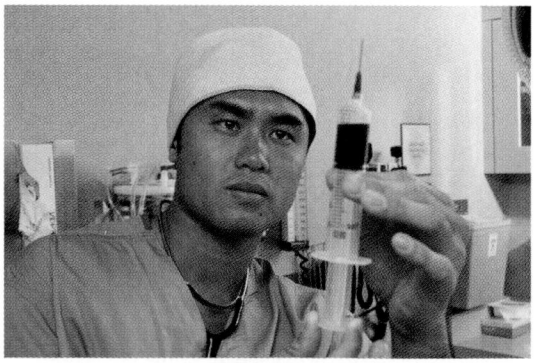

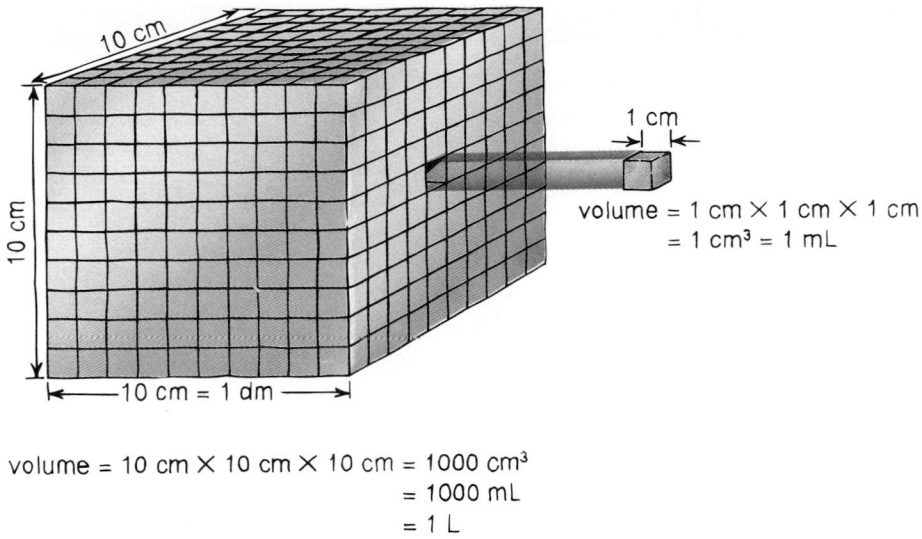

Figure 1.10 A cube measuring 10 cm on each side has a volume of 1000 cm^3, or 1 L; a cube measuring 1 cm on each side has a volume of 1 cm^3 or 1 mL.

When you see *1 cm,* you are reading about length; when you see *1 cc* or *1* cm^3 or *1 mL,* you are reading about volume. A comparison of units of volume is illustrated in Figure 1.10.

Measuring Mass

When you get a physical examination, your mass is recorded in kilograms, whereas the results of your laboratory tests are reported in grams, milligrams (mg), or micrograms (μg). A kilogram is equal to 1000 g. One gram represents the same mass as 1000 mg, and one mg equals 1000 μg.

Some Mass Equalities

1 kg = 1000 g
1 g = 1000 mg
1 mg = 1000 μg

S·A·M·P·L·E 1.10 P·R·O·B·L·E·M

Writing Metric Relationships

1. Identify the larger unit in each of the following pairs:
   a. centimeter or kilometer **b.** L or dL **c.** mg or μg
2. Complete the following list of metric equalities:
   a. 1 L = _____ dL **b.** 1 km = _____ m
   c. 1 m = _____ cm **d.** 1 cm^3 = _____ mL

Solution

1. **a.** kilometer **b.** L **c.** mg
2. **a.** 10 dL **b.** 1000 m **c.** 100 cm **d.** 1 mL

Study Check

Complete the following equalities:
a. 1 kg = _____ g **b.** 1 mL = _____ L

Q·U·E·S·T·I·O·N·S & P·R·O·B·L·E·M·S

SI and Metric Prefixes

1.37 What are some advantages and disadvantages of the United States converting completely to the metric (SI) system?

1.38 What parts of the U.S. currency system are similar to the units in the metric system?

1.7

1.39 The speedometer in the margin is marked in both km/h and mi/h or mph. What is the meaning of each abbreviation?

1.40 In Canada, a highway sign gives a speed limit as 80 km/h. According to the speedometer in problem 1.39, would you be exceeding the speed limit of 55 mph if you were in the United States?

1.41 How does the prefix *kilo* affect the gram unit in *kilogram*?

1.42 How does the prefix *centi* affect the meter unit in *centimeter*?

1.43 Write the abbreviation for each of the following units:
a. milligram **b.** deciliter **c.** kilometer
d. kilogram **e.** microliter

1.44 Write the complete name for each of the following units:
a. cm **b.** kg **c.** dL **d.** mm **e.** μg

1.45 Write the numerical values for each of the following prefixes:
a. centi **b.** kilo **c.** milli **d.** deci **e.** mega

1.46 Write the complete name (prefix + unit) for each of the following numerical values:
a. 0.10 g **b.** 0.000 001 g **c.** 1000 g
d. $\frac{1}{100}$ g **e.** 0.001 g

1.47 Complete the following metric relationships:
a. 1 m = _____ cm
b. 1 km = _____ m
c. 1 mm = _____ m
d. 1 L = _____ mL

1.48 Complete the following metric relationships:
a. 1 kg = _____ g
b. 1 mL = _____ L
c. 1 g = _____ kg
d. 1 g = _____ mg

1.49 For each of the following pairs, which is the larger unit?
a. milligram or kilogram
b. milliliter or microliter
c. cm or km

1.50 For each of the following pairs, which is the smaller unit?
a. mg or g
b. centimeter or millimeter
c. mm or μm

Study Pack

Learning Goal

Write a conversion factor for two units that describe the same quantity.

1.7 Conversion Factors

Many problems in chemistry and the health sciences require a change of units. You make changes in units every day. For example, suppose you spent 2.0 hours (hr) on your homework, and someone asked you how many minutes that was. You would answer 120 minutes (min). You knew how to change from hours to minutes because you knew an equality (1 hr = 60 min) that related the two units. To do the problem, the equality is written in the form of a fraction called a **conversion factor.** One of the quantities is the numerator, and the other is the denominator. Be sure to include the units when you write the conversion factors. Two factors are always possible from any equality.

Two Conversion Factors for the Equality 1 hr = 60 min

$$\frac{\text{Numerator}}{\text{Denominator}} \begin{matrix} \longrightarrow \\ \longrightarrow \end{matrix} \quad \frac{60\text{ min}}{1\text{ hr}} \quad \text{and} \quad \frac{1\text{ hr}}{60\text{ min}}$$

These factors are read as "60 minutes per 1 hour," and "1 hour per 60 minutes." The term *per* means "divide." Some common relationships and their corresponding conversion factors are given in Table 1.7. It is important that the equality you select to construct a conversion factor is a true relationship.

Metric–Metric Conversion Factors

We can write metric conversion factors for the metric relationships we have studied. For example, from the equality for meters and centimeters, we can write the following factors:

Metric Equality	Conversion Factors		
1 m = 100 cm	$\frac{100\text{ cm}}{1\text{ m}}$	and	$\frac{1\text{ m}}{100\text{ cm}}$

Both are proper conversion factors for the relationship; one is just the inverse of the other. The usefulness of conversion factors is enhanced by the fact that we can turn a conversion factor over and use its inverse. Table 1.8 lists some conversion factors from the metric relationships we have discussed.

Metric–U.S. System Conversion Factors

Suppose you need to convert from pounds, a unit in the U.S. system, to kilograms in the metric (or SI) system. A relationship you could use is

1 kg = 2.20 lb

Table 1.7 Some Common Equalities and Their Corresponding Conversion Factors

Equality	Conversion Factors		
1 yard = 3 feet	$\frac{1\text{ yd}}{3\text{ ft}}$	and	$\frac{3\text{ ft}}{1\text{ yd}}$
1 dollar = 100 cents	$\frac{1\text{ dollar}}{100\text{ cents}}$	and	$\frac{100\text{ cents}}{1\text{ dollar}}$
1 hour = 60 minutes	$\frac{1\text{ hr}}{60\text{ min}}$	and	$\frac{60\text{ min}}{1\text{ hr}}$
1 gallon = 4 quarts	$\frac{1\text{ gal}}{4\text{ qt}}$	and	$\frac{4\text{ qt}}{1\text{ gal}}$

1.7

Table 1.8 Some Useful Metric Conversion Factors

Metric Relationship (Equality)	Conversion Factors		
Length			
1 m = 1000 mm	$\frac{1\text{ m}}{1000\text{ mm}}$	and	$\frac{1000\text{ mm}}{1\text{ m}}$
1 cm = 10 mm	$\frac{1\text{ cm}}{10\text{ mm}}$	and	$\frac{10\text{ mm}}{1\text{ cm}}$
Volume			
1 L = 1000 mL	$\frac{1\text{ L}}{1000\text{ mL}}$	and	$\frac{1000\text{ mL}}{1\text{ L}}$
1 dL = 100 mL	$\frac{1\text{ dL}}{100\text{ mL}}$	and	$\frac{100\text{ mL}}{1\text{ dL}}$
Mass			
1 kg = 1000 g	$\frac{1\text{ kg}}{1000\text{ g}}$	and	$\frac{1000\text{ g}}{1\text{ kg}}$
1 g = 1000 mg	$\frac{1\text{ g}}{1000\text{ mg}}$	and	$\frac{1000\text{ mg}}{1\text{ g}}$

The corresponding conversion factors would be

$$\frac{2.20\text{ lb}}{1\text{ kg}} \quad \text{and} \quad \frac{1\text{ kg}}{2.20\text{ lb}}$$

Table 1.9 lists some useful metric–U.S. relationships, and Figure 1.11 illustrates the different units used to describe the contents of some packaged foods. Figure 1.12 shows the quantities of some common items in both metric and U.S. units.

Table 1.9 Some Metric–U.S. Relationships and Conversion Factors

Relationship	Conversion Factors		
Length			
2.54 cm = 1 in.	$\frac{2.54\text{ cm}}{1\text{ in.}}$	and	$\frac{1\text{ in.}}{2.54\text{ cm}}$
1 m = 39.4 in.	$\frac{1\text{ m}}{39.4\text{ in.}}$	and	$\frac{39.4\text{ in.}}{1\text{ m}}$
Volume			
946 mL = 1 qt	$\frac{946\text{ mL}}{1\text{ qt}}$	and	$\frac{1\text{ qt}}{946\text{ mL}}$
1 L = 1.06 qt	$\frac{1\text{ L}}{1.06\text{ qt}}$	and	$\frac{1.06\text{ qt}}{1\text{ L}}$
Mass			
454 g = 1 lb	$\frac{454\text{ g}}{1\text{ lb}}$	and	$\frac{1\text{ lb}}{454\text{ g}}$
1 kg = 2.20 lb	$\frac{1\text{ kg}}{2.20\text{ lb}}$	and	$\frac{2.20\text{ lb}}{1\text{ kg}}$

Figure 1.11 Metric and U.S. units describe the contents of some packaged foods.

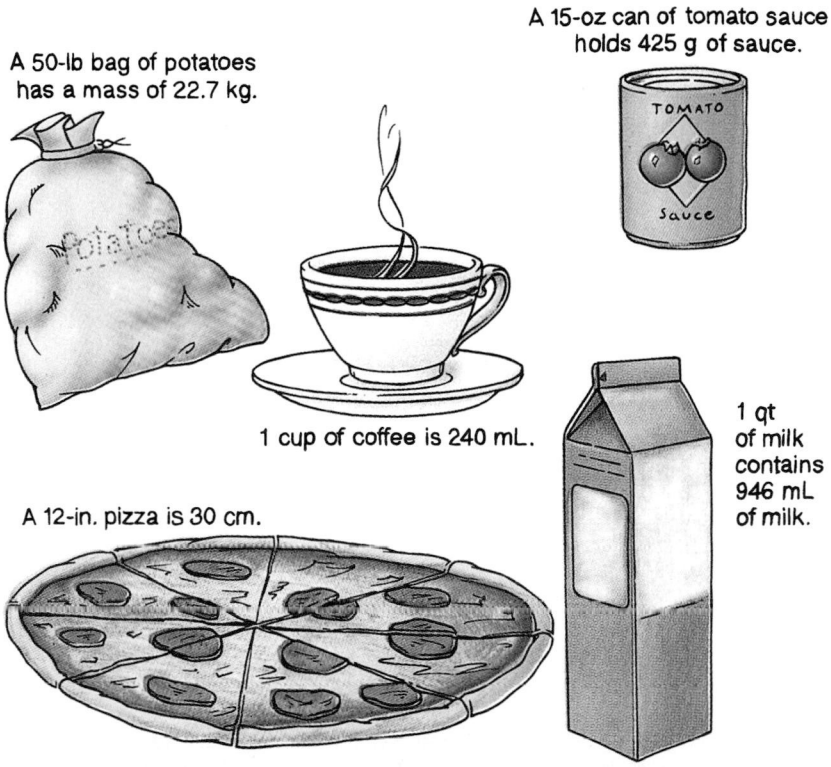

Figure 1.12 Metric and U.S. quantities for some common items.

EXPLORE YOUR WORLD

SI and Metric Equalities on Product Labels

Read the labels on some food products on your kitchen shelves and in your refrigerator. List the amount of product given in different units. Write a relationship for two of the amounts for the same product and container. Look for measurements of grams and pounds, or quarts and milliliters.

Questions

1. Use the stated measurement to derive a metric–U.S. conversion factor.
2. How do your results compare to the conversion factors we have described in this text?

1.7

S·A·M·P·L·E 1.11 P·R·O·B·L·E·M

Writing Conversion Factors for Equalities

Write conversion factors for the following equalities:

a. There are 1000 mg in 1 g.
b. One day has 24 hours.
c. There are 12 eggs in 1 dozen eggs.

Solution

Equality	Conversion Factors		
a. 1 g = 1000 mg	$\frac{1\text{g}}{1000\text{ mg}}$	and	$\frac{1000\text{ mg}}{1\text{g}}$
b. 1 day = 24 hr	$\frac{1\text{ day}}{24\text{ hr}}$	and	$\frac{24\text{ hr}}{1\text{ day}}$
c. 12 eggs = 1 dozen eggs	$\frac{12\text{ eggs}}{1\text{ dozen eggs}}$	and	$\frac{1\text{ dozen eggs}}{12\text{ eggs}}$

Study Check

Write the equality and conversion factors for a relationship between inches and centimeters.

Q·U·E·S·T·I·O·N·S & P·R·O·B·L·E·M·S

Conversion Factors

1.51 Why can two conversion factors be written for an equality such as 1 m = 100 cm?

1.52 How can you check that you have written the correct conversion factors for an equality?

1.53 In a cookbook, look up weights and measures and list some U.S. and metric equivalents. Rewrite them as conversion factors.

1.54 In a dictionary, look up measurement and list some U.S. and metric equivalents. Write each one as conversion factors.

1.55 For cooking spaghetti, the recipe says to add 1 teaspoon (tsp) salt to 2 quarts (qt) of water. How would you write this measurement in the form of conversion factors?

1.56 At the gas pump, the cost of 1 gallon (gal) of gas is $1.65. How would you write the cost per gallon of gas as conversion factors?

1.57 What equality is expressed by the conversion factor $\frac{1000\text{ g}}{1\text{ kg}}$?

1.58 What equality is expressed by the conversion factor $\frac{1\text{ m}}{100\text{ cm}}$?

1.59 Write a numerical relationship and conversion factors for each of the following statements:

a. One yard is 3 feet.
b. One mile is 5280 feet.
c. One minute is 60 seconds.
d. A car goes 27 miles on 1 gallon of gas.

1.60 Write a numerical relationship and conversion factors for each of the following statements:
a. One gallon is 4 quarts.
b. At the store, oranges are $1.29 per lb.
c. There are 7 days in 1 week.
d. One dollar has four quarters.

1.61 Write the numerical relationship and conversion factors for the following pairs of units:
a. centimeters and meters **b.** milligrams and grams
c. liters and milliliters **d.** deciliters and milliliters

1.62 Write the numerical relationship and corresponding conversion factors for the following pairs of units:
a. centimeters and inches **b.** pounds and kilograms
c. pounds and grams **d.** quarts and milliliters

1.8 Using Conversion Factors

Learning Goal

Use a conversion factor to change from one unit to another.

The process of problem solving in chemistry is much the same as the process you use when you work out everyday problems such as changing dollars to cents or hours to minutes. To illustrate this process, let's take a look at an everyday situation.

Suppose you go to the store to buy some apples. The sign at the fruit stand states that 1 lb of apples costs 48 cents as shown in Figure 1.13. When you weigh the apples, you have 2.0 lb. How much will you pay, in cents, for the apples? A close look at the way we think about this problem might show the following:

Step 1 The quantity 2.0 lb is the given or stated quantity.

Given: 2.0 lb apples

Figure 1.13 Converting units.

Step 2 It is helpful to decide on a unit plan. For this problem, our unit plan is to change from pounds to cents.

Unit Plan: lb ⟶ cents
Given — Unit for answer

Step 3 The relationship stated in the problem as 48 cents per pound of apples can be written in the form of conversion factors.

Relationship: 1 lb apples = 48 cents

Conversion Factors: $\frac{1 \text{ lb apples}}{48 \text{ cents}}$ and $\frac{48 \text{ cents}}{1 \text{ lb apples}}$

1.8

Step 4 Now we can write the setup for the problem using our unit plan (step 2) and the conversion factors (step 3). First, write down the given quantity, 2.0 lb. Then multiply it by the conversion factor that has the unit lb in the denominator because that will cancel out the given unit of lb in the numerator. The unit of cents in the numerator (top number) gives the correct unit for the answer.

Problem Setup:

Unit for answer goes here

$$2.0\cancel{\text{lb}} \times \frac{48 \text{ cents}}{1\cancel{\text{lb}}} = 96 \text{ cents}$$

Given (stated unit) — Conversion factor (cancels out given) — Answer (desired unit)

Take a look at the way the units cancel. The unit that you want in the answer is the one that remains after all the other units have canceled out. This is a helpful way to check a problem.

$$\frac{\text{Numerator} \longrightarrow}{\text{Denominator} \longrightarrow} \quad \cancel{\text{lb}} \times \frac{\text{cents}}{\cancel{\text{lb}}} = \text{cents} \quad \text{Answer unit}$$

Do the calculation on your calculator to give the following:

2.0 ⊠ 48 = 96

The number 96 combined with the desired unit, cents, gives the final answer of 96 cents. With few exceptions, answers to numerical problems must contain a number and a unit.

96 cents Answer

S·A·M·P·L·E 1.12 ·P·R·O·B·L·E·M

Problem Solving Using Conversion Factors

On a recent bicycle trip, Maria averaged 35 miles per day. How many days did it take her to cover 175 miles?

Solution

When a problem contains several pieces of information, it is necessary to determine which one is the given. The given or starting unit is usually the measurement with a single unit, which would be 175 miles. The other quantity, 35 mi/day, is a relationship of two units (mi/day), which is useful as a conversion factor, but it is not the starting unit.

Step 1 **Given:** 175 miles

Step 2 **Unit plan:** miles ⟶ days Unit for answer

Step 3 **Relationship (from problem):** 35 miles = 1 day

Conversion Factors: $\frac{1 \text{ day}}{35 \text{ miles}}$ and $\frac{35 \text{ miles}}{1 \text{ day}}$

Step 4 **Problem Setup:**

Unit for answer goes here

$$\underset{\text{3 SFs}}{175 \cancel{\text{miles}}} \times \underset{\text{2 SFs}}{\frac{1 \text{ day}}{35 \cancel{\text{miles}}}} = \underset{\text{2 SFs}}{5.0 \text{ days}}$$

Suppose the problem were set up in the following way. If the units do not cancel, you know there is an error in the setup.

$$175 \text{ miles} \times \frac{35 \text{ miles}}{\text{day}} = 6215 \textbf{ miles}^2\textbf{/day}$$ Wrong units!

Study Check

A recipe for shark fin soup calls for 3.0 quarts of chicken broth. If 1 quart contains 4 cups, how many cups of broth are needed?

S·A·M·P·L·E 1.13 ·P·R·O·B·L·E·M

Problem Solving Using Metric Factors

The Daily Value (DV) for sodium is 2400 mg. How many grams of sodium is that?

Solution

Step 1 **Given:** 2400 mg

Step 2 **Unit Plan:** mg ⟶ g Unit for answer

Step 3 **Metric Equality:** 1 g = 1000 mg

Conversion Factors: $\frac{1 \text{g}}{1000 \text{ mg}}$ and $\frac{1000 \text{ mg}}{1 \text{g}}$

Step 4 **Problem Setup:**

Unit for answer goes here

$$\underset{\text{Given}}{2400 \cancel{\text{mg}}} \times \underset{\text{Metric factor}}{\frac{1 \text{ g}}{1000 \cancel{\text{mg}}}} = \underset{\text{Answer (in grams)}}{2.4 \text{ g of sodium needed daily}}$$

Study Check

A can containing 473 mL of frozen orange juice is mixed with 1415 mL of water. How many liters of orange juice were prepared?

Using Two or More Conversion Factors

In many problems you will need to use two or more steps in your unit plan. Then two or more conversion factors will be required. These can be constructed from the equalities you have learned or those stated in the problem. In setting up the problem, one factor follows the other. Each factor is arranged to cancel the preceding unit until you obtain the desired unit.

S·A·M·P·L·E **1.14** ·P·R·O·B·L·E·M

1.8

Problem Solving Using Two Factors

A recipe for salsa requires 3.0 cups of tomato sauce. If only metric measures are available, how many milliliters of tomato sauce are needed? (There are 4 cups in 1 quart.)

Solution

You may not know a relationship between cups and milliliters. However, you do know how to change cups to quarts, and quarts to milliliters.

Step 1 **Given:** 3.0 cups

Step 2 **Unit Plan:** cups ⟶ quarts ⟶ milliliters Unit for answer

Step 3 **Relationships and Conversion Factors:**

1. 1 qt = 4 cups $\frac{1\text{ qt}}{4\text{ cups}}$ and $\frac{4\text{ cups}}{1\text{ qt}}$

2. 1 qt = 946 mL $\frac{1\text{ qt}}{946\text{ mL}}$ and $\frac{946\text{ mL}}{1\text{ qt}}$

Step 4 **Problem Setup:** Use factor 1 to convert from cups to quarts:

$$3.0\,\cancel{\text{cups}} \times \frac{1\text{ qt}}{4\,\cancel{\text{cups}}} = 0.75\text{ qt}$$

Then use factor 2 to convert from quarts to milliliters:

$$0.75\,\cancel{\text{qt}} \times \frac{946\text{ mL}}{1\,\cancel{\text{qt}}} = 710\text{ mL}$$

In a continuous setup all the factors are used one after the other. The unit in the denominator of each new factor always cancels the unit in the previous numerator.

$$\underset{\text{Given quantity}}{3.0\,\cancel{\text{cups}}} \times \underset{\text{U.S. factor}}{\frac{1\,\cancel{\text{qt}}}{4\,\cancel{\text{cups}}}} \times \underset{\text{Metric–U.S. factor}}{\frac{946\text{ mL}}{1\,\cancel{\text{qt}}}} = \underset{\text{Answer (in mililiters)}}{710\text{ mL}}$$

All the calculations are done in a sequence on the calculator. No intermediate values are written down, only the final answer.

3.0	÷	4	×	946	=	710
Two SFs		Exact		Three SFs		Two SFs

Study Check

One medium bran muffin contains 4.2 g of fiber. How many ounces (oz) of fiber are obtained by eating three medium bran muffins, if 1 lb = 16 oz? (*Hint:* Number of muffins ⟶ g of fiber ⟶ lb ⟶ oz)

Using a sequence of two or more conversion factors is a very efficient way to set up and solve problems, especially if you are using a calculator. Once you have the problem set up, the calculations can be done without writing out the intermediate values. This process is worth practicing until you understand unit cancellation and the mathematical calculations. A summary of steps in problem solving is given in Table 1.10.

Table 1.10 Steps in Problem Solving

Step 1	Identify the given quantity and unit.
Step 2	Write a unit plan to help you think about changing units from the given to the answer unit. Be sure you can supply a conversion factor for each change.
Step 3	Determine the equalities and corresponding conversion factors you will need to change from one unit to another.
Step 4	Set up the problem according to your unit plan. Arrange each conversion factor to cancel the preceding unit. Check that the units cancel to give the unit of the answer. Carry out the calculations and give a final answer with the correct number of significant figures and unit.

Clinical Calculations Using Conversion Factors

Conversion factors are also useful for calculating medications. For example, if an antibiotic is available in 5-mg tablets, the dosage can be written as a conversion factor, 5 mg/1 tablet. In many hospitals, the apothecary unit of grains (gr) is still in use; there are 65 mg in 1 gr. When you do a medication problem, you often start with a doctor's order that contains the quantity to give the patient. The medication dosage is used as a conversion factor.

S·A·M·P·L·E 1.15 ·P·R·O·B·L·E·M

Clinical Calculations with Factors

Dr. Alvarez orders 0.050 g of a medication for your patient. If each tablet contains 10 mg of medication, how many tablets are needed?

Solution

Step 1 **Given:** 0.050 g

Step 2 **Unit Plan:** g ⟶ mg ⟶ tablets Unit for answer

Step 3 **Relationships and Conversion Factors:**

1. 1 g = 1000 mg $\quad \dfrac{1\text{ g}}{1000\text{ mg}}$ and $\dfrac{1000\text{ mg}}{1\text{ g}}$

2. 1 tablet = 10 mg of medication $\quad \dfrac{1\text{ tablet}}{10\text{ mg}}$ and $\dfrac{10\text{ mg}}{1\text{ tablet}}$

Step 4 **Problem Setup:** g ⟶ mg ⟶ tablets

Unit for answer

$$\underset{\text{Given}}{0.050\ \cancel{\text{g}}} \times \underset{\text{Metric factor}}{\frac{1000\ \cancel{\text{mg}}}{1\ \cancel{\text{g}}}} \times \underset{\text{Dosage factor}}{\frac{1\text{ tablet}}{10\ \cancel{\text{mg}}}} = \underset{\text{Answer (in tablets)}}{5\text{ tablets}}$$

1.8

Study Check

An aspirin tablet contains 5.0 grains (gr) of aspirin. How many milligrams (mg) of aspirin are in two aspirin tablets, if 1 gr = 65 mg?

Using Percents as Conversion Factors

Sometimes percents are given in a problem. To work with a percent, it is convenient to write it as a conversion factor. To do this, we choose units from the problem to express the numerical relationship of the part to 100 parts of the whole. For example, an athlete might have 18% body fat by weight. We can use a choice of units for weight or mass to write the following conversion factors:

$$\frac{18\text{ kg body fat}}{100\text{ kg body mass}} \qquad \frac{18\text{ lb body fat}}{100\text{ lb body weight}}$$

If a question is asked about the number of kilograms of body fat in a marathon runner with a total body mass of 48 kg, we would state the units in the factor in kg body fat. Note that the units in the factor for the percent cancel the given.

$$\underset{\text{Given}}{48\ \cancel{\text{kg body mass}}} \times \underset{\text{Percent factor}}{\frac{18\text{ kg body fat}}{100\ \cancel{\text{kg body mass}}}} = 8.6\text{ kg body fat}$$

If the problem asks about the pounds of body fat in a dancer weighing 126 lb, we would state the factor in lb body fat.

$$126\ \cancel{\text{lb body weight}} \times \frac{18\text{ lb body fat}}{100\ \cancel{\text{lb body weight}}} = 23\text{ lb body fat}$$

S·A·M·P·L·E 1.16 P·R·O·B·L·E·M

Using Percent Factors

A wine contains 13% alcohol by volume. How many milliliters of alcohol are contained in a glass containing 125 mL of the wine?

Solution

The factor for 13% alcohol can be written using mL units as:

$$\frac{13 \text{ mL alcohol}}{100 \text{ mL wine}}$$

The factor is then used with the given to set up the problem.

$$125 \cancel{\text{mL wine}} \times \frac{13 \text{ mL alcohol}}{100 \cancel{\text{mL wine}}} = 16 \text{ mL alcohol}$$

Study Check

There are 85 students in a chemistry class. On the last test, 20.% received grades of A. How many A's were earned on the test?

QUESTIONS & PROBLEMS

Using Conversion Factors

1.63 How would you explain conversion factors to a friend?

1.64 What advice would you give your friend for converting from one unit to another?

1.65 When you convert one unit to another, how do you know which unit to place in the denominator?

1.66 When you convert one unit to another, how do you know which unit to place in the numerator?

1.67 (True or false) When you read a problem, you should:
- **a.** Start the problem with the first number in the problem.
- **b.** Identify the given quantity and unit.
- **c.** Determine a unit plan.
- **d.** Do all the work in your head.
- **e.** Drop all the units while you work on the problem.

1.68 When you set up a problem, the following are important (true or false):
- **a.** The unit to cancel appears in the denominator of the conversion factor.
- **b.** It doesn't matter how you write the conversion factor.
- **c.** The setup should lead to the unit you desire as an answer.
- **d.** You don't need to remember the metric prefixes.
- **e.** You should check each conversion factor one more time.

1.69 Use U.S. system conversion factors to solve the following problems:
- **a.** How many yards are in 24 ft?
- **b.** How many seconds are in 15 min?
- **c.** You need 3.5 qt of oil for your car. How many gallons is that?

1.70 Use U.S. system conversion factors to solve the following problems:
- **a.** A necklace contains 2.6 oz of gold. How many pounds is that?
- **b.** You ran a total of 2.4 miles today. How far is that in feet? (1 mi = 5280 ft)
- **c.** One game at the arcade requires one quarter. You have $3.50 in your pocket. How many games can you play?

1.71 Use metric conversion factors to solve the following problems:

a. The height of a student is 175 centimeters. How tall is the student in meters?

b. A cooler has a volume of 5500 mL. What is the capacity of the cooler in liters?

c. A hummingbird has a mass of 0.055 kg. What is the mass of the hummingbird in grams?

1.72 Use metric conversion factors to solve the following problems:

a. The Daily Value of phosphorus is 800 mg. How many grams of phosphorus are recommended?

1.8

b. A glass of orange juice contains 0.85 dL of juice. How many milliliters of orange juice is that?

c. A package of chocolate instant pudding contains 2840 mg of sodium. How many grams of sodium is that?

1.73 Solve the following problems using one or more conversion factors:

a. A container holds 0.750 qt of liquid. How many milliliters of lemonade will it hold?

b. What is the mass in kilograms of a person who weighs 165 lb?

c. The femur, or thighbone, is the longest bone in the body. In a 6-ft-tall person, the femur might be 19.5 in. long. What is the length of that person's femur in millimeters?

d. A person on a diet has been losing weight at the rate of 3.5 lb per week. If the person has been on the diet for 6 weeks, how many kilograms were lost?

1.74 Solve the following problems using one or more conversion factors:

a. You need 4.0 ounces of a steroid ointment. If there are 16 oz in 1 lb, how many grams of ointment does the pharmacist need to prepare?

b. During surgery, a patient receives 5.0 pints of plasma. How many milliliters of plasma were given? (1 quart = 2 pints)

c. A piece of plastic tubing measuring 560 mm in length is needed for an intravenous setup. How many feet of tubing are required?

d. Zippy the snail moves at the rate of 2.0 inches per hour. How many centimeters does Zippy travel in 4.0 hours?

1.75 Using conversion factors, solve the following clinical problems:

a. You have used 250 L of distilled water for a dialysis patient. How many gallons of water is that?

b. A patient needs 0.024 g of a sulfa drug. There are 8-mg tablets in stock. How many tablets should be given?

c. The daily dose of ampicillin for the treatment of an ear infection is 115 mg/kg of body weight. What is the daily dose for a 34-lb child?

1.76 Using conversion factors, solve the following clinical problems:

a. The physician has ordered 1.0 g of tetracycline to be given every 6 hours to a patient. If your stock on hand is 500-mg tablets, how many will you need for 1 day's treatment?

b. An intramuscular medication is given at 5.00 mg/kg of body weight. If you give 425 mg of medication to a patient, what is the patient's weight in pounds?

c. A physician has ordered 325 mg of atropine, intramuscularly. If atropine were available as 0.50 g/mL of solution, how many milliliters would you need to give?

1.77 Oxygen makes up 46.7% by mass of the earth's crust. How many grams of oxygen are present if a sample of the earth's crust has a mass of 325 g?

1.78 Magnesium makes up 2.1% by mass of the earth's crust. How many grams of magnesium are present if a sample of the earth's crust has a mass of 1.25 g?

1.79 Water is 11.2% by mass hydrogen. How many kilograms of water would contain 5.0 g of hydrogen?

1.80 Water is 88.8% by mass oxygen. How many grams of water would contain 2.25 kg of oxygen?

1.9 Density

Study Pack

Learning Goal

Calculate the density or specific gravity of a substance, and use the density or specific gravity to calculate the mass or volume of a substance.

Why do objects sink or float? In Figure 1.14, we see that a sample of lead sinks in water, but a cork floats. Why does this happen? We need to look at density. The density of lead is greater than the density of water, which makes the lead object sink. The cork floats because cork is less dense than water. Differences in density determine whether an object will sink or float.

The mass and volume of any object can be measured. However, the separate measurements do not tell us how tightly packed the substance might be. However, if we compare the mass of the object to its volume, we obtain a relationship called **density**.

$$\text{Density} = \frac{\text{mass of substance}}{\text{volume of substance}}$$

In the metric system, the densities of solids and liquids are usually expressed as grams per cubic centimeter (g/cm^3) or grams per milliliter (g/mL). The density of gases is usually stated as grams per liter (g/L). Table 1.11 gives the densities of some common substances.

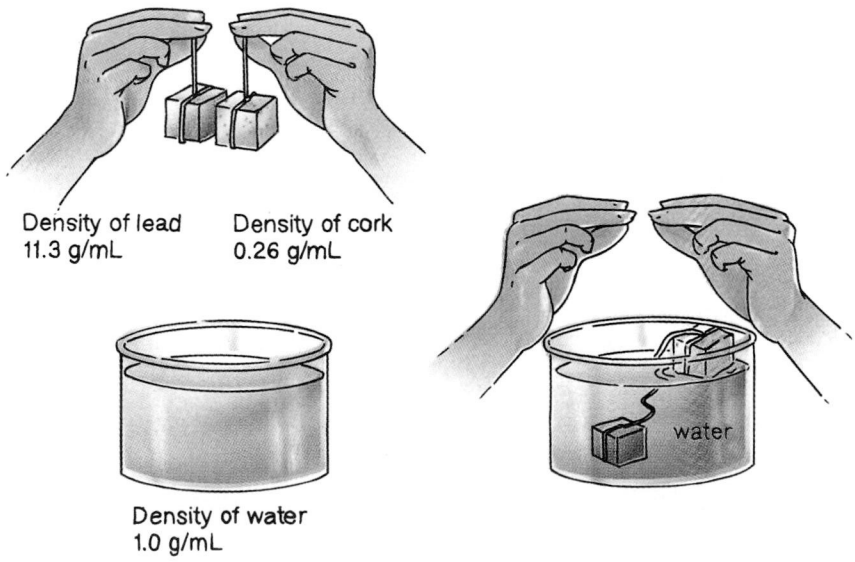

Figure 1.14 Differences in density determine whether an object will sink or float.

1.9

Table 1.11 Density of Some Common Substances

Solids (at 25°C)	Density (g/mL)	Liquids (at 25°C)	Density (g/mL)	Gases (at 0°C)	Density (g/L)
Cork	0.26	Gasoline	0.66	Hydrogen	0.090
Wood (maple)	0.75	Ethyl alcohol	0.79	Helium	0.179
Ice	0.92	Olive oil	0.92	Methane	0.714
Sugar	1.59	Water (at 4°C)	1.000	Neon	0.90
Bone	1.80	Plasma (blood)	1.03	Nitrogen	1.25
Aluminum	2.70	Urine	1.003–1.030	Air (dry)	1.29
Cement	3.00	Milk	1.04	Oxygen	1.45
Diamond	3.52	Mercury	13.6	Carbon dioxide	1.96
Silver	10.5				
Lead	11.3				
Gold	19.3				

Density of Solids

The density of a solid is calculated from its mass and volume. When a solid is completely submerged, it displaces a volume of water that is *equal to its own volume*. In Figure 1.15, the water level rises from 25 mL to 35 mL. This means that 10. mL of water is displaced and that the volume of the object is 10. mL.

Volume of water displaced

35 mL − 25 mL = 10. mL

$$\text{DENSITY OF SOLID} = \frac{\text{Mass}}{\text{Volume}} = \frac{32.0\text{ g}}{10.\text{ mL}} = 3.2\text{ g/mL}$$

Figure 1.15 Determining the density of a solid by using volume displacement.

EXPLORE YOUR WORLD

Sink or Float?

1. Fill a large container or bucket with water. Place a can of diet and a can of nondiet soft drink in the water. What happens? Using information on the label, how might you account for your observations?
2. Design an experiment to determine the substance that is the most dense in each of the following:
   a. water and vegetable oil
   b. water and ice
   c. rubbing alcohol and ice
   d. vegetable oil, water, and ice

S·A·M·P·L·E 1.17 P·R·O·B·L·E·M

Calculating Density

A 50.0-mL sample of buttermilk has a mass of 56.0 g. What is the density of the buttermilk?

Solution

To calculate density, substitute the mass (g) and volume (mL) of the buttermilk into the expression for density.

$$\text{Density} = \frac{\text{mass}}{\text{volume}} = \frac{56.0\text{ g}}{50.0\text{ mL}} = \frac{1.12\text{ g}}{1\text{ mL}} = 1.12\text{ g/mL}$$

Two units in this answer

Study Check

A copper sample has a mass of 44.65 g and a volume of 5.0 cm^3. What is the density of copper, expressed as grams per cubic centimeter (g/cm^3)?

S·A·M·P·L·E 1.18 P·R·O·B·L·E·M

Using Volume Displacement to Calculate Density

A lead weight used in the belt of a scuba diver has a mass of 226 g. When the weight is carefully placed in a graduated cylinder containing 200.0 mL of water, the water level rises to 220.0 mL. What is the density of the lead weight (g/mL)?

Solution

Both the mass and volume of the lead weight are needed to calculate its density. Its mass, 226 g, is given. The volume of the lead weight is equal to the volume of water displaced, which is calculated as follows:

Volume displaced

$$\begin{array}{r} 220.0\text{ mL} \\ -200.0\text{ mL} \\ \hline 20.0\text{ mL} \end{array}$$

Answer to first decimal place

Volume of the lead weight = 20.0 mL

$$\text{Density of lead} = \frac{\text{mass of lead}}{\text{volume of lead}} = \frac{\overset{\text{Three SFs}}{226\text{ g}}}{\underset{\text{Three SFs}}{20.0\text{ mL}}} = 11.3\text{ g/mL}$$

Three SFs in answer

In the expression for density, be sure to use the volume of water the object displaces and *not* the original volume of water.

Study Check

A total of 0.50 lb of glass marbles is added to 425 mL of water. The water level rises to a volume of 528 mL. What is the density (g/mL) of the glass marbles?

S·A·M·P·L·E 1.19 ·P·R·O·B·L·E·M

Sink or Float?

From their densities, determine whether each of the following substances will sink or float when placed in seawater, which has a density of 1.025 g/mL.

Substance	Density
Gasoline	0.66 g/mL
Asphalt	1.2 g/mL
Cardboard	0.69 g/cm^3

1.9

Solution

The gasoline and cardboard will float because they are less dense than seawater. Asphalt will sink because it is more dense than water.

Study Check

Using Table 1.11, explain why helium-filled balloons float in air.

Problem Solving Using Density

Because density is a comparison of two units, it can be used as a conversion factor. For example, if the mass and the density of a sample are known, the volume of the sample can be calculated.

S·A·M·P·L·E 1.20 ·P·R·O·B·L·E·M

Problem Solving Using Density

The cast on Malcolm's leg has a density of 2.32 g/cm^3. What is the volume of the cast if it weighs 4.25 lb?

Solution

Step 1 **Given:** 4.25 lb

Step 2 **Unit Plan:** lb $\longrightarrow$ g $\longrightarrow$ cm^3 Unit for answer

Step 3 **Relationships and Conversion Factors:**

Metric–U.S.: 1 lb = 454 g $\quad \dfrac{454\text{ g}}{1\text{ lb}}$ and $\dfrac{1\text{ lb}}{454\text{ g}}$

Density: 2.32 g = 1 cm^3 $\quad \dfrac{2.32\text{ g}}{1\text{ cm}^3}$ and $\dfrac{1\text{ cm}^3}{2.32\text{ g}}$

Step 4 **Problem Setup:**

$$4.25\cancel{\text{lb}} \times \frac{454\cancel{\text{g}}}{1\cancel{\text{lb}}} \times \frac{1\text{ cm}^3}{2.32\cancel{\text{g}}} = 832\text{ cm}^3$$

Given | Metric–U.S. Factor | Density factor | Answer to three SFs

Study Check

Milk has a density of 1.04 g/mL. What is the mass of milk in a 1.00-qt container, if 1 qt = 946 mL?

HEALTH NOTE

Determination of Percentage of Body Fat

Body mass is made up of protoplasm, extracellular fluid, bone, and adipose tissue. One way to determine the amount of adipose tissue is to measure the whole-body density. After the on-land mass of the body is determined, the underwater body mass is obtained by submerging the person in water. (See Figure 1.16.) Because water helps support the body by giving it buoyancy, the underwater body mass is less. A higher percentage of body fat will make a person more buoyant, causing the underwater mass to be even lower. This occurs because fat has a lower density than the rest of the body.

The difference between the on-land mass and underwater mass, known as the buoyant force, is used to determine the body volume. Then the mass and volume of the person are used to calculate body density. For example, suppose a 70.0-kg person has a body volume of 66.7 L. The body density is calculated as

$$\frac{\text{Body mass}}{\text{Body volume}} = \frac{70.0\ \text{kg}}{66.7\ \text{L}}$$

$$= 1.05\ \text{kg/L} \quad \text{or} \quad 1.05\ \text{g/mL}$$

When the body density is determined, it is compared with a chart that correlates the percentage of adipose tissue with body density. A person with a body density of 1.05 g/mL has 21% body fat, according to such a chart. This procedure is used by athletes to determine exercise and diet programs.

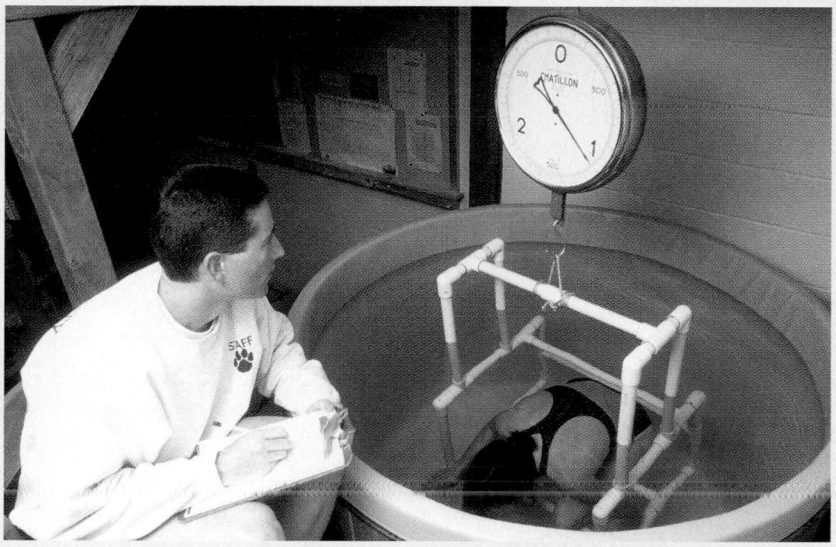

Figure 1.16 A person is submerged in a water tank to determine underwater body mass and to calculate percentage of body fat.

Specific Gravity

Specific gravity (sp gr) is a ratio between the density of a substance and the density of water. Specific gravity is calculated by dividing the density of a sample by the density of water. In this text, we will use the metric value of 1.00 g/mL for the density of water. A substance with a specific gravity of 1.00 has the same density as water. A substance with a specific gravity of 3.00 is three times as dense as water, whereas a substance with a specific gravity of 0.50 is just one-half as dense as water.

$$\text{Specific gravity} = \frac{\text{density of sample}}{\text{density of water}}$$

In the calculations for specific gravity, the units of density must match. Then, all units cancel to leave only a number. *Specific gravity is one of the few unitless values you will encounter in chemistry.*

An instrument called a hydrometer is often used to measure the specific gravity of fluids such as battery fluid or a sample of urine. In Figure 1.17, a hydrometer is used to measure the specific gravity of a fluid.

Figure 1.17 The specific gravity of a beer is measured with a hydrometer.

S·A·M·P·L·E **1.21** ·P·R·O·B·L·E·M

Specific Gravity

What is the specific gravity of coconut oil that has a density of 0.925 g/mL?

Solution

$$\text{sp gr oil} = \frac{\text{density of oil}}{\text{density of water}} = \frac{0.925\ \cancel{\text{g/mL}}}{1.00\ \cancel{\text{g/mL}}} = 0.925 \quad \text{No units}$$

Study Check

1.9

What is the specific gravity of ice if 35.0 g of ice has a volume of 38.2 mL?

S·A·M·P·L·E **1.22** ·P·R·O·B·L·E·M

Problem Solving with Specific Gravity

John took 2.0 teaspoons (tsp) of cough syrup (sp gr 1.20) for a persistent cough. If there are 5.0 mL in 1 tsp, what was the mass (in grams) of the cough syrup?

Solution

Step 1 **Given:** 2.0 tsp

Step 2 **Unit Plan:** tsp ⟶ mL ⟶ g Unit for answer

Step 3 **Relationships and Conversion Factors:**

$$1\ \text{tsp} = 5.0\ \text{mL} \qquad \frac{1\ \text{tsp}}{5.0\ \text{mL}} \quad \text{and} \quad \frac{5.0\ \text{mL}}{1\ \text{tsp}}$$

For problem solving, it is convenient to convert the specific gravity value to density.

$$\text{Density} = (\text{sp gr})\ \text{g/mL} = 1.20\ \text{g/mL}$$

Relationship: 1 mL = 1.20 g

Conversion Factors: $\frac{1\ \text{mL}}{1.20\ \text{g}}$ and $\frac{1.20\ \text{g}}{1\ \text{mL}}$

Step 4 **Problem Setup:**

$$2.0\ \cancel{\text{tsp}} \times \frac{5.0\ \cancel{\text{mL}}}{1\ \cancel{\text{tsp}}} \times \frac{1.20\ \text{g}}{1\ \cancel{\text{mL}}} = 12\ \text{g of syrup}$$

Study Check

An ebony carving has a mass of 275 g. If ebony has a specific gravity of 1.33, what is the volume of the carving?

HEALTH NOTE

Specific Gravity of Urine

The specific gravity of urine is often determined as part of a laboratory evaluation of the health of an individual. The specific gravity of urine is normally in the range 1.003–1.030. This is somewhat greater than the specific gravity of water because compounds such as salts and urea are dissolved in urine.

If the specific gravity of a person's urine is too low or too high, a physician might suspect a kidney problem. For example, if a urine sample shows a specific gravity of 1.001, significantly lower than normal, malfunctioning of the kidneys is a possibility. When water has been lost because of dehydration, the urine may have a specific gravity above normal.

QUESTIONS & PROBLEMS

Density

1.81 In an old trunk, you find a piece of metal that you think may be aluminum, silver, or lead. In lab you find it has a mass of 217 g and a volume of 19.2 cm^3. Using Table 1.11, what is the metal you have found?

1.82 Suppose you have two 100-mL graduated cylinders. In each cylinder there are 40.0 mL of water. You also have two cubes: one is lead and the other is aluminum. Each cube measures 2.0 cm on each side. After you carefully lower each into the water of its own cylinder, what will the new water level be in each of the cylinders?

1.83 What is the density (g/mL) of each of the following samples:

a. A 20.0-mL sample of a salt solution that has a mass of 24.0 g.

b. A solid object with a mass of 1.65 lb and a volume of 170 mL.

c. A gem has a mass of 45.0 g. When the gem is placed in a graduated cylinder containing 20.0 mL of water, the water level rises to 34.5 mL. 3.1

1.84 What is the density of each of the following samples?

a. A medication, if 3.00 mL has a mass of 3.85 g.

b. The fluid in a car battery, if it has a volume of 125 mL and a mass of 155 g.

c. A 5.00-mL urine sample from a patient suffering from symptoms resembling those of diabetes mellitus. The mass of the urine sample is 5.025 g.

1.85 Use the density value to solve the following problems:

a. What is the mass, in grams, of 150 mL of a liquid with a density of 1.4 g/mL?

b. What is the mass of a glucose solution that fills a 0.500-L intravenous bottle if the density of the glucose solution is 1.15 g/mL?

c. A sculptor has prepared a mold for casting a bronze figure. The figure has a volume of 225 mL. If bronze has a density of 7.8 g/mL, how many ounces of bronze are needed in the preparation of the bronze figure?

1.86 Use the density value to solve the following problems:

a. A graduated cylinder contains 18.0 mL of water. What is the new water level after 35.6 g of silver metal with a density of 10.5 g/mL is submerged in the water?

b. A thermometer containing 8.3 g of mercury has broken. If mercury has a density of 13.6 g/mL, what volume spilled?

c. A fish tank holds 35 gallons of water. Using the density of 1.0 g/mL for water, determine the number of pounds of water in the fish tank.

1.87 Solve the following specific gravity problems:

a. A urine sample has a density of 1.030 g/mL. What is the specific gravity of the sample?

b. A liquid has a volume of 40.0 mL and a mass of 45.0 g. What is the specific gravity of the liquid?

c. The specific gravity of a vegetable oil is 0.85. What is its density?

1.88 Solve the following specific gravity problems:

a. A 5% glucose solution has a specific gravity of 1.02. What is the mass of 500 mL of glucose solution?

b. A bottle containing 325 g of cleaning solution is used for carpets. If the cleaning solution has a specific gravity of 0.850, what volume of solution was used?

c. Butter has a specific gravity of 0.86. What is the mass, in grams, of 2.15 L of butter?

1.9

Chapter Review

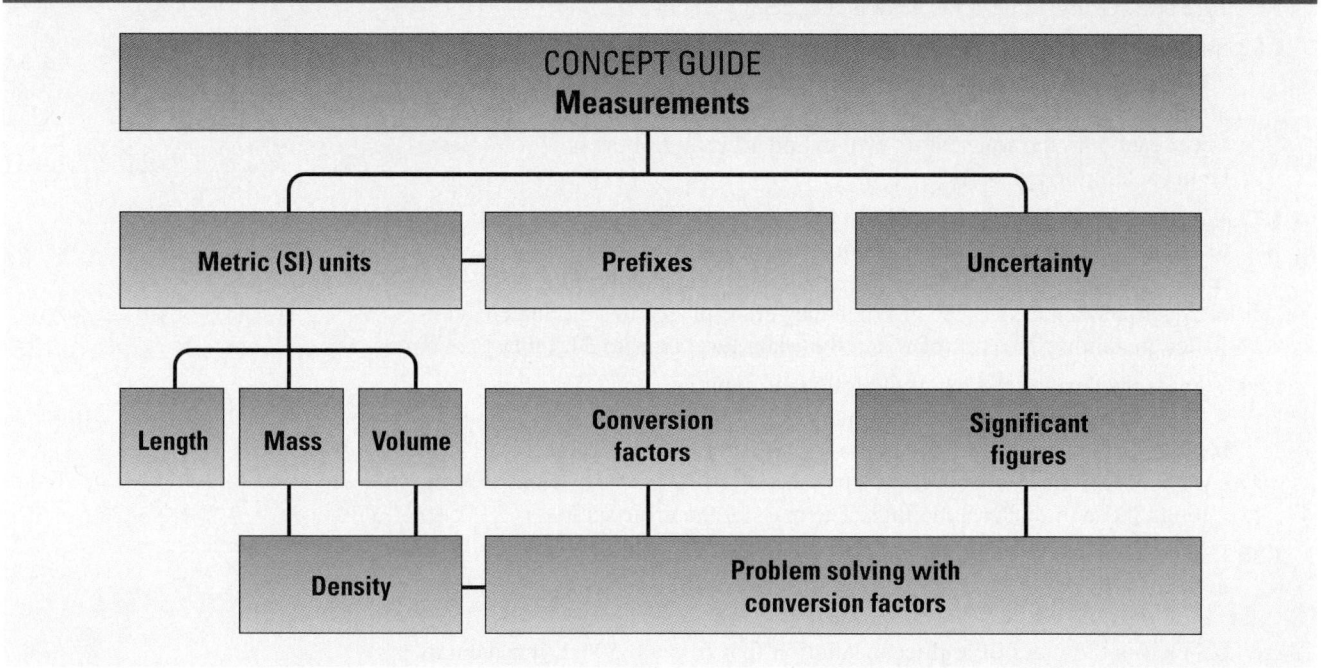

Chapter Summary

1.1 Units of Measurement

In science, physical quantities are described in units of the metric or International System (SI). Some important units are meter (m) for length, liter (L) for volume, gram (g) and kilogram (kg) for mass, and Celsius (°C) for temperature.

1.2 Scientific Notation

Large and small numbers can be written using scientific notation in which the decimal point is moved to give a coefficient between 1 and 10 and the number of decimal places moved shown as a power of 10. A large number will have a positive power of 10, while a small number will have a negative power of 10.

1.3 Measured and Exact Numbers

A measured number is any number derived from using a measuring device to determine a quantity. An exact number is obtained by counting items or from a definition; no measuring device is used.

1.4 Significant Figures

Significant figures are the numbers reported in a measurement including the last estimated digit. Zeros in front of a decimal number or at the end of a large number are not significant.

1.5 Significant Figures in Calculations

In multiplication or division, the measured number in the calculation that has the smallest number of significant figures determines the number of significant figures in an answer. In addition or subtraction, the final answer reflects the last place that all the measured numbers have in common.

1.6 SI and Metric Prefixes

Prefixes placed in front of a unit change the size of the unit by factors of 10. Prefixes such as *centi*, *milli*, and *micro* provide smaller units; prefixes such as *kilo* provide larger units. An equality relates two metric units that measure the same quantity of length, volume, or mass. Examples of metric equalities are 1 m = 100 cm; 1 L = 1000 mL; 1 kg = 1000 g.

1.7 Conversion Factors

Conversion factors are used to express a relationship in the form of a fraction. Two factors can be written for any relationship in the metric or U.S. system.

1.8 Using Conversion Factors

Factors are useful when changing a quantity expressed in one unit to one expressed in another unit.

1.9 Density

The density of a substance is a ratio of its mass to its volume, usually g/mL or g/cm^3. The units of density can be used as a factor to convert between the mass and volume of a substance. Specific gravity (sp gr) compares the density of a substance to the density of water, 1.00 g/mL.

Key Terms

accuracy How close a measurement is to the true value.

Celsius (°C) temperature scale A temperature scale on which water has a freezing point of 0°C and a boiling point of 100°C.

centimeter (cm) A unit of length in the metric system; there are 2.54 cm in 1 in.

conversion factor A ratio in which the numerator and denominator are quantities from an equality or given relationship. For example, the conversion factors for the relationship 1 kg = 2.20 lb are written as the following:

$$\frac{2.20\text{ lb}}{1\text{ kg}} \quad \text{and} \quad \frac{1\text{ kg}}{2.20\text{ lb}}$$

cubic centimeter (cm^3, cc) The volume of a cube that has 1-cm sides, equal to 1 mL.

density The relationship of the mass of an object to its volume expressed as grams per cubic centimeter (g/cm^3), grams per milliliter (g/mL), or grams per liter (g/L).

equality A relationship between two units that measure the same quantity.

exact number A number obtained by counting or definition.

gram (g) The metric unit used in measurements of mass.

Kelvin (K) temperature scale A temperature scale on which the lowest possible temperature is 0 K, which makes the freezing point of water 273 K and the boiling point of water 373 K.

kilogram (kg) A metric mass of 1000 g, equal to 2.20 lb. The kilogram is the SI standard unit of mass.

liter (L) The metric unit for volume that is slightly larger than a quart.

mass A measure of the quantity of material in an object.

measured number A number obtained when a quantity is determined by using a measuring device.

meter (m) The metric unit for length that is slightly longer than a yard. The meter is the SI standard unit of length.

metric system A system of measurement used by scientists and in most countries of the world.

milliliter (mL) A metric unit of volume equal to one-thousandth of a L (0.001 L).

precision Ability to repeat a measurement and obtain values close together.

prefix The part of the name of a metric unit that precedes the base unit and specifies the size of the measurement. All prefixes are related on a decimal scale.

scientific notation A form of writing large and small numbers using a coefficient between 1 and 10, followed by a power of 10.

SI units An International System of units that modifies the metric system.

significant figures The numbers recorded in a measurement.

specific gravity (sp gr) A relationship between the density of a substance and the density of water:

$$\text{sp gr} = \frac{\text{density of sample}}{\text{density of water}}$$

temperature An indicator of the hotness or coldness of an object.

volume The amount of space occupied by a substance.

Chapter Practice

This group of questions and problems is related to the topics in this chapter. However, they do not all follow the chapter order and they require you to combine concepts and skills from several sections. These problems will help you increase your critical thinking skills and help you prepare for your next exam.

1.89 During a workout at the gym, you set the treadmill at a pace of 55.0 meters per minute. How many minutes will you walk if you cover a distance of 7500 feet?

1.90 A fish company delivers 22 kg of salmon, 5.5 kg of crab, and 3.48 kg of oysters to your seafood restaurant.
a. What is the total mass, in kilograms, of the seafood?
b. What is the total number of pounds?

1.91 In France, grapes are 14 francs per kilogram. What is the cost of grapes in dollars per pound, if the exchange rate is 7.5 francs per dollar?

1.92 In Mexico, avocados are 125 pesos per kilogram. What is the cost in cents of an avocado that weighs 0.45 lb if the exchange rate is 7.5 pesos to the dollar?

1.93 Bill's recipe for onion soup calls for 4.0 lb of thinly sliced onions. If an onion has an average mass of 115 g, how many onions does Bill need?

1.94 The price of 1 pound (lb) of potatoes is $1.75. If all the potatoes sold today at the store bring in $1,420, how many kilograms (kg) of potatoes did grocery shoppers buy?

1.95 The following nutrition information is listed on a box of crackers:
Serving size 0.50 oz (6 crackers)
Fat 4 g per serving Sodium 140 mg per serving
a. If the box has a net weight (contents only) of 8.0 oz, about how many crackers are in the box?
b. If you ate 10 crackers, how many ounces of fat are you consuming?
c. How many grams of sodium are used to prepare 50 boxes of crackers?

1.96 A dialysis unit requires 75,000 mL of distilled water. How many gallons of water are needed? (1 gal = 4 qt)

1.97 To prevent bacterial infection, a doctor orders 4 tablets of amoxicillin per day for 10 days. If each tablet contains 250 mg of amoxicillin, how many ounces of the medication are given in 10 days?

1.98 Celeste's diet restricts her intake of protein to 24 g per day. If she eats an 8.0-oz burger that is 15.0% protein, has she exceeded her protein limit for the day? How many ounces of a burger would be allowed for Celeste?

1.99 A letter has arrived for you that reads as follows:

Dear Cousin:

How are you? I hear you are taking chemistry now. Maybe you can help me. I have been working in our mine and recently found a brilliant stone that looks like a diamond, but I am not sure. Do you know how I could determine its density? I read in a mineralogy magazine that diamond has a density of 3.51 g/cm^3. Please describe how I can determine the density of this stone, including the tools I need to buy, the measurements I need to make, and the necessary calculations. I hope you are enjoying your chemistry class.

Sincerely,

Cousin Emma

Write a letter to Emma using complete sentences.

1.100 An object has a mass of 3.15 oz. When it is submerged in a graduated cylinder initially containing 325.2 mL of water, the water level rises to 442.5 mL. What is the density (g/mL) of the object?

1.101 The density of lead is 11.3 g/mL. The water level in a graduated cylinder initially at 215 mL rises to 285 mL after a piece of lead is submerged. What is the mass in grams of the lead?

1.102 A graduated cylinder contains 155 mL of water. A 15.0-g piece of iron (density = 7.86 g/cm^3) and a 20.0-g piece of lead (density = 11.3 g/cm^3) are added. What is the new water level in the cylinder?

1.103 How many cubic centimeters (cm^3) of olive oil have the same mass as 1.00 L of gasoline? (See Table 1.11.)

1.104 Ethyl alcohol has a specific gravity of 0.79. What is the volume, in quarts, of 1.50 kg of alcohol?

1.105 A typical adult body contains 55% water. If a person weighs 175 lb, how many kilograms of water do they have in their body?

1.106 A typical adult body contains 55% water. If a person has a mass of 65 kg, how many pounds of water do they have in their body?

1.107 **a.** Some athletes have as little as 3.0% body fat. If such a person has a body mass of 45 kg, how much body fat does that person have?
b. In a process called liposuction, a doctor removes some adipose tissue from a person's body. If body fat has a density of 0.94 g/mL and 3.0 liters of fat are removed, how many pounds of fat were removed from the patient? (See the Health Note "Determination of Percentage of Body Fat.")

1.108 When a 50.0-kg person is immersed in a water tank for body fat testing, she has an underwater mass of 2.0 kg. This difference in body mass is equal to the mass of water displaced by the body. The density of water is 1.00 g/mL. (See the Health Note "Determination of Percentage of Body Fat.")
a. What volume of water is displaced?
b. What is the volume, in L, of the person?
c. What is the body density, in g/mL, of that person?

1.109 A urinometer is a hydrometer used to measure the specific gravity of a sample of urine. (See the Health Note "Specific Gravity of Urine.")
a. What is the density of a urine sample that has a specific gravity of 1.012?
b. What is the mass of a 5.00-mL urine sample with a specific gravity of 1.022?

1.110 A normal range for specific gravity of urine is 1.010 to 1.025. A 10.0-mL sample of urine has a mass of 10.31 g. What is the specific gravity of the urine? Is the urine considered normal? Why or why not? (See the Health Note "Specific Gravity of Urine.")

Answers

Answers to Study Checks

1.1 length

1.2 **a.** 4.25×10^5 m **b.** 8×10^{-7} g

1.3 exact

1.4 14 inches

1.5 **a.** two **b.** one **c.** five

1.6 **a.** 36 m **b.** 0.0026 L **c.** 3800 g **d.** 1.3 kg

1.7 **a.** 0.4924 **b.** 0.0080 or 8.0×10^{-3} **c.** 2.0

1.8 **a.** 83.70 mg **b.** 0.5 L

1.9 **a.** mega **b.** centi

1.10 **a.** 1000 **b.** 0.001 or $\frac{1}{1000}$

1.11 Equality: 1 in. = 2.54 cm

Conversion Factors: $\frac{1 \text{ in.}}{2.54 \text{ cm}}$ and $\frac{2.54 \text{ cm}}{1 \text{ in.}}$

1.12 12 cups

1.13 1.888 L

1.14 0.44 oz

1.15 650 mg

1.16 17 students got A's

1.17 8.9 g/cm^3

1.18 2.2 g/mL

1.19 Air has a density of 1.29 g/L. Because helium has a density of 0.179 g/L, a helium-filled balloon is less dense than air and floats in air.

1.20 984 g of milk

1.21 0.916

1.22 207 mL

Answers to Selected Questions and Problems

1.1 The metric system is the measurement system used in Mexico.

1.3 In the U.S., weight is measured in pounds (lb), height in feet and inches, gasoline in gallons, and temperature in Fahrenheit (°F). In France, weight is measured in kilograms, height in meters, gasoline in liters, and temperature in Celsius (°C).

1.5 In the metric system, the meter (m) is used in length measurements. In the U.S. system, several units are used, including inch (in.), feet (ft), and miles (mi).

1.7 **a.** meter; length
b. gram; mass
c. milliliter; volume
d. meter; length
e. Celsius; temperature

1.9 **a.** 5.5×10^4 m
b. 4.8×10^2 g
c. 5×10^{-6} cm
d. 1.4×10^{-4} s

1.11 **a.** 7.2×10^3
b. 3.2×10^{-2}
c. 1×10^4
d. 6.8×10^{-2}

1.13 **a.** 1×10^{-6}, 1×10^{-3}, 1×10^3
b. 3.5×10^{-5}, 0.035, 3.5×10^4
c. 7.8×10^{-3}, 3.5×10^3, 24 000

1.15 **a.** first decimal place (0.6)
b. second decimal place (0.05)
c. first decimal place (0.0)

1.17 The number four was determined by counting the number of chairs. No measuring device was used.

1.19 **a.** measured
b. exact
c. exact
d. measured

1.21 Significant figures represent the precision of a measurement. All the measured figures plus the estimated digits are significant digits.

1.23 **a.** 5 **b.** 2 **c.** 2
d. 3 **e.** 4 **f.** 3

1.25 **a.** 5.0×10^3 L
b. 3.0×10^4 g
c. 1.0×10^5 m

1.27 The precision of the answer is limited by the precision of the measurements used in the calculation.

1.29 An answer from multiplication or division can have only as many significant figures as the measurement that has the smallest number of significant figures.

1.31 **a.** 1.85
b. 184
c. 0.00474
d. 8810
e. 1.83

1.33 **a.** 1.6
b. 0.01
c. 27.6
d. 3.5

1.35 **a.** 53.54 cm
b. 127.6 g
c. 121.5 mL
d. 0.50 L

1.37 Changing to the metric system would give a uniform decimal system of measurement that would be used throughout the world. However, the change may be expensive and frustrating to many U.S. citizens.

1.39 km/h is kilometers per hour; mi/h is miles per hour

1.41 The prefix *kilo* means to multiply by 1000. One kg is the same mass as 1000 g.

1.43 **a.** mg
b. dL
c. km
d. kg
e. μL

1.45 **a.** 0.01
b. 1000
c. 0.001
d. 0.1
e. 1 000 000

1.47 **a.** 100 cm
b. 1000 m
c. 0.001 m
d. 1000 mL

1.49 **a.** kilogram
b. milliliter
c. km

1.51 A conversion factor can be inverted to give a second conversion factor.

1.53 3 teaspoons = 1 tablespoon; $\frac{3 \text{ tsp}}{1 \text{ tbs}}$ and $\frac{1 \text{ tbs}}{3 \text{ tsp}}$

2 cups = 1 pint; $\frac{2 \text{ cups}}{1 \text{ pint}}$ and $\frac{1 \text{ pint}}{2 \text{ cups}}$

1.55 $\frac{1 \text{ tsp salt}}{2 \text{ qt water}}$ and $\frac{2 \text{ qt water}}{1 \text{ tsp salt}}$

1.57 1 kg = 1000 g

1.59 **a.** 3 ft = 1 yd; $\frac{3 \text{ ft}}{1 \text{ yd}}$ and $\frac{1 \text{ yd}}{3 \text{ ft}}$

b. 1 mile = 5280 feet; $\frac{5280 \text{ ft}}{1 \text{ mi}}$ and $\frac{1 \text{ mi}}{5280 \text{ ft}}$

c. 1 min = 60 sec; $\frac{60 \text{ sec}}{1 \text{ min}}$ and $\frac{1 \text{ min}}{60 \text{ sec}}$

d. 1 gal = 27 mi; $\frac{1 \text{ gal}}{27 \text{ mi}}$ and $\frac{27 \text{ mi}}{1 \text{ gal}}$

1.61 **a.** 100 cm = 1 m; $\frac{100 \text{ cm}}{1 \text{ m}}$ and $\frac{1 \text{ m}}{100 \text{ cm}}$

b. 1000 mg = 1 g; $\frac{1000 \text{ mg}}{1 \text{ g}}$ and $\frac{1 \text{ g}}{1000 \text{ mg}}$

c. 1 L = 1000 mL; $\frac{1000 \text{ mL}}{1 \text{L}}$ and $\frac{1 \text{ L}}{1000 \text{ mL}}$

d. 1 dL = 100 mL; $\frac{100 \text{ mL}}{1 \text{ dL}}$ and $\frac{1 \text{ dL}}{100 \text{ mL}}$

1.63 Conversion factors are quantitative relationships written in a fraction form that are useful in solving numerical problems.

1.65 The unit in the denominator must cancel with the preceding unit.

1.67 **a.** F **b.** T **c.** T **d.** F **e.** F

1.69 **a.** 8.0 yd **b.** 9.0×10^2 s **c.** 0.88 gal

1.71 **a.** 1.75 m **b.** 5.5 L **c.** 55 g

1.73 **a.** 710. mL **b.** 75.0 kg **c.** 495 mm **d.** 9.5 kg

1.75 **a.** 66 gal **b.** 3 tablets **c.** 1800 mg

1.77 152 g of oxygen

1.79 0.045 kg water

1.81 lead; 11.3 g/mL

1.83 **a.** 1.20 g/mL **b.** 4.4 g/mL **c.** 3.10 g/mL

1.85 **a.** 210 g **b.** 575 g **c.** 62 oz

1.87 **a.** 1.030 **b.** 1.13 **c.** 0.85 g/mL

1.89 42 min

1.91 $0.85 per lb

1.93 16 onions

1.95 **a.** 96 crackers
b. 0.2 oz of fat
c. 110 g of sodium

1.97 0.35 oz

1.99 Dear Cousin,

If you want to find the density of a diamond you will need a balance and a graduated cylinder. You can measure the mass of the diamond on the balance. Since its shape is probably irregular, you would first place some water in the cylinder and record its level. Then place the diamond under water and record the new water level. Subtract the first volume measurement from the second to find the volume of the diamond. Don't forget the units! Density is calculated by dividing the mass of the diamond by its volume. Your answer will be in units of g/mL or g/cm^3. If you get 3.51g/cm^3, you probably have a real diamond and you should take it to a jewelry store. If the density doesn't match, it may just be a piece of glass.

Good luck,
Cousin Bill

1.101 790 g

1.103 720 cm^3

1.105 44 kg water

1.107 **a.** 1.4 kg (1400 g) body fat
b. 6.2 lb

1.109 **a.** 1.012 g/mL
b. 5.11 g

2

Atoms and Elements

Fireworks display vibrant colors that result when electrons in atoms of various elements used in the fireworks are excited by heat. Atoms of strontium and sodium produce the colors red and yellow.

Looking Ahead

2.1	Elements and Symbols
2.2	The Periodic Table
2.3	The Atom
2.4	Atomic Number and Mass Number
2.5	Isotopes and Atomic Mass
2.6	Electron Arrangement and the Periodic Law

www.chemplace.com/college/timberlake

Visit the web site above to test yourself, run simulations, and review key concepts from this chapter. For even more interactive review and quizzing, use the two CD-ROMs included with the book.

All of the matter that surrounds us is composed of primary substances called *elements.* There are different kinds of elements. The elements calcium and phosphorus build your teeth and bones. The hemoglobin that carries oxygen in your blood contains the element iron. The elements carbon, hydrogen, oxygen, and nitrogen derived from the digestion of food are used by the cells of your body to build proteins. Today, there are 112 different elements. Of these, 88 occur naturally and are found in different combinations, providing the great number of compounds that make up our world. The rest of the elements have been produced artificially and are not found in nature.

You have probably seen the element aluminum. Imagine that you are tearing a piece of aluminum foil into smaller and smaller pieces. Now imagine that you have a piece so small that you can no longer break it down further. Then you would have an atom of aluminum, the smallest particle of an element that still retains the characteristics of that element.

Study Pack

Chem of Life Atoms and Molecules: Atomic Structure 3

2.1 Elements and Symbols

Learning Goal

Given the name of an element, write its correct symbol; from the symbol, write the correct name.

Elements are primary substances from which all other things are built. They cannot be broken down into simpler substances. Many of the elements were named for planets, mythological figures, minerals, colors, geographic locations, and famous people. Some sources of the names of elements are listed in Table 2.1.

Chemical Symbols

Chemical symbols are one- or two-letter abbreviations for the names of the elements. Only the first letter of an element's symbol is capitalized; a second letter, if there is one, is lowercase. That way, we know when a different element is indicated. If both letters are capitalized, it represents the symbols of two different elements. For example, the element nickel has the symbol Ni. However, the two capital letters NI specify two elements, nitrogen (N) and iodine (I).

One-Letter Symbols		Two-Letter Symbols	
C	carbon	Co	cobalt
S	sulfur	Si	silicon
N	nitrogen	Ne	neon
I	iodine	Ni	nickel

Although most of the symbols use letters from the current names, some are derived from their ancient Latin or Greek names. For example, Na, the symbol for sodium, comes from the Latin word *natrium.* The symbol for iron, Fe, is derived from the Latin name *ferrum.* Table 2.2 lists the names and symbols of some common elements. Learning their names and symbols will greatly help your learning of chemistry. A complete list of all the elements and their symbols appears on the inside front cover of this text. Figure 2.1 shows examples of some common elements.

Table 2.1 Some Elements and Their Names

Element	Source of Name
Uranium	The planet Uranus
Titanium	Titans (mythology)
Chlorine	*Chloros,* "greenish yellow" (Greek)
Iodine	*Ioeides,* "violet" (Greek)
Magnesium	Magnesia, a mineral
Californium	California
Curium	Marie and Pierre Curie

2.1

Table 2.2 Names and Symbols of Some Common Elements

Name[a]	Symbol	Name[a]	Symbol
Aluminum	Al	Lead *(plumbum)*	Pb
Argon	Ar	Lithium	Li
Barium	Ba	Magnesium	Mg
Boron	B	Mercury *(hydrargyrum)*	Hg
Bromine	Br	Neon	Ne
Cadmium	Cd	Nickel	Ni
Calcium	Ca	Nitrogen	N
Carbon	C	Oxygen	O
Chlorine	Cl	Phosphorus	P
Cobalt	Co	Potassium *(kalium)*	K
Copper *(cuprum)*	Cu	Silicon	Si
Fluorine	F	Silver *(argentum)*	Ag
Gold *(aurum)*	Au	Sodium *(natrium)*	Na
Helium	He	Strontium	Sr
Hydrogen	H	Sulfur	S
Iodine	I	Tin *(stannum)*	Sn
Iron *(ferrum)*	Fe	Zinc	Zn

[a]Names given in parentheses are ancient Latin or Greek words from which the symbols are derived.

Figure 2.1 Some common elements and objects made of them: carbon (C) as graphite in a pencil; copper (Cu) in pennies; nickel (Ni) in nickels; aluminum (Al) in an aluminum can.

HEALTH NOTE

Elements Essential to Health

Several elements are essential for the well-being and survival of the human body. Some examples and the amounts present in a 60–kg person are listed in Table 2.3.

Table 2.3 Elements Essential to Health

Element	Symbol	Amount in a 60-kg Person	Where Found
Oxygen	O	39 kg	Water, carbohydrates, fats, proteins
Carbon	C	11 kg	Carbohydrates, fats, proteins
Hydrogen	H	6 kg	Water, carbohydrates, fats, proteins
Nitrogen	N	2 kg	Proteins, DNA, RNA
Calcium	Ca	1 kg	Bones, teeth
Phosphorus	P	0.6 kg	Bones, teeth, DNA, RNA
Potassium	K	0.2 kg	Inside cells (important in conduction of nerve impulses)
Sulfur	S	0.2 kg	Some amino acids
Sodium	Na	0.1 kg	Body fluids (important in nerve conduction and fluid balance)
Magnesium	Mg	0.1 kg	Inside cells (important in enzyme function)
Chlorine	Cl	0.1 kg	Outside of cells (major electrolyte)

HEALTH NOTE

Latin Names for Elements in Clinical Usage

In medicine, the Latin names for sodium and potassium are often used. The condition in which there is too much sodium in the body is called hypernatremia, and a low sodium level is called hyponatremia. In the case of potassium, both the modern name and the Latin name are used. For example, a high potassium level may be called hyperpotassemia or hyperkalemia; a below-normal potassium level may be called hypopotassemia or hypokalemia.

EXPLORE YOUR WORLD

Origins of Element Names

Use a reference such as a dictionary, the Internet (look for an interactive periodic table), or the CRC *Handbook of Chemistry and Physics* (Chemical Rubber Publishing Company). Look up an element and determine the origin of its name. For example, polonium is named for Poland and uranium is named for the planet Uranus. Make a list of the element names and their origins.

Questions

1. What elements are named for a country, a region, a state, or a city?
2. What elements are named for people or figures in mythology?
3. What other sources do you find for the names of elements?

Figure 2.2 Cake molds made of copper metal.

2.1

SAMPLE 2.1 PROBLEM

Writing Chemical Symbols

What are the chemical symbols for the following elements?

a. carbon **b.** nitrogen **c.** chlorine **d.** copper

Solution

a. C **b.** N **c.** Cl **d.** Cu

Study Check

What are the chemical symbols for the elements silicon, sulfur, and silver?

SAMPLE 2.2 PROBLEM

Naming Chemical Elements

Give the name of the element that corresponds to each of the following chemical symbols:

a. Zn **b.** K **c.** H **d.** Fe

Solution

a. zinc **b.** potassium **c.** hydrogen **d.** iron

Study Check

What are the names of the elements with the chemical symbols Mg, Al, and F?

Table 2.4 Some Physical Properties of Copper

Color	Reddish-orange
Odor	Odorless
Melting point	1083°C
Boiling point	2567°C
State at 25°C	Solid
Luster	Very shiny
Conduction of electricity	Excellent
Conduction of heat	Excellent

Physical Properties

Suppose a friend asks you to look into a mirror and describe what you see. You might tell her the color of your hair, eyes, and skin, or the length and texture of your hair. Perhaps you have freckles or dimples. These are descriptions of some of your physical properties. Every element has **physical properties** too, which are those characteristics that can be observed or measured without affecting the identity of an element. The physical properties of a substance include its shape, color, odor, taste, density, hardness, melting point, and boiling point. A sample of copper, shown in Figure 2.2, has the physical properties listed in Table 2.4.

EXPLORE YOUR WORLD

Physical Properties of Aluminum

Obtain a piece of aluminum foil. List all the physical properties of aluminum that you can observe. Use a reference manual such as the *Handbook of Chemistry and Physics* to determine more physical properties.

Questions

1. What is the chemical symbol for aluminum?
2. What is the color and shape of the aluminum sample?
3. What is its density?
4. What are the melting and boiling points of aluminum?

S·A·M·P·L·E 2.3 P·R·O·B·L·E·M

Physical Properties

What physical property could you use to distinguish between copper and silver?

Solution

Copper has a red-orange color and silver has a gray color.

Study Check

What are some physical properties of the mercury in a thermometer?

Q·U·E·S·T·I·O·N·S & P·R·O·B·L·E·M·S

Elements and Symbols

Answers for odd-numbered Questions and Problems are given at the end of Chapter 2. Complete solutions to odd-numbered Questions and Problems are in the *Study Guide*.

2.1 Write the symbols for the following elements:

a. copper **b.** silicon **c.** potassium **d.** nitrogen
e. iron **f.** barium **g.** lead **h.** strontium

2.2 Write the symbols for the following elements:

a. oxygen **b.** lithium **c.** sulfur **d.** aluminum
e. hydrogen **f.** neon **g.** tin **h.** gold

2.3 Write the name of the element for each symbol.

a. C **b.** Cl **c.** I **d.** Hg
e. F **f.** Ar **g.** Zn **h.** Ni

2.4 Write the name of the element for each symbol.

a. He **b.** P **c.** Na **d.** Mg
e. Ca **f.** Br **g.** Cd **h.** Si

2.5 What elements are in the following substances?

a. table salt, NaCl **b.** plaster casts, $CaSO_4$
c. Demerol, $C_{15}H_{22}ClNO_2$ **d.** antacid, $CaCO_3$

2.6 What elements are in the following substances?

a. water, H_2O **b.** baking soda, $NaHCO_3$
c. lye, NaOH **d.** sugar, $C_{12}H_{22}O_{11}$

2.7 Describe the color of the following elements shown in Figure 2.1.

a. carbon **b.** copper

2.8 Describe the color of the following elements shown in Figure 2.1.

a. nickel **b.** aluminum

2.2 The Periodic Table

Learning Goal

Use the periodic table to identify the group and the period of an element and whether it is a metal or a nonmetal.

By the late 1800s, scientists began to notice that certain elements looked alike and behaved much the same way in reactions. A Russian chemist, Dmitri Mendeleev, suggested that the elements could be arranged in groups or families according to their similar properties. Today, we call this arrangement of elements the **periodic table.**

Groups and Periods

A vertical column on the periodic table contains a **group** of elements, which have similar properties. The elements in the table are divided into **representative elements** and the **transition elements** as shown in Figure 2.3. We have identified the groups of the representative elements with the numbers 1–8 because most of our discussion in this text will concern them.

A single horizontal row in the periodic table is called a **period.** Each row is counted from the top of the table down as Period 1, Period 2, and so on. The first period contains only the elements hydrogen (H) and helium (He). The second period, which is the second row of elements, contains lithium (Li), beryllium (Be), boron (B), carbon (C), nitrogen (N), oxygen (O), fluorine (F), and neon (Ne). The third period begins with sodium (Na) and ends at argon (Ar). There are seven periods in the periodic table.

2.2

S·A·M·P·L·E 2.4 P·R·O·B·L·E·M

Groups and Periods in the Periodic Table

State whether each set represents elements in a group, a period, or neither:

a. F, Cl, Br, I **b.** Na, Al, P **c.** K, Al, O

Solution

a. The elements F, Cl, Br, and I are part of a group of elements; they all appear in the vertical column 7.

Figure 2.3 Groups and periods in the periodic table.

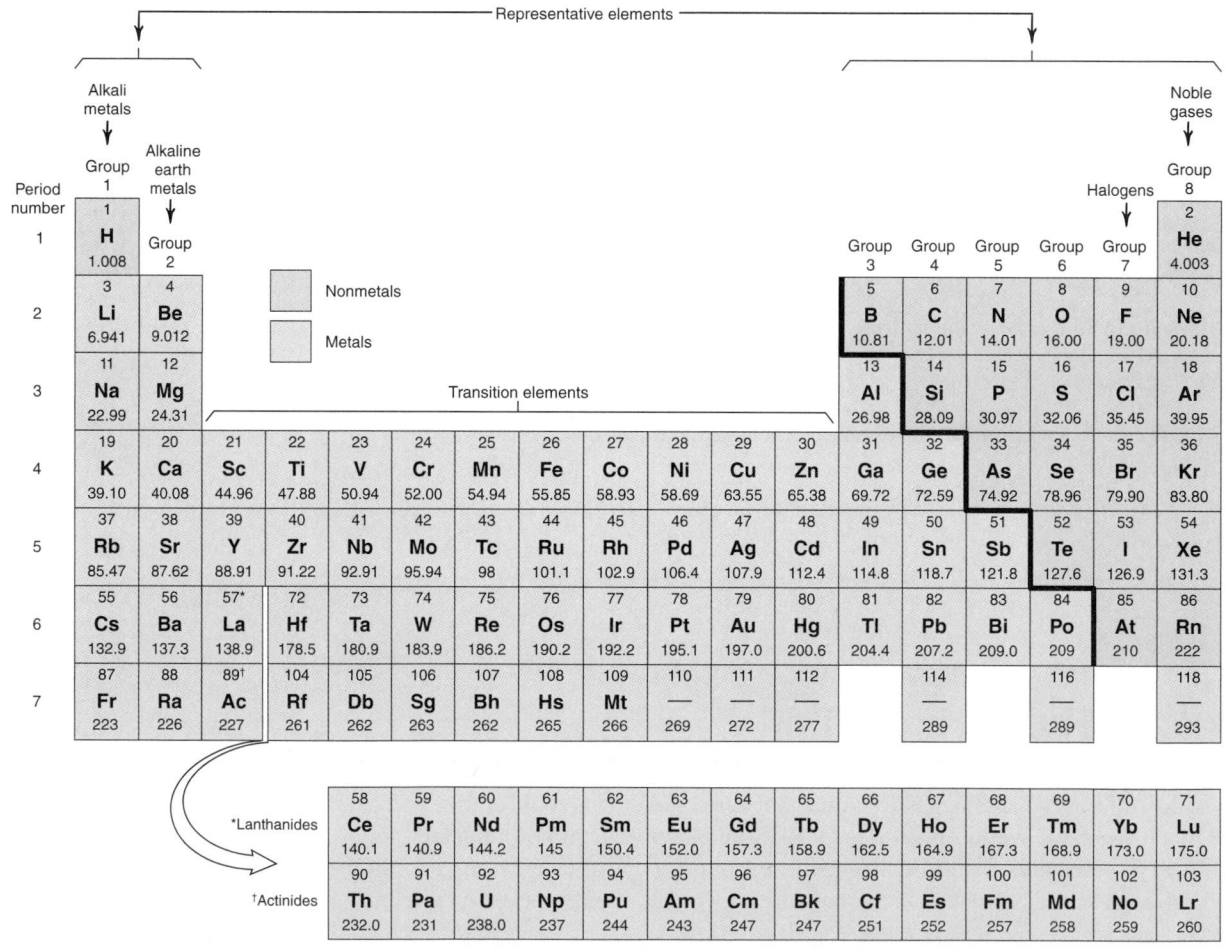

b. The elements Na, Al, and P all appear in the third row or third period in the periodic table.
c. Neither. The elements K, Al, and O are not part of the same group and they do not belong to the same period.

Study Check

a. What elements are found in Period 2?
b. What elements are found in Group 2?

S·A·M·P·L·E **2.5** P·R·O·B·L·E·M

Stating Group and Period Number

Identify the period number and the group number for the following elements:
a. calcium **b.** tin

Solution

The period is found by counting down the horizontal rows of the elements on the periodic table, and the group number is found at the top of the vertical column that contains that element.
a. Calcium (Ca) is in Period 4 and Group 2.
b. Tin (Sn) is in Period 5 and Group 4.

Study Check

Give the symbols of the elements that are represented by the following period and group numbers:
a. Period 3, Group 5
b. Period 6, Group 8

Classification of Groups

Several groups in the periodic table have special names, as Figure 2.3 shows. Group 1 elements, lithium (Li), sodium (Na), potassium (K), rubidium (Rb), cesium (Cs), and francium (Fr), are a family of elements known as the **alkali metals** (see Figure 2.4). The elements within this group are soft, shiny metals that are good conductors of heat and electricity and have relatively low melting points. Alkali metals react vigorously with water and form white products when they combine with oxygen.

Group 2 elements, beryllium (Be), magnesium (Mg), calcium (Ca), strontium (Sr), barium (Ba), and radium (Ra), are called the **alkaline earth metals.** They are also shiny metals like those in Group 1, but they are not quite so reactive.

The **halogens** are found on the right side of the periodic table in Group 7. They include the elements fluorine (F), chlorine (Cl), bromine (Br), and iodine (I), as shown in Figure 2.5.

Lithium (Li)

Sodium (Na)

Potassium (K)

Figure 2.4 Lithium (Li), sodium (Na), and potassium (K), some alkali metals from Group 1.

Figure 2.5 From left to right: Chlorine (Cl_2), bromine (Br_2), and iodine (I_2), some halogens from Group 7.

2.2

Group 8 contains the **noble gases,** helium (He), neon (Ne), argon (Ar), krypton (Kr), xenon (Xe), and radon (Rn). They are quite unreactive and are seldom found in combination with other elements.

HEALTH NOTE

Calcium and Strontium

Calcium (Ca) and strontium (Sr) are two elements of the alkaline earth group, Group 2. The chemical behavior of strontium is so similar to that of calcium that when strontium is ingested, it replaces some of the calcium in the bones and teeth. This similarity in behavior caused great concern among scientists during nuclear testing because radioactive strontium is a product of certain nuclear reactions. If the radioactive strontium were to drift to cattle-grazing lands, it could become a part of cow's milk and eventually find its way to the bones of young children. Once it is there, the effects of the radioactivity are detrimental to proper growth and development.

Metals and Nonmetals

Figure 2.6 Silver, a shiny gray solid, is a metal.

Another feature of the periodic table is the heavy zigzag line that separates the elements into the *metals* and the *nonmetals.* The metals are those elements on the left of the line *except for hydrogen,* and the nonmetals are the elements on the right.

In general, most **metals** are shiny solids. They can be shaped into wires (ductile) or hammered into a flat sheet (malleable). Metals are good conductors of heat and electricity. They usually melt at higher temperatures than nonmetals. All of the metals are solids at room temperature, except for mercury (Hg), which is a liquid. Some typical metals are sodium (Na), magnesium (Mg), copper (Cu), gold (Au), silver (Ag), iron (Fe), and tin (Sn).

Nonmetals are not very shiny, malleable, or ductile, and they are often poor conductors of heat and electricity. They typically have low melting points and low densities. You may have heard of nonmetals such as hydrogen (H), carbon (C), nitrogen (N), oxygen (O), chlorine (Cl), and sulfur (S). Table 2.5 compares some characteristics of silver, a metal, with those of sulfur, a nonmetal, as seen in Figures 2.6 and 2.7.

Figure 2.7 Sulfur, a yellow solid, is a nonmetal.

Table 2.5 Some Characteristics of Silver, a Metal, and Sulfur, a Nonmetal

Silver	Sulfur
A shiny metal	A dull, yellow nonmetal
Extremely ductile	Brittle
Can be hammered into sheets (malleable)	Shatters when hammered
Good conductor of heat and electricity	Poor conductor, good insulator
Used in coins, jewelry, tableware	Used in gunpowder, rubber, fungicides
Density 10.5 g/mL	Density 3.1 g/mL
Melting point 962°C	Melting point 113°C

S·A·M·P·L·E 2.6 P·R·O·B·L·E·M

Classifying Elements as Metals and Nonmetals

Using a periodic table, classify the following elements as metals or nonmetals:
a. Na **b.** Si **c.** Cl **d.** Cu

Solution

a. Metal; sodium is located to the left of the heavy zigzag line that separates metals and nonmetals.
b. Nonmetal; silicon is on the right side of the zigzag line.
c. Nonmetal; chlorine is on the right side of the zigzag line.
d. Metal; copper is located to the left of the zigzag line.

Study Check

Which of the following elements, nickel or nitrogen, would be a gas and a poor conductor of heat?

HEALTH NOTE

Some Important Trace Elements in the Body

Some metals and nonmetals known as trace elements are essential to the proper functioning of the body. Although they are required in very small amounts, their absence can disrupt major biological processes and cause illness. The trace elements listed in Table 2.6 are present in the body combined with other elements.

Table 2.6 Some Important Trace Elements in the Body

Element	Adult DV[a]	Biological Function	Deficiency Symptoms	Dietary Sources
Iron (Fe)	10 mg (males) 18 mg (females)	Formation of hemoglobin; enzymes	Dry skin, spoon nails, decreased hemoglobin count, anemia	Beef, kidneys, liver, egg yolk, oysters, spinach, beans, apricots, raisins, whole-wheat bread
Copper (Cu)	2.0–5.0 mg	Necessary in many enzyme systems; growth; aids formation of red blood cells and collagen	Uncommon; anemia, decreased white cell count; bone demineralization	Nuts, organ meats, whole-wheat grains, shellfish, eggs, poultry, leafy green vegetables
Zinc (Zn)	15 mg	Amino acid metabolism; enzyme systems; energy production; collagen	Retarded growth and bone formation; skin inflammation; loss of taste and smell; poor healing	Wheat germ, shellfish, milk, lima beans, fish, eggs, whole grains, turkey (dark meat), cheddar cheese
Manganese (Mn)	2.5–5.0 mg	Necessary for some enzyme systems; collagen formation; central nervous system; fat and carboyhydrate metabolism; blood clotting	Abnormal skeletal growth; impairment of central nervous system	Cereals, peas, beans, lettuce, wheat bran, meat, poultry, fish
Iodine (I)	150 μg	Necessary for activity of thyroid gland	Hypothyroidism; goiter; cretinism	Iodized table salt, seafood
Fluorine (F)	1.5–4.0 mg	Necessary for solid tooth formation and retention of calcium in bones with aging	Dental cavities	Tea, fish, milk, eggs, water in some areas, supplementary drops, toothpaste

[a]DV, daily value.

QUESTIONS & PROBLEMS

The Periodic Table

2.2

2.9 Identify the group or period number described by each of the following statements:
a. contains the elements C, N, and O
b. begins with helium
c. the alkali metals
d. ends with neon

2.10 Identify the group or period number described by each of the following statements:
a. contains Na, K, and Rb
b. the horizontal row that begins with Li
c. the noble gases
d. contains F, Cl, Br, and I

2.11 Classify the following as an alkali metal, alkaline earth metal, transition element, halogen, or noble gas:
a. Ca **b.** Fe **c.** Xe **d.** Na **e.** Cl

2.12 Classify the following as an alkali metal, alkaline earth metal, transition element, halogen, or noble gas:
a. Ne **b.** Mg **c.** Cu **d.** Br **e.** Ba

2.13 Give the symbol of the element described by the following:
a. Group 4, Period 2
b. a noble gas in Period 1
c. an alkali metal in Period 3
d. Group 2, Period 4
e. Group 3, Period 3

2.14 Give the symbol of the element described by the following:
a. an alkaline earth metal in Period 2
b. Group 5, Period 3
c. a noble gas in Period 4
d. a halogen in Period 5
e. Group 4, Period 4

2.15 Is each of the following elements most likely a metal or nonmetal?
a. calcium
b. sulfur
c. an element that is shiny
d. does not conduct heat
e. located in Group 7
f. phosphorus
g. magnesium
h. silver

2.16 Is each of the following elements most likely a metal or nonmetal?
a. located in Group 2
b. a good conductor of electricity
c. chlorine
d. iron
e. an element that is not shiny
f. silicon
g. nitrogen
h. aluminum

2.3 The Atom

Study Pack

Chem of Life Atoms and Molecules: Atomic Structure 1

Learning Goal

Describe the electrical charge and location in an atom for a proton, a neutron, and an electron.

2.3

All the elements you see listed on the periodic table are made up of small particles called atoms. An **atom** is the smallest particle of an element that retains the characteristics of that element.

Billions of atoms are packed together to build you and everything around you. The paper in this book contains atoms of carbon, hydrogen, and oxygen. The ink on this paper, even the dot over the letter *i,* contains huge numbers of atoms. There are as many atoms in that dot as there are seconds in 10 billion years.

The concept of the atom is relatively recent. Although the Greek philosophers in 500 B.C. reasoned that everything must contain minute particles they called *atomos,* the idea of atoms did not become a scientific theory until 1808. Then John Dalton (1766–1844) developed an atomic theory that proposed that atoms were responsible for the combinations of elements found in compounds.

Atomic Theory

1. All matter is made up of tiny particles called atoms.
2. All atoms of a given element are similar to one another and different from atoms of other elements.
3. Atoms of two or more different elements combine to form compounds. A particular compound is always made up of the same kinds of atoms and always has the same number of each kind of atom.
4. A chemical reaction involves the rearrangement, separation, or combination of atoms. Atoms are never created or destroyed during a chemical reaction.

Atoms are the building blocks of everything we see around us; yet we cannot see an atom or even a billion atoms with the naked eye. However, when billions and billions of atoms are packed together, the characteristics of each atom are added to those of the next until we can see the characteristics we associate with the element. For example, a small piece of the shiny, copper-colored element we call copper consists of many, many copper atoms. Through a special kind of microscope called a scanning tunneling microscope, we can now see images of individual atoms, such as the atoms of carbon in graphite shown in Figure 2.8.

Figure 2.8 Graphite, a form of carbon, magnified millions of times by a scanning tunneling microscope. This instrument generates an image of the atomic structure. The round yellow objects are atoms.

Figure 2.9 Attraction and repulsion of electrical charges.

2.3 Parts of an Atom

By the early part of the twentieth century, growing evidence indicated that atoms were not solid spheres, as Dalton had imagined. New experiments showed that atoms were composed of even smaller bits of matter called **subatomic particles.** The chemistry of an element depends upon the subatomic particles that are the building blocks of the atoms.

There are three subatomic particles of interest to us, the proton, neutron, and electron. Two of these carry electrical charges. The **proton** has a positive charge (+), and the **electron** carries a negative charge (−). The **neutron** has no electrical charge; it is neutral.

Like charges repel; they push away from each other. When you brush your hair on a dry day, electrical charges that are alike build up on the brush and in your hair; as a result your hair flies away from the brush. Opposite or unlike charges attract. The crackle of clothes taken from the clothes dryer indicates the presence of electrical charges. The clinginess of the clothing is due to the attraction of opposite, unlike charges, as shown in Figure 2.9.

EXPLORE YOUR WORLD

Repulsion and Attraction

1. Obtain a tape dispenser with clear tape, a hairbrush, comb, and a piece of paper. Tear off a piece of the tape that is about 20 cm long (the length of your hand). Stick the tape to the edge of a table leaving the end hanging down. Tear off a second piece of tape and slowly bring it close to the first one. What happens? Is there an attraction or repulsion?

   Slide your thumb and finger along the tape you are holding. Bring it close to the piece that is hanging from the table. What happens? Is there an attraction or repulsion? Attach the second tape to the edge of the table. Brush your hair and bring the brush close to each piece of tape hanging from the table. What do you observe?
2. Tear a small piece of paper into bits. Brush your hair several times and place the brush just above the bits of paper. Use your knowledge of electrical charges to give an explanation for your observations. Try the same experiment with a comb.

Questions

1. What happens when objects with like charges are placed close together?
2. What happens when objects with unlike charges are placed close together?

Mass

All of the subatomic particles are extremely small compared with the things you see around you. One proton has a mass of 1.7×10^{-24} g, and the neutron is about the same. The mass of the electron is even less. Because the mass of a subatomic particle is so minute, chemists use a small unit of mass called an **atomic mass unit (amu).** An amu is defined as one-twelfth of the mass of the carbon atom with 6 protons and 6 neutrons, a standard with which the mass of every other atom is compared. In biology, the atomic mass unit is called a dalton in honor of John Dalton. On the amu scale, the proton and neutron each have a mass of about 1 amu. Because the electron is so light, its mass is usually ignored in atomic mass calculations. Table 2.7 summarizes some information about the subatomic particles in an atom.

Table 2.7 Particles in the Atom

Subatomic Particle	Symbol	Electrical Charge	Approximate Mass	Location in Atom
Proton	p or p^+	1+	1 amu	Inside nucleus
Neutron	n or n^0	0	1 amu	Inside nucleus
Electron	e^-	1−	0	Outside nucleus

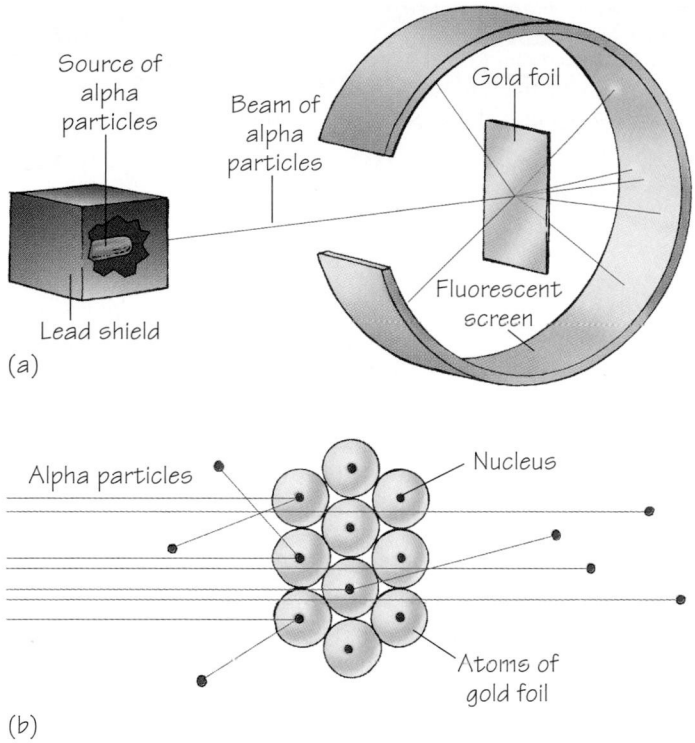

Figure 2.10 **(a)** Positive alpha particles are aimed at a piece of gold foil. **(b)** Alpha particles that come close to the atomic nucleus are deflected from their straight path.

Structure of the Atom

By the early 1900s, scientists knew that atoms contained subatomic particles, but they did not know how those particles were arranged. In 1911, an experiment led Ernest Rutherford to propose that most of the mass of the atom was contained in a small region at the center of the atom. In his gold-foil experiment, positively charged alpha particles were aimed at a very thin sheet of gold foil as seen in Figure 2.10. He thought that the heavy alpha particles would travel in straight paths and go right on through the gold foil. But much to his surprise, some of the alpha particles changed direction as they passed through the gold foil. Some were deflected so much that they seemed to go back in the direction of the source of the particles. According to Rutherford, it was as though he had shot a cannonball at a piece of tissue paper, and it bounced back at him.

Nucleus of the Atom

From the gold-foil experiment, Rutherford concluded that the alpha particles would change direction only if they came close to some type of dense, positively charged region in the atom. This dense core of an atom called the **nucleus,** where the protons and neutrons are located, has a positive charge and contains most of the mass of an atom. Most of an atom is empty space, which is occupied only by fast-moving electrons. (See Figure 2.11.) If an atom were the size of a football stadium, the nucleus would be about the size of a golf ball placed in the center of the field.

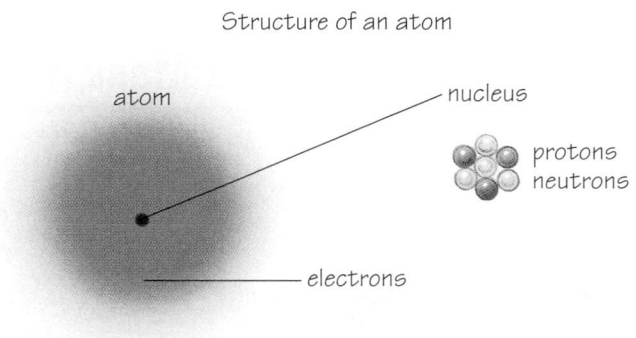

Figure 2.11 Arrangement of the subatomic particles in an atom. The protons and neutrons are located in the nucleus at the center of the atom; the electrons are located outside the nucleus.

S·A·M·P·L·E 2.7 P·R·O·B·L·E·M

Identifying Subatomic Particles

Complete the following table for subatomic particles:

Name	Symbol	Mass (amu)	Charge	Location in Atom
Electron	_____	_____	_____	_____
_____	_____	1	0	_____

Solution

Electron	e^-	0	1−	Outside nucleus
Neutron	n^0	1	0	Nucleus

Study Check

Give the symbol, mass, electrical charge, and location of a proton in the atom.

Q·U·E·S·T·I·O·N·S & P·R·O·B·L·E·M·S

The Atom

2.17 Is a proton, neutron, or electron described by each of the following?
a. has the smallest mass **b.** carries a positive charge
c. is found outside the nucleus **d.** is electrically neutral

2.18 Is a proton, neutron, or electron described by each of the following?
a. has a mass about the same as a proton's **b.** is found in the nucleus
c. is found in the larger part of the atom **d.** carries a negative charge

2.19 What did Rutherford's gold-foil experiment tell us about the organization of the subatomic particles in an atom?

2.20 Why does the nucleus in every atom have a positive charge?

2.21 Which of the following particles have opposite charges?
a. two protons **b.** a proton and an electron
c. two electrons **d.** a proton and a neutron

2.22 Which of the following pairs of particles have the same charges?
a. two protons **b.** a proton and an electron
c. two electrons **d.** an electron and a neutron

2.23 On a dry day, your hair flies away when you brush it. What types of charges are on the hair strands?

2.24 Sometimes clothes removed from the dryer cling together. What kinds of charges are on the clothes?

Atoms and Molecules: Atomic Structure 3

Learning Goal

Given the atomic number and the mass number of an atom, state the number of protons, neutrons, and electrons.

2.4 Atomic Number and Mass Number

All of the atoms of the same element always have the same number of protons. This feature distinguishes atoms of one element from atoms of all the other elements. An **atomic number,** which is equal to the number of protons in the nucleus of an atom, is used to identify each element.

Atomic number = number of protons in an atom

On the inside front cover of this text is a list of all the elements, their chemical symbols, and their atomic numbers. Next to it is a periodic table, which gives all of the elements in order of increasing atomic number. In the periodic table, the atomic number is the whole number that appears above the symbol for each element, as illustrated in Figure 2.12. For example, a hydrogen atom, with atomic number 1, has 1 proton; a helium atom, with atomic number 2, has 2 protons; an atom of oxygen, with atomic number 8, has 8 protons; and gold, with atomic number 79, has 79 protons.

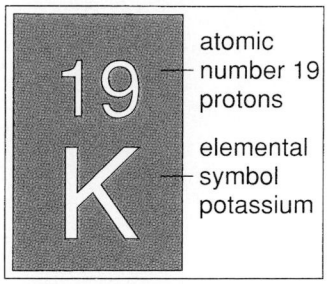

Figure 2.12 Atomic numbers appear above the symbols of the elements listed in the periodic table. The atomic number indicates the number of protons in an atom of that element.

Atoms Are Neutral

An atom is electrically neutral. That means that the number of protons in an atom is equal to the number of electrons. This electrical balance gives an atom an overall charge of zero. Thus, in every atom, the atomic number also gives the number of electrons. Table 2.8 lists the atomic number and corresponding number of protons and electrons for some atoms.

S·A·M·P·L·E· **2.8** ·P·R·O·B·L·E·M

Using Atomic Number to Find the Number of Protons and Electrons

Using the periodic table in Figure 2.3, state the atomic number, number of protons, and number of electrons for an atom of each of the following elements:
a. nitrogen **b.** magnesium **c.** bromine

Solution

a. atomic number 7; 7 protons and 7 electrons
b. atomic number 12; 12 protons and 12 electrons
c. atomic number 35; 35 protons and 35 electrons

Study Check

Consider an atom that has 26 electrons.
a. How many protons are in its nucleus? **b.** What is its atomic number?
c. What is its name, and what is its symbol?

Mass Number

We now know that the protons and neutrons determine the mass of the nucleus. For any atom, the **mass number** is the sum of the number of protons and neutrons in the nucleus.

Mass number = number of protons + number of neutrons

For example, an atom of oxygen that contains 8 protons and 8 neutrons has a mass number of 16. An atom of iron that contains 26 protons and 30 neutrons has a mass number of 56. Table 2.9 illustrates the relationship between atomic number, mass number, and the number of protons, neutrons, and electrons in some atoms of different elements.

Table 2.8 Number of Protons and Electrons in Some Atoms

Element	Symbol	Atomic Number	Number of Protons	Number of Electrons	Overall Charge
Hydrogen	H	1	1	1	0
Helium	He	2	2	2	0
Oxygen	O	8	8	8	0
Sodium	Na	11	11	11	0
Iron	Fe	26	26	26	0
Gold	Au	79	79	79	0

Table 2.9 Composition of Some Atoms of Different Elements

Element	Symbol	Atomic Number	Mass Number	Number of Protons	Number of Neutrons	Number of Electrons
Hydrogen	H	1	1	1	0	1
Nitrogen	N	7	14	7	7	7
Phosphorus	P	15	31	15	16	15
Chlorine	Cl	17	37	17	20	17
Iron	Fe	26	56	26	30	26

2.4

S · A · M · P · L · E · 2.9 · P · R · O · B · L · E · M

Calculating Mass Number

Calculate the mass number of an atom by using the information given:

a. 5 protons and 6 neutrons
b. 18 protons and 22 neutrons
c. atomic number 48 and 64 neutrons

Solution

a. Mass number = 5 + 6 = 11
b. Mass number = 18 + 22 = 40
c. Mass number = 48 + 64 = 112

Study Check

What is the mass number of a silver atom that has 60 neutrons?

S · A · M · P · L · E · 2.10 · P · R · O · B · L · E · M

Calculating Numbers of Protons and Neutrons

For an atom of phosphorus that has a mass number of 31, determine the following:

a. the number of protons
b. the number of neutrons
c. the number of electrons

Solution

a On the periodic table, the atomic number of phosphorus is 15. A phosphorus atom has 15 protons.

b. The number of neutrons in this atom is found by subtracting the atomic number from the mass number. The number of neutrons is 16.

$$\text{Mass number} - \text{atomic number} = \text{number of neutrons}$$
$$31 - 15 = 16$$

c. Because an atom is neutral, there is an electrical balance of protons and electrons. Because the number of electrons is equal to the number of protons, the phosphorus atom has 15 electrons.

Study Check

How many neutrons are in the nucleus of a bromine atom that has a mass number of 80?

QUESTIONS & PROBLEMS

Atomic Number and Mass Number

2.25 Would you use atomic number, mass number, or both to obtain the following?
a. number of protons in an atom
b. number of neutrons in an atom
c. number of particles in the nucleus
d. number of electrons in a neutral atom

2.26 What do you know about the subatomic particles from the following?
a. atomic number
b. mass number
c. mass number − atomic number
d. mass number + atomic number

2.27 Write the names and symbols of the elements with the following atomic numbers:
a. 3 **b.** 9 **c.** 20 **d.** 30
e. 10 **f.** 14 **g.** 53 **h.** 8

2.28 Write the names and symbols of the elements with the following atomic numbers:
a. 1 **b.** 11 **c.** 19 **d.** 26
e. 35 **f.** 47 **g.** 15 **h.** 2

2.29 How many protons and electrons are there in a neutral atom of the following?
a. magnesium **b.** zinc **c.** iodine **d.** potassium

2.30 How many protons and electrons are there in a neutral atom of the following?
a. carbon **b.** fluorine **c.** calcium **d.** sulfur

2.31 Complete the following table for neutral atoms.

Name of Element	Symbol	Atomic Number	Mass Number	Number of Protons	Number of Neutrons	Number of Electrons
	Al		27			
		12			12	
Potassium					20	
				16	15	
			56			26

2.32 Complete the following table for neutral atoms.

Name of Element	Symbol	Atomic Number	Mass Number	Number of Protons	Number of Neutrons	Number of Electrons
	N		15			
Calcium			42			
				38	50	
		14			16	
		56	138			

Study Pack

Chem of Life Atoms and Molecules: Atomic Structure 4

Learning Goal

Give the number of protons, electrons, and neutrons in the isotopes of an element.

2.5

2.5 Isotopes and Atomic Mass

We have seen that all atoms of the same element have the same number of protons and electrons. However, the atoms of any one element are not completely identical because they can have different numbers of neutrons. **Isotopes** are atoms of the same element that have different numbers of neutrons. For example, all atoms of the element magnesium (Mg) have 12 protons. However, some magnesium atoms have 12 neutrons, others have 13 neutrons, and still others have 14 neutrons. The differences in numbers of neutrons for these magnesium atoms cause their mass numbers to be different but not their chemical behavior. The three isotopes of magnesium have the same atomic number but different mass numbers.

To distinguish between the different isotopes of an element, we can write an **atomic symbol** that indicates the mass number and the atomic number of the atom.

Atomic Symbol for an Isotope of Magnesium

Mass number → 24
Symbol of element → Mg
Atomic number → 12

An isotope may be referred to by its name or symbol followed by the mass number, such as magnesium–24 or Mg–24. Magnesium has three naturally occurring isotopes, as shown in Table 2.10. Table 2.11 lists atomic symbols for the isotopes of some selected elements.

Table 2.10 Isotopes of Magnesium

	$12e^-$ $12p^+$ $12n^0$	$12e^-$ $12p^+$ $13n^0$	$12e^-$ $12p^+$ $14n^0$
Atomic symbol	$^{24}_{12}Mg$	$^{25}_{12}Mg$	$^{26}_{12}Mg$
Number of protons	12	12	12
Number of electrons	12	12	12
Mass number	**24**	**25**	**26**
Number of neutrons	**12**	**13**	**14**

Table 2.11 Isotopes and Average Atomic Mass of Some Elements

Element	Isotopes	Average Atomic Mass
Lithium	$^{6}_{3}Li$ $^{7}_{3}Li$	6.941 amu
Carbon	$^{12}_{6}C$ $^{13}_{6}C$ $^{14}_{6}C$	12.01 amu
Oxygen	$^{16}_{8}O$ $^{17}_{8}O$ $^{18}_{8}O$	16.00 amu
Sulfur	$^{32}_{16}S$ $^{33}_{16}S$ $^{34}_{16}S$ $^{36}_{16}S$	32.06 amu
Chlorine	$^{35}_{17}Cl$ $^{37}_{17}Cl$	35.45 amu
Copper	$^{63}_{29}Cu$ $^{65}_{29}Cu$	63.55 amu

S·A·M·P·L·E **2.11** ·P·R·O·B·L·E·M

Identifying Protons and Neutrons in Isotopes

State the number of protons and neutrons in the following isotopes of neon (Ne):
a. $^{20}_{10}Ne$ **b.** $^{21}_{10}Ne$ **c.** $^{22}_{10}Ne$

Solution

The atomic number of Ne is 10; each isotope has 10 protons. The number of neutrons in each isotope is found by subtracting the atomic number (10) from each mass number.
a. 10 protons; 10 neutrons (20 − 10)
b. 10 protons; 11 neutrons (21 − 10)
c. 10 protons; 12 neutrons (22 − 10)

Study Check

Write a symbol for the following isotopes:
a. a nitrogen atom with 8 neutrons
b. an atom with 20 protons and 22 neutrons
c. an atom with mass number 27 and 14 neutrons

Average Atomic Mass

In laboratory work, a scientist generally uses samples that contain many atoms of an element. Among those atoms are all of the various isotopes with their different masses. To obtain a convenient mass to work with, chemists use the mass of an "average atom" of each element. This average atom has an **atomic mass,** which is the weighted average mass of all of the naturally occurring isotopes of that element.

To calculate an atomic mass, the percent abundance of each isotope must be known. For example, in chlorine (Cl), 75.5% of the atoms have a mass number of 35; the other 24.5% have a mass number of 37. Using these values and the mass number of each isotope (which is quite close to each isotope's mass in amu), we can calculate the average atomic mass for the element.

Determination of the Average Atomic Mass of Chlorine

Isotope	Percent of Sample		Mass Number		Contribution to Average Atom
$^{35}_{17}Cl$	$\frac{75.5}{100}$	×	35	=	26.4 amu
$^{37}_{17}Cl$	$\frac{24.5}{100}$	×	37	=	9.1 amu
			Atomic mass Cl	=	35.5 amu

This calculation illustrates the way in which isotopes determine the overall atomic mass for an element. Most elements consist of several isotopes, and this is one reason atomic masses listed on the periodic table are seldom whole numbers.

On the periodic table, the atomic mass is given below the symbol of each element. The element chlorine is shown in Figure 2.13 as it appears in the periodic table.

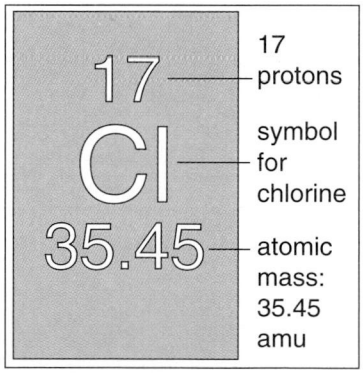

Figure 2.13 Information for chlorine in the periodic table.

SAMPLE 2.12 PROBLEM

Identifying Atomic Mass

Using the periodic table, give the atomic mass to one decimal place (tenths place) for each of the following elements.

a. hydrogen **b.** iron **c.** sulfur **d.** potassium

Solution

a. 1.0 amu **b.** 55.9 amu **c.** 32.1 amu **d.** 39.1 amu

2.5

Study Check

What is the name of the element with atomic mass 65.38 amu?

QUESTIONS & PROBLEMS

Isotopes and Atomic Mass

2.33 What are the number of protons, neutrons, and electrons in the following isotopes?

a. $^{27}_{13}Al$ **b.** $^{52}_{24}Cr$ **c.** $^{34}_{16}S$ **d.** $^{56}_{26}Fe$

2.34 What are the number of protons, neutrons, and electrons in the following isotopes?

a. $^{2}_{1}H$ **b.** $^{14}_{7}N$ **c.** $^{26}_{14}Si$ **d.** $^{70}_{30}Zn$

2.35 Write the atomic symbols for isotopes with the following:

a. 15 protons and 16 neutrons
b. 35 protons and 45 neutrons
c. 13 electrons and 14 neutrons
d. a chlorine atom with 18 neutrons

2.36 Write the atomic symbols for isotopes with the following:

a. an oxygen atom with 10 neutrons
b. 4 protons and 5 neutrons
c. 26 electrons and 30 neutrons
d. a mass number of 24 and 13 neutrons

2.37 There are four isotopes of sulfur with mass numbers 32, 33, 34, and 36.

a. Write the atomic symbol for each of these atoms.
b. How are these isotopes alike?
c. How are they different?
d. Why is the atomic mass of sulfur listed on the periodic table not a whole number?

2.38 There are four isotopes of strontium with mass numbers 84, 86, 87, 88.

a. Write the atomic symbol for each of these atoms.
b. How are these isotopes alike?
c. How are they different?
d. Why is the atomic mass of strontium listed on the periodic table not a whole number?

2.39 What information do you need about an element and its isotopes in order to calculate its average atomic mass?

2.40 Copper consists of two isotopes, copper-63 and copper-65. If the atomic mass for copper on the periodic table is 63.55, are there more atoms of Cu-63 or Cu-65 in a sample of copper?

2.6 Electron Arrangement and the Periodic Law

Study Pack

Chem of Life Atoms and Molecules: Atomic Structure 2

Learning Goal

Given the name or symbol of one of the first 20 elements in the periodic table, write the electron arrangement and use to explain the periodic law.

The chemical behavior of an element is primarily determined by the way the electrons are arranged about the nucleus. Every electron has a specific amount of energy. Electrons of similar energy are grouped in an energy level called a **shell.** The shells closest to the nucleus contain electrons with the lowest energies, whereas shells farther away from the nucleus contain electrons with higher energies.

Each electron shell holds a different number of electrons. As shown in Table 2.12, shell 1, the lowest energy level, can hold up to 2 electrons; shell 2 can hold up to 8 electrons; shell 3 can take 18 electrons; and shell 4 has room for 32 electrons.

There are additional energy levels or shells in an atom, but they are beyond our consideration in this text. In the atoms of the elements known today, electrons can occupy energy levels as high as shell 7.

Table 2.12 Capacity of Some Electron Shells

Electron Shell	Maximum Number of Electrons
1	2
2	8
3	18
4	32

Electron Arrangements for the First 20 Elements

The **electron arrangement** of an atom gives the number of electrons in each shell. We might imagine electron shells as floors in a hotel. The ground floor fills first, then the second floor, and so on. In Figure 2.14, 2 electrons are placed in shell 1, the lowest energy level. The next 8 electrons go into shell 2. Both shell 1 and shell 2 are now filled. Shell 3 initially takes 8 electrons; it then stops filling for a while, even though it is capable of holding more electrons. This break in filling is due to an overlap in the energy values of shell 3 and shell 4. At this point, shell 4 takes the next 2 electrons.

The electron arrangements for the first 20 elements can be written by placing electrons in shells beginning with the lowest energy. The single electron of hydrogen and the 2 electrons of helium can be placed in shell 1. When we wish to draw a simple diagram of the electron arrangement, we indicate the nucleus of the atom and draw curved lines to represent each of the occupied shells. In the illustrations that follow, one isotope has been chosen for each element shown. The electron configurations for hydrogen and helium are shown in the illustration in the margin.

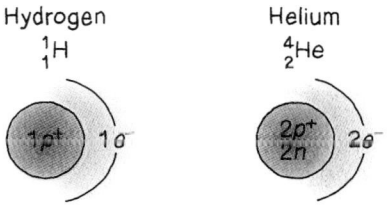

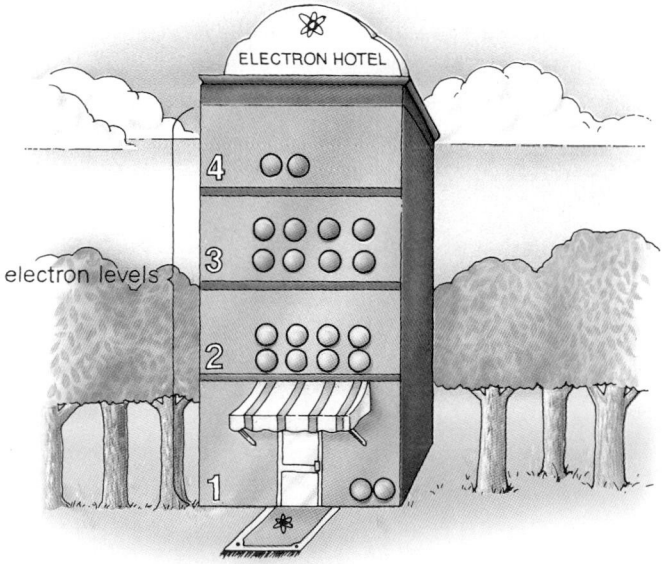

Figure 2.14 Electron occupancy for the first 20 electrons.

The elements of the second period (lithium, Li, to neon, Ne) have enough electrons to fill the first shell and part or all of the second shell. For example, lithium has 3 electrons. Two of those electrons complete shell 1. The remaining electron goes into the second shell. As we go across the period, more and more electrons enter the second shell. For example, an atom of carbon, with a total of 6 electrons, fills shell 1 with 2 electrons, and 4 remaining electrons enter the second shell. The last element in Period 2 is neon. The 10 electrons in an atom of neon completely fill the first and second shells.

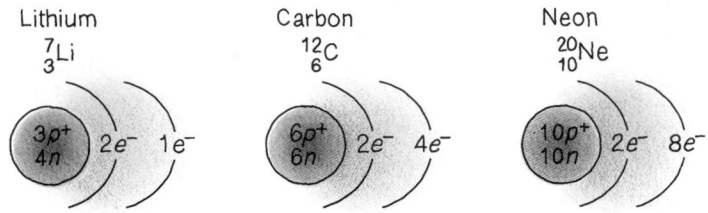

2.6

In an atom of sodium, atomic number 11, the first and second electron shells are filled and the last electron enters the third shell. The rest of the elements in the third period continue to add to the third shell. For example, a sulfur atom with 16 electrons has 2 electrons in the first shell, 8 electrons in the second shell, and 6 electrons in the third shell. At the end of the period, we find that argon has 8 electrons in the third shell.

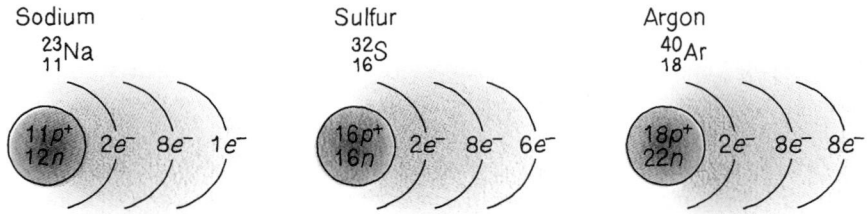

At this point, the third shell with 8 electrons stops filling for a while. The remaining electrons in potassium and calcium actually begin to fill the fourth shell. The third shell will continue filling in elements with higher atomic numbers. The electron arrangements for the first 20 elements are summarized in Table 2.13. Although there are many more elements in the periodic table, we will not consider the electron arrangements of elements beyond atomic number 20.

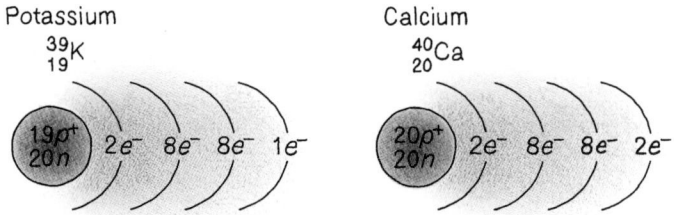

Table 2.13 Electron Arrangements for the First 20 Elements

Element	Symbol	Atomic Number	Number of Electrons in Shell 1	2	3	4
Hydrogen	H	1	1			
Helium	He	2	2			
Lithium	Li	3	2	1		
Beryllium	Be	4	2	2		
Boron	B	5	2	3		
Carbon	C	6	2	4		
Nitrogen	N	7	2	5		
Oxygen	O	8	2	6		
Fluorine	F	9	2	7		
Neon	Ne	10	2	8		
Sodium	Na	11	2	8	1	
Magnesium	Mg	12	2	8	2	
Aluminum	Al	13	2	8	3	
Silicon	Si	14	2	8	4	
Phosphorus	P	15	2	8	5	
Sulfur	S	16	2	8	6	
Chlorine	Cl	17	2	8	7	
Argon	Ar	18	2	8	8	
Potassium	K	19	2	8	8	1
Calcium	Ca	20	2	8	8	2

S·A·M·P·L·E 2.13 ·P·R·O·B·L·E·M

Writing Electron Arrangements

Write the electron arrangement for each of the following:

a. oxygen **b.** chlorine

Solution

a. Oxygen has an atomic number of 8. Therefore, there are 8 electrons in the electron arrangement:

$2e^-$ $6e^-$

b. An atom of chlorine has 17 protons and 17 electrons. The electrons are arranged as follows:

$2e^-$ $8e^-$ $7e^-$

Study Check

What element has the following electron arrangement?

$2e^-$ $8e^-$ $2e^-$

Energy-Level Changes

Whenever possible, electrons occupy the lowest energy levels. However, if a source of energy such as heat or light is available, an electron can absorb a certain amount of energy and jump to a higher energy level. Electrons in this higher energy state, however, are unstable. They drop back to one of the lower, more stable energy levels, releasing some of their energy, as illustrated in Figure 2.15. Neon lights are an example of how electrons gain energy from electricity and emit energy as light (see Figure 2.16).

The amount of energy released when electrons drop to lower energy levels can be large or small. High-energy emissions include X rays and gamma rays. Low-energy emissions include infrared rays (heat), radio waves, and microwaves. Figure 2.17 illustrates some forms of energy emitted when electrons drop to lower energy levels in atoms.

2.6

Only the energies in the visible region are detected by our eyes. For example, when sunlight passes through a prism or a raindrop, it is separated into the colors known as the visible spectrum, which we see in rainbows.

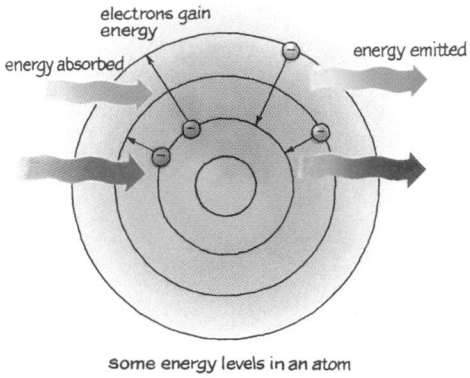

Figure 2.15 Electrons gain energy and jump to higher energy levels. When they drop back to lower energy levels, energy is emitted.

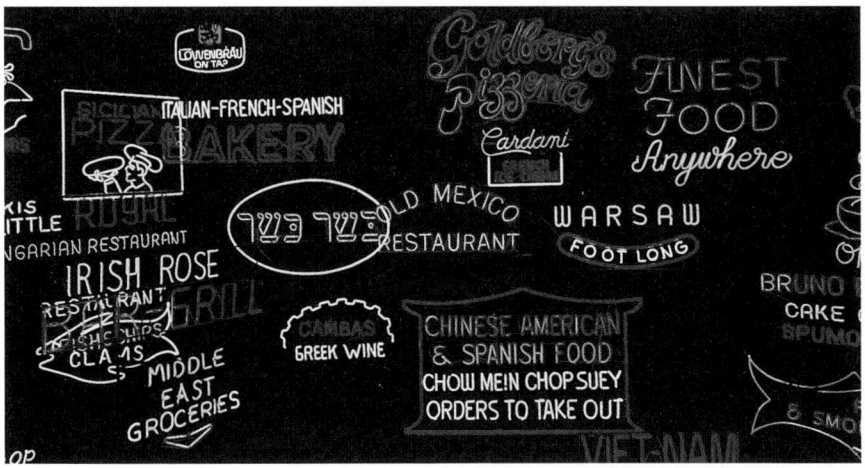

Figure 2.16 When an electric current flows through a gas-filled tube, electrons jump to higher energy levels and fall again, emitting light of different colors.

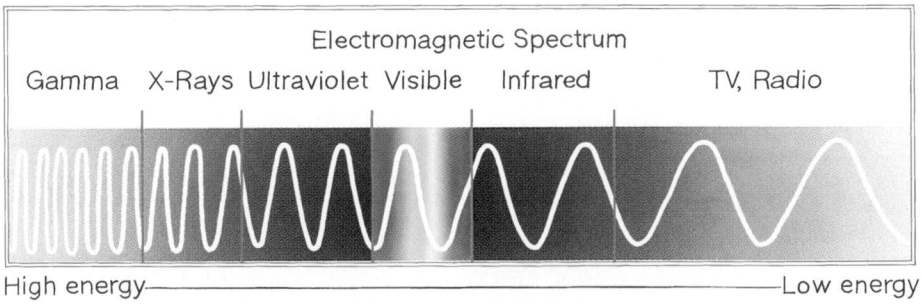

Figure 2.17 The electromagnetic spectrum shows a pattern of increasing and decreasing energy.

EXPLORE YOUR WORLD

Light Released from Candy

For this investigation, you will need some candies with wintergreen such as Lifesaver or other brands of wintergreen mints or Necco wafers. You also need pliers, a room that can be darkened, and some clear tape. Cover the end of the pliers with clear tape, break the candies in half, and darken the room. Wait 3–5 minutes so that your eyes can adjust to the darkness before you proceed. Place a piece of a candy in the pliers and crush it. Look for visible light to be emitted. If you have a friend or relative you can do this with, watch each other as you bite and chew a candy.

The crushing action introduces energy, which is absorbed by the electrons, causing them to jump to a higher energy level. When the electrons drop from a higher energy level back to a lower level, energy is released, but this time in the form of visible light energy that you can see.

Questions

1. How does the change in energy levels cause energy to be emitted as visible light?
2. What color(s) of light is emitted?
3. Are there differences in the brightness of the light?

ENVIRONMENTAL NOTE

Biological Reactions to UV Light

Our everyday life depends on sunlight, but exposure to sunlight can have damaging effects on living cells, and too much exposure can even cause their death. The light energy, especially ultraviolet (UV), excites electrons and may lead to unwanted chemical reactions. The list of damaging effects of sunlight includes sunburn; wrinkling; premature aging of the skin; changes in the DNA of the cells, which can lead to skin cancers and melanomas; inflammation of the eyes; and perhaps cataracts. Some drugs, like the acne medications Accutane and Retin-A, as well as antibiotics, diuretics, sulfonamides, and estrogens, make the skin extremely photosensitive and can cause undesirable changes in its reaction to sunlight. Using a sunscreen is now recommended by doctors to prevent the adverse effects of sun exposure.

High-energy radiation is the most damaging biologically. Most of the radiation in this range is absorbed in the epidermis of the skin. The degree to which radiation is absorbed depends on the thickness of the epidermis, the hydration of the skin, the amount of coloring pigments and proteins of the skin, and the arrangement of the blood vessels. In light-skinned people, 85–90% of the radiation is absorbed by the epidermis, with the rest reaching the dermis layer. In dark-skinned people, 90–95% of the radiation is absorbed by the epidermis, with a smaller percentage reaching the dermis.

However, medicine does take advantage of the beneficial effect of sunlight. Phototherapy can be used to treat certain skin conditions including psoriasis, eczema, and dermatitis. In the treatment of psoriasis, for example, oral drugs are given to make the skin more photosensitive; exposure to UV radiation follows. Low-energy radiation is used to break down bilirubin in neonatal jaundice. Sunlight is also a factor in stimulating the immune system. (See Figure 2.18.)

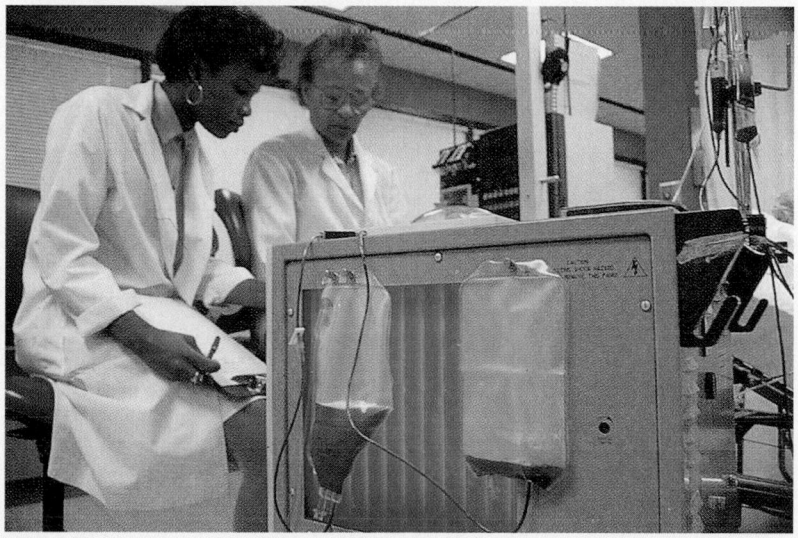

Figure 2.18 In cutaneous T-cell lymphoma, an abnormal increase in T cells causes painful ulceration of the skin. The skin is treated by photophoresis, in which the patient receives a photosensitive chemical, and then blood is removed from the body and exposed to ultraviolet light. The blood is returned to the patient, and the treated T cells stimulate the immune system to respond to the cancer cells.

Table 2.14 A Comparison of Electron Arrangements of Some Group 1 Elements

Element	Atomic Number	Number of Electrons in Shell 1	2	3	4	
Lithium	3	2	**1**			One electron in each outer shell
Sodium	11	2	8	**1**		
Potassium	19	2	8	8	**1**	

2.6 The Periodic Law

We have seen that the groups in the periodic table contain elements having similar properties. This repetition of properties with increasing atomic number is known as the **periodic law.** If we now observe the electron arrangements for the elements in a group, we find another similarity. All the elements in a group (also called a family) of elements have the same number of electrons in their outermost shells.

For example, the elements lithium, sodium, and potassium are part of Group 1. All of these elements have one electron in their outer shells, as seen in Table 2.14. The outer shell is the shell with the highest energy that contains one or more electrons. Thus, the similarity of properties among elements in a group can now be attributed to their having the same number of electrons in each of their outermost shells.

Group Number

The **group numbers** 1–8 appear at the top of the periodic table. Each group number is equal to the number of electrons in the outer shell of the elements in that column. All elements in Group 1 have 1 electron in their outer shells, elements in Group 2 have 2 electrons in their outer shells, elements in Group 3 have 3 electrons in their outer shells, and so on. We are most interested in the number of electrons in the outer shells because these electrons have the greatest effect on the way an atom forms compounds. Table 2.15 shows the electron arrangement by group for the first 20 elements.

S·A·M·P·L·E· 2.14 ·P·R·O·B·L·E·M

Using Group Numbers

Using the periodic table, write the group number and the number of electrons in the outer electron level of the following elements:

a. sodium **b.** sulfur **c.** aluminum

Solution

a. Sodium is in Group 1; sodium has 1 electron in the outer electron level.
b. Sulfur is in Group 6; sulfur has 6 electrons in the outer electron level.
c. Aluminum is in Group 3; aluminum has 3 electrons in the outer electron level.

Study Check

What is the group number, total number of electrons, and name of an atom that has 5 electrons in the third electron level?

Table 2.15 Electron Arrangements, by Group, for the First 20 Elements

		Number of Electrons in Shell			
Group Number	**Element**	**1**	**2**	**3**	**4**
1	Hydrogen	**1**			
	Lithium	2	**1**		
	Sodium	2	8	**1**	
	Potassium	2	8	8	**1**
2	Beryllium	2	**2**		
	Magnesium	2	8	**2**	
	Calcium	2	8	8	**2**
3	Boron	2	**3**		
	Aluminum	2	8	**3**	
4	Carbon	2	**4**		
	Silicon	2	8	**4**	
5	Nitrogen	2	**5**		
	Phosphorus	2	8	**5**	
6	Oxygen	2	**6**		
	Sulfur	2	8	**6**	
7	Fluorine	2	**7**		
	Chlorine	2	8	**7**	
8	Helium	**2**			
	Neon	2	**8**		
	Argon	2	8	**8**	

QUESTIONS & PROBLEMS

Electron Arrangement and the Periodic Law

2.41 Electrons exist in specific energy levels. Explain.

2.42 In what order do electrons fill the energy levels 1–4 for the first 20 elements on the periodic table?

2.43 How many electrons are in the second energy level of the following elements?

a. sodium **b.** nitrogen **c.** sulfur
d. helium **e.** chlorine

2.44 How many electrons are in the third energy level of the following elements?

a. oxygen **b.** sulfur **c.** phosphorus
d. calcium **e.** fluorine

2.45 Write the electron arrangement for each of the following elements:
Example: sodium 2, 8, 1

a. carbon
b. argon
c. sulfur
d. potassium
e. an atom with 13 protons and 14 neutrons
f. nitrogen

2.6

2.46 Write the electron arrangement for each of the following atoms:
Example: sodium 2, 8, 1
a. phosphorus
b. neon
c. oxygen
d. an atom with atomic number 18
e. aluminum
f. silicon

2.47 Identify the elements that have the following electron arrangements:

Energy Level:	*1*	*2*	*3*
a.	$2e^-$	$1e^-$	
b.	$2e^-$	$8e^-$	$2e^-$
c.	$1e^-$		
d.	$2e^-$	$8e^-$	$7e^-$
e.	$2e^-$	$6e^-$	

2.48 Identify the elements that have the following electron arrangements:

Energy Level:	*1*	*2*	*3*
a.	$2e^-$	$5e^-$	
b.	$2e^-$	$8e^-$	$6e^-$
c.	$2e^-$	$4e^-$	
d.	$2e^-$	$8e^-$	$8e^-$
e.	$2e^-$	$8e^-$	$3e^-$

2.49 Answer the following with *gain* or *emit*:
a. Electrons can jump to higher energy levels when they _____ a specific amount of energy.
b. When electrons drop to lower energy levels, they _____ a certain amount of energy.

2.50 **a.** Comparing X rays, infrared, and radio waves as shown in Figure 2.17, which has the highest radiation energy?
b. Why do we protect ourselves from X rays but not radio waves?

2.51 The elements boron and aluminum are in the same group on the periodic table.
a. Write the electron arrangements for B and Al.
b. How many electrons are in the outer energy level of each atom?
c. What is their group number?

2.52 The elements fluorine and chlorine are in the same group on the periodic table.
a. Write the electron arrangements for F and Cl.
b. How many electrons are in each of their outer energy levels?
c. What is their group number?

2.53 What is the number of electrons in the outer energy level and the group number for each of the following elements?
Example: fluorine $7e^-$; Group 7
a. magnesium **b.** chlorine **c.** oxygen **d.** nitrogen
e. barium **f.** bromine

2.54 What is the number of electrons in the outer energy level and the group number for each of the following elements?
Example: fluorine $7e^-$; Group 7
a. lithium **b.** silicon **c.** neon **d.** argon
e. tin **f.** cesium (Cs)

2.55 Would you expect Mg, Ca, and Sr to have similar properties? Why?

2.56 Name two elements that would exhibit physical and chemical behavior similar to chlorine.

Chapter Review

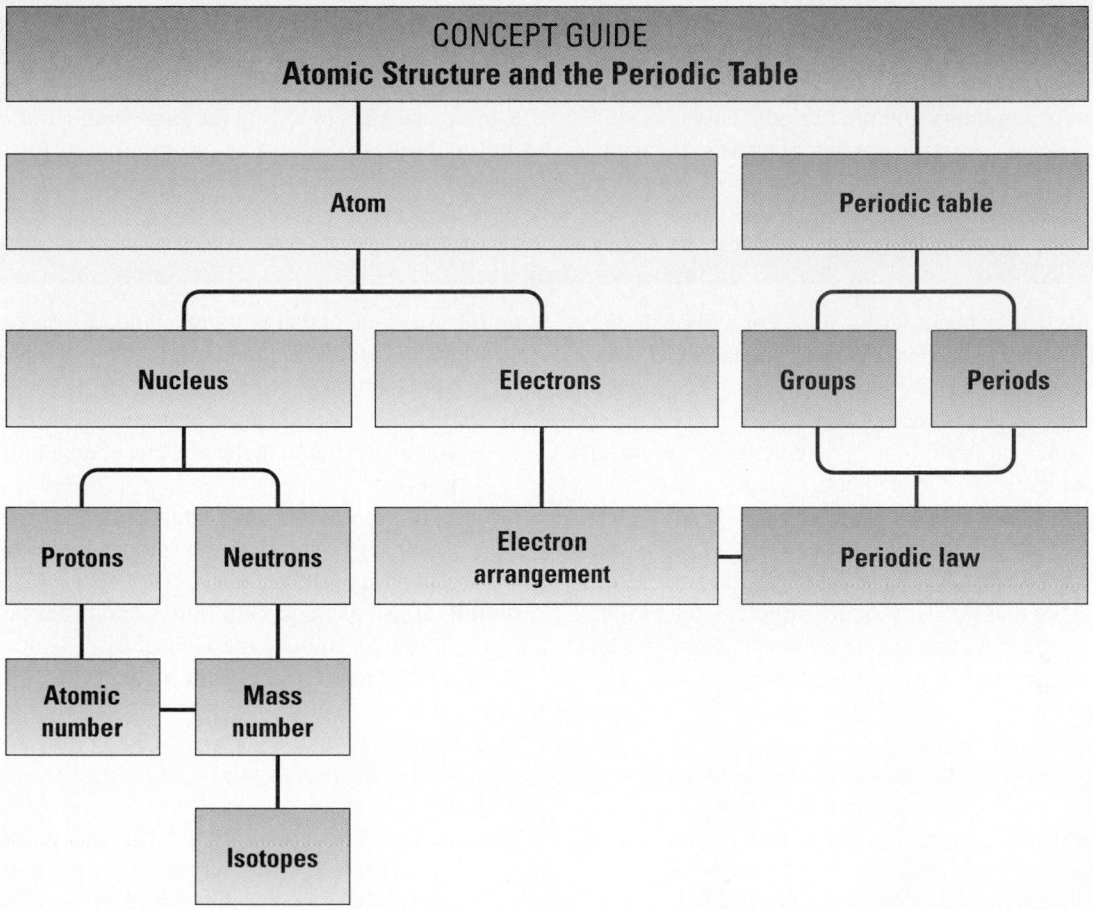

Chapter Summary

2.1 Elements and Symbols

Elements are the primary substances of matter. Chemical symbols are one- or two-letter abbreviations of the names of the elements.

2.2 The Periodic Table

The periodic table is an arrangement of the elements by increasing atomic number. A vertical column on the periodic table containing elements with similar properties is called a group. A horizontal row is called a period. Elements in Group 1 are called the alkali metals; Group 2, alkaline earth metals; Group 7, the halogens; and Group 8, the noble gases. On the periodic table, metals are located on the left of the heavy zigzag line, and nonmetals are to the right of the heavy zigzag line.

2.3 The Atom

An atom is the smallest particle that retains the characteristics of an element. Atoms are composed of three subatomic particles. Protons have a positive charge (+), electrons carry a negative charge (−), and neutrons are electrically neutral. The protons and neutrons, each with a mass of about 1 amu, are found in the tiny, dense nucleus. Electrons with a much smaller mass are located outside the nucleus.

2.4 Atomic Number and Mass Number

The atomic number gives the number of protons in all the atoms of the same element. In a neutral atom, there is an equal number of protons and electrons. The mass number is the total number of protons and neutrons in an atom.

2.5 Isotopes and Atomic Mass

Atoms that have the same number of protons but different numbers of neutrons are called isotopes. The atomic mass of an element is the weighted average mass of all the isotopes in a naturally occurring sample of that element.

2.6 Electron Arrangement and the Periodic Law

Every electron has a specific amount of energy. In an atom, the electrons of similar energy are grouped in specific energy levels or shells. The first level nearest the nucleus can hold 2 electrons, the second level can hold 8 electrons, the third level will take up to 18 electrons. The electron arrangement is written by placing the number of electrons in that atom in order from the lowest energy levels and filling to higher levels. The similarity of behavior for the elements in a group is related to having the same number of electrons in their outermost shells. The group number for an element gives the number of electrons in its outermost shell.

Key Terms

alkali metals Elements of Group 1 except hydrogen; these are soft, shiny metals with one outer shell electron.

alkaline earth metals Group 2 elements, which have 2 electrons in their outer shells.

atom The smallest particle of an element that retains the characteristics of the element.

atomic mass The weighted average mass of all the naturally occurring isotopes of an element.

atomic mass unit (amu) A small mass unit used to describe the mass of very small particles such as atoms and subatomic particles; 1 amu is equal to one-twelfth the mass of a carbon-12 atom.

atomic number A number that is equal to the number of protons in an atom.

atomic symbol An abbreviation used to indicate the mass number and atomic number of an isotope.

chemical symbol An abbreviation that represents the name of an element.

electron A negatively charged subatomic particle having a very small mass that is usually ignored in calculations; its symbol is e^-.

electron arrangement An organization of electrons within the atom by increasing energy shells.

element A primary substance that cannot be separated into any simpler substances.

group A vertical column in the periodic table that contains elements having similar physical and chemical properties.

group number A number that appears at the top of each vertical column (group) in the periodic table and indicates the number of electrons in the outermost shell.

halogen Group 7 elements of fluorine, chlorine, bromine, and iodine.

isotope An atom that differs only in mass number from another atom of the same element. Isotopes have the same atomic number (number of protons) but different numbers of neutrons.

mass number The total number of neutrons and protons in the nucleus of an atom.

metal An element that is shiny, malleable, and a good conductor of heat and electricity. The metals are located to the left of the zigzag line in the periodic table.

neutron A neutral subatomic particle having a mass of 1 amu and found in the nucleus of an atom; its symbol is n or n^0.

noble gas An element in Group 8 of the periodic table, generally unreactive and seldom found in combination with other elements.

nonmetal An element with little or no luster that is a poor conductor of heat and electricity. The nonmetals are located to the right of the zigzag line in the periodic table.

nucleus The compact, very dense center of an atom, containing the protons and neutrons of the atom.

period A horizontal row of elements in the periodic table.

periodic law The repetition of similar chemical and physical properties with increasing atomic number due to the reappearance of the same number of electrons in the outermost shells of atoms of those elements.

periodic table An arrangement of elements by increasing atomic number such that elements having similar chemical behavior are grouped in vertical columns.

physical property A characteristic that can be observed or measured without affecting the identity of an element, including shape, color, odor, taste, density, hardness, melting point, and boiling point.

proton A positively charged subatomic particle having a mass of 1 amu and found in the nucleus of an atom; its symbol is p or p^+.

representative element An element found in Groups 1 through 8 of the periodic table.

shell An energy level containing electrons of similar energies.

subatomic particle A particle within an atom; protons, neutrons, and electrons are subatomic particles.

transition element An element located between Groups 2 and 3 on the periodic table.

Chapter Practice

2.57 Why is Co the symbol for cobalt, but not CO?

2.58 Which of the following is correct? Write the correct symbol if needed.

a. copper, Co **b.** silicon, SI **c.** iron, Fe
d. fluorine, Fl **e.** potassium, P **f.** sodium, Na
g. gold, Au **h.** lead, PB

2.59 Give the symbol and name of the element found in the following group and period on the periodic table.

a. Group 2, Period 3 **b.** Group 7, Period 4
c. Group 3, Period 3 **d.** Group 6, Period 2

2.60 Give the group and period number for the following elements:

a. potassium **b.** phosphorus
c. carbon **d.** neon

2.61 The following statements are *false*. Reword the underlined word or phrase to make a *true* statement.

a. The proton is a <u>neutral</u> particle.
b. The electrons are found <u>in the nucleus</u>.
c. The nucleus is the <u>largest</u> part of the atom.
d. The <u>neutron</u> has a negative charge.
e. Most of the mass of an atom is due to its <u>electrons</u>.

2.62 State the number of protons and neutrons in the following atoms:

a. ${}^{2}_{1}H$ **b.** ${}^{37}_{17}Cl$ **c.** ${}^{106}_{48}Cd$ **d.** ${}^{209}_{83}Bi$

2.63 The most abundant isotope of iron is Fe-56.

a. How many protons, neutrons, and electrons are in this isotope?
b. What is the symbol of another isotope of iron with 25 neutrons?
c. What is the symbol of a different atom with the same mass number of 51 and 27 neutrons?

2.64 Cadmium, atomic number 48, consists of eight naturally occurring isotopes. Do you expect any of the isotopes to have the atomic mass listed on the periodic table for cadmium? Explain.

2.65 Distinguish between:

a. atoms and isotopes
b. atomic number and mass number

2.66 Distinguish between:

a. elemental symbol and atomic symbol
b. atomic number and atomic mass

2.67 Consider the following atoms in which the chemical symbol of the element is represented by X.

${}^{16}_{8}X$ ${}^{16}_{9}X$ ${}^{18}_{10}X$ ${}^{17}_{8}X$ ${}^{18}_{8}X$

a. What atoms have the same number of protons?
b. Which atoms are isotopes? Of what element?
c. Which atoms have the same mass number?
d. What atoms have the same number of neutrons?

2.68 Five isotopes of zinc are zinc-64, zinc-66, zinc-67, zinc-68, and zinc-70.

a. Write the atomic symbols, including atomic number and mass number for each of these atoms.
b. Give the number of protons, electrons, and neutrons for each of the zinc isotopes.

2.69 A sample of copper has two naturally occurring isotopes. The isotope Cu-63 (mass 62.9) makes up 69.2% of all the copper atoms, and Cu-65 (mass 64.9) makes up 30.8% of the sample. What average atomic mass would you calculate for copper?

2.70 In a sample of magnesium, Mg-24 (mass 24.0) is 79.0% of the sample, Mg-25 (mass 25.0) is 10.0%, and Mg-26 (mass 26.0) is 11.0%. What atomic mass would you calculate for magnesium?

2.71 If the diameter of a sodium atom is 3.14×10^{-8} cm, how many sodium atoms would fit along a line exactly 1 inch long?

2.72 A lead atom has a mass of 3.4×10^{-22} g. How many lead atoms are in a cube of lead that has a volume of 2.00 cm^3, if the density of lead is 11.3 g/cm^3?

2.73 Write the names of two representative elements that have the following number of electrons in their outer energy levels:

a. 2 electrons **b.** 6 electrons
c. 8 electrons **d.** 3 electrons

2.74 Explain the difference between the following:

a. atomic mass and average atomic mass
b. atom and nucleus

2.75 Write the names of two elements that are in the following groups:

a. halogens **b.** noble gases
c. alkali metals **d.** alkaline earth metals

2.76 The following are trace elements that have been found to be crucial to the biochemical and physiological processes in the body. Indicate whether each is a metal or nonmetal.

a. zinc **b.** cobalt **c.** manganese (Mn)
d. iodine **e.** copper **f.** selenium (Se)
g. nickel **h.** iron

Answers

Answers to Study Checks

2.1 Si, S, and Ag

2.2 magnesium, aluminum, and fluorine

2.3 Shiny, liquid, gray color, forms spherical drops if the thermometer breaks.

2.4 **a.** Period 2: Li, Be, B, C, N, O, F, Ne
b. Group 2: Be, Mg, Ca, Sr, Ba, Ra

2.5 **a.** P **b.** Rn

2.6 nitrogen

2.7 A proton, symbol p or p^+ with a mass of 1 amu and a charge of 1+, is found in the nucleus of an atom.

2.8 **a.** 26 **b.** 26 **c.** iron, Fe

2.9 Because the atomic number of silver is 47, it has 47 protons. The mass number is 107, which is the sum of 47 protons and 60 neutrons.

2.10 45 neutrons

2.11 **a.** $^{15}_{7}N$ **b.** $^{42}_{20}Ca$ **c.** $^{27}_{13}Al$

2.12 zinc

2.13 magnesium

2.14 5; 15; phosphorus

Answers to Selected Questions and Problems

2.1 **a.** Cu **b.** Si **c.** K **d.** N
e. Fe **f.** Ba **g.** Pb **h.** Sr

2.3 **a.** carbon **b.** chlorine **c.** iodine
d. mercury **e.** fluorine **f.** argon
g. zinc **h.** nickel

2.5 **a.** sodium, chlorine
b. calcium, sulfur, oxygen
c. carbon, hydrogen, chlorine, nitrogen, oxygen
d. calcium, carbon, oxygen

2.7 **a.** black **b.** orange

2.9 **a.** Period 2 **b.** Group 8
c. Group 1 **d.** Period 2

2.11 **a.** alkaline earth metal **b.** transition element
c. noble gas **d.** alkali metal
e. halogen

2.13 **a.** C **b.** He **c.** Na **d.** Ca **e.** Al

2.15 **a.** metal **b.** nonmetal **c.** metal
d. nonmetal **e.** nonmetal **f.** nonmetal
g. metal **h.** metal

2.17 **a.** electron **b.** proton **c.** electron **d.** neutron

2.19 Rutherford determined that an atom contains a small, compact nucleus that is positively charged.

2.21 **b.**

2.23 In the process of brushing hair, strands of hair become charged with like charges that repel each other.

2.25 **a.** atomic number **b.** both
c. mass number **d.** atomic number

2.27 **a.** lithium, Li **b.** fluorine, F **c.** calcium, Ca
d. zinc, Zn **e.** neon, Ne **f.** silicon, Si
g. iodine, I **h.** oxygen, O

2.29 **a.** 12 **b.** 30 **c.** 53 **d.** 19

2.31 *See Table 2.17*

Table 2.17

Name of Element	Symbol	Atomic Number	Mass Number	Number of Protons	Number of Neutrons	Number of Electrons
Aluminum	Al	13	27	13	14	13
Magnesium	Mg	12	24	12	12	12
Potassium	K	19	39	19	20	19
Sulfur	S	16	31	16	15	16
Iron	Fe	26	56	26	30	26

2.33 **a.** 13 protons, 14 neutrons, 13 electrons
b. 24 protons, 28 neutrons, 24 electrons
c. 16 protons, 18 neutrons, 16 electrons
d. 26 protons, 30 neutrons, 26 electrons

2.35 **a.** ${}^{31}_{15}P$ **b.** ${}^{80}_{35}Br$ **c.** ${}^{27}_{13}Al$ **d.** ${}^{35}_{17}Cl$

2.37 **a.** ${}^{32}_{16}S$ ${}^{33}_{16}S$ ${}^{34}_{16}S$ ${}^{36}_{16}S$
b. They all have the same number of protons and electrons.
c. They have different numbers of neutrons, which gives them different mass numbers.
d. The atomic mass of S listed on the periodic table is the average atomic mass of all the isotopes.

2.39 You need to know the percentage and the mass for each of the isotopes that make up a particular element.

2.41 The energy of electrons is of a specific quantity for each energy level. Electrons cannot have energies that are between the energy levels.

2.43 **a.** 8 **b.** 5 **c.** 8 **d.** 0 **e.** 8

2.45 **a.** 2, 4 **b.** 2, 8, 8 **c.** 2, 8, 6
d. 2, 8, 8, 1 **e.** 2, 8, 3 **f.** 2, 5

2.47 **a.** Li **b.** Mg **c.** H
d. Cl **e.** O

2.49 **a.** gain **b.** emit

2.51 **a.** B 2, 3 Al 2, 8, 3
b. 3
c. Group 3

2.53 **a.** $2e^-$, Group 2 **b.** $7e^-$, Group 7 **c.** $6e^-$, Group 6
d. $5e^-$, Group 5 **e.** $2e^-$, Group 2 **f.** $7e^-$, Group 7

2.55 Yes. They each have two electrons in their outer shells.

2.57 The first letter of a symbol is a capital, but a second letter is lowercase. The symbol Co is for cobalt, but the symbols in CO are for carbon and oxygen.

2.59 **a.** Mg, magnesium **b.** Br, bromine
c. Al, aluminum **d.** O, oxygen

2.61 **a.** The proton is a positive particle.
b. Electrons are found in the outer part of the atom.
c. The nucleus is a very small part of the atom.
d. The electron has a negative charge.
e. Most of the mass of the atom is due to its protons and neutrons.

2.63 **a.** 26 protons, 30 neutrons, 26 electrons
b. ${}^{51}_{26}Fe$ **c.** ${}^{51}_{24}Cr$

2.65 **a.** An isotope is one kind of atom that makes up a particular element. Most elements are mixtures of two or more isotopes.
b. Atomic number gives the number of protons in an atom; mass number gives the total number of protons and neutrons.

2.67 **a.** ${}^{16}_{8}X$, ${}^{17}_{8}X$, ${}^{18}_{8}X$ all have 8 protons
b. ${}^{16}_{8}X$, ${}^{17}_{8}X$, ${}^{18}_{8}X$ all isotopes of O
c. ${}^{16}_{8}X$ and ${}^{16}_{9}X$ have mass number 16, and ${}^{18}_{10}X$ and ${}^{18}_{8}X$ have a mass number of 18.
d. ${}^{16}_{8}X$ and ${}^{18}_{10}X$ both have 8 neutrons.

2.69 63.5 amu

2.71 8.09×10^7 sodium atoms

2.73 **a.** any two elements in Group 2 such as magnesium, calcium, and barium
b. any two elements in Group 6 such as oxygen and sulfur
c. any two elements in Group 8 except helium: neon, argon, krypton, xenon, or radon
d. any two elements in Group 3 such as boron and aluminum

2.75 **a.** any two elements in Group 7 such as fluorine, chlorine, bromine, or iodine
b. any two elements in Group 8 such as helium, neon, argon, krypton, xenon, or radon
c. any two elements in Group 1 such as lithium, sodium, and potassium, except hydrogen.
d. any two elements in Group 2 such as magnesium, calcium, and barium

3

Nuclear Radiation

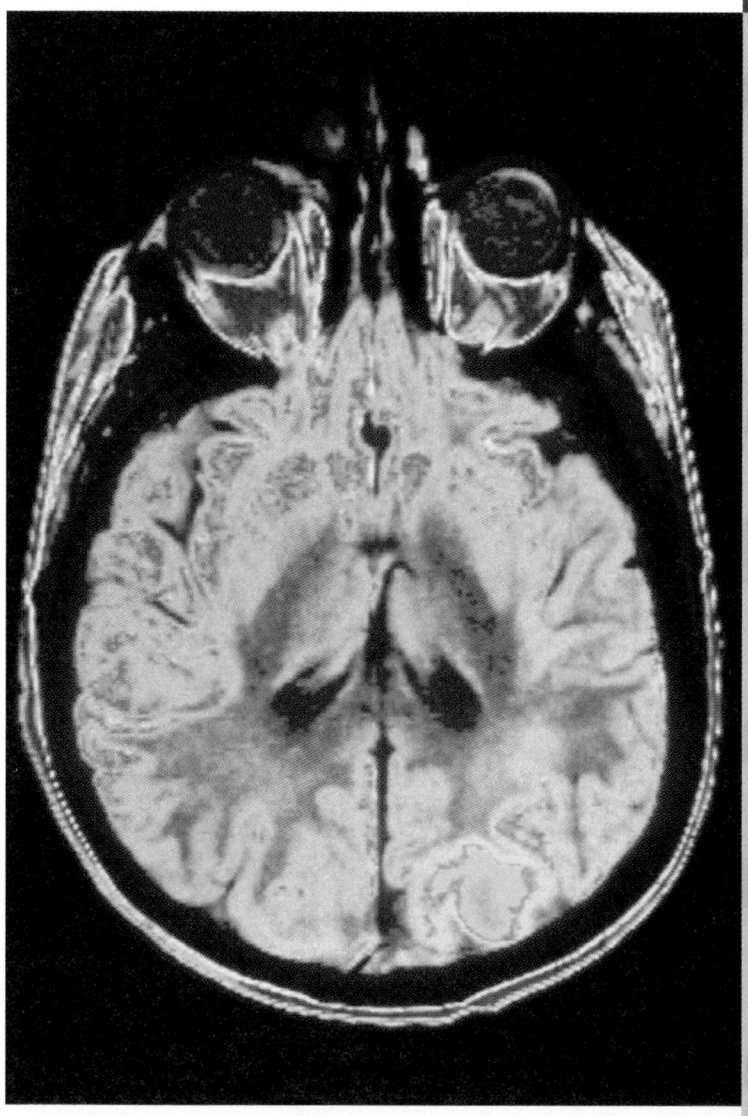

Using magnetic resonance imaging (MRI), the magnetic properties of the atoms in a brain produce this image of a brain with cancer.

Looking Ahead

3.1	Natural Radioactivity
3.2	Nuclear Equations
3.3	Producing Radioactive Isotopes
3.4	Radiation Detection and Measurement
3.5	Medical Applications Using Radioactivity
3.6	Half-Life of a Radioisotope
3.7	Nuclear Fission and Fusion

www.chemplace.com/college/timberlake

Visit the web site above to test yourself, run simulations, and review key concepts from this chapter. For even more interactive review and quizzing, use the two CD-ROMs included with the book.

In nuclear medicine, a radiologist uses radioactive isotopes to determine the size and shape of an organ, locate a tumor, or measure the metabolic activity of cells in the thyroid, brain, or liver. Radioactive isotopes are also used to measure the amounts of substances such as drugs and hormones in blood and urine samples.

The amount of radiation received by a patient is carefully monitored, because radiation can cause damage in the cells of the body. Medically, radiation is used to treat cancer and diseases of the blood or bone. The radiation penetrates the abnormal cells, inhibiting their growth. Nearby normal cells are also affected by the radiation, but they have a greater capacity to recover than do the abnormal cells.

3.1 Natural Radioactivity

Learning Goal

Describe alpha, beta, and gamma radiation.

Most naturally occurring isotopes of elements up to atomic number 19 have stable nuclei. Elements with higher atomic numbers (20 to 83) consist of a mixture of isotopes, some of which may have unstable nuclei. When the nucleus of an isotope is unstable, it is *radioactive,* which means that it will spontaneously emit energy to become more stable. This energy, called *radiation,* may take the form of particles such as alpha (α) particles or beta (β) particles or pure energy such as gamma (γ) rays. Elements with atomic numbers of 84 and higher consist only of radioactive isotopes. So many protons and neutrons are crowded together in their nuclei that the strong repulsions between the protons makes those nuclei unstable.

In Chapter 2, we wrote symbols to distinguish among the different isotopes of an element. Recall that an atom's mass number is equal to the sum of the protons and neutrons in the nucleus, and its atomic number is equal to the number of protons. In the symbol for an isotope, the mass number is written in the upper left corner of the symbol, and the atomic number is written in the lower left corner. For example, a radioactive isotope of iodine used in the diagnosis and treatment of thyroid conditions has a mass number of 131 and an atomic number of 53. We can write its symbol as

Mass number (protons and neutrons)
Element
Atomic number (protons)

$$^{131}_{53}\mathrm{I}$$

This isotope is also called iodine-131 or I-131. (Note that radioactive isotopes are named by writing the mass number after the element's name or symbol.) When necessary, we can obtain the atomic number from the periodic table. Table 3.1 compares some stable, nonradioactive isotopes with some radioactive isotopes.

Table 3.1 Stable and Radioactive Isotopes of Some Elements

Magnesium	Iodine	Uranium
Stable Isotopes		
$^{24}_{12}\mathrm{Mg}$	$^{127}_{53}\mathrm{I}$	None
Magnesium-24	Iodine-127	
Radioactive Isotopes		
$^{23}_{12}\mathrm{Mg}$	$^{125}_{53}\mathrm{I}$	$^{235}_{92}\mathrm{U}$
Magnesium-23	Iodine-125	Uranium-235
$^{27}_{12}\mathrm{Mg}$	$^{131}_{53}\mathrm{I}$	$^{238}_{92}\mathrm{U}$
Magnesium-27	Iodine-131	Uranium-238

Types of Radiation

We cannot feel, taste, or smell radiation, but high-energy radiation is capable of creating havoc within the cells of our bodies. Different forms of radiation are emitted from an unstable nucleus when a change takes place among its protons and neutrons. Energy is released, and a new, more stable nucleus is formed. One type of radiation consists of alpha particles. An **alpha particle** contains 2 protons and 2 neutrons, which gives it a mass number of 4 and an atomic number of 2. Because it has 2 protons, an alpha particle has a charge of 2+. That makes it identical to a helium nucleus. In equations, it is written as the Greek letter alpha (α) or as the symbol for helium.

3.1

α or ${}^{4}_{2}\text{He}$

Alpha particle

Another type of radiation occurs when a radioisotope emits **beta particles.** A beta particle, which is a high-energy electron, has a charge of 1− and a mass number of 0. It is represented by the Greek letter beta (β) or by the symbol for the electron (e^-). When the symbol e^- is used for a beta particle, the subscript of −1 is used to show its charge.

β or ${}^{0}_{-1}e$

Beta particle

Beta particles are produced from unstable nuclei when neutrons change into protons. The high-energy electrons were not any of the electrons of the atoms and they do not exist in the nucleus until there is a transformation of neutrons within the nucleus.

${}^{1}_{0}\text{n}$	$\longrightarrow$	${}^{1}_{1}\text{H}$	+	${}^{0}_{-1}e$
Neutron in the nucleus		New proton remains in the nucleus		Electron formed and emitted as a beta particle

Gamma rays are high-energy radiation similar to X rays released as an unstable nucleus undergoes a rearrangement to give a more stable, lower-energy nucleus. A gamma ray is shown as the Greek letter gamma (γ). Because gamma rays are energy only, there is no mass or charge associated with their symbol.

γ

Gamma ray

Table 3.2 summarizes the types of radiation we will use in nuclear equations.

Table 3.2 Some Common Forms of Radiation

Type of Radiation	Symbol	Mass Number	Atomic Number	Charge
Alpha particle	α, ${}^{4}_{2}\text{He}$	4	2	2+
Beta particle	β, ${}^{0}_{-1}e$	0	0	1−
Gamma ray, X ray	γ	0	0	0
Proton	${}^{1}_{1}\text{H}$, ${}^{1}_{1}\text{p}$	1	1	1+
Neutron	${}^{1}_{0}\text{n}$	1	0	0

S·A·M·P·L·E· **3.1** ·P·R·O·B·L·E·M

Writing Symbols for Radiation Particles

Write the symbol for an alpha particle.

Solution

The alpha particle contains 2 protons and 2 neutrons. It has a mass number of 4 and an atomic number of 2.

α or $^{4}_{2}He$

Study Check

What is the symbol used for beta radiation?

HEALTH NOTE

Biological Effects of Radiation

When high-energy radiation strikes molecules in its path, electrons may be knocked away. The result of this *ionizing radiation* is the formation of unstable ions or radicals. A free radical is a particle that has an unpaired electron. For example, when radiation passes through the human body, it may interact with water molecules, removing electrons and producing H_2O^+ ions or it may produce free radicals.

When ionizing radiation strikes the cells of the body, the unstable ions or free radicals that form can cause undesirable chemical reactions. The cells in the body most sensitive to radiation are the ones undergoing rapid division—those of the bone marrow, skin, reproductive organs, and intestinal lining, as well as all cells of growing children. Damaged cells may lose their ability to produce necessary materials. For example, if radiation damages cells of the bone marrow, red blood cells may no longer be produced. If sperm cells or ova or the cells of a fetus are damaged, birth defects may result. In contrast, cells of the nerves, muscles, liver, and adult bones are much less sensitive to radiation because they undergo little or no cellular division.

Cancer cells are another example of rapidly dividing cells. Because cancer cells are highly sensitive to radiation, large doses of radiation are used to destroy them. The surrounding normal tissue, dividing at a slower rate, shows a greater resistance to radiation and suffers less damage. In addition, normal tissue is able to repair itself more readily than cancerous tissue. However, this repair is not always complete. Long-term effects of ionizing radiation include a shortened life span, malignant tumors, leukemia, anemia, and genetic mutations.

Radiation Protection

Because many cells in the body are sensitive to radiation, it is important that the radiologist, doctor, and nurse working with radioactive isotopes use proper radiation protection. Proper **shielding** is necessary to prevent exposure. Alpha particles are the heaviest of the radiation particles; they travel only a few centimeters in the air before they collide with air molecules, acquire electrons, and become helium atoms. A piece of paper, clothing, and our skin are protection against alpha particles. Lab coats and gloves will also provide sufficient shielding. However, if ingested or inhaled, alpha particles can bring about serious internal damage because their heavy mass causes much ionization in a short distance.

Beta particles have a very small mass and move much faster and farther than alpha particles, traveling as much as several meters through air. They can pass through paper and penetrate as far as 4–5 mm into body tissue. External exposure to beta particles can burn the surface of the skin, but they are stopped before they can reach the internal organs. Heavy clothing such as lab coats and gloves are needed to protect the skin from beta particles. (See Figure 3.1.)

Gamma rays travel great distances through the air and pass through many materials, including body tissues. Only very dense shielding, such as lead or concrete, will stop them. Because they can penetrate so deeply, exposure to gamma rays can be extremely

Figure 3.1 Shielding material needed to protect a person from alpha, beta, and gamma radiation.

hazardous. Even the syringe used to give an injection of a gamma-emitting radioactive isotope is placed inside a special lead-glass cover.

When preparing radioactive materials, the radiologist wears special gloves and works behind leaded windows. Long tongs are used within the work area to pick up vials of radioactive material, keeping them away from the hands and body. (See Figure 3.2.) Table 3.3 summarizes the shielding materials required for the various types of radiation.

Try to keep the time you must spend in a radioactive area to a minimum. A certain amount of radiation is emitted every minute. Remaining in a radioactive area twice as long exposes a person to twice as much radiation.

Keep your distance! The greater the distance from the radioactive source, the lower the intensity of radiation received. If you double your distance from the radiation source, the intensity of radiation drops to $(\frac{1}{2})^2$ or one-fourth of its previous value, as Figure 3.3 shows. This is one of the reasons dentists and x-ray technicians leave the room and stand behind a shield or lead-lined wall to take your X rays. They are exposed to radiation every day and must minimize the amount of radiation they receive.

Figure 3.2 A technician in a radiation laboratory wears special protective clothing and stands behind a lead shield when using radioisotopes.

Table 3.3 Properties of Ionizing Radiation and Shielding Required

Property	Alpha	Beta	Gamma
Characteristics	Helium nucleus	Electron	High-energy rays
Symbols	α, $^{4}_{2}He$	β, $^{0}_{-1}e$	γ, $^{0}_{0}\gamma$
Mass Number	4	0	0
Charge	2+	1−	0
Travel Distance in Air	2–4 cm	200–300 cm	500 m
Tissue Depth	0.05 mm	4–5 mm	50 cm or more
Shielding	Paper, clothing	Heavy clothing, lab coats, gloves	Lead, thick concrete
Typical Source	Radium-226	Carbon-14	Technetium-99m

EXPLORE YOUR WORLD

Dental X Rays

Ask your dentist about dental X rays on your next office visit.

Question

1. What amount of radiation is used? How are you and the x-ray technician protected?

S·A·M·P·L·E 3.2 P·R·O·B·L·E·M

Radiation Protection

How does the type of shielding for alpha radiation differ from that used for gamma radiation?

Solution

Alpha radiation is stopped by paper and clothing. However, lead or concrete is needed for protection from gamma radiation.

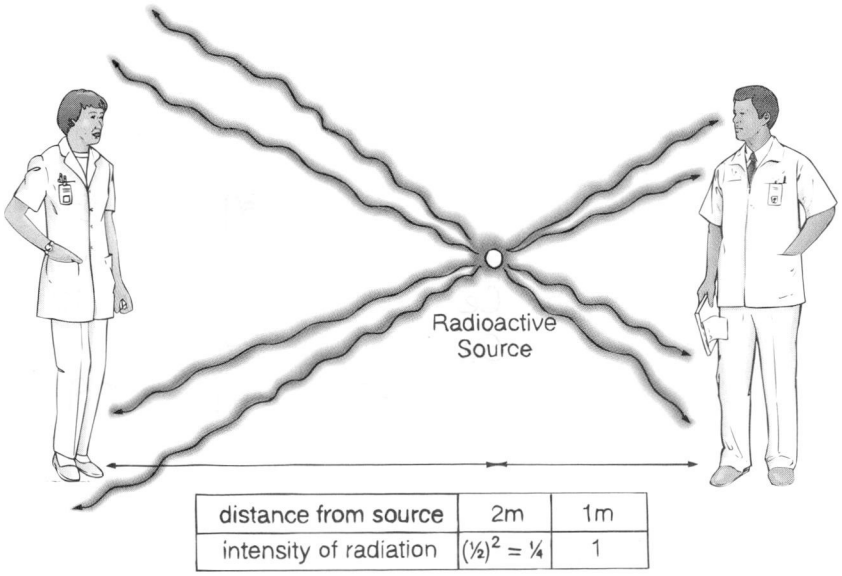

Figure 3.3 The amount of radiation a person receives is less when he or she is further away from the radioactive source.

Study Check

Besides shielding, what other methods help reduce exposure to radiation?

QUESTIONS & PROBLEMS

Natural Radioactivity

The answers for the odd-numbered problems throughout this chapter are given at the end of Chapter 3. The complete solutions are in the *Study Guide.*

3.1

3.1 **a.** How are an alpha particle and a helium nucleus similar? different?
b. What symbols are used for alpha particles?

3.2 **a.** How are a beta particle and an electron similar? different?
b. What symbols are used for beta particles?

3.3 Naturally occurring potassium consists of three isotopes, potassium-39, potassium-40, and radioactive potassium-41.
a. Write the atomic symbols for each isotope.
b. In what ways are the isotopes similar, and in what ways do they differ?

3.4 Naturally occurring iodine is iodine-127. Medically radioactive isotopes of iodine-125 and iodine-130 are used.
a. Write the atomic symbols for each isotope.
b. In what ways are the isotopes similar, and in what ways do they differ?

3.5 Supply the missing information in the following table:

Medical Use	Atomic Symbol	Mass Number	Number of Protons	Number of Neutrons
Heart imaging	$^{201}_{81}Tl$			
Radiation therapy		60	27	
Abdominal scan			31	36
Hyperthyroidism	$^{131}_{53}I$			
Leukemia treatment		32		17

3.6 Supply the missing information in the following table:

Medical Use	Atomic Symbol	Mass Number	Number of Protons	Number of Neutrons
Cancer treatment	$^{60}_{27}Co$	60	27	33
Brain scan	Tc	99	43	56
Blood flow	Ce	141	58	83
Bone scan	Sr	85	38	47
Lung function	$^{133}_{54}Xe$	133	54	79

3.7 Write a symbol for the following:
a. alpha particle **b.** neutron **c.** beta particle
d. nitrogen-15 **e.** iodine-125

3.8 Write a symbol for the following:
a. proton **b.** gamma ray **c.** electron
d. barium-131 **e.** cobalt-60

3.9 Identify the symbol for X in each of the following:
a. ${}^{0}_{-1}X$ **b.** ${}^{4}_{2}X$ **c.** ${}^{1}_{0}X$ **d.** ${}^{24}_{11}X$ **e.** ${}^{14}_{6}X$

3.10 Identify the symbol for X in each of the following:
a. ${}^{1}_{1}X$ **b.** ${}^{32}_{15}X$ **c.** ${}^{0}_{0}X$ **d.** ${}^{59}_{26}X$ **e.** ${}^{85}_{38}X$

3.11 **a.** Why does beta radiation penetrate further in solid material than alpha radiation?
b. How does ionizing radiation cause damage to cells of the body?
c. Why does the x-ray technician leave the room when you receive an X ray?
d. What is the purpose of wearing gloves when handling radioisotopes?

3.12 **a.** As a nurse in an oncology unit, you sometimes give an injection of a radioisotope. What are three ways you can minimize your exposure to radiation?
b. Why are cancer cells more sensitive to radiation than nerve cells?
c. What is the purpose of placing a lead apron on a patient who is receiving routine dental X rays?
d. Why are the walls in a radiology office built of thick concrete blocks?

Study Pack

Chem of Life Atoms and Molecules: Atomic Structure 4

3.2 Nuclear Equations

Learning Goal

Write an equation showing mass numbers and atomic numbers for radioactive decay.

When a nucleus spontaneously breaks down by emitting radiation, the process is called **radioactive decay**. It can be shown as a *nuclear equation* using the symbols for the original radioactive nucleus, the new nucleus, and the radiation emitted.

Radioactive nucleus $\longrightarrow$ new nucleus + radiation (α, β, γ)

A nuclear equation is balanced when the sum of the mass numbers and the sum of the atomic numbers of the particles and atoms on one side of the equation are equal to their counterparts on the other side.

Alpha Emitters

Alpha emitters are radioisotopes that decay by emitting alpha particles. For example, uranium-238 decays to thorium-234 by emitting an alpha particle, as shown in Figure 3.4.

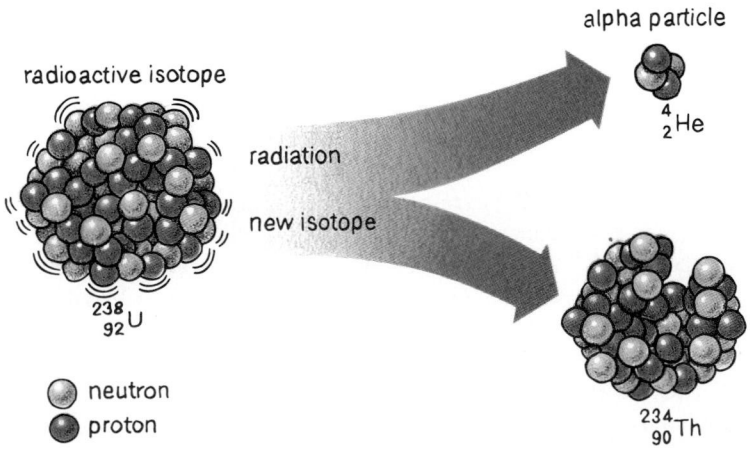

Figure 3.4 An unstable uranium-238 nucleus undergoes radioactive decay by emitting an alpha particle.

$$\underset{\text{Radioactive nucleus}}{{}^{238}_{92}\text{U}} \longrightarrow \underset{\text{New nucleus}}{{}^{234}_{90}\text{Th}} + \underset{\text{Alpha particle}}{{}^{4}_{2}\text{He}}$$

The alpha particle emitted contains 2 protons, which gives the new nucleus 2 fewer protons, or 90 protons. That means that the new nucleus has an atomic number of 90 and is therefore thorium (Th). Because the alpha particle has a mass number of 4, the mass number of the thorium isotope is 234, 4 less than that of the original uranium nucleus.

3.2 Completing a Nuclear Equation

In another example of radioactive decay, radium-226 emits an alpha particle to form a new isotope whose mass number, atomic number, and identity we must determine. We would first write an incomplete nuclear equation:

$${}^{226}_{88}\text{Ra} \longrightarrow \boxed{\ ?\ } + {}^{4}_{2}\text{He}$$

To complete the equation, we must make the mass number of the reactant equal the combined mass numbers of the products. Therefore, the sum of the mass number of the alpha particle and the mass number of the new isotope must equal 226, the mass number of radium. We can calculate the mass number of the new nucleus by subtracting:

$$226 - 4 = 222 \quad \text{Mass number of new nucleus}$$

The atomic number of radium (88) must equal the sum of the atomic numbers of the alpha particle and the new isotope. Therefore, we obtain the atomic number of the new nucleus by the following calculation:

$$88 - 2 = 86 \quad \text{Atomic number of new nucleus}$$

In the periodic table, the element that has an atomic number of 86 is radon (Rn). We complete the nuclear equation by writing the atomic symbol for radon-222:

$${}^{226}_{88}\text{Ra} \longrightarrow \boxed{{}^{222}_{86}\text{Rn}} + {}^{4}_{2}\text{He}$$

ENVIRONMENTAL NOTE

Radon in Our Homes

The presence of radon has become a much publicized environmental and health issue because of radiation danger. Radioactive isotopes that produce radon, such as radium-226 and uranium-238, are naturally present in many types of rocks and soils. Radium-226 emits an alpha particle and is converted into radon gas, which diffuses out of the rocks and soil.

$${}^{226}_{88}\text{Ra} \longrightarrow {}^{222}_{86}\text{Rn} + {}^{4}_{2}\text{He}$$

As uranium-238 decays, it also forms radium-226, which in turn produces radon. Uranium-238 has been found in particularly high levels in an area between Pennsylvania and New England.

Outdoors, radon gas poses little danger, because it dissipates in the air. However, if the source of radon is under a house or building, the gas can enter the house through cracks in the foundation or other openings. Then the radon is inhaled by those living or working there. Inside the lungs, radon emits alpha particles to form polonium-218, which is known to cause cancer when present in the lungs.

$${}^{222}_{86}\text{Rn} \longrightarrow {}^{218}_{84}\text{Po} + {}^{4}_{2}\text{He}$$

Some researchers have estimated that 10% of all lung cancer deaths in the United States are due to radon. The Environmental Protection Agency (EPA) recommends that the maximum level of radon not exceed 4 picocuries (pCi) per liter of air in a home. One (1) picocurie (pCi) is equal to 10^{-12} curies (Ci); curies are described in Section 3.4. In California, 1% of all the houses surveyed exceeded the EPA's recommended maximum radon level.

S·A·M·P·L·E **3.3** P·R·O·B·L·E·M

Writing an Equation for Alpha Decay

Smoke detectors, now required in many homes and apartments, contain an alpha emitter such as americium-241. The alpha particles ionize air molecules, producing a constant stream of electrical current. However, when smoke particles enter the detector, they interfere with the formation of ions in the air, and the electric current is interrupted. This causes the alarm to sound and warns the occupants of the danger of fire. Complete the following nuclear equation for the decay of americium-241:

$$^{241}_{95}\text{Am} \longrightarrow \boxed{\ ?\ } + {}^{4}_{2}\text{He}$$

Solution

The mass number of the new nucleus is 237, obtained by subtracting the mass number of the alpha particle (4) from the mass number of americium (241).

Mass number of Am	−	Mass number of alpha particle	=	Mass number of new nucleus
241	−	4	=	237

The atomic number of the new nucleus is obtained by subtracting the atomic number of the alpha particle (2) from the atomic number of americium (95).

Atomic number of Am	−	Atomic number of alpha particle	=	Atomic number of new nucleus
95	−	2	=	93

The element whose atomic number is 93 is neptunium, Np. The completed nuclear equation for the reaction is written as follows:

$$^{241}_{95}\text{Am} \longrightarrow {}^{237}_{93}\text{Np} + {}^{4}_{2}\text{He}$$

Study Check

Write a balanced nuclear equation for the alpha emitter polonium-214.

EXPLORE YOUR WORLD

Smoke Detectors

Read the label of contents on a smoke detector. What isotope is used? What are the products of radioactive decay for the isotope? (*Hint:* Use a chemistry handbook.)

Questions

1. What type of radiation is emitted?
2. Why would this be used in a home?
3. How does the alarm detect smoke?
4. What is the activity of the isotope?
5. How often should it be replaced? Why?

Beta Emitters

A beta emitter is a radioactive isotope that decays by emitting beta particles. To form a beta particle, the unstable nucleus converts a neutron into a proton. The newly formed proton adds to the number of protons already in the nucleus and increases the atomic

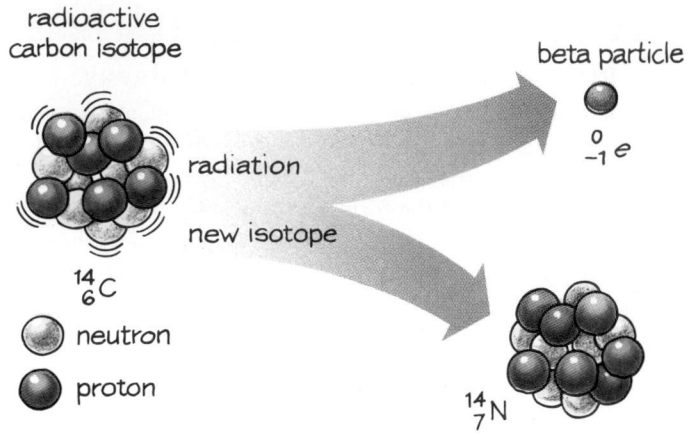

Figure 3.5 A carbon-14 nucleus emits beta radiation when a neutron changes to a proton and electron (beta particle).

3.2

number by 1. However, the mass number of the newly formed nucleus stays the same. For example, carbon-14 decays by emitting a beta particle and forming a nitrogen isotope. (See Figure 3.5.)

In the nuclear equation of a beta emitter, the mass number of the radioactive isotope and the mass number of the new nucleus are the same, and the atomic number of the new nucleus increases by 1, indicating a change of one element into another. This is the nuclear equation for the beta decay of carbon-14:

Mass number
is the same for
both nuclei

$$^{14}_{6}C \longrightarrow {}^{14}_{7}N + {}^{0}_{-1}e$$

Atomic number
of the new nucleus
increases by 1

S·A·M·P·L·E 3.4 P·R·O·B·L·E·M

Writing an Equation for Beta Decay

Cobalt-60, a radioactive isotope used in the treatment of cancer, decays by emitting a beta particle. Write the nuclear equation for its decay.

Solution

The radioactive isotope is a cobalt isotope that has a mass number of 60. Looking at the periodic table, we find that the atomic number of cobalt is 27. The products of the nuclear decay are a beta particle and a new nucleus:

$$^{60}_{27}Co \longrightarrow \boxed{\quad ? \quad} + {}^{0}_{-1}e$$

Because the mass number does not change in beta emission, we can assign a mass number of 60 to the new isotope. The atomic number of the new nucleus will be 1 more than that of cobalt, or 28.

$$\text{Atomic number of new isotope} = \text{Atomic number of Co} + 1$$
$$28 = 27 + 1$$

Because the element that has an atomic number of 28 is nickel (Ni), we can write the equation for the beta decay of the radioactive cobalt as

$$^{60}_{27}Co \longrightarrow {}^{60}_{28}Ni + {}^{0}_{-1}e$$

Study Check

Iodine-131, a beta emitter, is used to check thyroid function and to treat hyperthyroidism. Write its nuclear equation.

HEALTH NOTE

Beta Emitters in Medicine

The radioactive isotopes of several biologically important elements are beta emitters. When a radiologist wants to treat a malignancy within the body, a beta emitter may be used. The short range of penetration into the tissue by beta particles is advantageous for certain conditions. For example, some malignant tumors increase the fluid within the body tissues. A compound containing phosphorus-32, a beta emitter, is injected into the body cavity where the tumor is located. The beta particles travel only a few millimeters through the tissue, so only the malignancy and any tissue within that range are affected. The growth of the tumor is slowed or stopped, and the production of fluid decreases. Phosphorus-32 is also used to treat leukemia, polycythemia vera (excessive production of red blood cells), and lymphomas.

$$^{32}_{15}P \longrightarrow {}^{32}_{16}S + {}^{0}_{-1}e$$

Another beta emitter, iron-59, is used in blood tests to determine the level of iron in the blood, and the rate of production of red blood cells by the bone marrow.

$$^{59}_{26}Fe \longrightarrow {}^{59}_{27}Co + {}^{0}_{-1}e$$

Gamma Emitters

There are very few pure gamma emitters, although gamma radiation accompanies most alpha and beta radiation. In radiology, one of the most commonly used gamma emitters is technetium (Tc), a radioactive element that does not occur naturally on Earth. The excited state called metastable technetium may be written as technetium-99m, Tc-99m, or ^{99m}Tc. By emitting energy in the form of gamma rays, the excited nucleus becomes more stable.

$$^{99m}_{43}Tc \longrightarrow {}^{99}_{43}Tc + \gamma$$

Figure 3.6 summarizes the changes in the nucleus for alpha, beta, and gamma radiation.

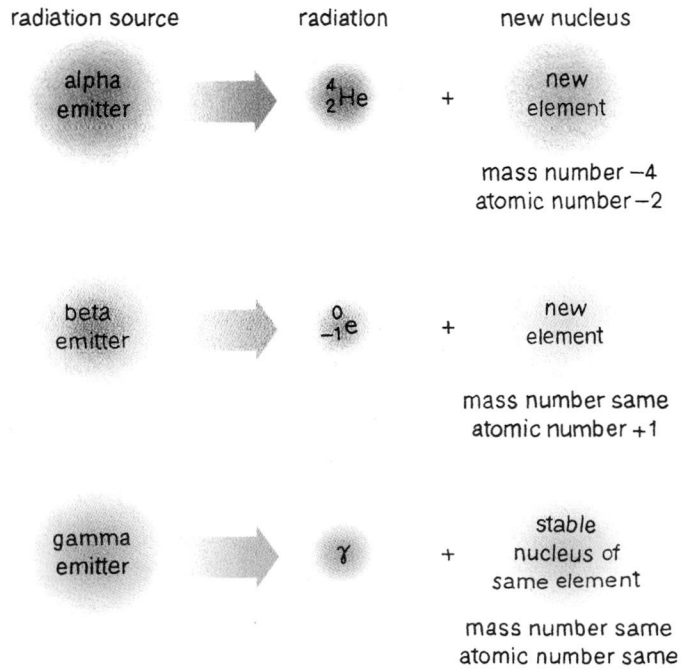

Figure 3.6 Changes in the nucleus for alpha, beta, and gamma radiation.

Q·U·E·S·T·I·O·N·S & P·R·O·B·L·E·M·S

Nuclear Equations

3.13 Write a balanced nuclear equation for the alpha decay of each of the following radioactive isotopes:

a. $^{208}_{84}Po$ **b.** $^{232}_{90}Th$ **c.** $^{251}_{102}No$ **d.** $^{220}_{86}Rn$

3.14 Write a balanced nuclear equation for the alpha decay of each of the following radioactive isotopes:

a. $^{243}_{96}Cm$ **b.** $^{252}_{99}Es$ **c.** $^{251}_{98}Cf$ **d.** $^{261}_{107}Bh$

3.3

3.15 Write a balanced nuclear equation for the beta decay of each of the following radioactive isotopes:

a. $^{25}_{11}Na$ **b.** $^{20}_{8}O$ **c.** $^{92}_{38}Sr$ **d.** $^{42}_{19}K$

3.16 Write a balanced nuclear equation for the beta decay of each of the following radioactive isotopes:

a. potassium-42 **b.** iron-59 **c.** iron-60 **d.** barium-141

3.17 Complete each of the following nuclear equations:

a. $^{28}_{13}Al \longrightarrow ? + {}^{0}_{-1}e$ **b.** $? \longrightarrow {}^{86}_{36}Kr + {}^{1}_{0}n$

c. $^{66}_{29}Cu \longrightarrow {}^{66}_{30}Zn + ?$ **d.** $? \longrightarrow {}^{4}_{2}He + {}^{234}_{90}Th$

3.18 Complete each of the following nuclear equations:

a. $^{11}_{6}C \longrightarrow {}^{7}_{4}Be + ?$ **b.** $^{35}_{16}S \longrightarrow ? + {}^{0}_{-1}e$

c. $? \longrightarrow {}^{90}_{39}Y + {}^{0}_{-1}e$ **d.** $^{210}_{83}Bi \longrightarrow ? + {}^{4}_{2}He$

Study Pack

3.3 Producing Radioactive Isotopes

Learning Goal

Write an equation for the formation of a radioactive isotope.

Today, more than 1500 radioisotopes are produced by converting stable, nonradioactive isotopes into radioactive ones. To do this, a stable atom is bombarded by fast-moving alpha particles, protons, or neutrons. When one of these particles is absorbed by a stable nucleus, the nucleus becomes unstable and the atom is now a radioactive isotope, or **radioisotope**.

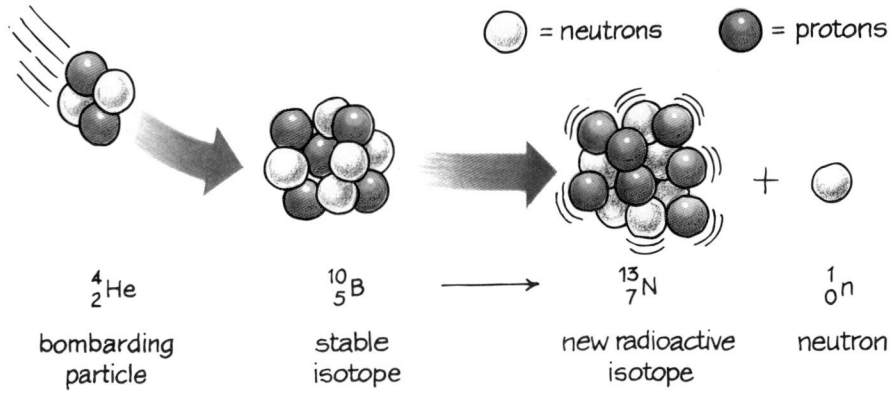

Figure 3.7 Transmutation: the formation of a radioactive isotope by nuclear bombardment.

When a nonradioactive isotope such as boron-10 is bombarded by an alpha particle, it is converted to nitrogen-13, a radioisotope, as shown in Figure 3.7. In this bombardment reaction, a neutron is emitted.

$$\underset{\text{Bombarding particle}}{{}^{4}_{2}\text{He}} + \underset{\text{Stable isotope}}{{}^{10}_{5}\text{B}} \longrightarrow \underset{\text{Radioactive isotope}}{{}^{13}_{7}\text{N}} + \underset{\text{Neutron}}{{}^{1}_{0}\text{n}}$$

The process of changing one element into another is called **transmutation.**

All of the known elements that have atomic numbers greater than 92 have been produced by bombardment; none of these elements occurs naturally. Most have been produced in only small amounts and exist for such a short time that it is difficult to study their properties. An example is element 105, dubnium, which is produced when californium-249 is bombarded with nitrogen-15.

$${}^{249}_{98}\text{Cf} + {}^{15}_{7}\text{N} \longrightarrow {}^{260}_{105}\text{Db} + 4\,{}^{1}_{0}\text{n}$$

Technetium-99m is a radioisotope used in nuclear medicine for several diagnostic procedures, including the detection of brain tumors and examinations of the liver and spleen. The source of technetium-99m is molybdenum-99, which is produced in a nuclear reactor by neutron bombardment of molybdenum-98.

$${}^{98}_{42}\text{Mo} + {}^{1}_{0}\text{n} \longrightarrow {}^{99}_{42}\text{Mo}$$

Many radiology laboratories have a small generator containing the radioactive molybdenum-99, which decays to give the technetium-99m radioisotope. (See Figure 3.8.)

$${}^{99}_{42}\text{Mo} \longrightarrow {}^{99\text{m}}_{43}\text{Tc} + {}^{0}_{-1}e$$

The technetium-99m radioisotope has a half-life of 6 hours and decays by emitting gamma rays. Gamma emission is most desirable for diagnostic work because the gamma rays pass through the body to the detection equipment.

$${}^{99\text{m}}_{43}\text{Tc} \longrightarrow {}^{99}_{43}\text{Tc} + \gamma$$

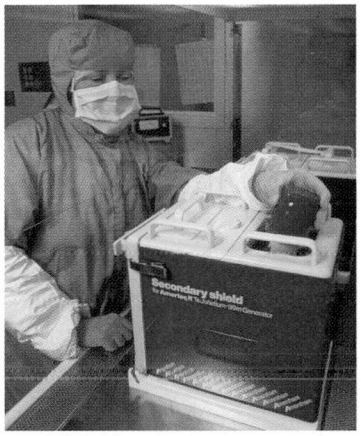

Figure 3.8 Medical physicist unloading technetium-99m (Tc-99m) from a hospital generator.

S·A·M·P·L·E **3.5** ·P·R·O·B·L·E·M

Writing Equations for Transmutations

Gallium-67 is used in the treatment of lymphomas. It is produced by the bombardment of zinc-66 by a proton. Write the equation for this nuclear bombardment.

Solution

The reactants in this nuclear equation are zinc-66 and a proton. The radioactive isotope formed is gallium-67.

$${}^{66}_{30}\text{Zn} + \underset{\text{Proton}}{{}^{1}_{1}\text{H}} \longrightarrow \underset{\text{New radioisotope}}{{}^{67}_{31}\text{Ga}}$$

Study Check

Write the equation for the bombardment of aluminum-27 by an alpha particle to produce the radioactive isotope phosphorus-30 and one neutron.

S·A·M·P·L·E 3.6 P·R·O·B·L·E·M

Completing a Nuclear Equation for a Bombardment Reaction

Write the missing symbol in the following bombardment reaction:

$$^{58}_{28}\text{Ni} + \underset{\text{Proton}}{^{1}_{1}\text{H}} \longrightarrow \boxed{\ ?\ } + {}^{4}_{2}\text{He}$$

3.3

Solution

The sum of the mass numbers for nickel and hydrogen is 59. Therefore, the mass number of the new isotope must be 59 minus 4, or 55. The sum of the atomic numbers is 29. The atomic number of the new isotope is 29 minus 2, or 27. The element that has atomic number 27 is cobalt (Co).

$$^{58}_{28}\text{Ni} + \underset{\text{Proton}}{^{1}_{1}\text{H}} \longrightarrow \underset{\text{New isotope}}{\boxed{^{55}_{27}\text{Co}}} + {}^{4}_{2}\text{He}$$

Study Check

Complete the following bombardment equation:

$$\boxed{} + {}^{4}_{2}\text{He} \longrightarrow {}^{17}_{8}\text{O} + {}^{1}_{1}\text{H}$$

Q·U·E·S·T·I·O·N·S & P·R·O·B·L·E·M·S

Producing Radioactive Isotopes

3.19 Complete each of the following bombardment reactions:

a. $^{9}_{4}\text{Be} + {}^{1}_{0}\text{n} \longrightarrow ?$

b. $^{32}_{16}\text{S} + ? \longrightarrow {}^{32}_{15}\text{P}$

c. $? + {}^{1}_{0}\text{n} \longrightarrow {}^{24}_{11}\text{Na} + {}^{4}_{2}\text{He}$

d. When Al-27 is bombarded by an alpha particle, it forms Si-30. What other particle is produced?

3.20 Complete each of the following bombardment reactions:

a. $^{40}_{18}\text{Ar} + ? \longrightarrow {}^{43}_{19}\text{K} + {}^{1}_{1}\text{H}$

b. $^{238}_{92}\text{U} + {}^{1}_{0}\text{n} \longrightarrow ?$

c. $? + {}^{1}_{0}\text{n} \longrightarrow {}^{14}_{6}\text{C} + {}^{1}_{1}\text{H}$

d. When an alpha particle bombards N-14, it forms a radioisotope and a proton. What is the radioisotope that is produced?

3.4 Radiation Detection and Measurement

Study Pack

Learning Goal

Describe the detection and measurement of radiation.

One of the most common instruments for detecting beta and gamma radiation is the Geiger counter, shown in Figure 3.9. It consists of a metal tube filled with a gas such as argon. Wires connect the metal tube to a battery. When radiation enters a window on the end of the tube, it produces ions in the gas; the ions produce an electrical current. Each burst of current is amplified to give a click and a readout on a meter.

$$\text{Ar} + \text{radiation} \longrightarrow \text{Ar}^+ + e^-$$

Radiation is measured in several different ways. We can measure the activity of a radioactive sample or determine the impact of radiation on biological tissue.

Activity

When a radiology laboratory obtains a radioisotope, the activity of the sample is measured in terms of the number of nuclear disintegrations per second. The **curie (Ci),** the original unit of activity, was defined as the number of disintegrations that occur in one second for 1 g of radium, which is equal to 3.7×10^{10} disintegrations per second. The curie was named for Marie Curie, a Polish scientist, who along with Pierre Curie discovered the radioactive elements radium and polonium. A newer unit of radiation activity is the **becquerel (Bq),** which is one disintegration per second.

Radiation Absorbed

After we measure the activity of a radioisotope, we often want to know how much radiation the tissues in the body absorb. The **rad (radiation absorbed dose)** is a unit that measures the amount of radiation absorbed by a gram of material such as body tissue. A newer unit for absorbed dose is the **gray (Gy),** which is equal to 100 rads.

Biological Damage (Equivalent Dose)

The **rem (radiation equivalent in humans)** measures the biological effects of different kinds of radiation. Alpha particles don't penetrate the skin. But if they should enter the body by some other route, they cause a lot of damage even though the particles travel a short distance in tissue. High-energy radiation such as beta particles and high-energy protons and neutrons that penetrate the skin and travel into tissue cause more damage. Gamma rays are damaging because they travel a long way through tissue and create a great deal of ionization.

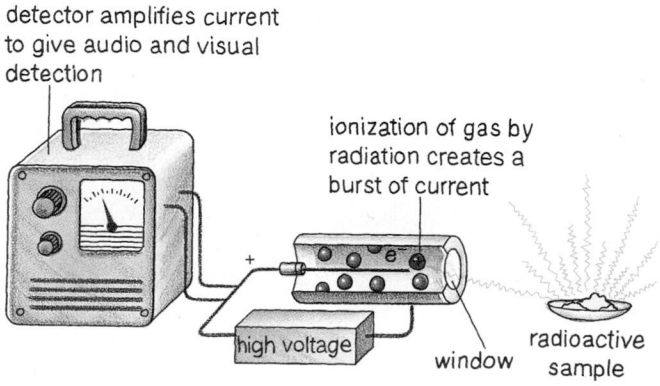

Figure 3.9 A Geiger counter detects an electrical current produced when radiation ionizes a gas.

To determine the equivalent or rem dose, the absorbed dose (rads) is multiplied by a factor that adjusts for biological damage caused by a particular form of radiation. For beta and gamma radiation, and X rays, the factor is 1, so the biological damage in rems is the same as the absorbed radiation (rads). For high-energy protons and neutrons, the factor is about 10, and for alpha particles it is 20.

$$\text{Biological damage (rem)} = \text{absorbed dose (rad)} \times \text{factor}$$

Often the measurement for an equivalent dose will be in units of **millirems (mrem)**. One rem is equal to 1000 mrem. A newer unit is the **sievert (Sv).** One sievert is equal to 100 rems. Table 3.4 summarizes the units used to measure radiation.

3.4

HEALTH NOTE

Radiation and Food

Foodborne illnesses caused by pathogenic bacteria such as *Salmonella*, *Listeria,* and *Escherichia coli* have become a major health concern in the United States. The Centers for Disease Control and Prevention estimates that each year *E. coli* in contaminated foods infects 20,000 people in the United States, and that 500 people die. A recent outbreak of *E. coli* O157:H7 resulted in more than 700 cases of illness and four deaths. This same *E. coli* has been responsible for at least nine outbreaks of illness from contaminated fruit juices, lettuce, and alfalfa sprouts.

The Food and Drug Administration (FDA) has approved the use of 0.3 kilogray (0.3 kGy) to 1 kGy of ionizing radiation produced by cobalt-60 or cesium-137 for the treatment of foods. The irradiation technology is much like that used to sterilize medical supplies. Cobalt pellets are placed in stainless steel tubes, which are arranged in racks. When food moves through the series of racks, the gamma rays pass through the food and kill the bacteria.

It is important for consumers to understand that when food is irradiated, it never comes in contact with the radioactive source. The gamma rays pass through the food to kill bacteria, but that does not make the food radioactive. The radiation kills bacteria because it stops their ability to divide and grow. We cook or heat food thoroughly for the same purpose. Radiation, as well as heat, has little effect on the food itself because its cells are no longer dividing or growing. Thus irradiated food is not harmed although a small amount of vitamin B_1 and C may be lost.

Currently, tomatoes, blueberries, and mushrooms are being irradiated to allow them to be harvested when completely ripe and extend their shelf life (see Figure 3.10). The FDA has also approved the irradiation of pork, poultry, and beef in order to decrease potential infections and to extend shelf life. Currently, irradiated vegetable and meat products are available in retail markets in South Africa. Apollo 17 astronauts ate irradiated foods on the moon, and some U.S. hospitals and nursing homes now use irradiated poultry to reduce the possibility of infections among patients. The extended shelf life of irradiated food also makes it useful for campers and military personnel. Soon consumers concerned about food safety will have a choice of irradiated meats, fruits, and vegetables at the market.

a

b

Figure 3.10 **(a)** The FDA requires this symbol to appear on irradiated retail foods. **(b)** After 2 weeks, the irradiated blueberries on the right show no spoilage. Mold is starting to grow on the nonirradiated ones on the left.

Table 3.4 Some Units of Radiation Measurement

Measurement	Common Unit	SI Unit	Relationship
Activity	curie (Ci) = 3.7×10^{10} disintegrations/s	becquerel (Bq) = 1 disintegration/s	1 Ci = 3.7×10^{10} Bq
Absorbed Dose	rad	gray (Gy)	1 Gy = 100 rad
Biological Damage	rem = rad × factor	1 sievert (Sv)	1 Sv = 100 rem

S · A · M · P · L · E · 3.7 · P · R · O · B · L · E · M

Measuring Activity

A treatment for leukemia uses phosphorus-32, a beta emmitter, which has an activity of 2 millicuries (mCi). How many beta particles are emitted in 1 s?

Solution

We can calculate the number of beta particles from the radioisotope's activity.

$$2\ \cancel{\text{mCi}} \times \frac{1\ \cancel{\text{Ci}}}{1000\ \cancel{\text{mCi}}} \times \frac{3.7 \times 10^{10}\ \beta \text{ particles}}{\cancel{1\ \text{Ci}}} \times 1\ \cancel{\text{s}}$$

$$= 7 \times 10^{7}\ \beta \text{ particles}$$

Study Check

An iodine-131 source has an activity of 0.25 Ci. How many radioactive atoms will disintegrate in 1.0 min?

Background Radiation

We are all exposed to low levels of radiation every day. Naturally occurring radioactive isotopes are part of the atoms of wood, brick, and concrete in our homes and the buildings where we work and go to school. This radioactivity, called background radiation, is present in the soil, in the food we eat, in the water we drink, and in the air we breathe. For example, one of the naturally occurring isotopes of potassium, potassium-40, is radioactive. It is found in the body because it is always present in any potassium-containing food. Other naturally occurring radioisotopes in air and food are carbon-14, radon-222, strontium-90, and iodine-131.

We are constantly exposed to radiation (cosmic rays) produced in space by the sun. At higher elevations, the amount of radiation from outer space is greater because there are fewer air molecules to absorb the radiation. People living at high altitudes or flying in an airplane receive more radiation from cosmic rays than those who live at sea level. For example, a person living in Denver receives about twice the cosmic radiation as a person living in Los Angeles.

Other Sources of Radiation

You may also receive some radiation from nuclear testing, although such testing has decreased in recent years. If nuclear testing occurs, radioactive isotopes enter the atmosphere and are carried around the world by wind currents. They reach the earth

Table 3.5 Average Annual Radiation Received by a Person in the United States

Source	Dose (mrem)
Natural	
The ground	15
Air, water, food	30
Cosmic rays	40
Wood, concrete, brick	50
Medical	
Chest X ray	50
Dental X ray	20
Upper gastrointestinal tract X ray	200
Other	
Television	2
Air travel	1
Radon	200[a]
Cigarette smoking	35

[a]Varies widely.

during rainstorms and other weather phenomena. A person living close to a nuclear power plant normally does not receive much additional radiation, perhaps 0.1 millirem (mrem) in 1 year. (One rem equals 1000 mrem.) However, in the accident at the Chernobyl nuclear power plant in 1986, it is estimated that people in a nearby town received as much as 1 rem/hr.

In addition to naturally occurring radiation from construction materials in our homes, we receive radiation from television. In the medical clinic, dental and chest X rays also add to our radiation exposure. Table 3.5 lists some common sources of radiation. The average person in the United States receives about 0.17 rem or 170 mrem of radiation annually.

3.4 Radiation Sickness

The larger the dose of radiation received at one time, the greater the effect on the body. Exposure to radiation under 25 rem usually cannot be detected. Whole-body exposure of 100 rem produces a temporary decrease in the number of white blood cells. If the exposure to radiation is 100 rem or higher, the person suffers the symptoms of radiation sickness: nausea, vomiting, fatigue, and a reduction in white cell count. A whole-body dosage greater than 300 rem can lower the white cell count to zero. The victim suffers diarrhea, hair loss, and infection.

Lethal Dose

Exposure to radiation of about 500 rem is expected to cause death in 50% of the people receiving that dose. This amount of radiation is called the lethal dose for one-half the population, or the LD_{50}. The LD_{50} varies for different life-forms, as Table 3.6 shows. Radiation dosages of about 600 rem would be fatal to all humans within a few weeks.

Table 3.6 Lethal Doses of Radiation for Some Life-Forms

Life-Form	LD_{50} (rem)
Insect	100,000
Bacterium	50,000
Rat	800
Human	500
Dog	300

Maximum Permissible Dose

Any person working with radiation, such as a radiologist, a radiation technician, or a nurse whose patient has received a radioisotope, must wear some type of detection apparatus to measure exposure to radiation. A standard for occupational exposure is the maximum permissible dose (MPD). If this dose is not exceeded, the probability of injury is minimized. The detection apparatus is usually a badge, ring, or pin containing a small piece of photographic film. The window that covers the film absorbs light and beta rays, but gamma rays penetrate the covering and expose the film. The darker the film, the more the person has been exposed to gamma radiation. These badges are checked periodically to prevent exposure to more than the maximum permissible dose. The maximum permissible dose for occupational exposure is 5 rem (5000 mrem) per year.

QUESTIONS & PROBLEMS

Radiation Detection and Measurement

3.21 **a.** How does a Geiger counter detect radiation?

b. What are the SI unit and the older unit that describe the activity of a radioactive sample?

c. What are the SI unit and the older unit that describe the radiation dose absorbed by tissue?

d. What is meant by the term kilogray?

3.22 **a.** What is background radiation?
b. What are the SI unit and the older unit that describe the biological effect of radiation?
c. What is meant by the terms mCi and mrem?
d. Why is a factor used to determine the dose equivalent?

3.23 **a.** A sample of iodine-131 has an activity of 3.0 Ci. How many disintegrations occur in the iodine-131 sample in 20. s?
b. The recommended dosage of iodine-131 is 4.20 μCi/kg of body weight. How many microcuries of iodine-131 are needed for a 70.0-kg patient with hyperthyroidism?

3.24 **a.** The dosage of technetium-99m for a lung scan is 20 μCi/kg of body weight. How many millicuries should be given to a 50.0-kg patient? (1 mCi = 1000 μCi)
b. A patient receives 50 mrads in a chest X ray. What is that amount in grays? What would be the dose equivalent in mrems?
c. Suppose a person absorbed 50 mrads of alpha radiation. What would be the dose equivalent in mrems? How does it compare with the mrems in part b?

3.25 Why would an airline pilot be exposed to more background radiation than the person who works at the ticket counter?

3.26 In radiation therapy, a patient receives high doses of radiation. What symptoms of radiation sickness might the patient exhibit?

3.5 Medical Applications Using Radioactivity

Learning Goal

Describe the use of radioisotopes in medicine.

Suppose a radiologist wants to determine the condition of an organ in the body. How is this done? In some cases, the patient is given a radioisotope that is known to concentrate in that organ. The cells in the body cannot differentiate between a nonradioactive atom and a radioactive one. All atoms of an element, including any radioactive isotopes, have the same electron arrangement and the same chemistry in the body. The difference is that radioactive atoms can be detected because they emit radiation and the nonradioactive atoms do not. Some radioisotopes used in nuclear medicine are listed in Table 3.7.

After a patient receives a radioisotope, the radiologist determines the level and location of radioactivity emitted by the radioisotope. An apparatus called a scanner is used to produce an image of the organ. The scanner moves slowly across the patient's body above the region where the organ containing the radioisotope is located. The gamma rays emitted from the radioisotope in the organ can be used to expose a photographic plate, producing a **scan** of the organ. On a scan, an area of decreased or increased radiation can indicate such conditions as a disease of the organ, a tumor, a blood clot, or edema.

A common method of determining thyroid function is the use of radioactive iodine uptake (RAIU). Taken orally, the radioisotope iodine-131 mixes with the iodine already present in the thyroid. Twenty-four hours later, the amount of iodine taken up by the thyroid is determined. A detection tube held up to the area of the thyroid gland detects the radiation coming from the iodine-131 that has located there. (See Figure 3.11.)

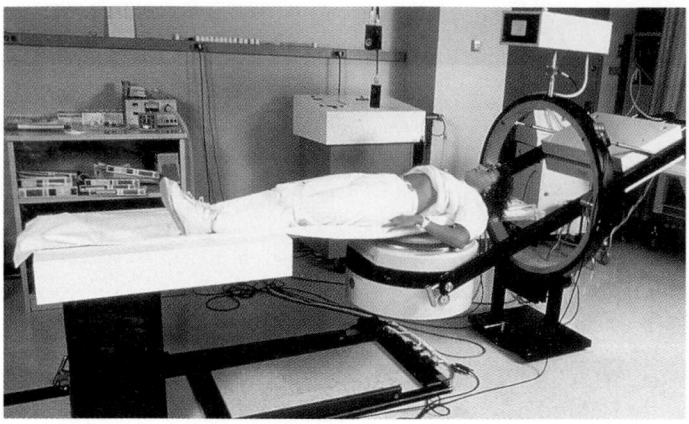

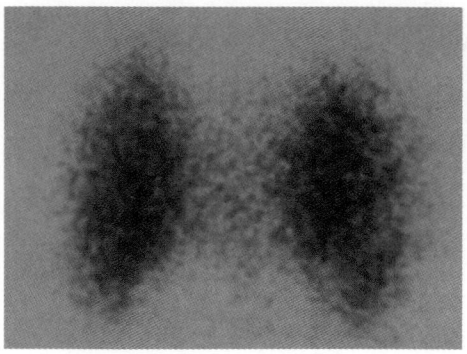

Figure 3.11 A scanner is used to detect radiation from radioisotopes in the organs of the chest (left) and thyroid (right).

The iodine uptake is directly proportional to the activity of the thyroid. A patient with a hyperactive thyroid will have a higher than normal level of radioactive iodine, whereas a patient with a hypoactive thyroid will record low values.

If the patient has hyperthyroidism, treatment is begun to lower the activity of the thyroid. One treatment involves giving the patient a therapeutic dosage of radioactive iodine, which has a higher radiation count than the diagnostic dose. The radioactive iodine goes to the thyroid where its radiation destroys some of the thyroid cells. The thyroid produces less thyroid hormone, bringing the hyperthyroid condition under control. Normally, radioactive isotopes do not enter brain cells because there is a protective blood-brain barrier. However, if a blood vessel has broken, or if there is a brain tumor drawing on the blood supply, radioisotopic uptake does occur within the brain and appears on a brain scan. (See Figure 3.12.)

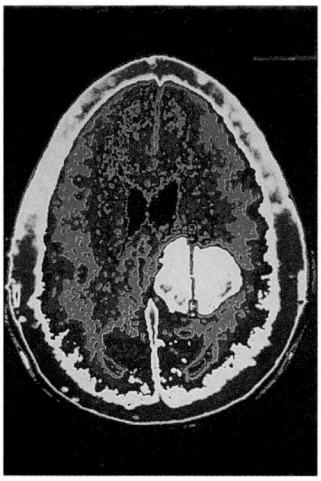

Figure 3.12 A CT scan showing a brain tumor (yellow area) in the center of the right side of the brain.

Table 3.7 Medical Applications of Radioisotopes

Isotope	Half-Life	Medical Application
Ce-141	32.5 days	Gastrointestinal tract diagnosis; measuring myocardial blood flow
Co-60	5.3 yr	External radiation therapy; sterilize surgical instruments and medicines
Ga-67	78 hr	Abdominal imaging; tumor detection
Ga-68	68 min	Detect pancreatic cancer
P-32	4.3 days	Treatment of leukemia, polycythemia vera (excess red blood cells), pancreatic cancer
I-123	13.1 hr	Imaging brain, thyroid, kidney, and heart; measuring cerebral blood flow; and detection of neurological disease
I-125	60 days	Treatment of brain cancer; osteoporosis detection
I-131	8 days	Imaging thyroid; treatment of Graves' disease, goiter, and hyperthyroidism; treatment of thyroid and prostate cancer
Sr-85	65 days	Detection of bone lesions; brain scans
Sr-89	50 days	Alleviation of bone cancer pain; treatment of prostate cancer
Tc-99m	6 hr	Imaging of skeleton and heart muscle, brain, liver, heart, lungs, bone, spleen, kidney, and thyroid; *most widely used radioisotope in nuclear medicine*
Th-201	73 hr	Heart imaging; diagnosis and location of myocardial infarction

HEALTH NOTE

Radiation Doses in Diagnostic and Therapeutic Procedures

We can compare the levels of radiation exposure commonly used during diagnostic and therapeutic procedures in nuclear medicine. In diagnostic procedures, the radiologist minimizes radiation damage by using only enough radioisotope to evaluate the condition of an organ or tissue. (See Table 3.8.)

The doses used in **radiation therapy** are much greater than those used for diagnostic procedures. For example, a therapeutic dose would be used to destroy the cells in a malignant tumor. Although there will be some damage to surrounding tissue, the healthy cells are more resistant to radiation and can repair themselves. (See Table 3.9.)

Table 3.8 Radiation Doses Used for Diagnostic Procedures

Organ	Dose (rem)
Liver	0.3
Thyroid	50.0
Lung	2.0

Table 3.9 Radiation Doses Used for Therapeutic Procedures

Condition	Dose (rem)
Lymphoma	4500
Skin cancer	5000–6000
Lung cancer	6000
Brain tumor	6000–7000

Positron Emission Tomography (PET)

Radioisotopes that emit a positron are used in an imaging method called positron emission tomography (PET). A positron is a particle emitted from the nucleus that has the same mass as an electron but has a positive charge.

$$\underset{\text{Positron}}{\beta^+} \quad \text{or} \quad {}_{+1}^{0}e$$

Carbon-11 is an example of a radioisotope that emits a positron when it decays.

$${}_{6}^{11}\text{C} \longrightarrow \underset{\text{Positron}}{{}_{+1}^{0}e} + {}_{5}^{11}\text{B}$$

A positron is produced when a proton changes to a neutron.

$${}_{1}^{1}\text{H} \longrightarrow {}_{+1}^{0}e + {}_{0}^{1}\text{n}$$

The new nucleus retains the same mass number but has a lower atomic number. The positron exists for only a moment before it collides with an electron producing gamma rays, which can be detected.

$$\underset{\text{Positron}}{{}_{+1}^{0}e} + \underset{\text{Electron in an atom}}{{}_{-1}^{0}e} \longrightarrow \underset{\text{Gamma rays produced}}{2\gamma}$$

Medically, positron emitters such as carbon-11, oxygen-15, and nitrogen-13 are used to diagnose conditions involving blood flow, metabolism, and particularly brain function. Glucose labeled with carbon-11 is used to detect damage in the brain from epilepsy, stroke, and Parkinson's disease. The gamma rays from the positrons emitted by carbon-11 are detected by computerized equipment to create a three-dimensional image of the organ. (See Figure 3.13.)

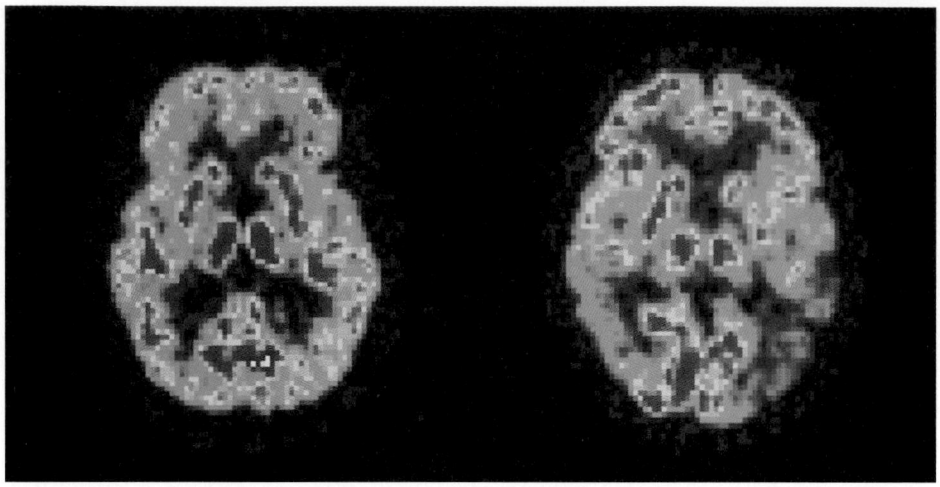

Figure 3.13 These PET scans of the brain show a normal brain on the left and a brain affected by Alzheimer's disease on the right.

3.5

HEALTH NOTE

Other Imaging Methods

Computed Tomography (CT)

Another imaging method used to detect changes within the body is computed tomography (CT). A computer monitors the degree of absorption of 30,000 x-ray beams directed at the brain at successive layers. The differences in absorption based upon the densities of the tissues and fluids in the brain provide a series of images of the brain. This technique is successful in the identification of brain hemorrhages, tumors, and atrophy.

Magnetic Resonance Imaging (MRI)

In another imaging technique called magnetic resonance imaging (MRI), no radiation is used. The atomic nuclei in atoms such as hydrogen emit small amounts of energy when they are excited by a strong magnetic field. Such low-energy changes are harmless to the patient, but they can be detected to create an image on a television monitor. (See Figure 3.14.) Magnetic resonance imaging is used in the detection of multiple sclerosis, abnormalities of the spine and brain, tumors, and birth malformations. In some cases, MRI is replacing the use of X rays and other scanning techniques, including the CT scanner.

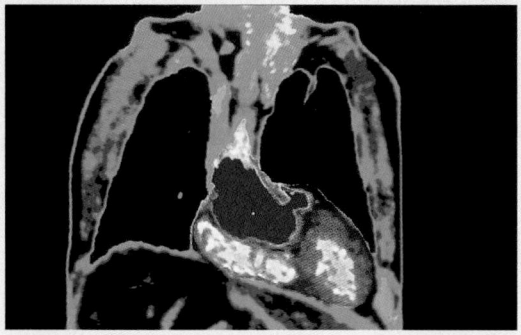

Figure 3.14 An MRI scan of the heart and lungs, with the left ventricle shown in red.

S·A·M·P·L·E 3.8 ·P·R·O·B·L·E·M

Medical Application of Radioactivity

In hyperthyroidism, the thyroid produces an overabundance of thyroid hormone. Why would a radioisotope such as iodine-131 with an activity of 10 μCi be used to image the thyroid?

Solution

The thyroid takes up the iodine-131 because the thyroid uses most of the iodine in the body. An activity level of 10 μCi is high enough to determine the level of thyroid function, but low enough to keep cell damage to a minimum.

Study Check

The evaluation of the thyroid image determined that the patient had an enlarged thyroid. A few weeks later, that same patient was given 100 mCi of iodine-131 intravenously. Why was the activity of the iodine-131 sample used for treatment so much greater than the activity of the iodine-131 used in the diagnostic procedure?

Q·U·E·S·T·I·O·N·S & P·R·O·B·L·E·M·S

Medical Applications Using Radioactivity

3.27 Bone and bony structures contain calcium and phosphorus.

a. Why would the radioisotopes of calcium-47 and phosphorus-32 be used in the diagnosis and treatment of bone diseases?

b. During nuclear tests, scientists were concerned that strontium-85, a radioactive product, would be harmful to the growth of bone in children. Explain.

3.28 **a.** Technetium-99m emits only gamma radiation. Why would this type of radiation be used in diagnostic imaging rather than an isotope that also emits beta or alpha radiation?

b. A patient with polycythemia vera (excess production of red blood cells) receives radioactive phosphorus-32. Why would this treatment reduce the production of red blood cells in the bone marrow of the patient?

3.29 In a diagnostic test for leukemia, a patient receives 4.0 mL of a solution containing selenium-75. If the activity of the selenium-75 is 45 μCi/mL, what is the dose received by the patient?

3.30 A vial contains radioactive iodine-131 with an activity of 2.0 mCi per milliliter. If the thyroid test requires 3.0 mCi in an "atomic cocktail," how many milliliters are used to prepare the iodine-131 solution?

3.6 Half-Life of a Radioisotope

Learning Goal

Given a half-life, calculate the amount of radioisotope remaining after one or more half-lives.

The **half-life** of a radioisotope is the time it takes for one-half of a sample to decay. For example, iodine-131, a radioisotope used in diagnosis and treatment of thyroid disorders, has a half-life of 8 days. If we begin with a sample containing 20 grams of iodine-131, there would be 10 g remaining after 8 days. In another 8 days, we would have 5 g of iodine-131 left. Thus, in each half-life of 8 days, one-half of the radioactive sample has disintegrated.

$$20\text{ g }^{131}\text{I} \xrightarrow[\text{10 g }^{131}\text{I decay}]{\text{8 days}} 10\text{ g }^{131}\text{I} \xrightarrow[\text{5 g }^{131}\text{I decay}]{\text{8 days}} 5\text{ g }^{131}\text{I}$$

This information can be summarized as follows:

Time elapsed	0	8 days	16 days	24 days
Number of half-lives	0	1	2	3
Quantity of iodine-131	20. g	10. g	5.0 g	2.5 g

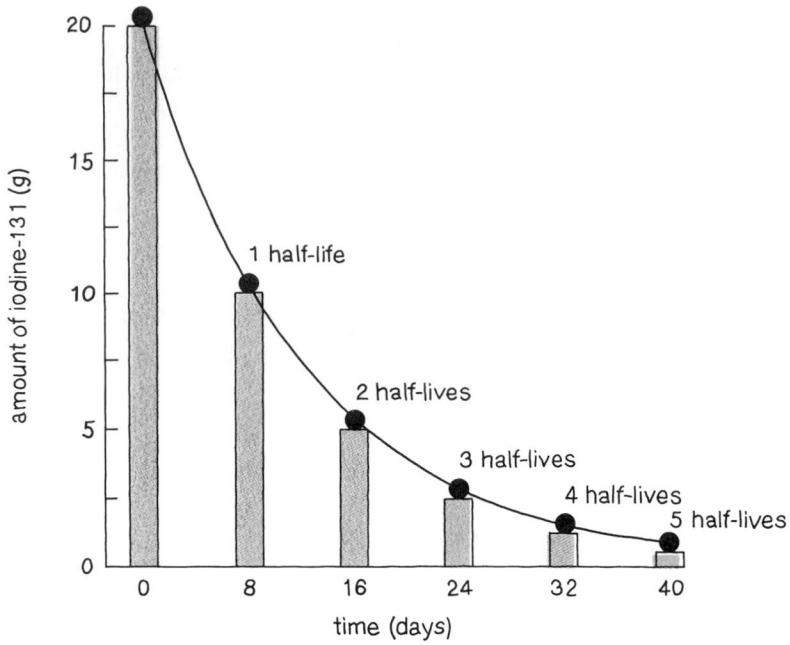

Figure 3.15 Decay curve for iodine-131. One-half of the sample decays with each half-life. Iodine-131 has a half-life of 8 days.

A decay curve is a diagram of the decay of a radioactive isotope. Figure 3.15 shows such a curve for the iodine-131 we have discussed.

Naturally occurring isotopes of the elements usually have long half-lives, as shown in Table 3.10. They disintegrate slowly and produce radiation over a long period of time, even hundreds or millions of years. In contrast, many of the radioisotopes used in nuclear medicine have much shorter half-lives. They disintegrate rapidly and produce almost all their radiation in a short period of time. For example, technetium-99m emits half of its radiation in the first 6 hr. This means that a small amount of the radioisotope given to a patient is essentially gone within 2 days. The decay products of technetium-99m are totally eliminated by the body.

Table 3.10 Half-Lives of Some Radioisotopes

Element	Radioisotope	Half-Life	Type of Radiation
Naturally Occurring Radioisotopes			
Carbon	^{14}C	5730 yr	β
Potassium	^{40}K	1.3×10^9 yr	β, γ
Radium	^{226}Ra	1600 yr	α, γ
Uranium	^{238}U	4.5×10^9 yr	α, γ
Some Medical Radioisotopes			
Carbon	^{11}C	20 min	β^+
Chromium	^{51}Cr	28 days	γ
Iodine	^{131}I	8 days	β, γ
Iodine	^{125}I	60 days	γ
Iron	^{59}Fe	46 days	β, γ
Phosphorus	^{32}P	14 days	β
Oxygen	^{15}O	2 min	β^+
Potassium	^{42}K	12 hr	β, γ
Sodium	^{24}Na	15 hr	β, γ
Strontium	^{85}Sr	64 days	γ
Technetium	^{99m}Tc	6.0 hr	γ

EXPLORE YOUR WORLD

Modeling Half-Lives

Model 1

Obtain a deck of cards. Place 32 cards faceup on the table. The cards facing up represent radioactive atoms. Using a clock with a second hand, begin timing. After 1 minute, turn half the cards facedown. They represent the number of atoms that emit radiation during the first half-life. How many radioactive atoms (faceup cards) are left? When the next minute passes, turn half of the remaining cards facedown. How many radioactive atoms (faceup cards) are left? For each of the next 3 minutes, continue to turn over half the radioactive atoms (faceup cards). Write your own definition of radioactive half-life.

Model 2

Obtain a licorice stick or celery stalk, and a piece of paper. Draw a vertical and a horizontal axis on the paper. Label the vertical axis as radioactive atoms and the horizontal axis as minutes. Place the licorice stick or celery against the vertical axis and mark its height for zero minutes. In the next minute, cut the licorice stick or celery in two. (You can eat the half if you are hungry.) Place the shortened licorice stick or celery at 1 minute on the horizontal axis and mark its height. Every minute cut the licorice stick or celery in half again and mark the shorter height at the corresponding time. Keep reducing the length by half until you cannot divide the licorice or celery in half any more. Connect the points you made for each minute. What does the curve look like? How does this curve represent the concept of a half-life for a radioisotope?

S·A·M·P·L·E 3.9 ·P·R·O·B·L·E·M

Calculating Quantity of Radioisotopes

Nitrogen-13, which has a half-life of 10 min, is used to image organs in the body. For the diagnostic procedure, the patient receives an injection of a compound containing the radioisotope. Originally, the nitrogen-13 has an activity of 40 microcuries (μCi). If the procedure requires 30 min, what is the remaining activity of the radioisotope?

Solution

$$\text{Number of half-lives} = 30\ \cancel{\text{min}} \times \frac{1\ \text{half-life}}{10\ \cancel{\text{min}}}$$

$$= 3$$

The activity of the radioisotope after 3 half-lives is

$$40\ \mu\text{Ci} \xrightarrow{10\ \text{min}} 20\ \mu\text{Ci} \xrightarrow{10\ \text{min}} 10\ \mu\text{Ci} \xrightarrow{10\ \text{min}} 5\ \mu\text{Ci}$$

Another way to calculate the activity of radioactive nitrogen-13 left in the sample is to construct a chart to show the number of half-lives, elapsed time, and the amount of radioactive isotope that is left in the sample.

Time elapsed	0	10 min	20 min	30 min
Number of half-lives elapsed	0	1	2	3
Activity of N-13 remaining	40 μCi	20 μCi	10 μCi	5 μCi

Study Check

Iron-59, used in the determination of bone marrow function, has a half-life of 46 days. If the laboratory receives a sample of 8.0 g of iron-59, how many grams are still active after 184 days?

ENVIRONMENTAL NOTE

Dating Ancient Objects

3.6

A technique known as radiological dating is used by geologists, archaeologists, and historians as a way to determine the age of ancient objects. The age of an object derived from plants or animals (such as wood, fiber, natural pigments, bone, and cotton or woolen clothing) is determined by measuring the amount of carbon-14, a naturally occurring radioactive form of carbon. In 1960, Willard Libby received the Nobel Prize for the work he did developing carbon-14 dating techniques during the 1940s. Carbon-14 is produced in the upper atmosphere by the bombardment of $^{14}_{7}N$ by high-energy neutrons from cosmic rays.

$$^{1}_{0}n + ^{14}_{7}N \longrightarrow$$

Neutron from cosmic rays — Nitrogen in atmosphere

$$^{14}_{6}C + ^{1}_{1}H$$

Radioactive carbon-14 — Proton

The carbon-14 reacts with oxygen to form radioactive carbon dioxide, $^{14}CO_2$. Because carbon dioxide is continuously absorbed by living plants during the process of photosynthesis, some carbon-14 will be taken into the plant. After the plant dies, no more carbon-14 is taken up, and the amount of carbon-14 contained in the plant decreases as it undergoes radioactive decay emitting β particles.

$$^{14}_{6}C \longrightarrow ^{14}_{7}N + ^{0}_{-1}e$$

Scientists use the half-life of carbon-14 (5730 years) to calculate the amount of time that has passed since the plant died, a process called **carbon dating.** The smaller the amount of carbon-14 remaining in the sample, the greater the number of half-lives that have passed. Thus, the approximate age of the sample can be determined. For example, a wooden beam found in an ancient Indian dwelling might have one-half of the carbon-14 found in living plants today. Thus, the dwelling was probably constructed about 5730 years ago, one half-life of carbon-14. (See Figure 3.16.) This technique is useful for dating samples that have ages up to 30,000 years. For older objects, carbon dating is not reliable.

A radiological dating method used for determining the age of rocks is based on the radioisotope uranium-238, which decays through a series of reactions to lead-206. The uranium-238 isotope has a very long half-life, about 4×10^9 (4 billion) years. Measurements of the amounts of uranium-238 and lead-206 enable geologists to determine the age of rock samples. The older rocks will have a higher percentage of lead-206 because more of the uranium-238 has decayed. The age of rocks brought back from the moon by the *Apollo* missions was determined using uranium-238. They were found to be about 4×10^9 years old, approximately the same age calculated for Earth.

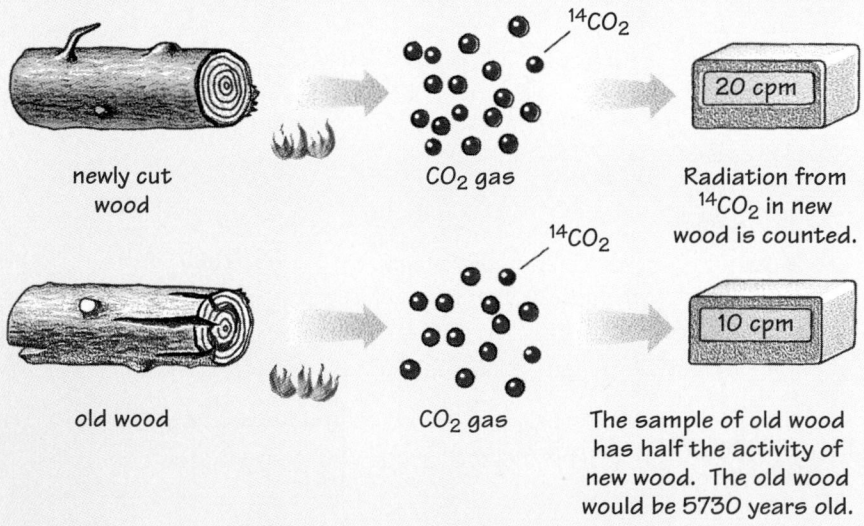

Figure 3.16 The age of an ancient boat or wood from a prehistoric village can be determined by carbon-14 dating.

S·A·M·P·L·E 3.10 P·R·O·B·L·E·M

Dating Using Half-Lives

In Los Angeles, the remains of ancient animals have been unearthed at La Brea tar pits. Suppose a bone sample from the tar pits is subjected to the carbon-14 dating method. If the sample shows that two half-lives have passed, when did the animal live?

Solution

We can calculate the age of the bone sample by using the half-life of carbon-14 (5730 years).

$$2\ \cancel{\text{half-lives}} \times \frac{5730\ \text{years}}{1\ \cancel{\text{half-life}}} = 11{,}000\ \text{years}$$

We would estimate that the animal lived 11,000 years ago, or about 9000 B.C.

Study Check

Suppose that a piece of wood found in a tomb had $\frac{1}{8}$ (3 half-lives) of its original C-14 activity. About how many years ago was the wood part of a living tree?

Q·U·E·S·T·I·O·N·S & P·R·O·B·L·E·M·S

Half-Life of a Radioisotope

3.31 What is meant by the term half-life?

3.32 Why do radioisotopes used for diagnosis in nuclear medicine have short half-lives?

3.33 Technetium-99m is an ideal radioisotope for scanning organs because it has a half-life of 6.0 hr and is a pure gamma emitter. Suppose that 80.0 mg were prepared in the technetium generator this morning. How many milligrams would remain after the following intervals?

a. one half-life **b.** two half-lives
c. 18 hr **d.** 24 hr

3.34 A sample of sodium-24 with an activity of 12 mCi is used to study the rate of blood flow in the circulatory system. If sodium-24 has a half-life of 15 hr, what is the activity of the sodium after $2\frac{1}{2}$ days?

3.35 Strontium-85, used for bone scans, has a half-life of 64 days. How long will it take for the radiation level of strontium-85 to drop to one-fourth of its original level? To one-eighth?

3.36 Fluorine-18, which has a half-life of 110 min, is used in PET scans. If 100 mg of fluorine-18 is shipped at 8:00 A.M., how many milligrams of the radioisotope are still active if the sample arrives at the radiology laboratory at 1:30 P.M.?

Study Pack

Learning Goal

Describe the processes of nuclear fission and fusion.

3.7

3.7 Nuclear Fission and Fusion

Nuclear Fission

In the 1930s, scientists bombarding uranium-235 with neutrons discovered that the U-235 nucleus splits into two medium-weight nuclei and a great amount of energy. This was the discovery of a new kind of nuclear reaction called nuclear **fission.** The energy generated by nuclear fission, splitting the atom, was called atomic energy. When uranium-235 absorbs a neutron, it breaks apart to form two smaller nuclei, several neutrons, and a great amount of energy. (See Figure 3.17.) A typical equation for nuclear fission is

$$^{235}_{92}\text{U} + ^{1}_{0}\text{n} \longrightarrow ^{139}_{56}\text{Ba} + ^{94}_{36}\text{Kr} + 3\,^{1}_{0}\text{n} + \text{energy}$$

If we could weigh these products with great accuracy, we would find that their total mass is slightly less than the mass of the starting materials. The missing mass has been converted into energy, consistent with the famous equation derived by Albert Einstein:

$$E = mc^2$$

E is the energy released, m is the mass lost, and c is the speed of light, 3×10^8 m/s. Even though the mass loss is very small, when it is multiplied by the speed of light squared the result is a large value for the energy released. The fission of 1 g of uranium-235 produces about as much energy as the burning of 3 tons of coal.

Chain Reaction

Fission begins when a neutron collides with the nucleus of a uranium atom. The resulting nucleus is unstable and splits into smaller nuclei. This fission process also releases several neutrons and large amounts of gamma radiation and energy. The neutrons emitted have high energies and bombard more uranium-235 nuclei. As fission continues, there is a rapid increase in the number of high-energy neutrons capable of splitting more uranium atoms, a process called a **chain reaction,** shown in Figure 3.18. To sustain a nuclear chain reaction, sufficient quantities of uranium-235 must be brought together to provide a critical mass in which almost all the neutrons immediately collide with more uranium-235 nuclei. So much heat and energy build up that an atomic explosion occurs.

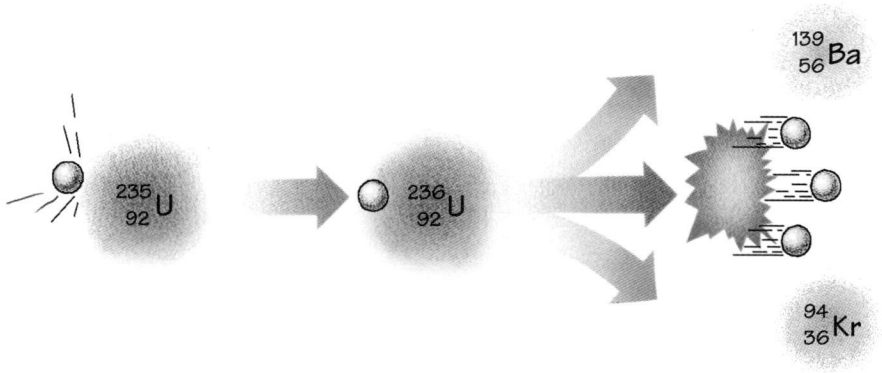

Figure 3.17 Diagram of nuclear fission. After absorbing a fast-moving neutron, a $^{235}_{92}$U nucleus undergoes fission to produce two smaller nuclei, 3 neutrons, and a great amount of energy.

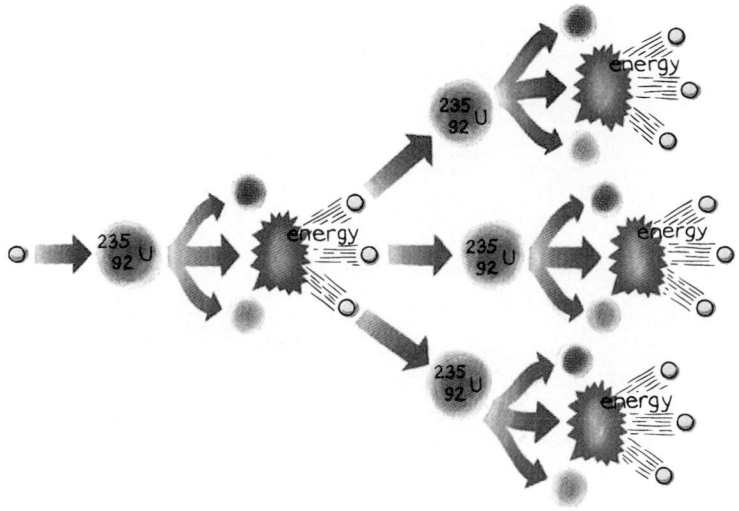

Figure 3.18 Nuclear chain reaction by fission of uranium-235.

Nuclear Power Plants

In a nuclear power plant, the quantity of uranium-235 is held below a critical mass, so it cannot sustain a chain reaction. The fission reactions are slowed by placing control rods, which absorb some of the fast-moving neutrons, among the uranium samples. In this way, less fission occurs, and there is a slower, controlled production of energy. The heat from the controlled fission is used to produce steam. The steam drives a generator, which produces electricity as shown in Figure 3.19. Approximately 10% of the electrical energy produced in the United States is generated in nuclear power plants.

Although nuclear power plants help meet some of our energy needs, there are major problems. One of the most serious problems is the production of radioactive by-products that have very long half-lives. It is essential that these waste products be stored safely for a very long time in a place where they do not contaminate the environment. Early in 1990, the Environmental Protection Agency gave its approval for the storage of radioactive hazardous wastes in chambers 2150 ft underground. In 1998 the Waste Isolation Pilot Plant (WIPP) repository site in New Mexico was ready to receive plutonium waste from former U.S. bomb factories. Although authorities claim the caverns are safe, other people are concerned with the safe transport of the radioactive waste by trucks on the highways.

Nuclear Fusion

In **fusion,** two small nuclei such as those in hydrogen combine to form a larger nucleus. Mass is lost, and a tremendous amount of energy is released, even more than the energy released from nuclear fission. However, a very high temperature (100,000,000°C) is required to overcome the repulsion of the hydrogen nuclei and cause them to undergo fusion. Fusion reactions occur continuously in the sun and other stars, providing us with heat and light. The huge amounts of energy produced by our sun come from the fusion of 6×10^{11} kilograms of hydrogen every second. The fusion reaction shown in Figure 3.20 involves the combination of two isotopes of hydrogen.

$$^{3}_{1}\text{H} + ^{2}_{1}\text{H} \longrightarrow ^{4}_{2}\text{He} + ^{1}_{0}\text{n} + \text{energy}$$

The fusion reaction has tremendous potential as a possible source for future energy needs. One of the advantages of fusion as an energy source is that hydrogen is plentiful in the oceans. Although scientists expect some radioactive waste from fusion reactors, the amount is expected to be much less than that from fission, and the waste products should have shorter half-lives. However, fusion is still in the experimental

Figure 3.19 Generating electricity at Indian Point, a nuclear power plant in Buchanan, New York.

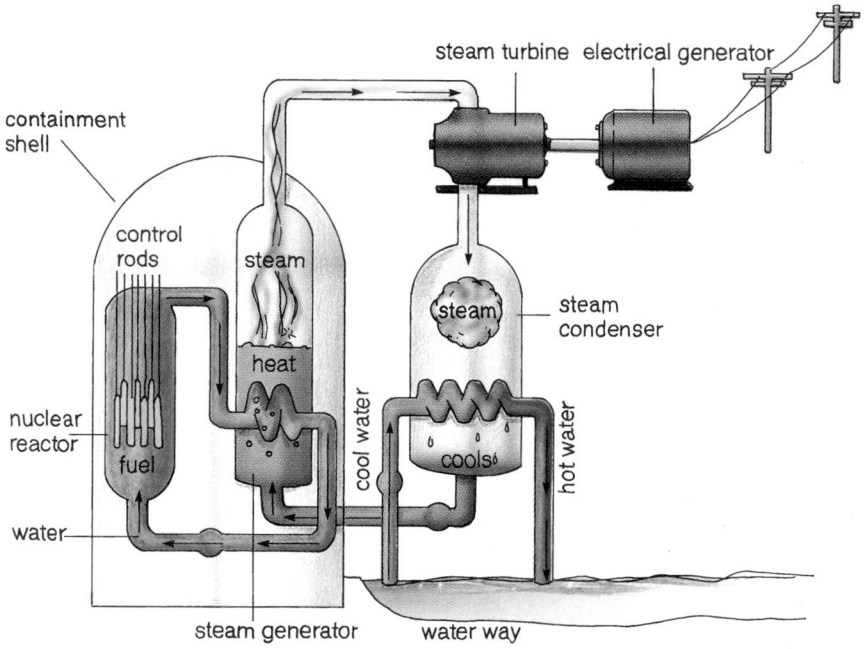

Figure 3.20 The fusion process: Energy is released when small masses combine to form a larger nucleus.

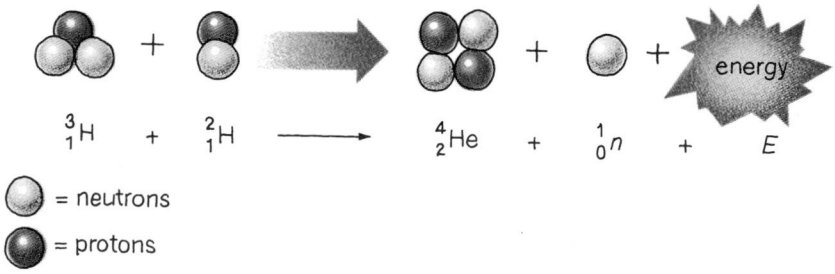

stage because the extremely high temperatures needed have been difficult to reach and even more difficult to maintain. Research groups around the world are attempting to develop the technology needed to make the fusion reaction a reality in our lifetime.

S·A·M·P·L·E 3.11 P·R·O·B·L·E·M

Identifying Fission and Fusion

Classify the following as pertaining to nuclear fission, nuclear fusion, or both:

a. Small nuclei combine to form larger nuclei.
b. Large amounts of energy are released.
c. Very high temperatures are needed for reaction.

Solution

a. fusion
b. both fusion and fission
c. fusion

Study Check

Would you expect the following reaction to be an example of a fission or fusion reaction?

$$^{2}_{1}H + ^{1}_{1}H \longrightarrow ^{3}_{2}He$$

Q·U·E·S·T·I·O·N·S & P·R·O·B·L·E·M·S

Nuclear Fission and Fusion

3.37 What is nuclear fission?

3.38 How does a chain reaction occur in nuclear fission?

3.39 Complete the following fission reaction:

$$^{235}_{92}U + ^{1}_{0}n \longrightarrow ^{131}_{50}Sn + ^{?}_{?} + 2\,^{1}_{0}n + \text{energy}$$

3.40 In another fission reaction, U-235 bombarded with a neutron produces Sr-94, another small nucleus, and 3 neutrons. Write the complete equation for the fission reaction.

3.41 Indicate whether each of the following are characteristic of the fission or fusion process or both:

a. Neutrons bombard a nucleus.
b. The nuclear process occurring in the sun.
c. A large nucleus splits into smaller nuclei.
d. Small nuclei combine to form larger nuclei.

3.42 Indicate whether each of the following are characteristic of the fission or fusion process or both:

a. Very high temperatures are required to initiate the reaction.
b. Less radioactive waste is produced.
c. Hydrogen nuclei are the reactants.
d. Large amounts of energy are released when the nuclear reaction occurs.

Chapter Review

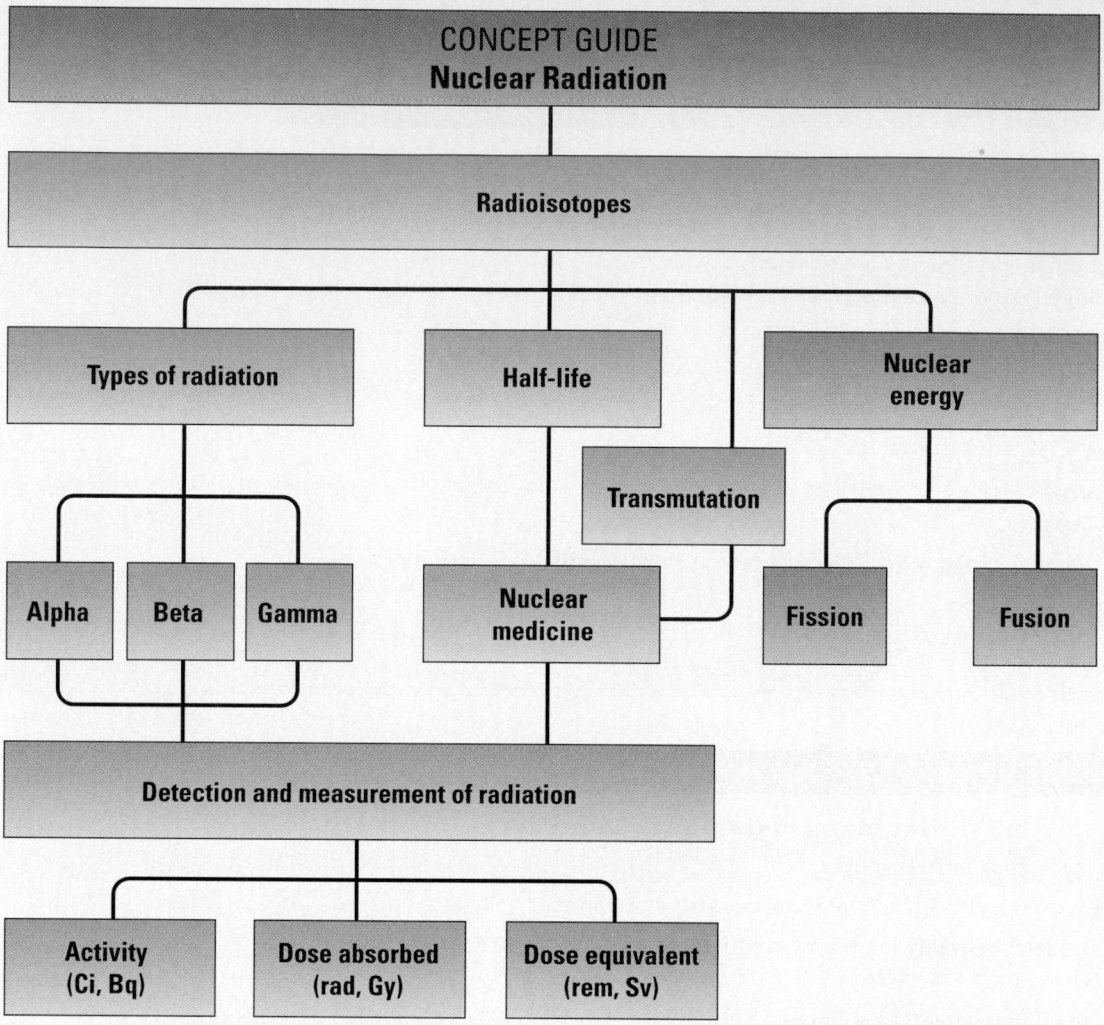

Chapter Summary

3.1 Natural Radioactivity

Radioactive isotopes have unstable nuclei that break down (decay), spontaneously emitting alpha (α), beta (β), and gamma (γ) radiation. Because radiation can damage the cells in the body, proper protection must be used: shielding, limiting the time of exposure, and distance.

3.2 Nuclear Equations

A balanced equation is used to represent the changes that take place in the nuclei of the reactants and products. The new isotopes and the type of radiation emitted can be determined from the symbols that show the mass numbers and atomic numbers of the isotopes in the nuclear reaction.

3.3 Producing Radioactive Isotopes

A radioisotope is produced artificially when a nonradioactive isotope is bombarded by a small particle such as a proton, or an alpha or beta particle. Many radioactive isotopes used in nuclear medicine are produced in this way.

3.4 Radiation Detection and Measurement

In a Geiger counter, radiation ionizes gas in a metal tube, which produces an electrical current. The curie (Ci) measures the number of nuclear transformation of a radioactive sample. Activity is also measured in becquerel (Bq) units. The amount of radiation absorbed by a substance is measured in rads or the gray (Gy). The rem and the sievert (Sv) are units used to determine the biological damage from the different types of radiation.

3.5 Medical Applications Using Radioactivity

In nuclear medicine, radioisotopes are given that go to specific sites in the body. By detecting the radiation they emit, an evaluation can be made about the location and extent of an injury, disease, tumor, or the level of function of a particular organ.

3.6 Half-Life of a Radioisotope

Every radioisotope has its own rate of emitting radiation. The time it takes for one-half of a radioactive sample to decay is called its half-life. For many medical radioisotopes, such as Tc-99m and I-131, half-lives are short; for others, usually naturally occurring ones such as C-14, Ra-226, and U-238, half-lives are extremely long.

3.7 Nuclear Fission and Fusion

In fission, a large nucleus breaks apart into smaller pieces, releasing one or more types of radiation and a great amount of energy. In fusion, small nuclei combine to form a larger nucleus while great amounts of energy are released.

Key Terms

alpha particle A nuclear particle identical to a helium ($^{4}_{2}He$, or α) nucleus (2 protons and 2 neutrons).

becquerel (Bq) A unit of the activity of a radioactive sample equal to one disintegration per second.

beta particle A particle identical to an electron ($^{0}_{-1}e$, or β) that forms in the nucleus when a neutron changes to a proton and an electron.

carbon dating A technique used to date ancient specimens that contain carbon. The age is determined by the amount of active C-14 that remains in the sample.

chain reaction A fission reaction that will continue once it has been initiated with a high-energy neutron bombarding a heavy nucleus such as U-235.

curie (Ci) A unit of radiation equal to 3.7×10^{10} disintegrations/s.

equivalent dose The measure of biological damage from an absorbed dose that has been adjusted for the type of radiation.

fission A process in which large nuclei are split into smaller pieces, releasing large amounts of energy.

fusion A reaction in which large amounts of energy are released when small nuclei combine to form larger nuclei.

gamma ray High-energy radiation (γ) emitted to make a nucleus more stable.

gray (Gy) A unit of absorbed dose equal to 100 rads.

half-life The length of time it takes for one-half of a radioactive sample to decay.

rad (radiation absorbed dose) A measure of an amount of radiation absorbed by the body.

radiation therapy The use of high doses of radiation to destroy harmful tissues in the body.

radioactive decay The process by which a radioactive nucleus breaks down with the release of high-energy radiation.

radioisotope A radioactive atom of an element.

rem (radiation equivalent in humans) A measure of the biological damage caused by the various kinds of radiation (rad $\times$ radiation biological factor).

scan The image of a site in the body created by the detection of radiation from radioactive isotopes that have accumulated in that site.

shielding Materials used to provide protection from radioactive sources.

sievert (Sv) A unit of biological damage (equivalent dose) equal to 100 rems.

transmutation The formation of a radioactive nucleus by bombarding a stable nucleus with fast-moving particles.

Chapter Practice

3.43 Carbon-12 is a nonradioactive isotope of carbon, and carbon-14 is a radioactive isotope. What is similar and different about the two carbon isotopes?

3.44 Give the number of protons, neutrons, and electrons in atoms of the following isotopes:

a. boron-10 **b.** zinc-72
c iron-59 **d.** gold-198

3.45 Describe alpha, beta, and gamma radiation in terms of:

a. type of radiation
b. symbols
c. depth of tissue penetration
d. type of shielding needed for protection

3.46 When you have dental X rays at the dentist's office, the technician places a heavy lead apron over you, and then leaves the room to take your X rays. What is the purpose of these activities?

3.47 Write the balanced nuclear equations for each of the following alpha emitters:

a. thorium-225 **b.** bismuth-210
c. platinum-190 **d.** radium-210

3.48 Write the balanced nuclear equations for each of the following beta emitters:

a. potassium-40 **b.** sulfur-35
c. cesium-137 **d.** tin-126

3.49 Complete each of the following nuclear equations:

a. $^{14}_{7}N + ^{4}_{2}He \longrightarrow$? $+ ^{1}_{1}H$

b. $^{27}_{13}Al + ^{4}_{2}He \longrightarrow ^{30}_{14}Si +$?

c. $^{235}_{92}U + ^{1}_{0}n \longrightarrow ^{90}_{38}Sr + 3^{1}_{0}n +$?

3.50 Complete each of the following nuclear equations:

a. $^{59}_{27}Co +$? $\longrightarrow ^{56}_{25}Mn + ^{4}_{2}He$

b. ? $\longrightarrow ^{14}_{7}N + ^{0}_{-1}e$

c. $^{76}_{36}Kr + ^{0}_{-1}e \longrightarrow$?

3.51 Write the atomic symbols and a complete nuclear equation for the following:

a. When two oxygen-16 atoms collide, one of the products is an alpha particle. What is the other product?

b. When californium-249 is bombarded by oxygen-18, a new element, seaborgium-263, and four neutrons are produced.

c. Radon-222 emits an alpha particle, and the product emits another alpha particle. Write the two nuclear reactions.

3.52 Write the symbols and complete the following nuclear equations:

a. When polonium-210 emits an alpha particle, the product is lead-206.

b. Bismuth-211 decays by emitting an alpha particle. The product is another radioisotope that emits a beta particle. Write equations for the two nuclear changes.

3.53 If the amount of radioactive phosphorus-32 in a sample decreases from 1.2 g to 0.30 g in 28 days, what is the half-life of phosphorus-32?

3.54 If the amount of radioactive iodine-123 in a sample decreases from 0.4 g to 0.1 g in 26.2 hours, what is the half-life of iodine-123?

3.55 Iodine-131, a beta emitter, has a half-life of 8.0 days.

a. Write the nuclear equation for the beta-decay of iodine-131.

b. How many grams of a 12.0-g sample of iodine-131 would remain after 40 days?

c. How many days have passed if 48 g of iodine-131 decayed to 3.0 grams of iodine-131?

3.56 Cesium-137, a beta emitter, has a half-life of 30 years.

a. Write the nuclear equation for cesium-137.

b. How many grams of a 16 g sample of cesium-137 would remain after 90 years?

c. How many years will be needed for 28 grams of cesium-137 to decay to 3.5 g of cesium-137?

3.57 A nurse was accidentally exposed to potassium-42 while doing some brain scans for possible tumors. The error was not discovered until 36 hours later when the activity of the potassium-42 sample was 2.0 μCi. If potassium-42 has a half-life of 12 hr, what was the activity of the sample at the time the nurse was exposed?

3.58 A wooden object from the site of an ancient temple has a carbon-14 activity of 10 counts per minute compared with a reference piece of wood cut today that has an activity of 40 counts per minute. If the half-life for carbon-14 is 5730 years, what is the age of the ancient wood object?

3.59 A 120-mg sample of technetium-99m is used for a diagnostic test. If technetium-99m has a half-life of 6.0 hr, how much of the technetium-99m sample remains 24 hr after the test?

3.60 The half-life of oxygen-15 is 124 seconds. If a sample of oxygen-15 has an activity of 4000 becquerels, how many minutes will elapse before it reaches an activity of 500 becquerels?

3.61 Why does a radiation technician wear a film badge?

3.62 When can exposure to radiation be detected? At about what exposure level will a person begin to suffer symptoms of radiation sickness?

3.63 What is the purpose of irradiating meats, fruits, and vegetables?

3.64 The irradiation of foods was approved in the United States in the 1980s.

a. Why have we not seen many irradiated products in our markets?

b. Would you buy foods that have been irradiated? Why or why not?

3.65 What is the difference between fission and fusion?

3.66 **a.** What are the products in the fission of uranium-235 that make possible a nuclear chain reaction?

b. What is the purpose of placing control rods among uranium samples in a nuclear reactor?

3.67 Where does fusion occur naturally?

3.68 Why are scientists continuing to try to build a fusion reaction even though very high temperatures have been difficult to reach and maintain?

Answers

Answers to Study Checks

3.1 β or ${}^{0}_{-1}e$

3.2 Limiting the time one spends near a radioactive source and staying as far away as possible will reduce exposure to radiation.

3.3 ${}^{214}_{84}\text{Po} \longrightarrow {}^{210}_{82}\text{Pb} + {}^{4}_{2}\text{He}$

3.4 ${}^{131}_{53}\text{I} \longrightarrow {}^{131}_{54}\text{Xe} + {}^{0}_{-1}e$

3.5 ${}^{27}_{13}\text{Al} + {}^{4}_{2}\text{He} \longrightarrow {}^{30}_{15}\text{P} + {}^{1}_{0}\text{n}$

3.6 ${}^{14}_{7}\text{N}$

3.7 5.6×10^{11} iodine-131 atoms

3.8 An iodine-131 sample with a higher activity is used in radiation therapy when radiation is needed to destroy some of the cells in the thyroid gland.

3.9 0.50 g

3.10 17,000 years

3.11 fusion

Answers to Selected Questions and Problems

3.1 **a.** Both an alpha particle and a helium nucleus have 2 protons and 2 neutrons. However, an α-particle is emitted from a nucleus during radioactive decay.
b. α, ${}^{4}_{2}\text{He}$

3.3 **a.** ${}^{39}_{19}\text{K}$, ${}^{40}_{19}\text{K}$, ${}^{41}_{19}\text{K}$
b. They all have 19 protons and 19 electrons, but they differ in the number of neutrons.

3.5

Medical Use	Isotope Symbol	Mass Number	Number of Protons	Number of Neutrons
Heart imaging	${}^{201}_{81}\text{Tl}$	201	81	120
Radiation therapy	${}^{60}_{27}\text{Co}$	60	27	33
Abdominal scan	${}^{67}_{31}\text{Ga}$	67	31	36
Hyperthyroidism	${}^{131}_{53}\text{I}$	131	53	78
Leukemia treatment	${}^{32}_{15}\text{P}$	32	15	17

3.7 **a.** α, ${}^{4}_{2}\text{He}$ **b.** ${}^{1}_{0}\text{n}$ **c.** β, ${}^{0}_{-1}e$
d. ${}^{15}_{7}\text{N}$ **e.** ${}^{125}_{53}\text{I}$

3.9 **a.** β or e^-
b. α or He
c. n
d. Na
e. C

3.11 **a.** Because β particles move faster than α particles, they can penetrate further into tissue.
b. Ionizing radiation breaks bonds and forms reactive species that cause undesirable reactions in the cells.
c. X-ray technicians leave the room to increase the distance between them and the radiation. Also a wall that contains lead shields them.
d. Wearing gloves shields the skin from α and β radiation.

3.13 **a.** ${}^{208}_{84}\text{Po} \longrightarrow {}^{204}_{82}\text{Pb} + {}^{4}_{2}\text{He}$
b. ${}^{232}_{90}\text{Th} \longrightarrow {}^{228}_{88}\text{Ra} + {}^{4}_{2}\text{He}$
c. ${}^{251}_{102}\text{No} \longrightarrow {}^{247}_{100}\text{Fm} + {}^{4}_{2}\text{He}$
d. ${}^{220}_{86}\text{Rn} \longrightarrow {}^{216}_{84}\text{Po} + {}^{4}_{2}\text{He}$

3.15 **a.** ${}^{25}_{11}\text{Na} \longrightarrow {}^{25}_{12}\text{Mg} + {}^{0}_{-1}e$
b. ${}^{20}_{8}\text{O} \longrightarrow {}^{20}_{9}\text{F} + {}^{0}_{-1}e$
c. ${}^{92}_{38}\text{Sr} \longrightarrow {}^{92}_{39}\text{Y} + {}^{0}_{-1}e$
d. ${}^{42}_{19}\text{K} \longrightarrow {}^{42}_{20}\text{Ca} + {}^{0}_{-1}e$

3.17 **a.** ${}^{28}_{14}\text{Si}$
b. ${}^{87}_{36}\text{Kr}$
c. ${}^{0}_{-1}e$
d. ${}^{238}_{92}\text{U}$

3.19 **a.** ${}^{10}_{4}\text{Be}$ **b.** ${}^{0}_{-1}e$
c. ${}^{27}_{13}\text{Al}$ **d.** ${}^{1}_{1}\text{H}$

3.21 **a.** When radiation enters the Geiger counter, it ionizes a gas in the detection tube, which produces a burst of current that is detected by the instrument.
b. becquerel (Bq), curie (Ci)
c. gray (Gy), rad
d. 1000 Gy

3.23 **a.** 2.2×10^{12} disintegrations
b. 294 μCi

3.25 When pilots are flying at high altitudes, there is less atmosphere to protect them from cosmic radiation.

3.27 **a.** Since the elements Ca and P are part of bone, their radioactive isotopes will also become part of the bony structures of the body. Once the radioactive isotopes are in the bone, their radiation can be used to diagnose or treat bone diseases.
b. Strontium (Sr) acts much like calcium (Ca) because both are Group 2 elements. The body will accumulate radioactive strontium in bones in the same way that it incorporates calcium. Radioactive strontium is harmful to children because the radiation it produces causes more damage in cells that are dividing rapidly.

3.29 180 μCi

3.31 A half-life is the time it takes for one-half of a radioactive sample to decay.

3.33 **a.** 40.0 mg **b.** 20.0 mg
c. 10.0 mg **d.** 5.00 mg

3.35 128 days, 192 days

3.37 Nuclear fission is the splitting of a large atom into smaller fragments with the release of large amounts of energy.

3.39 $^{103}_{42}Mo$

3.41 **a.** fission **b.** fusion **c.** fission **d.** fusion

3.43 Each has 6 protons and 6 electrons, but carbon-12 has 6 neutrons and carbon-14 has 8 neutrons. Carbon-12 is a stable isotope, but carbon-14 is radioactive and will emit radiation.

3.45 **a.** α radiation consists of a helium nucleus emitted from the nucleus of a radioisotope. β radiation is an electron and γ radiation is high energy emitted from the nucleus of a radioisotope.
b. α particle: $^{4}_{2}He$
β particle: $^{0}_{-1}e$
γ radiation: γ
c. α particles penetrate 0.05 mm into tissue, β particles 4–5 mm, and γ rays 50 cm or more.
d. Paper or clothing will shield you from α particles; heavy clothing, lab coats, and gloves will shield you from β particles; and lead and concrete are needed for shielding from γ radiation.

3.47 **a.** $^{225}_{90}Th \longrightarrow {}^{221}_{88}Ra + {}^{4}_{2}He$
b. $^{210}_{83}Bi \longrightarrow {}^{206}_{81}Tl + {}^{4}_{2}He$
c. $^{190}_{78}Pt \longrightarrow {}^{186}_{76}Os + {}^{4}_{2}He$
d. $^{210}_{88}Ra \longrightarrow {}^{206}_{86}Rn + {}^{4}_{2}He$

3.49 **a.** $^{17}_{8}O$ **b.** $^{1}_{1}H$ **c.** $^{143}_{54}Xe$

3.51 **a.** $^{16}_{8}O + {}^{16}_{8}O \longrightarrow {}^{4}_{2}He + {}^{28}_{14}Si$
b. $^{249}_{98}Cf + {}^{18}_{8}O \longrightarrow {}^{263}_{106}Sg + 4\,{}^{1}_{0}n$
c. $^{222}_{86}Rn \longrightarrow {}^{218}_{84}Po + {}^{4}_{2}He$
$^{218}_{84}Po \longrightarrow {}^{214}_{82}Pb + {}^{4}_{2}He$

3.53 14 days

3.55 **a.** $^{131}_{53}I \longrightarrow {}^{0}_{-1}e + {}^{131}_{54}Xe$
b. 0.375 g
c. 32 days

3.57 16 μCi

3.59 7.5 mg

3.61 A film badge detects the amount of radiation exposure. The badge is checked periodically to make sure that a technician has not received more than the maximum permissible radiation dose.

3.63 The irradiation of meats, fruits and vegetables kills bacteria such as *E. coli* that can cause foodborne illnesses. In addition, spoilage is deterred, and shelf life is extended.

3.65 In the fission process, an atom splits into smaller nuclei. In fusion, small nuclei combine (fuse) to form a larger nucleus.

3.67 Fusion occurs naturally in the sun and other stars.

4

Compounds and Their Bonds

The main component in beach sand, a compound called silicon dioxide (SiO_2), is made of silicon and oxygen. Water (H_2O) is a combination of hydrogen and oxygen. Seawater also contains salt (NaCl), which has the chemical name sodium chloride.

Looking Ahead

4.1	Valence Electrons
4.2	Ions and the Octet Rule
4.3	Ionic Compounds
4.4	Naming and Writing Ionic Formulas
4.5	Covalent Bonds
4.6	Covalent Compounds
4.7	Bond Polarity
4.8	Polyatomic Ions

www.chemplace.com/college/timberlake

Visit the web site above to test yourself, run simulations, and review key concepts from this chapter. For even more interactive review and quizzing, use the two CD-ROMs included with the book.

Most of the things you see and use every day are made of *compounds,* in which atoms of one element are combined with atoms of other elements. Although there are 112 elements known today, there are millions of different compounds because of the many different ways in which atoms may combine. In ionic compounds such as table salt (sodium chloride, NaCl), electrons are transferred from one atom to another to form positive and negative ions. Strong attractive forces called ionic bonds hold the ions together. In covalent compounds such as water (H_2O), electrons are shared by the atoms, an attraction called a covalent bond.

The substances necessary for life include ionic and covalent compounds. For example, the human body is about 60–65% water, a covalent compound made from the elements hydrogen and oxygen. Some other covalent compounds necessary for life are carbohydrates, fats, and proteins. All are made of the elements carbon, hydrogen, and oxygen; proteins contain nitrogen and sulfur also. These compounds, obtained from your diet, are needed to build and repair cells, provide energy, and direct metabolic processes. Ionic compounds such as sodium chloride and potassium chloride (KCl) are needed for the proper functioning of the heart muscle and nerve conduction.

4.1

4.1 Valence Electrons

Learning Goal

Using the periodic table, write the electron-dot structure for a representative element.

For the atoms of the representative elements in the periodic table, the electrons in the outer shell determine their chemical properties. These influential electrons, called **valence electrons,** are located in the valence shell, which is the outermost energy level of an atom. For example, all of the elements in Group 1, such as potassium, have 1 valence electron. Recall that in an atom of potassium, the electrons are arranged in the first four energy levels as 2, 8, 8, 1. Because the fourth shell is the outermost shell for potassium, the electron in the fourth shell is a valence electron. In an electron-dot structure, this valence electron is placed next to the symbol for the element, as shown in Figure 4.1.

The elements in Group 2 have 2 valence electrons; similarly, the elements in Group 3 have 3 valence electrons, elements in Group 5 have 5 valence electrons, and elements in Group 7 have 7 valence electrons. If we compare the electron arrangement and the group number of a representative element, we find that the number of valence electrons in an atom is equal to the group number. Some examples of the relationship between electron arrangement, group number, and number of valence electrons are shown in Table 4.1.

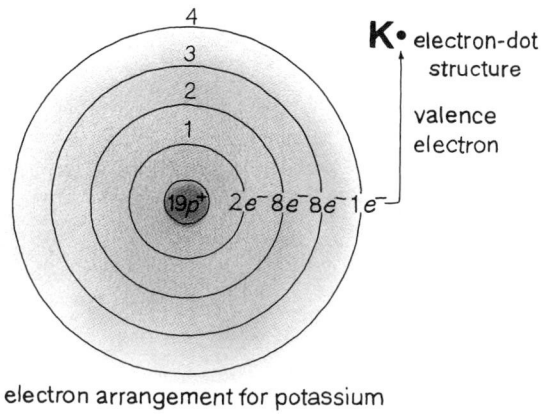

Figure 4.1 The electron arrangement for potassium and its electron-dot structure.

Table 4.1 Electron Arrangement

Element	Group Number	Number of Electrons in Energy Level 1	2	3	4	Number of Valence Electrons
Li	1	2	**1**			1
Ca	2	2	8	8	**2**	2
Al	3	2	8	**3**		3
C	4	2	**4**			4
N	5	2	**5**			5
S	6	2	8	**6**		6
Cl	7	2	8	**7**		7

S·A·M·P·L·E **4.1** ·P·R·O·B·L·E·M

Counting Valence Electrons

Using the electron arrangement, state the number of valence electrons in atoms of each of the following elements:

a. O **b.** Na

Solution

a. The element oxygen has 6 valence electrons, as shown in its electron arrangement.

2 **6**

b. The element sodium has 1 valence electron, because there is 1 electron in its highest energy level, shell 3.

2 8 **1**

Study Check

How many electrons are in the outer shell of sulfur?

Electron-Dot Structures

An **electron-dot structure** is a convenient way to represent the valence shell. Valence electrons are shown as dots placed on the sides, top, or bottom of the symbol for the element. It does not matter on which of the four sides you place the dots. However, 1 to 4 valence electrons are usually arranged as single dots. When there are more than 4 electrons, the electrons begin to pair up. Any of the following would be an acceptable electron-dot structure for magnesium, which has 2 valence electrons:

Possible Electron-Dot Structures for the 2 Valence Electrons in Magnesium

$\dot{\underset{\cdot}{M}}g\cdot$ $\quad$ $\dot{\underset{\cdot}{M}}g$ $\quad$ $\cdot\dot{M}g$ $\quad$ $\cdot Mg\cdot$ $\quad$ $\underset{\cdot}{M}g\cdot$ $\quad$ $\cdot\underset{\cdot}{M}g$

Electron-dot structures for the first 20 elements are given in Table 4.2. Note that although helium has just 2 valence electrons, it appears in Group 8 with the rest of the noble gases because shell 1, its outermost shell, is filled.

Table 4.2 Electron-Dot Structures for the First Twenty Elements

	Group Number							
	1	**2**	**3**	**4**	**5**	**6**	**7**	**8**
Number of Valence Electrons	$1e^-$	$2e^-$	$3e^-$	$4e^-$	$5e^-$	$6e^-$	$7e^-$	$8e^-$
	H·							He:
	Li·	$\dot{\text{Be}}$·	·$\dot{\text{B}}$·	·$\dot{\underset{\cdot}{\text{C}}}$·	·$\ddot{\underset{\cdot}{\text{N}}}$·	·$\ddot{\underset{\cdot}{\text{O}}}$:	·$\ddot{\underset{\cdot\cdot}{\text{F}}}$:	:$\ddot{\underset{\cdot\cdot}{\text{Ne}}}$:
	Na·	$\dot{\text{Mg}}$·	·$\dot{\text{Al}}$·	·$\dot{\underset{\cdot}{\text{Si}}}$·	·$\dot{\underset{\cdot}{\text{P}}}$·	·$\dot{\underset{\cdot}{\text{S}}}$:	·$\ddot{\underset{\cdot\cdot}{\text{Cl}}}$:	:$\ddot{\underset{\cdot\cdot}{\text{Ar}}}$:
	K·	$\dot{\text{Ca}}$·						

4.1

S·A·M·P·L·E 4.2 P·R·O·B·L·E·M

Writing Electron-Dot Structures

Write the electron-dot structure for each of the following elements:

a. chlorine **b.** aluminum **c.** argon

Solution

a. Because the group number for chlorine is 7, we can state that chlorine has 7 valence electrons.

·$\ddot{\underset{\cdot\cdot}{\text{Cl}}}$:

b. Aluminum, in Group 3, has 3 valence electrons.

·$\dot{\text{Al}}$·

c. Argon, in Group 8, has 8 valence electrons.

:$\ddot{\underset{\cdot\cdot}{\text{Ar}}}$:

Study Check

What is the electron-dot structure for the alkaline-earth metal in Period 4?

Q·U·E·S·T·I·O·N·S & P·R·O·B·L·E·M·S

Valence Electrons

4.1 Where are the valence electrons of an atom?

4.2 Why is the group number for a representative element the same as the number of valence electrons?

4.3 Write the electron arrangement and the number of valence electrons in an atom of each of the following elements:
a. N **b.** O **c.** Ar **d.** K **e.** S

4.4 Write the electron arrangement and the number of valence electrons in an atom of each of the following elements:
a. Na **b.** Al **c.** Cl **d.** Mg **e.** P

4.5 Write the group number and electron-dot structure for each element:
a. sulfur **b.** nitrogen **c.** calcium
d. sodium **e.** potassium

4.6 Write the group number and electron-dot structure for each element:
a. carbon **b.** oxygen **c.** fluorine
d. lithium **e.** chlorine

4.7 Using the symbol M for a metal atom, draw the electron dot structure for an atom of a metal in the following groups:
a. Group 1 **b.** Group 2

4.8 Using the symbol NM for a nonmetal atom, draw the electron dot structure for an atom of a nonmetal in the following groups:
a. Group 5 **b.** Group 7

4.9 As shown in Figure 2.4 of Chapter 2, the alkali metals are in the same family on the periodic table. What is their group number and how many valence electrons does each have?

4.10 As shown in Figure 2.5 of Chapter 2, the halogens are in the same family on the periodic table. What is their group number and how many valence electrons does each have?

Study Pack

Chem of Life Atoms and Molecules: Ions 1, 2

4.2 Ions and the Octet Rule

Learning Goal

Using the octet rule, write the symbols of the simple ions for the representative elements.

The noble gases do not combine readily with other elements. One explanation for their lack of reactivity is that 8 valence electrons make them particularly stable. All of the noble gases have an octet of 8 valence electrons, except for helium, which is stable with 2 electrons in its first shell. The electron arrangements and electron-dot structures for some noble gases are shown in Table 4.3.

Except for the noble gases, most elements tend to combine with other elements to form a *compound.* In most of the **compounds** formed by representative elements (Groups 1–7), each atom has a noble gas arrangement. The **octet rule** indicates that the atoms of the elements in a compound lose, gain, or share valence electrons in order to produce a stable, noble gas arrangement of eight electrons. Thus, the octet rule provides a basic key to understanding the ways in which atoms bond together in compounds.

Table 4.3 Electron Arrangement for Some Noble Gases

Noble Gas	Number of Electrons in Shell 1	2	3	4	Electron-Dot Structure
He	**2**				He:
Ne	2	**8**			:Ne:
Ar	2	8	**8**		:Ar:
Kr	2	8	18	**8**	:Kr:

S·A·M·P·L·E **4.3** P·R·O·B·L·E·M

Octet Rule

Write the electron-dot structures for neon (Ne), sodium (Na), and fluorine (F). Which atoms have octets and which do not?

Solution

:Ne: Na· ·F:

Neon, a noble gas, has an octet of 8 valence electrons. Sodium, however, has just 1 valence electron, and fluorine has 7. They would be expected to lose, gain, or share electrons to give each atom an octet of valence electrons.

Study Check

Which of the following elements have octets?
a. carbon **b.** calcium **c.** krypton (Kr)

HEALTH NOTE

Some Uses for Noble Gases

Noble gases may be used when it is necessary to have a substance that is unreactive. Scuba divers normally use a pressurized mixture of nitrogen and oxygen gases for breathing under water. However, when the air mixture is used at depths where pressure is high, the nitrogen gas is absorbed into the blood, where it can cause mental disorientation. To avoid this problem, a breathing mixture of oxygen and helium may be substituted. The diver still obtains the necessary oxygen, but the unreactive helium that dissolves in the blood does not cause mental disorientation. However, its lower density does change the vibrations of the vocal cords, and the diver will sound like Donald Duck.

Helium is also used to fill blimps and balloons. When dirigibles were first designed, they were filled with hydrogen, a very light gas. However, when they came in contact with any type of spark or heating source, they exploded violently because of the extreme reactivity of hydrogen gas with oxygen present in the air. Today blimps are filled with unreactive helium gas, which presents no danger of explosion. (See Figure 4.2.)

Lighting tubes are generally filled with a noble gas such as neon or argon. While the electrically heated filaments that produce the light get very hot, the surrounding noble gases do not react with the hot filament. If heated in air, the elements that constitute the filament would soon burn up.

Figure 4.2 The Goodyear blimp filled with helium gas is ready for takeoff.

4.2

Positive Ions

Metals lose their valence electrons in order to attain a noble gas arrangement. By losing electrons, they form **ions,** which are atoms that have a different number of electrons than protons. For example, when a sodium atom loses its outer electron, the next shell of electrons has a noble gas arrangement, as shown in Figure 4.3.

By losing an electron, sodium now has 10 electrons instead of 11. Because there are still 11 protons in its nucleus, the atom is no longer neutral. It has become a sodium ion and has an electrical charge, called an **ionic charge,** of 1+. In the symbol for the ion, the ionic charge is written in the upper right-hand corner, as in Na^+.

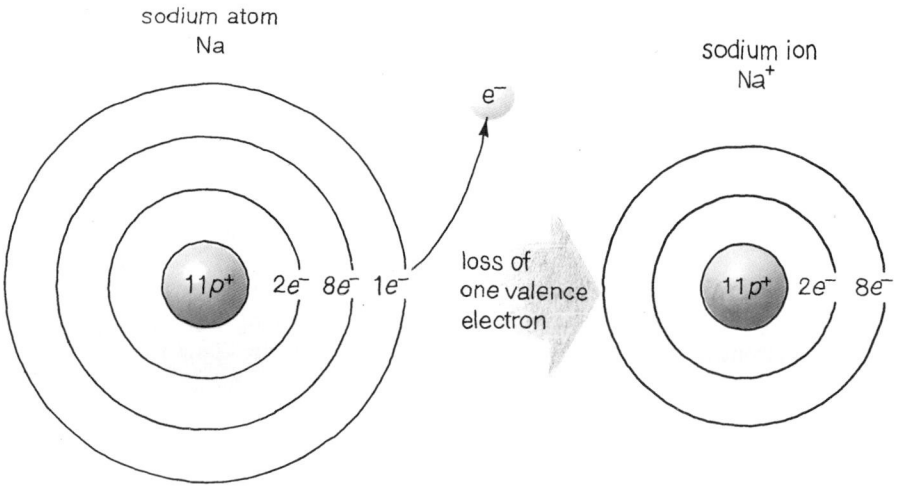

Figure 4.3 The loss of a valence electron changes a sodium atom to a sodium ion, which has a stable electron arrangement. What noble gas has the same electron arrangement as sodium ion?

Comparing the Charges in the Sodium Atom and the Sodium Ion

	Na Atom	Na^+ Ion
Charge from electrons	11−	10−
Charge from protons	11+	11+
Overall ionic charge	0	1+

In general, metals in ionic compounds have lost their valence electrons to form positively charged ions. Positive ions are also called **cations** (pronounced *cat'-ions*). Magnesium, a metal in Group 2, attains a noble gas arrangement by losing 2 valence electrons to form a positive ion with a 2+ ionic charge (see Figure 4.4). A metal ion is named by its element name.

Comparing the Charges in the Magnesium Atom and the Magnesium Ion

	Mg Atom	Mg^{2+} Ion
Charge from electrons	12−	10−
Charge from protons	12+	12+
Overall ionic charge	0	2+

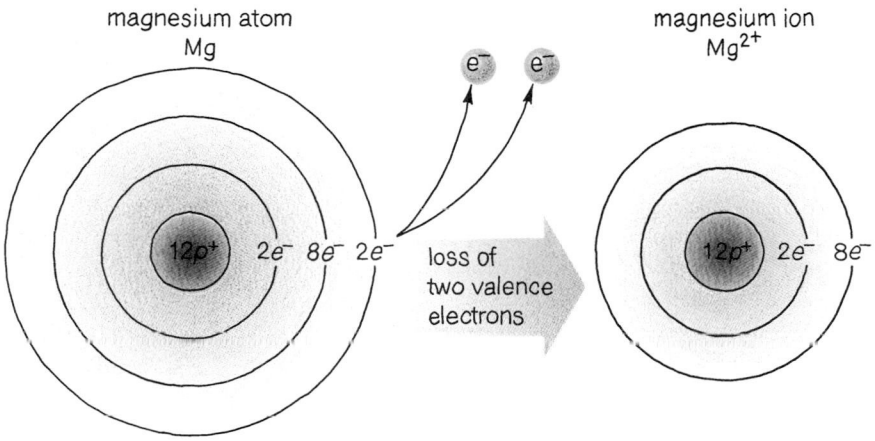

Figure 4.4 A magnesium ion has a stable electron arrangement after it loses 2 valence electrons. What noble gas has the same electron arrangement as magnesium ion?

Negative Ions

A nonmetal forms a negative ion when it gains one or more electrons to fill its valence shell. For example, when an atom of chlorine with 7 valence electrons obtains one more electron, it completes a noble gas arrangement, as shown in Figure 4.5. The resulting particle is a chloride ion having a negative ionic charge (Cl^-). Ions with negative ionic charges are called **anions** (pronounced *an'-ions*). Nonmetal ions are named by using *ide* at the end of their element names.

Comparing the Charges in the Chlorine Atom and the Chloride Ion

	Cl Atom	Cl^- Ion
Charge from electrons	17−	18−
Charge from protons	17+	17+
Overall ionic charge	0	1−

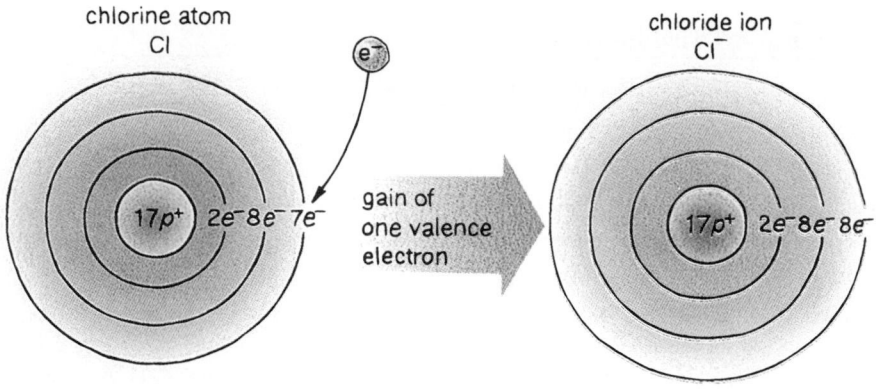

Figure 4.5 When a chlorine atom gains 1 valence electron, it becomes a chloride ion, which has a stable electron arrangement. What noble gas has the same electron arrangement as chloride ion?

4.2

S·A·M·P·L·E **4.4** ·P·R·O·B·L·E·M

Calculating Ionic Charge

Write the symbol and the ionic charge for each of the following ions:

a. a nitrogen ion that has 7 protons and 10 electrons
b. a calcium ion that has 20 protons and 18 electrons
c. a lithium ion that has 3 protons and 2 electrons

Solution

a. $7p^+$ and $10e^-$ = N^{3-} **b.** $20p^+$ and $18e^-$ = Ca^{2+}
c. $3p^+$ and $2e^-$ = Li^+

Study Check

How many protons and electrons are in a bromide ion, Br^-?

Ionic Charges from Group Numbers

Group numbers can be used to determine the ionic charges for most ions of the representative elements. We have seen that metals lose electrons to form positive ions. The elements in Groups 1, 2, and 3 lose the same number of electrons (1, 2, and 3, respectively). Group 1 metals form ions with 1+ charges, Group 2 metals form ions with 2+ charges, and Group 3 metals form ions with 3+ charges.

The nonmetals from Groups 5, 6, and 7 form negative ions. Group 5 nonmetals form ions with 3− charges, Group 6 nonmetals form ions with 2− charges, and Group 7 nonmetals form ions with 1− charges. The elements of Group 4 do not typically form ions. Table 4.4 lists the ionic charges for some common ions of representative elements.

Table 4.4 Ionic Charges for Representative Elements

Group Number	Number of Valence Electrons	Electron Change to Give an Octet	Ionic Charge	Examples
Metals				
1	1	Lose 1	1+	Li^+, Na^+, K^+
2	2	Lose 2	2+	Mg^{2+}, Ca^{2+}
3	3	Lose 3	3+	Al^{3+}
Nonmetals				
5	5	Gain 3	3−	N^{3-}, P^{3-}
6	6	Gain 2	2−	O^{2-}, S^{2-}
7	7	Gain 1	1−	F^-, Cl^-, Br^-, I^-

S·A·M·P·L·E **4.5** ·P·R·O·B·L·E·M

Writing Ions

Consider the elements aluminum and oxygen.

a. Identify each as a metal or a nonmetal.
b. State the number of valence electrons for each.
c. State the number of electrons that must be lost or gained for each to acquire an octet.
d. Write the symbol of each resulting ion, including its ionic charge.

Solution

	Aluminum	**Oxygen**
a.	metal	nonmetal
b.	3 valence electrons	6 valence electrons
c.	loses $3e^-$	gains $2e^-$
d.	Al^{3+}	O^{2-}

Study Check

What are the symbols for the ions formed by potassium and sulfur?

HEALTH NOTE

Some Important Ions in the Body

A number of ions in body fluids have important physiological and metabolic functions. Some of them are listed in Table 4.5.

Table 4.5 Ions in the Body

Ion	Occurrence	Function	Source	Result of Too Little	Result of Too Much
Na^+	Principal cation outside the cell	Regulation and control of body fluids	Salt, seafood, meat	Hyponatremia, anxiety, diarrhea, circulatory failure, decrease in body fluid	Hypernatremia, little urine, thirst, edema
K^+	Principal cation inside the cell	Regulation of body fluids and cellular functions	Bananas, orange juice, skim milk, prunes, meat	Hypokalemia (hypopotassemia), lethargy, muscle weakness, failure of neurological impulses	Hyperkalemia (hyperpotassemia), irritability, nausea, little urine, cardiac arrest
Ca^{2+}	Cation outside the cell; 90% of calcium in the body in bone as $Ca_3(PO_4)_2$ or $CaCO_3$	Major cation of bone; muscle smoothant	Milk, cheese, butter, meat, some vegetables	Hypocalcemia, tingling fingertips, muscle cramps, tetany	Hypercalcemia, relaxed muscles, kidney stones, deep bone pain,
Mg^{2+}	Cation outside the cell; 70% of magnesium in the body in bone structure	Essential for certain enzymes, muscles, and nerve control	Widely distributed (part of chlorophyll of all green plants), nuts, grains	Disorientation, hypertension, tremors, slow pulse	Drowsiness
Cl^-	Principal anion outside the cell	Gastric juice, regulation of body fluids	Salt, seafood, meat	Same as for Na^+	Same as for Na^+

4.2

QUESTIONS & PROBLEMS

Ions and the Octet Rule

4.11 **a.** In Figure 4.3, how does the octet rule explain the formation of a sodium ion?
b. What noble gas has the same electronic arrangement as the sodium ion?
c. Why do you think Group 1 and Group 2 elements are found in many compounds, but not Group 8 elements?

4.12 **a.** In Figure 4.5, how does the octet rule explain the formation of a chloride ion?
b. What noble gas has the same electronic arrangement as the chloride ion?
c. Why do you think Group 7 elements are found in many compounds, but not Group 8 elements?

4.13 Write the electron-dot arrangement for each of the following elements. Indicate which elements have a complete valence shell, and which do not.
a. neon **b.** oxygen **c.** lithium **d.** argon

4.14 Write the electron-dot arrangement for each of the following elements. Indicate which elements have a complete valence shell, and which do not.
a. magnesium **b.** nitrogen **c.** helium **d.** potassium

4.15 State the number of electrons that must be lost by atoms of each of the following elements to acquire a noble gas electron arrangement:
a. Li **b.** Mg **c.** Al **d.** Cs **e.** Ba

4.16 State the number of electrons that must be gained by atoms of each of the following elements to acquire a noble gas electron arrangement:
a. Cl **b.** O **c.** N **d.** I **e.** P

4.17 What noble gas has the same electronic arrangement as each of the following ions?
a. Na^+ **b.** Mg^{2+} **c.** K^+ **d.** O^{2-} **e.** F^-

4.18 What noble gas has the same electronic arrangement as each of the following ions?
a. Li^+ **b.** Sr^{2+} **c.** S^{2-} **d.** Al^{3+} **e.** Br^-

4.19 State the number of electrons lost or gained when the following elements form ions:
a. Mg **b.** P **c.** Group 7 **d.** Na **e.** Al

4.20 State the number of electrons lost or gained when the following elements form ions:
a. O **b.** Group 2 **c.** F **d.** Li **e.** N

4.21 Write the symbols of the ions with the following number of protons and electrons:
a. 3 protons, 2 electrons **b.** 9 protons, 10 electrons
c. 12 protons, 10 electrons **d.** 26 protons, 23 electrons
e. 30 protons, 28 electrons

4.22 How many protons and electrons are in the following ions?
a. O^{2-} **b.** K^+ **c.** Br^- **d.** S^{2-} **e.** Sr^{2+}

4.23 Write the symbol for the ion of each of the following:
a. chloride **b.** potassium **c.** oxide **d.** aluminum

4.24 Write the symbol for the ion of each of the following:
a. fluoride **b.** calcium **c.** sodium **d.** lithium

4.25 What is the name of each of the following ions?
a. K^+ **b.** S^{2-} **c.** Ca^{2+} **d.** N^{3-}

4.26 What is the name of each of the following ions?
a. Mg^{2+} **b.** Ba^{2+} **c.** I^- **d.** Cl^-

4.3 Ionic Compounds

Study Pack

Learning Goal

Using charge balance, write the correct formula for an ionic compound.

Ionic compounds consist of positive and negative ions. The ions are held together by strong electrical attractions between the opposite charges, called **ionic bonds.** Consider the ionic compound sodium chloride, NaCl. Its formula, NaCl, indicates that it consists of Na^+ ions and Cl^- ions. Figure 4.6 illustrates how NaCl is formed by the attraction between a sodium ion and a chloride ion.

Sodium atom	Chlorine atom		Sodium ion	Chloride ion		Sodium chloride, an ionic compound
Na·	·$\ddot{\underset{..}{Cl}}$:	⟶	Na^+	$\left[:\ddot{\underset{..}{Cl}}:\right]^-$	=	NaCl

Charge Balance in Ionic Compounds

The **formula** of an ionic compound indicates the number and kinds of ions that make up the ionic compound. The sum of the ionic charges in the formula is always zero. For example, the NaCl formula indicates that there is one sodium ion, Na^+, for every chloride ion, Cl^-, in the compound. Note that the ionic charges of the ions do not appear in the formula of the ionic compound.

Charge balance for the ions in the ionic compound NaCl

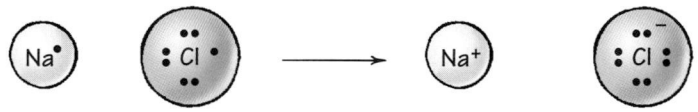

Loses $1e^-$	Gains $1e^-$	=	One sodium ion Na^+	One chloride ion Cl^-

[1+ + 1− = 0]

NaCl, sodium chloride

The physical and chemical properties of an ionic compound such as NaCl are very different from those of the original elements. For example, the original elements of NaCl were sodium, a soft, shiny metal, and chlorine, a yellow-green poisonous gas. Yet, as positive and negative ions, they form table salt, NaCl, a white, crystalline substance that is common in our diet. Scientists diagram the solid form of NaCl as alternating Na^+ and Cl^- ions packed together in a lattice structure, as shown in Figure 4.7.

Figure 4.6 In sodium chloride, NaCl, sodium loses its valence electron and chlorine gains an electron, forming a Na^+ cation and a Cl^- anion. The resulting ionic compound is held together by ionic bonds between the positive and negative ions.

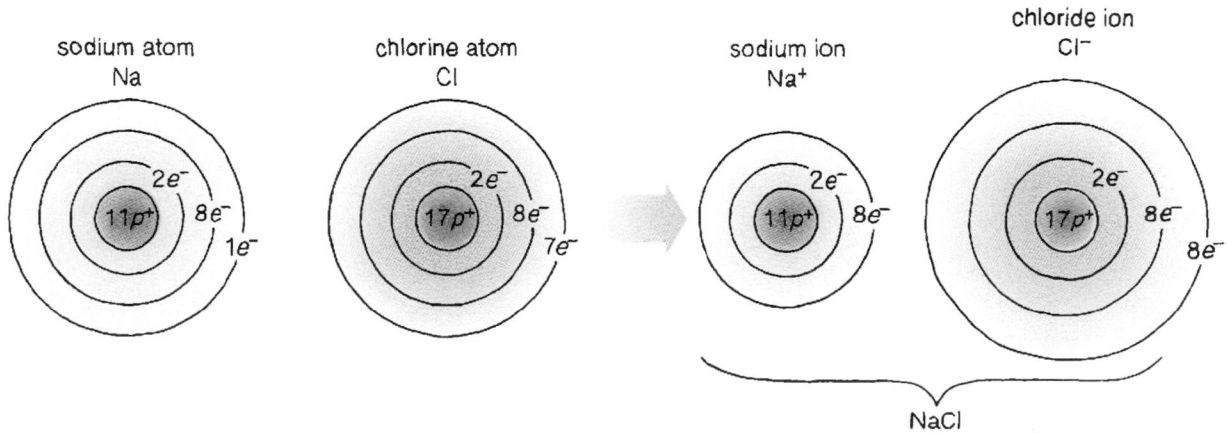

a b c

Figure 4.7 **(a)** Crystals of table salt, the ionic compound sodium chloride (NaCl). **(b)** Crystals of NaCl under magnification. **(c)** A diagram of the arrangement of Na^+ and Cl^- packed together in a NaCl crystal.

4.3 Subscripts in Formulas

The formula of any ionic compound has a zero overall charge. That means that the total amount of positive charge is equal to the total amount of negative charge. Consider a compound of magnesium and chlorine. To achieve an octet, a Mg atom loses its 2 valence electrons to form Mg^{2+}. Each Cl atom gains 1 electron to form Cl^-, which has a complete valence shell. In this example, two Cl^- ions are needed to balance the positive charge of Mg^{2+}. This gives the formula, $MgCl_2$, magnesium chloride, in which a subscript of 2 shows that two Cl^- were needed for charge balance.

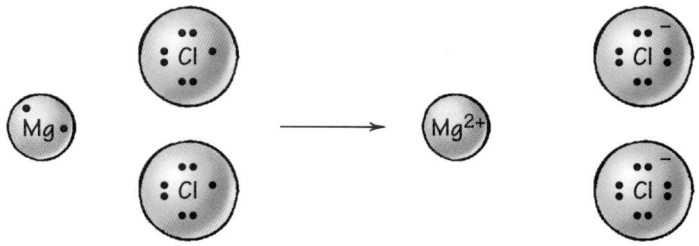

Loses $2e^-$ Each gains $1e^-$ One magnesium ion Mg^{2+} Two chloride ions $2\ Cl^-$

$[(2+) + 2(1-) = 0]$

$MgCl_2$, magnesium chloride

S·A·M·P·L·E· 4.6 ·P·R·O·B·L·E·M

Diagramming an Ionic Compound

Diagram the formation of the ionic compound aluminum fluoride, AlF_3.

Solution

In their electron-dot structures, aluminum has 3 valence electrons and fluorine has 7. The aluminum loses its 3 valence electrons, and each fluorine atom gains an electron, to give ions with noble gas arrangements in the ionic compound AlF_3.

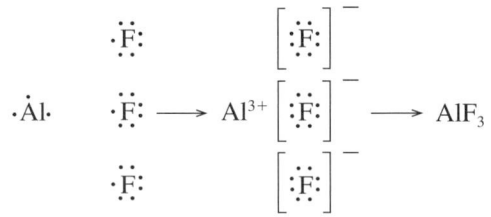

Loses $3e^-$ Gains $3e^-$ Al^{3+} Three F^-

$3(+) + 3(-) = 0$

Study Check

Diagram the formation of lithium sulfide, Li_2S.

Writing Ionic Formulas from Ionic Charges

The subscripts in the formula of an ionic compound represent the number of positive and negative ions that give an overall charge of zero. Thus, we can now write a formula directly from the ionic charges of the positive and negative ions. Suppose we wish to write the formula of the ionic compound containing Na^+ and S^{2-} ions. To balance the ionic charge of the S^{2-} ion, we will need to place two Na^+ ions in the formula. This gives the formula Na_2S, which has an overall charge of zero.

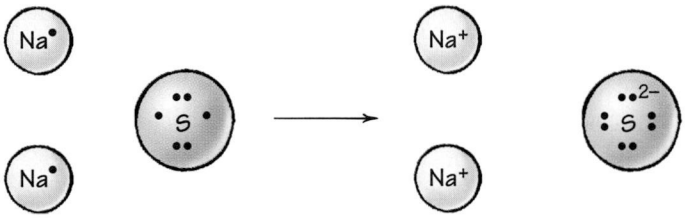

Each loses $1e^-$ Gains $2e^-$ Two sodium ions One sulfide ion

$[2(1+) + (2-) = 0]$

Na_2S, sodium sulfide

S·A·M·P·L·E 4.7 P·R·O·B·L·E·M

Writing Formulas from Ionic Charges

Use ionic charge balance to write the formula for the ionic compound containing Na^+ and N^{3-}.

Solution

Determine the number of each ion needed for charge balance. The charge for nitrogen (3−) is balanced by three Na^+ ions. Writing the positive ion first gives the formula Na_3N.

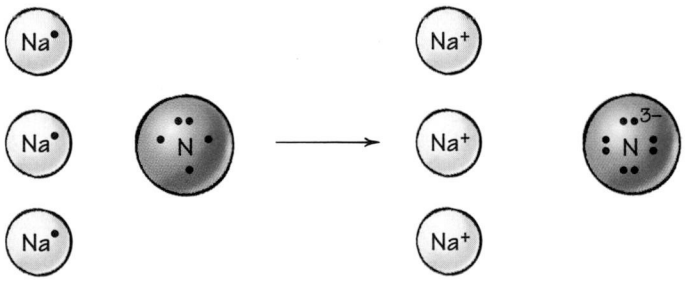

Each loses $1e^-$ Gains $3e^-$ Three sodium ions One nitride ion

Na^+ N^{3-}

$[3(1+) + (3-) = 0]$

Na_3N, sodium nitride

Study Check

Use ionic charges to determine the formula of the compound that would form when calcium and chlorine react.

Q·U·E·S·T·I·O·N·S & P·R·O·B·L·E·M·S

4.4

Ionic Compounds

4.27 Which of the following pairs of elements are likely to form an ionic compound?

a. lithium and chlorine **b.** oxygen and chlorine
c. potassium and oxygen **d.** sodium and neon
e. sodium and magnesium

4.28 Which of the following pairs of elements are likely to form an ionic compound?

a. helium and oxygen **b.** magnesium and chlorine
c. chlorine and bromine **d.** potassium and sulfur
e. sodium and potassium

4.29 Using electron-dot structures, diagram the formation of the following ionic compounds:

a. KCl **b.** $CaCl_2$ **c.** Na_3N

4.30 Using electron-dot structures, diagram the formation of the following ionic compounds:

a. MgS **b.** $AlCl_3$ **c.** Li_2O

4.31 Write the correct ionic formula for compounds formed between the following ions:

a. Na^+ and O^{2-} **b.** Al^{3+} and Br^- **c.** Ba^{2+} and O^{2-}
d. Mg^{2+} and Cl^- **e.** Al^{3+} and S^{2-}

4.32 Write the correct ionic formula for compounds formed between the following ions:

a. Al^{3+} and Cl^- **b.** Ca^{2+} and S^{2-} **c.** Li^+ and S^{2-}
d. K^+ and N^{3-} **e.** K^+ and I^-

4.33 Write the correct formula for ionic compounds formed by the following metals and nonmetals:

a. sodium and sulfur **b.** potassium and nitrogen
c. aluminum and iodine **d.** lithium and oxygen

4.34 Write the correct formula for ionic compounds formed by the following metals and nonmetals:

a. calcium and chlorine **b.** barium and bromine
c. sodium and phosphorus **d.** magnesium and oxygen

4.4 Naming and Writing Ionic Formulas

Learning Goal

Given the formula of an ionic compound, write the correct name.

The name of a metal ion is the same as its elemental name. The name of a nonmetal ion is obtained by replacing the end of its elemental name with *ide*. Table 4.6 lists the names of some important metal and nonmetal ions.

S·A·M·P·L·E **4.8** P·R·O·B·L·E·M

Naming Ions

Write the name of each of the following ions:
a. Al^{3+} **b.** S^{2-}

Solution

a. Aluminum ion; a metal ion uses the elemental name.
b. Sulfide ion: a nonmetal changes the ending of the name of the element to *ide*.

Study Check

Write the name of each of the following ions:
a. Ca^{2+} **b.** N^{3-} **c.** Br^{-}

Naming Ionic Compounds Containing Two Elements

In the name of an ionic compound made up of two elements, the metal ion is named followed by the name of the nonmetal ion. Subscripts are never mentioned; they are understood as a result of the charge balance of the ions in the compound.

Compound	Metal Ion	Nonmetal Ion	Name
NaF	Na^{+} Sodium	F^{-} Fluoride	Sodium fluoride
$MgBr_2$	Mg^{2+} Magnesium	Br^{-} Bromide	Magnesium bromide
Al_2O_3	Al^{3+} Aluminum	O^{2-} Oxide	Aluminum oxide

Table 4.7 gives more formulas and names of some ionic compounds.

Table 4.7 Formulas and Names of Some Ionic Compounds

	Ion		
Formula	Metal	Nonmetal	Name
MgS	Mg^{2+}	S^{2-}	Magnesium sulfide
K_2O	K^{+}	O^{2-}	Potassium oxide
$CaCl_2$	Ca^{2+}	Cl^{-}	Calcium chloride
Na_3N	Na^{+}	N^{3-}	Sodium nitride
Al_2O_3	Al^{3+}	O^{2-}	Aluminum oxide

S·A·M·P·L·E **4.9** P·R·O·B·L·E·M

Naming Ionic Compounds

Write the name of each of the following ionic compounds:
a. Na_2O **b.** Al_2S_3

Table 4.6 Formulas and Names of Some Common Ions

Group Number	Formula of Ion	Name of Ion
Metals		
1	Li^{+}	Lithium
	Na^{+}	Sodium
	K^{+}	Potassium
2	Mg^{2+}	Magnesium
	Ca^{2+}	Calcium
	Ba^{2+}	Barium
3	Al^{3+}	Aluminum
Nonmetals		
5	N^{3-}	Nitride
	P^{3-}	Phosphide
6	O^{2-}	Oxide
	S^{2-}	Sulfide
7	F^{-}	Fluoride
	Cl^{-}	Chloride
	Br^{-}	Bromide
	I^{-}	Iodide

Solution

Compound	Ions and Names		Name of Compound
a. Na_2O	Na^+	O^{2-}	
	Sodium	Oxide	Sodium oxide
b. Al_2S_3	Al^{3+}	S^{2-}	
	Aluminum	Sulfide	Aluminum sulfide

Study Check

Name the compound Ca_3P_2.

4.4

Metals That Form Two Positive Ions

Cations of Iron	Fe^{2+}	Fe^{3+}
Systematic Name	Iron(II)	Iron(III)
Older Name	Ferrous	Ferric

Transition Metal	Ion
Silver	Ag^+
Zinc	Zn^{2+}

Like all metals, the transition metals and the representative metals in Group 4 also form positive ions. However, these metals typically form two or more kinds of positive ions. For example, in some ionic compounds, iron is in the Fe^{2+} form, but in a different compound, it takes the Fe^{3+} form. Copper also forms two different ions: Cu^+ is present in some compounds and Cu^{2+} in others. (The reasons for these different ions are complex and will not be considered in this text.) When a metal can form two or more ions, it is not possible to predict the ionic charge from the group number. We say that it has a variable valence.

When two different ions are possible for a metal, a naming system is needed to identify the particular cation in a compound. To do this, a Roman numeral that matches the ionic charge is placed in parentheses after the elemental name of the metal. For the cations of iron, Fe^{2+} is named iron(II), and Fe^{3+} is named iron(III) as shown by the first table in the left margin of this page. An older system names Fe^{2+} as the *ferrous ion* and Fe^{3+} as the *ferric ion*.

Table 4.8 lists the ions of some common metals that produce two or more ions. Figure 4.8 shows the common ions we have discussed and their location on the periodic table. Typically, the transition metals form two or more positive ions. However, zinc and silver form only one ion. The ionic charges of silver and zinc are fixed like Group 1 and 2 metals, so their elemental names are sufficient when naming their ionic compounds. (See the second table in the margin.)

Figure 4.8 Location on the periodic table of ions formed by some metals and nonmetals.

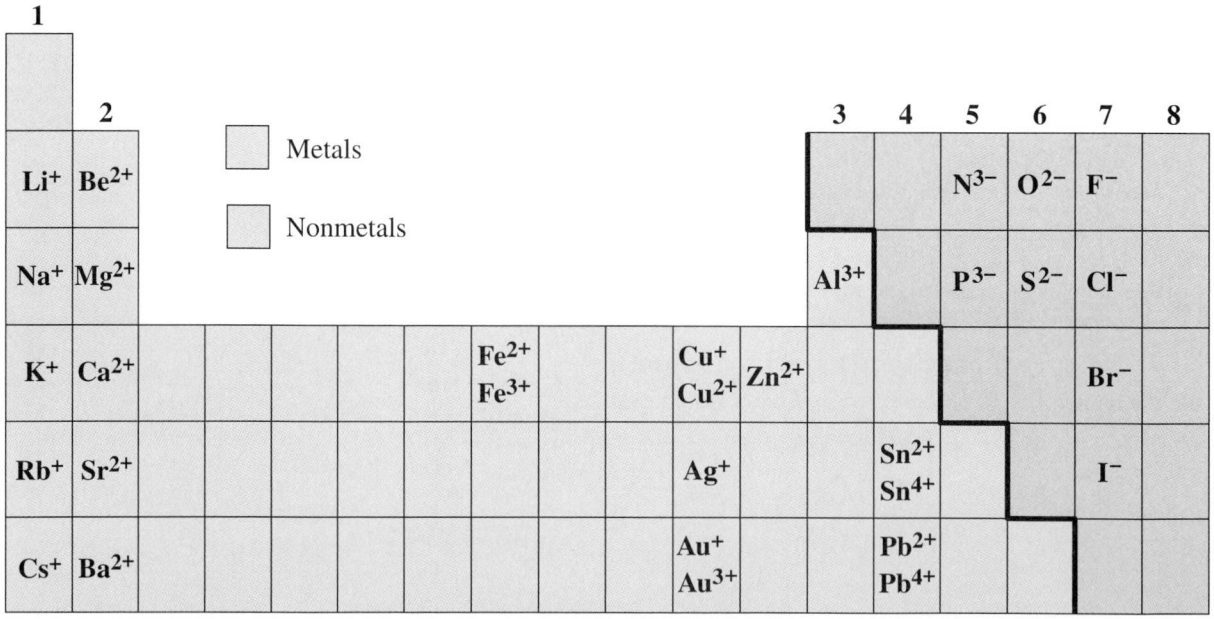

Table 4.8 Some Metals That Form Two Positive Ions

Element	Possible Ions	Systematic Name	Older Name
Iron	Fe^{2+} Fe^{3+}	Iron(II) Iron(III)	Ferrous Ferric
Copper	Cu^{+} Cu^{2+}	Copper(I) Copper(II)	Cuprous Cupric
Gold	Au^{+} Au^{3+}	Gold(I) Gold(III)	Aurous Auric
Tin	Sn^{2+} Sn^{4+}	Tin(II) Tin(IV)	Stannous Stannic
Lead	Pb^{2+} Pb^{4+}	Lead(II) Lead(IV)	Plumbous Plumbic

Table 4.9 Some Ionic Compounds of Metals That Form Two Kinds of Positive Ions

Compound	Systematic Name	Older Name
$FeCl_2$	Iron(II) chloride	Ferrous chloride
$FeCl_3$	Iron(III) chloride	Ferric chloride
Cu_2S	Copper(I) sulfide	Cuprous sulfide
$CuCl_2$	Copper(II) chloride	Cupric chloride
$SnCl_2$	Tin(II) chloride	Stannous chloride
$PbBr_4$	Lead(IV) bromide	Plumbic bromide

S·A·M·P·L·E 4.10 P·R·O·B·L·E·M

Writing Simple Ions

Write the symbol of the following ions:

a. copper(II) **b.** iron(III) **c.** zinc

Solution

a. Cu^{2+} **b.** Fe^{3+} **c.** Zn^{2+}

Study Check

What is the name of the Cu^{+} ion?

A summary of how ionic compounds are named follows. Table 4.9 lists names of some ionic compounds in which the metals form more than one type of positive ion.

Rules for Naming Ionic Compounds

Number of Ions Possible for Metal	Name of Metal Ion	Name of Nonmetal Ion
One positive ion: Groups 1, 2, 3, Zn, Ag	Use element name	Element name ends in *ide*
More than one positive ion: Transition metals and Group 4 metals	Use element name + Roman numeral in parentheses that is equal to the positive charge	Element name ends in *ide*

Naming Compounds That Include Transition Metals

When a metal forms more than one type of positive ion, the ionic charge of that metal in a particular formula must be included as a Roman numeral in paretheses when the metal is named. The selection of the correct Roman numeral depends upon the calculation of the ionic charge of the metal in the formula of a specific ionic compound. For example, we know that in the formula $CuCl_2$ the positive charge of the copper ion must balance the negative charge of two chloride ions. Because we know that chloride ions each have a 1− charge, there must be a total negative charge of 2−. Balancing the 2− by the positive charge gives a charge of 2+ for Cu (a Cu^{2+} ion):

$CuCl_2$

Cu charge	+	Cl^- charge	= 0
(?)	+	2(1−)	= 0
(2+)	+	2−	= 0

Because copper forms ions with two different charges, Cu^+ and Cu^{2+}, we need to use the Roman numeral system and place (II) after *copper* when naming the compound:

4.4

Copper(II) chloride

S·A·M·P·L·E· 4.11 ·P·R·O·B·L·E·M

Naming Ionic Compounds

Write the name for each of the following ionic compounds:

a. FeO **b.** $AlBr_3$ **c.** Cu_2S

Solution

a. Because iron is a transition metal that forms more than one ion, it is necessary to determine the ionic charge of Fe in FeO:

Fe charge	+	O^{2-} charge	= 0
?	+	(2−)	= 0
(2+)	+	(2−)	= 0

The charge on Fe is therefore 2+. So we place the Roman numeral II in parentheses after *iron:*

Iron(II) oxide

b. Because aluminum is in Group 3 and forms only Al^{3+}, the name of the metal is sufficient. A Roman numeral is not needed.

$AlBr_3$ Aluminum bromide

c. As a transition metal, copper forms more than one positive ion. Balancing the charge in Cu_2S, we determine the ionic charge of the copper ion:

2(Cu charge)	+	S^{2-} charge	= 0
2(?)	+	(2−)	= 0
2(1+)	+	(2−)	= 0

The charge on Cu is therefore 1+. So we place the Roman numeral I in parentheses after copper:

Copper(I) sulfide

Although the total positive charge is 2+, it is divided between two copper ions; therefore, each copper ion in the formula is indicated as Cu^+.

Study Check

Write the name of the compound whose formula is $AuCl_3$.

Writing Formulas from the Name of an Ionic Compound

The formula of an ionic compound can be written from its name because the first term in the name describes the metal ion and the second term specifies the nonmetal ion. Subscripts are added as needed to balance the charge. The steps for writing a formula from the name of an ionic compound follow.

Steps for Writing a Formula from the Name of an Ionic Compound

Step 1 Write the positive ion from the first term in the name, and the negative ion from the second term.

Step 2 Balance the ionic charges to give an overall charge of zero.

Step 3 Write the formula, placing the symbol for the metal first and the symbol for the nonmetal second. Use appropriate subscripts for charge balance.

S·A·M·P·L·E **4.12** P·R·O·B·L·E·M

Writing Formulas for Ionic Compounds

Write the correct formula for iron(III) chloride.

Solution

Step 1 Write the names of the metal ion and the nonmetal ion. Because the Roman numeral (III) indicates the ionic charge of the particular iron ion, we can then write Fe^{3+} for the positive ion. The chloride ion has an ionic charge of 1−.

Iron(III)	chloride
Fe^{3+}	Cl^-

Step 2 Balance the charges.

Ions needed to balance charge:	Fe^{3+}	Cl^-
		Cl^-
		Cl^-
Total charge:	(3+) +	3(1−) = 0

Step 3 Write the formula of the compound, metal first, using subscripts when needed.

$FeCl_3$

Study Check

What is the correct formula of the ionic compound aluminum sulfide?

Q·U·E·S·T·I·O·N·S & P·R·O·B·L·E·M·S

Naming and Writing Ionic Formulas

4.35 Write names for the following ionic compounds:

a. Al_2O_3 **b.** $CaCl_2$ **c.** Na_2O
d. Mg_3N_2 **e.** KI

4.36 Write names for the following ionic compounds:

a. $MgCl_2$ **b.** K_3P **c.** Li_2S
d. LiBr **e.** MgO

4.37 Why is a Roman numeral placed after the name of most transition metals ions?

4.38 The compound $CaCl_2$ is named calcium chloride; the compound $CuCl_2$ is named copper(II) chloride. Explain why a Roman numeral is used in one name but not the other.

4.39 Write the names of the following transition metal ions (include the Roman numeral when necessary):
a. Fe^{2+} **b.** Cu^{2+} **c.** Zn^{2+} **d.** Pb^{4+}

4.40 Write the names of the following transition metal ions (include the Roman numeral when necessary):
a. Ag^{+} **b.** Cu^{+} **c.** Fe^{3+} **d.** Sn^{2+}

4.41 Write names for the following ionic compounds:
a. $SnCl_2$ **b.** FeO **c.** Cu_2S **d.** CuS

4.42 Write names for the following ionic compounds:
a. Ag_3P **b.** PbS **c.** SnO_2 **d.** $AuCl_3$

4.43 Indicate the charge on the metal ion in each of the following formulas:
a. $AuCl_3$ **b.** Fe_2O_3 **c.** PbI_4 **d.** $SnCl_2$

4.44 Indicate the charge on the metal ion in each of the following formulas:
a. $FeCl_2$ **b.** CuO **c.** Fe_2S_3 **d.** AlP

4.45 Write formulas for the following ionic compounds:
a. magnesium chloride **b.** sodium sulfide
c. copper(I) oxide **d.** zinc phosphide
e. gold(III) nitride

4.46 Write formulas for the following ionic compounds:
a. iron(III) oxide **b.** barium fluoride
c. tin(IV) chloride **d.** silver sulfide
e. copper(II) chloride

Study Pack

Chem of Life Atoms and Molecules: Covalent Bonds 1–4

Learning Goal

Diagram the electron-dot structure for a covalent molecule.

4.5 Covalent Bonds

When atoms of two nonmetals combine, covalent compounds result. The atoms, which are similar or even identical, are held together by sharing electrons. This type of bond is called a **covalent bond.**

The simplest covalent molecule is hydrogen, H_2. A hydrogen atom attains a noble gas arrangement when its first shell is filled with 2 electrons. To accomplish this, two hydrogen atoms share their single electrons and form a covalent bond:

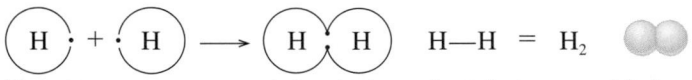

Electrons to share | A shared pair of electrons | A covalent bond | A hydrogen molecule

The covalent bond representing a shared pair of electrons can also be written as a line between the two atoms. Atoms that are bonded through covalent bonds produce a **molecule.**

In most covalent molecules, each of the atoms has an octet of 8 electrons. For example, a fluorine molecule, F_2, consists of two fluorine atoms. Each atom needs one more valence electron for a noble gas arrangement. An octet for each is achieved when the fluorine atoms share a pair of electrons in their outer shells.

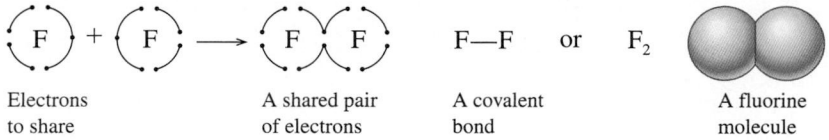

Hydrogen (H_2) and fluorine (F_2) are examples of nonmetal elements whose natural state is diatomic; that is, they contain two atoms. The elements that exist as diatomic molecules are listed in Table 4.10.

Table 4.10 Elements That Exist as Diatomic, Covalent Molecules

Element	Diatomic Molecule	Name
H	H_2	Hydrogen
N	N_2	Nitrogen
O	O_2	Oxygen
F	F_2	Fluorine
Cl	Cl_2	Chlorine
Br	Br_2	Bromine
I	I_2	Iodine

4.5

Sharing Electrons Between Atoms of Different Elements

The number of electrons that an atom shares and the number of covalent bonds it forms are usually equal to the number of electrons needed to acquire a noble gas arrangement. For example, carbon has 4 valence electrons. Because carbon needs to acquire 4 more electrons for an octet, it forms 4 covalent bonds by sharing its 4 valence electrons. Table 4.11 relates the group numbers of some nonmetals to the number of covalent bonds the elements typically form.

Methane, a component of natural gas is a compound made of carbon and hydrogen. To attain an octet, each carbon shares 4 electrons, and each hydrogen shares 1 electron. In a molecule, a carbon atom forms four covalent bonds with four hydrogen atoms. (See Figure 4.9.) The electron-dot structure for the molecule is written with the carbon atom in the center and the hydrogen atoms on the sides.

Table 4.12 gives the electron-dot structures for several covalent molecules.

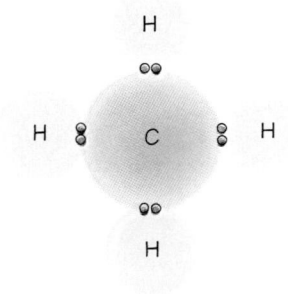

Figure 4.9 Diagram of a carbon atom sharing electrons with hydrogen atoms to form four covalent bonds. The carbon atom has 8 electrons in its valence shell, and each hydrogen has a filled outer shell of 2 electrons.

Table 4.11 Covalent Bonds Required by Various Nonmetals

Group Number	Nonmetal	Number of Valence Electrons	Number of Electrons Shared to Complete an Octet	Number of Covalent Bonds
1	II	1	1	1
4	C, Si	4	4	4
5	N, P	5	3	3
6	O, S	6	2	2
7	F, Cl, Br, I	7	1	1

S·A·M·P·L·E **4.13** ·P·R·O·B·L·E·M

Writing Electron-Dot Structures for Covalent Compounds

Draw an electron-dot structure for H_2S.

Solution

Sulfur, which has 6 valence electrons, shares 2 electrons to form two covalent bonds. Two atoms of hydrogen, each having 1 valence electron, will form two covalent bonds, one for each atom:

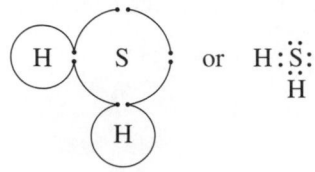

Electron-dot structure for H_2S

Study Check

Write the electron-dot structure for NH_3.

4.5

Table 4.12 Electron-Dot Structures for Some Covalent Compounds

CH_4	NH_3	H_2O
Structures Using Electron Dots Only		
H H:C:H H	H:N:H H	:O:H H
Structures Using Bonds and Electron Dots		
H H—C—H H	H—N—H H	:O—H H
Molecular Models		
Methane molecule	Ammonia molecule	Water molecule

Multiple Covalent Bonds

Up to now, we have looked at covalent bonding in molecules having only single bonds. In many covalent compounds, atoms share two or three pairs of electrons to complete their electron octets. A **double bond** is the sharing of two pairs of electrons, and in a **triple bond,** three pairs of electrons are shared.

Double and triple bonds are formed when single covalent bonds fail to complete the octets of all the atoms in the molecule. For example, in the electron-dot structure for the covalent compound N_2, an octet is achieved when each nitrogen atom shares 3 electrons. Thus, three covalent bonds, or a triple bond, will form.

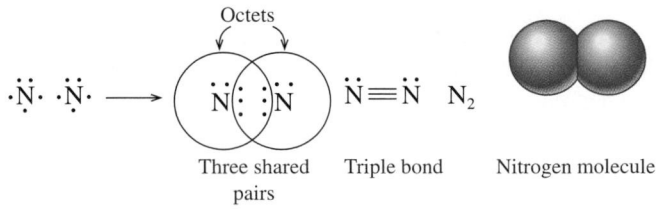

We use the steps in the following example to write an electron-dot structure for SO_2.

Steps for Writing Electron-Dot Structures

Step 1 Determine the central atom of the compound and place other atoms around it. A skeleton structure may also be given.

For SO_2, the central atom is S, which is bonded to the two oxygen atoms.

O S O

Step 2 Find the total number (sum) of valence electrons by adding the number of valence electrons for each atom in the compound.

$$\text{The total valence electrons for } SO_2 = \begin{array}{lrcl} S & 1 \times 6e^- & = & 6e^- \\ O & 2 \times 6e^- & = & \underline{12e^-} \\ & & & 18e^- \end{array}$$

Step 3 Place a pair of electrons (single bond) to attach each of the atoms to the central atom. Determine the total number of bonding electrons used.

O : S : O $4e^-$ used to connect atoms

Step 4 Subtract the number of bonding electrons used in step 3 from the total number of valence electrons. Place the remaining electrons as nonbonding pairs around the atoms and try to give each atom an octet. (Remember, hydrogen needs only 2 electrons, a single bond.)

$18e^- - 4e^-$ (step 3) $= 14e^-$ remaining

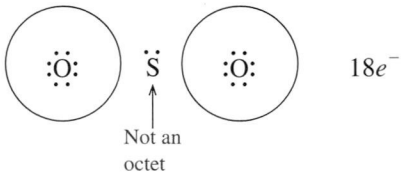

$18e^-$

Step 5 If octets for all the atoms were not completed in step 4, move a nonbonding pair of electrons from an adjacent atom to the atom that needs an octet to form a multiple bond.

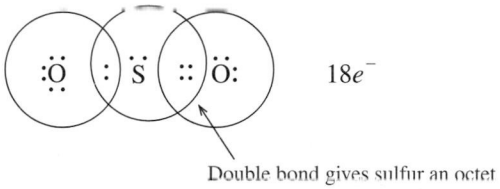

$18e^-$

:Ö — S̈ = Ö:

S·A·M·P·L·E **4.14** ·P·R·O·B·L·E·M

Drawing Electron-Dot Structures Having Multiple Bonds

Draw the electron-dot structure for the compound CO_2. Carbon is the central atom.

Step 1 Arrange the atoms for CO_2 with C as the central atom: O C O

Step 2 The 4 valence electrons of carbon and the 6 electrons of two oxygen atoms gives a total of 16 valence electrons.

$$\text{Total valence electrons for } CO_2 = \begin{array}{lrcl} C & 1 \times 4e^- & = & 4e^- \\ O & 2 \times 6e^- & = & \underline{12e^-} \\ & & & 16e^- \end{array}$$

4.5

Step 3 Use two pairs of electrons ($4e^-$) to attach each of the oxygen atoms to the central carbon atom.

O : C : O $4e^-$ used to connect atoms

Step 4 Place the remaining 12 valence electrons around the atoms:
$16e^- - 4e^- = 12e^-$

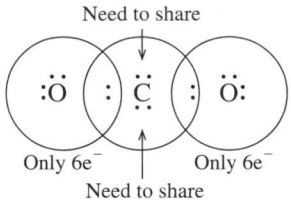

Step 5 Octets for the oxygen atoms are not complete. Move two nonbonding pairs of electrons on the carbon atom between the carbon atom and the oxygen atoms to complete the octets for both oxygens.

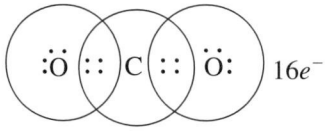

:Ö = C = Ö: CO_2

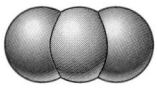

Molecule of carbon dioxide

Study Check

Determine the number of valence electrons for the compound HCN and arrange them in the electron-dot structure.

Q·U·E·S·T·I·O·N·S & P·R·O·B·L·E·M·S

Covalent Bonds

4.47 Write the electron-dot structure for each of the following covalent molecules:
a. Br_2 **b.** H_2 **c.** HF **d.** OF_2

4.48 Write the electron-dot structure for each of the following covalent molecules:
a. NCl_3 **b.** CCl_4 **c.** Cl_2 **d.** SiF_4

4.49 The following covalent compounds have double or triple bonds. Write their electron-dot structures. (The order of atoms is indicated in parentheses.)
a. C_2H_2 (H C C H)
b. CH_2O (O H C H)
c. SO_3 (O O S O)

4.50 The following covalent compounds have double or triple bonds. Write their electron-dot structures. (The order of atoms is indicated in parentheses.)
a. HCN **b.** H_2CCH_2 (H H H C C H) **c.** HNO_2 (H O N O)

4.6 Covalent Compounds

Study Pack

Learning Goal

Given the formula of a covalent compound, write its correct name; given the name of a covalent compound, write its formula.

4.6

Unlike the names of ionic compounds, the names of covalent compounds need prefixes because several different compounds can be formed from the same two nonmetals. For example, carbon and oxygen can form two different compounds, carbon monoxide, CO, and carbon dioxide, CO_2. Nitrogen and oxygen also form several different covalent molecules. We could not distinguish between them by using the name *nitrogen oxide.* Therefore, prefixes are used with molecular compounds composed of two nonmetals.

In the name of a covalent compound, the first nonmetal in the formula is named by its elemental name; the second nonmetal is named by its elemental name with the ending changed to *ide.* All subscripts indicating two or more atoms of an element are expressed as prefixes and attached in front of each name. Table 4.13 lists some prefixes used in naming covalent compounds.

Table 4.13 Prefixes Used in Naming Covalent Compounds

Number of Atoms	Prefix
1	Mono
2	Di
3	Tri
4	Tetra
5	Penta
6	Hexa
7	Hepta
8	Octa

Some Covalent Compounds Formed by Nitrogen and Oxygen

NO	Nitrogen monoxide
N_2O	Dinitrogen monoxide
N_2O_3	Dinitrogen trioxide
N_2O_4	Dinitrogen tetroxide
N_2O_5	Dinitrogen pentoxide

In the name of a covalent compound, the prefix *mono* is usually omitted for the first element. When the vowels *o* and *o* or *a* and *o* appear together, the first vowel is omitted. Table 4.14 lists the formulas, names, and commercial uses of some other covalent compounds.

Table 4.14 Some Common Covalent Compounds

Formula	Name	Commercial Uses
CS_2	Carbon disulfide	Manufacture of rayon
CO_2	Carbon dioxide	Carbonation of beverages; fire extinguishers, propellant in aerosols, dry ice
SiO_2	Silicon dioxide	Manufacture of glass
NCl_3	Nitrogen trichloride	Bleaching of flour in some countries (prohibited in U.S.)
SO_2	Sulfur dioxide	Preserving fruits, vegetables; disinfectant in breweries, bleaching textiles
SO_3	Sulfur trioxide	Manufacture of explosives
SF_6	Sulfur hexafluoride	Electrical circuits
ClO_2	Chlorine dioxide	Bleaching pulp (for making paper), flour, leather
ClF_3	Chlorine trifluoride	Rocket propellant

4.6

S·A·M·P·L·E 4.15 P·R·O·B·L·E·M

Naming Covalent Compounds

Name each of the following covalent compounds:

a. NCl_3 **b.** P_4O_6

Solution

a. In NCl_3, the first nonmetal is nitrogen. We could use the term *mononitrogen,* but the prefix *mono* is usually dropped from the name of the first element. The second nonmetal is chlorine, which is named *chloride.* The subscript indicating three chlorine atoms is shown as the prefix *tri* in front of the name of the second element, trichloride.

NCl_3

Nitrogen trichloride

b. In P_4O_6, the first nonmetal is phosphorus. Because there are four P atoms, it is named *tetraphosphorus.* The second nonmetal, consisting of six oxygen atoms, is named *hexoxide.* (When the vowels *a* and *o* appear together, as would happen in *hexa* plus *oxide,* the ending of the prefix is dropped.) Putting the names for the two nonmetals together, we have

P_4O_6

Tetraphosphorus hexoxide

Study Check

Scientists have recently combined some of the noble gases with other elements. One of the compounds produced was XeF_6. How would you name this compound?

S·A·M·P·L·E 4.16 P·R·O·B·L·E·M

Writing Formulas from Names of Covalent Compounds

Write the formulas of the following covalent compounds:

a. sulfur dichloride **b.** dinitrogen pentoxide

Solution

a. The first nonmetal is sulfur. Since there is no prefix given, we know that there is one atom of sulfur. The second nonmetal in the formula is chlorine. The prefix *di* indicates that there are two atoms of chlorine, which means that the subscript 2 appears with chlorine in the formula.

Sulfur dichloride

SCl_2

b. The first nonmetal in the name is nitrogen. The prefix *di* means that there are two atoms of nitrogen and therefore the subscript 2 is needed. The second nonmetal is oxygen. The prefix *pent(a)* means "five"—there are five atoms of oxygen so the subscript 5 is needed.

Dinitrogen pentoxide

N_2O_5

Study Check

What is the formula of iodine pentafluoride?

QUESTIONS & PROBLEMS

Covalent Compounds

4.51 Name the following covalent compounds:
a. PBr_3 **b.** CBr_4 **c.** SiO_2
d. HF **e.** NI_3

4.52 Name the following covalent compounds:
a. CS_2 **b.** P_2O_5 **c.** Cl_2O
d. PCl_3 **e.** N_2O_4

4.53 Name the following covalent compounds:
a. N_2O_3 **b.** NCl_3 **c.** $SiBr_4$
d. PCl_5 **e.** SO_3

4.54 Name the following covalent compounds:
a. SiF_4 **b.** IBr_3 **c.** CO_2
d. SO_2 **e.** N_2O

4.55 Write the formulas of the following covalent compounds:
a. carbon tetrachloride **b.** carbon monoxide
c. phosphorus trichloride **d.** dinitrogen tetroxide

4.56 Write the formulas of the following covalent compounds:
a. sulfur dioxide **b.** silicon tetrachloride
c. iodine pentafluoride **d.** dinitrogen monoxide

4.57 Write the formulas of the following covalent compounds:
a. oxygen difluoride **b.** boron trifluoride
c. dinitrogen trioxide **d.** sulfur hexafluoride

4.58 Write the formulas of the following covalent compounds:
a. sulfur dibromide **b.** carbon disulfide
c. tetraphosphorus hexoxide **d.** dinitrogen pentoxide

Chem of Life Atoms and Molecules: Electronegativity 1, 2

4.7 Bond Polarity

Learning Goal

Using electronegativity values, classify a bond as nonpolar covalent, polar covalent, or ionic.

In covalent bonds, as we have seen, electrons may be shared between atoms that are identical, as in Cl_2, or between atoms that are different, as in HCl. When electrons are shared between identical nonmetal atoms, they are shared equally. However, when they are shared between atoms of different nonmetals, they are shared unequally because one of the atoms in the covalent bond has a stronger attraction for the pair of electrons than the other. The ability of an atom to attract the shared electrons is called its electronegativity, and it is indicated by a number called an **electronegativity value.** The element with the greater electronegativity value pulls the shared electrons closer to its nucleus.

In the table of electronegativity values shown in Figure 4.10, fluorine, the most electronegative element, has the highest value of 4.0. By contrast, the metals in Groups 1 and 2 have low electronegativity values. The electronegativity values increase going from left to right across each period of the periodic table, and upward through a group. The nonmetals, which have high electronegativity values, have a greater attraction for electrons than do the metals.

The **polarity** of a bond is related to the difference in the electronegativities of the atoms in the bond. In a covalent bond between identical atoms, the electrons are shared equally, and the bond is a **nonpolar covalent bond.** In a covalent bond between atoms

4.7

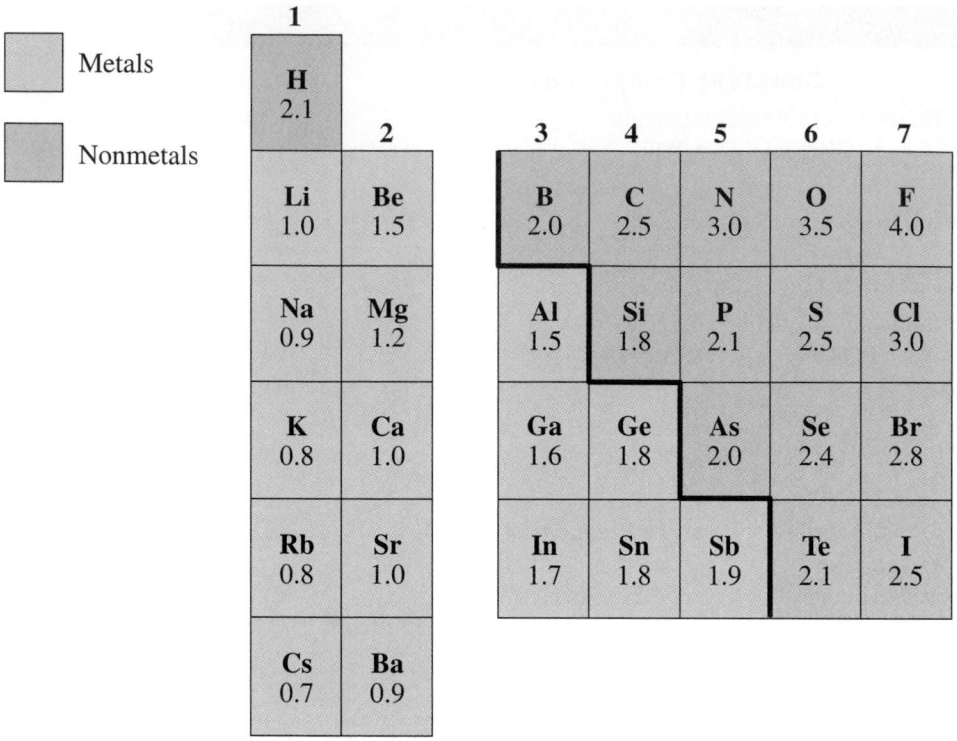

Figure 4.10 Electronegativity values for the representative elements in the periodic table.

with different electronegativity values, the shared electrons are pulled closer to the atom with the higher value. This type of bond is called a **polar covalent bond**. For example, in HCl the shared electron pair is pulled closer to the more electronegative chlorine atom, as shown in Figure 4.11.

In a polar covalent bond, the more electronegative atom may be shown with a partial negative charge, indicated by the lowercase Greek letter delta with a negative sign, δ^-. The atom with the lower electronegativity is farther from the electron pair and is shown with a partial positive charge, indicated by the Greek letter delta with a positive sign, δ^+.

Figure 4.11 Equal and unequal sharing of electrons in the nonpolar covalent bond of H_2 and the polar covalent bond of HCl.

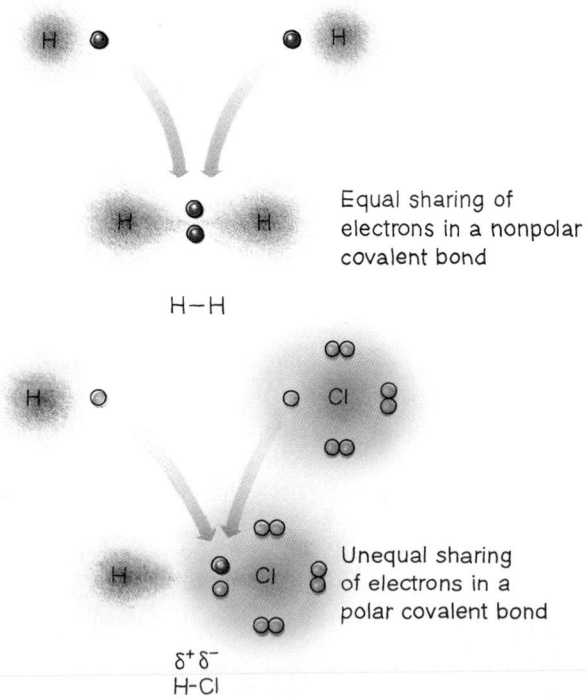

Table 4.15 Predicting Bond Type from Electronegativity Differences

Molecule		Type of Electron Sharing	Electronegativity Difference[a]	Bond Type
H_2	H—H	Shared equally	2.1 − 2.1 = 0	Nonpolar covalent
Cl_2	Cl—Cl	Shared equally	3.0 − 3.0 = 0	Nonpolar covalent
HI	$\overset{\delta+}{H}—\overset{\delta-}{I}$	Shared unequally	2.5 − 2.1 = 0.4	Polar covalent
HCl	$\overset{\delta+}{H}—\overset{\delta-}{Cl}$	Shared unequally	3.0 − 2.1 = 0.9	Polar covalent
NaCl	$Na^+ Cl^-$	Electron transfer	3.0 − 0.9 = 2.1	Ionic
MgO	$Mg^{2+} O^{2-}$	Electron transfer	3.5 − 1.2 = 2.3	Ionic

[a]Values are taken from Figure 4.10

Covalent bonds with electronegativity differences up to about 1.6 are considered to be polar. When the difference in electronegativity is greater than 1.6, the electrons are transferred from one atom to the other forming ions. Table 4.15 gives some examples of how bond type can be predicted from electronegativity differences.

S·A·M·P·L·E· 4.17 ·P·R·O·B·L·E·M

Identifying Bond Type

Indicate whether bonds between the following atoms would be nonpolar covalent, polar covalent, or ionic. If polar covalent, indicate the partial charges with δ^+ and δ^- notation.

a. carbon and chlorine **b.** fluorine and fluorine

Solution

a. The bond between the two nonmentals C and Cl has an electronegativity difference of 0.5. The bond is polar covalent. Because the chlorine atom has the greater electronegativity (3.0), the Cl is designated as partially negative, δ^-, in the polar bond.

$\overset{\delta+}{C}—\overset{\delta-}{Cl}$

b. The bond F—F occurs between identical atoms. It is a nonpolar covalent bond, as indicated by an electronegativity difference of zero (0).

Study Check

Determine if the bonds between the following would be nonpolar covalent, polar covalent, or ionic. If polar covalent, show the partial charges with δ^+ and δ^-.

a. hydrogen and hydrogen **b.** hydrogen and oxygen

A Review of Bonding

As we have seen, the types of bonds range from the nonpolar covalent bonds, in which electrons are shared equally, all the way to the ionic bond, in which electrons are completely transferred from one atom to another. This variation in bonding is continuous; there is no definite point where one type of bond stops and the next starts. However, for purposes of discussion, we can state some general rules (summarized in Table 4.16) that may be used to predict the type of bond formed between two elements. Examples of the different bonding types are given in Table 4.17.

4.7

Table 4.16 Types of Bonding

Typical Elements	Two Nonmetals (identical)	Two Nonmetals (different)	Metal and Nonmetal
Electron bonding	Shared equally	δ^+ Shared unequally δ^-	+ −
Electronegativity difference	0	1.6 or less	Over 1.6
Bond type	Nonpolar covalent	Polar covalent	Ionic
Examples	H—H	$H^{\delta+}$—$Br^{\delta-}$	K^+ Br^-

Table 4.17 Examples of Types of Bonding

Elements to Bond	Electron-Dot Structure	Type of Elements	Type of Bonding	Molecule or Ionic Unit	Formula
F and F	:F· :F·	Two nonmetals (the same element)	Nonpolar covalent	F—F	F_2
P and Cl	·P· :Cl·	Two nonmetals (different elements)	Polar covalent	Cl—P—Cl (Cl)	PCl_3
Na and F	Na· ·F:	Metal, nonmetal	Ionic	$Na^+ F^-$	NaF
Ca and O	Ca· :O·	Metal, nonmetal	Ionic	$Ca^{2+} O^{2-}$	CaO

Rules for Predicting Bond Type

1. A bond between two atoms of the same element or between atoms having the same electronegativity value is a nonpolar covalent bond.
2. A covalent bond between atoms of two different nonmetals having different electronegativities is a polar covalent bond.
3. A bond between a metal and a nonmetal is usually ionic.

S·A·M·P·L·E 4.18 P·R·O·B·L·E·M

Predicting Bond Polarity

Predict the type of bond between each of the following pairs of elements:

a. Ca and Cl **b.** P and S **c.** Br and Br

Solution

a. A bond between calcium, a metal, and chlorine, a nonmetal, would be ionic.
b. A bond between phosphorus and sulfur, two nonmetals with different electronegativity values, is polar covalent.
c. A bond between atoms of the same element would be nonpolar covalent.

Study Check

Predict the type of bond for the following:

a. S—O **b.** Na—O **c.** O—O

Q·U·E·S·T·I·O·N·S & P·R·O·B·L·E·M·S

Bond Polarity

4.59 Identify the bonding between the following pairs of elements as ionic, polar covalent, or nonpolar covalent. If no bond forms, write *none*.
a. Cl and Cl **b.** Ne and O **c.** F and F
d. Mg and F **e.** S and Cl

4.60 Identify the bonding between the following pairs of elements as ionic, polar covalent, or nonpolar covalent. If no bond forms, write *none*.
a. Na and Cl **b.** O and F **c.** K and S
d. H and H **e.** N and H

4.61 Predict the type of bonding (ionic, polar covalent, or nonpolar covalent) in each of the following:
a. SO_2 **b.** Na_2O **c.** PCl_3 **d.** Cl_2 **e.** NH_3

4.62 Predict the type of bonding (ionic, polar covalent, or nonpolar covalent) in each of the following:
a. $CaCl_2$ **b.** H_2 **c.** OF_2 **d.** N_2O **e.** CO

4.63 Identify the atom that has the higher electronegativity in each of the following pairs:
a. Na and F **b.** N and F **c.** S and Cl
d. K and Br **e.** Cl and Br

4.64 Identify the atom that has the higher electronegativity in each of the following pairs:
a. Li and Na **b.** Si and Cl **c.** S and O
d. N and F **e.** Ca and Br

4.65 Place the symbols for partially positive (δ^+) and partially negative (δ^-) above the appropriate atoms in the following polar covalent bonds:
a. H—F **b.** C—Cl **c.** N—O **d.** N—F

4.66 Place the symbols for partially positive (δ^+) and partially negative (δ^-) above the appropriate atoms in the following polar covalent bonds:
a. O—H **b.** O—S **c.** P—Cl **d.** S—Cl

4.8 Polyatomic Ions

Learning Goal

Write a formula of a compound containing a polyatomic ion.

In a **polyatomic ion,** a group of atoms has acquired an electrical charge. Most polyatomic ions consist of a nonmetal such as phosphorus, sulfur, carbon, or nitrogen covalently bonded to oxygen atoms. These oxygen-containing polyatomic ions have an ionic charge of 1−, 2−, or 3− because 1, 2, or 3 electrons were added to the atoms in the group to complete their octets. Only one of the common polyatomic ions, NH_4^+, is positively charged. Some models of polyatomic ions are shown in Figure 4.12.

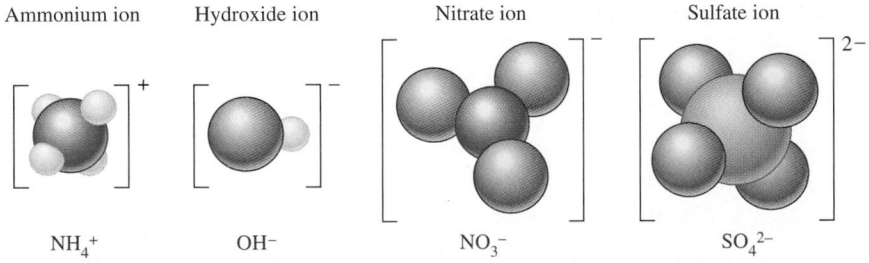

Figure 4.12 Some common polyatomic ions, which are groups of atoms that carry an ionic charge.

4.8

Naming Polyatomic Ions

The names of the most common polyatomic ions end in *ate*. The *ite* ending is used for the names of related ions that have one less oxygen atom. Recognizing these endings will help you identify polyatomic ions in the name of a compound. The hydroxide ion (OH^-) and cyanide ion (CN^-) are exceptions to this naming pattern. There is no easy way to learn polyatomic ions. You will need to memorize the number of oxygen atoms and the charge associated with each ion, as shown in Table 4.18. By memorizing the formulas and the names of the ions shown in the boxes, you can derive the related ions. For example, the sulfate ion is SO_4^{2-}. We can write the formula of the sulfite ion, which has one less oxygen atom, as SO_3^{2-}. The formula of hydrogen carbonate can be written by placing a hydrogen cation (H^+) in front of the formula for carbonate (CO_3^{2-}) and decreasing the charge from 2− to 1− to give HCO_3^-.

Writing Formulas for Compounds Containing Polyatomic Ions

No polyatomic ion exists by itself. Like any ion, a polyatomic ion must be associated with ions of opposite charge. The bonding between polyatomic ions and other ions is one of electrical attraction. For example, the compound sodium sulfate consists of sodium ions (Na^+) and sulfate ions (SO_4^{2-}) held together by ionic bonds. Figure 4.13 shows the formation of Na_2SO_4.

To write correct formulas for compounds containing polyatomic ions we follow the same rules of charge balance that we used for writing the formulas of simple ionic compounds. The total negative and positive charges must equal zero. For example, consider the formula for a compound containing calcium ions and carbonate ions. The ions are written as

Ca^{2+}	CO_3^{2-}
Calcium ion	Carbonate ion

Ionic charge: $(2+) + (2-) = 0$

Because one ion of each balances the charge, the formula is written as

$CaCO_3$
Calcium carbonate

Table 4.18 Names and Formulas of Some Common Polyatomic Ions

Nonmetal	Formula of Ion[a]	Name of Ion
Hydrogen	OH^-	Hydroxide
Nitrogen	NH_4^+	Ammonium
	$\boxed{NO_3^-}$	Nitrate
	NO_2^-	Nitrite
Chlorine	$\boxed{ClO_3^-}$	Chlorate
	ClO_2^-	Chlorite
Carbon	$\boxed{CO_3^{2-}}$	Carbonate
	HCO_3^-	Hydrogen carbonate (or bicarbonate)
	CN^-	Cyanide
Sulfur	$\boxed{SO_4^{2-}}$	Sulfate
	HSO_4^-	Hydrogen sulfate (or bisulfate)
	SO_3^{2-}	Sulfite
	HSO_3^-	Hydrogen sulfite (or bisulfite)
Phosphorus	$\boxed{PO_4^{3-}}$	Phosphate
	HPO_4^{2-}	Hydrogen phosphate
	$H_2PO_4^-$	Dihydrogen phosphate
	PO_3^{3-}	Phosphite

[a]Boxed formulas are the most common polyatomic ion for that element.

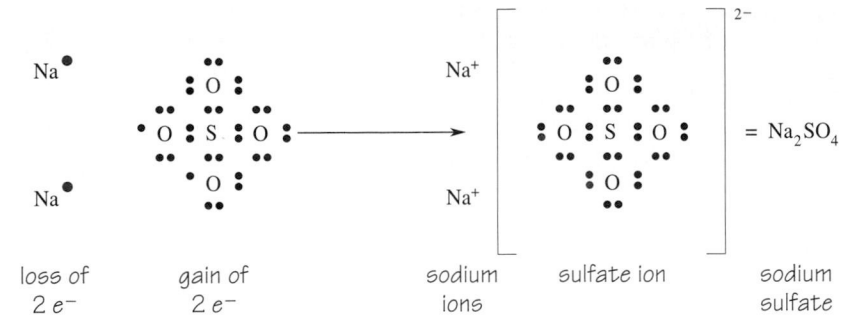

Figure 4.13 The ionic compound sodium sulfate contains a polyatomic ion.

When more than one polyatomic ion is needed for charge balance, parentheses are used to enclose the formula of the ion. A subscript is written outside the closing parenthesis. Consider the formula for magnesium nitrate. The ions in this compound are the magnesium ion and the nitrate ion, a polyatomic ion.

Mg^{2+} — Magnesium ion

NO_3^- — Nitrate ion

To balance the positive charge of 2+, two nitrate ions are needed. The formula, including the parentheses around the nitrate ion, is as follows:

Mg^{2+} NO_3^- NO_3^-

$(2+) + 2(1-) = 0$

Magnesium nitrate

$Mg(NO_3)_2$

Parentheses enclose the formula of the nitrate ions

Subscript outside the parentheses indicates the use of two nitrate ions

HEALTH NOTE

Polyatomic Ions in Bone and Teeth

Bone structure consists of two parts: a solid mineral material, and a second phase made up primarily of collagen proteins. The mineral substance is a compound called hydroxyapatite, a solid formed from calcium ions, phosphate ions, and hydroxide ions. This material is deposited in the web of collagen to form a very durable bone material. (See Figure 4.14).

$Ca_{10}(PO_4)_6(OH)_2$
Calcium hydroxyapatite

In most individuals, bone material is continuously being absorbed and re-formed. After age 40, more bone material may be lost than formed, a condition called osteoporosis. Bone mass reduction occurs at a faster rate in women than in men, and at different rates in different parts of the skeleton. The reduction in bone mass can be as much as 50% over a period of 30 to 40 years. It is recommended that persons over 35, especially women, include a daily calcium supplement in their diet.

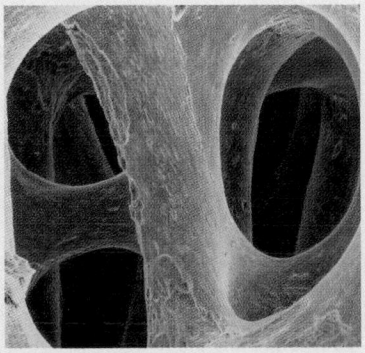

a

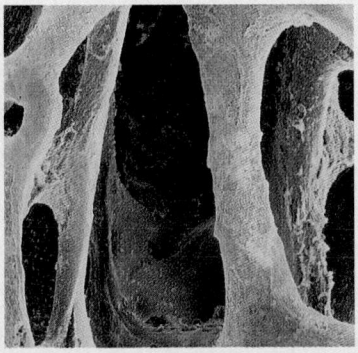

b

Figure 4.14 Scanning electron micrographs (SEMs) of **(a)** normal bone and **(b)** bone in osteoporosis due to calcium loss.

EXPLORE YOUR WORLD

Names of Compounds Listed on Products

Calcium carbonate is the active ingredient in some antacids. One of the inert ingredients is silicon dioxide. Are these compounds ionic or covalent compounds? What are their formulas? Look at the labels of other food and/or health products. See if you can recognize some of the names. What are some of the ionic compounds? What compounds have covalent bonds? Try to write some of their formulas. You may use a reference book such as the *Handbook of Chemistry and Physics* or the *Merck Index* to find or check some of the formulas.

4.8

S·A·M·P·L·E 4.19 ·P·R·O·B·L·E·M

Writing Formulas Having Polyatomic Ions

Write the formula for a compound containing aluminum ions and bicarbonate ions.

Solution

The positive ion in the compound is the aluminum ion, Al^{3+}, and the negative ion is the bicarbonate ion, HCO_3^-, which is a polyatomic ion.

Al^{3+}	HCO_3^-
Aluminum ion	Bicarbonate ion

Three HCO_3^- ions are required to balance the Al^{3+} charge.

Al^{3+} HCO_3^-
HCO_3^-
HCO_3^-

$$(3+) + 3 \times (1-) = 0$$

The formula for the compound is written by enclosing the formula of the bicarbonate ion, HCO_3^-, in parentheses and writing the subscript 3 just outside the last parenthesis:

$Al(HCO_3)_3$

Study Check

Write the formula for a compound containing ammonium ions and phosphate ions.

Naming Compounds Containing Polyatomic Ions

When naming ionic compounds containing polyatomic ions, we write the positive ion, usually a metal, first, and then we write the name of the polyatomic ion. It is important that you learn to recognize the polyatomic ion in the formula and name it correctly. As with other ionic compounds, no prefixes are used.

Na_2SO_4	$FePO_4$	$Al_2(CO_3)_3$
$Na_2\boxed{SO_4}$	$Fe\boxed{PO_4}$	$Al_2(\boxed{CO_3})_3$
Sodium sulfate	Iron(III) phosphate	Aluminum carbonate

Table 4.19 lists the formulas and names of some ionic compounds that include polyatomic ions, and also gives their uses in medicine and industry.

Table 4.19 Some Compounds That Contain Polyatomic Ions

Formula	Name	Use
$BaSO_4$	Barium sulfate	Radiopaque medium
$CaCO_3$	Calcium carbonate	Antacid, calcium supplement
$Ca_3(PO_4)_2$	Calcium phosphate	Calcium replenisher
$CaSO_3$	Calcium sulfite	Preservative in cider and fruit juices
$CaSO_4$	Calcium sulfate	Plaster casts
$AgNO_3$	Silver nitrate	Topical anti-infective
$NaHCO_3$	Sodium bicarbonate	Antacid
$Zn_3(PO_4)_2$	Zinc phosphate	Dental cements
$FePO_4$	Iron(III) phosphate	Food and bread enrichment
K_2CO_3	Potassium carbonate	Alkalizer, diuretic
$Al_2(SO_4)_3$	Aluminum sulfate	Antiperspirant, anti-infective
$AlPO_4$	Aluminum phosphate	Antacid
$MgSO_4$	Magnesium sulfate	Cathartic, Epsom salts

S·A·M·P·L·E· **4.20** ·P·R·O·B·L·E·M

Naming Compounds Containing Polyatomic Ions

Name the following ionic compounds:

a. $CaSO_4$ **b.** $Cu(NO_2)_2$

Solution

Compound Formula	Ions Present	Name of Compound
a. $CaSO_4$	Ca^{2+}, calcium SO_4^{2-}, sulfate	Calcium sulfate
b. $Cu(NO_2)_2$	Cu^{2+}, copper(II) NO_2^-, nitrite One less oxygen than nitrate, NO_3^-	Copper(II) nitrite

Study Check

What is the name of $Ca_3(PO_4)_2$?

Summary of Naming Compounds

Throughout this chapter we have examined strategies for naming ionic and covalent compounds. Now we can summarize the rules, as illustrated in Figure 4.15. In general, compounds having two elements are named by stating the first element, followed by the second element with an *ide* ending. If the first element is a metal, the compound is ionic: if the first element is a nonmetal, the compound is covalent. For ionic compounds, it is necessary to determine whether the metal can form more than one type of positive ion; if so, a Roman numeral following the name of the metal indicates the particular ionic charge. In naming covalent compounds having two elements, prefixes are necessary to indicate the number of atoms of each nonmetal as shown in that particular formula. Ionic compounds having three or more elements include some type of polyatomic ion. They are named by ionic rules, but have an *ate* or *ite* ending when the polyatomic ion has a negative charge. Table 4.20 summarizes some naming rules for elements, ionic compounds, and covalent compounds.

4.8

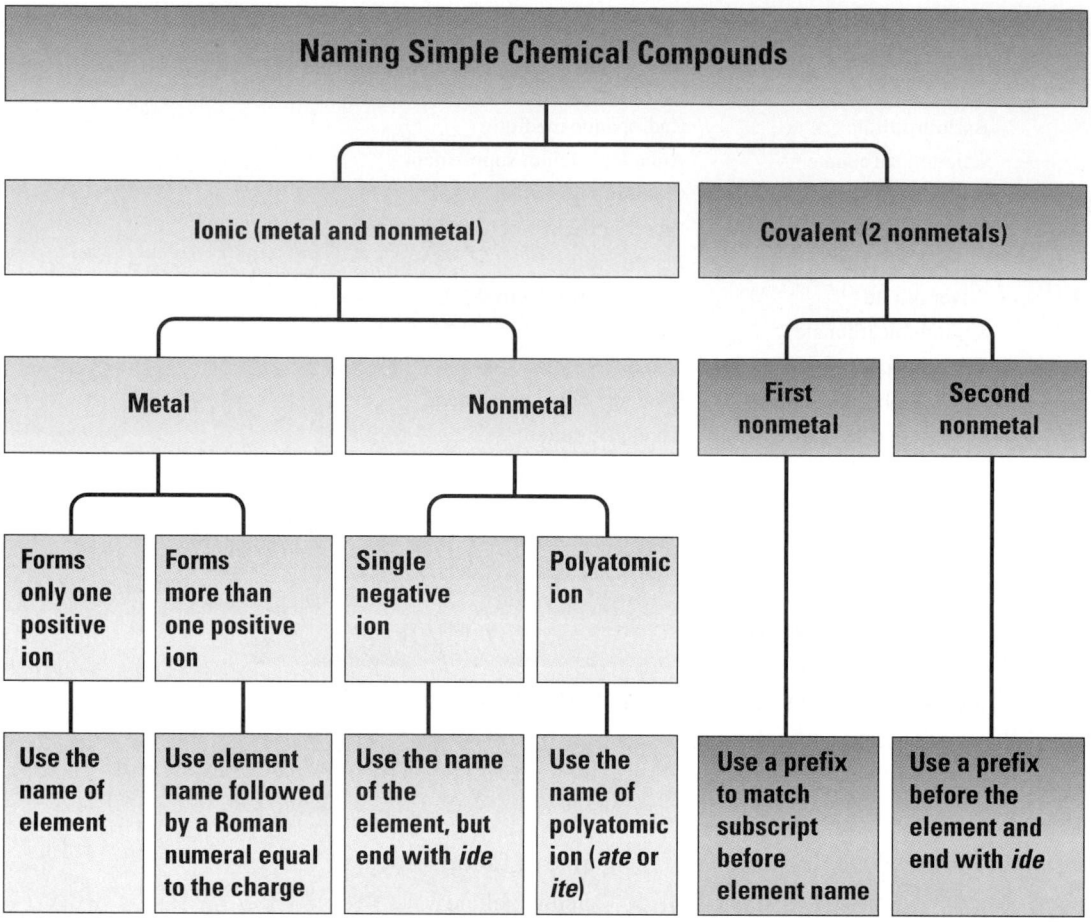

Figure 4.15 A flowchart of how ionic and covalent compounds are named.

S·A·M·P·L·E· **4.21** ·P·R·O·B·L·E·M

Naming Ionic and Covalent Compounds

Use Figure 4.15 to name the following compounds:

a. Na_3P **b.** $CuSO_4$ **c.** SO_3

Solution

a. Na_3P is an ionic compound. Na is a metal that forms a single ion, Na^+, which is named sodium. P^{3-}, a single negative ion, is named phosphide. The compound Na_3P is sodium phosphide.

b. $CuSO_4$ is an ionic compound. Cu is a transition metal that forms more than one ion. The Cu^{2+} is the positive ion that balances the charge on SO_4^{2-}. Roman numeral II specifies the charge in the name copper(II). SO_4^{2-} is a polyatomic ion named sulfate. The compound $CuSO_4$ is named copper(II) sulfate.

c. SO_3 is covalent because it consists of two nonmetals. The first element S is named sulfur (no prefix is needed). The subscript three uses the prefix *tri* and the element oxygen becomes oxide. The compound SO_3 is named sulfur trioxide.

Study Check

What is the name of $Fe(NO_3)_2$?

Table 4.20 Rules for Naming Elements, Ionic Compounds, and Covalent Compounds

Type	Formula Feature	Naming Procedure
Element	Symbol of element; may be diatomic. *Examples:* Ca Cl_2	Use name of element. *Examples:* Calcium Chlorine
Ionic compound (two elements)	Symbol of metal followed by symbol of nonmetal; subscripts used for charge balance. *Examples:* Na_2O Fe_2S_3	Use element name for metal; Roman numeral required if two or more positive ions possible. For nonmetal use element name with *ide* ending. *Examples:* Sodium oxide Iron(III) sulfide
Ionic compound (three elements)	Usually symbol of metal followed by a polyatomic ion composed of nometals; parentheses may enclose polyatomic ion for charge balance. *Examples:* $Mg(NO_3)_2$ $CuSO_4$ $(NH_4)_2CO_3$	Use element name for metal, with Roman numeral if needed, followed by name of polyatomic ion. *Examples:* Magnesium nitrate Copper(II) sulfate Ammonium carbonate
Covalent compound (two elements)	Symbols of two nonmetals; subscripts show number of atoms in a molecule. *Examples:* N_2O_3 CCl_4	Place prefixes before each element name if there are two or more atoms in the formula; add *ide* ending to second element. *Examples:* Dinitrogen trioxide Carbon tetrachloride

QUESTIONS & PROBLEMS

Polyatomic Ions

4.67 Write the formulas including the charge for the following polyatomic ions:
a. bicarbonate **b.** ammonium
c. phosphate **d.** hydrogen sulfate

4.68 Write the formulas including the charge for the following polyatomic ions:
a. nitrite **b.** sulfite **c.** hydroxide **d.** phosphite

4.69 Name the following polyatomic ions:
a. SO_4^{2-} **b.** CO_3^{2-} **c.** PO_4^{3-} **d.** NO_3^-

4.70 Name the following polyatomic ions:
a. OH^- **b.** HSO_3^- **c.** CN^- **d.** NO_2^-

4.71 Complete the following table with the formula of the compound:

	OH^-	NO_2^-	CO_3^{2-}	HSO_4^-	PO_4^{3-}
Li+	LiOH	$LiNO_2$	Li_2CO_3	$LiHSO_4$	Li_3PO_4
Cu^{2+}	$Cu(OH)_2$	$Cu(NO_2)_2$	$CuCO_3$	$Cu(HSO_4)_2$	$Cu_3(PO_4)_2$
Ba^{2+}	$Ba(OH)_2$	$Ba(NO_2)_2$	$BaCO_3$	$Ba(HSO_4)_2$	$Ba_3(PO_4)_2$

4.72 Complete the following table with the formula of the compound:

	OH^-	NO_3^-	HCO_3^-	SO_3^{2-}	PO_4^{3-}
NH_4^+					
Al^{3+}					
Pb^{4+}					

4.8

4.73 Write the formula for the polyatomic ion in each of the following and name each compound:

a. Na_2CO_3 **b.** NH_4Cl **c.** Li_3PO_4
d. $Cu(NO_2)_2$ **e.** $FeSO_3$

4.74 Write the formula for the polyatomic ion in each of the following and name each compound:

a. KOH **b.** $NaNO_3$ **c.** $CuCO_3$ **d.** $NaHCO_3$ **e.** $BaSO_4$

4.75 Write the correct formula for the following compounds:

a. barium hydroxide **b.** sodium sulfate **c.** iron(II) nitrate
d. zinc phosphate **e.** iron(III) carbonate

4.76 Write the correct formula for the following compounds:

a. aluminum chloride **b.** ammonium oxide
c. magnesium bicarbonate **d.** sodium nitrite
e. copper(I) sulfate

4.77 Name the compounds that are found in the following sources:

a. $Al_2(SO_4)_3$ antiperspirant
b. $CaCO_3$ antacid
c. N_2O "laughing gas," inhaled anesthetic
d. Na_3PO_4 cathartic
e. $(NH_4)_2SO_4$ fertilizer
f. Fe_2O_3 pigment

4.78 Name the compounds that are found in the following sources:

a. N_2 Earth's atmosphere
b. $Mg_3(PO_4)_2$ antacid
c. $FeSO_4$ iron supplement in vitamins
d. $MgSO_4$ Epsom salts
e. Cu_2O fungicide
f. SnF_2 dental caries prophylactic

Chapter Review

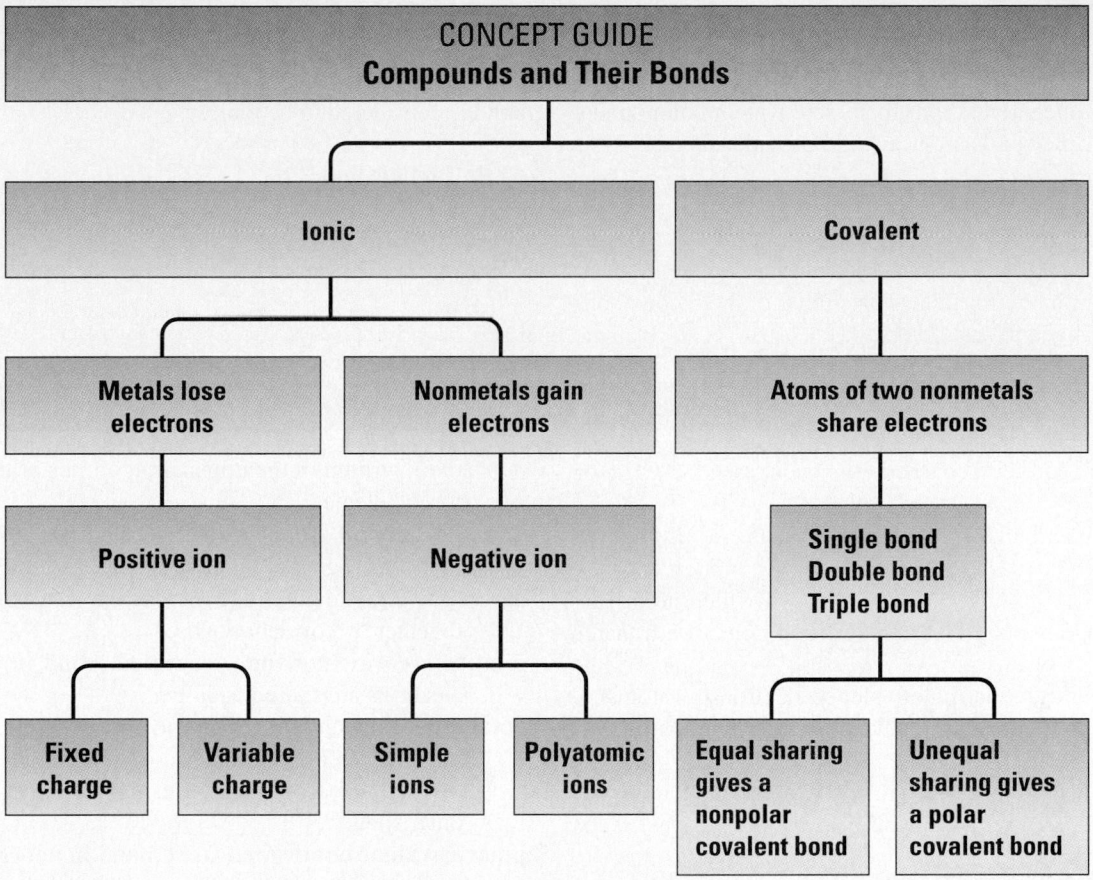

Chapter Summary

4.1 Valence Electrons

The electrons located in the outermost energy level of an atom play a major role in determining its chemical properties. An electron-dot structure represents the valence electrons in an atom.

4.2 Ions and the Octet Rule

The nonreactivity of the noble gases is associated with the stable arrangement of 8 electrons, an octet, in their valence shells; helium needs 2 electrons for stability. Atoms of elements other than the noble gases achieve stability by losing, gaining, or sharing their valence electrons with other atoms in the formation of compounds. Metals of the representative elements form an octet by losing their valence electrons to form positively charged cations; Group 1, 1+, Group 2, 2+, and Group 3, 3+. When they react with metals, nonmetals gain electrons to form octets in their valence shells. As anions, they have negative charges; Group 5, 3−, Group 6, 2−, Group 7, 1−.

4.3 Ionic Compounds

The total positive and negative ionic charge must balance in the formula of an ionic compound. Charge balance in a formula is achieved by using subscripts after each symbol so that the overall charge is zero.

4.4 Naming and Writing Ionic Formulas

In naming ionic compounds, the positive ion is given first, followed by the name of the negative ion. Ionic compounds containing two elements end with *ide*. When the metal can form more than one positive ion, its ionic charge is determined from the total negative charge in the formula. Typically transition metals form cations with two or more ionic charges. The charge is given as a Roman numeral in the name, such as iron(II) and iron(III) for the cations of iron with 2+ and 3+ ionic charges.

4.5 Covalent Bonds

In a covalent bond, electrons are shared by atoms of nonmetals. By sharing, each of the atoms achieves a noble gas arrangement. In a nonpolar covalent bond, electrons are shared equally by atoms. In a polar covalent bond, electrons are unequally shared because they are attracted by the more electronegative atom. In some covalent compounds, double or triple bonds are needed to provide an octet.

4.6 Covalent Compounds

Two nonmetals can form two or more different covalent compounds. In their names, prefixes are used to indicate the subscript in the formula. The ending of the second nonmetal is changed to *ide*.

4.7 Bond Polarity

Electronegativity values indicate the ability of atoms to attract electrons. In general, the electronegativity of a metal is low; a nonmetal is high. In a nonpolar covalent bond, the difference in electronegativities is zero; in a polar covalent bond, the difference can be up to 1.6. Greater than 1.6, the bond is considered to be ionic.

4.8 Polyatomic Ions

A group of nonmetal atoms that carries an electrical charge, for example, the carbonate ion has the formula CO_3^{2-}. Most polyatomic ions have names that end with *ate* or *ite*.

Key Terms

anion A negatively charged ion such as Cl^-, O^{2-}, or SO_4^{2-}.

cation A positively charged ion such as Na^+, Mg^{2+}, or Al^{3+}.

compound A combination of atoms in which noble gas arrangements are attained through electron transfer or electron sharing.

covalent bond A sharing of valence electrons by atoms.

double bond A sharing of two pairs of electrons by two atoms.

electron-dot structure The representation of an atom that shows each valence electron as a dot above, below, or beside the symbol of the element.

electronegativity value A number that indicates the relative ability of an element to attract electrons.

formula The group of symbols that represent the atoms or ions in a compound.

ion An atom or group of atoms having an electrical charge because of a loss or gain of electrons.

ionic bond The attraction between oppositely charged ions that results from the transfer of electrons.

ionic charge The difference between the number of protons (positive) and the number of electrons (negative), written in the upper right corner of the symbol for the element.

ionic compound A compound of positive and negative ions held together by ionic bonds.

molecule The smallest unit of two or more atoms held together by covalent bonds.

nonpolar covalent bond A covalent bond in which the electrons are shared equally.

octet rule Elements in Groups 1–7 react with other elements by forming ionic or covalent bonds to produce a noble gas arrangement, usually 8 electrons in the outer shell.

polar covalent bond A covalent bond in which the electrons are shared unequally.

polarity A measure of the unequal sharing of electrons, indicated by the difference in electronegativity values.

polyatomic ion A group of covalently bonded nonmetal atoms that has an overall electrical charge.

triple bond A sharing of three pairs of electrons by two atoms.

valence electrons The electrons present in the outermost shell of an atom, which are largely responsible for the chemical behavior of the element.

Chapter Practice

4.79 What are the elements in Period 3 represented by the following?

a. $\cdot\ddot{\underset{\cdot}{X}}\cdot$ **b.** $X\cdot$ **c.** $\cdot\dot{X}\cdot$ **d.** $\cdot\dot{\underset{\cdot}{X}}\cdot$

4.80 Write the electron arrangement of the following:

a. Ar **b.** Na^+ **c.** S^{2-}
d. Cl^- **e.** Ca^{2+}

4.81 Write the electron arrangement of the following:

a. N^{3-} **b.** Ne **c.** F^-
d. Al^{3+} **e.** Li^+

4.82 What noble gas has the same arrangement as the following ions?

a. Al^{3+} **b.** Cl^- **c.** Ca^{2+}
d. Na^+ **e.** N^{3-}

4.83 Consider an ion with the symbol X^{2+} formed from a representative element.

a. What is the group number of the element?
b. What is the electron-dot structure of the element?
c. If X is in Period 2, what is the element?

4.84 Consider the following electron-dot structures for representative elements X and Y:

X· ·$\ddot{\underset{\cdot\cdot}{Y}}$·

a. What are the group numbers of X and Y?
b. Will a compound of X and Y be ionic or covalent?
c. What ions would be formed by X and Y?
d. What would be the formula of a compound of X and Y?
e. What would be the formula of a compound of X and chlorine?
f. What would be the formula of a compound of Y and chlorine?

4.85 One of the ions of tin is tin(IV).
a. What is the symbol for this ion?
b. How many protons and electrons are in the ion?
c. What is the formula of tin(IV) oxide?
d. What is the formula of tin(IV) phosphate?

4.86 As shown in the Health Note "Polyatomic Ions in Bone and Teeth," the mineral portion of bone is composed of calcium hydroxyapatite, $Ca_{10}(PO_4)_6(OH)_2$. Name the ions in the formula.

4.87 Classify the following compounds as ionic or covalent.

a. Li_2O	**b.** N_2O	**c.** CF_4
d. Cl_2O	**e.** MgF_2	**f.** CO
g. $CaCl_2$	**h.** K_3N	

4.88 Hydrogen and helium are both very light gases. Why is helium gas, but not hydrogen gas, used to fill balloons and blimps?

4.89 Name the following compounds:

a. $FeCl_3$	**b.** Cl_2O_7	**c.** Br_2
d. $Ca_3(PO_4)_2$	**e.** PCl_3	**f.** $Al_2(CO_3)_3$
g. $PbCl_4$	**h.** $MgCO_3$	**i.** NO_2
j. $SnSO_4$	**k.** $Ba(NO_3)_2$	**l.** CuS

4.90 Write the formulas for the following compounds:

a. tin(II) carbonate	**b.** lithium phosphide
c. silicon tetrachloride	**d.** iron(III) sulfide
e. carbon dioxide	**f.** calcium bromide
g. sodium carbonate	**h.** nitrogen dioxide
i. aluminum nitrate	**j.** copper(I) nitride
k. potassium phosphate	**l.** lead(IV) sulfide

Answers

Answers to Study Checks

4.1 six electrons

4.2 ·Ca·

4.3 **c.** krypton (Kr)

4.4 $35p^+$ and $36e^-$

4.5 K^+ and S^{2-}

4.6

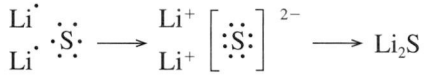

4.7 $CaCl_2$

4.8 **a.** calcium ion **b.** nitride ion
c. bromide ion

4.9 calcium phosphide

4.10 copper(I) ion; cuprous ion

4.11 gold(III) chloride; auric chloride

4.12 Al_2S_3

4.13 H :$\ddot{\underset{\cdot\cdot}{N}}$: H
H

4.14 Total number of valence electrons: $1e^- + 4e^- + 5e^- = 10\,e^-$
A triple bond gives octets to C and N. H : C ::: N :
or H—C≡N:

4.15 xenon hexafluoride

4.16 IF_5

4.17 **a.** nonpolar covalent
b. polar covalent $H^{\delta+}$—$O^{\delta-}$

4.18 **a.** polar covalent **b.** ionic
c. nonpolar covalent

4.19 $(NH_4)_3PO_4$

4.20 calcium phosphate

4.21 iron(II) nitrate

Answers to Selected Questions and Problems

4.1 The valence electrons are the electrons in the highest energy level of an atom.

4.3 **a.** 2, 5: 5 valence electrons
b. 2, 6: 6 valence electrons
c. 2, 8, 8: 8 valence electrons
d. 2, 8 ,8, 1: 1 valence electron
e. 2, 8, 6: 6 valence electrons

4.5 **a.** Group 6 :$\ddot{\underset{\cdot\cdot}{S}}$·
b. Group 5 ·$\ddot{\underset{\cdot}{N}}$·
c. Group 2 ·Ca·
d. Group 1 Na·
e. Group 1 K·

4.7 **a.** M· **b.** ·M·

4.9 They are all in Group 1; each has 1 valence electron.

4.11 **a.** By losing 1 valence electron from the third energy level, sodium achieves an octet in the second shell.
b. The sodium ion Na^+ has the same electron arrangement as Ne (2, 8).
c. Group 1 and 2 elements become stable by losing electrons to form compounds.Group 8 elements are stable.

4.13 **a.** :Ne: Stable, has a complete valence shell
b. :O· Needs 2 electrons to complete valence shell
c. Li· Needs to lose 1 electron to complete valence shell
d. :Ar: Stable, has an octet

4.15 **a.** 1 **b.** 2 **c.** 3 **d.** 1 **e.** 2

4.17 **a.** neon **b.** neon **c.** argon
d. neon **e.** neon

4.19 **a.** lose $2e^-$ **b.** gain $3e^-$ **c.** gain $1e^-$
d. lose $1e^-$ **e.** lose $3e^-$

4.21 **a.** Li^+ **b.** F^- **c.** Mg^{2+}
d. Fe^{3+} **e.** Zn^{2+}

4.23 **a.** Cl^- **b.** K^+ **c.** O^{2-} **d.** Al^{3+}

4.25 **a.** potassium **b.** sulfide
c. calcium **d.** nitride

4.27 **a, c**

4.29 **a.** K· ·Cl: ⟶ K^+ [:Cl:]$^-$ ⟶ KCl

b. ·Ca· ·Cl: ⟶ Ca^{2+} [:Cl:]$^-$ ⟶ $CaCl_2$
·Cl: [:Cl:]$^-$

c. Na· Na· Na· ·N· ⟶ Na^+ Na^+ Na^+ [:N:]$^{3-}$ ⟶ Na_3N

4.31 **a.** Na_2O **b.** $AlBr_3$ **c.** BaO
d. $MgCl_2$ **e.** Al_2S_3

4.33 **a.** Na_2S **b.** K_3N **c.** AlI_3 **d.** Li_2O

4.35 **a.** aluminum oxide **b.** calcium chloride
c. sodium oxide **d.** magnesium nitride
e. potassium iodide

4.37 Most of the transition metals form more than one positive ion. The specific ion is indicated in a name by writing a Roman numeral that is the same as the ionic charge. For example, iron forms Fe^{2+} and Fe^{3+} ions, which are named iron(II) and iron(III).

4.39 **a.** iron(II) or ferrous **b.** copper(II) or cupric
c. zinc **d.** lead (IV) or plumbic

4.41 **a.** tin(II) chloride, stannous chloride
b. iron(II) oxide, ferrous oxide
c. copper(I) sulfide; cuprous sulfide
d. copper(II) sulfide; cupric sulfide

4.43 **a.** Au^{3+} **b.** Fe^{3+} **c.** Pb^{4+} **d.** Sn^{2+}

4.45 **a.** $MgCl_2$ **b.** Na_2S **c.** Cu_2O
d. Zn_3P_2 **e.** AuN

4.47 **a.** :Br:Br:
b. H:H
c. H:F:
d. :O:F: :F:

4.49 **a.** H:C::C:H
b. :O: H:C:H
c. :O: :O:S::O:

4.51 **a.** phosphorus tribromide **b.** carbon tetrabromide
c. silicon dioxide **d.** hydrogen fluoride
e. nitrogen triiodide

4.53 **a.** dinitrogen trioxide
b. nitrogen trichloride
c. silicon tetrabromide
d. phosphorus pentachloride
e. sulfur trioxide

4.55 **a.** CCl_4 **b.** CO **c.** PCl_3 **d.** N_2O_4

4.57 **a.** OF_2 **b.** BF_3 **c.** N_2O_3 **d.** SF_6

4.59 **a.** nonpolar covalent **b.** none
c. nonpolar covalent **d.** ionic
e. polar covalent

4.61 **a.** polar covalent **b.** ionic
c. polar covalent **d.** nonpolar covalent
e. polar covalent

4.63 **a.** F **b.** F **c.** Cl **d.** Br **e.** Cl

4.65 **a.** $\overset{\delta^+}{H}—\overset{\delta^-}{F}$

b. $\overset{\delta^+}{C}—\overset{\delta^-}{Cl}$

c. $\overset{\delta^+}{N}—\overset{\delta^-}{O}$

d. $\overset{\delta^+}{N}—\overset{\delta^-}{F}$

4.67 **a.** HCO_3^- **b.** NH_4^+
c. PO_4^{3-} **d.** HSO_4^-

4.69 **a.** sulfate **b.** carbonate
c. phosphate **d.** nitrate

4.71

	OH^-	NO_2^-	CO_3^{2-}	HSO_4^-	PO_4^{3-}
Li^+	$LiOH$	$LiNO_2$	Li_2CO_3	$LiHSO_4$	Li_3PO_4
Cu^{2+}	$Cu(OH)_2$	$Cu(NO_2)_2$	$CuCO_3$	$Cu(HSO_4)_2$	$Cu_3(PO_4)_2$
Ba^{2+}	$Ba(OH)_2$	$Ba(NO_2)_2$	$BaCO_3$	$Ba(HSO_4)_2$	$Ba_3(PO_4)_2$

4.73 **a.** CO_3^{2-}, sodium carbonate
b. NH_4^+, ammonium chloride
c. PO_4^{3-}, lithium phosphate
d. NO_2^-, copper(II) nitrite
e. SO_3^{2-}, iron(II) sulfite

4.75 **a.** $Ba(OH)_2$ **b.** Na_2SO_4 **c.** $Fe(NO_3)_2$
d. $Zn_3(PO_4)_2$ **e.** $Fe_2(CO_3)_3$

4.77 **a.** aluminum sulfate **b.** calcium carbonate
c. dinitrogen monoxide **d.** sodium phosphate
e. ammonium sulfate **f.** iron(III) oxide

4.79 **a.** P **b.** Na **c.** Al **d.** Si

4.81 **a.** 2, 8 **b.** 2, 8 **c.** 2, 8
d. 2, 8 **e.** 2

4.83 **a.** Group 2
b. ·X·
c. Be

4.85 **a.** Sn^{4+} **b.** 50 protons, 46 electrons
c. SnO_2 **d.** $Sn_3(PO_4)_4$

4.87 **a.** ionic **b.** covalent **c.** covalent
d. covalent **e.** ionic **f.** covalent
g. ionic **h.** ionic

4.89 **a.** iron(III) chloride **b.** dichlorine heptoxide
c. bromine **d.** calcium phosphate
e. phosphorus trichloride **f.** aluminum carbonate
g. lead(IV) chloride **h.** magnesium carbonate
i. nitrogen dioxide **j.** tin(II) sulfate
k. barium nitrate **l.** copper(II) sulfide

5

Chemical Reactions and Quantities

When exposed to air and water, iron reacts with oxygen to form rust.

Looking Ahead

5.1	Chemical Changes
5.2	Chemical Equations
5.3	Balancing a Chemical Equation
5.4	Types of Reactions
5.5	A Closer Look: Oxidation-Reduction Reactions
5.6	The Mole
5.7	Calculations Using Molar Mass
5.8	Mole Relationships in Chemical Equations
5.9	A Closer Look: Mass Calculations for Reactions

www.chemplace.com/college/timberlake

Visit the web site above to test yourself, run simulations, and review key concepts from this chapter. For even more interactive review and quizzing, use the two CD-ROMs included with the book.

When we know the formulas of the substances in a chemical reaction, we can write the chemical equation, which shows the reactants and the products of the reaction. We do much the same thing at home when we use a recipe to make cookies or add the right amount of water to make soup. We also use an equation to determine the correct amounts of the reactants and calculate the amounts of product. At the automotive repair shop, a mechanic adjusts the carburetor or fuel injection system of an engine to allow the correct amounts of fuel and oxygen from the air to mix together so the engine will run properly. In medicine, the dosages of medications are calculated to give an amount that is appropriate for a person's body mass.

5.1 Chemical Changes

Study Pack

5.1

Learning Goal

Identify a change in a substance as a chemical or a physical change.

In a **physical change,** the appearance of a substance is altered, but not its composition. When liquid water becomes a gas, or freezes to a solid, it is still water. If we smash a rock or tear a piece of paper, only the size of the material changes. The smaller pieces are still made of rock or paper because there was no change in the composition of the substances.

In a **chemical change,** at least one of the reacting substances changes into a new substance with a different composition. A chemical change also involves physical changes such as a change in color, a change in temperature, or production of bubbles or formation of a solid. For example, if we burn a match, we see that light and heat are produced along with a charred matchstick. A chemical change produces new substances with new properties and compositions.

Every day, chemical changes are taking place around us. A **chemical reaction** takes place with atoms of the reacting substances forming new combinations with new physical properties. (See Figure 5.1.) When you see rust forming on a nail or piece of iron, the iron (Fe) is reacting with oxygen (O_2) to produce the new substance, rust (Fe_2O_3). Perhaps you have cleaned a silver dish or a piece of silver jewelry because the metal (Ag) has reacted with sulfur (S) to produce tarnish, which has a dull, gray-black color. You may have dropped an antacid tablet in a glass of water and noticed it fizzing and bubbling as sodium bicarbonate ($NaHCO_3$) reacted to form carbon dioxide (CO_2) gas. Table 5.1 gives some examples of some typical physical and chemical changes.

Figure 5.1 Examples of chemical change include a rusty nail, tarnish on a silver fork, and the fizzing (bubbling) of an antacid tablet in water.

S·A·M·P·L·E 5.1 ·P·R·O·B·L·E·M

Classifying Chemical and Physical Change

Classify each of the following changes as physical or chemical:

a. water freezing into an icicle **b.** burning a match
c. breaking a chocolate bar **d.** digesting a chocolate bar

Solution

a. Physical. Freezing water involves only a change from liquid water to ice. No change has occurred in the substance water.
b. Chemical. Burning a match causes the formation of heat, light, and ash that were not present before striking the match.
c. Physical. Breaking a chocolate bar does not affect its composition.
d. Chemical. The digestion of the chocolate bar converts its components into new substances.

Study Check

Classify the following changes as physical or chemical:

a. chopping a carrot **b.** developing a Polaroid picture
c. inflating a balloon

Table 5.1 Comparison of Some Chemical and Physical Changes

Chemical Changes	Physical Changes
Rusting nail	Melting ice
Bleaching a stain	Boiling water
Burning a log	Sawing a log in half
Tarnishing silver	Tearing paper
Fermenting grapes	Breaking a glass
Souring of milk	Pouring milk

5.1

EXPLORE YOUR WORLD

Evidence of a Chemical Change

1. Place $\frac{1}{2}$ cup of vinegar in a glass. Add a teaspoon of baking soda and observe. When the reaction ends, add another teaspoon of baking soda. Record all the changes in physical properties that you observe.
2. Light a candle and watch it burn for a while. Record all the changes in physical properties you observe.
3. Obtain a piece of silverware that is tarnished. List some of the physical properties of the tarnished silverware. Use a silver polish to clean part of the tarnished area on the silverware. Compare the polished area with the tarnished area. List the changes in physical properties.

Question

What evidence of chemical change did you observe in these trials?

QUESTIONS & PROBLEMS

Chemical Changes

5.1 Classify each of the following changes as chemical or physical:
a. chewing a gum drop
b. ignition of fuel in the space shuttle
c. drying clothes
d. neutralizing stomach acid with an antacid tablet
e. formation of snowflakes
f. an exploding dynamite stick

5.2 Classify each of the following changes as chemical or physical:
a. fogging the mirror during a shower
b. tarnishing of a silver bracelet
c. breaking a bone
d. mending a broken bone
e. burning paper
f. picking up the leaves in the yard

5.2 Chemical Equations

Study Pack

Chem of Life Reactions and Equilibrium: Chemical Equations 1, 2

Learning Goal

State a chemical equation in words and determine the number of atoms in the reactants and products.

Whenever you put together a bicycle, build a model airplane, prepare a new recipe, mix a medication, or clean a patient's teeth, you follow a set of directions. These directions tell you what materials to use, and the products you will obtain. In chemistry, a **chemical equation** tells us the materials we need and the products that will form in a chemical reaction.

Writing an Everyday Equation

When we breathe, digest our food, or cook, atoms in the starting materials are rearranged to form new substances. Suppose we decide to make some pancakes. The ingredients (reactants) in the recipe are milk, eggs, and pancake mix. Writing the recipe as an equation shows the ingredients to the left of an arrow—the arrow means "yield" or "react to form." The product (pancakes) is written to the right of the arrow as follows.

Milk + eggs + pancake mix $\longrightarrow$ pancakes

However, the equation in this form does not tell us how much of each reactant to use. We need to include the quantities of the reactants and product to know how much of each reactant to measure out and how many pancakes the recipe makes. Figure 5.2 illustrates the reactants and the products of our cooking experiment.

1 cup milk + 3 eggs + 2 cups pancake mix $\longrightarrow$ 16 pancakes

Writing a Chemical Equation

When sulfur is burned in oxygen, the product is sulfur dioxide. To write this as a chemical equation, the formulas of the **reactants** are written to the left of the arrow, and the formulas of the **products** are written to the right. When there are two or more formulas on the same side, they are separated by plus (+) signs.

Reactants	$\longrightarrow$	**Products**
$S(s) + O_2(g)$	$\longrightarrow$	$SO_2(g)$

Sometimes, as in this case, the formulas in an equation may include letters in parentheses, which give the physical state of the substances as solid (*s*), liquid (*l*), or gas (*g*). If a substance is dissolved in water, it is an aqueous (*aq*) solution. Table 5.2 summarizes some of the symbols used in equations.

When a reaction takes place, the bonds between the atoms of the reactants are broken and new bonds are formed to give the products. Atoms cannot be gained, lost, or changed into other types of atoms during a reaction. Therefore, a reaction must be

Table 5.2 Some Symbols Used in Writing Equations

Example equation:

$$C(s) + O_2(g) \xrightarrow{\Delta} CO_2(g)$$

Symbol	Meaning
+	Separates two or more formulas
$\longrightarrow$	Reacts to form products
Δ	The reactants are heated
(*s*)	Solid
(*l*)	Liquid
(*g*)	Gas
(*aq*)	Aqueous

Figure 5.2 In preparing pancakes, the proper amounts of milk, eggs, and pancake mix are mixed together. After the ingredients undergo a reaction, the pancakes are ready to eat.

written as a **balanced equation,** which shows the same number of atoms for each element on both sides of the arrow. For example, we can write an equation to show that carbon burns in oxygen gas and produces carbon dioxide as follows.

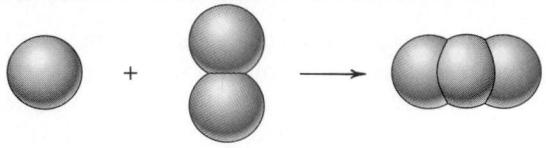

$$C(s) + O_2(g) \longrightarrow CO_2(g)$$

5.2

Reactants	**Product**
1 carbon atom	1 carbon atom
2 oxygen atoms	2 oxygen atoms

We say that the equation is *balanced* because there is one carbon atom and two oxygen atoms on each side of the arrow.

Now consider the reaction in which hydrogen reacts with oxygen to form water. First we write the formulas of the reactants and products:

$$H_2 + O_2 \longrightarrow H_2O$$

Counting the atoms, we see that there are two oxygen atoms on the reactant side, and only one oxygen atom on the product side. Because the equation does not have an equal number of atoms on both sides, the equation is *not balanced.* In the balanced equation for this reaction, a coefficient of 2 is written in front of the H_2O formula. Now the product has four hydrogen atoms, which means that a coefficient of 2 must be written in front of the formula H_2 in the reactants.

$$2H_2 + O_2 \longrightarrow 2H_2O$$

Now the number of hydrogen atoms and oxygen atoms are the same in the reactants as in the products. The equation is *balanced.* The small, whole numbers we placed in front of the formulas are called **coefficients.**

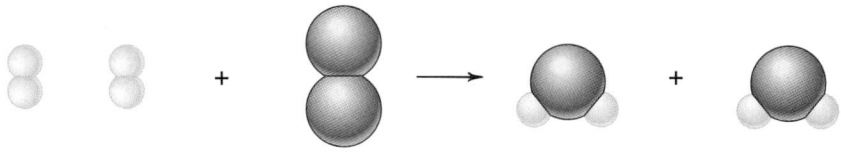

$$2H_2 + O_2 \longrightarrow 2H_2O$$

Reactants	**Products**
4 H atoms	4 H atoms
2 O atoms	2 O atoms

S·A·M·P·L·E· 5.2 ·P·R·O·B·L·E·M

Chemical Equations

Hydrogen and nitrogen react to form ammonia, NH_3.

$$3H_2 + N_2 \longrightarrow \underset{\text{Ammonia}}{2NH_3}$$

a. What are the coefficients in the equation?

b. How many atoms of each element are in the reactants and products of the equation?

Solution

a. The coefficients are three (3) in front of H_2, one (1) in front of N_2, which is understood, and two (2) in front of NH_3.

b. In the reactants there are 6 hydrogen atoms and 2 nitrogen atoms. There are also 6 hydrogen atoms and 2 nitrogen atoms in the product.

Study Check

When ethane (C_2H_6) burns in oxygen, the products are carbon dioxide and water. The balanced equation is as follows.

$$2C_2H_6 + 7O_2 \longrightarrow 4CO_2 + 6H_2O$$

State the total number of atoms of each element in the balanced equation.

QUESTIONS & PROBLEMS

Chemical Equations

5.3 State the number of atoms of each element on the reactant and on the product sides of the following equations:

a. $2NO + O_2 \longrightarrow 2NO_2$
b. $5C + 2SO_2 \longrightarrow CS_2 + 4CO$
c. $2C_2H_2 + 5O_2 \longrightarrow 4CO_2 + 2H_2O$
d. $N_2H_4 + 2H_2O_2 \longrightarrow N_2 + 4H_2O$

5.4 State the number of atoms of each element on the reactant and on the product sides of the following equations:

a. $CH_4 + 2O_2 \longrightarrow CO_2 + 2H_2O$
b. $4P + 5O_2 \longrightarrow P_4O_{10}$
c. $4NH_3 + 6NO \longrightarrow 5N_2 + 6H_2O$
d. $6CO_2 + 6H_2O \longrightarrow \underset{\text{Glucose}}{C_6H_{12}O_6} + 6O_2$

5.5 Determine whether each of the following equations is balanced or not balanced:

a. $S + O_2 \longrightarrow SO_3$
b. $2Al + 3Cl_2 \longrightarrow 2AlCl_3$
c. $H_2 + O_2 \longrightarrow H_2O$
d. $C_3H_8 + 5O_2 \longrightarrow 3CO_2 + 4H_2O$

5.6 Determine whether each of the following equations are balanced or not balanced:

a. $PCl_3 + Cl_2 \longrightarrow PCl_5$
b. $CO + 2H_2 \longrightarrow CH_3OH$
c. $2KClO_3 \longrightarrow 2KCl + O_2$
d. $Mg + N_2 \longrightarrow Mg_3N_2$

5.7 All of the following are balanced equations. State the number of atoms of each element in the reactants and in the products.

a. $2Na + Cl_2 \longrightarrow 2NaCl$
b. $PCl_3 + 3H_2 \longrightarrow PH_3 + 3HCl$
c. $P_4O_{10} + 6H_2O \longrightarrow 4H_3PO_4$

5.8 All of the following are balanced equations. State the number of atoms of each element in the reactants and in the products.

a. $2N_2 + 3O_2 \longrightarrow 2N_2O_3$
b. $Al_2O_3 + 6HCl \longrightarrow 2AlCl_3 + 3H_2O$
c. $C_5H_{12} + 8O_2 \longrightarrow 5CO_2 + 6H_2O$

Study Pack

Chem of Life Reactions and Equilibrium: Chemical Equations 3, 4

Learning Goal

Write a balanced chemical equation from the formulas of the reactants and products for a reaction.

5.3

5.3 Balancing a Chemical Equation

We have seen that a chemical equation must be balanced. In many cases, we use a method of trial and error. However, Table 5.3 gives some steps that might be useful. We can use these steps to balance a reaction in which hydrogen and chlorine react to form hydrogen chloride.

Step 1 Write the formulas of the reactants on the left side of the arrow and the products on the right.

$$H_2 + Cl_2 \longrightarrow HCl$$

Step 2 When we compare the number of hydrogen atoms and chlorine atoms on the reactant side with the number of these atoms on the product side, we find that they are not balanced.

Reactants		Products
$H_2 + Cl_2$	$\longrightarrow$	HCl
2 H		1 H
2 Cl		1 Cl

Step 3 To balance the equation, the coefficient 2 must be written in front of the HCl formula. Remember that the subscripts in the formulas are never changed as you balance an equation.

$$H_2 + Cl_2 \longrightarrow 2HCl$$

Step 4 This time when we compare the atoms in the equation, we find they are equal. The equation is balanced with the lowest possible whole numbers as coefficients.

Reactants		Products	
$H_2 + Cl_2$	$\longrightarrow$	2HCl	Balanced equation
2 H		~~1 H~~ 2 H	
2 Cl		~~1 Cl~~ 2 Cl	

Figure 5.3 illustrates the unbalanced and the balanced relationship for this equation.

In another reaction, the gas methane, CH_4, reacts with oxygen to produce carbon dioxide and water.

Step 1 The formulas are written as follows.

$$CH_4 + O_2 \longrightarrow CO_2 + H_2O$$

Table 5.3 Steps for Balancing Chemical Equations

Step 1	Use the correct formulas for the reactants and products.
Step 2	Count the number of atoms of each element on the reactant side and the product side. Count a polyatomic ion as a group if unchanged. Determine if the equation is balanced.
Step 3	Begin to balance the equation, one element at a time, placing coefficients in front of the formulas to equalize atoms. A typical starting place is with a formula that has subscripts. Balance O_2 and other diatomic elements last. Do not try to adjust or add to the subscripts of any formula as you proceed.
Step 4	Recheck the atoms on the reactant side and on the product side to make sure the equation is correctly balanced. Sometimes, balancing one element will undo the balance of another. If that happens, repeat the process. Make sure that the coefficients are the lowest possible whole numbers. If you can divide all the coefficients by 2 or some other number, then reduce them all to the lowest proportion.

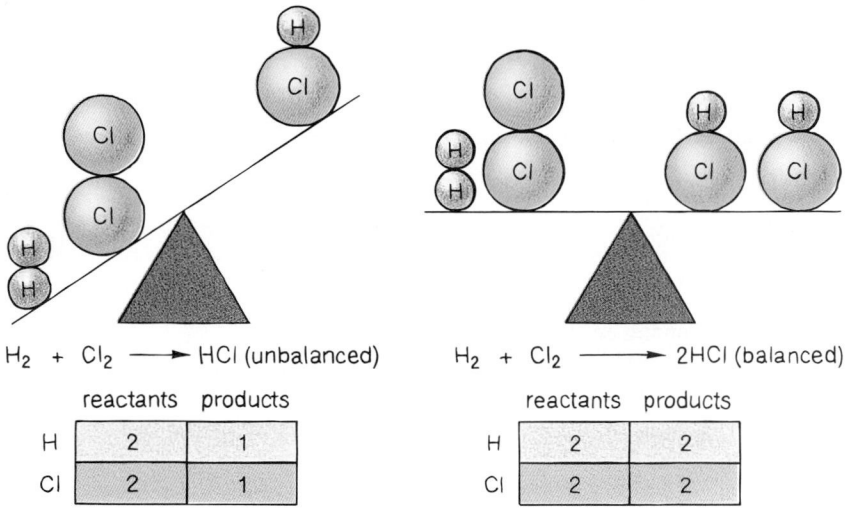

Figure 5.3 Unbalanced and balanced equations.

Step 2 When we compare the atoms on the reactant side with the atoms on the product side, we see that there are more hydrogen atoms on the left side and more oxygen atoms on the right.

Reactants	**Products**	
$CH_4 + O_2 \longrightarrow$	$CO_2 + H_2O$	
1 C atom	1 C atom	
4 H atoms	2 H atoms	Not balanced
2 O atoms	3 O atoms	Not balanced

Step 3 To balance the hydrogen, a coefficient of 2 is placed in front of the formula for water, H_2O.

$$CH_4 + O_2 \longrightarrow CO_2 + 2H_2O$$

The four oxygen atoms now in the products are balanced in the reactants by placing a 2 in front of the O_2 formula.

$$CH_4 + 2O_2 \longrightarrow CO_2 + 2H_2O$$

Step 4 Rechecking the atoms shows that the number of carbon, hydrogen, and oxygen atoms are now the same in both the reactants and the products. The equation is *balanced* using the lowest possible whole numbers as coefficients.

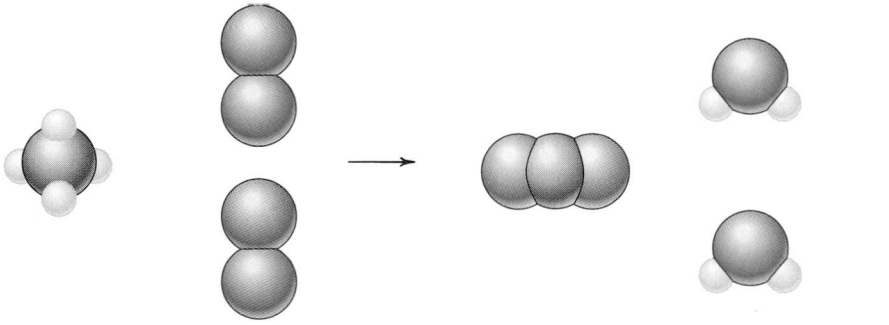

$CH_4 + 2O_2 \longrightarrow CO_2 + 2H_2O$ Balanced

Reactants	**Products**
1 C atom	1 C atom
4 H atoms	4 H atoms
4 O atoms	4 O atoms

Suppose you had written the equation as follows.

$$2CH_4 + 4O_2 \longrightarrow 2CO_2 + 4H_2O \quad \text{Incorrect}$$

Although there are equal numbers of atoms on both sides of the equation, this is not written correctly. All the coefficients must be divided by 2 to obtain the lowest possible whole numbers.

S·A·M·P·L·E 5.3 P·R·O·B·L·E·M

Balancing Equations

5.3

Balance the following equations:

a. $Al + Cl_2 \longrightarrow AlCl_3$

b. $Na_3PO_4 + MgCl_2 \longrightarrow Mg_3(PO_4)_2 + NaCl$

Solution

a. Step 1 The correct formulas are written in the equation.

$$Al + Cl_2 \longrightarrow AlCl_3$$

Step 2 When we compare the number of atoms on the reactant and product sides, we find that the chlorine atoms are not balanced.

Reactants		Product	
$Al + Cl_2$	$\longrightarrow$	$AlCl_3$	
1 Al		1 Al	
2 Cl		3 Cl	Not balanced

Step 3 In the formulas with chlorine, there is an odd-even relationship. A 2 placed in front of $AlCl_3$ will give an even number of 6 Cl atoms. Any number we use to multiply the Cl_2 will double the number of Cl atoms. To balance the 6 Cl in the product, we place a 3 in front of Cl_2. Finally, we must rebalance the 2 Al atoms that are now in the product by placing a 2 in front of the Al on the left side.

Step 4 When we recheck the balance of atoms, we see that the equation is balanced.

Reactants		Product	
$2Al + 3Cl_2$	$\longrightarrow$	$2AlCl_3$	Balanced
~~1 Al~~ 2 Al	=	~~1 Al~~ 2 Al	
~~2 Cl~~ 6 Cl	=	~~3 Cl~~ 6 Cl	

b. Step 1 In the equation, the correct formulas are written.

$$Na_3PO_4 + MgCl_2 \longrightarrow Mg_3(PO_4)_2 + NaCl$$

Step 2 When we compare the number of ions on the reactant and product sides, we find that the equation is not balanced. In this equation, it is more convenient to balance the phosphate polyatomic ion as a group instead of its individual atoms.

Reactants		Products	
$Na_3PO_4 + MgCl_2$	$\longrightarrow$	$Mg_3(PO_4)_2 + NaCl$	
3 Na^+		1 Na^+	Not balanced
1 PO_4^{3-}		2 PO_4^{3-}	Not balanced
1 Mg^{2+}		3 Mg^{2+}	Not balanced
2 Cl^-		1 Cl^-	Not balanced

Step 3 We begin with the formula of $Mg_3(PO_4)_2$ because there are several subscripts. A 3 in front of $MgCl_2$ will balance magnesium and a 2 in front of Na_3PO_4 will balance the phosphate ion. Looking again at each of the ions in the reactants and products, we see that the sodium and chloride ions are not yet equal. A 6 is needed in front of the NaCl to complete the balance.

Reactants		**Products**	
$2Na_3PO_4 + 3MgCl_2$	$\longrightarrow$	$Mg_3(PO_4)_2 + 6NaCl$	Balanced
$6\ Na^+$	=	$6\ Na^+$	
$2\ PO_4^{3-}$	=	$2\ PO_4^{3-}$	
$3\ Mg^{2+}$	=	$3\ Mg^{2+}$	
$6\ Cl^-$	=	$6\ Cl^-$	

5.3

Study Check

Balance the following equations:

a. $Fe + O_2 \longrightarrow Fe_3O_4$

b. $C_3H_8 + O_2 \longrightarrow CO_2 + H_2O$

Q·U·E·S·T·I·O·N·S & P·R·O·B·L·E·M·S

Balancing a Chemical Equation

5.9 Balance the following equations:

a. $N_2 + O_2 \longrightarrow NO$

b. $HgO \longrightarrow Hg + O_2$

c. $Fe + O_2 \longrightarrow Fe_2O_3$

d. $Na + Cl_2 \longrightarrow NaCl$

e. $Cu_2O + O_2 \longrightarrow CuO$

5.10 Balance the following equations:

a. $Al + Cl_2 \longrightarrow AlCl_3$

b. $P_4 + O_2 \longrightarrow P_4O_{10}$

c. $C_4H_8 + O_2 \longrightarrow CO_2 + H_2O$

d. $Sb_2S_3 + HCl \longrightarrow SbCl_3 + H_2S$

e. $Fe_2O_3 + C \longrightarrow Fe + CO$

5.11 Balance the following equations:

a. $Mg + AgNO_3 \longrightarrow Mg(NO_3)_2 + Ag$

b. $CuCO_3 \longrightarrow CuO + CO_2$

c. $Al + CuSO_4 \longrightarrow Cu + Al_2(SO_4)_3$

d. $Pb(NO_3)_2 + NaCl \longrightarrow PbCl_2 + NaNO_3$

e. $Al + HCl \longrightarrow AlCl_3 + H_2$

5.12 Balance the following equations:

a. $Zn + H_2SO_4 \longrightarrow ZnSO_4 + H_2$

b. $Al + H_2SO_4 \longrightarrow Al_2(SO_4)_3 + H_2$

c. $K_2SO_4 + BaCl_2 \longrightarrow BaSO_4 + KCl$

d. $CaCO_3 \longrightarrow CaO + CO_2$

e. $Al_2(SO_4)_3 + KOH \longrightarrow Al(OH)_3 + K_2SO_4$

Study Pack

Chem of Life Reactions and Equilibrium: Chemical Reactions 1, 2

Learning Goal

Identify a reaction as a combination, decomposition, replacement, or combustion reaction.

5.4

5.4 Types of Reactions

A great number of reactions occur in nature, in biological systems, and in the laboratory. However, there are some general patterns among all reactions that help us classify reactions. Some reactions may fit into two reaction types.

Combination Reactions

In a **combination reaction,** two or more elements or simple compounds bond to form one product. For example, sulfur and oxygen combine to form the product sulfur dioxide.

Combination reaction

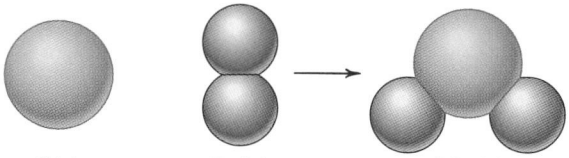

$$S(s) + O_2(g) \longrightarrow SO_2(g)$$

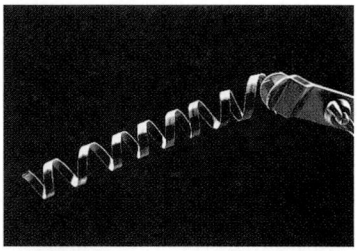

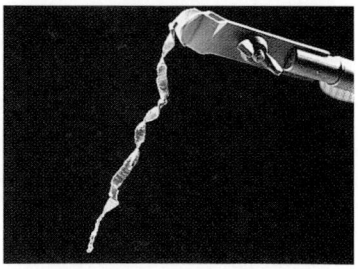

Figure 5.4 This combination reaction is: $2Mg(s) + O_2(g) \xrightarrow{\Delta} 2MgO(s)$

In Figure 5.4, the elements magnesium and oxygen combine to form a single product, magnesium oxide.

$$2Mg(s) + O_2(g) \longrightarrow 2MgO(s)$$

In other examples of combination reactions, elements or simple compounds combine to form a single product.

$$N_2(g) + 3H_2(g) \longrightarrow 2NH_3(g) \quad \text{Ammonia}$$

$$Cu(s) + S(s) \longrightarrow CuS(s)$$

$$MgO(s) + CO_2(g) \longrightarrow MgCO_3(s)$$

Decomposition Reactions

In a **decomposition reaction,** a reactant splits into two or more simpler products. For example, when mercury(II) oxide is heated, the products are mercury and oxygen.

$$2HgO(s) \longrightarrow 2Hg(l) + O_2(g)$$

In another example of a decomposition reaction, calcium carbonate breaks apart into simpler compounds of calcium oxide and carbon dioxide.

$$CaCO_3(s) \longrightarrow CaO(s) + CO_2(g)$$

Replacement Reactions

In replacement reactions, elements in compounds are replaced by other elements. In a **single replacement reaction,** a reacting element switches place with an element in the reacting compound. For example, in the reaction shown in Figure 5.5, zinc replaces copper in copper(II) sulfate.

$$Zn(s) + CuSO_4(aq) \longrightarrow ZnSO_4(aq) + Cu(s)$$

In another single replacement reaction shown in Figure 5.6, zinc replaces hydrogen in hydrochloric acid, $HCl(aq)$.

$$Zn(s) + 2HCl(aq) \longrightarrow ZnCl_2(aq) + H_2(g)$$

In the following single replacement reaction, the halogen (Group 7) chlorine replaces bromine in the compound potassium bromide.

$$Cl_2(g) + 2KBr(s) \longrightarrow 2KCl(s) + Br_2(l)$$

Figure 5.5 The equation for this single replacement reaction is: $Zn(s) + CuSO_4(aq) \longrightarrow Cu(s) + ZnSO_4(aq)$

Figure 5.6 This single replacement reaction is written: $Zn(s) + 2HCl(aq) \longrightarrow ZnCl_2(aq) + H_2(g)$

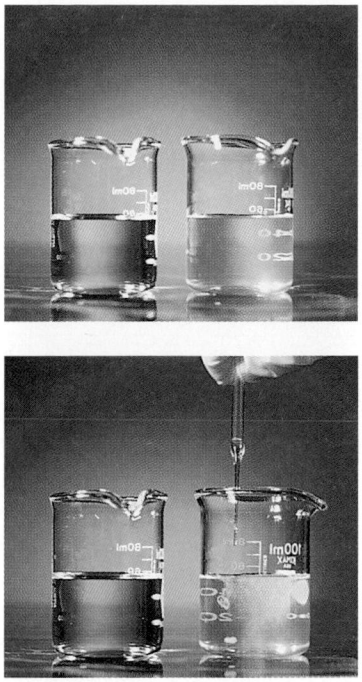

Figure 5.7 The equation for this double replacement reaction is: $Na_2SO_4(aq) + BaCl_2(aq) \longrightarrow BaSO_4(s) + 2NaCl(aq)$

In a **double replacement reaction,** the metal ions in the reacting compounds switch places. For example, in the reaction shown in Figure 5.7, barium ions change places with sodium ions in the reactants to form sodium chloride and a white solid precipitate of barium sulfate.

$$BaCl_2(aq) + Na_2SO_4(aq) \longrightarrow BaSO_4(s) + 2NaCl(aq)$$

When sodium hydroxide and hydrochloric acid (HCl) react, sodium and hydrogen switch places, forming sodium chloride and water.

$$NaOH(aq) + HCl(aq) \longrightarrow NaCl(aq) + HOH(l)$$

Combustion Reactions

The burning of a log in a fireplace and the burning of fuel in the engine of a car are examples of combustion reactions. In a **combustion reaction,** fuel and oxygen are required and often the reaction produces carbon dioxide, water, and heat energy. For example, methane gas (CH_4) reacts with oxygen to produce carbon dioxide and water. The heat energy produced by this combustion reaction cooks our food and heats our homes. (See Figure 5.8.)

$$CH_4(g) + 2O_2(g) \longrightarrow CO_2(g) + 2H_2O + \text{heat energy}$$

Figure 5.8 In a gas burner, methane (CH_4) reacts with oxygen to produce carbon dioxide (CO_2) and water (H_2O).

Combustion reactions also occur in the cells of the body in order to metabolize food, which provides building materials and energy for the activities we want to do. We absorb oxygen (O_2) from the air to burn glucose ($C_6H_{12}O_6$) from our food, and eventually our cells produce CO_2, H_2O, and energy.

$$C_6H_{12}O_6 + 6O_2 \longrightarrow 6CO_2 + 6H_2O + \text{energy}$$

When a combustion reaction adds oxygen to form a single product, it can also be categorized as a combination reaction. For example, in the following reactions, sulfur and aluminum are burned in oxygen, producing oxide products.

$$S + O_2 \longrightarrow SO_2$$

$$2Al + 3O_2 \longrightarrow 2Al_2O_3$$

Table 5.4 summarizes the reaction types and gives examples.

HEALTH NOTE

5.4

Smog and Health Concerns

There are two types of smog. One, photochemical smog, requires sunlight to initiate reactions that produce pollutants such as nitrogen oxides and ozone. The other type of smog, industrial or London smog, occurs in areas where coal containing sulfur is burned and the unwanted product sulfur dioxide is emitted.

Photochemical smog is most prevalent in cities where people are dependent on cars for transportation. On a typical day in Los Angeles, for example, nitrogen monoxide (NO) emissions from car exhausts increase as traffic increases on the roads. The nitrogen monoxide is formed when N_2 and O_2 react at high temperatures in car and truck engines.

$$N_2 + O_2 \xrightarrow{\text{Heat}} 2NO$$

Then, NO reacts with oxygen in the air to produce NO_2, a reddish brown gas that is irritating to the eyes and damaging to the respiratory tract. See Figure 5.9.

$$2NO + O_2 \longrightarrow 2NO_2$$

When NO_2 is exposed to sunlight, it is converted into NO and oxygen atoms.

$$NO_2 \xrightarrow{\text{Sunlight}} NO + \underset{\text{Oxygen atoms}}{O}$$

Oxygen atoms are so reactive that they combine with oxygen molecules in the atmosphere, forming ozone.

$$O + O_2 \longrightarrow \underset{\text{Ozone}}{O_3}$$

In the upper atmosphere (the stratosphere), ozone is beneficial because it protects us from harmful ultraviolet radiation that comes from the sun. However, in the lower atmosphere, ozone irritates the eyes and respiratory tract, where it causes coughing, decreased lung function, and fatigue. It also causes deterioration of fabrics, cracks rubber, and damages trees and crops.

Industrial smog is prevalent in areas where coal with a high sulfur content is burned to produce electricity. During combustion, the sulfur is converted to sulfur dioxide:

$$S + O_2 \longrightarrow SO_2$$

The SO_2 is damaging to plants, suppressing growth, and it is corrosive to metals such as steel. SO_2 is also damaging to humans and can cause lung impairment and respiratory difficulties. The SO_2 in the air reacts with more oxygen to form SO_3. SO_3 combines with water in the air to form sulfuric acid, which makes the rain acidic.

$$2SO_2 + O_2 \longrightarrow 2SO_3$$

$$SO_3 + H_2O \longrightarrow \underset{\text{Sulfuric acid}}{H_2SO_4}$$

The presence of sulfuric acid in rivers and lakes causes an increase in the acidity of the water, reducing the ability of animals and plants to survive.

Figure 5.9 Nitrogen dioxide, a reddish brown gas that irritates the respiratory tract, is one of the pollutants in photochemical smog.

Table 5.4 Summary of Reaction Types

Combination	
$A + B \longrightarrow AB$	$Ca(s) + Cl_2(g) \longrightarrow CaCl_2(s)$
Decomposition	
$AB \longrightarrow A + B$	$Fe_2S_3(s) \longrightarrow 2Fe(s) + 3S(s)$
Single Replacement	
$A + BC \longrightarrow AC + B$	$Cu(s) + 2AgNO_3(aq) \longrightarrow Cu(NO_3)_2(aq) + 2Ag(s)$
Double Replacement	
$AB + CD \longrightarrow AD + CB$	$BaCl_2(aq) + K_2SO_4(aq) \longrightarrow BaSO_4(s) + 2KCl(aq)$
Combustion	
$AH_4 + 2O_2 \longrightarrow AO_2 + 2H_2O$ + heat energy	$CH_4(g) + 2O_2(g) \longrightarrow CO_2(g) + 2H_2O(g)$ + heat energy

S·A·M·P·L·E 5.4 ·P·R·O·B·L·E·M

Classification of Reactions

Classify the following reactions as combination, decomposition, single replacement, double replacement, or combustion.

a. $2Fe_2O_3 + 3C \longrightarrow 3CO_2 + 4Fe$

b. $2Na + Cl_2 \longrightarrow 2NaCl$

c. $Fe_2S_3 \longrightarrow 2Fe + 3S$

d. $C_2H_4 + 3O_2 \longrightarrow 2CO_2 + 2H_2O$

e. $BaCl_2 + K_2SO_4 \longrightarrow BaSO_4 + 2KCl$

Solution

a. single replacement **b.** combination **c.** decomposition
d. combustion **e.** double replacement

Study Check

Nitrogen monoxide and oxygen react to form nitrogen dioxide. Write the balanced equation and identify the reaction type(s).

Q·U·E·S·T·I·O·N·S & P·R·O·B·L·E·M·S

Types of Reactions

5.13 a. Why is the following reaction called a decomposition reaction?

$$2Al_2O_3 \longrightarrow 4Al + 3O_2$$

b. Why is the following reaction called a single replacement reaction?

$$Br_2 + BaI_2 \longrightarrow BaBr_2 + I_2$$

5.14 a. Why is the following reaction called a combination reaction?

$$H_2 + Br_2 \longrightarrow 2HBr$$

b. Why is the following reaction called a double replacement reaction?

$$AgNO_3 + NaCl \longrightarrow AgCl + NaNO_3$$

5.15 Classify each of the following reactions as a combination, decomposition, single replacement, double replacement, or combustion reaction:

a. $4Fe + 3O_2 \longrightarrow 2Fe_2O_3$
b. $Mg + 2AgNO_3 \longrightarrow Mg(NO_3)_2 + 2Ag$
c. $CuCO_3 \longrightarrow CuO + CO_2$
d. $NaOH + HCl \longrightarrow NaCl + H_2O$
e. $C_4H_8 + 6O_2 \longrightarrow 4CO_2 + 4H_2O$
f. $ZnCO_3 \longrightarrow CO_2 + ZnO$
g. $Al_2(SO_4)_3 + 6KOH \longrightarrow 2Al(OH)_3 + 3K_2SO_4$
h. $Pb + O_2 \longrightarrow PbO_2$

5.5

5.16 Classify each of the following reactions as a combination, decomposition, single replacement, double replacement, or combustion reaction:

a. $CuO + 2HCl \longrightarrow CuCl_2 + H_2O$
b. $2Al + 3Br_2 \longrightarrow 2AlBr_3$
c. $2C_2H_2 + 5O_2 \longrightarrow 4CO_2 + 2H_2O$
d. $Pb(NO_3)_2 + 2NaCl \longrightarrow PbCl_2 + 2NaNO_3$
e. $2Mg + O_2 \longrightarrow 2MgO$
f. $Fe_2O_3 + 3C \longrightarrow 2Fe + 3CO$
g. $C_6H_{12}O_6 \longrightarrow 2C_2H_6O + 2CO_2$
h. $BaCl_2(aq) + K_2CO_3(aq) \longrightarrow BaCO_3(s) + 2KCl(aq)$

5.17 Try your hand at predicting the products that would result from the following types of reactions. Balance each equation you write:

a. combination: $Mg + Cl_2 \longrightarrow$
b. decomposition: $HBr \longrightarrow$
c. single replacement: $Mg + Zn(NO_3)_2 \longrightarrow$
d. double replacement: $K_2S + Pb(NO_3)_2 \longrightarrow$
e. combustion: $C_2H_6 + O_2 \longrightarrow$

5.18 Try your hand at predicting the products that would result from the reactions of the following:

a. combination: $Ca + S \longrightarrow$
b. decomposition: $PbO_2 \longrightarrow$
c. single replacement: $KI + Cl_2 \longrightarrow$
d. double replacement: $CuCl_2 + Na_2S \longrightarrow$
e. combustion: $C_2H_4 + O_2 \longrightarrow$

Study Pack

5.5 A Closer Look: Oxidation-Reduction Reactions

Learning Goal

Define the terms oxidation and reduction.

In the previous section, there were chemical changes that can also be described as oxidation and reduction reactions. The burning of methane gas to heat our homes, formation of rust on a nail, production of glucose in photosynthesis, or the extraction of energy from our food are reactions that involve oxidation and reduction. Oxidation must always occur together with a reduction.

Loss and Gain of Electrons

Oxidation-reduction reactions are defined as a loss and a gain of electrons among the reactants. In an oxidation reaction, one element loses electrons. Simultaneously, a reduction reaction occurs as another element gains electrons. You may like to use the shorthand phrase "LEO the lion goes GER" to remind yourself that a loss of electrons is oxidation (LEO), and a gain of electrons is a reduction (GER).

When we looked at the formation of ionic compounds, we also saw that metals lost electrons and nonmetals gained electrons. Now we can say that metals are oxidized and nonmetals are reduced.

Consider the elements and ions in the reaction where calcium loses electrons to form calcium ion and where sulfur gains electrons to form sulfide ion. The elemental forms of calcium and sulfur have no ionic charge; they are neutral.

$$Ca + S \longrightarrow Ca^{2+} + S^{2-} = CaS$$

If we write the formation of each ion of the compound, we find that there is an oxidation and a reduction. In every oxidation-reduction reaction, the number of electrons exchanged among the reactants must be equal.

$$Ca \longrightarrow Ca^{2+} + 2e^- \quad \text{Oxidation (LEO)}$$

$$S + 2e^- \longrightarrow S^{2-} \quad \text{Reduction (GER)}$$

The loss of electrons is an **oxidation.** We say that calcium was oxidized. At the same time, the sulfur atom gained electrons. This gain of electrons is called **reduction.** We say that the sulfur was reduced. The two parts must occur simultaneously and with the same total number of electrons. We can write the two parts as follows.

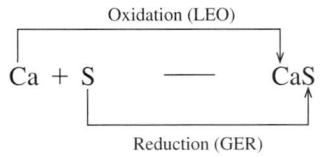

If we look at a single replacement reaction, we also find that it has two parts: an oxidation and a reduction. Consider the reaction of zinc and copper(II) sulfate.

$$Zn + CuSO_4 \longrightarrow ZnSO_4 + Cu$$

We can rewrite the equation to show the ions of the compounds.

$$Zn + Cu^{2+} + SO_4^{2-} \longrightarrow Zn^{2+} + SO_4^{2-} + Cu$$

Now we can show that zinc loses 2 electrons to form Zn^{2+}, and Cu^{2+} gains 2 electrons to form Cu.

$$Zn \longrightarrow Zn^{2+} + 2e^- \quad \text{Oxidation (LEO)}$$

$$Cu^{2+} + 2e^- \longrightarrow Cu \quad \text{Reduction (GER)}$$

In this single replacement reaction, zinc was oxidized and copper(II) ion was reduced.

S·A·M·P·L·E· 5.5 ·P·R·O·B·L·E·M

Oxidation-Reduction Reactions

In photographic film, the following decomposition reaction occurs in the presence of light. What is oxidized and what is reduced?

$$2AgBr \xrightarrow{\text{Light}} 2Ag + Br_2$$

Solution

In the AgBr compound, silver ion has an ionic charge of 1+ and the bromide ion has an ionic charge of 1−. We can write the reaction as follows.

$$2Ag^+ + 2Br^- \longrightarrow 2Ag + Br_2$$

We can determine the oxidation reaction and the reduction reaction by writing the changes for each ion separately.

$$2Br^- \longrightarrow Br_2 + 2e^- \quad \text{Oxidation (LEO)}$$

$$2Ag^+ + 2e^- \longrightarrow 2Ag \quad \text{Reduction (GER)}$$

In the reaction, bromide ion is oxidized, and silver ion is reduced.

Study Check

5.5

In the following combination reaction, what is oxidized and what is reduced?

$$2Li + F_2 \longrightarrow 2LiF$$

EXPLORE YOUR WORLD

Oxidation of Fruits and Vegetables

Freshly cut surfaces of fruits and vegetables discolor when exposed to oxygen in the air. Cut three slices of a fruit or vegetable such as apple, potato, avocado, or banana. Leave one piece on the kitchen counter (uncovered). Wrap one piece in plastic wrap and leave on the kitchen counter. Dip one piece in lemon juice (vitamin C, also called ascorbic acid, an antioxidant) and leave uncovered.

Questions

1. What changes take place in each sample after 1–2 hours?
2. Why would wrapping fruits and vegetables slow the rate of discoloration?
3. If lemon juice contains vitamin C (an antioxidant), why would dipping a fruit or vegetable in lemon juice affect the oxidation reaction on the surface of the fruit or vegetable?
4. Other kinds of antioxidants are vitamin E, citric acid, and BHT. Look for these antioxidants on the labels of cereals, potato chips, and other foods in your kitchen. Why are antioxidants added to food products that will be stored on our kitchen shelves?

Gain and Loss of Oxygen

Sometimes an oxidation-reduction is described in terms of an exchange of oxygen, or an exchange of hydrogen in a reaction.

An early definition described oxidation as a reaction that added oxygen to an element or compound. When you see rust forming on a car or a nail, the iron is being oxidized. Aluminum and sulfur are also oxidized when they add oxygen.

$$4Fe + 3O_2 \longrightarrow 2Fe_2O_3 \quad \text{Oxidation: add O}$$

$$2Al + 3O_2 \longrightarrow 2\,Al_2O_3 \quad \text{Oxidation: add O}$$

$$S + O_2 \longrightarrow SO_2 \quad \text{Oxidation: add O}$$

A reduction would be a reaction in which a compound loses oxygen.

$$2KClO_3 \longrightarrow 2KCl + 3O_2 \quad \text{Reduction: lose O}$$

Loss and Gain of Hydrogen in Biological Systems

In organic and biochemical systems, the oxidation of carbon compounds often involves the loss of hydrogen atoms. Therefore, in a reduction reaction, there is a gain of hydrogen atoms. For example, methyl alcohol (CH_3OH), a poisonous substance, is metabolized in the body by the following reactions.

$$\underset{\text{Methyl alcohol}}{CH_3OH} \longrightarrow \underset{\text{Formaldehyde}}{H_2CO} + 2H \quad \text{Oxidation: loss of H atoms}$$

The formaldehyde can be oxidized further, this time by the addition of oxygen, to produce formic acid.

$$\underset{\text{Formaldehyde}}{2H_2CO} + O_2 \longrightarrow \underset{\text{Formic acid}}{2H_2CO_2} \quad \text{Oxidation: add O}$$

Finally, formic acid is oxidized to carbon dioxide and water.

$$\underset{\text{Formic acid}}{2H_2CO_2} + O_2 \longrightarrow 2CO_2 + 2H_2O \quad \text{Oxidation: add O}$$

The intermediate products of the oxidation of methyl alcohol are quite toxic, causing blindness and possibly death as they interfere with key reactions in the cells of the body.

In summary, we find that the particular definition of oxidation and reduction we use depends on the process that occurs in the reaction. All of these definitions are summarized in Table 5.5. Oxidation always involves a loss of electrons, but it may also be seen as an addition of oxygen, or the loss of hydrogen atoms. A reduction always involves a gain of electrons, and may also be seen as the loss of oxygen, or the gain of hydrogen.

Table 5.5 Characteristics of Oxidation and Reduction

Always Involves	May Involve
Oxidation	
Loss of electrons (LEO)	Addition of oxygen Loss of hydrogen
Reduction	
Gain of electrons (GER)	Loss of oxygen Gain of hydrogen

SAMPLE 5.6 PROBLEM

Identification of Oxidation-Reduction Reactions

Determine if each of the following is an oxidation-reduction reaction.

a. $Zn + Cl_2 \longrightarrow ZnCl_2$

b. $C_2H_4 + 3O_2 \longrightarrow 2CO_2 + 2H_2O$

c. $BaCl_2 + Na_2CO_3 \longrightarrow BaCO_3 + 2NaCl$

Solution

a. Yes. Zinc loses electrons; it is oxidized. $Zn \longrightarrow Zn^{2+} + 2e^-$
Chlorine gains electrons; it is reduced. $Cl_2 + 2e^- \longrightarrow 2Cl^-$

b. Yes. The addition of oxygen indicates that an oxidation-reduction reaction has occurred.

c. No. There are no changes in ionic charges between reactants and products. For example, barium is Ba^{2+} in $BaCl_2$ and in $BaCO_3$. The sodium is Na^+ and chlorine is Cl^- in all the compounds. There is no transfer of electrons in this double replacement reaction.

Study Check

Indicate what is oxidized and what is reduced in the following equation:

$$2Fe + 3Br_2 \longrightarrow 2FeBr_3$$

Q·U·E·S·T·I·O·N·S & P·R·O·B·L·E·M·S

A Closer Look: Oxidation-Reduction Reactions

5.19 Indicate whether each of the following describes an oxidation or a reduction:
a. loss of hydrogen atoms **b.** gain of electrons
c. addition of oxygen atoms **d.** always accompanies oxidation

5.20 Indicate whether each of the following describes an oxidation or a reduction:
a. loss of electrons **b.** gain of hydrogen atoms
c. loss of oxygen atoms **d.** always accompanies reduction

5.6

5.21 In the following reactions, identify what is oxidized, and what is reduced.
a. $Zn + Cl_2 \longrightarrow ZnCl_2$ **b.** $Cl_2 + 2NaBr \longrightarrow 2NaCl + Br_2$
c. $2PbO \longrightarrow 2Pb + O_2$ **d.** $2Fe^{3+} + Sn^{2+} \longrightarrow 2Fe^{2+} + Sn^{4+}$

5.22 In the following reactions, identify what is oxidized, and what is reduced.
a. $2Li + F_2 \longrightarrow 2LiF$ **b.** $Cl_2 + 2KI \longrightarrow 2KCl + I_2$
c. $Zn + Cu^{2+} \longrightarrow Zn^{2+} + Cu$
d. $Fe + CuSO_4 \longrightarrow FeSO_4 + Cu$

5.23 In the mitochondria of human cells, energy for the production of ATP is provided by the oxidation and reduction reactions of the iron ions in the cytochromes of the electron transport chain. Identify each of the following reactions as oxidation or reduction.
a. $Fe^{3+} + e^- \longrightarrow Fe^{2+}$ **b.** $Fe^{2+} \longrightarrow Fe^{3+} + e^-$

5.24 Chlorine (Cl_2) is a strong germicide used to disinfect drinking water and to kill microbes in swimming pools. If the product is Cl^-, was the elemental chlorine oxidized or reduced?

5.25 When linoleic acid, an unsaturated fatty acid, reacts with hydrogen, it forms a saturated fatty acid.

$$C_{18}H_{32}O_2 + 2H_2 \longrightarrow C_{18}H_{36}O_2$$

Is linoleic acid oxidized or reduced in the hydrogenation reaction?

5.26 In one of the reactions in the citric acid cycle, which provides energy for ATP synthesis, succinic acid is converted to fumaric acid.

$$\underset{\text{Succinic acid}}{C_4H_6O_4} \longrightarrow \underset{\text{Fumaric acid}}{C_4H_4O_4} + 2H$$

The reaction is accompanied by a coenzyme, flavin adenine dinucleotide (FAD).

$$FAD + 2H \longrightarrow FADH_2$$

a. Is succinic acid oxidized or reduced?
b. Is FAD oxidized or reduced?
c. Why would the two reactions occur together?

Reactions and Equilibrium: Stoichiometry 1, 2

Learning Goal

Given the chemical formula of a substance, calculate its molar mass.

5.6 The Mole

At the store, you buy eggs by the dozen. In an office, pencils are ordered by the gross and paper by the ream. In a restaurant, soda is ordered by the case. In each of these examples, the terms *dozen, gross, ream,* and *case* count the number of items present. For example, when you buy a dozen eggs, you know you will get 12 eggs in the carton.

Table 5.6 Number of Items in Some Typical Collections

Collection	Number of Items
1 trio of singers	3 singers
1 dozen eggs	12 eggs
1 case of soda	24 cans of soda
1 gross of pencils	144 pencils
1 ream of paper	500 sheets of paper
1 mole of atoms	6.02×10^{23} atoms
1 mole of molecules	6.02×10^{23} molecules

(See Table 5.6 and Figure 5.10.) In chemistry, particles such as atoms and molecules are counted by the **mole,** a unit that contains 6.02×10^{23} items. This very large number, called **Avogadro's number** after Amedeo Avogadro, an eighteenth-century Italian physicist, looks like this when written with 3 significant figures:

Avogadro's Number

$$602\,000\,000\,000\,000\,000\,000\,000 = 6.02 \times 10^{23}$$

One mole of any element always has Avogadro's number of atoms. For example, 1 mole of carbon contains 6.02×10^{23} carbon atoms; 1 mole of aluminum contains 6.02×10^{23} aluminum atoms; 1 mole of sulfur contains 6.02×10^{23} sulfur atoms.

$$1 \text{ mole of an element} = 6.02 \times 10^{23} \text{ atoms of that element}$$

One mole of a compound also contains Avogadro's number of molecules or formula units. (Molecules are the particles of covalent compounds; formula units are the group of ions given by the formula of an ionic compound.) For example, 1 mole of CO_2, a covalent compound, contains 6.02×10^{23} molecules of CO_2. One mole of NaCl, an ionic compound, contains 6.02×10^{23} formula units of NaCl (Na^+, Cl^-). Table 5.7 gives examples of the number of particles in some 1-mole quantities.

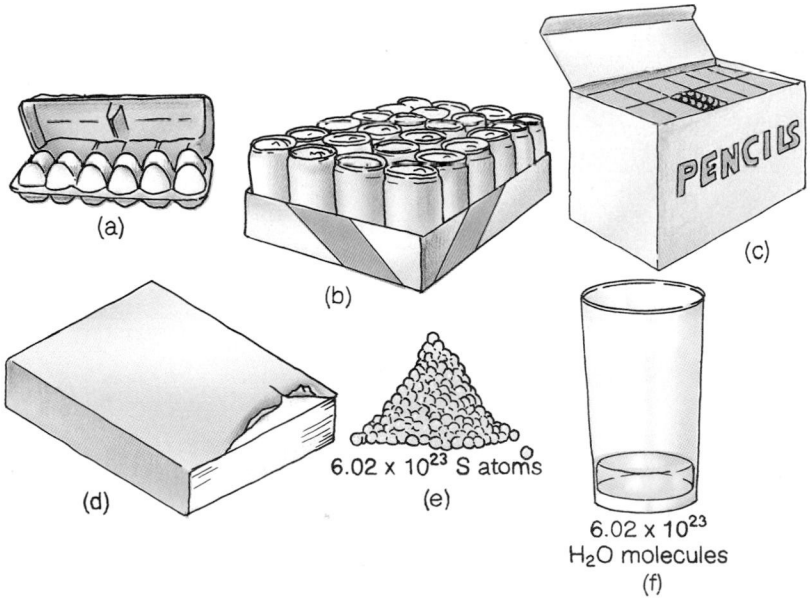

Figure 5.10 Some common collections and the number of items in them. **(a)** A dozen eggs is 12 eggs. **(b)** A case of soda contains 24 cans of soda. **(c)** A gross of pencils is 144 pencils. **(d)** A ream of paper contains 500 sheets of paper. **(e)** A mole of sulfur contains 6.02×10^{23} sulfur atoms. **(f)** A mole of water contains 6.02×10^{23} water molecules.

Table 5.7 Number of Particles in One-Mole Samples

Substance	Number and Type of Particles
1 mole carbon	6.02×10^{23} carbon atoms
1 mole aluminum	6.02×10^{23} aluminum atoms
1 mole sulfur	6.02×10^{23} sulfur atoms
1 mole water (H_2O)	6.02×10^{23} H_2O molecules
1 mole NaCl	6.02×10^{23} NaCl formula units
1 mole sucrose ($C_{12}H_{22}O_{11}$)	6.02×10^{23} sucrose molecules
1 mole vitamin C ($C_6H_8O_6$)	6.02×10^{23} vitamin C molecules

5.6

Moles of Elements in a Compound

In Chapter 4, we learned that the symbols and subscripts in a formula tell us the number of atoms of each type of element. For example, the formula for aspirin, $C_9H_8O_4$, indicates that there are nine carbon atoms, eight hydrogen atoms, and four oxygen atoms in each molecule. One mole of aspirin has the same number of moles of each element as the atoms in a molecule of aspirin: 9 moles of carbon (C), 8 moles of hydrogen (H), and 4 moles of oxygen (O).

In one molecule	C_9 Nine atoms of carbon	H_8 Eight atoms of hydrogen	O_4 Four atoms of oxygen
In one mole	C_9 Nine moles of carbon	H_8 Eight moles of hydrogen	O_4 Four moles of oxygen

S·A·M·P·L·E· 5.7 ·P·R·O·B·L·E·M

Counting the Moles of Elements in a Formula

State the number of moles for each element present in 1 mole of the following compounds:

a. SO_3 **b.** $Al(OH)_3$, antacid **c.** $C_8H_{10}N_4O_2$, caffeine

Solution

a. 1 mole of SO_3 = 1 mole of S and 3 moles of O

b. 1 mole of $Al(OH)_3$ = 1 mole of Al, 3 moles of O, and 3 moles of H
Be sure to multiply everything in the parentheses by the subscript 3.

c. 1 mole $C_8H_{10}N_4O_2$ = 8 moles of C, 10 moles of H, 4 moles of N, and 2 moles of O

Study Check

Vitamin C (ascorbic acid) has the formula $C_6H_8O_6$. How many moles of each element are in 1 mole of vitamin C?

Molar Mass

You will never weigh a single atom or molecule. They are much too small to weigh, even on the most accurate balance. Instead, you will work in a laboratory with samples that you can weigh in grams using a balance. If you need 1 mole of an element, you would weigh out the **molar mass,** which is the mass in grams equal to the numerical value of its atomic mass. For example, the atomic mass of carbon is 12.0. In this text, we will take the atomic mass of the elements on the periodic table to one decimal place (tenths). Therefore, 1 mole of carbon has a mass of 12.0 grams. In one molar mass (12.0 g) of carbon, there are 6.02×10^{23} carbon atoms.

$$1 \text{ atom C} = 12.0 \text{ amu} \quad \text{Atomic mass of carbon}$$

$$1 \text{ mole C} = 12.0 \text{ g C}$$

$$\text{Molar mass of C} = 12.0 \text{ g C}$$

To determine the molar mass of a compound, multiply the molar mass of each element by its subscript in the formula, and add the results. For example, the molar mass of sulfur trioxide, SO_3, is obtained by adding the molar masses of 1 mole of sulfur and 3 moles of oxygen.

Number of Moles in the Formula		Molar Mass		Total Mass for Each
1 mole S	×	$\frac{32.1 \text{ g S}}{1 \text{ mole S}}$	=	32.1 g S
3 moles O	×	$\frac{16.0 \text{ g O}}{1 \text{ mole O}}$	=	48.0 g O
		Molar mass of SO_3	=	80.1 g

Figure 5.11 shows some 1-mole quantities of substances. Table 5.8 lists the molar mass for several 1-mole samples.

Table 5.8 The Molar Mass and Number of Particles in One-Mole Quantities

Substance	Molar Mass	Number of Particles in One Mole
Carbon (C)	12.0 g	6.02×10^{23} C atoms
Sodium (Na)	23.0 g	6.02×10^{23} Na atoms
Iron (Fe)	55.9 g	6.02×10^{23} Fe atoms
NaF (preventative for dental caries)	42.0 g	6.02×10^{23} NaF formula units
$CaCO_3$ (antacid)	100.1 g	6.02×10^{23} $CaCO_3$ formula units
$C_6H_{12}O_6$ (glucose)	180.0 g	6.02×10^{23} glucose molecules
$C_8H_{10}N_4O_2$ (caffeine)	194.0 g	6.02×10^{23} caffeine molecules

a

b

Figure 5.11 One-mole samples of **(a)** elements and **(b)** compounds.
Top left: 12 g of carbon
Top right: 32.1 g of sulfur
Bottom left: 129.9 g of $CoCl_2$
Bottom right: 294.2 g of $K_2Cr_2O_7$

S·A·M·P·L·E 5.8 P·R·O·B·L·E·M

Calculating Molar Mass of Compounds

Find the molar mass of each compound.

a. $CaCO_3$, a compound used as an antacid and to replenish calcium in the diet

b. $Mg(NO_3)_2$ used in fireworks.

Solution

a. Number of Moles in the Formula		Molar Mass		Total Mass for Each Element
1 mole Ca	×	$\frac{40.1 \text{ g Ca}}{1 \text{ mole Ca}}$	=	40.1 g Ca
1 mole C	×	$\frac{12.0 \text{ g C}}{1 \text{ mole C}}$	=	12.0 g C
3 moles O	×	$\frac{16.0 \text{ g O}}{1 \text{ mole O}}$	=	48.0 g O
		Molar mass of $CaCO_3$	=	100.1 g

b. Number of Moles in the Formula		Molar Mass		Total Mass for Each Element
1 mole Mg	×	$\frac{24.3 \text{ g Mg}}{1 \text{ mole Mg}}$	=	24.3 g Mg
2 moles N	×	$\frac{14.0 \text{ g N}}{1 \text{ mole N}}$	=	28.0 g N
6 moles O	×	$\frac{16.0 \text{ g O}}{1 \text{ mole O}}$	=	96.0 g O
		Molar mass of $Mg(NO_3)_2$	=	148.3 g

5.6

Study Check

Calculate the molar mass of salicylic acid, $C_7H_6O_3$.

QUESTIONS & PROBLEMS

The Mole

5.27 What is a mole?

5.28 What is Avogadro's number?

5.29 **a.** How many carbon atoms are in 1 mole of carbon?
b. How many iron atoms are in 1 mole of iron?
c. Which contains more atoms: 1.0 mole of Ca or 1.0 mole of Mg?
d. Which contains more atoms: 1.0 mole of O_2 or 1.0 mole of O_3?

5.30 **a.** How many CO molecules are in 1 mole of CO?
b. How many CO_2 molecules are in 1 mole of CO_2?
c. Which contains more molecules: 1.0 mole of CO or 1.0 mole of CO_2?
d. Which contains more atoms: 1.0 mole of CO or 1.0 mole of CO_2?

5.31 How many moles of each element are in 1 mole of each of the following compounds?
a. $FeSO_4$ (iron supplement)
b. Al_2O_3 (absorbent and abrasive)
c. $Zn(C_2H_3O_2)_2$ (zinc dietary supplement)
d. $(NH_4)_2CO_3$ (baking powder)
e. $C_7H_5NO_3S$ (saccharin, a noncaloric sweetener)
f. $C_{22}H_{25}ClN_2O_8$ (tetracycline, an antibiotic)

5.32 How many moles of each element are in 1 mole of each of the following compounds?
a. Li_2CO_3 (antidepressant)
b. $Al_2(SO_4)_3$ (antiperspirant)
c. $Mg(OH)_2$ (antacid, laxative)
d. $(NH_4)_2SO_4$ (fertilizer)
e. $C_{14}H_{29}NO_4S$ (penicillin, an antibiotic)
f. $C_{20}H_{24}N_2O_2$ (quinine)

5.33 Calculate the molar mass of each compound in problem 5.31.

5.34 Calculate the molar mass of each compound in problem 5.32.

5.7 Calculations Using Molar Mass

Study Pack

Chem of Life Reactions and Equilibrium: Stoichiometry 3

Learning Goal

Given the number of moles of a substance, calculate the mass in grams; given the mass, calculate the number of moles.

5.7

The molar mass of an element or a compound is one of the most useful conversion factors in chemistry. Molar mass is used to change from moles of a substance to grams, or from grams to moles. To do these calculations, we use the molar mass as a conversion factor. For example, 1 mole of magnesium has a mass of 24.3 g. To express molar mass as an equality, we can write

$$1 \text{ mole Mg} = 24.3 \text{ g Mg}$$

From this equality, two conversion factors can be written.

$$\frac{24.3 \text{ g Mg}}{1 \text{ mole Mg}} \quad \text{and} \quad \frac{1 \text{ mole Mg}}{24.3 \text{ g Mg}}$$

Conversion factors are written for compounds in the same way. For example, the molar mass of the compound H_2O is 18.0 g.

$$1 \text{ mole } H_2O = 18.0 \text{ g } H_2O$$

The conversion factors from the molar mass of H_2O are written as

$$\frac{18.0 \text{ g } H_2O}{1 \text{ mole } H_2O} \quad \text{and} \quad \frac{1 \text{ mole } H_2O}{18.0 \text{ g } H_2O}$$

We can now change from moles to grams, or grams to moles, using the conversion factors derived from the molar mass. (Remember, you must calculate the molar mass of the substance first.)

S·A·M·P·L·E 5.9 ·P·R·O·B·L·E·M

Converting Moles of an Element to Grams

Silver metal is used in the manufacture of tableware, mirrors, jewelry, and dental alloys. If the design for a piece of jewelry requires 0.750 mole of silver, how many grams of silver would be needed?

Solution

Using the atomic mass, we write the molar mass of silver as an equality.

$$1 \text{ mole Ag} = 107.9 \text{ g}$$

From the equality, we can write conversion factors that relate grams and moles.

$$\frac{107.9 \text{ g Ag}}{1 \text{ mole Ag}} \quad \text{and} \quad \frac{1 \text{ mole Ag}}{107.9 \text{ g Ag}}$$

Unit plan: 0.750 mole Ag ⟶ g Ag

The problem is set up by multiplying the given (0.750 mole Ag) by the conversion factor that cancels "mole Ag." For this problem, we use the first conversion factor.

$$\underset{\text{Given}}{0.750 \cancel{\text{ mole Ag}}} \times \underset{\text{Molar mass conversion factor}}{\frac{107.9 \text{ g Ag}}{1 \cancel{\text{ mole Ag}}}} = \underset{\text{Answer (rounded to three significant figures)}}{80.9 \text{ g Ag}}$$

Study Check

Calculate the number of grams of gold (Au) present in 0.124 mole of gold.

5.7

S·A·M·P·L·E 5.10 P·R·O·B·L·E·M

Converting Moles of a Compound to Grams

Salicylic acid has a formula of $C_7H_6O_3$. How many grams of salicylic acid are in 2.25 mole of $C_7H_6O_3$?

Solution

The molar mass of salicylic acid is 138.0 g. The molar mass gives the equality

$$1 \text{ mole } C_7H_6O_3 = 138.0 \text{ g } C_7H_6O_3$$

From the molar mass, we can write two conversion factors:

$$\frac{138.0 \text{ g salicylic acid}}{1 \text{ mole salicylic acid}} \quad \text{and} \quad \frac{1 \text{ mole salicylic acid}}{138.0 \text{ g salicylic acid}}$$

Unit plan: $2.25 \text{ mole } C_7H_6O_3 \longrightarrow \text{g } C_7H_6O_3$

In the problem setup, the given, 2.25 mole of salicylic acid, is multiplied by the first factor to cancel the unit "mole salicylic acid."

$$\underset{\text{Given}}{2.25 \cancel{\text{mole salicylic acid}}} \times \underset{\text{Molar mass conversion factor}}{\frac{138.0 \text{ g salicylic acid}}{1 \cancel{\text{mole salicylic acid}}}} = \underset{\text{Answer (rounded to 3 SFs)}}{311 \text{ g salicylic acid}}$$

Study Check

Calculate the mass (grams) of 0.65 mole of ammonium carbonate, $(NH_4)_2CO_3$, a substance found in baking powders.

S·A·M·P·L·E 5.11 P·R·O·B·L·E·M

Converting Mass of an Element to Moles

The total bone mass of a typical 70.0-kg person contains 115 g of phosphorus. How many moles of phosphorus are in the bone mass?

Solution

Phosphorus has a molar mass of 31.0 g.

$$1 \text{ mole P} = 31.0 \text{ g P}$$

The two conversion factors for this molar mass are

$$\frac{31.0 \text{ g P}}{1 \text{ mole P}} \quad \text{and} \quad \frac{1 \text{ mole P}}{31.0 \text{ g P}}$$

Unit plan: $115 \text{ g P} \longrightarrow \text{mole P}$

In the problem setup, the given, 115 g P, is multiplied by the conversion factor that cancels "g P."

$$\underset{\text{Given}}{115 \cancel{\text{g P}}} \times \underset{\text{Molar mass conversion factor}}{\frac{1 \text{ mole P}}{31.0 \cancel{\text{g P}}}} = \underset{\text{Answer (rounded to 3 SFs)}}{3.71 \text{ mole P}}$$

Study Check

How many moles of iron are contained in 255 g of iron (Fe) metal?

S·A·M·P·L·E **5.12** P·R·O·B·L·E·M

Converting Mass of a Compound to Moles

Iodized salt provides sufficient iodine in a person's diet to prevent goiter, a condition that occurs in the thyroid when iodine levels are too low. Iodized salt is prepared by adding 0.010% by mass potassium iodide (KI) to table salt. If 0.0737 g of KI are added to each box of salt (737 g), how many moles of KI are required in the preparation of 250 boxes of salt?

Solution

The molar mass of KI is

1 mole KI = 166.0 g KI

Written in the form of conversion factors, this molar mass gives

$$\frac{166.0 \text{ g KI}}{1 \text{ mole KI}} \quad \text{and} \quad \frac{1 \text{ mole KI}}{166.0 \text{ g KI}}$$

Unit plan: boxes of salt ⟶ g KI ⟶ mole KI

The problem is set up with the given, 250 boxes of salt, multiplied by a factor of 0.0737 g KI per box. The final conversion factor cancels "g KI."

$$\underset{\text{Given (exact)}}{250 \cancel{\text{boxes of salt}}} \times \underset{\text{Conversion factor}}{\frac{0.0737 \cancel{\text{g KI}}}{1 \cancel{\text{box of salt}}}} \times \underset{\text{Molar mass conversion factor}}{\frac{1 \text{ mole KI}}{166.0 \cancel{\text{g KI}}}} = \underset{\text{Answer (rounded to 3 SFs)}}{0.111 \text{ mole KI}}$$

Study Check

One gelcap of an antacid contains 311 mg $CaCO_3$ and 232 mg $MgCO_3$. In a recommended dosage of two gelcaps, how many moles each of $CaCO_3$ and $MgCO_3$ are present?

Q·U·E·S·T·I·O·N·S & P·R·O·B·L·E·M·S

Calculations Using Molar Mass

5.35 Calculate the number of grams in each of the following:
a. 2.00 moles Na **b.** 2.80 moles Ca
c. 0.125 mole Sn **d.** 3.00 moles S
e. 0.040 mole Cu

5.36 Calculate the number of grams in each of the following:
a. 1.50 moles of K **b.** 10.0 moles Al
c. 2.5 moles of C **d.** 0.25 mole P
e. 0.018 mole Ag

5.37 Calculate the number of grams in each of the following:
a. 0.500 mole NaCl **b.** 1.75 moles Na_2O
c. 0.225 mole H_2O **d.** 4.00 moles CO_2
e. 0.015 mole H_3PO_4

5.38 Calculate the number of grams in each of the following:
a. 2.0 moles $MgCl_2$ **b.** 0.100 mole NO_2
c. 3.5 moles C_3H_8 **d.** 5.00 moles C_2H_6O
e. 0.400 mole $(NH_4)_2O$

5.39 How many grams are in 0.500 mole of each of the following?
a. K **b.** Cl_2 **c.** Na_2CO_3 **d.** He **e.** C_3H_8O

5.40 How many grams are in 2.25 moles of each of the following?
a. Zn **b.** N_2 **c.** NaBr **d.** H_2O **e.** C_6H_{14}

5.41 **a.** The compound $MgSO_4$ is called Epsom salts. How many grams will you need to prepare a bath containing 5.00 moles of Epsom salts?
b. In a bottle of soda, there is 0.25 mole CO_2. How many grams of CO_2 are in the bottle?

5.42 **a.** Cyclopropane, C_3H_6, is an anesthetic given by inhalation. How many grams are in 0.25 mole of cyclopropane?
b. The sedative Demerol has the formula $C_{15}H_{22}ClNO_2$. How many grams are in 0.025 mole of Demerol?

5.8

5.43 How many moles are contained in each of the following?
a. 50.0 g Ag **b.** 40.0 g Cu **c.** 0.200 g C
d. 15.0 g NH_3 **e.** 75.0 g SO_2

5.44 How many moles are contained in each of the following?
a. 50.0 g Fe **b.** 25.0 g Ca **c.** 5.00 g S
d. 40.0 g H_2O **e.** 100.0 g O_2

5.45 How many moles are in 25.0 g of each of the following compounds?
a. CO_2 **b.** $Al(OH)_3$ **c.** $MgCl_2$
d. C_2H_6O **e.** NH_4Cl

5.46 How many moles are in 4.00 g of each of the following compounds?
a. NH_3 **b.** $Ca(NO_3)_2$ **c.** SO_3 **d.** H_2SO_4 **e.** $FeCl_3$

5.47 A can of Drāno contains 480 g of NaOH. How many moles of NaOH are in the can of Drāno?

5.48 A gold nugget weighs 35.0 g. How many moles of gold are in the nugget?

Study Pack

Chem of Life Reactions and Equilibrium: Stoichiometry 4, 5

Learning Goal

Given a quantity in moles of reactant or product, use mole–mole conversion factors to calculate the quantity of another substance in the reaction.

5.8 Mole Relationships in Chemical Equations

In the first part of this chapter, we saw that equations are balanced in terms of the numbers of atoms in a chemical reaction. However, when we do experiments in the laboratory, prepare medications in the hospital, or follow a recipe for pancakes, we work with samples measured in grams that contain billions of atoms and molecules.

For any balanced equation the total amount of matter in the reactants must be equal to that of the products. Let us go back to our pancake recipe and see what this means. If we weighed all of the ingredients, we would find that their total mass was equal to the mass of the pancakes. After those substances undergo a chemical reaction, their mass becomes part of the mass of the pancakes. The conservation of mass is illustrated in Figure 5.12.

Figure 5.12 Mass is conserved in a recipe. The pancakes produced are equal in mass to the ingredients.

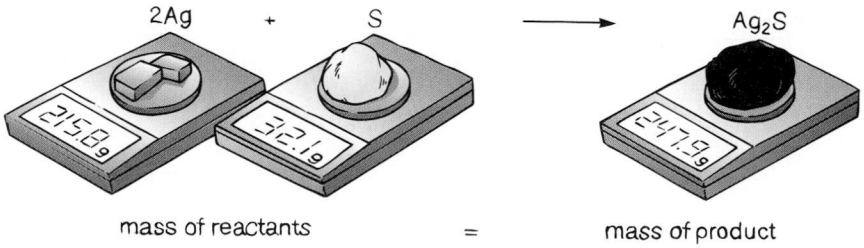

Figure 5.13 In the reaction of silver and sulfur to form silver sulfide (Ag_2S), mass is conserved: the mass of the reactants on the laboratory balance is equal to the mass of the product.

1 cup milk + 3 eggs + 2 cups pancake mix ⟶ 16 pancakes
250 g + 150 g + 300 g = 700 g

Conservation of Mass

Mass of the reactants = mass of the products
700 g = 700 g

Conservation of Mass in an Equation

When tarnish forms, silver reacts with sulfur to form silver sulfide.

$$2Ag + S \longrightarrow Ag_2S$$

Earlier we interpreted equations in terms of the particles of the reactants and products. In this reaction, we see that the number of silver atoms that react is two times the number of sulfur atoms. If we had 200 silver atoms, we would need 100 sulfur atoms for a complete reaction. Now that we have learned the mole concept, we can interpret the equation in terms of moles. This equation tells us that 2 moles of silver is needed for every 1 mole of sulfur that reacts. Using the molar mass of each substance, we can also determine the mass in grams of the reactants and products as seen in Figure 5.13. Let us look at all the ways we can now interpret a chemical equation:

$2Ag + S \longrightarrow Ag_2S$

Number of atoms:

2 Ag atoms + 1 S atom ⟶ 1 Ag_2S formula unit (ions)
200 Ag atoms + 100 S atoms ⟶ 100 Ag_2S formula units
$2(6.02 \times 10^{23})$ + 6.02×10^{23} ⟶ 6.02×10^{23}
Ag atoms S atoms Ag_2S formula units

Number of moles:

2 moles Ag + 1 mole S ⟶ 1 mole Ag_2S

Number of grams:

2(107.9g) + 32.1 g S ⟶ 247.9 g Ag_2S
247.9 g reactants ⟶ 247.9 g product

S·A·M·P·L·E **5.13** ·P·R·O·B·L·E·M

Conservation of Mass

Calculate the total mass of reactants and the total mass of products in the following equation:

$$CH_4 + 2O_2 \longrightarrow CO_2 + 2H_2O$$

Solution

Multiplying the number of moles (coefficients) in the equation by their molar masses gives the mass of the reactants and products.

	Reactants			Products		
	CH_4	+	$2O_2$ $\longrightarrow$	CO_2	+	$2H_2O$
Moles:	1 mole CH_4		2 mole O_2	1 mole CO_2		2 mole H_2O
Molar mass:	× 16.0 g/mole		× 32.0 g/mole	× 44.0 g/mole		× 18.0 g/mole
Mass:	16.0 g	+	64.0 g $\longrightarrow$	44.0 g	+	36.0 g
Total mass:	80.0 g reactants			=	80.0 g products	

Study Check

For the following equation, compare the total mass of the reactants and products.

$$4NH_3 + 3O_2 \longrightarrow 2N_2 + 6H_2O$$

5.8 Mole–Mole Factors

When iron reacts with sulfur, the product is iron(III) sulfide.

$$2Fe(s) + 3S(s) \longrightarrow Fe_2S_3(s)$$

Because the equation is balanced, we know the proportions of iron and sulfur in the reaction. For this reaction, we see that 2 moles of iron reacts with 3 moles of sulfur to form 1 mole of iron(III) sulfide. From the proportions given by the coefficients, we can write **mole–mole conversion factors** between reactants and product:

Fe and S: $\dfrac{2 \text{ moles Fe}}{3 \text{ moles S}}$ and $\dfrac{3 \text{ moles S}}{2 \text{ moles Fe}}$

Fe and Fe_2S_3: $\dfrac{2 \text{ moles Fe}}{1 \text{ mole } Fe_2S_3}$ and $\dfrac{1 \text{ mole } Fe_2S_3}{2 \text{ moles Fe}}$

S and Fe_2S_3: $\dfrac{3 \text{ moles S}}{1 \text{ mole } Fe_2S_3}$ and $\dfrac{1 \text{ mole } Fe_2S_3}{3 \text{ moles S}}$

SAMPLE 5.14 PROBLEM

Writing Mole Conversion Factors

Consider the following balanced equation:

$$4Na + O_2 \longrightarrow 2Na_2O$$

Write the mole–mole conversion factors for

a. Na and O_2 **b.** Na and Na_2O

Solution

a. $\dfrac{4 \text{ moles Na}}{1 \text{ mole } O_2}$ and $\dfrac{1 \text{ mole } O_2}{4 \text{ moles Na}}$

b. $\dfrac{4 \text{ moles Na}}{2 \text{ moles } Na_2O}$ and $\dfrac{2 \text{ moles } Na_2O}{4 \text{ moles Na}}$

Study Check

Using the equation in Sample Problem 5.14, write the mole–mole conversion factors for O_2 and Na_2O.

Using Mole–Mole Factors in Calculations

Whenever you prepare a recipe, adjust a carburetor for the proper mixture of fuel and air, or prepare medicines in a pharmaceutical laboratory, you need to know the proper amounts of reactants to use and how much of the product will form. Earlier we wrote all the possible conversion factors that can be obtained from the balanced equation: $2Fe + 3S \longrightarrow Fe_2S_3$. Now we will show how mole–mole factors are used in chemical calculations.

5.8

S·A·M·P·L·E 5.15 ·P·R·O·B·L·E·M

Calculating Moles of a Reactant

In the reaction of iron and sulfur, how many moles of sulfur are needed to react with 6.0 moles of iron?

$$2Fe + 3S \longrightarrow Fe_2S_3$$

Solution

In this problem, we need to find the relationship between the moles of Fe and the moles of S.

mole Fe $\longrightarrow$ mole S

Using the coefficients of Fe and S in the balanced equation, two mole–mole factors can be written:

$$\frac{2 \text{ moles Fe}}{3 \text{ moles S}} \quad \text{and} \quad \frac{3 \text{ moles S}}{2 \text{ moles Fe}}$$

To solve the problem, we write the given quantity of 6.0 moles of Fe and convert to moles of S by using the mole–mole factor.

$$\underset{\text{Given}}{6.0 \cancel{\text{moles Fe}}} \times \underset{\text{Mole mole factor}}{\frac{3 \text{ moles S}}{2 \cancel{\text{moles Fe}}}} = \underset{\text{Answer}}{9.0 \text{ moles of S}}$$

Study Check

Using the equation in Sample Problem 5.15, calculate the number of moles of iron needed to completely react with 2.7 moles of sulfur.

S·A·M·P·L·E 5.16 ·P·R·O·B·L·E·M

Finding the Moles of Product

In a combustion reaction, propane reacts with oxygen. How many moles of CO_2 can be produced when 6.0 moles of propane reacts?

$$C_3H_8 + 5O_2 \longrightarrow 3CO_2 + 4H_2O$$

5.8

Solution

We are given the moles of C_3H_8 and need to find the moles of CO_2.

$$\text{mole } C_3H_8 \longrightarrow \text{mole } CO_2$$

The coefficients in the equation tell us that 1 mole of C_3H_8 will produce 3 moles of CO_2, which can be written as two mole–mole factors as follows.

$$\frac{1 \text{ mole } C_3H_8}{3 \text{ moles } CO_2} \quad \text{and} \quad \frac{3 \text{ moles } CO_2}{1 \text{ mole } C_3H_8}$$

Now we can write the given of 6.0 moles C_3H_8 and multiply by the mole–mole factor that gives the moles of CO_2.

$$\underset{\text{Given}}{6.0 \cancel{\text{ moles } C_3H_8}} \times \underset{\substack{\text{Mole–mole factor} \\ \text{(coefficients)}}}{\frac{3 \text{ moles } CO_2}{1 \cancel{\text{ mole } C_3H_8}}} = \underset{\text{Answer}}{18.0 \text{ moles } CO_2}$$

Study Check

Using the equation in Sample Problem 5.16, calculate how many moles of oxygen must react to produce 2.0 moles of water.

QUESTIONS & PROBLEMS

Mole Relationships in Chemical Equations

5.49 Give an interpretation of the following equations in terms of (1) number of particles and (2) number of moles:

a. $2SO_2 + O_2 \longrightarrow 2SO_3$

b. $4P + 5O_2 \longrightarrow 2P_2O_5$

5.50 Give an interpretation of the following equations in terms of (1) number of particles and (2) number of moles:

a. $2Al + 3Cl_2 \longrightarrow 2AlCl_3$

b. $4HCl + O_2 \longrightarrow 2Cl_2 + 2H_2O$

5.51 Calculate the total masses of the reactants and the products in each of the equations of problem 5.49.

5.52 Calculate the total masses of the reactants and the products in each of the balanced equations of problem 5.50.

5.53 Write all of the mole factors for the equations listed in problem 5.49.

5.54 Write all of the mole factors for the equations listed in problem 5.50.

5.55 The reaction of hydrogen with oxygen produces water.

$$2H_2(g) + O_2(g) \longrightarrow 2H_2O(g)$$

a. How many moles of O_2 are required to react with 2.0 moles of H_2?

b. If you have 5.0 moles of O_2, how many moles of H_2 are needed for the reaction?

c. How many moles of H_2O form when 2.5 moles of O_2 reacts?

5.56 Ammonia is produced by the reaction of hydrogen and nitrogen.

$$N_2(g) + 3H_2(g) \longrightarrow \underset{\text{Ammonia}}{2NH_3(g)}$$

a. How many moles of H_2 are needed to react with 1.0 mole of N_2?
b. How many moles of N_2 reacted if 0.60 mole of NH_3 is produced?
c. How many moles of NH_3 are produced when 1.4 moles of H_2 reacts?

5.57 Carbon disulfide and carbon monoxide are produced when carbon is heated with sulfur dioxide.

$$5C + 2SO_2 \longrightarrow CS_2 + 4CO$$

a. How many moles of C are needed to react with 0.500 mole of SO_2?
b. How many moles of CO are produced when 1.2 moles of C reacts?
c. How many moles of SO_2 are required to produce 0.50 mole CS_2?
d. How many moles of CS_2 are produced when 2.5 moles of C reacts?

5.58 In the acetylene torch, acetylene gas (C_2H_2) burns in oxygen to produce carbon dioxide and water.

$$2C_2H_2(g) + 5O_2(g) \longrightarrow 4CO_2(g) + 2H_2O(g)$$

a. How many moles of O_2 are needed to react with 2.00 moles of C_2H_2?
b. How many moles of CO_2 are produced when 3.5 moles of C_2H_2 reacts?
c. How many moles of C_2H_2 are required to produce 0.50 mole H_2O?
d. How many moles of CO_2 are produced from 0.100 mole O_2?

5.9 A Closer Look: Mass Calculations for Reactions

Study Pack

Chem of Life Reactions and Equilibrium: Stoichiometry 6

Learning Goal

Given the quantity in grams of a substance in a reaction, calculate the quantity of another substance in the reaction.

Often, when you perform a chemistry experiment, you need to use a laboratory balance to weigh out a certain mass of reactant. Once you know its mass you can convert grams to moles. Then, using mole–mole factors from the balanced equation, you can determine the moles and grams of other substances that react or are produced in the reaction.

S·A·M·P·L·E **5.17** ·P·R·O·B·L·E·M

Calculating the Mass of a Product

In the formation of smog, nitrogen reacts with oxygen to produce nitrogen monoxide. Calculate the grams of NO produced when 1.5 moles of O_2 reacts.

$$N_2(g) + O_2(g) \longrightarrow 2NO(g)$$

Solution

From the moles of O_2, you need to determine the mass, in grams, of NO.

$$\text{mole } O_2 \longrightarrow \text{g NO}$$

From the equation we know that 1 mole of N_2 reacts with 1 mole of O_2 to form 2 moles of NO. First, we can change the given, 1.5 moles O_2, to moles of NO using the mole–mole factors derived from the coefficients of NO and O_2 in the balanced equation.

$$1.50\ \cancel{\text{moles } O_2} \times \frac{2 \text{ moles NO}}{1\ \cancel{\text{mole } O_2}} = 3.00 \text{ moles NO}$$

Now the moles of NO_2 can be converted to grams of NO using the molar mass.

$$3.00\ \cancel{\text{moles NO}} \times \frac{30.0 \text{ g NO}}{1\ \cancel{\text{mole NO}}} = 90.0 \text{ g NO}$$

These two steps can be combined to give a sequence that calculates the grams of NO from the moles of O_2.

5.9

mole O_2 ⟶ mole NO ⟶ g NO

The combined solution shows both conversion factors in the setup.

$$\underset{\text{Given}}{1.50\ \cancel{\text{mole } O_2}} \times \underset{\text{Mole factor}}{\frac{2\ \cancel{\text{moles NO}}}{1\ \cancel{\text{mole } O_2}}} \times \underset{\text{Molar mass NO}}{\frac{30.0 \text{ g NO}}{1\ \cancel{\text{mole NO}}}} = 90.0 \text{ g NO}$$

Study Check

Using the equation in Sample Problem 5.17, calculate how many grams of NO can be produced when 0.25 mole of N_2 reacts?

S·A·M·P·L·E 5.18 ·P·R·O·B·L·E·M

Calculating the Mass of a Product

In a combustion reaction, acetylene (C_2H_2) burns with oxygen.

$$2C_2H_2 + 5O_2 \longrightarrow 4CO_2 + 2H_2O$$

How many grams of carbon dioxide are produced when 22.0 g of C_2H_2 are burned?

Solution

The question asks us to convert grams of C_2H_2 to grams of CO_2 produced.

g C_2H_2 ⟶ g CO_2

First we convert g C_2H_2 to moles of C_2H_2. Then we change moles of C_2H_2 to moles of CO_2 by using the relationship in the balanced equation that shows that 4 moles of CO_2 is produced when 2 moles of C_2H_2 reacts. Finally, the moles of CO_2 can be changed to g CO_2.

g C_2H_2 ⟶ mole C_2H_2 ⟶ mole CO_2 ⟶ g CO_2

$$\underset{\text{Given}}{22.0\ \cancel{\text{g } C_2H_2}} \times \underset{\text{Molar mass } C_2H_2}{\frac{1\ \cancel{\text{mole } C_2H_2}}{26.0\ \cancel{\text{g } C_2H_2}}} \times \underset{\text{Mole factor}}{\frac{4\ \cancel{\text{moles } CO_2}}{2\ \cancel{\text{moles } C_2H_2}}} \times \underset{\text{Molar mass } CO_2}{\frac{44.0 \text{ g } CO_2}{1\ \cancel{\text{mole } CO_2}}} = 74.5 \text{ g } CO_2$$

Study Check

Using the equation in Sample Problem 5.18, calculate how many grams of CO_2 can be produced when 25.0 g of O_2 reacts.

Summary of Calculations in Chemical Reactions

In all of these calculations using equations, we have used the molar masses of the substances and their mole factors from the coefficients in the balanced equation. We use the following steps to set up and solve problems that involve calculations and equations.

Steps in Finding the Moles or Masses of Substances in a Chemical Reaction

1. If necessary, convert the mass of substance A to moles using its molar mass.

   grams of A ⟶ moles of A

2. Convert the moles of A to moles of the desired substance B using the mole factor derived from the coefficients of the balanced equation.

   moles of A ⟶ moles of B

3. If needed, convert the moles of B to grams using its molar mass.

   moles of B ⟶ grams of B

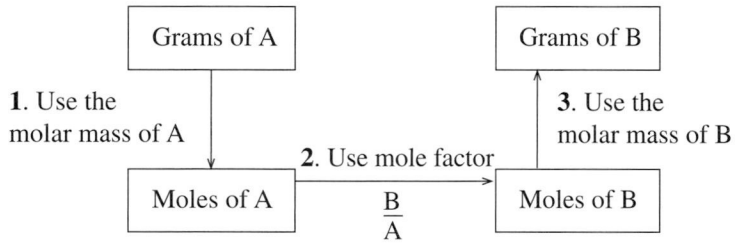

Think of the preceding diagram as a map. Locate the initial and final units for a problem and follow the steps to obtain your answer.

S·A·M·P·L·E· 5.19 ·P·R·O·B·L·E·M

Calculating the Mass of a Reactant

Propane gas, C_3H_8, a fuel for camp stoves and some specially equipped automobiles, reacts with oxygen to produce carbon dioxide and water. How many grams of O_2 are required to react with 22.0 g of C_3H_8?

$$C_3H_8(g) + 5O_2(g) \longrightarrow 3CO_2(g) + 4H_2O(g)$$

Solution

The equation indicates that 1 mole of C_3H_8 reacts with 5 moles of O_2 to produce 3 moles of CO_2 and 4 moles of H_2O. The 22.0 g of C_3H_8 are first changed to moles of C_3H_8. Then the moles of C_3H_8 are changed to moles of O_2, which can be converted to grams of O_2.

$$\text{g } C_3H_8 \longrightarrow \text{mole } C_3H_8 \longrightarrow \text{mole } O_2 \longrightarrow \text{g } O_2$$

$$\underset{\text{Given}}{22.0\ \cancel{\text{g } C_3H_8}} \times \underset{\text{Molar mass } C_3H_8}{\frac{1\ \cancel{\text{mole } C_3H_8}}{44.0\ \cancel{\text{g } C_3H_8}}} \times \underset{\text{Mole factor}}{\frac{5\ \cancel{\text{moles } O_2}}{1\ \cancel{\text{mole } C_3H_8}}} \times \underset{\text{Molar mass } O_2}{\frac{32.0\text{ g } O_2}{1\ \cancel{\text{mole } O_2}}} = 80.0\text{ g } O_2$$

Study Check

Using the equation in Sample Problem 5.19, calculate how many grams of C_3H_8 are needed to produce 15.0 g of H_2O.

QUESTIONS & PROBLEMS

A Closer Look: Mass Calculations for Reactions

5.9

5.59 Sodium reacts with oxygen to produce sodium oxide.

$$4Na(s) + O_2(g) \longrightarrow 2Na_2O(s)$$

a. How many grams of Na_2O are produced when 2.5 moles of Na react?
b. If you have 18.0 g of Na, how many grams of O_2 are required for reaction?
c. How many grams of O_2 are needed in a reaction that produces 75.0 g of Na_2O?

5.60 Nitrogen gas reacts with hydrogen gas to produce ammonia by the following equation:

$$N_2(g) + 3H_2(g) \longrightarrow 2NH_3(g)$$

a. If you have 1.8 moles of H_2, how many grams of NH_3 can be produced?
b. How many grams of H_2 are needed to react with 2.80 g of N_2?
c. How many grams of NH_3 can be produced from 12 g of H_2?

5.61 Ammonia and oxygen react to form nitrogen and water.

$$\underset{\text{Ammonia}}{4NH_3} + 3O_2 \longrightarrow 2N_2 + 6H_2O$$

a. How many grams of O_2 are needed to react with 8.0 moles of NH_3?
b. How many grams of N_2 can be produced when 6.50 g of O_2 reacts?
c. How many grams of water are formed from the reaction of 34 g of NH_3?

5.62 Iron(III) oxide reacts with carbon to give iron and carbon monoxide.

$$Fe_2O_3 + 3C \longrightarrow 2Fe + 3CO$$

a. How many grams of C are required to react with 2.50 moles of Fe_2O_3?
b. How many grams of CO are produced when 36.0 g of C reacts?
c. How many grams of Fe can be produced when 6.00 g of Fe_2O_3 reacts?

Chapter Review

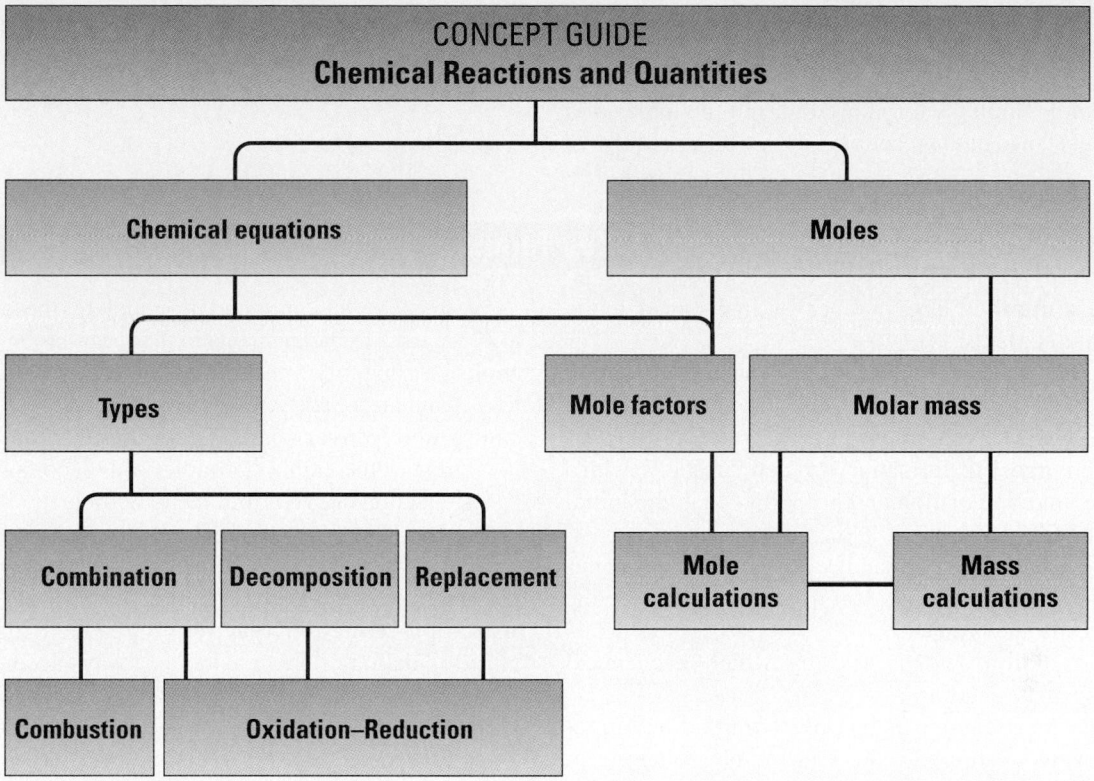

Chapter Summary

5.1 Chemical Changes

A chemical change occurs when the atoms of the initial substances rearrange to form new substances. When new substances form, a chemical reaction has taken place.

5.2 Chemical Equations

A chemical equation shows the formulas of the substances that react on the left of a reaction arrow, and the products that form on the right side of the reaction arrow.

5.3 Balancing a Chemical Equation

An equation is balanced by writing the smallest whole numbers (coefficients) in front of formulas to equalize the atoms of each element in the reactants and the products.

5.4 Types of Reactions

Many chemical reactions can be organized by reaction type: combination, decomposition, single replacement, double replacement, and/or combustion.

5.5 A Closer Look: Oxidation-Reduction Reactions

When electrons are transferred in a reaction, it is an oxidation-reduction reaction. One reactant loses electrons, and another reactant gains electrons. Overall, the number of electrons lost and gained is equal.

5.6 The Mole

One mole of an element contains 6.02×10^{23} atoms; a mole of a compound contains 6.02×10^{23} molecules or formula units. The molar mass (g/mole) of any substance is the mass in grams equal numerically to its atomic mass, or the sum of the atomic masses, which have been multiplied by their subscripts in a formula.

5.7 Calculations Using Molar Mass

The molar mass is the mass in grams of 1 mole of an element or a compound. It becomes an important conversion factor when it is used to change a quantity in grams to moles, or to change a given number of moles to grams.

5.8 Mole Relationships in Chemical Equations

In a balanced equation, the total mass of the reactants is equal to the total mass of the products. The coefficients in an equation describing the relationship between the moles of any two components are used to write mole–mole factors. When the number of moles for one substance is known, a mole–mole factor is used to find the moles of a different substance in the reaction.

5.9 A Closer Look: Mass Calculations for Reactions

In calculations using equations, the molar masses of the substances and their mole factors are used to change the number of grams of one substance to the corresponding grams of a different substance.

Key Terms

Avogadro's number The number of items in a mole, equal to 6.02×10^{23}.

balanced equation The final form of a chemical reaction that shows the same number of atoms of each element in the reactants and products.

chemical change The formation of a new substance with a different composition and properties than the initial substance.

chemical equation A shorthand way to represent a chemical reaction using chemical formulas to indicate the reactants and products.

chemical reaction The process by which a chemical change takes place.

coefficients Whole numbers placed in front of the formulas in an equation to balance the number of atoms or moles of atoms of each element in an equation.

combination reaction A reaction in which reactants combine to form a single product.

combustion reaction A reaction in which an element or a compound reacts with oxygen to form oxide products.

decomposition reaction A reaction in which a single reactant splits into two or more simpler substances.

double replacement reaction A reaction in which parts of two different reactants exchange places.

molar mass The mass in grams of 1 mole of an element equal numerically to its atomic mass. The molar mass of a compound is equal to the sum of the masses of the elements in the formula.

mole A group of atoms, molecules, or formula units that contains 6.02×10^{23} of these items.

mole–mole conversion factor A conversion factor that relates the number of moles of two compounds in an equation derived from their coefficients.

oxidation The loss of electrons by a substance. Biological oxidation may involve the addition of oxygen, or the loss of hydrogen.

oxidation-reduction reaction A reaction in which the oxidation of one reactant is always accompanied by the reduction of another reactant.

physical change A change in which the physical properties change but the chemical composition does not change.

products The substances formed as a result of a chemical reaction.

reactants The initial substances that undergo change in a chemical reaction.

reduction The gain of electrons by a substance. Biological reduction may involve the loss of oxygen, or the gain of hydrogen.

single replacement reaction An element replaces a different element in a compound.

Chapter Practice

5.63 Balance each of the following equations and identify the type of reaction.

a. $Al + Cl_2 \longrightarrow AlCl_3$
b. $Fe + H_2SO_4 \longrightarrow Fe_2(SO_4)_3 + H_2$
c. $AgNO_3 + H_2S \longrightarrow Ag_2S + HNO_3$
d. $Cl_2 + KI \longrightarrow KCl + I_2$

5.64 Balance each of the following equations and identify the type of reaction.

a. $NI_3 \longrightarrow N_2 + I_2$
b. $Mg + N_2 \longrightarrow Mg_3N_2$
c. $Na_2SO_4 + Al(OH)_3 \longrightarrow Al_2(SO_4)_3 + NaOH$
d. $Fe + HCl \longrightarrow FeCl_3 + H_2$

5.65 Complete and balance each of the following combustion reactions.

a. $C_4H_8 + O_2 \longrightarrow$
b. $C_5H_{12} + O_2 \longrightarrow$
c. $C_6H_{12}O_6 + O_2 \longrightarrow$

5.66 Complete and balance each of the following combustion reactions.

a. $C_2H_6 + O_2 \longrightarrow$
b. $C_6H_6 + O_2 \longrightarrow$
c. $C_3H_8O_3 + O_2 \longrightarrow$

5.67 Identify each of the following as an oxidation or a reduction reaction.

a. $Zn^{2+} + 2e^- \longrightarrow Zn$
b. $Al \longrightarrow Al^{3+} + 3e^-$
c. $Pb \longrightarrow Pb^{2+} + 2e^-$
d. $Cl_2 + 2e^- \longrightarrow 2Cl^-$

5.68 Write a balanced chemical equation for each of the following oxidation-reduction reactions.

a. Sulfur reacts with molecular chlorine to form sulfur dichloride.
b. Molecular chlorine and sodium bromide react to form molecular bromine and sodium chloride.
c. Aluminum metal and iron(III) oxide react to produce aluminum oxide and elemental iron.
d. Copper(II) oxide reacts with elemental C to form elemental copper and carbon dioxide.

5.69 In problem 5.68, show how electrons are lost and gained. Identify which element is oxidized and which element is reduced in each reaction.

5.70 You have collected 20.0 lb of aluminum cans. How many moles of aluminum do you have?

5.71 A typical human body contains 60.0% water by mass. If a person weighs 70.0 kg, how many moles of water are present in that person's body?

5.72 A blood level of 400. mg of alcohol per 100. mL of blood can cause coma and possibly be fatal. How many moles of alcohol (C_2H_6O) are present in 10.0 mL of blood?

5.73 At a winery, glucose ($C_6H_{12}O_6$) in grapes undergoes fermentation to produce ethanol (C_2H_6O) and carbon dioxide.

$$\underset{\text{Glucose}}{C_6H_{12}O_6} \longrightarrow \underset{\text{Ethanol}}{2C_2H_6O} + 2CO_2$$

a. How many moles of glucose are required to form 124 g of ethanol?
b. How many grams of ethanol would be formed from the reaction of 0.240 kg of glucose?

5.74 Gasohol is a fuel containing ethanol (C_2H_6O) that burns in oxygen (O_2) to give carbon dioxide and water.

a. State the reactants and products for this reaction in the form of a balanced equation.
b. How many moles of O_2 are needed to completely react with 4.0 moles of C_2H_6O?
c. If a car produces 88 g of CO_2, how many grams of O_2 are used up in the reaction?
d. If you add 125 g of C_2H_6O to your gas, how many grams of CO_2 and H_2O can be produced?

5.75 Balance the following equation:

$$NH_3 + F_2 \longrightarrow N_2F_4 + HF$$

a. How many moles of each reactant are needed to produce 4.00 moles of HF?
b. How many grams of F_2 are required to react with 1.50 moles of NH_3?
c. How many grams of N_2F_4 can be produced when 3.40 g of NH_3 reacts?

Answers

Answers to Study Checks

5.1 **a.** physical **b.** chemical **c.** physical

5.2 There are 4 carbon atoms, 12 hydrogen atoms, and 14 oxygen atoms on the reactant side of the equation, and on the product side.

5.3 **a.** $3Fe + 2O_2 \longrightarrow Fe_3O_4$
b. $C_3H_8 + 5O_2 \longrightarrow 3CO_2 + 4H_2O$

5.4 $2NO + O_2 \longrightarrow 2NO_2$ Combination reaction

5.5 Lithium is oxidized: $2Li \longrightarrow 2Li^+ + 2e^-$
Fluorine is reduced: $F_2 + 2e^- \longrightarrow 2F^-$

5.6 Each iron is oxidized: $Fe \longrightarrow Fe^{3+} + 3e^-$
Each bromine is reduced: $Br_2 + 2e^- \longrightarrow 2Br^-$

5.7 6 moles of C, 8 moles of H, 6 moles of O

5.8 138.0 g

5.9 24.4 g Au

5.10 62 g $(NH_4)_2CO_3$

5.11 4.56 moles of Fe

5.12 0.00621 mole of $CaCO_3$, 0.00550 mole of $MgCO_3$

5.13 The total mass of the reactants is 164.0 g (68.0 g NH_3 + 96.0 g O_2), which equals the total mass of the products of 164.0 g (56.0 g N_2 + 108.0 g H_2O).

5.14 $\dfrac{2 \text{ moles } Na_2O}{1 \text{ mole } O_2}$ and $\dfrac{1 \text{ mole } O_2}{2 \text{ moles } Na_2O}$

5.15 1.8 moles of Fe

5.16 2.5 moles of O_2

5.17 15 g NO

5.18 27.5 g CO_2

5.19 9.17 g C_3H_8

Answers to Selected Questions and Problems

5.1 **a.** physical **b.** chemical **c.** physical **d.** chemical **e.** physical **f.** chemical

5.3 **a.** Reactant side: 2 N atoms, 4 O atoms
Product side: 2 N atoms, 4 O atoms
b. Reactant side: 5 C atoms, 2 S atoms, 4 O atoms
Product side: 5 C atom, 2 S atoms, 4 O atom
c. Reactant side: 4 C atoms, 4 H atoms, 10 O atoms
Product side: 4 C atoms, 4 H atoms, 10 O atoms
d. Reactant side: 2 N atoms, 8 H atoms, 4 O atoms
Product side: 2 N atoms, 8 H atoms, 4 O atoms

5.5 **a.** not balanced **b.** balanced **c.** not balanced **d.** balanced

5.7 **a.** 2 Na atoms, 2 Cl atoms
b. 1 P atom, 3 Cl atoms, 6 H atoms
c. 4 P atoms, 16 O atoms, 12 H atoms

5.9 **a.** $N_2 + O_2 \longrightarrow 2NO$
b. $2HgO \longrightarrow 2Hg + O_2$
c. $4Fe + 3O_2 \longrightarrow 2Fe_2O_3$
d. $2Na + Cl_2 \longrightarrow 2NaCl$
e. $2Cu_2O + O_2 \longrightarrow 4CuO$

5.11 **a.** $Mg + 2AgNO_3 \longrightarrow Mg(NO_3)_2 + 2Ag$
b. $CuCO_3 \longrightarrow CuO + CO_2$
c. $2Al + 3CuSO_4 \longrightarrow 3Cu + Al_2(SO_4)_3$
d. $Pb(NO_3)_2 + 2NaCl \longrightarrow PbCl_2 + 2NaNO_3$
e. $2Al + 6HCl \longrightarrow 2AlCl_3 + 3H_2$

5.13 **a.** A single reactant splits into two simpler substances (elements).
b. One element in the reacting compound is replaced by the other reactant.

5.15 **a.** combination; combustion
b. single replacement
c. decomposition
d. double replacement
e. combustion
f. decomposition
g. double replacement
h. combination; combustion

5.17 **a.** $Mg + Cl_2 \longrightarrow MgCl_2$
b. $2HBr \longrightarrow H_2 + Br_2$
c. $Mg + Zn(NO_3)_2 \longrightarrow Mg(NO_3)_2 + Zn$
d. $K_2S + Pb(NO_3)_2 \longrightarrow PbS + 2KNO_3$
e. $2C_2H_6 + 7O_2 \longrightarrow 4CO_2 + 6H_2O$

5.19 **a.** oxidation **b.** reduction **c.** oxidation **d.** reduction

5.21 **a.** Zn is oxidized; Cl_2 is reduced.
b. Br^- in NaBr is oxidized; Cl_2 is reduced.
c. The O^{2-} in PbO is oxidized; the Pb^{2+} is reduced.
d. Sn^{2+} is oxidized; Fe^{3+} is reduced.

5.23 **a.** reduction **b.** oxidation

5.25 Linoleic acid gains hydrogen atoms and is reduced.

5.27 One mole contains 6.02×10^{23} atoms of an element, molecules of a covalent substance, or formula units of an ionic substance.

5.29 **a.** There are 6.02×10^{23} C atoms in one mole of C.
b. There are 6.02×10^{23} Fe atoms in one mole of Fe.
c. 1.0 mole of Ca and 1.0 mole of Mg both contain the same number of atoms; 6.02×10^{23} Ca atoms and 6.02×10^{23} Mg atoms.
d. 1.0 mole of O_2 contains $2(6.02 \times 10^{23}$ O atoms) or 1.2×10^{24} O atoms. 1.0 mole of O_3 molecules contains $3(6.02 \times 10^{23}$ O atoms), or 1.8×10^{24} O atoms.

5.31 **a.** 1 mole of Fe, 1 mole of S, and 4 moles of O
b. 2 moles of Al and 3 moles of O
c. 1 mole Zn, 4 moles C, 6 moles H, and 4 moles O
d. 2 moles N, 8 moles H, 1 mole C, and 3 moles O
e. 7 moles C, 5 moles H, 1 mole N, 3 moles O, and 1 mole S
f. 22 moles C, 25 moles H, 1 mole Cl, 2 moles N, and 8 moles O

5.33 **a.** 152.0 g **b.** 102.0 g **c.** 183.4 g **d.** 96.0 g **e.** 183.1 g **f.** 481 g

5.35 **a.** 46.0 g **b.** 112 g **c.** 14.8 g **d.** 96.3 g **e.** 2.5 g

5.37 **a.** 29.3 g **b.** 109 g **c.** 4.05 g **d.** 176 g **e.** 1.5 g

5.39 **a.** 19.6 g **b.** 35.5 g **c.** 53.0 g **d.** 2.00 g **e.** 30.0 g

5.41 **a.** 602 g **b.** 11g

5.43 **a.** 0.463 mole **b.** 0.629 mole **c.** 0.0167 mole **d.** 0.882 mole **e.** 1.17 moles

5.45 **a.** 0.568 mole **b.** 0.321 mole **c.** 0.262 mole **d.** 0.543 mole **e.** 0.467 mole

5.47 12 moles

5.49 **a.** **(1)** Two molecules of sulfur dioxide react with 1 molecule of oxygen to produce 2 molecules of sulfur trioxide.
a. **(2)** Two moles of sulfur dioxide react with 1 mole of oxygen to produce 2 moles of sulfur trioxide.
b. **(1)** Four atoms of phosphorus react with 5 molecules of oxygen to produce 2 molecules of diphosphorus pentoxide.
b. **(2)** Four moles of phosphorus react with 5 moles of oxygen to produce 2 moles of diphosphorus pentoxide.

5.51 **a.** 160.2 g reactants = 160.2 g products
b. 284 g reactants = 284 g products

5.53 **a.** $\frac{2 \text{ moles } SO_2}{1 \text{ mole } O_2}$ and $\frac{1 \text{ mole } O_2}{2 \text{ moles } SO_2}$

$\frac{2 \text{ moles } SO_2}{2 \text{ moles } SO_3}$ and $\frac{2 \text{ moles } SO_3}{2 \text{ moles } SO_2}$

$\frac{2 \text{ moles } SO_3}{1 \text{ mole } O_2}$ and $\frac{1 \text{ mole } O_2}{2 \text{ moles } SO_3}$

b. $\frac{4 \text{ moles P}}{5 \text{ moles } O_2}$ and $\frac{5 \text{ moles } O_2}{4 \text{ moles P}}$

$\frac{4 \text{ moles P}}{2 \text{ moles } P_2O_5}$ and $\frac{2 \text{ moles } P_2O_5}{4 \text{ moles P}}$

$\frac{5 \text{ moles } O_2}{2 \text{ moles } P_2O_5}$ and $\frac{2 \text{ moles } P_2O_5}{5 \text{ moles } O_2}$

5.55 **a.** 1.0 mole O_2 **b.** 10. moles of H_2
c. 5.0 moles H_2O

5.57 **a.** 1.25 moles C **b.** 0.96 mole CO
c. 1.0 mole SO_2 **d.** 0.50 mole CS_2

5.59 **a.** 78 g Na_2O **b.** 6.26 g O_2 **c.** 19.4 g O_2

5.61 **a.** 190 g O_2 **b.** 3.79 g N_2 **c.** 54 g H_2O

5.63 **a.** $2Al + 3Cl_2 \longrightarrow 2AlCl_3$ Combination
b. $2Fe + 3H_2SO_4 \longrightarrow Fe_2(SO_4)_3 + 3H_2$
Single replacement
c. $2AgNO_3 + H_2S \longrightarrow Ag_2S + 2HNO_3$
Double replacement
d. $Cl_2 + 2KI \longrightarrow 2KCl + I_2$
Single replacement

5.65 **a.** $C_4H_8 + 6O_2 \longrightarrow 4CO_2 + 4H_2O$
b. $C_5H_{12} + 8O_2 \longrightarrow 5CO_2 + 6H_2O$
c. $C_6H_{12}O_6 + 6O_2 \longrightarrow 6CO_2 + 6H_2O$

5.67 **a.** reduction **b.** oxidation
c. oxidation **d.** reduction

5.69 **a.** S is oxidized: $S \longrightarrow S^{2+} + 2e^-$
Cl is reduced: $Cl_2 + 2e^- \longrightarrow 2Cl^-$
b. Br^- is oxidized; $2Br^- \longrightarrow Br_2 + 2e^-$
Cl is reduced: $Cl_2 + 2e^- \longrightarrow 2Cl^-$
c. Al is oxidized: $Al \longrightarrow Al^{3+} + 3e^-$
Fe^{3+} is reduced: $Fe^{3+} + 3e^- \longrightarrow Fe$
d. Cu^{2+} is reduced: $Cu^{2+} + 2e^- \longrightarrow Cu$
C is oxidized: $C \longrightarrow C^{4+} + 4e^-$
(in CO_2, $2 \times O^{2-}$ means that C is 4+)

5.71 2330 moles of water

5.73 **a.** 1.35 moles of glucose
b. 123 g of ethanol

5.75 $2NH_3 + 5F_2 \longrightarrow N_2F_4 + 6HF$
a. 1.33 moles of NH_3 and 3.33 moles of F_2
b. 143 g F_2
c. 10.4 g of N_2F_4

6

Energy and States of Matter

The three states of matter for water are ice, liquid, and gas.

Looking Ahead

6.1	Energy
6.2	Measuring Temperature
6.3	Measuring Heat Energy
6.4	A Closer Look: Energy and Nutrition
6.5	States of Matter
6.6	Melting and Freezing
6.7	Boiling and Condensation
6.8	Heating and Cooling Curves
6.9	Energy in Chemical Reactions
6.10	A Closer Look: Chemical Equilibrium

www.chemplace.com/college/timberlake

Visit the web site above to test yourself, run simulations, and review key concepts from this chapter. For even more interactive review and quizzing, use the two CD-ROMs included with the book.

Energy makes it possible to do work. You use energy to do the work of running, playing tennis, studying, and breathing. You obtain that energy from the foods you eat. Energy takes many forms, but you are probably most familiar with heat, also called thermal energy. In your home, you use thermal energy to cook food, heat water, warm the room, and dry your hair.

Matter is everything around us that has mass and occupies space. Matter can be a solid, liquid, or gas. For example, the water in an ice cube, an ice rink, or an iceberg is solid. Water from a faucet or in a pool is a liquid. Water becomes a gas when it evaporates from wet clothes or boils in a pan on the stove. Many gases are invisible to us, but some are detected by a characteristic odor or color. For example, you know when a bottle of perfume or ammonia is opened because you smell the gas as it fills the room.

The state of matter can be changed by adding or removing heat. When you heat water in a tea kettle, the water gets hotter. Perhaps it boils and changes to a gas. You make ice cubes by using a freezer to remove heat from the water in an ice cube tray.

A loss or a gain of heat energy also occurs in chemical reactions. If a reaction mixture becomes warmer, heat is given off. We say that an exothermic reaction has taken place. If the reaction mixture becomes cooler, heat is absorbed. We say that the reaction is endothermic. The amount of heat that is lost or gained is written in a chemical equation as the heat of reaction.

The rate at which a reaction takes place is measured by how fast the products form. It depends on the number of reacting particles, the size of the particles, and the amount of energy required to make the product. Many reactions are reversible, which means that interactions between the products can change them back into the original reactants. Chemical equilibrium is reached when reactants change into products at the same rate that products change back into reactants.

6.1 Energy

Learning Goal

Describe some forms of energy.

When you are running, walking, dancing, or thinking, you are using energy to do **work**. In fact, **energy** is defined as the ability to do work. Suppose you are climbing a steep hill. Perhaps you become too tired to go on. We could say that you do not have sufficient energy to do any more work. Now, suppose you sit down and have lunch. In a while you will have obtained some energy from the food, and you will be able to do more work and complete the climb.

Potential and Kinetic Energy

All energy is potential or kinetic. **Potential energy** is stored energy, whereas **kinetic energy** is the energy of motion. A boulder resting on top of a mountain has potential energy because of its location. If the boulder rolls down the mountain, the potential energy becomes kinetic energy. Water stored in a reservoir has potential energy. When the water goes over the dam, the stored energy becomes kinetic energy. Even the food you eat has potential energy stored in its chemical bonds. When you digest food, you convert its stored energy to kinetic energy to do biological work.

SAMPLE 6.1 PROBLEM

Forms of Energy

Identify the energy in each of the following as potential or kinetic:

a. gasoline **b.** skating **c.** a candy bar

Solution

a. potential energy (stored)
b. kinetic energy (motion)
c. potential energy (stored)

Study Check

Would the energy in a stretched rubber band be kinetic or potential?

6.1

Forms of Energy

Energy from the sun is stored as chemical energy during the formation of trees, plants, and fuels. When wood is burned, chemical energy is released as heat or thermal energy, and light, which is radiant energy. **Heat** is associated with the motion of particles in a substance. A frozen pizza feels cold because the particles in the pizza have little heat, or thermal energy, and are moving very slowly. As heat is added, the motions of the particles increase, and the pizza becomes warm. Eventually the particles have enough thermal energy to make the pizza hot and ready to eat.

In an electric power plant, thermal energy from the burning of fossil fuels such as natural gas or coal is used to produce steam, which turns the turbines and generates electrical energy. In your home, electrical energy is converted to radiant energy when you switch on a light bulb, to mechanical energy when you use a mixer or a washing machine, and to thermal energy when you use a hair dryer or a toaster (see Figure 6.1). In your body, chemical energy from the food you eat is converted to mechanical energy when you use muscles to pedal a bicycle, run a marathon, or mow the lawn.

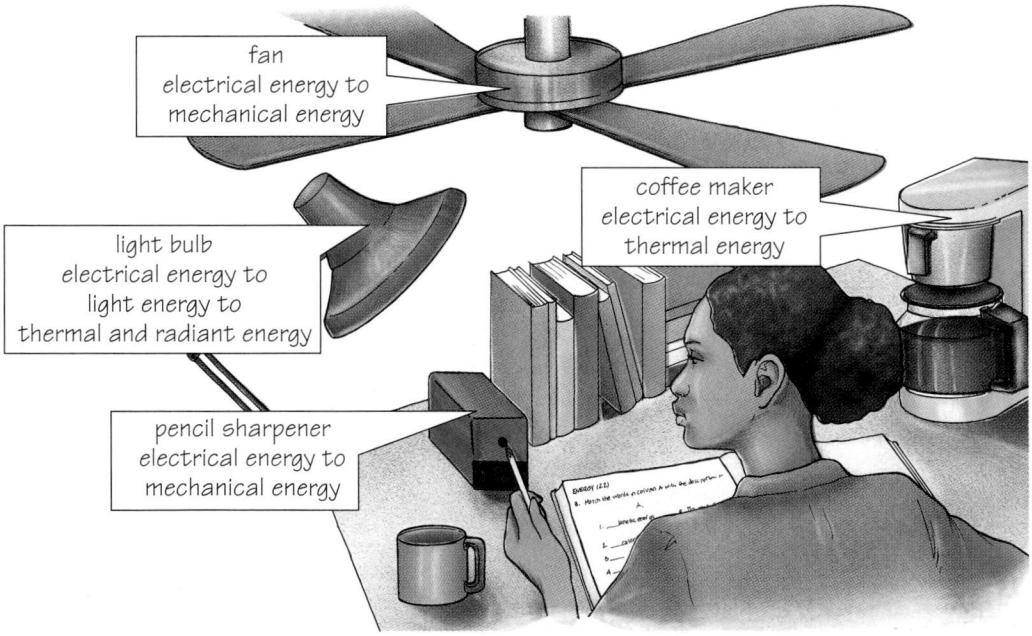

Figure 6.1 Examples of electrical energy converted into light energy, mechanical energy, and thermal energy.

ENVIRONMENTAL NOTE

Global Warming

The amount of carbon dioxide (CO_2) gas in our atmosphere is on the increase as we burn more gasoline, coal, and natural gas. The oceans and forests normally absorb carbon dioxide, but they cannot keep up with the continued increase. The cutting of trees in the rain forests (deforestation) reduces the amount of carbon dioxide removed from the atmosphere. Many of the trees are also burned as land is cleared. It has been estimated that deforestation may account for 15–30% of the carbon dioxide that goes into the atmosphere each year.

The carbon dioxide in the atmosphere acts like the glass in a greenhouse. When sunlight warms the earth's surface, some of the heat is trapped by carbon dioxide as shown in Figure 6.2. As CO_2 levels rise, more heat is trapped. It is not yet clear how severe the effects of global warming might be. Some scientists estimate that by around the year 2030, the atmospheric level of carbon dioxide could double and cause the temperature of Earth's atmosphere to rise by 2–5°C. If that should happen, it would have a profound impact on Earth's climate. For example, an increase in the melting of snow and ice could raise the ocean levels by as much as 2 m, which is enough to flood many cities located on the ocean shorelines.

Worldwide efforts are being made to reduce fossil fuel use and to slow or stop deforestation. It will require cooperation throughout the world to avoid the bleak future that some scientists predict should global warming continue unchecked.

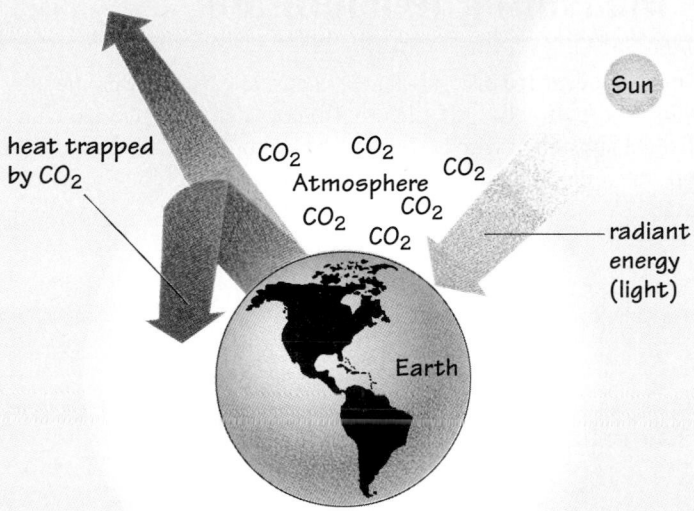

Figure 6.2 As the level of carbon dioxide (chemical formula CO_2) increases in the atmosphere, more heat is retained, causing an increase in Earth's temperature.

QUESTIONS & PROBLEMS

Energy

6.1 Discuss the changes in the potential and kinetic energy of a roller-coaster ride as the roller-coaster climbs up a hill and goes down the other side.

6.2 Discuss the changes in the potential and kinetic energy of a ski jumper taking the elevator to the top of the jump, and going down the ramp.

6.3 Indicate whether each statement describes potential or kinetic energy:
a. water at the top of a waterfall **b.** kicking a ball
c. the energy in a lump of coal **d.** a skier at the top of a hill

6.4 Indicate whether each statement describes potential or kinetic energy:
a. the energy in your food **b.** a tightly wound spring
c. an earthquake **d.** a car speeding down the freeway

6.5 What are the forms of energy involved in the following examples?
Example: Striking a match converts chemical energy into heat and light
a. using a hair dryer **b.** using an electric fan
c. burning gasoline in a car engine
d. sunlight falling on a solar water heater

6.2

6.6 What are the forms of energy involved in the following examples?
a. turning on a light switch **b.** cooking food on a gas stove
c. using a microwave oven **d.** burning a log in the fireplace

---

6.2 Measuring Temperature

Learning Goal

Given a temperature, calculate a corresponding temperature on another scale.

When we take the **temperature** of a substance, we are measuring the intensity of heat. Temperature tells us the direction that heat will flow between two substances. Heat always flows from a substance with a higher temperature to a substance with a lower temperature until the temperatures of both are the same. When you drink hot coffee or touch a hot stove, heat flows to your mouth or hand, which is at a lower temperature. When you touch an ice cube, it feels cold because heat flows from your hand to the colder ice cube.

We use thermometers to determine how "hot" or "cold" a substance is. When a thermometer is placed in a substance, the heat flow causes the liquid in the thermometer—alcohol or mercury—to expand or contract.

Celsius and Fahrenheit Temperatures

Temperatures in science, and in most of the world, are measured and reported in *Celsius (°C)* units. In the United States, everyday temperatures are commonly reported in *Fahrenheit (°F)* units. A typical room temperature of 22°C would be the same as 72°F. A normal body temperature of 37.0°C is 98.6°F.

On the Celsius and Fahrenheit scales, the temperatures of melting ice and boiling water are used as reference points. On the **Celsius scale,** the freezing point of water is defined as 0°C, and the boiling point as 100°C. On a **Fahrenheit scale,** water freezes at 32°F and boils at 212°F. On each scale, the temperature difference between freezing and boiling is divided into smaller units, or degrees. On the Celsius scale, there are 100 units between the freezing and boiling of water compared with 180 units on the Fahrenheit scale. That makes a Celsius unit almost twice the size of a Fahrenheit degree: 1°C = 1.8°F. (See Figure 6.3.)

$$180 \text{ Fahrenheit units} = 100 \text{ Celsius units}$$

$$\frac{180 \text{ Fahrenheit units}}{100 \text{ Celsius units}} \quad \text{or} \quad \frac{1.8 \text{ Fahrenheit units}}{1 \text{ Celsius unit}}$$

In a chemistry laboratory or a hospital, you will measure temperatures in Celsius units. If you do need to convert to Fahrenheit, multiply the Celsius temperature by the above conversion factor to change to Fahrenheit units, and add 32 degrees. The 32 you

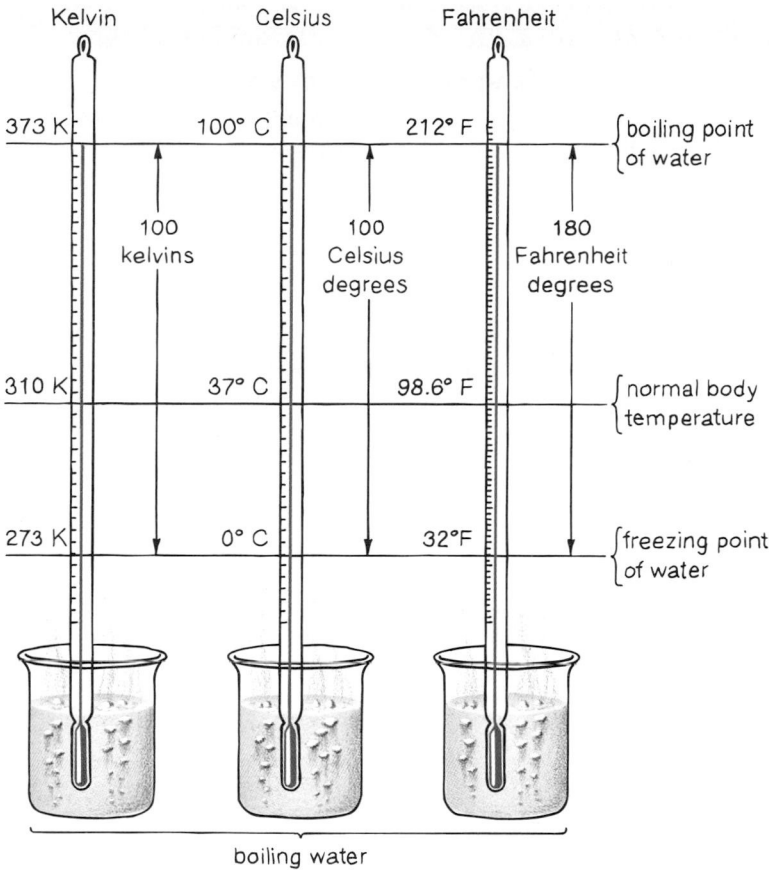

Figure 6.3 A comparison of the Fahrenheit, Celsius, and Kelvin temperature scales.

add adjusts the freezing point on the Celsius scale of 0°C to 32°F on the Fahrenheit scale.

$$°F = \underbrace{\frac{1.8°F}{1°C}(°C)}_{\text{Changes °C to °F}} + \underbrace{32°}_{\text{Adjusts freezing points}} \longrightarrow °F = 1.8\,(°C) + 32°$$

HEALTH NOTE

Variation in Body Temperature

Normal body temperature is considered to be 37.0°C, although it varies throughout the day and from person to person. Oral temperatures of 36.1°C are common in the morning and climb to a high of 37.2°C between 6 P.M. and 10 P.M. Temperatures above 37.2°C for a person at rest are usually an indication of disease. Individuals involved in prolonged exercise may also experience elevated temperatures. Body temperatures of marathon runners can range from 39°C to 41°C as heat production during exercise exceeds the body's ability to lose heat.

Changes of more than 3.5°C from the normal body temperature begin to interfere with bodily functions. Temperatures above 41°C can lead to convulsions, particularly in children, which may cause permanent brain damage. Heatstroke occurs above 41.1°C. Sweat production stops and the skin becomes hot and dry. The pulse rate is elevated, and respiration becomes weak and rapid. The person can become lethargic and lapse into a coma. Damage to internal organs is a major concern, and treatment, which must be immediate, may include immersing the person in an ice-water bath.

In hypothermia, body temperature can drop as low as 28.5°C. The person may appear cold and pale and have an irregular heartbeat. Unconsciousness can occur if the body temperature drops below 26.7°C. Respiration becomes slow and shallow, and oxygenation of the tissues decreases. Treatment involves providing oxygen and increasing blood volume with glucose and saline fluids. Internal temperature may be restored by injecting warm fluids (37.0°C) into the peritoneal cavity.

6.2

S·A·M·P·L·E 6.2 P·R·O·B·L·E·M

Temperature Conversion

While traveling in China, you discover that your temperature is 38.2°C. What is your temperature measured with a Fahrenheit thermometer?

Solution

Using the temperature conversion equation, we write:

$$°F = 1.8(°C) + 32°$$

In the equation, *the values of 1.8 and 32 are exact numbers.* Therefore, the answer is reported to the same decimal place as the initial temperature.

$$°F = 1.8(38.2) + 32$$

$$= 68.8 + 32 \quad \text{Exact}$$

$$= 100.8°F \quad \text{Answer to first decimal place}$$

Study Check

When making ice cream, you use rock salt to chill the mixture. If the temperature drops to −11°C, what is it in °F?

S·A·M·P·L·E 6.3 P·R·O·B·L·E·M

Converting to Celsius Temperature

You are going to cook a turkey at 325°F. If you use an oven with Celsius settings, at what temperature should you set it?

Solution

To convert from Fahrenheit to Celsius, we rearrange the temperature equation.

$$°F = 1.8\,(°C) + 32$$

$$°F - 32 = 1.8\,(°C)$$

$$\frac{(°F - 32)}{1.8} = °C$$

Entering the temperature of 325°F into the equation gives:

$$°C = \frac{(325 - 32)}{1.8} = \frac{\overset{\text{Three SFs}}{293}}{\underset{\text{Exact}}{1.8}} = 163°C$$

On your calculator, first subtract (325 − 32), and then divide by 1.8.

Study Check

Your patient has a temperature of 103.6°F. What is this temperature on a Celsius thermometer?

Kelvin Temperature Scale

Scientists tell us that the coldest temperature possible is −273°C (more precisely, −273.16°C). On the **Kelvin scale,** this temperature, called **absolute zero,** has the value of 0 Kelvin (0 K). Units on the Kelvin scale are called kelvins (K); no degree symbol is used. Because there are no lower temperatures, the Kelvin scale has no negative numbers. Between the freezing and boiling points of water, there are 100 kelvins, which makes a kelvin equal in size to a Celsius unit.

$$1\ \text{K} = 1°\text{C}$$

To calculate a Kelvin temperature, add 273 to the Celsius temperature:

$$\text{K} = °\text{C} + 273$$

Table 6.1 gives a comparison of some temperatures on the three scales.

SAMPLE 6.4 PROBLEM

Converting Celsius Temperature to Kelvins

What is a normal body temperature of 37°C on the Kelvin scale?

Solution

To convert from Celsius temperature to a Kelvin temperature, we use the equation

$$\text{K} = °\text{C} + 273$$
$$\text{K} = 37 + 273$$
$$= 310.\ \text{K}$$

Study Check

You are cooking a pizza at 375°F. What is that temperature in kelvins? (*Hint:* Convert to Celsius temperature first.)

SAMPLE 6.5 PROBLEM

Converting from Kelvin to Celsius Temperature

In a cryogenic laboratory, a technician freezes a sample using liquid nitrogen at 77 K. What is the temperature in °C?

Solution

To find the Celsius temperature, we rearrange the Kelvin equation:

$$\text{K} = °\text{C} + 273$$
$$\text{K} - 273 = °\text{C}$$

Using our given temperature of 77 K, we calculate the Celsius temperature for the freezing of the sample.

$$°\text{C} = 77\ \text{K} - 273 = -196°\text{C}$$

Study Check

On the planet Mercury, the average night temperature is 13 K, and the average day temperature is 683 K. What are these temperatures in Celsius degrees?

6.2

Table 6.1 A Comparison of Temperatures

Example	Fahrenheit (°F)	Celsius (°C)	Kelvin (K)
Sun	9937	5503	5776
A hot oven	450	232	505
A desert	120	49	322
A high fever	104	40	313
Room temperature	72	22	295
Water freezes	32	0	273
A northern winter	−76	−60	213
Helium boils	−452	−269	4
Absolute zero	−459	−273	0

QUESTIONS & PROBLEMS

Measuring Temperature

6.7 True or false:

a. When you heat water, the water molecules move faster.
b. When molecules move faster, the substance gets hotter.
c. When a substance gets hotter, its temperature goes down.
d. Energy always flows from a hotter substance to a cooler one.

6.8 True or false:

a. When water cools in a refrigerator, the water molecules move slower.
b. When water molecules move slower, the water gets colder.
c. When a substance cools, its temperature goes down.
d. Cooling involves a loss of energy.

6.9 Your friend who is visiting from France just took her temperature. When she reads 99.8, she becomes concerned that she is quite ill. How would you explain the temperature to your friend?

6.10 You have a friend who is using a recipe for flan from a Mexican cookbook. You notice that he set your oven temperature at 175°F. What would you advise him to do?

6.11 Solve the following temperature conversions:

a. 37.0°C = ____ °F
b. 65.3°F = ____ °C
c. −27°C = ____ K
d. 62°C = ____ K
e. 114°F = ____ °C
f. 72°F = ____ K

6.12 Solve the following temperature conversions:

a. 25°C = ____ °F
b. 155°C = ____ °F
c. −25°F = ____ °C
d. 224 K = ____ °C
e. 545 K = ____ °C
f. 875 K = ____ °F

6.13 **a.** A patient with heat stroke has a temperature of 106°F. What does this read on a Celsius thermometer?

b. Because high fevers can cause convulsions in children, the doctor wants to be called if the child's temperature goes over 40.0°C. Should the doctor be called if a child has a temperature of 103°F?

6.14 a. Hot compresses are being prepared for the patient in room 32B. The water is heated to 145°F. What is the temperature of the hot water in °C?

b. During extreme hypothermia, a young woman's temperature dropped to 20.6°C. What was her temperature on the Fahrenheit scale?

Study Pack

Chem of Life Water: Properties of Water 5

6.3

Learning Goal

Given the mass of a sample, and the temperature change, calculate the heat lost or gained.

6.3 Measuring Heat Energy

In chemistry and the health sciences, heat is commonly measured in calories, from the Latin *caloric* meaning "heat." A **calorie (cal)** is the amount of heat needed to raise the temperature of exactly 1 g of water by exactly 1°C. A **kilocalorie (kcal),** which is also the nutritional "big" *Calorie,* is used for larger amounts of heat. The **joule (J),** pronounced "jewel," is the SI unit of energy. One calorie is the same amount of energy as 4.18 joules. One kilojoule is 1000 joules.

1 kcal = 1000 calories

1 cal = 4.18 J

1 kJ = 1000 J

EXPLORE YOUR WORLD

Loss and Gain of Heat

Obtain three matching cups or glasses. Place 2 cups of cold tap water in the first one, and 2 cups of hot tap water in the second one. If you have a candy thermometer, determine the temperature of each. Otherwise, use your finger to quickly test each water sample. Pour about one-half of each water sample into a third cup or glass. Compare the temperature of the mixed water sample with the initial two water samples.

Questions

1. How does the temperature of the mixed water sample compare to the original hot water sample and cold water sample?
2. What happened to some of the heat energy of the hot water in the mixed sample?
3. Why did the temperature of the cold sample change in the mixed sample?

Specific Heat

All substances absorb energy, but some need only a small amount of heat to become hot and others need more. The amount of heat that raises the temperature of 1 g of a substance by 1°C is known as its **specific heat.**

$$\text{Specific heat} = \frac{\text{heat (cal or J)}}{1\text{ g} \times 1°\text{C}}$$

On a hot day at the beach, the ocean water is cool to the touch, the air is warm; but the sand is so hot, it can burn your feet. Sand has a lower specific heat than water. Therefore, sand reaches a higher temperature than water when it absorbs the same amount of heat. Because water has a high specific heat, one of its main functions in the body is to absorb large amounts of heat without causing fluctuations in body temperature.

When you cook, you may use aluminum or copper pans. If you compare specific heats, you find that the amount of heat that raises 1 g of water 1°C increases the temperature of 1 g of aluminum by about 5°C, and 1 g of copper by 10°C. Materials such as aluminum and copper can heat quickly to cook food. Table 6.2 lists the specific heats of a variety of materials.

Table 6.2 Specific Heats of Some Substances

Substance	Specific Heat (cal/g °C)	Specific Heat (J/g °C)
Water	1.00	4.18
Alcohol	0.58	2.4
Wood	0.42	1.8
Aluminum	0.22	0.90
Sand	0.19	0.79
Iron	0.11	0.46
Copper	0.093	0.39
Silver	0.057	0.24
Gold	0.031	0.13

EXPLORE YOUR WORLD

Specific Heat

Obtain two identical containers, such as bowls or pie tins, and two thermometers. Place water in one container and an equal amount of sand in the other. Set the containers in the sun and place a thermometer in each to the same depth. Record the temperature initially and every 10 minutes after that for a total of 60 minutes.

Questions

1. Which sample had the higher final temperature?
2. How would you graph this experiment?
3. Why does one substance heat up more quickly than another?

6.3

Calculations Using Specific Heat

The heat gained or lost by a substance is calculated by multiplying its mass, temperature change, and specific heat. The temperature change often noted as ΔT is the difference in the temperatures of the sample before and after heat is added or removed. (The Greek letter Delta, symbol Δ, means "change in"; read ΔT as "delta T.")

$$\text{Heat gain or loss} = (\text{mass}) \times \begin{pmatrix}\text{temperature}\\ \text{change}(\Delta T)\end{pmatrix} \times (\text{specific heat})$$

$$\text{Heat (calories)} = \text{g} \times {}^\circ\text{C} \times \frac{\text{cal}}{\text{g}\ {}^\circ\text{C}}$$

$$\text{Heat (Joules)} = \text{g} \times {}^\circ\text{C} \times \frac{\text{J}}{\text{g}\ {}^\circ\text{C}}$$

S·A·M·P·L·E 6.6 P·R·O·B·L·E·M

Calculating Heat Energy

How many calories are needed to warm 15 g of water from 12°C to 36°C?

Solution

The temperature change (ΔT) is the difference between the final and initial temperatures.

$$\Delta T = 36^\circ\text{C} - 12^\circ\text{C} = 24^\circ\text{C}$$

The heat is calculated by multiplying the mass of the water, the temperature change, and the specific heat of water (1.00 cal/g °C).

$$\underset{\text{Mass}}{15\,\cancel{\text{g}}} \times \underset{\Delta T}{24\,\cancel{^\circ\text{C}}} \times \underset{\text{Specific heat}}{\frac{1.00\ \text{cal}}{\cancel{\text{g}}\,\cancel{^\circ\text{C}}}} = \underset{\text{Heat energy}}{360\ \text{cal}}$$

Study Check

How many calories are needed to heat 30.0 g of copper from 75°C to 125°C if the specific heat of copper is 0.093 cal/g °C?

S·A·M·P·L·E **6.7** P·R·O·B·L·E·M

Calculating Heat Loss in Joules

A 225-g sample of hot tea cools from 95.0°C to 23.0°C. What is the energy loss in joules assuming that tea has the same specific heat as water (4.18 J/g °C)?

Solution

$$\Delta T = 95.0°C - 23.0°C = 72.0°C$$

The number of joules lost during cooling is calculated as follows:

$$\underset{\text{Mass}}{225\,\cancel{g}} \times \underset{\Delta T}{72.0\,\cancel{°C}} \times \underset{\text{Specific heat}}{\frac{4.18\text{ J}}{\cancel{g}\,\cancel{°C}}} = \underset{\text{Heat released}}{67{,}700\text{ J}} \quad \text{or} \quad 6.77 \times 10^4\text{ J}$$

Study Check

How much heat in joules is released when 15 g of gold cools from 215°C to 35°C? (See Table 6.2.)

Q·U·E·S·T·I·O·N·S & P·R·O·B·L·E·M·S

Measuring Heat Energy

6.15 Why are cooking pans made of aluminum, iron, or copper?

6.16 Substances A and B with the same mass and the same initial temperature absorb the same amount of heat energy. The final temperature of A is 55°C higher than the temperature of B. What does this tell you about the specific heats of A and B?

6.17 Convert each of the following:
a. 3500 cal to kcal **b.** 28 cal to joules
c. 425 J to cal **d.** 4.5 kJ to cal

6.18 Convert each of the following:
a. 8.1 kcal to cal **b.** 325 J to kJ
c. 2550 cal to kJ **d.** 2.50 kcal to J

6.19 What is the amount of heat required in each of the following?
a. calories to heat 25 g of water from 15°C to 25°C
b. joules to heat 150 g of water from 0°C to 75°C
c. joules to heat 10.0 g of copper from 25°C to 275°C
d. kilojoules to heat 150 g of water in a kettle from 15°C to 77°C

6.20 What is the amount of heat involved in each of the following?
a. calories given off when 85 g of water cools from 45°C to 25°C
b. joules given off when 0.50 kg of water cools from 80.°C to 60.°C
c. kilocalories given off when 250 g of sand cools from 155°C to 42°C
d. joules given off when 25 g of water cools from 86°C to 61°C

6.21 How many joules are needed to raise the temperature of 10.0 g of silver from 15°C to 237°C?

6.22 An electric power plant releases warm water into a nearby stream. The temperature of 5.0 kg of streamwater increases from 22°C to 28°C. How many kilocalories of heat were absorbed by the water?

Study Pack

Learning Goal

Calculate the energy of a sample using calorimetry data.

6.4 A Closer Look: Energy and Nutrition

When you are watching your food intake, the Calories you are counting are actually kilocalories. In the field of nutrition, it is common to use the **Calorie, Cal** (with an uppercase C) to mean 1000 cal, or 1 kcal. Nutritional values may also be given in kilojoules (kJ).

6.4

Nutritional Calories

1 Cal = 1 kcal

1 kcal = 1000 cal

1 kcal = 4.18 kJ

The number of Calories in a food is determined by using an apparatus called a **calorimeter,** shown in Figure 6.4. A sample is placed in a steel container within the calorimeter and water is added to fill a surrounding chamber. When the food burns (combustion), the heat emitted causes an increase in the temperature of the water. By calculating the kilocalories (or Calories) absorbed by the water, we can determine the caloric content of the food as shown in Sample Problem 6.8.

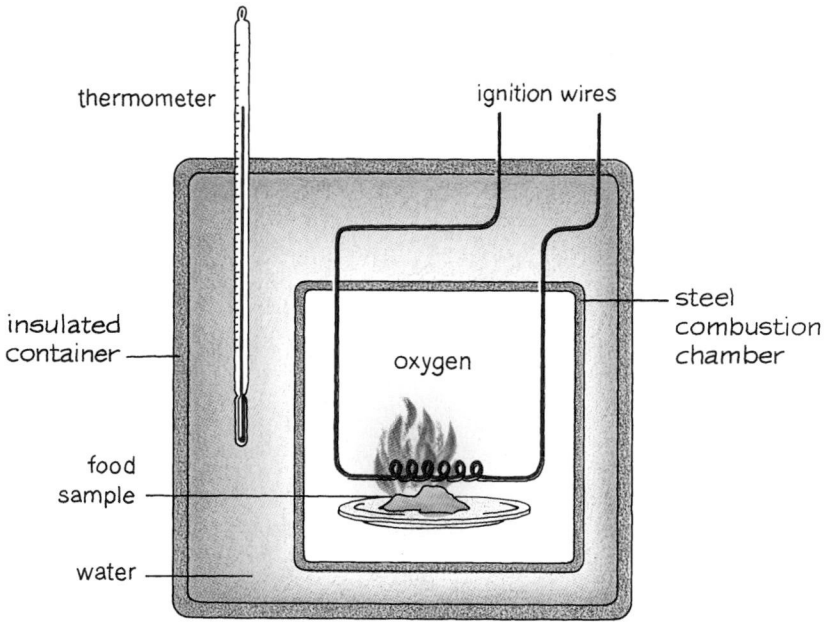

a

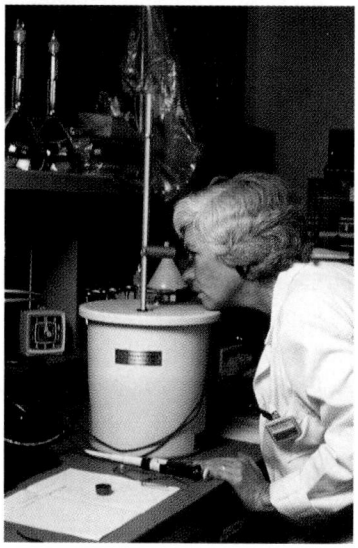

b

Figure 6.4 **(a)** A calorimeter. The energy released by the combustion of a food sample is determined by measuring the mass of the food sample, the mass of water in the calorimeter, and the change in temperature of the water. **(b)** The temperature change is observed during the combustion of a food sample in a calorimeter.

S·A·M·P·L·E **6.8** ·P·R·O·B·L·E·M

Calculating Caloric Values

A 2.3-g sample of butter, a fat, is placed in a calorimeter that is surrounded by 600. g of water at an initial temperature of 17°C. After combustion of the butter, the water has a temperature of 52°C. What is the caloric value (kcal/g) of butter?

Solution

The heat absorbed by the water is calculated as follows:

$$\underbrace{600.\,\cancel{g}}_{\text{Mass of water}} \times \underbrace{35\,\cancel{°C}}_{\Delta T} \times \underbrace{\frac{1.00\text{ cal}}{\cancel{g}\,\cancel{°C}}}_{\text{Specific heat}} = 21{,}000\text{ cal}$$

$$21{,}000\,\cancel{\text{cal}} \times \frac{1\text{ kcal}}{1000\,\cancel{\text{cal}}} = 21\text{ kcal}$$

Because the fat provided the 21 kcal of heat, the caloric value of butter is calculated as follows:

$$\frac{21\text{ kcal}}{2.3\text{ g fat}} = 9.1\text{ kcal/g fat}$$

Study Check

A 5.4-g sample of glucose, a carbohydrate, is placed in a calorimeter. The water in the container has a mass of 1000. g, and an initial temperature of 24°C. After all the glucose is burned, the water temperature is 46°C. What is the caloric value for glucose?

Caloric Food Values

The **caloric values** are the kilocalories per gram of the three types of food: carbohydrates, fats, and proteins. These values, which we will use as exact numbers, are listed in Table 6.3. If the composition of a food is known in terms of the mass of each food type, the caloric content (the total number of Calories) can be calculated.

$$\text{Kilocalories} = \underbrace{\cancel{g}}_{\text{Mass of carbohydrate, fat, or protein}} \times \underbrace{\frac{\text{kcal}}{\cancel{g}}}_{\text{Caloric value}}$$

The caloric content of a packaged food is listed in the nutritional information on the package, usually in terms of the number of Calories in a serving, as shown in Figure 6.5. The general composition and caloric content of some foods are given in Table 6.4.

Figure 6.5 The nutritional facts label on a food package lists the Calories and the grams of each food type per serving.

Table 6.3 Caloric Values for the Three Food Types

Food Type	Carbohydrate	Fat (lipid)	Protein
Caloric value	$\frac{4\text{ kcal}}{1\text{ g}}$	$\frac{9\text{ kcal}}{1\text{ g}}$	$\frac{4\text{ kcal}}{1\text{ g}}$

6.4

Table 6.4 General Composition and Caloric Content of Some Foods

Food	Protein (g)	Fat (g)	Carbohydrate (g)	Energy (kcal[a])
Banana, 1 medium	1	trace	26	110
Beans, red kidney, 1 cup	15	1	42	240
Beef, lean, 3 oz	22	5	trace	130
Carrots, raw, 1 cup	1	trace	11	50
Chicken, no skin, 3 oz	20	3	0	110
Egg, 1 large	6	6	trace	80
Milk, 4% fat, 1 cup	9	9	12	165
Milk, nonfat, 1 cup	9	trace	12	85
Oil, olive, 1 tbs.	0	14	0	130
Potato, baked	3	trace	23	105
Salmon, 3 oz	17	5	0	115
Steak, 3 oz	20	27	0	320
Yogurt, lowfat, 1 cup	8	4	13	120

[a] Values in kcal are rounded to nearest five.

S·A·M·P·L·E 6.9 ·P·R·O·B·L·E·M

Caloric Content for a Food

How many kilocalories are in a piece of chocolate cake that contains 35 g of carbohydrate, 10 g of fat, and 5 g of protein? Report the answer to the nearest tens place.

Solution

Using the caloric values for carbohydrate, fat, and protein (Table 6.3), we can calculate the total number of kcal:

$$\text{Carbohydrate:} \quad 35\,\cancel{g} \times \frac{4\text{ kcal}}{1\,\cancel{g}} = 140\text{ kcal}$$

$$\text{Fat:} \quad 10\,\cancel{g} \times \frac{9\text{ kcal}}{1\,\cancel{g}} = 90\text{ kcal}$$

$$\text{Protein:} \quad 5\,\cancel{g} \times \frac{4\text{ kcal}}{1\,\cancel{g}} = 20\text{ kcal}$$

$$\text{Total caloric content:} \quad 250\text{ kcal}$$

Study Check

A 1-oz (28 g) serving of oat bran hot cereal with half a cup of whole milk contains 22 g of carbohydrate, 7 g of fat, and 10 g of protein. If you eat two servings of the oat bran for breakfast, how many kilocalories will you obtain?

EXPLORE YOUR WORLD

Counting Calories

Obtain a food item that has a nutrition label. From the information on the label, determine the number of grams of carbohydrate, fat, and protein in one serving. Using the caloric values, calculate the total Calories for one serving. (Most products round off to the tens place.)

Question

How does your total for the Calories in one serving compare to the Calories stated on the label for a single serving?

HEALTH NOTE

Losing and Gaining Weight

The number of kilocalories needed in your daily diet depends on your age, sex, and physical activity. Some general levels of energy needs are given in Table 6.5.

When food intake exceeds energy output, a person's body weight increases. Food intake is usually regulated by the hunger center in the hypothalamus, located in the brain. The regulation of food intake is normally proportional to the nutrient stores in the body. If these nutrient stores are low, you feel hungry; if they are high, you do not feel like eating.

Weight reduction occurs when food intake is less than energy output. Many diet products contain cellulose, which has no nutritive value but provides bulk and makes you feel full. Some diet drugs depress the hunger center and must be used with caution because they excite the nervous system and can elevate blood pressure. Because muscular exercise is an important way to expend energy, an increase in daily exercise aids weight loss. Table 6.6 lists some activities and the amount of energy they require.

Table 6.5 Typical Energy Requirements

Age (yr)	Weight (lb)	Mass (kg)	Energy (kcal)
Young Adult			
Female (13–19)	115	52	2400
Male (13–19)	130	59	2980
Adult			
Female	121	55	2200
Male	143	65	3000

Table 6.6 Energy Expended by a 70-kg (155-lb) Person

Activity	Energy Expended (kcal/hr)
Sleeping	60
Sitting	100
Walking	200
Swimming	500
Running	550

EXPLORE YOUR WORLD

Calories in One Day

For one day, list the quantity and caloric content of all the foods you eat as well as your activities. Using a nutritional reference, look up the various food items and determine the Calories (kilocalories) you consume in one day. If you did some exercise, use Table 6.6 to determine the number of Calories you expend in one day.

Questions

1. If your caloric intake exceeds your caloric output, what will happen?
2. If you want to lose weight (or gain weight), how would the number of Calories you take in compare to the number of Calories you use in one day?

Q·U·E·S·T·I·O·N·S & P·R·O·B·L·E·M·S

A Closer Look: Energy and Nutrition

6.5

6.23 Use the following data for foods burned in a calorimeter to calculate the number of Calories (kilocalories) for each food:

Food	Mass (g) of Water in a Calorimeter	Temperature (°C) Initial	Temperature (°C) Final
a. celery (1 stalk)	500.	25	35
b. waffle (7-in. diameter)	5000.	20.	62
c. popcorn (1 cup, no oil)	1000.	25	50.

6.24 A 0.50-g sample of octane, a component in gasoline, is burned in a calorimeter. If the water in the calorimeter has a mass of 1200 g and an initial temperature of 22°C, what is the amount of heat in kilojoules per gram produced by the octane if the final temperature is 26°C?

6.25 Using the caloric values for foods, determine each of the following:
- **a.** The total Calories for 1 cup of orange juice that has 26 g of carbohydrate, no fat, and 2 g of protein.
- **b.** The grams of carbohydrate in 1 apple if the apple has no fat and no protein and provides 72 kcal of energy.
- **c.** The number of Calories in 1 tablespoon of vegetable oil, which contains 14 g of fat and no carbohydrate and no protein.
- **d.** How many Calories are in 1 breakfast roll if it has 30. g of carbohydrates, 15 g of fat, and 5 g of protein?

6.26 Using the caloric values for foods, determine each of the following:
- **a.** The total Calories for 2 tablespoons of crunchy peanut butter that contains 6 g of carbohydrate, 16 g of fat, and 7 g of protein.
- **b.** The grams of protein in 1 cup of soup that has 110 Cal with 7 g of fat and 9 g of carbohydrate.
- **c.** How many grams of sugar (carbohydrate) are in 1 can of cola if there are 140 Cal and no fat and no protein?
- **d.** How many grams of fat are in 1 avocado if there are 405 Calories, 13 g of carbohydrate, and 5 g of protein?

6.27 One cup of clam chowder contains 9 g of protein, 12 g of fat, and 16 g of carbohydrate. How many kilocalories are in the clam chowder? In Europe, the energy value on the label would be given as kJ. How many kJ should be on the label?

6.28 A high-protein diet contains 70. g of carbohydrate, 150 g of protein, and 5.0 g of fat. How many kilocalories does this diet provide? How many kilojoules is that?

6.5 States of Matter

Learning Goal

Identify the physical state of a substance as a solid, liquid, or gas.

Matter is anything that occupies space and has mass. This book, the food you eat, the water you drink, your cat or dog, and the air you breathe are just a few examples of matter. On Earth, matter exists in one of three *physical states:* solid, liquid, or gas. All matter is made up of tiny particles. In a **solid,** very strong attractive forces hold the particles close together. They are arranged in such a rigid pattern that they can only vibrate slowly in their fixed positions. This gives a solid a definite shape and volume. For many solids, this rigid structure produces a crystal as seen in sugar, quartz, and amethyst (see Figure 6.6).

Figure 6.6 The solid states of **(a)** sugar crystals, **(b)** quartz, and **(c)** amethyst, a purple form of quartz.

In a **liquid,** the particles have enough energy to move freely in random directions. They are still close to each other with strong attractions to maintain a definite volume, but there is no rigid structure. Thus, when oil, water, or vinegar is poured from one container to another, the liquid maintains its own volume but takes the shape of the new container (see Figure 6.7).

The air you breathe is made of gases, mostly nitrogen and oxygen. In a **gas,** the particles move so fast that the attractions are no longer as important. As a result, they move far apart, creating great distances between each gas particle. This behavior allows gases to expand to fill their container or to compress if the container size is reduced. Gases have no definite shape or volume of their own; they take the shape and volume of their container, as shown in Figure 6.8. Table 6.7 compares some of the properties of the three states of matter.

Figure 6.7 A liquid with a volume of 100 mL takes the shape of its container.

Table 6.7 Some Properties of Solids, Liquids, and Gases

Property	Solid	Liquid	Gas
	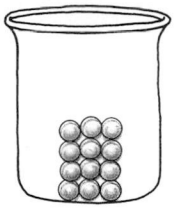		
Shape	Has a definite shape	Takes the shape of the container	Takes the shape of its container
Volume	Has a definite volume	Has a definite volume	Fills the volume of the container
Arrangement of particles	Fixed, very close	Random, close	Random, far apart
Interaction between particles	Very strong	Strong	Essentially none
Movement of particles	Very slow	Moderate	Very fast
Examples	Ice, salt, iron	Water, oil, vinegar	Water vapor, helium, air

Figure 6.8 A gas takes the shape and volume of its container.

EXPLORE YOUR WORLD

States of Matter

1. Place an object such as an eraser or marble in a glass. Transfer the object to another container and observe the shape and volume of the object in the new container.
2. Remove the object and place one cup of water in the container. Transfer all the water to another container. Measure the volume of the water again.
3. Open a bottle of household ammonia or perfume and stand a few feet away. Measure the time it takes for you to detect its odor.

6.5

Questions

1. What can you say about the interaction between and arrangement of the particles in the solid object?
2. **a.** What happens to the shape of the water when you place it in a different container?
   b. Did the volume of the water change when you poured it into a different container?
   c. What can you say about the arrangement and attraction between the particles in a liquid?
3. **a.** Could you smell the ammonia (or perfume) right away? After a few minutes?
   b. How did the ammonia (or perfume) molecules get from the bottle to your nose?
   c. What can you say about the attraction and relationship between gas particles?
   d. Summarize the properties of solids, liquids, and gases.

SAMPLE 6.10 PROBLEM

States of Matter

Identify each of the following as a solid, a liquid, or a gas:
a. oxygen in the air
b. a substance with a fixed arrangement of particles
c. safflower oil
d. a substance that has great distances between its particles

Solution

a. gas **b.** solid **c.** liquid **d.** gas

Study Check

What state of water has a definite volume but takes the shape of its container?

QUESTIONS & PROBLEMS

States of Matter

6.29 Indicate whether each of the following describes a gas, a liquid, or a solid:
a. This substance has no definite volume or shape.
b. The particles in this substance are not attracted to each other.
c. The particles of this substance are held in a definite structure.

6.30 Indicate whether each of the following describes a gas, a liquid, or a solid:
a. The substance has a definite volume but takes the shape of the container.
b. The particles of this substance are very far apart.
c. This substance occupies the entire volume of the container.

6.6 Melting and Freezing

Study Pack

Chem of Life Water: Properties of Water 8

Learning Goal

Describe the change of state between solid and liquid and calculate the energy involved.

6.6

Matter undergoes a **change of state** when it is converted from one state to another state. When an ice cube melts in a drink, a change of state has occurred.

When we heat a solid, its particles become so lively that they overcome the forces holding them together. At a temperature called the **melting point (mp),** the particles of the solid separate and move about in random patterns. The substance is **melting,** changing from a solid to a liquid.

If the temperature is lowered, the reverse process takes place. Kinetic energy is lost, the particles slow down, and attractive forces pull the particles close together. The substance is **freezing.** A liquid changes to a solid at the **freezing point (fp),** which is the same temperature as the melting point. Every substance has its own freezing (melting) point: water freezes at 0°C; ice melts at 0°C; gold freezes (melts) at 1064°C; nitrogen freezes (melts) at −210°C.

Suppose we have a glass containing ice and water. At warm temperatures, the ice melts, forming more liquid. At cold temperatures, heat is lost and the liquid freezes. However, at the melting (freezing) point of 0°C, ice melts at the same rate that water freezes. The reversible processes of melting and freezing are in balance, shown by writing double arrows in opposite directions. We say that the ice–liquid system has reached dynamic equilibrium.

$$\text{Solid} \underset{\text{Freezing}}{\overset{\text{Melting}}{\rightleftharpoons}} \text{liquid}$$

Heat of Fusion

During melting, energy called the **heat of fusion** is needed to separate the particles of a solid. For example, 80. calories of heat are needed to melt exactly 1 g of ice.

Heat of Fusion for Water

$$\frac{80.\ \text{cal}}{1\ \text{g ice}}$$

The heat of fusion (80. cal/g) is also the heat that must be removed to freeze 1 g of water. Water is sometimes sprayed in fruit orchards during very cold weather. If the air temperature drops to 0°C, the water begins to freeze. Heat is released in this change of state, which warms the air and protects the fruit.

To determine the heat needed to melt a sample of ice, multiply the mass of the ice by its heat of fusion. There is no temperature change in the calculation because temperature remains constant as long as the ice is melting.

Calculating Heat to Melt (or Freeze) Water

$$\text{Heat} = \text{mass} \times \text{heat of fusion}$$
$$\text{cal} = \text{g} \times 80.\ \text{cal/g}$$

S·A·M·P·L·E **6.11** ·P·R·O·B·L·E·M

Heat of Fusion

Ice cubes at 0°C with a mass of 26 g are added to your soft drink.

a. How much heat (calories) will be absorbed to completely melt the ice at 0°C?

b. What happens to the temperature of your drink? Why?

Solution

a. The heat in calories required to melt the ice is calculated as follows:

$$\underset{\text{Mass}}{26\,g} \times \underset{\text{Heat of fusion}}{\frac{80.\ \text{cal}}{1\,g}} = 2100\ \text{cal}$$

b. The drink will be colder because heat from the drink is providing the energy to melt the ice.

Study Check

6.6

In a freezer, 120 g of water at 0°C is placed in an ice cube tray. How many kilojoules of heat must be removed to form ice cubes?

Sublimation

In a process called **sublimation,** the particles on the surface of a solid absorb enough heat to change directly to a gas with no temperature change. For example, dry ice, which is solid carbon dioxide, sublimes at −78°C. It is called "dry" because it does not form a liquid. You may have noticed that mothballs seem to disappear at room temperature. They have sublimed, a process we can detect by the odor of their vapor. In very cold areas, snow does not melt but sublimes directly to vapor. In a frost-free refrigerator, ice on the walls of the freezer sublimes when warm air is circulated through the compartment during the defrost cycle.

$$\text{Solid} \underset{\text{Deposition}}{\overset{\text{Sublimation}}{\rightleftharpoons}} \text{gas}$$

Figure 6.9 Freeze-dried foods are produced by sublimation.

Freeze-dried foods prepared by sublimation are convenient for long storage, and for camping and hiking. (See Figure 6.9.) A food that has been frozen is placed in a vacuum chamber where it dries as the ice sublimes. The dried food retains all of its nutritional value and needs only water to be edible. A food that is freeze-dried does not need refrigeration because bacteria cannot grow without moisture. The reverse process of sublimation is called **deposition.**

QUESTIONS & PROBLEMS

Melting and Freezing

6.31 Which of the following is a change of state?
- **a.** An iceberg is solid water.
- **b.** Water on the street freezes during a cold wintry night.
- **c.** A wood log is chopped to make kindling.
- **d.** Butter is melted to put on popcorn.

6.32 Which of the following is a change of state?
- **a.** Apple chips are made by freeze-drying apple slices.
- **b.** Dough is used to make different pasta shapes.
- **c.** Solid nitrogen melts at 63 K.
- **d.** Salt is dissolved in water to make a gargle for the throat.

6.33 Identify each of the following changes of state as melting, freezing, or sublimation.
- **a.** The solid structure of a substance breaks down as liquid forms.
- **b.** Heat is added to silver at its melting point of 962°C.
- **c.** Coffee is freeze-dried.
- **d.** Water on the street turns to ice during a cold wintry night.

6.34 Identify each of the following changes of state as melting, freezing, or sublimation.
a. Dry ice in an ice cream cart disappears.
b. Snow on the ground turns to liquid water.
c. Heat is removed from 125 g of liquid water at 0°C.
d. On a cold wintry day, rain turns to snow.

6.35 Calculate the heat needed at 0°C to make each of the following changes of state. Indicate whether heat was absorbed or released.
a. calories to melt 65 g of ice
b. joules to melt 17.0 g of ice
c. kilocalories to freeze 225 g of water
d. kilojoules to freeze 50.0 g of water

6.36 Calculate the heat needed at 0°C to make each of the following changes of state. Indicate whether heat was absorbed or released.
a. calories to freeze 35 g of water
b. joules to freeze 250 g of water
c. kilocalories to melt 140 g of ice
d. kilojoules to melt 5.0 kg of ice

Water:
Properties of Water 6, 7

6.7 Boiling and Condensation

Learning Goal

Describe the change of state between gas and liquid and calculate the energy involved.

Water in a mud hole disappears, unwrapped food dries out, and clothes hung on a line dry. Water **evaporates** when fast-moving water molecules have enough energy to escape from the surface. (See Figure 6.10a.) The loss of the "hot" water molecules removes heat and leaves the liquid cooler.

As water vapor particles cool, they lose kinetic energy and slow down. In **condensation,** the water molecules form liquid again as attractive forces pull them together, a process that is the reverse of evaporation. You may have noticed that condensation forms on the mirror when you take a shower. Because heat is lost as a liquid condenses, it is a warming process. That is why, when a rain storm is approaching, we notice a warming of the air as gaseous water molecules condense to rain.

EXPLORE YOUR WORLD

Evaporation, Boiling, and Condensation

1. Place some water in a teakettle or small pan on the stove. Hold a small mirror a few inches from the spout or over the water. What happens? Heat the water in the teakettle or pan. As it warms, hold the mirror a few inches from the spout or the surface of the water. When the water boils, *very carefully* hold the mirror near the spout or surface. Wear a kitchen mitt or use a hot pad. *Keep your fingers away! Steam burns!* In old Western movies, a mirror was held near a person's mouth or nose to see if that person was still alive. Explain.
2. Seal some slices of a fruit or vegetable such as celery or carrots in a plastic bag and place the same number in an open container. Place both in a frost-free freezer. Every few days check the food.

Questions

1. Why does steam change state when water vapor hits a mirror?
2. **a.** Compare the appearance of the food left in an open container with that sealed in a plastic bag. How would you account for any difference?
   b. Why should you tightly wrap food before putting it in a freezer?

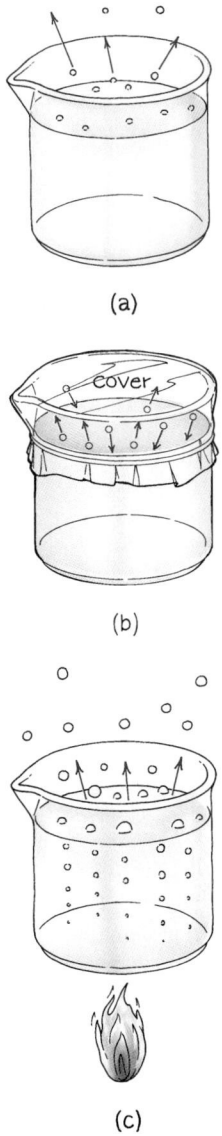

Figure 6.10 **(a)** Evaporation occurs at the surface of a liquid. **(b)** Evaporation and condensation reach equilibrium in a closed container. **(c)** Boiling occurs as bubbles of gas form throughout the liquid.

6.7

Liquid–Gas Equilibrium

With enough time, water in an open container will all evaporate. However, if a tight-fitting cover is placed on the container, the water level will go down just a small amount. At first, the water molecules evaporate from the surface. Then, some of the vapor molecules begin to condense and return to liquid. Eventually, the number of evaporating molecules is equal to the number condensing. (See Figure 6.10b.) The reverse processes of evaporation and condensation have equalized. As a result of the liquid–gas equilibrium, the water level does not go any lower.

$$\text{Liquid} \underset{\text{Condensation}}{\overset{\text{Evaporation}}{\rightleftharpoons}} \text{gas (vapor)}$$

Boiling

As the heat added to a liquid increases, more and more particles evaporate. At the **boiling point,** the particles throughout the liquid have the energy needed to change into a gas. The **boiling** of the liquid occurs as gas bubbles form throughout the liquid, then rise to the surface and escape. (See Figure 6.10c.) All the changes of state are illustrated in Figure 6.11.

S·A·M·P·L·E 6.12 P·R·O·B·L·E·M

Identifying States and Changes of States

Give the state or change of state described in the following:

a. particles on the surface of a liquid escaping to form vapor
b. a liquid changing to a solid
c. gas bubbles forming throughout a liquid

Solution

a. evaporation **b.** freezing **c.** boiling

Study Check

What is the phase change that occurs in condensation?

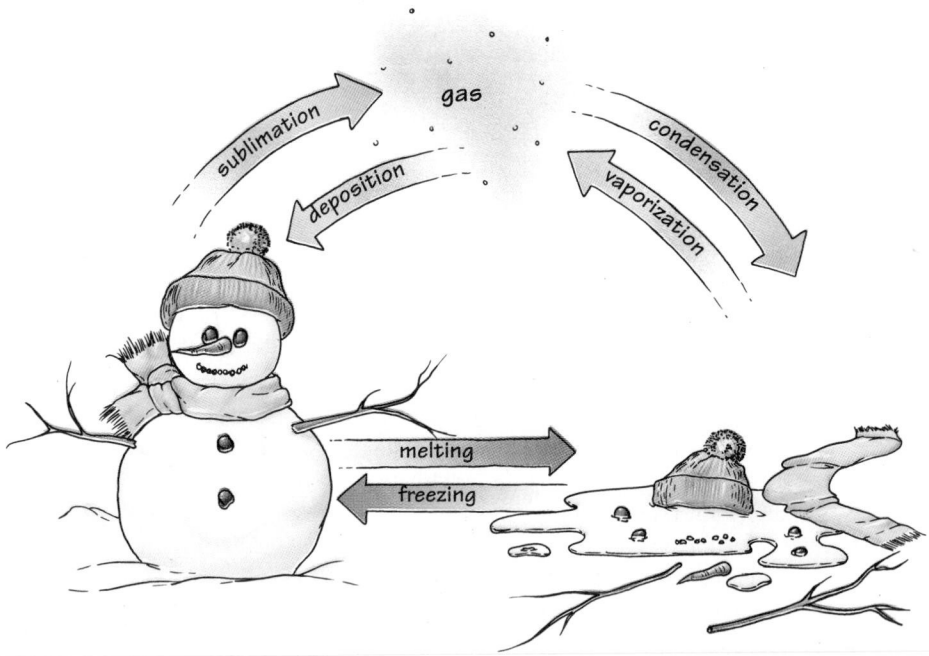

Figure 6.11 A summary of the changes of state.

Heat of Vaporization

The energy needed to vaporize exactly 1 g of liquid to gas is called the **heat of vaporization.** For water, 540 calories, a great amount of heat, is needed to convert 1 g of water to vapor.

Heat of Vaporization for Water

$$\frac{540 \text{ cal}}{1 \text{ g water (liquid)}}$$

When 1 g of water condenses, the heat of vaporization, 540 calories, is the amount of energy that must be removed. Therefore, 540 cal/g is also the *heat of condensation* of water. To calculate the amount of heat needed to vaporize (or condense) a sample of water, multiply the mass of the sample by the heat of vaporization. As before, no temperature change occurs for a change of state.

Calculating Heat to Vaporize (or Condense) Water

$$\text{Heat} = \text{mass} \times \text{heat of vaporization}$$
$$\text{cal} = \text{g} \times 540 \text{ cal/g}$$

SAMPLE 6.13 PROBLEM

Calculating Heat for Vaporization

In a sauna, 150 g of water is converted to steam at 100°C. How many kilocalories of heat are needed?

Solution

The heat in calories to vaporize 150 g of water at 100°C is calculated by multiplying its mass by the heat of vaporization of water. To convert calories to kilocalories, we use the equality: 1 kcal = 1000 cal

$$150\,\cancel{g} \times \frac{540\,\cancel{cal}}{1\,\cancel{g}} \times \frac{1 \text{ kcal}}{1000\,\cancel{cal}} = 81 \text{ kcal}$$

Study Check

When steam from a pan of boiling water reaches a cool window, it condenses. How much heat in kilocalories (kcal) is released when 25 g of steam condenses at 100°C?

HEALTH NOTE

Steam Burns

Hot water at 100°C will cause burns and damage to the skin. However, getting steam on the skin is even more dangerous. Let us consider 25 g of hot water at 100°C. If this water falls on a person's skin, the temperature of the water will drop to body temperature, 37°C. The heat released during cooling burns the skin. The amount of heat can be calculated from the temperature change of 63°C:

$$25\,\cancel{g} \times 63\,\cancel{°C} \times \frac{1.00 \text{ cal}}{\cancel{g}\,\cancel{°C}} = 1{,}600 \text{ cal heat}$$

For comparison, we can calculate the amount of heat released when 25 g of steam at 100°C hits the skin. First, the steam condenses to water (liquid) at 100°C:

$$25\,\cancel{g} \times \frac{540 \text{ cal}}{1\,\cancel{g}} = 14{,}000 \text{ cal heat released}$$

Now the temperature of the 25 g of water drops from 100°C to 37°C, releasing still more heat—1600 cal, as we saw earlier. Now we can calculate the total amount of heat released from the condensation of the steam as follows:

Condensation (100°C)	=	14,000 cal
Cooling (100°C to 37°C)	=	1,600 cal
Heat released	=	16,000 cal (rounded)

The amount of heat released from steam is 10 times greater than the heat from the same amount of hot water.

QUESTIONS & PROBLEMS

Boiling and Condensation

6.37 Identify each of the following changes of state as evaporation, boiling, or condensation:
a. The water vapor in the clouds changes to rain.
b. Wet clothes dry on a clothesline.
c. Lava flows into the ocean and forms steam.
d. After a hot shower, your bathroom mirror is covered with water.

6.8

6.38 Identify each of the following changes of state as evaporation, boiling, or condensation:
a. At 100°C, the water in a pan changes to steam.
b. On a cool morning, the windows in your car fog up.
c. A shallow pond dries up in the summer.
d. Your teakettle whistles when the water is ready for tea.

6.39 **a.** How does perspiration during heavy exercise cool the body?
b. Why do clothes dry more quickly on a hot summer day than on a cold winter day?
c. Why do wet clothes stay wet in a plastic bag?

6.40 **a.** When a sports injury occurs during a game, a spray such as ethyl chloride may be used to numb an area of the skin. Explain how a substance like ethyl chloride that evaporates quickly can numb the skin.
b. Why does water in a wide, flat, shallow dish evaporate more quickly than the same amount of water in a tall, narrow glass?
c. Which will dry out faster: a sandwich on a plate or a sandwich in plastic wrap?

6.41 Calculate the heat change at 100°C in each of the following problems. Indicate whether heat was absorbed or released.
a. calories to vaporize 10.0 g of water
b. joules to vaporize 50.0 g of water
c. kilocalories to condense 8.0 kg of steam
d. kilojoules to condense of 170 g of steam

6.42 Calculate the heat change at 100°C in each of the following problems. Indicate whether heat was absorbed or released.
a. calories to condense 10.0 g of steam
b. joules to condense 75 g of steam
c. kilocalories to vaporize 44 g of water
d. kilojoules to vaporize 5.0 kg of water

6.8 Heating and Cooling Curves

Learning Goal

Draw a heating or cooling curve for a substance and calculate the heat involved.

Diagrams for Changes of State

When heat is added to a solid, it begins to melt, forming a liquid. As more heat is added, the liquid warms until it boils and forms a gas. Each of these changes of state can be illustrated visually as a diagram called a **heating curve.** (See Figure 6.12.) In a heating curve, the temperature is shown on the vertical axis. The addition of heat is shown on the horizontal axis. Each of the segments represents a warming step or a change of state.

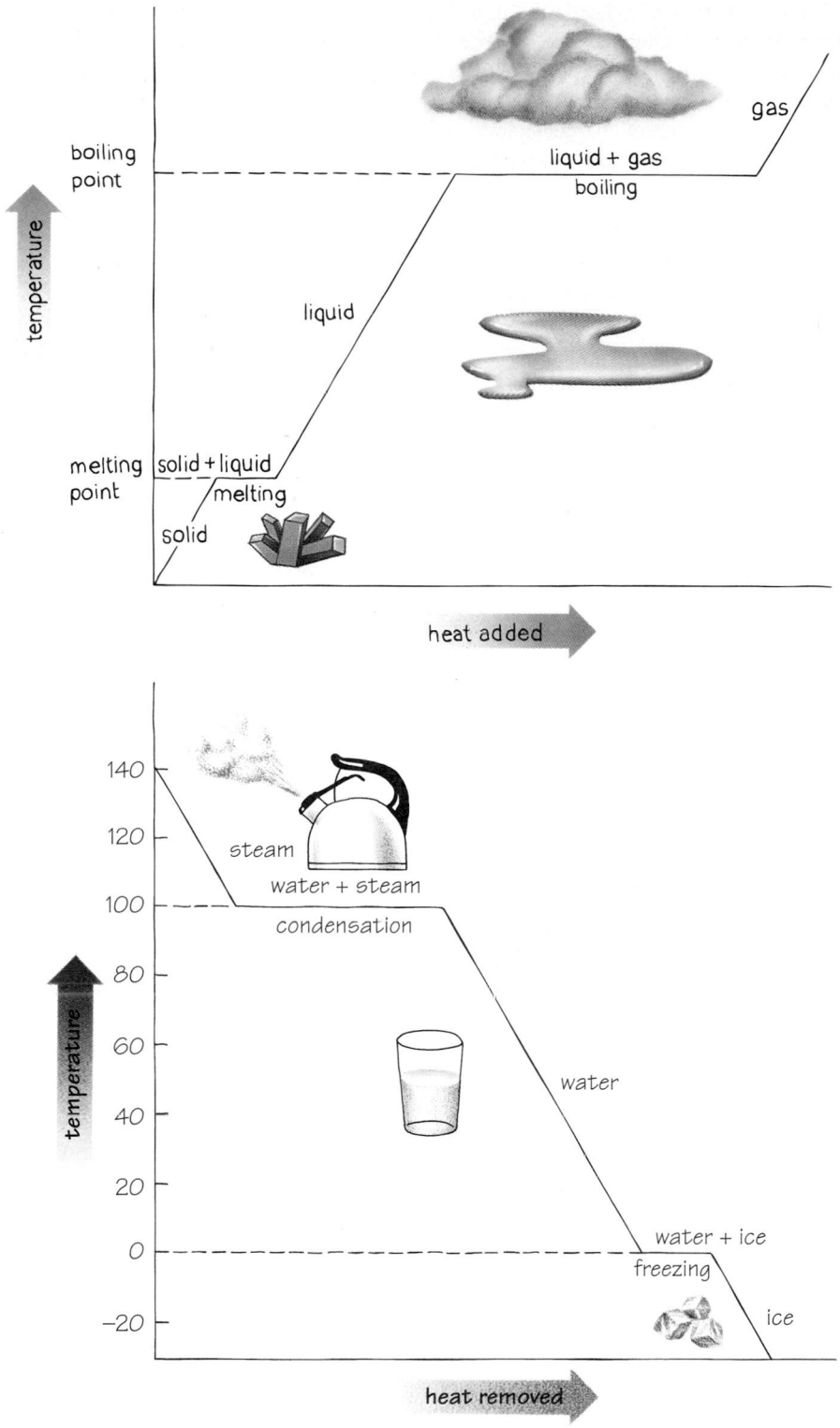

Figure 6.12 *Top graph:* A heating curve diagrams the temperature increases and changes in state as heat is added. *Bottom graph:* A cooling curve for water.

Steps on a Heating Curve

The first diagonal line indicates a warming of a solid as heat is added. When the melting temperature is reached, solid begins to change to a liquid. There is no change in temperature as the solid changes to liquid. This is shown as a flat line, or plateau, at the melting point on the heating curve. During the melting process, all of the heat is used to break apart the particles of the solid while temperature remains constant.

After all the particles are in the liquid state, the temperature begins to rise again as shown by the next diagonal line. As more heat is added, the liquid warms and the particles move about more vigorously until the boiling point is reached. Now the liquid begins to boil, indicated as another flat region on the diagram at its boiling point. Again there is no temperature change as long as liquid is changing into gas. After all the liquid has changed to gas, additional heat will cause the temperature to rise again.

Steps on a Cooling Curve

6.8

When heat is removed from a substance such as water vapor, the gas cools and condenses. A **cooling curve** is a diagram of this cooling process. (See Figure 6.12.) On the cooling curve, the temperature is plotted on the vertical axis and the removal of heat on the horizontal axis. If a gas is cooled, heat is lost and the temperature drops. At the condensation point (same as the boiling point), the gas begins to condense, forming liquid. This process is indicated by a flat line (plateau) on the cooling curve at the condensation point.

After all of the gas has changed into liquid, the particles within the liquid cool as indicated by the downward sloping line that shows the temperature decrease. At the freezing point, the particles in the liquid slow so much that solid begins to form. A second flat line at the freezing point indicates the change of state from liquid to solid (freezing). Once all of the substance is frozen, more heat can be lost, which lowers the solid's temperature below its freezing point.

Combining Energy Calculations

Up to now, we have calculated one step in a heating or cooling curve. However, many problems require a combination of steps that include a temperature change as well as a change of state. The heat is calculated for each step separately and then added together to find the total energy as seen in Sample Problem 6.14.

S·A·M·P·L·E 6.14 ·P·R·O·B·L·E·M

Combined Heat Calculations

Calculate the calories needed to convert 15.0 g of water at 25°C to steam at 100°C. You will need to use the specific heat of liquid water, 1.00 cal/g °C, and the heat of vaporization for water, 540 cal/g.

Solution

If we sketched a heating curve for water, we would find that two steps are needed for the calculation of heat. One step is the heat needed to warm the liquid from 25°C to 100°C, and the second step is the heat for the vaporization of the water to gas.

Step 1 Heat to warm the water (liquid) from 25°C to 100°C:

$$\underset{\text{Mass}}{15.0\,\cancel{g}} \times \underset{\Delta T}{75\,\cancel{°C}} \times \underset{\text{Specific heat of water}}{\frac{1.00\text{ cal}}{\cancel{g}\,\cancel{°C}}} = 1100\text{ cal}$$

Step 2 Heat to vaporize the water at 100°C:

$$\underset{\text{Mass}}{15.0\,\cancel{g}} \times \underset{\text{Heat of vaporization of water}}{\frac{540\text{ cal}}{1\,\cancel{g}}} = 8100\text{ cal}$$

We calculate the total heat by adding step 1 and step 2:

1. Heating water	1100 cal
2. Changing liquid to steam	8100 cal
Total heat needed:	9200 cal

Study Check

How many calories are released when 25.0 g of steam condenses at 100°C, cools to 0°C, and freezes? (*Hint:* The solution will require three energy steps.)

QUESTIONS & PROBLEMS

Heating and Cooling Curves

6.43 Draw a heating curve for a sample of ice that is heated from −20°C to 140°C. Indicate the segment of the graph that corresponds to each of the following:
a. solid **b.** melting point **c.** liquid
d. boiling point **e.** gas

6.44 Draw a cooling curve for a sample of steam that cools from 110°C to −10°C. Indicate the segment of the graph that corresponds to each of the following:
a. solid **b.** freezing point **c.** liquid
d. condensation point (boiling point) **e.** gas

6.45 Using the values for the heat of fusion, specific heat of water, and/or heat of vaporization, calculate the amount of heat energy in each of the following:
a. calories needed to warm 20.0 g of water at 15°C to 72°C (one step)
b. calories needed to melt 50.0 g of ice at 0°C, and to warm the liquid to 65°C (two steps)
c. kilojoules given off when 15 g of steam condenses at 100°C, and the liquid cools to 0°C (two steps)
d. kilocalories needed to melt 24 g of ice at 0°C, to warm the liquid to 100°C, and to vaporize it at 100°C (three steps)

6.46 Using heat of fusion, specific heat of water, and/or heat of vaporization, calculate the amount of heat energy in each of the following:
a. calories to condense 125 g of steam at 100°C, and to cool the liquid to 15°C (two steps)
b. joules to melt a 525-g ice cube at 0°C, and warm the liquid to 15°C (two steps)
c. kilocalories to condense 85 g of steam at 100°C, cool the liquid, and freeze it at 0°C (three steps)
d. calories to warm 55 mL of water (density = 1.0 g/mL) from 10.°C to 100°C, and vaporize it at 100°C (two steps)

6.47 A bag containing 250 g of ice at 0°C was placed on a burn on a patient's hand. When the bag was removed, the ice had melted, and the liquid water had come to a temperature of 24°C. How many kilocalories of heat were absorbed by the ice and cool water? How many kilojoules is that?

6.48 A 115-g sample of steam at 100°C is emitted from a volcano. It condenses, cools, and falls as snow at 0°C. How many kilocalories of heat are released? How many kilojoules is that?

Study Pack

Chem of Life Reactions and Equilibrium: Chemical Reactions 3–5

Learning Goal

Describe the energy in exothermic and endothermic reactions; list factors that change the rate of a reaction.

6.9

6.9 Energy in Chemical Reactions

For a chemical reaction to take place, the molecules of the reactants must come in contact with each other. If there are no collisions between the molecules (or atoms), no reaction is possible. When the molecules collide, bonds between atoms are broken and new bonds can form. The energy needed to break apart those bonds is called the **activation energy.** If the energy of a collision is less than the activation energy, the molecules bounce apart without reacting.

The concept of activation energy is analogous to climbing over a hill. To reach a destination on the other side, we must expend energy to climb to the top of the hill. Once we are at the top, we can easily roll down the other side. The energy needed to get us from our starting point to the top of the hill would be our activation energy.

Exothermic and Endothermic Reactions

As the bonds of the products form, energy is released into the reaction system. However, that energy may be greater or less than the activation energy of the reaction. The **heat of reaction** is the energy difference between the reactants and the products. In **exothermic reactions,** the energy of the products is lower than that of the reactants and heat is given off. In **endothermic reactions,** the energy of the products is higher than the energy of the reactants and heat must be absorbed for products to form. (See Figure 6.13.)

For example, the combustion of methane (CH_4) is an exothermic reaction. Heat (213 kcal per mole of CH_4 burned) is released along with the products CO_2 and H_2O.

Exothermic Reaction

$$\underset{\text{Methane}}{CH_4} + 2O_2 \longrightarrow CO_2 + 2H_2O + \underset{\text{Heat released}}{213 \text{ kcal of heat}}$$

The formation of hydrogen iodide is an endothermic reaction. In the overall energy change, heat is absorbed when the hydrogen and iodine undergo reaction.

Endothermic Reaction

$$H_2 + I_2 + \underset{\text{Heat absorbed}}{12 \text{ kcal of heat}} \longrightarrow 2HI$$

Reaction	Energy Change	Heat in the Equation
Exothermic	Heat released	Product side
Endothermic	Heat absorbed	Reactant side

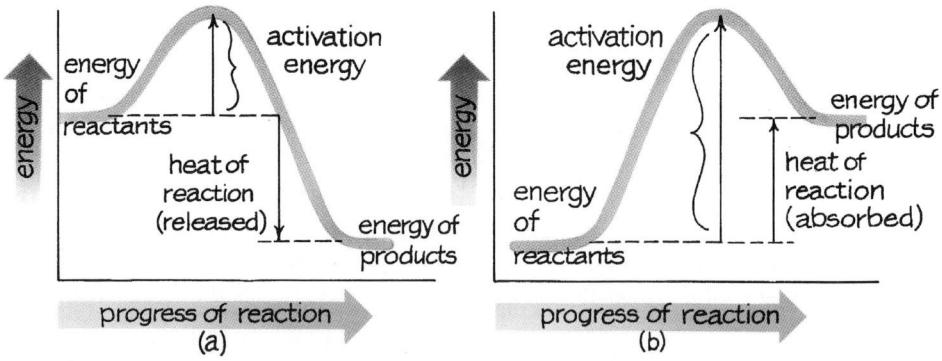

Figure 6.13 **(a)** In an exothermic reaction, the energy of the products is lower than the energy of the reactants; energy is released. **(b)** In an endothermic reaction, the energy of the products is higher than the energy of the reactants; energy is absorbed.

S·A·M·P·L·E 6.15 P·R·O·B·L·E·M

Exothermic and Endothermic Reactions

In the reaction of one mole of solid carbon with oxygen gas, the energy of the carbon dioxide product is 94 kcal lower than the energy of the reactants.

a. Is the reaction exothermic or endothermic?

b. What is the equation for the reaction?

Solution

a. exothermic **b.** $C + O_2 \longrightarrow CO_2 + 94\text{ kcal}$

Study Check

When two moles of ethanol (C_2H_5OH) reacts with oxygen (O_2), the products are carbon dioxide, water, and 326 kcal of heat. Write a balanced equation including the heat of reaction.

6.9

HEALTH NOTE

Hot Packs and Cold Packs

In a hospital, at a first-aid station, or at an athletic event, a *cold pack* may be used to reduce swelling from an injury, remove heat from inflammation, or decrease capillary size to lessen the effect of hemorrhaging. Inside the plastic container of a cold pack, there is a compartment containing solid ammonium nitrate (NH_4NO_3) that is separated from a compartment containing water (see Figure 6.14). The pack is activated when it is hit or squeezed hard enough to break the walls between the compartments and cause the ammonium nitrate to mix with the water (shown as H_2O over the reaction arrow). In an endothermic process, each gram of NH_4NO_3 that dissolves absorbs 79 cal of heat from the water. The temperature drops and the pack becomes cold and ready to use:

Endothermic Reaction in a Cold Pack

$$6.3\text{ kcal} + NH_4NO_3(s) \xrightarrow{H_2O} NH_4NO_3(aq)$$

Hot packs are used to relax muscles, lessen aches and cramps, and increase circulation by expanding capillary size. Constructed in the same way as cold packs, a hot pack may contain the salt $CaCl_2$. The dissolving of the salt in water is exothermic releasing 160 cal per gram of salt. The temperature rises and the pack becomes hot and ready to use:

Exothermic Reaction in a Hot Pack

$$CaCl_2(s) \xrightarrow{H_2O} CaCl_2(aq) + 18\text{ kcal}$$

Figure 6.14 In a cold pack, an endothermic reaction lowers the temperature of the pack. In a hot pack, an exothermic reaction raises the temperature.

EXPLORE YOUR WORLD

Hotter or Colder?

Place 1 cup of water in a glass and test the temperature with your finger. You can use a candy thermometer if you have one. Add a tablespoon of baking soda and stir.

Questions

1. What happened to the temperature of water when you added baking soda?
2. Is the reaction exothermic or endothermic?
3. Would the heat for the reaction be written on the reactant or the product side of an equation for the reaction?
4. Predict what will happen if you add another tablespoon of baking soda. Test your prediction.

6.9

Rate of Reaction

The **rate** (or speed) **of a reaction** is measured by the amount of reactant used up, or the amount of product formed, in a certain period of time. Reactions with low activation energies go faster than reactions with high activation energies. The rate of a reaction can be affected by the temperature, the amounts of reactants in the container, and catalysts.

At higher temperatures, the increase in kinetic energy makes the reactants move faster and collide more often, and provides more collisions with the required energy of activation. Reactions go faster at higher temperatures. For every approximately 10°C increase in temperature, most reactions rates double. If we want food to cook faster, we raise the temperature. When body temperature rises, there is an increase in the pulse rate, rate of breathing, and metabolic rate. On the other hand, we slow down a reaction by lowering the temperature. We refrigerate perishable foods to make them last longer. In some cardiac surgeries, body temperature is lowered to 28°C so the heart can be stopped when less oxygen is required by the brain. This is also the reason that some people have survived submersion in icy lakes for long periods of time.

The rate of a reaction also increases when reactants are added. Then there are more collisions between the reactants and the reaction goes faster. For example, a patient having difficulty breathing may be given a breathing mixture with a higher oxygen content than the atmosphere. The increase in the number of oxygen molecules in the lungs increases the rate at which oxygen combines with hemoglobin. The increased rate of oxygenation of the blood means that the patient can breathe more easily.

$$\underset{\text{Hemoglobin}}{Hb} + \underset{\text{Oxygen}}{O_2} \longrightarrow \underset{\text{Oxyhemoglobin}}{HbO_2}$$

Another way to speed up a reaction is to lower the energy of activation. This can be done by adding a **catalyst,** which speeds up the reaction but is not itself changed or used up. (See Figure 6.15.) Earlier, we discussed the energy required to climb a hill. If instead we found a tunnel through the hill, we won't need as much energy to get to the other side. A catalyst acts by providing an alternate pathway with a lower energy requirement. That makes it possible for more collisions to form product successfully. Catalysts have found many uses in industry. In the production of margarine, the reaction of hydrogen with vegetable oils is normally very slow. However, when finely divided platinum is present, the reaction occurs rapidly. In the body, biocatalysts called enzymes make most metabolic reactions go at the rates necessary for proper cellular activity.

A summary of the factors affecting reaction rate is given in Table 6.8.

Table 6.8 Factors That Increase Reaction Rate

Factor	Reason
More reactants	More collisions
Higher temperature	More collisions, more collisions with energy of activation
Adding a catalyst	Lowers energy of activation

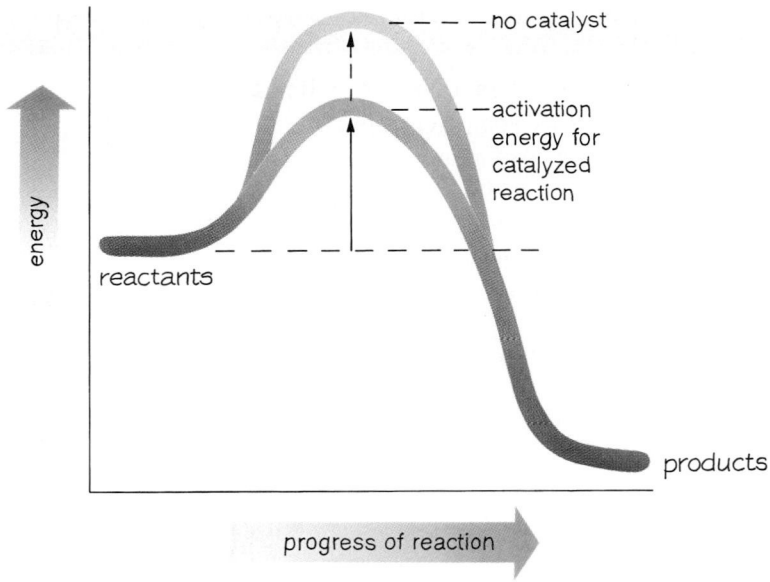

Figure 6.15 The rate of reaction increases when a catalyst is used to lower the energy of activation.

S·A·M·P·L·E· 6.16 ·P·R·O·B·L·E·M

Factors That Affect the Rate of Reaction

Indicate whether the following changes will increase, decrease, or have no effect upon the rate of reaction:

a. increase in temperature
b. decrease in the number of reactants
c. adding a catalyst

Solution

a. increase **b.** decrease **c.** increase

Study Check

How does the lowering of temperature affect the rate of reaction?

EXPLORE YOUR WORLD

How Fast Can It Go?

Obtain three matching glasses. Half fill one with hot water, one with room temperature water, and one with cold (ice) water. Add a few drops of food coloring or ink to the water in each glass. Do not disturb.

Questions

1. In which glass does the color spread most rapidly?
2. In which water sample is the spread of color the slowest?
3. How does the temperature of each water sample affect the rate at which the color spreads throughout the water?

QUESTIONS & PROBLEMS

Energy in Chemical Reactions

6.9

6.49 **a.** Why do chemical reactions require an energy of activation?
b. What is the function of a catalyst?
c. In an exothermic reaction, is the energy of the products higher or lower?
d. Draw an energy diagram for an exothermic reaction.

6.50 **a.** What is measured by the heat of reaction?
b. How does the heat of reaction differ in exothermic and endothermic reactions?
c. In an endothermic reaction, is the energy of the products higher or lower than the reactants?
d. Draw an energy diagram for an endothermic reaction.

6.51 Classify the following as exothermic or endothermic reactions:
a. 125 kcal are released.
b. In the energy diagram, the energy level of the products is higher than the reactants.
c. The metabolism of glucose in the body provides energy.

6.52 Classify the following as exothermic or endothermic reactions:
a. In the energy diagram, the energy level of the products is lower than the reactants.
b. In the body, the synthesis of proteins requires energy.
c. 30 kcal are absorbed.

6.53 Classify the following as exothermic or endothermic reactions:
a. gas burning in a Bunsen burner:

$$CH_4(g) + 2O_2(g) \longrightarrow CO_2(g) + H_2O(g) + 213 \text{ kcal}$$

b. dehydrating limestone:

$$Ca(OH)_2 + 15.6 \text{ kcal} \longrightarrow CaO + H_2O$$

c. formation of aluminum oxide and iron from aluminum and iron(III) oxide:

$$2Al + Fe_2O_3 \longrightarrow Al_2O_3 + 2Fe + 204 \text{ kcal}$$

6.54 Classify the following as exothermic or endothermic reactions:
a. the combustion of propane:

$$C_3H_8 + 5O_2 \longrightarrow 3CO_2 + 4H_2O + 531 \text{ kcal}$$

b. the formation of "table" salt:

$$2Na + Cl_2 \longrightarrow 2NaCl + 196 \text{ kcal}$$

c. decomposition of phosphorus pentachloride:

$$PCl_5 + 16 \text{ kcal} \longrightarrow PCl_3 + Cl_2$$

6.55 **a.** What is meant by the rate of a reaction?
b. Why does bread grow mold more quickly at room temperature than in the refrigerator?

6.56 **a.** How does a catalyst affect the activation energy?
b. Why is pure oxygen used in respiratory distress?

6.57 How would each of the following change the rate of the reaction shown here?

$$2SO_2 + O_2 \longrightarrow 2SO_3$$

a. adding SO_2
b. raising the temperature
c. adding a catalyst
d. removing some SO_2

6.58 How would each of the following change the rate of the reaction shown here?

$$2NO(g) + 2H_2(g) \longrightarrow N_2(g) + 2H_2O(g)$$

a. adding more NO
b. lowering the temperature
c. removing some H_2
d. adding a catalyst

6.10 A Closer Look: Chemical Equilibrium

Reactions and Equilibrium: Direction of Change 1–3

Learning Goal

Use the concept of reversible reactions to explain chemical equilibrium.

In Chapter 5, we looked at reactions as a change in reactants to form products. However, in many reactions the products can also interact and change back into the reactants. When a reaction goes in both directions, it is called a **reversible reaction.** For example, SO_2 and O_2 react to produce SO_3. Initially, this reaction will go in the forward direction to form SO_3. However, as SO_3 accumulates, collisions also occur between SO_3 molecules to form the reactants again. A reversible reaction is really two reactions that occur at the same time.

$$2SO_2(g) + O_2(g) \longrightarrow 2SO_3(g) \quad \text{Forward reaction}$$
$$2SO_2(g) + O_2(g) \longleftarrow 2SO_3(g) \quad \text{Reverse reaction}$$

Chemical Equilibrium

Eventually the rate at which the products change back to reactants becomes equal to the rate of the forward reaction. Then we say that the reaction has reached **chemical equilibrium,** which means that the reactants are changing into products at the same rate that the products are forming reactants. When the reaction reaches equilibrium, there is no further change in the relative amounts of the reactants and the products.

To show that a reaction is reversible, we write the forward and reverse reactions as a single equation using a double arrow.

$$2SO_2(g) + O_2(g) \rightleftharpoons 2SO_3(g)$$

Some reversible reactions produce a high proportion of products when equilibrium is reached. Then we say that the equilibrium favors the formation of products. However, if equilibrium occurs when there is still a high concentration of reactants and a low amount of products, we say that the equilibrium favors the reactants.

EXPLORE YOUR WORLD

Modeling an Equilibrium System

Obtain four or five tennis balls. Sit on the floor with another person sitting several feet away. Person 1 (the reactant) takes all the tennis balls. Begin the forward reaction by rolling the tennis balls, one by one, from person 1 to person 2 (the product). When person 2 (product) receives two of the tennis balls, he or she can begin to roll the tennis balls back (reverse reaction).

Questions

1. What is the only direction possible at the beginning of this reaction (forward or reverse)?
2. When does your reaction reach equilibrium?
3. What happens to the equilibrium if you give two additional tennis balls to the reactant person?
4. What happens to the equilibrium if you take one or two of the product tennis balls out?

6.10

Effect of Concentration on Equilibrium

We can change the rate of the forward or reverse reaction by changing the ratio of products to reactants. When we alter a reaction that is at equilibrium, we say that there is a stress on the equilibrium. Then, according to **LeChâtelier's principle,** the rates of forward and reverse reactions will change to relieve that stress.

Consider the reaction of nitrogen and oxygen to form nitrogen monoxide. At equilibrium, the rates of the forward and reverse reactions are equal.

$$N_2(g) + O_2(g) \rightleftharpoons 2NO(g)$$

If O_2 is added to the system, it places a stress on the system. To remove this stress, the rate of the forward reaction increases in order to form more product. On the other hand, if some O_2 is removed from the equilibrium system, the rate of the reverse reaction increases to replenish O_2.

$$\begin{array}{rcl} & & \text{Stress: Add } O_2 \\ & & \text{Shift: } \longrightarrow \\ N_2(g) + O_2(g) & \rightleftharpoons & 2NO(g) \\ & & \text{Stress: Remove } O_2 \\ & & \text{Shift: } \longleftarrow \end{array}$$

Suppose that some NO is added to the equilibrium system. With more NO molecules, the reverse reaction speeds up. The reaction shifts toward the reactants. If some NO is removed, the forward reaction will try to replace the NO. The reaction shifts towards the products.

$$\begin{array}{rcl} & & \text{Stress: Remove NO} \\ & & \text{Shift: } \longrightarrow \\ N_2(g) + O_2(g) & \rightleftharpoons & 2NO(g) \\ & & \text{Stress: Add NO} \\ & & \text{Shift: } \longleftarrow \end{array}$$

In general, we can think of adding or removing one of the substances in a reaction as having the effect of *pushing* the reaction towards the products, or *pulling* the reaction back to the reactants. Adding more reactant *pushes* the forward reaction in the direction of the products. Adding more product *pushes* the reverse reaction back toward the reactants. If some product is removed, the reaction is *pulled* in the direction of the products. If a reactant is removed, the reaction is *pulled* in the direction of the reactants.

HEALTH NOTE

Oxygen, Hemoglobin, and Carbon Monoxide Poisoning

When we take in air, the O_2 moves through our bronchial passages and diffuses through the air sacs in the lungs into the bloodstream. In the blood, a large molecule called hemoglobin (Hb) picks up the O_2 in a reversible reaction. At the air sacs in the lungs, the high concentration of O_2 favors the forward reaction to produce HbO_2.

$$Hb + O_2 \rightleftharpoons HbO_2$$

When a person is placed in a hyperbaric chamber where O_2 is at a higher concentration, the forward reaction favors the formation of more oxygenated hemoglobin. The increased amount of HbO_2 carries more O_2 to the tissues of the body, and helps to fight infections and cancer. In the body tissues, O_2 concentration is low, so the reverse reaction is favored and the HbO_2 releases the O_2.

Carbon monoxide (CO) is produced in an incomplete combustion reaction where ventilation is not adequate, such as in a furnace that isn't working properly. CO is attracted to Hb 140 times more strongly than O_2 is, and it attaches to the same binding site on the Hb as the O_2. When a molecule of HbCO forms, the Hb has no sites available to pick up O_2. In carbon monoxide poisoning, high concentrations of CO push the reaction toward the formation of HbCO in the blood, which can lead to dizziness and even death.

$$Hb + CO \rightleftharpoons HbCO$$

One way to treat carbon monoxide poisoning is to place a person in an atmosphere of pure oxygen. (Air is only 20% oxygen.) A higher concentration of O_2 will push the following reaction toward the products, which include the release of CO from the hemoglobin and a return to HbO_2.

$$HbCO + O_2 \rightleftharpoons HbO_2 + CO$$

6.10

Effect of Temperature on Equilibrium

When the temperature of an equilibrium system is increased, the reaction that is favored is the one that removes the heat. For example, when heat is added to an exothermic reaction, the equilibrium favors the reactants because the reverse reaction requires heat. If heat is removed (temperature decreases) from an exothermic reaction, the forward reaction is favored because it produces heat.

$$\begin{array}{rl} \text{Stress:} & \text{Remove heat} \\ \text{Shift:} & \longrightarrow \\ 2SO_2(g) + O_2(g) & \rightleftharpoons 2SO_3(g) + \text{heat} \\ \text{Stress:} & \text{Add heat} \\ \text{Shift:} & \longleftarrow \end{array}$$

When heat is added to an endothermic reaction, the equilibrium favors the forward reaction because it absorbs heat.

$$\begin{array}{rl} \text{Stress:} & \text{Add heat} \\ \text{Shift:} & \longrightarrow \\ N_2(g) + O_2(g) + \text{heat} & \rightleftharpoons 2NO(g) \\ \text{Stress:} & \text{Remove heat} \\ \text{Shift:} & \longleftarrow \end{array}$$

Table 6.9 summarizes the changes in equilibrium when changes are made in the amounts of products or reactants or temperature.

Table 6.9 Effects of Changes on Equilibrium

Factor	Change (stress)	Reaction Favored to Remove Stress
Concentration	Add more reactant	Forward
	Remove reactant	Reverse
	Add product	Reverse
	Remove product	Forward
Temperature	Raise	Endothermic direction to absorb heat
	Lower	Exothermic direction to produce heat

HEALTH NOTE

Homeostasis: Regulation of Body Temperature

In a physiological system of equilibrium called homeostasis, changes in our environment are balanced by changes in our bodies. It is crucial to our survival that we balance heat gain with heat loss. If we do not lose enough heat, our body temperature rises. At high temperatures, the body can no longer regulate our metabolic reactions. If we lose too much heat, body temperature drops. At low temperatures, essential functions proceed too slowly.

The skin plays an important role in the maintenance of body temperature. When the outside temperature rises, receptors in the skin send signals to the brain. The temperature-regulating part of the brain stimulates the sweat glands to produce perspiration. As perspiration evaporates from the skin, heat is removed and the body temperature is lowered.

In cold temperatures, epinephrine is released, causing an increase in metabolic rate, which increases the production of heat. Receptors on the skin signal the brain to contract the blood vessels. Less blood flows through the skin, and heat is conserved. The production of perspiration stops to lessen the heat lost by evaporation. A summary of the body's temperature regulation is illustrated in Figure 6.16.

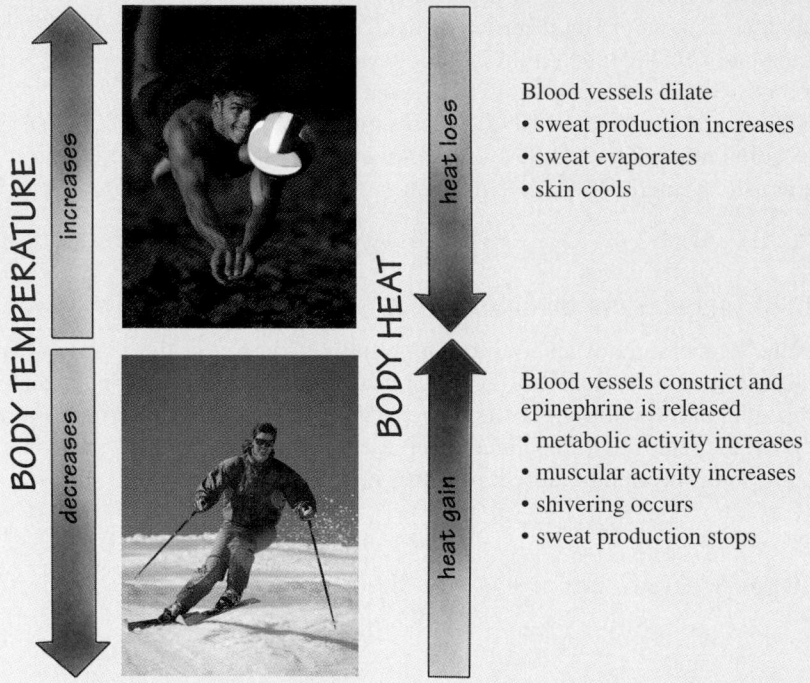

Figure 6.16 In homeostasis, a physiological equilibrium maintains a constant internal body temperature.

S·A·M·P·L·E **6.17** ·P·R·O·B·L·E·M

Factors Affecting Equilibrium

How do each of the following affect the equilibrium of the reaction shown here?

$$N_2(g) + 3H_2(g) \rightleftharpoons 2NH_3(g) + \text{heat}$$

a. add N_2 **b.** lower the temperature **c.** add NH_3

Solution

a. The addition of N_2 pushes the reaction toward products; the equilibrium shifts toward the products.

b. A lower temperature removes heat, which favors the exothermic direction; the equilibrium shifts toward the products.

c. The addition of NH_3 favors the reverse reaction toward reactants; the equilibrium shifts toward the reactants.

Study Check

If some H_2 is removed from the reaction in the problem, how will the equilibrium be affected?

QUESTIONS & PROBLEMS

A Closer Look: Chemical Equilibrium

6.59 What is meant by the term "reversible reaction"?

6.60 When does a reversible reaction reach equilibrium?

6.61 Hydrogen chloride can be made by reacting hydrogen gas and chlorine gas.

$$H_2(g) + Cl_2(g) \rightleftharpoons 2HCl(g) + 44 \text{ kcal}$$

What effect does each of the following changes have on equilibrium?

a. add H_2 **b.** add heat

c. remove HCl **d.** remove Cl_2

6.62 Ammonia is produced by reacting nitrogen gas and hydrogen gas.

$$N_2(g) + 3H_2(g) \rightleftharpoons 2NH_3(g) + 22 \text{ kcal}$$

What is the effect of each of the following changes on equilibrium?

a. remove N_2 **b.** remove heat

c. add NH_3 **d.** add H_2

6.63 In the lower atmosphere, oxygen is converted to ozone (O_3) by the energy provided from lightning.

$$3O_2(g) + \text{heat} \rightleftharpoons 2O_3(g)$$

What is the effect of each of the following changes on equilibrium?

a. add O_2 **b.** add O_3

c. add heat **d.** remove heat

6.64 When heated, carbon reacts with water to produce carbon monoxide and hydrogen.

$$C(s) + H_2O(g) + \text{heat} \rightleftharpoons CO(g) + H_2(g)$$

What is the effect of each of the following changes on equilibrium?

a. add heat **b.** remove heat

c. remove CO **d.** add H_2O

6.65 Give an explanation for each of the following:
When you open a bottle of a soft drink, the drink eventually goes flat. How does LeChâtelier's principle explain this?

$$H_2CO_3(aq) \rightleftharpoons CO_2(g) + H_2O(l)$$

6.66 When you exercise, energy is produced by increasing the rate of the following reaction involving glucose. Why do you breathe faster when you exercise?

$$\underset{\text{glucose}}{C_6H_{12}O_6(aq)} + 6O_2(g) \rightleftharpoons 6CO_2(g) + 6H_2O(l) + \text{energy}$$

Chapter Review

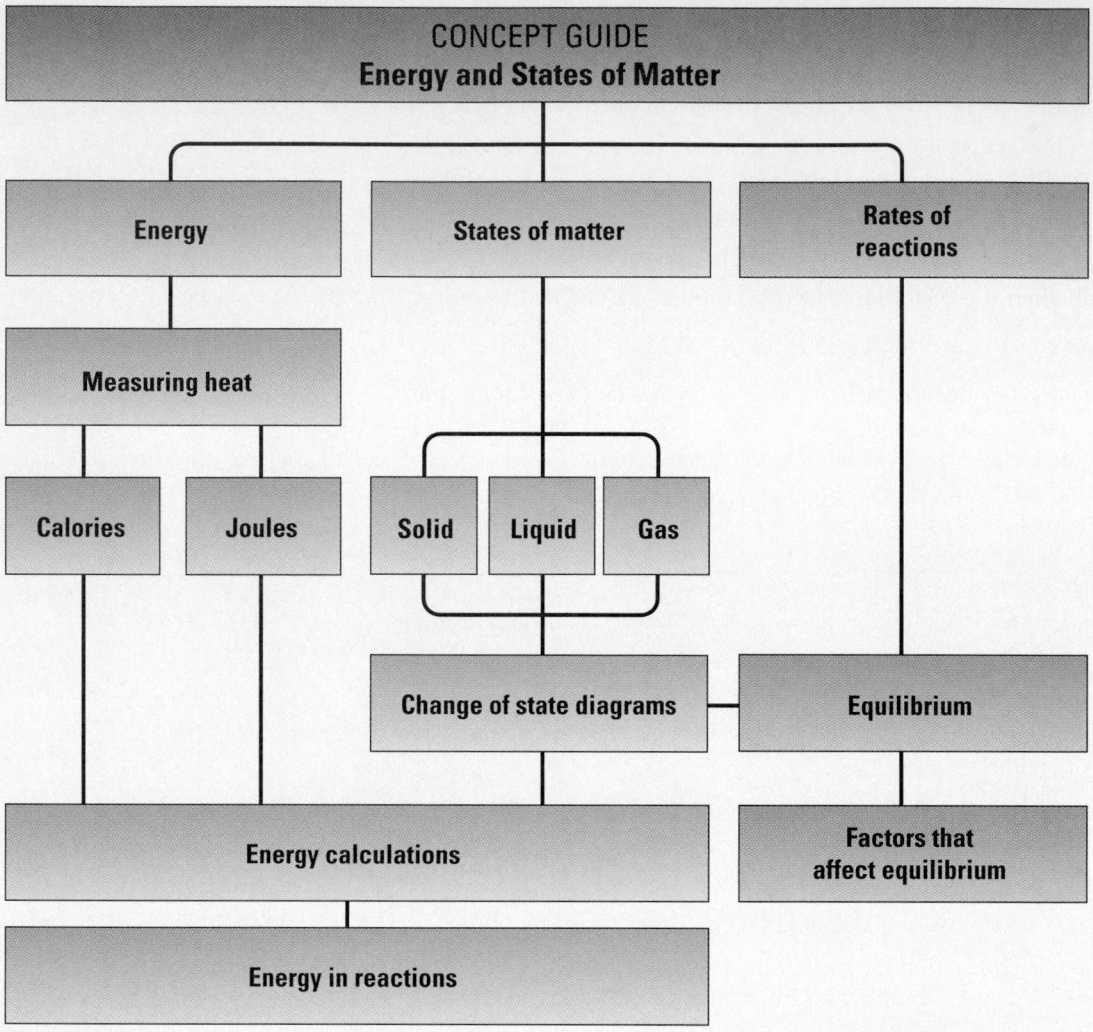

Chapter Summary

6.1 Energy

Energy is the ability to do work. Potential energy is stored energy; kinetic energy is the energy of motion.

6.2 Measuring Temperature

In science, temperature is measured in Celsius units, °C, or kelvins, K. In the United States, the Fahrenheit scale, °F, is still in use.

6.3 Measuring Heat Energy

Specific heat is the amount of energy required to raise the temperature of exactly 1 g of a substance by exactly 1°C. The heat lost or gained by a substance is determined by multiplying its mass (g), the temperature change (ΔT), and its specific heat (cal/g °C).

6.4 A Closer Look: Energy and Nutrition

The nutritional Calorie is the same amount of energy as 1 kcal or 1000 calories. The caloric content of a food is the sum of calories from carbohydrate, fat, and protein.

6.5 States of Matter

Matter is anything that has mass and occupies space. Physical properties such as shape, state, or color can be observed without affecting the identity of a substance. The three states of matter are solid, liquid, and gas.

6.6 Melting and Freezing

A substance undergoes a physical change when its shape, size, or state changes, but its identity does not change. Melting occurs when the particles in a solid absorb enough energy to break apart and form a liquid. The amount of energy required to convert exactly 1 g of solid to liquid is called the heat of fusion. For water, 80. cal are needed to melt 1 g of ice, or must be removed to freeze 1 g of water. Sublimation is a special process whereby a solid changes directly to a gas.

6.7 Boiling and Condensation

Evaporation occurs when particles in a liquid state absorb enough energy to break apart and form gaseous particles. Boiling is the vaporization of liquid at its boiling point. The heat of vaporization is the amount of heat needed to convert exactly 1 g of liquid to vapor. For water, 540 cal are needed to vaporize 1 g of water, or removed to condense 1 g of steam.

6.8 Heating and Cooling Curves

A heating or cooling curve illustrates the changes in temperature and state as heat is added to or removed from a substance. Plateaus on the graph indicate changes of state. The total heat absorbed or removed from a substance undergoing temperature changes and changes of state is the sum of energy calculations for change(s) of state and change(s) in temperature.

6.9 Energy in Chemical Reactions

In a reaction, the reacting particles must collide with energy equal to or greater than the energy of activation. The heat of reaction is the energy difference between the initial energy of the reactants and the final energy of the products. In exothermic reactions, energy is released. In endothermic reactions, energy is absorbed. The rate of a reaction, which is the speed at which the reactants are converted to products, can be increased by adding more reactants, raising the temperature, or adding a catalyst.

6.10 A Closer Look: Chemical Equilibrium

Chemical equilibrium occurs in a reversible reaction when the rate of the forward reaction is equal to the rate of the reverse reaction. The addition of reactants or removal of products favors the forward reaction. Removal of reactants or addition of products favors the reverse reaction. Adding or removing heat shifts the equilibrium of exothermic and endothermic reactions.

Key Terms

absolute zero The lowest possible temperature that matter can attain.

activation energy The energy needed upon collision to break apart the bonds of the reacting molecules.

boiling The formation of bubbles of gas throughout a liquid.

boiling point (bp) The temperature at which a substance exists as a liquid and gas; liquid changes to gas (boils), and gas changes to liquid (condenses).

caloric value The kilocalories obtained per gram of the food types: carbohydrate, fat, and protein.

calorie (cal) The amount of heat energy that raises the temperature of exactly 1 g of water exactly 1°C.

Calorie (Cal) The dietary unit of energy, equal to 1000 cal, or 1 kcal.

calorimeter An instrument used to measure the amount of heat released by burning a sample.

catalyst A substance that increases the rate of reaction by lowering the activation energy.

Celsius scale A temperature scale on which the freezing point of water is 0°C and the boiling point is 100°C.

change of state The transformation of one state of matter to another; for example, solid to liquid, liquid to solid, liquid to gas.

chemical equilibrium The point at which the forward and reverse reactions take place at the same rate so that there is no further change in amounts of reactants and products.

condensation The change of state of a gas to a liquid.

cooling curve A diagram that illustrates temperature changes and changes of states for a substance as heat is removed.

deposition The change of a gas directly to a solid; the reverse of sublimation.

endothermic reaction A reaction that requires heat; the energy of the products is higher than the energy of the reactants.

energy The ability to do work.

evaporation The formation of a gas (vapor) by the escape of high-energy molecules from the surface of a liquid.

exothermic reaction A reaction that releases heat; the energy of the products is lower than the energy of the reactants.

Fahrenheit scale A temperature scale on which ice melts at 32°F and water boils at 212°F.

freezing A change of state from liquid to solid.

freezing point (fp) The temperature at which the solid and liquid forms of a substance are in equilibrium; a liquid changes to a solid (freezes), a solid changes to a liquid (melts).

gas A state of matter characterized by no definite shape or volume. Particles in a gas move rapidly.

heat The energy associated with the motion of particles in a substance.

heat of fusion The energy required to melt exactly 1 g of a substance. For water, 80. cal are needed to melt 1 g of ice; 80. cal are released when 1 g of water freezes.

heat of reaction The energy released or absorbed during a chemical reaction equal to the energy difference between the reactants and products.

heat of vaporization The energy required to vaporize 1 g of a substance. For water, 540 calories are needed to vaporize exactly 1 g of liquid: 1 g of steam gives off 540 cal when it condenses.

heating curve A diagram that shows the temperature changes and changes of state of a substance as it is heated.

joule The SI unit of heat energy; 4.18 J = 1 cal.

Kelvin scale An SI temperature scale that assigns zero to the lowest possible temperature (absolute zero). A kelvin (K) represents a unit of temperature on the Kelvin scale: K = °C + 273.

kinetic energy A type of energy that is required for actively doing work; energy of motion.

kilocalorie An amount of heat energy equal to 1000 calories.

LeChâtelier's principle When a stress is placed on a system at equilibrium, the reaction shifts to relieve that stress.

liquid A state of matter that takes the shape of its container but has a definite volume.

matter Anything that has mass and occupies space.

melting The conversion of a solid to a liquid.

melting point (mp) The temperature at which a solid becomes a liquid (melts). It is the same temperature as the freezing point.

potential energy An inactive type of energy that is stored for use in the future.

rate of reaction The speed at which reactants are used to form product(s).

reversible reaction A reaction in which a forward reaction occurs from reactants to products, and a reverse reaction occurs from products back to reactants.

solid A state of matter that has its own shape and volume.

specific heat A quantity of heat that changes the temperature of exactly 1 g of a substance by exactly 1°C.

sublimation The change of state in which a solid is transformed directly to a gas without forming a liquid.

temperature An indication of the direction of heat flow, high to low.

work An activity that requires energy.

Chapter Practice

6.67 On a hot day, the beach sand gets hot, but the water stays cool. Explain.

6.68 Why do drops of liquid water form on a glass of iced tea?

6.69 When it rains or snows, the air temperature seems warmer. Explain.

6.70 **a.** Water is sprayed on the ground of an orchard when temperatures are near freezing to keep the fruit from freezing. Explain.

b. How many kilocalories of heat are released if 5.0 kg of water at 15°C is sprayed on the ground and cools and freezes at 0°C?

6.71 If you used the 2000 Calories you expend in energy in one day to heat 50,000 g of water at 20°C, what would be its new temperature?

6.72 A 0.50-g sample of vegetable oil is placed in a calorimeter. When the sample is burned, 18.9 kJ are given off. What is the caloric value (kcal/g) of the oil?

6.73 A typical diet in the United States provides 15% of the calories from protein, 45% from carbohydrates, and the remainder from fats. Calculate the grams of protein, carbohydrate, and fat to be included each day in diets having the following caloric requirements:

a. 1200 kcal **b.** 1900 kcal **c.** 2600 kcal

6.74 You have just eaten a quarter-pound cheeseburger, French fries, and chocolate shake. How many hours will a 70.0-kg person need to run to "burn off" the kilocalories in this meal? See Table 6.6 (assume you are a 70-kg person).

Item	Protein (g)	Fat (g)	Carbohydrate (g)
Cheeseburger	31	29	34
French fries	3	11	26
Chocolate shake	11	9	60.

6.75 If you want to lose 1 pound of "fat," which is 15% water, how many kilocalories do you need to expend? How many kilojoules is that?

6.76 Calculate the Cal (kcal) in 1 cup of whole milk: 12 g of carbohydrate, 9 g of fat, and 9 g of protein.

6.77 A hot-water bottle contains 725 g of water at 65°C. If the water cools to body temperature (37°C), how many kilojoules of heat could be transferred to sore muscles?

6.78 When 1.0 g of gasoline burns, 11,500 calories of energy are given off. If the density of gasoline is 0.74 g/mL, how many kilocalories of energy are obtained from 1.5 gallons of gasoline?

6.79 An ice cube tray holds 325 g of water. If the water initially has a temperature of 25°C, how many kilojoules of heat must be removed to cool and freeze the water at 0°C?

6.80 A Fahrenheit thermometer reads −40°F. What would the temperature be on the Celsius scale?

6.81 A 3.0-kg block of lead is taken from a furnace at 300.°C and placed on a large block of ice at 0°C. The specific heat of lead is 0.028 cal/g °C. If all the heat given up by the lead is used to melt ice, how much ice is melted after the temperature of the lead drops to 0°C?

6.82 The melting point of benzene is 5.5°C and its boiling point is 80.1°C. Sketch a heating curve for benzene from 0°C to 100°C.

a. What is the state of benzene at 15°C?
b. What happens on the curve at 5.5°C?
c. What is the state of benzene at 63°C?
d. What is the state of benzene at 98°C?
e. At what temperature will both liquid and gas be present?

6.83 A sample of 50.0 g of ice at 0°C is added to 50.0 g of water (liquid) at 0°C. How many kcal are needed to melt the ice, warm all the liquid water to 100°C, and change it to steam at 100°C?

6.84 When a gas barbecue is turned on, there is no reaction. Why is a match or lighter required to start burning the fuel?

6.85 In an endothermic reaction, nitrogen gas reacts with oxygen gas to produce nitrogen dioxide gas. Write the balanced equation for this reaction at equilibrium. Determine the effects on the equilibrium of each of the following:

a. add heat **b.** remove N_2
c. add O_2 **d.** remove NO_2
e. remove heat

6.86 The reaction between iron and oxygen that forms Fe_2O_3 also produces a lot of heat. However, the reaction is very slow. Would this indicate that the forward reaction has a high or low energy of activation? Explain.

6.87 At high altitudes, there is a lower concentration of O_2 in the air and less O_2 to combine with hemoglobin (Hb).

$$Hb + O_2 \rightleftharpoons HbO_2$$

People who live at high altitudes produce more hemoglobin molecules. How will an increase in Hb molecules affect the Hb + O_2 equilibrium?

6.88 Gasohol is a fuel containing ethyl alcohol (C_2H_6O) that burns in oxygen to give carbon dioxide, water, and heat. Write the balanced equation for the reaction. What is the effect of the following on the rate of the reaction?

a. add oxygen **b.** add a catalyst

Answers

Answers to Study Checks

6.1 potential energy

6.2 12°F

6.3 39.8°C

6.4 464 K

6.5 night, −260.°C; day, 410.°C

6.6 140 cal

6.7 350 J

6.8 4.1 kcal/g

6.9 380 kcal

6.10 liquid

6.11 40. kJ

6.12 Gas changes to liquid.

6.13 14 kcal

6.14 18,000 cal

6.15 $2C_2H_5OH + 6O_2 \longrightarrow 4CO_2 + 6H_2O + 326\text{ kcal}$

6.16 The rate of reaction will decrease because the number of collisions between reacting particles will be less, and a smaller number of the collisions that do occur will have sufficient energy for activation.

6.17 Removing H_2 favors the reactants; the equilibrium shifts to the reactants.

Answers to Selected Problems

6.1 When the car is at the top of a hill, it has its maximum potential energy. As it descends, potential energy changes to kinetic energy. At the bottom of the hill, all the energy is kinetic.

6.3 **a.** potential **b.** kinetic **c.** potential **d.** potential

6.5 **a.** Electrical energy changes to heat energy.
b. Electrical energy changes to kinetic energy.
c. Chemical energy changes to kinetic energy.
d. Radiant energy changes to heat energy.

6.7 **a.** true **b.** true **c.** false **d.** true

6.9 In the United States, we still use the Fahrenheit temperature scale. In °F, normal body temperature is 98.6. On the Celsius scale, her temperature would be 37.7°C, a mild fever.

6.11 **a.** 98.6°F **b.** 18.5°C **c.** 246 K **d.** 335 K **e.** 46°C **f.** 295 K

6.13 **a.** 41°C
b. No. The temperature is equivalent to 39°C.

6.15 Metals (aluminum, iron, or copper) conduct heat easily and get hot quickly.

6.17 **a.** 3.5 kcal **b.** 120 J **c.** 102 cal **d.** 1100 cal

6.19 **a.** 250 cal **b.** 47,000 J **c.** 980 J **d.** 39 kJ

6.21 530 J

6.23 **a.** 5. Cal **b.** 210 Cal **c.** 25 Cal

6.25 **a.** 110 Cal **b.** 18 g **c.** 130 Cal **d.** 280 Cal

6.27 210 kcal; 880 kJ

6.29 **a.** gas **b.** gas **c.** solid

6.31 **b.** freezing **d.** melting

6.33 **a.** melting **b.** melting **c.** sublimation **d.** freezing

6.35 **a.** 5200 cal absorbed **b.** 5700 J absorbed **c.** 18 kcal released **d.** 17 kJ released

6.37 **a.** condensation **b.** evaporation **c.** boiling **d.** condensation

6.39 **a.** The heat from the skin is used to evaporate the water (perspiration). Therefore the skin is cooled.
b. On a hot day, there are more molecules with sufficient energy to become water vapor.
c. In a closed bag, some molecules evaporate but they cannot escape and will condense back to liquid; the clothes will not dry.

6.41 **a.** 5400 cal absorbed **b.** 110,000 J absorbed **c.** 4300 kcal released **d.** 380 kJ released

6.43

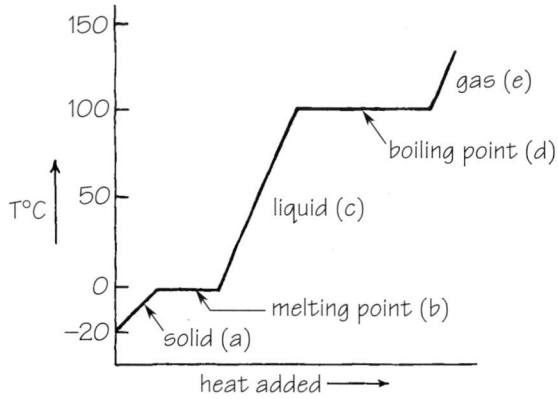

6.45 **a.** 1100 cal **b.** 7300 cal **c.** 40. kJ **d.** 17 kcal

6.47 26 kcal; 110 kJ

6.49 **a.** The energy of activation is the energy required to break the bonds of the reacting molecules.
b. A catalyst provides a pathway that lowers the activation energy and speeds up a reaction.
c. In exothermic reactions, the energy of the products is lower than the reactants.
d.

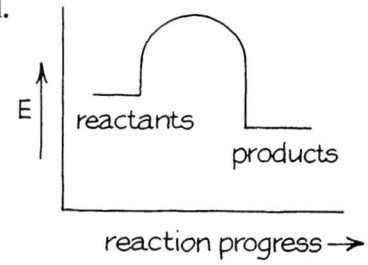

6.51 **a.** exothermic **b.** endothermic **c.** exothermic

6.53 **a.** exothermic **b.** endothermic **c.** exothermic

6.55 **a.** The rate of a reaction tells how fast the products are formed.
b. Reactions go faster at higher temperatures.

6.57 **a.** increase **b.** increase **c.** increase **d.** decrease

6.59 Reversible reactions are reactions that proceed in both the forward and reverse directions.

6.61 **a.** Products are favored.
b. Reactants are favored.
c. Products are favored.
d. Reactants are favored.

6.63 **a.** Products are favored.
b. Reactants are favored.
c. Products are favored.
d. Reactants are favored.

6.65 The escape of CO_2 gas removes CO_2; equilibrium favors the formation of products. The soda no longer bubbles.

6.67 In Table 6.2, sand has a lower specific heat than water. When both substances absorb the same amount of heat, the final temperature of the sand will be higher than that of water.

6.69 When water vapor condenses or liquid water freezes, heat is released, which warms the air.

6.71 60°C

6.73 **a.** 45 g protein, 140 g carbohydrate, 53 g fat
b. 71 g protein, 210 g carbohydrate, 84 g fat
c. 98 g protein, 290 g carbohydrate, 120 g fat

6.75 3500 kcal, 15,000 kJ

6.77 85 kJ

6.79 140 kJ

6.81 320 g of ice melts

6.83 68 kcal

6.85 $N_2(g) + 2O_2(g) + \text{heat} \rightleftharpoons 2NO_2(g)$
a. Products are favored.
b. Reactants are favored.
c. Products are favored.
d. Products are favored.
e. Reactants are favored.

6.87 An increase in hemoglobin molecules favors the products, which provides more oxygenated hemoglobin in the blood.

7

Gases

The nitrogen in air breathed under pressure at depths of 100 to 300 feet can be harmful to divers. At these depths, the oxygen in scuba tanks is mixed with helium.

Looking Ahead

7.1	Properties of Gases
7.2	Gas Pressure
7.3	Pressure and Volume (Boyle's Law)
7.4	Temperature and Volume (Charles' Law)
7.5	Temperature and Pressure (Gay-Lussac's Law)
7.6	The Combined Gas Law
7.7	Volume and Moles (Avogadro's Law)
7.8	Partial Pressures (Dalton's Law)

www.chemplace.com/college/timberlake

Visit the web site above to test yourself, run simulations, and review key concepts from this chapter. For even more interactive review and quizzing, use the two CD-ROMs included with the book.

We all live at the bottom of a sea of gases called the atmosphere. The most important of these gases is oxygen, which constitutes about 21% of the atmosphere. Without oxygen life on this planet would be impossible because oxygen is vital to all life processes of plants and animals. Ozone (O_3), formed in the upper atmosphere by the interaction of oxygen with ultraviolet light, absorbs some of the harmful radiation before it can strike the earth's surface. The other gases in the atmosphere include nitrogen (78% of the atmosphere), argon, carbon dioxide (CO_2), and water vapor. Carbon dioxide gas, a product of human cellular metabolism, is used by plants in photosynthesis, which produces the oxygen that is essential for humans and animals.

The atmosphere has become a dumping ground for other gases, such as methane, chlorofluorohydrocarbons (CFCs), sulfur dioxide, and nitrogen oxides. The chemical reactions of these gases with sunlight and oxygen in the air are contributing to air pollution, ozone depletion, global warming, and acid rain. Such chemical changes can seriously affect our health and the way all of us live. A knowledge of the behavior of gases and some of the laws that govern gas behavior can help us understand the nature of matter and allow us to make decisions concerning the important environmental and health issues we face today.

7.1 Properties of Gases

Learning Goal

Describe the kinetic theory of gases.

In Chapter 6, we saw that the behavior of gases is quite different from that of liquids and solids. Gas particles are far apart, whereas particles of both liquids and solids are held close together because of strong attractive forces. This means that a gas has no definite shape or volume and will completely fill any container. Because there are great distances between its particles, a gas is less dense than a solid or liquid and can be compressed. A model for the behavior of a gas, called the **kinetic theory,** helps us understand gas behavior.

Kinetic Theory of Gases

1. A gas is composed of very small particles (molecules or atoms).
2. The particles of a gas are very far apart. Thus, a gas is mostly empty space, and we assume that the volume of a container of gas is the same as the volume of the gas.
3. Gas particles move rapidly, colliding with other gas particles and hitting against the walls of the container.
4. Gas particles do not attract or repel one another.
5. Gas particles move faster when the temperature increases.

The kinetic theory helps explain some of the characteristics of gases. For example, we can quickly smell perfume from a bottle that is opened on the other side of a room, because its particles move rapidly in all directions. They move faster at higher temperatures, and more slowly at lower temperatures. Sometimes tires and gas-filled containers explode when temperatures are too high. From the kinetic theory, we know that gas particles move faster when heated, hit the walls of a container with more force, and cause a buildup of pressure inside a container.

S·A·M·P·L·E· 7.1 ·P·R·O·B·L·E·M

Kinetic Theory of Gases

How would you explain the following characteristics of a gas using the kinetic theory of gases?

a. The particles of gas are not attracted to each other.

b. Gas particles move faster when the temperature is increased.

Solution

a. The distance between the particles of a gas is so great that the particles are not attracted to each other.

b. Temperature increases the kinetic energy of the gas particles, which makes them move faster.

Study Check

Why can you smell food that is cooking in the kitchen when you are in a different part of the house?

7.1

When we talk about a gas, we describe it in terms of four properties: pressure, volume, temperature, and the amount of gas.

Pressure (*P*) Gas particles are extremely small. However, when billions and billions of gas particles hit against the walls of a container, they exert a great force known as the **pressure** of the gas. The more molecules that strike the wall, the greater the pressure. If we heat the container to make the molecules move faster, they smash into the walls of the container more often and with increased force, which increases the pressure. The gas molecules of oxygen and nitrogen in the air around us are exerting pressure on us all the time. We call the pressure exerted by the air atmospheric pressure. As you go to higher altitudes the atmospheric pressure is less because there are fewer molecules of oxygen and nitrogen in the air. The most common units used for gas measurement are the atmosphere (atm) and millimeters of mercury (mm Hg). On the TV weather report, you may hear or see the atmospheric pressure given in inches of mercury, or kilopascals in countries other than the United States.

Volume (*V*) The volume of gas equals the size of the container in which the gas is placed. When you inflate a tire or a basketball, you are adding more gas particles, which increases the number of particles hitting the walls of the tire or basketball, and its volume increases. Sometimes, on a cool morning, a tire looks flat. The volume of the tire has decreased because a lower temperature decreases the speed of the molecules, which reduces the force of their impacts on the walls of the tire. The most common units for volume measurement are liters (L) and milliliters (mL).

Temperature (*T*) The temperature of a gas is related to the kinetic energy of its particles. For example, if we have a gas at 200 K in a rigid container and heat it to a temperature of 400 K, the gas particles will have twice the kinetic energy that they did at 200 K. That also means that the gas at 400 K exerts twice the pressure of the gas at 200 K. Although you measure gas temperature using a Celsius thermometer, all comparisons of gas behavior and all calculations related to temperature must use the Kelvin temperature. No one has quite created the conditions for absolute zero (K), but we predict that the particles will have zero kinetic energy, and the gas will exert zero pressure.

Amount of Gas (*n*) When you add air to a bicycle tire, you are increasing the amount of gas, which results in a higher pressure in the tire. Usually we measure the amount of gas by its mass (grams). In gas laws, we need to change the grams of gas to moles.

A summary of the four properties of a gas are given in Table 7.1.

Table 7.1 Properties That Describe a Gas

Property	Description	Unit(s) of Measurement
Pressure (P)	The force exerted by gas against the walls of the container	atmosphere (atm); mm Hg
Volume (V)	The space occupied by the gas	liter (L); milliliter (mL)
Temperature (T)	Determines the kinetic energy and motion of the gas particles	Celsius (°C); Kelvin (K) *required in calculations*
Amount (n)	The quantity of gas present in a container	grams (g); moles (n) *required in calculations*

S·A·M·P·L·E 7.2 P·R·O·B·L·E·M

Properties of Gases

Identify the property of a gas that is described by each of the following:

a. increases the kinetic energy of gas particles
b. the force of the gas particles hitting the walls of the container
c. the space that is occupied by a gas

Solution

a. temperature **b.** pressure **c.** volume

Study Check

When helium is added to a balloon, the number of grams of gas increases. What property of a gas is described?

EXPLORE YOUR WORLD

Forming a Gas

Obtain a jar or a plastic bottle, and baking soda. You will also need an elastic glove that fits over the mouth of the jar, or a balloon that fits snugly over the top of the plastic bottle. Place a cup of vinegar in the jar or bottle. Sprinkle some baking soda into the fingertips of the glove or into the balloon. Carefully fit the glove or balloon over the top of the jar or bottle. Slowly lift the fingers of the glove or the balloon so that the baking soda falls into the vinegar. Watch what happens. Squeeze the glove or balloon.

Question

1. Describe the properties of gas that you observe as the reaction takes place between vinegar and baking soda.
2. How do you know that a gas was formed?

Q·U·E·S·T·I·O·N·S & P·R·O·B·L·E·M·S

Properties of Gases

7.1 Use the kinetic theory of gases to explain each of the following:
a. Gases move faster at higher temperatures.
b. Gases can be compressed much more than liquids or solids.
c. Gases have low densities.

7.2 Use the kinetic theory of gases to explain each of the following:
a. A container of non-stick cooking spray explodes when thrown into a fire.
b. The air in a hot-air balloon is heated to make the balloon rise.
c. The atmospheric pressure on top of Mt. Everest is lower than atmospheric pressure at sea level.

7.3 Identify the property of a gas that is measured in each of the following:
a. 350 K **b.** space occupied by a gas **c.** 2.00 g of O_2
d. force of gas particles striking the walls of the container

7.4 Identify the property of a gas that is measured in each of the following measurements:
a. determines the kinetic energy of the gas particles
b. 1.0 atm **c** 10.0 L **d.** 0.50 mole of He

7.2 Gas Pressure

Learning Goal

Describe the units of measurement used for pressure and change from one unit to another.

When water boils in a pan covered with a lid, the collisions of the molecules of steam (gas) lift up the lid. The gas molecules exert **pressure,** which is defined as a force acting on a certain area.

$$\text{Pressure } (P) = \frac{\text{force}}{\text{area}}$$

The air that covers the surface of the earth, the atmosphere, contains vast numbers of gas particles. Because the air particles have mass, they are pulled toward the earth by gravity, where they exert an **atmospheric pressure.**

The atmospheric pressure can be measured using a **barometer,** as shown in Figure 7.1. A long glass tube is closed on one end and filled with mercury. Then it is inverted and its open end is placed in a dish of mercury. The weight of the mercury begins to push the mercury out of the bottom of the tube. However, the pressure of the atmosphere against the mercury in the dish will push the mercury back up the tube.

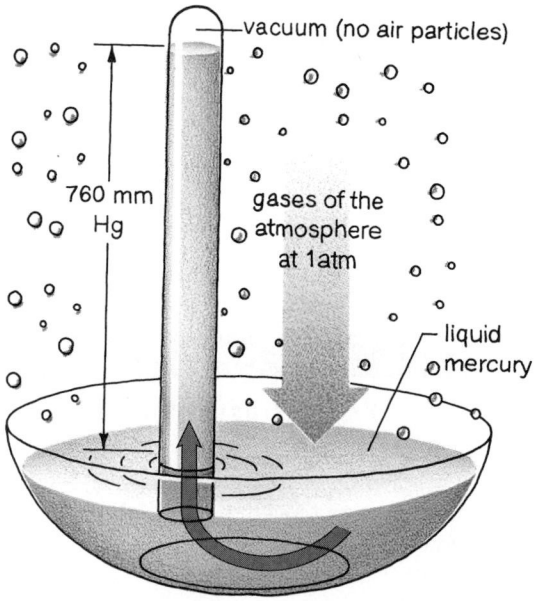

Figure 7.1 A barometer: A column of mercury 760 mm high is supported by the pressure exerted by the gases in the atmosphere at sea level.

Eventually, the mercury in the tube reaches a level where its weight is equal to the atmospheric pressure. At a pressure of exactly 1 atmosphere (atm), the mercury column would be exactly 760 mm high. We say that the atmospheric pressure is 760 mm Hg (millimeters of mercury), which is the same as a standard **atmosphere (atm).**

HEALTH NOTE

Measuring Blood Pressure

The measurement of your blood pressure is one of the important measurements a doctor or nurse makes during a physical examination. Acting like a pump, the heart contracts to create the pressure that pushes blood through the circulatory system. During contraction, the blood pressure is called systolic and is at its highest. When the heart muscles relax, the blood pressure is called diastolic and falls. Normal range for systolic pressure is 100–120 mm Hg, and for diastolic pressure, 60–80 mm Hg, usually expressed as a ratio such as 100/80. These values are somewhat higher in older people. When blood pressures are elevated, especially the systolic pressure, such as 140/90, there is a greater risk of stroke, heart attack, or kidney damage. Low blood pressure prevents the brain from receiving adequate oxygen, causing dizziness and fainting.

The blood pressures are measured by a sphygmomanometer, an instrument consisting of a stethoscope and an inflatable cuff connected to a tube of mercury called a manometer. (See Figure 7.2.) After the cuff is wrapped around the upper arm, it is pumped up with air until it cuts off the flow of blood through the arm. With the stethoscope over the artery, the air is slowly released from the cuff. When the pressure equals the systolic pressure, blood starts to flow again, and the noise it makes is heard. As air continues to be released, the cuff deflates until no sound in the artery is heard. That second pressure reading is noted as the diastolic pressure, the pressure when the heart is not contracting.

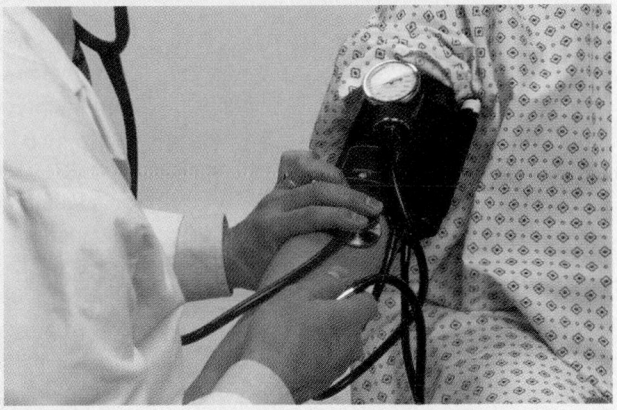

Figure 7.2 A doctor uses a sphygmomanometer to measure blood pressure.

Atmospheric Pressure and Altitude

At sea level, where there are more air particles, the atmospheric pressure is about 1 atm. In Los Angeles, atmospheric pressure can fluctuate between 730 and 760 mm Hg, depending on the weather. At higher altitudes, there are fewer air particles and the atmospheric pressure is less than 1 atm. In Denver (1.6 km), a typical atmospheric pressure might be 630 mm Hg, and if you were climbing Mount Everest (9.3 km) it might get as low as 270 mm Hg.

Deep-sea divers must be concerned about increasing pressures on their ears and lungs when they dive below the surface of the ocean. Because water is denser than air, the pressure on a diver increases rapidly as the diver descends. At a depth of 33 ft below the surface of the ocean, an additional 1 atmosphere of pressure is exerted by the water on a diver, for a total of 2 atm. At 100 ft down, there is a total of 4 atm on a diver. The air tanks a diver carries continuously adjust the pressure of the breathing mixture to match the increase of pressure.

One atmosphere of pressure may also be expressed as 760 **torr,** a pressure unit named to honor Evangelista Torricelli, the inventor of the barometer. Because they are equal, units of torr and mm Hg are used interchangeably.

$$1 \text{ atm} = 760 \text{ mm Hg} = 760 \text{ torr}$$

$$1 \text{ mm Hg} = 1 \text{ torr}$$

Table 7.2 Units for Measuring Pressure

Unit	Abbreviation	Unit Equivalent to 1 atm
Atmosphere	atm	1 atm
Millimeters of Hg	mm Hg	760 mm Hg
Torr	torr	760 torr
Inches of Hg	in. Hg	29.9 in. Hg
Pounds per square inch	lb/in.^{2} (psi)	14.7 lb/in.^{2}
Pascal	Pa	101,325 Pa

7.2

In SI units, pressure is measured in pascals (Pa); 1 atm is equal to 101,000 Pa. Because a pascal is a very small unit, it is likely that pressures would be reported in kilopascals. However, the unit is not yet in common use.

$$1 \text{ atm} = 1.01 \times 10^5 \text{ Pa} = 101 \text{ kPa}$$

If you have a barometer in your home, it probably gives pressure in inches of mercury. One atmosphere is also equal to the pressure of a column of mercury that is 29.9 inches high. Weather reports in the United States are given in inches of mercury. A lowering of pressure often indicates rain or snow, whereas an increase in pressure (referred to as a high-pressure system) usually brings dry and sunny weather.

The American equivalent of 1 atm is 14.7 pounds per square inch (lb/in.^{2}). When you use a pressure gauge to check the air pressure in the tires of a car it may read 30–35 psi. Table 7.2 summarizes the various units used in the measurement of pressure.

S·A·M·P·L·E 7.3 P·R·O·B·L·E·M

Units of Pressure Measurement

A sample of neon gas has a pressure of 0.50 atm. Give the pressure of the neon in:
a. millimeters of Hg **b.** inches of Hg

Solution

a. The equality 1 atm = 760 mm Hg can be written as conversion factors:

$$\frac{760 \text{ mm Hg}}{1 \text{ atm}} \quad \text{or} \quad \frac{1 \text{ atm}}{760 \text{ mm Hg}}$$

Using the appropriate conversion factor, the problem is set up as

$$0.50 \cancel{\text{atm}} \times \frac{760 \text{ mm Hg}}{1 \cancel{\text{atm}}} = 380 \text{ mm Hg}$$

b. One atm is equal to 29.9 in. Hg. Using this equality as a conversion factor in the problem setup, we obtain

$$0.50 \cancel{\text{atm}} \times \frac{29.9 \text{ in. Hg}}{1 \cancel{\text{atm}}} = 15 \text{ in. Hg}$$

Study Check

What is the pressure in atmospheres for a gas that has a pressure of 655 torr?

Vapor Pressure

In Chapter 6, we learned that liquid molecules with sufficient kinetic energy can break away from the surface of the liquid as they become gas particles or vapor. In an open container, all the liquid will eventually evaporate. In a closed container, the vapor accumulates and creates pressure called **vapor pressure.** Each liquid exerts its own vapor pressure at a given temperature. As temperature increases, more vapor forms, and vapor pressure increases. Table 7.3 lists the vapor pressure of water at various temperatures.

EXPLORE YOUR WORLD

Vapor Pressure

Measure 1 cup of water and pour it into a wide, flat dish and 1 cup of water in a tall, narrow glass. Mark the water level in each using a pen or piece of tape. Set both containers on the counter. After 1 week, check the level of water in each container. Use the measuring cup to determine the amount of water remaining in each container.

Questions

1. In which container did evaporation occur more rapidly?
2. Explain why evaporation occurs at a different rate in the two types of containers.
3. How did the amount of water remaining in each container after one week confirm your answer in question 1?

Vapor Pressure and Boiling Point

A liquid reaches its boiling point when its vapor pressure becomes equal to the external pressure. As boiling occurs, bubbles of the gas form within the liquid and quickly rise to the surface. For example, at an atmospheric pressure of 760 mm Hg, water will boil at 100°C, the temperature at which its vapor pressure reaches 760 mm Hg. (See Figure 7.3.)

At higher altitudes, atmospheric pressures are lower and the boiling point of water is less than 100°C. Earlier, we saw that the typical atmospheric pressure in Denver is 630 mm Hg. This means that water in Denver needs a vapor pressure of 630 mm Hg to boil. Because

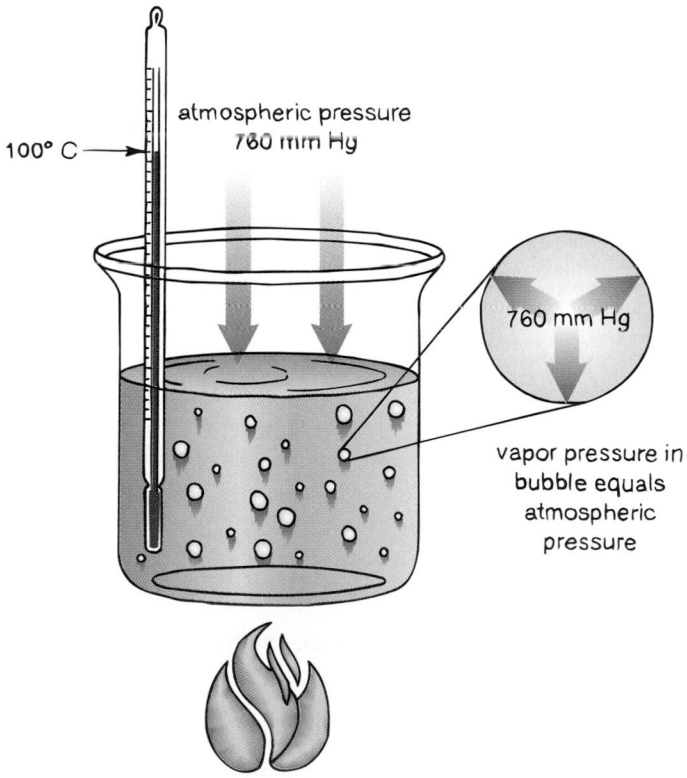

Figure 7.3 Vapor pressure and boiling point for water at 1 atm pressure (760 mm Hg).

Table 7.3 Vapor Pressure of Water

Temperature (C°)	Vapor Pressure (mm Hg)
0	5
10	9
20	18
30	32
37	47[a]
40	55
50	93
60	149
70	234
80	355
90	528
100	760

[a]At body temperature.

7.2

Table 7.4 Altitude, Atmospheric Pressure, and the Boiling Point of Water

Location	Altitude (km)	Atmospheric Pressure (mm Hg)	Boiling Point (°C)
Sea level	0	760	100
Los Angeles	0.09	752	99
Las Vegas	0.70	700	98
Denver	1.60	630	95
Mount Whitney	4.00	467	87
Mount Everest	9.30	270	70

water has a vapor pressure of 630 mm Hg at 95°C, water boils at 95°C in Denver. Water boils at even lower temperatures at still lower atmospheric pressures, as Table 7.4 shows.

People who live at high altitudes often use pressure cookers to obtain higher temperatures when preparing food. When the external pressure is greater than 1 atm, a temperature higher than 100°C is needed to boil water. Laboratories and hospitals use devices called autoclaves to sterilize laboratory and surgical equipment. An autoclave, like a pressure cooker, is a closed container that increases the total pressure above the liquid so it will boil at higher temperatures. Table 7.5 shows how the boiling point of water increases as pressure increases.

Table 7.5 Pressure and the Boiling Point of Water

Pressure (mm Hg)	Boiling Point (°C)
760	100
800	100.4
1075	110
1520 (2 atm)	120
2026	130
7600 (10 atm)	180

QUESTIONS & PROBLEMS

Gas Pressure

7.5 What units are used to describe the pressure of gas?

7.6 Which of the following statement(s) describes the pressure of a gas?
a. the force of the gas particles on the walls of the container
b. the number of gas particles in a container
c. the volume of the container
d. 3.00 atm
e. 750 torr

7.7 An oxygen tank contains oxygen (O_2) at a pressure of 2.00 atm. What is the pressure in the tank of terms of the following units?
a. torr **b.** lb/in.^{2} **c.** mm Hg

7.8 On a climb up Mt. Whitney, the atmospheric pressure is 467 mm Hg. What is the pressure in terms of the following units?
a. atm **b.** torr **c.** in. Hg

7.9 Match the terms *vapor pressure, atmospheric pressure,* and *boiling point* to the following descriptions.
a. the temperature at which bubbles of vapor appear within the liquid
b. the pressure exerted by a gas above the surface of its liquid
c. the pressure exerted on the earth by the particles in the air
d. the temperature at which the vapor pressure of a liquid becomes equal to the external pressure

7.10 In which pair(s) would boiling occur?

	Atmospheric Pressure	Vapor Pressure
a.	760 mm Hg	700 mm Hg
b.	480 torr	480 mm Hg
c.	1.2 atm	912 mm Hg
d.	1020 mm Hg	760 mm Hg
e.	740 torr	1.0 atm

7.11 Give an explanation for the following observations:
a. Water boils at 87°C on the top of Mt. Whitney.
b. Food cooks more quickly in a pressure cooker than in an open pan.

7.12 Give an explanation for the following observations:
a. A wet towel dries faster on a windy day.
b. Water used to sterilize surgical equipment is heated to 120°C at 2.0 atm in an autoclave.

7.3 Pressure and Volume (Boyle's Law)

Study Pack

Learning Goal

Use the pressure–volume relationship (Boyle's law) to determine the new pressure or volume of a certain amount of gas at a constant temperature.

Imagine that you can see air particles hitting the walls inside a bicycle tire pump. What happens to the pressure inside the pump as we push down on the handle? As the air is compressed, the air particles are crowded together. In the smaller volume, more collisions occur, and the air pressure increases.

When a change in one property (in this case, volume) causes a change in another property (in this case, pressure), those properties are related. Furthermore, when one change causes a change in the opposite direction, such as an increase in pressure causing a decrease in volume, the properties have an **inverse relationship.** The relationship between the pressure and volume of a gas is known as **Boyle's law.** The law states that the volume (V) of a sample of gas changes inversely with the pressure (P) of the gas as long as there has been no change in the temperature (T) or amount of gas (n), as illustrated in Figure 7.4.

S·A·M·P·L·E 7.4 ·P·R·O·B·L·E·M

Relationship of Pressure and Volume

Complete the information for the pressure and volume of a gas sample at constant temperature.

	Pressure	Volume
a.	increases	____
b.	____	increases

Solution

a. Pressure and volume are inversely related. When the pressure of a gas increases, the volume decreases.
b. If the volume of the gas sample increases, the pressure of the gas decreases.

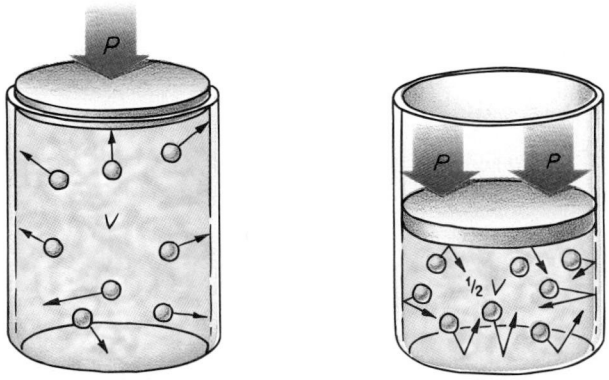

Figure 7.4 When the volume of a gas decreases, its pressure increases as long as there is no change in the temperature or the amount of the gas.

Study Check

If the pressure of a gas decreases, what happens to its volume if no change has occurred in the temperature or in the amount of the gas?

7.3

If we change the volume or pressure of a gas sample without any change occurring in the temperature or in the amount of the gas, the new pressure and volume will give the same product as the initial pressure and volume. Because the PV product has the same value under both conditions, we can set the initial and final PV values equal to each other.

$$P_iV_i = P_fV_f \quad \text{No change in number of moles and temperature}$$

HEALTH NOTE

Pressure–Volume Relationship in Breathing

The importance of Boyle's law becomes more apparent when you consider the mechanics of breathing. Our lungs are elastic, balloon-like structures contained within an airtight chamber called the thoracic cavity. The diaphragm, a muscle, forms the flexible floor of the cavity.

Inspiration

The process of taking a breath of air begins when the diaphragm flattens, and the rib cage expands, causing an increase in the volume of the thoracic cavity. The elasticity of the lungs allows them to expand when the thoracic cavity expands. According to Boyle's law, the pressure inside the lungs will decrease when their volume increases. This causes the pressure inside the lungs to fall below the pressure of the atmosphere. This difference in pressures produces a **pressure gradient** between the lungs and the atmosphere. In a pressure gradient, molecules flow from an area of greater pressure to an area of lower pressure, a process called **diffusion.** Thus, we inhale as air flows into the lungs *(inspiration)*, until the pressure within the lungs becomes equal to the pressure of the atmosphere.

Expiration

Expiration, or the exhalation phase of breathing, occurs when the diaphragm relaxes and moves back up into the thoracic cavity to its resting position. This reduces the volume of the thoracic cavity, which squeezes the lungs and decreases their volume. Now the pressure in the lungs is greater than the pressure of the atmosphere, so air flows out of the lungs. Thus, breathing is a process in which pressure gradients are continuously created between the lungs and the environment as a result of the changes in the volume and pressure. (See Figure 7.5.)

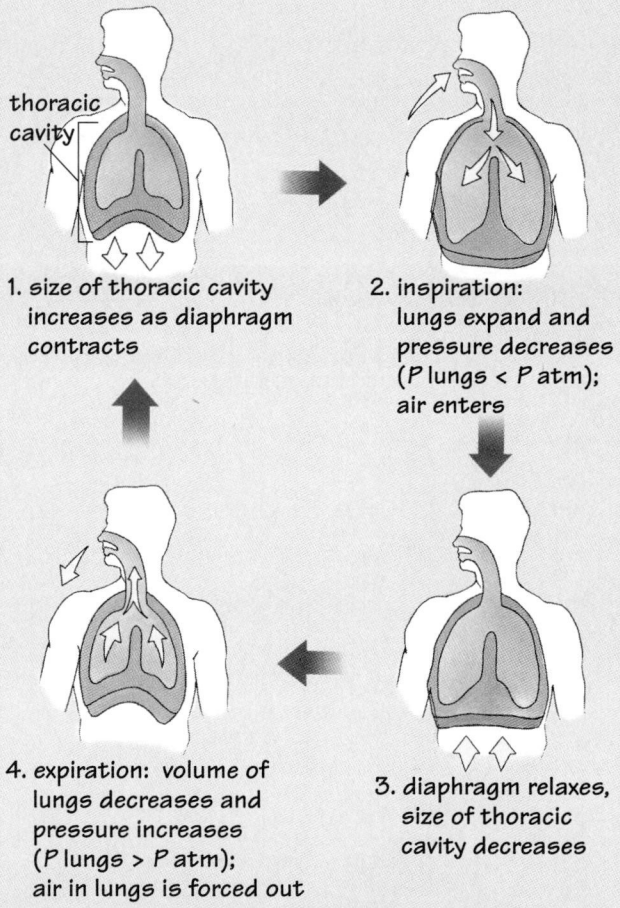

Figure 7.5 The mechanics of breathing are based on a pressure–volume relationship.

S·A·M·P·L·E **7.5** P·R·O·B·L·E·M

Calculating Pressure When Volume Changes

A sample of hydrogen gas (H_2) has a volume of 4.0 L and a pressure of 1.0 atm. What is the new pressure if the volume is decreased to 2.0 L?

Solution

In calculations with gas laws, it is helpful to organize the data in a table of initial and final conditions.

Initial Conditions	Final Conditions	Change
P_i = 1.0 atm	P_f = ?	P must increase
V_i = 4.0 L	V_f = 2.0 L	V decreases

In this problem, we want to know the final pressure (P_f) for the change in volume.

The problem setup begins with the initial pressure of 1.0 atm followed by a volume factor using the initial and final volumes because their units will cancel. There are two possible volume factors; one will decrease pressure, and the other will increase pressure.

Volume Factors

$$\frac{2.0\text{ L}}{4.0\text{ L}} \quad \text{and} \quad \frac{4.0\text{ L}}{2.0\text{ L}}$$

This factor will decrease pressure — This factor will increase pressure

In this problem the volume has decreased. Because pressure and volume are inversely related, the pressure must increase. Therefore, we will use the volume factor that increases pressure.

$$\underset{\text{New pressure}}{P_f} = \underset{\text{Initial pressure}}{1.0\text{ atm}} \times \underset{\text{Volume factor that increases pressure}}{\frac{4.0\cancel{\text{L}}}{2.0\cancel{\text{L}}}} = 2.0\text{ atm}$$

The pressure of the hydrogen gas increases from 1.0 atm to 2.0 atm. This is the change expected, because the volume decreased.

Alternatively, the pressure–volume equation can be solved for the final pressure by dividing both sides by V_f.

$$\frac{P_iV_i}{V_f} = \frac{P_f\cancel{V_f}}{\cancel{V_f}}$$

$$P_f = \frac{P_iV_i}{V_f}$$

Substituting the values listed in the table of information into the equation for P_f gives

$$P_f = \frac{(1.0\text{ atm})(4.0\cancel{\text{L}})}{(2.0\cancel{\text{L}})} = 2.0\text{ atm}$$

Study Check

A sample of helium gas has a volume of 250 mL at 800 torr. If the volume is changed to 500 mL, what is the new pressure, assuming no change in temperature or number of moles?

EXPLORE YOUR WORLD

Medicine Dropper "Diver"

You can observe the effect of pressure change on the ability of a medicine dropper to sink or float in water. Obtain a plastic bottle with a cap and a medicine dropper. A 2-liter soda bottle or a 1-liter water bottle works well. The dropper needs to fit through the top of the bottle. Take up water in the medicine dropper until it is one-half to three-quarters full. The rest of the space is air (gas). Place the medicine dropper in a glass of water. It should just float at the surface. If not, adjust the water level.

7.3

Fill the plastic bottle with water up to about 1 to $1\frac{1}{2}$ inches from the top. Put the dropper with water in the bottle. Tightly cap the bottle. When you squeeze the sides of the bottle, the dropper should sink. If you cannot make the dropper sink, change the amount of water you take into the dropper, or the amount of water in the bottle. You may have to adjust several times.

Questions

1. What does squeezing the sides of the bottle do to the pressure of the gas in the bottle and in the dropper?
2. What happens to the volume of gas in the dropper when you squeeze the bottle?
3. How does this cause the dropper to sink?
4. Why does the dropper rise when you release the sides of the bottle?
5. What is the relationship between the pressure and volume of a gas?

S·A·M·P·L·E 7.6 ·P·R·O·B·L·E·M

Calculating Volume When Pressure Changes

In the hospital respiratory unit, the gauge on a 10.0-L tank of compressed oxygen reads 4500 mm Hg. How many liters of oxygen can be delivered from the tank at a pressure of 750 mm Hg?

Solution

Placing our information in a table gives the following:

Initial Conditions	Final Conditions	Change
P_i = 4500 mm Hg	P_f = 750 mm Hg	P decreases
V_i = 10.0 L	V_f = ?	V must increase

According to Boyle's law, a decrease in the pressure will cause an increase in the volume. The initial volume (V_i) must be multiplied by a pressure factor that gives a greater volume.

$$V_f = V_i \times \frac{P_i}{P_f}$$

$$V_f = 10.0\text{ L} \times \frac{4500\ \cancel{\text{torr}}}{750\ \cancel{\text{torr}}} = 60.\text{ L}$$

New volume — Initial volume — Pressure factor that increases volume

Study Check

A sample of methane gas (CH_4) has a volume of 125 mL at 0.600 atm pressure and 25°C. How many milliliters will it occupy at a pressure of 1.50 atm and 25°C?

QUESTIONS & PROBLEMS

Pressure and Volume (Boyle's Law)

7.13 Why do scuba divers need to exhale air when they ascend to the surface of the water?

7.14 Why does a sealed bag of chips expand when you take it to a higher altitude?

7.15 What happens to the volume of your lungs during expiration when you get rid of some CO_2?

7.16 How do respirators (or CPR) help a person obtain oxygen and expel carbon dioxide?

7.17 The air in a cylinder with a piston has a volume of 220 mL and a pressure of 650 mm Hg.

a. If a change results in a higher pressure inside the cylinder, does cylinder A or B represent the final volume? Explain your choice.

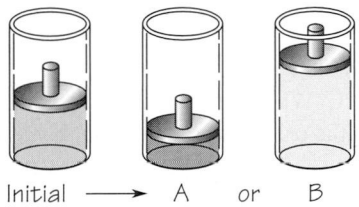

b. If the pressure inside the cylinder increases to 1.2 atm, what is the final volume of the cylinder? Complete the following data table:

Property	Initial	Final
Pressure (P)		
Volume (V)		

7.18 A balloon is filled with helium gas. When the following changes are made at constant temperature, which of these diagrams (A, B, or C) shows the new volume of the balloon?

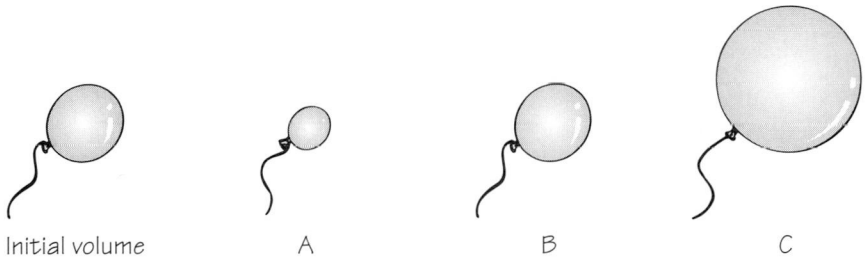

a. The balloon floats to a higher altitude where the outside pressure is lower.

b. The balloon is taken inside the house, but the atmospheric pressure remains the same.

c. The balloon is put in a hyperbaric chamber in which the pressure is increased.

7.19 Complete the following using the pressure–volume relationship.

Pressure	Volume
a. decreases	_____
b. increases	_____

7.20 Complete the following using the pressure–volume relationship:

Pressure	Volume
a. _____	decreases
b. _____	increases

7.21 A 10.0-L balloon contains He gas at a pressure of 655 mm Hg. What is the new pressure of the He gas at each of the following volumes if there is no change in temperature?
a. 20.0 L **b.** 5.00 L **c.** 1500 mL **d.** 100.0 mL

7.3

7.22 The air in a 5.00-L tank has a pressure of 1.20 atm. What is the new pressure of the air when the air is placed in tanks that have the following volumes, if there is no change in temperature?
a. 1.00 L **b.** 2500 mL **c.** 750 mL **d.** 20.00 L

7.23 A sample of nitrogen (N_2) gas has a volume of 4.5 L at a pressure of 760 mm Hg. What is the new pressure if the gas sample is compressed to a volume of 1.5 L if there is no change in temperature?

7.24 An emergency tank of oxygen holds 20.0 L of oxygen (O_2) at a pressure of 15.0 atm and 22°C. When the gas is released at 22°C, it provides 300.0 L of oxygen. What is the pressure at which the oxygen is released?

7.25 A sample of nitrogen (N_2) has a volume of 50.0 L at a pressure of 760. mm Hg. What is the volume of the gas at each of the following pressures if there is no change in temperature?
a. 1500 mm Hg **b.** 2.0 atm **c.** 0.500 atm **d.** 850 torr

7.26 A sample of methane (CH_4) has a volume of 25 mL at a pressure of 0.80 atm. What is the volume of the gas at each of the following pressures if there is no change in temperature?
a. 0.40 atm **b.** 2.00 atm **c.** 2500 mm Hg **d.** 80.0 torr

7.27 Cyclopropane, C_3H_6, is a general anesthetic. A 5.0-L sample has a pressure of 5.0 atm. What is the volume of the anesthetic given to a patient at a pressure of 1.0 atm?

7.28 The volume of air in a person's lungs is 615 mL at a pressure of 760 mm Hg. Inhalation occurs as the pressure in the lungs drops to 752 mm Hg. To what volume did the lungs expand?

7.29 Use the words *inspiration* and *expiration* to describe the part of the breathing cycle that occurs as a result of each of the following:
a. The diaphragm contracts (flattens out).
b. The volume of the lungs decreases.
c. The pressure within the lungs is less than the atmosphere.

7.30 Use the words *inspiration* and *expiration* to describe the part of the breathing cycle that occurs as a result of each of the following:
a. The diaphragm relaxes, moving up into the thoracic cavity.
b. The volume of the lungs expands.
c. The pressure within the lungs is greater than the atmosphere.

Figure 7.6 As the air inside a hot-air balloon is heated, its volume expands and its density decreases. The hot-air balloon rises and floats in the air.

7.4 Temperature and Volume (Charles' Law)

Study Pack

Learning Goal

Use the temperature–volume relationship (Charles' law) to determine the new temperature or volume of a certain amount of gas at a constant pressure.

When preparing a hot-air balloon for a flight, the air in the balloon is heated with a small propane heater. As the air warms, its volume increases. The resulting decrease in density allows the balloon to rise. (See Figure 7.6.)

To study the effect of changing temperature on the volume of a gas, we must not change the pressure or the amount of the gas. Suppose we increase the Kelvin temperature of a gas sample. The kinetic theory shows that the activity (kinetic energy) of the gas will also increase. To keep pressure constant, the volume of the container must increase. (See Figure 7.7.) By contrast, if the temperature of the gas is lowered, the volume of the container must be reduced to maintain the same pressure.

Suppose that you are going to take a ride in a hot air balloon. The captain turns on a propane burner to heat the air inside the balloon. As the temperature rises, the air particles move faster and spread out, causing the volume of the balloon to increase. Eventually, the air in the balloon becomes less dense than the air outside, and the balloon and its passengers rise. In fact, it was in 1787 that Jacques Charles, a balloonist as well as a physicist, proposed that the volume of a gas is related to the temperature. This became **Charles' law,** which states that the volume (V) of a gas is directly related to the temperature (K) when there is no change in the pressure (P) or amount (n) of gas. A

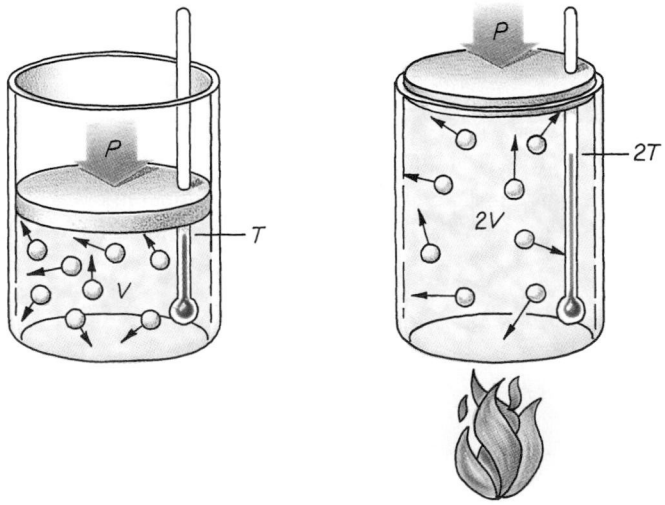

Figure 7.7 The Kelvin temperature of a gas is directly related to the volume of the gas when there is no change in pressure or amount.

direct relationship is one in which the related properties increase or decrease together. For two conditions, initial and final, we can write Charles' law as follows.

$$\frac{V_i}{T_i} = \frac{V_f}{T_f}$$

Remember that all temperatures used in gas law calculations must be converted to their corresponding Kelvin (K) temperatures.

7.4

S·A·M·P·L·E· 7.7 ·P·R·O·B·L·E·M

Relationship of Volume and Temperature

Complete the information for the volume and temperature of a gas when the amount of gas and its pressure remain constant.

Volume	Temperature
a. _____	decreases
b. increases	_____

Solution

a. If the temperature of a gas decreases, its volume also decreases.
b. When the volume of a gas increases, the temperature also increases.

Study Check

If the volume of a sample of gas decreases, how must the temperature change if the pressure is to remain constant?

EXPLORE YOUR WORLD

Temperature and Volume

Obtain three balloons of the same type and size. Partially inflate (blow up) each balloon so that their volumes are about equal. Place one balloon in the sun, or a warm place. Place the other balloon in the freezer compartment of your refrigerator. Leave the third balloon in the room. Compare the volumes of the three balloons after 15 minutes, 30 minutes, and 1 hour.

Questions

1. Which balloon has the greatest volume?
2. Which balloon has the smallest volume?
3. How would you explain the effect of the changes in temperature on the volumes of the balloons?

S·A·M·P·L·E· 7.8 ·P·R·O·B·L·E·M

Calculating Volume When Temperature Changes

A sample of neon gas has a volume of 5.0 L and a temperature of 17°C. Find the new volume of the gas after the temperature has been increased to 47°C.

Solution

When the temperatures are given in degrees Celsius, they must be changed to Kelvins.

$$T_i = 17°C + 273 = 290.\ K$$

$$T_f = 47°C + 273 = 320.\ K$$

Initial Conditions	Final Conditions	Change
$T_i = 290.\ \text{K}$	$T_f = 320.\ \text{K}$	T increases
$V_i = 5.0\ \text{L}$	$V_f = ?$	V must increase

Because an increase in temperature increases the volume, multiply the initial volume by the factor that gives a greater final volume.

$$V_f = V_i \times \text{temperature factor}$$

$$\underset{\text{New volume}}{V_f} = \underset{\text{Initial volume}}{5.0\ \text{L}} \times \underset{\text{Temperature factor that increases volume}}{\frac{320.\ \cancel{\text{K}}}{290.\ \cancel{\text{K}}}} = 5.5\ \text{L}$$

7.4

Study Check

A mountain climber inhales 500. mL of air at a temperature of −10.°C. What volume will the air occupy in the lungs if the climber's body temperature is 37°C?

QUESTIONS & PROBLEMS

Temperature and Volume (Charles' Law)

7.31 Select the diagram that shows the new volume of a balloon when the following changes are made at constant pressure.

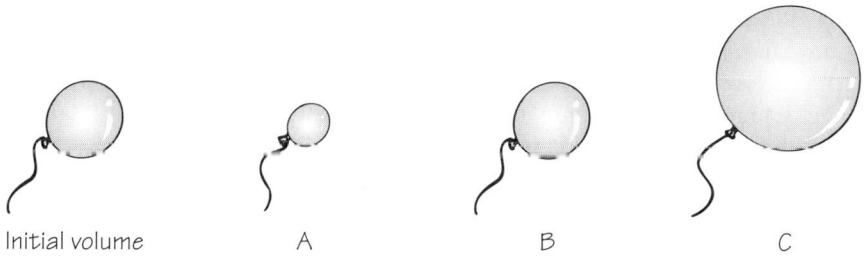

a. The temperature is changed from 150 K to 300 K.
b. The balloon is placed in a freezer.
c. The balloon is first warmed, and then returned to its starting temperature.

7.32 Indicate whether the final volume of gas in each of the following is the same, larger, or smaller than the initial volume:
a. A volume of 500 mL of air on a cold winter day at 5°C is breathed into the lungs, where body temperature is 37°C.
b. The heater used to heat 1400 L of air in a hot air balloon is turned off.
c. A balloon filled with helium at the amusement park is left in a car on a hot day.

7.33 State the change that occurs for volume or temperature when P and n are not changed.

Volume	Temperature
a. _____	decreases
b. increases	_____

7.34 State the change that occurs for volume or temperature when P and n are not changed.

Volume	Temperature
a. decreases	_____
b. _____	increases

7.5

7.35 A balloon contains 2500 mL of helium gas at 75°C. What is the new volume of the gas when the temperature changes to the following, if n and P are not changed?

a. 55°C **b.** 680 K **c.** −25°C **d.** 240 K

7.36 A gas has a volume of 4.00 L at 0°C. What final temperature in degrees Celsius is needed to cause the volume of the gas to change to the following, if n and P are not changed?

a. 10.0 L **b.** 1200 mL **c.** 2.50 L **d.** 50.0 mL

7.5 Temperature and Pressure (Gay-Lussac's Law)

Learning Goal

Use the temperature–pressure relationship (Gay-Lussac's law) to determine the new temperature or pressure of a certain amount of gas at a constant volume.

If we could watch the molecules of a gas as the temperature rises, we would notice that they move faster and hit the sides of the container more often and with greater force. If we keep the volume of the container the same, we would observe an increase in the pressure. A temperature–pressure relationship, also known as **Gay-Lussac's law,** states that the pressure of a gas is directly related to its Kelvin temperature. This means that an increase in temperature increases the pressure of a gas, and a decrease in temperature decreases the pressure of the gas, provided the volume and number of moles of the gas remain the same. (See Figure 7.8.)

The ratio of pressure (P) to temperature (T) is the same under all conditions as long as volume (V) and amount of gas (n) do not change.

$$\frac{P_i}{T_i} = \frac{P_f}{T_f}$$

S·A·M·P·L·E 7.9 ·P·R·O·B·L·E·M

Relationship of Temperature and Pressure

Complete the information for the temperature and pressure of a gas when volume is kept constant.

Temperature	Pressure
a. increases	_____
b. _____	decreases

Solution

a. If the temperature increases, the pressure of the gas also increases.
b. If the pressure decreases, the temperature of the gas also decreases.

Study Check

If the temperature of the gas in a tire decreases, how will pressure change?

S·A·M·P·L·E **7.10** P·R·O·B·L·E·M

Calculating Pressure When Temperature Changes

Aerosol cans can be dangerous if they are heated because they can explode. Suppose a can of hair spray with a pressure of 4.0 atm at a room temperature of 27°C is thrown into a fire. If the temperature of the gas inside the aerosol can reaches 402°C, what will be its pressure? The aerosol can may explode if the pressure inside exceeds 8.0 atm. Would you expect the can to explode?

Solution

We must first change the temperatures to kelvins.

$$T_i = 27°C = 273 = 300. \text{ K}$$

$$T_f = 402°C + 273 = 675 \text{ K}$$

Initial Conditions	Final Conditions	Change
$P_i = 4.0$ atm	$P_f = ?$	P must increase
$T_i = 300.$ K	$T_f = 675$ K	T increases

Because the temperature of the gas has increased, the final pressure must increase. In the calculation, we use a factor that gives a greater final pressure.

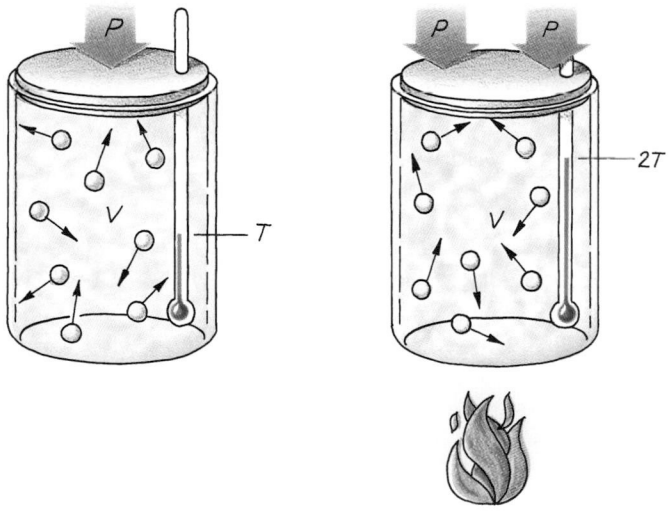

Figure 7.8 The Kelvin temperature of a gas is directly related to the pressure of the gas when there is no change in volume or amount.

$$P_f = 4.0 \text{ atm} \times \frac{675\,\cancel{K}}{300.\,\cancel{K}} = 9.0 \text{ atm}$$

New pressure — Initial pressure — Temperature factor that increases pressure

Because the calculated pressure exceeds 8.0 atm, we expect the can to explode.

Study Check

7.5 In a storage area where the temperature has reached 55°C, the pressure of oxygen gas in a 15.0-L steel cylinder is 965 torr. To what temperature would the gas have to be cooled to reduce the pressure to 850. torr?

QUESTIONS & PROBLEMS

Temperature and Pressure (Gay-Lussac's Law)

7.37 Why do aerosol cans explode if heated?

7.38 How can the tires on a car look flat on a cold, winter morning, but have a blowout when the car is driven on hot pavement in the desert?

7.39 State the expected change for temperature and pressure according to Gay-Lussac's Law, with no change in n and V.

Pressure	Temperature
a. _____	increases
b. _____	decreases

7.40 State the expected change for temperature and pressure according to Gay Lussac's Law, with no change in n and V.

Pressure	Temperature
a. decreases	_____
b. increases	_____

7.41 Solve for the new pressure when each of the following temperature changes occurs, with n and V constant:

a. A gas sample has a pressure of 1200 torr at 155°C. What is the final pressure of the gas after the temperature has dropped to 0°C?

b. An aerosol can has a pressure of 1.40 atm at 12°C. What is the final pressure in the aerosol can if it is used in a room where the temperature is 35°C?

7.42. Solve for the new temperature in degrees Celsius when pressure is changed.

a. A 10.0-L container of helium gas has a pressure of 250 torr at 0°C. To what Celsius temperature does the sample need to be heated to obtain a pressure of 1500 torr?

b. A 500.0-mL sample of air at −10.°C is inhaled. What is the new volume when the temperature of the air sample rises to body temperature of 37°C?

7.6 The Combined Gas Law

Study Pack

Learning Goal

Use the combined gas law to find the new pressure, volume, or temperature of a gas when changes in two of these properties are given.

7.6

All of the pressure–volume–temperature relationships for gases that we have studied may be combined into a single relationship called the combined gas law. This expression is useful for studying the effect of changes in two of these variables on the third as long as the amount of gas (number of moles) remains constant.

$$\frac{P_iV_i}{T_i} = \frac{P_fV_f}{T_f} \quad \text{No change in amount of gas}$$

S·A·M·P·L·E 7.11 P·R·O·B·L·E·M

Using the Combined Gas Law

A 25.0-mL bubble is released from a diver's air tank at a pressure of 4.00 atm and a temperature of 11°C. What is the volume of the bubble when it reaches the ocean surface, where the pressure is 1.00 atm and the temperature is 18°C?

Solution

We must first change the temperature to kelvins.

$$T_i = 11°C + 273 = 284 \text{ K}$$
$$T_f = 18°C + 273 = 291 \text{ K}$$

Initial Conditions	Final Conditions	Change
$P_i = 4.00$ atm	$P_f = 1.00$ atm	P decreases
$V_i = 25.0$ mL	$V_f = ?$	
$T_i = 284$ K	$T_f = 291$ K	T increases

To set up the calculation, we will first consider the effect of the pressure change on the volume. Because the pressure change will increase volume (Boyle's law), we multiply the initial volume by a pressure factor that gives a greater final volume.

$$V_f = 25.0 \text{ mL} \times \frac{4.00 \cancel{\text{atm}}}{1.00 \cancel{\text{atm}}} \times ?$$

New volume = Initial volume × Pressure factor that increases volume

Now we will consider the effect of the temperature increase, which would cause an increase in volume (Charles' law). We need to multiply by a temperature factor that increases the volume.

$$V_f = 25.0 \text{ mL} \times \frac{4.00 \cancel{\text{atm}}}{1.00 \cancel{\text{atm}}} \times \frac{291 \cancel{\text{K}}}{284 \cancel{\text{K}}} = 102 \text{ mL}$$

New volume = Initial volume × Pressure factor that increases volume × Temperature factor that increases volume

The combined effect of the pressure decrease and temperature increase is an increase in the volume.

Study Check

A weather balloon is filled with 15.0 L of helium at a temperature of 25°C and a pressure of 685 mm Hg. What is the pressure of the helium in the balloon in the upper atmosphere when the temperature is −35°C and the volume becomes 34.0 L?

Q·U·E·S·T·I·O·N·S & P·R·O·B·L·E·M·S

7.7

The Combined Gas Law

7.43 Write the expression for the combined gas law. What gas laws are combined to make the combined gas law?

7.44 Rearrange the variables in the combined gas law to give an expression for the following:
a. V_f **b.** P_f

7.45 A sample of Cl_2 gas has a volume of 10.0 L at a pressure of 2.00 atm and a temperature of 15°C. What is the pressure of the gas when the volume and temperature of the gas are changed to the following?
a. 5.00 L and 125°C
b. 20.0 L and 0°C
c. 760 mL and 250.°C
d. 50.0 L and 350.°F

7.46 A sample of argon gas has a volume of 750 mL at a pressure of 1.00 atm and a temperature of 127°C. What is the volume of the gas when the pressure and temperature of the gas are changed to the following?
a. 2.50 atm and 250.°C
b. 0.58 atm and 75°C
c. 4.00 atm and −23°C
d. 0.85 atm and 400.°C

7.47 A 100.0-mL bubble of hot gases at 225°C and 1.80 atm escapes from an active volcano. What is the new volume of the bubble outside the volcano where the temperature is −25°C and the pressure is 0.80 atm?

7.48 A scuba diver 40 ft below the ocean surface inhales 50.0 mL of compressed air in a scuba tank at a pressure of 3.00 atm and a temperature of 8°C. What is the pressure of air in the lungs if the gas expands to 150.0 mL at a body temperature of 37°C?

7.7 Volume and Moles (Avogadro's Law)

Learning Goal

Describe the relationship between the amount of a gas and its volume and use this relationship in calculations.

In our study of the gas laws, we have looked at changes in properties for a specified amount (n) of gas. Now we will consider how the properties of a gas change when there is a change in number of moles or grams. For example, when you blow up a balloon, its volume increases because you add more air molecules. If a basketball gets a hole in it, and some of the air leaks out, its volume decreases. In 1811, Amedeo Avogadro stated that the volume of a gas is directly related to the number of moles of a gas when temperature and pressure are not changed. We refer to this statement as **Avogadro's law.** If the number of moles of a gas are doubled, then the volume will double as long

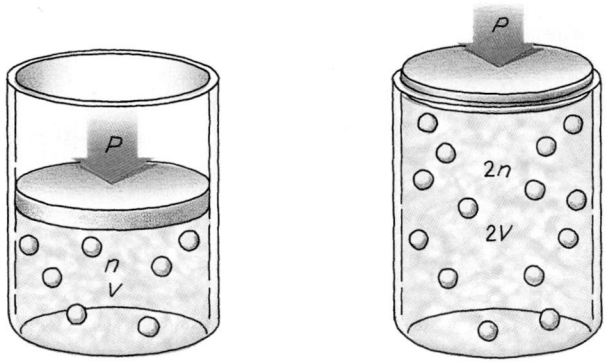

Figure 7.9 If the amount (number of moles) of a gas is doubled, the volume of the gas doubles when there is no change in pressure or temperature.

as we do not change the pressure or the temperature. (See Figure 7.9.) For two conditions, we can write

$$\frac{V_i}{n_i} = \frac{V_f}{n_f}$$

S·A·M·P·L·E· **7.12** ·P·R·O·B·L·E·M

Calculating Volume for a Change in Moles

A balloon with a volume of 220 mL is filled with 2.0 moles of helium. To what volume will the balloon expand if 3.0 moles of helium are added, to give a total of 5.0 moles of helium, and the pressure and temperature do not change?

Solution

A data table for our given information can be set up as follows:

Initial Conditions	Final Conditions	Change
V_i = 220 mL	V_f = ?	V must increase
n_i = 2.0 moles	n_f = 5.0 moles	n increases

Because the amount of gas (number of moles) has increased, the volume of the balloon must increase according to Avogadro's law.

$$\underset{\text{New volume}}{V_f} = \underset{\text{Initial volume}}{220\text{ mL}} \times \underset{\text{Mole factor that increases volume}}{\frac{5.0\ \cancel{\text{moles}}}{2.0\ \cancel{\text{moles}}}} = 550\text{ mL}$$

Study Check

At a certain temperature and pressure, 8.00 g of oxygen has a volume of 5.00 L. What is the volume after 4.00 g of oxygen is added to the balloon?

STP and Molar Volume

Scientists have chosen standard temperature and pressure **(STP)** conditions of 0°C (273 K) and exactly 1 atm (760 mm Hg).

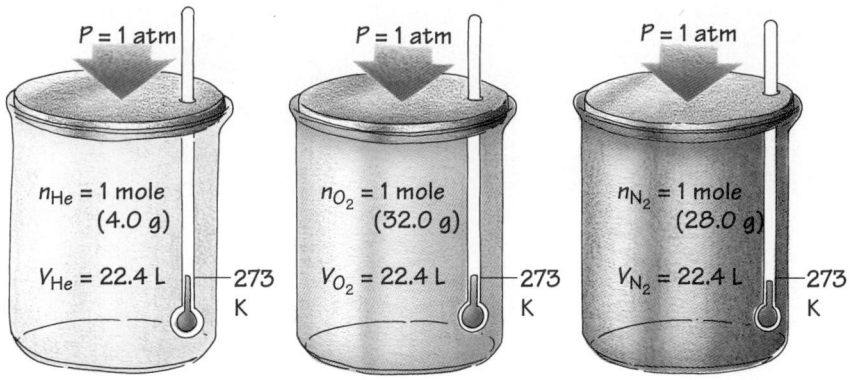

Figure 7.10 Molar volume: One mole of a gas at STP has a volume of 22.4 L.

7.7

STP Conditions

Standard temperature is 0°C (273 K)

Standard pressure is 1 atm (760 mm Hg)

At STP, 1 mole of any gas occupies a volume of 22.4 L. This value is known as the *molar volume* of a gas. (See Figure 7.10.) Suppose we have three containers, one filled with 1 mole of oxygen gas (O_2), another filled with 1 mole of nitrogen gas (N_2), and one filled with 1 mole of helium gas (He). When the gases are at STP conditions (1 atm and 273 K), each has a volume of 22.4 L. Thus, the volume of 1 mole of any gas at STP is 22.4 L, the **molar volume** of a gas.

Molar Volume

The volume of 1 mole of gas at STP = 22.4 L

As long as a gas is at STP conditions (0°C and 1 atm), its molar volume can be used as a conversion factor to convert between the number of moles of gas and its volume.

Molar Volume Conversion Factors

$$\frac{1 \text{ mole gas (STP)}}{22.4 \text{ L}} \quad \text{and} \quad \frac{22.4 \text{ L}}{1 \text{ mole gas (STP)}}$$

S·A·M·P·L·E **7.13** ·P·R·O·B·L·E·M

Calculations Using Molar Volume

What is the volume of 64.0 g of O_2 gas at STP?

Solution

The mass of the gas is changed to moles by its molar mass, which is 32.0 g/mole. Because the gas is at STP, we can use the molar volume to convert from moles to liters of gas.

$$64.0 \cancel{\text{g } O_2} \times \frac{1 \cancel{\text{mole } O_2}}{32.0 \cancel{\text{g } O_2}} \times \frac{22.4 \text{ L } O_2}{1 \cancel{\text{mole } O_2}} = 44.8 \text{ L } O_2$$

Molar mass of O_2 — Molar volume of a gas

Study Check

How many grams of nitrogen (N_2) gas are in 5.6 L of the gas at STP?

Q·U·E·S·T·I·O·N·S & P·R·O·B·L·E·M·S

Volume and Moles (Avogadro's Law)

7.49 What happens to the volume of a bicycle tire or a basketball when you use an air pump to add air?

7.50 Sometimes when you blow up a balloon and release it, it flies around the room. What is happening to the air that was in the balloon and its volume?

7.51 Solve the following problems when no changes occur in pressure and temperature:

a. A sample of 4.00 moles of argon gas has a volume of 10.0 L. A small leak causes one-half of the molecules to escape. What is the new volume of the gas?

b. A balloon containing 1.00 mole of helium has a volume of 440 mL. What is the new volume after 2.00 more moles of helium is added to the balloon?

c. A 1500 mL sample of SO_2 contains 6.00 moles SO_2. How many moles of SO_2 are lost when the volume is reduced to 750 mL?

7.52 Solve the following problems when no changes occur in pressure and temperature:

a. A 250-mL sample of neon gas contains 1.0 mole Ne. What is the new volume if 3.0 moles of Ne are added to the container?

b. A 10.0-L sample of helium gas contains 4.00 g of He. If 2.00 g of helium are added to the initial sample, what is the new volume of the gas at the same pressure and temperature?

c. A 2500-mL sample of O_2 contains 5.00 moles of O_2. If oxygen is released until the volume drops to 1500 mL, how many moles of O_2 were lost?

7.53 Use the molar volume of a gas to solve the following at STP:

a. the number of moles of O_2 in 44.8 L of O_2 gas

b. the number of moles of CO_2 in 4.00 L of CO_2 gas

c. the volume (L) of 6.40 g of O_2

d. the volume (mL) occupied by 50.0 g of neon

7.54 Use molar volume to solve the following problems at STP:

a. the volume (L) occupied by 2.5 moles of N_2

b. the volume (mL) occupied by 0.420 mole of He

c. the number of grams of neon contained in 11.2 L of Ne gas

d. the number of moles of H_2 in 1600 mL of H_2 gas

7.8 Partial Pressures (Dalton's Law)

Learning Goal

Use partial pressures to calculate the total pressure of a mixture of gases.

Many gas samples are a mixture of gases. For example, the air you breathe is a mixture of mostly oxygen and nitrogen gases. For gas mixtures, we use the gas laws we have studied, because particles of all gases behave in the same way. Therefore, the total pressure of the gases in a mixture is a result of the collisions of the gas particles regardless of what type of gas they are.

In a gas mixture, each gas exerts its **partial pressure,** which is the pressure it would exert if it were the only gas in the container. **Dalton's law** states that the total pressure

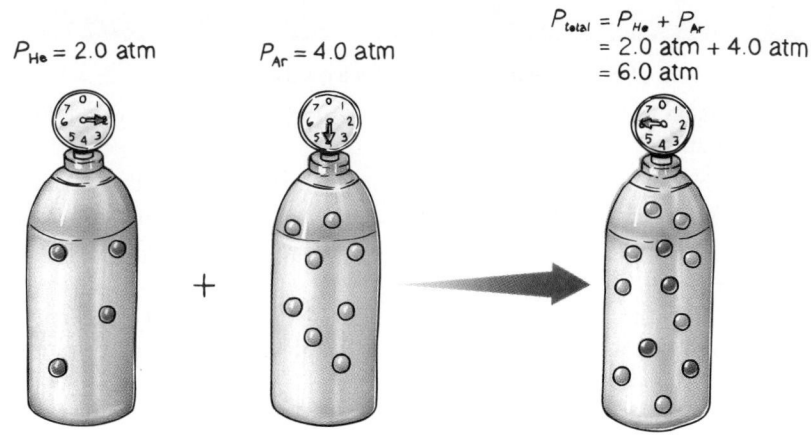

Figure 7.11 The total pressure of a gas mixture, which depends on the total number of gas particles present, is the sum of the partial pressures of the individual gases.

7.8

of a gas mixture is the sum of the partial pressures of the gases in the mixture. (See Figure 7.11.)

$$P_{total} = P_1 + P_2 + P_3 + \cdots$$

Total pressure of a gas mixture = Sum of the partial pressures of the gases in the mixture

S·A·M·P·L·E· 7.14 ·P·R·O·B·L·E·M

Calculating the Total Pressure of a Gas Mixture

A 10-L gas tank contains propane (C_3H_8) gas at a pressure of 300 torr. Another 10-L gas tank contains methane (CH_4) gas at a pressure of 500 torr. In preparing a gas fuel mixture, the gases from both tanks are combined in a 10-L container at the same temperature. What is the pressure of the gas mixture?

Solution

Using Dalton's law of partial pressure, we find that the total pressure of the gas mixture is the sum of the partial pressures of the gases in the mixture.

$$\begin{aligned} P_{total} &= P_{propane} + P_{methane} \\ &= 300 \text{ torr} + 500 \text{ torr} \\ &= 800 \text{ torr} \end{aligned}$$

Therefore, when both propane and methane are placed in the same container, the total pressure of the mixture is 800 torr.

Study Check

A gas mixture consists of helium with a partial pressure of 315 mm Hg, nitrogen with a partial pressure of 204 mm Hg, and argon with a partial pressure of 422 mm Hg. What is the total pressure in atmospheres?

Air Is a Gas Mixture

The air you breathe is a mixture of gases. What we call the atmospheric pressure is actually the sum of the partial pressures of the gases in the air. Table 7.6 lists partial pressures for the gases in air on a typical day.

Table 7.6 Composition of Air

Gas	Pressure (mm Hg)	Percentage (%)
Nitrogen, N_2	594.0	78
Oxygen, O_2	160.0	21
Carbon dioxide, CO_2	0.3	1
Water vapor, H_2O	5.7	
Total air	760.0	100

S·A·M·P·L·E 7.15 ·P·R·O·B·L·E·M

Partial Pressure of a Gas in a Mixture

A mixture of oxygen and helium is prepared for a scuba diver who is going to descend 200 ft below the ocean surface. At that depth, the diver breathes a gas mixture that has a total pressure of 7.0 atm. If the partial pressure of the oxygen in the tank at that depth is 1.5 atm, what is the partial pressure of the helium?

Solution

From Dalton's law of partial pressures, we know that the total pressure is equal to the sum of the partial pressures:

$$P_{total} = P_{O_2} + P_{He}$$

To solve for the partial pressure of helium (P_{He}), we rearrange the expression to give the following:

$$P_{He} = P_{total} - P_{O_2}$$
$$P_{He} = 7.0 \text{ atm} - 1.5 \text{ atm}$$
$$= 5.5 \text{ atm}$$

Thus, in the gas mixture that the diver breathes, the partial pressure of the helium is 5.5 atm.

Study Check

An anesthetic consists of a mixture of cyclopropane gas, C_3H_6, and oxygen gas, O_2. If the mixture has a total pressure of 825 torr, and the partial pressure of the cyclopropane is 73 torr, what is the partial pressure of the oxygen in the anesthetic?

Amount of Gas Dissolved in a Liquid

Henry's law states that the amount of gas that will dissolve in a liquid at a particular temperature is directly related to the pressure of that gas above the liquid. (See Figure 7.12.)

A can of soda contains dissolved carbon dioxide gas. When the can is opened, bubbling occurs. Sometimes the amount of escaping gas is so great that the liquid sprays out of the container. (See Figure 7.13.) When the can was originally filled and sealed, the pressure of the CO_2 gas in it was high, to make sure that sufficient CO_2 dissolved in the liquid. When the can is opened, the CO_2 pressure drops, less CO_2 is soluble, and bubbles of gas rapidly escape from the beverage.

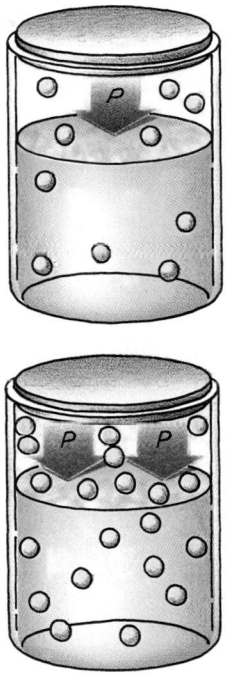

Figure 7.12 Henry's law: The amount of gas dissolved in a liquid is related to the pressure of that gas above the liquid. When the pressure of the gas increases, the number of gas molecules that enter the liquid increases.

Figure 7.13 When a can of soda is opened, the CO_2 gas in the liquid rapidly escapes.

7.8

EXPLORE YOUR WORLD

Oxygen in the Air

For this activity, you will need some candles of different heights, a shallow bowl, a glass or jar tall enough to fit over the tallest candle, and a watch with a second hand. Using some melted wax from a burning candle, stick one candle to the bottom of a shallow bowl. Light the candle and place a jar over the candle. Observe the candle flame. Measure the time that the candle continues to burn.

Stick another candle that has a different height to the bottom of the bowl. You can try three candles if you wish. Light all the candles and carefully lower the jar over all the candles. The opening of the jar should be sitting on the bottom of the bowl. Measure the time that the candles continue to burn. Watch the order in which they are extinguished.

Remove the jar and place some water around the candles. Light the candles again, and place the jar over them. Observe the water level inside the jar as the candles go out.

Questions

1. What happens to the flame when you cover the candle with a jar? Explain.
2. How does the time for one candle's flame to go out compare to the time for two or three candles? Explain.
3. In what order were the candles extinguished? Explain.
4. What happens to the water level inside the jar after the candle flames go out?
5. Approximately how far up the jar did the water rise?
6. If you express the change in water level in the jar as a percentage, and the atmospheric pressure is 760 mm Hg, what would you calculate for the partial pressure of oxygen in the air?

Blood Gases

Our cells continuously use oxygen and produce carbon dioxide. Both gases move in and out of the lungs through the membranes of the alveoli, the tiny air sacs at the ends of the airways in the lungs. An exchange of gases occurs in which oxygen from the air diffuses into the lungs and into the blood, while carbon dioxide produced in the cells is carried to the lungs to be exhaled. In Table 7.7, partial pressures are given for the gases in air that we inhale (inspired air), air in the alveoli, and the air that we exhale (expired air). The partial pressure of water vapor increases within the lungs because the vapor pressure of water is 47 mm Hg at body temperature.

Table 7.7 Partial Pressures of Gases During Breathing

	Partial Pressure (mm Hg)		
Gas	**Inspired Air**	**Alveolar Air**	**Expired Air**
Nitrogen, N_2	594.0	573	569
Oxygen, O_2	160.0	100	116
Carbon dioxide, CO_2	0.3	40	28
Water vapor, H_2O	5.7	47	47
Total	760.0	760	760

At sea level, oxygen normally has a partial pressure of 100 mm Hg in the alveoli of the lungs. Because the partial pressure of oxygen in venous blood is 40 mm Hg, oxygen diffuses from the alveoli into the bloodstream. The oxygen combines with hemoglobin, which carries it to the tissues of the body where the partial pressure of oxygen can be very low, less than 30 mm Hg. Oxygen diffuses from the blood where the partial pressure of O_2 is high into the tissues where O_2 pressure is low.

As oxygen is used in the cells of the body during metabolic processes, carbon dioxide is produced, so the partial pressure of CO_2 may be as high as 50 mm Hg or more. Carbon dioxide diffuses from the tissues into the bloodstream and is carried to the lungs. There it diffuses out of the blood, where CO_2 has a partial pressure of 46 mm Hg, into the alveoli, where the CO_2 is at 40 mm Hg and is exhaled. (See Figure 7.14.) Table 7.8 gives the partial pressures of blood gases in the tissues and in oxygenated and deoxygenated blood.

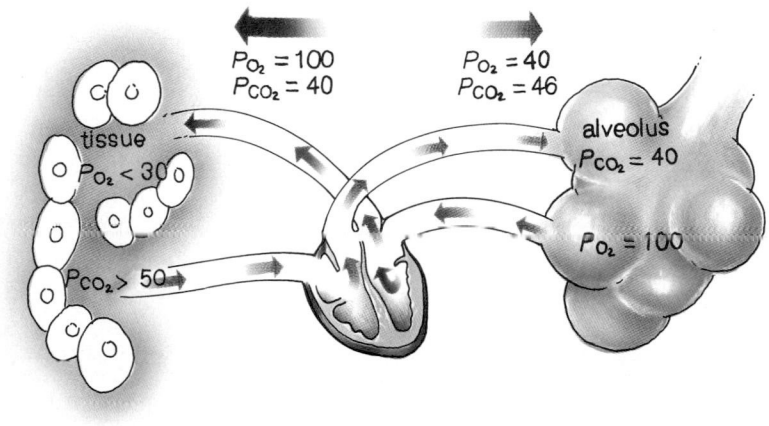

Figure 7.14 The diffusion of oxygen and carbon dioxide across the membranes of the alveoli and tissues during the exchange of blood gases.

Table 7.8 Partial Pressures of Oxygen and Carbon Dioxide in Blood and Tissues

	Partial Pressure (mm Hg)		
Gas	**Oxygenated Blood**	**Deoxygenated Blood**	**Tissues**
O_2	100	40	30 or less
CO_2	40	46	50 or greater

HEALTH NOTE

Hyperbaric Chambers

7.8

A burn patient may undergo treatment for burns and infections in a hyperbaric chamber, a device in which pressures can be obtained that are two to three times greater than atmospheric pressure. A greater oxygen pressure increases the level of dissolved oxygen in the blood and tissues, where it fights bacterial infections. High levels of oxygen are toxic to many strains of bacteria. The hyperbaric chamber may also be used during surgery, to help counteract carbon monoxide (CO) poisoning, and to treat some cancers. (See Figure 7.15.)

The blood is normally capable of dissolving up to 95% of the oxygen. Thus, if the partial pressure of the oxygen is 2280 mm Hg (3 atm), 95% of that or 2160 mm Hg of oxygen can dissolve in the blood where it saturates the tissues. In the case of carbon monoxide poisoning, this oxygen can replace the carbon monoxide that has attached to the hemoglobin.

A patient undergoing treatment in a hyperbaric chamber must also undergo decompression (reduction of pressure) at a rate that slowly reduces the concentration of dissolved oxygen in the blood. If decompression is too rapid, the oxygen dissolved in the blood may form gas bubbles in the circulatory system.

If divers do not decompress slowly, they suffer a similar condition called the bends. While below the surface of the ocean, divers breathe air at higher pressures. At such high pressures, nitrogen gas will dissolve in their blood. If they ascend to the surface too quickly, the dissolved nitrogen forms bubbles in the blood that can produce life-threatening blood clots. The gas bubbles can also appear in the joints and tissues of the body and be quite painful. A diver suffering from the bends is placed immediately in a decompression chamber where pressure is first increased and then slowly decreased. The dissolved nitrogen can then diffuse through the lungs until atmospheric pressure is reached.

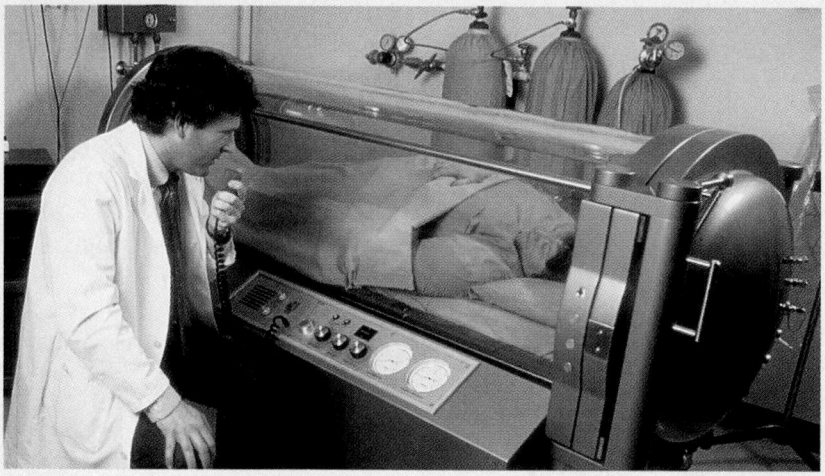

Figure 7.15 A hyperbaric chamber exposes a patient to oxygen at a pressure that is two to three times atmospheric pressure.

S·A·M·P·L·E· 7.16 ·P·R·O·B·L·E·M

Blood Gases

Use partial pressures to explain why carbon dioxide passes through the alveolar membrane from the blood into the lungs.

Solution

Oxygen diffuses from a higher partial pressure of 100 mm Hg in the air to a lower partial pressure of 40 mm Hg in the blood at the alveoli.

Study Check

Muscles and tissues in the body need oxygen to continue to function and carry out metabolic reactions. Why does oxygen flow from the blood into the body tissues?

QUESTIONS & PROBLEMS

Partial Pressures (Dalton's Law)

7.55 A typical air sample in the lungs contains oxygen at 100 mm Hg, nitrogen at 573 mm Hg, carbon dioxide at 40 mm Hg, and water vapor at 47 mm Hg. Why are these pressures called partial pressures?

7.56 Suppose a mixture contains helium and oxygen gases. If the partial pressure of helium is the same as the partial pressure of oxygen, what do you know about the number of helium atoms compared to the number of oxygen molecules? Explain.

7.57 In a gas mixture, the partial pressures are nitrogen 425 torr, oxygen 115 torr, and helium 225 torr. What is the total pressure (torr) exerted by the gas mixture?

7.58 In a gas mixture, the partial pressures are argon 415 mm Hg, neon 75 mm Hg, and nitrogen 125 mm Hg. What is the total pressure (atm) exerted by the gas mixture?

7.59 A gas mixture containing oxygen, nitrogen, and helium exerts a total pressure of 925 torr. If the partial pressures are oxygen 425 torr and helium 75 torr, what is the partial pressure (torr) of the nitrogen in the mixture?

7.60 A gas mixture containing oxygen, nitrogen, and neon exerts a total pressure of 1.20 atm. If helium added to the mixture increases the pressure to 1.50 atm, what is the partial pressure (atm) of the helium?

7.61 In certain lung ailments such as emphysema, there is a decrease in the ability of oxygen to diffuse into the blood.

a. How would the partial pressure of oxygen in the blood change?

b. Why does a person with severe emphysema sometimes use a portable oxygen tank?

7.62 An accident to the head can affect the ability of a person to ventilate (breathe in and out), and so can certain drugs.

a. What would happen to the partial pressures of oxygen and carbon dioxide in the blood if a person cannot properly ventilate?

b. When a person with hypoventilation is placed on a ventilator, an air mixture is delivered at pressures that are alternately above the air pressure in the person's lung, and then below. How will this move oxygen gas into the lungs, and carbon dioxide out?

Chapter Review

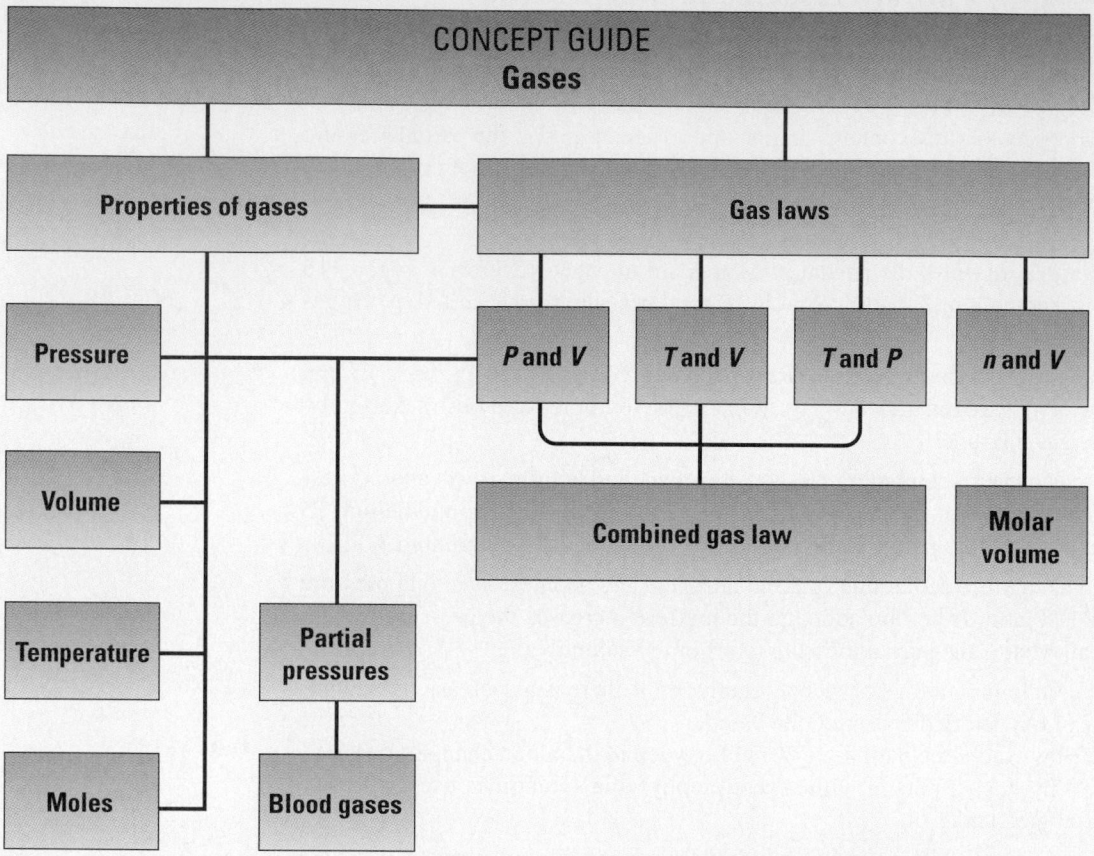

Chapter Summary

7.1 Properties of Gases

In a gas, particles are so far apart and moving so fast that their attractions are unimportant. A gas is described by the physical properties of pressure (P), volume (V), temperature (T), and amount in moles (n).

7.2 Gas Pressure

A gas exerts pressure, the force of the gas particles on the surface of a container. Gas pressure is measured in units of torr, mm Hg, atm, and pascal. Vapor pressure is the pressure of the gas that forms when a liquid evaporates. At the boiling point of a liquid, the vapor pressure equals the external pressure.

7.3 Pressure and Volume (Boyle's Law)

The volume (V) of a gas changes inversely with the pressure (P) of the gas if there is no change in the amount and temperature: $P_iV_i = P_fV_f$. This means that the pressure increases if volume decreases; pressure decreases if volume increases.

7.4 Temperature and Volume (Charles' Law)

The volume (V) of a gas is directly related to its Kelvin temperature (T) when there is no change in the amount and pressure of the gas:

$$\frac{V_i}{T_i} = \frac{V_f}{T_f}$$

Therefore, if temperature increases, the volume of the gas increases; if temperature decreases, volume decreases.

7.5 Temperature and Pressure (Gay-Lussac's Law)

The pressure (P) of a gas is directly related to its Kelvin temperature (T).

$$\frac{P_i}{T_i} = \frac{P_f}{T_f}$$

This means that an increase in temperature (T) increases the pressure of a gas, or a decrease in temperature decreases the pressure, as long as the amount and volume stays constant.

7.6 The Combined Gas Law

Gas laws combine into a relationship of pressure (P), volume (V), and temperature (T).

$$\frac{P_iV_i}{T_i} = \frac{P_fV_f}{T_f}$$

This expression is used to determine the effect of changes in two of the variables on the third.

7.7 Volume and Moles (Avogadro's Law)

The volume (V) of a gas is directly related to the number of moles (n) of the gas when the pressure and temperature of the gas do not change.

$$\frac{V_i}{n_i} = \frac{V_f}{n_f}$$

If the moles of gas are increased, the volume must increase; or if the moles of gas are decreased, the volume decreases.

7.8 Partial Pressures (Dalton's Law)

In a mixture of two or more gases, the total pressure is the sum of the partial pressures of the individual gases.

$$P_{\text{total}} = P_1 + P_2 + P_3 + \cdots$$

The partial pressure of a gas in a mixture is the pressure it would exert if it were the only gas in the container.

Key Terms

atmosphere (atm) The pressure exerted by a column of mercury 760 mm high.

atmospheric pressure The pressure exerted by the atmosphere.

Avogadro's law A gas law that states that the volume of gas is directly related to the number of moles of gas in the sample when pressure and temperature do not change.

barometer An instrument used to measure atmospheric pressure.

Boyle's law A gas law stating that the pressure of a gas is inversely related to the volume, when temperature and moles of the gas do not change; that is, if volume decreases, pressure increases.

Charles' law A gas law stating that the volume of a gas changes directly with a change in Kelvin temperature, when pressure and moles of the gas do not change.

combined gas law A relationship that combines several gas laws relating pressure, volume, and temperature.

$$\frac{P_iV_i}{T_i} = \frac{P_fV_f}{T_f}$$

Dalton's Law A gas law stating that the total pressure exerted by a mixture of gases in a container is the sum of the partial pressures that each gas would exert alone.

diffusion The movement of gas particles from a more concentrated area (high pressure) to a less concentrated area (low pressure).

direct relationship A relationship in which two properties increase or decrease together.

Gay-Lussac's law A gas law stating that the pressure of a gas changes directly with a change in temperature when the number of moles of a gas and its volume are held constant.

Henry's Law A gas law stating that the amount of gas dissolved in a liquid at a certain temperature is directly related to the pressure of the gas above the liquid.

inverse relationship A relationship in which two properties change in opposite directions.

kinetic theory A model used to explain the behavior of gases.

molar volume A volume of 22.4 L occupied by 1 mole of a gas at STP conditions of 0°C (273 K) and 1 atm.

partial pressure The pressure exerted by a single gas in a gas mixture.

pressure The force exerted by gas particles that hit the walls of a container.

pressure gradient A difference in pressure causing gas particles to move from the area of high pressure to the area of low pressure.

STP Standard conditions of 0°C (273 K) temperature and 1 atm pressure used for the comparison of gases.

torr A unit of pressure equal to 1 mm Hg; 760 torr = 1 atm.

vapor pressure The pressure exerted by the particles of vapor above a liquid.

Chapter Practice

7.63 When a gas is heated at constant volume, the pressure increases. Use ideas from the kinetic molecular theory to explain.

7.64 When you drink through a straw, you breathe in to reduce the pressure in your mouth. What pressure is exerted on the top of the liquid you are drinking? Why does the liquid move up the straw?

7.65 At a restaurant, a customer chokes on a piece of food. You put your arms around the person's waist and use your fists to push up on the person's abdomen, an action called the Heimlich maneuver.

a. How would this action change the volume of the chest and lungs?

b. Why does it cause the person to expel the food item from the airway?

7.66 An airplane is pressurized to 650 mm Hg, which is the atmospheric pressure at a ski resort at 13,000 ft altitude.

a. If air is 21% oxygen, what is the partial pressure of oxygen on the plane?

b. If the partial pressure of oxygen drops below 100 mm Hg, passengers become drowsy. If this happens, oxygen masks are released. What is the total cabin pressure at which oxygen masks are dropped?

7.67 A fire extinguisher has a pressure of 150 lb/in.^{2} at 25°C. What is the pressure in atmospheres if the fire extinguisher is used at a temperature of 75°C?

7.68 A weather balloon has a volume of 750 L when filled with helium at 8°C at a pressure of 380 torr. What is the new volume of the balloon, where the pressure is 0.20 atm and the temperature is −45°C?

7.69 A sample of hydrogen (H_2) gas at 127°C has a pressure of 2.00 atm. At what temperature (°C) will the pressure of the H_2 decrease to 0.25 atm?

7.70 Nitrogen (N_2) is prepared and a sample of 250 mL is collected over water at 30°C and a total pressure of 745 mm Hg.

a. Using the vapor pressure of water in Table 7.3, what is the partial pressure of the nitrogen?

b. What is the volume of the nitrogen at STP?

7.71 A gas mixture with a total pressure of 2400 torr is used by a scuba diver. If the mixture contains 2.0 moles of helium and 6.0 moles of oxygen, what is the partial pressure of each gas in the sample?

7.72 What is the total pressure in mm Hg of a gas mixture containing argon gas at 0.25 atm, helium gas at 350 mm Hg, and nitrogen gas at 360 torr?

7.73 A gas mixture contains oxygen and argon at partial pressures of 0.60 atm and 425 mm Hg. If nitrogen gas added to the sample increases the total pressure to 1250 torr, what is the partial pressure in torr of the nitrogen added?

7.74 A gas mixture contains helium and oxygen at partial pressures of 255 torr and 0.450 atm. What is the total pressure in mm Hg of the mixture after it is placed in a container one-half the volume of the original container?

7.75 When you open a bottle of a carbonated beverage, why is there a burst of bubbles?

7.76 At a party, a helium-filled balloon floats close to some hot lights and bursts. What might be a reason for this?

7.77 Compare the partial pressures of the following respiratory gases:

a. oxygen in the lungs and in the blood coming to the alveoli

b. oxygen in arterial blood and venous blood

c. carbon dioxide in the tissues and arterial blood

d. carbon dioxide in the venous blood and the lungs

7.78 For each of the comparisons in problem 7.77, describe the direction of diffusion for each gas.

7.79 When heated, calcium carbonate decomposes to give calcium oxide and carbon dioxide gas.

$$CaCO_3(s) \longrightarrow CaO(s) + CO_2(g)$$

If 2.00 moles of $CaCO_3$ react, how many liters of CO_2 gas are produced at STP?

7.80 In the reaction

$$2H_2(g) + O_2(g) \longrightarrow 2H_2O(g)$$

a. If 16.00 g of O_2 react at 0.800 atm and 127°C, what volume of oxygen was used?

b. How many moles of water are produced in part a?

c. If the water is produced at 760 mm Hg and 115°C, what volume of water was collected?

Answers

Answers to Study Checks

7.1 The gas molecules that carry the odor of the food move throughout the house until they reach the room that you are in.

7.2 The mass in grams gives the amount of gas.

7.3 0.862 atm

7.4 The volume increases.

7.5 400 torr

7.6 50.0 mL

7.7 Temperature must decrease.

7.8 589 mL

7.9 Pressure will decrease.

7.10 16°C

7.11 241 mm Hg

7.12 7.50 L

7.13 7.0 g N_2

7.14 1.24 atm

7.15 752 torr

7.16 Oxygen diffuses from a higher partial pressure of 100 mm Hg in the arterial blood to a lower partial pressure of 30 mm Hg in the muscles and body tissues where oxygen has been depleted.

Answers to Selected Questions and Problems

7.1 **a.** At a higher temperature, gas particles have greater kinetic energy, which makes them move faster.
b. Because there are great distances between the particles of a gas, they can be pushed closer together and still remain a gas.
c. Gas particles are very far apart, which means that the mass of a gas in a certain volume is very small, resulting in a low density.

7.3 **a.** temperature **b.** volume **c.** amount **d.** pressure

7.5 atmospheres (atm), mm Hg, torr, lb/in.^{2}, pascals

7.7 **a.** 1520 torr **b.** 29.4 lb/in.^{2} **c.** 1520 mm Hg

7.9 **a.** boiling point **b.** vapor pressure **c.** atmospheric pressure **d.** boiling point

7.11 **a.** On top of a mountain, water boils below 100°C because the atmospheric (external) pressure is less than 1 atm.
b. Because the pressure inside a pressure cooker is greater than 1 atm, water boils above 100°C. At a higher temperature, food cooks faster.

7.13 As a diver ascends to the surface, external pressure decreases. If the air in the lungs were not exhaled, its volume would expand and severely damage the lungs. The pressure in the lungs must adjust to changes in the external pressure.

7.15 In order to exhale (expiration) air containing CO_2, the pressure in the lungs must increase, which is done when the volume of the lungs is decreased as the diaphragm relaxes.

7.17 **a.** The pressure is greater in cylinder A. According to Boyle's law, a decrease in volume pushes the gas particles closer together, which will cause an increase in the pressure.
b.

Property	Initial	Final
Pressure (P)	650 mm Hg	1.2 atm
Volume (V)	220 mL	160 mL

7.19 **a.** increases **b.** decreases

7.21 **a.** 328 mm Hg **b.** 1310 mm Hg **c.** 4400 mm Hg **d.** 65 500 mm Hg

7.23 2300 mm Hg

7.25 **a.** 25 L **b.** 25 L **c.** 100. L **d.** 45 L

7.27 25 L

7.29 **a.** inspiration **b.** expiration **c.** inspiration

7.31 **a.** C **b.** A **c.** B

7.33 **a.** decreases **b.** increases

7.35 **a.** 2400 mL **b.** 4900 mL **c.** 1800 mL **d.** 1700 mL

7.37 An increase in temperature increases the pressure inside the can. When the pressure exceeds the pressure limit of the can, it explodes.

7.39 **a.** increases **b.** decreases

7.41 **a.** 770 torr **b.** 1.51 atm

7.43 $\frac{P_iV_i}{T_i} = \frac{P_fV_f}{T_f}$

Boyle's, Charles', and Gay-Lussac's laws are combined to make this law.

7.45 **a.** 5.53 atm **b.** 0.948 atm **c.** 48 atm **d.** 0.625 atm

7.47 110 mL

7.49 The volume increases because the amount of gas particles is increased.

7.51 **a.** 5.00 L **b.** 1300 mL **c.** 3.0 moles of SO_2

7.53 **a.** 2.00 moles of O_2 **b.** 0.179 mole of CO_2 **c.** 4.48 L **d.** 55 400 mL

7.55 In a gas mixture, the pressure that each gas exerts as part of the total pressure is called the partial pressure of that gas. Because the air sample is a mixture of gases, the total pressure is the sum of the partial pressures of each gas in the sample.

7.57 765 torr

7.59 425 torr

7.61 **a.** The partial pressure of oxygen will be lower than normal.

b. Breathing a higher concentration of oxygen will help to increase the diffusion of oxygen into the lungs and blood and raise the partial pressure of oxygen in the blood.

7.63 The increase in temperature increases the motion of the gas particles. They hit the walls with more force, which creates more pressure.

7.65 **a.** The volume of the chest and lungs is decreased.

b. The decrease in volume increases the pressure, which can dislodge the food in the trachea.

7.67 12 atm

7.69 −223°C

7.71 He 600 torr, O_2 1800 torr

7.73 370 torr

7.75 According to Henry's law, the CO_2 has a high pressure in the closed bottle. As the bottle is opened, CO_2 molecules escape as a burst of bubbles because they are less soluble in the liquid when the external pressure of CO_2 is decreased.

7.77 **a.** The P_{O_2} is high in the lungs and low in the blood coming to the lungs.

b. The P_{O_2} is high in arterial blood and low in venous blood.

c. The P_{CO_2} is high in the tissues and low in arterial blood.

d. The P_{CO_2} is high in the venous blood and low in the lungs.

7.79 44.8 L CO_2

8

Solutions

A solution consists of a solute and a solvent that are uniformly mixed.

Looking Ahead

8.1	Properties of Water
8.2	Solutions
8.3	Electrolytes and Nonelectrolytes
8.4	Solubility
8.5	Percent Concentration
8.6	Colloids and Suspensions
8.7	A Closer Look: Osmosis and Dialysis
8.8	Molarity

www.chemplace.com/college/timberlake

Visit the web site above to test yourself, run simulations, and review key concepts from this chapter. For even more interactive review and quizzing, use the two CD-ROMs included with the book.

Solutions are everywhere around us. Each consists of a mixture of at least one substance dissolved in another. The air we breathe is a solution of oxygen and nitrogen gases. Carbon dioxide gas dissolved in water makes carbonated drinks. When we make solutions of coffee or tea, we use hot water to dissolve substances from coffee beans or tea leaves. The ocean is also a solution, consisting of many salts such as sodium chloride dissolved in water. In a hospital, the antiseptic tincture of iodine is a solution of iodine dissolved in alcohol.

In body fluids, there are ions such as K^+, Na^+, Cl^-, H^+, and HCO_3^-, along with molecules of glucose and urea. Significant changes in their concentrations can indicate illness or injury. Therefore, the measurement of their concentrations is a valuable diagnostic tool.

8.1

Another type of mixture, called a colloid, contains large particles. Mayonnaise, whipped cream, and gelatin are all colloids. In a third type of mixture, called a suspension, the particles are so large they settle out. You may have used suspensions such as liquid penicillin or calamine lotion, which had to be shaken before use.

In the processes of osmosis and dialysis, we will see how water, essential nutrients, and waste products enter and leave the cells of the body. In osmosis, water flows in and out of the cells of the body. In dialysis, small particles in solution as well as water diffuse through semipermeable membranes.

Water:
Water Molecule 1, 2

Learning Goal

Describe hydrogen bonding and surface tension in water.

8.1 Properties of Water

Water is one of the most common substances in nature. In the H_2O molecule, an oxygen atom shares electrons with two hydrogen atoms. Because the oxygen atom is much more electronegative, the O—H bonds are strongly polar. In each of the polar bonds, the oxygen atom has a slight negative (δ^-) charge, and the hydrogen atom has a slight positive (δ^+) charge. Because of the arrangement of the polar bonds, water is a *polar substance.*

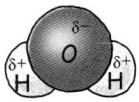

Water, a polar molecule

S·A·M·P·L·E 8.1 ·P·R·O·B·L·E·M

Properties of Water

Why are the O—H bonds in water polar?

Solution

The electron pair shared between each oxygen atom and hydrogen atom is attracted to the more electronegative oxygen, which gives the oxygen atom a slightly negative charge and the hydrogen atom a slightly positive charge.

Study Check

Why do the molecules at the surface of liquid water behave differently than the water molecules within the liquid?

HEALTH NOTE

Water in the Body

The average adult contains about 60% water by weight, and the average infant about 75%. About 60% of the body's water is contained within the cells as intracellular fluids; the other 40% makes up extracellular fluids, which include the interstitial fluid in tissues, and the plasma in the blood. These external fluids carry nutrients and waste materials between the cells and the circulatory system.

Every day you lose between 1500 and 3000 mL of water from the kidneys as urine, from the skin as perspiration, from the lungs as you exhale, and from the gastrointestinal tract. Serious dehydration can occur in an adult if there is a 10% net loss in total body fluid, and a 20% loss of fluid can be fatal. An infant suffers severe dehydration with a 5–10% loss in body fluid.

Water loss is continually replaced by the liquids and foods in the diet and from metabolic processes that produce water in the cells of the body. (See Figure 8.1.) Table 8.1 lists the amounts of water contained in some foods.

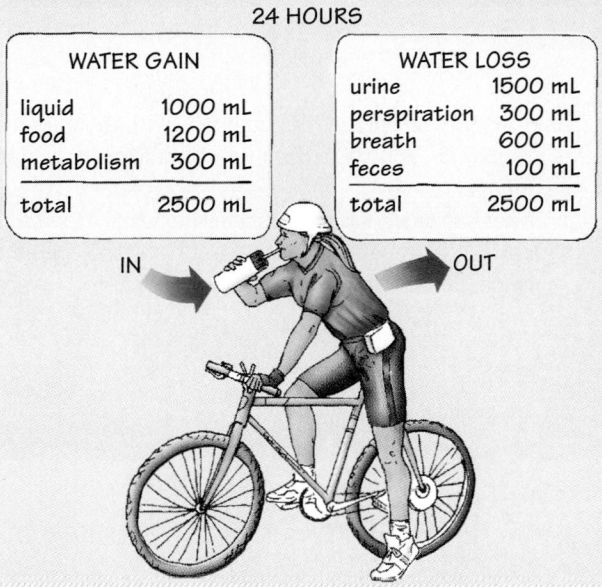

Figure 8.1 The balance of water loss and gain in the body

Table 8.1 Percentage of Water in Some Foods

Food	Water (%)	Food	Water (%)
Vegetables		**Meats/Fish**	
Carrot	88	Chicken, cooked	71
Celery	94	Hamburger, broiled	60
Cucumber	96	Salmon	71
Tomato	94	**Grains**	
Fruits		Cake	34
Apple	85	French bread	31
Banana	76	Noodles, cooked	70
Cantaloupe	91	**Milk Products**	
Grapefruit	89	Cottage cheese	78
Orange	86	Milk, whole	87
Strawberry	90	Yogurt	88
Watermelon	93		

Hydrogen Bonds in Water

A **hydrogen bond** is an attraction between the oxygen atom of one water molecule and a hydrogen atom in another water molecule. In Figure 8.2, hydrogen bonds are shown as dots between the water molecules. Hydrogen bonds occur between molecules where hydrogen is bonded to the very electronegative atoms of O, N, or F. Although hydrogen bonds are much weaker than covalent or ionic bonds, there are many of them linking molecules together. As a result hydrogen bonding plays an important role in the properties of water, biological compounds such as proteins, and DNA.

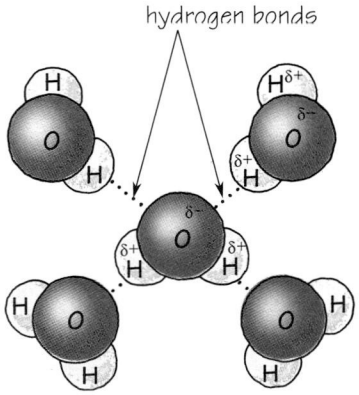

Figure 8.2 Hydrogen bonding links water molecules together.

Figure 8.3 The surface tension of water forms a resting place for this water bug.

Surface Tension

8.1

Imagine that you have filled a glass up to its rim with water. Carefully adding a few more drops of water or dropping in some pennies does not cause the water to overflow. Instead the water seems to adhere to itself, forming a dome that rises slightly above the rim of the glass. This effect is the result of the polarity of water. Throughout the liquid in the glass, water molecules are attracted in all directions by surrounding water molecules. However, the water molecules on the surface are pulled like a skin toward the rest of the water in the glass. As a result, the water molecules on the surface become more tightly packed, a feature called **surface tension.** Because of surface tension, a needle floats on the top of water, certain water bugs can travel across the surface of a pond or lake, and drops of water are spherical. (See Figure 8.3.)

EXPLORE YOUR WORLD

Surface Tension I

Obtain a glass or jar that is smooth around the top. Fill it with water as full as you can without running over. Obtain several pennies. Predict the number of pennies you can add before water runs down the side. Holding a penny vertically, carefully slide it into the glass of water. Keep adding pennies until you see water run down the outside of the glass. Observe the surface of the water. Repeat the test with containers of different shapes and sizes.

Questions

1. How did your prediction compare with the number of pennies you were able to add before the water ran over?
2. What happened to the surface of the water? Explain.
3. How does the number of pennies added compare for containers of different shapes or sizes?

EXPLORE YOUR WORLD

Surface Tension II

Place some water in a small bowl. Carefully place a sewing needle on the surface. Push the needle below the surface of the water. Pick up the needle and float it on the water surface again. This time add a few drops of a liquid detergent or touch the water surface with the tip of a bar of soap.

Using a new water sample in the bowl, sprinkle pepper on the surface of the water. Observe. Add a few drops of liquid detergent or touch the surface with the tip of a bar of soap. Soaps and detergents contain surfactants, which means they decrease the surface tension.

Questions

1. Why does a needle float on the surface of the water?
2. What makes the needle sink?
3. How did the detergent or soap change the appearance of the pepper on the surface of the water?

QUESTIONS & PROBLEMS

Properties of Water

8.1 **a.** Draw a water molecule and describe its polarity.
b. Show how hydrogen bonding occurs between two water molecules.
c. Why does HF exhibit hydrogen bonding?
d. Would you expect CH_4 molecules to hydrogen bond? Explain.

8.2 Match the following statements with *hydrogen bonding, surface tension,* or *polar substance:*

a. the pulling together of water molecules at the surface to form a thick layer that will support the weight of small bugs
b. attractive forces between water molecules
c. behavior of water in forming solutions with ionic compounds
d. formation of beads of water on a waxed surface

Study Pack

Chem of Life Water: Solutions 1, 2

8.2

8.2 Solutions

Learning Goal

Define solute and solvent; describe the formation of a solution.

A **solution** is a mixture in which one substance called the **solute** is dispersed uniformly in another substance called the **solvent.** Because the solute and the solvent do not react with each other, they can be mixed in varying proportions. A little salt dissolved in water tastes slightly salty. When more salt dissolves, the water tastes very salty. Usually, the solute (in this case, salt) is the substance present in the smaller amount, whereas the solvent (in this case, water) is the larger amount. A $CuSO_4$ solution forms when the particles of the solute become evenly dispersed among the molecules of the solvent (water). (See Figure 8.4.)

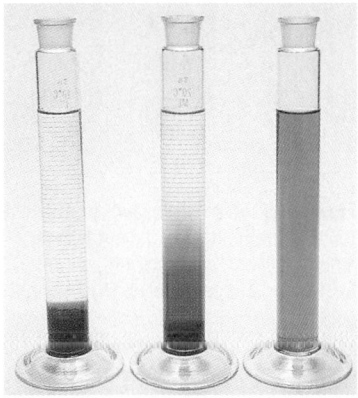

Figure 8.4 A solution of copper (II) sulfate ($CuSO_4$) is formed when particles of solute become evenly dispersed among the solvent particles.

S·A·M·P·L·E **8.2** ·P·R·O·B·L·E·M

Identifying a Solute and a Solvent

Identify the solute and the solvent in each of the following solutions:

a. 1.0 g of sugar dissolved in 100 g of water
b. 50 mL of water mixed with 20 mL of isopropyl alcohol (rubbing alcohol)

Solution

a. Sugar is the smaller quantity that is dissolving. It is the solute; water is the solvent.
b. Because both water and isopropyl alcohol are liquids, the one with the smaller volume, isopropyl alcohol, is the solute. Water is the solvent.

Study Check

A tincture of iodine is prepared with 0.10 g of I_2 and 10.0 mL of ethyl alcohol. What is the solute, and what is the solvent?

Types of Solutes and Solvents

Solutes and solvents may be solids, liquids, or gases. The solution that forms has the same physical state as the solvent. When sugar is dissolved in a glass of water, a liquid sugar solution forms. Sugar is the solute, and water is the solvent. Soda water and soft drinks are prepared by dissolving CO_2 gas in water. The CO_2 gas is the solute, and water is the solvent. Table 8.2 lists some solutes and solvents and their solutions.

Like Dissolves Like

Gases form solutions easily because their particles are moving rapidly, are far apart, and are not attracted to the other gas particles. When solids or liquids form solutions,

8.2

Table 8.2 Some Examples of Solutions

Type	Example	Solute	Solvent
Gas Solutions			
Gas in a gas	Air	Oxygen (gas)	Nitrogen (gas)
Liquid Solutions			
Gas in a liquid	Soda water Household ammonia	Carbon dioxide (gas) Ammonia (gas)	Water (liquid) Water (liquid)
Liquid in a liquid	Vinegar	Acetic acid (liquid)	Water (liquid)
Solid in a liquid	Seawater Tincture of iodine	Sodium chloride (solid) Iodine (solid)	Water (liquid) Alcohol (liquid)
Solid Solutions			
Liquid in a solid	Dental amalgam	Mercury (liquid)	Silver (solid)
Solid in a solid	Brass Steel	Zinc (solid) Carbon (solid)	Copper (solid) Iron (solid)

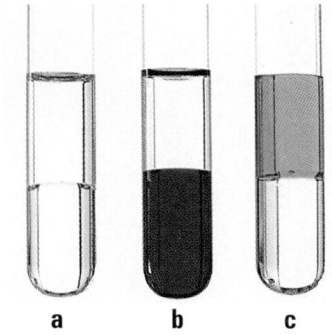

Figure 8.5 Like dissolves like. **(a)** Test tubes contain an upper layer of water, a polar solvent, and a lower layer of CH_2Cl_2, a nonpolar solvent. **(b)** The nonpolar solute I_2 dissolves in the nonpolar solvent. **(c)** The ionic solute $Ni(NO_3)_2$ dissolves in the polar solvent.

there must be an attraction between the solute particles and the solvent particles. Then the particles of the solute and solvent will mix together. If there is no attraction between a solute and a solvent, their particles do not mix and no solution forms.

A salt such as NaCl will form a solution with water because the Na^+ and Cl^- ions in the salt are attracted to the positive and negative parts of the individual water molecules. You can dissolve table sugar, $C_{12}H_{22}O_{11}$, in water because the molecule has many polar —OH groups that attract water molecules.

However, compounds containing nonpolar molecules such as iodine (I_2), oil, or grease do not dissolve in water because water is polar. Nonpolar solutes require nonpolar solvents for a solution to form. The expression "like dissolves like" is a way of saying that the polarities of a solute and a solvent must be similar in order to form a solution. Figure 8.5 illustrates the formation of some polar and nonpolar solutions.

Like Dissolves Like

Polar solutes dissolve in polar solvents;

nonpolar solutes dissolve in nonpolar solvents.

S·A·M·P·L·E 8.3 P·R·O·B·L·E·M

Polar and Nonpolar Solutes

Indicate whether each of the following substances will dissolve in water, a polar solvent. Explain.

a. KCl **b.** Cl_2 **c.** octane, C_8H_{18}, a compound in gasoline
d. ethanol, C_2H_6O, a substance in mouthwash

Solution

a. Yes. KCl is an ionic compound.
b. No. Cl_2 is a nonpolar molecule.
c. No. C_8H_{18} is a nonpolar substance.
d. Yes. C_2H_6O is a polar substance.

Study Check

Will oil, a nonpolar substance, dissolve in hexane, a nonpolar solvent?

EXPLORE YOUR WORLD

Like Dissolves Like

Mix together small amounts of the following substances:

a. oil and water
b. water and vinegar
c. salt and water
d. sugar and water
e. salt and oil

Questions

1. Which of the mixtures formed a solution? Which did not?
2. Why do some mixtures form solutions, but others do not?

8.2

Dissolving a Solute

The formation of a solution depends on similar polarities of the solute and solvent particles. An ionic compound, such as sodium chloride (NaCl), dissolves in water because water is a polar solvent. As the water molecules bombard the surface of the salt, as shown in Figure 8.6, the Cl^- ions are attracted to the partially positive hydrogen atoms of the water molecules. The Na^+ ions are attracted to the partially negative oxygen atoms. These attractions pull the ions into the water, where they undergo hydration: They become surrounded by water molecules. A sodium chloride solution has formed.

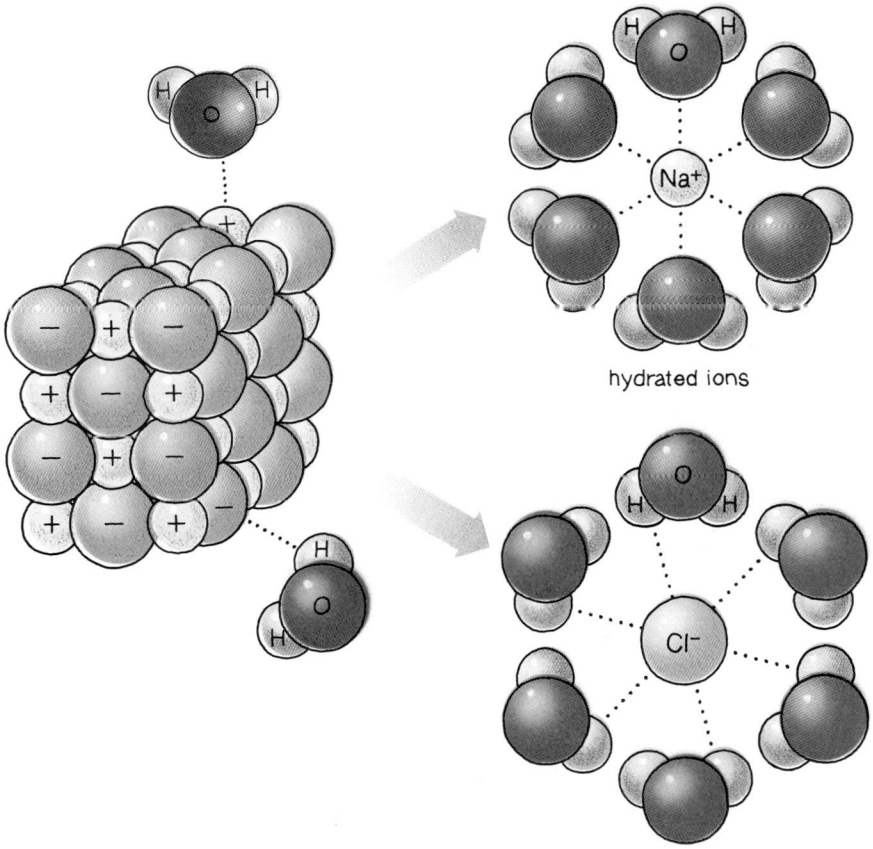

Figure 8.6 The dissolving of NaCl, an ionic solute, in water. Attracted by polar water molecules, Na^+ and Cl^- become part of the solution, where they are hydrated.

S·A·M·P·L·E· 8.4 ·P·R·O·B·L·E·M

Dissolving an Ionic Solute

What happens to the ions in solid KBr when it dissolves in water?

Solution

The ions, K^+ and Br^-, are pulled away from the solid as polar water molecules attract and separate them from the salt. Once dispersed in water, the ions are hydrated.

8.2

Study Check

What happens to the ions in solid $CaCl_2$ when it dissolves in water?

Factors Affecting Rate of Solution Formation

The time it takes to form a solution depends on how fast the solute particles spread throughout the solvent. If a solid is broken down into smaller particles, it will form a solution faster because its surface area is greater. Note that we are not changing the amount of dissolved solute—only the time it takes to form the solution. For example, in a glass of water, a spoonful of powdered sugar dissolves faster than the same amount of sugar in a sugar cube. The more compact sugar cube has a smaller surface area, so there are relatively fewer molecules on the surface to interact with water. Eventually the final composition of the solutions will be identical.

Another way to increase the rate of solution is by stirring or agitating the solution. If the solute particles are dispersed more quickly, more solvent can interact with the remaining undissolved solute.

A higher temperature increases the rate of solution by increasing the rate at which the solvent bombards the solute and moves the solute into the solution.

The rate of solution increases when:

1. The solute is crushed.
2. The mixture is stirred.
3. The mixture is heated.

S·A·M·P·L·E· 8.5 ·P·R·O·B·L·E·M

Rate of Solution Formation

You are going to dissolve some salt in a glass of water. Indicate whether each of the following conditions will *increase* or *decrease* the rate of solution for the salt.

a. using hot water **b.** crushing the salt

Solution

a. An increase in the temperature of the water increases the rate of solution.

b. Increasing the surface area of the salt that comes in contact with the water molecules will increase the rate of solution.

Study Check

How will placing some salt in a glass of water and cooling the mixture affect the rate of solution formation?

QUESTIONS & PROBLEMS

Solutions

8.3 Identify the solute and the solvent in each solution composed of the following:
a. 10.0 g of NaCl and 100.0 g of H_2O
b. 50.0 mL of ethanol, $C_2H_5OH(l)$, and 10.0 mL of H_2O
c. 0.20 L of O_2 and 0.80 L of N_2

8.4 Identify the solute and the solvent in each solution composed of the following:
a. 50.0 g of silver and 4.0 g of mercury
b. 100.0 mL of water and 5.0 g of sugar
c. 1.0 g of I_2 and 50.0 mL of alcohol

8.5 Water is a polar solvent; CCl_4 is a nonpolar solvent. In which solvent are each of the following more likely to be soluble?
a. KCl, ionic **b.** I_2, nonpolar
c. sugar, polar **d.** gasoline, nonpolar

8.6 Water is a polar solvent; hexane is a nonpolar solvent. In which solvent are each of the following more likely to be soluble?
a. vegetable oil, nonpolar **b.** benzene, nonpolar
c. $LiNO_3$, ionic **d.** Na_2SO_4, ionic

8.7 Match the following statements with *oxygen atom, hydrogen atom,* or *hydration* to describe the formation of a NaBr solution:
a. the portion of the water molecule attracted to Na^+ ions
b. the surrounding of Na^+ ions in solution by water molecules
c. the portion of the water molecule attracted to the Br^- ions

8.8 Match the following statements with *oxygen atom, hydrogen atom,* or *hydration* to describe the formation of a $CaCl_2$ solution:
a. the portion of the water molecule attracted to the Cl^- ions
b. the portion of the water molecule attracted to Ca^{2+} ions
c. the surrounding of Cl^- ions in solution by water molecules

8.9 Describe the formation of a KI solution.

8.10 Describe the formation of a LiBr solution.

8.11 Indicate whether each of the following actions will increase or decrease the rate of solution of KI in water:
a. stirring the mixture
b. using large chunks of KI
c. placing the beaker of KI and water in ice
d. heating the beaker containing the KI and water

8.12 Indicate whether each of the following actions will increase or decrease the rate of solution of sugar (sucrose) in water:
a. crushing the sugar into a fine powder
b. placing the beaker of sugar and water in ice
c. stirring the mixture
d. heating the sugar and water

Study Pack

8.3 Electrolytes and Nonelectrolytes

Learning Goal

Describe the solution process for electrolytes and nonelectrolytes in water.

8.3

Solutions containing ions are conductors of electricity. The ions come from substances called **electrolytes.** To test for electrolytes, we can use an apparatus that consists of a battery and a pair of electrodes connected by wires to a light bulb. (See Figure 8.7.) The light bulb glows when current flows through it, and that only happens when there are ions moving about the solution.

In water, a **strong electrolyte** exists only as ions in solution. When there are many ions in solution, the light bulb glows brightly. However, in a solution containing **weak electrolytes,** there are only a few ions. With a small number of ions in solution, the light bulb glows dimly. When the electrodes are placed in pure water or a sugar solution, the light bulb does not glow at all. Substances that dissolve in water as covalent molecules and do not conduct electricity are **nonelectrolytes.**

Strong Electrolytes

Sodium chloride (NaCl), a strong electrolyte, consists of sodium ions and chloride ions held together by strong ionic bonds. In water, the ions are attracted to water molecules and separate from the solid, a process called dissociation. A solution forms that looks like water but tastes salty because it contains dissolved sodium ions and chloride ions. (See Figure 8.8.) We write an equation for the dissociation of NaCl in water to show that solid NaCl separates into hydrated ions of $Na^+(aq)$ and $Cl^-(aq)$:

$$\underset{}{NaCl(s)} \xrightarrow[\text{100\% dissociation}]{H_2O} \underset{\text{Ions in solution}}{Na^+(aq) + Cl^-(aq)}$$

or simplified:

$$NaCl \xrightarrow{H_2O} Na^+ + Cl^-$$

Other soluble ionic compounds dissolve in a similar way. When we write an ionic equation, we balance the electrical charges of the ions. The total positive charge must equal the total negative charge. For example, magnesium chloride, $MgCl_2$, dissolves in water to give one magnesium ion for every two chloride ions in solution.

$$MgCl_2 \xrightarrow{H_2O} \underset{\text{Charge balance for ions in solution}}{Mg^{2+}(aq) + 2Cl^-(aq)}$$

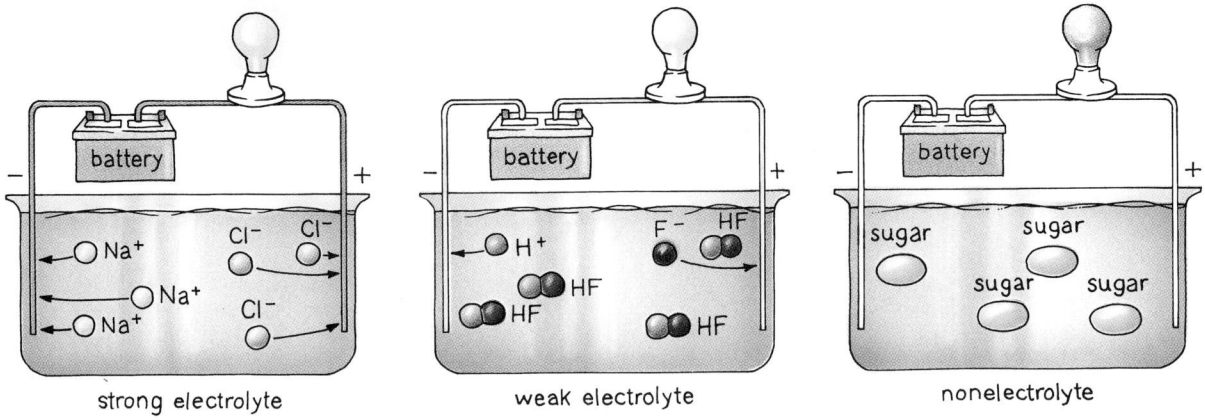

Figure 8.7 In solution a strong electrolyte produces ions and conducts an electrical current; a weak electrolyte produces only a few ions and conducts a weak electrical current. No ions, only molecules, exist in a solution of a nonelectrolyte; there is no electrical current.

S·A·M·P·L·E 8.6 P·R·O·B·L·E·M

Writing Equations for Ionic Compounds

Write the equation for the formation of an aqueous solution of $Mg(NO_3)_2$.

Solution

The formula indicates one magnesium ion and two nitrate ions. The equation for the formation of its solution in water is

$$Mg(NO_3)_2 \xrightarrow{H_2O} Mg^{2+}(aq) + 2NO_3^-(aq)$$

Study Check

Write an equation for the formation of an aqueous solution of Na_2SO_4.

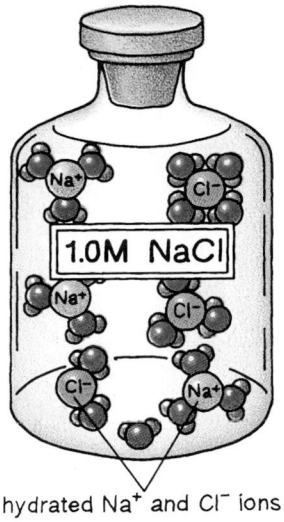

Figure 8.8 A solution of NaCl, a strong electrolyte, contains sodium ions (Na^+) and chloride ions (Cl^-).

Weak Electrolytes

A weak electrolyte is typically a polar covalent compound that dissolves in water mostly as nondissociated molecules. However, a few of the molecules ionize to produce a small number of ions in solution. For example, most HF molecules that dissolve in water remain as molecules. However, a few HF molecules ionize, producing a small number of H^+ and F^- ions. Thus, in water, weak electrolytes are only slightly ionized, which means that most of the dissolved substance is in the molecular form. There are really two reactions taking place in this solution. One is the ionization of HF molecules written as

$$HF \xrightarrow{\text{Ionization}} H^+ + F^-$$

As the number of H^+ and F^- build up in solution, some of the ions recombine to give HF molecules.

$$HF \xleftarrow{\text{Recombination}} H^+ + F^-$$

Eventually the reverse reaction goes as fast as the forward reaction. Then the concentrations of HF, H^+, and F^- no longer change. An equilibrium exists between the HF molecules and its ions. The forward and reverse reactions are shown in the equation by using a double arrow. (We used a single arrow for strong electrolytes that produced only ions.) The formation of the HF solution (and other weak electrolytes) may be written as follows:

$$\underset{\text{A few ionize}}{HF \rightleftharpoons H^+ + F^-}$$

Nonelectrolytes

Nonelectrolytes are polar covalent compounds that dissolve in water as molecules. They produce no ions and their solutions do not conduct electricity. For example, sucrose (sugar) is a nonelectrolyte that dissolves in water to produce a solution of sugar molecules. (See Figure 8.9.)

$$\underset{\text{Sucrose}}{C_{12}H_{22}O_{11}(s)} \xrightarrow{H_2O} \underset{\text{Solution of sucrose molecules}}{C_{12}H_{22}O_{11}(aq)}$$

In an aqueous solution:

A strong electrolyte is completely (100%) ionized.

A weak electrolyte is slightly ionized.

A nonelectrolyte forms no ions, only molecules.

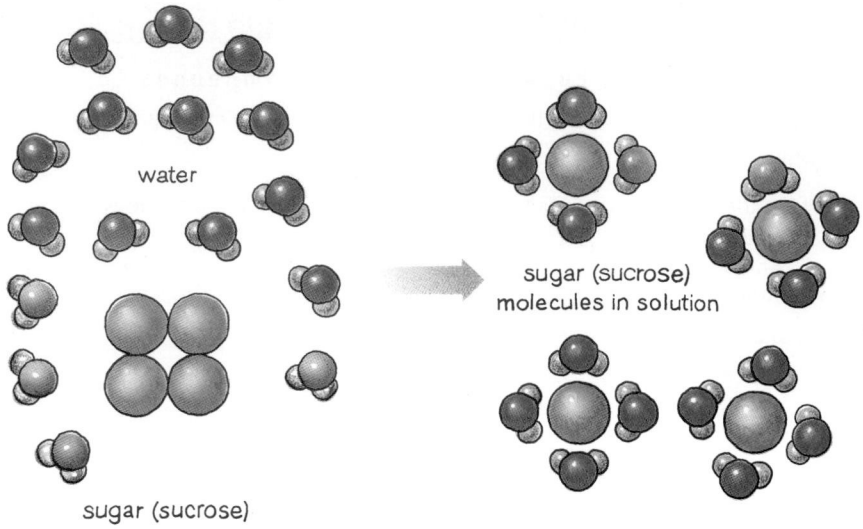

Figure 8.9 The formation of a sugar solution as polar sucrose molecules ($C_{12}H_{22}O_{11}$) are attracted by polar water molecules.

S·A·M·P·L·E **8.7** ·P·R·O·B·L·E·M

Solutions of Electrolytes and Nonelectrolytes

Indicate whether solutions of each of the following contain ions, molecules, or both:

a. Na_2SO_4, a soluble salt **b.** CH_3OH, a nonelectrolyte

Solution

a. A solution of Na_2SO_4 contains the ions of the salt, Na^+ and SO_4^{2-}.
b. A nonelectrolyte such as CH_3OH dissolves in water as molecules.

Study Check

Boric acid, H_3BO_3, is a weak electrolyte. Would you expect a boric acid solution to contain ions, molecules, or both?

Equivalents

We can describe the amounts of electrolytes, such as Na^+, Cl^-, K^+, and Ca^{2+}, in terms of equivalents. An **equivalent (Eq)** is the amount of ion that carries 1 mole of electrical charge. For example, 1 mole of Na^+ ions and 1 mole of Cl^- ions are each 1 equivalent because they each contain 1 mole of charge. This is the same for ions such as K^+, H^+, OH^-, and HCO_3^-. For an ion with a charge of 2+ or 2−, there are 2 equivalents for each mole. Some examples of ions and equivalents are shown in Table 8.3.

Table 8.3 Equivalents of Electrolytes

Ion	Electrical Charge	Number of Equivalents in 1 mole
Na^+	1+	1 Eq
Ca^{2+}	2+	2 Eq
Fe^{3+}	3+	3 Eq
Cl^-	1−	1 Eq
SO_4^{2-}	2−	2 Eq

Because equivalents describe the amount of positive or negative charge per mole of ion or compound, we can use the number of equivalents in 1 mole as a conversion factor. For example, Ca^{2+} contains 2 equivalents in 1 mole of ions, which can be written as two possible conversion factors.

$$\frac{2 \text{ Eq } Ca^{2+}}{1 \text{ mole } Ca^{2+}} \quad \text{and} \quad \frac{1 \text{ mole } Ca^{2+}}{2 \text{ Eq } Ca^{2+}}$$

S·A·M·P·L·E 8.8 P·R·O·B·L·E·M

Calculating Equivalents

How many equivalents are in 6.0 g of Ca^{2+}?

Solution

The conversion factor of molar mass converts the grams to moles. Because there are 2 equivalents in 1 mole of calcium ion, we calculate the number of equivalents in our sample.

$$6.0 \cancel{\text{g } Ca^{2+}} \times \frac{1 \cancel{\text{mole } Ca^{2+}}}{40.1 \cancel{\text{g } Ca^{2+}}} \times \frac{2 \text{ Eq } Ca^{2+}}{1 \cancel{\text{mole } Ca^{2+}}} = 0.30 \text{ Eq } Ca^{2+}$$

Study Check

How many grams of Cl^- are present in 2.0 Eq of Cl^-?

S·A·M·P·L·E 8.9 P·R·O·B·L·E·M

Electrolyte Concentration

In body fluids, concentrations of electrolytes are often expressed as milliequivalent (mEq) per liter. A typical concentration for Na^+ in the blood is 138 mEq/L. How many grams of sodium ion are in 1.00 L of blood?

Solution

Using the volume and the electrolyte concentration (in mEq/L) we can find the number of equivalents in 1.00 L of blood.

$$1.00 \cancel{\text{L}} \times \frac{138 \cancel{\text{mEq}}}{1 \cancel{\text{L}}} \times \frac{1 \text{ Eq}}{1000 \cancel{\text{mEq}}} = 0.138 \text{ Eq } Na^+$$

We can then convert equivalents to moles (for Na^+ there is 1 Eq per mole), and moles to grams using molar mass.

$$0.138 \cancel{\text{Eq } Na^+} \times \frac{1 \cancel{\text{mole } Na^+}}{1 \cancel{\text{Eq } Na^+}} \times \frac{23.0 \text{ g } Na^+}{1 \cancel{\text{mole } Na^+}} = 3.17 \text{ g } Na^+$$

Study Check

A Ringer's solution for intravenous fluid replacement contains 155 mEq Cl^- per liter of solution. If a patient receives 500. mL of Ringer's solution, how many grams of chloride were given?

HEALTH NOTE

Electrolytes in Body Fluids

The concentrations of electrolytes present in body fluids and in intravenous fluids given to a patient are often expressed in milliequivalents per liter (mEq/L) of solution.

$$1 \text{ Eq} = 1000 \text{ mEq}$$

Table 8.4 gives the concentrations of some typical electrolytes in blood plasma. There is a charge balance, because the total number of positive charges is equal to the total number of negative charges. The use of a specific intravenous solution depends on the nutritional, electrolyte, and fluid needs of the individual patient. Examples of various types of solutions are given in Table 8.5.

8.3

Table 8.4 Some Typical Concentrations of Electrolytes in Blood Plasma

Electrolyte	Concentration (mEq/L)
Cations (Positive Ions)	
Na^+	138
K^+	5
Mg^{2+}	3
Ca^{2+}	4
Total	150
Anions (Negative Ions)	
Cl^-	110
HCO_3^-	30
HPO_4^{2-}	4
Proteins	6
Total	150

Table 8.5 Electrolyte Concentrations in Intravenous Replacement Solutions

Solution	Electrolytes (mEq/L)	Use
Sodium chloride (0.9%)	Na^+ 154, Cl^- 154	Replacement of fluid loss
Potassium chloride with 5% dextrose	K^+ 40, Cl^- 40	Treatment of malnutrition (low potassium levels)
Ringer's solution	Na^+ 147, K^+ 4, Ca^{2+} 4, Cl^- 155	Replacement of fluids and electrolytes lost through dehydration
Maintenance solution with 5% dextrose	Na^+ 40, K^+ 35, Cl^- 40, lactate$^-$ 20, HPO_4^{2-} 15	Maintenance of fluid and electrolyte levels
Replacement solution (extracellular)	Na^+ 140, K^+ 10, Ca^{2+} 5, Mg^{2+} 3, Cl^- 103, acetate$^-$ 47, citrate^{3-} 8	Replacement of electrolytes in extracellular fluids

Q·U·E·S·T·I·O·N·S & P·R·O·B·L·E·M·S

Electrolytes and Nonelectrolytes

8.13 KF is a strong electrolyte, and HF is a weak electrolyte. How are they different?

8.14 NaOH is a strong electrolyte, and CH_3OH is a nonelectrolyte. How are they different?

8.15 The following soluble salts are strong electrolytes. Write a balanced equation for their dissociation in water:

a. KCl **b.** $CaCl_2$ **c.** K_3PO_4 **d.** $Fe(NO_3)_3$

8.16 The following salts are strong electrolytes. Write a balanced equation for their dissociation in water:

a. LiBr **b.** $NaNO_3$ **c.** $FeCl_3$ **d.** $Mg(NO_3)_2$

8.17 Indicate whether aqueous solutions of the following will contain ions only, molecules only, or molecules and some ions:

a. acetic acid ($HC_2H_3O_2$), found in vinegar, a weak electrolyte

b. NaBr, a salt

c. fructose ($C_6H_{12}O_6$), a nonelectrolyte

8.18 Indicate whether aqueous solutions of the following will contain ions only, molecules only, or molecules and some ions:
a. Na_2SO_4, a salt
b. ethanol, C_2H_5OH, a nonelectrolyte
c. HCN, hydrocyanic acid, a weak electrolyte

8.19 Write equations for the formation of aqueous solutions of each substance in problem 8.17.

8.20 Write equations for the formation of aqueous solutions of each substance in problem 8.18.

8.21 Indicate the type of electrolyte represented in the following equations:

a. $K_2SO_4 \xrightarrow{H_2O} 2K^+ + SO_4^{2-}$

b. $NH_4OH \xrightleftharpoons{H_2O} NH_4^+ + OH^-$

c. $C_6H_{12}O_6(s) \xrightarrow{H_2O} C_6H_{12}O_6(aq)$

8.22 Indicate the type of electrolyte represented in the following equations:

a. $CH_3OH \xrightarrow{H_2O} CH_3OH(aq)$

b. $MgCl_2 \xrightarrow{H_2O} Mg^{2+} + 2Cl^-$

c. $HClO \xrightleftharpoons{H_2O} H^+ + ClO^-$

8.23 Indicate the number of equivalents in each of the following:
a. 1 mole K^+ **b.** 2 moles OH^-
c. 1 mole Ca^{2+} **d.** 3 moles CO_3^{2-}

8.24 Indicate the number of equivalents in each of the following:
a. 1 mole Mg^{2+} **b.** 0.5 mole H^+
c. 4 moles Cl^- **d.** 2 moles Fe^{3+}

8.25 Calculate the number of equivalents in each of the following:
a. 25.0 g Cl^- **b.** 15.0 g Fe^{3+}
c. 4.0 g Ca^{2+} **d.** 1.0 g H^+

8.26 Calculate the number of equivalents in each of the following:
a. 10.0 g Na^+ **b.** 8.0 g OH^-
c. 20.0 g Mg^{2+} **d.** 4.0 g Al^{3+}

8.27 If blood plasma typically contains 3.0 mEq/L of Mg^{2+}, how many grams of Mg^{2+} would be in 5.0 L of plasma?

8.28 If blood plasma typically contains 110 mEq/L of Cl^-, how many grams of Cl^- would be in 5.0 L of plasma?

8.29 A physiological saline solution that is 0.90% (m/v) NaCl contains 154 mEq/L each of Na^+ and Cl^-. How many grams each of Na^+ and Cl^- are in 1.0 L of the saline solution?

8.30 A solution to replace potassium loss contains 40. mEq/L each of K^+ and Cl^-. How many grams each of K^+ and Cl^- are in 1.5 L of the solution?

8.31 A solution contains 40. mEq/L of Cl^- and 15 mEq/L of HPO_4^{2-}. If Na^+ is the only cation in the solution, what is the Na^+ concentration in milliequivalents per liter?

8.32 A sample of Ringer's solution contains the following concentrations (mEq/L) of cations: Na^+ 147, K^+ 4, and Ca^{2+} 4. If Cl^- is the only anion in the solution, what is the Cl^- concentration in milliequivalents per liter?

Study Pack

Learning Goal

Define solubility; distinguish between an unsaturated and a saturated solution. Identify an insoluble salt.

8.4

8.4 Solubility

The term **solubility** is used to describe the amount of a solute that can dissolve in a given amount of solvent. Many factors such as the type of solute, the type of solvent, and temperature affect a solute's solubility. Solubility, usually expressed in grams of solute in exactly 100 grams of solvent, is the maximum amount of solute that can be dissolved at a certain temperature. If a solute readily dissolves when added to the solvent, the solution does not contain the maximum amount of solute. We call the solution an **unsaturated solution.** When a solution contains all the solute that can dissolve, it is **saturated.** If we try to add more solute, undissolved solute will remain on the bottom of the container. For example, 36 g of NaCl can dissolve in 100 grams of water at 20°C. (See Figure 8.10.) If we add 15 g of NaCl to 100 g of water at 20°C, it all dissolves; the solution is unsaturated. The NaCl solution becomes saturated when we dissolve a total of 36 g of NaCl.

S·A·M·P·L·E **8.10** P·R·O·B·L·E·M

Saturated Solutions

At 70°C, the solubility of KCl is 50. g/100 g of water. In the laboratory, a student mixes 75 g of KCl with 100 g of water at a temperature of 70°C.

a. How much of the KCl will dissolve?

b. Is the solution saturated?

c. What is the mass in grams of the undissolved KCl that forms on the bottom of the container?

Solution

a. A total of 50. g of the KCl will dissolve because that is its solubility and, therefore, the maximum amount of KCl in solution at 70°C.

b. Yes, the solution is saturated.

c. The mass of the undissolved KCl is 25 g.

Study Check

At 30°C, the solubility of $CuSO_4$ is 25 g/100 g of water. How many grams of $CuSO_4$ are needed to make a saturated $CuSO_4$ solution with 300. g of water at 30°C?

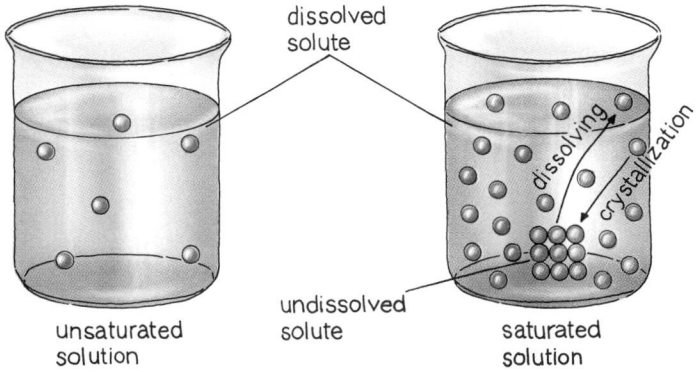

Figure 8.10 In an unsaturated solution, all of the solute dissolves. In a saturated solution, the maximum amount of solute for a given temperature has dissolved.

HEALTH NOTE

Gout: A Problem of Saturation in Body Fluids

Gout is a disease that affects adults, primarily men, over the age of 40. Attacks of gout occur when the concentration of uric acid in blood plasma exceeds its solubility, and undissolved uric acid is deposited in the joints, causing severe pain. At body temperature, 37°C, the solubility of uric acid is 7 mg/100 mL of plasma. Typically, uric acid concentrations range from 2 mg to 6 mg/100 mL of plasma. If the uric acid concentration exceeds 7 mg/100 mL of plasma, solid uric acid crystals can appear in the cartilage, tendons, and soft tissues, as well as in the tissues of the kidneys, where they can cause renal damage.

Foods in the diet that contribute to high levels of uric acid are often rich in protein and include certain meats, sardines, mushrooms, and asparagus. Certain enzyme defects can also cause an overproduction of uric acid. Treatment for gout includes a reduction in protein intake and the restriction of certain foods. Drugs such as colchicine may be used to reduce uric acid production or to increase the excretion of uric acid.

EXPLORE YOUR WORLD

Preparing Solutions

In your kitchen, place $\frac{1}{4}$ or $\frac{1}{2}$ cup of cold water in a glass. Begin adding 1 tablespoon of sugar at a time and stir thoroughly. Take a sip of the liquid in the glass as you proceed. As the sugar solution becomes more concentrated, you may need to stir for a few minutes until all the sugar dissolves. Each time observe the solution after several minutes to determine when it is saturated.

Repeat the above activity with $\frac{1}{4}$ or $\frac{1}{2}$ cup of warm water. Count the number of tablespoons of sugar you need to add to form a saturated solution.

Questions

1. What do you notice about how sweet each sugar solution tastes?
2. How did you know when you obtained a saturated solution?
3. How much sugar dissolved in the warm water compared to the cold water?

Effect of Temperature

Most solids become more soluble in water as temperature increases. For example, if you add sugar to ice tea, a layer of undissolved sugar forms on the bottom of the glass when saturation is reached. However, if the tea is heated, more sugar dissolves because the solubility of sugar is greater in hot water. Figure 8.11 illustrates the effect of higher temperatures on the solubility of some solids.

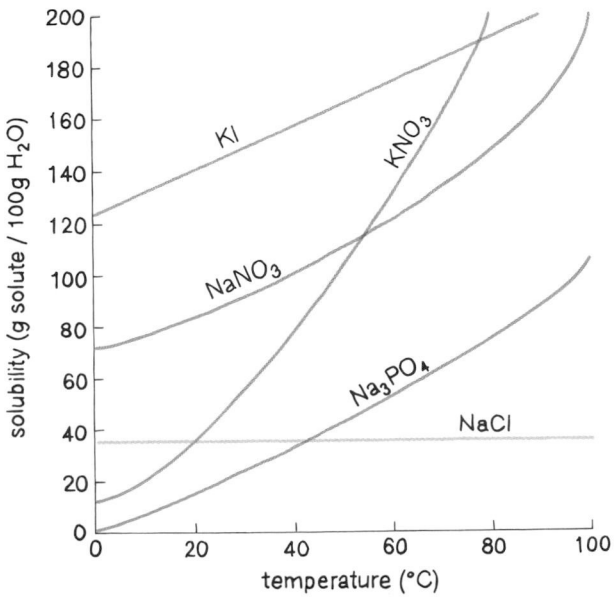

Figure 8.11 Most solids are more soluble in water at higher temperatures.

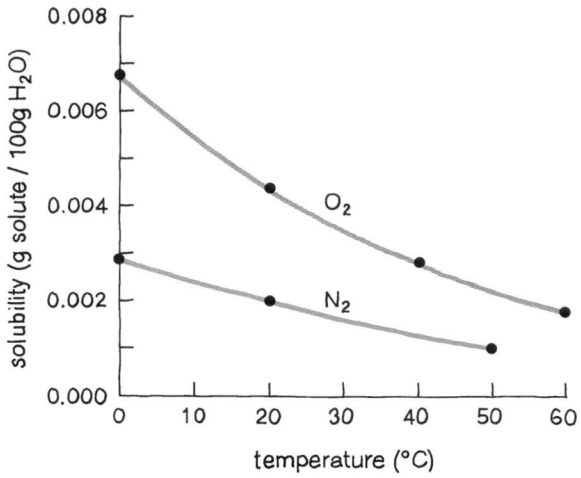

Figure 8.12 The solubility of gases decreases at higher temperatures.

8.4

In contrast, gases are less soluble in water at higher temperature. (See Figure 8.12.) As the temperature increases, increasing numbers of gas molecules escape from the solution. At high temperatures, bottles containing carbonated solutions may burst, as more gas molecules leave the solution and increase the gas pressure inside the bottle. Biologists have found that increased temperatures in rivers and lakes cause the amount of dissolved oxygen to decrease until the warm water can no longer support a biological community. Electric generating plants are required to have their own ponds to use with their cooling towers to lessen the threat of thermal pollution.

EXPLORE YOUR WORLD

Rock Candy: A Saturated Sugar Solution

A recipe for rock candy is a recipe for a saturated sugar solution. This activity will involve heating a sugar solution to the boiling point. Be very careful when you handle the hot solution, especially if children are watching.

Materials and Ingredients

You will need $2\frac{1}{2}$ cups of sugar, 1 cup of water, a saucepan, a spoon, a metal container, a candy thermometer if available, and a piece of string.

Procedure

1. Place the sugar and water in the saucepan and heat slowly, stirring constantly, until all of the sugar dissolves.
2. Stop stirring and let the sugar solution come to a boil. Cook until the temperature reaches 260°F (the hardball stage on a candy thermometer). To determine when this temperature is reached without a thermometer, drop a small amount of the solution into very cold water. If it forms a ball hard enough to hold its shape, but does not form threads, then it is at the hard ball stage.
3. Carefully pour the hot liquid into a metal container. Don't use a glass container; it can be cracked by the hot liquid. Tie a piece of string to a spoon or pencil and place it across the top of the container so that the string hangs down into the solution. You can attach a paper clip on the end of the string to hold it down. Let the solution cool. Set it aside without disturbing it for several days. In 5–7 days, lift up the spoon or pencil and the string should be covered with sugar crystals.

Questions

a. Why do sugar crystals form when the solution cools?

b. What other process occurs as the solution sits that affects the solubility of the sugar?

S·A·M·P·L·E 8.11 P·R·O·B·L·E·M

Factors Affecting Solubility

Indicate whether the solubility of the solute will increase or decrease in each of the following situations:

a. using 80°C water instead of 25°C water to dissolve sugar
b. effect on dissolved O_2 in a lake as it warms

Solution

a. An increase in the temperature increases the solubility of the sugar.
b. An increase in the temperature decreases the solubility of O_2 gas.

Study Check

At 10°C, the solubility of KNO_3 is 20 g/100 g H_2O. Would the value of 5 g/100 g H_2O or 65 g/100 g H_2O be the more likely solubility at 40°C? Explain.

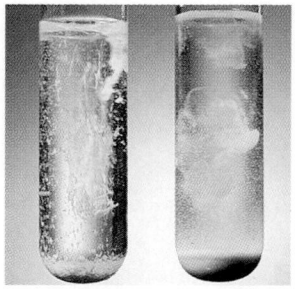

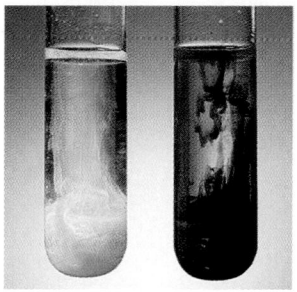

Figure 8.13 Some insoluble salts in water.

Soluble and Insoluble Salts

Up to now, we have considered ionic compounds that dissolve in water; they are **soluble salts.** However, some ionic compounds do not separate into ions in water. They are **insoluble salts** that remain as solids even in contact with water. (See Figure 8.13.)

Salts that are always soluble in water must contain at least one of the following ions: Li^+, Na^+, K^+, NH_4^+, or NO_3^-. Most salts containing Cl^- are soluble, but AgCl, $PbCl_2$, or Hg_2Cl_2 are not; they are insoluble chloride salts. Similarly, most salts containing SO_4^{2-} are soluble, but a few are insoluble as shown in Table 8.6. Most other salts are insoluble salts that do not dissolve in water. In an insoluble salt, attractions between its positive and negative ions are too strong for the polar water molecules to break. We use solubility rules to predict whether a salt (a solid ionic compound) would be expected to dissolve in water. Table 8.7 illustrates the use of these rules.

In medicine, the insoluble salt $BaSO_4$ is used as an opaque substance to enhance X rays of the gastrointestinal tract. (See Figure 8.14.) $BaSO_4$ is so insoluble that it does not dissolve in gastric fluids. Other barium salts cannot be used, because they dissolve in water releasing Ba^{2+}, which is poisonous.

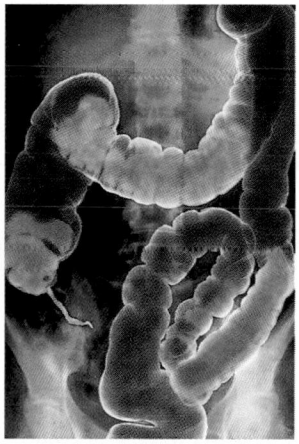

Figure 8.14 Barium sulfate–enhanced X ray of the abdomen showing the large intestine.

Table 8.6 Solubility Rules for Ionic Compounds

Ability to Dissolve in Water	Salts That Contain These Ions	But Not These Salts
Always soluble	Li^+, Na^+, K^+, NH_4^+, NO_3^-	
Mostly soluble	Cl^-	AgCl, $PbCl_2$, and Hg_2Cl_2
	SO_4^{2-}	$BaSO_4$, $PbSO_4$, and $CaSO_4$
Insoluble (do not dissolve or will form a precipitate)	OH^-, CO_3^{2-}, S^{2-}, PO_4^{3-}	Any salt that contains Li^+, Na^+, K^+, NH_4^+, NO_3^-

Table 8.7 Using Solubility Rules

Ionic Compound	Solubility in Water	Reasoning
K_2S	Soluble	Contains K^+
$Ca(NO_3)_2$	Soluble	Contains NO_3^-
$PbCl_2$	Insoluble	Is an insoluble chloride
NaOH	Soluble	Contains Na^+
$AlPO_4$	Insoluble	Contains no soluble ions

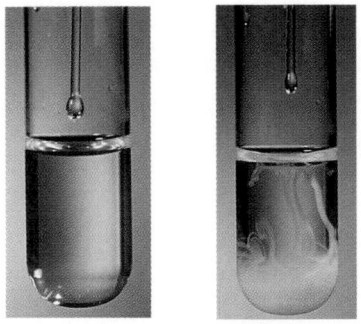

Figure 8.15 A precipitate of AgCl forms when solutions of $AgNO_3$ and NaCl are mixed. The soluble salt, $NaNO_3$, remains in solution.

8.4

Formation of a Solid

Sometimes, when we mix two solutions, a solid called a *precipitate* forms and drops to the bottom of the container. A solid forms when ions of an insoluble salt come in contact with one another. For example, when a solution of $AgNO_3$ (Ag^+ and NO_3^-) is mixed with a solution of NaCl (Na^+ and Cl^-), the white insoluble salt AgCl is produced. (See Figure 8.15.) We can write the reaction as a double replacement equation.

$$AgNO_3(aq) + NaCl(aq) \longrightarrow AgCl(s) + NaNO_3(aq)$$

In a *net ionic equation,* we write only the ions that form the insoluble salt AgCl, because Na^+ and NO_3^- ions remain in solution.

$$Ag^+ + Cl^- \longrightarrow AgCl(s)$$

S·A·M·P·L·E· **8.12** ·P·R·O·B·L·E·M

Soluble and Insoluble Salts

1. Predict whether each of the following salts is soluble in water:
   a. Na_3PO_4 **b.** $CaCO_3$
2. Solutions of $BaCl_2$ and K_2SO_4 are mixed and a white solid forms.
   a. Write the equation (double replacement) for the reaction.
   b. What is the white solid that forms?

Solution

1. **a.** Soluble. All salts with Na^+ are soluble.
   b. Insoluble. Most salts with CO_3^{2-} are insoluble.
2. **a.** $BaCl_2(aq) + K_2SO_4(aq) \longrightarrow BaSO_4(s) + 2KCl(aq)$
   b. $BaSO_4$ is the white solid.

Study Check

Would you expect the following salts to be soluble in water? Why?
a. $PbCl_2$ **b.** Na_3PO_4 **c.** $FeCO_3$

HEALTH NOTE

Kidney Stones Are Insoluble Salts

The kidneys continuously excrete substances that have low solubilities. If the urine becomes saturated with these substances, they crystallize and form kidney stones. One type formed in the urinary tract, renal calculus, is composed of calcium phosphate and calcium oxalate. The excessive ingestion of mineral salts and insufficient water intake leads to the saturation of the urine with minerals and consequently to kidney stones. Typically, one stone is formed every 2–3 years, and the tendency to form stones is strongly familial.

When a kidney stone passes through the urinary tract, it can cause considerable pain and discomfort, necessitating the use of painkillers and surgery. Sometimes ultrasound is used to break up large kidney stones in the kidneys. Persons prone to kidney stones are advised to drink six to eight glasses of water every day to prevent saturation levels of minerals in the urine.

QUESTIONS & PROBLEMS

Solubility

8.33 State whether each of the following refers to a saturated or unsaturated solution:

a. A crystal added to a solution does not change in size.

b. A sugar cube completely dissolves when added to a cup of coffee.

8.34 State whether each of the following refers to a saturated or unsaturated solution:

a. A spoonful of salt added to boiling water dissolves.

b. A layer of sugar forms on the bottom of a glass of tea as ice is added.

8.35 Use this table to determine whether each of the solutions that follows is saturated or unsaturated:

Solubility (g/100 g H_2O)		
T(°C)	**KBr**	**KI**
20	65	145
40	80	160
60	90	175
80	100	190
100	110	210

a. 75 g of KBr in 100 g of H_2O at 40°C

b. 140 g of KI in 50 g of H_2O at 40°C

c. 100 g of KBr in 100 g of H_2O at 60°C

d. 100 g of KI in 200 g of H_2O at 20°C

8.36 Use the table in problem 8.35 to determine whether each of the following is a saturated or unsaturated solution:

a. 100 g of KBr in 100 g of H_2O at 100°C

b. 50 g of KI in 50 g of H_2O at 60°C

c. 250 g of KI in 100 g of H_2O at 100°C

d. 40 g KBr in 200 g of H_2O at 80°C

8.37 A solution containing 100 g of KBr in 100 g of H_2O at 100°C is cooled to 40°C. Use the table in problem 8.35 to answer the following:

a. Is the solution saturated or unsaturated at 100°C?

b. Is the solution saturated or unsaturated at 40°C?

c. How much solid KBr, if any, formed during the cooling process?

8.38 A solution containing 150 g of KI in 100 g of water is heated to 80°C and cooled to 20°C. Use the table in problem 8.35 to answer the following:

a. Is the solution saturated or unsaturated at 80°C?

b. Is the solution saturated or unsaturated at 20°C?

c. How much solid KI, if any, formed during the cooling process?

8.39 Explain the following observations:

a. More sugar dissolves in hot tea than in ice tea.

b. Champagne in a warm room goes flat.

c. A warm can of soda has more spray when opened than a cold one.

8.40 Explain the following observations:

a. An open can of soda loses its "fizz" quicker at room temperature than in the refrigerator.

b. Chlorine gas in tap water escapes as the sample warms to room temperature.

c. Less sugar dissolves in iced coffee than in hot coffee.

8.41 Predict whether each of the following ionic compounds is soluble in water:
a. LiCl **b.** AgCl **c.** $BaCO_3$ **d.** K_2O **e.** $Fe(NO_3)_3$

8.42 Predict whether each of the following ionic compounds is soluble in water:
a. PbS **b.** NaI **c.** Na_2S **d.** Ag_2O **e.** $CaSO_4$

8.43 Determine whether a solid forms when solutions containing the following salts are mixed. If so, write the equation (double replacement) for the reaction, and identify the insoluble salt.
a. KCl and Na_2S **b.** $AgNO_3$ and K_2S **c.** $CaCl_2$ and Na_2SO_4

8.44 Determine whether a solid forms when solutions containing the following salts are mixed. If so, write the equation (double replacement) for the reaction, and identify the insoluble salt.
a. Na_3PO_4 and $AgNO_3$
b. K_2SO_4 and Na_2CO_3
c. $Pb(NO_3)_2$ and Na_2CO_3

8.5

Study Pack

8.5 Percent Concentration

Learning Goal

Calculate the percent concentration of a solute in a solution; use percent concentration to calculate the amount of solute or solution.

The amount of solute dissolved in a certain amount of solution is called the **concentration** of the solution. Although there are many ways to express a concentration, they all specify a certain amount of solute in a given amount of solution.

$$\text{Concentration of a solution} = \frac{\text{amount of solute}}{\text{amount of solution}}$$

Mass/Mass Percent

The **mass/mass percent,** % (m/m), of a solution is the percent by mass of solute in the solution. To calculate a mass/mass percent divide the grams of solute by the grams of the solution, and multiply by 100.

$$\text{mass/mass \%} = \frac{\text{grams of solute}}{\text{grams of solution}} \times 100$$

The mass of the solution is the sum of the grams of solute and the grams of solvent.

$$\text{g solution} = \text{g solute} + \text{g solvent}$$

Suppose we prepared a solution from 10.0 g of KCl (solute) and 90.0 g of water (solvent). The mass of the solution is 100.0 g of solution (10.0 g + 90.0 g). The percent concentration by mass of KCl in the solution is 10.0%.

$$\text{\% KCl (m/m)} = \frac{10.0 \text{ g KCl (solute)}}{100.0 \text{ g solution}} \times 100 = 10.0\% \text{ KCl}$$

10.0 g KCl + 90.0 g H_2O

S·A·M·P·L·E 8.13 P·R·O·B·L·E·M

Percent Concentration by Mass

If 15 g of NaOH are dissolved in 285 g of water, what is the mass percent of NaOH in the solution?

Solution

We find the mass of a solution by adding the mass of the solute and the solvent.

$$15 \text{ g NaOH (solute)} + 285 \text{ g } H_2O \text{ (solvent)} = 300. \text{ g solution}$$

Placing grams of solute and grams of solution in the mass percent expression gives

$$\frac{15 \text{ g NaOH}}{300. \text{ g solution}} \times 100 = 5.0\% \text{ (m/m) NaOH solution}$$

Study Check

What is the mass percent of NaCl in a solution made by dissolving 2.0 g of NaCl in 56 g of water?

Volume/Volume Percent

Because the volumes of liquids or gases can be easily measured, the concentrations of their solutions are often expressed as **volume/volume percent,** % (v/v). The units of volume used in the ratio must be the same, for example, both in milliliters or both in liters.

$$\text{volume/volume } \% = \frac{\text{volume (mL) solute}}{\text{volume (mL) solution}} \times 100$$

↑ Units must cancel

We interpret a volume/volume percent as the volume of solute in exactly 100 mL of solution. In the wine industry, a label that reads 12% (v/v) by volume means 12 mL of alcohol in 100 mL of wine.

Mass/Volume Percent

A **mass/volume percent,** % (m/v), commonly used in health fields, is calculated by dividing the grams of the solute by the milliliters of solution and multiplying by 100. Widely used in hospitals and pharmacies, the preparation of intravenous solutions and medicines involves the mass/volume percent even though the units do not cancel.

$$\text{mass/volume } \% = \frac{\text{grams of solute}}{\text{milliliters of solution}} \times 100$$

The mass/volume percent, % (m/v), indicates the grams of a substance that are contained in exactly 100 mL of a solution. For example, in every 100 mL of a 5% (m/v) glucose solution, there are 5 g of glucose.

8.5

S·A·M·P·L·E 8.14 P·R·O·B·L·E·M

Calculating Percent Concentration

A student prepared a solution by dissolving 5 g of KI with enough water to give a final volume of 250 mL. (See Figure 8.16.) What is the mass/volume percent of the KI solution?

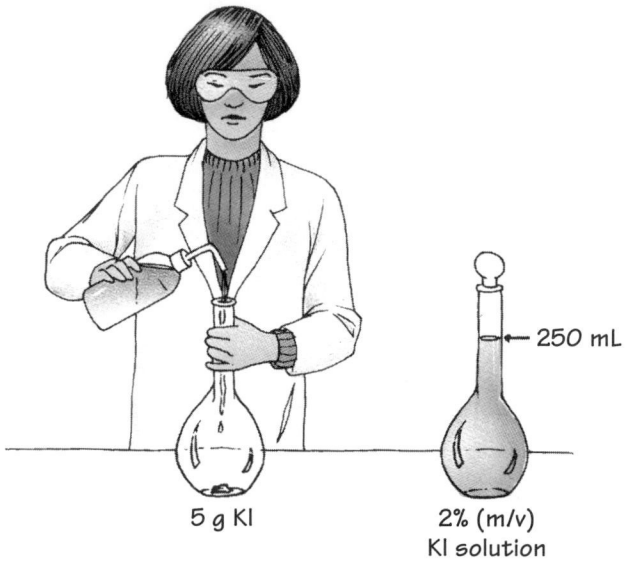

Figure 8.16 A 2% (m/v) KI solution is prepared by placing 5 g of KI in a 250-mL flask and filling the flask to the 250-mL mark.

Solution

Placing grams of solute and milliliters of solution in the expression for percent (m/v) concentration gives

$$\text{mass/volume \%} = \frac{\overset{\text{Mass of solute}}{5\text{ g KI}}}{\underset{\text{Volume of solution}}{250\text{ mL solution}}} \times 100 = 2\%\text{ (m/v) KI}$$

Study Check

What is the mass/volume percent, % (m/v), of Br_2 in a solution prepared by dissolving 12 g of bromine (Br_2) in enough carbon tetrachloride to make 250 mL of solution?

Percent Concentrations as Conversion Factors

In the preparation of solutions, we often need to calculate the amount of solute or solution. Then the percent concentration is useful as a conversion factor. Some examples of percent concentrations, their meanings, and possible conversion factors are given in Table 8.8.

Table 8.8 Conversion Factors from Percent Concentrations

Percent Concentration	Meaning	Conversion Factors		
10% (m/m) KCl	There are 10 g of KCl in exactly 100 g of solution.	$\frac{10\text{ g KCl}}{100\text{ g solution}}$	and	$\frac{100\text{ g solution}}{10\text{ g KCl}}$
5% (m/v) glucose	There are 5 g of glucose in exactly 100 mL of solution.	$\frac{5\text{ g glucose}}{100\text{ mL solution}}$	and	$\frac{100\text{ mL solution}}{5\text{ g glucose}}$

SAMPLE 8.15 PROBLEM

Using Mass/Volume Percent

A laboratory technician must prepare 2.00 L of a 5% (m/v) glucose solution. How many grams of glucose must be measured out?

Solution

Using the 5% (m/v) concentration as a conversion factor, we calculate the grams of glucose to weigh out for the solution as

$$\underset{\text{Given}}{2.00\,\cancel{\text{L}}} \times \underset{\text{Metric factor}}{\frac{1000\,\cancel{\text{mL}}}{1\,\cancel{\text{L}}}} \times \underset{\text{Percent factor}}{\frac{5\text{ g glucose}}{100\,\cancel{\text{mL}}}} = 100\text{ g glucose}$$

Study Check

How many grams of NaCl are needed to prepare 3.0 L of a 1% (m/v) NaCl solution?

8.5

QUESTIONS & PROBLEMS

Percent Concentration

8.45 What is the difference between a 5% (m/m) glucose solution and a 5% (m/v) glucose solution?

8.46 What is the difference between a 10% (v/v) methyl alcohol (CH_3OH) solution and a 10% (m/m) methyl alcohol solution?

8.47 Calculate the mass/mass percent of the solute in each of the following solutions:
a. 20.0 g of KCl in 150. g of solution
b. 40.0 g of $CaCl_2$ in 250. g of solution
c. 50.0 g of $NaNO_3$ in 400. g of H_2O

8.48 Calculate the mass/mass percent of the solute in each of the following solutions:
a. 8.0 g of NaBr in 100. g of H_2O
b. 75.0 g of KCl in 500.0 g of solution
c. 2.0 g of KOH in 20.0 g of H_2O

8.49 Calculate the mass/volume percent of the solute in each of the following solutions:
a. 75.0 g of Na_2SO_4 in 250 mL of solution
b. 1.5 g of Na_2CO_3 in 500. mL of solution
c. 50.0 g of glucose ($C_6H_{12}O_6$) in 0.50 L of solution

8.50 Calculate the mass/volume percent of the solute in each of the following solutions:
a. 2 g of sucrose ($C_{12}H_{22}O_{11}$) in 100 mL of solution
b. 0.80 g of $MgCl_2$ in 20.0 mL of solution
c. 60.0 g of KOH in 150 mL of solution

8.51 Write two conversion factors for the following solution concentrations:
a. 10% (m/m) NaOH **b.** 0.5% (m/v) NaCl **c.** 15% (v/v) CH_3OH

8.52 Write two conversion factors for the following solution concentrations:
a. 4% (m/v) KCl **b.** 1.0% (m/m) K_2CO_3
c. 15% (v/v) isopropyl alcohol solution

8.53 Calculate the amount of solute (g or mL) needed to prepare the following solutions:
a. 50.0 mL of a 5.0% (m/v) KCl solution
b. 225 g of a 10.0% (m/m) K_2CO_3 solution

c. 1250 mL of a 4.0% (m/v) NH_4Cl solution
d. 250 mL of a 10.0% (v/v) acetic acid solution

8.54 Calculate the amount of solute (g or mL) needed to prepare the following solutions:
a. 75 g of a 25% (m/m) NaOH solution
b. 150 mL of a 40.0% (m/v) $LiNO_3$ solution
c. 450 mL of a 2.0% (m/v) KCl solution
d. 225 mL of a 15% (v/v) isopropyl alcohol solution

8.55 A mouthwash contains 22.5% alcohol by volume. If the bottle of mouthwash contains 355 mL, what is the volume in milliliters of the alcohol?

8.6

8.56 A bottle of champagne is 11% alcohol by volume. If there are 750 mL of champagne in the bottle, how many milliliters of alcohol are present?

8.57 A patient receives 100 mL of 20% (m/v) mannitol solution every hour.
a. How many grams of mannitol are given in 1 hour?
b. How many grams of mannitol does the patient receive in 15 hours?

8.58 A patient receives 250 mL of a 4.0% (m/v) amino acid solution twice a day.
a. How many grams of amino acids are in 250 mL of solution?
b. How many grams of amino acids does the patient receive in 1 day?

8.59 Calculate the amount of solution (g or mL) that contains each of the following amounts of solute.
a. 10.0 g of HCl from a 4.0% (m/m) HCl solution
b. 5.0 g of $LiNO_3$ from a 25% (m/v) $LiNO_3$ solution
c. 40.0 g of KOH from a 10.0% (m/v) KOH solution
d. 2.0 mL of acetic acid from a 10.0% (v/v) acetic acid solution

8.60 Calculate the amount of solution (g or mL) that contains each of the following amounts of solute.
a. 7.50 g of NaCl from a 2.0% (m/v) NaCl solution
b. 4.0 g of NaOH from a 25% (m/v) NaOH solution
c. 20.0 g of KBr from a 8.0% (m/v) KBr solution
d. 40.0 mL of isopropyl alcohol from a 15% (v/v) isopropyl alcohol solution

8.61 A patient needs 100 g of glucose in the next 12 hours. How many liters of a 5% (m/v) glucose solution must be given?

8.62 A patient receives 2.0 g of NaCl in 8 hours. How many milliliters of a 0.90% (m/v) NaCl (saline) solution were delivered?

8.6 Colloids and Suspensions

Learning Goal

From its properties, identify a mixture as a solution, a colloid, or a suspension.

The solute particles in a solution play an important role in determining the properties of that solution. In most of the solutions discussed so far, the solute is dissolved as small particles that are uniformly dispersed throughout the solvent to give a homogeneous solution. When you observe a solution, such as salt water, you cannot visually distinguish the solute from the solvent. The solution appears transparent even when a light shines through it because the particles are too small to scatter the light. The particles are so small that they go through filters and through semipermeable membranes. A **semipermeable membrane** allows solvent molecules such as water and very small solute particles to pass through, but not large solute molecules.

Colloids

The particles in colloidal dispersions, or **colloids,** are much larger than solute particles in a solution. Colloidal particles are large molecules, such as proteins, or groups of molecules or ions. Colloids are homogeneous mixtures that do not separate or settle out. Colloidal particles are small enough to pass through filters but too large to pass through semipermeable membranes.

When a beam of light shines through a colloid, it produces the **Tyndall effect,** in which the large solute particles scatter the light and make the light beam visible, as shown in Figure 8.17. The Tyndall effect causes a light beam to be visible in fog because the water droplets in the air scatter the light. When there is smoke or smog in the air, the blues and greens of sunlight are scattered more, so that we observe the oranges and reds of a beautiful sunset.

Figure 8.17 The Tyndall effect: When light is reflected by colloids, it becomes visible.

8.6

EXPLORE YOUR WORLD

Tyndall Effect

If you have a small flashlight, you can observe the Tyndall effect in various solutions. Obtain solutions such as skim milk, soft drinks, vinegar, tea, salt water, and cleaning solutions. Shine a beam of light through each solution. Observe how the beam of light appears in the solution.

Questions

1. Which of the solutions contain colloids?
2. Why don't some solutions scatter light?

Types of Colloids

Colloids are classified by the type of solute and the dispersing medium. There are four types of colloids: aerosols, foams, emulsions, and sols. *Aerosols* consist of liquid or solid particles dispersed in a gas. *Foams* are dispersions of gases in liquids or solids. An *emulsion* is a liquid dispersed in another liquid or a solid. For example, when milk is not homogenized, the suspended particles of cream separate. The process of homogenization breaks the cream particles into colloids that remain dispersed within the milk solution. In a *sol,* particles of a solid are dispersed in a liquid or solid to give a more rigid solution. Table 8.9 lists several examples of colloids.

Table 8.9 Colloids

Examples	Type	Substance Dispersed	Dispersing Medium
Fog, clouds, sprays	Aerosol	Liquid	Gas
Dust, smoke	Aerosol	Solid	Gas
Shaving cream, whipped cream, soapsuds	Foam	Gas	Liquid
Styrofoam, marshmallows	Foam	Gas	Solid
Mayonnaise, butter, homogenized milk, hand lotions	Emulsion	Liquid	Liquid
Cheese, butter	Emulsion	Liquid	Solid
Blood plasma, paints (latex), gelatin	Sol	Solid	Liquid
Cement, pearls	Sol	Solid	Solid

Suspensions

Suspensions are heterogeneous, nonuniform mixtures that are very different from solutions or colloids. The particles of a suspension are so large that they can often be seen with the naked eye. They are trapped by filters and semipermeable membranes.

The weight of the suspended solute particles causes them to settle out soon after mixing. If you stir muddy water, it mixes but then quickly separates as the suspended particles settle to the bottom and leave clear liquid at the top. You can find suspensions among the medications in a hospital or in your medicine cabinet. These include Kaopectate, calamine lotion, antacid mixtures, and liquid penicillin. It is important to "shake well before using" to suspend all the particles before giving a medication that is a suspension.

8.6

Water-treatment plants make use of the properties of suspensions to purify water. When coagulants such as aluminum sulfate or ferric sulfate are added to untreated water, they react with impurities to form large suspended particles called floc. In the water-treatment plant, a system of filters traps the suspended particles but clean water passes through.

Figure 8.18 illustrates some properties of solutions, colloids, and suspensions, and Table 8.10 compares the different types of solutions.

S·A·M·P·L·E 8.16 ·P·R·O·B·L·E·M

Classifying Types of Mixtures

Classify each of the following as a solution, colloid, or suspension:

a. a homogeneous mixture in which a beam of light is visible
b. a mixture that settles rapidly upon standing
c. a mixture whose solute particles pass through both filters and membranes

Solution

a. colloid **b.** suspension **c.** solution

Study Check

Enzymes are large molecules of protein that catalyze chemical reactions inside the cells of the body. If they cannot pass through the cell membrane, are they solutions or colloids?

Table 8.10 Comparison of Solutions, Colloids, and Suspensions

Type of Mixture	Type of Particle	Effect of Light	Settling	Separation
Solution	Small particles such as single atoms, ions, or molecules	Transparent	Particles do not settle	Particles cannot be separated by filters or semipermeable membranes
Colloid	Larger molecules or groups of molecules or ions	Tyndall effect occurs	Particles do not settle	Particles can be separated by semipermeable membranes but not by filters
Suspension	Very large particles that may be visible	Opaque (not transparent)	Particles settle rapidly	Particles can be separated by filters

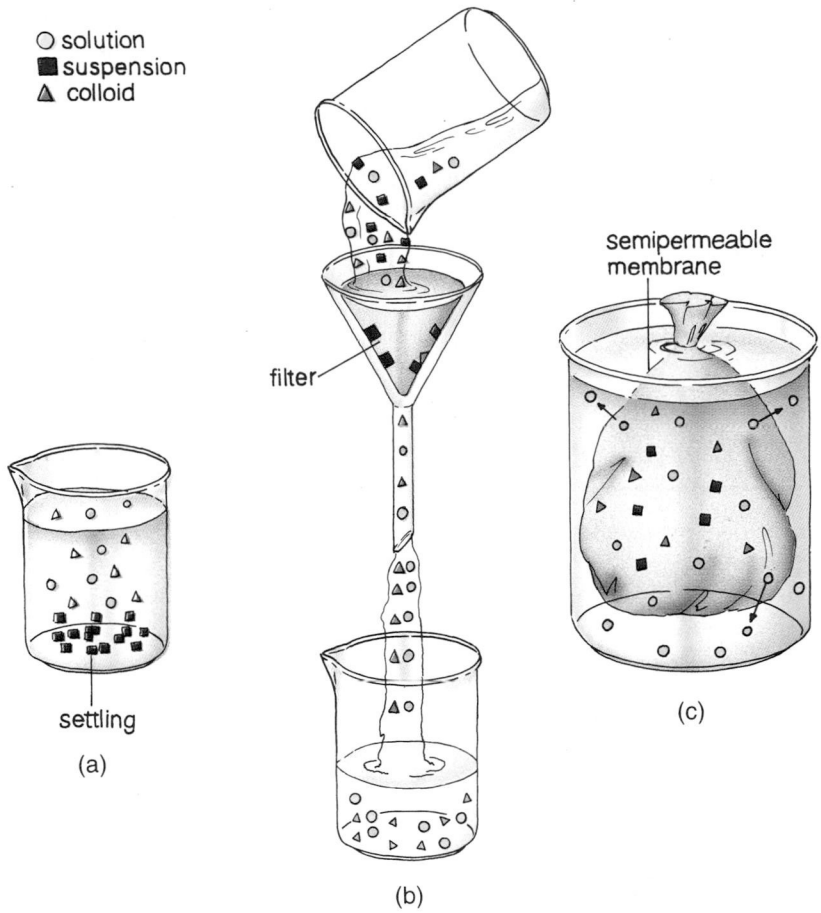

Figure 8.18 Properties of different types of solutions: **(a)** Suspensions settle out; **(b)** suspensions are separated by a filter; **(c)** solution particles go through a semipermeable membrane, but colloids and suspensions do not.

HEALTH NOTE

Colloids and Solutions in the Body

Colloids in the body are isolated by semipermeable membranes. For example, the intestinal lining allows solution particles to pass into the blood and lymph circulatory systems. However, the colloids from foods are too large to pass through the membrane, and they remain in the intestinal tract. Digestion breaks down large colloidal particles, such as starch and protein, into smaller particles, such as glucose and amino acids, that can pass through the intestinal membrane and enter the circulatory system. Certain foods, such as bran, a fiber, cannot be broken down by human digestive processes, and they move through the intestine intact.

Large proteins, such as enzymes, are formed inside cells. Because they are colloids, they remain inside the cell. However, many of the nutrients that must be obtained by cells, such as oxygen, amino acids, electrolytes, glucose, and minerals, can pass through cellular membranes. Waste products, such as urea and carbon dioxide pass out of the cell to be excreted.

QUESTIONS & PROBLEMS

Colloids and Suspensions

8.63 Identify the following as characteristic of a solution, colloid, or suspension:

a. a transparent mixture that cannot be separated by a semipermeable membrane

b. a mixture that appears cloudy when a beam of light is passed through it

c. a mixture that settles out upon standing

8.64 Identify the following as characteristic of a solution, colloid, or suspension:
a. Particles of this mixture remain inside a semipermeable membrane, but pass through filters.
b. A beam of light is not visible in this solution.
c. The particles of solute in this solution are very large and visible.

8.65 What is an emulsion?

8.66 What is an aerosol?

8.7

Study Pack

8.7 A Closer Look: Osmosis and Dialysis

Learning Goal

Describe the changes in concentration of solute and solvent in the processes of osmosis and dialysis.

The movement of water into and out of the cells of plants as well as our own bodies is an important biological process. In a process called **osmosis,** the solvent water moves through a semipermeable membrane from a solution that has a lower concentration of solute into a solution where the solute concentration is higher. (See Figure 8.19.) This movement, or osmosis, of water happens in a direction that equalizes (or attempts to equalize) the concentrations on both sides of the membrane. For example, when water is separated from sucrose (sugar) solution by a semipermeable membrane, water molecules, but not the larger sucrose molecules, move through the membrane to dilute the sucrose solution. As more water flows across the membrane, the level of the sugar solution rises as the level of water on the other side decreases.

At some point, the levels of the liquids on each side of the semipermeable membrane stop changing. Because the liquid level of the sugar solution is higher than the solvent side, the weight of the higher level of solution exerts a downward pressure called osmotic pressure. **Osmotic pressure** is the pressure that prevents the flow of additional water into the more concentrated solution. The water now moves equally between the two compartments as the system comes to equilibrium. The osmotic pressure of a solution depends on the number of solute particles in the solution. Pure water has no osmotic pressure. The greater the number of particles dissolved in a solution, the higher its osmotic pressure.

If a pressure greater than the osmotic pressure is applied to a solution, osmosis is reversed and solvent flows out of the solution and the level in the solvent compartment increases. This process, known as reverse osmosis, is used in desalinization plants in some parts of the world to produce drinking water from sea (salt) water.

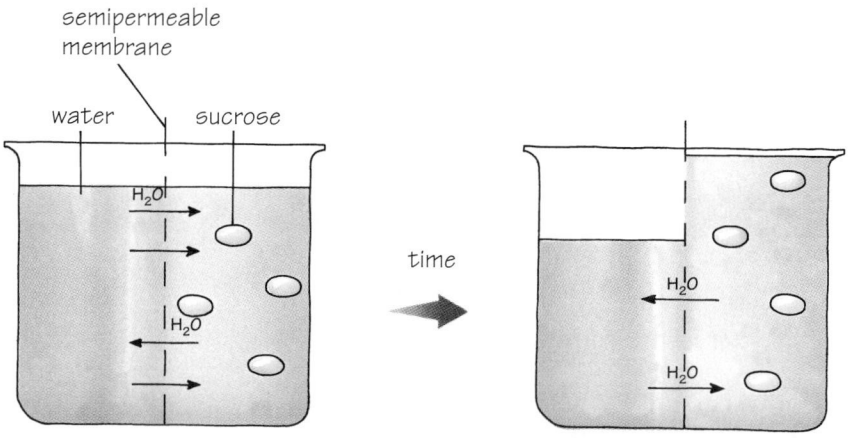

Figure 8.19 In osmosis, water flows through a semipermeable membrane from the solution of lower solute concentration into the solution of higher solute concentration.

S·A·M·P·L·E 8.17 ·P·R·O·B·L·E·M

Osmotic Pressure

A 2% sucrose solution and an 8% sucrose solution are separated by a semipermeable membrane.

a. Which sucrose solution exerts the greater osmotic pressure?
b. In what direction does water flow initially?
c. Which solution will have the higher level of liquid at equilibrium?

Solution

a. The 8% sucrose solution has the higher solute concentration, more solute particles, and the greater osmotic pressure.
b. Initially, water will flow out of the 2% solution into the more concentrated 8% solution.
c. The level of the 8% solution will be higher.

Study Check

If a 10% glucose solution is separated from a 5% glucose solution by a semipermeable membrane, which solution will decrease in volume?

Isotonic Solutions

Because the cell membranes in biological systems are semipermeable, osmosis is an ongoing process. The solutes in body solutions such as blood, tissue fluids, lymph, and plasma all exert osmotic pressure. Most intravenous solutions are **isotonic solutions,** which exert the same osmotic pressure as body fluids. *Iso* means "equal to," and *tonic* refers to the osmotic pressure of the solution in the cell. In the hospital, isotonic solutions or **physiological solutions** include 0.90% (m/v) NaCl and 5% (m/v) glucose solution. Although they do not contain the same particles as the body fluids, they exert the same osmotic pressure.

Hypotonic and Hypertonic Solutions

A red blood cell placed in an isotonic solution retains its normal volume because there is an equal flow of water into and out of the cell. (See Figure 8.20a.) However, if a red blood cell is placed in a solution that is not isotonic, the differences in osmotic pressure inside and outside the cell can drastically alter the volume of the cell. When a red blood cell is placed in pure water, a **hypotonic solution** (*hypo* means "lower than"), water flows into the cell by osmosis. (See Figure 8.20b.) The increase in fluid causes the cell to swell, and possibly burst, a process called **hemolysis.** A similar process occurs when you place dehydrated food, such as raisins or dried fruit, in water. The water enters the cells and the food becomes plump and smooth.

If a red blood cell is placed in a **hypertonic solution,** which has a higher solute concentration (*hyper* means "greater than"), water leaves the cell by osmosis. Suppose a red blood cell is placed in a 10% (m/v) NaCl solution. Because the osmotic pressure in the red blood cell is equal to that of a 0.90% (m/v) NaCl solution, the 10% (m/v) NaCl solution has a much greater osmotic pressure. As water is lost, the cell shrinks, a process called **crenation.** (See Figure 8.20c.) A similar process occurs when making pickles, which uses a hypertonic salt solution that causes the cucumbers to shrivel as they lose water.

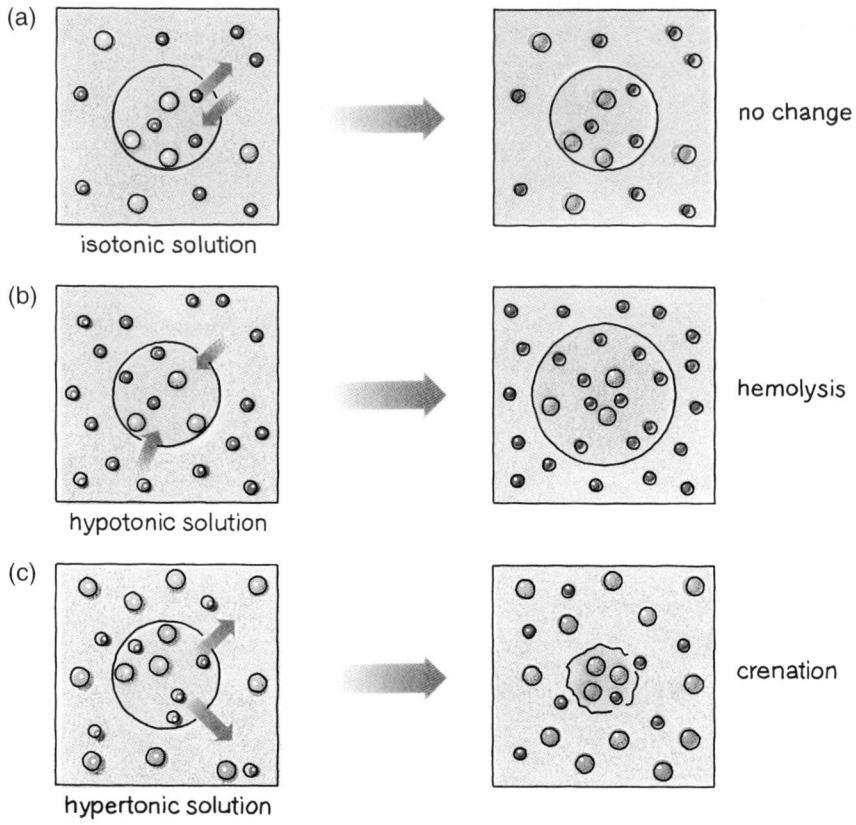

Figure 8.20 **(a)** In an isotonic solution, a red blood cell retains its normal volume.

(b) In a hypotonic solution, water flows into a red blood cell, causing it to swell and perhaps burst (hemolysis).

(c) In a hypertonic solution, water leaves the red blood cell, causing it to shrink (crenation).

S·A·M·P·L·E· **8.18** ·P·R·O·B·L·E·M

Isotonic, Hypotonic, and Hypertonic Solutions

Describe each of the following solutions as isotonic, hypotonic, or hypertonic. Indicate whether a red blood cell placed in each solution will undergo hemolysis, crenation, or no change.

a. a 5% (m/v) glucose solution **b.** a 0.2% (m/v) NaCl solution

Solution

a. A 5% (m/v) glucose solution is isotonic. A red blood cell will not undergo any change.

b. A 0.2% (m/v) NaCl solution is hypotonic. A red blood cell will undergo hemolysis.

Study Check

What is the effect of a 10% (m/v) glucose solution on a red blood cell?

EXPLORE YOUR WORLD

Everyday Osmosis

1. Place a few pieces of dry fruit such as raisins, prunes, or banana chips in water. Observe them after 1 hour or more. Look at them again the next day.
2. Place some grapes in a concentrated salt-water solution. Observe them after 1 hour or more. Look at them again the next day.
3. Place one potato slice in water and another slice in a concentrated salt-water solution. After 1–2 hours, observe the shapes and size of the slices. Look at them again the next day.

Questions

1. How did the shape of the dried fruit change after being in water? Explain.
2. How did the appearance of the grapes change after being in a concentrated salt solution? Explain.
3. How does the appearance of the potato slice that was placed in water compare to the appearance of the potato slice placed in salt water? Explain.
4. At the grocery store, why are sprinklers used to spray water on fresh produce such as lettuce, carrots, and cucumbers?

Dialysis

Dialysis is a process that is similar to osmosis. In dialysis, a semipermeable membrane, called a dialyzing membrane, permits small solute molecules and ions as well as solvent water molecules to pass through, but it retains large particles, such as colloids. Dialysis is a way to separate solution particles from colloids.

Suppose we fill a cellophane bag with a solution containing NaCl, glucose, starch, and protein and place it in pure water. Cellophane is a dialyzing membrane, and the sodium ions, chloride ions, and glucose molecules will pass through it into the surrounding water. However, starch and protein remain inside because they are colloids. Water molecules will flow by osmosis into the cellophane bag. Eventually, the concentrations of sodium ions, chloride ions, and glucose molecules inside and outside the dialysis bag become equal. To remove more NaCl or glucose, the cellophane bag must be placed in a fresh sample of pure water. (See Figure 8.21.)

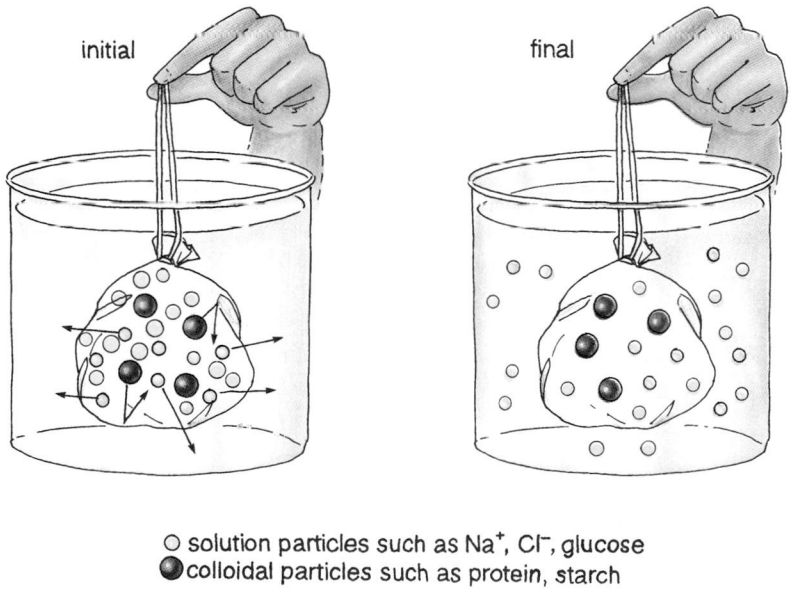

Figure 8.21 Dialysis, the separation of solutions from colloids in water.

HEALTH NOTE

Dialysis by the Kidneys and the Artificial Kidney

The fluids of the body undergo dialysis by the membranes of the kidneys, which remove waste materials, excess salts, and water. In an adult, each kidney contains about 2 million nephrons. (See Figure 8.22.) At the top of each nephron, there is a network of arterial capillaries called the glomerulus.

As blood flows into the glomerulus, small particles, such as amino acids, glucose, urea, water, and certain ions, will move through the capillary membranes into the nephron. As this solution moves through the nephron, substances still of value to the body (such as amino acids, glucose, certain ions, and 99% of the water) are reabsorbed. The major waste product, urea, is excreted in the urine.

8.7

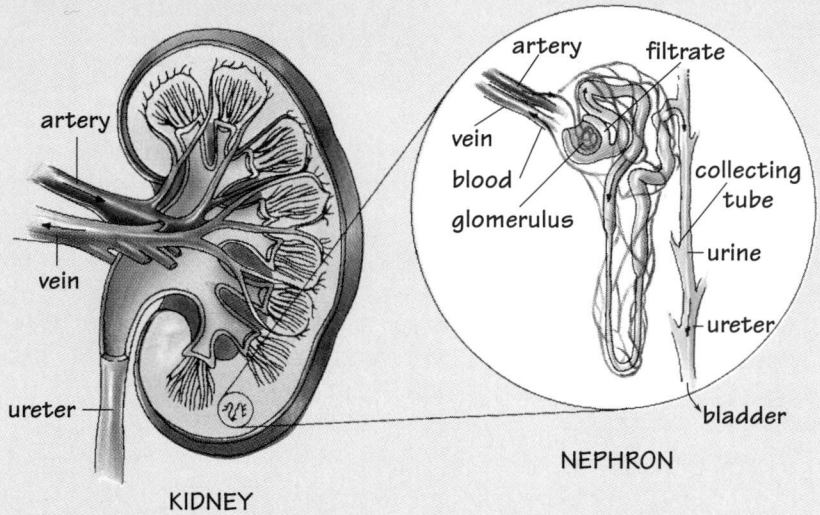

Figure 8.22 The nephron, the working unit of the kidney. Water and small particles dialyze out of the blood across the membrane of the glomerulus. Most of the reusable substances, including water, glucose, and amino acids, are reabsorbed from the filtrate, leaving urea and other waste products to be excreted in the urine.

Hemodialysis

If the kidneys fail to dialyze waste products, increased levels of urea can become life-threatening in a relatively short time. A person with kidney failure must use an artificial kidney, which cleanses the blood by **hemodialysis,** as shown in Figure 8.23.

A typical artificial kidney machine contains a large tank filled with about 100 L of water containing selected electrolytes. In the center of this dialyzing bath (dialysate), there is a dialyzing coil or membrane made of cellulose tubing. As the patient's blood flows through the dialyzing coil, the highly concentrated waste products dialyze out of the blood. No blood is lost because the membrane is not permeable to large particles such as red blood cells.

Dialysis patients do not produce much urine. As a result, they retain large amounts of water between dialysis treatments, which produces a strain on the heart. The intake of fluids for a dialysis patient may be restricted to as little as a few teaspoons of water a day. In the dialysis procedure, the pressure of the blood is increased as it circulates through the dialyzing coil so water can be squeezed out of the blood. For some dialysis patients, 2–10 L of water may be removed during one treatment. Dialysis patients have from two to three treatments a week, each treatment requiring about 5–7 hr. Some of the newer treatments require less time. For many patients, dialysis is done at home with a home dialysis unit.

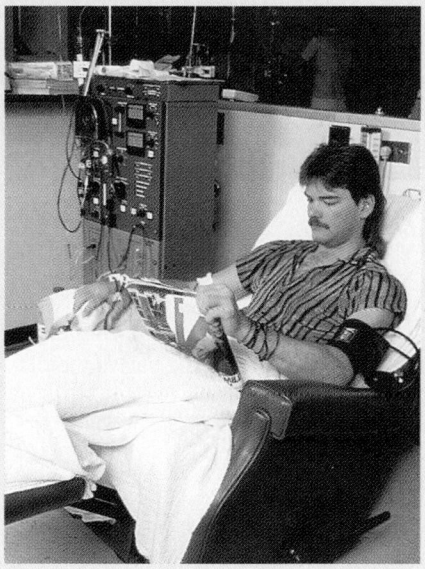

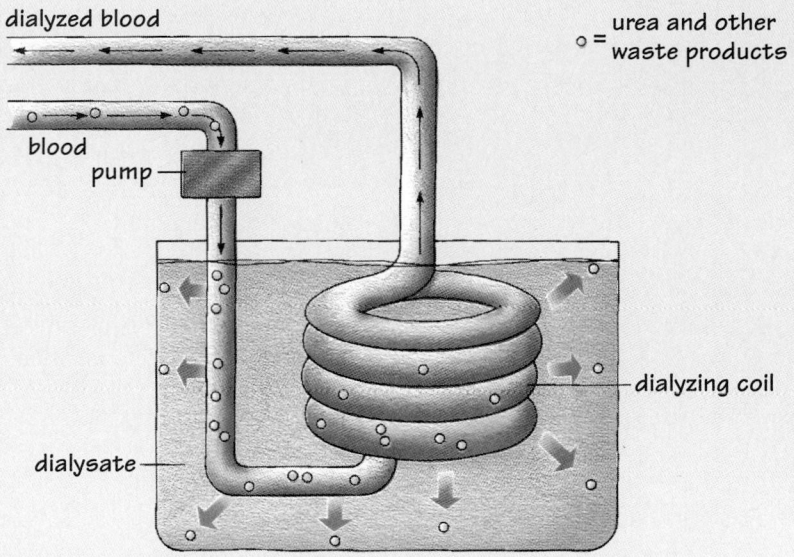

Figure 8.23 Hemodialysis, the dialysis of the blood by an artificial kidney. The initial dialysate consists of water and electrolytes. As blood flows through the dialyzing coil, urea and other waste products dialyze out of the coil and into the dialysate.

QUESTIONS & PROBLEMS

A Closer Look: Osmosis and Dialysis

8.67 **a.** How do plants obtain water from the ground?
b. How does a pickle get all shriveled up?

8.68 **a.** Why can't you drink large quantities of seawater?
b. Why does salt preserve foods?

8.69 A 10% (m/v) starch solution is separated from pure water by an osmotic membrane.
a. Which solution has the higher osmotic pressure?
b. In which direction will water flow initially?
c. In which compartment will the volume level rise?

8.70 Two solutions, a 0.1% (m/v) albumin solution and a 2% (m/v) albumin solution, are separated by a semipermeable membrane. (Albumins are colloidal proteins.)
a. Which compartment has the higher osmotic pressure?
b. In which direction will water flow initially?
c. In which compartment will the volume level rise?

8.71 Indicate the compartment (A or B) that will increase in volume for each of the following pairs of solutions separated by semipermeable membranes:

	A	*B*
a.	5% (m/v) glucose	10% (m/v) glucose
b.	4% (m/v) albumin	8% (m/v) albumin
c.	0.1% (m/v) NaCl	10% (m/v) NaCl

8.72 Indicate the compartment (A or B) that will increase in volume for each of the following pairs of solutions separated by semipermeable membranes:

	A	*B*
a.	20% (m/v) glucose	10% (m/v) glucose
b.	10% (m/v) albumin	2% (m/v) albumin
c.	0.5% (m/v) NaCl	5% (m/v) NaCl

8.73 Are the following solutions isotonic, hypotonic, or hypertonic compared with a red blood cell?

a. distilled H_2O **b.** 1% (m/v) glucose
c. 0.90% (m/v) NaCl **d.** 5% (m/v) glucose
e. 10% (m/v) NaCl

8.74 Will a red blood cell undergo crenation, hemolysis, or no change in each of the following solutions?

a. 1% (m/v) glucose **b.** 2% (m/v) NaCl
c. 5% (m/v) NaCl **d.** 0.1% (m/v) NaCl
e. 10% (m/v) glucose

8.8

8.75 Each of the following mixtures is placed in a dialyzing bag and immersed in distilled water. Which substances will be found outside the bag in the distilled water?

a. NaCl solution
b. starch (colloid) and alanine (amino acid solution)
c. NaCl solution and starch (colloid)
d. urea solution

8.76 Each of the following mixtures is placed in a dialyzing bag and immersed in distilled water. Which substances will be found outside the bag in the distilled water?

a. KCl solution and glucose solutions
b. an albumin solution (colloid)
c. an albumin solution (colloid), KCl solution, and glucose solution
d. urea solution and NaCl solution

8.8 Molarity

Learning Goal

Calculate the molarity of a solution; use molarity as a conversion factor to calculate the moles of solute or the volume needed to prepare a solution.

When the solutes of solutions take part in reactions, chemists are interested in the number of reacting particles. For this purpose, chemists use **molarity (M),** a concentration that states the number of moles of solute in exactly 1 liter of solution. The molarity of a solution can be calculated knowing the moles of solute and the volume of solution.

$$\text{Molarity (M)} = \frac{\text{moles of solute}}{\text{1 liter (1 L) of solution}}$$

For example, if 6.0 moles of NaOH were dissolved in enough water to prepare 3.0 L of solution, the resulting NaOH solution has a molarity of 2.0 M. The abbreviation M indicates the units of moles per liter (moles/L).

$$\text{M} = \frac{\text{moles of solute}}{\text{liters of solution}} = \frac{\text{6.0 moles NaOH}}{\text{3.0 L of solution}} = \frac{\text{2.0 moles NaOH}}{\text{1 L}} = \text{2.0 M NaOH}$$

S·A·M·P·L·E· 8.19 ·P·R·O·B·L·E·M

Calculating Molarity

What is the molarity (M) of each of the following solutions?

a. 2.0 moles of calcium chloride ($CaCl_2$) in 500 mL of solution
b. 60.0 g of NaOH in 0.250 L of solution

Solution

a. Because molarity requires liters (L), the volume in liters must be determined.

$$500\ \cancel{mL} \times \frac{1\ L}{1000\ \cancel{mL}} = 0.5\ L \text{ of solution}$$

Placing the moles of solute and the volume in liters into the concentration expression, we calculate the molarity as

$$\text{molarity (M)} = \frac{\text{moles } CaCl_2}{\text{liters of solution}} = \frac{2.0 \text{ moles } CaCl_2}{0.5\ L} = 4\ M$$

b. Because molarity requires moles of solute, we must change from grams of NaOH to moles of NaOH using the molar mass NaOH (40.0).

$$\underset{\text{Grams of NaOH}}{60.0\ \cancel{g\ NaOH}} \times \underset{\text{Molar mass}}{\frac{1 \text{ mole NaOH}}{40.0\ \cancel{g\ NaOH}}} = 1.50 \text{ moles NaOH}$$

Now we use the moles of NaOH and volume of the solution in liters to calculate the molarity of the solution.

$$M = \frac{\overset{\text{Moles of solute}}{1.50 \text{ moles NaOH}}}{\underset{\text{Liters of solution}}{0.250\ L}} = 6.00\ M \text{ solution}$$

Study Check

What is the molarity of a solution that contains 75 g of KNO_3 dissolved in 350 mL of solution?

Molarity as a Conversion Factor

When we need to calculate the moles of solute or the volume of solution, the molarity is used as a conversion factor. Examples of conversion factors from molarity are given in Table 8.11.

Table 8.11 Some Examples of Molar Solutions

Molarity	Meaning	Conversion Factors		
6.0 M HCl	6.0 moles of HCl in 1 liter of solution	$\frac{6.0 \text{ moles HCl}}{1\ L}$	and	$\frac{1\ L}{6.0 \text{ moles HCl}}$
0.20 M NaOH	0.20 mole of NaOH in 1 liter of solution	$\frac{0.20 \text{ mole NaOH}}{1\ L}$	and	$\frac{1\ L}{0.20 \text{ mole NaOH}}$

S·A·M·P·L·E· 8.20 ·P·R·O·B·L·E·M

Molarity as a Conversion Factor

How many moles of NaCl are present in 4.0 L of a 2.0 M NaCl solution?

Solution

The molarity (2.0 M) of the solution gives two possible conversion factors:

$$\frac{2.0 \text{ moles NaCl}}{1\ L} \quad \text{and} \quad \frac{1\ L}{2.0 \text{ moles NaCl}}$$

Using the factor that cancels volume (L), we can calculate the moles of NaCl.

$$\underset{\text{Given volume}}{4.0\cancel{L}} \times \underset{\text{Molarity as a conversion factor}}{\frac{2.0 \text{ moles NaCl}}{1\cancel{L}}} = 8.0 \text{ moles NaCl}$$

Study Check

How many moles of HCl are present in 750 mL of a 6.0 M HCl solution?

8.8

To prepare a solution, we must convert the number of moles of solute needed into grams. Using the volume and the molarity of the solution with the molar mass of the solute, we can calculate the number of grams of solute necessary. This type of calculation is illustrated in Sample Problem 8.21.

S·A·M·P·L·E· 8.21 ·P·R·O·B·L·E·M

Using Molarity

How many grams of KCl would you need to weigh out to prepare 0.250 L of a 2.00 M KCl solution?

Solution

The number of moles of KCl is determined by using the given volume of solution and the molarity as a conversion factor.

$$\underset{\text{Given volume}}{0.250\cancel{\text{L solution}}} \times \underset{\text{Molarity factor}}{\frac{2.00 \text{ moles KCl}}{1\cancel{\text{L solution}}}} = 0.500 \text{ mole KCl}$$

Knowing the moles of solute needed, the number of grams of KCl is calculated using the molar mass of KCl (74.6 g/mol).

$$\underset{\text{Moles KCl}}{0.500\cancel{\text{mole KCl}}} \times \underset{\text{Molar mass KCl}}{\frac{74.6 \text{ g KCl}}{1\cancel{\text{mole KCl}}}} = 37.3 \text{ g KCl}$$

Study Check:

How many grams of baking soda, $NaHCO_3$, are in 325 mL of a 4.50 M $NaHCO_3$ solution?

The volume of a solution can be calculated if the number of moles of solute and the molarity of the solution are given.

S·A·M·P·L·E· 8.22 ·P·R·O·B·L·E·M

Calculating Volume of a Molar Solution

What volume, in liters, of 1.5 M HCl solution is needed to provide 6.0 moles of HCl?

Solution

Using the molarity as a conversion factor, we set up the following calculation:

$$\underset{\text{Given}}{6.0\ \cancel{\text{moles HCl}}} \times \underset{\text{Molarity factor}}{\frac{1\ \text{L solution}}{1.5\ \cancel{\text{moles HCl}}}} = 4.0\ \text{L HCl}$$

Study Check

How many milliliters of 2.0 M NaOH solution will provide 20.0 g of NaOH? (*Hint:* Convert grams of NaOH to moles of NaOH first.)

8.8

HEALTH NOTE

Osmolarity of Physiological Solutions and Body Fluids

Every molecule and every ion of solute that is dissolved in a solution contributes to its osmotic pressure. Osmolarity is a concentration term that describes the total number of moles of all solute particles in a liter of solution. For example, 1 mole of NaCl dissolves in a liter of water and produces 1 mole of sodium ions (Na^+) and 1 mole of chloride ions (Cl^-) for a total of 2 moles (2 osmoles) of particles. Thus, one mole of NaCl in 1 liter of solution gives a 2 osmolar (Osm) solution.

$$\begin{aligned} 1\ \text{mole NaCl/L} &= (1\ \text{mole Na}^+ + 1\ \text{mole Cl}^-)/\text{L} \\ &= 2\ \text{moles ions/L} \\ &= 2\ \text{osmoles/L} \\ &= 2\ \text{Osm} \end{aligned}$$

The osmolarity of a 0.10 M K_2SO_4 solution would be 0.30 Osm because three ions form when K_2SO_4 dissolves.

In hospitals, the osmolarity of a 0.9% (m/v) NaCl solution and a 5% (m/v) glucose solution must be similar to the osmolarity of the body fluids to prevent damage to blood cells. Normal serum and urine each have an osmolarity of about 0.3 Osm, or 300 milliosmoles/liter (300 mOsm). A 0.9% NaCl solution is 0.3 Osm or 300 milliosmoles per liter (300 mOsm), which is calculated as follows:

$$\frac{0.9\ \text{g NaCl}}{100\ \text{mL}} \times \frac{1\ \text{mole NaCl}}{58.5\ \text{g NaCl}} \times \frac{2\ \text{osmoles}}{1\ \text{mole NaCl}} \times \frac{1000\ \text{mL}}{1\ \text{L}} = 0.3\ \text{Osm} = 300\ \text{mOsm}$$

We can do a similar calculation to illustrate that a 5% (m/v) glucose solution has the same osmotic pressure as body fluids.

$$5\%\ \text{glucose} = \frac{5\ \text{g glucose}}{100\ \text{mL}}$$

$$\text{Molar mass of glucose} = 180\ \text{g/mole}$$

Because glucose, $C_6H_{12}O_6$, dissolves as molecules, one mole of glucose yields one osmole solute in solution.

$$\frac{5\ \text{g glucose}}{100\ \text{mL}} \times \frac{1000\ \text{mL}}{1\ \text{L}} \times \frac{1\ \text{mole glucose}}{180\ \text{g glucose}} \times \frac{1\ \text{osmole}}{1\ \text{mole glucose}} = 0.3\ \text{Osm} = 300\ \text{mOsm}$$

Therefore, the osmolarities of 0.9% NaCl and 5% glucose solutions are the same as the osmolarities of normal body fluids and will exert the same osmotic pressures as the body fluids.

QUESTIONS & PROBLEMS

Molarity

8.77 Calculate the molarity (M) of the following solutions:
a. 2.0 moles of glucose in 4.0 L of solution
b. 0.10 mole of NaCl in 40.0 mL of solution
c. 4.0 g of KOH in 2.0 L of solution
d. 20.0 g of NaOH in 0.500 L of solution

8.78 Calculate the molarity (M) of the following solutions:
a. 0.50 mole glucose in 0.200 L of solution
b. 0.300 mole LiBr in 300. mL of solution
c. 36.5 g of HCl in 1.0 L of solution
d. 8.0 g of NaOH in 250 mL of solution

8.8

8.79 Write two conversion factors for each of the following concentrations:
a. 6.00 M HCl **b.** 0.250 M $NaHCO_3$ **c.** 1.0 M H_2SO_4

8.80 Write two conversion factors for each of the following concentrations:
a. 2.0 M NaOH **b.** 0.10 M KBr **c.** 3.0 M Li_2CO_3

8.81 Calculate the moles of solute needed to prepare each of the following:
a. 1.0 L of a 3.0 M NaCl solution
b. 0.400 L of a 1.0 M KBr solution
c. 200.0 mL of a 4.0 M NaCl solution

8.82 Calculate the moles of solute needed to prepare each of the following:
a. 5.0 L of a 2.0 M $CaCl_2$ solution
b. 10.0 L of a 10.0 M NaOH solution
c. 25 mL of a 1.0 M KBr solution

8.83 Calculate the grams of solute needed to prepare each of the following solutions:
a. 1.0 L of a 1.0 M NaOH solution
b. 4.0 L of a 0.20 M KCl solution
c. 50.0 mL of a 1.0 M NaCl solution

8.84 Calculate the grams of solute needed to prepare each of the following solutions:
a. 2.0 L of a 6.0 M NaOH solution
b. 5.0 L of a 0.10 M $CaCl_2$ solution
c. 1500 mL of a 2.0 M NaCl solution

8.85 What volume in liters provides the following amounts of solute?
a. 2.0 moles of NaOH from a 2.0 M NaOH solution
b. 10 moles of NaCl from a 1 M NaCl solution
c. 80.0 g of NaOH from a 1 M NaOH solution

8.86 What volume in liters provides the following amounts of solute?
a. 0.100 mole of KCl from a 4.0 M KCl solution
b. 125 g of NaOH from a 1.0 M NaOH solution
c. 5.0 moles of HCl from a 6.0 M HCl solution

Chapter Review

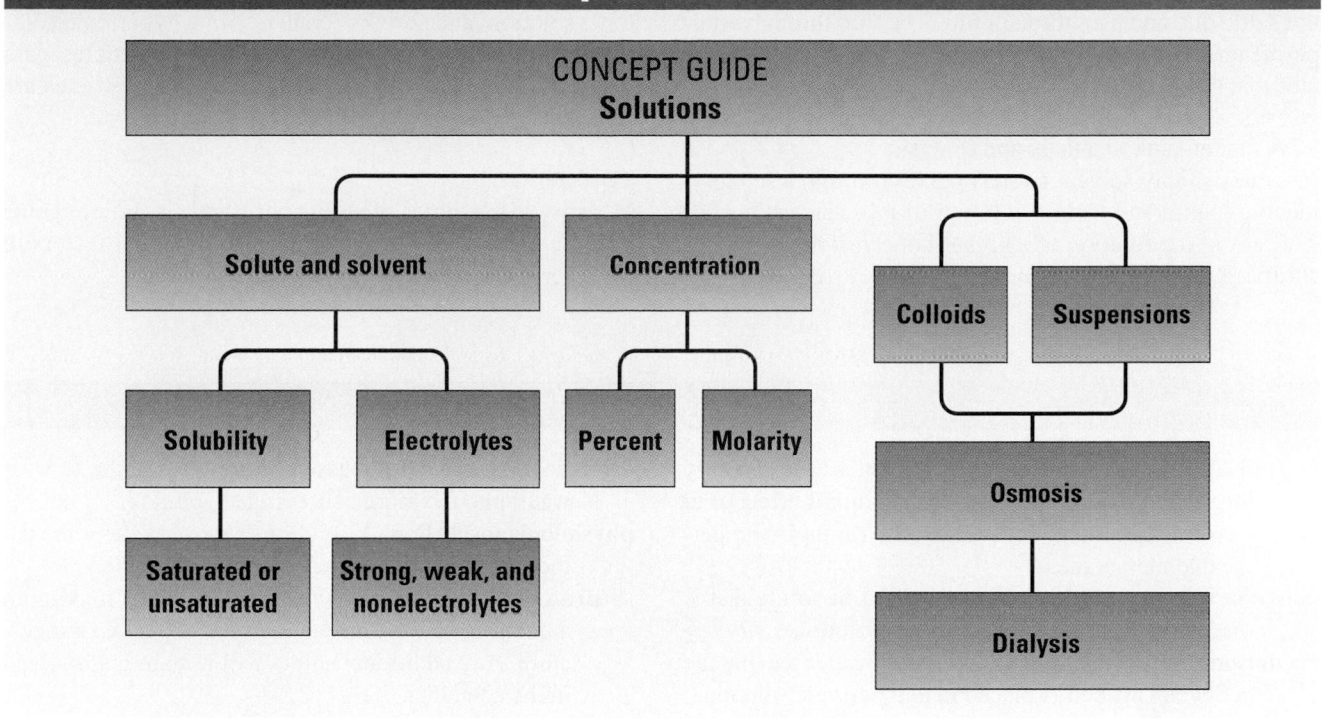

Chapter Summary

8.1 Properties of Water

The polar O — H bond permits hydrogen bonding between water molecules. Surface tension is a result of the hydrogen bonds in water.

8.2 Solutions

A solution forms when a solute dissolves in a solvent. The expression "like dissolves like" means that a polar or ionic solute dissolves in a polar solvent, and a nonpolar solute requires a nonpolar solvent. A solution will form faster if the solute is crushed or if the solution is stirred or heated.

8.3 Electrolytes and Nonelectrolytes

An ionic solute dissolves in water, a polar solvent, because the polar water molecules attract and pull the ions into solution where they become hydrated. Substances that release ions are called electrolytes because the solution will conduct an electrical current. Strong electrolytes are completely ionized, whereas weak electrolytes are only partially ionized. Nonelectrolytes are substances that dissolve in water to produce molecules and cannot conduct electrical currents.

8.4 Solubility

A solution that contains the maximum amount of dissolved solute is a saturated solution. The solubility of a solute is the maximum amount of a solute that can dissolve in 100 g of solvent. A solution containing less than the maximum amount of dissolved solute is unsaturated. An increase in temperature increases the solubility of most solids in water, but decreases the solubility of gases in water. Salts that are soluble in water usually contain Li^+, Na^+, K^+, NH_4^+ or NO_3^-.

8.5 Percent Concentration

The concentration of a solution is the amount of solute dissolved in a certain amount of solution. Mass percent expresses the ratio of the mass of solute to the mass of solution multiplied by 100. Percent concentration is also expressed as volume/volume and mass/volume ratios. In calculations of grams or milliliters of solute or solution, the percent concentration is used as a conversion factor.

8.6 Colloids and Suspensions

Colloids contain particles that show the Tyndall effect, do not settle out, and pass through filters but not through semipermeable membranes. Suspensions have very large particles that settle out of solution.

8.7 A Closer Look: Osmosis and Dialysis

In osmosis, only solvent (water) passes through a semipermeable membrane from a solution of a lower solute concentration to a solution of a higher concentration. Isotonic solutions have osmotic pressures equal to that of body fluids. A red blood cell maintains its volume in an isotonic solution, but swells and may burst (hemolysis) in a hypotonic solution, and shrinks (crenates) in a hypertonic solution. In dialysis, water and small solute particles pass through a dialyzing membrane, while larger particles are retained.

8.8 Molarity

Molarity is the moles of solute per liter of solution. Units of molarity, moles/liter, are used in conversion factors to solve for moles of solute or volume of solution.

Key Terms

colloid A mixture having particles that are moderately large. Colloids scatter light (the Tyndall effect) and pass through filters but cannot pass through semipermeable membranes.

concentration A measure of the amount of solute that is dissolved in a specified amount of solution.

crenation The shriveling of a cell due to water leaving the cell when the cell is placed in a hypertonic solution.

dialysis A process in which water and small solute particles pass through a semipermeable membrane.

electrolyte A substance that produces ions when dissolved in water; its solution conducts electricity.

equivalent (Eq) The amount of a positive or negative ion that supplies 1 mole of electrical charge.

hemodialysis A mechanical cleansing of the blood by an artificial kidney using the principle of dialysis.

hemolysis A swelling and bursting of red blood cells in a hypotonic solution due to an increase in fluid volume.

hydrogen bond The attraction between a partially positive hydrogen atom in one water molecule and the partially negative oxygen atom of another.

hypertonic solution A solution that has a higher osmotic pressure than the red blood cells of the body.

hypotonic solution A solution that has a lower osmotic pressure than the red blood cells of the body.

insoluble salt An ionic compound that does not dissolve in water.

isotonic solution A solution that has the same osmotic pressure as that of the red blood cells of the body.

mass/mass percent The grams of solute in exactly 100 grams of solution.

mass/volume percent The grams of solute in exactly 100 ml of solution.

molarity (M) The number of moles of solute in exactly 1 L of solution.

nonelectrolyte A substance that dissolves in water as molecules; its solution will not conduct an electrical current.

osmosis The flow of a solvent, usually water, through a semipermeable membrane into a solution of higher solute concentration.

osmotic pressure The pressure that prevents the flow of water into the more concentrated solution.

physiological solution A solution that exerts the same osmotic pressure as normal body fluids.

saturated solution A solution containing the maximum amount of solute that can dissolve at a given temperature. Any additional solute will remain undissolved in the container.

semipermeable membrane A membrane that permits the passage of certain substances while blocking or retaining others.

solubility The maximum amount of solute that can dissolve in exactly 100 g of solvent, usually water, at a given temperature.

soluble salt An ionic compound that dissolves in water.

solute The component in a solution that changes state upon dissolving; if no change in state occurs, it is the component present in the smaller quantity.

solution A homogeneous mixture in which the solute is made up of small particles (ions or molecules) that can pass through filters and semipermeable membranes.

solvent The substance in which the solute dissolves; usually the component present in greatest amount.

strong electrolyte An ionic compound that ionizes completely when it dissolves in water. Its solution is a good conductor of electricity.

surface tension A characteristic of water in which the hydrogen bonding between water molecules pulls its surface molecules together so tightly that they form a "skin."

suspension A mixture in which the solute particles are large enough and heavy enough to settle out and be retained by both filters and semipermeable membranes.

Tyndall effect The opaque path of a light beam through a colloid as light is scattered by the colloidal particles.

unsaturated solution A solution that contains less solute than can be dissolved.

weak electrolyte A substance that produces only a few ions along with many molecules when it dissolves in water. Its solution is a weak conductor of electricity.

Chapter Practice

8.87 A patient receives all her nutrition from fluids given through the vena cava. Every 12 hours, 750 mL of a solution that is 4% (m/v) amino acids from protein and 25% (m/v) glucose is given along with 500 mL of a 10% (m/v) lipid (fat).
a. In 1 day, how many grams of amino acids, glucose, and lipid are given to the patient?
b. How many kilocalories does she obtain in 1 day?

8.88 An 80-proof brandy is 40.0% (v/v) ethyl alcohol. The "proof" is twice the percent concentration of alcohol in the beverage. How many milliliters of alcohol are present in 750 mL of brandy?

8.89 A solution is prepared with 70.0 g of HNO_3 and 130.0 g of H_2O. It has a density of 1.21 g/mL.
a. What is the mass percent of the HNO_3 solution?
b. What is the total volume of the solution?
c. What is the mass/volume percent?
d. What is its molarity (M)?

8.90 Why would a dialysis unit (artificial kidney) use isotonic concentrations of NaCl, KCl, $NaHCO_3$, and glucose in the dialysate?

8.91 Why would solutions with high salt content be used to prepare dried flowers?

8.92 A patient on dialysis has a high level of urea, a high level of sodium, and a low level of potassium in the blood. Why is the dialyzing solution prepared with a high level of potassium but no sodium or urea?

8.93 Why is it dangerous to drink seawater even if you are stranded on a desert island?

8.94 In some countries, pure water is obtained from seawater by reverse osmosis. What do you think such a process involves?

8.95 Calculate the osmolarity of each of the following solutions:
a. 0.20 M NaCl **b.** 0.050 M K_2CO_3
c. 0.2 M $CaCl_2$ **d.** 0.1 M Na_2SO_4

8.96 In 1 liter of Ringer's solution there are 0.147 mole of NaCl, 0.004 mole KCl, and 0.004 mole of $CaCl_2$. What is the osmolarity of Ringer's solution? Why is it used as a physiological solution?

Answers

Answers to Study Checks

8.1 Within liquid water, the water molecules form hydrogen bonds in all directions with surrounding water molecules. However, the water molecules on the surface can only hydrogen bond in a downward direction, which pulls them more tightly together.

8.2 Iodine is the solute, and ethyl alcohol is the solvent.

8.3 Yes. Both the solute and solvent are nonpolar substances; "like dissolves like."

8.4 The ions Ca^{2+} and Cl^- are pulled away from the surface of the solid by the polar water molecules. In solution, water molecules surround the ions in a process called hydration.

8.5 Lowering the temperature will slow down the formation of the solution.

8.6 $Na_2SO_4(s) \xrightarrow{H_2O} 2Na^+(aq) + SO_4^{2-}(aq)$

8.7 A solution of a weak electrolyte will contain both molecules and ions.

8.8 71 g Cl^-

8.9 2.75 g Cl^-

8.10 75 g $CuSO_4$

8.11 The value of 65 g/100 g H_2O is more likely because the solubility of most solids increases when the temperature increases.

8.12 **a.** No; $PbCl_2$ is an insoluble chloride
b. Yes; a salt containing Na^+ ion is soluble
c. No; $FeCO_3$ is insoluble

8.13 3.4% (m/m) NaCl solution

8.14 4.8% (m/v) Br_2 in CCl_4

8.15 30 g NaCl

8.16 colloids

8.17 5% glucose

8.18 The red blood cell will shrink (crenate).

8.19 2.1 M KNO_3

8.20 4.5 moles of HCl

8.21 123 g KNO_3

8.22 250 mL

Answers to Selected Questions and Problems

8.1 **a.**

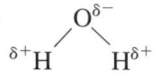

The oxygen in the water molecule has a partial negative charge and the hydrogen atoms have partial positive charges.

b.

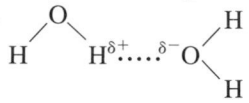

c. HF is polarized, with a partial positive charge on H and partial negative on F.

$H^{\delta+} — F^{\delta-}$

d. CH_4 would not form hydrogen bonds because the C — H bond is not sufficiently polar.

8.3 **a.** NaCl, solute; water, solvent
b. water, solute; ethanol, solvent
c. oxygen, solute; nitrogen, solvent

8.5 **a.** water **b.** CCl_4
c. water **d.** CCl_4

8.7 **a.** oxygen atom **b.** hydration
c. hydrogen atom

8.9 The polar water molecules pull the K^+ and I^- ions away from the solid and into solution, where they are hydrated.

8.11 **a.** increase **b.** decrease
c. decrease **d.** increase

8.13 In a solution of KF, only the ions of K^+ and F^- are present in the solvent. In an HF solution, there are a few ions of H^+ and F^- present but mostly dissolved HF molecules.

8.15 **a.** $KCl \xrightarrow{H_2O} K^+ + Cl^-$

b. $CaCl_2 \xrightarrow{H_2O} Ca^{2+} + 2Cl^-$

c. $K_3PO_4 \xrightarrow{H_2O} 3K^+ + PO_4^{3-}$

d. $Fe(NO_3)_3 \xrightarrow{H_2O} Fe^{3+} + 3NO_3^-$

8.17 **a.** mostly molecules and a few ions
b. ions only
c. molecules only

8.19 **a.** $HC_2H_3O_2 \overset{H_2O}{\rightleftharpoons} H^+ + C_2H_3O_2^-$

b. $NaBr \xrightarrow{H_2O} Na^+ + Br^-$

c. $C_6H_{12}O_6(s) \xrightarrow{H_2O} C_6H_{12}O_6(aq)$

8.21 **a.** strong electrolyte
b. weak electrolyte
c. nonelectrolyte

8.23 **a.** 1 Eq **b.** 2 Eq **c.** 2 Eq **d.** 6 Eq

8.25 **a.** 0.704 Eq **b.** 0.805 Eq
c. 0.20 Eq **d.** 1.0 Eq

8.27 0.18 g Mg^{2+}

8.29 3.5 g Na^+, 5.5 g Cl^-

8.31 55 mEq/L

8.33 **a.** saturated **b.** unsaturated

8.35 **a.** unsaturated **b.** saturated
c. saturated **d.** unsaturated

8.37 **a.** unsaturated **b.** saturated **c.** 20 g KBr

8.39 **a.** The solubility of solid solutes typically increases as temperature increases.
b. The solubility of a gas is less at a higher temperature.
c. Gas solubility is less at a higher temperature and the CO_2 pressure in the can is increased.

8.41 **a.** soluble **b.** insoluble **c.** insoluble
d. soluble **e.** soluble

8.43 **a.** no solid forms
b. Ag_2S is insoluble; $2AgNO_3 + K_2S \longrightarrow Ag_2S(s) + 2KNO_3$
c. $CaSO_4$ is insoluble; $CaCl_2 + Na_2SO_4 \longrightarrow CaSO_4(s) + 2NaCl$

8.45 5% (m/m) is 5 g of glucose in 100 g of solution, whereas 5% (m/v) is 5 g of glucose in 100 mL of solution.

8.47 **a.** 13.3% **b.** 16.0% **c.** 11.1%

8.49 **a.** 30.% **b.** 0.30% **c.** 10.%

8.51 **a.** $\frac{10\text{g NaOH}}{100\text{ g solution}}$ and $\frac{100\text{ g solution}}{10\text{ g NaOH}}$

b. $\frac{0.5\text{ g NaCl}}{100\text{ mL solution}}$ and $\frac{100\text{ mL solution}}{0.5\text{ g NaCl}}$

c. $\frac{15\text{ mL } CH_3OH}{100\text{ mL solution}}$ and $\frac{100\text{ mL solution}}{15\text{ mL } CH_3OH}$

8.53 **a.** 2.5 g KCl **b.** 22.5 g K_2CO_3
c. 50. g NH_4Cl **d.** 25 mL acetic acid

8.55 79.9 mL of alcohol

8.57 **a.** 20 g mannitol **b.** 300 g mannitol

8.59 **a.** 250 g **b.** 20. mL
c. 400. mL **d.** 20. mL

8.61 2 L

8.63 **a.** solution **b.** colloid **c.** suspension

8.65 An emulsion is a colloid in which a liquid is dispersed in another liquid or a solid.

8.67 **a.** Water flows from wet soil through plant membranes toward the higher solute concentration in the plants.
b. Because a brine solution (salt water) has a higher solute concentration, water flows from the cucumber to the brine and the cucumber becomes a pickle.

8.69 **a.** starch
b. from pure water into the starch
c. starch

8.71 **a.** B 10% glucose **b.** B 8% albumin
c. B 10% NaCl

8.73 **a.** hypotonic **b.** hypotonic **c.** isotonic
d. isotonic **e.** hypertonic

8.75 **a.** NaCl **b.** alanine **c.** NaCl **d.** urea

8.77 **a.** 0.50 M glucose **b.** 2.5 M NaCl
c. 0.036 M KOH **d.** 1.00 M NaOH

8.79 **a.** $\frac{6.00 \text{ moles HCl}}{1 \text{ L solution}}$ and $\frac{1 \text{ L solution}}{6.00 \text{ moles HCl}}$

b. $\frac{0.250 \text{ mole } NaHCO_3}{1 \text{ L solution}}$ and $\frac{1 \text{ L solution}}{0.250 \text{ mole } NaHCO_3}$

c. $\frac{1.0 \text{ mole } H_2SO_4}{1 \text{ L solution}}$ and $\frac{1 \text{ L solution}}{1.0 \text{ mole } H_2SO_4}$

8.81 **a.** 3.0 moles NaCl **b.** 0.40 mole KBr
c. 0.80 mole NaCl

8.83 **a.** 40. g NaOH **b.** 60. g KCl
c. 2.9 g NaCl

8.85 **a.** 1.0 L **b.** 10 L **c.** 2 L

8.87 **a.** 60 g of amino acids, 380 g of glucose and 100 g of lipids
b. 2600 kcal

8.89 **a.** 35.0% HNO_3 **b.** 165 mL
c. 42.4% **d.** 6.72 M

8.91 The solution will dehydrate the flowers because water will flow out of the cells of the flowers into the more concentrated salt solution.

8.93 Drinking seawater will cause water to flow out of the body cells and further dehydrate a person.

8.95 **a.** 0.40 Osm **b.** 0.15 Osm
c. 0.6 Osm **d.** 0.3 Osm

9

Acids and Bases

Citrus fruits taste sour because they contain weak acids such as citric acid.

Looking Ahead

9.1	Acids and Bases
9.2	Ionization of Water
9.3	The pH Scale
9.4	Strengths of Acids and Bases
9.5	Making Dilutions
9.6	Acid–Base Neutralization
9.7	Buffers
9.8	A Closer Look: Acid–Base Titration

www.chemplace.com/college/timberlake

Visit the web site above to test yourself, run simulations, and review key concepts from this chapter. For even more interactive review and quizzing, use the two CD-ROMs included with the book.

Lemons, grapefruits, and vinegar taste sour because they contain acidic solutions. We have acid in our stomach that helps us digest food; we produce lactic acid in our muscles when we exercise. Acid from bacteria turns milk sour to make cottage cheese or yogurt. Bases are solutions that neutralize acids. Sometimes we take antacids such as milk of magnesia to offset the effects of too much stomach acid.

The pH of a solution describes its acidity. The pH of body fluids, including blood and urine, is regulated primarily by the lungs and the kidneys. Major changes in the pH of the body fluids can severely affect biological activities within the cells. Buffers are present to prevent large fluctuations in pH.

Study Pack

Chem of Life Acids, Bases, and pH: Water and Acidity 1

9.1

9.1 Acids and Bases

Learning Goal

Describe acids and bases using the Arrhenius and the Brønsted–Lowry concepts.

The term *acid* comes from the Latin word *acidus,* which means "sour." We are familiar with the sour tastes of vinegar and lemons and other common acids in foods as illustrated in Figure 9.1.

In the nineteenth century, Arrhenius was the first to describe **acids** as substances that produce hydrogen ions (H^+) when they dissolve in water. For example, hydrogen chloride ionizes in water to give hydrogen ions, H^+, and chloride ions, Cl^-. For polar covalent compounds such as HCl, the formation of ions in water is called an ionization reaction. Using the Arrhenius concept, we write the ionization of HCl in water as

$$\underset{\text{Polar covalent compound}}{HCl(g)} \xrightarrow[\text{Ionization in water}]{H_2O} H^+ + Cl^-$$

The hydrogen ions, H^+, give all acids a sour taste, change blue litmus to red, and corrode some metals.

Figure 9.1 Some common household items that contain acids include vinegar, lemons, carbonated beverages, and vitamin C.

Naming Acids

When an acid dissolves in water to produce hydrogen ion and a simple nonmetal ion, the prefix *hydro-* is used before the name of the nonmetal and its *–ide* ending is changed to *–ic acid*. For example, hydrogen chloride (HCl) dissolves in water to form HCl (*aq*), which is named hydrochloric acid.

When an acid contains a polyatomic ion, the name of the acid comes from the name of the polyatomic ion. The *–ate* in the name is replaced with *–ic acid.* For example, HNO_3, which contains the nitrate ion (NO_3^-), is named nitric acid. If the acid contains a polyatomic ion with an *–ite* ending, its name ends with *–ous acid.* You can determine the *–ous acid* formula by using one oxygen less that the polyatomic ion of the *–ic acid.* The names of some common acids and their anions are listed in Table 9.1. For example, HNO_2, which contains the nitrite ion (NO_2^-), is named nitrous acid.

Acid	Anion	Name of Anion	Acid Name
HCl (aq)	Cl^-	chloride	**hydro**chlor**ic acid**
HNO_3 (aq)	NO_3^-	nitrate	nitr**ic acid**
HNO_2 (aq)	NO_2^-	nitrite	nitr**ous acid**

Figure 9.2 Some common bases include detergents, antacids, drain openers, and oven and window cleaners.

Bases

You may be familiar with some of the bases illustrated in Figure 9.2 such as detergents, antacids, drain openers, and oven and window cleaners. According to the Arrhenius theory, **bases** are ionic compounds that separate into a metal ion and hydroxide ions

9.1

Table 9.1 Naming Common Acids

Acid	Name of Acid	Anion	Name of Anion
HCl	**hydro**chloric **acid**	Cl^-	chlor**ide**
HBr	**hydro**bromic **acid**	Br^-	brom**ide**
HNO_3	nitr**ic acid**	NO_3^-	nitr**ate**
HNO_2	nitr**ous acid**	NO_2^-	nitr**ite**
H_2SO_4	sulfur**ic acid**	SO_4^{2-}	sulf**ate**
H_2SO_3	sulfur**ous acid**	SO_3^{2-}	sulf**ite**
H_2CO_3	carbon**ic acid**	CO_3^{2-}	carbon**ate**
H_3PO_4	phosphor**ic acid**	PO_4^{3-}	phosph**ate**

(OH^-) when they dissolve in water. For example, sodium hydroxide is an Arrhenius base that dissociates in water to give sodium ions, Na^+, and hydroxide ions, OH^-.

$$\underset{\text{Ionic compound}}{NaOH} \xrightarrow[\text{Dissociation}]{H_2O} Na^+ + \underset{\text{Hydroxide ion}}{OH^-}$$

Most bases are from Groups 1 and 2, such as KOH, LiOH, and $Ca(OH)_2$. There are other bases such as $Al(OH)_3$ and $Fe(OH)_3$, but because they are fairly insoluble, only a few OH^- ions are released into solution. The hydroxide ions (OH^-) give Arrhenius bases common characteristics such as a bitter taste and soapy feel.

Table 9.2 compares some characteristics of acids and bases.

Table 9.2 Characteristics of Acids and Bases

Characteristic	Acids	Bases
Ions in water	H^+ (H_3O^+)	OH^- (Arrhenius)
Behavior in water	H^+ donor	H^+ acceptor
Electrolytes	Yes	Yes
Taste	Sour	Bitter, chalky
Feel	May sting	Soapy, slippery
pH	Under 7	Over 7
Litmus	Red	Blue
Phenolphthalein	Clear	Red
Neutralization	Neutralizes bases	Neutralizes acids

Naming Bases

Typical Arrhenius bases are named as hydroxides.

Bases	Name
NaOH	sodium **hydroxide**
KOH	potassium **hydroxide**
$Ca(OH)_2$	calcium **hydroxide**
$Al(OH)_3$	aluminum **hydroxide**

S·A·M·P·L·E **9.1** ·P·R·O·B·L·E·M

Arrhenius Acids and Bases

Identify each of the following as characteristic of an Arrhenius acid or base:
a. is named magnesium hydroxide
b. tastes sour
c. turns litmus red

Solution

a. A base contains hydroxide ions.
b. An acid tastes sour.
c. An acid turns litmus red.

Study Check

Detergents have a soapy feel. Are they acids or bases?

Brønsted–Lowry Acids and Bases

Early in the twentieth century, Brønsted and Lowry expanded the definition of acids and bases. A **Brønsted–Lowry acid** donates a proton (hydrogen ion, H^+) to another substance, and a **Brønsted–Lowry base** accepts a proton.

A Brønsted–Lowry acid is a proton (H^+) donor.

A Brønsted–Lowry base is a proton (H^+) acceptor.

A proton (H^+) does not actually exist in water. Its attraction to polar water molecules is so strong that the proton bonds to the water molecule and forms a **hydronium ion, H_3O^+**. At times, we use the H^+ symbol for convenience, but understand that it is the H_3O^+ that is present in acidic aqueous solutions.

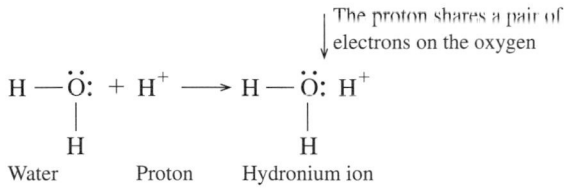

We can now write the formation of a hydrochloric acid solution as a transfer of a proton from hydrogen chloride to water. (See Figure 9.3.) By accepting a proton in the reaction, water is acting as a base according to the Brønsted–Lowry concept.

$$\underset{\substack{\text{Acid}\\(H^+\text{ donor})}}{HCl} + \underset{\substack{\text{Base}\\(H^+\text{ acceptor})}}{H_2O} \longrightarrow \underset{\substack{\text{Hydronium ion}\\\text{(acidic solution)}}}{H_3O^+} + Cl^-$$

In another reaction ammonia (NH_3) reacts with water. Because the nitrogen atom of NH_3 has a stronger attraction for a proton, water acts as an acid by donating a proton. (See Figure 9.4.)

$$\underset{\substack{\text{Base}\\(H^+\text{ acceptor})}}{NH_3} + \underset{\substack{\text{Acid}\\(H^+\text{ donor})}}{H_2O} \longrightarrow \underset{\substack{\text{Ammonium}\\\text{ion}}}{NH_4^+} + \underset{\substack{\text{Hydroxide ion}\\\text{(basic solution)}}}{OH^-}$$

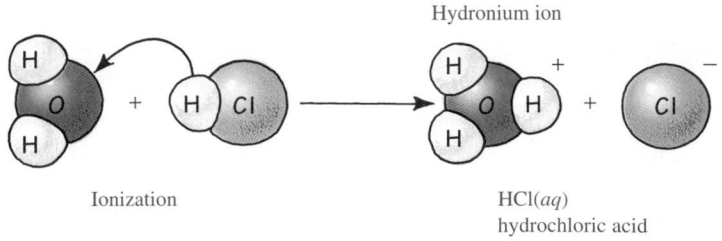

Figure 9.3 Polar covalent hydrogen chloride (HCl) ionizes in water to produce hydrochloric acid, which exists as hydronium ions (H_3O^+) and chloride ions (Cl^-).

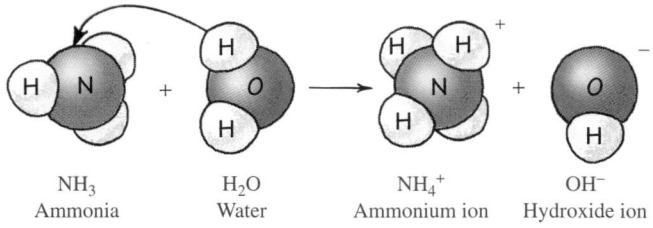

Figure 9.4 In water, ammonia (NH_3) acts as a Brønsted–Lowry base by accepting a proton (H^+) from water.

EXPLORE YOUR WORLD

Acids and Bases in Products at Home

A variety of acids and bases are in many of the foods we eat. Citric acid and ascorbic acid (vitamin C) give citrus fruits their sour taste. Vinegar is a 5% acetic acid solution. Many skin-care products now include acids from fruits. Antacids contain bases such as aluminum hydroxide and calcium hydroxide as well as calcium carbonate, which is a Brønsted–Lowry base because the carbonate ion accepts protons.

Read the labels of several products you use in your home, such as soft drinks, antacids, cereals, skin-care products, vitamins, vinegar, and juices. Look for the names of acids and bases, such as citric acid, acetic acid, lactic acid, folic acid, aluminum hydroxide, sodium hydroxide, and magnesium hydroxide. Make a list of the various products and the acids and bases each contains. Write out as many of the formulas of the acids and bases as you can. To find the formulas of some acids and bases, use a chemistry reference book such as the *Merck Index*. You could also do a search on the Internet.

S·A·M·P·L·E 9.2 P·R·O·B·L·E·M

Acids and Bases

In each of the following equations, identify the reactant that is an acid (H^+ donor) and the reactant that is a base (H^+ acceptor).

a. $HBr + H_2O \longrightarrow H_3O^+ + Br^-$ **b.** $H_2O + CN^- \longrightarrow HCN + OH^-$

Solution

a. HBr, acid; H_2O, base **b.** H_2O, acid; CN^-, base

Study Check

When HNO_3 reacts with water, water acts as a base (H^+ acceptor). Write the equation for the reaction.

Q·U·E·S·T·I·O·N·S & P·R·O·B·L·E·M·S

Acids and Bases

9.1 Indicate whether each of the following statements is characteristic of an acid or a base:

a. has a sour taste
b. neutralizes bases
c. produces H_3O^+ ions in water
d. is named potassium hydroxide
e. is a proton (H^+) acceptor

9.2 Indicate whether each of the following statements is characteristic of an acid or a base:
a. neutralizes acids **b.** produces OH^- in water
c. is a proton (H^+) donor **d.** has a soapy feel
e. turns litmus red

9.3 Identify each of the following compounds as an acid or base:
a. HCl **b.** $Ca(OH)_2$ **c.** H_2CO_3
d. HNO_3 **e.** H_2SO_3

9.4 Identify each of the following compounds as an acid or base:
a. $Al(OH)_3$ **b.** HBr **c.** H_2SO_4
d. KOH **e.** HNO_2

9.5 Give the names of the acids and bases in problem 9.3.

9.6 Give the names of the acids and bases in problem 9.4.

9.7 Write formulas for the following acids and bases:
a. magnesium hydroxide **b.** hydrofluoric acid
c. phosphoric acid **d.** lithium hydroxide
e. copper(II) hydroxide

9.8 Write formulas for the following acids and bases:
a. barium hydroxide **b.** hydriodic acid **c.** nitric acid
d. iron(III) hydroxide **e.** sodium hydroxide

9.9 In each of the following equations, identify the acid (proton donor) and base (proton acceptor).
a. $HI + H_2O \longrightarrow H_3O^+ + I^-$
b. $F^- + H_2O \longrightarrow HF + OH^-$

9.10 In each of the following equations, identify the acid (proton donor) and base (proton acceptor).
a. $CO_3^{2-} + H_2O \longrightarrow HCO_3^- + OH^-$
b. $H_2SO_4 + H_2O \longrightarrow H_3O^+ + HSO_4^-$

Study Pack

Chem of Life Acids, Bases, and pH: Water and Acidity 2, 3

9.2 Ionization of Water

Learning Goal

Use the ion product of water to calculate the $[H_3O^+]$ and $[OH^-]$ in a solution.

Most of the time we think of pure water as a nonelectrolyte that contains only molecules of water. However, water molecules are polar and occasionally, two water molecules collide in such a way that one donates a proton (H^+) to the other. One water molecule is acting as an acid and the other as a base. The products of this ionization are one hydronium ion (H_3O^+) and one hydroxide ion (OH^-).

$$H_2O + H_2O \longrightarrow H_3O^+ + OH^-$$

Equilibrium is reached quickly as the H_3O^+ and OH^- ions react in the reverse reaction to produce water molecules again.

$$H_2O + H_2O \longleftarrow H_3O^+ + OH^-$$

We write the ionization of water using a double arrow to show that it is reversible. The small arrow in the forward direction indicates that very few of the water molecules are ionized at any time.

$$\underset{\text{Water}}{H_2O} + \underset{\text{Water}}{H_2O} \rightleftharpoons \underset{\text{Hydronium ion}}{H_3O^+} + \underset{\text{Hydroxide ion}}{OH^-}$$

Sometimes, the reaction is written in a simplified form as a simple dissociation.

$$H_2O \rightleftharpoons H^+ + OH^-$$

Pure water produces equal amounts of H_3O^+ and OH^-. Therefore, pure water has no acidic or basic properties; it is **neutral.** At 25°C, the concentrations of H_3O^+ and OH^- are each 1×10^{-7} M. Square brackets written around a symbol indicate the concentration in moles per liter (M).

Pure Water Concentrations of H_3O^+ and OH^- at 25°C

$[H_3O^+] = 1 \times 10^{-7}$ M $\qquad [OH^-] = 1 \times 10^{-7}$ M

9.2

Ion Product for Water

If we multiply the concentrations of the H_3O^+ and OH^- ions in water, we obtain K_w, the **ion-product constant of water.**

Ion Product of Water

$$\begin{aligned} K_w &= [H_3O^+][OH^-] \\ &= (1 \times 10^{-7})(1 \times 10^{-7}) \\ &= 1 \times 10^{-14} \end{aligned}$$

The numerical value 1×10^{-14} is constant for all aqueous solutions even when acid or base is added.

Because $[H_3O^+]$ and $[OH^-]$ are equal in pure water, it is **neutral**. In an *acidic solution,* the $[H_3O^+]$ is greater than the $[OH^-]$ and maintains the $[H_3O^+][OH^-]$ value at 1×10^{-14}. The same idea applies to basic solutions. Figure 9.5 illustrates the relationship between $[H_3O^+]$ and $[OH^-]$. A comparison of $[H_3O^+]$ and $[OH^-]$ in some solutions is given in Table 9.3.

Hydronium–Hydroxide Balance in Acids, Bases, and Neutral Solutions

In acids, $[H_3O^+]$ is greater than $[OH^-]$.

In bases, $[OH^-]$ is greater than $[H_3O^+]$.

In neutral solutions, $[H_3O^+]$ equals $[OH^-]$.

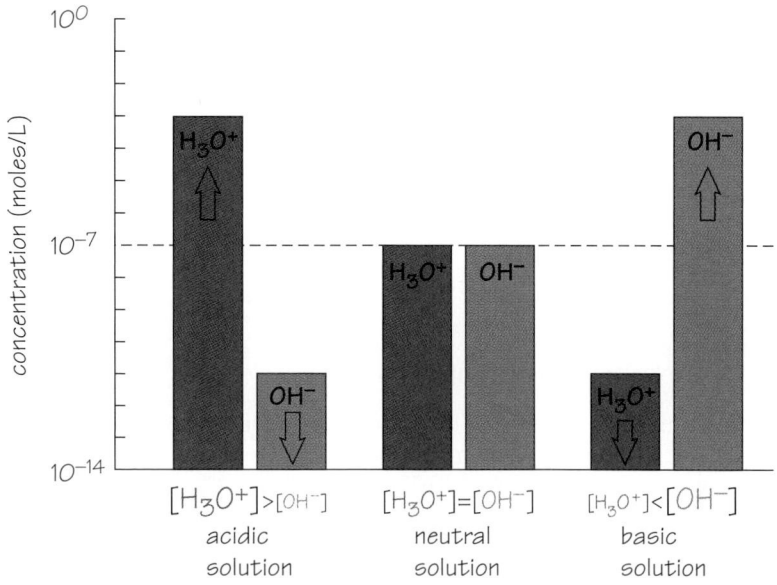

Figure 9.5 In a neutral solution such as pure water, $[H_3O^+]$ and $[OH^-]$ are equal. In acidic solutions, the $[H_3O^+]$ goes up as the $[OH^-]$ goes down. In basic solutions, the $[OH^-]$ goes up as the $[H_3O^+]$ goes down.

Table 9.3 Examples of $[H_3O^+]$ and $[OH^-]$ in Neutral, Acidic, and Basic Solutions

Type of Solution	$[H_3O^+]$	$[OH^-]$	Ion Product, K_w
Neutral	1×10^{-7}	1×10^{-7}	1×10^{-14}
Acidic	1×10^{-2}	1×10^{-12}	1×10^{-14}
	1×10^{-5}	1×10^{-9}	1×10^{-14}
Basic	1×10^{-8}	1×10^{-6}	1×10^{-14}
	1×10^{-11}	1×10^{-3}	1×10^{-14}

If we know the concentration of either the H_3O^+ or OH^- ion, we can calculate the concentration of the other ion by using the ion product expression.

$$[H_3O^+][OH^-] = 1 \times 10^{-14}$$

$$[H_3O^+] = \frac{1 \times 10^{-14}}{[OH^-]} \qquad [OH^-] = \frac{1 \times 10^{-14}}{[H_3O^+]}$$

S·A·M·P·L·E· 9.3 ·P·R·O·B·L·E·M

Calculating $[H_3O^+]$ in Solution

A solution of a window cleaner has a $[OH^-]$ of 1×10^{-4} M. What is the $[H_3O^+]$ in the cleaner?

Solution

We can solve for the $[H_3O^+]$ by using the ion product of water.

$$[H_3O^+][OH^-] = 1 \times 10^{-14}$$

$$[H_3O^+] = \frac{1 \times 10^{-14}}{[OH^-]}$$

The value for $[OH^-]$ of 1×10^{-4} can be substituted to solve the expression for $[H_3O^+]$.

$$[H_3O^+] = \frac{1 \times 10^{-14}}{1 \times 10^{-4}} = 1 \times 10^{-10}\ \text{M}$$

Study Check

What is the value for $[H_3O^+]$ of orange juice if its $[OH^-]$ is 1×10^{-11} M? Is the solution acidic, basic, or neutral?

S·A·M·P·L·E· 9.4 ·P·R·O·B·L·E·M

Calculating $[OH^-]$ in Solution

A vinegar solution has a $[H_3O^+]$ of 2×10^{-3} M. What is the $[OH^-]$ of this solution?

Solution

Rearranging the ion product expression for $[OH^-]$ gives:

$$[H_3O^+][OH^-] = 1 \times 10^{-14}$$

$$[OH^-] = \frac{1 \times 10^{-14}}{[H_3O^+]}$$

Substituting the given value of 2×10^{-3} for the $[H_3O^+]$ completes the calculation.

$$[OH^-] = \frac{1 \times 10^{-14}}{2 \times 10^{-3}}$$
$$= 5 \times 10^{-12}\ M$$

Study Check

A sample of gastric juice (HCl) has a $[H_3O^+]$ of 5×10^{-2} M. What is the $[OH^-]$ of the sample?

9.3

QUESTIONS & PROBLEMS

Ionization of Water

9.11 Why are the concentrations of H_3O^+ and OH^- equal in pure water?

9.12 In an acid solution, how does the concentration of the H_3O^+ compare to the concentration of OH^-?

9.13 In the following solutions, is the H_3O^+ concentration greater than, less than, or equal to the OH^- concentration?

a. acidic **b.** $[H_3O^+] = 1 \times 10^{-4}$ M
c. neutral **d.** $[OH^-] = 1 \times 10^{-5}$ M

9.14 In the following solutions, is the OH^- concentration greater than, less than, or equal to the H_3O^+ concentration?

a. basic **b.** $[H_3O^+] = 1 \times 10^{-7}$ M
c. $[H_3O^+] = 1 \times 10^{-10}$ M **d.** $[OH^-] = 1 \times 10^{-11}$ M

9.15 Calculate the $[H_3O^+]$ in each solution if the $[OH^-]$ has the following values:

a. 1×10^{-8} M **b.** 1×10^{-2} M
c. 1×10^{-11} M **d.** 2×10^{-5} M

9.16 Calculate the $[OH^-]$ in each solution if the $[H_3O^+]$ has the following values:

a. 1×10^{-1} M **b.** 1×10^{-7} M
c. 1×10^{-10} M **d.** 5×10^{-4} M

9.17 Indicate whether the following are acidic, basic, or neutral solutions:

a. $[OH^-] = 1 \times 10^{-6}$ M **b.** $[H_3O^+] = 1 \times 10^{-10}$ M
c. $[H_3O^+] = 1 \times 10^{-7}$ M **d.** $[H_3O^+] = 1 \times 10^{-2}$ M
e. $[OH^-] = 4 \times 10^{-4}$ M **f.** $[OH^-] = 8 \times 10^{-12}$ M

9.18 Indicate whether the following are acidic, basic, or neutral solutions:

a. $[OH^-] = 1 \times 10^{-11}$ M **b.** $[H_3O^+] = 1 \times 10^{-5}$ M
c. $[OH^-] = 1 \times 10^{-10}$ M **d.** $[H_3O^+] = 1 \times 10^{-8}$ M
e. $[OH^-] = 3 \times 10^{-3}$ M **f.** $[H_3O^+] = 6 \times 10^{-12}$ M

Acid, Bases, and pH: pH 1–3

Learning Goal

Calculate pH from $[H_3O^+]$; given the pH, calculate $[H_3O^+]$ and $[OH^-]$ of a solution.

9.3 The pH Scale

In acids and bases, the H_3O^+ concentrations can typically range from 1 M to 1×10^{-14} M. In clinical situations, a simpler way of describing acidity, the **pH** scale, is used. On the pH scale, the pH values range from 0 to 14 as shown in Figure 9.6. A pH value below 7 corresponds to an acidic solution, whereas solutions with pH values

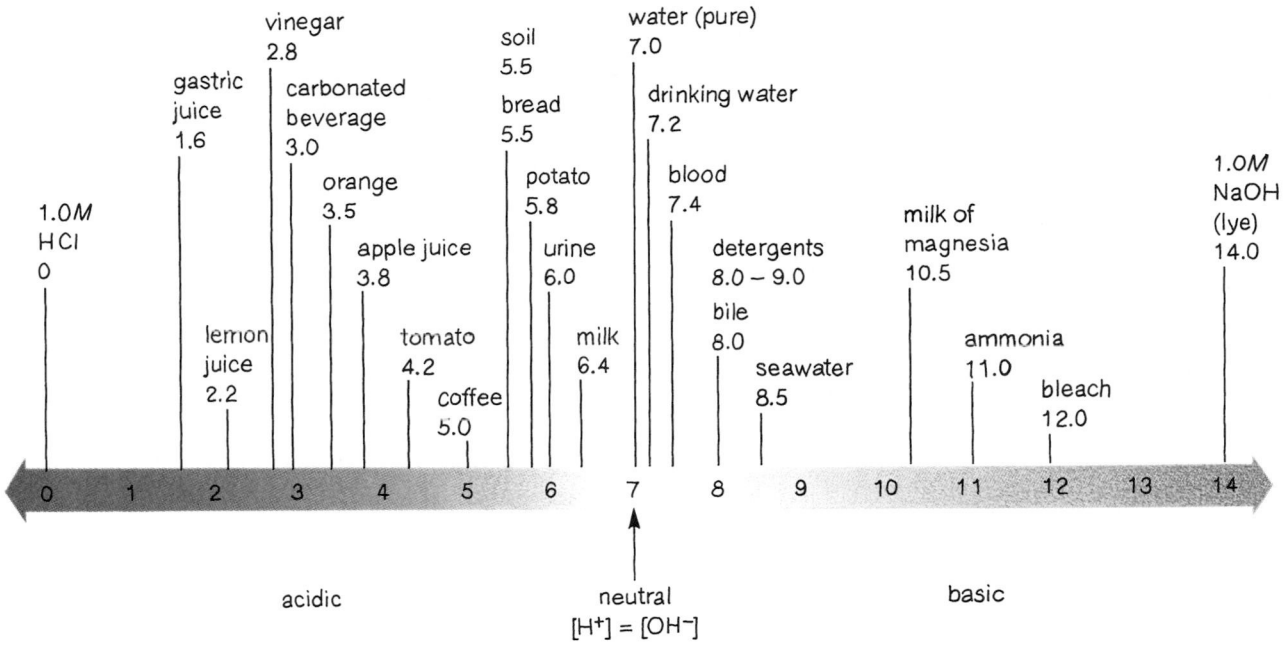

Figure 9.6 On the pH scale, values below 7 are acidic, a value of 7 is neutral, and values above 7 are basic.

above 7 are basic. A solution with a pH of 7, the midpoint of the pH scale, would be neutral. Table 9.4 lists the pH values for some common substances.

Neutral solution	pH = 7
Acidic solution	pH below 7
Basic solution	pH above 7

In the laboratory, the pH of a solution is determined by using a pH meter or an indicator that shows specific color changes at certain pH values as seen in Figure 9.7.

Table 9.4 pH Values of Some Common Substances

Solution	pH
1 M HCl	0
Gastric juice	1.6
Lemon juice	2.2
Vinegar	2.8
Carbonated drinks	3.0
Coffee	5.0
Urine	6.0
Rainwater	6.2
Water, pure	7.0
Blood	7.4
Bile	8.0
Detergents	8.0–9.0
Milk of magnesia	10.5
Ammonia solution	11.0
Bleach	12.0
1 M NaOH	14.0

S·A·M·P·L·E **9.5** ·P·R·O·B·L·E·M

Using pH to Identify Acidic and Basic Solutions

Classify the following as acidic, basic, or neutral solutions:

a. tomato juice, pH 4.2
b. saliva, pH 6.6
c. antacid, pH 8.5

Solution

a. Tomato juice is acidic because a pH of 4.2 is below 7.
b. Saliva is acidic because a pH value of 6.6 is below 7.
c. The antacid is basic. A pH of 8.5 is above 7.

Study Check

An NaCl solution has a pH of 7. Is it acidic, basic, or neutral?

Figure 9.7 The dye in red cabbage juice acts as a pH indicator by turning different colors that correspond to different pH values.

9.3

EXPLORE YOUR WORLD

Using Vegetables and Flowers as pH Indicators

Many flowers and vegetables with strong color, especially reds and purples, contain compounds that change color with changes in pH. Some examples are red cabbage, cranberry juice, and cranberry drinks.

Materials Needed

Red cabbage, water, and a saucepan; or cranberry juice or drinks

Several glasses or small glass containers and some tape and a pen or pencil to mark the containers

Several colorless household solutions such as vinegar, lemon juice, other fruit juices, baking soda, antacids, aspirin, window cleaners, soaps, shampoos, and detergents

Procedure

1. Obtain a bottle of cranberry juice or cranberry drink, or use a red cabbage to prepare the red cabbage pH indicator, as follows: Tear up several red cabbage leaves and place them in a saucepan and cover with water. Boil for about 5 minutes. Cool, strain, and collect the purple indicator solution.
2. Place small amounts of each household solution into separate, clear glass containers and mark what each one is. If the sample is a solid or a thick liquid, add a small amount of water. Add some cranberry juice or some red cabbage indicator until you obtain a color.
3. Observe the colors of the various samples. The colors that indicate acidic solutions are the pink and orange colors (pH 1–4) and the pink to lavender colors (5–6). A neutral solution has about the same purple color as the indicator. Bases will give blue to green color (pH 8–11) or a yellow color (pH 12–13).
4. Arrange your samples by color and pH. Classify each of the solutions as acidic (1–6), neutral (7), or basic (8–13).
5. Try to make an indicator using other colorful fruits or flowers.

Questions

1. Which products that tested acidic listed acids on their labels?
2. Which products that tested basic listed a base on their labels?
3. How many products were neutral?
4. Which flowers or vegetables behaved as indicators?

HEALTH NOTE

Stomach Acid, HCl

When a person sees, smells, thinks about, and/or tastes food, the gastric glands in the stomach begin to secrete a HCl solution that is strongly acidic. In a single day, a person may secrete as much as 2000 mL of gastric juice.

The HCl in the gastric juice activates a digestive enzyme called pepsin that breaks down proteins in food entering the stomach. The secretion of HCl continues until the stomach has a pH of about 2, which is the optimum pH for activating the digestive enzymes without ulcerating the stomach lining. Normally, large quantities of viscous mucus are secreted within the stomach to protect its lining from acid and enzyme damage.

When we compare pH values with their corresponding H_3O^+ concentrations, we find an acidic solution has a low pH value and a high $[H_3O^+]$. In a basic solution the pH value is high and the $[H_3O^+]$ is low.

Type of Solution	pH	$[H_3O^+]$
Neutral	7	1×10^{-7} M
Acidic	Below 7	More than 1×10^{-7} M
Basic	Above 7	Less than 1×10^{-7} M

The pH value is derived from the power of ten of the hydronium ion concentration. For example, if lemon juice has a $[H_3O^+]$ equal to 1×10^{-2} M, the pH is 2. We have taken the power of 10 (−2) and multiplied by a negative 1 to give $-(-2) = 2$. Mathematically, this is *finding the logarithm (log)*. In mathematical terms pH is defined as the negative value of the log of the molar hydronium ion concentration.

$$pH = -\log[H_3O^+]$$

For example, an acidic solution has a hydronium ion concentration of 1×10^{-6} M. The log of $[H_3O^+]$ is −6 (the power of 10). After we multiply the log by a negative sign $-(-6)$, we obtain a pH value of 6. For $[H_3O^+]$ with coefficients of 1, the pH is obtained simply by dropping the negative sign of the power of 10. *Significant figures in pH values:* In reporting a pH value, the number of digits after the decimal point is equal to the number of significant figures given in the coefficient of the concentration. Thus, a basic solution with a $[H_3O^+]$ of 1×10^{-11} has a pH of 11.0. Table 9.5 gives a comparison of some $[H_3O^+]$, $[OH^-]$, and pH values.

Table 9.5 A Comparison of $[H_3O^+]$, $[OH^-]$, and Corresponding pH Values at 25°C

pH	$[H_3O^+]$	$[OH^-]$	
0	10^0	10^{-14}	Acidic
1	10^{-1}	10^{-13}	
2	10^{-2}	10^{-12}	
3	10^{-3}	10^{-11}	
4	10^{-4}	10^{-10}	
5	10^{-5}	10^{-9}	
6	10^{-6}	10^{-8}	
7	10^{-7}	10^{-7}	Neutral
8	10^{-8}	10^{-6}	Basic
9	10^{-9}	10^{-5}	
10	10^{-10}	10^{-4}	
11	10^{-11}	10^{-3}	
12	10^{-12}	10^{-2}	
13	10^{-13}	10^{-1}	
14	10^{-14}	10^0	

S·A·M·P·L·E 9.6 ·P·R·O·B·L·E·M

Calculating pH

Determine the pH of the following solutions.

a. $[H_3O^+] = 1 \times 10^{-5}$ M **b.** $[H_3O^+] = 6 \times 10^{-8}$ M

c. $[OH^-] = 1 \times 10^{-3}$ M

9.3

Solution

a. Because the coefficient in the $[H_3O^+]$ is 1, the pH of the solution is the power of 10 without the negative sign, followed by a zero in the first decimal place.

$$[H_3O^+] = 1 \times 10^{-5} \qquad pH = 5.0$$

b. When the coefficient of the $[H_3O^+]$ is not 1, we must include it when we take the log of the concentration. Enter the concentration into your calculator, press the *log* key, and change signs as follows:

6 [EE or EXP] 8 [+/–] [log] [+/–] = 7.22184875

In this problem, the coefficient 6 has one significant figure. Therefore, the calculator answer is rounded to one decimal place and the pH value is reported as 7.2.

c. Because $[OH^-]$ is given, we have to calculate the $[H_3O^+]$ using the ion product of water.

$$[H_3O^+][OH^-] = 1 \times 10^{-14}$$

$$[H_3O^+] = \frac{1 \times 10^{-14}}{1 \times 10^{-3}}$$

$$= 1 \times 10^{-11}$$

$$pH = 11.0$$

Study Check

What is the pH of a bleach whose $[OH^-]$ is 1×10^{-4} M? (*Hint:* Calculate the $[H_3O^+]$ first.)

Sometimes, we need to determine the $[H_3O^+]$ from the pH. For whole number pH values, we can use the negative pH value as the power of 10 in the H_3O^+ concentration.

$$[H_3O^+] = 1 \times 10^{-pH}$$

S·A·M·P·L·E· 9.7 ·P·R·O·B·L·E·M

Calculating $[H_3O^+]$ from pH

Determine $[H_3O^+]$ for solutions having the following pH values:

a. pH = 3.0 **b.** pH = 12.0

Solution

For pH values that are whole numbers, the $[H_3O^+]$ can be written 1×10^{-pH}.

a. $[H_3O^+] = 1 \times 10^{-3}$ M (pH) **b.** $[H_3O^+] = 1 \times 10^{-12}$ M (pH)

Study Check

What is the $[H_3O^+]$ and $[OH^-]$ of beer if it has a pH of 5.0?

ENVIRONMENTAL NOTE

Acid Rain

Rain typically has a pH of 6.2. It is slightly acidic because carbon dioxide in the air combines with water to form carbonic acid. However, in many parts of the world, rain has become considerably more acidic, with pH values as low as 3 being reported. One cause of acid rain is the sulfur dioxide (SO_2) gas produced when coal that contains sulfur is burned.

In the air, the SO_2 gas reacts with oxygen to produce SO_3, which then combines with water to form sulfuric acid, H_2SO_4, a strong acid.

$$S + O_2 \longrightarrow SO_2$$

$$2SO_2 + O_2 \longrightarrow 2SO_3$$

$$SO_3 + H_2O \longrightarrow H_2SO_4$$

In parts of the United States, acid rain has made lakes so acidic they are no longer able to support fish and plant life. Limestone ($CaCO_3$) is sometimes added to these lakes to neutralize the acid. In Eastern Europe, acid rain has brought about an environmental disaster. Nearly 40% of the forests in Poland have been severely damaged, and some parts of the land are so acidic that crops will not grow. Throughout Europe and the United States, monuments made of marble (a form of $CaCO_3$) are deteriorating as acid rain dissolves the marble. (See Figure 9.8.)

$$2H^+ + CaCO_3 \longrightarrow Ca^{2+} + H_2O + CO_2$$

Efforts to slow or stop the damaging effects of acid rain include the reduction of sulfur emissions. This will require installation of expensive equipment in coal-burning plants to absorb more of the SO_2 gases before they are emitted. In some outdated plants, this may be impossible, and they will need to be closed. It is a difficult problem for engineers and scientists, but one that must be solved.

a

b

Figure 9.8 **(a)** Detail of the marble that is part of the Washington Square Arch in Washington Square Park, New York City, completed on July 17, 1935. **(b)** The destructive effect of acid rain on the same marble, June 1994.

QUESTIONS & PROBLEMS

The pH Scale

9.19 Why does a neutral solution have a pH of 7?

9.20 If you know the $[OH^-]$, how would you determine the pH of a solution?

9.21 State whether each of the following solutions is acidic, basic, or neutral:
a. blood, pH 7.4 **b.** vinegar, pH 2.8 **c.** drain cleaner, pH 11.2
d. coffee, pH 5.5 **e.** milk, pH 7.0

9.22 State whether each of the following solutions is acidic, basic, or neutral:
a. soda, pH 3.2 **b.** shampoo, pH 6.0
c. hot tub water, pH 7.8 **d.** acid rain, pH 6.2
e. laundry detergent, pH 9.5

9.3

9.23 Arrange the pH values of the following groups in order from the most acidic to the most basic:

a. 14, 5, 10, 7 **b.** 8, 10, 1, 13
c. 5.5, 1.6, 2.4, 8.5 **d.** 3.5, 9.7, 8.8, 11.4

9.24 Arrange the pH values of the following groups in order from the most acidic to the most basic:

a. 5, 3, 2, 1 **b.** 11, 4, 7, 9
c. 2.4, 1.8, 5.4, 6.3 **d.** 11.4, 7.6, 5.2, 10.8

9.25 Calculate the pH of each of the following solutions:

a. $[H_3O^+] = 1 \times 10^{-4}$ M **b.** $[H_3O^+] = 1 \times 10^{-9}$ M
c. $[H_3O^+] = 1 \times 10^{-12}$ M **d.** $[OH^-] = 1 \times 10^{-5}$ M
e. $[OH^-] = 1 \times 10^{-11}$ M

9.26 Calculate the pH of each of the following solutions:

a. $[H_3O^+] = 1 \times 10^{-1}$ M **b.** $[H_3O^+] = 1 \times 10^{-9}$ M
c. $[H_3O^+] = 1 \times 10^{-7}$ M **d.** $[OH^-] = 1 \times 10^{-2}$ M
e. $[OH^-] = 1 \times 10^{-6}$ M

9.27 Use a calculator to determine the pH of each of the following solutions:

a. $[H_3O^+] = 4 \times 10^{-2}$ M **b.** $[H_3O^+] = 2 \times 10^{-8}$ M
c. $[OH^-] = 5 \times 10^{-11}$ M

9.28 Use a calculator to determine the pH of each of the following solutions:

a. $[H_3O^+] = 8 \times 10^{-10}$ M **b.** $[H_3O^+] = 6 \times 10^{-3}$ M
c. $[OH^-] = 2 \times 10^{-6}$ M

9.29 Determine the $[H_3O^+]$ and $[OH^-]$ for each solution with the following pH values:

a. pH = 2.0 **b.** pH = 7.0
c. pH = 13.0 **d.** pH = 5.0

9.30 Determine the $[H_3O^+]$ and $[OH^-]$ for each solution with the following pH values:

a. pH = 10.0 **b.** pH = 4.0
c. pH = 1.0 **d.** pH = 8.0

9.31 Complete the following table of pH, $[H_3O^+]$, and $[OH^-]$:

$[H_3O^+]$	$[OH^-]$	pH	Acidic, Basic, or Neutral?
	1×10^{-6}		
		2.0	
1×10^{-5}			
1×10^{-7}			

9.32 Complete the following table for pH, $[H_3O^+]$, and $[OH^-]$:

$[H_3O^+]$	$[OH^-]$	pH	Acidic, Basic, or Neutral?
		10.0	
	1×10^{-7}		
1×10^{-12}			
	1×10^{-8}		

9.4 Strengths of Acids and Bases

Study Pack

Chem of Life Acids, Bases, and pH: Acids, Bases 1, 2

Learning Goal

Write an equation for the ionization of strong and weak acids.

9.4

Only a few acids, called **strong acids,** are strong electrolytes that ionize completely to produce only ions in solution. Three important strong acids are HCl, HNO_3, and H_2SO_4. In the ionization equation for nitric acid, a single arrow is used to indicate complete ionization.

$$HNO_3 + H_2O \longrightarrow \mathbf{H_3O^+ + NO_3^-}$$

Most acids are **weak acids** (weak electrolytes) that do not ionize very much. Even at high concentrations, weak acids contain only a small number of hydronium ions. In fact, you use some weak acids at home, such as citric acid, found in lemon or grapefruit juice, and vinegar, a 5% acetic acid solution. In carbonated soft drinks, CO_2 dissolves in water to form carbonic acid, H_2CO_3, a weak acid. In a weak acid, equilibrium is reached after a few of the dissolved molecules form ions. Using a double arrow, we show the reverse reaction that keeps the concentration of molecules of acetic acid high and that of the ions low.

$$\underset{\text{Acetic acid}}{\mathbf{HC_2H_3O_2}} + \mathbf{H_2O} \rightleftharpoons H_3O^+ + \underset{\text{Acetate ion}}{C_2H_3O_2^-}$$

Figure 9.9 compares the components of a solution of a strong acid (HCl) and a weak acid (HF).

S·A·M·P·L·E **9.8** ·P·R·O·B·L·E·M

Writing Equations for Ionization of Weak Acids

Write an equation for the ionization of formic acid ($HCHO_2$), a weak acid found in ant stings.

Solution

As a weak acid, formic acid exists in solution mostly as molecules with a few ions. For weak acids, the first hydrogen in the formula is the only hydrogen that ionizes in water.

$$HCHO_2 + H_2O \rightleftharpoons H_3O^+ + CHO_2^-$$

Study Check

Hypochlorous acid, HClO, is a weak acid that forms when chlorine reacts with water. Write an equation for its ionization in water.

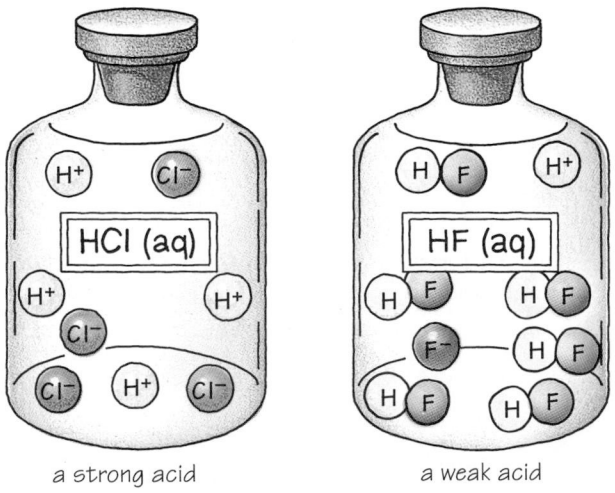

Figure 9.9 A strong acid, HCl, ionizes completely; a small amount of weak acid molecules (HF) ionizes to give a solution of HF molecules and ions.

9.4

Strong and Weak Bases

In water, a **strong base** (strong electrolyte) separates completely into metal ions and hydroxide ions (OH^-). The metal hydroxides of Group 1 as well as $Ba(OH)_2$ in Group 2 are all strong bases.

Sodium hydroxide $NaOH(s) \xrightarrow{H_2O} Na^+ + OH^-$

Potassium hydroxide $KOH(s) \xrightarrow{H_2O} K^+ + OH^-$

Barium hydroxide $Ba(OH)_2(s) \xrightarrow{H_2O} Ba^{2+} + 2OH^-$

Strong bases, such as NaOH (also known as lye), are used in household products to dissolve grease in ovens and to clean drains. Because hydroxide ions cause severe damage to the skin and eyes, the use of such products in the home should be carefully supervised. Most Group 2 hydroxides are not very soluble in water, although the amount that does dissolve dissociates completely. One such hydroxide is magnesium hydroxide, $Mg(OH)_2$, an antacid known as milk of magnesia that is ingested to counteract the effects of too much stomach acid.

$$Mg(OH)_2 \xrightarrow{H_2O} Mg^{2+} + 2OH^-$$

A **weak base** (weak electrolyte) produces only a few hydroxide ions when it reacts with water. The most common weak base is ammonia (NH_3), found in several household window cleaners. Ammonia reacts with water by accepting a proton. The resulting hydroxide ion makes the solution basic. The double arrow in the equation indicates an equilibrium in which only a few ammonia molecules have reacted at any given time.

$$\underset{\text{Ammonia}}{NH_3(g)} + \underset{\text{Water}}{H_2O} \rightleftharpoons \underset{\text{Ammonium ion}}{NH_4^+} + \underset{\text{Hydroxide ion}}{OH^-}$$

It is important to know when you are using acids or bases in the laboratory because they can burn the skin severely and damage the eyes. If you spill an acid or a base on your skin or get some in your eyes, be sure to flood the area immediately with water to lessen any damage.

S·A·M·P·L·E 9.9 ·P·R·O·B·L·E·M

Dissociation of a Base in Water

Write an equation for the dissociation of LiOH in water.

Solution

LiOH is a hydroxide of a Group 1 element. It dissociates in water to give a basic solution of lithium ions (Li^+) and hydroxide ions (OH^-).

$$LiOH \xrightarrow{H_2O} Li^+ + OH^-$$

Study Check

Calcium hydroxide is used in some antacids to counteract excess stomach acid. Write an equation for the formation of a calcium hydroxide solution.

Q·U·E·S·T·I·O·N·S & P·R·O·B·L·E·M·S

Strengths of Acids and Bases

9.33 What is the difference between a strong and weak acid?

9.34 What is the difference between a strong and weak base?

9.35 What are the formulas and names of three common strong acids?

9.36 What are the formulas and names of four common strong bases?

9.37 Write a balanced equation for the ionization of each of the following in water:

a. HI, a strong acid **b.** $Ca(OH)_2$
c. KOH **d.** HNO_2, a weak acid

9.38 Write a balanced equation for the ionization of each of the following in water:

a. HCN, a weak acid **b.** $Mg(OH)_2$
c. NH_3, a weak base **d.** HNO_3, a strong acid

9.5 Making Dilutions

Learning Goal

Describe the preparation of a dilute solution of an acid or base given the molarity or percent concentration of the solution.

In chemistry, we often need to prepare a dilute solution from a more concentrated solution. In a process called **dilution,** we add water to a solution to make a larger volume. For example, you might prepare some orange juice by adding three cans of water to the original can of concentrated orange juice. (See Figure 9.10.)

When more water is added, the solution volume increases, causing a decrease in the concentration. However, *the amount of solute does not change.* Therefore the initial amount of solute before dilution is equal to the final amount of solute in the diluted solution. (See Figure 9.11.) By diluting the solution, we decrease the concentration.

Amount of solute in the concentrated solution = amount of solute in the diluted solution

Figure 9.10 Dilution of orange juice at breakfast time.

9.5

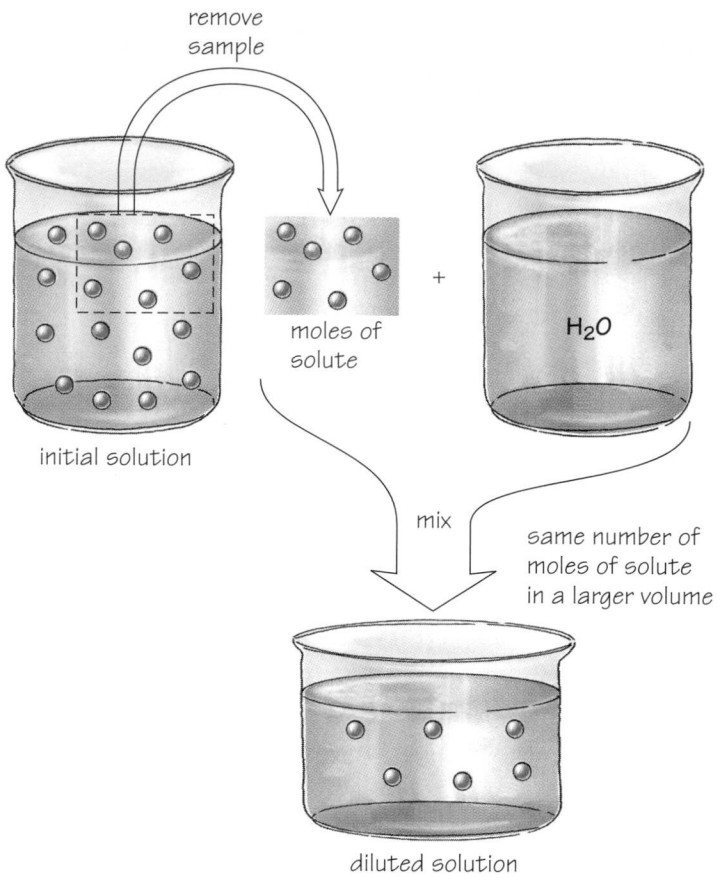

Figure 9.11 The number of moles of solute in the sample is equal to the number of moles of solute in the diluted solution.

When we need to calculate the new concentration of a diluted solution, we first determine the amount of the solute in moles (or grams).

$$\text{Moles of solute} = \text{original volume (L)} \times \text{molarity} \left(\frac{\text{moles}}{\text{liter}}\right)$$

or

$$\text{Grams of solute} = \text{original volume (mL)} \times \%\ \text{(m/v)} \left(\frac{\text{grams}}{100\ \text{mL}}\right)$$

The diluted concentration is calculated by placing the amount of solute and the new, diluted volume in the concentration expression.

$$\text{Molarity of diluted solution} = \frac{\text{moles of solute}}{\text{diluted volume (L)}}$$

or

$$\text{Percent (m/v) of diluted solution} = \frac{\text{grams of solute}}{\text{diluted volume (mL)}} \times 100$$

S·A·M·P·L·E· **9.10** ·P·R·O·B·L·E·M

Diluting a Base

You are going to prepare a diluted KOH solution by placing 75 mL of a 20.0% (m/v) KOH solution in a volumetric flask and adding water to make 500. mL of diluted KOH solution. (See Figure 9.12.) What is the final concentration of the diluted solution?

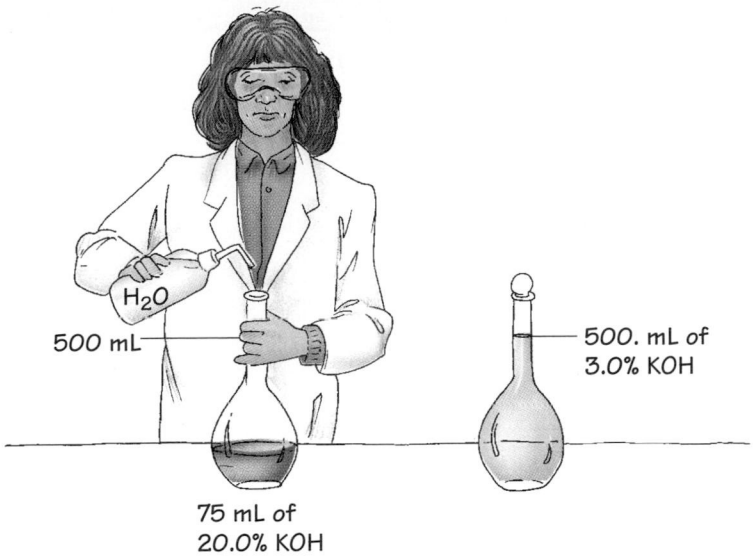

Figure 9.12 Water is added to 75 mL of a 20.0% (m/v) KOH solution to give a final volume of 500. mL. The final concentration of the diluted solution is 3.0% (m/v) KOH.

Solution

$$\text{Grams of KOH} = 75\ \cancel{\text{mL}} \times \frac{20.0\ \text{g KOH}}{100\ \cancel{\text{mL}}} = 15\ \text{g KOH}$$

$$\text{New volume} = 500.\ \text{mL}$$

$$\%\ (\text{m/v})\ \text{of diluted solution} = \frac{15\ \text{g KOH}}{500.\ \text{mL}} \times 100 = 3.0\%\ (\text{m/v})\ \text{KOH}$$

Study Check

A 25-mL sample of a 10.0% (m/v) NaOH solution is diluted with water to make a final volume of 100.0 mL. What is the concentration of the diluted NaOH solution?

S·A·M·P·L·E 9.11 ·P·R·O·B·L·E·M

Diluting an Acid Solution

When diluting acids, the acid is *always* added to the water needed for the dilution. If water is added to a concentrated acid, a great amount of heat is generated and acid may splatter out of the container. Calculate the molarity of a diluted HCl solution made by adding 25 mL of 12 M HCl to water making 250. mL of a diluted solution.

Solution

$$\text{Moles of HCl} = 0.025\ \cancel{\text{L}} \times \frac{12\ \text{moles HCl}}{1\ \cancel{\text{L}}} = 0.30\ \text{mole HCl}$$

$$\text{Diluted volume} = 25\ \text{mL (original)} + 225\ \text{mL of water}$$
$$= 250.\ \text{mL} = 0.250\ \text{L}$$

$$\text{Molarity of diluted solution} = \frac{0.30\ \text{mole HCl}}{0.250\ \text{L}} = 1.2\ \text{M}$$

Study Check

What is the molarity of the diluted solution after 75 mL of a 6.0 M HNO_3 solution is diluted to make a final volume of 150 mL?

9.5

S·A·M·P·L·E 9.12 P·R·O·B·L·E·M

Finding the Final Volume

What is the volume (mL) of a 0.20 M HCl that can be prepared from diluting 50.0 mL of a 1.0 M HCl solution?

Solution

Because the moles of solutes are unchanged in the dilution, we can set the initial conditions equal to the final conditions of

$$\text{Moles solute (concentrated)} = \text{moles solute (diluted)}$$

$$M_c V_c = M_d V_d$$

Solving for the unknown diluted volume (V_d) we obtain

$$V_d = \frac{M_c V_c}{M_d}$$

$$= \frac{(1.0\,\cancel{M})\,(0.050\text{ L})}{(0.20\,\cancel{M})}$$

$$= 0.25\text{ L} = 250\text{ mL (diluted volume)}$$

Study Check

You are going to prepare a 0.25 M NaOH by diluting 1.0 L of a 1.0 M NaOH solution. What is the final volume (L) of the diluted solution?

Q·U·E·S·T·I·O·N·S & P·R·O·B·L·E·M·S

Making Dilutions

9.39 After you open a can of condensed tomato soup, you add one can of water. Why is this a dilution?

9.40 A can of frozen lemonade calls for the addition of 3 cans of water to make a pitcher of the beverage. Why is this a dilution?

9.41 Calculate the final concentration of each of the following diluted solutions:

a. 2.0 L of a 6.0 M HCl solution is added to water so that the final volume is 6.0 L.

b. Water is added to 0.50 L of a 12 M NaOH solution to make 3.0 L of a diluted NaOH solution.

c. A 10.0-mL sample of 25% (m/v) KOH solution is diluted with water so that the final volume is 100.0 mL.

d. A 50.0-mL sample of 15% (m/v) H_2SO_4 solution is added to water to give a final volume of 250 mL.

9.42 Calculate the final concentration of each of the following diluted solutions:

a. 1.0 L of a 4.0 M HNO_3 solution is added to water so that the final volume is 8.0 L.

b. Water is added to 0.25 L of a 6.0 M KOH solution to make 2.0 L of a diluted KOH solution.

c. A 50.0-mL sample of an 8.0% (m/v) NaOH is diluted with water so that the final volume is 200.0 mL.

d. A 5.0-mL sample of 50.0% (m/v) acetic acid ($HC_2H_3O_2$) solution is added to water to give a final volume of 25 mL.

9.43 What is the final volume in liters needed to prepare each of the following diluted solutions?

a. a 0.20 M HCl solution prepared from 20.0 mL of a 6.0 M HCl solution

b. a 2.0% (m/v) NaOH solution prepared from 50.0 mL of a 10.0% (m/v) NaOH solution

c. a 0.50 M H_3PO_4 solution prepared from 0.500 L of a 6.0 M H_3PO_4 solution

9.44 What is the final volume (L) needed to prepare each of the following diluted solutions?

a. a 1.0 % (m/v) KOH solution prepared from 10.0 mL of a 20.0% KOH solution

b. a 0.10 M HCl solution prepared from 25 mL of a 6.0 M HCl solution

c. a 1.0 M NaOH solution prepared from 50.0 mL of a 12 M NaOH solution

9.6 Acid–Base Neutralization

Learning Goal

Write a balanced equation for the neutralization reaction between an acid and a base.

In a reaction called **neutralization**, acid and base form water and salt. A **salt** is an ionic compound that contains a metal ion and a nonmetal or polyatomic ion. For example, the neutralization reaction of HCl and NaOH is written as

$$\underset{\text{Acid}}{\mathbf{HCl}} + \underset{\text{Base}}{\mathrm{Na}\mathbf{OH}} \longrightarrow \underset{\text{Salt}}{\mathrm{NaCl}} + \underset{\text{Water}}{\mathbf{H_2O}}$$

To understand the process that occurs in neutralization, we need to write the acid, base, and salt in their ionic forms.

Acid	Base	React to Give	Salt	Water
$\mathbf{H^+} + \mathrm{Cl^-}$	$\mathrm{Na^+} + \mathbf{OH^-}$	$\longrightarrow$	$\mathrm{Na^+} + \mathrm{Cl^-}$	$\mathbf{H_2O}$

In solution, HCl exists as H^+ (or H_3O^+) and Cl^-, and NaOH exists as Na^+ and OH^-. In the neutralization, the H^+ reacts with OH^- to form water, leaving Na^+ and Cl^- ions in solution. By omitting the unreacted metal and nonmetal ions from the overall equation, we write a net ionic equation for any acid–base *neutralization* as

$$\underset{\text{From an acid}}{\mathbf{H^+}} + \underset{\text{From a base}}{\mathbf{OH^-}} \longrightarrow \underset{\text{Water forms}}{\mathbf{H_2O}}$$

This reaction occurs when a solution of any acid and a solution of a base, strong or weak, are neutralized.

Balancing an Acid–Base Neutralization Equation

In another neutralization reaction, sulfuric acid, H_2SO_4, reacts with sodium hydroxide, NaOH. We write the unbalanced equation as

$$\underset{\text{Acid}}{\mathrm{H_2SO_4}} + \underset{\text{Base}}{\mathrm{NaOH}} \longrightarrow \underset{\text{Salt}}{\text{salt}} + \underset{\text{Water}}{\mathrm{H_2O}}$$

In any neutralization reaction, the amounts of H^+ and the OH^- must be equal. Because the acid H_2SO_4 donates two H^+ ions, it requires two OH^- ions from the base. Place a coefficient of 2 in front of NaOH. The neutralization of two H^+ ions and two OH^- ions forms two molecules of H_2O.

$$\mathrm{H_2SO_4} + 2\mathrm{NaOH} \longrightarrow \text{salt} + 2\mathrm{H_2O}$$

Two Na^+ ions from the base and SO_4^{2-} ion from the acid give the formula of the salt, Na_2SO_4.

$$\underset{\text{Acid}}{H_2SO_4} + \underset{\text{Base}}{2NaOH} \longrightarrow \underset{\text{Salt}}{Na_2SO_4} + \underset{\text{Water}}{2H_2O}$$

Here are more examples of balanced neutralization reactions between acids and bases.

Some Acid–Base Neutralization Reactions

Acid	Base		Salt	Water
HNO_3	+ KOH	$\longrightarrow$	KNO_3	+ H_2O
2HCl	+ $Ca(OH)_2$	$\longrightarrow$	$CaCl_2$	+ $2H_2O$
H_3PO_4	+ 3NaOH	$\longrightarrow$	Na_3PO_4	+ $3H_2O$

9.6

S·A·M·P·L·E· 9.13 ·P·R·O·B·L·E·M

Writing a Neutralization Equation

Write the balanced equation for the neutralization of HCl by $Ba(OH)_2$.

Solution

In a neutralization of HCl (acid) and $Ba(OH)_2$ (base), the products are a salt and water. The unbalanced reaction is

$$HCl + Ba(OH)_2 \longrightarrow \text{salt} + H_2O$$

A 2 is placed in front of the HCl to provide two H^+ ions to neutralize the two OH^- ions from the base. The product of the neutralization is $2H_2O$.

$$2HCl + Ba(OH)_2 \longrightarrow \text{salt} + 2H_2O$$

The ions that remain in solution are the Ba^{2+} ion from the base and two Cl^- ions from the acid. The formula of the salt is $BaCl_2$.

$$2HCl + Ba(OH)_2 \longrightarrow BaCl_2 + 2H_2O$$

Study Check

Write a balanced equation for the neutralization of H_3PO_4 with KOH.

EXPLORE YOUR WORLD

Everyday Neutralization Reactions

Baking soda, $NaHCO_3$, can also be used as an antacid. When $NaHCO_3$ reacts with acetic acid in vinegar, the products are $CO_2(g)$, H_2O, and $NaC_2H_3O_2$ (salt).

$$NaHCO_3 + HC_2H_3O_2 \longrightarrow CO_2(g) + H_2O + NaC_2H_3O_2$$

For this exploration, you will need: baking soda, a glass, vinegar, and a pH indicator such as red cabbage juice (if available) or cranberry juice.

Place 1 tablespoon of baking soda in the glass. Add some pH indicator to give a color. Add one teaspoon of vinegar and observe the reaction. Continue to add more vinegar, one teaspoon at a time, and observe the reaction and color changes.

Try the neutralization using an antacid instead of baking soda. From the package label, identify the base(s) in the antacid product.

Questions

1. How does the pH change as you add more acid to the baking soda (base) or antacid (base)?
2. How do you know the base (baking soda or antacid) is being neutralized?
3. When do you know that the base has been neutralized?
4. What are the formulas of some bases that are typically used to prepare antacids?
5. What is the purpose of a base in an antacid?

HEALTH NOTE

Antacids

Antacids are substances used to neutralize excess stomach acid (HCl). Some antacids are mixtures of aluminum hydroxide and magnesium hydroxide. These hydroxides are not very soluble in water, so the levels of available OH^- are not damaging to the intestinal tract. However, aluminum hydroxide has the side effects of producing constipation and binding phosphate in the intestinal tract, which may cause weakness and loss of appetite. Magnesium hydroxide has a laxative effect. These side effects are less likely when a combination of the antacids is used.

$$Al(OH)_3 + 3HCl \longrightarrow AlCl_3 + 3H_2O$$

$$Mg(OH)_2 + 2HCl \longrightarrow MgCl_2 + 2H_2O$$

Some antacids use calcium carbonate to neutralize excess stomach acid. About 10% of the calcium is absorbed into the bloodstream, where it elevates the levels of serum calcium. Calcium carbonate is not recommended for patients who have peptic ulcers or a tendency to form kidney stones.

$$CaCO_3 + 2HCl \longrightarrow H_2O + CO_2(g) + CaCl_2$$

Still other antacids contain sodium bicarbonate. This type of antacid has a tendency to increase blood pH and elevate sodium levels in the body fluids. It also is not recommended in the treatment of peptic ulcers.

$$NaHCO_3 + HCl \longrightarrow NaCl + CO_2(g) + H_2O$$

The neutralizing substances in some antacid preparations are given in Table 9.6.

Table 9.6 Basic Compounds in Some Antacids

Antacid	Base(s)
Amphojel	$Al(OH)_3$
Milk of magnesia	$Mg(OH)_2$
Mylanta, Maalox, Di-Gel, Gelusil, Riopan	$Mg(OH)_2$, $Al(OH)_3$
Bisodol	$CaCO_3$, $Mg(OH)_2$
Titralac, Tums, Pepto-Bismol	$CaCO_3$
Alka-Seltzer	$NaHCO_3$, $KHCO_3$

QUESTIONS & PROBLEMS

Acid–Base Neutralization

9.45 Which of the following are neutralization reactions?

a. $SO_3 + H_2O \longrightarrow H_2SO_4$

b. $HNO_3 + KOH \longrightarrow KNO_3 + H_2O$

c. $2Mg + O_2 \longrightarrow 2MgO$

d. $H_3PO_4 + 3NaOH \longrightarrow Na_3PO_4 + 3H_2O$

9.46 Which of the following are neutralization reactions?

a. $2HCl + Ca(OH)_2 \longrightarrow CaCl_2 + 2H_2O$

b. $H_2SO_4 + Pb(NO_3)_2 \longrightarrow PbSO_4 + 2HNO_3$

c. $Mg + 2HCl \longrightarrow MgCl_2 + H_2$

d. $HClO_2 + NaOH \longrightarrow NaClO_2 + H_2O$

9.47 Balance each of the following neutralization reactions:

a. $HCl + Mg(OH)_2 \longrightarrow MgCl_2 + H_2O$

b. $H_3PO_4 + LiOH \longrightarrow Li_3PO_4 + H_2O$

9.48 Balance each of the following neutralization reactions:

a. $HNO_3 + Ba(OH)_2 \longrightarrow Ba(NO_3)_2 + H_2O$

b. $H_2SO_4 + Al(OH)_3 \longrightarrow Al_2(SO_4)_3 + H_2O$

9.49 Write a balanced equation for the neutralization of each of the following:

a. H_2SO_4 and NaOH **b.** HCl and $Fe(OH)_3$

c. H_2CO_3 and $Mg(OH)_2$

9.50 Write a balanced equation for the neutralization of each of the following:

a. H_3PO_4 and NaOH **b.** HI and LiOH

c. HNO_3 and $Ca(OH)_2$

9.7

Study Pack

Chem of Life

Acids, Bases, and pH: Buffers 1-4

Learning Goal

Describe the role of buffers in maintaining the pH of a solution.

9.7 Buffers

When a small amount of acid or base is added to pure water, the pH changes drastically. However, if a solution is buffered, there is little change in pH. A **buffer solution** resists a change in pH when small amounts of acid or base are added. For example, blood is a buffer solution in the body that maintains a pH of about 7.4. If the pH of the blood goes even slightly above or below this value, changes in our uptake of oxygen and our metabolic process can be drastic enough to cause death. Even though we are constantly obtaining acids and bases from foods and biological processes, the buffers in the body so effectively absorb those compounds that blood pH remains unchanged. (See Figure 9.13.)

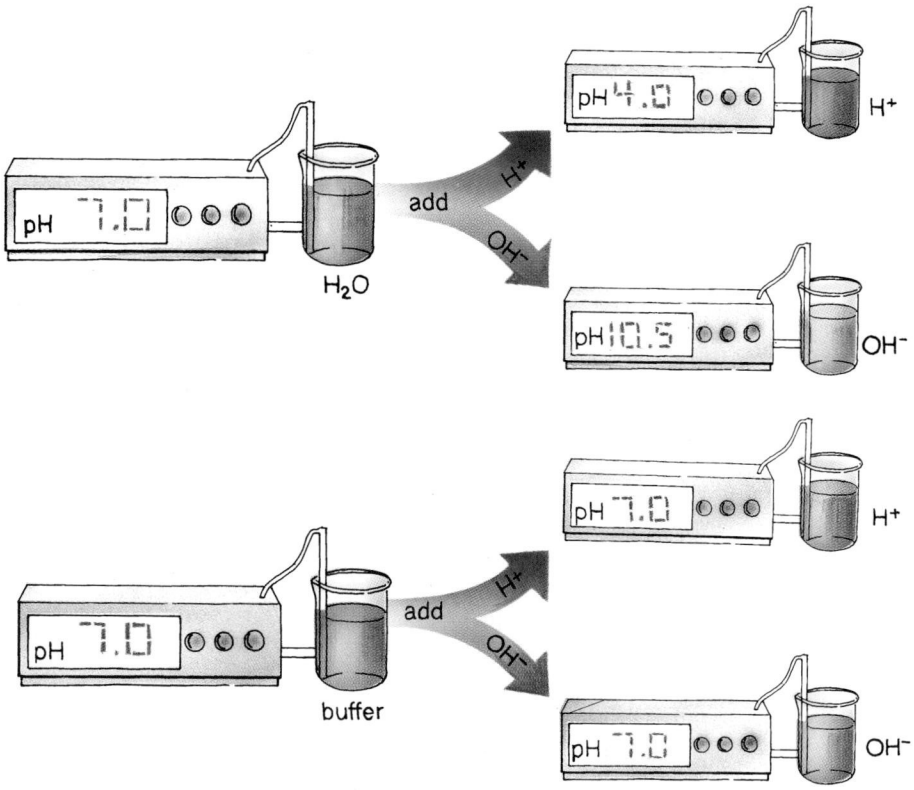

Figure 9.13 Adding an acid or a base to water changes the pH drastically, but a buffer resists pH change when an acid or base is added.

A buffer solution must contain two substances, a weak acid and a salt of that acid or a weak base and a salt of that base. For example, an important buffer in the blood consists of carbonic acid, a weak acid, and its salt, sodium bicarbonate, $NaHCO_3$. The buffering action is the result of the weak acid (or base) in equilibrium with its ions in the solution.

$$\underset{\text{Carbonic acid}}{H_2CO_3} + H_2O \rightleftharpoons H_3O^+ + \underset{\text{Bicarbonate ion}}{HCO_3^-}$$

If we add a salt of the weak acid, $NaHCO_3$, it will dissociate and increase the amount of bicarbonate ion, HCO_3^- in the solution.

$$\underset{\substack{\text{Sodium bicarbonate}\\ \text{(salt of the weak acid)}}}{NaHCO_3} \longrightarrow Na^+ + \underset{\text{More bicarbonate ion}}{HCO_3^-}$$

Although the weak acid itself does not produce very many bicarbonate ions, the HCO_3^- concentration is large because of the added salt. Thus, this buffer has high levels of both carbonic acid and bicarbonate ion.

Now let us see how the buffer works to resist pH changes when some H_3O^+ or OH^- is added. Suppose that some OH^- enters the blood. The weak acid available in the buffer reacts with this OH^- and neutralizes it:

$$\underset{\substack{\text{Carbonic}\\ \text{acid}}}{H_2CO_3} + \underset{\text{Hydroxide ion}}{OH^-} \longrightarrow H_2O + \underset{\text{Bicarbonate ion}}{HCO_3^-}$$

When H_3O^+ enters the buffer system, it reacts with the bicarbonate ion to form carbonic acid.

$$\underset{\substack{\text{Bicarbonate}\\ \text{ion}}}{HCO_3^-} + H_3O^+ \longrightarrow \underset{\text{Carbonic acid}}{H_2CO_3} + H_2O$$

In summary, when OH^- ions are added to a buffer solution, they react with the weak acid of the buffer. On the other hand, if H_3O^+ ions enter the buffer solution, they are tied up by the negative ions. Both components are necessary for a buffer solution to maintain pH. The equilibrium between these buffering processes is shown in Figure 9.14.

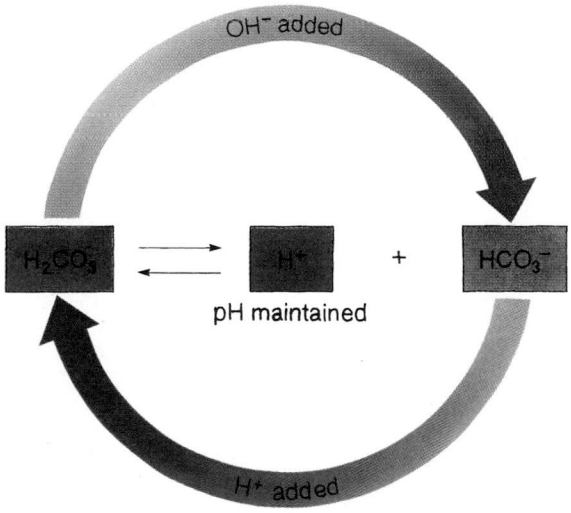

Figure 9.14 The equilibrium between carbonic acid, a weak acid, and the bicarbonate ion in the blood buffer system.

S·A·M·P·L·E· 9.14 ·P·R·O·B·L·E·M

Identifying Buffer Solutions

Indicate whether each of the following would make a buffer solution:

a. HCl, a strong acid, and NaCl
b. H_3PO_4, a weak acid
c. HF, a weak acid, and NaF

Solution

a. No. A solution of a strong acid and its salt is completely ionized.
b. No. A weak acid is not sufficient for a buffer; the salt of the weak acid is also needed.
c. Yes. This mixture contains a weak acid and its salt.

Study Check

Will a mixture of NaCl and Na_2CO_3 make a buffer solution? Explain.

9.7

HEALTH NOTE

CO_2 and the Bicarbonate Buffer

In our cells, carbon dioxide is continually produced as an end product of cellular metabolism. Some CO_2 is carried to the lungs for elimination, and the rest dissolves in body fluids, forming carbonic acid and bicarbonate ion.

Bicarbonate Buffer System

$$CO_2 + H_2O \rightleftharpoons H_2CO_3 \rightleftharpoons H^+ + HCO_3^-$$

Carbon dioxide — Carbonic acid — Bicarbonate ion

The partial pressure of CO_2 gas, P_{CO_2}, determines the concentration of carbonic acid, which in turn affects the pH of the blood. Table 9.7 lists the normal values for the components of this buffer in arterial blood.

A higher level of CO_2 produces more H_2CO_3 in the blood, which produces more H^+, and lowers the pH. This condition is called **acidosis.** When the origin of acidosis is respiratory, it is called respiratory acidosis. Poor respiration or below-normal gas exchange (hypoventilation) causes an increase in the CO_2 in the blood. This can happen in emphysema when CO_2 does not diffuse properly through the lung membrane; it can also happen when the respiratory center in the medulla of the brain is affected by an accident or by depressive drugs. Acidosis caused by conditions that are not respiratory is called metabolic acidosis.

A lowering of the CO_2 level leads to a higher blood pH, a condition called **alkalosis.** Excitement, trauma, or a high temperature may cause a person to breathe rapidly (hyperventilation). If sufficient CO_2 is expelled, its partial pressure in the blood falls below normal. The blood buffer restores the CO_2 level by converting some H_2CO_3 to CO_2 and H_2O. The decrease in H_2CO_3 causes some H^+ to combine with HCO_3^-. The H^+ drops and blood pH rises, a condition called respiratory alkalosis.

Although the amount of CO_2 in the blood depends on the respiratory processes, the kidneys control the amount of H^+ excreted in the urine. The pH of urine varies considerably and can range from 4.5 to 8.2. In severe conditions, even these mechanisms fail, and life-threatening changes in blood pH occur. Table 9.8 lists some of the conditions that lead to changes in the blood pH, and some possible treatments.

Table 9.7 Normal Values for Blood Buffer in Arterial Blood

P_{CO_2}	40 mm Hg
H_2CO_3	1.2 mEq/L of plasma
HCO_3^-	24 mEq/L of plasma
pH	7.35–7.45

Table 9.8 Acidosis and Alkalosis: Symptoms, Causes, and Treatments

Respiratory Acidosis: CO_2 ↑ pH ↓	
Symptoms:	Failure to ventilate, suppression of breathing, disorientation, weakness, coma
Causes:	Lung disease blocking gas diffusion (e.g., emphysema, pneumonia, bronchitis, and asthma); depression of respiratory center by drugs, cardiopulmonary arrest, stroke, poliomyelitis, or nervous system disorders
Treatment:	Correction of disorder, infusion of bicarbonate
Respiratory Alkalosis: CO_2 ↓ pH ↑	
Symptoms:	Increased rate and depth of breathing, numbness, light-headedness, tetany
Causes:	Hyperventilation due to anxiety, hysteria, fever, exercise; reaction to drugs such as salicylate, quinine, and antihistamines; conditions causing hypoxia (e.g., pneumonia, pulmonary edema, and heart disease)
Treatment:	Elimination of anxiety-producing state, rebreathing into a paper bag
Metabolic Acidosis: H^+ ↑ pH ↓	
Symptoms:	Increased ventilation, fatigue, confusion
Causes:	Renal disease, including hepatitis and cirrhosis; increased acid production in diabetes mellitus, hyperthyroidism, alcoholism, and starvation; loss of alkali in diarrhea; acid retention in renal failure
Treatment:	Sodium bicarbonate given orally, dialysis for renal failure, insulin treatment for diabetic ketosis
Metabolic Alkalosis: H^+ ↓ pH ↑	
Symptoms:	Depressed breathing, apathy, confusion
Causes:	Vomiting, diseases of the adrenal glands, ingestion of excess alkali
Treatment:	Infusion of saline solution, treatment of underlying diseases

QUESTIONS & PROBLEMS

Buffers

9.51 Which of the following represent a buffer system? Explain.

a. NaOH and NaCl **b.** H_2CO_3 and $NaHCO_3$
c. HF and KF **d.** KCl and NaCl

9.52 Which of the following represent a buffer system? Explain.

a. H_3PO_4 **b.** $NaNO_3$
c. $HC_2H_3O_2$ and $NaC_2H_3O_2$ **d.** HCl and NaOH

9.53 Consider the buffer system of hydrofluoric acid, HF, and its salt, NaF.

$$HF \rightleftharpoons H^+ + F^-$$

a. What is the purpose of the buffer system?
b. Why is a salt of the acid needed?
c. How does the buffer react when some H^+ is added?
d. How does the buffer react when some OH^- is added?

9.54 Consider the buffer system of nitrous acid, HNO_2, and its salt, $NaNO_2$.

$$HNO_2 \rightleftharpoons H^+ + NO_2^-$$

a. What is the purpose of a buffer system?
b. What is the purpose of $NaNO_2$ in the buffer?
c. How does the buffer react when some H^+ is added?
d. How does the buffer react when some OH^- is added?

9.55 Chronic lung diseases such as emphysema and asthma or injury to the brain respiratory center can cause hypoventilation and an elevation in the CO_2 level of the blood.

a. Would the blood become more acidic or basic?

b. Would you expect to see a blood pH above or below 7.45?

c. Why would a sodium bicarbonate solution be given intravenously?

9.56 A 22-year-old female is admitted with a history of vomiting and diarrhea. Her blood pH is 7.48 and HCO_3^- is 33 mEq/L.

a. Is the condition acidosis or alkalosis?

b. Is the condition metabolic or respiratory?

c. How will treatment to stop vomiting and diarrhea return the blood pH to normal?

9.8

Study Pack

9.8 A Closer Look: Acid–Base Titration

Learning Goal

Calculate the molarity or volume of an acid or base from titration information.

Suppose we need to find the molarity of an HCl solution of unknown concentration. We can do this by a laboratory procedure called **titration** in which we neutralize an acid sample with a known amount of base. In our titration, we first place a measured volume of the acid in a flask, and add a few drops of an **indicator** such as phenolphthalein. In an acidic solution, phenolpthalein is colorless. Then we fill a buret with a NaOH solution of known molarity and carefully add NaOH to the acid in the flask, as shown in Figure 9.15.

In the titration, we neutralize the acid by adding a volume of base that contains a matching number of moles of OH^-. We know that neutralization has taken place when the phenolphthalein in the solution changes from colorless to pink. This is called the neutralization endpoint. From the volume and molarity of the NaOH, we can calculate the number of moles of NaOH and then the concentration of the acid.

S·A·M·P·L·E· 9.15 ·P·R·O·B·L·E·M

Titration of an Acid

A 10.0-mL sample of HCl is titrated with 1.0 M NaOH. The equation for the reaction is

$$NaOH + HCl \longrightarrow NaCl + H_2O$$

If 20.0 mL of base were required to neutralize the HCl, what is the molarity of the acid?

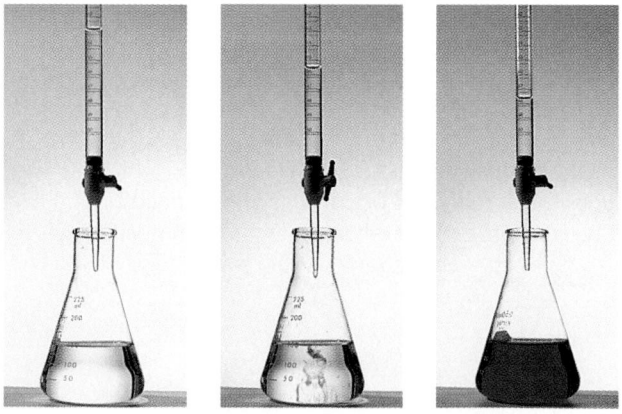

Figure 9.15 The titration of an acid. A known volume of an acid is placed in a flask with an indicator and titrated with a measured volume of NaOH to the neutralization point.

Solution

Step 1 **Finding the Reacting Moles of Base**
From the volume and the molarity of the NaOH, calculate the moles of NaOH used in the titration.

$$20.0\,\cancel{\text{mL}} \times \frac{1\,\cancel{\text{L}}}{1000\,\cancel{\text{mL}}} \times \underbrace{\frac{1.0 \text{ mole NaOH}}{1\,\cancel{\text{L}}}}_{\text{Molarity of NaOH}} = 0.020 \text{ mole NaOH}$$

Step 2 **Finding the Reacting Moles of Acid**
In the neutralization equation, 1 mole of NaOH reacts with 1 mole of HCl. The moles of HCl are calculated as follows:

$$0.020\,\cancel{\text{mole NaOH}} \times \frac{1 \text{ mole HCl}}{1\,\cancel{\text{mole NaOH}}} = 0.020 \text{ mole HCl}$$

Step 3 **Calculating the Molarity of the Acid**
To calculate the molarity of the acid, place the moles of acid and its initial volume in the expression for molarity.

$$\text{Molarity (M) of acid} = \frac{\text{moles of acid}}{\text{volume of acid}} = \frac{0.020 \text{ mole HCl}}{\underbrace{0.0100 \text{ L}}_{\text{10.0-mL HCl sample expressed in liters}}} = 2.0 \text{ M HCl}$$

Study Check

In a titration, a 25.0-mL HCl sample required 50.0 mL of a 0.100 M NaOH solution for neutralization. What is the molarity of the HCl sample?

S·A·M·P·L·E· 9.16 ·P·R·O·B·L·E·M

Titration of an Acid

A 5.00-mL sample of H_2SO_4 requires 40.0 mL of 2.0 M NaOH for complete neutralization. What is the molarity of the acid? The equation for the neutralization is

$$H_2SO_4 + 2NaOH \longrightarrow Na_2SO_4 + 2H_2O$$

Solution

Using the steps for titration calculations we find the following:

Step 1 From the volume and the molarity of the NaOH, calculate the moles of NaOH.

$$40.0\,\cancel{\text{mL}} \times \frac{1\,\cancel{\text{L}}}{1000\,\cancel{\text{mL}}} \times \underbrace{\frac{2.0 \text{ moles NaOH}}{1\,\cancel{\text{L}}}}_{\text{Molarity of NaOH}} = 0.080 \text{ mole NaOH}$$

Step 2 From the neutralization equation, we know that 2 moles of NaOH react with 1 mole of H_2SO_4. The moles of H_2SO_4 in the sample are calculated as follows:

$$0.080\,\cancel{\text{mole NaOH}} \times \frac{1 \text{ mole } H_2SO_4}{2\,\cancel{\text{moles NaOH}}} = 0.040 \text{ mole } H_2SO_4$$

Step 3 To calculate the molarity of the acid, place the number of moles of acid and its initial volume in the expression for molarity.

$$\text{Molarity (M) of acid} = \frac{\text{moles of acid}}{\text{volume of acid}} = \frac{0.040 \text{ mole } H_2SO_4}{\underset{\substack{\text{5.0-mL } H_2SO_4 \text{ sample} \\ \text{expressed in liters}}}{0.0050 \text{ L}}} = 8.0 \text{ M } H_2SO_4$$

Study Check

What is the volume (mL) of 0.100 M NaOH needed to completely neutralize 5.00 mL of 0.200 M H_3PO_4? (*Hint:* Write the balanced equation for neutralization.)

9.8

QUESTIONS & PROBLEMS

A Closer Look: Acid–Base Titration

9.57 If you need to determine the concentration of a solution of formic acid, $HCHO_2$, how would you proceed?

9.58 If you need to determine the concentration of a sample of acetic acid, $HC_2H_3O_2$, how would you proceed?

9.59 What is the molarity of the acid in each of the following titrations?

a. a 5.0-mL sample of HCl solution that is titrated with 22.0 mL of 2.0 M NaOH solution

$$HCl + NaOH \longrightarrow NaCl + H_2O$$

b. a 10.0-mL sample of sulfuric acid solution that is titrated with 15.0 mL of 1.0 M NaOH solution

$$H_2SO_4 + 2NaOH \longrightarrow Na_2SO_4 + 2H_2O$$

c. a 10.0-mL sample of phosphoric acid solution that is titrated with 16.0 mL of 1.0 M NaOH solution

$$H_3PO_4 + 3NaOH \longrightarrow Na_3PO_4 + 3H_2O$$

9.60 What is the molarity of the acid in each of the following titrations?

a. a 20.0-mL sample of HCl solution that is titrated with 8.0 mL of 6.0 M NaOH solution

$$HCl + NaOH \longrightarrow NaCl + H_2O$$

b. a 10.0-mL sample of sulfuric acid solution that is titrated with 25.0 mL of 0.50 M NaOH solution

$$H_2SO_4 + 2NaOH \longrightarrow Na_2SO_4 + 2H_2O$$

c. a 30.0-mL sample of phosphoric acid solution that is titrated with 18.0 mL of 3.0 M NaOH solution

$$H_3PO_4 + 3NaOH \longrightarrow Na_3PO_4 + 3H_2O$$

Chapter Review

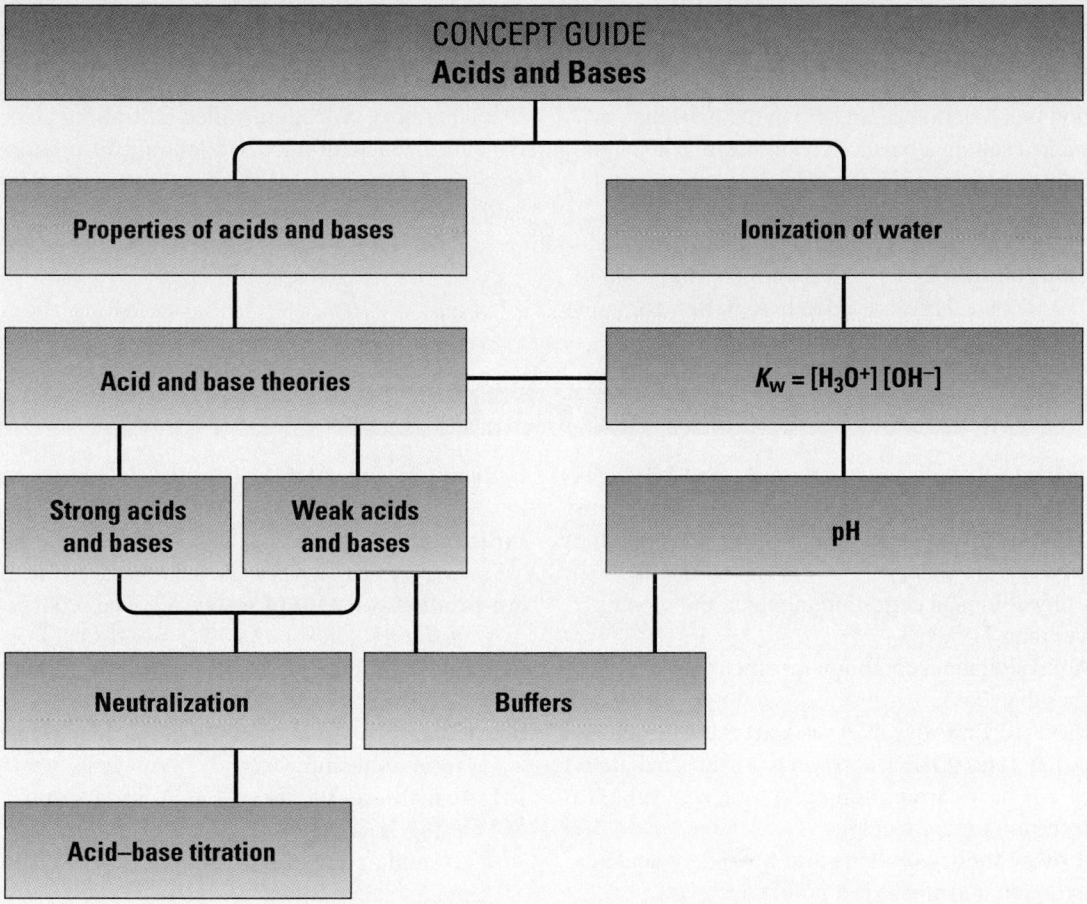

Chapter Summary

9.1 Acids and Bases

In water, an Arrhenius acid produces H^+, and an Arrhenius base produces OH^-. According to the Brønsted–Lowry theory, acids are proton (H^+) donors, and bases are proton acceptors. Protons form hydronium ions, H_3O^+, in water because they bond to polar water molecules.

9.2 Ionization of Water

In pure water, a few molecules transfer protons to other water molecules, producing small, but equal, amounts of $[H_3O^+]$ and $[OH^-]$ such that each has a concentration of 1×10^{-7} mole/L. The ion product, K_w, $[H_3O^+][OH^-] = 1 \times 10^{-14}$, applies to all aqueous solutions. In acidic solutions, the $[H_3O^+]$ is greater than the $[OH^-]$. In basic solutions, the $[OH^-]$ is greater than the $[H_3O^+]$.

9.3 The pH Scale

The pH scale is a range of numbers from 0 to 14 related to the $[H_3O^+]$ of the solution. A neutral solution has a pH of 7. In acidic solutions, the pH is below 7, and in basic solutions the pH is above 7. Mathematically, pH is the negative logarithm of the hydronium ion concentration ($-\log[H_3O^+]$).

9.4 Strengths of Acids and Bases

In strong acids, all the H^+ in the acid is donated to H_2O; in a weak acid, only a small percentage of acid molecules produce H_3O^+. Strong bases are hydroxides of Groups 1 and 2 that dissociate completely in water. An important weak base is ammonia, NH_3.

9.5 Making Dilutions

When water is added to a solution, the volume increases. The solute is now distributed throughout a larger volume, which dilutes the solution and decreases the concentration.

9.6 Acid–Base Neutralization

Neutralization is a reaction of an acid and a base that produces water and a salt. In a balanced neutralization equation, an equal number of moles of H^+ and OH^- must be reacted.

9.7 Buffers

A buffer solution resists changes in pH when small amounts of acid or a base are added. A buffer contains either a weak acid and its salt or a weak base and its salt. The weak acid picks up added OH^-, and the anion of the salt picks up added H^+. Buffers are important in maintaining the pH of the blood.

9.8 A Closer Look: Acid-Base Titration

In a laboratory procedure called a titration, an acid sample is neutralized with a known amount of a base. From the volume and molarity of the base, the concentration of the acid is calculated.

Key Terms

acid A substance that dissolves in water and produces hydrogen ions (H^+), according to the Arrhenius theory. All acids are proton donors, according to the Brønsted–Lowry theory.

acidosis A physiological condition in which the blood pH is lower than 7.35.

alkalosis A physiological condition in which the blood pH is higher than 7.45.

base A substance that dissolves in water and produces hydroxide ions (OH^-) according to the Arrhenius theory. All bases are proton acceptors, according to the Brønsted–Lowry theory.

Brønsted–Lowry theory of acids and bases An acid is a proton donor, and a base is a proton acceptor.

buffer solution A mixture of a weak acid or a weak base and its salt that resists changes in pH when small amounts of an acid or base are added.

dilution A process by which water (solvent) is added to a solution to increase the volume and decrease (dilute) the concentration of the solute.

hydronium ion, H_3O^+ The H_3O^+ ion formed by the attraction of a proton (H^+) to an H_2O molecule.

indicator A substance added to a titration sample that changes color when the acid or base is neutralized.

ion-product constant of water, K_w The product of $[H_3O^+]$ and $[OH^-]$ in solution; $K_w = [H_3O^+][OH^-]$.

neutral A solution with equal concentrations of $[H_3O^+]$ and $[OH^-]$.

neutralization A reaction between an acid and a base to form a salt and water.

pH A measure of the $[H_3O^+]$ in a solution; $pH = -\log[H_3O^+]$.

salt An ionic compound that contains a metal ion and a nonmetal or polyatomic ion.

strong acid An acid that completely ionizes in water.

strong base A base that completely ionizes in water.

titration The addition of base to an acid sample to determine the concentration of the acid.

weak acid An acid that ionizes only slightly in solution.

weak base A base that ionizes only slightly in solution.

Chapter Practice

9.61 Name each of the following and identify each as an Arrhenius acid or base.

a. H_2SO_4 **b.** KOH **c.** $Ca(OH)_2$
d. HCl **e.** HNO_2

9.62 For the reactants in each of the following, identify the Brønsted–Lowry acid and base:

a. $HBr + H_2O \longrightarrow H_3O^+ + Br^-$
b. $NH_3 + H_2O \rightleftharpoons NH_4^+ + OH^-$
c. $HSO_4^- + H_2O \rightleftharpoons SO_4^{2-} + H_3O^+$

9.63 What is meant when "pH balanced" is written on a shampoo bottle?

9.64 Are the following examples of body fluids acidic, basic, or neutral?

a. saliva, pH 6.8 **b.** urine, pH 5.9
c. pancreatic juice, pH 8.0 **d.** bile, pH 8.4
e. blood, 7.45

9.65 Why is it important to check the pH of an aquarium or swimming pool?

9.66 What is the pH of pure water?

9.67 What are some similarities and differences between strong and weak acids?

9.68 What are some ingredients found in antacids? What do they do?

9.69 One ingredient in some antacids is $Mg(OH)_2$.
a. If the base is not very soluble in water, why is it considered a strong base?
b. What is the neutralization reaction of $Mg(OH)_2$ with stomach acid, HCl?

9.70 Acetic acid, which is the acid in vinegar, is a weak acid. Why?

9.71 What is the pH of each of the following solutions?
a. $[H_3O^+] = 1 \times 10^{-8}$ M
b. $[H_3O^+] = 5 \times 10^{-2}$ M
c. $[OH^-] = 3 \times 10^{-12}$ M
d. $[H_3O^+] = 0.001$ M
e. $[OH^-] = 0.01$ M

9.72 Are the solutions in problem 9.71 acidic or basic?

9.73 What are the $[H_3O^+]$ and $[OH^-]$ for a solution with the following pH values?
a. 3.0 **b.** 6.0 **c.** 8.0 **d.** 11.0

9.74 Solution A has a pH of 4.0 and solution B has a pH of 6.0.
a. Which solution is more acidic?
b. What is the $[H_3O^+]$ in each?
c. What is the $[OH^-]$ in each?

9.75 What is the value of $[OH^-]$ in a solution that contains 0.20 g of NaOH in 0.25 L of solution?

9.76 What is the value of $[H_3O^+]$ in a solution that contains 1.5 g of HNO_3 in 0.50 L of solution?

9.77 Toothpaste contains SnF_4, which is a source of fluoride ion, F^-. HF is a weak acid.

$$HF \rightleftharpoons H^+ + F^-$$

a. In the stomach, where the pH is between 1 and 3, what direction of the reaction will be favored?
b. In the small intestine, where the pH rises to 8, what will be the more prevalent form, HF or F^-?

9.78 In the blood plasma, pH is maintained by the carbonic acid–bicarbonate buffer system.
a. How is pH maintained when acid is added to the buffer system?
b. How is pH maintained when base is added to the buffer system?

9.79 A person with untreated diabetes has an increase in acidic compounds in the blood. When admitted, this person has an increase in pulse and respiratory rates.
a. Why does the body respond by hyperventilating?
b. What treatment might be used to correct the condition?

9.80 How does ventilation in the respiration process play a role in respiratory acidosis and respiratory alkalosis?
a. What condition may be present with a blood pH value above 7.45 and a P_{CO_2} below 35 mm Hg?
b. What condition may be present with a blood pH value below 7.35 and a P_{CO_2} above 45 mm Hg?

9.81 A patient is admitted with a high temperature who is hyperventilating. The blood pH is 7.52.
a. Is the condition acidosis or alkalosis?
b. Is the condition metabolic or respiratory?
c. Why would the nurse help the patient breathe into a paper bag?

9.82 How many milliliters of 1.0 M NaOH solution are needed to completely neutralize 25.0 mL of 2.0 M H_2SO_4 solution?

9.83 Zinc reacts with hydrochloric acid solution to produce zinc chloride and hydrogen gas.

$$Zn(s) + 2HCl(aq) \longrightarrow ZnCl_2(aq) + H_2(g)$$

a. If 3.50 g of Zn reacts with 50.0 mL of HCl solution, what is the molarity of the HCl solution?
b. What volume (L) of hydrogen is produced at STP?

Answers

Answers to Study Checks

9.1 bases

9.2 $HNO_3 + H_2O \longrightarrow H_3O^+ + NO_3^-$

9.3 $[H_3O^+] = 1 \times 10^{-3}$ M; acidic

9.4 $[OH^-] = 2 \times 10^{-13}$ M

9.5 neutral

9.6 $[H_3O^+] = 1 \times 10^{-10}$ M; pH = 10.0

9.7 $[H_3O^+] = 1 \times 10^{-5}$ M; $[OH^-] = 1 \times 10^{-9}$ M

9.8 $HClO + H_2O \rightleftharpoons H_3O^+ + ClO^-$

9.9 $Ca(OH)_2 \xrightarrow{H_2O} Ca^{2+} + 2OH^-$

9.10 2.5 % (m/v) NaOH

9.11 3.0 M HNO_3

9.12 4.0 L

9.13 $H_3PO_4 + 3KOH \longrightarrow K_3PO_4 + 3H_2O$

9.14 No. Both substances are salts; the mixture has no weak acid present.

9.15 0.200 M HCl

9.16 30.0 mL

Answers to Selected Questions and Problems

9.1 **a.** acid **b.** acid **c.** acid **d.** base **e.** base

9.3 **a.** acid **b.** base **c.** acid **d.** acid **e.** acid

9.5 **a.** hydrochloric acid **b.** calcium hydroxide **c.** carbonic acid **d.** nitric acid **e.** sulfurous acid

9.7 **a.** $Mg(OH)_2$ **b.** HF **c.** H_3PO_4 **d.** LiOH **e.** $Cu(OH)_2$

9.9 **a.** HI is the acid (proton donor) and H_2O is the base (proton acceptor).
b. H_2O is the acid (proton donor) and F^- is the base (proton acceptor).

9.11 When water dissociates, equal amounts of H_3O^+ and OH^- are produced.

9.13 **a.** greater **b.** greater **c.** equal **d.** less

9.15 **a.** $[H_3O^+] = 1 \times 10^{-6}$ M
b. $[H_3O^+] = 1 \times 10^{-12}$ M
c. $[H_3O^+] = 1 \times 10^{-3}$ M
d. $[H_3O^+] = 5 \times 10^{-10}$ M

9.17 **a.** basic **b.** basic **c.** neutral **d.** acidic **e.** basic **f.** acidic

9.19 In a neutral solution, the $[H_3O^+]$ is 1×10^{-7} M and the pH is 7, which is the negative value of the power of 10.

9.21 **a.** basic **b.** acidic **c.** basic **d.** acidic **e.** neutral

9.23 **a.** 5, 7, 10, 14 **b.** 1, 8, 10, 13 **c.** 1.6, 2.4, 5.5, 8.5 **d.** 3.5, 8.8, 9.7, 11.4

9.25 **a.** 4.0 **b.** 9.0 **c.** 12.0 **d.** 9.0 **e.** 3.0

9.27 **a.** 1.4 **b.** 7.7 **c.** 3.7

9.29

	$[H_3O^+]$	$[OH^-]$
a.	1×10^{-2} M	1×10^{-12} M
b.	1×10^{-7} M	1×10^{-7} M
c.	1×10^{-13} M	1×10^{-1} M
d.	1×10^{-5} M	1×10^{-9} M

9.31

$[H_3O^+]$	$[OH^-]$	pH	Acidic, Basic, Neutral?
1×10^{-8} M	1×10^{-6} M	8.0	basic
1×10^{-2} M	1×10^{-12} M	2.0	acidic
1×10^{-5} M	1×10^{-9} M	5.0	acidic
1×10^{-7} M	1×10^{-7} M	7.0	neutral

9.33 A strong acid is completely ionized in solution, but a weak acid is only partially ionized.

9.35 HCl, (hydrochloric acid), HNO_3 (nitric acid), H_2SO_4 (sulfuric acid)

9.37 **a.** $HI + H_2O \xrightarrow{H_2O} H_3O^+ + I^-$

b. $Ca(OH)_2 \xrightarrow{H_2O} Ca^{2+} + 2OH^-$

c. $KOH \xrightarrow{H_2O} K^+ + OH^-$

d. $HNO_2 + H_2O \overset{H_2O}{\rightleftharpoons} H_3O^+ + NO_2^-$

9.39 Adding water (solvent) to the soup increases the volume and dilutes the tomato concentration.

9.41 **a.** 2.0 M HCl **b.** 2.0 M NaOH **c.** 2.5% (m/v) KOH **d.** 3.0% (m/v) H_2SO_4

9.43 **a.** 0.60 L **b.** 0.25 L **c.** 6.0 L

9.45 (b) and (d) are neutralization reactions; (a) and (c) are not.

9.47 **a.** $2HCl + Mg(OH)_2 \longrightarrow MgCl_2 + 2H_2O$
b. $H_3PO_4 + 3LiOH \longrightarrow Li_3PO_4 + 3H_2O$

9.49 **a.** $H_2SO_4 + 2NaOH \longrightarrow Na_2SO_4 + 2H_2O$
b. $3HCl + Fe(OH)_3 \longrightarrow FeCl_3 + 3H_2O$
c. $H_2CO_3 + Mg(OH)_2 \longrightarrow MgCO_3 + 2H_2O$

9.51 (b) and (c) are buffer systems. (b) contains the weak acid H_2CO_3 and its salt $NaHCO_3$. (c) contains HF, a weak acid, and its salt KF.

9.53 **a.** A buffer system keeps the pH constant.
b. To neutralize any H^+ added.
c. The added H^+ reacts with F^- from NaF.
d. The added OH^- is neutralized by the HF.

9.55 **a.** acidic
b. below
c. $NaHCO_3$ neutralizes excess acid.

9.57 To a known volume of formic acid, add a few drops of indicator. Place a NaOH solution of known molarity in a buret. Add base to acid until one drop changes the color of the solution. Use the volume and molarity of NaOH to calculate the concentration of the formic acid in the sample.

9.59 **a.** 8.8 M HCl
b. 0.75 M H_2SO_4
c. 0.53 M H_3PO_4

9.61 **a.** sulfuric acid, acid
b. potassium hydroxide, base
c. calcium hydroxide, base
d. hydrochloric acid, acid
e. nitrous acid, acid

9.63 A "pH balanced" shampoo has a pH that matches the pH of hair.

9.65 A high or low pH can kill the fish in an aquarium, or irritate the skin and eyes in a pool.

9.67 Both strong and weak acids produce H_3O^+ in water. Weak acids are only slightly ionized; whereas a strong acid exists as ions in solution.

9.69 **a.** The $Mg(OH)_2$ that dissolves is completely ionized, making it a strong base.
b. $Mg(OH)_2 + 2HCl \longrightarrow MgCl_2 + 2H_2O$

9.71 **a.** 8.0 **b.** 1.3 **c.** 2.5 **d.** 3.0 **e.** 12.0

9.73 **a.** $[H_3O^+] = 1 \times 10^{-3}$ M; $[OH^-] = 1 \times 10^{-11}$ M
b. $[H_3O^+] = 1 \times 10^{-6}$ M; $[OH^-] = 1 \times 10^{-8}$ M
c. $[H_3O^+] = 1 \times 10^{-8}$ M; $[OH^-] = 1 \times 10^{-6}$ M
d. $[H_3O^+] = 1 \times 10^{-11}$ M; $[OH^-] = 1 \times 10^{-3}$ M

9.75 $[OH^-] = 0.020$ M

9.77 **a.** The reaction shifts left toward more HF. **b.** F^-

9.79 **a.** Hyperventilation will lower the CO_2 level in the blood and shift the equilibrium of the carbonate buffer toward bicarbonate, lowering the H_2CO_3 concentration, which decreases the H_3O^+.
b. Treat the diabetes condition to reduce the production of acidic compounds; perhaps an infusion of bicarbonate.

9.81 **a.** alkalosis **b.** respiratory
c. Rebreathing CO_2 raises the CO_2 level in the blood. The equilibrium will shift toward more H_2CO_3, which increases the H^+ and lowers the pH to normal.

9.83 **a.** 2.14 M **b.** 1.20 L H_2

10

Alkanes and Aromatic Hydrocarbons

Crude oil is being pumped from below the ocean floor in an off-shore oil well.

Looking Ahead

10.1	Organic Compounds
10.2	Bonding in Hydrocarbons
10.3	Naming Alkanes
10.4	Properties of Alkanes
10.5	Branched Alkanes
10.6	Structural Formulas
10.7	Cycloalkanes
10.8	Aromatic Hydrocarbons

www.chemplace.com/college/timberlake

Visit the web site above to test yourself, run simulations, and review key concepts from this chapter. For even more interactive review and quizzing, use the two CD-ROMs included with the book.

The element carbon has a special role in chemistry because it has the capability to form so many different compounds, many of which are found in biological systems. Gasoline, coal, medicines, shampoos, plastic bottles, and perfumes are all made of organic compounds. The fabrics in your clothes may be naturally occurring organic compounds such as cotton and silk, or they may be synthetic organic compounds such as polyester or nylon. The food we eat is composed of many different organic compounds that supply us with fuel for energy and the carbon atoms needed for building and repairing the cells of our bodies. What we learn about organic molecules will be used to explain the behavior of the very large biomolecules that are important in health and medicine.

Before 1800, scientists used the term *organic* for compounds they thought could only be produced within the cells of living systems. They believed that a "vital force," found only in living organisms, was required to synthesize an organic compound. By the early 1800s, urea, a product of protein metabolism in animals, was prepared from the inorganic compound ammonium cyanate. Since then, millions of organic compounds have been prepared in laboratories and the idea of a vital force has become a part of chemical history.

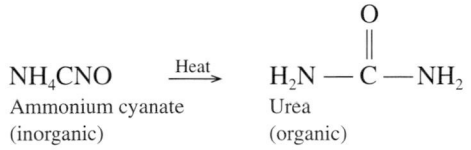

10.1

Study Pack

Chem of Life Organic Molecules: General Structure 1

10.1 Organic Compounds

Learning Goal

From its properties, classify a compound as organic or inorganic.

Organic chemistry is the study of compounds containing carbon. The location of carbon in the middle of the periodic table and its low atomic mass make it ideal as the major element for biological compounds. In organic compounds, carbon is bonded mostly to hydrogen and sometimes to oxygen, nitrogen, sulfur, phosphorus, and halogens. In general, **organic compounds** have covalent bonds, have low melting and boiling points, burn vigorously, and are soluble in nonpolar solvents. There are a great number of organic compounds because carbon atoms have a unique ability to bond to other carbon atoms to form large molecules.

Until now, most of the chemistry we have studied involved inorganic compounds that were ionic or had polar covalent bonds. In general, these compounds have high melting and boiling points, do not burn in air, and are soluble in water. Although there are variations among organic and inorganic compounds, it is helpful to compare some of their properties as listed in Table 10.1. Table 10.2 contrasts the properties of butane, an organic compound, and sodium chloride, an inorganic compound.

Table 10.1 Properties of Most Organic and Inorganic Compounds

Organic	Inorganic
All are carbon compounds.	Contain metals in salts and oxides.
All have covalent bonds.	All have ionic or polar bonds.
All have low melting points.	All have high melting points.
All have low boiling points.	All have high boiling points.
Most burn in air.	Few burn in air.
Most are soluble in nonpolar solvents.	Most are soluble in polar solvents.
Most are nonelectrolytes.	Many are electrolytes.
All can be large molecules with many atoms.	All are usually small with few atoms.

Table 10.2 A Comparison of an Organic and an Inorganic Compound

Compound	Butane	Sodium chloride
Formula	C_4H_{10}	NaCl
Molar mass	58.1 g/mole	58.5 g/mole
Bonds	Covalent	Ionic
State (room temperature)	Gas	Solid
Melting point	−138°C	801°C
Boiling point	0°C	1413°C
Solubility in water	Low	High
Solubility in a nonpolar solvent	High	Low
Electrolyte	No	Yes
Burns in air	Yes	No

10.1

Hydrocarbons

Today chemists know of millions of organic compounds. To study this vast array of structures, the compounds have been organized into *families* according to common features. A list of the families we will discuss is found on the inside back cover. We will study the features (functional groups) of each family and relate them to their chemical behavior.

All compounds classified as **hydrocarbons** contain only the elements carbon and hydrogen. The **alkanes** are hydrocarbons that have only single bonds (C—C). They are known as **saturated hydrocarbons** because every carbon atom is bonded to four atoms, the most a carbon atom can have. In Chapter 11, we will study the *unsaturated hydrocarbons,* the alkenes that contain a double bond (C=C), and alkynes with a triple bond (C≡C). In this chapter we will also look at the aromatic compounds, a group of hydrocarbons that contain benzene rings.

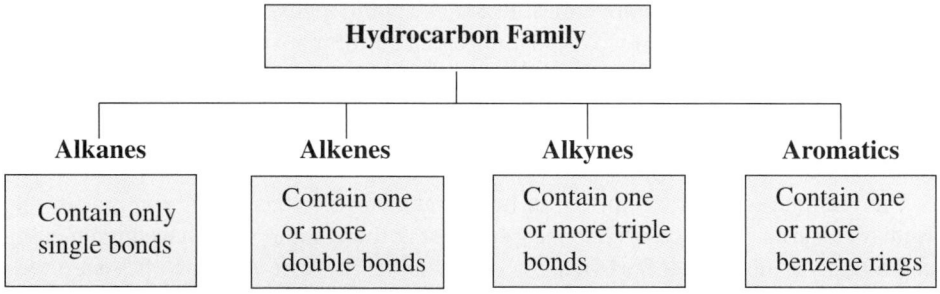

S·A·M·P·L·E 10.1 P·R·O·B·L·E·M

Classifying Properties of Organic Compounds

State whether each of the following properties is typical of an organic or an inorganic compound:

a. soluble in water **b.** low boiling point **c.** burns in air

Solution

a. Inorganic; most inorganic compounds are ionic or polar and soluble in water.
b. Organic; most organic compounds have low boiling points.
c. Organic; organic compounds are often very flammable.

Study Check

Would a compound that dissociates into ions in water most likely be an organic or inorganic compound?

QUESTIONS & PROBLEMS

Organic Compounds

10.1 What do vegetable oil, gasoline, a plastic raincoat, and sugar all have in common?

10.2 What do you mean when you use the term *organic*?

10.3 How does the historical view of organic chemistry differ from the modern definition of *organic chemistry?*

10.4 Distinguish between the following terms:
a. hydrocarbon and alkane **b.** alkane and alkene

10.5 Classify the following compounds as organic or inorganic.
a. $CaCl_2$ **b.** CH_3CH_2Cl **c.** C_4H_{10}
d. $CoCl_3$ **e.** CH_4

10.6 Classify the following compounds as organic or inorganic.
a. CH_3CH_3 **b.** KCl **c.** Cr_2O_3
d. Na_2SO_4 **e.** $C_{10}H_{22}$

10.7 Match the following physical and chemical properties with the compounds propane, C_3H_8, or potassium chloride, KCl.
a. melts at −190°C
b. burns vigorously
c. melts at 770°C
d. conducts electricity when it dissolves in water
e. is a gas at room temperature

10.8 Match the following physical and chemical properties with the compounds cyclohexane, C_6H_{12}, or calcium nitrate, $Ca(NO_3)_2$.
a. floats on the surface of water (density = 0.78 g/mL)
b. melts above 500°C
c. insoluble in water
d. density of 2.5 g/mL
e. produces ions in water

Study Pack

Chem of Life Organic Molecules: Carbon Skeletons 1

10.2 Bonding in Hydrocarbons

Learning Goal

Draw the complete and the condensed structural formula for an alkane.

In the simplest hydrocarbon, methane (CH_4), the four valence electrons of carbon are shared with four hydrogen atoms. A hydrogen atom only forms one bond in compounds. In the electron-dot structure, each shared pair of electrons represents a single bond. A **complete structural formula** is written when we show all the bonds between all of the atoms.

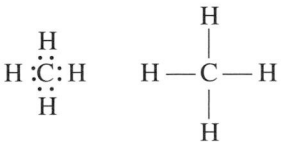

Although our drawing of methane is two-dimensional, methane actually has a three-dimensional shape (shown in Figure 10.1) called a *tetrahedral structure.* The carbon atom is in the center and bonded to four hydrogen atoms, which are as far apart from each other as possible and occupy the corners of the tetrahedron. All carbon atoms in

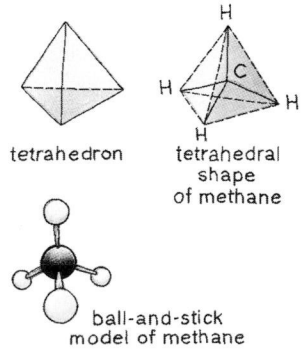

Figure 10.1 The representations of the atoms in a methane (CH_4) molecule.

alkanes exhibit the tetrahedral shape, even though we often write the flat formula for convenience.

The alkane with two atoms of carbon is ethane, which has the molecular formula C_2H_6. Each carbon is bonded to the other carbon atom and to three hydrogen atoms to give a total of four bonds to each carbon atom. A ball-and-stick model of ethane is shown in Figure 10.2.

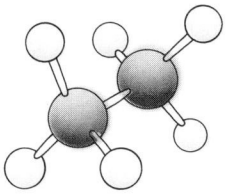

Figure 10.2 A ball-and-stick model of ethane, C_2H_6.

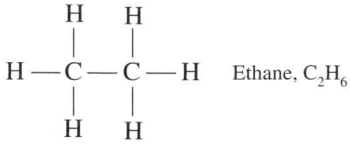

Ethane, C_2H_6

10.2

To write the complete structural formula of propane, C_3H_8, an alkane with three carbon atoms, we first connect the carbon atoms with single bonds.

C—C—C

Eight hydrogen atoms are attached to complete the four bonds to each carbon atom. A ball-and-stick model of propane is shown in Figure 10.3.

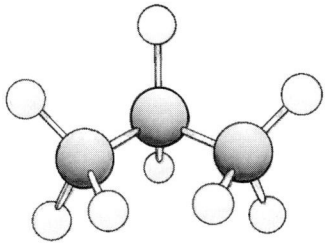

Figure 10.3 A ball-and-stick model of propane, C_3H_8.

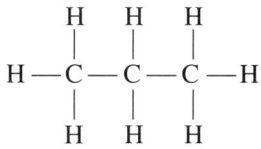

Propane, C_3H_8

Condensed Structural Formulas

Although complete structural formulas are helpful, we will not always draw all of the individual C—H and C—C bonds. In a **condensed structural formula,** we still show the arrangement or structure, but we write each carbon atom and its attached hydrogen atoms as a group. A subscript in each group indicates the number of hydrogen atoms bonded to that carbon atom. The C—H single bonds are understood but not written individually.

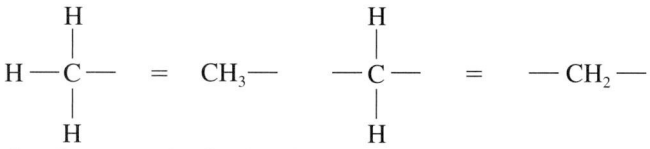

Complete structural Condensed Complete structural Condensed

We can simplify further by omitting the line drawn for the carbon–carbon (C—C) bonds. Both the complete structural and condensed structural formulas show the arrangement of the carbon atoms in a molecule. The molecular formulas give the total number of each kind of atom, but they indicate nothing about the arrangement of the atoms in the molecule. Here, for example, are the different formulas used to represent propane:

C_3H_8 — **Molecular formula** shows the number of atoms, but not their order

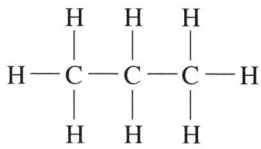

Complete structural formula shows arrangement of all bonds

$\boxed{CH_3}$ $\boxed{CH_2}$ $\boxed{CH_3}$ — Groups of carbon and hydrogen atoms

$CH_3—CH_2—CH_3$
or $CH_3CH_2CH_3$ — **Condensed structural formula** shows arrangement, but not single C—H bonds

S·A·M·P·L·E· **10.2** ·P·R·O·B·L·E·M

Drawing Structural and Condensed Structural Formulas for Alkanes

A molecule of butane has four carbon atoms in a row. What are its complete structural, condensed structural, and molecular formulas?

Solution

In the complete structural formula, four carbon atoms are joined to hydrogen atoms using single bonds to give each carbon atom a total of four bonds. In the condensed structural formula, each carbon atom and its attached hydrogen atoms are written as CH_3—, or —CH_2—.

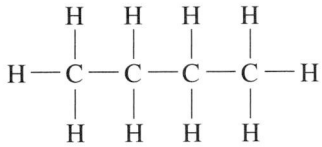

Complete structural formula

$CH_3—CH_2—CH_2—CH_3$

or

$CH_3CH_2CH_2CH_3$

Condensed formula

C_4H_{10}

Molecular formula

The molecular formula, C_4H_{10}, indicates only the carbon and hydrogen atoms in the molecule.

Study Check

Write the structural, condensed, and molecular formulas for an alkane that has five carbon atoms in a row.

S·A·M·P·L·E· **10.3** ·P·R·O·B·L·E·M

Formulas for Branched-Chain Alkanes

Draw a condensed structural formula for the following branched-chain alkane:

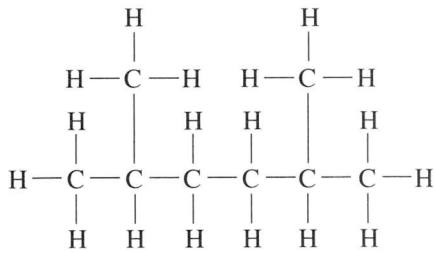

Solution

In the condensed structural formula, each carbon atom and its surrounding hydrogen atoms are shown as a group with a subscript to indicate the number of attached hydrogen atoms.

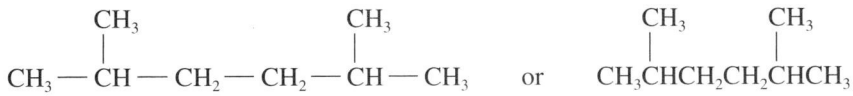

10.2

Study Check

Write a condensed structural formula for the following compound:

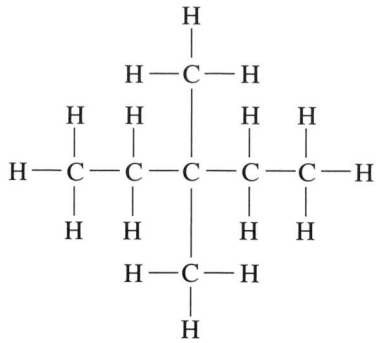

Q·U·E·S·T·I·O·N·S & P·R·O·B·L·E·M·S

Bonding in Hydrocarbons

10.9 The hydrocarbon ethane is composed of two carbon atoms and six hydrogen atoms, C_2H_6. For ethane, write each of the following:
a. complete structural formula
b. condensed structural formula

10.10 The hydrocarbon propane is composed of three carbon atoms and eight hydrogen atoms, C_3H_8. For propane, write each of the following:
a. complete structural formula
b. condensed structural formula

10.11 Give the condensed structural formula for each of the following complete structural formulas:

a.

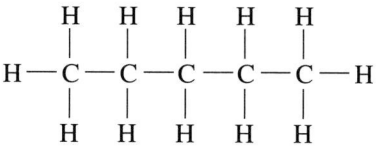

b.

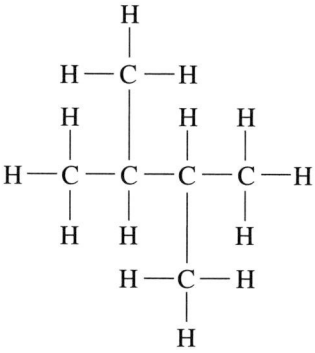

10.12 Give the condensed structural formula for each of the following complete structural formulas:

a.

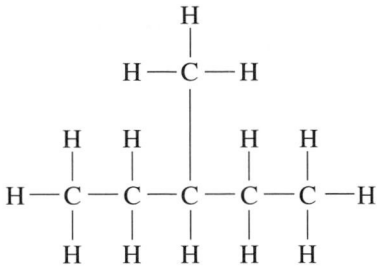

b.

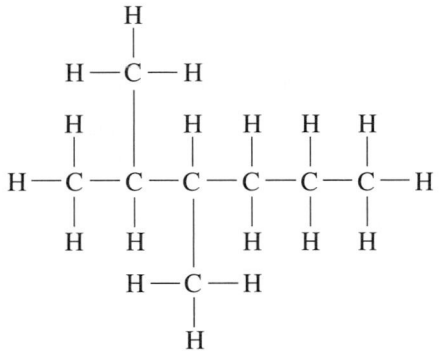

10.13 Write the complete structural formula and the condensed structural formula by adding hydrogen atoms to each of the following alkanes:

a. C—C

b.

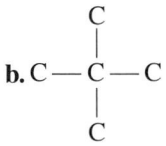

c. C—C—C—C

d.

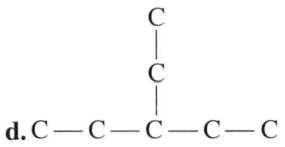

10.14 Write the complete structural formula and the condensed structural formula by adding hydrogen atoms to each of the following alkanes:

a. C—C—C

b.

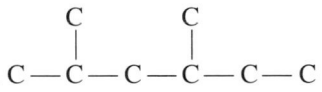

c.

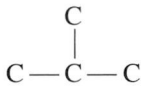

d.

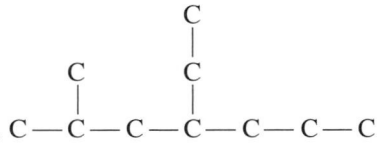

Study Pack

10.3 Naming Alkanes

Learning Goal

Write the names and condensed and structural formulas of the first ten unbranched alkanes.

10.3

When relatively few organic compounds were known, names were assigned in a random fashion. Although some of those *common names* are still in use, we now use a naming system known as the *IUPAC rules* (International Union of Pure and Applied Chemistry). Using the IUPAC rules, the stem of a name identifies the number of carbon atoms in the longest chain of a compound, and the suffix describes the family of the compound. The names given in Table 10.3 give the stems for unbranched alkanes with one to ten carbon atoms followed by the suffix *ane*, which indicates the alkane family. It is important to learn these stems because they are used in the names of all the families of organic compounds.

Table 10.3 IUPAC Names (Stem + *ane*) for the First Ten Unbranched Alkanes

Name	Number of Carbon Atoms	Molecular Formula	Condensed Structural Formula
Methane	1	CH_4	CH_4
Ethane	2	C_2H_6	CH_3CH_3
Propane	3	C_3H_8	$CH_3CH_2CH_3$
Butane	4	C_4H_{10}	$CH_3CH_2CH_2CH_3$
Pentane	5	C_5H_{12}	$CH_3CH_2CH_2CH_2CH_3$
Hexane	6	C_6H_{14}	$CH_3CH_2CH_2CH_2CH_2CH_3$
Heptane	7	C_7H_{16}	$CH_3CH_2CH_2CH_2CH_2CH_2CH_3$
Octane	8	C_8H_{18}	$CH_3CH_2CH_2CH_2CH_2CH_2CH_2CH_3$
Nonane	9	C_9H_{20}	$CH_3CH_2CH_2CH_2CH_2CH_2CH_2CH_2CH_3$
Decane	10	$C_{10}H_{22}$	$CH_3CH_2CH_2CH_2CH_2CH_2CH_2CH_2CH_2CH_3$

ENVIRONMENTAL NOTE

Petroleum Products

Mineral oil is a mixture of liquid hydrocarbons and is used as a laxative and a lubricant. Motor oil is a mixture of high-molecular-weight liquid hydrocarbons and is used to lubricate the internal compartments of engines. Petrolatum, or Vaseline, is a colloidal system in which liquid hydrocarbons having high boiling points are encapsulated in solid hydrocarbons. It is used in ointments and cosmetics and as a lubricant and a solvent. Applying a hydrocarbon mixture such as Vaseline softens the skin because the water in the skin is retained by the nonpolar compounds. Figure 10.4 shows some examples of mixtures of hydrocarbons.

Figure 10.4 Mixtures of hydrocarbons are found in products like these.

Crude oil contains a wide variety of hydrocarbons whose boiling points increase with molar mass. At an oil refinery, the crude oil components are separated by increasing the temperature (see Figure 10.5). Some of the products obtained from crude oil are listed in Table 10.4.

Figure 10.5 At an oil refinery, crude oil is separated into many useful organic products.

Table 10.4 Hydrocarbons in Crude Oil

Product	Number of Carbon Atoms	Temperature Range for Distillation (°C)
Natural gas	1–4	Below 30
Gasoline	5–12	30–200
Kerosene, heating oil	12–15	200–300
Diesel fuel, lubricating oil	15–25	300–400
Asphalt, tar	Over 25	Nonvolatile residue

Writing Formulas for Unbranched Alkanes

In an **unbranched alkane,** the carbon atoms are connected in a row, one after the other. You could trace a line through the carbon symbols without lifting the pencil off the paper.

Examples of Unbranched Alkanes

$CH_3—CH_2—CH_3$ $\quad$ $CH_3—CH_2—CH_2—CH_3$

Actually, alkane structures are not really flat or straight molecules. The tetrahedral shape of carbon makes the bonds between carbon atoms appear at angles to each other in a zigzag fashion. However, when we use condensed formulas, we only show the way the carbon atoms are connected. As a result, the formula appears flatter than it really is. Also, one compound can be represented by several diagrams. In Table 10.5, some possible condensed formulas for butane show a chain of four carbon atoms, but the four carbons don't need to be drawn in a straight line. However, if you trace the carbon atoms in each formula from one end to the other, you will see that they are all connected in a row.

Table 10.5 Some Possible Formulas for Butane

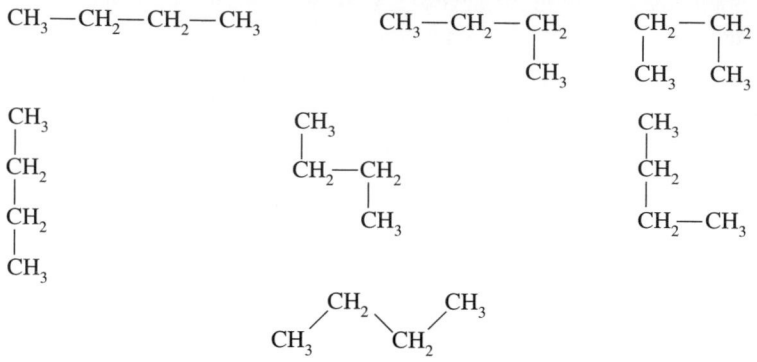

10.3

S·A·M·P·L·E **10.4** ·P·R·O·B·L·E·M

IUPAC Names

Give the IUPAC names for the following alkanes:

a. $CH_3CH_2CH_3$

b. 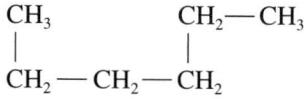

Solution

a. propane (three carbon atoms)
b. hexane (six carbon atoms)

Study Check

Write the condensed structural formula and the name of an unbranched alkane having seven carbon atoms.

Q·U·E·S·T·I·O·N·S & P·R·O·B·L·E·M·S

Naming Alkanes

10.15 What is meant by the term *unbranched alkane*?

10.16 Why are the carbon atoms in an unbranched alkane not really arranged in a straight line?

10.17 Write the condensed structural formulas of each of the following alkanes:
a. methane **b.** ethane **c.** pentane **d.** hexane

10.18 Write the condensed structural formulas of each of the following alkanes:
a. propane **b.** butane **c.** heptane **d.** octane

10.19 Give the IUPAC name for each of the following alkanes:

a. 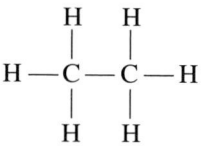

b. $CH_3 — CH_2 — CH_2 — CH_3$

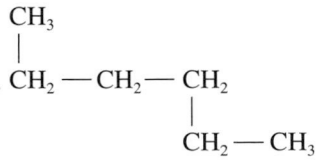

c.

d. $CH_3CH_2CH_2CH_2CH_2CH_2CH_2CH_3$

10.20 Give the IUPAC name for each of the following alkanes:

a.

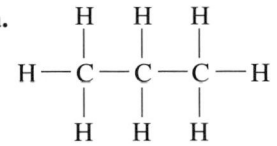

b. $CH_3—CH_2—CH_2—CH_2—CH_3$

c.

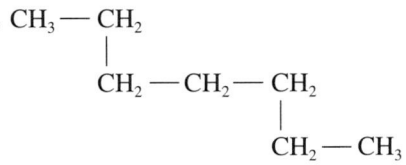

d. $CH_3CH_2CH_2CH_2CH_2CH_2CH_2CH_2CH_2CH_3$

10.4 Properties of Alkanes

Study Pack

Learning Goal

Describe some properties of alkanes.

The first four alkanes, methane, ethane, propane, and butane, are gases at room temperature and pressure and act as anesthetics if inhaled. They are widely used as heating fuels. Alkanes having 5 to 17 carbon atoms are liquids at room temperature; alkanes having 18 or more carbon atoms are waxy solids.

The bonds of the alkanes, C—C and C—H, are nonpolar, which means that alkanes are nonpolar. Thus, alkanes are soluble in nonpolar solvents such as other alkanes and gasoline. The liquid alkanes float on water because they are less dense and insoluble in water. In oil spills, most of the alkanes remain on the surface, where they spread over a large area.

The alkanes are among the least reactive of any of the organic compounds. The single covalent bonds of alkanes are difficult to break, a feature that makes them unreactive. However, alkanes do burn readily in oxygen.

Combustion

The growth of civilization has been strongly influenced by the discovery of fire, or combustion. Energy produced by burning wood was used to make pottery and glass, extract metals, make weapons, and forge tools. At high temperatures, an organic compound undergoes **combustion** when it reacts rapidly with oxygen to produce carbon dioxide, water, and heat. The methane gas we use to cook our foods and heat our homes, the propane in portable heaters and gas barbecues, and the gasoline that powers our cars are all alkanes that undergo combustion. (See Figure 10.6.) In combustion, a great amount of heat is also released. We can write the following equations for the combustion of methane and propane:

Figure 10.6 The propane fuel in the tank undergoes combustion to cook the food on the barbecue.

Combustion

$$\textit{Alkane} + O_2 \longrightarrow CO_2 + H_2O + \textit{heat}$$

Examples of Combustion

$$CH_4 + 2O_2 \longrightarrow CO_2 + 2H_2O + \text{heat}$$
$$C_3H_8 + 5O_2 \longrightarrow 3CO_2 + 4H_2O + \text{heat}$$

10.4

EXPLORE YOUR WORLD

Solubility of Alkanes

Obtain a small bowl and add water. Add a few drops of some mineral oil or a small amount of petroleum jelly (such as Vaseline) and observe. Stir.

Questions

1. What do your observations tell you about the solubility of alkanes in water?
2. If water is a polar solvent, how would you classify the solutes?
3. From your observations, are these hydrocarbons more or less dense than water? Explain.

EXPLORE YOUR WORLD

Combustion

In this exploration, we will look at the behavior of the products of combustion. You will need one or two candles, a Pyrex glass such as a measuring cup, and some matches or wooden splints.

Hold a Pyrex cup upside down, and place a burning match inside it. The match should continue to burn as long as oxygen is available. Light a candle and hold the inverted Pyrex cup above it for 15–20 seconds. Remove the cup from the candle and immediately place a burning match into it. The CO_2 accumulated from the combustion of the candle should extinguish the match.

Add some water and a lot of ice to the same Pyrex cup. It should become cold to the touch. Wipe the bottom of the cup and carefully hold the Pyrex cup over a burning candle. Look for condensation as water is formed from the combustion reaction. This may be more noticeable if two candles are used or the Pyrex cup is held for a few seconds over a gas flame on a cook top.

Questions

1. What are the products of combustion?
2. What was the evidence for the production of CO_2?
3. What observations gave evidence of the production of H_2O during combustion?

S·A·M·P·L·E 10.5 P·R·O·B·L·E·M

Writing an Equation for Combustion

Gasoline is a mixture of liquid alkanes including pentane, C_5H_{12}. Write and balance the equation for the combustion of pentane in the engine of a car.

Solution

The reactants of this combustion reaction are pentane and oxygen. The products of the combustion are carbon dioxide and water. The unbalanced equation can be written

$$C_5H_{12} + O_2 \longrightarrow CO_2 + H_2O$$

To write a balanced equation, we match the atoms of C, H, and O in the reactants with those in the products.

$$C_5H_{12} + 8O_2 \longrightarrow 5CO_2 + 6H_2O$$

Study Check

Write a balanced equation for the complete combustion of heptane.

HEALTH NOTE

Incomplete Combustion

You may already know that it is dangerous to burn natural gas, oil, or wood in a closed room where ventilation and fresh air is not adequate. A gas heater, fireplace, or wood stove must have proper ventilation. If the supply of oxygen is limited, incomplete combustion produces carbon monoxide. The incomplete combustion of methane in natural gas is written as:

$$2CH_4(g) + \underset{\text{Limited oxygen supply}}{3O_2(g)} \longrightarrow \underset{\text{Carbon monoxide}}{2CO(g)} + 4H_2O(g) + \text{heat}$$

Carbon monoxide (CO) is a colorless, odorless, poisonous gas. When inhaled, CO passes into the bloodstream, where it attaches to hemoglobin (as you saw in the Health Note on Oxygen, Hemoglobin and Carbon Monoxide Poisoning in Chapter 6). When CO binds to the hemoglobin, it reduces the amount of oxygen (O_2) reaching the organs and cells. As a result, a healthy person can experience a reduction in exercise capability, visual perception, and manual dexterity.

When the amount of hemoglobin bound to CO (COHb) is 10% or less, a person may experience shortness of breath, mild headache, and drowsiness, which are symptoms that may be mistaken for the flu. Heavy smokers can have as high as 9% COHb in their blood. When as much as 30% of the hemoglobin is bound to CO, a person may experience more severe symptoms including dizziness, mental confusion, severe headache, and nausea. If 50% or more of the hemoglobin is bound to CO, a person could become unconscious and die if not treated immediately with oxygen.

QUESTIONS & PROBLEMS

Properties of Alkanes

10.21 Write balanced equations for the complete combustion of each of the following:

a. methane, CH_4 **b.** propane, C_3H_8 **c.** butane, C_4H_{10}

10.22 Write balanced equations for the complete combustion of each of the following:

a. hexane, C_6H_{14} **b.** butane, C_4H_{10} **c.** octane, C_8H_{18}

10.23 Heptane, C_7H_{16}, is an unbranched alkane. It has a density of 0.68 g/mL, melts at −91°C, and boils at 98°C.

a. What is the structural formula of heptane?

b. Is it a solid, liquid, or gas at room temperature?

c. Is it soluble in water?

d. Will it float on the surface of water?

e. Heptane is a component of gasoline. Write the balanced equation for the combustion of heptane.

10.24 Nonane, C_9H_{20}, is an unbranched alkane. It has a density of 0.79 g/mL, melts at −51°C, and boils at 151°C.

a. What is the condensed structural formula of nonane?
b. Is it a solid, liquid, or gas at room temperature?
c. Is it soluble in water?
d. Will it float on the surface of water?
e. Nonane is a component of gasoline. Write the balanced equation for the complete combustion of nonane.

10.5

Study Pack

10.5 Branched Alkanes

Learning Goal

Use the IUPAC system to write the names of branched-chain alkanes.

When there are four or more carbon atoms in an alkane, the atoms can be arranged so that a **branch,** a side group of carbon atoms also called a *substituent*, is attached to a longer carbon chain. An alkane with at least one carbon branch is called a **branched** (or branched-chain) **alkane.** In the formulas of branched-chain compounds, the branches usually are drawn above or below the main carbon chain.

Examples of Some Branched-Chain Alkanes

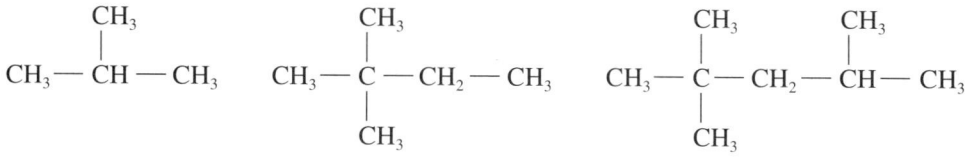

Alkyl Groups in Branched Alkanes

In the names of branched alkanes, the carbon groups attached to a longer carbon chain are named as alkyl groups. An **alkyl group** is derived from an alkane by removing one hydrogen. The *ane* ending of the alkane is changed to *yl* to indicate the alkyl group. Alkyl groups do not exist on their own; they must be attached to something such as a carbon chain. Some of the most common alkyl groups are illustrated in Table 10.6. The ball-and-stick models of the alkyl groups are shown in Figure 10.7.

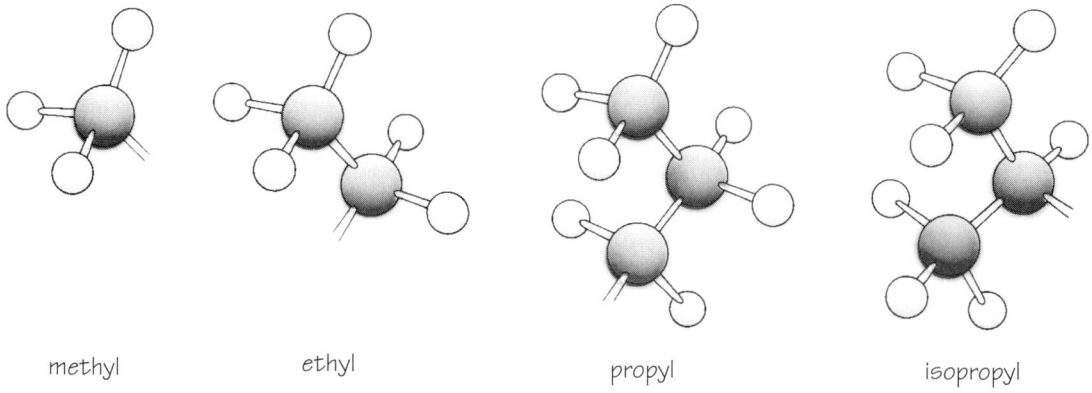

Figure 10.7 Ball-and-stick models of the alkyl groups derived from methane, ethane, and propane.

Table 10.6 Some Common Alkyl Groups

Alkane	Corresponding Alkyl Group	Condensed Structural Formula	Name
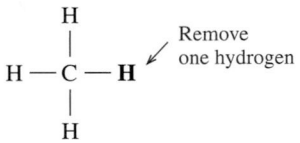 Methane	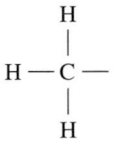	$CH_3—$	Methyl
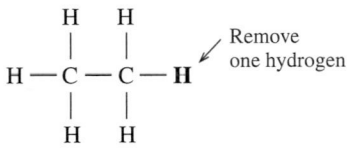 Ethane	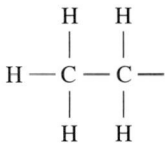	$CH_3CH_2—$	Ethyl
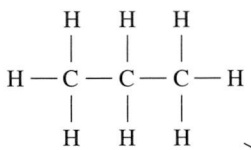 Propane	Remove end hydrogen 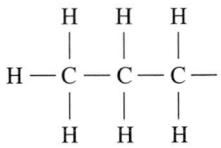	$CH_3CH_2CH_2—$	Propyl
	Remove center hydrogen 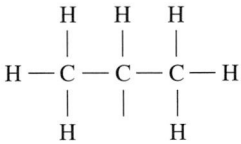	$CH_3CH(\vert)CH_3$	Isopropyl

Naming Branched-Chain Alkanes

Branched alkanes are named using the parent name and the names of the attached alkyl groups. We can now use the IUPAC rules to name the following branched-chain alkane.

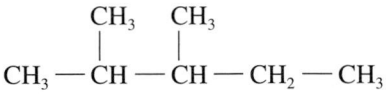

1. *Find the longest unbranched chain of carbon atoms and name it as the* ***parent chain****.* In our example, the parent chain is named *pentane* because it has an unbranched chain of five carbon atoms.

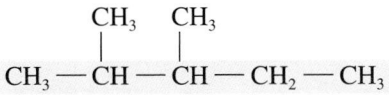

Pentane (longest chain = parent chain)

2. *Number the carbon atoms in the parent chain from the end nearest the first alkyl group.* Choose the direction of numbering that gives the first alkyl group the lowest possible number. If the first alkyl group is nearer the left end of the chain, number from *left to right.* If the first alkyl group is nearer the right end, number the chain from *right to left.*

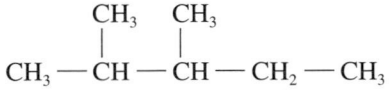

1 2 3 4 5 Numbering from left to right

3. *Determine the location and name for each alkyl group(s) attached to the main chain.* Use a hyphen to separate the number and the name of the alkyl group. In this example, we have a 2-methyl group and a 3-methyl group.

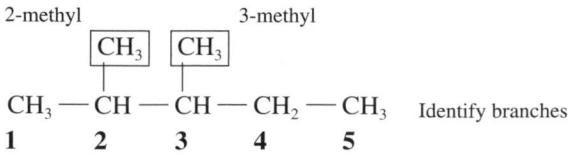

10.5

4. *Place the number(s) and name of the alkyl group(s) in* front *of the parent name, using a prefix if the same group appears more than once.* If the same alkyl group appears more than once, prefixes such as *di* (two), *tri* (three), and *tetra* (four) are used. In our example, we write 2,3-*dimethyl* in front of the parent name to indicate that there are two methyl groups, one on carbon 2 and another on carbon 3.

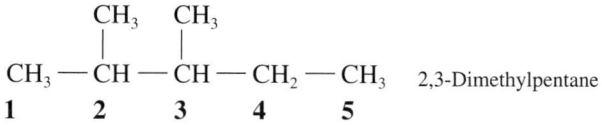

All the identical branches must have numbers, even if they are attached to the same carbon. Number the chain from the end of the carbon chain that gives the lowest combination of numbers, using commas to separate the numbers.

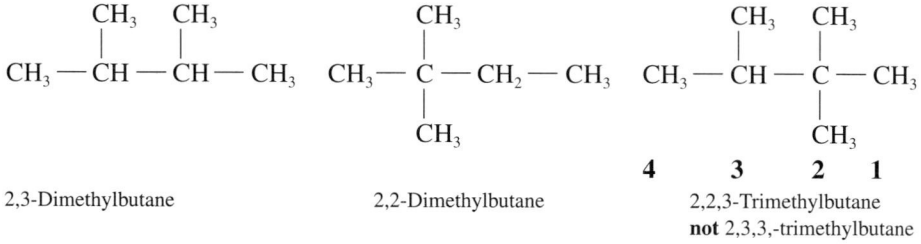

Once you start numbering a carbon chain with two or more branches, be sure to continue in that same direction to the other end of the chain.

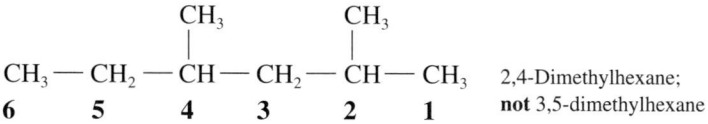

5. *List the names of two or more different substituents in alphabetical order with hyphens separating each number and name.* Prefixes for identical branches (di, tri, and so on) are not used in deciding alphabetical order.

Examples of Alphabetical Order of Alkyl Substituents

Ethyl, methyl

Triethyl, dimethyl

Ethyl, isopropyl, dimethyl

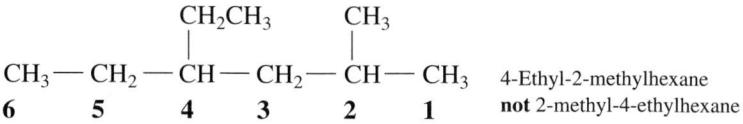

6. If two different branches can be given the same number, assign the lower number to the branch that is first alphabetically.

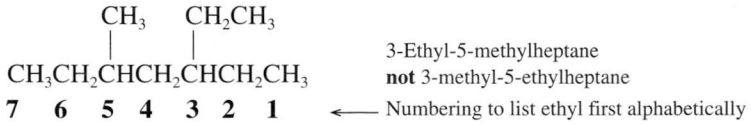

Be sure that you always find the longest continuous carbon chain. Watch out: The longest chain need not be the most obvious horizontal one. Look at the following structure:

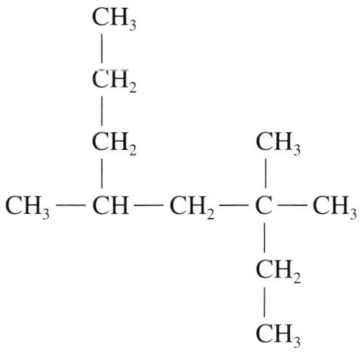

How many carbon atoms are in the parent chain? If you said five, look again. There is a chain of eight carbon atoms that goes around some corners!

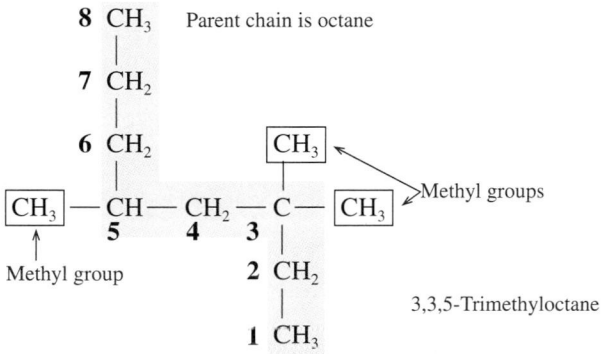

S·A·M·P·L·E· 10.6 ·P·R·O·B·L·E·M

Writing IUPAC Names for Branched-Chain Alkanes

Name the following compounds using the IUPAC system:

a. 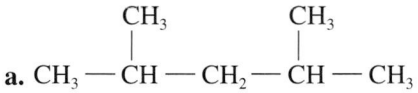

b. 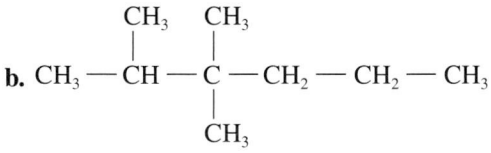

c. 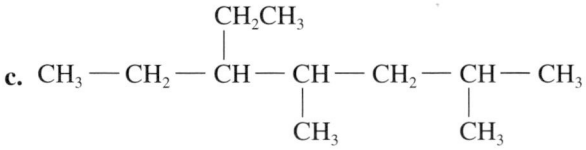

10.5

Solution

a. The longest chain of five carbon atoms is named *pentane.* There are two methyl groups on carbons 2 and 4. A prefix of *di* indicates the two methyl branches. The compound is named 2,4-dimethylpentane.

b. The longest chain of six carbon atoms is named *hexane.* There are three methyl groups indicated by the prefix *tri*, one on the second carbon and two on the third carbon. The compound is named 2,3,3-trimethylhexane.

c. The longest chain of seven carbon atoms is named *heptane.* Counting from *right to left,* there are two *(di)* methyl groups on carbons 2 and 4, and one ethyl group on carbon 5. Listing the alkyl groups in alphabetical order gives the name 5-ethyl-2,4-dimethylheptane.

Study Check

Name the following compound using the IUPAC system:

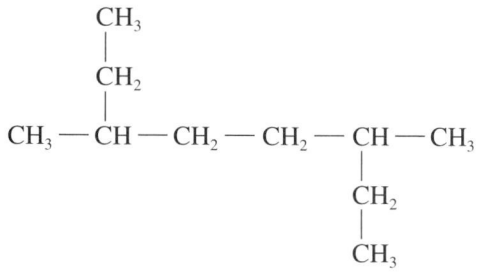

Q·U·E·S·T·I·O·N·S & P·R·O·B·L·E·M·S

Branched Alkanes

10.25 Draw the structure of the following alkyl groups:
a. methyl **b.** propyl

10.26 Draw the structure of the following alkyl groups:
a. ethyl **b.** isopropyl

10.27 Distinguish between the following pairs of terms:
a. methane and methyl group **b.** ethane and ethyl group

10.28 Distinguish between the following pairs of terms:
a. methyl group and ethyl group **b.** propyl group and isopropyl group

10.29 Give the IUPAC name for each of the following alkanes:

a. $CH_3-CH(CH_3)-CH_3$

b. $CH_3-CH_2-CH_2-CH(CH_3)-CH_3$

c. $CH_3-CH(CH_3)-CH(CH_3)-CH_2-CH_3$

d. $CH_2(CH_3)-CH(CH_3)-CH_2-CH(CH_3)-CH_3$

e.

```
CH3 — CH — CH3
      |
CH2 — CH — CH2 — CH2 — CH3
      |
      CH3
```

f.

```
                CH3
                |
CH3 — CH2 — CH — CH — CH3
            |
            CH2
            |
            CH3
```

g.

```
                CH3
                |
                CH2
                |
                CH2
                |
CH3 — CH2 — CH2 — CH — CH2
                |
                CH2
                |
                CH3
```

10.30 Give the IUPAC name for each of the following alkanes:

a.

```
        CH3   CH3
        |     |
CH3 — CH — CH — CH3
```

b.

```
                CH2 — CH3
                |
CH3 — CH2 — CH2 — CH — CH2 — CH3
```

c.

```
        CH3           CH3
        |             |
CH3 — CH — CH2 — CH — CH3
```

d.

```
        CH3   CH3             CH3
        |     |               |
CH3 — CH — CH — CH2 — C — CH3
                        |
                        CH3
```

e.

```
CH3 — CH — CH2 — CH3
      |
CH2 — CH2 — CH2 — CH — CH3
                       |
                       CH3
```

f.

```
                CH3           CH3
                |             |
CH3 — CH2 — CH — CH — CH2 — CH — CH3
            |
            CH2
            |
            CH3
```

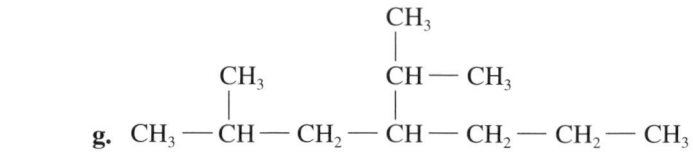

g.

Study Pack

10.6

Learning Goal

Write the structural formulas of alkanes and their isomers.

10.6 Structural Formulas

Suppose you are given the name 2,3-dimethylhexane and asked to write the structural formula or condensed structural formula of that compound. The name of the compound tells us the number of carbon atoms in the longest chain, the type of bonding, the names of any substituents, and where they are attached to the parent chain. We can break down the name 2,3-dimethylhexane in the following way.

2,3-Dimethylhexane

2,3-	di	methyl	hex	ane
Groups on carbons 2 and 3	Two identical groups	CH_3—	6 carbon atoms in parent chain	Single bonds

To draw the structural formula for the compound, we proceed as follows:

Step 1 Write the carbon atoms in the parent chain. The name hexane tells us that we need to draw a carbon chain containing six carbon atoms.

C—C—C—C—C—C **Hexane**

Step 2 Number the chain and attach the branches indicated. The numbers 2,3 in the name indicate that methyl groups, CH_3— , are attached to carbon 2 and carbon 3.

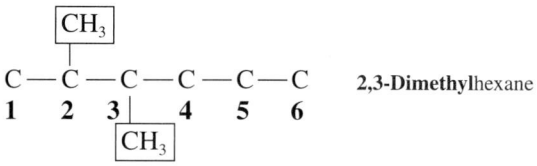

2,3-Dimethylhexane

Step 3 Add hydrogen and write a condensed structural formula.

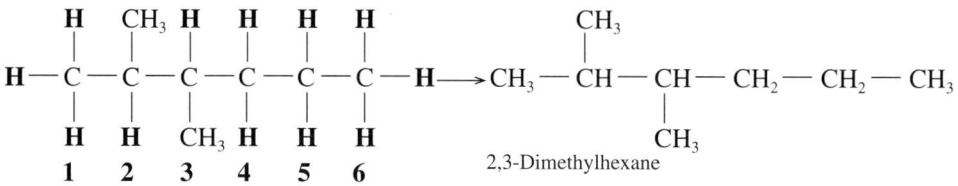

2,3-Dimethylhexane

S·A·M·P·L·E· **10.7** ·P·R·O·B·L·E·M

Writing a Condensed Structural Formula

Draw the condensed structural formula for 3-ethyl-5-methylheptane.

Solution

Because the parent name is heptane, we start by drawing a chain of seven carbon atoms.

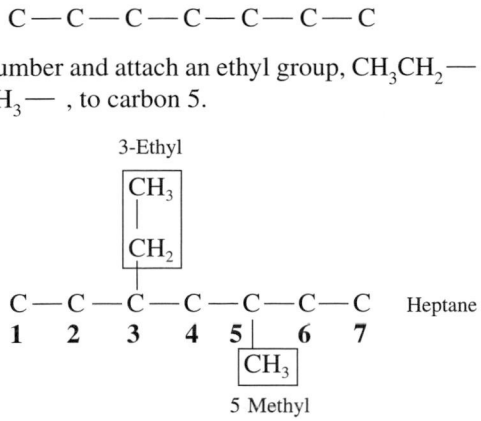

Add hydrogen and write the condensed structural formula.

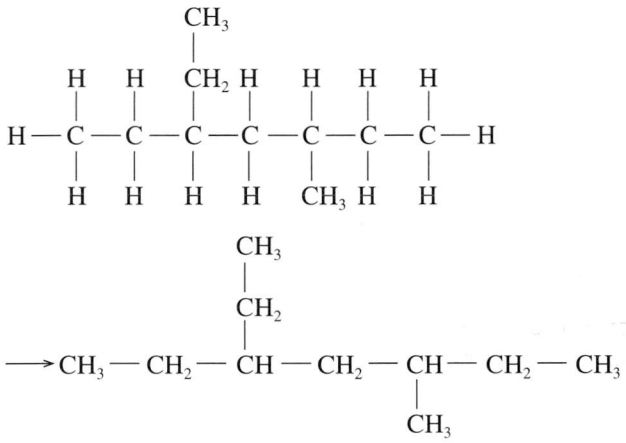

Study Check

Write the condensed structural formula of 2,2,4,4-tetramethylhexane.

Structural Isomers

Structural isomers are compounds that have the same molecular formula but different arrangements of atoms and therefore different structural formulas. For example, there are two different alkanes with the molecular formula C_4H_{10}. One is the unbranched chain; the other is a branched-chain. The ball-and-stick models of these two isomers are shown in Figure 10.8.

Condensed Structural Isomers for C_4H_{10}

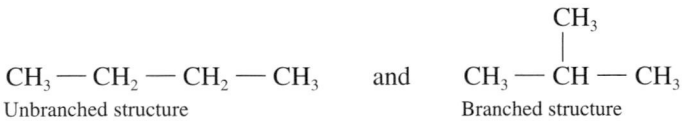

Although these two isomers contain the same type and number of atoms, the atoms are connected in different ways. The different structures of the isomers result in different physical and chemical properties such as boiling point, solubility, and reactivity as shown in Table 10.7.

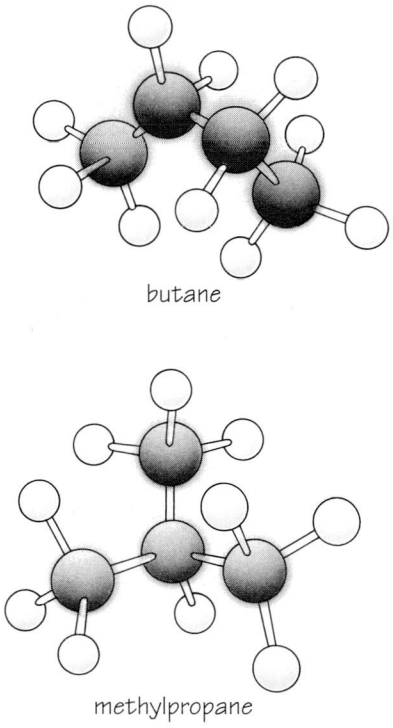

Figure 10.8 Ball-and-stick models of the isomers of C_4H_{10}.

Table 10.7 Some Properties of the Structural Isomers of C_4H_{10}

Condensed structural formulas	$CH_3—CH_2—CH_2—CH_3$	$CH_3—CH(CH_3)—CH_2$
Molar mass	58.1 g/mole	58.1 g/mole
Melting point	−138°C	−159°C
Boiling point	0.5°C	−12°C

10.6

Drawing Structural Isomers

There are three structural isomers of the molecular formula C_5H_{12}. Beginning with the unbranched-chain isomer, we write a carbon chain of five carbon atoms and fill in the appropriate number of hydrogen atoms:

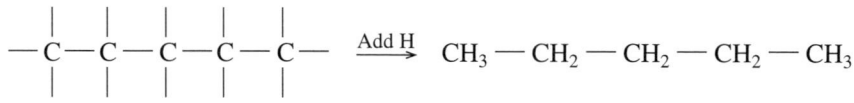

Five-carbon chain — Unbranched-chain isomer of C_5H_{12}

Both the other isomers of C_5H_{12} are branched-chain alkanes, but each must still have a total of five carbon atoms. We write a second isomer by writing a shorter chain of four carbon atoms and attaching the fifth carbon as a $CH_3—$ branch. Be sure to attach the $CH_3—$ to a carbon atom inside the chain. If the carbon group is added to an end of the carbon chain, you would just have a bent five-carbon chain.

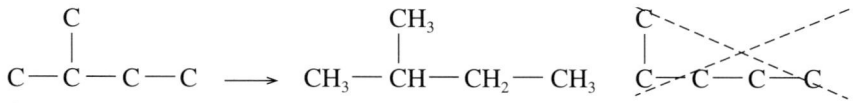

Carbon branch attached — A second isomer of C_5H_{12} — Not a branched isomer

If we move the $CH_3—$ to the other inside carbon in the chain, we get the same branched compound in reverse, not another isomer.

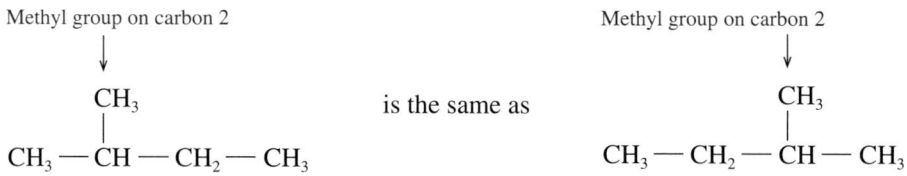

Chain is numbered from the left — Chain is numbered from the right

A third isomer is written by drawing a shorter chain of three carbon atoms and attaching two carbon atoms as $CH_3—$ branches to the center carbon. These are the only isomers we can draw with C_5H_{12}; any other arrangements will repeat one of the previous structures. The three isomers for C_5H_{12} are illustrated in Figure 10.9.

Attach two carbon branches — A third isomer of C_5H_{12}

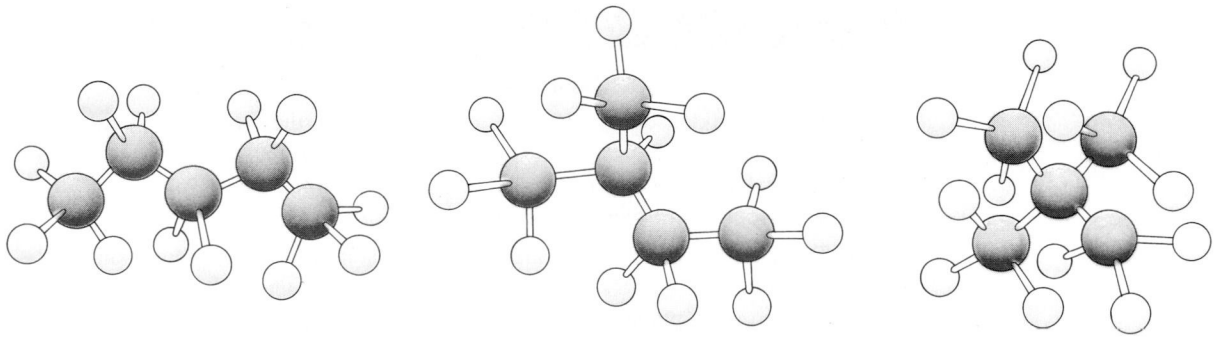

pentane 2-methylbutane 2,2-dimethylpropane

Figure 10.9 Ball-and-stick models of the structural isomers of C_5H_{12}.

S · A · M · P · L · E · **10.8** · P · R · O · B · L · E · M

Drawing Structural Isomers

There are five isomers that have the molecular formula C_6H_{14}. Draw their condensed structural formulas.

Solution

The first isomer is the unbranched-chain alkane with six carbon atoms.

$$CH_3 — CH_2 — CH_2 — CH_2 — CH_2 — CH_3$$

In the next two isomers, we write a five-carbon chain and attach a CH_3— to one of the carbon atoms inside the chain. In one isomer, the CH_3— group is attached to carbon 2 in one isomer, and to carbon 3 in the other isomer.

$$\begin{array}{ccccccccccc} & & CH_3 & & & & & & CH_3 & & \\ & & | & & & & & & | & & \\ CH_3 & — & CH & — & CH_2 & — & CH_2 & — & CH_3 & & \\ \mathbf{1} & & \mathbf{2} & & \mathbf{3} & & \mathbf{4} & & \mathbf{5} & & \end{array} \qquad \begin{array}{ccccccccccc} & & & & CH_3 & & & & & & \\ & & & & | & & & & & & \\ CH_3 & — & CH_2 & — & CH & — & CH_2 & — & CH_3 & & \\ \mathbf{1} & & \mathbf{2} & & \mathbf{3} & & \mathbf{4} & & \mathbf{5} & & \end{array}$$

Two more isomers are drawn using a four-carbon chain and attaching two CH_3 groups to carbon atoms within the chain.

$$\begin{array}{ccccccccc} & & CH_3 & & CH_3 & & & & \\ & & | & & | & & & & \\ CH_3 & — & CH & — & CH & — & CH_3 & & \end{array} \qquad \begin{array}{ccccccc} & & CH_3 & & & & \\ & & | & & & & \\ CH_3 & — & C & — & CH_2 & — & CH_3 \\ & & | & & & & \\ & & CH_3 & & & & \end{array}$$

Any other combination of atoms will repeat one of the structural isomers already drawn.

Study Check

Are the following pairs of compounds isomers or identical structures?

a. $\begin{array}{ccc} CH_3 & — & CH_2 \\ & & | \\ & & CH_2 — CH_3 \end{array}$ and $\begin{array}{ccc} CH_3 & & CH_3 \\ | & & | \\ CH_2 & — & CH_2 \end{array}$

b. $\begin{array}{c} CH_3 \\ | \\ CH_2 — CH_2 — CH_3 \end{array}$ and $\begin{array}{c} CH_3 \\ | \\ CH — CH_3 \\ | \\ CH_3 \end{array}$

QUESTIONS & PROBLEMS

Structural Formulas

10.31 Draw a condensed structural formula for each of the following alkanes:

a. butane
b. 3,3-dimethylpentane
c. 2,2,4-trimethyloctane
d. 3,3-dimethylhexane
e. 4-isopropyl-2-methylheptane

10.32 Draw a condensed structural formula for each of the following alkanes:

a. pentane
b. 3-ethyl-2-methylpentane
c. 4-propylheptane
d. 2,2,3,5-tetramethylhexane
e. 4-ethyl-2,2-dimethyloctane

10.6

10.33 What are structural isomers?

10.34 How would you determine whether two condensed structural formulas represented structural isomers?

10.35 Which of the following pairs of formulas represent structural isomers?

a. $CH_3-CH_2-CH_3$ and $CH_3-CH(CH_3)-CH_3$

b. $CH_3-CH_2-CH_2-CH_3$ and $CH_3-CH(CH_3)-CH_3$

c. $CH_3-CH(CH_3)-CH_2-CH_3$ and $CH_3-CH(CH_3)-CH_2-CH(CH_3)-CH_3$

d. $CH_3-CH_2-C(CH_3)_2-CH_3$ and $CH_3-CH_2-CH_2-CH_2-CH(CH_3)-CH_3$

10.36 Which of the following pairs of formulas represent structural isomers?

a. $CH_3-C(CH_3)_2-CH_3$ and $CH_3-CH(CH_3)-CH_3$

b. $CH_3-CH(CH_3)-CH(CH_3)-CH_2-CH_3$

and $CH_3-CH(CH_3)-CH_2-CH(CH_3)-CH_3$

c. $CH_3-CH_2-CH(CH_3)-CH_3$ and $CH_3-CH(CH_3)-CH_2-CH_3$

d. $CH_3-CH_2-CH_2-CH_3$ and $CH_3-C(CH_3)_2-CH_3$

10.37 Indicate whether each of the following pairs of alkanes are structural isomers:
a. propane and 2-methylpropane
b. butane and 2-methylpropane
c. 3-methylpentane and 2,2-dimethylpropane
d. hexane and 3,3-dimethylpentane

10.38 Indicate whether each of the following pairs of alkanes are structural isomers:
a. 2-methylpentane and 3-methylpentane
b. 3-methylpentane and 2,2,3-trimethylbutane
c. 3-methylhexane and 3,3-dimethylpentane
d. 3-ethylhexane and 2,2,3-trimethylpentane

10.7 Cycloalkanes

Study Pack

Learning Goal

Give the name for a cycloalkane; draw the condensed structural and geometric formulas.

Up to now, we have only looked at carbon atoms arranged in chains. However, hydrocarbons can also form simple rings or cyclic structures called **cycloalkanes.** The simplest cycloalkane, cyclopropane, consists of a ring of three carbon atoms. (See Figure 10.10). A cycloalkane has two fewer hydrogen atoms than the corresponding alkane because there are no end carbons. Thus, there are eight hydrogens in the open chain of propane but only six hydrogens in the ring of cyclopropane.

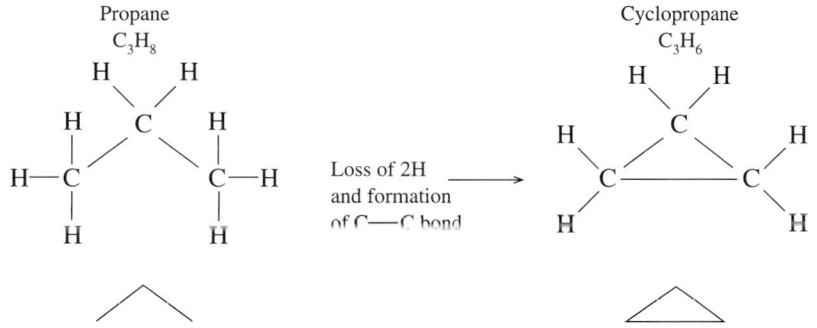

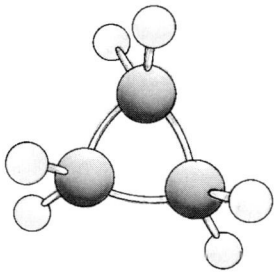

Figure 10.10 A ball-and-stick model of the ring structure of cyclopropane.

The names of cycloalkanes are similar to the names of the straight-chain alkanes with the same number of carbon atoms except that the prefix *cyclo* appears in front of the name. The structures of cycloalkanes are often drawn as simple geometric shapes. No hydrogen atoms are shown, although it is understood that they exist in the actual compound. For example, cyclopropane can be depicted as a triangle. Each corner of the triangle represents a carbon atom that is attached to two other carbon atoms and two hydrogen atoms. More examples of the geometric shapes for cycloalkanes are given in Table 10.8.

Naming Cycloalkanes with Alkyl Groups

Cycloalkanes with branches are named in a way similar to the alkanes. The prefix *cyclo* is placed in front of the alkane name for the same number of carbon atoms in the ring. The name of an attached alkyl group is placed in front of the cycloalkane name. The ring is not numbered for one side group. If there are two or more side groups, the carbon ring is numbered in the direction that gives the lowest numbers to the branches starting

10.7

Table 10.8 Formulas of Some Common Cycloalkanes

Condensed Structural Formula	Geometric Formula	Name
CH_2 / H_2C — CH_2		Cyclopropane
H_2C — CH_2 / H_2C — CH_2		Cyclobutane
CH_2 / H_2C CH_2 / H_2C — CH_2		Cyclopentane
CH_2 / H_2C CH_2 / H_2C CH_2 / CH_2		Cyclohexane

at one of the substituents. You may count clockwise or counterclockwise around the ring, whichever gives the lowest combination of numbers to the substituents.

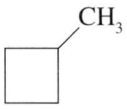

Methylcyclobutane

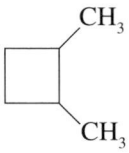

1,2-Dimethylcyclobutane

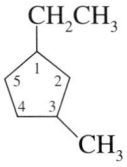

1-Ethyl-3-methylcyclopentane

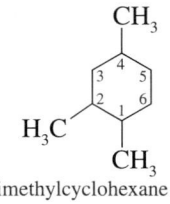

1,2,4- Trimethylcyclohexane
not 1,4,5-Trimethylcyclohexane
not 1,4,6-Trimethylcyclohexane

S·A·M·P·L·E· 10.9 ·P·R·O·B·L·E·M

Naming Cycloalkanes

Give the IUPAC name for each of the following cycloalkanes:

a. 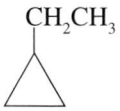

b. 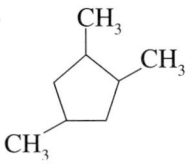

Solution

a. The triangle indicates a cyclic alkane with three carbon atoms, *cyclopropane.* The side group, $—CH_2CH_3$, is an *ethyl* group. The compound is named *ethylcyclopropane.*

b. The ring of five carbon atoms is *cyclopentane.* The ring is numbered to give the lowest possible numbers to the *methyl* groups. It is named as 1,2,4-trimethylcyclopentane.

Study Check

What is the IUPAC name for the following cyclic alkane?

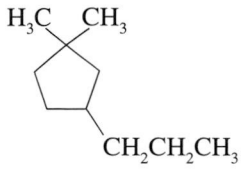

S·A·M·P·L·E· 10.10 ·P·R·O·B·L·E·M

Writing Condensed Structural Formulas of Cycloalkanes

Write the structure for 1,4-dimethylcyclohexane.

Solution

The parent name, *cyclohexane,* tells us that the parent chain is a ring of six carbon atoms. The rest of the name tells us there are methyl groups on carbons 1 and 4.

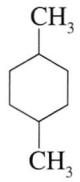

Study Check

What is the structural formula for 1-ethyl-2-methylcyclopentane?

Q·U·E·S·T·I·O·N·S & P·R·O·B·L·E·M·S

Cycloalkanes

10.39 How does butane differ from cyclobutane?

10.40 How does 2-methylpentane differ from methylcyclopentane?

10.41 Give the IUPAC name for each of the following cycloalkanes:

a.

b.

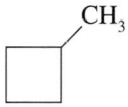

c.

d.

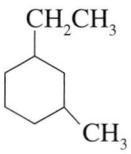

e.

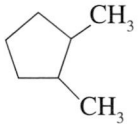

10.42 Give the IUPAC name for each of the following cycloalkanes:

a.

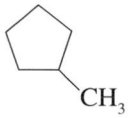

b.

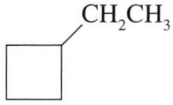

c.

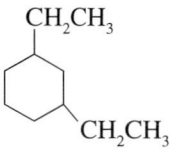

d. 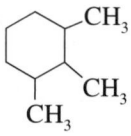

e. 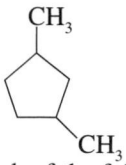

10.43 Draw the structure for each of the following cycloalkanes, using the geometric formula:

a. cyclopropane **b.** 1,2-dimethylcyclobutane
c. propylcyclohexane **d.** ethylcyclopentane
e. 3-ethyl-1,1-dimethylcyclohexane

10.7

10.44 Draw a structure for each of the following cycloalkanes, using the geometric formula:

a. cyclobutane **b.** methylcyclopropane
c. 1,1-dimethylcyclopentane **d.** 1-ethyl-4-methylcyclohexane
e. 1,2,4-trimethylcyclopentane

10.45 Are the following pairs of condensed structural formulas isomers, identical, or neither?

a. 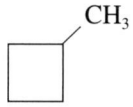

and 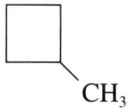

b. 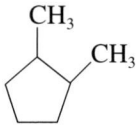

and 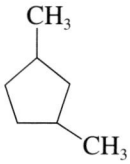

c. 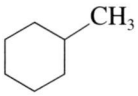

and 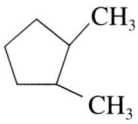

10.46 Are the following pairs of condensed structural formulas isomers, identical, or neither?

a. 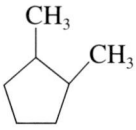

and 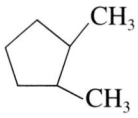

b. 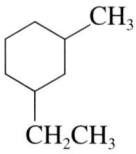

and 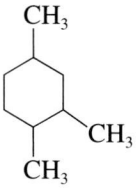

c. 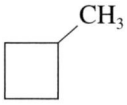

and 

10.8 Aromatic Hydrocarbons

Study Pack

Learning Goal

Describe the bonding in benzene; name aromatic compounds, and write their structural formulas.

10.8

In 1825, Michael Faraday isolated a hydrocarbon called benzene, which had a molecular formula C_6H_6. Because many compounds containing benzene had fragrant odors, the family of benzene compounds became known as **aromatic compounds.** A molecule of **benzene** consists of a ring of six carbon atoms with one hydrogen atom attached to each carbon. Each carbon atom uses 3 valence electrons to attach to the hydrogen atom and two adjacent carbons. That leaves 1 valence electron to share in a double bond with an adjacent carbon. In 1865, August Kekulé dreamed that the carbon atoms were arranged in a flat ring with alternating single and double bonds between the carbon atoms. This idea led to two ways of writing the benzene structure, as follows:

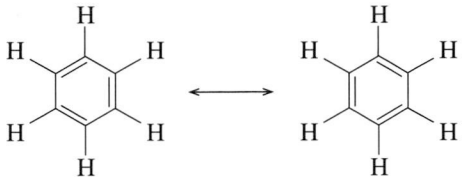

Structures for benzene

However scientists knew that the bonds in benzene are equal in length and behave more like a blend of a single and a double bond. Benzene is also represented as a hexagon with a circle in the center.

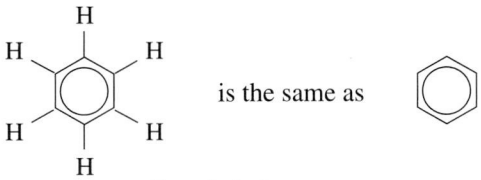

Formula for benzene

HEALTH NOTE

Aromatic Compounds in Health and Medicine

Aromatic compounds are common in nature and in medicine. Toluene is used as a starting reactant to make drugs, dyes, and explosives such as TNT (trinitrotoluene). The benzene ring is found in some amino acids (the building blocks of proteins), in pain relievers such as aspirin, acetaminophen, and ibuprofen, and flavorings such as vanillin.

Toluene

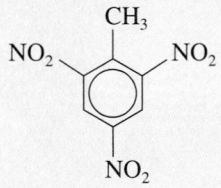

TNT (2,4,6-trinitrotoluene)

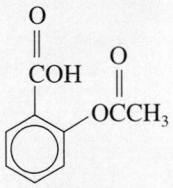

Aspirin

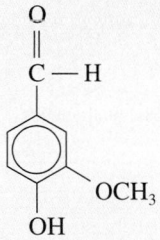

Vanillin

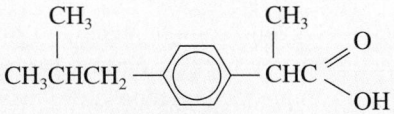

Ibuprofen

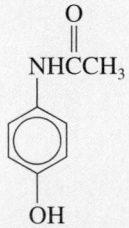

Acetaminophen

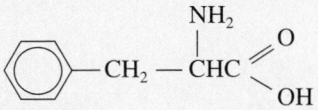

Phenylalanine, an amino acid

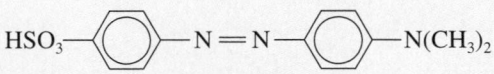

Methyl orange, textile dye

10.8 Naming Aromatic Compounds

In most aromatic compounds, benzene is named as the parent chain. Alkyl groups may be attached by taking the place of a hydrogen on one of the carbon atoms. A halogen atom may also replace a hydrogen on benzene. Then the halogen is named as a halo group, such as chloro (Cl) and bromo (Br).

Methylbenzene (toluene)

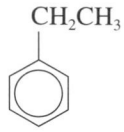

Ethylbenzene

Chlorobenzene

When two groups are attached to the benzene ring, the ring is numbered to give the lowest numbers to the attached groups. Sometimes, the prefixes *ortho, meta,* and *para* are used to indicate the relative position of two groups. The prefix *ortho (o)* indicates a 1,2 arrangement of groups, *meta (m)* is a 1,3 arrangement, and *para (p)* is used for a 1,4 arrangement.

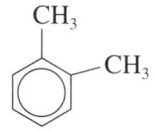

1,2-dimethylbenzene
(*o*-dimethylbenzene)

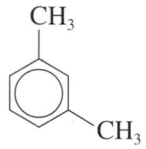

1,3-dimethylbenzene
(*m*-dimethylbenzene)

1,4-dimethylbenzene)
(*p*-dimethylbenzene)

HEALTH NOTE

Polycyclic Aromatic Hydrocarbons

Large aromatic compounds known as polycyclic aromatic hydrocarbons are formed by fusing together two or more benzene rings edge-to-edge. In a fused ring compound, neighboring benzene rings share two or more carbon atoms. Naphthalene with two benzene rings is well known for its use in mothballs. Anthracene with three rings is used in the manufacture of dyes.

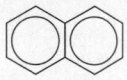

Naphthalene

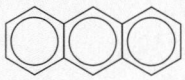

Anthracene

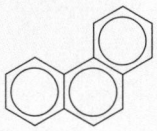

Phenanthrene

When a polycyclic compound contains phenanthrene, it may act as a carcinogen, a substance known to cause cancer. For example, benzpyrene, a product of combustion, has been identified in coal tar, tobacco smoke, barbecued meats, and automobile exhaust.

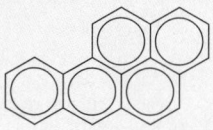

Benzpyrene

Compounds containing five or more fused benzene rings such as benzpyrene are potent carcinogens. The molecules interact with the DNA in the cells, causing abnormal cell growth and cancer. Increased exposure to carcinogens increases the chance of DNA alterations in the cells. (See Figure 10.11.)

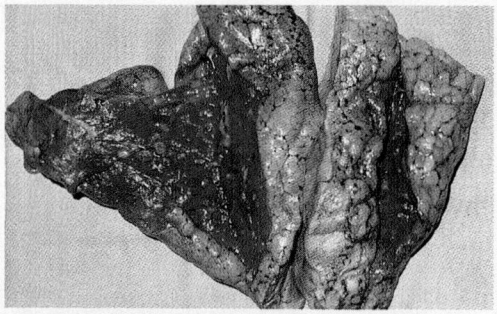

Figure 10.11 Some aromatic compounds in cigarette smoke cause cancer, as seen in the lung tissue of a heavy smoker.

S·A·M·P·L·E 10.11 · P·R·O·B·L·E·M

Naming Aromatic Compounds

Name the following compounds.

a. 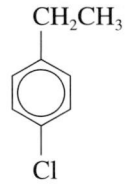

b. 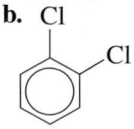

Solution

a. 1-chloro-4-ethylbenzene; *p*-chloroethylbenzene
b. 1,2-dichlorobenzene; *o*-dichlorobenzene

Study Check

Name the following compound.

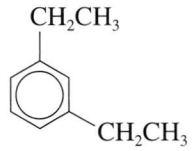

Q·U·E·S·T·I·O·N·S & P·R·O·B·L·E·M·S

Aromatic Hydrocarbons

10.47 Cyclohexane and benzene each have six carbon atoms. How are they different?

10.48 In the Health Note "Aromatic Compounds in Health and Medicine," what part of each molecule is the aromatic portion?

10.49 Name the following aromatic compounds:

a. 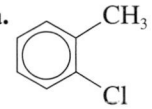

b. 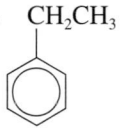

c. 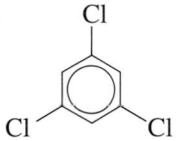

10.50 Name the following aromatic compounds:

a. 

b. 

c. 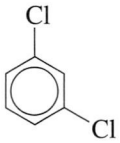

10.51 Draw the structural formulas for each of the following compounds:

a. methylbenzene
b. *m*-dichlorobenzene
c. 1-ethyl-4-methylbenzene
d. *p*-chlorotoluene

10.52 **a.** benzene
b. *o*-chloromethylbenzene
c. propylbenzene
d. 1,2,4-trichlorobenzene

Chapter Review

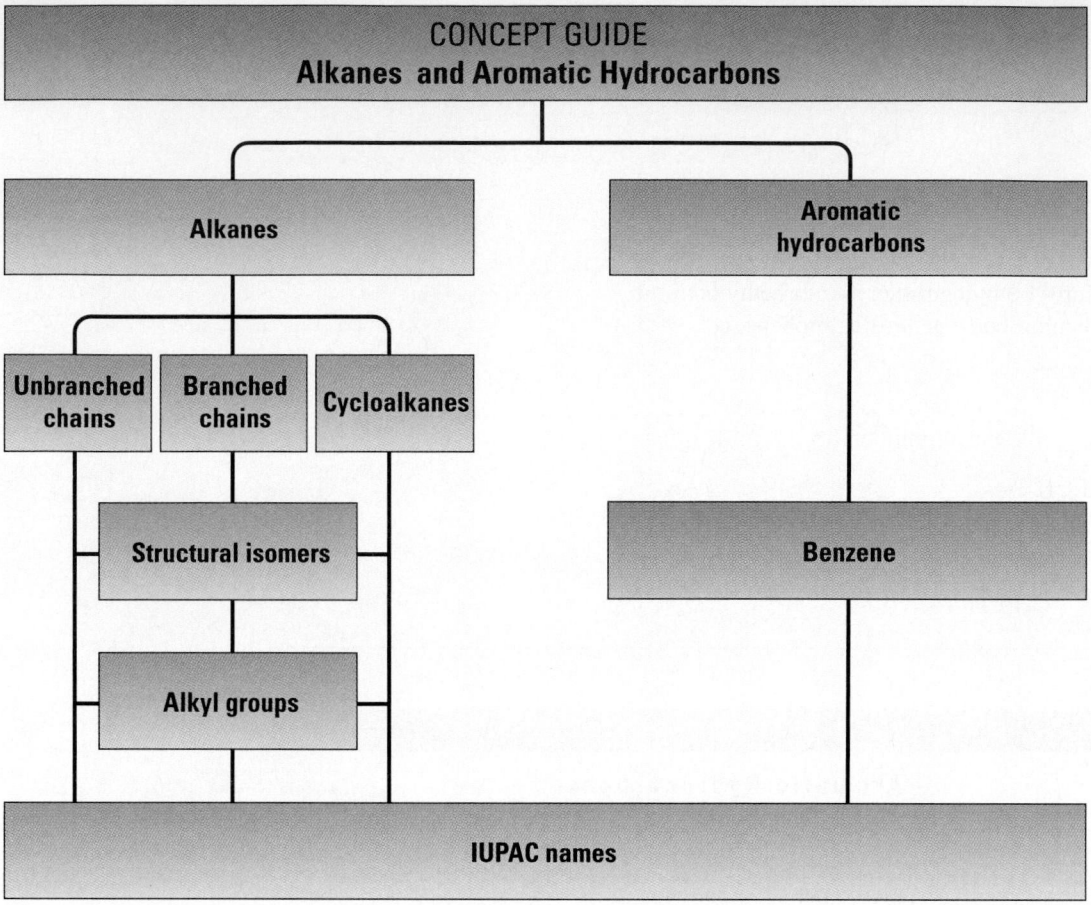

Chapter Summary

10.1 Organic Compounds

Organic compounds are compounds of carbon that typically have covalent bonds, have low melting and boiling points, burn vigorously, are nonelectrolytes, and are soluble in nonpolar solvents. The hydrocarbon family consists of compounds composed of carbon and hydrogen.

10.2 Bonding in Hydrocarbons

Alkanes are hydrocarbons that have only single bonds, C—C and C—H. Each carbon in an alkane always has four bonds arranged so that the bonded atoms are in the corners of a tetrahedron. A complete structural formula shows a separate line for every bonded atom; a condensed structural formula depicts each carbon atom and its attached hydrogen atoms as groups. A molecular formula shows only the number of each type of atom in the compound but not the arrangement of atoms.

Complete Structural Formula	*Condensed Structural Formula*	*Molecular Formula*
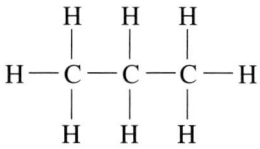	$CH_3—CH_2—CH_3$ or $CH_3CH_2CH_3$	C_3H_8

10.3 Naming Alkanes

The IUPAC system is a set of rules used to name organic compounds in a systematic manner. The name contains a stem that indicates the number of carbon atoms, and an ending that shows the family of the compound. For example, in the name *propane,* the stem *prop* indicates a chain of three carbon atoms and the ending *ane* indicates single bonds (alkane).

10.4 Properties of Alkanes

The alkanes are nonpolar, less dense than water, and mostly chemically unreactive, except they burn vigorously. Thus, they are widely used as fuels such as natural gas, methane, gasoline, and diesel fuels. In combustion, an alkane at a high temperature reacts rapidly with oxygen to produce carbon dioxide, water, and a great amount of heat.

10.5 Branched Alkanes

Branches of carbon atoms attached to a parent chain are named as alkyl groups. In a branched alkane, the longest continuous chain is named as the parent alkane. The alkyl groups attached to the chain are numbered and listed alphabetically in front of the name of the parent chain.

10.6 Structural Formulas

In an unbranched alkane, the carbon atoms are connected one after the other in a chain. A branched-chain alkane contains at least one alkyl branch or substituent. Structural isomers have the same molecular formula but differ in the sequence of atoms in each of their structural formulas.

10.7 Cycloalkanes

In cycloalkanes, the carbon atoms of alkanes form a cyclic structure. They are named by placing the prefix *cyclo* in front of the name of the alkane with the same number of carbon atoms. Branches are named as alkyl groups and must be numbered if there are two or more on a ring.

10.8 Aromatic Hydrocarbons

Most aromatic compounds contain benzene, a cyclic structure containing six CH units. The structure of benzene is represented as a hexagon with a circle in the center. Many aromatic compounds use the parent name benzene, although common names such as toluene are retained. The benzene ring is numbered and the branches are listed in alphabetical order. For two branches, the positions are often shown by the prefixes *ortho* (1,2-), *meta* (1,3-), and *para* (1,4-).

Summary of Naming

Type	Example	Characteristic	Structure
Alkane	Propane	Single C—C, C—H bonds	$CH_3—CH_2—CH_3$
Cycloalkane	Cyclopropane	Carbon ring	
Aromatic	Benzene	Ring of six CH units	
	Methylbenzene or toluene		
	1,4-dichlorobenzene or *para*-dichlorobenzene		

Summary of Reactions

Combustion

$$\text{Alkane} + O_2 \longrightarrow CO_2 + H_2O + \text{heat}$$

$$CH_4 + 2O_2 \longrightarrow CO_2 + 2H_2O + \text{heat}$$

Key Terms

alkane A hydrocarbon that has only single carbon-carbon bonds.

alkyl group A group derived from an alkane by removing one hydrogen atom. The ending of the alkane name is changed to *yl* to name it as an alkyl branch.

aromatic compound A compound that contains the ring structure of benzene.

benzene A ring of six carbon atoms each of which is attached to a hydrogen atom, C_6H_6.

branch An alkyl group attached to a parent carbon chain or ring.

branched alkane A hydrocarbon containing at least one alkyl group attached to the parent chain.

combustion A chemical reaction in which an alkane reacts with oxygen to produce CO_2, H_2O, and heat.

complete structural formula A type of formula in which each bond in the compound is shown as a line, C—H or C—C, to give the arrangement of all the atoms in the compound.

condensed structural formula A type of formula that illustrates the arrangement of the carbon atoms in an organic compound by grouping each carbon atom and its hydrogen atoms.

cycloalkane An alkane that exists as a ring or cyclic structure.

hydrocarbon An organic compound consisting of only carbon and hydrogen atoms.

organic compounds Compounds that contain carbon in covalent bonds. Most are nonpolar with low melting and boiling points, insoluble in water, soluble in nonpolar solvents, and flammable.

parent chain The longest unbranched chain of carbon atoms in a formula.

saturated hydrocarbon A compound of carbon and hydrogen that has only single carbon–carbon bonds (C—C).

structural isomer An organic compound that has the same molecular formula as another organic compound but a different structure.

Chapter Practice

10.53 Compare organic and inorganic compounds in terms of:

a. types of bonds **b.** solubility in water
c. melting points **d.** flammability

10.54 Identify each of the following as an organic or inorganic compound.

a. Na_2SO_4 **b.** CH_3CH_3
c. a compound that melts at 875°C
d. a compound that burns vigorously
e. methylcyclopentane
f. a compound that floats on water

10.55 Name the following alkanes:

a. $CH_3CH_2CH_3$

b. $CH_3CH_2CH_2CH_3$

c.
```
        CH3   CH3
         |     |
CH3 — CH — CH — CH3
```

d.
```
        CH3     CH3
         |       |
CH3CH2CHCH2CHCH3
```

10.56 Draw the structural formulas of the following alkanes:

a. octane
b. 3-methylpentane
c. 2,2-dimethylbutane
d. 2,3,4-trimethylhexane
e. 2,2,3,4-tetramethylheptane

10.57 In an automobile engine, "knocking" occurs when the combustion of gasoline occurs too rapidly. The octane number of gasoline represents the ability of a gasoline mixture to reduce knocking. A sample of gasoline is compared with heptane, rated 0 because it reacts with severe knocking, and 2,2,4-trimethylpentane (isooctane), which has a rating of 100 because of its low knocking. Write the condensed structural formula, molecular formula, and equation for the complete combustion of isooctane.

10.58 The density of pentane, a component of gasoline, is 0.63 g/mL. The heat of combustion for pentane is 845 kcal per mole (3536 kJ/mole).

a. Write an equation for the complete combustion of pentane.
b. What is the molar mass?
c. How much heat is produced when 1 gallon of pentane is burned? (1 gallon = 3.78 liters)
d. How many liters of CO_2 at STP are produced from the complete combustion of 1 gallon of pentane?

10.59 Name two structural isomers of each of the following:

a. hexane
b. 3-ethylpentane
c. 2,2-dimethylbutane
d. 1,2-dimethylcyclohexane

10.60 The following names are incorrect. Why? Write the correct IUPAC name.
a. 3-methylbutane
b. 3,4-dimethylpentane
c. 2,4-dimethylcyclohexane
d. 1,4-dimethylbutane

10.61 Identify each of the following formulas as molecular, complete structural, or condensed structural.
a. C_4H_{10}
b. $CH_3CH_2CH_2CH_2CH_3$

c.
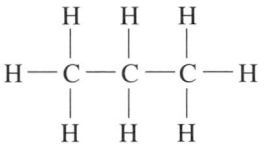

d. $CH_3CH_2CH_3$
e. C_6H_6

10.62 Using geometric formulas, draw the structure of the following cycloalkanes:
a. ethylcyclopropane
b. 1,3-dimethylcyclohexane
c. methylcyclopentane
d. 1,1-dimethylcyclopentane
e. 1-ethyl-3-methylcyclobutane

10.63 Write the IUPAC name of each of the following cycloalkanes:

a.

b.
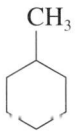

c.
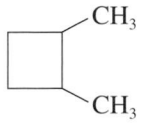

d.
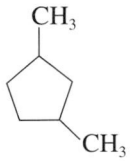

e.
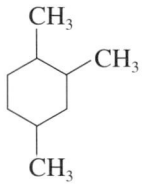

10.64 Write a balanced equation for the combustion of each of the following:
a. propane **b.** C_5H_{12}
c. benzene **d.** octane

10.65 Write four isomers of dimethylcyclohexane.

10.66 Write the structural formulas for each of the following:
a. ethylbenzene
b. *m*-dichlorobenzene
c. 1,2,4-trimethylbenzene
d. *p*-dimethylbenzene

10.67 Name each of the following aromatic compounds:

a.

b.
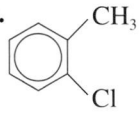

c.
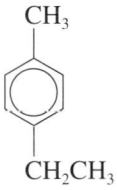

d.
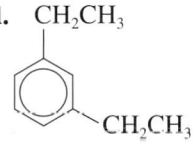

Answers

Answers to Study Checks

10.1 inorganic compound

10.2
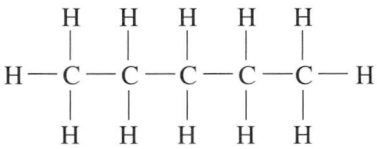

$CH_3CH_2CH_2CH_2CH_3$

C_5H_{12}

10.3
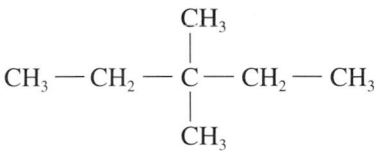

10.4 $CH_3CH_2CH_2CH_2CH_2CH_2CH_3$, heptane

10.5 $C_7H_{16} + 11O_2 \longrightarrow 7CO_2 + 8H_2O$

10.6 3,6-dimethyloctane

10.7

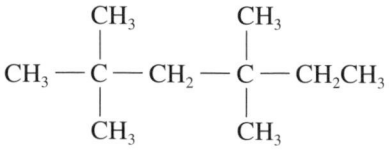

10.8 **a.** identical **b.** isomers

10.9 1,1-dimethyl-3-propylcyclopentane

10.10

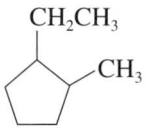

10.11 1,3-diethylbenzene; *m*-diethylbenzene

Answers to Selected Questions and Problems

10.1 Vegetable oil, gasoline, a plastic raincoat, and sugar are all composed of organic compounds.

10.3 Historically, organic chemistry involved the idea of a vital force, a force where a living organism was involved. Now, organic chemistry is the study of compounds containing carbon and hydrogen.

10.5 **a.** inorganic **b.** organic **c.** organic **d.** inorganic **e.** organic

10.7 **a.** propane **b.** propane **c.** KCl **d.** KCl **e.** propane

10.9 **a.**

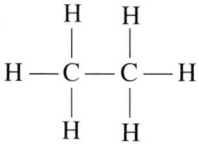

b. CH_3CH_3

10.11 **a.** $CH_3CH_2CH_2CH_2CH_3$

b.

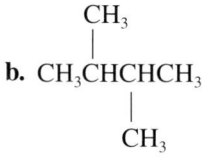

10.13

a.

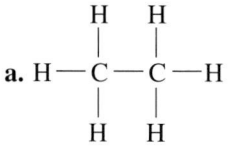

CH_3CH_3

b.

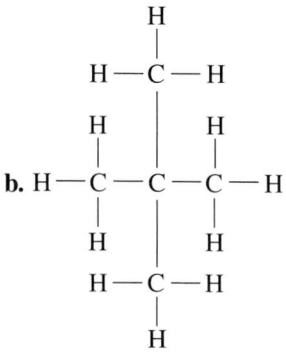

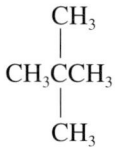

c.

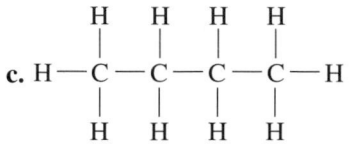

$CH_3CH_2CH_2CH_3$

d.

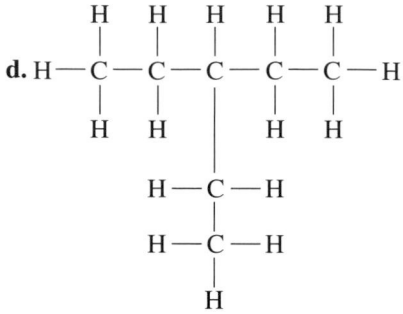

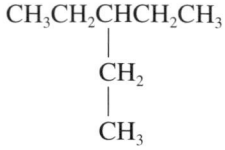

10.15 In an unbranched alkane, the carbon atoms are connected one after the other. There are no branches.

10.17 **a.** CH_4
b. CH_3CH_3
c. $CH_3CH_2CH_2CH_2CH_3$
d. $CH_3CH_2CH_2CH_2CH_2CH_3$

10.19 **a.** ethane **b.** butane **c.** hexane **d.** octane

10.21 **a.** $CH_4 + 2O_2 \longrightarrow CO_2 + 2H_2O$
b. $C_3H_8 + 5O_2 \longrightarrow 3CO_2 + 4H_2O$
c. $2C_4H_{10} + 13O_2 \longrightarrow 8CO_2 + 10H_2O$

10.23 **a.** $CH_3CH_2CH_2CH_2CH_2CH_2CH_3$
b. liquid
c. No.
d. Yes.
e. $C_7H_{16} + 11O_2 \longrightarrow 7CO_2 + 8H_2O$

10.25 **a.** $-CH_3$ **b.** $-CH_2CH_2CH_3$

10.27 **a.** Methane is a compound with the formula CH_4; a methyl group is an alkyl group with a formula $-CH_3$.
b. Ethane is an alkane with the formula CH_3CH_3. The ethyl group is an alkyl group that has a formula of $-CH_2CH_3$.

10.29 **a.** 2-methylpropane
b. 2-methylpentane
c. 2,3-dimethylpentane
d. 2,4-dimethylhexane
e. 2,4-dimethylheptane
f. 3-ethyl-2-methylpentane
g. 4-propylheptane

10.31 **a.** $CH_3CH_2CH_2CH_3$

b. $CH_3CH_2C(CH_3)_2CH_2CH_3$

c. $CH_3C(CH_3)_2CH_2CH(CH_3)CH_2CH_2CH_2CH_3$

d. $CH_3CH_2C(CH_3)_2CH_2CH_2CH_3$

e. $CH_3CH(CH_3)CH_2CH(CH_3)CH_2CH_2CH_3$

10.33 Structural isomers have the same molecular formula but a different sequence of atoms.

10.35 **a.** not structural isomers
b. structural isomers
c. not structural isomers
d. structural isomers

10.37 **a.** These are not isomers.
b. These are isomers.
c. These are not isomers.
d. These are not isomers.

10.39 Butane has four carbon atoms in an unbranched chain and a formula of C_4H_{10}. Cyclobutane has four carbon atoms in a cyclic structure and a formula of C_4H_8.

10.41 **a.** cyclobutane
b. methylcyclobutane
c. cyclopentane
d. 1-ethyl-3-methylcyclohexane
e. 1,2-dimethylcyclopentane

10.43 **a.**

b.

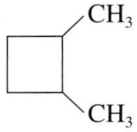

c.

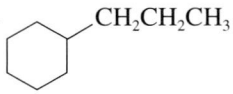

d.

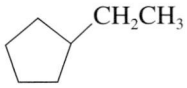

e.

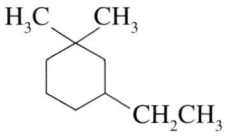

10.45 **a.** identical **b.** isomers **c.** isomers

10.47 Cyclohexane, C_6H_{12}, is a cycloalkane in which six carbon atoms in a ring are linked by single bonds. In benzene, C_6H_6, alternating single and double bonds link the six carbon atoms.

10.49 **a.** 1-chloro-2-methylbenzene; *o*-chloromethylbenzene; *o*-chlorotoluene
b. ethylbenzene
c. 1,3,5-trichlorobenzene

10.51 **a.**

b.

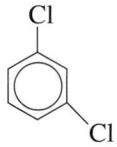

c.

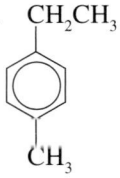

d.

10.53 **a.** Organic compounds contain covalent bonds: inorganic compounds have ionic and polar bonds.
b. Most organic compounds are not soluble in water: most inorganic compounds are soluble in water.
c. Organic compounds have low melting points, inorganic compounds have higher melting points.
d. Organic compounds burn easily; inorganic compounds do not.

10.55 **a.** propane **b.** butane
c. 2,3-dimethylbutane **d.** 2,4-dimethylhexane

10.57 Isooctane has the formula

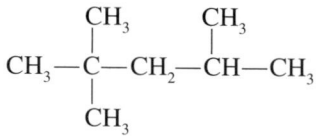

and a molecular formula of C_8H_{18}. The combustion reaction is $2C_8H_{18} + 25O_2 \longrightarrow 16CO_2 + 18H_2O$.

10.59 There are many isomers. Here are some possibilities.
a. 2-methylpentane; 3-methylpentane; 2,2-dimethylbutane, 2,3-dimethylbutane
b. 2-methylhexane; 3-methylhexane, 2,3-dimethylpentane, 2,2- dimethylpentane, 3,3-dimethylpentane, 2,2,3-trimethylbutane
c. 2-methylpentane, 3-methylpentane, 2,3-dimethylbutane
d. 1,1-dimethylcyclohexane, 1,3-dimethylcyclohexane, 1,4-dimethylcyclohexane, 1,2,3-trimethylcyclopentane, 1,2,4-trimethylcyclopentane, 1-ethyl-3-methylcyclopentane

10.61 **a.** molecular **b.** condensed structural
c. complete structural **d.** condensed structural
e. molecular

10.63 **a.** cyclopentane
b. methylcyclohexane
c. 1,2-dimethylcyclobutane
d. 1,3-dimethylcyclopentane
e. 1,2,4-trimethylcyclohexane

10.65

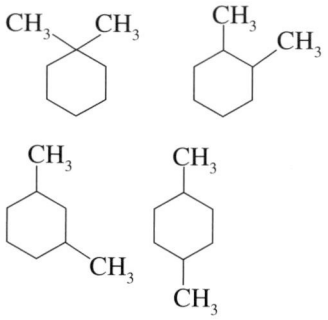

10.67 **a.** methylbenzene; toluene
b. 1-chloro-2-methylbenzene; *o*-chlorotoluene
c. 1-ethyl-4-methylbenzene; *p*-ethylmethylbenzene; *p*-ethyltoluene
d. 1,3-diethylbenzene; *m*-diethylbenzene

11

Haloalkanes, Alkenes, and Alkynes

Bubble wrap is made from two sheets of polypropylene, a type of plastic. The first sheet passes over a surface perforated by thousands of small holes that form pockets of air in the sheet, which are then sealed shut by the second sheet.

Looking Ahead

11.1	Functional Groups
11.2	Haloalkanes
11.3	Alkenes and Alkynes
11.4	Geometric Isomers of Alkenes
11.5	Addition Reactions
11.6	Polymers of Alkenes

www.chemplace.com/college/timberlake

Visit the web site above to test yourself, run simulations, and review key concepts from this chapter. For even more interactive review and quizzing, use the two CD-ROMs included with the book.

In this chapter, we will investigate the haloalkanes and unsaturated hydrocarbons, alkenes (double bonds), and alkynes (triple bonds). We have seen that the single bonds in the alkanes are not very reactive. In contrast, unsaturated hydrocarbons are highly reactive because the electrons involved in a double or triple bond are much more available for reactions. As a result, different atoms or groups of atoms can be added to alkenes and alkynes to prepare other kinds of organic compounds.

Study Pack

Chem of Life Organic Molecules: Functional Groups 1–6

11.1 Learning Goal

Classify organic compounds according to their functional groups.

11.1 Functional Groups

Organic compounds number in the millions, and more are synthesized every day. The task of learning organic chemistry seems overwhelming until we discover that each family of compounds has a characteristic structural feature that sets it apart from the other families. In Chapter 10 we learned that the alkanes, cycloalkanes, and aromatic hydrocarbons burn in oxygen to give carbon dioxide and water. However, they are generally unreactive compared to most other organic compounds.

A reactive organic compound has one section within its molecule where chemical reactions typically take place. This special feature, called a **functional group,** determines the family name and chemical reactivity of all the compounds with that functional group. A functional group is a specific grouping of atoms that will always react in a predictable way. It may be the carbon–carbon double bond in an alkene, an atom of oxygen in an ether, a hydroxyl group (—OH) in alcohols, or a nitrogen group such as $—NH_2$ in amines. Some typical functional groups are listed in Table 11.1. When we identify the functional group or groups in an organic compound, even a complex one, we can predict many of its chemical properties.

S·A·M·P·L·E 11.1 ·P·R·O·B·L·E·M

Functional Groups

Identify the family of each of the following organic compounds according to their functional groups:

a. $CH_3CH_2—OH$
b. $CH_3CH{=}CHCH_2CH_3$
c. $CH_3CH_2CH_2—NH_2$

Solution

a. alcohol
b. alkene
c. amine

Study Check

What is the organic family of the following compound?

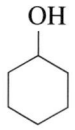

Table 11.1 Some Families of Organic Compounds

Family	Functional Group	Example
Alkane	$-\overset{\vert}{\underset{\vert}{C}}-\overset{\vert}{\underset{\vert}{C}}-$	H_3C-CH_3
Alkene	$-\overset{\vert}{C}=\overset{\vert}{C}-$	$H_2C=CH_2$
Alkyne	$-C\equiv C-$	$HC\equiv CH$
Aromatic		
Haloalkane	$-F$, $-Cl$, $-Br$, or $-I$	$CH_3-\mathbf{Cl}$
Alcohol	$-OH$	$CH_3-\mathbf{OH}$
Ether	$-O-$	$CH_3-\mathbf{O}-CH_3$
Thiol	$-SH$	$CH_3-\mathbf{SH}$
Aldehyde	$-\overset{O}{\overset{\Vert}{C}}H$	$CH_3-\overset{\mathbf{O}}{\overset{\Vert}{\mathbf{C}}}\mathbf{H}$
Ketone	$-\overset{O}{\overset{\Vert}{C}}-$	$CH_3-\overset{\mathbf{O}}{\overset{\Vert}{\mathbf{C}}}-CH_3$
Carboxylic acid	$-\overset{O}{\overset{\Vert}{C}}-OH$	$CH_3-\overset{\mathbf{O}}{\overset{\Vert}{\mathbf{C}}}-\mathbf{OH}$
Ester	$-\overset{O}{\overset{\Vert}{C}}-O-$	$CH_3-\overset{\mathbf{O}}{\overset{\Vert}{\mathbf{C}}}-\mathbf{O}-CH_3$
Amine	$-NH_2$	$CH_3-\mathbf{NH_2}$
Amide	$-\overset{O}{\overset{\Vert}{C}}-NH_2$	$CH_3-\overset{\mathbf{O}}{\overset{\Vert}{\mathbf{C}}}-\mathbf{NH_2}$

QUESTIONS & PROBLEMS

Functional Groups

11.1 Use Table 11.1 to write the functional group that would classify an organic compound as each of the following:
a. alcohol **b.** haloalkane **c.** aldehyde **d.** amine

11.2 Use Table 11.1 to write the functional group that would classify an organic compound as each of the following:
a. ketone **b.** alkene **c.** thiol **d.** alkyne

11.3 Use Table 11.1 to match these organic structures with the family names that follow.

A. $CH_3CH{=}CHCH_3$ **B.** $CH_3\overset{\overset{\displaystyle OH}{|}}{C}HCH_3$ **C.** $CH_3CH_2{-}Br$

D. $CH_3CH_2{-}O{-}CH_3$ **E.** $CH_3\overset{\overset{\displaystyle O}{\|}}{C}CH_2CH_3$

a. ketone **b.** ether **c.** alkene **d.** alcohol **e.** haloalkane

11.2

11.4 Use Table 11.1 to match these organic structures with the family names that follow.

A. $CH_3{-}\overset{\overset{\displaystyle O}{\|}}{C}{-}OH$ **B.** $CH_3CH_2CH_2{-}SH$ **C.** $CH_3CH_2{-}NH_2$

D. $CH_3C{\equiv}CCH_3$ **E.** $CH_3\overset{\overset{\displaystyle O}{\|}}{C}{-}O{-}CH_3$

a. ester **b.** alkyne **c.** carboxylic acid **d.** thiol **e.** amine

Study Pack

11.2 Haloalkanes

Learning Goal

Give the name for a haloalkane or draw the structural formula from the name.

The **haloalkanes** are alkanes that contain one or more atoms of fluorine, chlorine, bromine, and/or iodine. Not many haloalkanes are found in nature, but compounds containing halogens have been used to make many of the anesthetics and pesticides we use today. Many scientists now associate a group of haloalkanes known as chlorofluorocarbons (CFCs) with the breakdown of our protective ozone layer.

In the IUPAC system, the halogen atoms are named and located by numbering the parent alkane chain. In simple haloalkanes, the parent chain can also be named with a *yl* ending followed by the name of the specific halide.

Naming Haloalkanes

Functional group:	halogen atom: F (fluoro), Cl (chloro), Br (bromo), I (iodo)
IUPAC name:	haloalkane
Common name:	alkyl halide

In the following examples, the common names are given in parentheses.

$CH_3{-}Cl$
Chloromethane
(methyl chloride)

$CH_3CH_2{-}Br$
Bromoethane
(ethyl bromide)

$CH_3CH_2CH_2Cl$
1-Chloropropane
(propyl chloride)

$CH_3\overset{\overset{\displaystyle F}{|}}{C}CH_3$
$\underset{\underset{\displaystyle F}{|}}{}$
2,2-Difluoropropane

$CH_3\overset{\overset{\displaystyle Br}{|}}{C}HCH_2\overset{\overset{\displaystyle Cl}{|}}{C}HCH_3$
2-Bromo-4-chloropentane

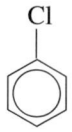
Chlorobenzene

HEALTH NOTE

Haloalkanes as Anesthetics

General anesthetics are compounds that are inhaled or injected to cause a loss of sensation so that surgery or other procedures can be done without causing pain to the patient. As nonpolar compounds they are soluble in the fats (also nonpolar) of the body including the nerve membranes, where they decrease the ability of the nerve cells to conduct the sensation of pain. One of the early anesthetics, carbon tetrachloride, CCl_4, was also used as a cleaning agent to remove oils and grease. However, it was found to be toxic to the liver and its use was discontinued.

One of the most widely used general anesthetics today is halothane, also called Fluothane. (See Figure 11.1.) It has a pleasant odor, is nonexplosive, has few side effects, undergoes few reactions within the body, and is eliminated quickly.

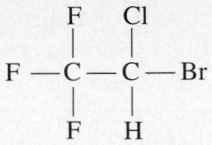

Halothane (Fluothane)

For minor surgeries, a local anesthetic such as chloroethane (ethyl chloride) may be applied to an area of the skin where it evaporates quickly to cool the skin and causes a loss of sensation.

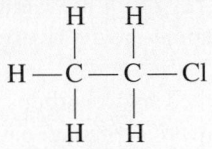

Chloroethane (ethyl chloride)

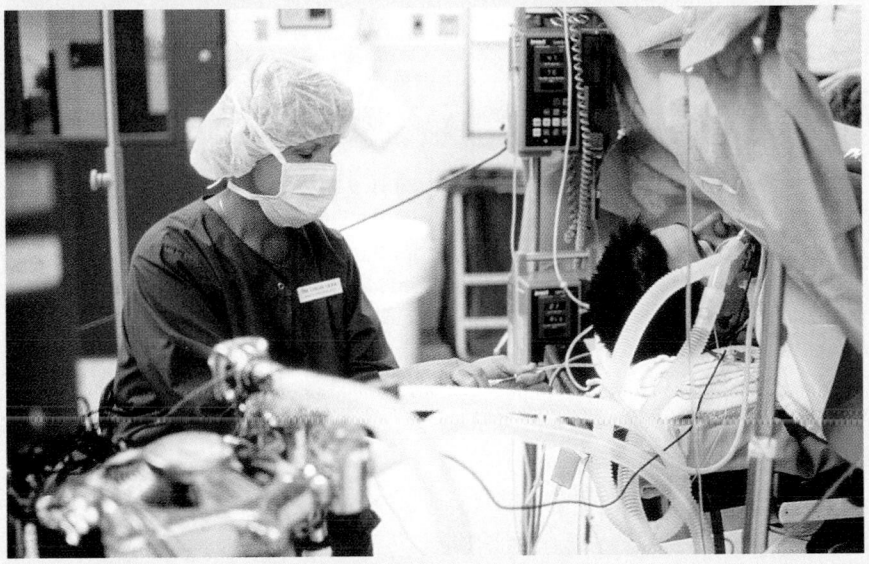

Figure 11.1 Anesthetics used during surgery produce a loss of all sensation.

ENVIRONMENTAL NOTE

Chlorinated Cyclic Compounds as Pesticides

Many insecticides and pesticides are chlorinated cyclic hydrocarbons with commercial names such as aldrin, chlordane, DDT, dieldrin, and lindane. (See Figure 11.2.) They have been used successfully in protecting crops from destruction by insects. In the 1940s, DDT was hailed as a miracle for its control of the mosquito and tsetse fly that spread malaria, typhus, and sleeping sickness.

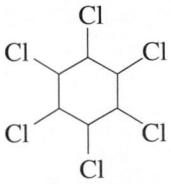

Lindane

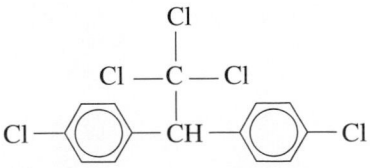

DDT (dichlorodiphenyltrichloroethane)

11.2

Because DDT is a nonpolar compound, it is soluble in nonpolar substances. In lakes and ponds where spraying of DDT occurs, small amounts of the pesticide are ingested and stored in the nonpolar lipids (fats) of microorganisms. Because these microorganisms serve as food for larger organisms and small fish, DDT becomes more concentrated in the body fat of larger creatures. Predatory birds and animals high in the food chain accumulate the greatest amounts of DDT. Eventually elevated DDT levels affect their reproduction and have contributed to the decline and near extinction of some species of hawks, eagles, falcons, and brown pelicans. Many of the chlorinated hydrocarbons, including DDT, have now been banned from use in widespread spraying in the United States, and the populations of some nearly extinct species have begun to rise again.

A major difficulty with chlorinated hydrocarbons is that they are nonbiodegradable; there are no natural pathways to break them down. They can be stored in human adipose tissue (body fat) and have been found in human breast milk. When they are absorbed by the human body, they act as poisons. Symptoms of pesticide poisoning in humans include dizziness, weakness, kidney damage, and nervous system excitability. Respiratory failure, coma, and death may follow.

Figure 11.2 Spraying a field of tulips.

S·A·M·P·L·E **11.2** ·P·R·O·B·L·E·M

Naming Haloalkanes

Freon 11 and Freon 12 are compounds that have been widely used as refrigerants in home and car air conditioners and as propellants in aerosol products. They are also known as CFCs or chlorofluorocarbons, a group of compounds known to destroy the ozone in the upper atmosphere. By 2000, many countries including the United States will stop production of CFCs. What are the IUPAC names for Freon 11 and Freon 12?

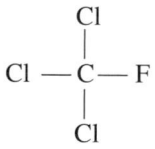

Freon 11

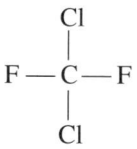

Freon 12

Solution

Freon 11, trichlorofluoromethane; Freon 12, dichlorodifluoromethane.

Study Check

As mentioned in the health note earlier in this chapter, halothane or Fluothane is the commercial name of a haloalkane used as an anesthetic. How would it be named using the IUPAC system?

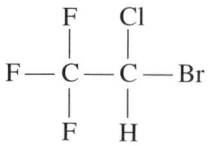

Halothane

ENVIRONMENTAL NOTE

CFCs and Ozone Depletion

From 10 to 30 miles above the earth, ultraviolet (UV) radiation produces highly reactive oxygen atoms that react with other oxygen atoms, producing ozone (O_3). The ozone in the upper atmosphere absorbs a large portion of the UV rays of the sun and acts as a protective shield for plants and animals on earth. The ultraviolet radiation that does get through can produce sunburn, cataracts, and skin cancers.

Normally, there is a balance between the formation and decomposition of ozone in the atmosphere. However, since the early 1970s scientists have become concerned that certain compounds entering the atmosphere are threatening the stability of the ozone layer. These compounds, called chlorofluorocarbons (CFCs), are the halogenated alkanes that are used as propellants for hair sprays, paints, deodorants, and refrigerants in air conditioners. Although they appear to be unreactive in the lower atmosphere, it has become evident that in the upper atmosphere CFCs decompose in the presence of UV light to produce highly reactive chlorine atoms and other products.

$$\underset{\text{Freon (CFC)}}{CFCl_3} \xrightarrow{\text{Sunlight (UV)}} Cl + \text{other products}$$

The reactive chlorine atoms catalyze the breakdown of ozone molecules. It has been estimated that one chlorine atom can destroy as many as 100,000 ozone molecules.

$$2O_3 \xrightarrow{Cl} 3O_2 \quad \text{Ozone destroyed}$$

Since 1985, researchers have observed the thinning of the ozone layer over the South Pole. In some areas as much as 50% of the ozone has been depleted, and at certain times of the year an ozone hole appears. There is recent evidence of a hole in the ozone layer over the Arctic as well. The concern of scientists is that a thinning of the ozone layer over more populated areas will increase the amount of unfiltered UV radiation received by plants and animals, with adverse effects. Already there are reports of an increase in cases of UV retinal damage in people living in the southern regions. (See Figure 11.3.)

By the early 1980s, CFCs were banned by the United States for use in spray containers. Many nations have agreed to ban CFCs. Chemical companies are working on substitutes to CFCs that are not damaging to the ozone, but so far they appear to be costly. However, that may be the price for retaining the life-protecting ozone layer.

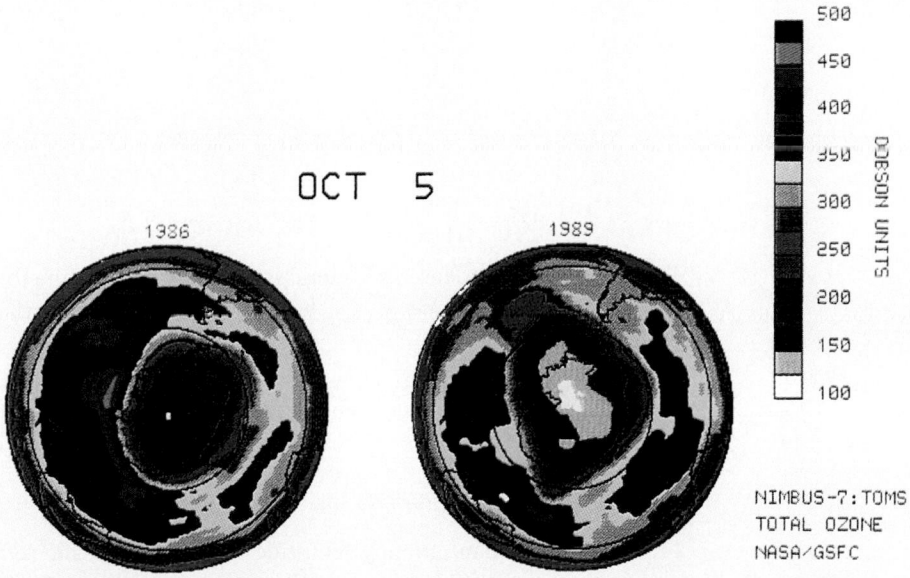

Figure 11.3 At certain times of the year, an ozone hole appears in an area over the South Pole where depletion of the ozone layer by CFCs is most severe.

QUESTIONS & PROBLEMS

Haloalkanes

11.2

11.5 Write the IUPAC name for each of the following haloalkanes:

a. 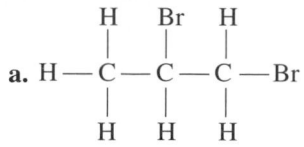

b. 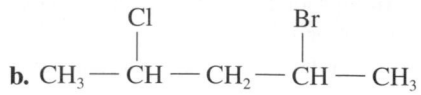

c. 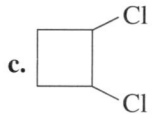

d. 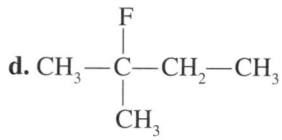

e. 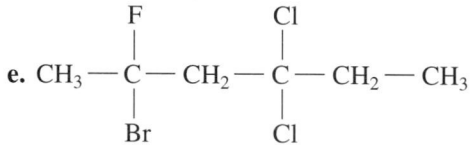

11.6 Write the IUPAC name for each of the following haloalkanes:

a. 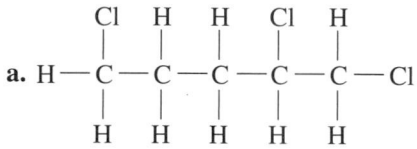

b. 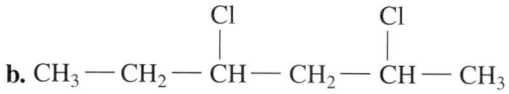

c. 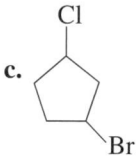

d. 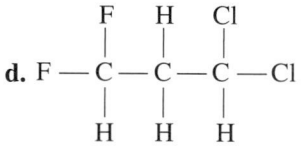

e. 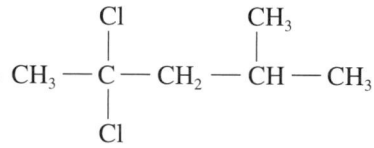

11.7 Write the condensed structural formula for each of the following:

a. 2-chloropropane **b.** 1-bromo-2-chloropropane
c. methyl bromide **d.** 1-chloro-1-fluoroethane
e. 1,2-difluorocyclopentane

11.8 Write the condensed structural formula for each of the following:

a. 1-chlorobutane **b.** 1-bromo-2-chlorocyclopropane
c. 3,3-dibromo-2-methylpentane **d.** 2,4-dichloro-3-methylhexane
e. 1,1,3,5-tetrachlorocyclohexane

11.9 Chloromethane (methyl chloride) and chloroethane (ethyl chloride) are used as topical anesthetics. When sprayed on the skin, they produce local anesthesia by lowering the temperature in the treated area to freeze the nerve endings. This makes local surgery near the surface of the skin painless. What are the structures of chloromethane and chloroethane?

11.10 The isomers of $C_2H_4Cl_2$ are used as industrial solvents. Write the condensed structural formulas for the two isomers and give their IUPAC names.

11.11 Write the four isomers of C_4H_9Cl and give the IUPAC name for each.

11.12 Write the four isomers of $C_3H_6F_2$ and give the IUPAC name for each.

11.3 Alkenes and Alkynes

Study Pack

Learning Goal

Write the IUPAC names for the structural formulas of alkenes, cycloalkenes, and alkynes, or write their structural formulas from their names.

11.3

Alkenes and **alkynes** are families of hydrocarbons that contain double or triple bonds. They are called **unsaturated compounds** because they have fewer hydrogen atoms attached to the carbon chain than alkanes do. Alkanes are called **saturated compounds** because they have the maximum number of hydrogen atoms attached to each carbon atom in the compound.

In an alkene, the double bond occurs when two adjacent carbon atoms in a chain share two pairs rather than only one pair of valence electrons. (See Figure 11.4.) The double bond makes an alkene molecule flat (planar): the carbon and hydrogen atoms in the double bond all lie in the same plane.

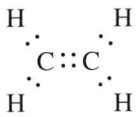

Double bond in electron-dot formula

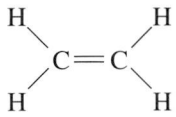

Complete structural formula

$H_2C{=}CH_2$ or $CH_2{=}CH_2$

Condensed structural formula

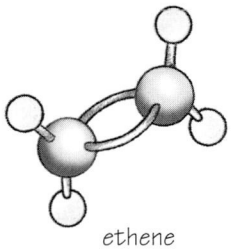

Figure 11.4 A ball-and-stick model of ethene, C_2H_4, the simplest alkene.

Cycloalkenes are similar to the cycloalkanes, except that there is a double bond between two of the carbon atoms in the ring.

Examples of Condensed Structural Formulas of Cycloalkenes

In an alkyne, a triple bond occurs when two adjacent carbon atoms share three pairs of valence electrons. (See Figure 11.5.)

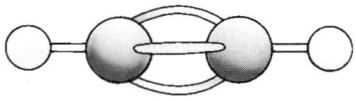

Figure 11.5 A ball-and-stick model of acetylene (IUPAC name *ethyne*).

$H{:}C{:::}C{:}H$

Triple bond in electron-dot formula

$H{-}C{\equiv}C{-}H$

Complete structural formula

$HC{\equiv}CH$

Condensed structural formula

S·A·M·P·L·E 11.3 ·P·R·O·B·L·E·M

Identifying Alkenes

Classify the following structural formulas as an alkane, alkene, or alkyne.

a. $CH_3{-}C{\equiv}C{-}CH_3$ **b.** $CH_3{-}CH_2{-}CH_3$

c. $CH_3CH_2C(CH_3){=}CHCH_2CH_3$

Solution

The structural formula with a double bond is an alkene, and the one with a triple bond is an alkyne.

a. alkyne **b.** alkane **c.** alkene

Study Check

Is the following structure a cycloalkane or cycloalkene?

Figure 11.6 Unripe fruit that is harvested and stored will be treated with ethene gas to produce ripening when the fruit is ready to go to market.

11.3

EXPLORE YOUR WORLD

Ripening Fruit

Obtain two unripe, green bananas. Place one of the green bananas in a plastic bag and seal it. Leave the two bananas on the counter next to each other. Check the bananas twice a day to see if there is a difference in the rate that they ripen. What are some possible reasons for any difference in the rate at which they ripen? If you wish to ripen an avocado, what procedure could you use?

Naming Alkenes and Alkynes

The IUPAC system is used to name alkenes and alkynes in a similar way to naming alkanes with an emphasis on the unsaturated functional group.

1. The parent carbon chain is the longest carbon chain *that contains both carbon atoms of the double or triple bond.* Replace the *ane* ending of the corresponding alkane with *ene* (double bond) or *yne* (triple bond).
2. Number the parent chain from the end nearest the double or triple bond. Place the lower number of the carbon atom that is part of the double or triple bond in front of the parent name.
3. Number any branches and name them as we did with the alkanes.

The simplest alkene, ethene (common name, ethylene), C_2H_4, is used to accelerate the ripening of fruit. Commercially grown fruit, such as avocados, bananas, tomatoes, apples, and melon, is picked before it is ripe. When the fruit is ready for market, it is placed in containers and ethene is added to complete the ripening process. (See Figure 11.6.)

Naming Alkenes and Alkynes

Functional group:	$>C{=}C<$	$-C{\equiv}C-$
IUPAC name:	alkene	alkyne
H_3C-CH_3 Ethane	$H_2C{=}CH_2$ Ethene (ethylene)	$HC{\equiv}CH$ Ethyne (acetylene)
$CH_3-CH_2-CH_3$ Propane	$H_2C{=}CH-CH_3$ Propene	$HC{\equiv}C-CH_3$ Propyne
$CH_3-CH_2-CH_2-CH_3$ Butane	$H_2C{=}CH-CH_2-CH_3$ 1-Butene	$HC{\equiv}C-CH_2CH_3$ 1-Butyne
	$CH_3-CH{=}CH-CH_3$ 2-Butene	$CH_3-C{\equiv}C-CH_3$ 2-Butyne

Figure 11.7 Acetylene is one of the gases used to obtain the high temperatures in a welder's torch.

The simplest alkyne, ethyne (common name, acetylene), C_2H_2, is used in acetylene torches to obtain high temperatures for welding. (See Figure 11.7.)

Naming Cycloalkenes

If a cyclic compound contains a double bond, it is named as a cycloalkene.

Cyclobutene

Cyclopentene

Cyclohexene

If a substituent is attached to the cycloalkene, the carbons in the double bond itself are always numbered 1 and 2. Then the ring is numbered in the direction that gives the next lower number to the attached group.

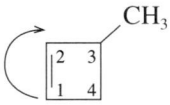

3-Methylcyclobutene

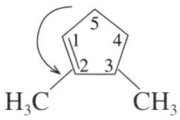

2,3-Dimethylcyclopentene

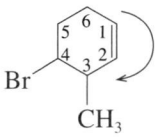

4-Bromo-3-methylcyclohexene

Some other examples of naming unsaturated compounds follow:

$CH_3CH_2CH_2-CH=CH_2$
1-Pentene
(**not** 4-pentene)

$CH_3-CH=CH-CH(CH_3)-CH_3$
4-Methyl-2-pentene

$CH_3CH_2CH(CH=CH-CH_3)CH_2CH_3$
4-Ethyl-2-hexene

Cyclohexene

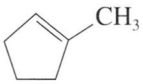

1-Methylcyclopentene

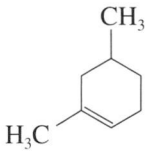

2,4-Dimethylcyclohexene

Compounds with two double bonds are named as **dienes**.

$CH_2=CH-CH=CH_2$
1,3-Butadiene

$CH_2=C(CH_3)-CH=CH-CH_3$
2-Methyl-1,3-pentadiene

ENVIRONMENTAL NOTE

Fragrant Alkenes

The odors you associate with cloves and peppermint, or perfumes from roses and lavender, are due to volatile oils that are synthesized by the plants. Unsaturated compounds known as terpenes, composed of two or more units of isoprene, are responsible for many of these fragrances.

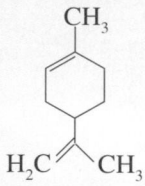

Limonene, found in oils of lemon, orange, and dill

11.3

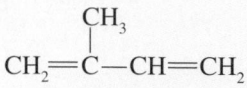

Isoprene

2-Methyl-1,3-butadiene

The structures with Figure 11.8 give examples of joined isoprene units.

Connected isoprene units

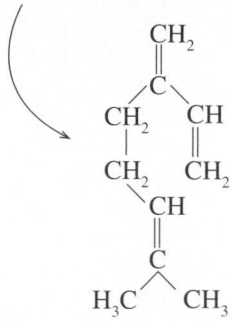

Myrcene, found in oils of bay leaves and verbena

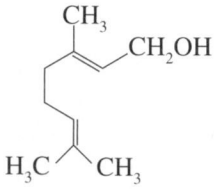

Geraniol, found in oils of rose and citronella

Figure 11.8 The fragrance of a rose is due to the unsaturated substances in the oils of the flower.

S·A·M·P·L·E 11.4 ·P·R·O·B·L·E·M

Alkenes and Alkynes

Write the IUPAC name for each of the following unsaturated compounds:

a. $CH_3CH_2CH{=}CHCH_3$

b.

c.

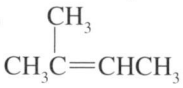

d. $HC{\equiv}CCH_2CH_2CH_3$

e.

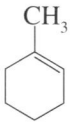

Solutions

a. 2-pentene **b.** cyclohexene **c.** 2-methyl-2-butene
d. 1-pentyne **e.** 1-methylcyclohexene

Study Check

Draw the structural formulas for each of the following:
a. 2-pentyne **b.** 3-chlorocyclohexene

Q·U·E·S·T·I·O·N·S & P·R·O·B·L·E·M·S

Alkenes and Alkynes

11.13 Compare the structural formulas of the following:
a. propene and propyne **b.** cyclohexane and cyclohexene

11.14 Compare the structural formulas of the following:
a. 1-butyne and 2-butyne
b. 1-methylcyclohexene and 3-methylcyclohexene

11.15 Give the IUPAC name for each of the following:

a. $CH_2{=}CH_2$ **b.** $CH_3C(CH_3){=}CH_2$ **c.** $CH_3CH_2C{\equiv}CCH_3$

d. 

e. 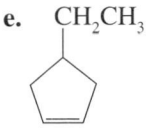

11.16 Give the IUPAC name for each of the following:

a. $CH_2{=}CHCH_2CH_2CH_2CH_3$ **b.** $CH_3C{\equiv}CCH_2CH_2CH(CH_3)CH_3$

c. 

d. 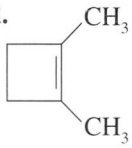

e. $CH_3CH(CH_3)CH_2CH(CH_3)CH_2CH{=}CH_2$

11.17 Draw the structural formula for each of the following compounds:
a. propene **b.** 1-pentene **c.** 2-methyl-1-butene
d. 3-methylcyclohexene **e.** 3-hexyne

11.18 Draw the structural formula for each of the following compounds:
a. 1-methylcyclopentene **b.** 3-methyl-1-butyne
c. 3, 4-dimethyl-1-pentene **d.** 4-ethyl-1-methylcyclohexene
e. 1,2-dimethylcyclopentene

11.4

Study Pack

Chem of Life Organic Molecules: Isomers 1

11.4 Geometric Isomers of Alkenes

Learning Goal

Write the names and structural formula of alkenes that have cis–trans isomers.

Alkenes can have structural isomers like the alkanes do. For example, here are three alkene structures with the formula C_4H_8:

Isomers of Alkenes with Formula C_4H_8

$CH_2{=}CH{-}CH_2{-}CH_3$	$CH_3{-}CH{=}CH{-}CH_3$	$CH_2{=}C(CH_3){-}CH_3$
1-Butene	2-Butene	2-Methyl-1-propene

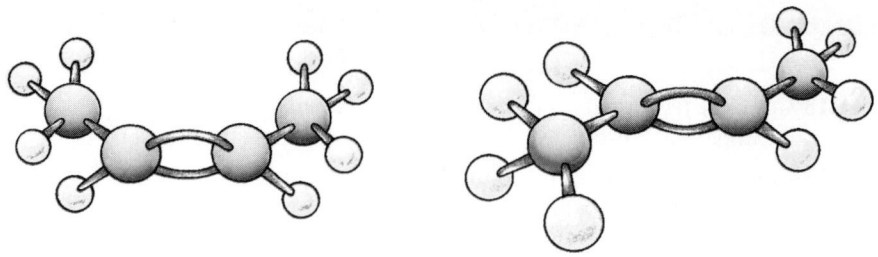

Figure 11.9 Ball-and-stick models of the cis and trans isomers of 2-butene.

11.4

Some alkenes can also have **geometric isomers,** which are two distinct arrangements of the groups attached to the double bond. In an alkene, the carbon atoms in the double bond are rigid and cannot rotate. As a result, the groups attached to the double bonds remain fixed on one side of the double bond or the other. In a **cis isomer,** the attached groups are on the *same side* of the double bond. In a **trans isomer,** the attached groups are on *opposite sides* of the double bond.

For example, we can write the structural isomers of 2-butene as geometric isomers. (See Figure 11.9.) As with any pair of cis–trans isomers, the cis and trans isomers of 2-butene are different compounds with different physical and chemical properties as noted below the following structures.

Geometric Isomers of 2-Butene

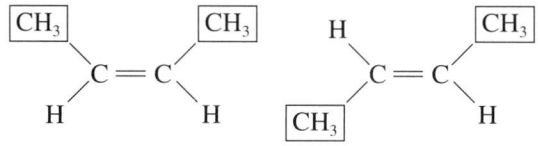

cis-2-Butene (mp −139°C; bp 3.7°C)

trans-2-Butene (mp −106°C; bp 0.3°C)

The other structural isomers of butene, 1-butene and 2-methylpropene, do not have cis–trans isomers because they have identical groups on at least one of the carbon atoms in the double bond.

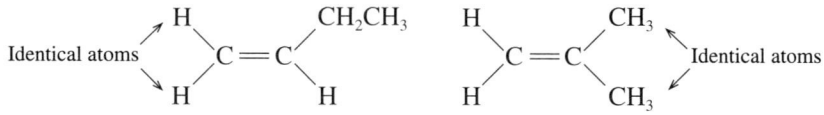

1-Butene 2-Methylpropene

Other examples of cis–trans isomers include the following:

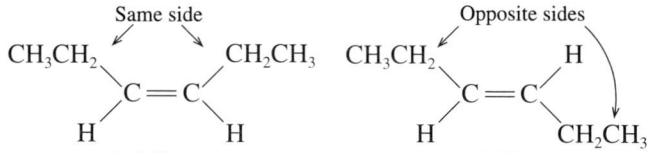

cis-3-Hexene *trans*-3-Hexene

S·A·M·P·L·E **11.5** ·P·R·O·B·L·E·M

Writing and Naming Cis–Trans Isomers

Give the condensed structural formulas and names for the geometric isomers of 2-pentene:

$$CH_3—CH{=}CH—CH_2CH_3$$

Solution

We can write the cis isomer by placing the alkyl groups on the same side of the double bond, and the trans isomer by placing them on opposite sides.

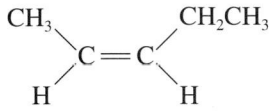

cis-2-Pentene

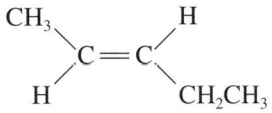

trans-2-Pentene

Study Check

Draw the condensed structural formula for *cis*- and *trans*-3-heptene.

11.4

EXPLORE YOUR WORLD

Modeling Cis–Trans Isomers

Because cis–trans isomerism is not easy to imagine, here are some things you can do to understand the difference in rotation around a single bond compared to a double bond and how it affects groups that are attached to the carbon atoms in the double bond.

Using Your Hands and Fingers as Single and Double Bonds

Put the fingertips of your index fingers together. This is a model of a single bond. Consider the index fingers as a pair of carbon atoms and think of your thumbs and other fingers as other parts of a carbon chain. While your index fingers are touching, twist your hands and change the position of the thumbs relative to each. Notice how the relationship of your other fingers changes.

Now place the tips of your index fingers and middle fingers together in a model of a double bond. As you did before, twist your hands to move the thumbs apart. What happens? Can you change the location of your thumbs relative to each other without breaking the double bond? The difficulty of moving your hands with two fingers touching represents the lack of rotation about a double bond. You have made a model of a cis isomer when both thumbs are on the same side. If you turn one hand over so one thumb points down, you have made a model of a trans isomer.

Using Toothpicks and Gumdrops to Form Single and Double Bonds

Use toothpicks for bonds and gumdrops for atoms. Place a toothpick between two gray gumdrops (carbon atoms). To each carbon atom attach three toothpicks with a yellow gumdrop on the other end. Rotate one of the gumdrop carbon atoms to show the movement of the attached atoms.

Remove a toothpick and yellow gumdrop from each carbon atom. Place a second toothpick between the carbon atoms. Each carbon atom should now be attached to two noncarbon atoms. Make those two atoms different colors, such as yellow and red. First place the red gumdrops on the same side of the double bond, which makes a model of a cis isomer. Try to twist the double toothpicks. Can you do it? If you turn the whole molecule upside down, are the red gumdrops still on the same side?

Now switch one of the red gumdrops with the yellow gumdrop on the same carbon. The red gumdrops should be opposite or trans to each other. Twist the double bond again. Can you move the red gumdrops to the same side of the double bond? It should not happen without breaking the double bond. How does this illustrate two isomers of a double bond molecule when the atoms in the double bond are attached to different groups of atoms? If you attach two yellow or two red gumdrops to the same carbon atom, can you make cis–trans isomers? Why or why not?

ENVIRONMENTAL NOTE

Pheromones in Insect Communication

Insects emit minute quantities of chemicals called pheromones that communicate with other members of the same species. Some pheromones warn of danger, others call for defense, mark a trail, or are sex attractants. The effectiveness of many of these compounds, containing 10 to 17 carbon atoms, depends on the cis or trans configuration of the double bond in the molecule. A certain species will respond to one isomer but not the other.

11.4

Scientists are interested in synthesizing pheromones to use as nontoxic alternatives to pesticides. For example, when a very small amount of the sex attractant of a certain insect is placed in a trap, the insects are unable to breed successfully. As a result the insects are removed from an agricultural area without affecting a wide range of other organisms. (See Figure 11.10.)

Figure 11.10 A trap in a crab apple tree is baited with codling moth pheromone.

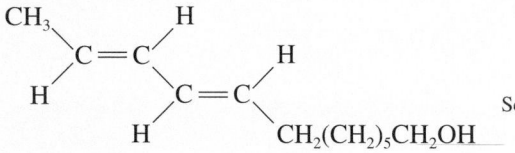

Sex attractant for the codling moth

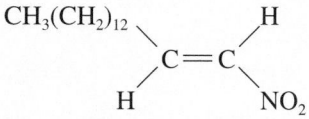

Defense signal for the termite

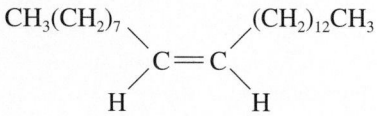

Sex attractant for the female housefly

HEALTH NOTE

Cis–Trans Isomers for Night Vision

The retinas of the eyes consist of two types of cells, rods and cones. The rods on the edge of the retina allow us to see in dim light, and the cones, in the center, produce our vision in bright light. In the rods, there is a substance called rhodopsin that absorbs light. Rhodopsin is composed of *cis*-11-retinal, an unsaturated compound, attached to a protein. When rhodopsin absorbs light, the *cis*-11-retinal isomer is converted to its trans isomer, which changes its shape. The trans form no longer fits the protein and it separates from the protein. The change from the cis to trans isomer and the separation from the protein generates an electrical signal that the brain perceives as light.

An enzyme (isomerase) converts the trans isomer back to the *cis*-11-retinal isomer and the rhodopsin re-forms. If there is a deficiency of rhodopsin in the rods of the retina, night blindness may occur. One common cause is a lack of vitamin A in the diet, because retinal is formed from vitamin A. Without a sufficient quantity of retinal, not enough rhodopsin is produced to enable us to see adequately in dim light.

Cis–trans isomers of retinal

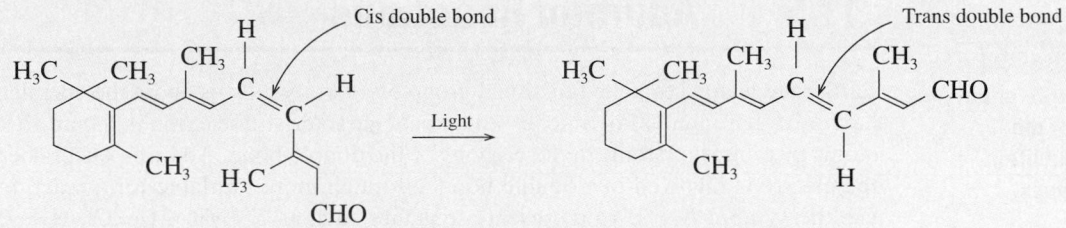

11.4

QUESTIONS & PROBLEMS

Geometric Isomers of Alkenes

11.19 What is the difference between a structural isomer and a geometric isomer?

11.20 How does *cis*-2-butene differ from *trans*-2-butene?

11.21 Which of the following can be written as cis–trans isomers?

a. $CH_2{=}CHCH_3$ **b.** $CH_3CH_2CH{=}CHCH_3$

c.

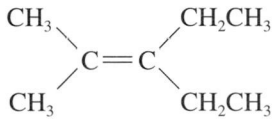

11.22 Which of the following do not have cis–trans isomers?

a.

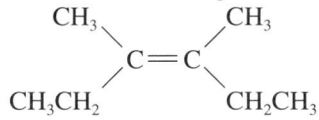

b. $CH_3CH_2CH_2CH{=}CH_2$

c.

11.23 Write the IUPAC name of each of the following isomers:

a.

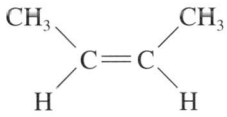

b.

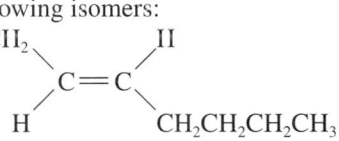

c.

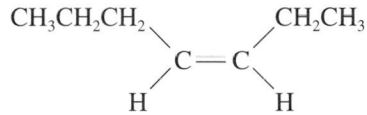

11.24 Write the IUPAC name of each of the following isomers.

a.

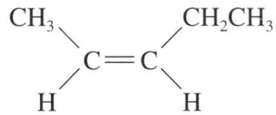

b.

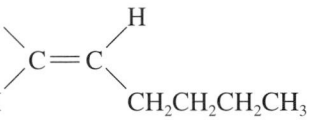

c.

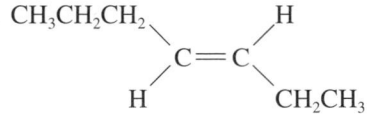

11.25 Draw the structural formula for each of the following geometric isomers:

a. *trans*-2-butene **b.** *cis*-2-pentene **c.** *trans*-3-heptene

11.26 Draw the structural formula for each of the following geometric isomers:

a. *cis*-3-hexene **b.** *trans*-2-pentene **c.** *cis*-4-octene

Study Pack

Learning Goal

Write the structural formulas and names for the products of the addition reactions of alkenes.

11.5

11.5 Addition Reactions

Earlier we learned that the functional group of a family tells us about the chemical behavior of that family. For alkenes, the most characteristic reaction is the **addition** of atoms or groups of atoms to the carbons of the double bond. Addition occurs because the electrons involved in a double bond are much more available for a reaction. We use the symbol A—B to represent a reactant such as H—H, Cl—Cl, H—Cl, or H—OH. The general equation for an addition reaction of an alkene is written as follows:

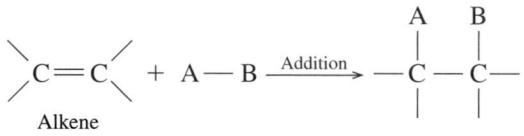

Table 11.2 Reactants and Addition Reactions

Reactant Added	Name of Addition Reaction
H_2	Hydrogenation
Cl_2, Br_2	Halogenation
HCl, HBr, HI	Hydrohalogenation
HOH	Hydration

The addition reactions have different names that depend on the type of reactant we add to the alkene, as Table 11.2 shows.

Adding Hydrogen

In the reaction called **hydrogenation,** two atoms of hydrogen are added across a double bond. The hydrogenation of the double bond in an alkene (unsaturated) converts it to a saturated bond of an alkane. A catalyst such as platinum (Pt) or nickel (Ni) is usually added to speed up the rate of the reaction. The symbol for the catalyst is shown over the arrow in the equation because a catalyst speeds up the reaction but does not appear in the products. The general equation for hydrogenation can be written as follows:

Hydrogenation of Alkenes

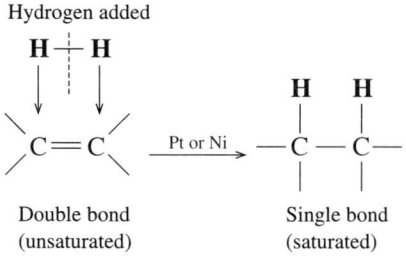

Some examples of the hydrogenation of alkenes follow:

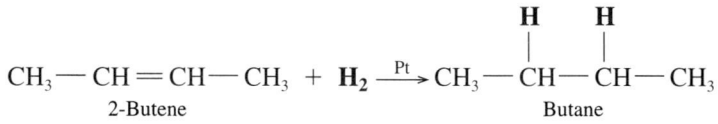

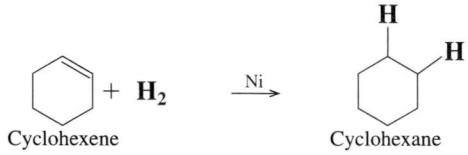

S·A·M·P·L·E **11.6** ·P·R·O·B·L·E·M

Writing Equations for Hydrogenation

Write the structural formula for the product of the following hydrogenation reactions:

a. $CH_3—CH{=}CH_2 + H_2 \xrightarrow{Pt}$

b.  $+ H_2 \xrightarrow{Pt}$

Solution

In an addition reaction, hydrogen adds to the double bond to give an alkane.

a. $CH_3—CH_2—CH_3$

b. 

Study Check

Draw the structural formula of the product of the hydrogenation of 2-methyl-1-butene using a platinum catalyst.

11.5

HEALTH NOTE

Hydrogenation of Unsaturated Fats

Vegetable oils such as corn oil or safflower oil are unsaturated fats composed of fatty acids that contain double bonds. The process of hydrogenation is used commercially to convert the double bonds in the unsaturated fats in vegetable oils to saturated fats such as margarine, which are more solid. Adjusting the amount of added hydrogen produces partially hydrogenated fats such as soft margarine, solid margarine in sticks, and shortenings, which are used in cooking. (See Figure 11.11.) For example, oleic acid is a typical unsaturated fatty acid in olive oil and has a cis-double bond at carbon 9. When oleic acid is hydrogenated, it is converted to stearic acid, a saturated fatty acid.

Figure 11.11 The vegetable oils in peanut butter, shortening, and margarine have been hydrogenated to form solid products.

$$CH_3(CH_2)_7—CH{=}CH—(CH_2)_7\overset{\overset{\displaystyle O}{\|}}{C}OH + H_2 \xrightarrow{Pt} CH_3(CH_2)_7—CH_2—CH_2—(CH_2)_7\overset{\overset{\displaystyle O}{\|}}{C}OH$$

Oleic acid (The cis isomer is found in olive oil and other unsaturated fats)

Stearic acid (found in saturated fats)

EXPLORE YOUR WORLD

Unsaturation in Fats and Oils

Read the labels on some containers of vegetable oils, margarine, peanut butter, and shortenings.

Questions

1. What terms on the label tell you that the compounds contain double bonds?
2. A label on a bottle of canola oil lists saturated, polyunsaturated, and monounsaturated fats. What do these terms tell you about the type of bonding in the fats?
3. A peanut butter label states that it contains partially hydrogenated vegetable oils or completely hydrogenated vegetable oils. What does this tell you about the type of reaction that took place in preparing the peanut butter?

11.5

Adding the Halogens Br_2 and Cl_2

Alkenes typically react with chlorine or bromine (the addition of fluorine is explosive and iodine is slow). The reaction, called **halogenation,** occurs readily, without the use of any catalyst, adding the atoms of the halogen to the double bond to yield a dihaloalkane product. In the equation for the general reaction, the symbol X—X or X_2 is used for Br_2 or Cl_2.

Halogenation of Alkenes

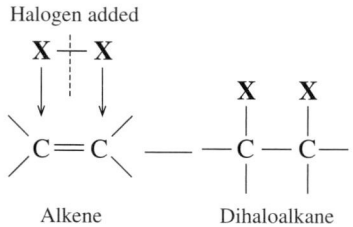

Here are some examples of adding Cl_2 or Br_2 to alkenes:

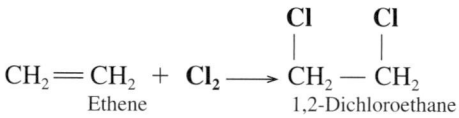

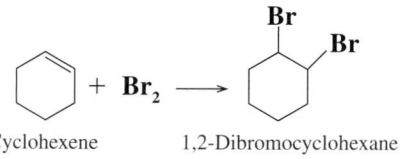

The addition reaction of bromine is sometimes used to test for the presence of double and triple bonds, as shown in Figure 11.12.

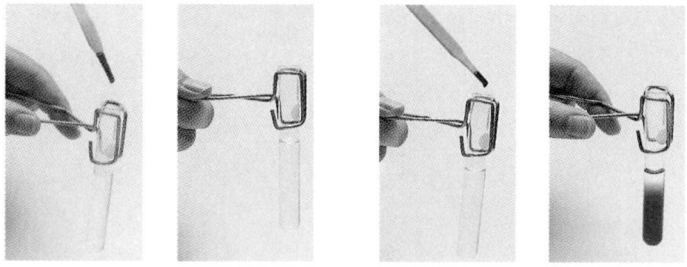

Figure 11.12 **(a)** When bromine is added to the alkene in the first test tube, the red color of the bromine is lost immediately as the bromine atoms add to the double bond. **(b)** In the second test tube, the red color remains because bromine does not react, or reacts only slowly, with the alkane.

S·A·M·P·L·E· 11.7 ·P·R·O·B·L·E·M

Writing Products of Halogenation

Write the condensed structural formula of the product of the following reaction:

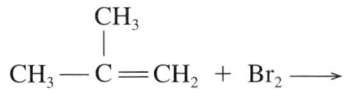

Solution

The addition of bromine to an alkene places a bromine atom on each of the carbon atoms of the double bond.

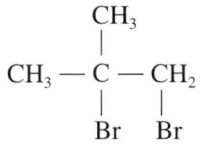

Study Check

What is the name of the product formed when chlorine is added to 1-butene?

11.5

Adding Hydrogen Halides

In the reaction called **hydrohalogenation,** a hydrogen halide (HCl, HBr, or HI) adds to an alkene to yield a haloalkane. The hydrogen atom bonds to one carbon of the double bond, and the halogen atom adds to the other carbon. The general reaction, in which HX represents HCl, HBr, or HI, can be written as follows:

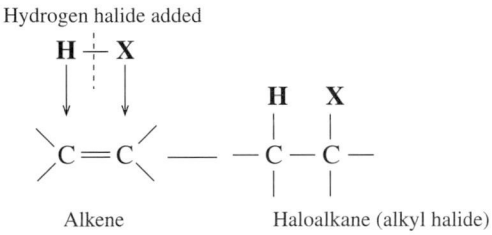

Two examples of hydrohalogenation follow:

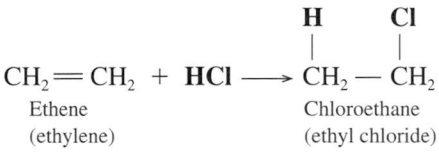

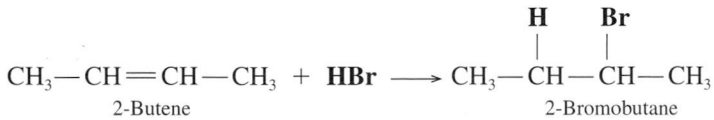

Markovnikov's Rule

When adding HX to some alkenes, you may find you have a choice of where to place the atoms on the double bond. We solve this problem by using **Markovnikov's rule,** which tells us that the hydrogen in HX adds to the carbon that already has the most hydrogen atoms, and the halogen adds to the carbon with the fewer hydrogen atoms. Using this rule, the hydrogen is added to the end carbon, and the chlorine to the second carbon in 1-propene.

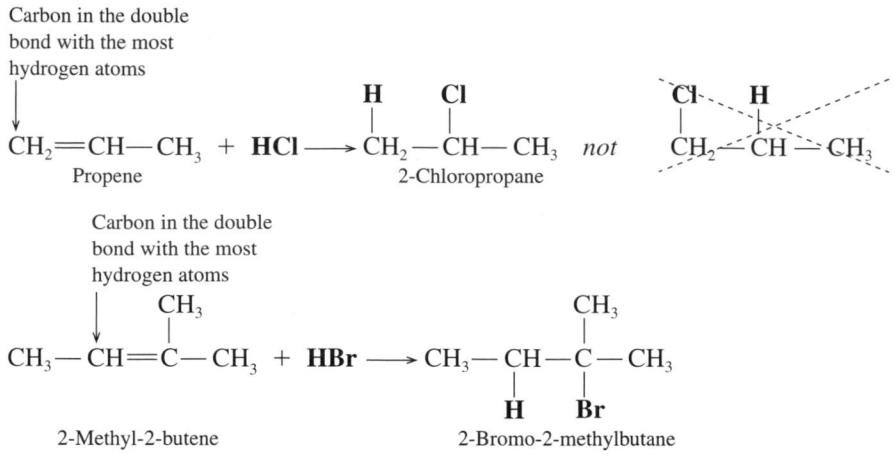

11.5

S·A·M·P·L·E **11.8** ·P·R·O·B·L·E·M

Addition Products by Using Markovnikov's Rule

Draw the structural formula for the product of the following reaction:

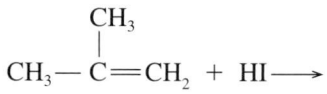

Solution

In the parent chain, carbon 1 of the double bond has more hydrogen atoms than carbon 2. Using Markovnikov's rule, we add the hydrogen atom from HI to carbon 1. The iodine atom adds to carbon 2 in the double bond.

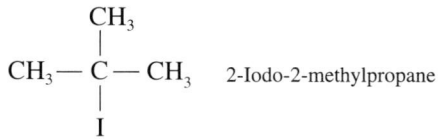

Study Check

Give the structural formula and name of the product obtained when HBr is added to 1-methylcyclopentene.

S·A·M·P·L·E **11.9** ·P·R·O·B·L·E·M

Writing a Synthesis by Using an Alkene

What alkene would you start with to prepare the following compound?

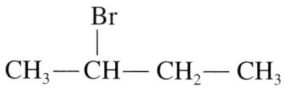

Solution

The starting alkene could be 1-butene or 2-butene.

Study Check

Write an equation using an alkene and HCl to prepare 2-chloropentane.

Adding Water

Alkenes react with water (HOH) when the reaction is catalyzed by an acid such as H_2SO_4 or H_3PO_4. In this reaction called **hydration,** H— attaches to one of the carbon atoms in the double bond, and —OH group to the other carbon. Hydration is used to prepare alcohols, which have the functional group —OH. In the general equation the acid is represented by H^+.

Hydration of Alkenes

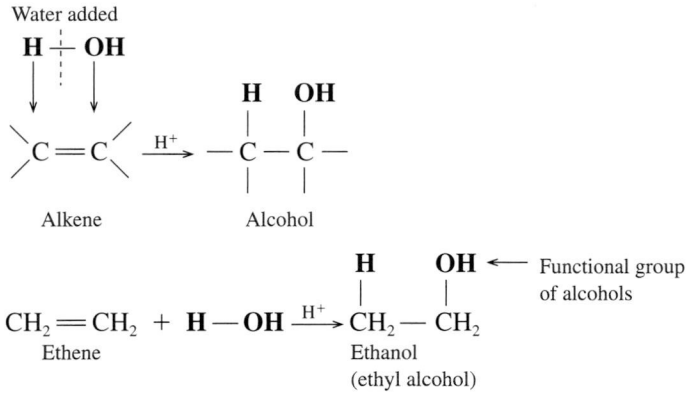

The addition of water to a double bond in which the carbon atoms are attached to different groups follows Markovnikov's rule.

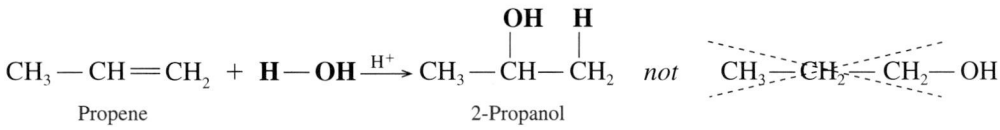

S·A·M·P·L·E 11.10 ·P·R·O·B·L·E·M

Writing Products of Hydration

Write the structural formulas for the products that form in the following hydration reactions:

a. $CH_3CH_2CH_2—CH=CH_2 + HOH \xrightarrow{H^+}$

b.  + HOH $\xrightarrow{H^+}$

Solution

a. Water adds H— and —OH to the double bond. We use Markovnikov's rule to add the H— to the CH_2 in the double bond, and the —OH to the CH.

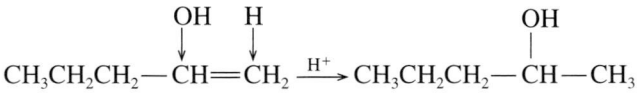

b. In cyclobutene, the H— adds to one side of the double bond, and the —OH adds to the other side.

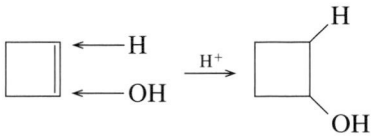

Study Check

Draw the structural formula for the alcohol obtained by the hydration of 2-methyl-2-butene.

QUESTIONS & PROBLEMS

Addition Reactions

11.5

11.27 Give the condensed structural formulas and names of the products in each of the following reactions:

a. $CH_3—CH{=}CH_2 + H_2 \xrightarrow{Pt}$

b. $CH_2{=}CH—\overset{\overset{CH_3}{|}}{C}H—CH_3 + Cl_2 \longrightarrow$

c.  $+ Br_2 \longrightarrow$

d. cyclopentene $+ H_2 \xrightarrow{Pt}$

e. 2-methyl-2-butene $+ Cl_2 \longrightarrow$

11.28 Give the condensed structural formulas and names of the products in each of the following reactions:

a. $CH_3—CH_2—CH{=}CH_2 + Br_2 \longrightarrow$

b. cyclohexene $+ H_2 \xrightarrow{Pt}$

c. *cis*-2-butene $+ H_2 \xrightarrow{Pt}$

d. $CH_3—\overset{\overset{CH_3}{|}}{C}{=}CH—CH_2—CH_3 + Cl_2 \longrightarrow$

e. 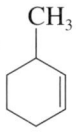 $+ Br_2 \longrightarrow$

11.29 Give the condensed structural formulas of the products in each of the following reactions using Markovnikov's rule when necessary:

a. $CH_3—CH{=}CH—CH_3 + HBr \longrightarrow$

b. cyclopentene $+ HOH \xrightarrow{H^+}$

c. $CH_2{=}CH—CH_2—CH_3 + HCl \longrightarrow$

d. $CH_3—\overset{\overset{CH_3}{|}}{C}{=}\underset{\underset{CH_3}{|}}{C}—CH_3 + HI \longrightarrow$

e. $CH_3CH_2—\overset{\overset{CH_3}{|}}{C}{=}CHCH_3 + HBr$

11.30 Give the condensed structural formulas of the products in each of the following reactions using Markovnikov's rule when necessary:

a. $CH_3-\overset{\overset{\displaystyle CH_3}{|}}{C}=CH-CH_3 + HCl \longrightarrow$

b. $CH_3CH_2-CH=CH-CH_2CH_3 + HOH \xrightarrow{H+}$

c. $CH_3-\overset{\overset{\displaystyle CH_3}{|}}{C}=CH_2 + HBr \longrightarrow$

d. 1-methylcyclohexene + HOH $\xrightarrow{H+}$

e.  + HBr $\longrightarrow$

11.31 Write an equation including any catalysts for the following reactions:
a. hydrogenation of 2-methylpropene
b. addition of hydrogen chloride to cyclopentene
c. addition of bromine to 2-pentene
d. hydration of propene

11.32 Write an equation including any catalysts for the following reactions:
a. hydration of 1-methylcyclobutene
b. hydrogenation of 3-hexene
c. addition of hydrogen bromide to 2-methyl-2-butene
d. addition of chlorine to 2,3-dimethyl-2-pentene

Study Pack

Chem of Life Organic Molecules: Polymers 1, 2

11.6 Polymers of Alkenes

Learning Goal

Describe the process of forming polymers using alkenes.

Polymers are long-chain molecules that consist of many repeating units of smaller carbon molecules called **monomers.** Many important polymers exist in nature. In biological systems, enzymes link the monomers together at everyday temperatures and pressures. Cellulose and starch are carbohydrate polymers made from many thousands of glucose molecules. Silk and wool are protein polymers made from amino acids. DNA, the molecule that carries genetic information, and RNA, another nucleic acid that makes proteins using DNA's information, are polymers of nucleotides.

Over the past hundred years, the plastics industry has made polymers that are in many of the materials we use every day, such as carpeting, plastic wrap, nonstick pans, nylon, Dacron, and other synthetic fibers. In medicine, biopolymers are used to replace diseased or damaged body parts such as hip joints, teeth, heart valves, and blood vessels. (See Figure 11.13.) Many of the synthetic polymers are made by addition reactions between monomers that are small alkenes. The conditions for many polymerization reactions require high temperatures and very high pressures (over 1000 atm).

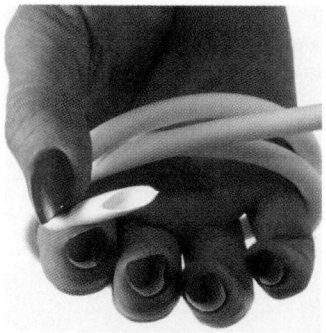

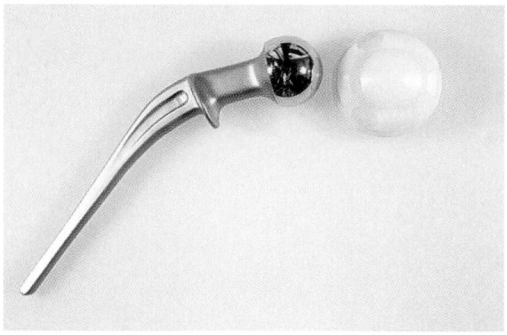

Figure 11.13 Synthetic polymers are finding extensive use in replacing diseased body parts such as an artery or hip joint.

11.6

Figure 11.14 Plastic containers are used to hold various products including soft drinks, juices, salad oils and dressings, tennis balls, and mouthwash.

Polyethylene is a common polymer used in the manufacturing of many home products such as plastic bottles, film, and plastic dinnerware. (See Figure 11.14.) Polyethylene is made from molecules of ethene ($CH_2{=}CH_2$). During polymerization, the double bonds of the ethene monomers open up and join with the next monomer. Eventually 100 to 1000 of these monomers are knit together to form a long carbon chain of a synthetic polymer.

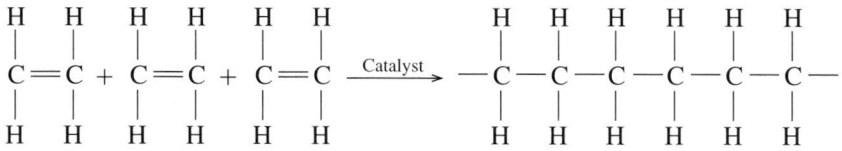

Ethylene molecules — Polyethylene

Another polymer known as polyvinyl chloride, PVC, is used as "vinyl" plastic to make such things as plastic covers, plastic pipes and tubing, body replacement parts, and garden hoses.

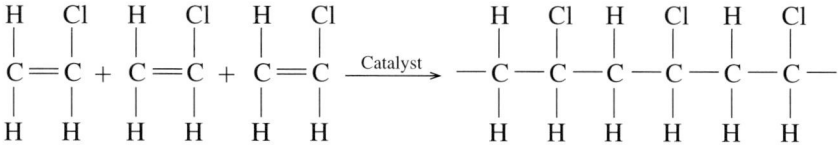

Chloroethene (vinyl chloride) — Polychloroethene (polyvinyl chloride, PVC)

Figure 11.15 Recycling information on a plastic bottle includes a triangle with a code number that identifies the type of polymer used to produce the plastic.

Table 11.3 lists several alkene monomers that are used to produce specific synthetic polymers and shows examples of each. The alkane-like nature of these plastic synthetic polymers makes them unreactive. Thus, they do not decompose easily (they are nonbiodegradable) and have become contributors to pollution. Efforts are being made to make them more degradable. It becomes increasingly important to recycle plastic material, rather than add to our growing landfills. You can identify the type of polymer used to manufacture a plastic item by looking for the recycle symbol (arrows in a triangle) found on the label or on the bottom of the plastic container as shown in Figure 11.15. For example, either the number 5 or the letters PP inside the triangle is a code for a polypropylene plastic.

1	2	3	4	5	6
PETE	HDPE	PV	LDPE	PP	PS
Polyethylene terephthalate	High-density polyethylene	Polyvinyl chloride	Low-density polyethylene	Polypropylene	Polystyrene

Table 11.3 Some Alkenes and Their Polymers

Monomer	Polymer	Uses for Polymer	Example
$H_2C{=}CH_2$ Ethene (ethylene)	$-CH_2-CH_2-CH_2-CH_2-$ Polyethylene	Plastic bottles, film, insulation material	
$H_2C{=}CH(Cl)$ Chloroethene (vinyl chloride)	$-CH_2-CH(Cl)-CH_2-CH(Cl)-$ Polyvinyl chloride (PVC)	Plastic pipes and tubing, garden hoses, garbage bags	
$CH_3CH{=}CH_2$ Propene (propylene)	$-CH_2-CH(CH_3)-CH_2-CH(CH_3)-$ Polypropylene	Ski and hiking clothing, carpets, artificial joints	
$F-C(F){=}C(F)-F$ Tetrafluoroethene	$-C(F)(F)-C(F)(F)-C(F)(F)-C(F)(F)-$ Teflon	Nonstick coatings	
$H_2C-C(Cl)-Cl$ 1,1-Dichloroethene	$-CH_2-C(Cl)(Cl)-CH_2-C(Cl)(Cl)-$ Saran	Plastic film and wrap	
$H_2C{=}CH(C_6H_5)$ Phenylethene (styrene)	$-CH_2-CH(C_6H_5)-CH_2-CH(C_6H_5)-$ Polystyrene	Plastic coffee cups and cartons, insulation	

EXPLORE YOUR WORLD

Polymers and Recycling Plastics

1. Make a list of the items you use or have in your room or home that are made of polymers.
2. Recycling information on the bottom or side of a plastic bottle includes a triangle with a code number that identifies the type of polymer used to make the plastic. Make a collection of several different kinds of plastic bottles. Try to find plastic items with each type of polymer.

Questions

1. What are the most common types of plastics among the plastic containers in your collection?
2. What are the monomer units of some of the plastics you looked at?

11.6

S·A·M·P·L·E **11.11** ·P·R·O·B·L·E·M

Polymers

What are the starting monomers for the following polymers?

a. polypropylene

b. Saran

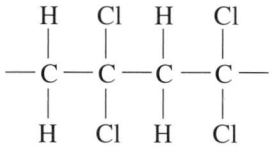

Solution

a. propene (propylene)

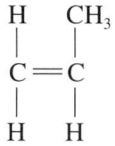

b. 1,1-dichloroethene,

Study Check

What is the monomer for PVC?

Q·U·E·S·T·I·O·N·S & P·R·O·B·L·E·M·S

Addition Polymers of Alkenes

11.33 What is a polymer?

11.34 What is a monomer?

11.35 Write an equation that represents the formation of a part of the Teflon polymer from two of the monomer units.

11.36 Write an equation that represents the formation of a part of the polystyrene polymer from two of the monomer units.

11.37 A plastic called polyvinylidene difluoride, PVDF, is made from monomers of 1,1-difluoroethene. Write the structure of the polymer formed from the addition of two monomers of the 1,1-difluoroethene.

11.38 An alkene called acrylonitrile is the monomer used to form the polymer used in the fabric material called Orlon. Write an equation that represents the formation of a part of the polyacrylonitrile polymer from two of the monomer units.

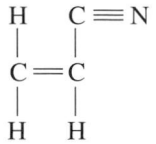

Acrylonitrile

Chapter Review

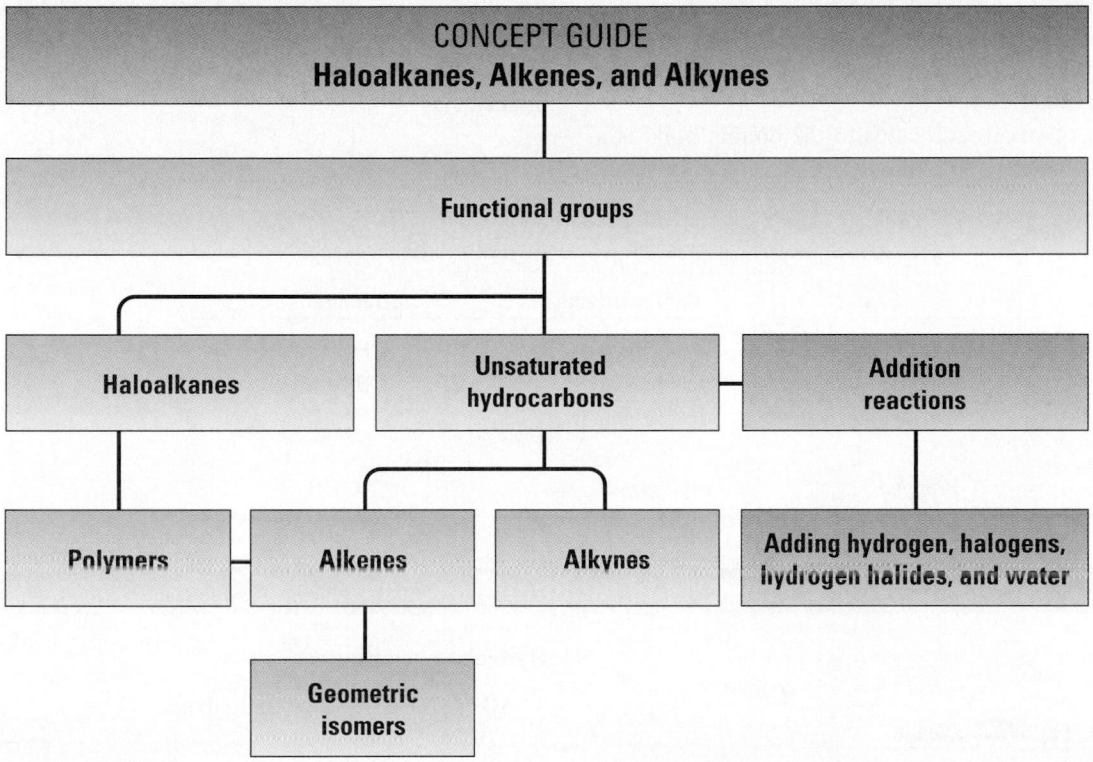

Chapter Summary

11.1 Functional Groups

An organic molecule contains a region that determines the molecule's family name and chemical reactivity. This functional group is a specific atom or grouping of atoms that will always react in a very predictable way. Examples of functional groups are a carbon–carbon double bond in an alkene, an atom of oxygen in an ether, a hydroxyl group, —OH, in alcohols, and a nitrogen group such as $—NH_2$ in amines.

11.2 Haloalkanes

When a hydrogen atom in an alkane is replaced by a halogen atom, F, Cl, Br, or I, the compound is a haloalkane. The halogen atoms are named as substituents (fluoro, chloro, bromo, iodo) attached to the alkane chain.

11.3 Alkenes and Alkynes

Alkenes are unsaturated hydrocarbons that contain at least one double bond (C═C) as the functional group. Alkenes and cycloalkenes use IUPAC rules for naming similar to the alkanes but use an *ene* ending. In alkenes, the longest carbon chain containing the double bond is numbered from the end nearest the double bond. If there are any substituents on a cycloalkene, the double bond is assigned the positions of 1 and 2, and the ring is numbered to give the next lowest numbers to the substituents. The alkynes are a family of unsaturated hydrocarbons that contain at least one triple bond. They use naming rules similar to the alkenes, but the parent chain ends with *yne.*

11.4 Geometric Isomers of Alkenes

Geometric isomers of alkenes occur when the carbon atoms in the double bond are connected to different atoms or groups. In the cis isomer the attached groups are on the same side of the double bond, whereas in the trans isomer they are connected on the opposite sides of the double bond.

11.5 Addition Reactions

The addition of small molecules to the double bond is a characteristic reaction of alkenes. Hydrogenation adds hydrogen atoms to the double bond of an alkene to yield an alkane. Halogenation adds bromine or chlorine atoms to produce dihaloalkanes. Hydrogen halides and water can also add to a double bond. When there are a different number of groups attached to the carbons in the double bond, Markovnikov's rule tells us to add the H from the adding reactant (HX or H—OH) to the carbon with the greater number of hydrogen atoms.

11.6 Polymers of Alkenes

Polymers are long-chain molecules that consist of many repeating units of smaller carbon molecules called **monomers.** In nature, cellulose and starch are polymers of glucose and proteins are polymers of amino acids. Many materials that we use every day are synthetic polymers, including carpeting, plastic wrap, nonstick pans, and nylon. These synthetic materials are often made by addition reactions in which a catalyst links the carbon atoms from various kinds of alkene molecules.

Summary of Naming

Type	Example	Characteristic	Structure
Haloalkane	1-Bromopropane	halogen atom	$CH_3—CH_2—CH_2—Br$
Alkene	Propene (propylene)	double bond	$CH_3—CH{=}CH_2$
Cycloalkene	Cyclopropene	double bond in a carbon ring	
Alkyne	Propyne	triple bond	$CH_3—C{\equiv}CH$

Summary of Addition Reactions

Hydrogenation

$$\text{Alkene} + H_2 \xrightarrow{Pt} \text{alkane}$$

$$CH_2{=}CH—CH_3 + H_2 \xrightarrow{Pt} CH_3—CH_2—CH_3$$

Halogenation

$$\text{Alkene} + Cl_2\ (\text{or } Br_2) \longrightarrow \text{dihaloalkane}$$

$$CH_2{=}CH—CH_3 + Cl_2 \longrightarrow CH_2(Cl)—CH(Cl)—CH_3$$

Hydrohalogenation

$$\text{Alkene} + HX \longrightarrow \text{haloalkane}$$

$$CH_2{=}CH—CH_3 + HCl \longrightarrow CH_3—CH(Cl)—CH_3$$

Markovnikov's rule

Hydration of Alkenes

$$\text{Alkene} + H—OH \xrightarrow{H^+} \text{alcohol}$$

$$CH_2{=}CH—CH_3 + H—OH \xrightarrow{H^+} CH_3—CH(OH)—CH_3$$

Markovnikov's rule

Key Terms

addition A reaction in which atoms or groups of atoms bond to a double bond. Addition reactions include the addition of hydrogen (hydrogenation), halogens (halogenation), hydrogen halides (hydrohalogenation), or water (hydration).

alkene An unsaturated hydrocarbon containing a carbon–carbon double bond.

alkyne An unsaturated hydrocarbon containing a carbon–carbon triple bond.

cis isomer A geometric isomer in which similar groups are connected on the same side of the double bond.

cycloalkene A cyclic hydrocarbon that contains a double bond in the ring.

functional group An atom or group of atoms that has a strong influence on the overall chemical behavior of an organic compound.

geometric isomers Two distinct arrangements of unsymmetrical alkenes made possible by the lack of rotation around the double bond. In a cis isomer, the groups are attached to the same side of the double bond; in the trans isomer, they appear on opposite sides.

haloalkane A type of alkane in which one or more hydrogen atoms has been replaced by a halogen atom.

halogenation The addition of Cl_2 or Br_2 to an alkene to form halogen-containing compounds.

hydration An addition reaction in which the components of water, H— and —OH, bond to the carbon–carbon double bond to form an alcohol.

hydrogenation The addition of hydrogen (H_2) to the double bond of alkenes to yield alkanes.

hydrohalogenation The addition of a hydrogen halide such as HCl or HBr to a double bond.

Markovnikov's rule When adding HX or HOH to alkenes with different numbers of groups attached to the double bonds, the H adds to the carbon that has the greater number of hydrogen atoms.

monomer The small organic molecule that is repeated many times in a polymer.

polymer A very large molecule that is composed of many small, repeating structural units that are identical.

saturated compounds A compound of carbon and hydrogen in which the carbon chain consists of only single carbon–carbon bonds ($H_3C—CH_3$).

trans isomer A geometric isomer in which similar groups are connected to opposite sides of the double bond in an alkene.

unsaturated compounds A compound of carbon and hydrogen in which the carbon chain contains at least one double (alkene) or triple carbon–carbon bond (alkyne). An unsaturated compound is capable of an addition reaction with hydrogen, which converts the double or triple bonds to single carbon–carbon bonds.

Chapter Practice

11.39 Compare the formulas and bonding in propane, cyclopropane, propene, and propyne.

11.40 Compare the formulas and bonding in butane, cyclobutane, cyclobutene, and 2-butyne.

11.41 Using Table 11.1, identify the family of each of the following organic compounds according to the functional group.

a. $CH_3CH_2CH_2Cl$

b.

c. CH_3CH_2OH

d. $CH_3C{\equiv}CH$

e.

f. $CH_3\overset{\overset{\displaystyle O}{\|}}{C}CH_2CH_3$

g. $CH_3CH_2CH_3$

11.42 Using Table 11.1, identify the family of each of the following organic compounds according to the functional group.

a. $CH_3CH_2CH{=}CHCH_3$

b.

c. $CH_3CH_2NH_2$

d. $CH_3\overset{\overset{\displaystyle O}{\|}}{C}OCH_3$

e. $CH_3CH_2CH_2CH_2CH_3$

f. $CH_3OCH_2CH_3$

g. $CH_3CH_2CH_2{-}SH$

11.43 Give the IUPAC name for each of the following compounds:

a.

b. $CH_2\underset{\displaystyle Cl}{\underset{|}{C}}HCH_2\underset{\displaystyle CH_3}{\underset{|}{C}}HCH_3$

c. $CH_2{=}\overset{\overset{\displaystyle CH_3}{|}}{C}CH_2CH_2CH_3$

d. $CH_3CH_2C{\equiv}CCH_3$

e.

f.

g.

11.44 Write the condensed structures of each of the following compounds:

a. 1,2-dibromocyclopentane

b. 2-pentyne

c. *cis*-2-heptene
d. 3,3-dichloro-2-methylpentane
e. *trans*-3-hexene
f. 2-bromo-3-chlorocyclohexene
g. 2,3-dichloro-1-butene

11.45 Indicate if the following pairs of structures represent structural isomers, geometric isomers, or identical compounds.

a.

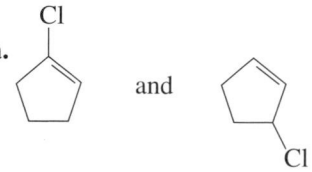

b.

c.

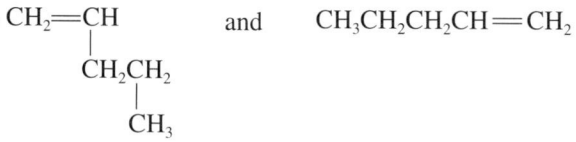

d.

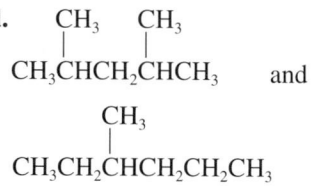

11.46 Draw the condensed structures and give the names for all the isomers of C_4H_8 including cyclic and cis–trans isomers.

11.47 Methylcyclopentane is formed by four different alkenes that react with hydrogen (H_2) in the presence of a Ni catalyst. Draw the condensed structures of each of these alkenes.

11.48 What role do cis–trans isomers play in night vision?

11.49 Write the cis and trans isomers for each of the following:

a. 2-pentene **b.** 3-hexene
c. 2-butene **d.** 2-hexene

11.50 Write the structures of the products, if any, for the following.

a. $CH_3CH{=}CHCH_3 + H_2 \xrightarrow{Ni}$

b.

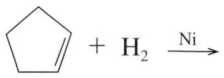

c. $CH_3CH{=}CHCH_3 + HBr \longrightarrow$

d.

e. $CH_3CH{=}CHCH_3 + Cl_2 \longrightarrow$

f. $CH_3CH_2CH_3 + H_2 \xrightarrow{Ni}$

g. $CH_3CH{=}CHCH_3 + HOH \xrightarrow{H^+}$

h.

i.

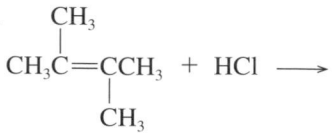

11.51 What is the condensed structural formula of the organic compound needed to prepare each of the following products?

a.

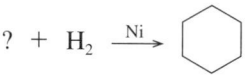

b.

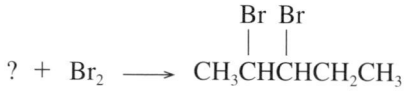

c.

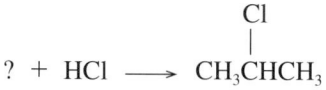

d.

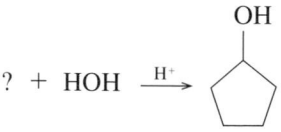

11.52 Write the combustion reaction for the acetylene used in a welder's torch.

11.53 Draw at least six structural isomers of $C_5H_{11}Cl$.

11.54 Some halogenated compounds used as refrigerants have the following names. Draw their structures.

Freon 14	tetrafluoromethane
Freon 114	1,2-dichloro-1,1,2,2-tetrafluoroethane
Freon C318	octafluorocyclobutane

11.55 What is the IUPAC name of lindane, a pesticide?

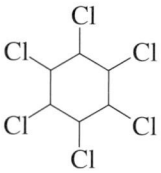

11.56 Lucite or Plexiglas is a polymer of methylmethacrylate. Write the part of the polymer that is made from the addition of two of these monomers.

11.57 Copolymers contain more than one type of monomer. One copolymer used in medicine is made of alternating units of styrene and acrylonitrile. Write a section of the copolymer that would have these alternating units. For structures of the monomers, see Polymers of Alkenes (Section 11.6).

Answers

Answers to Study Checks

11.1 alcohol

11.2 1-bromo-1-chloro-2,2,2-trifluoroethane

11.3 cycloalkene

11.4 **a.** $CH_3C{\equiv}CCH_2CH_3$ **b.**

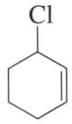

11.5 $CH_3CH_2CH{=}CHCH_2CH_2CH_3$, the structural formula for 3-heptene, can be written as the following cis–trans isomers:

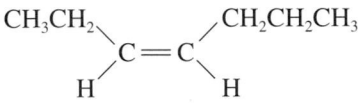

cis-3-Heptene

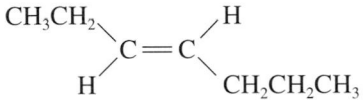

trans-3-Heptene

11.6

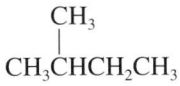

11.7 1,2-dichlorobutane

11.8

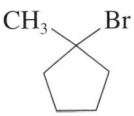

1-bromo-1-methylcyclopentane

11.9

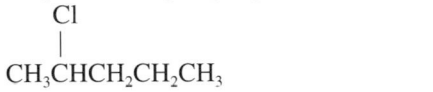

11.10

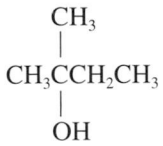

11.11 The monomer of PVC, polyvinyl chloride, is chloroethene:

Answers to Selected Questions and Problems

11.1 **a.** —OH **b.** —F, —Cl, —Br, or —I

c. $-\overset{\overset{\displaystyle O}{\|}}{C}H$ **d.** $-NH_2$

11.3 **a.** E **b.** D **c.** A **d.** B e. C

11.5 **a.** 1,2-dibromopropane
b. 2-bromo-4-chloropentane
c. 1,2-dichlorocyclobutane
d. 2-fluoro-2-methylbutane
e. 2-bromo-4,4-dichloro-2-fluorohexane

11.7 **a.**

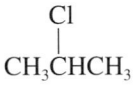

b.

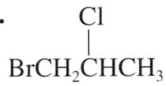

c. CH_3Br

d.

e.

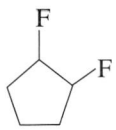

11.9 chloromethane, CH_3Cl; choroethane, CH_3CH_2Cl

11.11 $CH_3CH_2CH_2CH_2Cl$ 1-chlorobutane

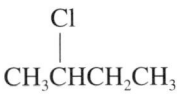

2-chlorobutane

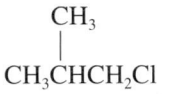

1-chloro-2-methylpropane

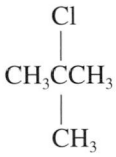

2-chloro-2-methylpropane

11.13 **a.** Propene contains a double bond, and propyne has a triple bond.
b. Cyclohexane is a saturated cycloalkane, and cyclohexene contains a double bond and is unsaturated.

11.15 **a.** ethene **b.** 2-methylpropene
c. 2-pentyne **d.** cyclobutene
e. 4-ethylcyclopentene

11.17 **a.** $CH_3CH{=}CH_2$ **b.** $CH_2{=}CHCH_2CH_2CH_3$

c. $CH_2{=}C(CH_3)CH_2CH_3$ **d.**

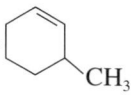

e. $CH_3CH_2C{\equiv}CCH_2CH_3$

11.19 Structural isomers have the same molecular formula, but different arrangements of the atoms. Geometric isomers in alkenes have different geometric relationship of the groups attached to a double bond.

11.21 **a.** There are no cis–trans isomers.
b. This alkene has cis–trans isomers.
c. There are no cis–trans isomers.

11.23 **a.** *cis*-2-butene **b.** *trans*-3-octene
c. *cis*-3-heptene

11.25 **a.**

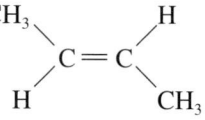

b.

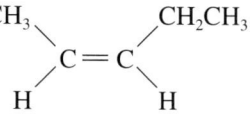

c.

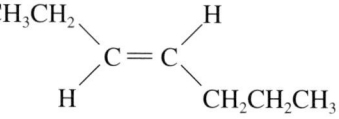

11.27 **a.** $CH_3CH_2CH_3$ propane

b.

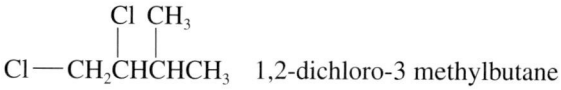

1,2-dichloro-3 methylbutane

c.

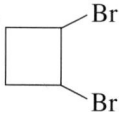

1,2-dibromocyclobutane

d.

cyclopentane

e.

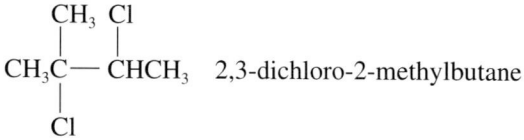

2,3-dichloro-2-methylbutane

11.29 **a.**

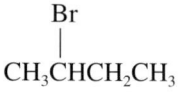

b.

c.

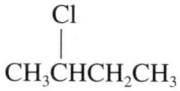

d.

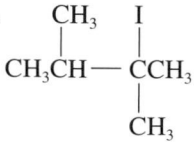

e.

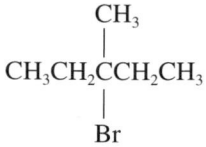

11.31 **a.**

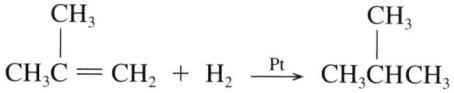

b.

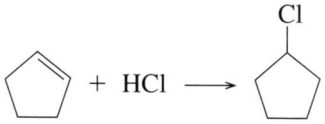

c.

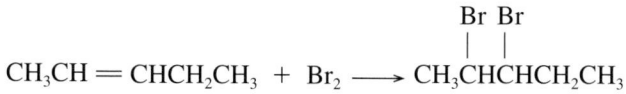

d.

11.33 A polymer is a very large molecule composed of small units that are repeated many times.

11.35

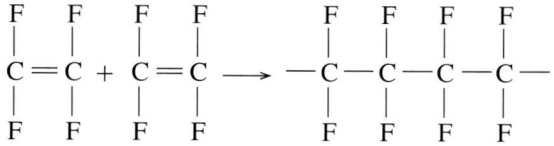

11.37

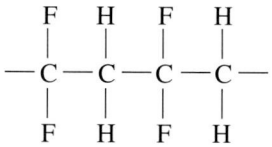

11.39 All the compounds have three carbon atoms. Propane is a saturated alkane, and cyclopropane is a saturated cyclic hydrocarbon. Both propene and propyne are unsaturated hydrocarbons, but propene has a double bond and propyne has a triple bond.

11.41 **a.** haloalkane **b.** cycloalkene **c.** alcohol
d. alkyne **e.** aromatic **f.** ketone
g. alkane

11.43 **a.** chlorocyclopentane
b. 2-chloro-4-methylpentane
c. 2-methyl-1-pentene
d. 2-pentyne
e. 1-chlorocyclopentene
f. *trans*-2-pentene
g. 1,3-dichlorocyclohexene

11.45 **a.** structural isomers **b.** geometric isomers
c. identical **d.** structural isomers

11.47

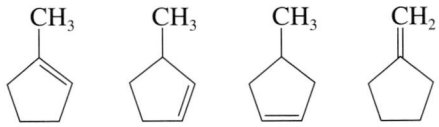

11.49 **a.**

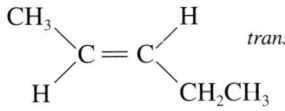

trans-2-Pentene

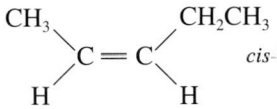

cis-2 Pentene

b.

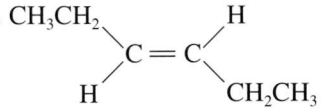

trans-3-Hexene

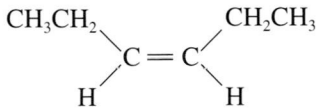

cis-3-Hexene

c.

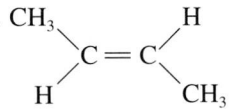

trans-2-Butene

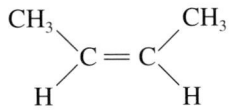

cis-2-Butene

d.

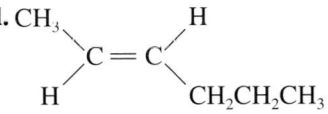

trans-2 Hexene

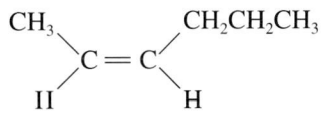

cis-2-Hexene

11.51 **a.**

b. $CH_3CH{=}CHCH_2CH_3$

c. $CH_2{=}CHCH_3$ **d.**

11.53

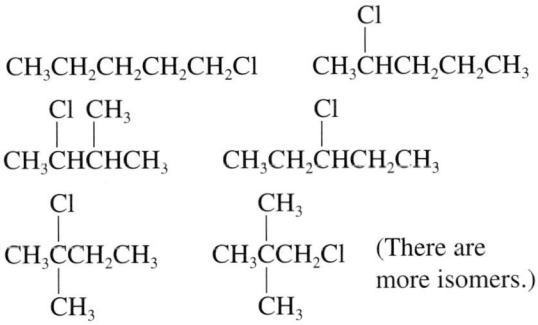

(There are more isomers.)

11.55 1,2,3,4,5,6-hexachlorocyclohexane

11.57

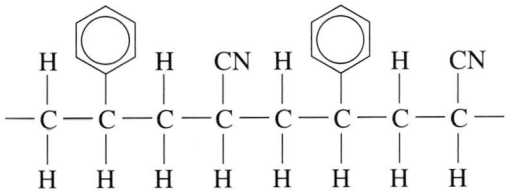

12

Alcohols, Phenols, Ethers, Aldehydes, and Ketones

In winemaking, the sugars in grapes are converted to ethanol in a process called fermentation, which is one of the oldest known chemical reactions.

Looking Ahead

12.1	Alcohols, Phenols, and Thiols
12.2	Ethers
12.3	Reactions of Alcohols
12.4	Aldehydes and Ketones
12.5	Oxidation and Reduction
12.6	Aldehydes and Ketones React with Alcohols

www.chemplace.com/college/timberlake

Visit the web site above to test yourself, run simulations, and review key concepts from this chapter. For even more interactive review and quizzing, use the two CD-ROMs included with the book.

In this chapter, we will look at families of organic compounds that contain oxygen or sulfur. Several of these families are important in biochemical pathways. Alcohols, phenols, and ethers contain one or more oxygen atoms connected by a single bond to a carbon atom. In an alcohol or phenol, the functional group is a hydroxyl group (—OH). For centuries, grains, vegetables, and fruits have been fermented to produce the ethyl alcohol in alcoholic beverages. Hydroxyl groups are important groups in sugars and starches as well as steroids such as cholesterol and estradiol. Menthol is an alcohol used in cough drops, shaving creams, and ointments. In a phenol, the hydroxyl group is attached to a benzene ring. Thiols are a family of compounds related to alcohols in which a sulfur group (—SH) replaces the hydroxyl group. The compounds in the ether family contain an oxygen atom connected to two carbon atoms (—O—). Ethyl ether was used for general anesthesia for more than a hundred years.

Aldehydes and ketones contain functional groups that have an oxygen atom with a double bond to a carbon atom. The simplest aldehyde, formaldehyde, is used for preserving tissues. Naturally occurring aldehydes such as those found in vanilla beans and almonds are used by the food industry as fragrances and flavorings. Ketones such as propanone (acetone) are commonly found in paint and polish removers.

Study Pack

Chem of Life Organic Molecules: Functional Groups 1, 2, 6

12.1 Alcohols, Phenols, and Thiols

Learning Goal

Name simple alcohols, phenols, and thiols, and draw their condensed structural formulas. Describe their solubility in water.

In the family of organic compounds known as **alcohols,** a **hydroxyl group (—OH)** has taken the place of a hydrogen atom in an alkane. In a **phenol,** the hydroxyl group is attached to a carbon atom of a benzene ring. We can write the general structures of alcohols and phenols by showing a —OH group connected to an alkyl (R) group or a benzene ring in an aromatic (Ar) compound:

General Formula

R—OH	Ar—OH
Alcohol	Phenol

Examples:

CH_3—OH

CH_3CH_2—OH

Naming Alcohols

Common names, widely used for the smaller alcohols, consist of the name of the alkyl group (R) followed by the family name *alcohol*. To name alcohols using the IUPAC rules, select the parent chain and change the *e* of the corresponding alkane name to *ol*. For example, methane becomes methan*ol* and ethan*e* becomes ethan*ol*. (See Figure 12.1.)

Naming Alcohols	R—OH
Functional group:	hydroxyl, —OH
IUPAC name:	alkanol
Common name:	alkyl alcohol

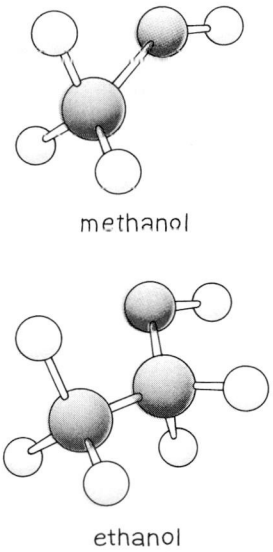

Figure 12.1 Ball-and-stick models of methanol and ethanol.

Some examples of the IUPAC names of alcohols and their common names (in parentheses) follow:

CH_3OH	CH_3CH_2OH	$CH_3CH_2CH_2OH$	$CH_3CH(OH)CH_3$
Methanol (methyl alcohol)	Ethanol (ethyl alcohol)	1-Propanol (propyl alcohol)	2-Propanol (isopropyl alcohol)

Suppose you want to name the following alcohol using the IUPAC rules.

12.1

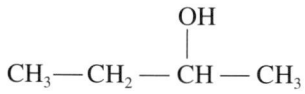

Step 1 Select the longest carbon chain containing the —OH group as the parent chain and replace the *e* in its name by *ol.*

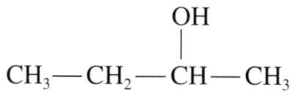

Butanol (replace *e* of butane with *ol*)

Step 2 Number the parent chain starting at the end closest to the —OH group.

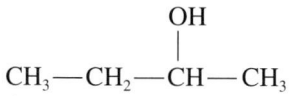

4 3 2 1 ⟵ Number chain from the right.

Step 3 Give the number for the position of the hydroxyl group in front of the parent name. In this example, the —OH group is on carbon 2 of the parent chain.

2-Butanol

Step 4 Give the names and numbers of any substituents.

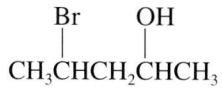

4-Bromo-2-pentanol

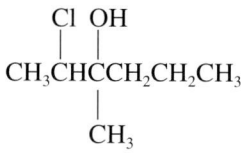

2-Chloro-3-methyl-3-hexanol

A cyclic alcohol is named as a *cycloalkanol.* The carbon ring is numbered so that the carbon atom bonded to the —OH group is carbon 1. Any other side groups are shown by numbering from carbon 1. Alcohols containing two —OH groups are named *diols;* three —OH groups, *triols*. Each of the hydroxyl groups is numbered.

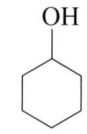

Cyclohexanol

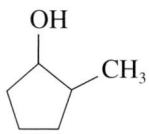

2-Methylcyclopentanol

1,2-Ethanediol (ethylene glycol)

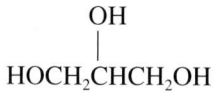

1,2,3- Propanetriol (glycerol)

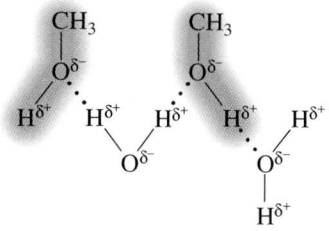

Figure 12.2 Hydrogen bonding occurs between polar hydroxyl groups of methanol molecules and water.

Solubility of Alcohols in Water

In Chapters 10 and 11 we learned that the hydrocarbons including alkanes, alkenes, alkynes, and aromatic compounds are composed of only carbon and hydrogen, which makes them nonpolar and not soluble in water. However, the hydroxyl group, —OH, of alcohols is a very polar functional group and forms hydrogen bonds with water. (See Figure 12.2.) Alcohols with one to four carbons are very soluble in water. However, when alcohols contain five or more carbon atoms, the nonpolar carbon chain causes them to behave more like alkanes. As a result, alcohols with longer carbon chains are not very soluble in water. For example, methanol with one carbon and ethanol with two carbon atoms are completely soluble in water, whereas octanol with eight carbon atoms is not soluble.

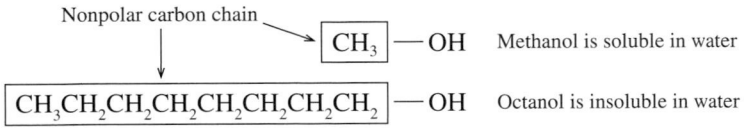

Naming Phenols

12.1

Aromatic compounds with —OH groups are named *phenols.* When there is a second group on the benzene ring, the ring is numbered starting from the carbon 1 bonded to the —OH group, or the prefixes *ortho, meta,* and *para* may be used.

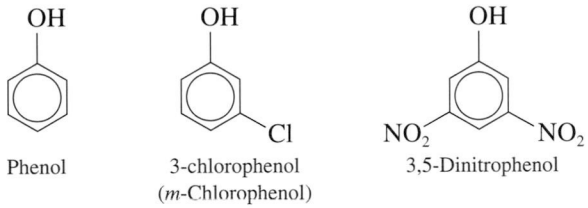

Properties of Phenols

The behavior of phenols differs considerably from that of alkyl alcohols. In water, phenol ionizes slightly as a weak acid. In fact, an early name for phenol was *carbolic acid.*

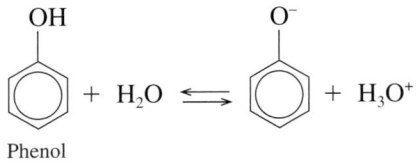

A concentrated solution of phenol is very corrosive and highly irritating to the skin, causing severe burns; ingestion can be fatal. Solutions containing phenol should be used with care. Dilute solutions of phenol were previously used in hospitals as antiseptics, but they have generally been replaced.

S·A·M·P·L·E 12.1 P·R·O·B·L·E·M

Naming Alcohols and Phenols

Name each of the following using IUPAC names:

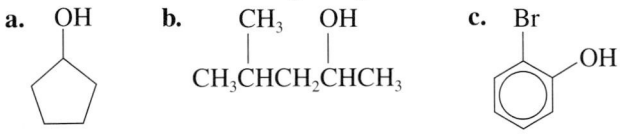

Solution

a. The —OH is attached to a cyclopentane ring. Changing the *e* to *ol* gives the IUPAC name of *cyclopentanol.*

b. The parent chain is pentane; the alcohol is named *pentanol.* The carbon chain is numbered to give the position of the —OH group on carbon 2 and the methyl group on carbon 4. The compound is named *4-methyl-2-pentanol.*

c. The compound is a *phenol* because the —OH is attached to a benzene ring. The ring is numbered with carbon 1 attached to the —OH in the direction that gives the bromine the lower number. The compound is named *2-bromophenol* or *o-bromophenol.*

Study Check

Name the following using IUPAC rules.

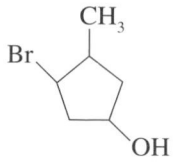

12.1

S·A·M·P·L·E 12.2 ·P·R·O·B·L·E·M

Writing Condensed Structural Formulas of Alcohols and Phenols

a. The alcohol 3,5,5-trimethyl-1-hexanol is used as a plasticizer, a substance added to plastics to keep them pliable. Draw its condensed structural formula.

b. Thymol, used as an antiseptic in mouthwashes, has an IUPAC name of 5-methyl-2- isopropylphenol. Draw its condensed structural formula.

Solution

a. The parent chain of six carbon atoms has an —OH group on carbon 1. Methyl groups are bonded to carbon 3 and to carbon 5.

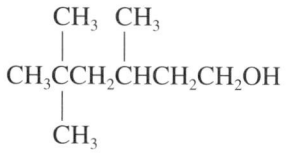

b. The phenol part of the name tells us to draw a benzene ring with an —OH group on carbon 1. Then the alkyl groups are attached, methyl to carbon 5 and isopropyl to carbon 2.

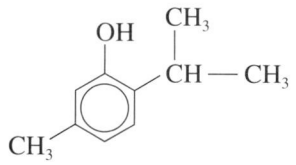

Study Check

Draw the condensed structural formula of 5-chloro-4-methyl-3-heptanol.

EXPLORE YOUR WORLD

Alcohols in Household Products

Read the labels on household products such as mouthwashes, cold remedies, alcoholic beverages, rubbing alcohol, and flavoring extracts. Look for names of alcohols such as ethyl alcohol, isopropyl alcohol, thymol, and menthol.

Questions

1. What part of the name tells you that it is an alcohol?
2. What alcohol is usually meant by the term "alcohol"?
3. What is the percent of alcohol in the products?
4. Write out the structures of the alcohols you find listed on the labels. You may need to use a reference book for some structures.

HEALTH NOTE

Alcohols in Health and Medicine

Methanol (CH_3OH), the simplest alcohol, is found in many solvents and paint removers. It is sometimes called *wood alcohol* because methanol is produced when wood decomposes at high temperatures in the absence of air. Methanol, a colorless and odorous liquid, is extremely poisonous and, if ingested, oxidizes to formaldehyde, a substance that can cause headaches, fatigue, blindness, and death.

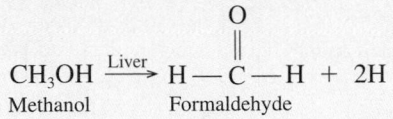

Ethanol (CH_3CH_2OH), or ethyl alcohol, is often used as a solvent for perfumes, varnishes, and some medicines, such as tincture of iodine. An antiseptic solution that is 70% by volume ethanol and 30% water is used to prepare an area of skin for injection and to sterilize equipment. It destroys bacteria by coagulating protein.

Isopropyl alcohol (2-propanol), commonly referred to as *rubbing alcohol*, is used as an astringent because it evaporates rapidly and cools the skin, reducing the size of blood vessels near the surface and decreasing pore size. It is also used to clean the skin before giving an injection or taking a blood sample.

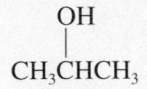

2-Propanol
(isopropyl alcohol)

Ethylene glycol (1,2-ethanediol) is commonly used as an antifreeze in heating and cooling systems, as a solvent for paints, inks, and plastics, and in the production of synthetic fibers such as Dacron. (See Figure 12.3.) If ingested, it is extremely toxic and causes drowsiness, convulsions, renal damage, coma, and death.

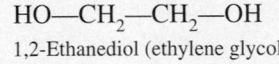

1,2-Ethanediol (ethylene glycol)

Glycerol (glycerin or 1,2,3-propanetriol), a trihydroxy alcohol, is a viscous liquid obtained from oils and fats during the production of soaps. The presence of several polar —OH groups makes it strongly attracted to water, a feature that makes glycerin useful as a skin softener in products such as skin lotions, cosmetics, shaving creams, and liquid soaps. It is also used as an antifreeze and in fluid shock absorbers. Used in the manufacture of dynamite, a mixture of glycerin and a strong oxidizer such as potassium chlorate can be explosive.

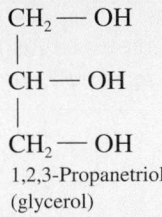

1,2,3-Propanetriol
(glycerol)

Menthol is a cyclic alcohol with a peppermint taste and odor that is used in candy, throat lozenges, and nasal inhalers. It causes the mucous membranes to increase their secretions and soothes the respiratory tract.

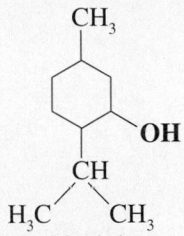

Menthol

Cholesterol is the major steroid found in the body, especially in the brain and nerve tissue. It is the starting compound for several hormones, vitamin D, and the bile salts, and it is the principal component of gallstones.

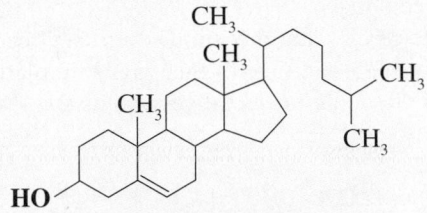

Cholesterol

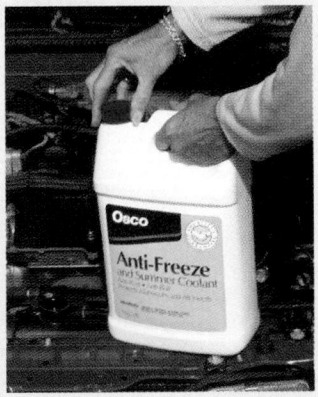

Figure 12.3 The diol ethylene glycol is the major component of many antifreeze fluids.

HEALTH NOTE

Derivatives of Phenol Are Found in Antiseptics, Essential Oils, and Antioxidants

Phenol and its derivatives are used as antiseptics in throat lozenges and mouthwashes. In household disinfectant sprays such as Lysol, the active ingredient is *o*-phenylphenol.

12.1

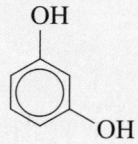

Resorcinol
(antiseptic)

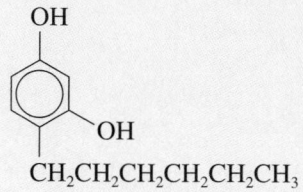

4-Hexylresorcinol
(antiseptic)

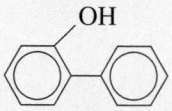

o-Phenylphenol
(disinfectant)

Several of the essential oils of plants that produce the odor or flavor of the plant are derivatives of phenol. Eugenol is found in cloves, vanillin in the vanilla bean, isoeugenol in nutmeg, and thymol in thyme and mint. Thymol has a pleasant, minty taste and is also used in mouthwashes and by dentists to disinfect a cavity before adding a filling compound. (See Figure 12.4.)

Foods such as cereals and oils spoil when they react with the oxygen in the air. One of the ways to prolong their shelf life is to add a preservative such as BHA or BHT to slow food spoilage.

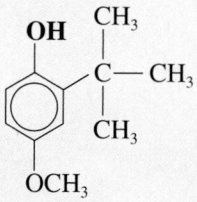

BHA (**b**utylated **h**ydroxy**a**nisole)

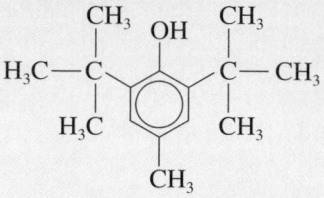

BHT (**b**utylated **h**ydroxy**t**oluene)

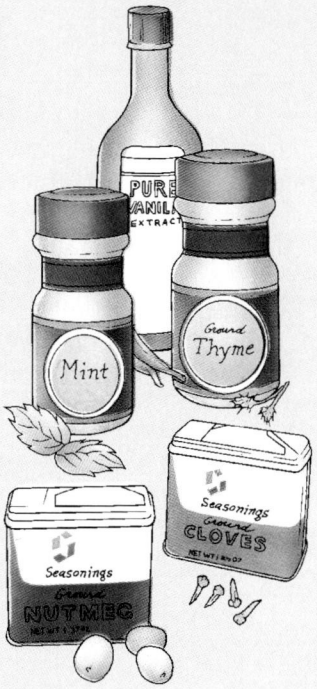

Figure 12.4 Derivatives of phenol are found in the essential oils of vanilla, cloves, nutmeg, thyme, and mint.

Thiols

Thiols are a family of sulfur-containing organic compounds similar to alcohols but having an —SH functional group in place of the —OH. Many short-chain thiols have strong, disagreeable odors. The smell of natural gas is not that of the methane, which is odorless, but of a small amount of ethanethiol that has been added to help detect a gas leak. Methanethiol is the characteristic odor of oysters, cheddar cheese, and "bad breath."

In the IUPAC system, thiols are named by adding *thiol* to the name of the longest carbon chain attached to the —SH group. An older term for thiol is mercaptan. The location of the —SH group is indicated by numbering the chain. Some examples of thiols follow:

Naming Thiols R—SH

Functional group: —SH

IUPAC name: alkanethiol

$CH_3—SH$	$CH_3CH_2—SH$	$CH_3\overset{\overset{\displaystyle SH}{\vert}}{C}HCH_3$
Methanethiol (oysters and cheese)	Ethanethiol (gas detection)	2-Propanethiol

There are thiols in the spray emitted when a skunk senses danger. The characteristic odor of onions is due to 1-propanethiol, which is also a lachrymator, a substance that will make your eyes tear. Garlic contains thiols including 2-propene-1-thiol. (See Figure 12.5.) We can break this name down as follows.

2-	prop	ene	-1-	thiol
Carbon 2 has C=C	3 carbons in chain	Alkene	On carbon 1	—SH group

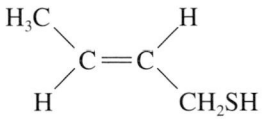

trans-2-Butene-1-thiol (in skunk spray)

$CH_3CH_2CH_2—SH$ 1-Propanethiol (in onions)

$CH_2=CH—CH_2—SH$ 2-Propene-1-thiol (in garlic)

Figure 12.5 Some common substances that have characteristic odors due to thiols.

12.1

S·A·M·P·L·E **12.3** ·P·R·O·B·L·E·M

Thiols

Draw the condensed structural formula of the following:

a. 1-butanethiol **b.** cyclohexanethiol

Solution

a. This compound has a —SH group on the first carbon of a butane chain.

$CH_3CH_2CH_2CH_2—SH$

b. This compound has a —SH group on a cyclic ring of hexane.

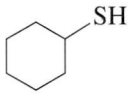

Study Check

What is the condensed structural formula of ethanethiol?

Q·U·E·S·T·I·O·N·S & P·R·O·B·L·E·M·S

Alcohols, Phenols, and Thiols

12.1 Why is ethanol soluble in water, but ethane is not?

12.2 Why is propanol more soluble in water than hexanol?

12.1

12.3 Give the IUPAC name for each of the following alcohols:

a. CH_3CH_2OH

b. $CH_3CH_2CH(OH)CH_3$

c. $CH_3CH_2CH(OH)CH_2CH_2CH_3$

d. $CH_3CH(CH_3)CH_2CH_2OH$

e.

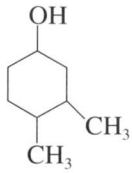

f. $CH_3-CH_2-C(CH_3)_2-CH_2-CH(CH_3)-CH_2-CH_2-OH$

12.4 Give the IUPAC name for each of the following alcohols:

a.

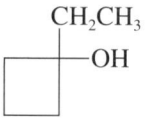

b. $CH_3-CH(Cl)-CH(CH_3)-CH_2-OH$

c. $CH_3CH_2CH(CH_3)CH(CH_3)CH_2OH$

d. $Cl-CH-CH_2-CH_2-CH(OH)-CH_3$

e.

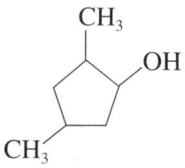

f. $CH_3-CH_2-CH(CH_2CH_2OH)-CH_2-CH_3$

12.5 Write the condensed structural formula of each of the following alcohols:

a. 1-propanol **b.** methyl alcohol **c.** 3-pentanol
d. 2-methyl-2-butanol **e.** cyclohexanol **f.** 1,4-butanediol

12.6 Write the condensed structural formula of each of the following alcohols:

a. ethyl alcohol **b.** 3-methyl-1-butanol
c. 2,4-dichlorocyclohexanol **d.** propyl alcohol
e. 1,3-cyclopentanediol **f.** 2,2,4-trimethyl-3-hexanol

12.7 Name each of the following phenols:

a.

b.

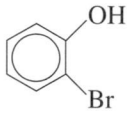

c.

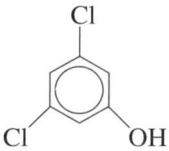

d.

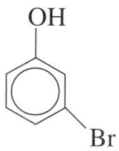

12.8 Name each of the following phenols:

a. 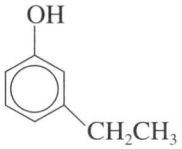

b. 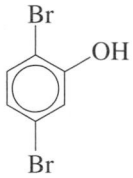

c. 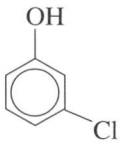

d. 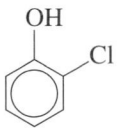

12.9 Write the condensed structural formula of each of the following phenols:
a. *m*-bromophenol **b.** *p*-chlorophenol **c.** 2,5-dichlorophenol
d. *o*-phenylphenol **e.** 4-ethylphenol

12.10 Write the condensed structural formula of each of the following phenols:
a. *o*-ethylphenol **b.** 2,4-dichlorophenol
c. 2,4-dimethylphenol **d.** 2-ethyl-5-methylphenol
e. *m*-phenylphenol

12.11 Give the IUPAC name for each of the following thiols:

a. $CH_3—SH$

b. $CH_3—\overset{\overset{SH}{|}}{CH}—CH_3$

c. $CH_3—\overset{\overset{CH_3}{|}}{CH}—\overset{\overset{CH_3}{|}}{CH}—CH_2—SH$

d. 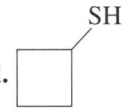

12.12 Give the IUPAC name for each of the following thiols:

a. $CH_3CH_2CH_2—SH$

b. $CH_3—CH_2—CH_2—\overset{\overset{SH}{|}}{CH}—CH_3$

c. $CH_3—\underset{\underset{CH_3}{|}}{\overset{\overset{CH_3}{|}}{C}}—CH_2—SH$

d. 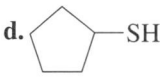

12.2 Ethers

Study Pack

Learning Goal

Give the name of an ether; write the condensed structural formula from the name.

An **ether** contains an oxygen atom attached by single bonds to two alkyl or aromatic groups. (See Figure 12.6.) The attached groups may be alkyl groups, aromatic, or one of each.

$CH_3—O—CH_3$ Dimethyl ether

$CH_3—O—CH_2CH_3$ Ethyl methyl ether

$CH_3—O—$  Methyl phenyl ether

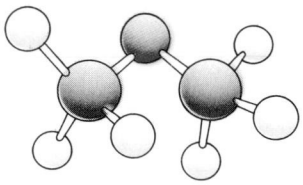

Figure 12.6 Ball-and-stick model of dimethyl ether.

12.2

Naming Ethers

The simple ethers are generally referred to by their common names. Each of the groups (alkyl or aromatic) attached to the oxygen atom is listed in alphabetical order, followed by the name *ether.* If the groups are identical, the prefix *di* may be used.

$CH_3—O—CH_2—CH_2—CH_3$ Common name: *methyl propyl ether*

Methyl Propyl

Naming Ethers	**R—O—R**
Functional group:	—O—
IUPAC name:	alkoxyalkane
Common name:	dialkyl ether alkyl[1] alkyl[2] ether alkyl phenyl ether

A complex ether is named according to the IUPAC system, in which the smaller group and the oxygen are named as an *alkoxy (**alk**yl + **oxy**gen)* group attached to the longer alkane chain. Suppose you want to name methyl propyl ether by its IUPAC name.

Step 1 Select the longer carbon chain and write its alkane name.

Step 2 Name the shorter carbon group and oxygen (oxy) as an alkoxy group and number the parent chain beginning at the end nearest to the alkoxy group.

Step 3 Give the location of the alkoxy group on the parent chain.

$CH_3—O—\underset{1}{CH_2}—\underset{2}{CH_2}—\underset{3}{CH_3}$ 1-Methoxypropane

Some examples of naming ethers follow:

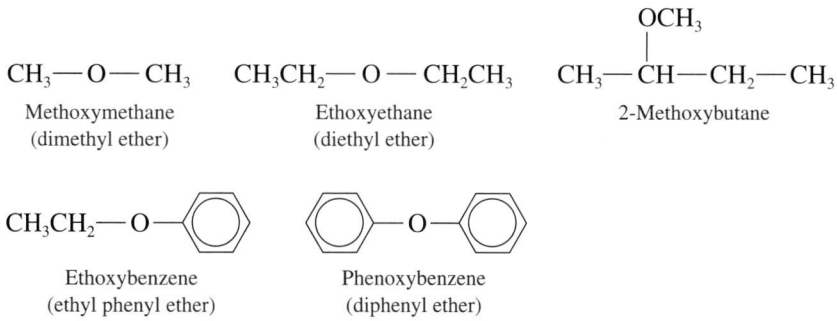

S·A·M·P·L·E 12.4 ·P·R·O·B·L·E·M

Ethers

Assign a common name and the IUPAC name to the following ethers:

a. $CH_3CH_2—O—CH_2CH_2CH_3$

b. 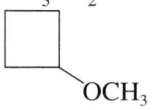

Solution

a. The groups attached to the oxygen are an ethyl group and a propyl group. The common name is *ethyl propyl ether.* Naming the shorter alkyl group and the oxygen as ethoxy gives the IUPAC name of *1-ethoxypropane.*

b. The groups attached to the oxygen are a methyl group and a cyclobutyl group. The common name is *cyclobutyl methyl ether.* In the IUPAC name, the CH_3O— is named methoxy; the cyclobutane is the longer carbon chain, to give the name *methoxycyclobutane.*

Study Check

What is the common name of ethoxybenzene?

Solubility of Ethers

12.2

Ethers are less soluble in water than alcohols but more soluble than alkanes. The oxygen atom in ethers makes them slightly polar but not as polar as alcohols. An O—H bond is more polar than an O—R bond. The shorter-chain ethers such as dimethyl ether and diethyl ether are somewhat soluble in water because the polar oxygen in the ether can form a hydrogen bond with water. However, ethers that contain longer carbon chains are not soluble in water.

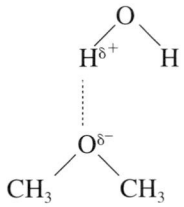

Hydrogen bond between water and ether

Uses of Ethers

The slight polarity of ethers, diethyl ether in particular, makes them very useful as solvents. They are also less reactive than many other organic materials and do not react with the substances they dissolve. However, ether vapors are highly flammable and react explosively with oxygen in the air. The utmost care must be taken when working with ethers.

HEALTH NOTE

Ethers as Anesthetics

Anesthesia is the loss of all sensation and consciousness. A general anesthetic is a substance that blocks signals to the awareness centers in the brain so the person has a loss of memory, a loss of feeling pain, and an artificial sleep. The term *ether* has been associated with anesthesia because diethyl ether was the most widely used anesthetic for more than a hundred years. Although it is easy to administer, ether is very volatile and highly flammable. A small spark in the operating room could cause an explosion. Since the 1950s, anesthetics such as Forane (isoflurane), Ethrane (enflurane), and Penthrane (methoxyflurane) have been developed that are not as flammable and do not cause nausea. (See Figure 12.7.) Most of these anesthetics retain the ether group, but the addition of many halogen atoms reduces the volatility and flammability of the ethers. More recently, they have been replaced by halothane (1-bromo-1-chloro-2,2,2-trifluoroethane), discussed in Chapter 11, because of the side effects of the ether-type inhalation anesthetics.

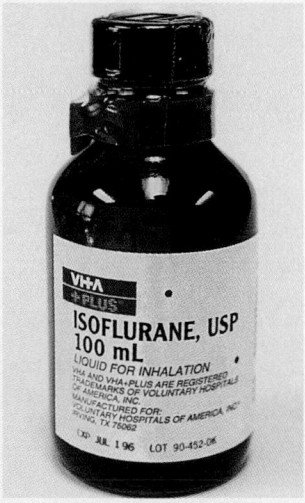

Figure 12.7 Isoflurane is an inhalation anesthetic used for the induction and maintenance of general anesthesia.

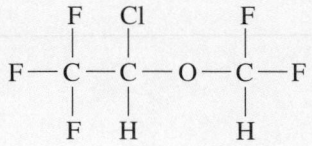

Forane
(isoflurane)

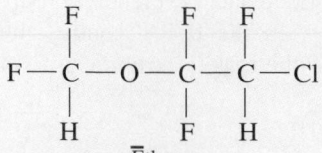

Ethrane
(enflurane)

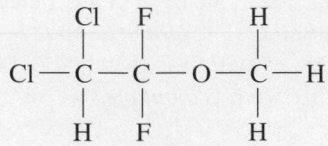

Penthrane
(methoxyflurane)

12.3

Q·U·E·S·T·I·O·N·S & P·R·O·B·L·E·M·S

Ethers

12.13 Two isomers of C_2H_6O are an alcohol and ether.
a. What are the formulas of the isomers?
b. Which one would be more soluble in water? Explain.

12.14 **a.** What is the function of a general anesthetic?
b. Why was the anesthetic diethyl ether replaced?
c. How do the structures of some anesthetics used today compare to the structure of diethyl ether?

12.15 Give the IUPAC name and a common name for each of the following ethers:
a. $CH_3—O—CH_2CH_3$
b. 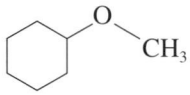
c. 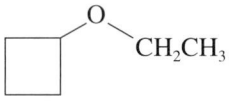
d. $CH_3—O—CH_2CH_2CH_3$

12.16 Give the IUPAC name and a common name for each of the following ethers:
a. $CH_3CH_2—O—CH_2CH_2CH_3$
b. 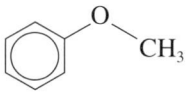
c. 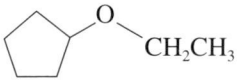
d. $CH_3—O—CH_3$

12.17 Write the condensed structural formula for each of the following ethers:
a. ethyl propyl ether
b. ethyl cyclopropyl ether
c. methoxycyclopentane
d. 1-ethoxy-2-methylbutane
e. 2,3-dimethoxypentane

12.18 Write the condensed structural formula for each of the following ethers:
a. diethyl ether
b. diphenyl ether
c. ethoxycyclohexane
d. 2-methoxy-2,3-dimethylbutane
e. 1,2-dimethoxybenzene

Study Pack

12.3 Reactions of Alcohols

Learning Goal

Write equations for the combustion and dehydration of an alcohol.

In Chapter 10, we learned that hydrocarbons undergo combustion in the presence of oxygen. Alcohols burn with oxygen too. For example, in a restaurant, a dessert is prepared by pouring alcohol on fruit and lighting it. (See Figure 12.8.) The combustion of ethanol proceeds as follows:

$$CH_3CH_2—OH + 3O_2 \xrightarrow{\text{Heat}} 2CO_2 + 3H_2O + \text{energy}$$

Figure 12.8 The burning of ethanol produces a flaming dessert.

12.3

EXPLORE YOUR WORLD

Combustion of Alcohol

Find a recipe for a dessert that is flamed, such as cherries jubilee, bananas Foster, or bananas flambé. Get a group of chemistry friends together and prepare your dessert. Explain to your friends how this dramatic presentation works.

Questions

1. What reaction does an alcohol undergo when you have crêpes suzette or bananas flambé?
2. In the Sterno, also called canned heat and used to heat fondue dishes, methanol is prepared as a gel or a liquid. Methanol is a highly toxic alcohol and can cause blindness if just 4 mL of methanol is ingested. What is the combustion reaction for methanol?

Dehydration of Alcohols

Earlier we saw that alkenes can add water to yield alcohols. In a reverse reaction, most alcohols can *dehydrate* (lose a water molecule) when they are heated with an acid catalyst such as H_2SO_4. During the **dehydration** of an alcohol, a H— and —OH are removed from *adjacent carbon atoms of the same alcohol* to produce a water molecule. A double bond forms between the same two carbon atoms to produce an alkene product.

Dehydration of an Alcohol (at High Temperatures) Forms an Alkene

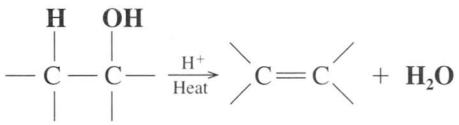

Examples:

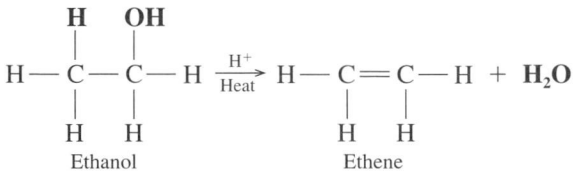

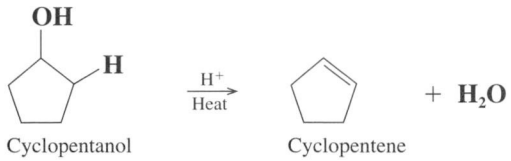

12.3

The dehydration of a secondary alcohol molecule can result in the formation of either of two products. The most prevalent product (major product) is the one that results when the hydrogen (H—) is removed from the carbon atom with the smallest number of hydrogen atoms.

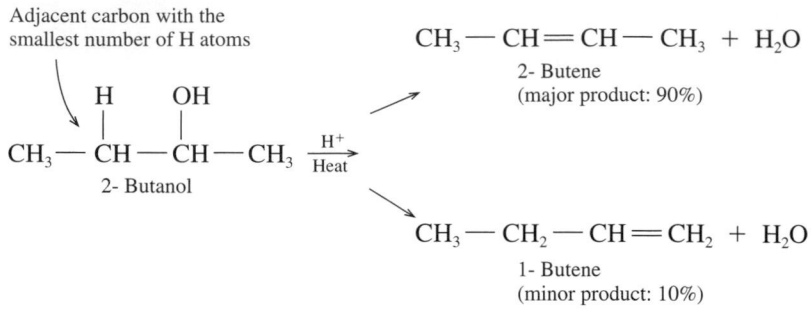

S·A·M·P·L·E· 12.5 ·P·R·O·B·L·E·M

Dehydration of Alcohols

Draw the condensed structural formulas for the alkenes produced by the dehydration of the following alcohols:

a.

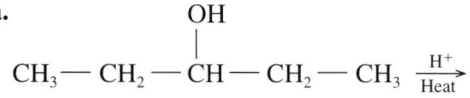

b.

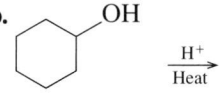

Solution

a. CH_3—CH_2—CH=CH—CH_3 + H_2O

b. The —OH of this alcohol is removed along with an H from an adjacent carbon. Remember that the hydrogens are not drawn in this type of geometric figure.

Study Check

What is the name of the alkene produced by the dehydration of cyclopentanol?

Formation of Ethers

Ethers form when the dehydration of alcohols occurs at lower temperatures in the presence of an acid catalyst. Then the components of water are removed from two separate molecules: an H— from one alcohol and the —OH from another. When the remaining portions of the two alcohols join, an ether is produced.

Dehydration of Two Alcohols Forms an Ether

Example:

$$\underset{\text{Methanol}}{CH_3{-}OH} + \underset{\text{Methanol}}{HO{-}CH_3} \xrightarrow[\text{Heat}]{H^+} \underset{\text{Dimethyl ether}}{CH_3{-}O{-}CH_3} + H_2O$$

S·A·M·P·L·E· 12.6 ·P·R·O·B·L·E·M

Ether Formation

Write the structural formula and name of the ether produced by the following reaction:

$$2CH_3CH_2OH \xrightarrow[\text{Heat}]{H^+}$$

Solution

The structural formula of the ether product is written by attaching the alkyl groups of the reacting alcohols to an oxygen atom.

$CH_3CH_2{-}O{-}CH_2CH_3$ Diethyl ether

Study Check

What alcohol is needed to form the ether named dicyclohexyl ether?

Q·U·E·S·T·I·O·N·S & P·R·O·B·L·E·M·S

Reactions of Alcohols

12.19 Draw the condensed structural formula of the alkene produced by each of the following dehydration reactions:

a. $CH_3{-}CH_2{-}CH_2{-}CH_2{-}OH \xrightarrow[\text{Heat}]{H^+}$

b. 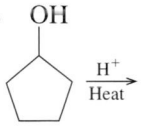

c. 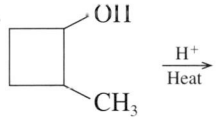

d.

$$CH_3{-}CH_2{-}CH_2{-}\overset{\overset{OH}{|}}{CH}{-}CH_3 \xrightarrow[\text{Heat}]{H^+}$$

12.20 Draw the condensed structural formula of the alkene produced by each of the following dehydration reactions:

a.

$$CH_3{-}\overset{\overset{CH_3}{|}}{CH}{-}CH_2{-}OH \xrightarrow[\text{Heat}]{H^+}$$

b.

$$CH_3{-}\overset{\overset{OH}{|}}{CH}{-}\overset{\overset{CH_3}{|}}{CH}{-}CH_2CH_3 \xrightarrow[\text{Heat}]{H^+}$$

12.3

c. 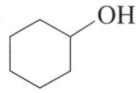

d. 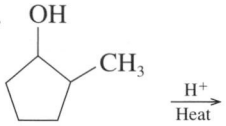

12.21 Write the ether product from the reaction of each of the following:

a. $2CH_3OH \xrightarrow[\text{Heat}]{H^+}$ **b.** $2CH_3CH_2CH_2OH \xrightarrow[\text{Heat}]{H^+}$

12.22 Write the ether product from the reaction of each of the following:

12.4

a. $2CH_3CH_2OH \xrightarrow[\text{Heat}]{H^+}$ **b.** $2CH_3CH(CH_3)CH_2OH \xrightarrow[\text{Heat}]{H^+}$

12.23 What alcohol(s) could be used to produce each of the following compounds?

a. $CH_2{=}CH_2$ **b.** $CH_3{-}O{-}CH_2CH_3$ **c.** 

12.24 What alcohol(s) could be used to produce each of the following compounds?

a. $CH_3CH_2{-}O{-}CH_2CH_3$ **b.** $CH_3{-}C(CH_3){=}CHCH_3$ **c.** 

Study Pack

Chem of Life

Organic Molecules: Functional Groups 3

Learning Goal

Name aldehydes and ketones; draw the condensed structural formula from the name.

12.4 Aldehydes and Ketones

Many important organic and biochemical compounds contain a **carbonyl group,** which is a carbon atom joined to an oxygen atom by a double bond.

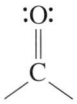

Carbonyl group

In an **aldehyde,** the carbon of the carbonyl group is attached to at least one hydrogen atom. The carbonyl group may be attached to a carbon in an alkyl group, a cycloalkyl group, or an aromatic group. (See Figure 12.9.)

In a **ketone,** the carbonyl group is attached to two carbon atoms. Therefore, the carbonyl group of a ketone is found within the carbon chain.

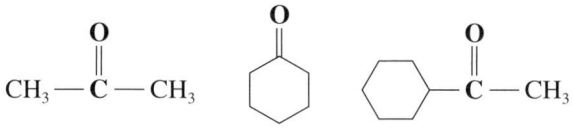

Solubility of Aldehydes and Ketones in Water

In a carbonyl group, the highly electronegative oxygen atom attracts the electrons of the double bond. As a result the carbonyl bond is strongly polar having an oxygen that has a partial negative charge and a carbon that is partially positive.

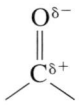

Polar carbonyl group

The polarity of the carbonyl bond determines the solubility of the aldehydes and ketones. Compounds with one to four carbons are soluble in water because the oxygen forms hydrogen bonds with water and pulls the small alkane chain into solution.

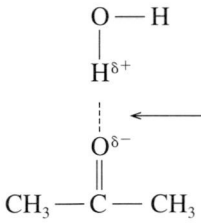

Hydrogen bonding makes ketones with 1–4 carbon atoms soluble in water

Dimethyl ketone

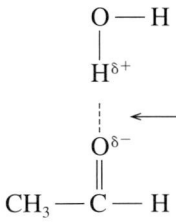

Hydrogen bonding makes aldehydes with 1–4 carbon atoms soluble in water

Acetaldehyde

The polarity of the carbonyl group makes ketones good solvents for many organic compounds. (See Figure 12.10.) Propanone (also called acetone or dimethyl ketone) is found in paint removers and fingernail polish remover. In the body, acetone may be produced in uncontrolled diabetes, fasting, and high-protein diets when large amounts of fats are metabolized for energy.

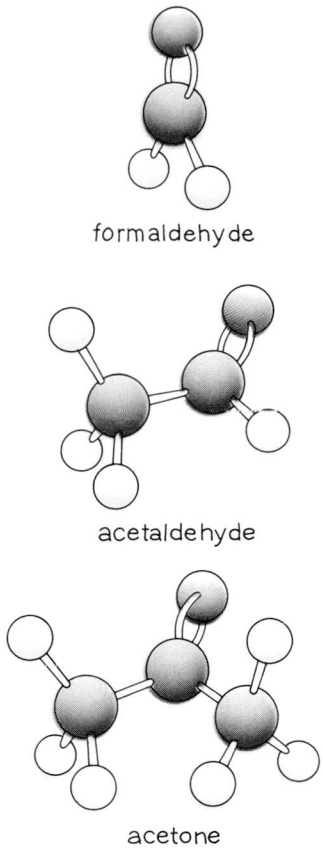

Figure 12.9 Ball-and-stick models of aldehydes and ketones: formaldehyde, acetaldehyde, and acetone.

Naming Aldehydes

To name an aldehyde using the IUPAC system, the longest carbon chain containing the aldehyde group is identified. Then the *e* ending of the name of that carbon chain is replaced with *al*. Any groups attached to the parent chain are named and numbered counting from the carbonyl carbon of the aldehyde group as carbon 1. For unbranched aldehydes with one to four carbons, the common names, which are one word, are most often used.

Naming Aldehydes

Functional group:	—C(=O)—H or —CHO
IUPAC name:	alkan**al,** benz**aldehyde**
Common name:	(form-, acet-, propion-, or butyr-)**aldehyde**

Figure 12.10 Butanone (commercially called methyl ethyl ketone, MEK), used to remove paint, is a hazardous chemical that must be used safely.

Some examples of IUPAC and common names for aldehydes follow:

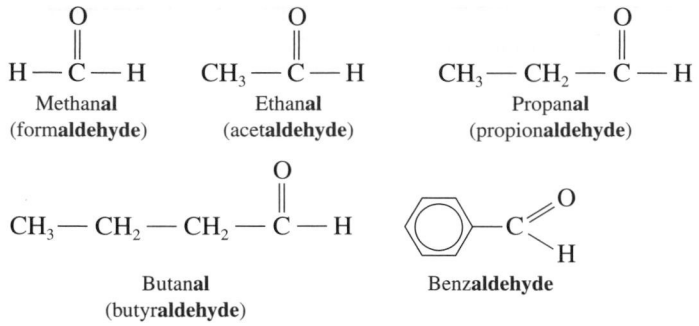

12.4

In an aldehyde, the carbonyl group is always the first carbon in the chain; no number is needed to indicate its position. In the IUPAC system, branches are located by numbering from the carbonyl carbon (carbon 1). In common names, the position of substituents is indicated by the lowercase Greek letters α (alpha), β (beta), and γ (gamma) starting with the carbon next to the carbonyl group.

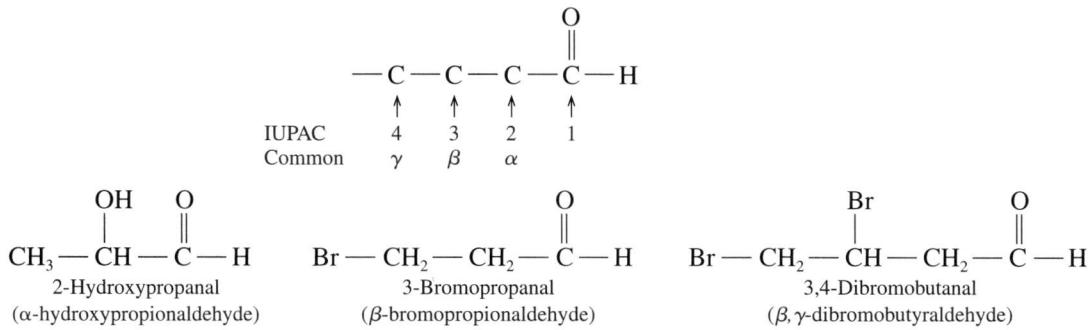

S·A·M·P·L·E· **12.7** ·P·R·O·B·L·E·M

Naming Aldehydes

Give the IUPAC names for the following aldehydes:

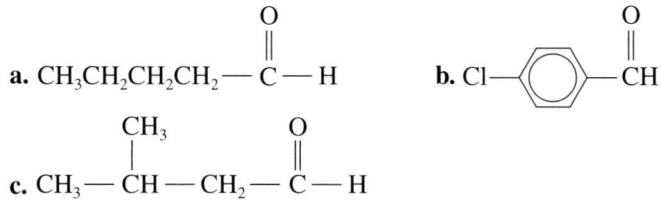

Solution

a. pentanal
b. 4-chlorobenzaldehyde, *p*-chlorobenzaldehyde
c. The longest unbranched chain has four atoms with a methyl group on the third carbon. The IUPAC name is *3-methylbutanal.*

Study Check

What are the IUPAC and common names of the aldehyde with three carbon atoms?

Naming Ketones

In the IUPAC system, a ketone is named by replacing the *e* with *one* in the name of the longest carbon chain containing the carbonyl group. The chain is numbered starting at the end nearest to the carbonyl group. (Propanone and butanone do not require numbers.) In a cyclic ketone, the carbonyl carbon is designated as carbon 1. In the common names of the simple ketones, which are widely used, the alkyl groups on either side of the carbonyl group are listed in alphabetical order followed by *ketone.*

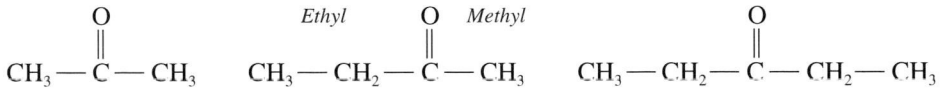

Propanone (dimethyl ketone) — Butanone (ethyl methyl ketone) — 3-Pentanone (diethyl ketone)

Cyclohexanone — Phenylethanone (methyl phenyl ketone)

12.4

S·A·M·P·L·E 12.8 P·R·O·B·L·E·M

Names of Ketones

Give the IUPAC name for the following ketone:

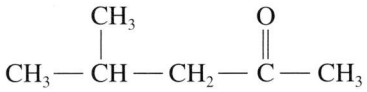

Solution

The longest chain is five carbon atoms. Counting from the right, the carbonyl group is on carbon 2 and a methyl group is on carbon 4. The IUPAC name is *4-methyl-2-pentanone.*

Study Check

What is the common name of 3-hexanone?

HEALTH NOTE

Aldehydes and Ketones in Health and Medicine

Several naturally occurring aromatic aldehydes and ketones are responsible for the odor and taste of food flavorings. Benzaldehyde is found in almonds, vanillin in vanilla beans, and cinnamaldehyde in cinnamon.

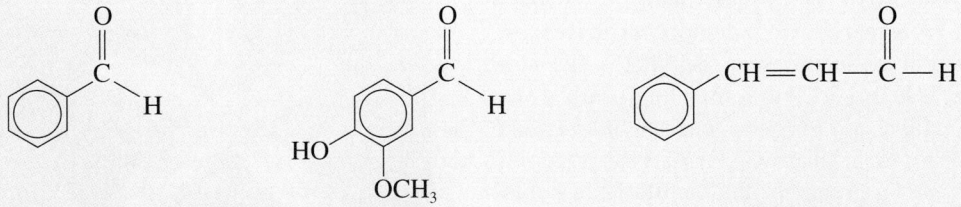

Benzaldehyde (oil of almonds) — Vanillin (vanilla bean) — Cinnamaldehyde (oil of cinnamon)

The flavor of butter or margarine is from butanedione, muscone is used to make musk perfumes, and oil of spearmint contains carvone.

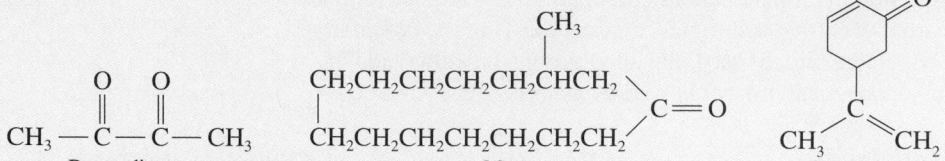

Butanedione (butter flavor) Muscone (musk) Carvone (spearmint oil)

12.4

Important hormones of the steroid family such as cortisone, testosterone, and progesterone contain a carbonyl group.

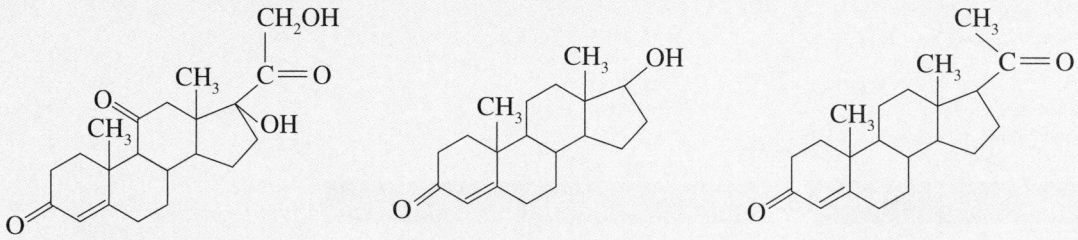

Cortisone (protein metabolism; inflammation reducer) Testosterone (male sex hormone) Progesterone (female sex hormone)

ENVIRONMENTAL NOTE

Vanilla

Vanilla has been used as a flavoring for over a thousand years. After drinking a beverage made from powdered vanilla and cocoa beans with Emperor Montezuma in Mexico, Cortez took vanilla back to Europe where it became popular for flavoring, and for scenting perfumes and tobacco. Thomas Jefferson introduced vanilla to the United States in the late 1700s. Today much of the vanilla we use in the world is grown in Mexico, Madagascar, Réunion, Seychelles, Tahiti, Ceylon, Java, the Philippines, and Africa.

The vanilla plant is a member of the orchid family. (See Figure 12.11.) There are many species of *Vanilla,* but *Vanilla planifolia* (or *V. fragrans*) is considered to produce the best flavor. The vanilla plant grows like a vine, and can grow to 100 feet in length. Its flowers are hand-pollinated to produce a green fruit that is picked in 8 or 9 months. It is sun-dried to form a long, dark brown pod, which is called "vanilla bean" because it looks like a string bean.

The flavor and fragrance of the vanilla bean comes from the black seeds found inside the dried vanilla bean; the seeds and pod are used to flavor desserts such as custards and ice cream. An extract of vanilla is made by chopping up vanilla beans and mixing them with a 35% alcohol–water mixture. The liquid, which contains the aldehyde vanillin, is drained from the bean residue and used for flavoring.

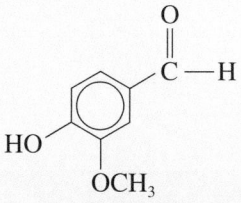

Vanillin

Figure 12.11 The seeds of the dried fruit of this orchid contain vanillin, a popular flavoring compound.

QUESTIONS & PROBLEMS

Aldehydes and Ketones

12.25 Write the IUPAC and common name (if any) for each of the following:

a. $CH_3-\overset{\overset{\displaystyle O}{\|}}{C}-H$

b. $CH_3-\overset{\overset{\displaystyle O}{\|}}{C}-CH_2-CH_3$

c. $CH_3-\overset{\overset{\displaystyle O}{\|}}{C}-CH_2-\underset{\underset{\displaystyle CH_3}{|}}{CH}-CH_3$

d. 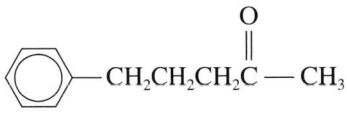

e. 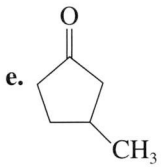

12.26 Write the IUPAC and common name (if any) for each of the following:

a. $CH_3CH_2CH_2CH_2CH_2\overset{\overset{\displaystyle O}{\|}}{C}H$

b. $CH_3-\underset{\underset{\displaystyle Br}{|}}{CH}-CH_2-\overset{\overset{\displaystyle O}{\|}}{C}-CH_3$

c. $CH_3\overset{\overset{\displaystyle Cl}{|}}{C}H\underset{\underset{\displaystyle CH_3}{|}}{C}HCH_2-\overset{\overset{\displaystyle O}{\|}}{C}-CH_2CH_2CH_3$

d. 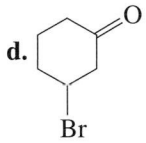

e. 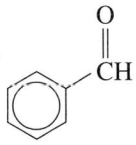

12.27 Write the condensed structural formula for each of the following compounds:

a. propionaldehyde
b. formaldehyde
c. 3-methyl-2-pentanone
d. β-chlorobutyraldehyde
e. benzaldehyde
f. ethyl phenyl ketone

12.28 Write the condensed structural formulas for each of the following compounds:

a. 3-methylpentanal
b. acetaldehyde
c. α-chlorobutyraldehyde
d. 4-bromocyclohexanone
e. 3-hydroxypropanal
f. isopropyl methyl ketone

12.29 Vanillin, a naturally occurring compound in vanilla beans and potato parings, has the following structural formula. Name all the functional groups in vanillin.

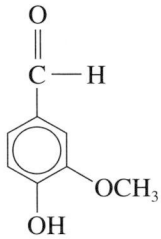

12.30 Veratraldehyde is prepared by adding a methyl group to the hydroxyl group of vanillin. Name all the functional groups in veratraldehyde.

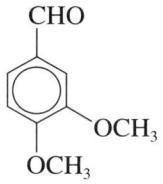

12.31 Which compound in each of the following pairs would be more soluble in water? Explain.

12.5

a. $CH_3CH_2CH_3$ or $CH_3\overset{O}{\overset{\|}{C}}CH_3$

b. $CH_3\overset{O}{\overset{\|}{C}}H$ or $CH_3CH_2CH_2CH_2\overset{O}{\overset{\|}{C}}H$

12.32 Which compound in each of the following pairs would be more soluble in water? Explain.

a. $CH_3-\overset{O}{\overset{\|}{C}}-CH_2CH_3$ or $CH_3-\overset{O}{\overset{\|}{C}}-\overset{O}{\overset{\|}{C}}-CH_3$

b. $CH_3-\overset{O}{\overset{\|}{C}}-CH_2CH_3$ or $CH_3-\overset{OH}{\overset{|}{C}H}-CH_2CH_3$

Study Pack

12.5 Oxidation and Reduction

Learning Goal

Write the reactions and/or products for the oxidation and reduction of alcohols, aldehydes, and ketones.

The aldehydes and ketones we discussed in the last section are produced from alcohols, which are easily oxidized. In Chapter 5, we learned that an **oxidation** was the loss of two hydrogen atoms. In some reactions, oxidation also means that a compound gains an oxygen atom. When two hydrogen atoms are removed from a carbon–carbon bond, a double bond (alkene) is produced, which can be oxidized further to a triple bond. In an alcohol such as ethanol, the removal of two hydrogen atoms oxidizes the ethanol to an aldehyde. The aldehyde is oxidized further by the gain of an oxygen atom to form a carboxylic acid called acetic acid. We can compare the various levels of oxidation as follows. The symbol [O] indicates that an oxidation takes place, which may be the loss of 2 H or the addition of an O.

Less oxidized → *More oxidized*

$$\underset{\text{Alkane}}{CH_3-CH_3} \xrightarrow{[O]} \underset{\text{Alkene}}{CH_2=CH_2} \xrightarrow{[O]} \underset{\text{Alkyne}}{HC\equiv CH}$$

$$\underset{\text{Alcohol}}{CH_3CH_2\overset{OH}{|}} \xrightarrow{[O]} \underset{\text{Aldehyde}}{CH_3-\overset{O}{\overset{\|}{C}}-H} \xrightarrow{[O]} \underset{\text{Carboxylic acid}}{CH_3-\overset{O}{\overset{\|}{C}}-OH}$$

Classification of Alcohols

Alcohols are classified according to the number of alkyl groups attached to the *carbon atom* bonded to the —OH group. In a **primary (1°) alcohol**, there is one alkyl group attached to the carbon bonded to the —OH group. Methanol is considered a primary alcohol. In a **secondary (2°) alcohol,** there are two alkyl groups attached. When there are three alkyl groups, the alcohol is classified as a **tertiary (3°) alcohol.**

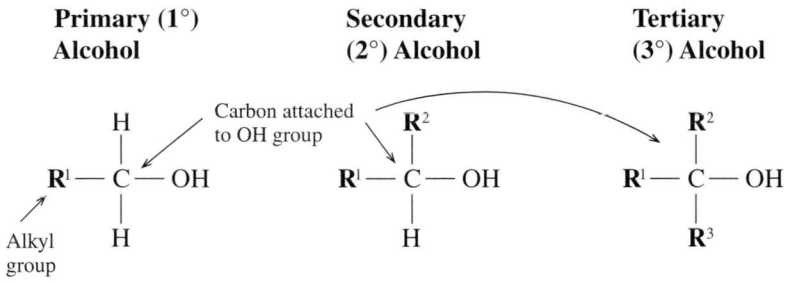

Examples:

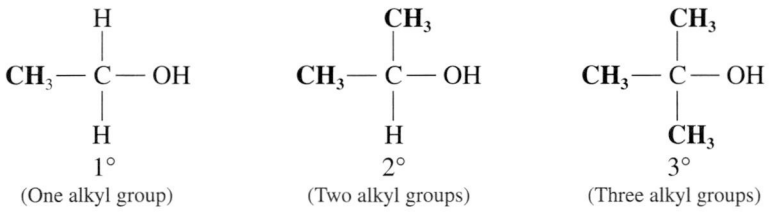

S·A·M·P·L·E 12.9 ·P·R·O·B·L·E·M

Classifying Alcohols

Classify each of the following alcohols as primary, secondary, or tertiary.

a. $CH_3—CH_2—CH_2—OH$

b. $CH_3—CH_2—C(OH)(CH_3)—CH_3$

c. 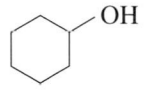

Solution

a. The carbon atom bonded to the —OH group is also attached to one alkyl group. This is a primary (1°) alcohol.

b. The carbon bonded to the —OH is attached to three alkyl groups. This is a tertiary (3°) alcohol.

c. The carbon bonded to the —OH group is bonded to two carbon atoms in the cyclic structure. This is a secondary (2°) alcohol.

Study Check

Classify the following alcohol as 1°, 2°, or 3°:

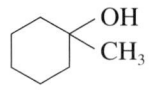

12.5

Oxidation of Primary Alcohols

In the oxidation of a primary alcohol, one hydrogen is removed from the carbon attached to the —OH, and the other hydrogen is taken from the —OH itself. The aldehyde product contains the carbon–oxygen double bond attached to a hydrogen. The other product, water, forms by combining the two "lost" hydrogen atoms with oxygen.

Oxidation of Primary Alcohols to Aldehydes

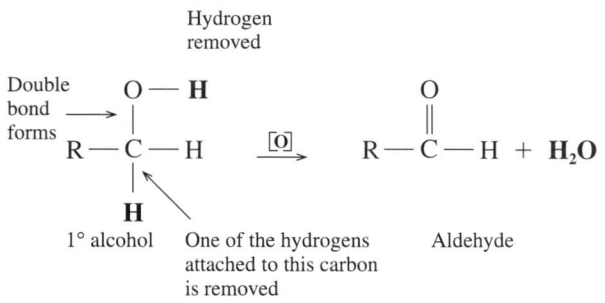

Examples:

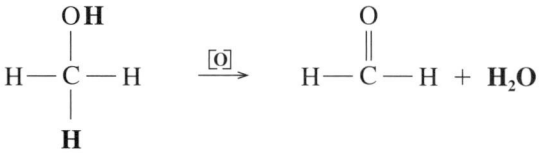

Methanol (methyl alcohol) → Methanal (formaldehyde)

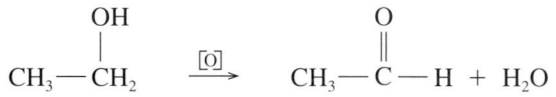

Ethanol (ethyl alcohol) → Ethanal (acetaldehyde)

Aldehydes oxidize further, this time by the addition of another oxygen to form a carboxylic acid. This step occurs so readily that it is often difficult to isolate the aldehyde product during oxidation. We will learn more about carboxylic acids in Chapter 14.

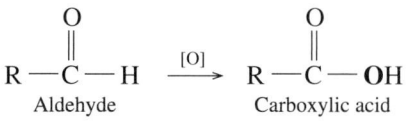

Aldehyde → Carboxylic acid

Example:

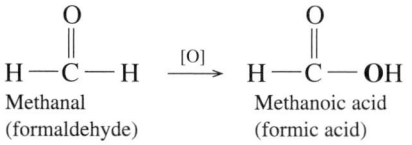

Methanal (formaldehyde) → Methanoic acid (formic acid)

Oxidation of Alcohol in the Body

The term *alcohol* commonly refers to the ethyl alcohol produced by fermentation when yeast is added to sugars in foods such as potatoes, grapes, rice, barley, and rye.

$$C_6H_{12}O_6 \xrightarrow{\text{Yeast}} 2CH_3CH_2OH + 2CO_2$$

Table 12.1 Typical Behaviors Exhibited by a Person Consuming Alcohol

Number of Beers (12 oz) or Glasses of Wine (5 oz)	Blood Alcohol Level (%)	Typical Behavior
1	0.025	Slightly dizzy, agreeable talker
2	0.05	Euphoria, loud talking, and laughing
4	0.10	Loss of inhibition, loss of coordination, drowsiness, legally drunk in most states
8	0.20	Intoxicated, quick to anger, exaggerated emotions
12	0.30	Unconscious
16–20	0.40–0.50	Coma and death

Ethanol is the most commonly abused drug in the United States. When ingested in small amounts, ethanol may produce a feeling of euphoria in the body although it is classified as a depressant. In the liver, enzymes such as alcohol dehydrogenases oxidize ethanol to acetaldehyde, a substance that impairs mental and physical coordination. If the blood alcohol concentration exceeds 0.4%, coma or death may occur. Table 12.1 gives some of the typical behaviors exhibited at various levels of blood alcohol.

$$\underset{\text{Ethanol (ethyl alcohol)}}{CH_3CH_2OH} \xrightarrow{[O]} \underset{\text{Ethanal (acetaldehyde)}}{CH_3\overset{\overset{\displaystyle O}{\|}}{C}H} \xrightarrow{[O]} 2CO_2 + H_2O$$

The acetaldehyde produced from ethanol in the liver is further oxidized to acetic acid, which is converted to carbon dioxide and water in the citric acid (Krebs) cycle. Thus, the enzymes in the liver can eventually break down ethanol, but the aldehyde and carboxylic acid intermediates can cause considerable damage while they are present within the cells of the liver.

A person weighing 150 lb requires about one hour to completely metabolize 10 ounces of beer. However, the rate of metabolism of ethanol varies between nondrinkers and drinkers. Typically, nondrinkers and social drinkers can metabolize 12–15 mg of ethanol/dL of blood in one hour, but an alcoholic can metabolize as much as 30 mg of ethanol/dL in one hour. Some effects of alcohol metabolism include an increase in liver lipids (fatty liver), an increase in serum triglycerides, gastritis, pancreatitis, ketoacidosis, alcoholic hepatitis, and psychological disturbances. Ethanol also exaggerates the effect of narcotics and sedatives, which can produce a combination that may lead to respiratory arrest, coma, and death.

When the Breathalyzer test is used for suspected drunk drivers, the driver exhales a volume of breath into a solution containing the orange Cr^{6+} ion. If there is ethyl alcohol present in the exhaled air, the alcohol is oxidized, and the Cr^{6+} is reduced to give a green solution of Cr^{3+}. (See Figure 12.12.)

$$\underset{\text{Ethanol}}{CH_3CH_2OH} + \underset{\text{Orange}}{Cr^{6+}} \longrightarrow \underset{\text{Acetic acid}}{CH_3\overset{\overset{\displaystyle O}{\|}}{C}OH} + \underset{\text{Green}}{Cr^{3+}}$$

Sometimes, alcoholics are treated with a drug called Antabuse (disulfiram), which prevents the oxidation of acetaldehyde to acetic acid. As a result, acetaldehyde accumulates in the blood, which causes nausea, profuse sweating, headache, dizziness, vomiting, and respiratory difficulties.

Figure 12.12 The Breathalyzer test determines the blood alcohol level in a possible drunk driver.

12.5

Oxidation of Methanol

Methanol, "wood alcohol," is a highly toxic alcohol present in products such as antifreeze, windshield washer fluid, Sterno, and paint strippers. Methanol is rapidly absorbed in the gastrointestinal tract. In the liver, it is metabolized to formaldehyde and then formic acid, a substance that causes nausea, severe abdominal pain, and blurred vision. Blindness can occur because the intermediate products destroy the retina of the eye. As little as 4 mL of methanol can produce blindness. The formic acid, which is not readily eliminated from the body, lowers blood pH so severely that just 30 mL of methanol can lead to coma and death.

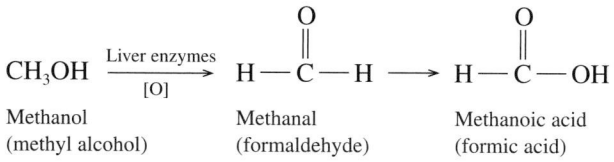

The treatment for methanol poisoning involves giving sodium bicarbonate to neutralize the formic acid in the blood. In some cases, ethanol is given intravenously to the patient. The enzymes in the liver pick up ethanol molecules to oxidize instead of methanol molecules, which gives more time for the methanol to be eliminated via the lungs without the formation of its dangerous oxidation products.

Oxidation of Secondary Alcohols

The oxidation of the secondary alcohols is similar to that of primary alcohols. One hydrogen is removed from the —OH and another from the carbon bonded to the —OH group. The result is a ketone that has the carbon–oxygen double bond attached to alkyl groups on both sides. Because there are no hydrogen atoms attached to the double bond in a ketone, no further oxidation is possible.

Oxidation of Secondary Alcohols to Ketones

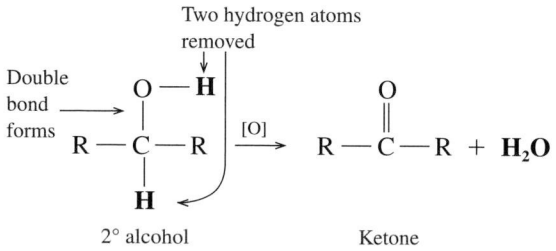

Examples:

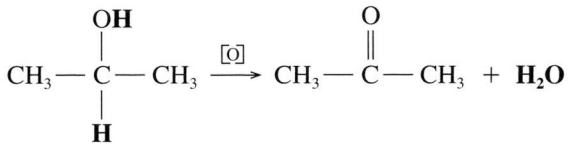

2-Propanol (isopropyl alcohol) — Propanone (dimethyl ketone; acetone)

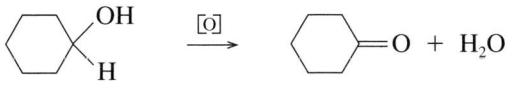

Cyclohexanol — Cyclohexanone

During vigorous exercise, lactic acid accumulates in the muscles and causes fatigue. When the activity level is decreased, oxygen enters the muscles and oxidizes the secondary —OH group in lactic acid to a ketone group in pyruvic acid. Pyruvic acid is metabolized further until it is completely oxidized to CO_2 and H_2O. The muscles in highly trained athletes are capable of taking up greater quantities of oxygen so that vigorous exercise can be maintained for longer periods of time.

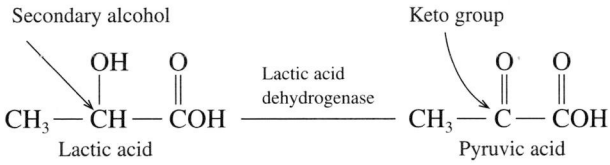

Tertiary alcohols do not oxidize because there are no hydrogen atoms to remove from the alcohol carbon. Therefore, tertiary alcohols do not oxidize.

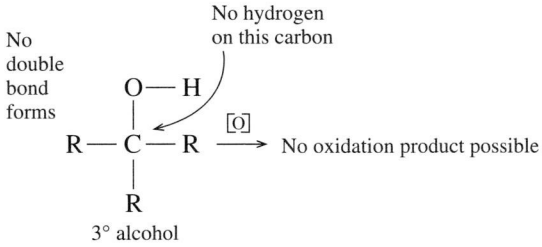

S·A·M·P·L·E 12.10 ·P·R·O·B·L·E·M

Oxidation of Alcohols

Draw the structural formulas of the organic products that form when the following alcohols undergo oxidation.

a. $CH_3CH_2\overset{\overset{\displaystyle OH}{|}}{C}HCH_3 \xrightarrow{[O]}$

b. $CH_3CH_2CH_2OH \xrightarrow{[O]}$

Solution

a. Because the reactant is a secondary alcohol, oxidation produces a ketone and water.

$$CH_3CH_2\overset{\overset{\displaystyle O}{\|}}{C}CH_3$$

b. This primary alcohol can oxidize to an aldehyde, and further oxidize to a carboxylic acid.

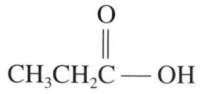

Study Check

When propene is hydrated in the presence of an acid cataylst, the product forms a ketone when it is oxidized. Explain using equations.

12.5

Reduction of Aldehydes and Ketones

Aldehydes and ketones are converted to alcohols when they gain hydrogen. **Reduction** is the addition of hydrogen to a compound or the loss of oxygen. When hydrogen is added, usually in the presence of a platinum or nickel catalyst, aldehydes are reduced to primary alcohols; ketones are reduced to secondary alcohols.

Reduction

Aldehydes Are Reduced to Primary Alcohols

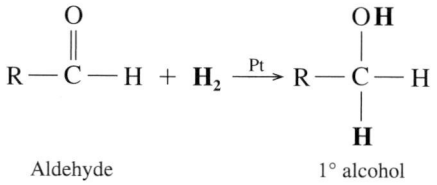

Ketones Are Reduced to Secondary Alcohols

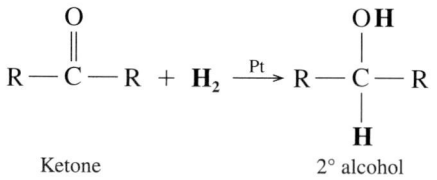

Examples:

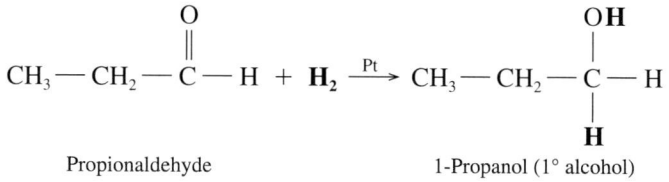

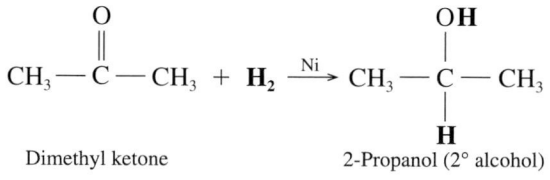

In biological systems, the reduction of aldehydes and ketones is an important process. In an enzyme-catalyzed reaction, hydrogen atoms are provided by a coenzyme known as NADH (nicotinamide adenine dinucleotide). In this reaction, NADH is oxidized to NAD^+, which is available to oxidize other biological molecules.

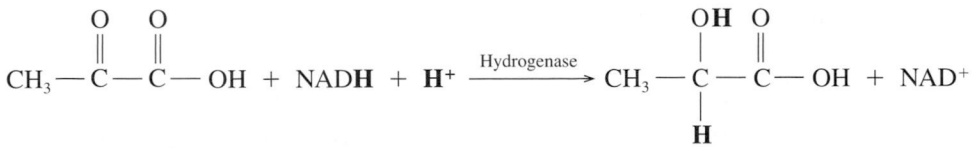

S·A·M·P·L·E· **12.11** ·P·R·O·B·L·E·M

Reduction of Carbonyl Groups

Write an equation for the reduction of cyclopentanone in the presence of a nickel catalyst.

Solution

The reacting molecule is a cyclic ketone that has five carbon atoms. Hydrogen atoms will add to the carbon and oxygen in the carbonyl group to form the corresponding secondary alcohol.

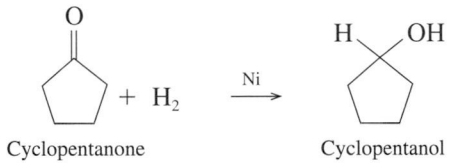

Study Check

What is the name of the product obtained from the hydrogenation of propionaldehyde?

Q·U·E·S·T·I·O·N·S & P·R·O·B·L·E·M·S

Oxidation and Reduction

12.33 Draw the condensed structural formula of the alcohol needed to give each of the following oxidation products:

a. formaldehyde **b.** cyclopentanone **c.** butanone
d. benzaldehyde **e.** 3-methylcyclohexanone

12.34 Draw the condensed structural formula of the alcohol needed to give each of the following oxidation products:

a. acetaldehyde **b.** 3-methylbutanone **c.** cyclohexanone
d. propionaldehyde **e.** 3-methylbutanal

12.35 Draw the condensed structural formula of the organic product when each of the following alcohols is oxidized [O] (if no reaction, write *none*):

a. $CH_3CH_2CH_2CH_2CH_2OH$ **b.** $CH_3CH_2CH(OH)CH_3$

c. 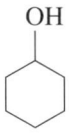

d. $CH_3-CH(OH)-CH_2-CH(CH_3)-CH_3$

e. $CH_3-CH(CH_3)-CH_2-CH_2-OH$

12.36 Draw the condensed structural formula of the organic product when each of the following alcohols is oxidized [O] (if no reaction, write *none*):

a. 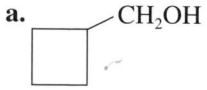

b. 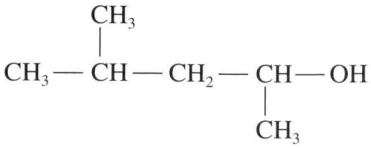

c. 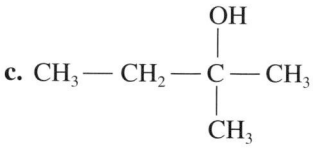

d. 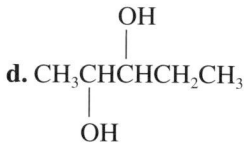

e. 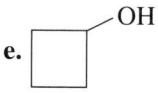

12.37 Give the condensed structural formula of the organic product formed when each of the following is reduced by hydrogen in the presence of a nickel catalyst:

a. butyraldehyde **b.** acetone
c. 3-bromohexanal **d.** 2-methyl-3-pentanone

12.38 Give the condensed structural formula of the organic product formed when each of the following is reduced by hydrogen in the presence of a nickel catalyst:

a. ethyl propyl ketone **b.** formaldehyde
c. 3-chlorocyclopentanone **d.** 2-pentanone

12.6

Study Pack

12.6 Aldehydes and Ketones React with Alcohols

Learning Goal

Draw the structures of the products formed when an alcohol adds to an aldehyde or a ketone.

The carbon–oxygen double bond of the carbonyl group in aldehydes and ketones readily undergoes addition reactions. An important example of addition is the addition of a molecule of an alcohol to the carbonyl group of an aldehyde or a ketone.

When a molecule of an alcohol adds to an aldehyde or a ketone, the hydrogen atom from the alcohol bonds to the oxygen atom of the carbonyl group to form a new hydroxyl group. The remaining portion of the alcohol molecule adds as an ether group to the carbon atom of the carbonyl group. The product of adding an alcohol to an aldehyde is called a **hemiacetal.** When an alcohol adds to a ketone, the product is a **hemiketal.**

Formation of a hemiacetal

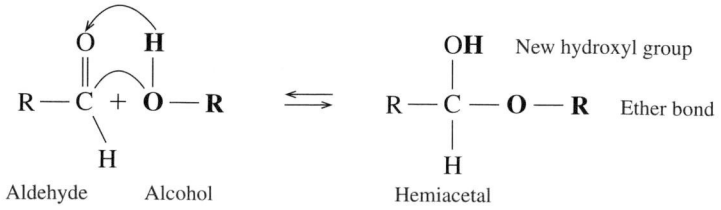

Example:

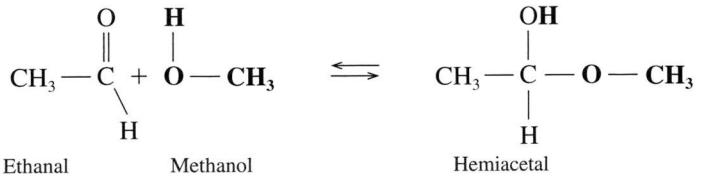

Formation of a hemiketal

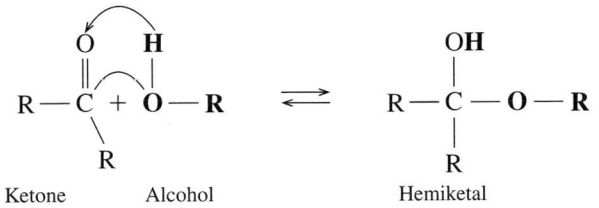

Example:

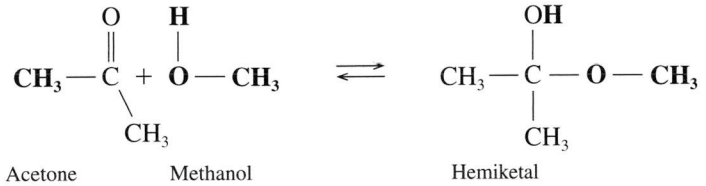

In a reaction mixture, the equilibrium favors the aldehyde or ketone, which makes it difficult to obtain a hemiacetal or hemiketal product. One very important exception is the formation of a cyclic hemiacetal or hemiketal, which can occur when both the —CHO group and the —OH group are in the *same* molecule. In Chapter 13, we see that the cyclic hemiacetal structure is the most stable form of sugars such as glucose. We can write this in a simplified form as follows:

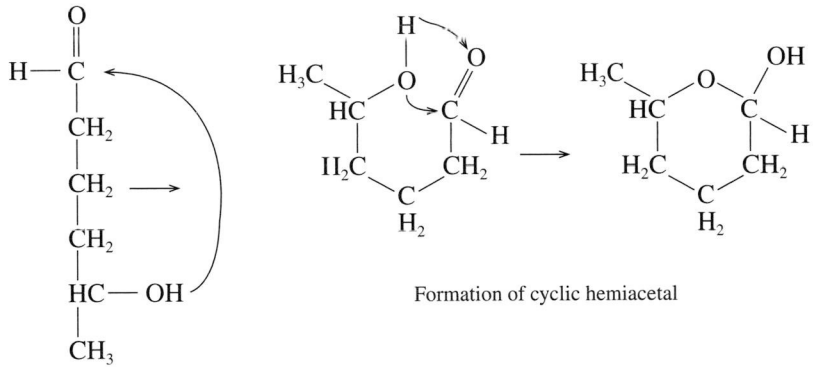

Formation of cyclic hemiacetal

Acetals and Ketals

If an acid catalyst is added to the reaction mixture, the hemiacetal or hemiketal will react with another molecule of alcohol. The term *hemi* means the compound is halfway to a complete addition of two molecules of an alcohol. When a second molecule of alcohol is added, the product is called an **acetal** or **ketal.** In each of the products, the carbon atom that was originally part of the carbonyl group is now attached to two ether groups. However, the acetal has a hydrogen atom attached to the carbonyl carbon, where the ketal has none.

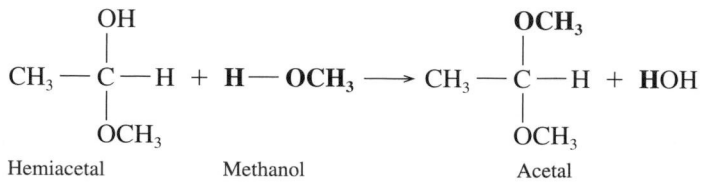

Unlike hemiacetal, the acetals are very stable and can be isolated from the reaction mixture. When we study carbohydrates in the next chapter, we will see that acetal bonds are used to link glucose molecules together in complex carbohydrates such as starch and cellulose.

12.6

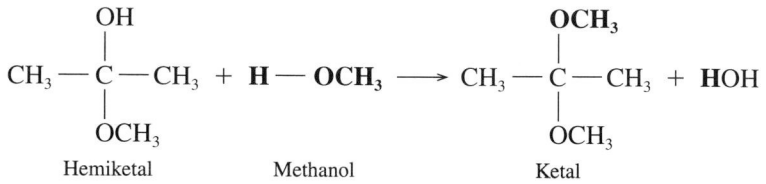

S·A·M·P·L·E 12.12 ·P·R·O·B·L·E·M

Acetals and Ketals

Write the structural formula of the hemiacetal and acetal products when methanol adds to propionaldehyde.

Solution

To form the hemiacetal, the hydrogen from the alcohol adds to the oxygen of the carbonyl group to form a new hydroxyl group and the remaining part of the alcohol adds to the carbon atom in the carbonyl group. The acetal forms when a second molecule of methanol is added to the carbonyl carbon atom.

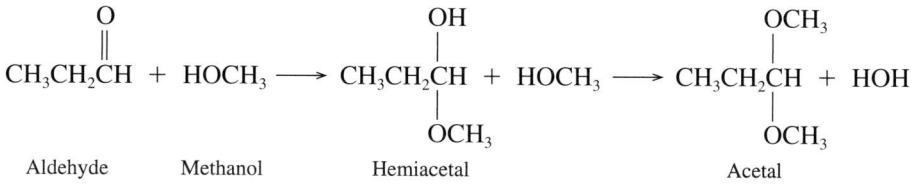

Study Check

What is the structural formula of the hemiketal produced when methanol adds to propanone?

Q·U·E·S·T·I·O·N·S & P·R·O·B·L·E·M·S

Aldehydes and Ketones React with Alcohols

12.39 Write the hemiacetal or hemiketal formed in each of the following reactions:

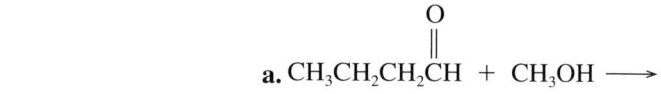

b. $CH_3\overset{\overset{\large O}{\|}}{C}CH_3 + CH_3CH_2OH \longrightarrow$

12.40 Write the hemiacetal or hemiketal products formed in each of the following reactions:

a. $CH_3\overset{\overset{\large O}{\|}}{C}CH_2CH_3 + CH_3OH \longrightarrow$

b. $CH_3\overset{\overset{\large O}{\|}}{C}H + CH_3CH_2OH \longrightarrow$

12.41 Write the structural formulas for the acetals or ketals produced when a second molecule of alcohol adds to each of the "hemi" products in problem 12.39.

12.42 Write the structural formulas for the acetals or ketals produced when a second molecule of alcohol adds to each of the "hemi" products in problem 12.40.

Chapter Review

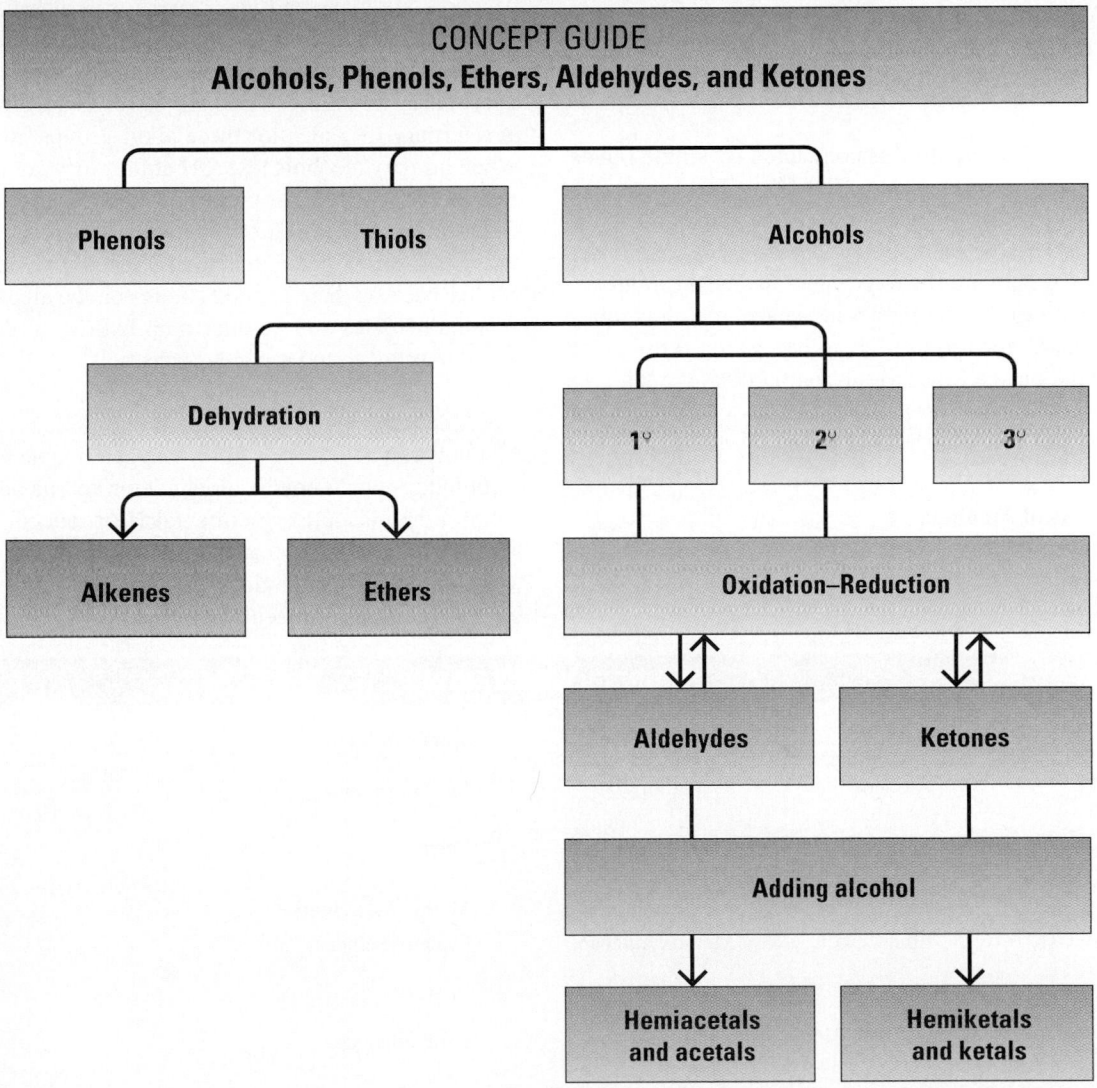

Chapter Summary

12.1 Alcohols, Phenols, and Thiols

The functional group of an alcohol is the hydroxyl group —OH attached to a carbon chain, R—OH. In a phenol, the hydroxyl group is attached to a benzene ring, Ar—OH. In the IUPAC system, the names of alcohols have *ol* endings, and the location of the —OH group is given by numbering the carbon chain. Simple alcohols are generally named by their common names with the alkyl name preceding the term *alcohol.* A cyclic alcohol is named as a cycloalkanol. An aromatic alcohol is named as a phenol.

The electronegative oxygen atom makes the —OH functional group very polar. Smaller alcohols (up to four carbons) are soluble in water because the —OH group can hydrogen bond with water molecules. In water, a phenol acts as a weak acid.

Thiols are similar to alcohols, except they have an —SH functional group in place of the —OH group. Thiols are named by the alkane chain followed by the word *thiol.*

12.2 Ethers

In an ether, an oxygen atom is connected by single bonds to two alkyl or aromatic groups, R—O—R. In the common names of ethers, the alkyl groups are listed alphabetically followed by the name *ether.* In the IUPAC name, the smaller alkyl group and the oxygen are named as an *alkoxy group* attached to the longer alkane chain, which is numbered to give the location of the alkoxy group. Less polar than alcohols, only a few shorter-chain ethers are slightly soluble in water. Ethers are widely used as solvents but can be explosive because their vapors are highly flammable.

12.3 Reactions of Alcohols

At high temperatures, alcohols dehydrate in the presence of an acid to yield alkenes. At lower temperatures, two molecules of alcohol lose H and OH to produce an ether.

12.4 Aldehydes and Ketones

Aldehydes and ketones contain a carbonyl group $\left(\diagup\!\!\diagdown C{=}O\right)$ which consists of a double bond between a carbon in the chain and an oxygen atom. In an aldehyde, the carbonyl group appears at the end of a carbon chain attached to at least one hydrogen atom. In a ketone, the carbonyl group occurs between carbon groups and has no hydrogens attached to it.

In the IUPAC system, aldehydes and ketones are named by replacing the *e* in the longest chain containing the carbonyl group with *al* for aldehydes, and *one* for ketones. The location of the carbonyl group in a ketone is given if there are more than four carbon atoms in the chain. The polarity of the carbonyl group makes aldehydes and ketones of one to four carbon atoms soluble in water.

12.5 Oxidation and Reduction

Alcohols are classified according to the number of alkyl groups attached to the hydroxyl carbon. In a primary (1°) alcohol one alkyl group is attached to the hydroxyl carbon. In a secondary (2°) alcohol two alkyl groups are attached, and in a tertiary (3°) alcohol three alkyl groups are attached. When primary alcohols lose 2 H atoms, they are oxidized to aldehydes, which can oxidize further to carboxylic acids when they gain an O atom. Secondary alcohols lose 2 H atoms and are oxidized to ketones. Tertiary alcohols do not oxidize because there are no H atoms on the alcohol carbon. When aldehydes and ketones gain hydrogen, they are reduced to primary and secondary alcohols.

12.6 Aldehydes and Ketones React with Alcohols

Alcohols can add to the carbon–oxygen double bond of the carbonyl group. When an aldehyde or ketone adds one alcohol, a hemiacetal or hemiketal is produced. When the hemiacetal or hemiketal adds a second molecule of alcohol to the carbon of the original carbonyl group, the more stable acetal or ketal molecule is formed.

Summary of Naming

Structure	Family	IUPAC Name	Common Name
CH_3—**OH**	Alcohol	Methanol	Methyl alcohol
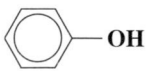	Phenol	Phenol	Phenol
CH_3—**SH**	Thiol	Methanethiol	Methyl mercaptan
CH_3—**O**—CH_3	Ether	Methoxymethane	Dimethyl ether
H—**C**(=**O**)—**H**	Aldehyde	Methanal	Formaldehyde
CH_3—**C**(=**O**)—CH_3	Ketone	Propanone	Dimethyl ketone (acetone)

Summary of Reactions

Dehydration of Alcohols to Form Alkenes

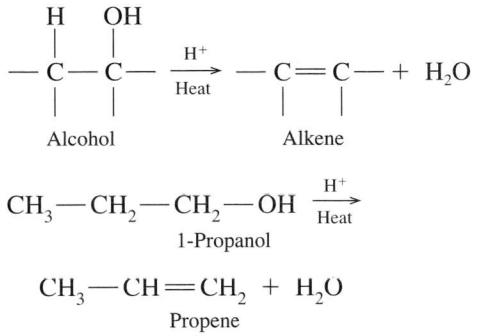

Dehydration of Alcohols to Form Ethers

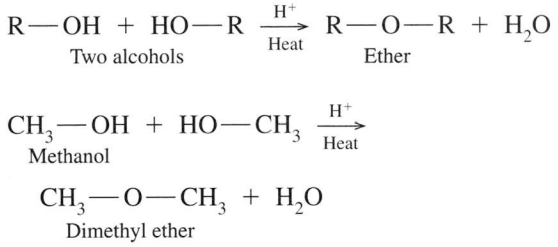

Oxidation of Primary Alcohols to Form Aldehydes and Carboxylic Acids

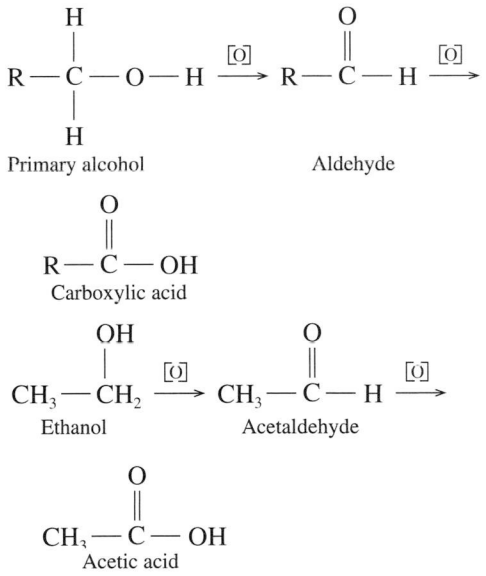

Oxidation of Secondary Alcohols to Form Ketones

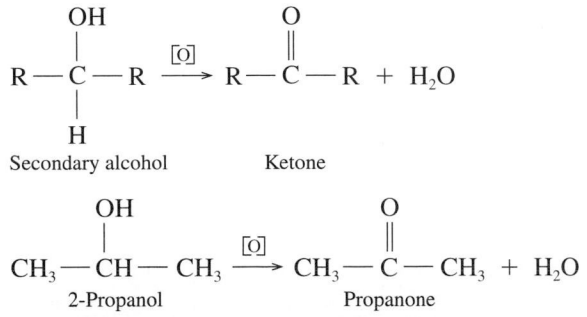

Reduction of Aldehydes to Primary Alcohols

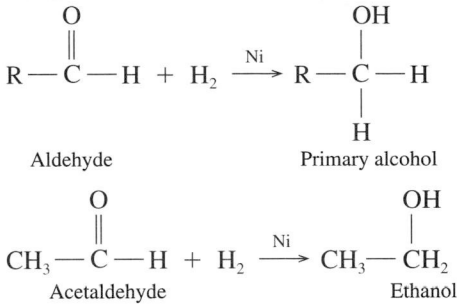

Reduction of Ketones to Secondary Alcohols

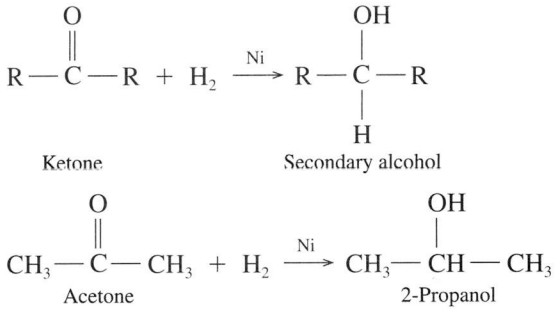

Formation of Hemiacetals and Hemiketals

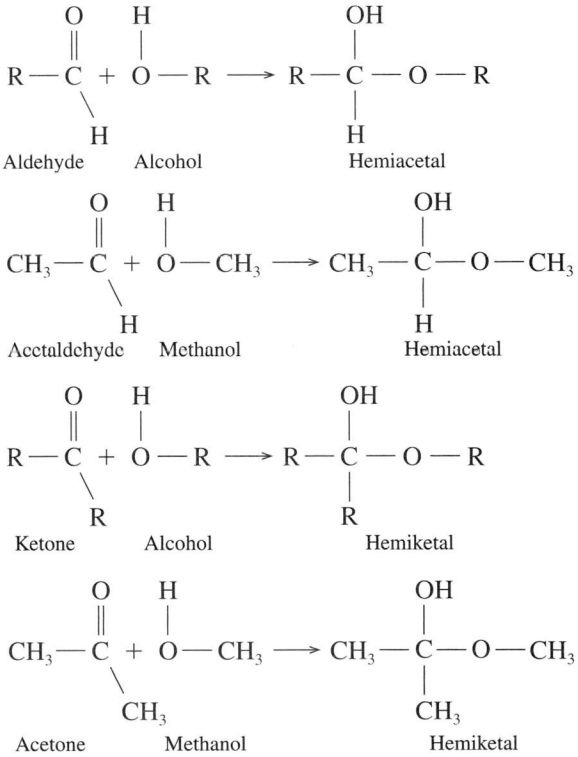

Formation of Acetals and Ketals

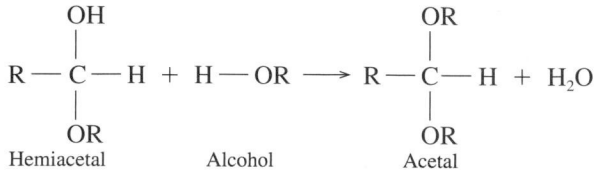

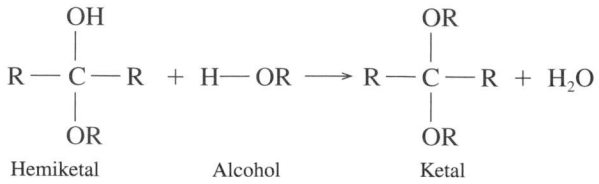

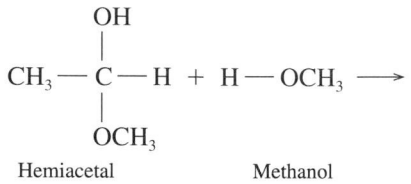

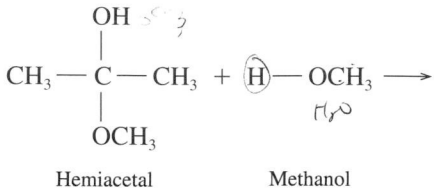

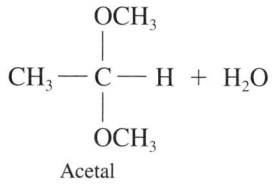

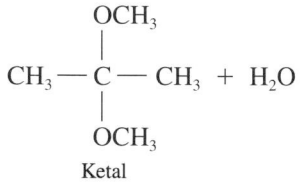

Key Terms

acetal The product of the addition of a second molecule of alcohol to a hemiacetal.

alcohols Organic compounds that contain the hydroxyl (—OH) functional group, attached to a carbon chain, R—OH.

aldehyde An organic compound with a carbonyl functional group and a hydrogen on the end carbon of the chain,

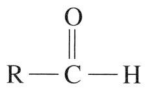

carbonyl group A functional group that contains a carbon–oxygen double bond $>C=O$

dehydration A reaction that removes water from an alcohol in the presence of an acid to form alkenes at high temperature, or ethers at lower temperatures.

ether An organic compound, R—O—R, in which an oxygen atom is bonded to two alkyl or two aromatic groups, or a mix of the two.

hemiacetal The product of the addition of an alcohol to the double bond of the carbonyl group in an aldehyde.

hemiketal The product of the addition of an alcohol to the double bond of the carbonyl group in a ketone.

hydroxyl group The —OH functional group.

ketal The product of the addition of a second molecule of alcohol to a hemiketal.

ketone An organic compound, R—C(=O)—R, in which the carbonyl functional group is bonded to two alkyl groups.

oxidation The loss of two hydrogen atoms from a reactant to give a more oxidized compound, e.g., primary alcohols oxidize to aldehydes, secondary alcohols oxidize to ketones. An oxidation can also be the addition of an oxygen atom as in the oxidation of aldehydes to carboxylic acids.

phenol An organic compound that has an —OH group attached to a benzene ring.

primary (1°) alcohol An alcohol that has one alkyl group attached to the alcohol carbon atom, R—CH_2OH.

reduction The addition of hydrogen to a carbonyl bond. Aldehydes are reduced to primary alcohols; ketones to secondary alcohols. Reduction can also be the loss of an oxygen atom.

secondary (2°) alcohol An alcohol that has two alkyl groups attached to the carbon atom bonded to the —OH group.

tertiary (3°) alcohol An alcohol that has three alkyl groups attached to the carbon atom attached to the —OH.

thiols Organic compounds that contain a thiol group (—SH) in place of the —OH group of an alcohol.

Chapter Practice

The following groups of problems include functional groups, physical and chemical properties, and reactions from all of the organic chapters.

12.43 Give the IUPAC and common names (if any) for each of the following compounds:

a. $CH_3CH(CH_3)CH_2CH_2CH(CH_3)CH_3$

b. $CH_3CH(Br)CH_2CH(OH)CH_2CH_3$

c. $CH_3CH_2C(=O)H$

d. $(CH_3)(H)C=C(CH_2CH_3)(H)$

e. 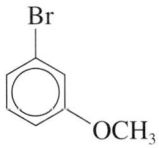

f. 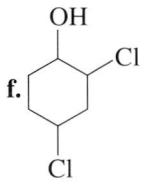

g. 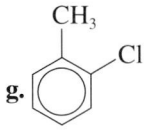

12.44 Give the IUPAC and common names (if any) for each of the following compounds:

a. $CH_3—O—CH_2CH_2CH_3$

b. $CH_3—CH_2—CH(SH)—CH_2—CH_3$

c. 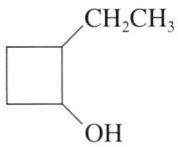

d. $CH_3CH_2—C\equiv C—CH_3$

e. 

f. 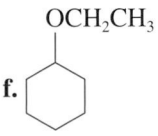

g. $CH_3C(=O)CH_2CH_2CH_2CH_2CH_2CH_3$

12.45 Draw the condensed structural formula of each of the following:

a. 3-methylcyclopentanone
b. *p*-dichlorobenzene
c. β-chloropropionaldehyde
d. *m*-bromophenol
e. diethyl ether

12.46. Draw the structural formula of each of the following:

a. methanethiol
b. β-chlorobutyraldehyde
c. 3-methyl-1-butanol
d. 3-methoxypentanal
e. *m*-bromotoluene

12.47 Which of the following compounds are soluble in water?

a. $CH_3CH_2CH_2CH_2CH_3$
b. CH_3OH
c. $CH_3CH_2—O—CH_2CH_3$
d. $CH_3C(=O)CH_3$
e. CH_3CH_2Cl
f. $CH_3CH_2CH_2OH$
g. $CH_3CH_2CH(OH)CH_2CH_2CH_3$

12.48 Which of the following compounds are soluble in water?

a. $HO—CH_2CH_2CH_2—OH$
b. $CH_3CH_2CH_2CH(CH_3)CH_3$
c. $CH_2—CH—CH_3$
d. $CH_3CH_2C(=O)CH_3$
e. $CH_3CH_2C(=O)H$
f. 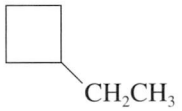
g. $CH_2(OH)—CH(OH)—CH_2(OH)$

12.49 Draw the structural formulas of all the alcohols with a molecular formula $C_4H_{10}O$.

12.50 Draw the structural formulas of all the ethers with a molecular formula $C_5H_{12}O$.

12.51 Draw the structural formulas of an alcohol, an ether, an aldehyde, and a ketone, all with the formula $C_6H_{12}O$.

12.52 Draw the structural formulas of two ketones and two aldehydes with the formula $C_5H_{10}O$.

12.53 Sometimes, several steps are needed to prepare a compound. Using a combination of the reactions we have studied, indicate how you might prepare the following from the starting substance given. (For example, 2-propanol could be prepared from 1-propanol by first dehydrating the alcohol to give propene and then hydrating it again to give 2-propanol according to Markovnikov's rule.)

$$CH_3CH_2CH_2OH \xrightarrow[\text{Heat}]{H^+} CH_3CH{=}CH_2 + H_2O \xrightarrow{H^+} CH_3\overset{\overset{\displaystyle OH}{|}}{C}HCH_3$$

a. acetaldehyde from ethene
b. 2-methylpropane from 2-methyl-2-propanol
c. butanone from 1-butene

12.54 As in problem 12.53, indicate how you might prepare the following from the starting substance given:
a. 1-pentene from pentanal
b. butanone from butanal
c. cyclopentene from cyclopentanone
d. 1,2-dibromobutane from butanal

12.55 Identify the functional groups in the following molecules:

a.

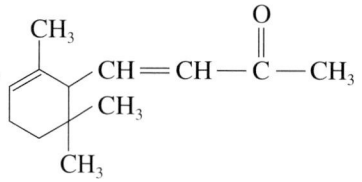

Ionone (odor of violets in perfumery)

b.

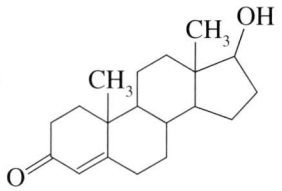

Testosterone

12.56 Identify the functional groups in the following molecules:

a.

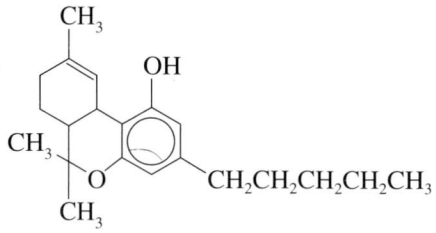

Tetrahydrocannabinol (THC)

b.

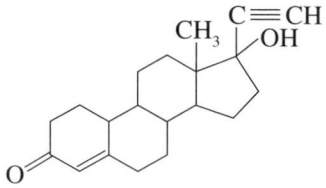

Norlutin (birth control drug)

12.57 Hexylresorcinol, an antiseptic ingredient used in mouthwashes and throat lozenges, has the IUPAC name of 4-hexyl-1,3-benzenediol. Draw its condensed structural formula.

12.58 Menthol, which has a minty flavor, is used in throat sprays and lozenges. Thymol is used as a topical antiseptic to destroy mold. Give each of their IUPAC names. What is similar and what is different about their structures?

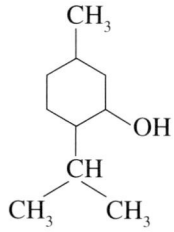

Menthol

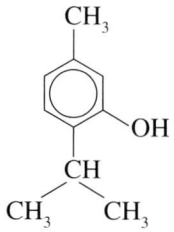

Thymol

12.59 Write the condensed structural formulas for each of the following naturally occurring compounds:
a. 2-heptanone, an alarm pheromone of bees
b. *trans*-2-hexenal, an alarm pheromone of a myrmicine ant
c. 2,5-dichlorophenol, a defense pheromone of a grasshopper
d. pinacolone, which is 3,3-dimethyl-2-butanone, odor of peppermint or camphor

12.60 Dimethyl ether and ethyl alcohol both have the molecular formula C_2H_6O. One has a boiling point of –24°C, and the other, 79°C. Draw the condensed structural formulas of each compound. Decide which boiling point goes with which compound and explain. Check the boiling points in a chemistry handbook.

Answers

Answers to Study Checks

12.1 3-bromo-4-methylcyclopentanol

12.2

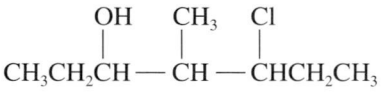

12.3 $CH_3CH_2—SH$

12.4 ethyl phenyl ether

12.5 cyclopentene

12.6 cyclohexyl alcohol

12.7 propanal (IUPAC), propionaldehyde (common)

12.8 ethyl propyl ketone

12.9 3°; The carbon attached to the hydroxyl group is bonded to three carbon atoms.

12.10 Propene adds water in the presence of an acid to form 2-propanol, which loses two hydrogen atoms to form propanone (acetone).

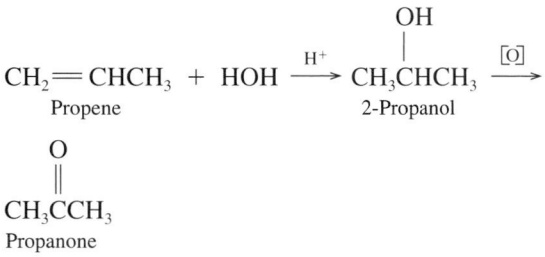

12.11 1-propanol

12.12 $CH_3C(OH)(OCH_3)CH_3$

Answers to Selected Questions and Problems

12.1 Ethanol has a polar —OH group that forms hydrogen bonds with the polar oxygen and hydrogen of water to dissolve ethanol molecules. Ethane is a saturated hydrocarbon that is nonpolar. Because it does not hydrogen bond to water, it does not dissolve in water.

12.3 **a.** ethanol **b.** 2-butanol
c. 3-hexanol **d.** 3-methyl-1-butanol
e. 3,4-dimethylcyclohexanol
f. 3,5,5-trimethyl-1-heptanol

12.5 **a.** $CH_3CH_2CH_2OH$ **b.** CH_3OH
c. $CH_3CH_2CH(OH)CH_2CH_3$ **d.** $CH_3C(OH)(CH_3)CH_2CH_3$

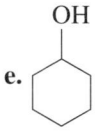

e.

f. $HOCH_2CH_2CH_2CH_2OH$

12.7 **a.** phenol
b. 2-bromophenol, *o*-bromophenol
c. 3,5-dichlorophenol
d. 3-bromophenol, *m*-bromophenol

12.9

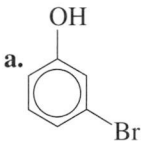

a.

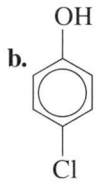

b.

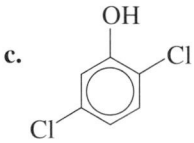

c.

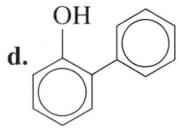

d.

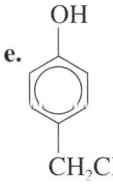

e.

12.11 **a.** methanethiol **b.** 2-propanethiol
c. 2,3-dimethyl-1-butanethiol
d. cyclobutanethiol

12.13 **a.** CH_3CH_2OH; $CH_3—O—CH_3$
b. Because the hydroxyl group of the alcohol can hydrogen bond with water, the short-chain alcohol is more soluble in water.

12.15 **a.** methoxyethane, ethyl methyl ether
b. methoxycyclohexane, cyclohexyl methyl ether
c. ethoxycyclobutane, cyclobutyl ethyl ether
d. 1-methoxypropane, methyl propyl ether

12.17 **a.** $CH_3CH_2—O—CH_2CH_2CH_3$

b. $CH_3CH_2—O—$ 

c. 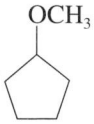

d. $CH_3CH_2—O—CH_2\overset{\overset{CH_3}{|}}{C}HCH_2CH_3$

e. $CH_3\overset{\overset{OCH_3}{|}}{C}H\underset{\underset{OCH_3}{|}}{C}HCH_2CH_3$

12.19 **a.** $CH_3CH_2CH{=}CH_2$ **b.** 

c. 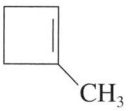

d. $CH_3CH_2CH{=}CHCH_3$

12.21 **a.** CH_3OCH_3 **b.** $CH_3CH_2CH_2OCH_2CH_2CH_3$

12.23 **a.** CH_3CH_2OH **b.** $CH_3OH + CH_3CH_2OH$

c. 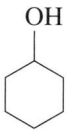

12.25 **a.** ethanal, acetaldehyde
b. butanone, ethyl methyl ketone
c. 4-methyl-2-pentanone
d. 5-phenyl-2-pentanone
e. 3-methylcyclopentanone

12.27 **a.** $CH_3CH_2\overset{\overset{O}{\|}}{C}H$ **b.** $\overset{\overset{O}{\|}}{C}HCH$

c. $CH_3\overset{\overset{O}{\|}}{C}\underset{\underset{CH_3}{|}}{C}HCH_2CH_3$ **d.** $CH_3\overset{\overset{Cl}{|}}{C}H\overset{\overset{O}{\|}}{C}H$

e. 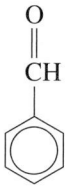

f. 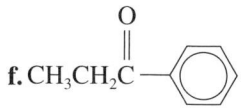

12.29 The functional groups are aldehyde, phenol, and ether.

12.31 **a.** $CH_3\overset{\overset{O}{\|}}{C}CH_3$ Water hydrogen bonds to the polar oxygen atom of the carbonyl group, which makes this compound more soluble.

b. $CH_3\overset{\overset{O}{\|}}{C}H$ The alkyl group in this compound has fewer carbon atoms, which makes it more water soluble.

12.33 **a.** CH_3OH **b.** 

c. $CH_3CH_2\overset{\overset{OH}{|}}{C}HCH_3$ **d.** 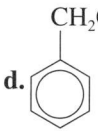

e. 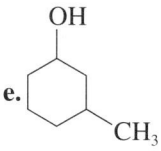

12.35 **a.** $CH_3CH_2CH_2CH_2\overset{\overset{O}{\|}}{C}H$ **b.** $CH_3CH_2\overset{\overset{O}{\|}}{C}CH_3$

c. 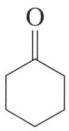

d. $CH_3\overset{\overset{O}{\|}}{C}CH_2\overset{\overset{CH_3}{|}}{C}HCH_3$

e. $CH_3\overset{\overset{CH_3}{|}}{C}HCH_2\overset{\overset{O}{\|}}{C}H$

12.37 **a.** $CH_3CH_2CH_2CH_2OH$ **b.** $CH_3\overset{\overset{OH}{|}}{C}HCH_3$

c. $CH_3CH_2CH_2\overset{\overset{Br}{|}}{C}HCH_2CH_2OH$

d. $CH_3\overset{\overset{CH_3}{|}}{C}H\underset{\underset{OH}{|}}{C}HCH_2CH_3$

12.39 **a.** $CH_3CH_2CH_2CH(OH)(OCH_3)$ **b.** $CH_3C(OH)(OCH_2CH_3)CH_3$

12.41 **a.** $CH_3CH_2CH_2CH(OCH_3)(OCH_3)$ **b.** $CH_3C(OCH_2CH_3)(OCH_2CH_3)CH_3$

12.43 **a.** 2,2,5-trimethylhexane
b. 5-bromo-3-hexanol
c. propanal, propionaldehyde
d. *cis*-2-pentene
e. 1-bromo-3-methoxybenzene, *m*-methoxybromobenzene
f. 2,4-dichlorocyclohexanol
g. 2-chlorotoluene, *o*-chlorotoluene

12.45 **a.** 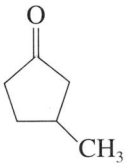
b. 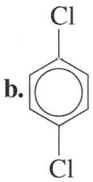
c. $CH_2CH_2CH(Cl)C(=O)$
d. 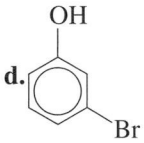
e. $CH_3CH_2—O—CH_2CH_3$

12.47 **a.** insoluble **b.** soluble **c.** insoluble
d. soluble **e.** insoluble **f.** soluble
g. insoluble

12.49 $CH_3CH_2CH_2CH_2OH$ $CH_3CH(OH)CH_2CH_3$
$CH_3C(OH)(CH_3)CH_3$ $CH_3CH(CH_3)CH_2OH$

12.51 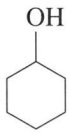
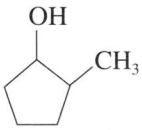
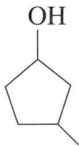
Many other alcohols are possible.

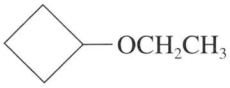
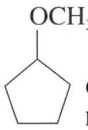
Other ethers are possible.

$CH_3CH=CHCH_2OCH_2CH_3$

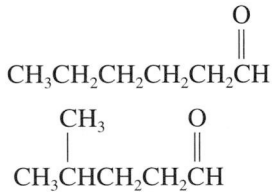
There are other aldehydes

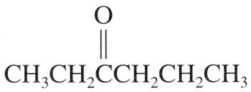
There are other ketones.

12.53 **a.** $CH_2=CH_2 + H_2O \xrightarrow{H^+} CH_3CH_2OH \xrightarrow{[O]} CH_3CH(=O)$

b. $CH_3C(OH)(CH_3)CH_3 \xrightarrow[\text{Heat}]{H^+} CH_3C(CH_3)=CH_2 + H_2 \xrightarrow{Pt} CH_3CH(CH_3)CH_3$

c. $CH_2=CHCH_2CH_3 + H_2O \xrightarrow{H^+}$
$CH_3CH(OH)CH_2CH_3 \xrightarrow{[O]} CH_3C(=O)CH_2CH_3$

12.55 **a.** alkene, cycloalkene, ketone
b. cycloalkene, ketone, alcohol

12.57 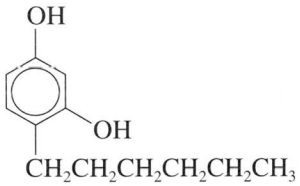

12.59 **a.** $CH_3C(=O)CH_2CH_2CH_2CH_2CH_3$

b. 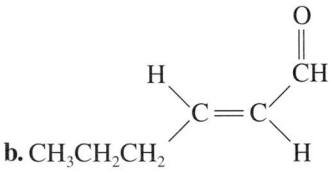

c. 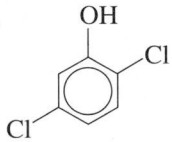

d. $CH_3C(=O)—C(CH_3)_2CH_3$

13

Carbohydrates

Complex carbohydrates, found in grains such as wheat, serve as a major source of glucose and energy for the body.

Looking Ahead

13.1	Classification of Carbohydrates
13.2	Monosaccharides
13.3	Chiral Carbon Atoms
13.4	Structures of Some Important Monosaccharides
13.5	Cyclic Structures of Monosaccharides
13.6	Disaccharides
13.7	Polysaccharides

www.chemplace.com/college/timberlake

Visit the web site above to test yourself, run simulations, and review key concepts from this chapter. For even more interactive review and quizzing, use the two CD-ROMs included with the book.

Bread, pasta, potatoes, rice, and many other foods in our diets contain carbohydrates. A carbohydrate is also called a **saccharide,** a word that comes from the Latin term *saccharum*, "sugar." Digestion breaks the bonds of the polysaccharides in these foods to give monosaccharides, primarily glucose, a carbohydrate that provides most of our cellular energy. Cellulose is a polysaccharide in plants that forms rigid cell walls. Although cellulose is not digestible by humans, it does play an important role in our diets by providing fiber. Our diets may also include sucrose (table sugar) and lactose (milk sugar).

The National Academy of Sciences has recommended that one-half of our calories be obtained from complex carbohydrates found in foods such as potatoes, corn, rice, and wheat. Because complex carbohydrates are digested slowly, a constant level of glucose is maintained in the blood over a longer period of time.

13.1 Classification of Carbohydrates

Learning Goal

Classify carbohydrates as monosaccharides, disaccharides, and polysaccharides.

Carbohydrates such as table sugar, lactose, and cellulose are made of carbon, hydrogen, and oxygen. Simple sugars, which have formulas of $C_n(H_2O)_n$, were once thought to be hydrates of carbon, thus the name *carbohydrate.* In the process of photosynthesis, carbohydrates are produced in the green leaves of plants including algae in lakes and oceans. Energy from sunlight is used to combine carbon dioxide (CO_2) and water into simple carbohydrates such as glucose.

Carbon cycle

$$6CO_2 + 6H_2O + \text{energy} \underset{\text{Respiration}}{\overset{\text{Photosynthesis}}{\rightleftharpoons}} \underset{\text{Glucose}}{C_6H_{12}O_6} + 6O_2$$

In our body tissues, glucose is oxidized in metabolic reactions known as respiration and chemical energy is released to do work in the cells. Carbon dioxide and water are formed. The combination of photosynthesis and respiration is called the carbon cycle, in which energy from the sun is stored by photosynthesis and made available to our cells when carbohydrates are metabolized. (See Figure 13.1.)

Types of Carbohydrates

The simplest carbohydrates are the **monosaccharides. Disaccharides** consist of two monosaccharide units joined together, whereas **polysaccharides** are complex

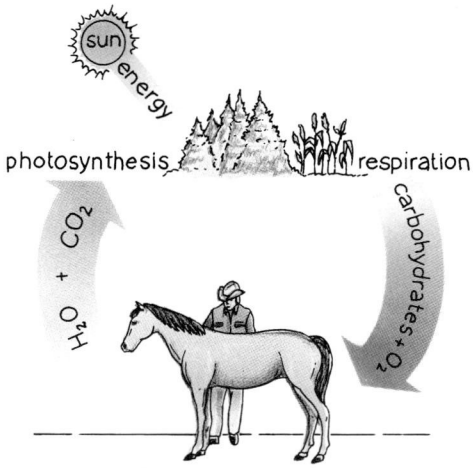

Figure 13.1 The carbon cycle in nature depicts the interdependence of photosynthesis and respiration.

EXPLORE YOUR WORLD

Photosynthesis

Carefully wrap four or five leaves of a houseplant with aluminum foil and masking tape. Place the plant in the sun for 1 week. Be sure to water. After 1 week unwrap the leaves.

Questions

1. What differences do you observe between the leaves that were wrapped and the unwrapped leaves?
2. What explanation would you give?
3. What are the requirements for photosynthesis?

13.1

carbohydrates that may contain thousands of monosaccharide units. When disaccharides or polysaccharides react with water, they can be completely hydrolyzed to yield monosaccharides.

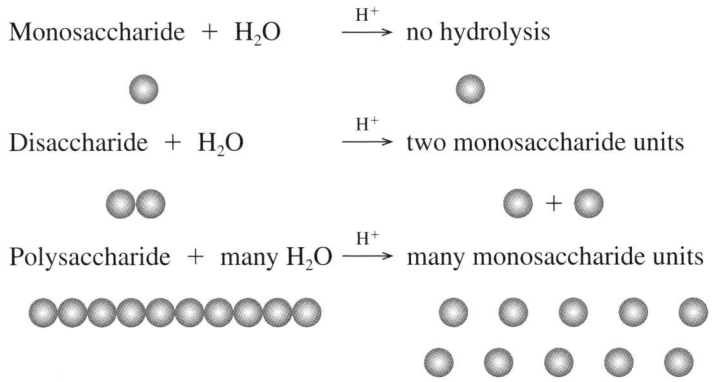

S·A·M·P·L·E· 13.1 ·P·R·O·B·L·E·M

Types of Carbohydrates

Classify the following carbohydrates as mono-, di-, or polysaccharides:

a. When lactose, milk sugar, is hydrolyzed, two monosaccharide units are produced.

b. Cellulose, a carbohydrate in cotton, yields thousands of monosaccharide units when completely hydrolyzed.

Solution

a. A disaccharide contains two monosaccharide units.

b. A polysaccharide contains many monosaccharide units.

Study Check

Fructose found in fruits does not undergo hydrolysis. What type of carbohydrate is it?

Q·U·E·S·T·I·O·N·S & P·R·O·B·L·E·M·S

Classification of Carbohydrates

13.1 What reactants are needed for photosynthesis and respiration?

13.2 What is the relationship between photosynthesis and respiration?

13.3 What is a monosaccharide? A disaccharide?

13.4 What is a polysaccharide?

Study Pack

Chem of Life Carbohydrates: Monosaccharides 1

13.2

13.2 Monosaccharides

Learning Goal

Classify a monosaccharide as an aldose or ketose and indicate the number of carbon atoms.

Monosaccharides are simple sugars that have an unbranched chain of three to eight carbon atoms, one of them in a carbonyl group and the rest attached to hydroxyl groups. In a polyhydroxy aldehyde called an **aldose,** the carbonyl group is on the first carbon (—CHO); a **ketose** contains the carbonyl group on the second carbon atom as a ketone (C=O).

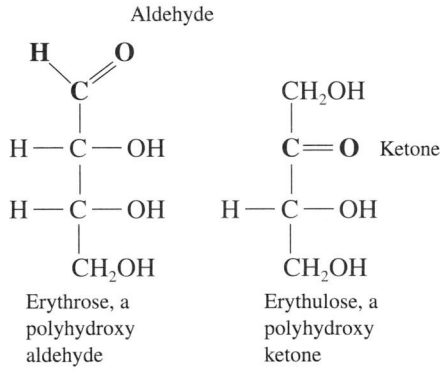

A monosaccharide with three carbon atoms is a *triose,* one with four carbon atoms is a *tetrose,* a *pentose* has five carbons, and a *hexose* contains six carbons. We can use both classification systems to indicate the type of carbonyl group and the number of carbon atoms. An aldopentose is a five-carbon monosaccharide that is an aldehyde; a ketohexose would be a six-carbon monosaccharide that is a ketone. Some examples:

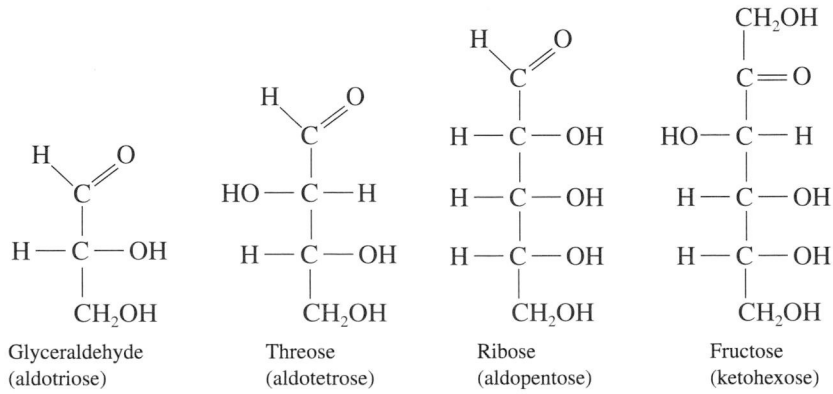

13.2

S·A·M·P·L·E 13.2 P·R·O·B·L·E·M

Monosaccharides

Classify each of the following monosaccharides to indicate their carbonyl group and number of carbon atoms:

a.

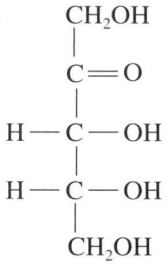

Ribulose

b.

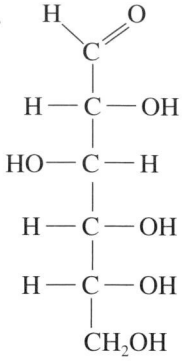

Glucose

Solution

a. The structural formula has a ketone group; ribulose is a ketose. Because there are five carbon atoms, it is a pentose. Combining these classifications makes it a ketopentose.

b. The structural formula has an aldehyde group; glucose is an aldose. Because there are six carbon atoms, it is an aldohexose.

Study Check

The simplest ketose is a triose named dihydroxyacetone. Draw its structural formula.

Q·U·E·S·T·I·O·N·S & P·R·O·B·L·E·M·S

Monosaccharides

13.5 What functional groups are found in all monosaccharides?

13.6 What is the difference between an aldose and a ketose?

13.7 What are the functional groups and number of carbons in a ketopentose?

13.8 What are the functional groups and number of carbons in an aldohexose?

13.9 Classify each of the following monosaccharides as an aldose or ketose.

$$(-\text{CHO} = -\overset{\overset{\displaystyle O}{\|}}{\text{C}}-\text{H})$$

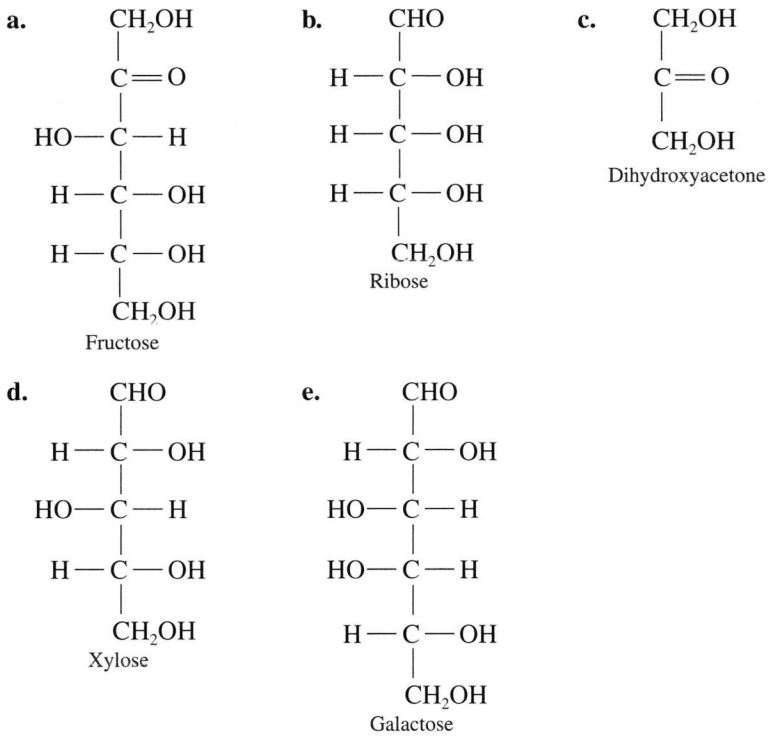

13.10 Classify each of the monosaccharides in problem 13.9 according to the number of carbon atoms in the chain.

13.3

13.3 Chiral Carbon Atoms

Study Pack

Learning Goal

Draw and identify Fischer projections for carbohydrate molecules.

If you look at your hands, each has a palm, a back, and the same number of fingers and thumbs. When the palms are up, your thumbs are on opposite sides. This is the reason you cannot put a left-handed glove on your right hand. If you turn your palms toward each other, they are mirror images. The left hand is the mirror image of the right hand. (See Figure 13.2.) We say that our hands are nonsuperimposable. Objects like hands that are nonsuperimposable mirror images are **chiral** objects. Left and right shoes are chiral; left- and right-handed golf clubs are chiral. They are mirror images of each other

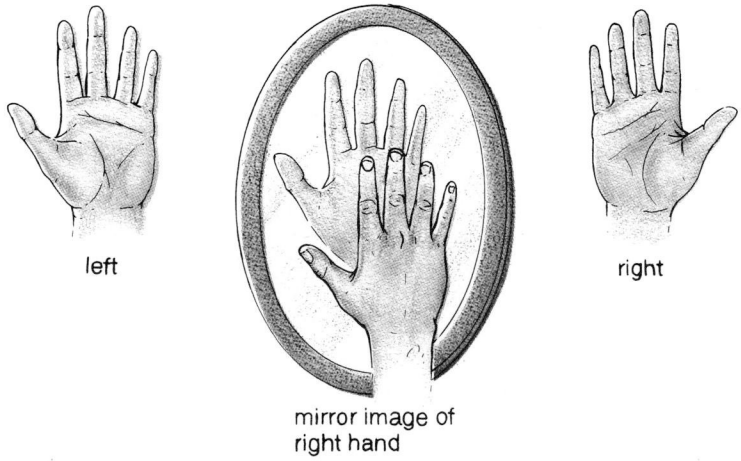

Figure 13.2 A pair of hands are chiral because they are mirror images of each other that are not superimposable.

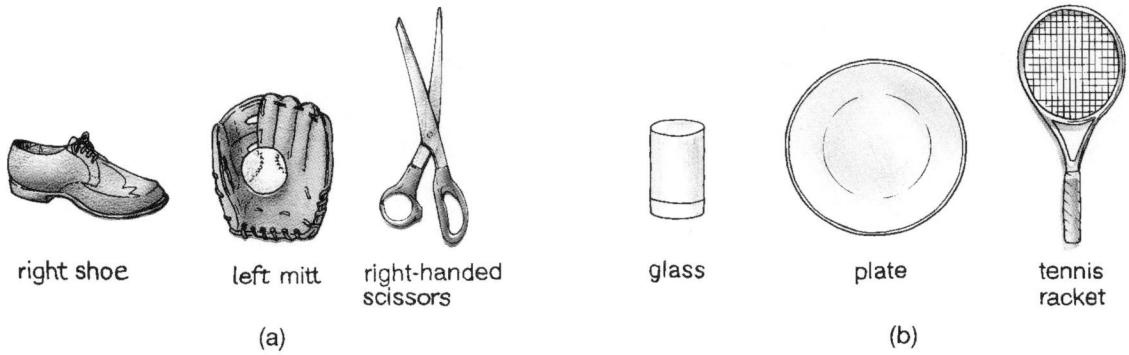

13.3

Figure 13.3 **(a)** Objects that are chiral. **(b)** Objects that are achiral.

that cannot be superimposed. Objects such as a glass or a ball whose mirror images can be superimposed on each other are achiral. (See Figure 13.3.)

S·A·M·P·L·E **13.3** ·P·R·O·B·L·E·M

Chiral and Achiral Objects

Classify each of the following objects as chiral or achiral:

a. left ear **b.** a Styrofoam coffee cup **c.** a left-handed mitt

Solution

a. Chiral; the left ear cannot be superimposed on the right ear.
b. Achiral; the mirror images of a cup can be superimposed on each other.
c. Chiral; a left-handed mitt cannot be superimposed on a right-handed mitt.

Study Check

Would a bowling pin be chiral or achiral?

Stereoisomers

A carbon compound is chiral if it has a carbon atom (a chiral carbon) attached to four different atoms or groups. It is chiral because four different atoms or groups can be attached to a carbon atom in two different ways. The resulting structures are mirror images and nonsuperimposable. Compounds that have the same structural formula but different spatial arrangements of the attached atoms are called **stereoisomers.** When stereoisomers are mirror images of each other, they are **enantiomers.** In Figure 13.4, we have drawn a chiral molecule and its mirror image. If we line up the hydrogen and iodine atoms, the bromine and chlorine atoms appear on opposite sides.

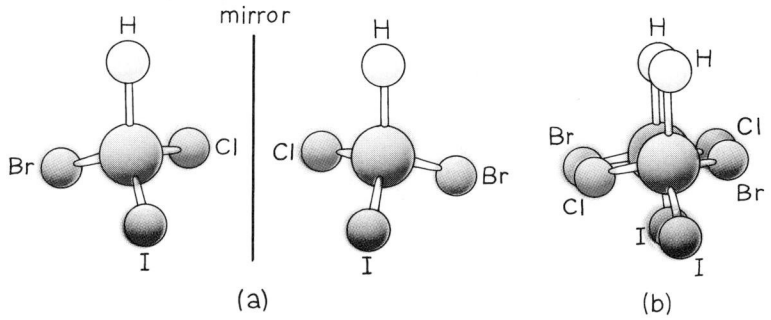

Figure 13.4 **(a)** The enantiomers of a chiral molecule are mirror images. **(b)** The enantiomers of a chiral molecule cannot be superimposed on each other.

S·A·M·P·L·E **13.4** ·P·R·O·B·L·E·M

Chiral Carbon Atoms

Indicate whether the central carbon atom in each of the following compounds is chiral or achiral; if chiral draw its mirror image.

a.

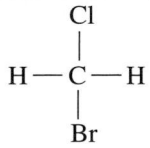

b.

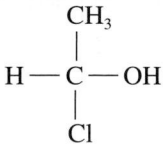

c.

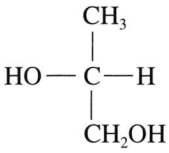

Solution

a. This molecule is achiral because two of the atoms attached to the carbon atom are the same (two hydrogens).

b. This molecule is chiral because there are four different atoms or groups attached to the central carbon. The mirror image would be written as:

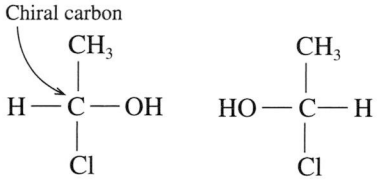

Its mirror image

c. The central carbon atom is chiral because there are four different groups attached. Its mirror image is written as:

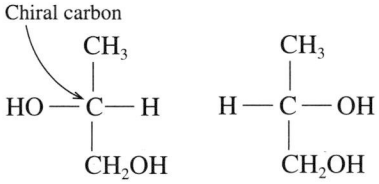

Its mirror image

Study Check

Identify the chiral carbon atom in 2-butanol.

Fischer Projections

In a three-dimensional view of glyceraldehyde, the simplest sugar, the bonds that extend back from the chiral carbon are shown as dotted wedges; bonds that extend to the front are solid wedges. In a two-dimensional view called a **Fischer projection,** the bonds that extend to the front are drawn as horizontal lines to the left and right sides of the chiral atoms. The bonds that extend back are drawn as vertical lines. The chiral carbon is the intersection of the horizontal and vertical lines.

Three-dimensional view

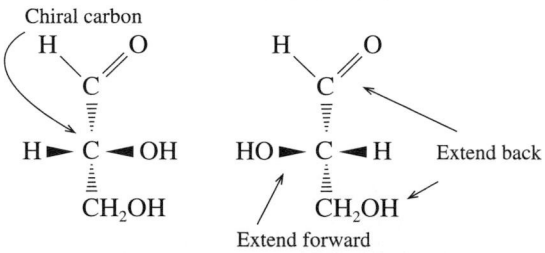

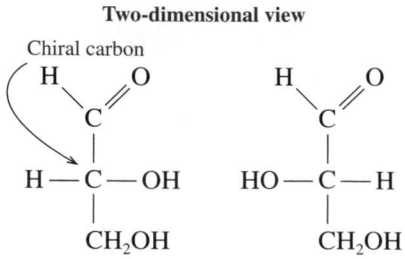

13.3

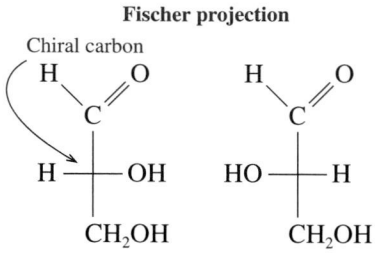

D and L Notation for Enantiomers

One way to distinguish between enantiomers is to label the molecules as the left- or right-handed form. By convention, the Fischer projection of glyceraldehyde is written with the carbonyl group at the top (carbon 1) and the $—CH_2OH$ group at the bottom. The letter L is assigned to the stereoisomer if the —OH group is on the left of the chiral carbon. In D-glyceraldehyde, the —OH is on the right.

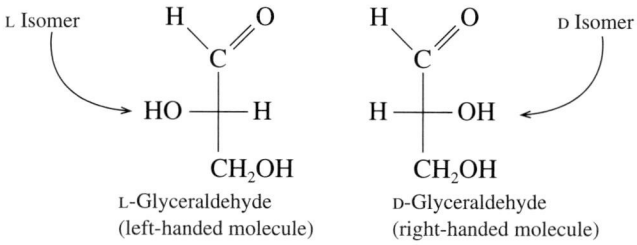

Most carbohydrates have longer carbon chains and several chiral carbon atoms. Then the chiral carbon furthest from the carbonyl group determines the D or L isomer. For example, in glucose, carbon 5 is the reference chiral carbon in the chain (carbon 6 in $—CH_2OH$ is not chiral).

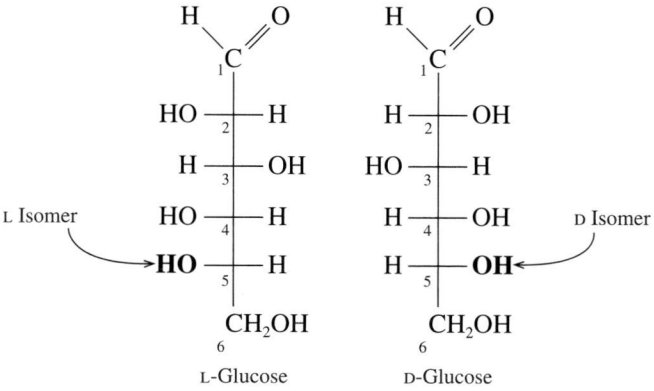

HEALTH NOTE

Chiral Compounds in Nature

Most stereoisomers that are active in biological systems consist of only one of the enantiomers. Rarely are both the D and L forms of a biomolecule biologically active. This happens because our enzymes and cell surface receptors on which metabolic reactions take place also have "handedness." Thus, only one of the enantiomers of a reactant or a drug interacts with those enzymes or receptors; the other is inactive. For example, only the D isomers of the monosaccharides of glucose, galactose, and fructose are active. Among the amino acids, only the L isomers such as L-alanine or L-lysine are used to make proteins.

A substance called carvone exists as D and L isomers. The L-carvone isomer gives the flavor to spearmint oil, and its enantiomer D-carvone is the flavor of caraway seed oil. Thus, our senses of taste and smell are sensitive to the chirality of molecules. In the brain, D-LSD causes hallucinations because it affects the production of serotonin, a chemical that is important in sensory perception. However, its enantiomer L-LSD produces little effect in the brain.

In the past, a drug contained both enantiomers, even though one was not useful, or might even have been toxic. For example, in the late 1950s, a drug called thalidomide was prescribed as a tranquilizer for pregnant women in some countries (not the United States). Its tragic effects on developing embryos was soon apparent as babies were born with deformed limbs. Scientists have determined that only one enantiomer of thalidomide produced the birth defects, but its mirror image was useful as a tranquilizer.

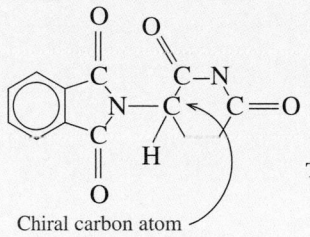

Thalidomide

Today, drug researchers are beginning to apply *chiral technology* to produce the active enantiomers of chiral drugs. The benefits of producing the active enantiomer include reducing the dose, enhancing activity, and reducing interactions with other drugs and harmful side effects. A substance used to treat Parkinson's disease is L-dopamine; its enantiomer D-dopamine has no biological effect. Even the active enantiomer of the popular analgesic ibuprofen, the active ingredient in Advil, Motrin, and Nuprin, is expected to appear on the market soon.

13.3

S·A·M·P·L·E· **13.5** ·P·R·O·B·L·E·M

Identifying D and L Isomers of Sugars

Is the following structure of ribose the D or L enantiomer?

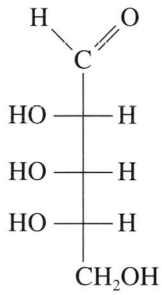

Solution

In ribose, carbon 4 is the chiral atom furthest from the carbonyl group. Because the hydroxyl group on carbon 4 is on the left, this enantiomer is L-ribose.

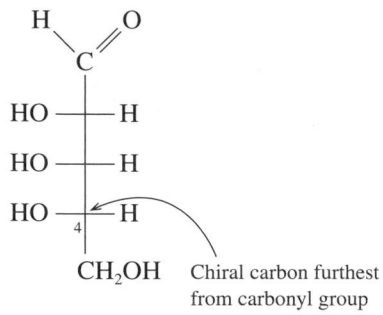

13.3

Study Check

Draw the Fischer projection for D-ribose.

QUESTIONS & PROBLEMS

Chiral Carbon Atoms

13.11 Indicate whether the following objects are chiral or not chiral. Consider each object to be without markings or printing.

a. a Ping-Pong ball for table tennis **b.** a tennis shoe
c. an unsharpened pencil **d.** a new piece of chalk

13.12 Indicate whether the following objects are chiral or not chiral. Consider each object to be without markings or printing.

a. a boxing glove **b.** a blank sheet of typing paper
c. a golf club **d.** a golf tee

13.13 Indicate whether the starred carbon atom in each of the following compounds is a chiral carbon or not:

a.
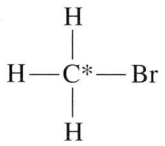

b.
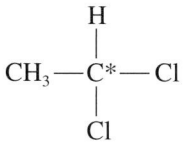

c.
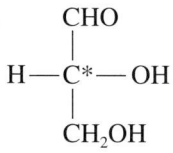

d.
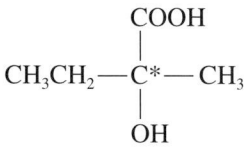

13.14 Indicate whether the central (starred) carbon atom in each of the following compounds is a chiral carbon or not:

a.
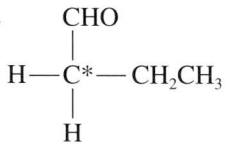

b.
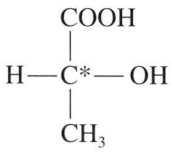

Lactic acid

c.
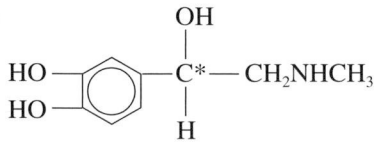

Epinephrine

d.
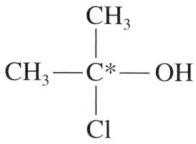

13.15 What is a Fischer projection?

13.16 Write the projection formula for D-glyceraldehyde and L-glyceraldehyde.

13.17 State whether each of the following sugars is the D or L isomer:

a.

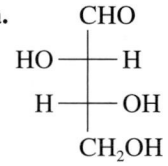

Threose

b.

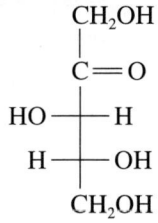

Xylulose

c.

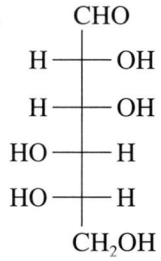

Mannose

d.

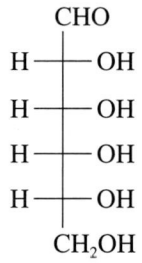

Allose

13.18 State whether each of the following sugars is the D or L isomer:

a.

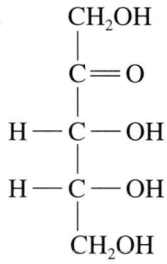

Ribulose

b.

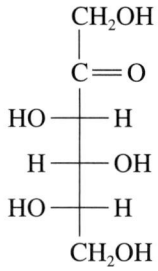

Sorbose

c.

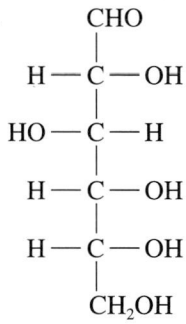

Glucose

d.

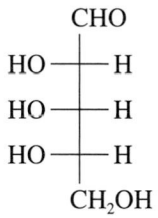

Ribose

Study Pack

13.4 Structures of Some Important Monosaccharides

Learning Goal

Draw the open-chain structures for D-glucose, D-galactose, and D-fructose.

13.4

The hexoses glucose, galactose, and fructose are important monosaccharides. Although structural formulas for both the D and L isomers can be written for these hexoses, the D isomers are the more common form of carbohydrates found in nature.

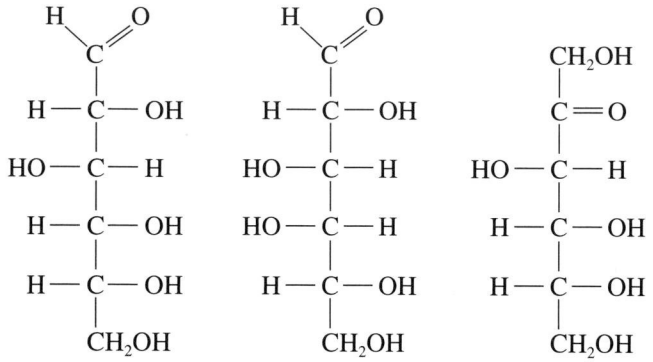

D-Glucose

The most common hexose is D-glucose, which has the molecular formula $C_6H_{12}O_6$. Also known as dextrose, **glucose** is found in fruits, vegetables, corn syrup, and honey. It is a building block of the disaccharides sucrose, lactose, and maltose and is the monosaccharide unit in polysaccharides such as starch, cellulose, and glycogen.

In the body, glucose normally occurs at a concentration of 70–90 mg/dL (100 mL) of blood. However, the amount of glucose depends on the time that has passed since eating. In the first hour after a meal, the level of glucose rises to about 130 mg/dL (100 mL) of blood, and then decreases over the next 2–3 hours as it is used in the tissues. Some glucose is converted to glycogen and stored in the liver and muscle.

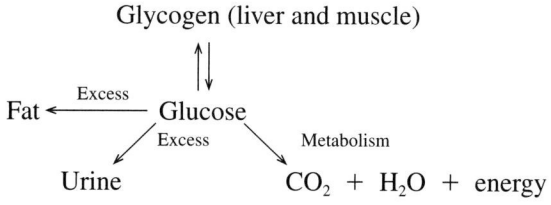

HEALTH NOTE

Hyperglycemia and Hypoglycemia

A doctor may order a glucose tolerance test to evaluate the body's ability to return to normal glucose concentration in response to the ingestion of a specified amount of glucose. The patient fasts for 12 hours and then drinks a solution containing 100 g of glucose. A blood sample is taken immediately, followed by more blood samples each half-hour for 2 hours, and then every hour for a total of 5 hours. If the blood glucose exceeds 140 mg/dL (1 dL = 100 ml) plasma and remains high, hyperglycemia may be indicated. An example of a disease that can cause hyperglycemia is diabetes mellitus, which occurs when the pancreas is unable to produce sufficient quantities of insulin. As a result, glucose levels in the body fluids can rise as high as 350 mg/dL plasma. Symptoms of diabetes in people under the age of 40 include thirst, excessive urination, increased appetite, and weight loss. In older persons, diabetes is sometimes a consequence of excessive weight gain.

When a person is hypoglycemic, the blood glucose level rises and then decreases rapidly to levels as low as 40 mg/dL plasma. In some cases, hypoglycemia is caused by overproduction of insulin by the pancreas. Low blood glucose can cause dizziness, general weakness, and muscle tremors. A diet may be prescribed that consists of several small meals high in protein and low in carbohydrate. Some hypoglycemic patients are finding success with diets that include more complex carbohydrates rather than simple sugars. (See Figure 13.5.)

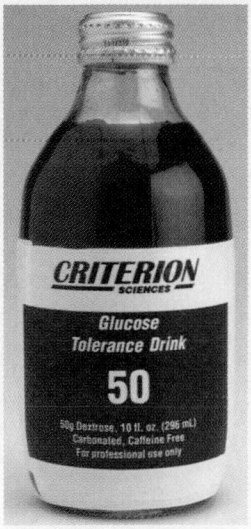

a

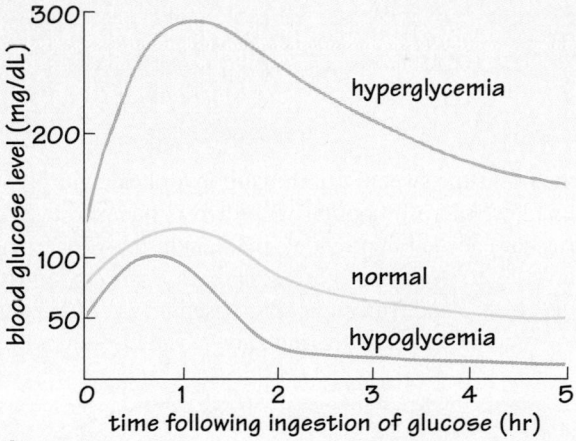

b

Figure 13.5 **(a)** In a glucose tolerance test, 100 g of glucose is given to a patient. **(b)** Typical blood glucose levels for normal, hyperglycemic, and hypoglycemic conditions.

13.4

D-Galactose

Galactose is an aldohexose that does not occur free in nature. It is obtained as a hydrolysis product of the disaccharide lactose, a sugar found in milk and milk products. Galactose is an important monosaccharide that is found in the cellular membranes of the brain and nervous system. The only difference in the structures of D-glucose and D-galactose is the arrangement of the —OH group on carbon 4.

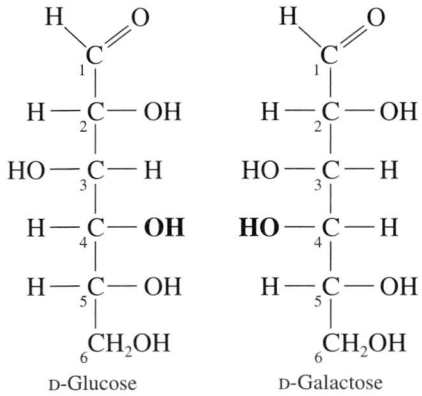

In a condition called *galactosemia,* an enzyme needed to convert galactose to glucose is missing. The accumulation of galactose in the blood and tissues can lead to cataracts, mental retardation, and cirrhosis. The treatment for galactosemia is the removal of all galactose-containing foods, mainly milk and milk products, from the diet. If this is done for an infant immediately after birth, the damaging effects of galactose can be avoided.

13.4

D-Fructose

In contrast to glucose and galactose, **fructose** is a ketohexose. The structure of fructose differs from glucose at carbons 1 and 2 by the location of the carbonyl group.

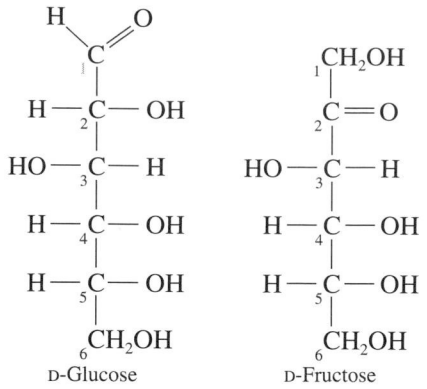

Fructose is the sweetest of the carbohydrates, twice as sweet as sucrose (table sugar). This makes fructose popular with dieters because less fructose, and therefore fewer calories, are needed to provide a pleasant taste. After fructose enters the bloodstream, it is converted to its isomer, glucose. Fructose is found in fruit juices and honey; it is also called levulose and fruit sugar. Fructose is also obtained as one of the hydrolysis products of sucrose, the disaccharide known as table sugar. (See Figure 13.6.)

Figure 13.6 Some common carbohydrates in our diets are fructose in fruit and honey, glucose in corn syrup, lactose in milk and milk products, and sucrose in table sugar.

QUESTIONS & PROBLEMS

Structures of Some Important Monosaccharides

13.19 Draw the open-chain structure of D-glucose and L-glucose.

13.20 Draw the open-chain structure of D-fructose and L-fructose.

13.21 How does the open-chain structure of D-galactose differ from D-glucose?

13.22 How does the open-chain structure of D-fructose differ from D-glucose?

13.23 Identify a monosaccharide that fits each of the following descriptions:
a. also called blood sugar
b. not metabolized in galactosemia
c. also called fruit sugar

13.24 Identify a monosaccharide that fits each of the following descriptions:
a. high blood levels in diabetes
b. obtained as a hydrolysis product of lactose
c. the sweetest of the monosaccharides

Study Pack

Chem of Life Carbohydrates: Monosaccharides 2, 3

13.5

13.5 Cyclic Structures of Monosaccharides

Learning Goal

Draw and identify the cyclic structures of monosaccharides.

Up to now in this chapter, we have represented monosaccharides by their open-chain linear structural formulas. However, most of the time, chains of five- or six-carbon atoms exist as stable ring structures. In an aldohexose, the hydroxyl group on carbon 5 reacts easily with the aldehyde group at the other end of the molecule. The product is a cyclic hemiacetal structure with six atoms, the most prevalent form of the monosaccharides. See Section 12.6 for a discussion of hemiacetals and hemiketals. Let us look at the formation of the cyclic structures of glucose.

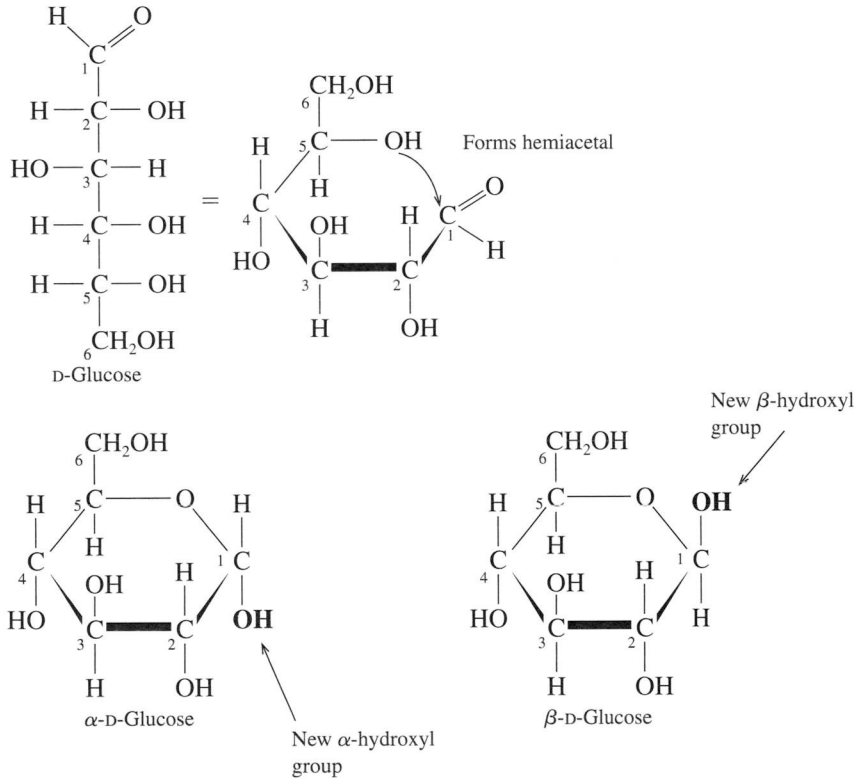

The structure at the left is the open-chain linear form of D-glucose. In the second structure, the open chain folds into a hexagon to bring the hydroxyl group on carbon 5 close to carbon 1 in the aldehyde group. In this form, called a **Haworth structure,** the groups on the right of the linear form appear below the ring. The intramolecular reaction between the alcohol group on carbon 5 and the aldehyde group forms the cyclic hemiacetal. Sometimes, the cyclic structure is simplified to show only the position of the hydroxyl groups on the six-atom ring structure.

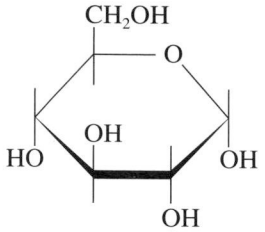

α-D-Glucose (simplified structure)

13.5

α and β Anomers

In the cyclic hemiacetal structures, a new hydroxyl group appears on carbon number 1. Because the new —OH at carbon 1 can be up or down, two isomers called **anomers** are formed. When the —OH on carbon 1 is *down,* the isomer is the α form of the sugar. In the β anomer, the —OH on carbon 1 is *up.* Such differences in structural forms may seem trivial. However, we can digest starch products such as pasta to obtain glucose because the polysaccharide contains the α isomers of glucose. We cannot digest paper or wood because cellulose consists of only β-D-glucose units. Humans have an α-amylase, an enzyme needed for the digestion of starches, but not a β-amylase for the digestion of cellulose.

Mutarotation

In solution, the α-glucose is in equilibrium with β-glucose. In a process called **mutarotation,** the hemiacetal of each isomer converts from the closed ring to the open chain and back again. As the ring opens and closes, the bond between carbons 1 and 2 can rotate, which allows the hydroxyl (—OH) group on carbon 1 to shift between the α and the β position. Although the open chain is an essential part of mutarotation, only a small amount of open chain is present at any given time.

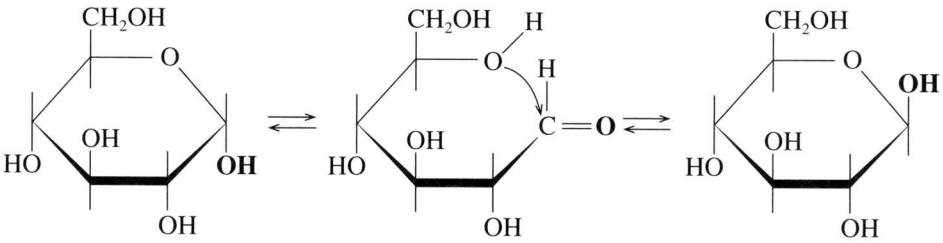

α-D-Glucose (36%) ⇌ D-Glucose open chain (trace) ⇌ β-D-Glucose (64%)

Cyclic Structures of Galactose

Galactose is an aldohexose like glucose, differing only in the arrangement of the —OH group on carbon 4. Thus, its cyclic structure is also similar to glucose, except that in galactose the —OH on carbon 4 is up. With the formation of a new hydroxyl group on carbon 1, galactose also exists as α and β anomers and undergoes mutarotation with the open-chain form in solution.

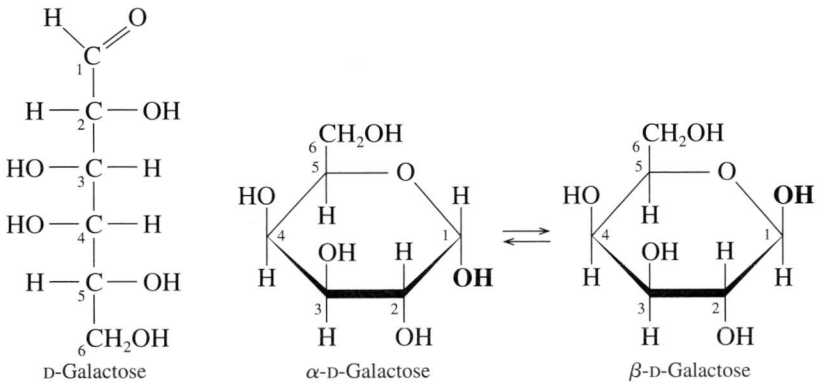

Cyclic Structures of Fructose

In contrast to glucose and galactose, fructose is a ketohexose. It forms a hemiketal when a hydroxyl group on carbon 5 reacts with the ketone group. The cyclic structure for fructose is a five-atom ring with carbon 2 at the right corner. A new hydroxyl group is on carbon 2 along with the carbon 1 in the CH_2OH group. There are also α and β anomers of fructose that undergo mutarotation in solution.

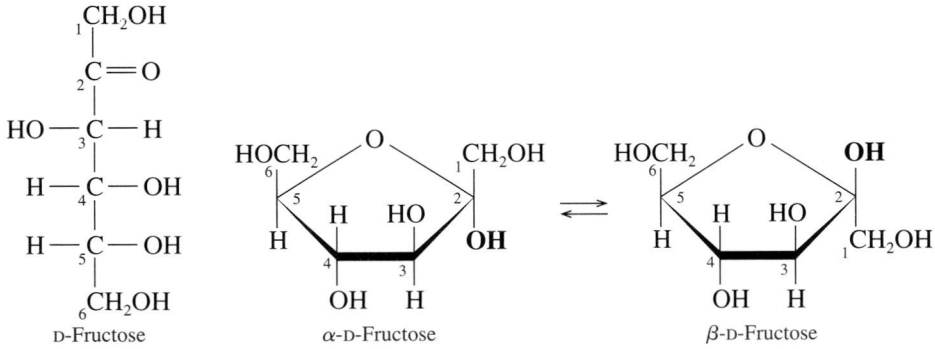

S·A·M·P·L·E 13.6 ·P·R·O·B·L·E·M

Drawing Cyclic Structures for Sugars

β-D-Mannose, a carbohydrate found in immunoglobulins, has the following open-chain structure. Draw the cyclic structure for β-D-mannose.

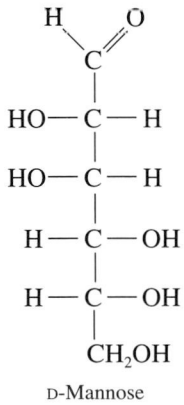

13.5

Solution

First, number the carbon atoms in the open chain starting at the aldehyde group.

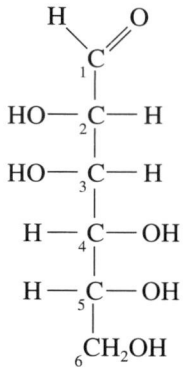

D-Mannose

Draw a six-atom hexagon containing carbon atoms 1–5 and an oxygen atom. Number the carbon atoms in the cyclic structure from the right corner in clockwise order. Write the CH_2OH group (carbon 6) above carbon 5.

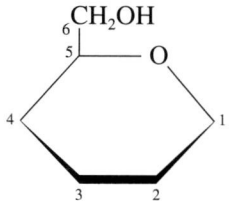

Draw the —OH groups on the left of the open chain above the ring, and the —OH groups on the right below. At carbon 1, write the new hydroxyl group upward to make the β-D-mannose anomer.

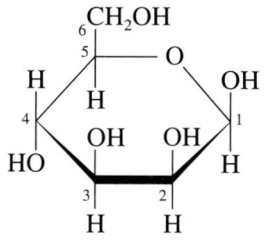

β-D-Mannose

Study Check

Draw the cyclic structure for α-D-glucose.

Oxidation of Monosaccharides

Monosaccharides are easily oxidized to yield carboxylic acids. Although they exist primarily in the cyclic forms, the aldehyde group of the open chain can add an O atom. Because another substance is reduced, monosaccharides are called **reducing sugars.** When a reagent called Benedict's solution is used, Cu^{2+} ions are reduced to Cu^{+}, which forms a precipitate of Cu_2O.

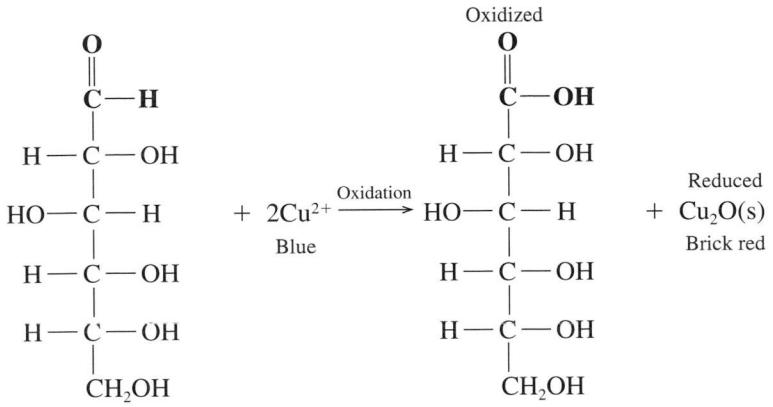

Fructose, a hemiketal, is also a reducing sugar. In the open-chain form, a rearrangement between the hydroxyl group on carbon 1 and the ketone group provides an aldehyde group that can be oxidized.

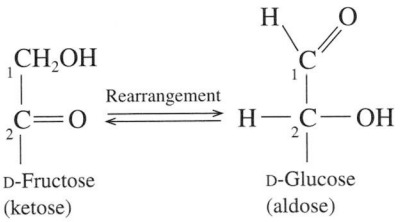

HEALTH NOTE

Testing for Glucose in Urine

Normally, blood glucose flows through the kidneys and is reabsorbed into the bloodstream. However, if the blood level exceeds about 160 mg of glucose/dL (100 mL) of blood, the kidneys cannot reabsorb it all, and glucose spills over into the urine, a condition known as glucosuria. A symptom of diabetes mellitus is a high level of glucose in the urine.

Benedict's test can be used to determine the presence of glucose in urine. The amount of cuprous oxide (Cu_2O) formed is proportional to the amount of reducing sugar present in the urine. Low to moderate levels of reducing sugar turn the solution green; solutions with high glucose levels turn Benedict's yellow or bright orange. Table 13.1 lists some colors associated with the concentration of glucose in the urine.

In another clinical test that is more specific for glucose, the enzyme glucose oxidase is used. The oxidase enzyme converts glucose to gluconic acid and hydrogen peroxide, H_2O_2. The peroxide produced reacts with a dye in the urine test strip to give different colors. The level of glucose present in the urine is found by matching the color produced to a color chart on the container. (See Figure 13.7.)

Table 13.1 Glucose Test Results

	Glucose Present in Urine	
Color	**%**	**mg/dL**
Blue	0	0
Blue-green	0.25	250
Green	0.50	500
Yellow	1.00	1000
Orange	2.00	2000

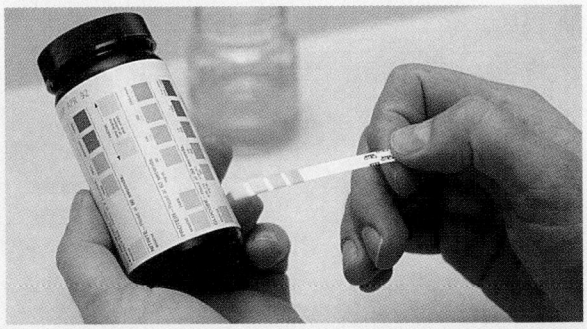

Figure 13.7 Glucose test: urine glucose level is determined by comparing the color of a urine test strip with the color chart on the container.

13.5

S·A·M·P·L·E 13.7 P·R·O·B·L·E·M

Reducing Sugars

Why is D-glucose called a *reducing sugar*?

Solution

D-glucose is easily oxidized by Benedict's reagent. A carbohydrate that reduces Cu^{2+} to Cu^{+} is called a reducing sugar.

Study Check

A test using Benedict's reagent turns orange with a urine sample. According to Table 13.1, what might this result indicate?

Q·U·E·S·T·I·O·N·S & P·R·O·B·L·E·M·S

Cyclic Structures of Monosaccharides

13.25 What are the kind and number of atoms in the ring portion of the cyclic structure of glucose?

13.26 What are the kind and number of atoms in the ring portion of the cyclic structure of fructose?

13.27 Draw the cyclic structures for the α and β anomers of D-glucose.

13.28 Draw the cyclic structures for the α and β anomers of D-fructose.

13.29 Identify each of the following cyclic structures as a hemiacetal or a hemiketal:

a.

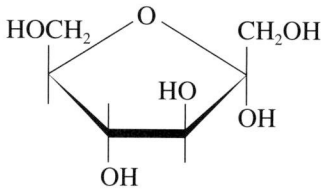

b.

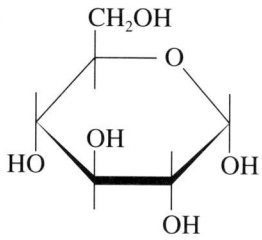

c.

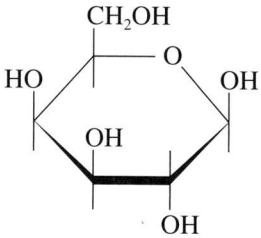

d.

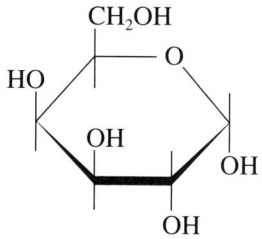

13.30 Identify each of the cyclic structures in problem 13.29 as the α or β anomer.

13.31 What is a reducing sugar?

13.32 What are the products when D-galactose reacts with Benedict's reagent?

13.6 Disaccharides

Study Pack

Chem of Life Carbohydrates: Disaccharides 1

Learning Goal

Describe the monosaccharide units and linkages in disaccharides.

13.6

A disaccharide is composed of two monosaccharides linked together. The most common disaccharides are maltose, lactose, and sucrose. Their hydrolysis gives the following monosaccharides.

$$\text{Maltose} + H_2O \longrightarrow \text{glucose} + \text{glucose}$$
$$\text{Lactose} + H_2O \longrightarrow \text{glucose} + \text{galactose}$$
$$\text{Sucrose} + H_2O \longrightarrow \text{glucose} + \text{fructose}$$

Glycosides (Acetals)

In their cyclic forms, the hydroxyl group on the anomeric carbon of a monosaccharide reacts with an alcohol to give an acetal. These acetal products of monosaccharides are called **glycosides,** and the acetal linkage is called a **glycosidic bond.** The glycosidic bond can be in the α or β position, depending on the anomer that reacted.

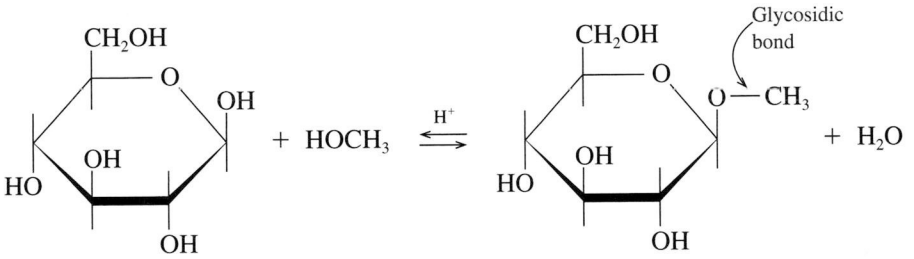

β-D-Glucose Methanol Methyl β-D-glucoside

In the acetal, the glycosidic bond does not open. Thus, neither mutarotation nor oxidation can occur.

Maltose

Maltose, or malt sugar, is a disaccharide obtained from the hydrolysis of starch. When maltose in barley and other grains is hydrolyzed by yeast enzymes, glucose is obtained that can undergo fermentation to give ethanol. Maltose is used in cereals, candies, and the brewing of beverages.

In maltose, the hydroxyl group on carbon 1 of one glucose molecule reacts with the hydroxyl group on carbon 4 of the second glucose molecule. A glycosidic bond joins the two glucose molecules and a molecule of water is eliminated. The glycosidic bond is designated as an α-1,4- to show that —OH on carbon 1 of the α anomer is joined to the —OH on carbon 4 of the second glucose. Because the second glucose molecule has a free —OH on the anomeric carbon, there are α and β anomers of maltose. This anomeric carbon also opens up to give a free aldehyde group to oxidize, which means that maltose is a reducing sugar.

13.6

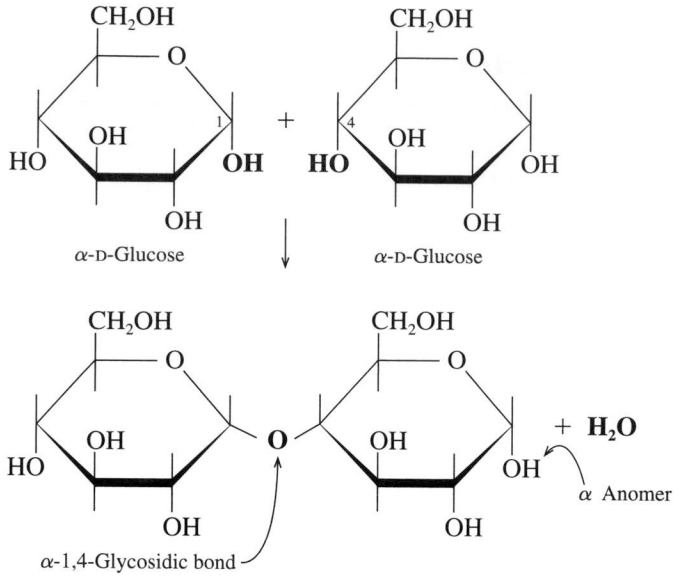

α-Maltose, a disaccharide

Lactose

Lactose, milk sugar, is a disaccharide found in milk and milk products. It makes up 6–8% of human milk and about 4–5% of cow's milk and is used in products that attempt to duplicate mother's milk. Some people do not produce sufficient quantities of the enzyme needed to hydrolyze lactose, and the sugar remains undigested, causing abdominal cramps and diarrhea. In some commercial milk products, an enzyme called lactase is added to break down lactose.

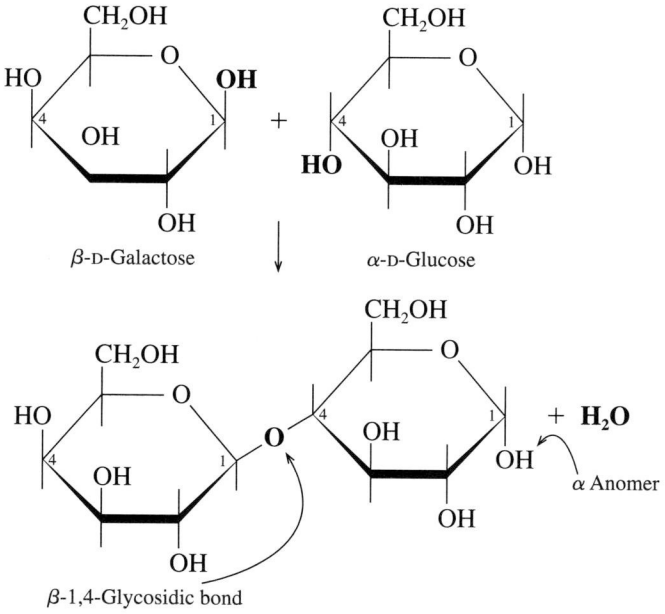

α-Lactose, a disaccharide

Lactose consists of β-D-galactose and D-glucose joined by a β-1,4-glycosidic bond. The free hydroxyl group on the anomeric carbon of the glucose gives α- and β-lactose. Although the glycosidic bond ties up the anomeric carbon in galactose, the anomeric carbon of the glucose can open to give an aldehyde group. Thus, lactose is a reducing sugar.

Sucrose

The disaccharide **sucrose,** table sugar, is extracted from sugar cane and sugar beets. See Figure 13.8. Some estimates indicate that each person in the United States consumes an average of 100 lb of sucrose every year either by itself or in a variety of food products.

Sucrose consists of α-D-glucose and β-D-fructose molecule joined by an α,β-1,2-glycosidic bond. The structure of sucrose differs from the other disaccharides because the glycosidic bond ties up the anomeric carbons of both monosaccharide units. There are no isomers for sucrose. Without a free aldehyde group, sucrose cannot react with Benedict's reagent; sucrose is not a reducing sugar.

Figure 13.8 Sucrose, one of the most abundant carbohydrates, is obtained from sugar beets.

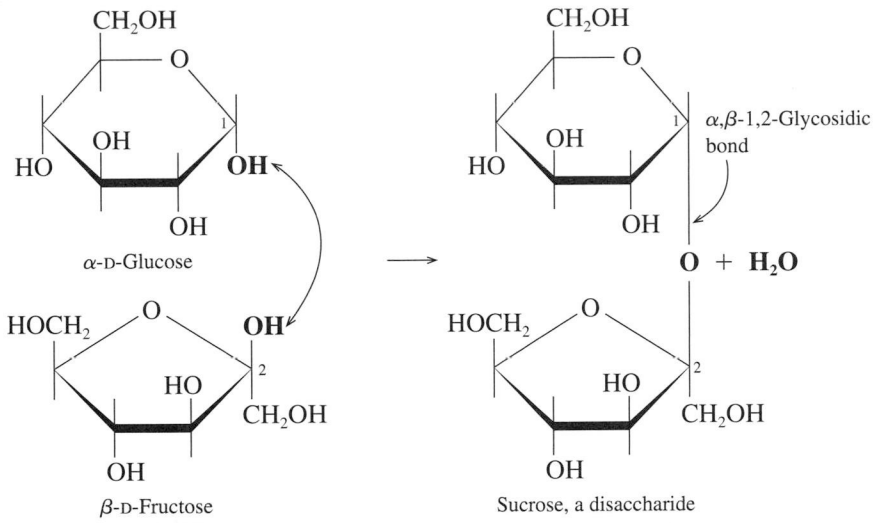

13.6

EXPLORE YOUR WORLD

Sugar and Sweeteners

Add a tablespoon of sugar to a glass of water and stir. Taste. Add more tablespoons of sugar, stir, and taste. If you have other carbohydrates such as fructose, honey, cornstarch, arrowroot or flour, add some of each to separate glasses of water and stir. If you have some artificial sweeteners, add a few drops of the sweetener or a package, if solid, to a glass of water. Taste each.

Questions

1. Which substance is the most soluble in water?
2. Place the substances in order from the one that tastes least sweet to the sweetest.
3. How does your list compare to the values in Table 13.2?
4. How does the sweetness of sucrose compare with the artificial sweeteners?
5. Check the labels of food products in your kitchen. Look for sugars such as sucrose, fructose, or artificial sweeteners such as aspartame or sucralose on the label. How many grams of sugar are in a serving of the food?

Fermentation

When yeast is added, the hexoses glucose and fructose (but not galactose) will undergo **fermentation** to produce ethanol and carbon dioxide gas.

$$\underset{\text{Glucose or fructose}}{C_6H_{12}O_6} \xrightarrow{\text{Yeast enzymes}} \underset{\text{Ethanol}}{2C_2H_5OH} + 2CO_2(g)$$

The disaccharides maltose and sucrose can also undergo fermentation because yeast contains enzymes for their hydrolysis to give glucose and fructose. However, lactose will not ferment because the enzyme lactase required for its hydrolysis is not present in yeast. Fermentation does not occur with polysaccharides.

HEALTH NOTE

How Sweet Is My Sweetener?

Although many of the monosaccharides and disaccharides taste sweet, they differ considerably in their degree of sweetness. Dietetic foods contain sweeteners that are noncarbohydrate or carbohydrates that are sweeter. Some examples of sweeteners compared with sucrose are shown in Table 13.2.

13.6

Table 13.2 Relative Sweetness of Sugars and Artificial Sweeteners

	Sweetness Relative to Sucrose (= 100)
Monosaccharides	
Galactose	30
Glucose	75
Fructose	175
Disaccharides	
Lactose	16
Maltose	33
Sucrose	100 ← reference standard
Artificial Sweeteners (Noncarbohydrate)	
Sucralose	600
Aspartame	18,000
Saccharin	45,000

S·A·M·P·L·E 13.8 ·P·R·O·B·L·E·M

Glycosidic Bonds in Disaccharides

Melebiose is a disaccharide that has a sweetness of about 30 compared with sucrose (= 100).

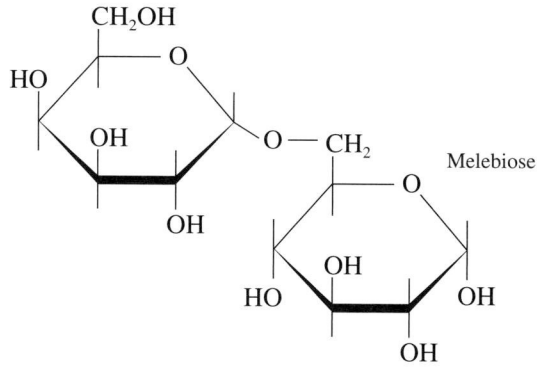

a. What are the monosaccharide units in melebiose?
b. How is the glycosidic bond designated?
c. Is the compound α- or β-melebiose?

Solution

a. The monosaccharide on the left side is α-D-galactose; on the right is α-D-glucose.
b. The monosaccharide units are linked by an α-1,6-glycosidic bond.
c. The downward position of the anomeric OH makes it α-melebiose.

Study Check

Cellobiose is a disaccharide composed of two glucose molecules linked by a β-1,4-glycosidic linkage. Draw a structural formula for β-cellobiose.

HEALTH NOTE

Blood Types and Carbohydrates

Every individual's blood can be typed as one of four blood groups, A, B, AB, and O. Although there is some variation among ethnic groups in the United States, the incidence of blood types in the general population is about 43% O, 40% A, 12% B, and 5% AB.

The blood types are determined by three or four monosaccharides that are attached to the red blood cells. All the blood types include *N*-acetylglucosamine, galactose, and fucose. In type A blood, a fourth monosaccharide, *N*-acetylgalactosamine, is bonded to the galactose. The structures of these monosaccharides are as follows:

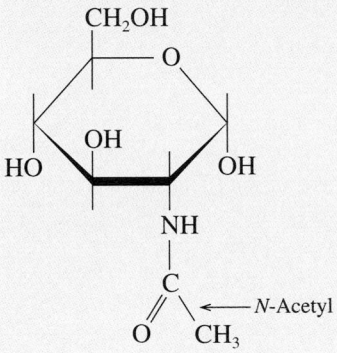

N-Acetylglucosamine (*N*-AcGlu)

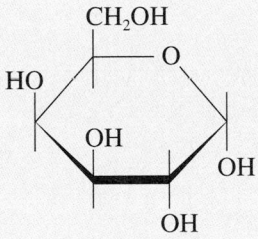

D-Galactose (Gal)

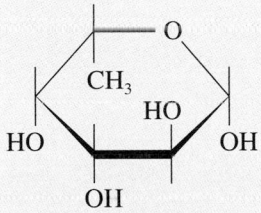

L-Fucose (Fuc)

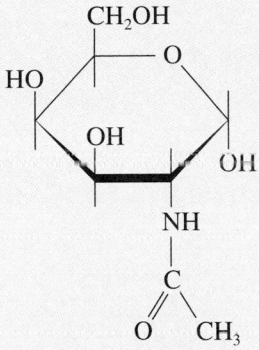

N-Acetylgalactosamine (*N*-AcGal)

Table 13.3 Compatibility of Blood Groups

Blood Group	Can Receive Blood Types	Cannot Receive Blood Types
A	A, O	B, AB
B	B, O	A, AB
AB	A, B, AB, O	None
O	O	A, B, AB

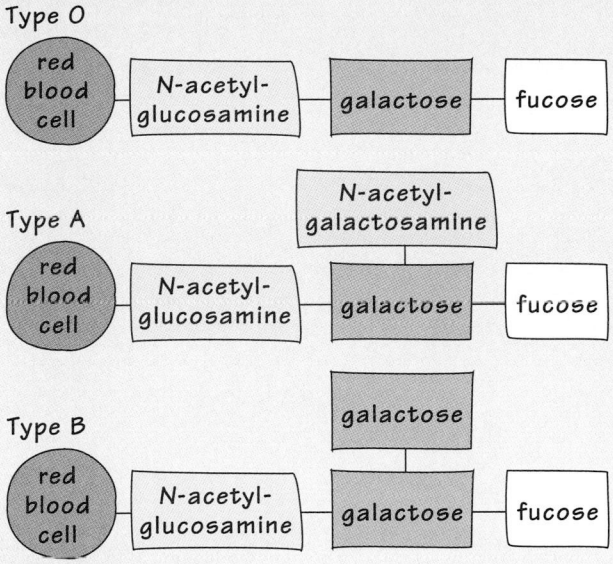

Figure 13.9 Specific carbohydrates attached to red blood cells determine a person's blood type.

In type B blood, there is a second molecule of galactose. Type AB blood contains both A and B blood types. (See Figure 13.9.) A person with type A blood produces antibodies against type B, whereas a person with type B blood produces antibodies against A. Type AB blood produces no antibodies, whereas type O produces both. Thus, if a person with type A blood receives a transfusion of type B blood, factors in the recipient's blood will agglutinate the donor's red blood cells. If you become a blood donor, your blood is screened carefully to make sure that an exact match is made with the blood type of the recipient. (See Figure 13.10.) Table 13.3 summarizes the compatibility of blood groups for transfusion.

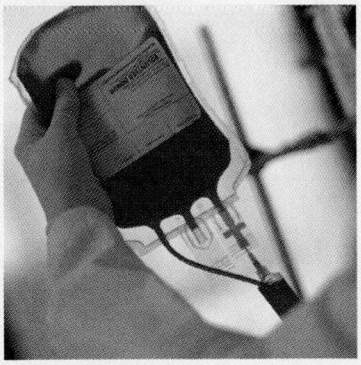

Figure 13.10 After blood is donated, it is typed by blood group and labeled.

13.6

QUESTIONS & PROBLEMS

Disaccharides

13.33 For each of the following disaccharides, give the monosaccharide units produced by hydrolysis, the type of glycosidic bond, and the identity of the disaccharide including the α or β anomer:

a.

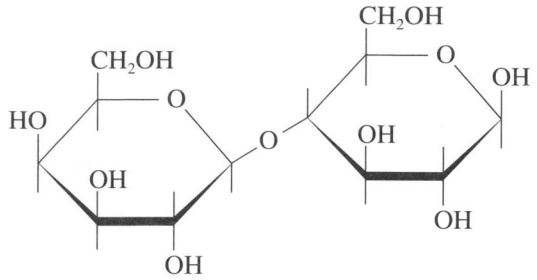

b.

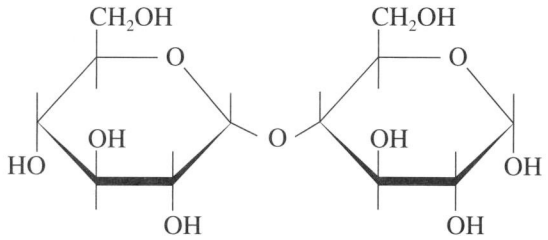

13.34 For each of the following disaccharides, give the monosaccharide units produced by hydrolysis, the type of glycosidic bond, and the identity of the disaccharide including the α or β anomer:

a.

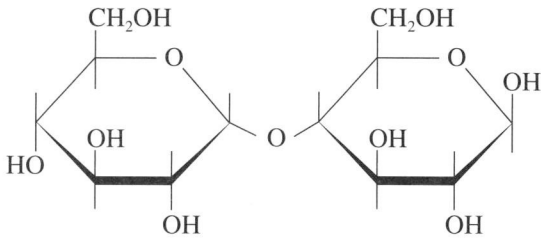

b.

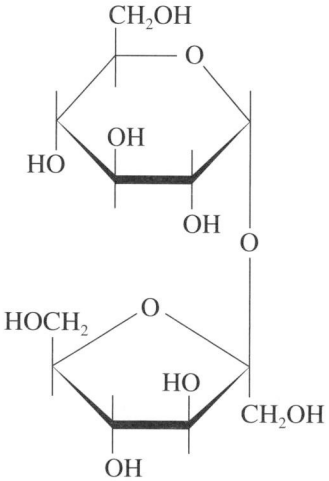

13.35 Identify disaccharides that fit each of the following descriptions:

a. Ordinary table sugar. **b.** Found in milk and milk products.
c. Also called *malt sugar.* **d.** Hydrolysis gives galactose and glucose.

13.36 Identify disaccharides that fit each of the following descriptions:

a. Not a reducing sugar. **b.** Composed of two glucose units.
c. Also called *milk sugar.* **d.** Hydrolysis gives glucose and fructose.

Study Pack

Chem of Life Carbohydrates: Polysaccharides 1

13.7

13.7 Polysaccharides

Learning Goal

Describe the structural features of amylose, amylopectin, glycogen, and cellulose.

A polysaccharide is a polymer of many monosaccharides joined together. Three important polysaccharides—starch, cellulose, and glycogen—are all polymers of D-glucose, which differ only in the type of glycosidic bonds and the amount of branching in the molecule.

Plant Starch: Amylose and Amylopectin

Starch, a storage form of glucose in plants, is found as insoluble granules in rice, wheat, potatoes, beans, and cereals. Starch is composed of two kinds of polysaccharides, amylose and amylopectin. **Amylose,** which makes up about 20% of starch, consists of α-D-glucose molecules connected by α-1,4-glycosidic bonds in a continuous chain. A typical polymer of amylose may contain from 250 to 4000 glucose units. Sometimes called a straight-chain polymer, polymers of amylose are actually coiled in helical fashion.

Amylopectin, which makes up as much as 80% of plant starch, is a branched-chain polysaccharide. Like amylose, the glucose molecules are connected by α-1,4-glycosidic bonds. However, at about every 25 glucose units, there is a branch of glucose molecules attached by an α-1,6-glycosidic bond between carbon 1 of the branch and carbon 6 in the main chain. See Figure 13.11.

Starches hydrolyze easily in water and acid to give smaller saccharides called dextrins, which then hydrolyze to maltose and finally glucose. In our bodies, these complex carbohydrates are digested by the enzymes amylase and maltase. The glucose obtained provides about 50% of our nutritional calories.

$$\text{Amylose, amylopectin} \xrightarrow{H^+ \text{ or amylase}} \text{dextrins} \xrightarrow{H^+ \text{ or amylase}} \text{maltose} \xrightarrow{H^+ \text{ or maltase}} \text{many D-glucose units}$$

Animal Starch: Glycogen

Glycogen, or animal starch, is a polymer of glucose that is stored in the liver and muscle of animals. It is hydrolyzed in our cells at a rate that maintains the blood level of glucose and provides energy between meals. The structure of glycogen is very similar to that of amylopectin found in plants except that glycogen is more highly branched. In glycogen, the glucose units are joined by α-1,4-glycosidic bonds, and branches occurring about every 10–15 glucose units are attached by α-1,6-glycosidic bonds.

Structural Polysaccharide: Cellulose

Cellulose is the major structural material of wood and plants. Cotton is almost pure cellulose. In **cellulose,** glucose molecules form a long unbranched chain similar to that of amylose. However, the glucose units in cellulose are linked by β-1,4-glycosidic bonds. The β isomers do not form coils like the α isomers but are aligned in parallel rows that are held in place by hydrogen bonds between hydroxyl groups in adjacent chains. This gives a rigid structure to the cell walls in wood and fiber that is more resistant to hydrolysis than the starches. See Figure 13.12.

13.7

Figure 13.11 The structures of **(a)** amylose, a straight-chain polysaccharide of glucose units, and **(b)** amylopectin, a branched chain.

(a) unbranched chain of amylose

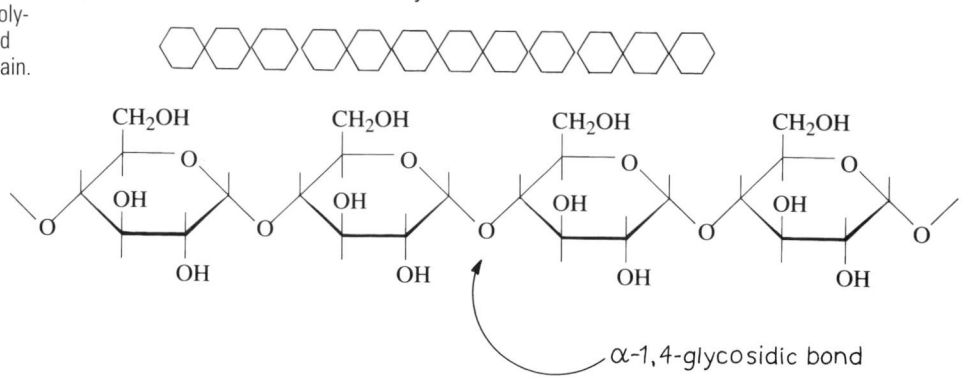

(b) branched-chain of amylopectin

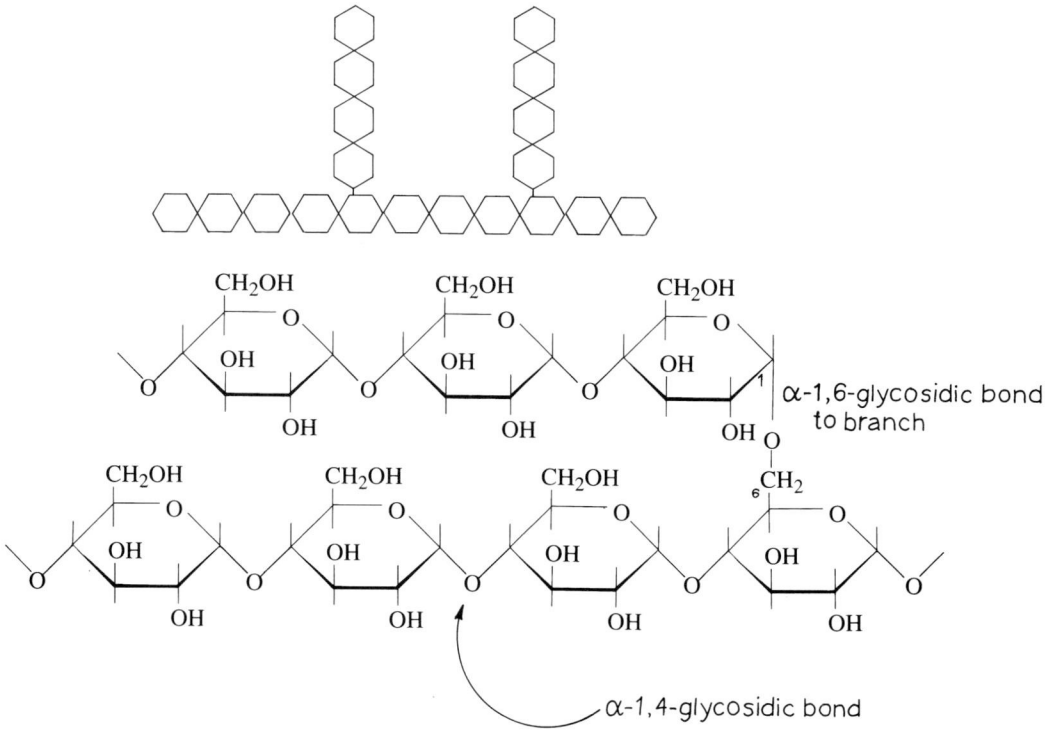

Figure 13.12 The polysaccharide cellulose, composed of β-1,4-glycosidic bonds, builds cell walls in wood and plants and makes up nearly 100% of cotton.

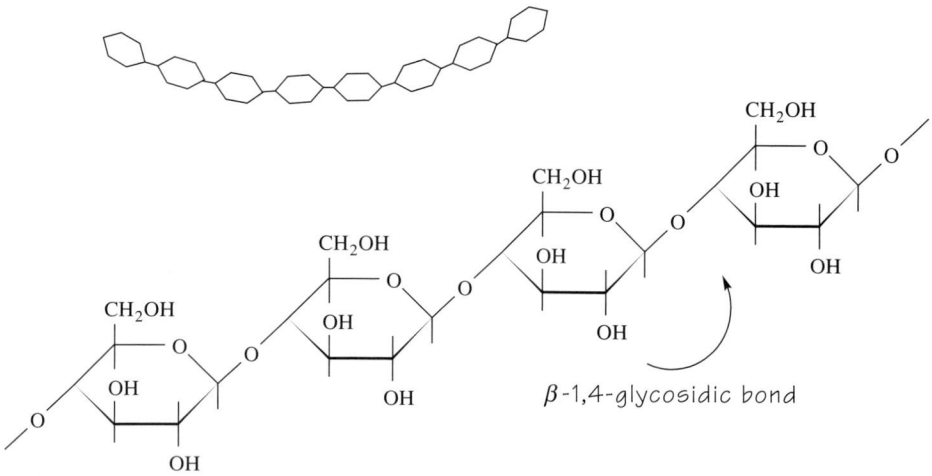

Cellulose

Enzymes in our saliva and pancreatic juices hydrolyze the α-1,4-glycosidic bonds of the starches. However, there are no enzymes in humans that are able to hydrolyze the β-1,4-glycosidic bonds of cellulose; we cannot digest cellulose. Some animals such as goats and cows and insects like termites are able to obtain glucose from cellulose. Their digestive systems contain bacteria and protozoa with enzymes that can hydrolyze β-1,4-glycosidic bonds.

Iodine Test

In the **iodine test,** iodine (I_2) is used to test for the presence of starch. The unbranched helical shape of the polysaccharide amylose in starch reacts strongly with iodine to form a deep blue-black complex. Amylopectin, cellulose, and glycogen produce reddish-purple and brown colors. Such colors do not develop when iodine is added to samples of mono- or disaccharides.

HEALTH NOTE

Fiber in the Diet

The term *dietary fiber* includes all plant material that is not digestible by humans, including cellulose. Food sources of dietary fiber include whole grains, bran, fruits, and vegetables. (See Figure 13.13.)

In the large intestine, the retention of water by dietary fiber has a laxative effect. As a result, wastes pass through the intestinal tract faster, which lessens the time that bacteria and any carcinogens come in contact with the intestine. Some studies indicate that a high-fiber diet may lower the incidence of colon cancer, diverticulitis (inflammation of the colon), heart disease, and hemorrhoids.

The absorptive effects of fiber may also be beneficial in weight maintenance. Fiber increases the bulk in the stomach and intestines without contributing to caloric intake. Fiber may also absorb some of the carbohydrate and cholesterol from the diet, thus decreasing the quantity that diffuses through the intestinal walls.

Figure 13.13 High-fiber foods include cucumbers, cabbage, broccoli, lettuce, potatoes, carrots, and many fruits.

EXPLORE YOUR WORLD

Iodine Test for Starch

If you have some iodine solution for cuts, you can use it to test for starch. Collect samples of food such as crackers, bread, pasta, cereals, and candy. You may also test aspirin tablets, antacids, and the glue on an envelope. Place a few drops of iodine on each of the items. If starch is present, a deep blue-black color will develop. After recording your observations, throw all your samples away! Iodine is toxic.

13.7

Questions

1. Which of the foods test positive for starch?
2. What other items contain starch?
3. What type of carbohydrate is starch?
4. Why do you think starch is used in preparing tablets such as aspirin?
5. Why is starch present in the glue of the envelope flap?

EXPLORE YOUR WORLD

Polysaccharides

Read the nutrition label on a box of crackers, cereal, bread, chips, or pasta. The major ingredient in crackers is flour, a starch. Chew on a single cracker for 4–5 minutes. An enzyme (amylase) in your saliva breaks apart the bonds in starch.

Questions

1. How are carbohydrates listed?
2. What is dietary fiber?
3. What other carbohydrates are listed?
4. How does the taste of the cracker change during the time that you chewed it?
5. What happened to the starches in the cracker as the amylase enzyme in your saliva reacted with the amylose or amylopectin?

S·A·M·P·L·E 13.9 ·P·R·O·B·L·E·M

Structures of Polysaccharides

Identify the polysaccharide described by each of the following statements:

a. A polysaccharide that is stored in the liver and muscle tissues
b. An unbranched polysaccharide containing β-1,4-glycosidic bonds
c. A starch containing α-1,4- and α-1,6-glycosidic bonds

Solution

a. glycogen **b.** cellulose **c.** amylopectin, glycogen

Study Check

Cellulose and amylose are both unbranched glucose polymers. How do they differ?

ENVIRONMENTAL NOTE

Some Other Polysaccharides

When carbon 6 of glucose is oxidized, it forms a carboxylic acid called glucuronic acid. In glucosamine and *N*-acetylglucosamine, a hydroxyl group on carbon 2 of glucose has been replaced by an amino or *N*-acetylamino group. The oxidation of glucose or the addition of an amino or amide group gives monosaccharide units that form polysaccharides that are important in cellular structure, connective tissues, anticoagulants, and antibiotics.

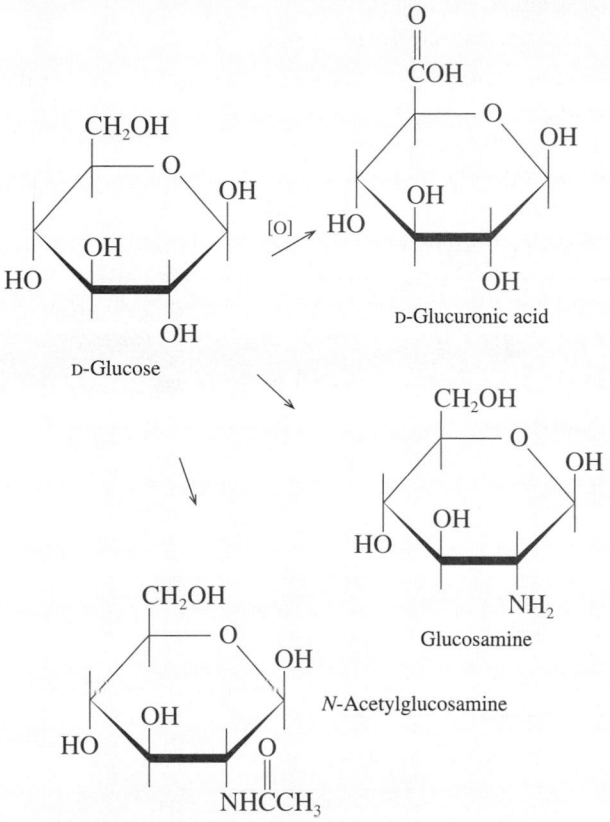

Chitin (pronounced *kite'-in*) is the polysaccharide that gives the rigid structure to the skeletons of insects, and forms the shells of crabs, lobsters, and shrimp. (See Figure 13.14.) Similar to cellulose, chitin is an unbranched polymer of *N*-acetylglucosamine units joined by β-1,4-glycosidic bonds.

The gel-like viscous material that bathes connective tissues, lubricates bone joints, and maintains the shape of the eyeball contains mucopolysaccharides such as hyaluronic acid. Hyaluronic acid is a polysaccharide consisting of alternating units of glucuronic acid and *N*-acetylglucosamine.

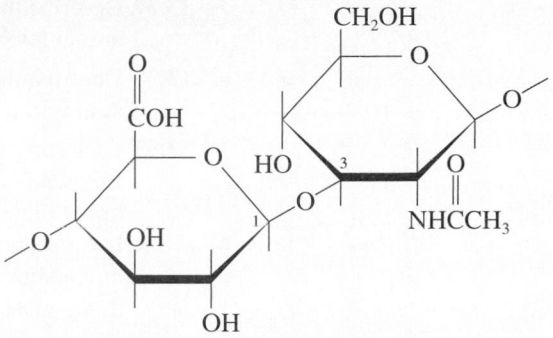

Hyaluronic acid, a repeating disaccharide unit in bone joints and connective tissues

Heparin, an acidic polysaccharide, is an anticoagulant. It is often used in open-heart surgery to prevent clotting of the blood. Heparin consists of repeating units of glucuronic acid and *N*-acetylglucosamine joined by α-1,4-glycosidic bonds.

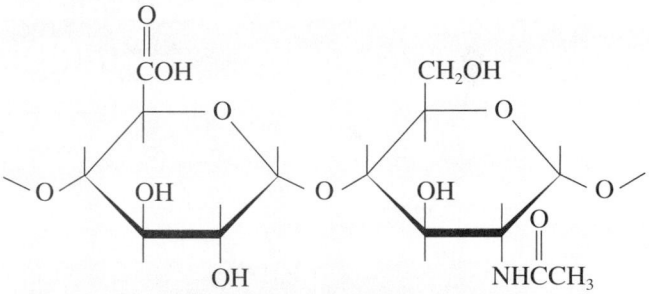

Repeating section of heparin

Figure 13.14 Chitin, a polymer of *N*-acetylglucosamine, is the structural polysaccharide in the shells of shrimp, crab, and lobsters.

QUESTIONS & PROBLEMS

Polysaccharides

13.37 Describe the similarities and differences in the following polysaccharides:
a. amylose and amylopectin **b.** amylopectin and glycogen

13.38 Describe the similarities and differences in the following polysaccharides:
a. amylose and cellulose **b.** cellulose and chitin

13.7

13.39 Give the name of one or more polysaccharides that matches each of the following descriptions:
a. not digestible by humans
b. the storage form of carbohydrates in plants
c. contains only α-1,4-glycosidic bonds
d. the most highly branched polysaccharide

13.40 Give the name of one or more polysaccharides that matches each of the following descriptions:
a. the storage form of carbohydrates in animals
b. contains only β-1,4-glycosidic bonds
c. contains both α-1,4- and α-1,6-glycosidic bonds
d. produces maltose during digestion

Chapter Review

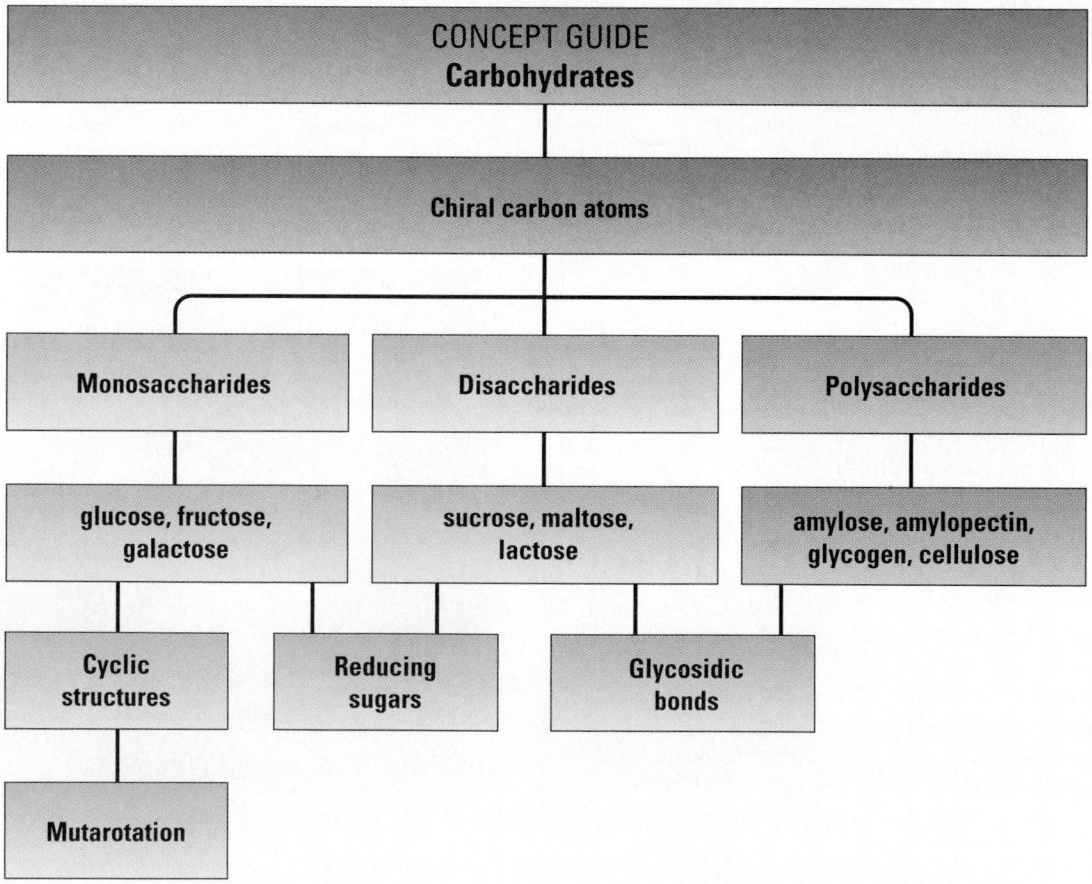

Chapter Summary

13.1 Classification of Carbohydrates

Carbohydrates are classified as monosaccharides (simple sugars), disaccharides (two monosaccharide units), and polysaccharides (many monosaccharide units).

13.2 Monosaccharides

Monosaccharides are polyhydroxy aldehydes (aldoses) or ketones (ketoses). Monosaccharides are also classified by their number of carbon atoms: *triose, tetrose, pentose,* or *hexose.*

13.3 Chiral Carbon Atoms

Chiral molecules have mirror images that cannot be superimposed on each other. In a chiral molecule, one or more carbon atoms are attached to four different atoms or groups. The mirror images of a chiral molecule represent two different molecules called enantiomers. In a Fischer projection (straight chain), the prefixes D- and L- are used to distinguish between the mirror images. In D-glyceraldehyde, the —OH is on the right of the chiral carbon; it is on the left in L-glyceraldehyde.

13.4 Structures of Some Important Monosaccharides

Important monosaccharides are the aldohexoses glucose and galactose and the ketohexose fructose.

13.5 Cyclic Structures of Monosaccharides

The predominant form of monosaccharides is the cyclic form of five or six atoms. The cyclic structure forms by a reaction between an OH (usually the one on carbon 5 in hexoses) with the carbonyl group of the same molecule. The formation of a new hydroxyl group on carbon 1 (or 2 in fructose) gives α and β anomers of the cyclic monosaccharide. Because the molecule opens and closes continuously (mutarotation) while in solution, both anomers are present. Monosaccharides are reducing sugars because the aldehyde group (also available in ketoses) is oxidized by a metal ion such as Cu^{2+}, which is reduced.

13.6 Disaccharides

Disaccharides are glycosides of two monosaccharide units joined together by a glycosidic bond. In the most common disaccharides, maltose, lactose, and sucrose, there is at least one glucose unit.

13.7 Polysaccharides

Polysaccharides are polymers of monosaccharide units. Starches consist of amylose, an unbranched chain of glucose, and amylopectin, a branched polymer of glucose. Glycogen, the storage form of glucose in animals, is similar to amylopectin with more branching. Cellulose is also a polymer of glucose, but in cellulose the glycosidic bonds are β bonds rather than α bonds as in the starches. Humans can digest starches, but not cellulose, to obtain energy. However, cellulose is important as a source of fiber in our diets.

Summary of Carbohydrates

Carbohydrate	Food Sources	Monosaccharides
Monosaccharides		
Glucose	Fruit juices, honey, corn syrup	Glucose
Galactose	Lactose hydrolysis	Galactose
Fructose	Fruit juices, honey, sucrose hydrolysis	Fructose
Disaccharides		
Maltose	Germinating grains, starch hydrolysis	Glucose + glucose
Lactose	Milk, yogurt, ice cream	Glucose + galactose
Sucrose	Sugar cane, sugar beets	Glucose + fructose
Polysaccharides		
Amylose	Rice, wheat, grains, cereals	Unbranched polymer of glucose joined by α-1,4-glycosidic bonds
Amylopectin	Rice, wheat, grains, cereals	Branched polymer of glucose joined by α-1,4- and α-1,6-glycosidic bonds
Glycogen	Liver, muscles	Highly branched polymer of glucose joined by α-1,4- and α-1,6-glycosidic bonds
Cellulose	Plant fiber, bran, beans, celery	Unbranched polymer of glucose joined by β-1,4-glycosidic bonds

Summary of Reactions

Glycoside (Acetal) Formation

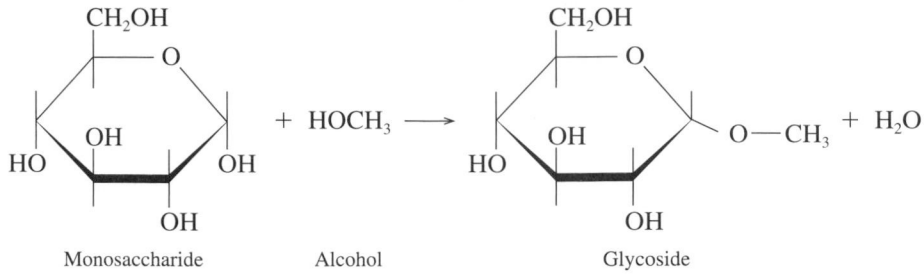

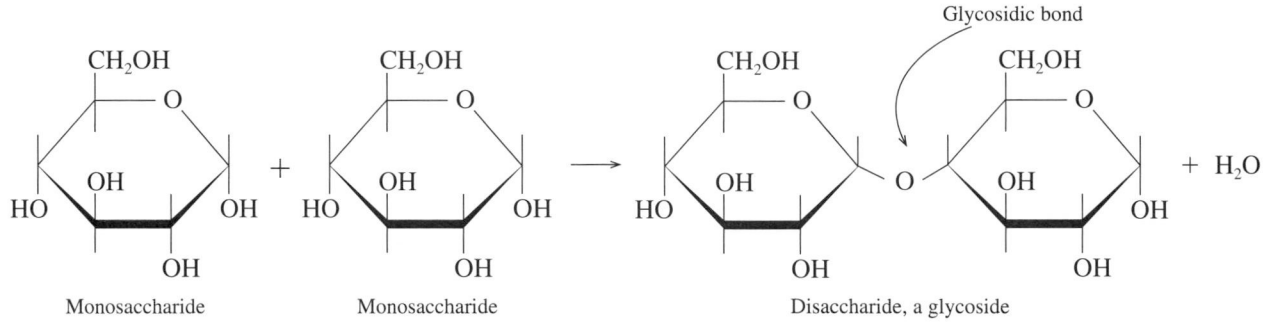

Hydrolysis of Disaccharides

Sucrose + H_2O ⟶ glucose + fructose

Lactose + H_2O ⟶ glucose + galactose

Maltose + H_2O ⟶ glucose + glucose

Key Terms

aldose Monosaccharides that contain an aldehyde group.

amylopectin A branched-chain polymer of starch composed of glucose units joined by α-1,4- and α-1,6-glycosidic bonds.

amylose An unbranched polymer of starch composed of glucose units joined by α-1,4-glycosidic bonds.

anomers The isomers that occur when the formation of a hemiacetal or hemiketal of monosaccharides produces a new hydroxyl group on carbon 1 (or carbon 2). In the α anomer, the OH is drawn downward; in the β isomer the OH is up.

carbohydrate A simple or complex sugar composed of carbon, hydrogen, and oxygen.

cellulose An unbranched polysaccharide composed of glucose units linked by β-1,4-glycosidic bonds that cannot be hydrolyzed by the human digestive system.

chiral A carbon atom that is attached to four different atoms or groups of atoms; the mirror images of a chiral molecule are not superimposable.

disaccharides Carbohydrates composed of two monosaccharides joined by a glycosidic bond.

enantiomers Stereoisomers that are mirror images of each other.

fermentation A reaction of glucose, fructose, maltose, or sucrose, in which the sugar reacts with enzymes in yeast to give ethanol and carbon dioxide gas.

Fischer projection A two-dimensional straight-chain representation of a molecule.

fructose A monosaccharide found in honey and fruit juices; it is combined with glucose in sucrose. Also called levulose and fruit sugar.

galactose A monosaccharide that occurs combined with glucose in lactose.

glucose The most prevalent monosaccharide in the diet. An aldohexose that is found in fruits, vegetables, corn syrup, and honey. Also known as blood sugar and dextrose. Combines in glycosidic bonds to form most of the polysaccharides.

glycogen A polysaccharide formed in the liver and muscles for the storage of glucose as an energy reserve. It is composed of glucose in a highly branched polymer joined by α-1,4- and α-1,6-glycosidic bonds.

glycosides Acetal products of a monosaccharide reacting with an alcohol or another sugar.

glycosidic bond The acetal bond that forms when an alcohol or a hydroxyl group of a monosaccharide adds to a hemiacetal. It is the type of bond that links monosaccharide units in di- or polysaccharides.

Haworth structure The cyclic structure that represents the closed chain of a monosaccharide.

iodine test A test for amylose that forms a blue-black color after iodine is added to the sample.

ketose A monosaccharide that contains a ketone group.

lactose A disaccharide consisting of glucose and galactose found in milk and milk products.

maltose A disaccharide consisting of two glucose units; it is obtained from the hydrolysis of starch and in germinating grains.

monosaccharide A polyhydroxy compound that contains an aldehyde or ketone group.

mutarotation The conversion between α and β anomers.

polysaccharides Polymers of many monosaccharide units, usually glucose. Polysaccharides differ in the types of glycosidic bonds and the amount of branching in the polymer.

reducing sugar A carbohydrate with a free aldehyde group capable of reducing the Cu^{2+} in Benedict's reagent.

saccharide A term from the Latin word *saccharum,* meaning "sugar"; it is used to describe the carbohydrate family.

stereoisomers Molecules with the same structural formulas and attached atoms but having different three-dimensional arrangements of the atoms.

sucrose A disaccharide composed of glucose and fructose; a nonreducing sugar, commonly called table sugar or "sugar."

Chapter Practice

13.41 D-Sorbitol, a sweetener found in seaweed and berries, contains only hydroxyl functional groups. When D-sorbitol is oxidized, it forms D-glucose. What is the structural formula of D-sorbitol?

13.42 Raffinose is a trisaccharide found in Australian manna and in cottonseed meal. It is composed of three different monosaccharides and is not a reducing sugar. Identify the monosaccharides in raffinose.

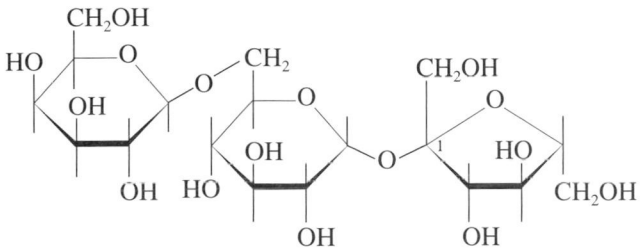

13.43 If α-galactose is dissolved in water, β-galactose is eventually present. Explain how this occurs.

13.44 Why are lactose and maltose reducing sugars, but sucrose is not?

13.45 β-Cellobiose is a disaccharide obtained from the hydrolysis of cellulose. It is quite similar to maltose except it has a β-1,4-glycosidic bond. What is the structure of β-cellobiose?

13.46 The disaccharide trehalose found in mushrooms is composed of two α-D-glucose molecules joined by an α-1,1-glycosidic bond. Draw the structure of trehalose.

Answers

Answers to Study Checks

13.1 a monosaccharide

13.2

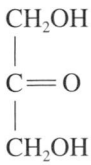

13.3 achiral

13.4

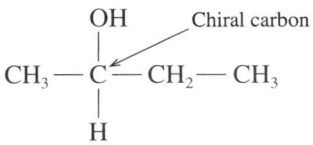

13.5

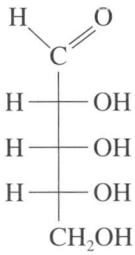

13.6

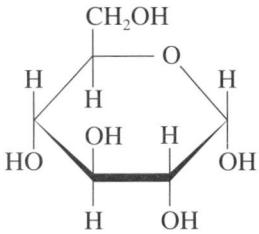

13.7 This indicates a high level of reducing sugar (probably glucose) in the urine. One common cause of this condition is diabetes mellitus.

13.8

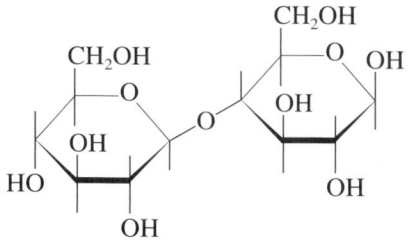

13.9 Cellulose contains glucose units connected by β-1,4-glycosidic bonds, whereas the glucose units in amylose are connected by α-1,4-glycosidic bonds.

Answers to Selected Questions and Problems

13.1 Photosynthesis requires CO_2, H_2O, and the energy from the sun. Respiration requires O_2 from the air and glucose from our foods.

13.3 A monosaccharide is a simple sugar composed of three to six carbon atoms. A disaccharide is composed of two monosaccharide units.

13.5 Hydroxyl groups are found in all monosaccharides along with a carbonyl on the first or second carbon.

13.7 A ketopentose contains hydroxyl and ketone functional groups and has five carbon atoms.

13.9 **a.** ketose **b.** aldose **c.** ketose **d.** aldose **e.** aldose

13.11 **a.** achiral **b.** chiral **c.** achiral **d.** achiral

13.13 **a.** achiral **b.** achiral **c.** chiral **d.** chiral

13.15 A Fischer projection is a two-dimensional representation of the three-dimensional structure of a molecule.

13.17 **a.** D **b.** D **c.** L **d.** D

13.19

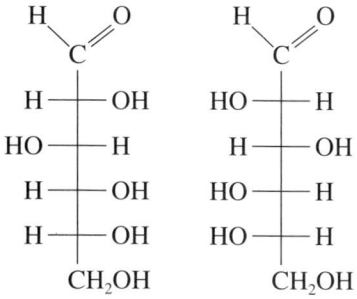

D-Glucose L-Glucose

13.21 In D-galactose the hydroxyl on carbon four extends to the left. In glucose this hydroxyl goes to the right.

13.23 **a.** glucose **b.** galactose **c.** fructose

13.25 In the cyclic structure of glucose, there are five carbon atoms and an oxygen.

13.27

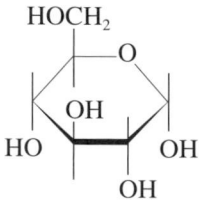

α-D-Glucose

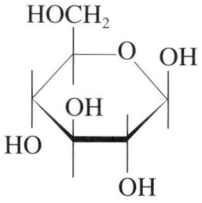

β-D-Glucose

13.29 **a.** hemiketal **b.** hemiacetal **c.** hemiacetal **d.** hemiacetal

13.31 A reducing sugar is one that reacts with Benedict's reagent.

13.33 **a.** galactose and glucose; β-1,4 bond; β-lactose
b. glucose; α-1,4 bond; α-maltose

13.35 **a.** sucrose **b.** lactose **c.** maltose **d.** lactose

13.37 **a.** Amylose is an unbranched polymer of glucose units joined by α-1,4 bonds; amylopectin is a branched polymer of glucose joined by α-1,4 and α-1,6 bonds.
b. Amylopectin, which is produced in plants, is a branched polymer of glucose, joined by α-1,4 and α-1,6 bonds. Glycogen, which is produced in animals, is a highly branched polymer of glucose, joined by α-1,4 and α-1,6 bonds.

13.39 **a.** cellulose **b.** amylose, amylopectin **c.** amylose **d.** glycogen

13.41

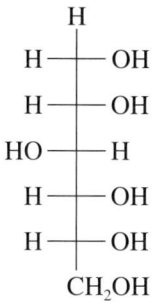

13.43 The α-galactose forms an open-chain structure, and when the chain closes, it can form both α- and β-galactose.

13.45

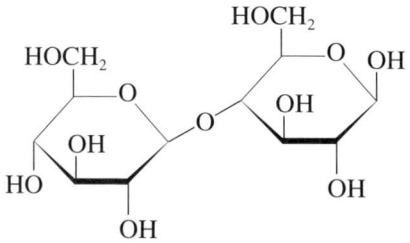

14

Carboxylic Acids, Esters, Amines, and Amides

The caffeine from these coffee beans is an alkaloid, which is a nitrogen-containing compound that has physiological effects on the body.

Looking Ahead

14.1	Carboxylic Acids
14.2	Properties of Carboxylic Acids
14.3	Esters
14.4	Esterification and Hydrolysis
14.5	Amines
14.6	Reactions of Amines
14.7	Amides
14.8	Reactions of Amides

www.chemplace.com/college/timberlake

Visit the web site above to test yourself, run simulations, and review key concepts from this chapter. For even more interactive review and quizzing, use the two CD-ROMs included with the book.

Carboxylic acids are similar to the inorganic acids we studied in Chapter 9. They have a sour or tart taste, produce hydronium ions in water, and neutralize bases. Many carboxylic acids are naturally occurring. A solution of acetic acid and water is the vinegar you use in your salad. The tartness of citrus fruits is due to citric acid. The burning sting of a red ant is due to the carboxylic acid called formic acid. The lactic acid produced in our muscles when we exercise is converted to pyruvic acid as oxygen returns to the muscle. When a carboxylic acid combines with an alcohol, an ester is produced. Aspirin is an ester. Esters give pleasant aromas and flavors to fruits and flowers.

Organic compounds that contain nitrogen include the amines and amides. Many, like adrenaline and amphetamine, are physiologically active. Alkaloids, such as curare, belladonna, and digitalis, are nitrogen-containing compounds obtained from plants. When a carboxylic acid combines with an amine, an amide is produced. Acetaminophen (Tylenol) is an amide. Some amides, such as phenobarbital, are used as sedatives and in anticonvulsant medications. Biologically, proteins are polymers held together by amide linkages between the carboxyl and amine groups of amino acids.

14.1

Study Pack

Chem of Life Organic Molecules: Functional Groups 3, 4

Learning Goal

Give the common names, IUPAC names, and condensed structural formulas of carboxylic acids.

14.1 Carboxylic Acids

In Chapter 12, we described the carbonyl group (C═O) as the functional group in aldehydes and ketones. In a **carboxylic acid,** a hydroxyl group is attached to the carbonyl group, forming a **carboxyl group.** The carboxyl functional group may be attached to an alkyl (R) group or an aromatic (Ar) group.

Carboxyl group

The carboxyl group can be written in several different ways. For example, the condensed structural formula for the carboxylic acid known as acetic acid can be written as follows:

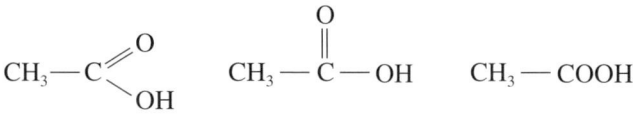

Some condensed structural formulas for acetic acid

Names of Carboxylic Acids

Using the IUPAC rules, carboxylic acids are named by replacing the *e* in the alkane parent name with *oic acid.* However, the common names of the simple carboxylic acids, which are derived from their natural sources, are used extensively. In Chapter 12, we named aldehydes using the prefixes from the common names of the carboxylic acids.

Naming Carboxylic Acids $R-\overset{\overset{\displaystyle O}{\|}}{C}-OH$

Functional group:	$-\overset{\overset{\displaystyle O}{\|}}{C}-OH$
IUPAC name:	alkanoic acid, benzoic acid
Common names:	formic acid, acetic acid, propionic acid, butyric acid

Table 14.1 Names and Natural Sources of Carboxylic Acids

Condensed Structural Formulas	IUPAC Name	Common Name	Occurs In
$HC(=O)OH$	Methanoic acid	Formic acid	Ant and bee stings (Latin *formica,* "ant")
$CH_3C(=O)OH$	Ethanoic acid	Acetic acid	Vinegar (Latin *acetum,* "sour")
$CH_3CH_2C(=O)OH$	Propanoic acid	Propionic acid	Dairy products (Greek, *pro,* "first," *pion,* "fat")
$CH_3CH_2CH_2C(=O)OH$	Butanoic acid	Butyric acid	Rancid butter (Latin *butyrum,* "butter")

Figure 14.1 Formic acid is the substance in the sting of a bee that irritates the skin.

Formic acid is injected under the skin from bee or red ant stings and other insect bites. (See Figure 14.1.) Acetic acid is the oxidation product of the ethanol in wines and apple cider. The resulting solution of acetic acid and water is known as vinegar. Butyric acid gives the foul odor to rancid butter. (See Table 14.1.) Some ball-and-stick models of carboxylic acids are shown in Figure 14.2.

Suppose you want to write the IUPAC and common name for the following carboxylic acid:

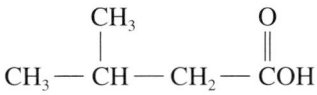

Step 1 Identify the longest carbon chain containing the carboxyl group and replace the *e* of the alkane name by *oic acid;* smaller carboxylic acids are often named by their common names.

Step 2 Number the carbon chain beginning with the carboxyl carbon as carbon 1. When using the common names, the Greek letters alpha (α), beta (β), and gamma (γ) are assigned to the carbons adjacent to the carboxyl carbon.

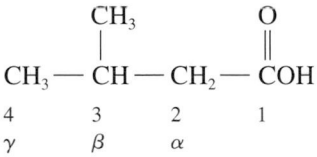

Step 3 Give the location and names of substituents on the parent chain. Use numbers in the IUPAC name and Greek letters in the common name.

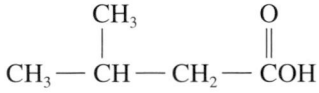

3-Methylbutanoic acid (β-methylbutyric acid)

Some more examples of naming carboxylic acids follow:

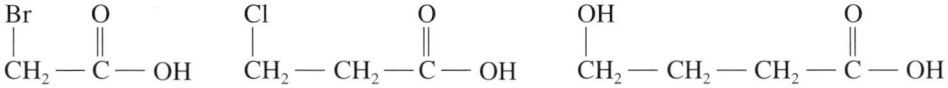

2-Bromoethanoic acid (α-bromoacetic acid)

3-Chloropropanoic acid (β-chloropropionic acid)

4-Hydroxybutanoic acid (γ-hydroxybutyric acid)

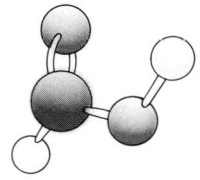

formic acid

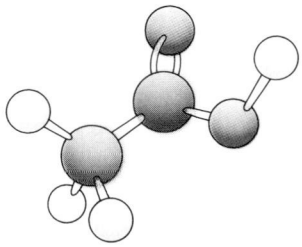

acetic acid

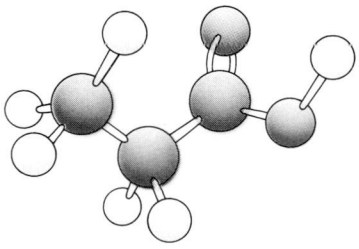

propionic acid

Figure 14.2 Ball-and-stick models of formic acid, acetic acid, and propionic acid.

The aromatic carboxylic acid is called benzoic acid. The carbon atoms in the benzene ring are numbered beginning with carbon 1, which is the carbon attached to the carboxyl group.

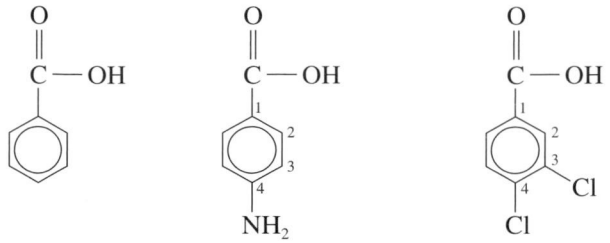

Benzoic acid | 4-Aminobenzoic acid (*p*-aminobenzoic acid) | 3,4-Dichlorobenzoic acid

14.1

HEALTH NOTE

Alpha Hydroxy Acids

Alpha hydroxy acids (AHAs) are naturally occurring carboxylic acids found in fruits, milk, and sugarcane. Cleopatra reportedly bathed in sour milk to smooth her skin. Dermatologists have been using products with a high concentration of AHAs to remove acne scars and reduce irregular pigmentation and age spots. Now lower concentrations (8–10%) of AHAs have been added to skin care products for the purpose of smoothing fine lines, improving skin texture, and cleansing pores. (See Figure 14.3.) Several alpha hydroxy acids may be found in skin care products singly or in combination. Glycolic acid and lactic acid are most frequently used.

Recent studies indicate that products with AHAs increase sensitivity of the skin to sun and UV radiation. It is recommended that a sunscreen with a sun protection factor (SPF) of at least 15 be used when treating the skin with products that include AHAs. Products containing AHAs at concentrations under 10% and pH values greater than 3.5 are generally considered safe. However, the Food and Drug Administration has reports of AHAs

Alpha Hydroxy Acid (Source)	Structure
Glycolic acid (Sugarcane, sugar beet)	$HO-CH_2-C(=O)OH$
Lactic acid (Sour milk)	$CH_3-CH(OH)-C(=O)OH$
Tartaric acid (Grapes)	$HOC(=O)-CH(OH)-CH(OH)-C(=O)OH$
Malic acid (Apples, grapes)	$HOC(=O)-CH_2-CH(OH)-C(=O)OH$
Citric acid (Citrus fruits: lemons, oranges, grapefruit)	$HO-C(CH_2-COOH)_2-COOH$

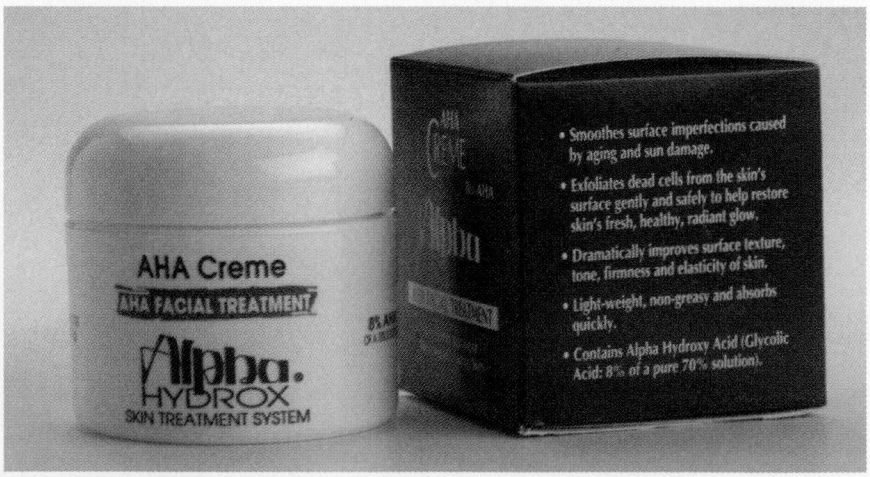

Figure 14.3 Alpha hydroxy acids, including glycolic, lactic, malic, tartaric, and citric acid, are used at concentrations of 8–10% in skin care products to soften skin cells and lessen wrinkles.

causing skin irritation including blisters, rashes, and discoloration of the skin. The FDA does not require product safety reports from cosmetic manufacturers, although they are responsible for marketing safe products. The FDA advises that you test any product containing AHAs on a small area of skin before you use it on a large area.

S·A·M·P·L·E **14.1** ·P·R·O·B·L·E·M

Naming Carboxylic Acids

Give the IUPAC and common name for each of the following carboxylic acids:

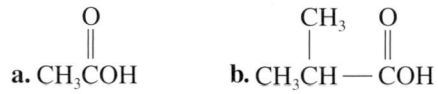

a. $CH_3\overset{\overset{\displaystyle O}{\|}}{C}OH$ **b.** $CH_3\overset{\overset{\displaystyle CH_3}{|}}{C}H-\overset{\overset{\displaystyle O}{\|}}{C}OH$

Solution

a. This carboxylic acid has two carbon atoms. In the IUPAC system, the *e* in ethane is replaced by *oic acid,* to give the name, *ethanoic acid.* Its common name is *acetic acid.*

b. This carboxylic acid has a methyl group on the second carbon. It has the IUPAC name *2-methylpropanoic acid.* In the common name, the Greek letter α specifies the carbon atom next to the carboxyl carbon, *α-methylpropionic acid.*

Study Check

Write the condensed structural formula of 3-phenylpropanoic acid.

Preparation of Carboxylic Acids

Carboxylic acids can be prepared from primary alcohols or aldehydes. As we saw in Chapter 12, the loss of two hydrogens from a primary alcohol produces an aldehyde. Oxidation continues easily as another oxygen is added to yield a carboxylic acid. For example, when ethyl alcohol in wine comes in contact with the oxygen in the air, vinegar is produced. The oxidation process converts the ethyl alcohol (primary alcohol) to acetaldehyde, and then to acetic acid, the carboxylic acid in vinegar. (See Figure 14.4.)

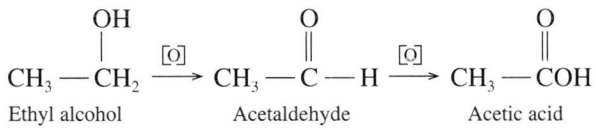

$$\underset{\text{Ethyl alcohol}}{CH_3-\overset{\overset{\displaystyle OH}{|}}{C}H_2} \xrightarrow{[O]} \underset{\text{Acetaldehyde}}{CH_3-\overset{\overset{\displaystyle O}{\|}}{C}-H} \xrightarrow{[O]} \underset{\text{Acetic acid}}{CH_3-\overset{\overset{\displaystyle O}{\|}}{C}OH}$$

Figure 14.4 Vinegar is a 5% solution of acetic acid and water.

S·A·M·P·L·E **14.2** ·P·R·O·B·L·E·M

Preparation of Carboxylic Acids

Write an equation for the oxidation of 1-propanol and name each product.

Solution

A primary alcohol will oxidize to an aldehyde, which can oxidize further to a carboxylic acid.

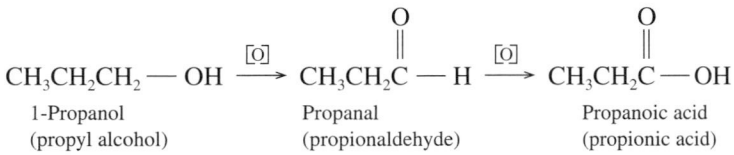

$$\underset{\substack{\text{1-Propanol}\\ \text{(propyl alcohol)}}}{CH_3CH_2CH_2-OH} \xrightarrow{[O]} \underset{\substack{\text{Propanal}\\ \text{(propionaldehyde)}}}{CH_3CH_2\overset{\overset{\displaystyle O}{\|}}{C}-H} \xrightarrow{[O]} \underset{\substack{\text{Propanoic acid}\\ \text{(propionic acid)}}}{CH_3CH_2\overset{\overset{\displaystyle O}{\|}}{C}-OH}$$

Study Check

Write the condensed structural formula of the carboxylic acid produced by the oxidation of 1-butanol.

QUESTIONS & PROBLEMS

Carboxylic Acids

14.1

14.1 What carboxylic acid is responsible for the pain of an ant sting?

14.2 What carboxylic acid is found in a solution of vinegar?

14.3 Explain the differences in the condensed structural formulas of propanal and propanoic acid.

14.4 Explain the differences in the condensed structural formulas of benzaldehyde and benzoic acid.

14.5 Give the IUPAC and common names (if any) for the following carboxylic acids:

a. $CH_3\overset{\overset{\displaystyle O}{\|}}{C}OH$

b. $CH_3CH_2CH_2\overset{\overset{\displaystyle O}{\|}}{C}OH$

c. $CH_3\overset{\overset{\displaystyle Cl}{|}}{C}H-\overset{\overset{\displaystyle O}{\|}}{C}OH$

d. $CH_3CH_2CH_2\overset{\overset{\displaystyle CH_3}{|}}{C}HCH_2\overset{\overset{\displaystyle O}{\|}}{C}OH$

14.6 Give the IUPAC and common names (if any) for the following carboxylic acids:

a. $H\overset{\overset{\displaystyle O}{\|}}{C}OH$

b. $CH_3CH_2CH_2\overset{\overset{\displaystyle Br}{|}}{C}H-\overset{\overset{\displaystyle O}{\|}}{C}OH$

c. 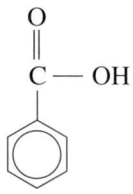

d. 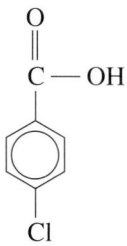

14.7 Draw the condensed structural formulas of each of the following carboxylic acids:

a. propionic acid
b. benzoic acid
c. 2-chloroethanoic acid
d. 3-hydroxypropanoic acid
e. β-methylbutyric acid
f. 3,5-dibromoheptanoic acid

14.8 Draw the condensed structural formulas of each of the following carboxylic acids:

a. butyric acid
b. 3-ethylbenzoic acid
c. α-hydroxyacetic acid
d. 2,4-dibromobutanoic acid
e. *m*-methylbenzoic acid
f. 4,4-dibromohexanoic acid

14.9 Draw the condensed structural formulas of the carboxylic acids formed by the oxidation of each of the following:

a. CH_3OH

b. $CH_3\overset{\overset{\displaystyle O}{\|}}{C}H$

c. $CH_3\overset{\overset{\displaystyle CH_3}{|}}{C}HCH_2CH_2OH$

d. 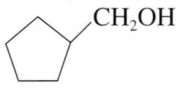

14.10 Draw the condensed structural formulas of the carboxylic acids formed by the oxidation of each of the following:

a. $CH_3CH_2CH_2CH_2CH_2CH_2OH$

b. $CH_3CH_2CH_2CH_2\overset{\overset{\displaystyle O}{\|}}{C}H$

c. $CH_3\overset{\overset{\displaystyle CH_3}{|}}{C}H\overset{\overset{\displaystyle O}{\|}}{C}H_2$

d. 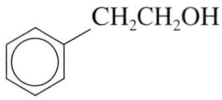

14.2

14.2 Properties of Carboxylic Acids

Study Pack

Learning Goal

Describe the solubility of carboxylic acids in water. Write equations for the ionization and neutralization of carboxylic acids.

Carboxylic acids are among the most polar organic compounds because they contain two polar groups: a hydroxyl (—OH) group and a carbonyl (C═O) group. The smaller carboxylic acids (one to four carbons) are very soluble in water because the carboxyl group can hydrogen bond with several water molecules. However, as the length of the carbon chain increases, the nonpolar hydrocarbon portion of the molecules reduces solubility. Carboxylic acids having five or more carbon atoms are only slightly soluble in water.

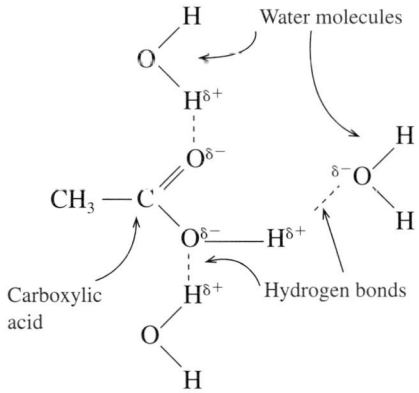

Hydrogen bonds form between the carboxylic acid and water molecules.

Carboxylic Acids Are Weak Acids

In water, carboxylic acids ionize by donating a proton to water to form hydronium ions, H_3O^+. However, carboxylic acids are weak acids, although they are more acidic than phenols. In solution, only a small percentage of the acid molecules are ionized; most of the acid is not ionized. The ion that forms is called a *carboxylate ion,* named by replacing the *ic acid* ending with *ate.*

14.2

Ionization of a Carboxylic Acid in Water

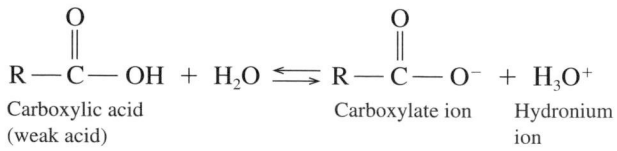

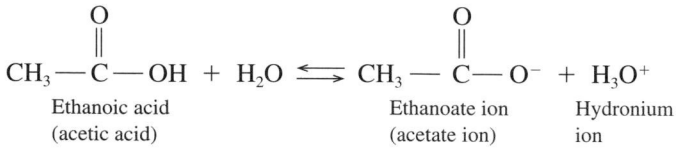

S·A·M·P·L·E 14.3 ·P·R·O·B·L·E·M

Ionization of Carboxylic Acids in Water

Write the equation for the ionization of propionic acid in water.

Solution

The ionization of propionic acid produces a carboxylate ion and a hydronium ion.

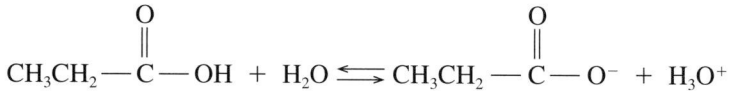

Study Check

Write an equation for the ionization of formic acid in water.

Neutralization of Carboxylic Acids

When a carboxylic acid is neutralized by a base, the products are the salt of the carboxylic acid and water. The salt is named by giving the name of the metal ion followed by the name of the carboxylate ion. Because salts of carboxylic acids are ionic, they are more soluble in water than the corresponding acids.

Neutralization of a Carboxylic Acid

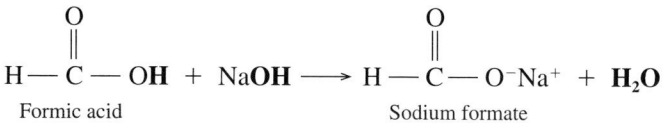

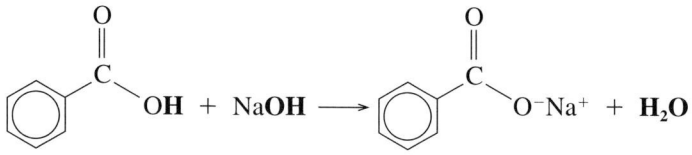

Sodium propionate, a preservative, is added to bread, cheeses, and bakery items to inhibit the spoilage of the food by microorganisms. Sodium benzoate, an inhibitor of mold and bacteria, is added to juices, margarine, relishes, salads, and jams. Monosodium glutamate (MSG) has been added to meats, fish, vegetables, and bakery items to enhance flavor, although it causes headaches in some people. (See Figure 14.5.)

Figure 14.5 Preservatives and flavor enhancers in soups and seasonings are carboxylic acids or their salts.

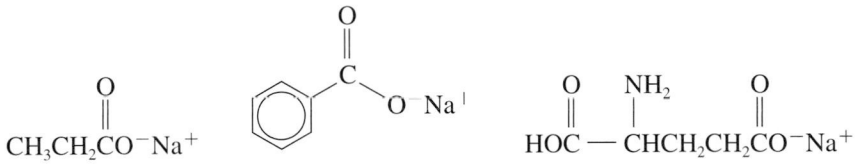

Sodium propionate — Sodium benzoate — Monosodium glutamate

S·A·M·P·L·E 14.4 ·P·R·O·B·L·E·M

Neutralization of a Carboxylic Acid

Write the equation for the neutralization of propionic acid with sodium hydroxide.

Solution

The neutralization of an acid with a base produces the salt of the acid and water.

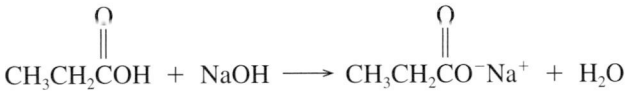

Propionic acid — Sodium propionate

Study Check

What carboxylic acid will give potassium butyrate when it is neutralized by KOH?

Q·U·E·S·T·I·O·N·S & P·R·O·B·L·E·M·S

Properties of Carboxylic Acids

14.11 In the following pairs, which compound would be more soluble in water? Explain.

a. $CH_3CH_2CH_3$ or $CH_3CH_2\overset{O}{\overset{\|}{C}}OH$

b. $CH_3CH_2CH_2CH_2OH$ or $CH_3CH_2CH_2\overset{O}{\overset{\|}{C}}OH$

c. $CH_3\overset{O}{\overset{\|}{C}}OH$ or $CH_3CH_2CH_2CH_2\overset{O}{\overset{\|}{C}}OH$

14.12 In the following pairs, which compound would be more soluble in water? Explain.

a. $CH_3CH_2CH_2OCH_3$ or $CH_3CH_2CH_2\overset{O}{\overset{\|}{C}}OH$

14.3

b. $CH_3CH_2CH_2CH_2CH_2\overset{O}{\overset{\|}{C}}OH$ or $CH_3CH_2\overset{O}{\overset{\|}{C}}OH$

c. $CH_3CH_2CH_2CH_2CH_2\overset{O}{\overset{\|}{C}}O^-Na^+$ or $CH_3CH_2CH_2CH_2CH_2\overset{O}{\overset{\|}{C}}OH$

14.13 Write equations for the ionization of each of the following carboxylic acids in water:

a. $H-\overset{O}{\overset{\|}{C}}-OH$ **b.** $CH_3CH_2\overset{O}{\overset{\|}{C}}OH$ **c.** acetic acid

14.14 Write equations for the ionization of each of the following carboxylic acids in water:

a. $CH_3\underset{\underset{CH_3}{|}}{C}H\overset{O}{\overset{\|}{C}}OH$ **b.** α-hydroxyacetic acid **c.** butanoic acid

14.15 Write equations for the reaction of each of the following carboxylic acids with NaOH:

a. formic acid **b.** propanoic acid **c.** benzoic acid

14.16 Write equations for the reaction of each of the following carboxylic acids with KOH:

a. acetic acid **b.** 2-methylbutanoic acid **c.** *p*-chlorobenzoic acid

Study Pack

14.3 Esters

Learning Goal

Give the common names, IUPAC names, and condensed structural formulas of esters.

In an **ester** the functional group looks like a carboxylic acid, except the carboxyl group is between two alkyl or aromatic groups.

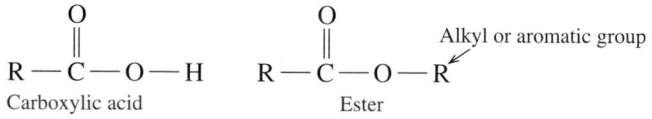

Naming Esters $\mathbf{R-\overset{O}{\overset{\|}{C}}-OR}$

Functional group:	$-\overset{O}{\overset{\|}{C}}-O-$
IUPAC name:	alkyl alkanoate (IUPAC name of acid)
Common name:	alkyl carboxylate (common name of acid)

The name of an ester consists of two parts, one from the alcohol and the other from the carboxylic acid.

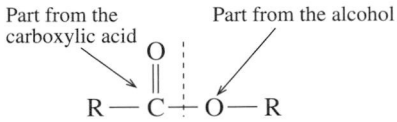

The first part of the ester name is from the alkyl group of the alcohol. The second part is from the name of the carboxylic acid with the ending of *ic acid* replaced by *ate*. In the common names of esters, the common names of the carboxylic acids are used. (See Figure 14.6.)

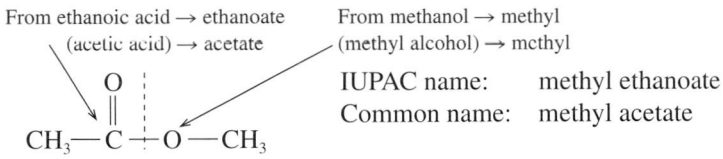

Examples:

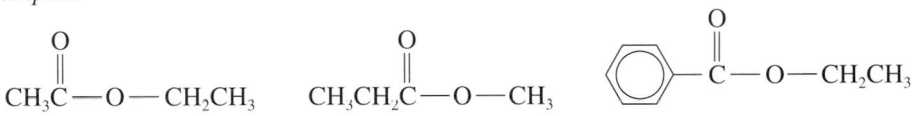

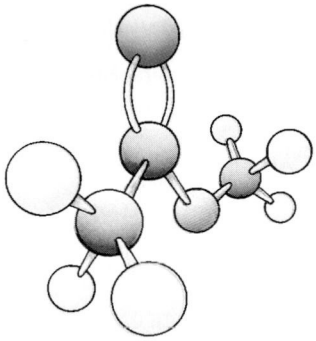

Figure 14.6 A ball-and-stick model of methyl acetate, an ester.

14.3

S · A · M · P · L · E · 14.5 · P · R · O · B · L · E · M

Naming Esters

Write the IUPAC and common names of the following ester:

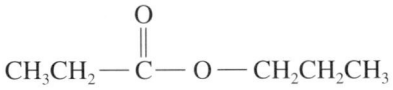

Solution

The alcohol part of the ester is propyl, and the carboxylic acid part is propanoic (propionic) acid.

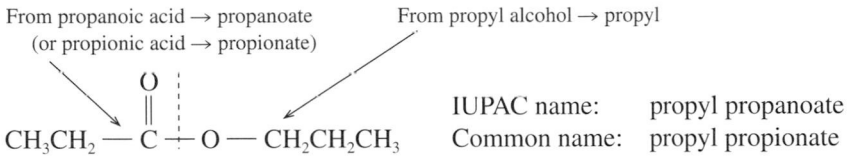

Study Check

Draw the condensed structural formula of pentyl acetate.

Esters in Plants

Many of the fragrances of perfumes and flowers and the flavors of fruits are due to esters. Small esters are volatile so we can smell them and soluble in water so we can taste them. (See Figure 14.7.) Several of these are listed in Table 14.2.

Figure 14.7 Esters are responsible for part of the odor and flavor of oranges, bananas, pears, pineapples, and strawberries.

Table 14.2 Some Naturally Occurring Esters in Fruits and Flavorings

Condensed Structural Formula	Name	Flavor/Odor
$HC(=O)-O-CH_2CH_3$	Ethyl methanoate (ethyl formate)	Rum
$HC(=O)-O-CH_2CH(CH_3)$	Isobutyl methanoate (isobutyl formate)	Raspberries
$CH_3C(=O)-O-CH_2CH_2CH_3$	Propyl ethanoate (propyl acetate)	Pears
$CH_3C(=O)-O-CH_2CH_2CH_2CH_2CH_3$	Pentyl ethanoate (pentyl acetate)	Bananas
$CH_3C(=O)-O-CH_2CH_2CH_2CH_2CH_2CH_2CH_2CH_3$	Octyl ethanoate (octyl acetate)	Oranges
$CH_3CH_2CH_2C(=O)-O-CH_2CH_3$	Ethyl butanoate (ethyl butyrate)	Pineapples
$CH_3CH_2CH_2C(=O)-O-CH_2CH_2CH_2CH_2CH_3$	Pentyl butanoate (pentyl butyrate)	Apricots
$CH_3CH_2CH_2C(=O)-SCH_3$	Methyl thiobutanoate (methyl thiobutyrate)	Strawberries

14.3

EXPLORE YOUR WORLD

Carboxylic Acids and Esters in Foods and Cosmetics

Smell some fruits such as a banana, strawberry, pear, and orange, and taste them if you like.

Look for the names of carboxylic acids and esters on the labels of food products, flavorings, and cosmetics you may use. List their names and write the formulas as you can. For complex names, use a *Merck Index* to look up the formulas of the substances whose names you read on the product labels.

Questions

1. For each fruit that you smell or taste, write the formula of the ester that is responsible for its characteristic smell and flavor.
2. What are the carboxylic acid or ester formulas found in some food products, flavorings, and cosmetics?
3. What are the formulas of some of the carboxylic acids and alcohols used in the esters?

QUESTIONS & PROBLEMS

Esters

14.17 What is the ester responsible for the flavor and odor of the following fruit?
a. banana **b.** orange **c.** apricot **d.** raspberry

14.18 What flavor would you notice if you smelled or tasted the following?
a. ethyl butanoate **b.** propyl acetate
c. methyl thiobutanoate **d.** ethyl formate

14.19 Identify the following as aldehydes, ketones, carboxylic acids or esters:

a. $CH_3\overset{\overset{\displaystyle O}{\|}}{C}H$ **b.** $CH_3\overset{\overset{\displaystyle O}{\|}}{C}-OCH_3$

c. $CH_3CH_2CH_2\overset{\overset{\displaystyle O}{\|}}{C}CH_3$ **d.** $CH_3\overset{\overset{\displaystyle CH_3}{|}}{C}HCH_2\overset{\overset{\displaystyle O}{\|}}{C}-OH$

14.20 Identify the following as aldehydes, ketones, carboxylic acids, or esters:

a. $CH_3\overset{\overset{\displaystyle O}{\|}}{C}OH$ **b.** $CH_3\overset{\overset{\displaystyle O}{\|}}{C}CH_3$

c. $CH_3CH_2CH_2\overset{\overset{\displaystyle O}{\|}}{C}H$ **d.** $CH_3\overset{\overset{\displaystyle CH_3}{|}}{C}HCH_2\overset{\overset{\displaystyle O}{\|}}{C}-OCH_2CH_3$

14.21 Name each of the following esters:

a. $H\overset{\overset{\displaystyle O}{\|}}{C}-OCH_3$ **b.** $CH_3\overset{\overset{\displaystyle O}{\|}}{C}-OCH_3$

c. $CH_3CH_2CH_2\overset{\overset{\displaystyle O}{\|}}{C}OCH_3$ **d.** $CH_3\overset{\overset{\displaystyle CH_3}{|}}{C}HCH_2\overset{\overset{\displaystyle O}{\|}}{C}-OCH_2CH_3$

14.22 Name each of the following esters:

a. $CH_3CH_2\overset{\overset{\displaystyle O}{\|}}{C}-OCH_2CH_3$ **b.** $CH_3CH_2CH_2CH_2CH_2\overset{\overset{\displaystyle O}{\|}}{C}-OCH_3$

c. $CH_3CH_2\underset{\underset{\displaystyle CH_3}{|}}{C}H\overset{\overset{\displaystyle O}{\|}}{C}OCH_3$ **d.** $CH_3CH_2\overset{\overset{\displaystyle O}{\|}}{C}-OCH_2CH_2CH_2CH_3$

14.23 Draw the condensed structural formulas of each of the following esters:

a. methyl acetate **b.** butyl formate
c. ethyl pentanoate **d.** 2-bromopropyl propanoate

14.24 Draw the condensed structural formulas of each of the following esters:

a. hexyl acetate **b.** propyl propionate
c. ethyl-2-hydroxybutanoate **d.** methyl benzoate

14.4

14.4 Esterification and Hydrolysis

Study Pack

Learning Goal

Write equations for the esterification, hydrolysis, and saponification of esters.

In a reaction called **esterification**, a carboxylic acid reacts with an alcohol when heated in the presence of an acid catalyst (usually H_2SO_4). In the reaction, water is formed by the —OH from the carboxylic acid and —H from the alcohol.

Esterification: A Reaction of a Carboxylic Acid and an Alcohol

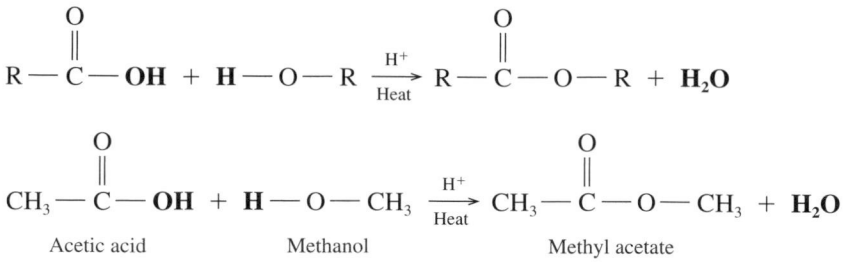

14.4

If we use acetic acid and 1-propanol, we can write an equation for the formation of the ester that is responsible for the flavor and odor of pears.

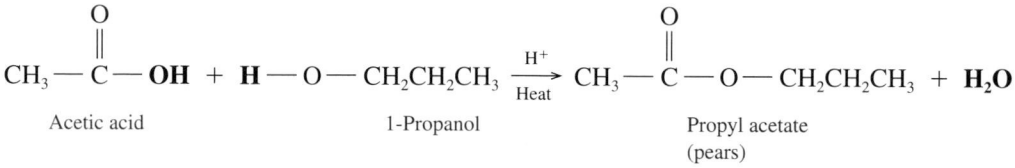

HEALTH NOTE

Salicylic Acid and Aspirin

Chewing on a piece of willow bark was a way of relieving pain for many centuries. By the 1800s, chemists discovered that salicylic acid was the agent in the bark responsible for the relief of pain. However, salicylic acid, which has both a carboxylic group and a hydroxyl group, irritates the stomach lining. An ester of salicylic acid and acetic acid called acetylsalicylic acid or "aspirin" that is less irritating was prepared 1899 by the Bayer chemical company in Germany. In some aspirin preparations, a buffer is added to neutralize the carboxylic acid group and lessen its irritation of the stomach. Aspirin is used as an analgesic (pain reliever), antipyretic (fever reducer), and anti-inflammatory agent.

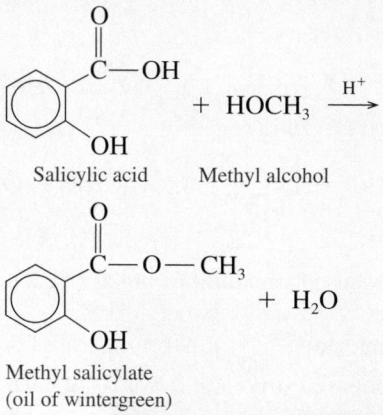

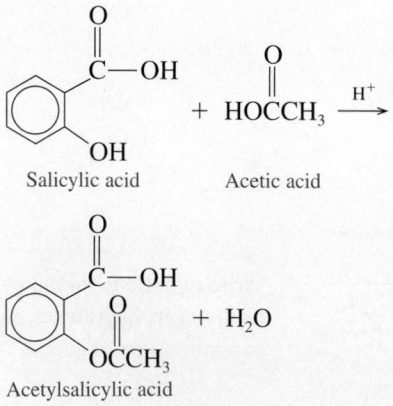

Oil of wintergreen, or methyl salicylate, has a spearmint odor and flavor. Because it can pass through the skin, methyl salicylate is used in skin ointments where it acts as a counterirritant, producing heat to soothe sore muscles. (See Figure 14.8.)

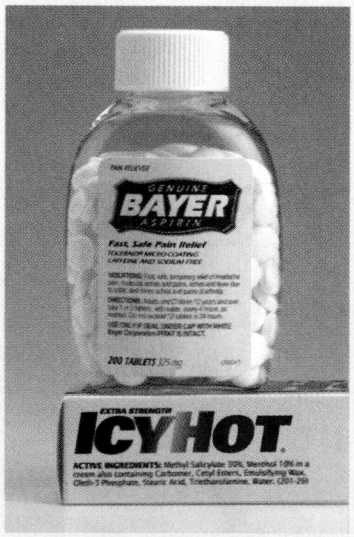

Figure 14.8 Products to relieve pain and fever made from salicylic acid.

S·A·M·P·L·E· 14.6 ·P·R·O·B·L·E·M

Writing Esterification Equations

The ester that gives the flavor and odor of apples can be synthesized from butyric acid and methyl alcohol. What is the equation for the formation of the ester in apples?

Solution

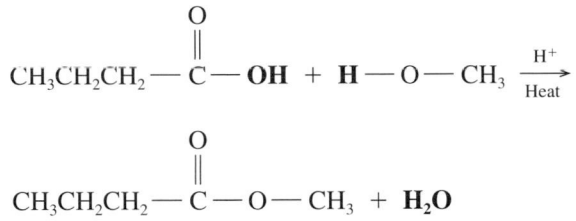

Study Check

What carboxylic acid and alcohol are needed to form the following ester, which gives the flavor and odor to apricots?

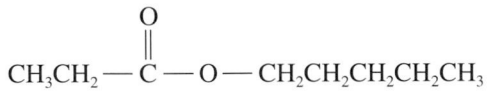

14.4

ENVIRONMENTAL NOTE

Plastics

Terephthalic acid (an acid with two carboxyl groups) is produced in large quantities for the manufacture of polyesters such as Dacron, and plastics. When terephthalic acid reacts with ethylene glycol, ester bonds can form on both ends of the molecules, allowing many molecules to combine until they have formed a long polymer known as a *polyester.*

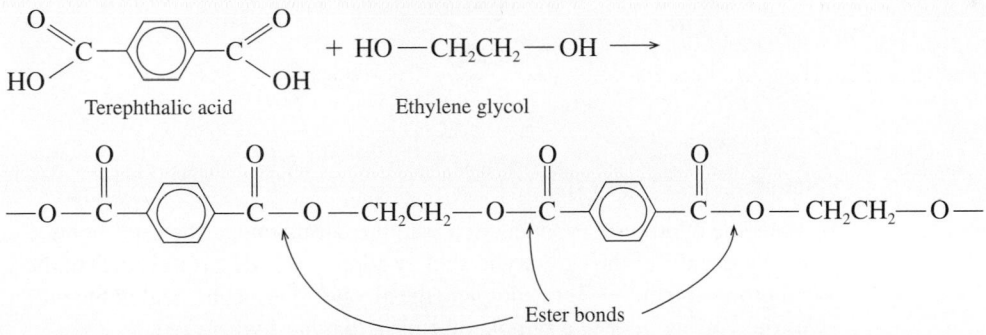

A section of the polyester Dacron

Dacron polyester is used to make permanent press fabrics, carpets, and clothes. In medicine, artificial blood vessels and valves are made of Dacron, which is biologically inert and does not clot the blood. The polyester can also be made as a film called Mylar and as a plastic known as PETE (**p**olyethylene**t**erphthalate). PETE is used for plastic soft drink bottles as well as for containers of salad dressings, shampoos, and dishwashing liquids.

Today PETE is the most widely recycled of all the plastics. In 1992, 365 million pounds (166 million kilograms) of PETE were recycled. After it is separated from other plastics, PETE can be changed into other useful items including polyester fabric for T-shirts and coats, fill for sleeping bags, door mats, and tennis ball containers.

Hydrolysis

In **hydrolysis,** esters are split apart by a reaction with water in the presence of an acid or base catalyst. The products of ester hydrolysis are the carboxylic acid and alcohol that combined to form the ester. Hydrolysis is an important reaction in the body too. For example, during digestion, the ester bonds in our dietary fats are hydrolyzed (split apart) by enzymes (biological catalysts called hydrolases) to produce glycerol and fatty acids.

The hydrolysis of an ester using a strong acid such as H_2SO_4 or HCl and heat, or an esterase enzyme, is the reverse of the reaction that formed the ester. The H of the water molecule becomes part of the alcohol, and the —OH part forms the carboxyl group in the acid.

14.4

Acid Hydrolysis of an Ester

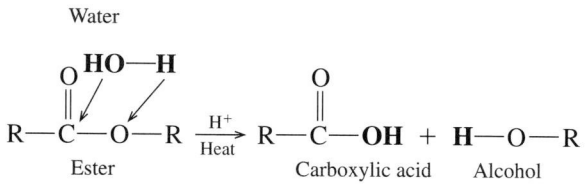

Example:

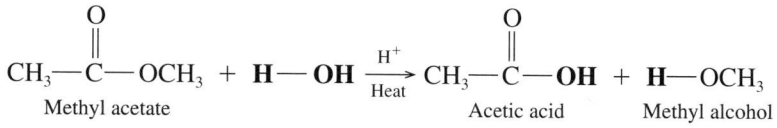

S·A·M·P·L·E 14.7 ·P·R·O·B·L·E·M

Acid Hydrolysis of Esters

Aspirin that has been stored for a long time may undergo hydrolysis in the presence of water and heat. What are the hydrolysis products of aspirin? Why does a bottle of old aspirin smell like vinegar?

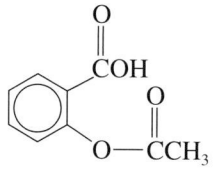

Aspirin

Solution

To write the hydrolysis products, separate the compound at the ester bond. Complete the formula of the carboxylic acid by adding —OH (from water) to the carbonyl group and the —H to complete the alcohol. The acetic acid in the products gives the vinegar odor to a sample of aspirin that has hydrolyzed.

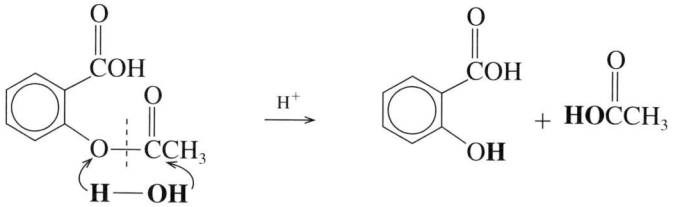

Aspirin Salicylic acid Acetic acid

Study Check

What are the names of the products from the acid hydrolysis of ethyl propionate?

Saponification is the hydrolysis of an ester using a strong base, NaOH or KOH, and heat to give the salt of the carboxylic acid and an alcohol.

Base Hydrolysis of an Ester

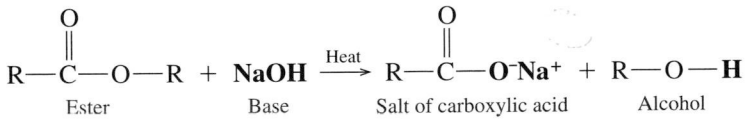

Example:

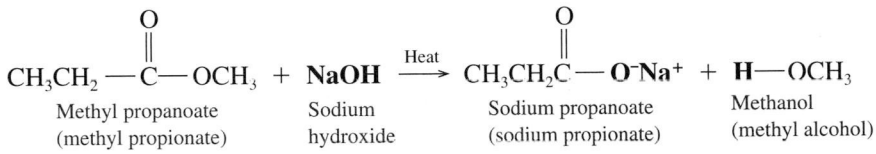

14.4

S·A·M·P·L·E 14.8 ·P·R·O·B·L·E·M

Base Hydrolysis of Esters

Ethyl acetate is a solvent widely used for fingernail polish, plastics, and lacquers. Write the equation of the hydrolysis of ethyl acetate by NaOH.

Solution

The hydrolysis of ethyl acetate by NaOH gives the salt of acetic acid and ethyl alcohol.

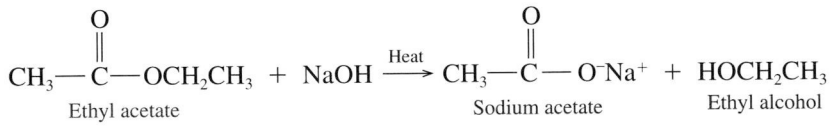

Study Check

Write the condensed structural formulas of the products from the hydrolysis of methyl benzoate by KOH.

Q·U·E·S·T·I·O·N·S & P·R·O·B·L·E·M·S

Esterification and Hydrolysis

14.25 Give the names of the carboxylic acids used to produce each of the following esters:

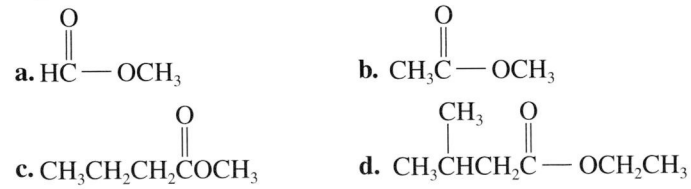

14.26 Give the names of the carboxylic acids used to produce the following esters.

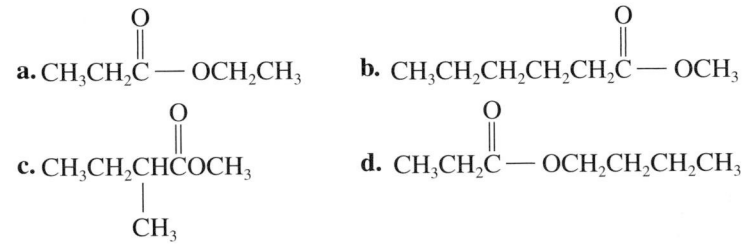

14.27 Write the condensed structural formulas of the esters formed when the following carboxylic acids react with methyl alcohol:
a. acetic acid **b.** butyric acid **c.** benzoic acid

14.28 Write the condensed structural formulas of the esters formed when each of the following carboxylic acids reacts with ethyl alcohol:
a. formic acid **b.** propionic acid **c.** 2-methylpentanoic acid

14.29 Draw the condensed structural formulas of the esters formed when each of the following carboxylic acids and alcohols react:

14.4

a. $CH_3\overset{\overset{\displaystyle O}{\|}}{C}-OH + HOCH_2CH_2CH_3 \xrightarrow{H^+}$

b. $CH_3CH_2CH_2CH_2\overset{\overset{\displaystyle O}{\|}}{C}OH + HO\overset{\overset{\displaystyle CH_3}{|}}{C}HCH_3 \xrightarrow{H^+}$

14.30 Draw the condensed structural formulas of the esters formed when each of the following carboxylic acids and alcohols react:

a. $CH_3-CH_2-\overset{\overset{\displaystyle O}{\|}}{C}-OH + HOCH_3 \xrightarrow{H^+}$

b. 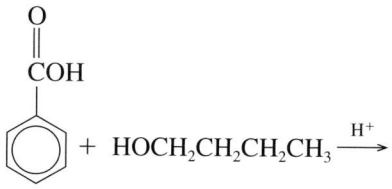

14.31 What are the products of the acid hydrolysis of an ester?

14.32 What are the products of the base hydrolysis of an ester?

14.33 Draw the condensed structural formulas of the products from the acid- or base-catalyzed hydrolysis of each of the following compounds:

a. $CH_3CH_2-\overset{\overset{\displaystyle O}{\|}}{C}-O-CH_3 + NaOH \longrightarrow$

b. $CH_3-\overset{\overset{\displaystyle O}{\|}}{C}-O-CH_2CH_2CH_3 + H_2O \xrightarrow{H^+}$

c. $CH_3CH_2CH_2-\overset{\overset{\displaystyle O}{\|}}{C}-O-CH_2CH_3 + H_2O \xrightarrow{H^+}$

d. 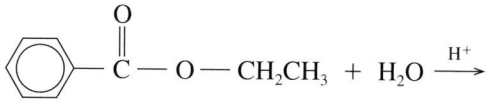

e. 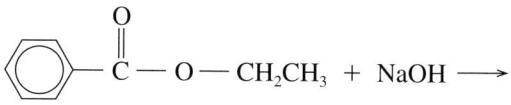

14.34 Draw the condensed structural formulas of the products from the acid- or base-catalyzed hydrolysis of each of the following compounds:

a. $CH_3CH_2-\overset{\overset{\large O}{\|}}{C}-O-CH_2CH_2CH_2CH_3 + H_2O \xrightarrow{H^+}$

b. $H-\overset{\overset{\large O}{\|}}{C}-O-CH_2CH_3 + NaOH \longrightarrow$

c. $CH_3CH_2-\overset{\overset{\large O}{\|}}{C}-O-CH_3 + H_2O \xrightarrow{H^+}$

d. $CH_3-CH_2-\overset{\overset{\large O}{\|}}{C}-O-$  $+ H_2O \xrightarrow{H^+}$

e. 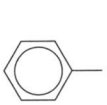 $-CH_2-\overset{\overset{\large O}{\|}}{C}-OCH_2CH_3 + NaOH \longrightarrow$

14.5

Study Pack

Chem of Life Organic Molecules: Functional Groups 5

14.5 Amines

Learning Goal

Classify amines; give their names and condensed structural formulas.

The element nitrogen plays an important role in the physiological activity of compounds such as antibiotics, amphetamines, alkaloids, and tranquilizers. **Amines** are organic compounds in which a nitrogen (N) atom is attached to one, two, or three alkyl or aromatic groups. (See Figure 14.9.)

Amines are classified as *primary (1°)* amines if a nitrogen atom is attached to one alkyl group. An amine is a *secondary (2°)* amine if two carbon groups are attached to nitrogen and a *tertiary (3°)* amine when there are three groups attached.

Primary amines	Secondary amines	Tertiary amines
Alkyl or aromatic group; Primary nitrogen atom		
$R^1-\underset{\underset{H}{\vert}}{N}-H$	$R^1-\underset{\underset{R^2}{\vert}}{N}-H$	$R^1-\underset{\underset{R^2}{\vert}}{N}-R^3$
Examples:		
$CH_3-\underset{\underset{H}{\vert}}{N}-H$	$CH_3-\underset{\underset{CH_3}{\vert}}{N}-H$	$CH_3-\underset{\underset{CH_3}{\vert}}{N}-CH_3$

Solubility of Amines

The electronegativity of the nitrogen atom in an amine makes an N—H bond polar. In primary and secondary amines, the N—H bond can hydrogen bond with water. In all amines, the lone pair of electrons on the nitrogen atom can also form hydrogen bonds with water, making amines with one to five carbon atoms water soluble.

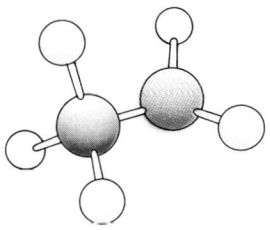

methylamine

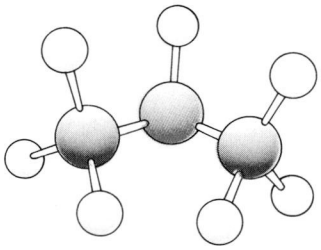

dimethylamine

Figure 14.9 Ball-and-stick models of amines: methylamine and dimethylamine.

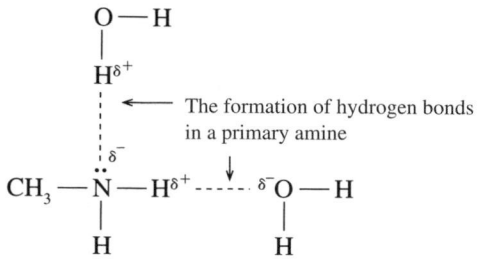

14.5

S·A·M·P·L·E 14.9 ·P·R·O·B·L·E·M

Classifying Amines

Classify the following amines as primary, secondary, or tertiary:

a. 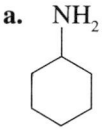

b. 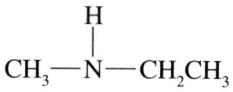

Solution

a. This is a primary amine because there is one alkyl group (cyclohexyl) attached to a nitrogen atom.

b. This is a secondary amine. There are two alkyl groups (methyl and ethyl) attached to the nitrogen atom.

Study Check

Classify the following amine as a primary, secondary, or tertiary amine:

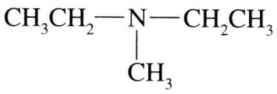

Naming Amines

Amines are named by both their common name and IUPAC name. Simple amines are most often named as one word using the alkyl group followed by the suffix *amine*. In the IUPAC system, the NH_2 group is named an *amino* group on an alkane chain.

Naming Amines	**R—NH$_2$**
Functional group:	—N—
IUPAC name:	aminoalkane
Common name:	alkylamine

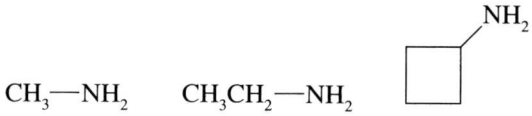

CH_3—NH_2 Aminomethane (methylamine)

CH_3CH_2—NH_2 Aminoethane (ethylamine)

Aminocyclobutane (Cyclobutylamine)

In more complex amines, the parent alkane chain is numbered and the substituents including the amino group are named.

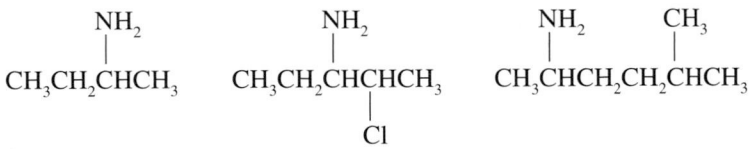

2-Aminobutane 2-Chloro-3-aminopentane 5-Methyl-2-aminohexane

Amines are among the substances responsible for the fishy odors of seafood and the putrid odors of the decay products of animal proteins.

1,4-Diaminobutane (putrescine) 1,5-Diaminopentane (cadaverine)

Naming Substituents on the Amino Group

In the common names for amines, the prefixes *di* and *tri* are used to indicate two or three identical alkyl groups attached to the N atom. In the IUPAC system, alkyl groups attached to the N atom of the amino group NH_2 are indicated by the prefix *N*-. If the alkyl groups are not identical, they are listed in alphabetical order.

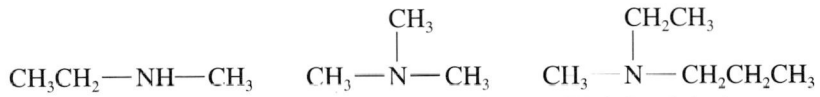

N-Methylaminoethane (ethylmethylamine) *N,N*-Dimethylaminomethane (trimethylamine) *N*-Ethyl-*N*-methylaminopropane (ethylmethylpropylamine)

Aromatic Amines. The name *aniline* has been approved as the IUPAC name of the amine of benzene.

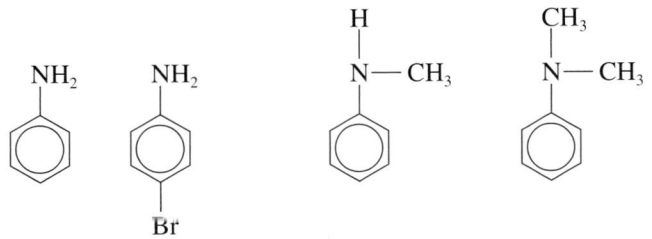

Aniline 4-Bromoaniline (*p*-bromoaniline) *N*-Methylaniline *N,N*-Dimethylaniline

S·A·M·P·L·E· 14.10 ·P·R·O·B·L·E·M

Naming Amines

Give the common name for each of the following amines:

a. $CH_3CH_2—NH_2$ **b.** 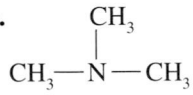 **c.** 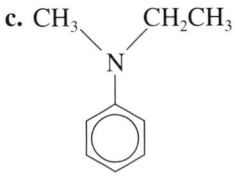

Solution

a. This amine has one ethyl group attached to the nitrogen atom; its name is *ethylamine.*

b. This amine has three methyl groups attached to the nitrogen atom; its name is *trimethylamine.*

c. This amine is a derivative of aniline with a methyl and an ethyl group attached to the nitrogen atom; its name is *N-ethyl-N-methylaniline.*

Study Check

Draw the structure of ethylpropylamine.

HEALTH NOTE

Amines in Health and Medicine

14.5

In response to allergic reactions or injury to cells, the body increases the production of histamine, which causes blood vessels to dilate and increases the permeability of the cells. Redness and swelling occur in the area. Administering an antihistamine such as diphenylhydramine helps block the effects of histamine.

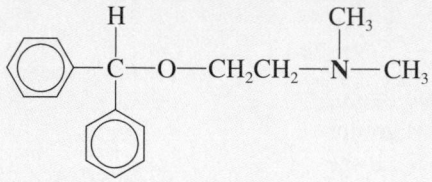

Diphenylhydramine

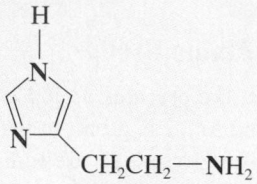

Histamine

In the body, hormones called biogenic amines carry messages between the central nervous system and nerve cells. Epinephrine (adrenaline) and norepinephrine (noradrenaline) are released by the adrenal medulla in "fight or flight" situations to raise the blood glucose level and move the blood to the muscles. Used in remedies for colds, hay fever, and asthma, the norepinephrine contracts the capillaries in the mucous membranes of the respiratory passages. Parkinson's disease is a result of a deficiency in another biogenic amine called dopamine.

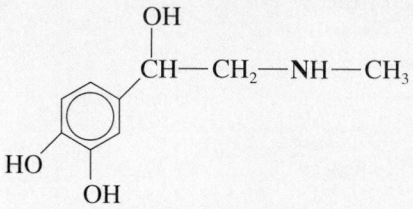

Epinephrine (adrenaline)

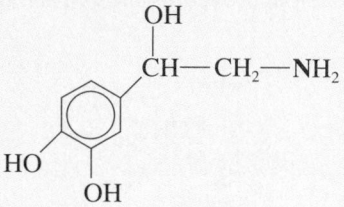

Norepinephrine (noradrenaline)

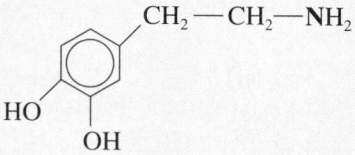

Dopamine

Produced synthetically, amphetamines (known as "uppers") are stimulants of the central nervous system much like adrenaline, but they also increase cardiovascular activity and depress the appetite. They are sometimes used to bring about weight loss, but they can cause chemical dependency. Benzedrine and Neo-Synephrine (phenylephrine) are used in medications to reduce respiratory congestion from colds, hay fever, and asthma. Sometimes, Benzedrine is taken internally to combat the desire to sleep, but it has side effects. Methedrine is used to treat depression and in the illegal form is known as "speed" or "crank."

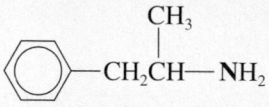

Benzedrine (amphetamine)

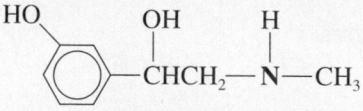

Neo-Synephrine (phenylephrine)

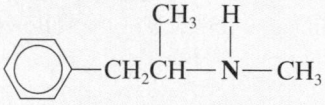

Methamphetamine (methedrine)

QUESTIONS & PROBLEMS

Amines

14.35 What is a primary amine?

14.36 What is a tertiary amine?

14.37 Classify the following amines as primary, secondary, or tertiary:

a. $CH_3CH_2CH_2CH_2—NH_2$

b. $CH_3—\overset{\overset{\displaystyle CH_3}{|}}{N}—CH_2CH_3$

c. 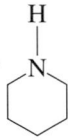

d. 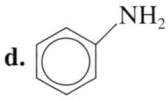

14.38 Classify the following amines as primary, secondary, or tertiary:

a. $CH_3-C(CH_3)_2-CH_2-NH_2$

b. $CH_3-CH_2-N(CH_3)-CH_3$

c. 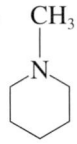

d. 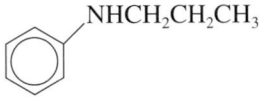

14.39 Give the common and IUPAC names for each of the following amines:

a. $CH_3CH_2-NH_2$

b. $CH_3-NH-CH_2CH_2CH_3$

c. $CH_3CH_2-N(CH_2CH_3)-CH_2CH_3$

d. 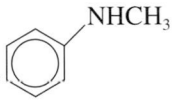

14.40 Give the common and IUPAC names for each of the following amines:

a. $CH_3-CH(CH_3)-CH_2-NH_2$

b. $CH_3CH_2CH_2CH_2CH_2-NH_2$

c. 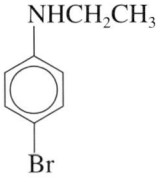

d. 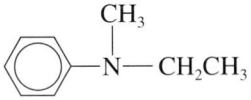

14.41 Draw the condensed structural formulas for each of the following amines:

a. ethylamine
b. *N*-methylaniline
c. butylpropylamine
d. 2-aminopentane

14.42 Draw the condensed structural formulas for each of the following amines:

a. dimethylamine
b. *p*-chloroaniline
c. *N,N*-diethylaniline
d. ethyldimethylamine

14.6 Reactions of Amines

Study Pack

Learning Goal

Write equations for the reactions of amines.

In Chapter 9, we saw that ammonia (NH_3) is a weak base that accepts a proton (H^+) from water to produce ammonium ion and hydroxide ion (OH^-).

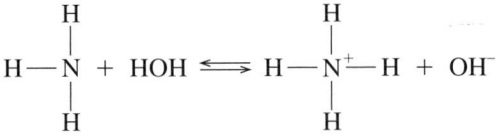

Ammonia Ammonium hydroxide

In water, amines are weak bases because the pair of electrons on the nitrogen atom can bond with a proton (H^+). The products of this ionization are an ammonium ion and a hydroxide ion.

Ionization of an Amine in Water

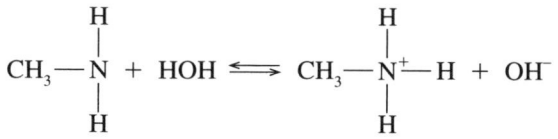

Methylamine Methylammonium hydroxide

14.6

When you squeeze lemon juice on fish, you counteract the fishy odor of the basic amines in the fish by neutralizing them with citric acid. The proton (H^+) from the acid bonds with the unshared pair of electrons on the nitrogen atom. The only product of the neutralization of an amine is an ammonium salt; no water is formed.

Neutralization of an Amine

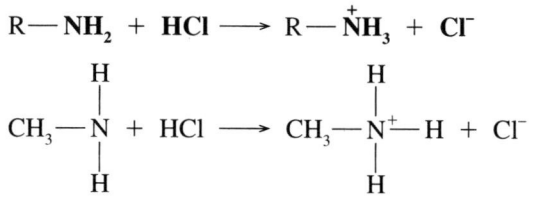

Methylamine Methylammonium chloride

S·A·M·P·L·E 14.11 P·R·O·B·L·E·M

Reactions of Amines

Write an equation that shows ethylamine

a. as a weak base in water **b.** reacting with HCl

Solution

a. In water, ethylamine acts as a weak base by accepting a proton from water to produce ethylammonium hydroxide.

$$CH_3CH_2{-}NH_2 + H{-}OH \rightleftharpoons CH_3CH_2{-}NH_3^+ + OH^-$$

b. $CH_3CH_2{-}NH_2 + HCl \longrightarrow CH_3CH_2{-}NH_3^+ + Cl^-$

Study Check

What is the condensed structural formula of the salt formed by the reaction of trimethylamine and HCl?

Amidation

When a carboxylic acid reacts with ammonia or an amine, an amide is produced. In **amidation,** a molecule of water is eliminated, and the fragments of the carboxylic acid and amine molecules join to form the amide, much like ester formation. Because a hydrogen must be lost from the amines, only primary and secondary amines will react.

Amidation

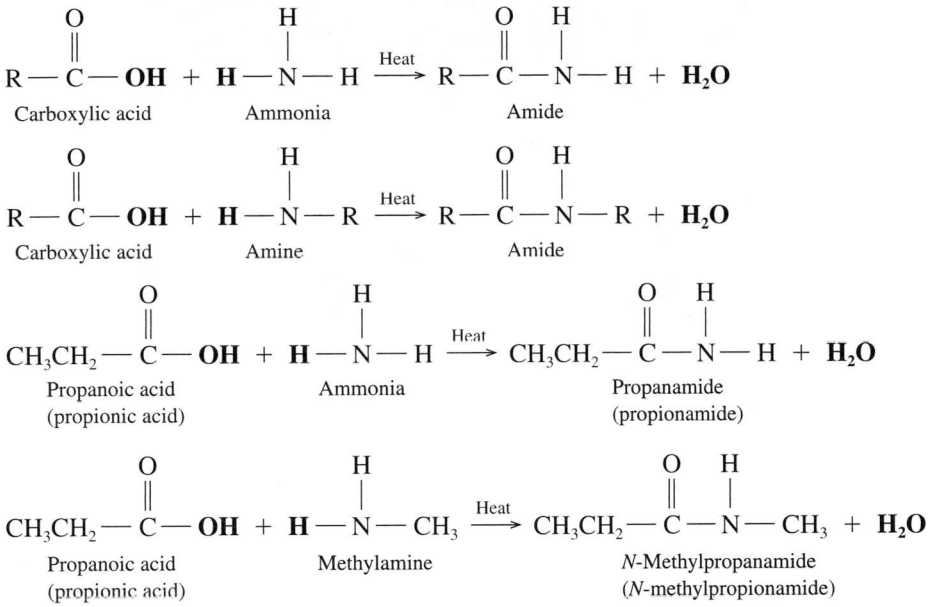

S·A·M·P·L·E 14.12 ·P·R·O·B·L·E·M

Amidation Products

Give the structural formula of the amide product in each of the following reactions:

a.

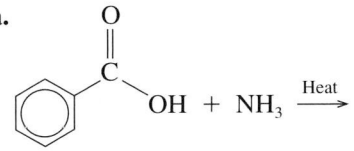

b.

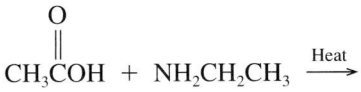

Solution

a. The structural formula of the amide product can be written by attaching the carbonyl group from the acid to the nitrogen atom of the amine. —OH is removed from the acid and —H from the amine to form water.

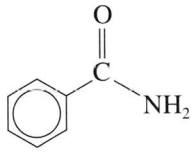

b.

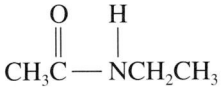

Study Check

What are the condensed structural formulas of the carboxylic acid and amine needed to prepare the following amide?

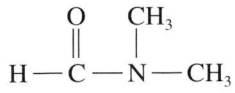

Figure 14.10 Nicotine and caffeine are alkaloids, nitrogen-containing compounds obtained from plants having physiological activity.

14.6

Alkaloids: Amines in Plants

Alkaloids are physiologically active nitrogen-containing compounds produced by plants. The term *alkaloid* refers to the basic (alkali-like) characteristics of amines. Certain alkaloids are used in anesthetics, in antidepressants, and as stimulants, although many are habit forming. (See Figure 14.10.)

Quinine obtained from the bark of the cinchona tree has been used in the treatment of malaria since the 1600s:

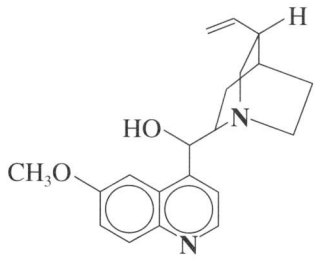

Quinine

Nicotine, from the leaves of the tobacco plant, and coniine, from hemlock, are extremely toxic alkaloids. Caffeine, a compound found in coffee beans and tea, is a stimulant of the central nervous system.

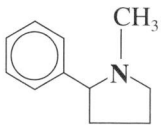

Nicotine

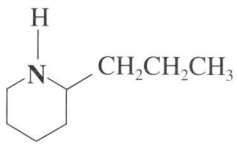

Coniine

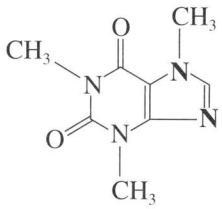

Caffeine

Atropine from belladonna and cocaine from the coca plant are used in low concentrations as anesthetics for eye and sinus procedures. However, higher doses produce euphoria followed by depression and a desire for additional quantities of the drug.

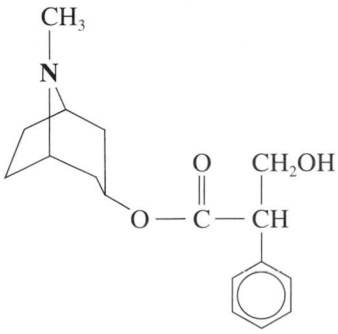

Atropine

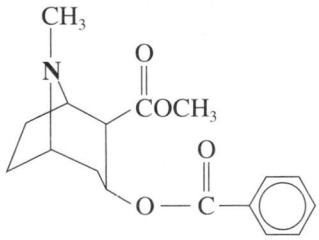

Cocaine

Chemists have altered the structures of substances such as atropine and cocaine to develop synthetic alkaloids such as procaine and lidocaine, which are used for local anesthesia. The synthetic products retain the anesthetic qualities of the natural alkaloid without the addictive side effects.

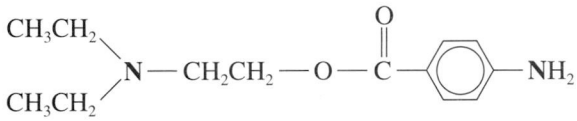

Procaine (novocaine)

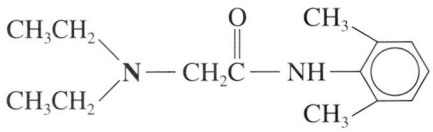

Lidocaine (Xylocaine)

For many centuries, morphine, an alkaloid found in the oriental poppy plant, has been used as a painkiller. However, it also has strong hallucinogenic and addictive side effects. Codeine, which is similar to morphine, is used in prescription painkillers. Heroin, an opiate similar to morphine, is strongly addicting and not used medically. One area of research in pharmacology is the synthesis of morphine-like compounds that can be used safely as effective painkillers with no side effects. A synthetic alkaloid, meperidine, or Demerol, was developed as a painkiller with a chemical structure similar to morphine but with reduced side effects.

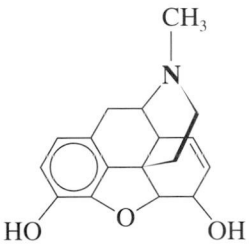

Morphine
(opium)

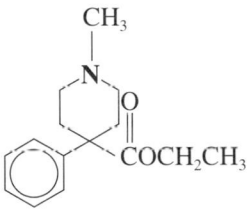

Meperidine
(Demerol)

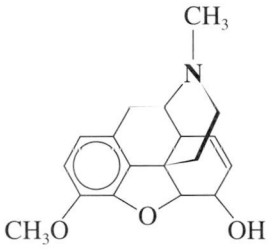

Codeine

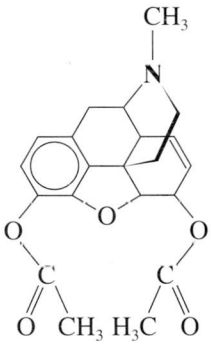

Heroin

Alkaloids are prevalent among the compounds known as hallucinogens. Examples include mescaline, from the peyote cactus, and LSD (lysergic acid diethylamide), prepared from lysergic acid that is produced by a fungus that grows on rye.

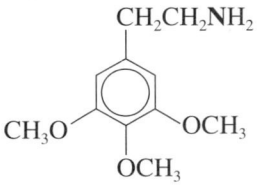

Mescaline

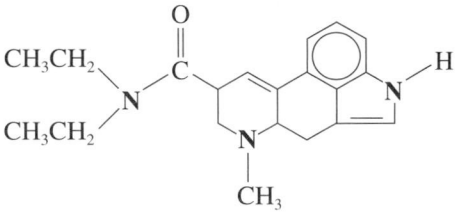

Lysergic acid diethylamide
(LSD)

14.6

Other heterocyclic amines act as tranquilizers by reducing the transmission of nerve impulses to the brain. Low levels of serotonin in the brain appear to be associated with depressed states. Reserpine has been used as a sedative for psychotic patients, and chlorpromazine has been effective as a medication for schizophrenia.

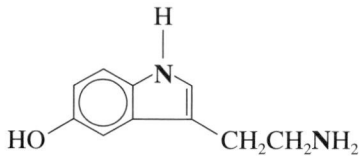

Serotonin

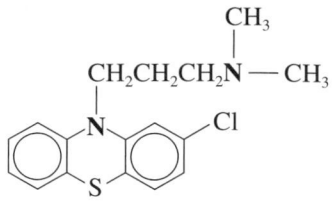

Chlorpromazine
(Thorazine)

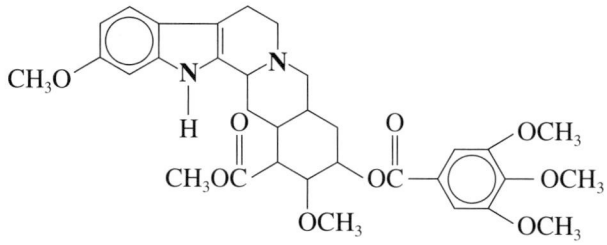

Reserpine

Q·U·E·S·T·I·O·N·S & P·R·O·B·L·E·M·S

Reactions of Amines

14.43 Write an equation for the ionization of each of the following amines in water:
a. methylamine **b.** dimethylamine **c.** aniline

14.44 Write an equation for the ionization of each of the following amines in water:
a. ethylamine **b.** propylamine **c.** *N*-methylaniline

14.45 Write the condensed structural formula of the salts obtained from the reactions of each of the amines in problem 14.43 with HCl.

14.46 Write the condensed structural formula of the salts obtained from the reactions of each of the amines in problem 14.44 with HBr.

14.47 Draw the condensed structural formulas of the amides formed in each of the following reactions:

a. $CH_3\overset{O}{\overset{\|}{C}}OH + NH_3 \xrightarrow{\text{Heat}}$

b. $CH_3\overset{O}{\overset{\|}{C}}OH + NH_2CH_2CH_3 \xrightarrow{\text{Heat}}$

c. 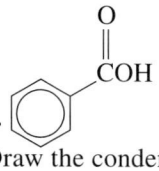 $+ NH_2CH_2CH_2CH_3 \xrightarrow{\text{Heat}}$

14.48 Draw the condensed structural formulas of the amides formed in each of the following reactions:

a. $CH_3CH_2CH_2CH_2\overset{O}{\overset{\|}{C}}OH + NH_3 \xrightarrow{\text{Heat}}$

b. $CH_3CH(CH_3)CH_2C(=O)OH + NH_2CH_2CH_2CH_3 \xrightarrow{\text{Heat}}$

c. $CH_3CH_2C(=O)OH +$ 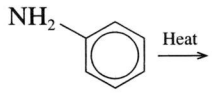

14.7 Amides

Study Pack

14.7

Learning Goal

Write the common names, IUPAC names, and condensed structural formulas of amides.

The **amides** are derivatives of carboxylic acids in which the hydroxyl group is replaced by an amino group. (See Figure 14.11.)

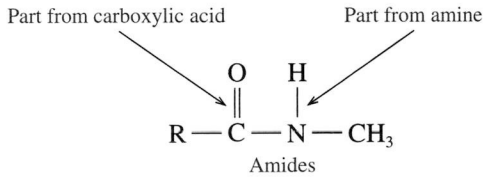

Naming Amides

In both the common and IUPAC names, amides are named by dropping the *ic acid* or *oic acid* from the carboxylic acid names (IUPAC or common) and adding the suffix *amide.*

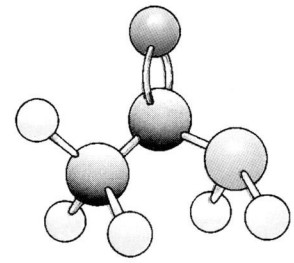

Figure 14.11 A ball-and-stick model of ethanamide (acetamide), an amide.

Naming Amides $R-C(=O)-NH_2$ **or** $R-C(=O)-NR_2$

Functional group: $-C(=O)-N-$

IUPAC name: alkanamide (using IUPAC name for carboxylic acid)

Common name: alkanamide (using common name for carboxylic acid)

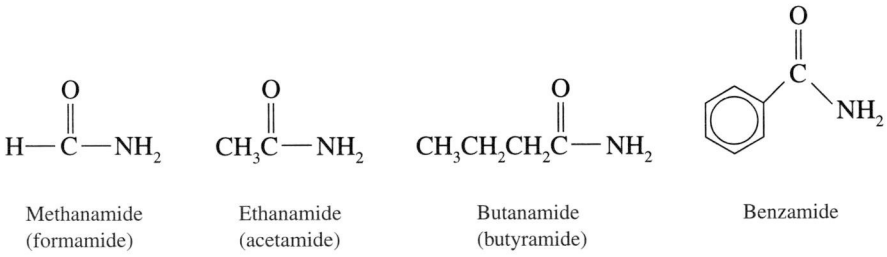

When alkyl groups are attached to the nitrogen atom, the name of an amide is preceded by *N*- or *N,N*- depending on whether there are one or two groups.

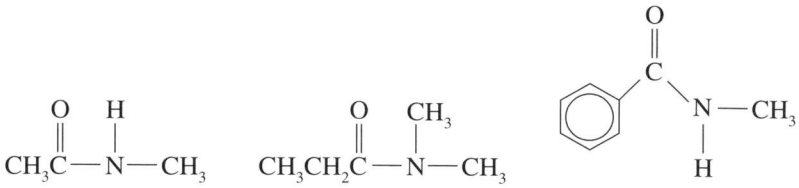

N-Methylethanamide (*N*-methylacetamide)

N,N-Dimethylpropanamide (*N,N*-dimethylpropionamide)

N-Methylbenzamide

14.7

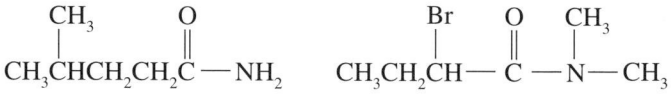

4-Methylpentanamide

N,N-Dimethyl-2-bromobutanamide (*N,N*-dimethyl-α-bromobutyramide)

S·A·M·P·L·E 14.13 ·P·R·O·B·L·E·M

Naming Amides

Give the common and IUPAC names for each of the following amides:

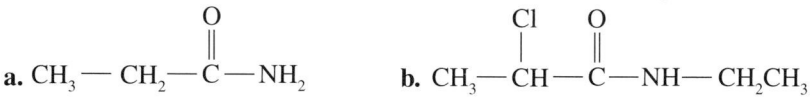

a. $CH_3-CH_2-\overset{\overset{\displaystyle O}{\|}}{C}-NH_2$ **b.** $CH_3-\overset{\overset{\displaystyle Cl}{|}}{CH}-\overset{\overset{\displaystyle O}{\|}}{C}-NH-CH_2CH_3$

Solution

a. The IUPAC name of the carboxylic acid is propanoic acid; the common name is propionic acid. Replacing the *oic acid* or *ic acid* ending with *amide* gives the IUPAC name of *propanamide* and common name of *propionamide.*

b. Using the same carboxylic acid as in part a, the ethyl group attached to the nitrogen atom is named *N-ethyl.* The amide is named *N-ethyl-2-chloropropanamide* (IUPAC), and *N-ethyl-α-chloropropionamide* (common).

Study Check

Draw the condensed structural formula of *N,N*-dimethyl-*p*-chlorobenzamide.

HEALTH NOTE

Amides in Health and Medicine

The simplest natural amide is urea, an end product of protein metabolism in the body. The kidneys remove urea from the blood and provide for its excretion in urine. If the kidneys malfunction, urea is not removed and builds to a toxic level, a condition called uremia. Urea is also used as a component of fertilizer, to increase nitrogen in the soil.

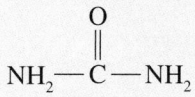

Urea

Synthetic amides are used as substitutes for sugar and aspirin. Saccharin is a very powerful sweetener and is used as a sugar substitute.

Aspirin substitutes contain phenacetin or acetaminophen, which is used in Tylenol. Like aspirin, acetaminophen is an analgesic and antipyretic, but it has little anti-inflammatory effect. (See Figure 14.12.)

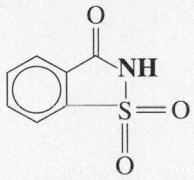

Saccharin

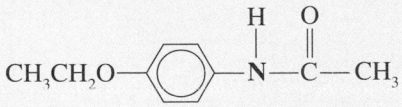

Phenacetin

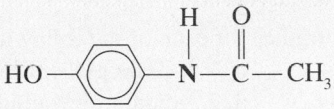

Acetaminophen

Many barbiturates are cyclic amides of barbituric acid that act as sedatives in small dosages or sleep inducers in larger dosages. They are often habit forming. Barbiturate drugs include phenobarbital (Luminal), pentobarbital (Nembutal), and secobarbital (Seconal).

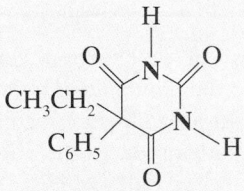

Luminal (phenobarbital)

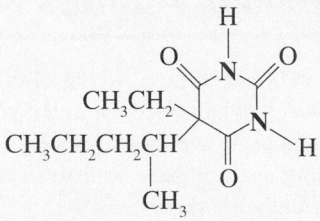

Nembutal (pentobarbital)

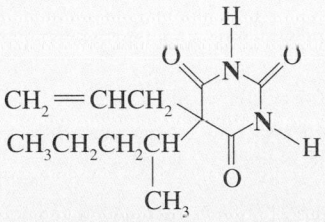

Seconal (secobarbital)

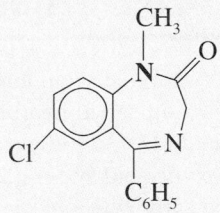

Valium (diazepam)

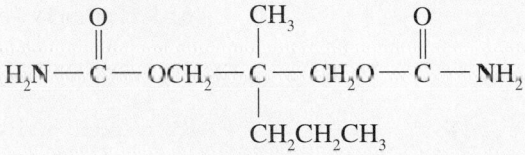

Equanil (meprobamate)

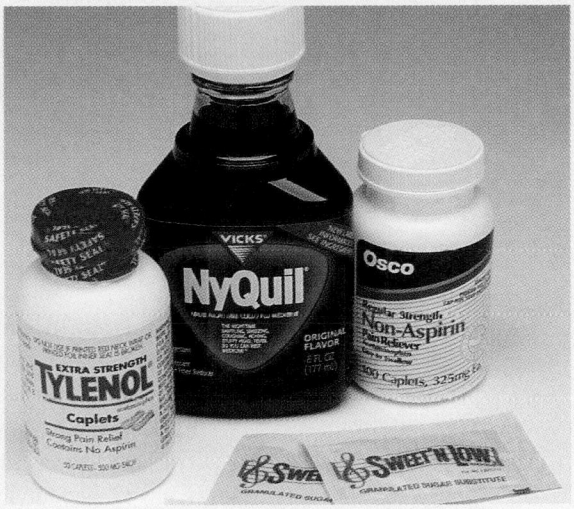

Figure 14.12 Sugar and aspirin substitutes contain amide compounds.

Another group of amides such as diazepam (Valium) and meprobamate (Equanil) are used to reduce anxiety.

14.7

QUESTIONS & PROBLEMS

Amides

14.49 Give the IUPAC and common names (if any) for each of the following amides:

a. $CH_3-\overset{O}{\overset{\|}{C}}-NH-CH_3$

b. $CH_3CH_2CH_2-\overset{O}{\overset{\|}{C}}-NH_2$

c. $H-\overset{O}{\overset{\|}{C}}-NH_2$

d. 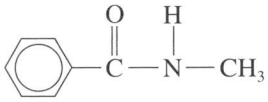

14.50 Give the IUPAC and common names (if any) for each of the following amides:

a. $CH_3CH_2\overset{\overset{\displaystyle O}{\|}}{C}-\overset{\overset{\displaystyle H}{|}}{N}-CH_2CH_3$

b. $CH_3CH_2CH_2CH_2CH_2\overset{\overset{\displaystyle O}{\|}}{C}-NH_2$

c. $CH_3\overset{\overset{\displaystyle O}{\|}}{C}-\overset{\overset{\displaystyle CH_3}{|}}{N}-CH_2CH_2CH_3$

d. 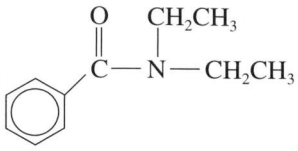

14.8

14.51 Draw the condensed structural formulas for each of the following amides:
a. propionamide **b.** 2-methylpentamide **c.** methanamide
d. *N*-ethylbenzamide **e.** *N*-ethylbutyramide

14.52 Draw the condensed structural formulas for each of the following amides:
a. formamide **b.** *N,N*-dimethylbenzamide
c. 3-methylbutyramide **d.** 2,2-dichlorohexamide
e. *N*-propyl-3-chloropentamide

Study Pack

14.8 Reactions of Amides

Learning Goal

Write equations for the hydrolysis of amides.

As we have seen, amide bonds are formed by the elimination of water. The reverse reaction called hydrolysis occurs when water is added back to the amide bond to split the molecule. When an acid is used, the hydrolysis products of an amide are the carboxylic acid and the ammonium salt. In base hydrolysis, the amide produces the salt of the carboxylic acid and ammonia or amine.

Acid Hydrolysis of Amides

$$\underset{\text{Amide}}{RC(=O)-NH_2} + HOH + HCl \longrightarrow \underset{\text{Carboxylic acid}}{RC(=O)-OH} + \underset{\text{Ammonium chloride, a salt}}{NH_4^+Cl^-}$$

Example:

$$\underset{\text{Ethanamide (acetamide)}}{CH_3C(=O)-NH_2} + HOH + HCl \longrightarrow \underset{\text{Ethanoic acid (acetic acid)}}{CH_3C(=O)-OH} + \underset{\text{Ammonium chloride}}{NH_4^+Cl^-}$$

Base Hydrolysis of Amides

$$\underset{\text{Amide}}{R-C(=O)-NH-R} + NaOH \longrightarrow \underset{\text{Sodium carboxylate, a salt}}{R-C(=O)-O^-Na^+} + \underset{\text{Alkyl amine}}{R-NH_2}$$

Example:

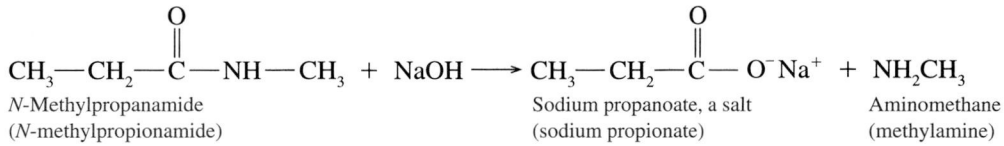

S·A·M·P·L·E **14.14** ·P·R·O·B·L·E·M

Hydrolysis of Amides

Write the structural formulas for the products for the hydrolysis of *N*-methylpentamide with NaOH.

Solution

Hydrolysis of the amide with a base produces a carboxylate salt (sodium pentanoate) and the corresponding amine (methylamine).

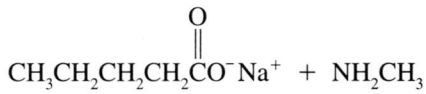

Study Check

What are the structures of the products from the hydrolysis of *N*-methylbutyramide with HBr?

HEALTH NOTE

Amide Hydrolysis in Nature

In biological systems, the peptide bonds of proteins are hydrolyzed by enzymes called proteases during the digestion of proteins or when they are broken down in cells. For example, the hydrolysis of a dipeptide to produce amino acids can be shown as follows:

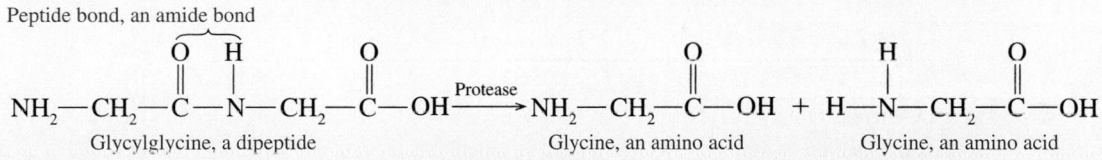

In 1989, a controversy arose concerning a hormone, Alar, that was being applied to apples to keep them on the trees longer and allow larger fruit to develop. Upon hydrolysis, the Alar produces an amine, UDMH, suspected of being a carcinogen. UDMH is the abbreviation for unsymmetrical dimethylhydrazine. Of particular concern was apple juice because high temperatures used during the processing of the apples increase the hydrolysis of Alar. Eventually, sale of Alar was discontinued in the United States.

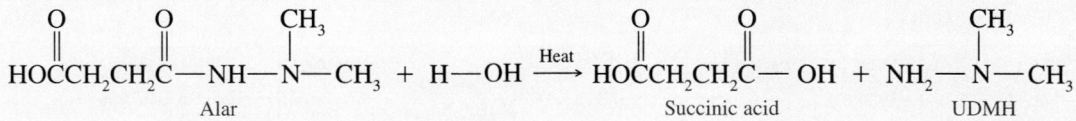

QUESTIONS & PROBLEMS

14.8

14.53 Write the condensed structural formulas for the products of the acid hydrolysis of each of the following amides with HCl:

a. $CH_3\overset{\overset{\large O}{\|}}{C}-NH_2$

b. $CH_3CH_2\overset{\overset{\large O}{\|}}{C}-NH_2$

c. $CH_3CH_2CH_2\overset{\overset{\large O}{\|}}{C}-NH-CH_3$

d. 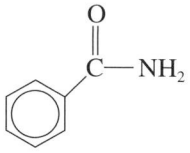

e. *N*-ethylpentanamide

14.54 Write the condensed structural formulas for the products of the base hydrolysis of each of the following amides with NaOH:

a. $CH_3CH_2\overset{\overset{\large CH_3}{|}}{CH}-\overset{\overset{\large O}{\|}}{C}-NH_2$

b. $CH_3CH_2CH_2\overset{\overset{\large O}{\|}}{C}-\overset{\overset{\large CH_2CH_3}{|}}{N}-CH_2CH_3$

c. 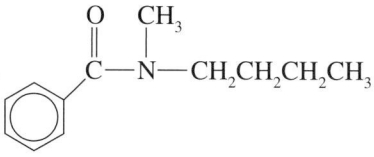

d. $\overset{\overset{\large Cl}{|}}{CH_3CH}-\overset{\overset{\large O}{\|}}{C}-\overset{\overset{\large CH_3}{|}}{N}-CH_2CH_3$

e. *N*-propyl benzamide

Chapter Review

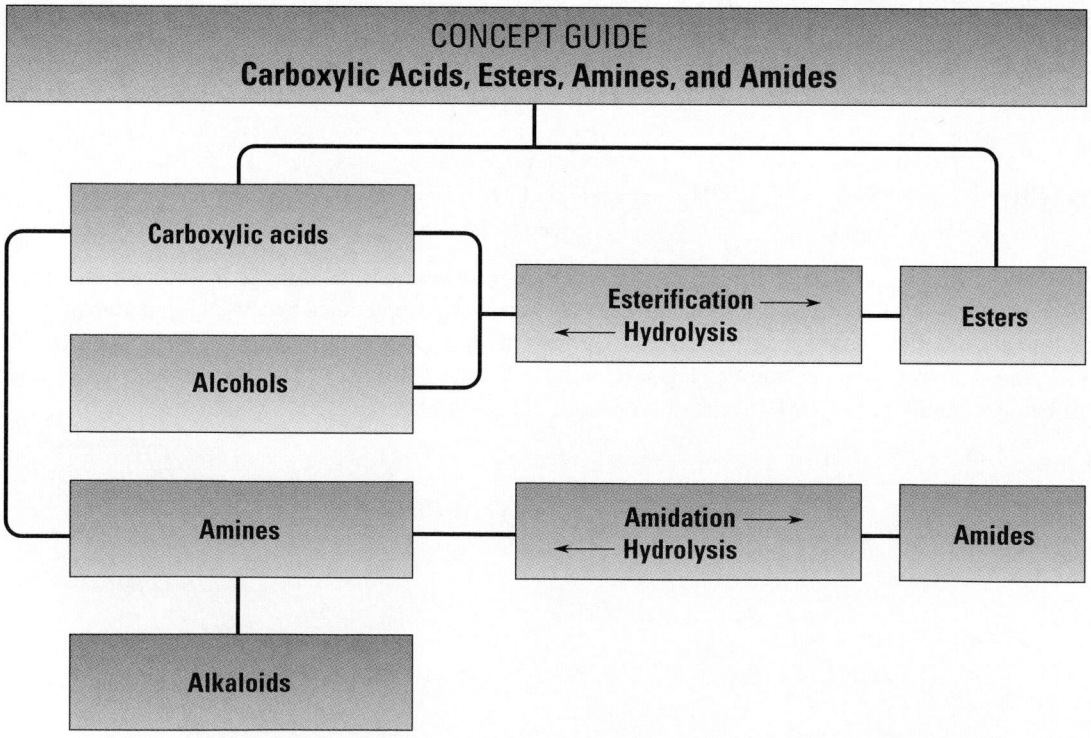

Chapter Summary

14.1 Carboxylic Acids

A carboxylic acid contains the carboxyl functional group, which is a hydroxyl group connected to the carbonyl group.

14.2 Properties of Carboxylic Acids

The carboxyl group contains polar bonds of O—H and C=O, which makes a carboxylic acid with one to four carbon atoms very soluble in water. As weak acids, carboxylic acids ionize slightly by donating a proton to water to form carboxylate and hydronium ions. Carboxylic acids are neutralized by a base producing the carboxylate salt and water.

14.3 Esters

In an ester, an alkyl or aromatic group has replaced the H of the hydroxyl group of a carboxylic acid. The names of esters consist of two words, one from the alcohol and the other from the carboxylic acid with the *ic* ending replaced by *ate*.

14.4 Esterification and Hydrolysis

In the presence of a strong acid, a carboxylic acid reacts with an alcohol to produce an ester. A molecule of water is removed: —OH from the carboxylic acid and —H from the alcohol molecule. Esters undergo acid hydrolysis by adding water to yield the carboxylic acid and alcohol (or phenol). Base hydrolysis or saponification of an ester produces the carboxylate salt and an alcohol.

14.5 Amines

A nitrogen atom attached to one, two, or three alkyl or aromatic groups forms a primary, secondary, or tertiary amine. Amines are prevalent in compounds, synthetic and naturally occurring, that have physiological activity. Amines are usually named by common names in which the names of the alkyl group are listed alphabetically preceding the suffix *amine*. In the IUPAC system, the word *amino* is used to give the position of the NH_2 group on an alkane chain.

14.6 Reactions of Amines

In water, amines act as weak bases by accepting protons from water to produce ammonium and hydroxide ions. Amines react with acids to produce ammonium salts. When a carboxylic acid reacts with ammonia or an amine, an amide is produced.

14.7 Amides

Amides are derivatives of carboxylic acids in which the hydroxyl group is replaced by an amino or *N*-alkyl (NH_2 or NHR) group. They are named by replacing the *ic acid* or *oic acid* ending with *amide*.

14.8 Reactions of Amides

Amides undergo hydrolysis (in acid or base) to produce the carboxylic acid and the amine (or ammonia) or their salts.

Summary of Naming

Family	Condensed Structural Formula	IUPAC Name	Common Name
Carboxylic acid	$CH_3\overset{\overset{\large O}{\|}}{C}—OH$	Ethanoic acid	Acetic acid
Carboxylate salt	$CH_3\overset{\overset{\large O}{\|}}{C}—O^-Na^+$	Sodium ethanoate	Sodium acetate
Ester	$CH_3\overset{\overset{\large O}{\|}}{C}—OCH_3$	Methyl ethanoate	Methyl acetate
Amine	$CH_3CH_2—NH_2$	Aminoethane	Ethylamine
Amide	$CH_3\overset{\overset{\large O}{\|}}{C}—NH_2$	Ethanamide	Acetamide

Summary of Reactions

Ionization of a Carboxylic Acid in Water

$$R-\overset{\overset{\displaystyle O}{\|}}{C}-OH + H_2O \rightleftharpoons R-\overset{\overset{\displaystyle O}{\|}}{C}-O^- + H_3O^+$$

Carboxylic acid (weak acid) · Water (weak base) · Carboxylate ion · Hydronium ion

$$CH_3-\overset{\overset{\displaystyle O}{\|}}{C}-OH + H_2O \rightleftharpoons CH_3-\overset{\overset{\displaystyle O}{\|}}{C}-O^- + H_3O^+$$

Ethanoic acid (acetic acid) · Ethanoate ion (acetate ion) · Hydronium ion

Neutralization of a Carboxylic Acid

$$R-\overset{\overset{\displaystyle O}{\|}}{C}-OH + NaOH \longrightarrow H-\overset{\overset{\displaystyle O}{\|}}{C}-O^-Na^+ + H_2O$$

Carboxylic acid · Base · Carboxylate salt

$$CH_3CH_2-\overset{\overset{\displaystyle O}{\|}}{C}-OH + NaOH \longrightarrow CH_3CH_2-\overset{\overset{\displaystyle O}{\|}}{C}-O^-Na^+ + H_2O$$

Propanoic acid (propionic acid) · Sodium hydroxide · Sodium propanoate (sodium propionate)

Esterification: Carboxylic Acid and an Alcohol

$$R-\overset{\overset{\displaystyle O}{\|}}{C}-OH + HO-R \xrightarrow{H^+} R-\overset{\overset{\displaystyle O}{\|}}{C}-O-R + H_2O$$

Carboxylic acid · Alcohol · Ester

$$CH_3-\overset{\overset{\displaystyle O}{\|}}{C}-OH + HO-CH_3 \xrightarrow{H^+} CH_3-\overset{\overset{\displaystyle O}{\|}}{C}-O-CH_3 + H_2O$$

Ethanoic acid (acetic acid) · Methanol (methyl alcohol) · Methyl ethanoate (methyl acetate)

Acid Hydrolysis of an Ester

$$R-\overset{\overset{\displaystyle O}{\|}}{C}-O-R + H-OH \xrightarrow[\text{Heat}]{H^+} R-\overset{\overset{\displaystyle O}{\|}}{C}-OH + R-OH$$

Ester · Carboxylic acid · Alcohol

$$CH_3-\overset{\overset{\displaystyle O}{\|}}{C}-OCH_3 + H-OH \xrightarrow[\text{Heat}]{H^+} CH_3-\overset{\overset{\displaystyle O}{\|}}{C}-OH + H-OCH_3$$

Methyl ethanoate (methyl acetate) · Ethanoic acid (acetic acid) · Methanol (methyl alcohol)

Base Hydrolysis of an Ester

$$R-\overset{\overset{\displaystyle O}{\|}}{C}-O-R + NaOH \xrightarrow{\text{Heat}} R-\overset{\overset{\displaystyle O}{\|}}{C}-O^-Na^+ + R-OH$$

Ester · Base · Carboxylate salt · Alcohol

$$CH_3CH_2-\overset{\overset{\displaystyle O}{\|}}{C}-OCH_3 + NaOH \xrightarrow{\text{Heat}} CH_3CH_2\overset{\overset{\displaystyle O}{\|}}{C}-O^-Na^+ + HOCH_3$$

Methyl propanoate (methyl propionate) · Sodium hydroxide · Sodium propanoate (sodium propionate) · Methanol (methyl alcohol)

Ionization of an Amine in Water

$$\underset{\text{Amine}}{R{-}NH_2} + HOH \rightleftharpoons \underset{\text{Alkyl ammonium hydroxide}}{R{-}NH_3^+} + OH^-$$

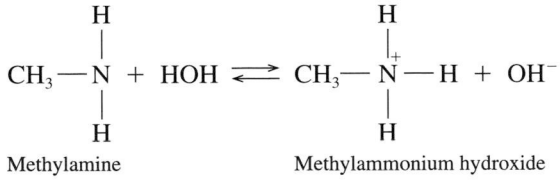

Neutralization of an Amine

$$\underset{\text{Amine}}{R{-}NH_2} + HCl \longrightarrow \underset{\text{Ammonium salt}}{R{-}NH_3^+\ Cl^-}$$

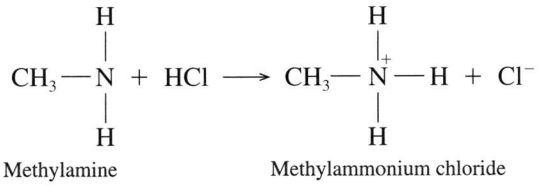

Amidation

$$\underset{\text{Carboxylic acid}}{R{-}C(=O){-}OH} + \underset{\text{Ammonia}}{H{-}N(H){-}H} \xrightarrow{\text{Heat}} \underset{\text{Amide}}{R{-}C(=O){-}N(H){-}H} + H_2O$$

$$\underset{\text{Carboxylic acid}}{R{-}C(=O){-}OH} + \underset{\text{Amine}}{H{-}N(H){-}R} \xrightarrow{\text{Heat}} \underset{\text{Amide}}{R{-}C(=O){-}N(H){-}R} + H_2O$$

$$\underset{\substack{\text{Propanoic acid}\\ \text{(propionic acid)}}}{CH_3CH_2{-}C(=O){-}OH} + \underset{\text{Ammonia}}{H{-}N(H){-}H} \xrightarrow{\text{Heat}} \underset{\substack{\text{Propanamide}\\ \text{(propionamide)}}}{CH_3CH_2{-}C(=O){-}N(H){-}H} + H_2O$$

$$\underset{\substack{\text{Propanoic acid}\\ \text{(propionic acid)}}}{CH_3CH_2{-}C(=O){-}OH} + \underset{\substack{\text{Aminomethane}\\ \text{(methylamine)}}}{H{-}N(H){-}CH_3} \xrightarrow{\text{Heat}} \underset{\substack{\textit{N}\text{-Methylpropanamide}\\ \text{(}\textit{N}\text{-methylpropionamide)}}}{CH_3CH_2{-}C(=O){-}N(H){-}CH_3} + H_2O$$

Hydrolysis of Amides

$$\underset{\text{Amide}}{RC(=O){-}NH_2} + H_2O + HCl \longrightarrow \underset{\substack{\text{Carboxylic}\\ \text{acid}}}{RC(=O){-}OH} + \underset{\substack{\text{Ammonium}\\ \text{chloride}}}{NH_4^+Cl^-}$$

$$\underset{\text{Amide}}{RC(=O){-}NH_2} + NaOH \longrightarrow \underset{\text{Carboxylate salt}}{RC(=O){-}O^-Na^+} + \underset{\text{Ammonia}}{NH_3}$$

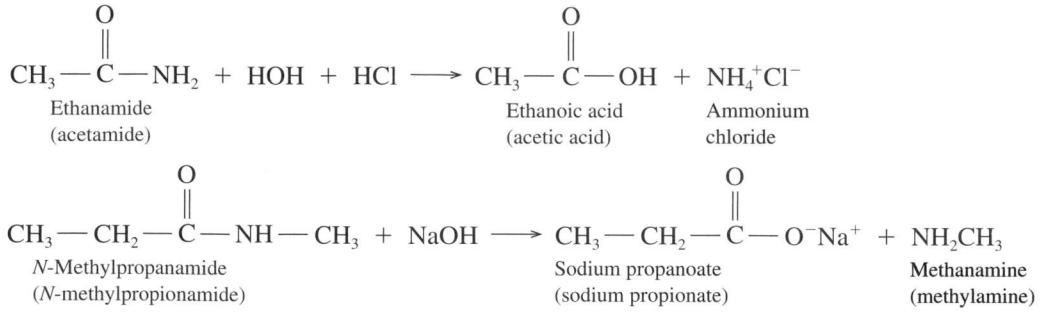

Key Terms

alkaloids Amines having physiological activity that are produced in plants.

amidation The formation of an amide from a carboxylic acid and ammonia or an amine.

amides Organic compounds containing the carbonyl group attached to an amino group or a substituted nitrogen atom:

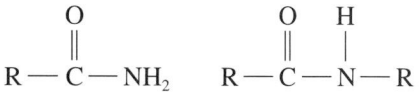

amines Organic compounds containing a nitrogen atom attached to one, two, or three hydrocarbon groups.

carboxyl group A functional group found in carboxylic acids composed of carbonyl and hydroxyl groups.

carboxylic acids A family of organic compounds containing the carboxyl group.

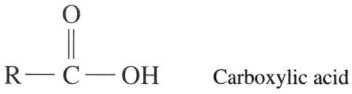

esterification The formation of an ester from a carboxylic acid and an alcohol with the elimination of a molecule of water in the presence of an acid catalyst.

esters A family of organic compounds in which an alkyl group replaces the hydrogen atom in a carboxylic acid.

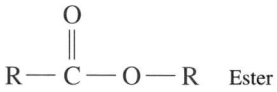

hydrolysis The splitting of a molecule by the addition of water. Esters hydrolyze to produce a carboxylic acid and an alcohol; amides to yield the corresponding carboxylic acid and amine or their salts.

saponification The hydrolysis of an ester with a strong base to produce a salt of the carboxylic acid and an alcohol.

Chapter Practice

14.55 Methyl benzoate is not soluble in water; but when it is heated with KOH, it dissolves. Write an equation for the reaction and explain what happens. When HCl is added to the solution, a white solid forms. What is the solid?

14.56 Hexanoic acid is soluble in NaOH solution, but hexanal is not. Explain.

14.57 Salicylic acid could be named *o*-hydroxybenzoic acid.
- **a.** What two reactive functional groups are present?
- **b.** Draw the structure of the ester product that forms when the hydroxyl group of salicylic acid reacts with acetic acid.
- **c.** Draw the structure of methyl salicylate, oil of wintergreen, formed when salicylic acid forms an ester with methyl alcohol.

14.58 What volume of 0.100 M NaOH is needed to neutralize 3.00 g of solid benzoic acid?

14.59 Draw the condensed structural formulas of the products produced by the hydrolysis of each of the following compounds:

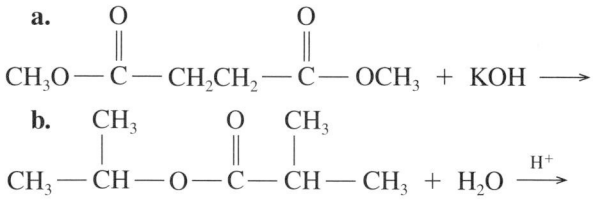

14.60 There are four amine isomers with the molecular formula C_3H_9N. Draw their condensed structural formulas. Name and classify each as a primary, secondary, or tertiary amine.

14.61 Starting with 1-propanol as your only organic compound, write reactions to convert the alcohol into each of the following:
- **a.** propionic acid
- **b.** 1-propyl propanoate
- **c.** propane
- **d.** 2-propanol
- **e.** acetone (propanone)
- **f.** propanamide

14.62 Toradol is used in dentistry to relieve pain. Name the functional groups in this molecule.

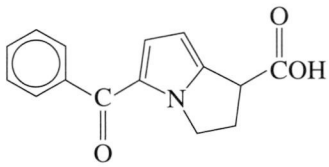

14.63 Voltaren is indicated for acute and chronic treatment of the symptoms of rheumatoid arthritis. Name the functional groups in this molecule.

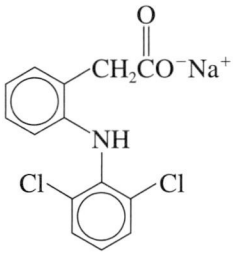

14.64 Many amine-containing drugs are given to patients in their salt form, such as hydrochloride or sulfate. What might be the reason?

14.65 Using a reference book such as the *Merck Index* or *Physicians' Desk Reference,* look up the structural formula of the following medicinal drugs and list the functional groups in the compounds. You may need to refer to the cross-index of names in the back of the reference book.

a. Keflex, an antibiotic
b. Inderal, a β-channel blocker used to treat heart irregularities
c. ibuprofen, an anti-inflammatory agent
d. Aldomet (methyldopa)
e. Percodan, a narcotic pain reliever
f. triamterene, a diuretic

Answers

Answers to Study Checks

14.1 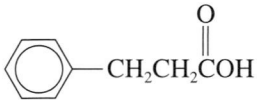

14.2 $CH_3CH_2CH_2\overset{O}{\overset{\|}{C}}OH$

14.3 $H\overset{O}{\overset{\|}{C}}OH + H_2O \rightleftharpoons H\overset{O}{\overset{\|}{C}}O^- + H_3O^+$

14.4 butanoic acid, butyric acid

14.5 $CH_3\overset{O}{\overset{\|}{C}}—OCH_2CH_2CH_2CH_2CH_3$

14.6 propanoic (propionic) acid and 1-pentanol

14.7 propanoic (propionic) acid and ethanol

14.8 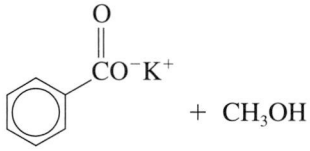

14.9 tertiary (3°)

14.10 $CH_3CH_2NHCH_2CH_2CH_3$

14.11 $CH_3—\overset{CH_3}{\underset{CH_3}{\overset{+}{N}}}—H + Cl^-$

14.12 $H\overset{O}{\overset{\|}{C}}OH$ and $H\overset{CH_3}{\overset{|}{N}}CH_3$

14.13 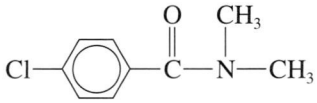

14.14 $CH_3CH_2CH_2\overset{O}{\overset{\|}{C}}OH$ and $CH_3NH_3^+Br^-$

Answers to Selected Problems

14.1 methanoic acid (formic acid)

14.3 Each compound contains three carbon atoms. They differ because propanal, an aldehyde, contains a carbonyl group bonded to a hydrogen. In propanoic acid, the carbonyl group connects to a hydroxyl group.

14.5 **a.** ethanoic acid (acetic acid)
b. butanoic acid (butyric acid)
c. 2-chloropropanoic acid (α-chloropropionic acid)
d. 3-methylhexanoic acid

14.7 **a.** $CH_3CH_2\overset{O}{\overset{\|}{C}}—OH$ **b.** 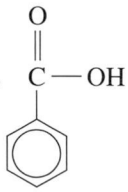

c. $Cl—CH_2\overset{O}{\overset{\|}{C}}—OH$ **d.** $HO—CH_2CH_2\overset{O}{\overset{\|}{C}}—OH$

e. $CH_3\overset{CH_3}{\overset{|}{C}}HCH_2\overset{O}{\overset{\|}{C}}—OH$

f. $CH_3CH_2\overset{Br}{\overset{|}{C}}HCH_2\overset{Br}{\overset{|}{C}}HCH_2\overset{O}{\overset{\|}{C}}OH$

14.9 **a.** $HC(=O)-OH$ **b.** $CH_3C(=O)-OH$

c. $CH_3CH(CH_3)CH_2C(=O)-OH$ **d.** 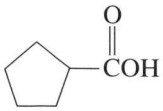

14.11 **a.** CH_3CH_2COOH has a polar carbon–oxygen double bond and a polar —OH; both polar groups hydrogen bond to water.
b. Both the carbonyl and the hydroxyl groups in $CH_3CH_2CH_2COOH$ are strongly polar and hydrogen bond to water.
c. CH_3COOH has a smaller carbon chain and will be more soluble.

14.13 **a.** $HC(=O)OH + H_2O \rightleftharpoons HC(=O)O^- + H_3O^+$

b. $CH_3CH_2C(=O)OH + H_2O \rightleftharpoons CH_3CH_2C(=O)O^- + H_3O^+$

c. $CH_3C(=O)OH + H_2O \rightleftharpoons CH_3C(=O)O^- + H_3O^+$

14.15 **a.** $HC(=O)OH + NaOH \longrightarrow HC(=O)O^-Na^+ + H_2O$

b. $CH_3CH_2C(=O)OH + NaOH \longrightarrow$

$CH_3CH_2C(=O)O^-Na^+ + H_2O$

c. 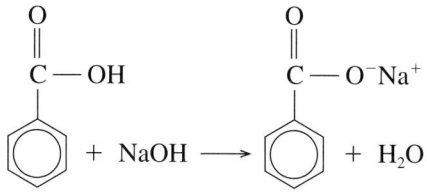

14.17 **a.** pentyl ethanoate (pentyl acetate)
b. octyl ethanoate (octyl acetate)
c. pentyl butanoate (pentyl butyrate)
d. isobutyl methanoate (isobutyl formate)

14.19 **a.** aldehyde **b.** ester
c. ketone **d.** carboxylic acid

14.21 **a.** methyl methanoate (methyl formate)
b. methyl ethanoate (methyl acetate)
c. methyl butanoate (methyl butyrate)
d. ethyl-3-methyl butanoate (ethyl-β-methyl butyrate)

14.23 **a.** $CH_3C(=O)-OCH_3$ **b.** $HC(=O)-CH_2CH_2CH_2CH_3$

c. $CH_3CH_2CH_2CH_2C(=O)-OCH_2CH_3$

d. $CH_3CH_2C(=O)-OCH_2CH(Br)CH_3$

14.25 **a.** methanoic acid (formic acid)
b. ethanoic acid (acetic acid)
c. butanoic acid (butyric acid)
d. 3-methylbutanoic acid (β-methylbutyric acid)

14.27 **a.** $CH_3C(=O)-OCH_3$ **b.** $CH_3CH_2CH_2C(=O)-OCH_3$

c. 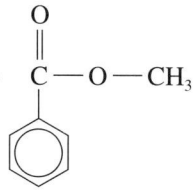

14.29 **a.** $CH_3C(=O)-O-CH_2CH_2CH_3$

b. $CH_3CH_2CH_2CH_2C(=O)-O-CH(CH_3)CH_3$

14.31 The products of the acid hydrolysis of an ester are an alcohol and a carboxylic acid.

14.33 **a.** $CH_3CH_2C(=O)O^-Na^+$ and CH_3OH

b. $CH_3C(=O)OH$ and $CH_3CH_2CH_2OH$

c. $CH_3CH_2CH_2C(=O)OH$ and CH_3CH_2OH

d. 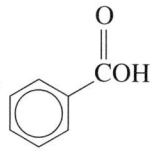 and CH_3CH_2OH

e. 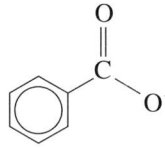 and CH_3CH_2OH

14.35 In a primary amine, there is one alkyl group (and two hydrogens) attached to a nitrogen atom.

14.37 **a.** primary **b.** tertiary
c. secondary **d.** primary

14.39 **a.** aminoethane (ethylamine)
b. *N*-methyl-1-aminopropane (methylpropylamine)
c. *N,N*-diethylaminoethane (triethylamine)
d. *N*-methylaniline

14.41 **a.** $CH_3CH_2NH_2$

b.

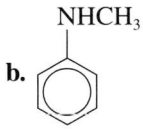

c. $CH_3CH_2CH_2CH_2-\overset{\overset{H}{|}}{N}-CH_2CH_2CH_3$

d. $CH_3\overset{\overset{NH_2}{|}}{C}HCH_2CH_2CH_3$

14.43 **a.** $CH_3NH_2 + H_2O \rightleftharpoons CH_3NH_3^+ + OH^-$

b. $CH_3-\overset{\overset{CH_3}{|}}{N}H + H_2O \rightleftharpoons CH_3-\overset{\overset{CH_3}{|}}{N}H_2^+ + OH^-$

c.

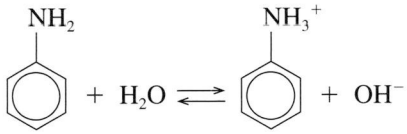

14.45 **a.** $CH_3NH_2 + HCl \longrightarrow CH_3NH_3^+Cl^-$

b. $CH_3\overset{\overset{CH_3}{|}}{N}H + HCl \longrightarrow CH_3\overset{\overset{CH_3}{|}}{N}H_2^+Cl^-$

c.

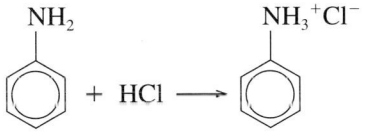

14.47 **a.** $CH_3\overset{\overset{O}{\|}}{C}-NH_2$

b. $CH_3\overset{\overset{O}{\|}}{C}-NHCH_2CH_3$

c.

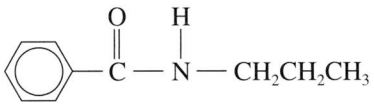

14.49 **a.** *N*-methylethanamide (*N*-methylacetamide)
b. butanamide (butyramide)
c. methanamide (formamide)
d. *N*-methylbenzamide

14.51 **a.** $CH_3CH_2\overset{\overset{O}{\|}}{C}-NH_2$

b. $CH_3CH_2CH_2\overset{\overset{CH_3}{|}}{C}H-\overset{\overset{O}{\|}}{C}-NH_2$

c. $H\overset{\overset{O}{\|}}{C}NH_2$

d.

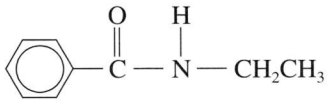

e. $CH_3CH_2CH_2\overset{\overset{O}{\|}}{C}-\overset{\overset{H}{|}}{N}-CH_2CH_3$

14.53 **a.** $CH_3\overset{\overset{O}{\|}}{C}OH + NH_4^+Cl^-$

b. $CH_3CH_2\overset{\overset{O}{\|}}{C}OH + NH_4^+Cl^-$

c. $CH_3CH_2CH_2\overset{\overset{O}{\|}}{C}OH + CH_3NH_3^+Cl^-$

d.

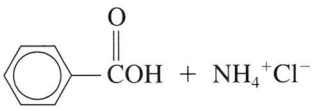

e. $CH_3CH_2CH_2CH_2\overset{\overset{O}{\|}}{C}OH + CH_3CH_2-NH_3^+Cl^-$

14.55

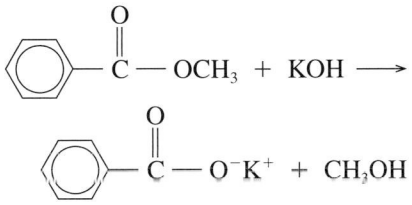

A soluble salt, potassium benzoate, is formed. When acid is added, the salt is converted to insoluble benzoic acid.

14.57 **a.** hydroxyl and carboxylic acid

b.

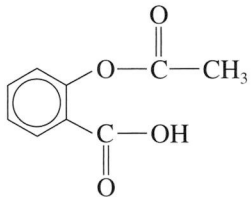

c.

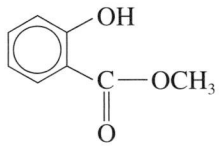

14.59

a. $K^+{}^-O\overset{\overset{O}{\|}}{C}CH_2CH_2\overset{\overset{O}{\|}}{C}O^-K^+ + 2CH_3OH$

b. $CH_3\overset{\overset{CH_3}{|}}{C}H-\overset{\overset{O}{\|}}{C}OH + CH_3\overset{\overset{CH_3}{|}}{C}H-OH$

14.61

a. $CH_3CH_2CH_2OH \xrightarrow{[O]}$

$$CH_3CH_2\overset{\overset{\displaystyle O}{\|}}{C}H \xrightarrow{[O]} CH_3CH_2\overset{\overset{\displaystyle O}{\|}}{C}OH$$

b. $CH_3CH_2CH_2OH \xrightarrow{[O]}$

$$CH_3CH_2\overset{\overset{\displaystyle O}{\|}}{C}H \xrightarrow{[O]} CH_3CH_2\overset{\overset{\displaystyle O}{\|}}{C}OH$$

$$CH_3CH_2\overset{\overset{\displaystyle O}{\|}}{C}OH + CH_3CH_2CH_2OH \xrightarrow{H^+}$$

$$CH_3CH_2\overset{\overset{\displaystyle O}{\|}}{C}-OCH_2CH_2CH_3$$

c. $CH_3CH_2CH_2OH \xrightarrow[\text{Heat}]{H^+}$

$$CH_3CH{=}CH_2 + H_2 \xrightarrow{Pt} CH_3CH_2CH_3$$

d. $CH_3CH_2CH_2OH \xrightarrow[\text{Heat}]{H^+}$

$$CH_3CH{=}CH_2 + H_2O \xrightarrow{H^+} CH_3\overset{\overset{\displaystyle OH}{|}}{C}HCH_3$$

e. $CH_3CH_2CH_2OH \xrightarrow[\text{Heat}]{H^+} CH_3CH{=}CH_2 + H_2O \xrightarrow{H^+}$

$$CH_3\overset{\overset{\displaystyle OH}{|}}{C}HCH_3 \xrightarrow{[O]} CH_3-\overset{\overset{\displaystyle O}{\|}}{C}-CH_3$$

f. $CH_3CH_2CH_2OH \xrightarrow{[O]} CH_3CH_2\overset{\overset{\displaystyle O}{\|}}{C}-H \xrightarrow{[O]}$

$$CH_3CH_2\overset{\overset{\displaystyle O}{\|}}{C}-OH + NH_3 \xrightarrow{\text{Heat}} CH_3CH_2\overset{\overset{\displaystyle O}{\|}}{C}-NH_2 + H_2O$$

14.63 carboxylate salt, aromatic, amine, haloaromatic

14.65 **a.** aromatic, amine, amide, carboxylic acid, cycloalkene
b. aromatic, ether, alcohol, amine
c. aromatic, carboxylic acid
d. phenol, amine, carboxylic acid
e. aromatic, ether, alcohol, amine, ketone
f. aromatic, amine

15

Lipids

The seeds of sunflowers yield an edible oil that is 90% unsaturated fat.

Looking Ahead

15.1	Types of Lipids
15.2	Fatty Acids
15.3	Waxes, Fats, and Oils
15.4	Chemical Properties of Triglycerides
15.5	Phospholipids and Glycolipids
15.6	Steroids and Cholesterol
15.7	Plasma (Cell) Membranes

www.chemplace.com/college/timberlake

Visit the web site above to test yourself, run simulations, and review key concepts from this chapter. For even more interactive review and quizzing, use the two CD-ROMs included with the book.

When we talk of fats, waxes, vegetable oils, steroids, or fat-soluble vitamins, we are discussing lipids. In the body, that ever-present adipose tissue known as body fat is a storage form of energy that provides more than twice as much energy (9 kcal/g) as carbohydrates (4 kcal/g). Fats also provide protection and insulation for internal organs.

Many people are concerned about the amounts of saturated fats and cholesterol in our diet. Researchers suggest that saturated fats and cholesterol are associated with diseases such as diabetes, cancers of the breast, pancreas, and colon, and arteriosclerosis, a condition in which deposits of lipid materials accumulate in the coronary blood vessels. These plaques restrict the flow of blood to the tissue, causing necrosis (death) of the tissue. In the heart, this could result in a *myocardial infarction* (heart attack).

The American Institute of Cancer Research has recommended that our diet contain more fiber and starch by adding more vegetables, fruits, and whole grains with moderate amounts of foods with low levels of fat and cholesterol such as fish, poultry, lean meats, and low-fat dairy products. They also suggest that we limit our intake of foods high in fat and cholesterol such as eggs, nuts, fatty or organ meats, cheeses, butter, and coconut and palm oil.

15.1

Study Pack

Chem of Life Lipids: Functions of Lipids 1

Learning Goal

Describe the classes of lipids.

15.1 Types of Lipids

Lipids are a family of biomolecules that have the common property of being soluble in organic solvents but not in water. The word "lipid" comes from the Greek word *lipos,* meaning "fat" or "lard." Typically, the lipid content of a plant cell can be extracted using an organic solvent such as ether, chloroform, or acetone. Lipids are an important feature in cell membranes, fat-soluble vitamins, and steroid hormones.

EXPLORE YOUR WORLD

Solubility of Fats and Oils

Place some water in a small bowl. Add a drop of a vegetable oil. Then add a few more drops of the oil. Now add a few drops of liquid soap and mix. Record your observations.

Place a small amount of fat such as margarine, butter, shortening, or vegetable oil on a dish or plate. Run water over it. Record your observations. Mix some soap with the fat substance and run water over it again. Record your observations.

Questions

1. Do the drops of oil in the water separate or do they come together? Explain.
2. How does the soap affect the oil layer?
3. Why don't the fats on the dish or plate wash off with water?
4. In general what is the solubility of lipids in water?
5. Why does soap help to wash the fats off the plate?

Within the lipid family, there are distinct structures that distinguish the different types of lipids, as shown in Table 15.1. Lipids such as waxes, fats, oils, and phospholipids are esters that can be hydrolyzed to give fatty acids along with other products including an alcohol. Sphingolipids contain an alcohol called sphingosine, and glycolipids contain a carbohydrate. Steroids are characterized by the steroid nucleus of four fused carbon rings. They do not contain fatty acids and cannot be hydrolyzed. Figure 15.1 illustrates the general structure of some classes of lipids.

Table 15.1 Classes of Lipid Molecules

Lipids	Composition
Waxes	Fatty acid and long-chain alcohol
Fats and oils (triglycerides)	Fatty acids and glycerol
Phospholipids	Fatty acids, glycerol, phosphate, amino alcohol
Sphingolipids	Fatty acids, sphingosine, phosphate, amino alcohol
Glycolipids	Fatty acids, glycerol or sphingosine, one or more monosaccharides
Steroids	A fused structure of three cyclohexanes and a cyclopentane

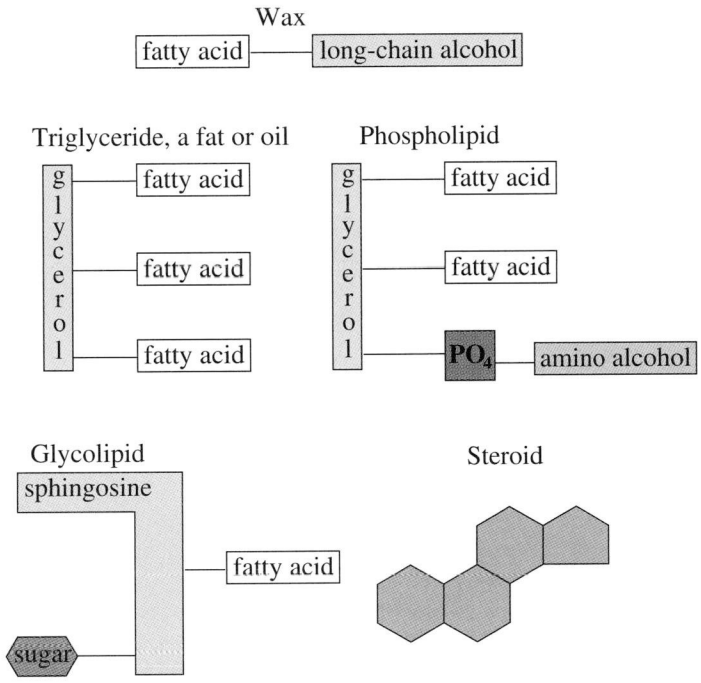

Figure 15.1 The general structure of some classes of lipids.

S·A·M·P·L·E **15.1** ·P·R·O·B·L·E·M

Classes of Lipids

What type of lipid does not contain fatty acids?

Solution

The steroids are a group of lipids with no fatty acids.

Study Check

What type of lipid contains a carbohydrate?

Q·U·E·S·T·I·O·N·S & P·R·O·B·L·E·M·S

Types of Lipids

15.1 What are some functions of lipids in the body?

15.2 What are some of the different kinds of lipids?

15.3 Lipids are not soluble in water. Are lipids polar or nonpolar molecules?

15.4 Which of the following solvents might be used to dissolve an oil stain?

a. water **b.** CCl_4
c. diethyl ether **d.** benzene
e. NaCl solution

15.2

Study Pack

15.2 Fatty Acids

Learning Goal

Identify a fatty acid as saturated or unsaturated.

Fatty acids are long-chain carboxylic acids that are insoluble in water. The fatty acids in fats and oils typically have an even number of carbon atoms, 12–18 carbon atoms being the most common in biological systems. An example is lauric acid, a 12-carbon acid found in coconut oil. The structural formula of a fatty acid may also be written in a zigzag form or in the condensed form as follows.

Some Formulas for Lauric Acid

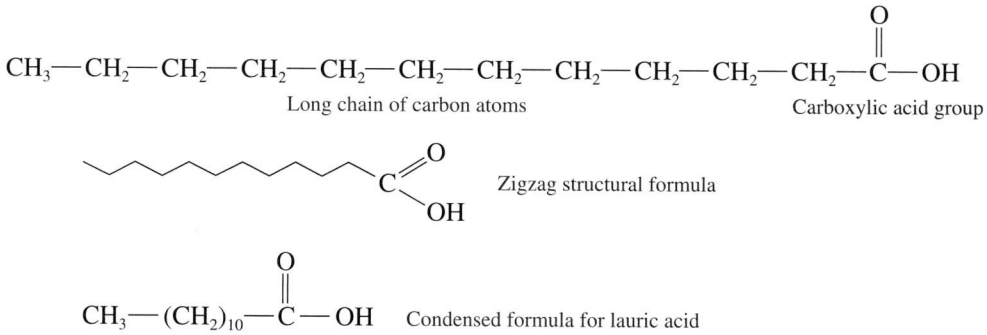

Saturated fatty acids such as lauric acid contain only single bonds between carbons. **Monounsaturated fatty acids** have one double bond in the carbon chain, and **polyunsaturated fatty acids** have two or more double bonds. Table 15.2 lists some of the typical fatty acids in lipids. An example of the structural, zigzag, and condensed formulas of oleic acid, a monounsaturated fatty acid found in olives and corn, is shown

Table 15.2 Properties of Common Fatty Acids

Common Name	Carbon Atoms	Structural Formula	Melting Point (°C)	Source
Saturated				
Lauric	12	$CH_3(CH_2)_{10}COOH$	43	Coconut
Myristic	14	$CH_3(CH_2)_{12}COOH$	54	Nutmeg
Palmitic	16	$CH_3(CH_2)_{14}COOH$	62	Palm
Stearic	18	$CH_3(CH_2)_{16}COOH$	69	Animal fat

Continued

Table 15.2 Properties of Common Fatty Acids—continued

Common Name	Carbon Atoms	Structural Formula	Melting Point (°C)	Source
Monounsaturated				
Palmitoleic	16	$CH_3(CH_2)_5CH{=}CH(CH_2)_7COOH$	0	Butter
Oleic	18	$CH_3(CH_2)_7CH{=}CH(CH_2)_7COOH$	13	Olives, corn
Polyunsaturated				
Linoleic	18	$CH_3(CH_2)_4CH{=}CHCH_2CH{=}CH(CH_2)_7COOH$	−9	Soybean, safflower, sunflower
Linolenic	18	$CH_3CH_2CH{=}CHCH_2CH{=}CHCH_2CH{=}CH(CH_2)_7COOH$	−17	Corn
Arachidonic	20	$CH_3(CH_2)_4(CH{=}CHCH_2)_4\,(CH_2)_2COOH$	−50	

15.2

next. In the complete structure, the bend in the carbon chain is caused by the double bond that has a cis configuration. The double bonds in most unsaturated fatty acids are cis, causing one or more bends in the carbon chain.

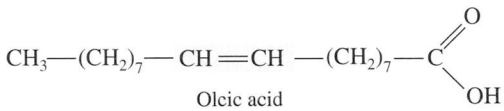

Olcic acid

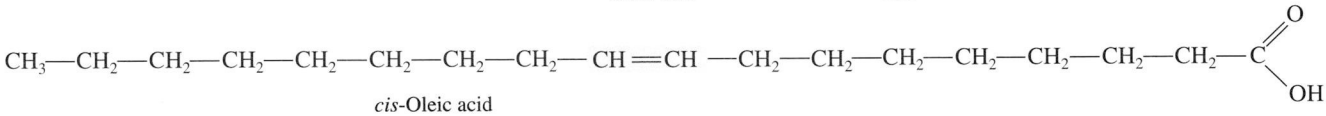

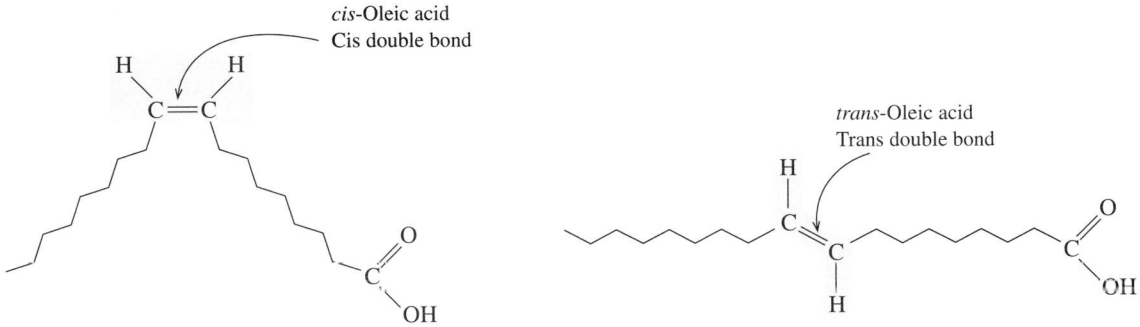

The human body is capable of synthesizing most fatty acids from carbohydrates or other fatty acids. However, humans and other mammals do not synthesize sufficient amounts of fatty acids that have more than one double bond, such as linoleic and linolenic acid. These fatty acids are called *essential* fatty acids because they must be provided by the diet. A deficiency of essential fatty acids can cause skin dermatitis in infants. However, the role of fatty acids in adult nutrition is not well understood. Adults do not usually have a deficiency of essential fatty acids.

Physical Properties of Fatty Acids

The uniform structure of the carbon chain in saturated fatty acids allows the molecules of saturated fatty acids to fit close together in a regular pattern. The close pattern of the saturated fatty acids allows strong attractions to occur between the carbon chains. (See Figure 15.2.) As a result, a significant amount of energy and high temperatures are required to separate the fatty acids and melt the fat. As the length of the carbon chain increases, more interactions occur between the carbon chains, requiring higher melting points. Saturated fatty acids are usually solids at room temperature.

In unsaturated fatty acids, the cis double bonds cause the carbon chain to bend, which gives the molecules an irregular shape. As a result, the molecules of an unsaturated fatty acid cannot stack together in a regular pattern, so fewer interactions occur between carbon chains. Consequently, the melting points of unsaturated fats are lower than those of saturated fats. Most unsaturated fats are liquids at room temperature.

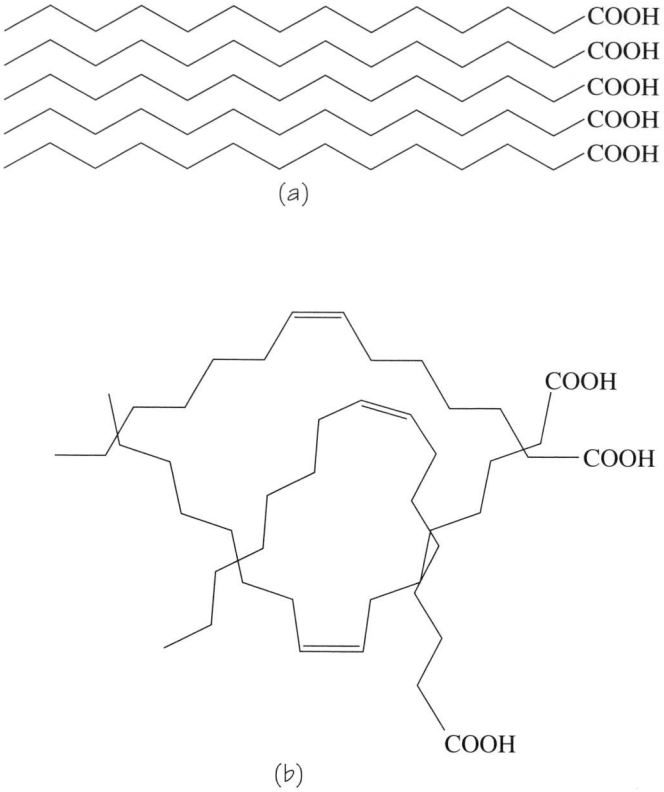

Figure 15.2 **(a)** A saturated fatty acid in which molecules fit closely together has a high melting point. **(b)** An unsaturated fatty acid in which molecules are further apart has a low melting point.

HEALTH NOTE

Omega-3 Fatty Acids in Fish Oils

Over the past several decades, Americans have been changing their diets to include more polyunsaturated fats and fewer saturated fats. This change is a response to research that indicates that atherosclerosis and heart disease are associated with high levels of saturated fats in the diet. However, this association does not seem to be correct for the Inuit people of Alaska, who have a high-fat diet and high levels of blood cholesterol, but a very low occurrence of atherosclerosis and heart attacks. The fats in the Inuit diet are primarily from fish rather than from land animals, as in many other people's diets.

Both fish and vegetable oils have high levels of polyunsaturated fats. The fatty acids in vegetable oils are omega-6 acids, which means that the first double bond occurs at carbon 6 counting from the methyl group. Two common omega-6 acids are linoleic acid and arachidonic acid. However, the fatty acids in the fish oils are mostly the omega-3 type, in which the first double bond occurs at the third carbon counting from the methyl group. Three common omega-3 fatty acids in fish are linolenic acid, eicosapentaenoic acid (EPA), and docosahexaenoic acid (DHA).

In atherosclerosis and heart disease, cholesterol forms plaques that adhere to the walls of the blood vessels. Blood pressure rises as blood has to squeeze through a smaller opening in the blood vessel. As more plaques form, there is also a possibility of blood clots blocking the blood vessels and causing a heart attack. Omega-3 fatty acids lower the tendency of blood platelets to stick together, thereby reducing the possibility of blood clots. However, high levels of omega-3 fatty acids can increase bleeding time if the ability of the platelets to form blood clots is reduced too much. It does seem that a diet that includes fish such as salmon, tuna, and herring can provide higher amounts of the omega-3 fatty acids, which help lessen the possibility of developing heart disease.

Linolenic acid, 18 carbon atoms

$$\underset{1}{CH_3}\underset{2}{CH_2}\underset{3}{CH}{=}CHCH_2CH{=}CHCH_2CH{=}CH(CH_2)_7COOH$$

$$CH_3CH_2(CH{=}CHCH_2)_5(CH_2)_2COOH$$ EPA, 20 carbon atoms

$$CH_3CH_2(CH{=}CHCH_2)_6CH_2COOH$$ DHA, 22 carbon atoms

HEALTH NOTE

Prostaglandins

Prostaglandins are hormone-like substances found in very low amounts in most tissues and fluids of the body. The parent structure, prostanoic acid, which has 20 carbon atoms, is formed from an unsaturated fatty acid called arachidonic acid. Small changes in the structure of prostanoic acid produce the different prostaglandins.

Prostaglandins produced in the tissues enter the blood, where they regulate several physiological activities. Some increase blood pressure as vasoconstrictors, and others lower blood pressure as vasodilators. There are also prostaglandins that stimulate contraction and relaxation of smooth muscle. Still others play a role in reproduction, respiration, fat metabolism, and nerve impulse transmission. Some also cause inflammation, pain, and fever. These effects are inhibited by anti-inflammatory agents such as aspirin and acetaminophen, which act by preventing the formation of the prostaglandins in the tissues.

The prostaglandins PGE_2 and $PGF_{2\alpha}$ are used medically to treat hypertension, to increase uterine contractions, and to relieve bronchial asthma.

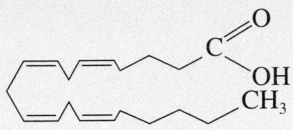

Arachidonic acid

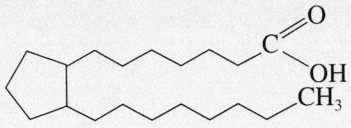

Prostanoic acid

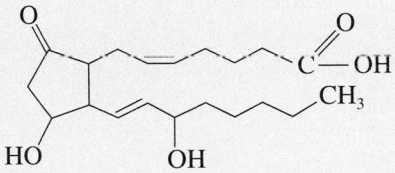

PGE_2

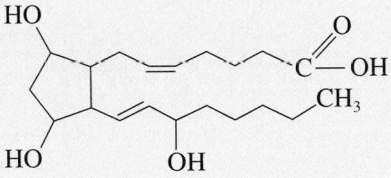

$PGF_{2\alpha}$

S·A·M·P·L·E 15.2 ·P·R·O·B·L·E·M

Structures and Properties of Fatty Acids

Consider the structural formula of oleic acid.

$$CH_3(CH_2)_7CH{=}CH(CH_2)_7\overset{\overset{\displaystyle O}{\|}}{C}OH$$

a. Why is the substance called an acid?
b. How many carbon atoms are in oleic acid?
c. Is it a saturated or unsaturated fatty acid?
d. Is it most likely to be solid or liquid at room temperature?
e. Would it be soluble in water?

Solution

a Oleic acid contains a carboxylic acid group.
b. It contains 18 carbon atoms.
c. It is an unsaturated fatty acid.
d. It is liquid at room temperature.
e. No, its long hydrocarbon chain makes it insoluble in water.

Study Check

Palmitoleic acid is a fatty acid with the following formula:

$$CH_3(CH_2)_5CH{=}CH(CH_2)_7\overset{\overset{\displaystyle O}{\|}}{C}OH$$

a. How many carbon atoms are in palmitoleic acid?
b. Is it a saturated or unsaturated fatty acid?
c. Is it most likely to be solid or liquid at room temperature?

15.3

Q·U·E·S·T·I·O·N·S & P·R·O·B·L·E·M·S

Fatty Acids

15.5 Describe some similarities and differences in the structures of a saturated fatty acid and an unsaturated fatty acid.

15.6 Stearic acid and linoleic acid both have 18 carbon atoms. Why does stearic acid melt at 69°C, but linoleic acid melts at −9°C?

15.7 Write the structure of the following fatty acids:
a. palmitic acid **b.** oleic acid

15.8 Write the structure of the following fatty acids:
a. stearic acid **b.** linoleic acid

15.9 Which of the following fatty acids are saturated, and which are unsaturated?
a. lauric acid **b.** linolenic acid
c. palmitoleic acid **d.** stearic acid

15.10 Which of the following fatty acids are saturated, and which are unsaturated?
a. linolenic acid **b.** palmitic acid
c. myristic acid **d.** oleic acid

15.11 How does the structure of a fatty acid with a cis double bond differ from the structure of a fatty acid with a trans double bond?

15.12 In each pair, identify the fatty acid with the lower melting point. Explain.
a. myristic acid and stearic acid
b. stearic acid and linoleic acid
c. oleic acid and linolenic acid

15.13 Describe the position of the first double bond in an omega-3 and an omega-6 fatty acid.

15.14 **a.** What are some sources of omega-3 and omega-6 fatty acids?
b. How may omega-3 fatty acids help in lowering the risk of heart disease?

Study Pack

Chem of Life Lipids: Introduction to Fats 1–6

Learning Goal

Write the structural formula of a wax, fat, or oil produced by the reaction of a fatty acid and an alcohol or glycerol.

15.3 Waxes, Fats, and Oils

Waxes are found in many plants and animals. Wax coatings on fruits and the leaves and stems of plants help to prevent loss of water and damage from pests. Waxes on the skin, fur, and feathers of animals and birds provide a waterproof coating. A **wax** is an ester of a saturated fatty acid and a long-chain alcohol, each containing from 14 to 30 carbon atoms.

Ester bond
↓
fatty acid — long-chain alcohol
Wax

Table 15.3 Some Typical Waxes

Type	Structural Formula	Source	Uses
Beeswax	$CH_3(CH_2)_{14}-C(=O)-O-(CH_2)_{29}CH_3$	Honeycomb	Candles, shoe polish, wax paper
Carnauba wax	$CH_3(CH_2)_{24}-C(=O)-O-(CH_2)_{29}CH_3$	Brazilian palm tree	Waxes for furniture, cars, floors, shoes
Jojoba wax	$CH_3(CH_2)_{18}-C(=O)-O-(CH_2)_{19}CH_3$	Jojoba	Candles, soaps, cosmetics

15.3

The formulas of some common waxes are given in Table 15.3. Beeswax obtained from honeycombs, and carnauba wax obtained from palm trees, are used to give a protective coating to furniture, cars, and floors. (See Figure 15.3.) Jojoba wax is used in making candles and cosmetics such as lipstick. Lanolin, a mixture of waxes obtained from wool, is used in hand and facial lotions to aid retention of water, which softens the skin.

Figure 15.3 Honeycomb is made of beeswax, which is composed of esters of long-chain fatty acids and long-chain alcohols.

Fats and Oils

Fats and *oils* are the most prevalent forms of lipids. Also known as **triglycerides** or *triacylglycerols,* fats and oils are esters of glycerol and three fatty acids. Glycerol is a trihydroxy alcohol that can form ester links with carboxyl groups of up to three fatty acids.

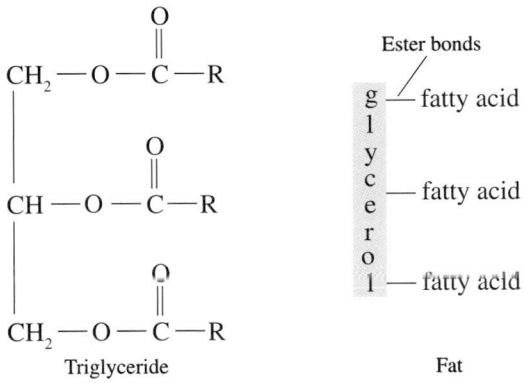

For example, three stearic acid molecules form ester bonds with glycerol to form the triglyceride tristearin.

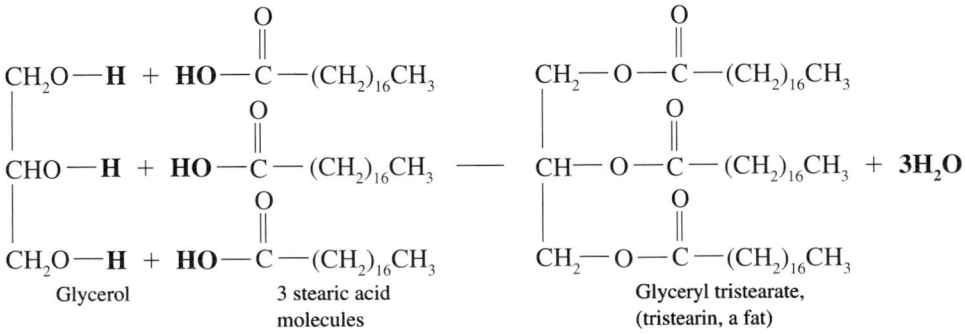

15.3

However, most fats and oils are **mixed triglycerides** that contain two or three different fatty acids. For example, a mixed triglyceride might be made from lauric acid, myristic acid, and palmitic acid. One possible structure for the mixed triglyceride follows:

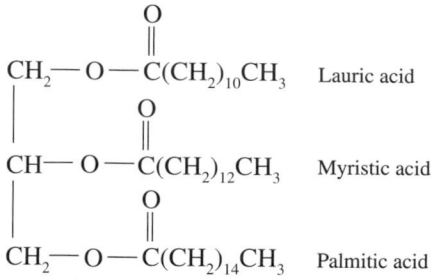

A mixed triglyceride

S·A·M·P·L·E 15.3 ·P·R·O·B·L·E·M

Writing Structures for Triglycerides

Draw the structural formula of triolein, a simple triglyceride.

Solution

Triolein is the triglyceride of glycerol and three oleic acid molecules. Each fatty acid is attached by an ester bond to one of the hydroxyl groups in glycerol.

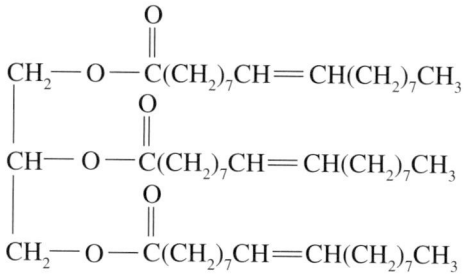

Glyceryl trioleate (triolein)

Study Check

Write the structure of trimyristin.

Melting Points of Fats and Oils

A **fat** is a triglyceride that is solid at room temperature, such as fats in meat, whole milk, butter, and cheese. Most fats come from animal sources.

An **oil** is a triglyceride, and most are liquid at room temperature. The most commonly used oils come from plant sources. Olive oil and peanut oil are called monounsaturated because they contain large amounts of oleic acid. Oils from corn, cottonseed, safflower, and sunflower are polyunsaturated because they contain large amounts of fatty acids with two or more double bonds. (See Figure 15.4.) However, a few oils such as palm oil and coconut oil are solid at room temperature because they consist mostly of saturated fatty acids.

The amounts of saturated, monounsaturated, and polyunsaturated fatty acids in some typical fats and oils are shown in Figure 15.5. Saturated fatty acids have higher melting points than unsaturated fatty acids. Animal fats usually contain more saturated fatty acids than do vegetable oils. Therefore the melting points of animal fats are higher than those of vegetable oils.

Figure 15.4 Vegetable oils such as olive oil, corn oil, and safflower oil contain unsaturated fats.

Figure 15.5 The melting points and polyunsaturation in some fats and oils.

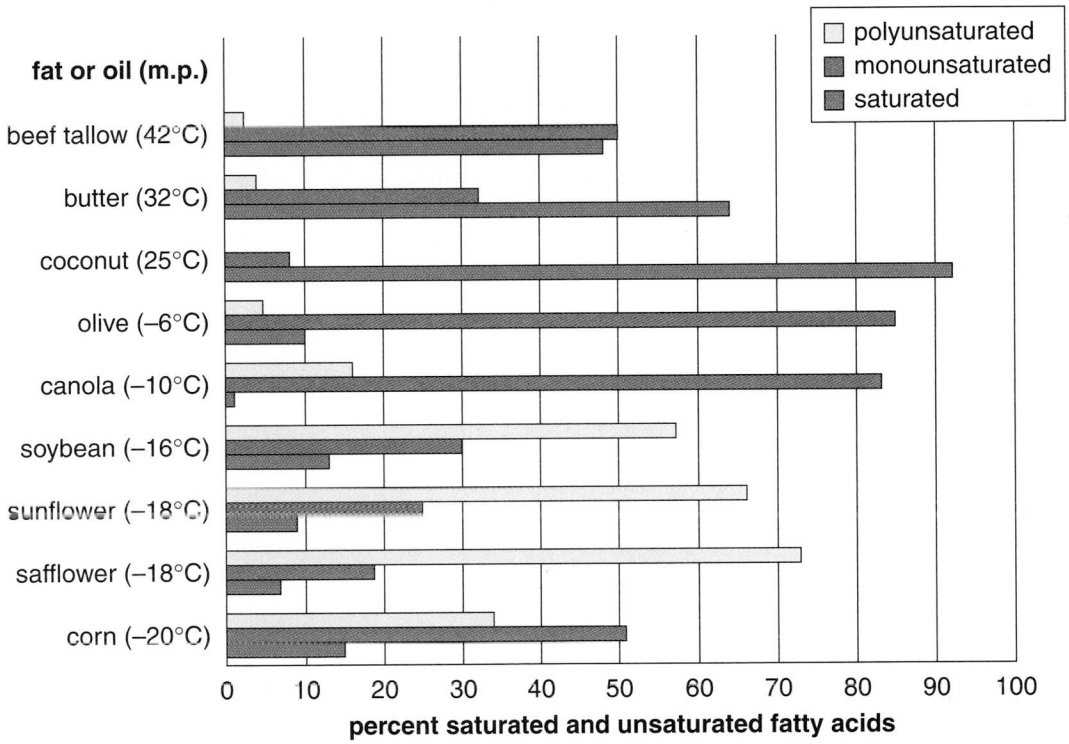

EXPLORE YOUR WORLD

Fats and Oils

1. Place some nuts such as walnuts, pecans, peanuts, or almonds in a plastic bag or on a piece of paper that is folded over. Carefully pound the nuts with a hammer or rolling pin. The oil from the nuts should appear as a liquid as it is pressed out of the nuts. This is an example of how peanut oil or walnut oil is obtained.
2. In mayonnaise, egg yolks are used to mix the oil and vinegar together so they won't separate. This process is called emulsification. The lipids in egg yolk coat the oil particles, which prevents them from coming back together in a separate layer. To make mayonnaise, place 1 egg yolk, $\frac{1}{4}$ teaspoon salt (if desired), and $\frac{1}{2}$ teaspoon of vinegar in a bowl and mix with an electric beater. Add $\frac{1}{2}$ cup of oil, slowly at first, drop by drop, to the mixture, and faster as the mixture thickens. This allows small amounts of oil to be coated by the emulsifying

agent in the egg yolk. Add a teaspoon of vinegar. The product should be yellow and shiny with no separate layers. You can use it, but because it has no preservatives it may spoil quickly.

3. Read the nutrition label on some packaged food such as cereals, potato chips, and milk.

Questions

1. How would you obtain oil from olives?
2. In preparing mayonnaise, a uniform mixture of oil and vinegar is prepared. How is this achieved when oil is not soluble in water?
3. For each packaged food, determine the following:
   a. How many grams of fat are in one serving of the food?
   b. Using the caloric value for fat (9 kcal/gram of fat), how many calories (kilocalories) come from the fat in one serving?
   c. What is the percentage of fat in one serving?

15.3

HEALTH NOTE

Olestra: A Fat Substitute

In 1968, food scientists designed an artificial fat called *olestra* as a source of nutrition for premature babies. However, olestra could not be digested and was never used for that purpose. Then scientists realized that olestra had the flavor and texture of a fat without the calories.

Olestra is manufactured by obtaining the fatty acids from the fats in cottonseed or soybean oils and bonding the fatty acids with the hydroxyl groups on sucrose. Chemically, olestra is composed of six to eight long-chain fatty acids attached by ester links to a sugar (sucrose) rather than to a glycerol molecule found in fats. This makes olestra a very large molecule, which cannot be absorbed through the intestinal walls. The enzymes and bacteria in the intestinal tract are unable to break down the olestra molecule and it travels through the intestinal tract undigested.

In 1996 the Food and Drug Administration approved olestra for use in potato chips, tortilla chips, crackers and fried snacks. In 1996, olestra snack products were test marketed in parts of Iowa, Wisconsin, Indiana, and Ohio. By 1997, there were reports of some adverse reactions, including diarrhea, abdominal cramps, and anal leakage, indicating that olestra may act as a laxative in some people. However, the manufacturers contend there is no direct proof that olestra is the cause of those effects.

The large molecule of olestra also combines with fat-soluble vitamins (A, D, E, and K) as well as the carotenoids from the foods we eat before they can be absorbed through the intestinal wall. Carotenoids are plant pigments in fruits and vegetables that protect against cancer, heart disease, and macular degeneration, a form of blindness in the elderly. The FDA now requires manufacturers to add the four vitamins, but not the carotenoids, to olestra products. The label on an olestra product must state the following: "This product contains olestra. Olestra may cause abdominal cramping and loose stools. Olestra inhibits the absorption of some vitamins and other nutrients. Vitamins A, D, E, and K have been added." Some snack foods made with olestra are now in supermarkets nationwide. Since there are already low-fat snacks on the market, it remains to be seen whether olestra will have any significant effect on reducing the problem of obesity.

QUESTIONS & PROBLEMS

Waxes, Fats, and Oils

15.15 Draw the structure of a component of beeswax that hydrolyzes to yield myricyl alcohol, $CH_3(CH_2)_{29}OH$, and palmitic acid.

15.16 Draw the structure of a component of jojoba wax that hydrolyzes to yield arachidic acid, a 20-carbon saturated fatty acid, and 1-docosanol, $CH_3(CH_2)_{21}OH$.

15.17 Draw the structure of a triglyceride that forms stearic acid and glycerol upon hydrolysis.

15.18 A mixed triglyceride contains two palmitic acid molecules to every one oleic acid molecule. Write two possible structures (isomers) for the compound.

15.19 Draw the structure of tripalmitin.

15.20 Draw the structure of triolein.

15.21 Safflower oil is called a polyunsaturated oil, whereas olive oil is a monounsaturated oil. Explain.

15.22 Why does olive oil have a lower melting point than butter fat?

15.23 Why does coconut oil, a vegetable oil, have a melting point similar to fats from animal sources?

15.24 A label on a bottle of 100% sunflower seed oil states that it is lower in saturated fats than all the leading oils.

a. How does the percentage of saturated fats in sunflower seed oil compare to that of safflower, corn, and canola oil? (See Figure 15.5.)

b. Is the claim valid?

15.4 Chemical Properties of Triglycerides

Study Pack

Learning Goal

Draw the structure of the product from the reaction of a triglyceride with hydrogen, an acid or base, or an oxidizing agent.

The chemical reactions of the triglycerides (fats and oils) are the same as we discussed for alkenes, carboxylic acids, and esters. We will look at the addition of hydrogen to the double bonds, the oxidation of double bonds, and the acid and base hydrolysis of the ester bonds of fats and oils.

Hydrogenation

The **hydrogenation** of unsaturated fats converts carbon–carbon double bonds to single bonds. The hydrogen gas is bubbled through the heated oil in the presence of a nickel catalyst.

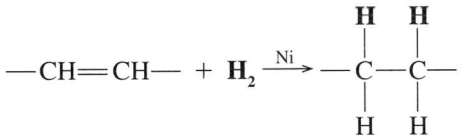

For example, when hydrogen adds to the double bonds of triolein, the product is the saturated fat tristearin.

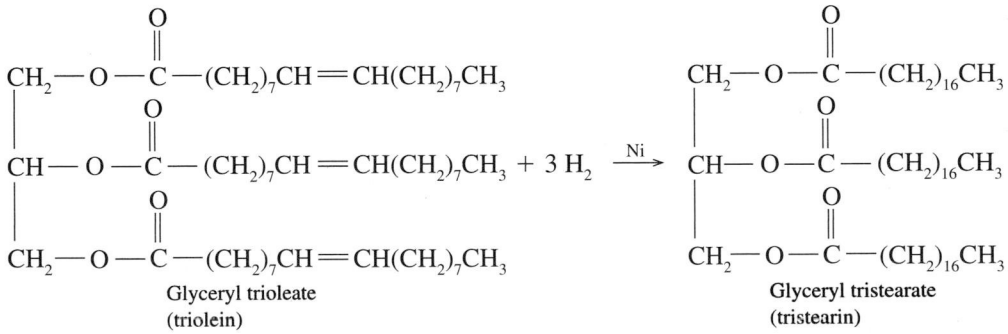

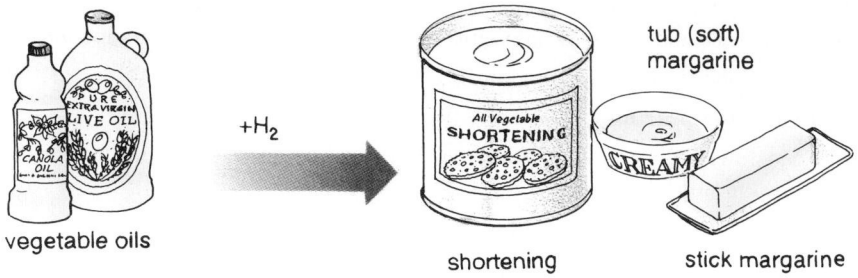

Figure 15.6 Soft margarines, stick margarines, and solid shortenings are produced by hydrogenating vegetable oils.

15.4

In commercial hydrogenation, the addition of hydrogen is stopped before all the double bonds in an oil become completely saturated. Complete hydrogenation gives a very brittle product, whereas the partial hydrogenation of a liquid vegetable oil changes it to a soft, semisolid fat. As the oil becomes more saturated, the melting point increases. Control of the degree of hydrogenation gives the various types of partially hydrogenated vegetable oil products on the market today—soft margarines, solid stick margarines, and solid shortenings. (See Figure 15.6.) Although these products now contain more saturated fatty acids than the original oils, they contain no cholesterol, unlike similar products from animal sources, such as butter and lard.

Figure 15.7 After strenuous exercise, increased body temperature and perspiration promote the oxidation of oils that accumulate on the surface of the skin.

Oxidation

A fat or oil becomes rancid when its double bonds are oxidized in the presence of oxygen and microorganisms. The products are short-chain fatty acids and aldehydes that have very disagreeable odors.

$$\underset{\text{Unsaturated fatty acids}}{-CH{=}CH-} \xrightarrow{[O]} \underset{\text{Short-chain aldehydes}}{-\overset{O}{\overset{\|}{C}}-H + H-\overset{O}{\overset{\|}{C}}-} \xrightarrow{[O]} \underset{\text{Short-chain carboxylic acids}}{-\overset{O}{\overset{\|}{C}}-OH + HO-\overset{O}{\overset{\|}{C}}-}$$

If a vegetable oil does not contain an antioxidant, it will oxidize rather easily. You can detect an oil that has become rancid by its unpleasant odor. If an oil is covered tightly and stored in a refrigerator, the process of oxidation can be slowed down and the oil will last longer.

Oxidation also occurs in the oils that accumulate on the surface of the skin during heavy exercise. (See Figure 15.7.) At body temperature, microorganisms on the skin promote rapid oxidation of the oils as they are exposed to oxygen and water. The resulting short-chain aldehydes and fatty acids account for the body odor associated with a workout and heavy perspiration.

EXPLORE YOUR WORLD

Types of Fats

Read the labels on food products that contain fats such as butter, margarine, vegetable oils, peanut butter, and potato chips. Look for terms such as saturated, monounsaturated, polyunsaturated, and partially or fully hydrogenated.

Questions

1. How many grams of saturated, monounsaturated, and polyunsaturated fat are in one serving of the product?
2. What type(s) of fats or oils are in the product?
3. What percent of the total fat is saturated fat? Unsaturated fat?
4. If the product is a vegetable oil, what information is given about how to store it? Why?
5. The label on a container of peanut butter states that the cottonseed and canola oils used to make the peanut butter have been fully hydrogenated. What are the typical products that would form when hydrogen is added?

HEALTH NOTE

Trans Fatty Acids and Hydrogenation

In the early 1900s, margarine became a popular replacement for the highly saturated fats such as butter and lard. Margarine is produced by partially hydrogenating the unsaturated fats in vegetable oils such as safflower oil, corn oil, canola oil, cottonseed oil, and sunflower oil. Fats that are more saturated are more resistant to oxidation.

In vegetable oils, the unsaturated fats usually contain cis double bonds. As hydrogenation occurs, double bonds are converted to single bonds. However, some of the cis double bonds are converted to trans double bonds, which causes a change in the overall structure of the fatty acids. If the label on a product states that the oils have been "partially" or "fully hydrogenated," that product will also contain trans fatty acids. In the United States, it is estimated that 2–4% of our total calories is in the form of trans fatty acids.

The concern about trans fatty acids is that their altered structure may make them behave like saturated fatty acids in the body. In the 1980s, research indicated that trans fatty acids have an effect on blood cholesterol similar to that of saturated fats, although study results vary. Several studies reported that trans fatty acids raise the levels of LDL-cholesterol, low-density lipoproteins containing cholesterol that can accumulate in the arteries. (LDLs and HDLs are described in the Health Note on lipoproteins later in the chapter.) Some studies also report that trans fatty acids lower HDL-cholesterol, high-density lipoproteins that carry cholesterol to the liver to be excreted. But other studies did not report any decrease in HDL-cholesterol. In some American and European studies, an increased risk of breast cancer was associated with increased intake of trans fatty acids. However, these studies are not conclusive and not all studies have supported such findings. Current evidence does not yet indicate that the intake of trans fatty acids is a significant risk factor for heart disease. The trans fatty acids controversy will continue to be debated as more research is done.

Foods containing trans fatty acids include milk, bread, fried foods, ground beef, baked goods, stick margarine, butter, soft margarine, cookies, crackers, and vegetable shortening. The American Heart Association recommends that margarine should have no more than 2 grams of saturated fat per tablespoon and a liquid vegetable oil should be the first ingredient. They also recommend the use of soft margarine, which is lower in trans fatty acids because soft margarine is only slightly hydrogenated, and diet margarine because it has less fat and therefore fewer trans fatty acids.

Many health organizations agree that fat should account for less than 30% of daily calories (the current average for Americans is 34%) and saturated fat should be less than 10% of total calories. Lowering the overall fat intake would also decrease the amount of trans fatty acids. The Food and Drug Administration and the U.S. Department of Agriculture are encouraging the use of new food labels to inform consumers of the fat content of food. The best advice may be to reduce total fat in the diet by using fats and oils sparingly, cooking with little or no fat, substituting olive oil or canola oil for other oils, and limiting the use of coconut oil and palm oil, which are high in saturated fatty acids.

Hydrolysis

Triglycerides are hydrolyzed (split by water) in the presence of strong acids or digestive enzymes called *lipases*. The products of hydrolysis of the ester bonds are glycerol and three fatty acids. The polar glycerol is soluble in water, but the fatty acids with their long hydrocarbon chains are not.

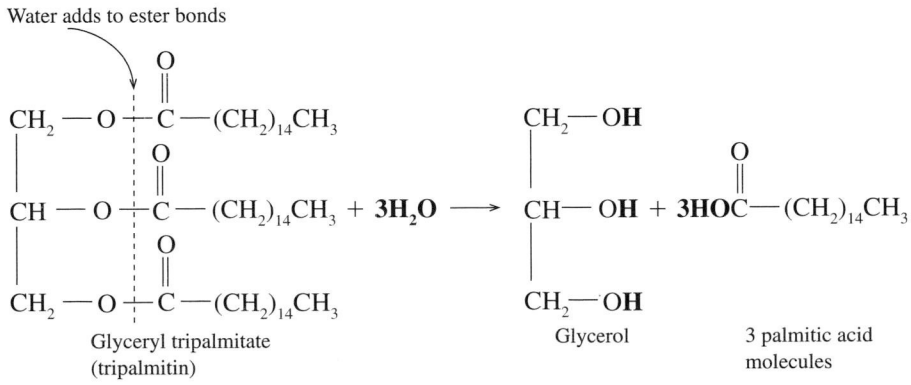

Saponification

When a fat is heated with a strong base such as sodium hydroxide, saponification of the fat gives glycerol and the sodium salts of the fatty acids, which are soaps. When NaOH is used, a solid soap is produced that can be molded into a desired shape; KOH produces a softer, liquid soap. Oils that are polyunsaturated produce softer soaps. Names like "coconut" or "avocado shampoo" tell you the sources of the oil used in the hydrolysis reaction.

Fat or oil + strong base ⟶ glycerol + salts of fatty acids (soaps)

15.4

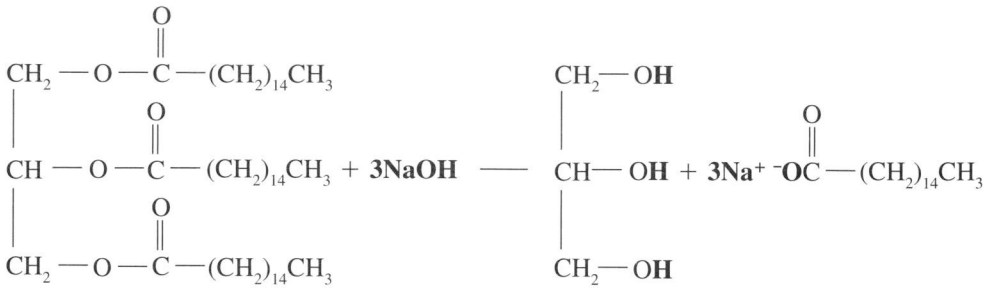

S·A·M·P·L·E· 15.4 ·P·R·O·B·L·E·M

Reactions of Lipids

Write the equation for the reaction catalyzed by the enzyme lipase that hydrolyzes trilaurin during the digestion process.

Solution

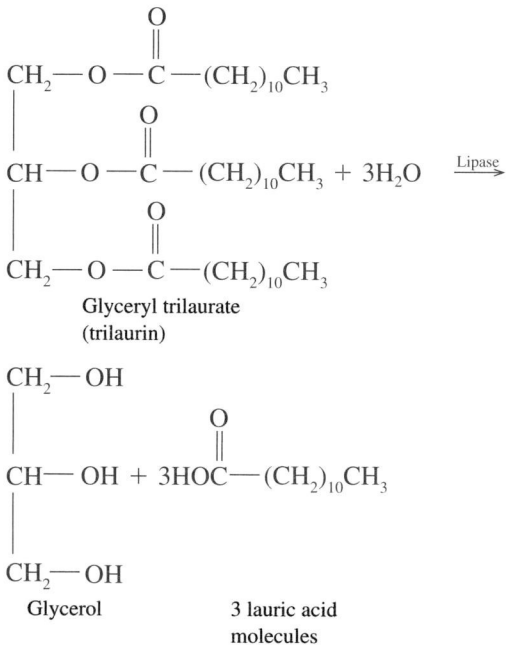

Study Check

What is the name of the product formed when a triglyceride containing oleic acid and linoleic acid is completely hydrogenated?

ENVIRONMENTAL NOTE

Cleaning Action of Soaps

For many centuries, soaps were made by heating a mixture of animal fats (tallow) with lye, a basic solution obtained from wood ashes. Today soaps are also prepared from fats such as coconut oil. Perfumes are added to give a pleasant-smelling soap. Because a soap is the salt of a long-chain fatty acid, the two ends of a soap molecule have different polarities. The long carbon chain end is nonpolar and *hydrophobic* (water-fearing). It is soluble in nonpolar substances such as oil or grease; but it is not soluble in water. The carboxylate salt end is ionic and *hydrophilic* (water-loving). It is very soluble in water but not in oils or grease. (See Figure 15.8.)

Figure 15.8 Soaps are sodium or potassium salts of long-chain carboxylic acids.

15.4

The Dual Polarity of a Soap (Salt of a Fatty Acid)

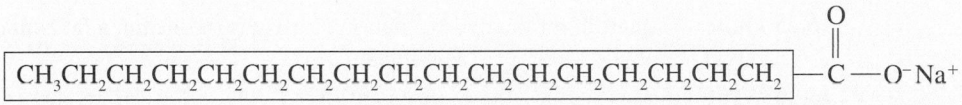

Nonpolar tail (hydrophobic) — Polar head (hydrophilic)

When a soap is used to clean grease or oil, the nonpolar ends of the soap molecules dissolve in nonpolar fats and oils that accompany dirt. The water-loving salt ends of the soap molecules extend outside where they can dissolve in water. The soap molecules coat the oil or grease, forming clusters called *micelles*. The ionic ends of the soap molecules provide polarity to the micelles, which makes them soluble in water. As a result, small globules of oil and fat coated with soap molecules are pulled into the water layer and rinsed away. (See Figure 15.9.)

One of the problems of using soaps is that the carboxylate end reacts with ions in water such as Ca^{2+} and Mg^{2+} and forms insoluble substances.

$$2CH_3(CH_2)_{16}COO^- + Mg^{2+} \longrightarrow$$

Stearate ion — Magnesium ion

$$[CH_3(CH_2)_{16}COO^-]_2\,Mg^{2+}$$

Magnesium stearate (insoluble)

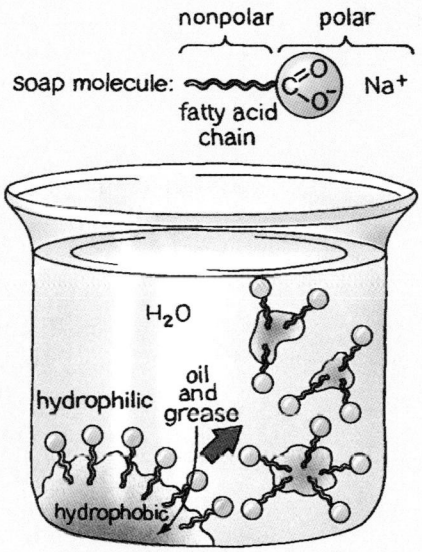

Figure 15.9 The cleaning action of soap. The nonpolar portion of the soap molecule dissolves in the grease and oil accompanying dirt on clothing and dishes. The polar ends of the soap molecules are attracted to water and pull the grease or oil into the aqueous solution.

These insoluble products produce a dull coating or scum on fabrics and dishes. To solve this problem, detergents that do not form insoluble products with ions in water were developed. One type of detergent found in shampoos consists of long nonpolar hydrocarbon chains from fatty acids such as lauric acid attached to sulfate. The calcium and magnesium salts of these detergents are soluble in water. They are also biodegradable, which means they break down into nonpolluting products in the environment as do soaps.

Sodium lauryl sulfate, a detergent

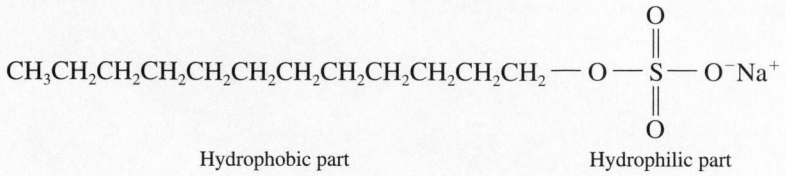

15.5

Q·U·E·S·T·I·O·N·S & P·R·O·B·L·E·M·S

Chemical Properties of Triglycerides

15.25 Write an equation for the hydrogenation of glyceryl trioleate, a fat containing glycerol and three oleic acid units.

15.26 Write an equation for the hydrogenation of glyceryl trilinolenate, a fat containing glycerol and three linolenic acid units.

15.27 A label on a container of margarine states that it contains partially hydrogenated corn oil.
a. How has the liquid corn oil been changed?
b. Why is the margarine product solid?

15.28 Why should a bottle of vegetable oil that has no preservatives be tightly covered and refrigerated?

15.29 **a.** Write an equation for the acid hydrolysis of glyceryl trimyristate (trimyristin).
b. Write an equation for the NaOH saponification of glyceryl trimyristate (trimyristin).

15.30 **a.** Write an equation for the acid hydrolysis of glyceryl trioleate (triolein).
b. Write an equation for the NaOH saponification of glyceryl trioleate (triolein).

15.31 Compare the structure of a triglyceride to the structure of olestra.

15.32 An oil is partially hydrogenated.
a. Are all or just some of the double bonds converted to single bonds?
b. What happens to many of the cis double bonds during hydrogenation?
c. How can you reduce the amount of trans fatty acids in your diet?

Study Pack

Chem of Life Lipids: Phospholipids 1–3, 5

Learning Goal

Describe the components of the phosphoglycerides, sphingolipids, and glycolipids.

15.5 Phospholipids and Glycolipids

The phospholipids are a family of lipids containing glycerol, fatty acids, and a phosphate group. Unlike most other lipids, phospholipids have polar and nonpolar regions, which enables them to play a role in the transport of less soluble lipids through the bloodstream. They are also major components in the structure of cell membranes.

Phosphoglycerides

Phosphoglycerides are the most abundant lipids in cell membranes, where they play an important role in cellular permeability. They make up much of the myelin sheath that protects nerve cells. In the body fluids, they combine with the less polar triglycerides and cholesterol to make them more soluble as they are transported in the body.

Like the triglycerides, **phosphoglycerides** contain glycerol and fatty acids, both saturated and unsaturated. They also contain a phosphate group and an amino alcohol that ionize. The resulting polar section makes phospholipids more soluble in water than most lipids.

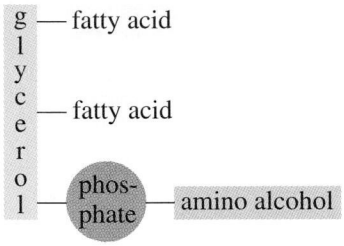

Phosphoglyceride

Three of the amino alcohols found in phosphoglycerides are as follows:

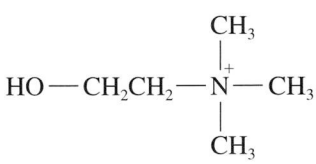

Choline

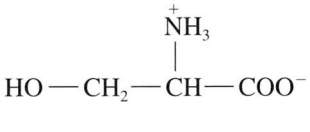

Serine

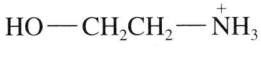

Ethanolamine

Lecithins and **cephalins** are two types of phosphoglycerides that are particularly abundant in brain and nerve tissues as well as in egg yolks, wheat germ, and yeast. Lecithins contain choline, and cephalins contain ethanolamine and sometimes serine. In the following structural formulas, palmitic acid is used as an example of a fatty acid.

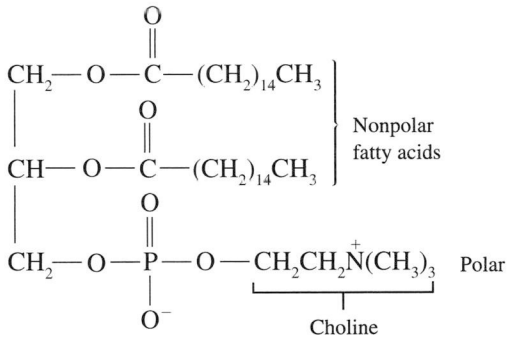

A lecithin

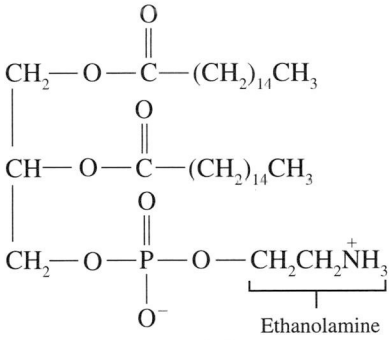

A cephalin

Phospholipids contain both polar and nonpolar regions, which allows them to interact with both polar and nonpolar substances. The ionized alcohol and phosphate portion called "the head" is polar and can hydrogen bond with water. The two fatty acids connected to the glycerol molecule represent the nonpolar "tails" of the phospholipid. The hydrocarbon chains that make up the "tails" are only soluble in other nonpolar substances, mostly lipids. (See Figure 15.10.)

15.5

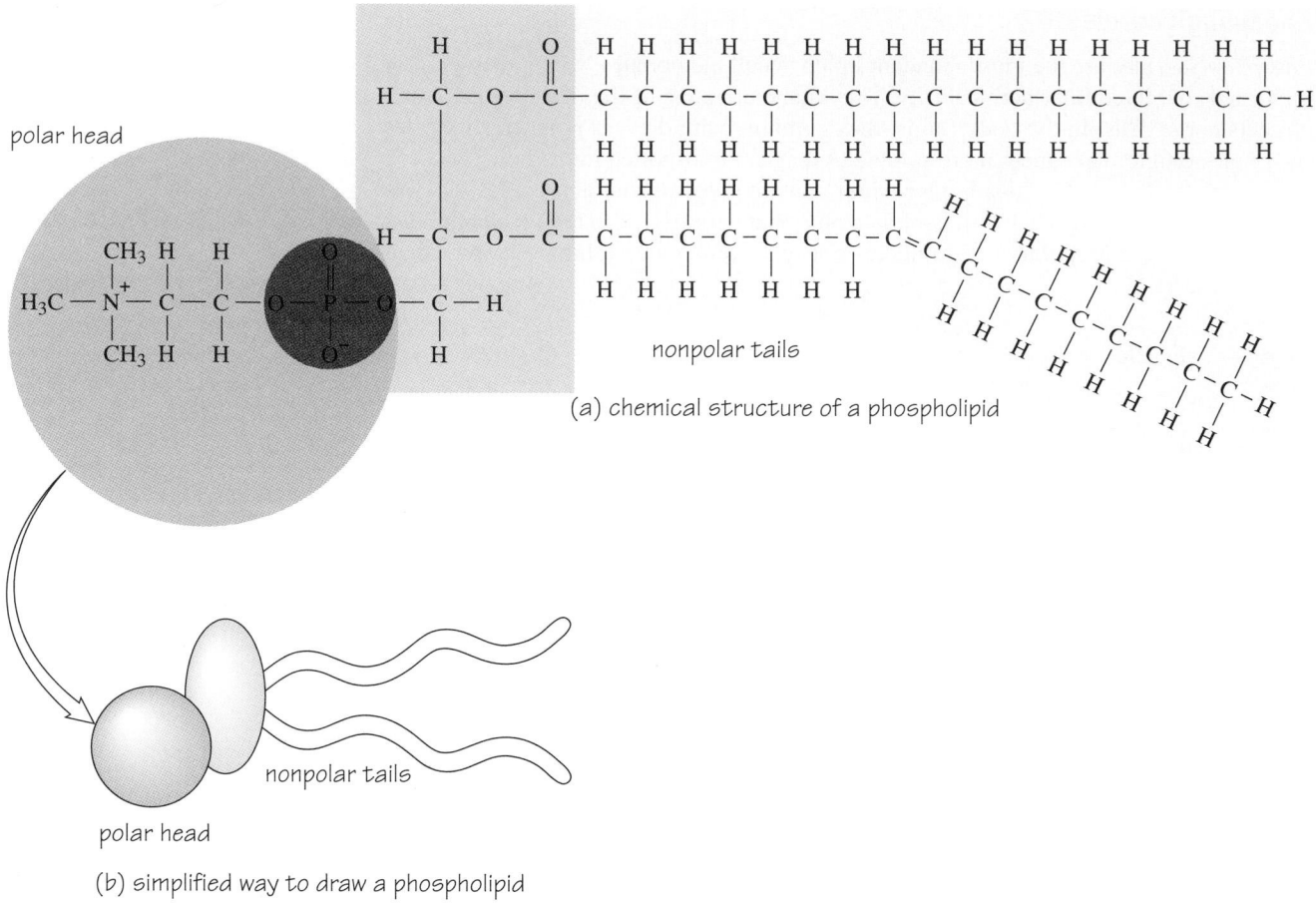

Figure 15.10 **(a)** In the structure of a phospholipid, a polar "head" contains the alcohol and phosphate groups, while the two fatty acids make up the nonpolar "tails." **(b)** A simplified drawing of a phospholipid indicates the polar region and the nonpolar region.

S · A · M · P · L · E · 15.5 · P · R · O · B · L · E · M

Drawing Phospholipid Structures

Draw the structure of cephalin, using stearic acid for the fatty acids and serine for the amino alcohol. Describe each of the components in the phosphoglyceride.

Solution

In general, phosphoglycerides are composed of a glycerol molecule in which two carbon atoms are attached to fatty acids such as stearic acid. The third carbon atom is attached in an ester bond to phosphate linked to an amino alcohol. In this example, the amino alcohol is serine.

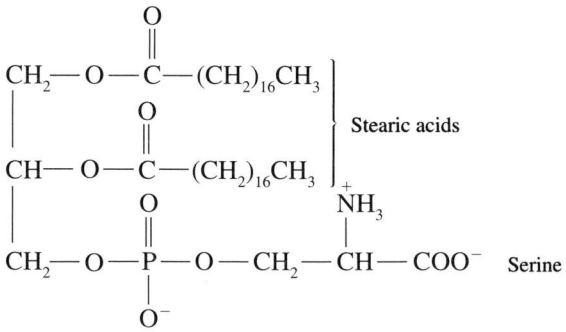

Study Check

How do phospholipids differ structurally from triglycerides?

Sphingolipids

The *sphingolipids* are another group of phospholipids that are also abundant in brain and nerve tissues. The **sphingolipids** are esters of an 18-carbon alcohol called sphingosine instead of glycerol.

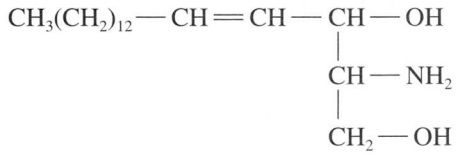

Sphingosine

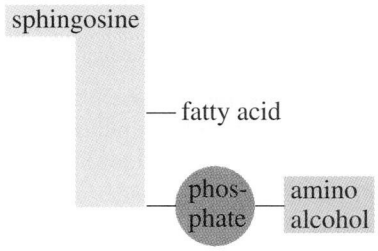

Sphingolipid

One of the most abundant sphingolipids is sphingomyelin. It is the white matter of the myelin sheath, a coating surrounding the nerve cells that increases the speed of nerve impulses and insulates and protects the nerve cells. In sphingomyelin, the sphingosine is linked by an amide bond to a fatty acid, and it is linked to a phosphate ester of choline, an amino alcohol.

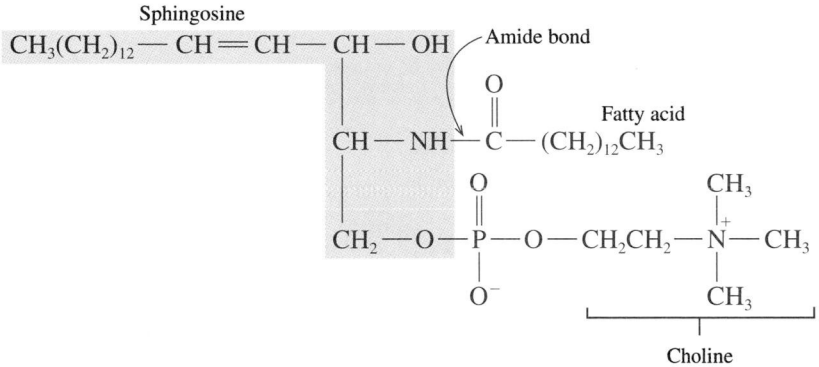

Sphingomyelin, a sphingolipid

Glycolipids

Glycolipids are abundant in the brain and in the myelin sheaths of nerves. Some **glycolipids** contain glycerol attached to two fatty acids and a monosaccharide, usually galactose.

15.5

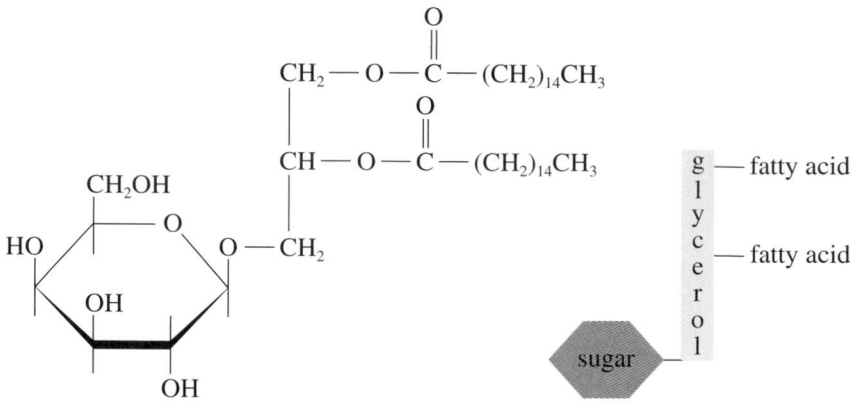

Galactose

Glycolipid

Cerebrosides, another type of glycolipid similar to the sphingolipids contain sphingosine, a fatty acid with 24 carbon atoms, and a monosaccharide, which is usually galactose and sometimes glucose. **Gangliosides** are similar to cerebrosides but contain two or more monosaccharides, such as glucose and galactose. Gangliosides are important in the membranes of neurons and act as receptors for hormones, viruses, and several drugs.

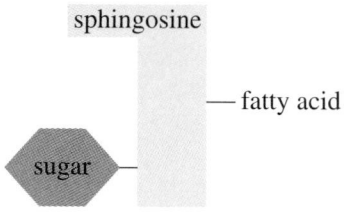

Cerebroside

Sphingosine

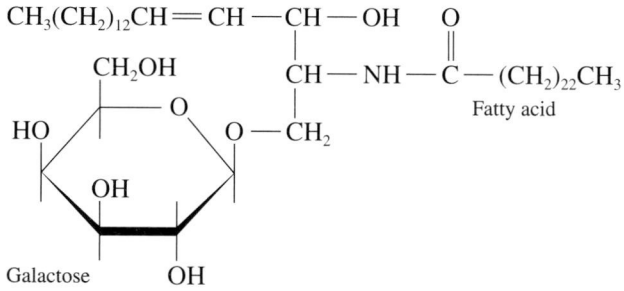

Galactocerebroside

HEALTH NOTE

Lipid Diseases

Many lipid diseases **(lipidoses)** involve the excessive accumulation of a sphingolipid or glycolipid because an enzyme needed for its breakdown is deficient or absent. The accumulation of these glycolipids may enlarge the spleen, liver, and bone marrow cells (Gaucher's disease) and cause mental retardation, seizures, blindness, and death in early infancy. Some lipid storage diseases are listed in Table 15.4.

In multiple sclerosis, sphingomyelins are lost from the myelin sheath, which is the protective membrane surrounding the neurons in the brain and spinal cord. (See Figure 15.11.) As the disease progresses, the myelin sheath deteriorates. Scars form on the neurons and impair the transmission of nerve signals. The symptoms of multiple sclerosis include various levels of muscle weakness and loss of coordination and vision depending on the amount of damage. The cause of multiple sclerosis is not yet known, although some researchers suggest that a virus is involved.

Table 15.4 Lipid Diseases

Names of Disease	Lipid Stored	Type	Enzyme Absent
Fabry's	Gal-gal-glucosylceramide	Ganglioside	α-Galactosidase
Gaucher's	Glucosylceramide	Cerebroside	β-Glucosidase
Niemann-Pick	Sphingomyelin	Sphingolipid	Sphingomyelinase
Tay-Sachs	G_{m2} ganglioside	Ganglioside	Hexosaminidase A

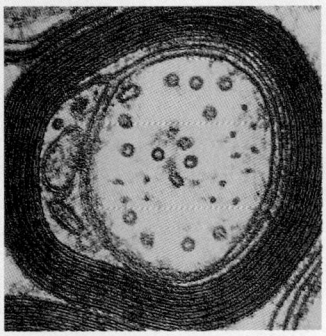

Figure 15.11 The myelin sheath with its high lipid content forms a protective membrane around the nerve axons in the brain and spinal cord.

15.5

S·A·M·P·L·E· **15.6** ·P·R·O·B·L·E·M

Glycolipids

In Fabry's disease, the trihexosylceramide shown here accumulates due to a deficiency of α-galactosidase. Identify the components A–E in this glycolipid.

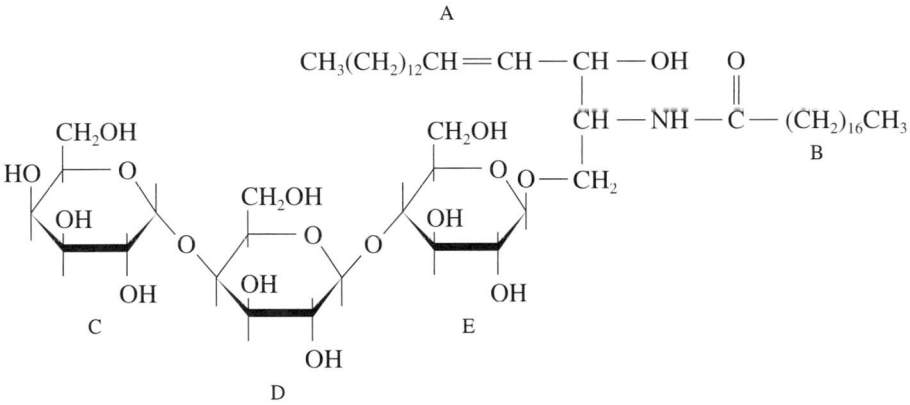

Solution

In this glycolipid, the components are sphingosine (A); stearic acid (B), an 18-carbon fatty acid; two galactose units (C, D); and one glucose (E).

Study Check

How do we know that this glycolipid is a breakdown product of a ganglioside rather than a cerebroside?

QUESTIONS & PROBLEMS

Phospholipids and Glycolipids

15.33 Describe the differences between the following pairs:
a. a fat and a phosphoglyceride
b. a phosphoglyceride and a sphingolipid

15.34 Describe the differences between the following pairs:
a. a lecithin and a cephalin **b.** a cerebroside and a ganglioside

15.6

15.35 Draw the structure of a phosphoglyceride containing two molecules of palmitic acid, and ethanolamine. What is another name for this type of phospholipid?

15.36 What amino alcohol is found in sphingomyelin? Draw the structure of a sphingomyelin containing palmitic acid.

15.37 Draw the structure of a cerebroside containing palmitic acid and galactose.

15.38 Draw the structure of a phospholipid that contains choline and palmitic acids.

15.39 Identify the following phospholipid and list its components:

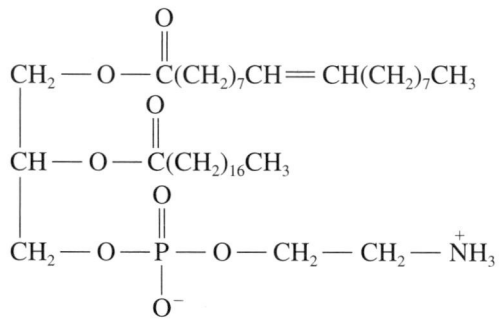

15.40 Identify the following phospholipid and list its components:

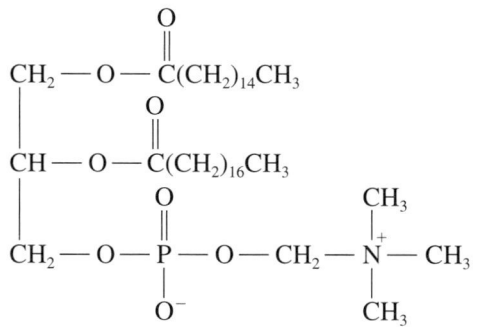

Study Pack

Chem of Life Lipids: Steroids 1, 2

Learning Goal

Describe the structure of a steroid and cholesterol.

15.6 Steroids and Cholesterol

The word *steroid* comes from the Latin word *stereos,* meaning "solid." **Steroids** are compounds containing the steroid nucleus, which consists of three cyclohexane rings and a cyclopentane ring fused together.

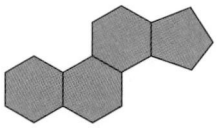

Steroid nucleus

When various groups attach to the steroid nucleus, many steroid compounds are formed that are of medical and biochemical importance. These include cholesterol, bile salts, steroid hormones, male and female sex hormones, and anabolic steroids.

Cholesterol

Cholesterol is the most important and abundant steroid compound in the body. It is called a *sterol* because it contains an alcohol group (—OH) on the first ring. In cholesterol, methyl groups and a branched chain of carbon atoms have been added to the steroid nucleus.

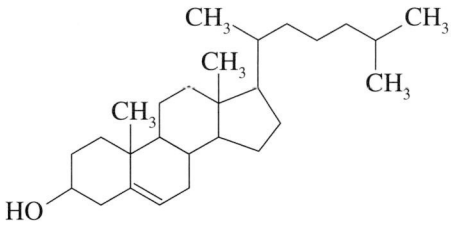

Cholesterol

S·A·M·P·L·E· 15.7 ·P·R·O·B·L·E·M

Cholesterol

Observe the structure of cholesterol for the following questions:

a. What part of cholesterol is the steroid nucleus?
b. What features have been added to the steroid nucleus in cholesterol?
c. What classifies cholesterol as a sterol?

Solution

a. The four fused rings form the steroid nucleus.
b. The cholesterol molecule contains an alcohol group (—OH) on the first ring, one double bond in the second ring, two methyl groups (—CH_3), and a branched carbon chain.
c. The alcohol group determines the sterol classification.

Study Check

Why is cholesterol in the lipid family?

HEALTH NOTE

Cholesterol in the Body

Cholesterol is a component of cellular membranes, myelin sheath, and brain and nerve tissue. It is also found in the liver, bile salts, and skin, where it forms vitamin D. In the adrenal gland, it is used to synthesize steroid hormones. Cholesterol in the body is obtained from eating meats, milk, and eggs, and it is also synthesized by the liver from fats, carbohydrates, and proteins. There is no cholesterol in vegetable and plant products.

If a diet is high in cholesterol, the liver produces less. A typical daily American diet includes 400–500 mg of cholesterol, one of the highest in the world. The American Heart Association has recommended that we consume no more than 300 mg of cholesterol a day. The cholesterol contents of some typical foods are listed in Table 15.5.

When cholesterol exceeds its saturation level in the bile, gallstones may form. Gallstones are composed of almost 100% cholesterol with some calcium salts, fatty acids, and phospholipids. High levels of cholesterol are also associated with the accumulation of lipid deposits (plaques) that line and narrow the coronary arteries. (See Figure 15.12.) Clinically, cholesterol levels are considered elevated if the total plasma cholesterol level exceeds 200–220 mg/dL.

Some research indicates that saturated fats in the diet may stimulate the production of cholesterol by the liver. A diet that is low in foods containing cholesterol and saturated fats appears to be helpful in reducing the serum cholesterol level. Other factors that may also increase the risk of heart disease are family history, lack of exercise, smoking, obesity, diabetes, gender, and age.

Table 15.5 Cholesterol Content of Some Foods

Food	Serving Size	Cholesterol (mg)
Liver (beef)	3 oz	370
Egg	1	250
Lobster	3 oz	175
Fried chicken	$3\frac{1}{2}$ oz	130
Hamburger	3 oz	85
Chicken (no skin)	3 oz	75
Fish (salmon)	3 oz	40
Butter	1 tablespoon	30
Whole milk	1 cup	35
Skim milk	1 cup	5
Margarine	1 tablespoon	0

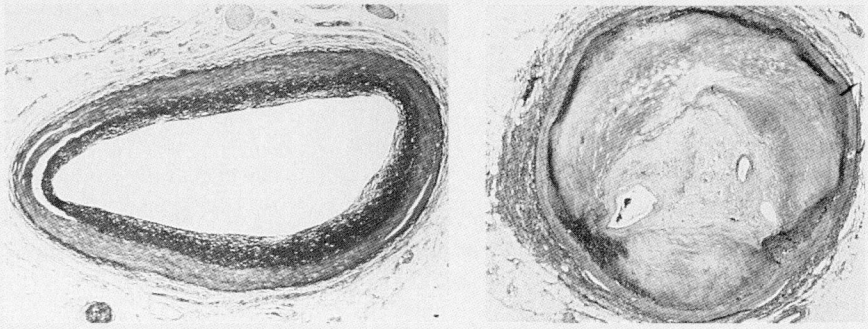

Figure 15.12 Excess cholesterol forms plaque that eventually can block an artery, resulting in a heart attack. **(a)** A normal, open artery shows no buildup of plaque. **(b)** An artery that is almost completely clogged by atherosclerotic plaque.

HEALTH NOTE

Lipoproteins: Transporting Lipids in Solution

Because trigylcerides and cholesterol are nonpolar, they are insoluble in blood and are not transported easily. To improve their solubility, they attach to proteins and phospholipids, forming a polar complex called a **lipoprotein.** (See Figure 15.13.)

There are four types of lipoproteins of varying densities and lipid composition. The chylomicrons and the VLDLs (very-low-density lipoproteins) are the major carriers of triglycerides between the intestines or the liver and the cells in the body where they are used for energy or stored as adipose tissue.

The LDLs (low-density lipoproteins) carry cholesterol to the tissues to be used for the synthesis of cell membranes and steroid hormones. If the need for cholesterol is exceeded by the LDLs, excess cholesterol can accumulate in the cells and the arteries. The HDLs (high-density

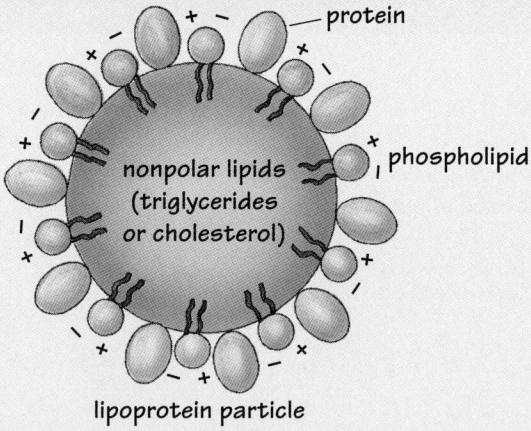

Figure 15.13 A lipoprotein particle; a form in which nonpolar lipids are carried through the bloodstream.

lipoproteins) carry cholesterol from the tissues to the liver where it produces bile salts, which are excreted. (See Table 15.6.)

Because high cholesterol levels have been associated with the onset of atherosclerosis and heart disease, the measurement of the serum levels of LDL and HDL is of much interest to medical researchers. When the LDL level is high, there is an increase in the levels of triglycerides and cholesterol in the blood serum. The excess cholesterol can deposit in the blood vessels, restricting blood flow and increasing the risk of developing heart disease and myocardial infarctions (heart attacks).

Low levels of LDL and high levels of HDL are associated with decreased risk of myocardial infarctions because HDLs remove cholesterol from the cells. A lower level of serum cholesterol decreases the risk of plaque formation within the blood vessels and, therefore, heart disease. Higher HDL levels are found in people who exercise regularly and eat less saturated fats. They are also higher in women whose estrogen levels are adequate.

Table 15.6 Some Properties of Plasma Lipoproteins

Type	Density (g/mL)	Protein (%)	Phospholipid (%)	Other Lipid[a] (%)
Chylomicrons	0.94	2	9	84
VLDL	0.940–1.006	10	18	72
LDL	1.006–1.063	25	20	55
HDL	1.063–1.210	55	25	20

[a] Other lipid includes cholesterol and triglycerides.

Bile Salts

The *bile* salts are synthesized in the liver from cholesterol and stored in the gallbladder. When bile is secreted into the small intestine, the bile salts mix with the water-insoluble fats and oils in our diets. The bile salts with their nonpolar and polar regions act much like soaps, breaking apart and emulsifying large globules of fat. The emulsions that form have a larger surface area for the lipases, enzymes that digest fat. The bile salts also help in the absorption of cholesterol into the intestinal mucosa.

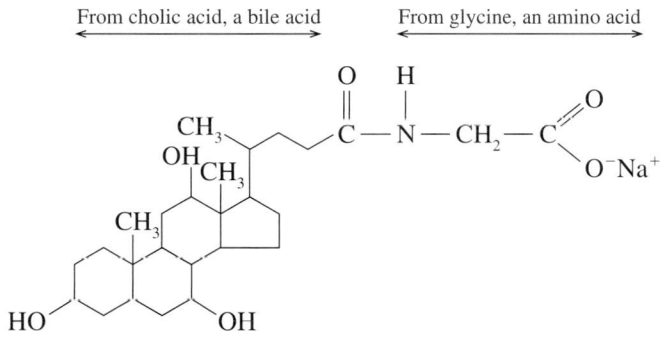

Sodium glycocholate, a bile salt

Fat-Soluble Vitamins

The fat-soluble vitamins A, D, E, and K are all nonpolar, which makes them soluble in body fat, where they are stored. An excess of a fat-soluble vitamin can result in hypervitaminosis and damage to the body. Vitamin D, a fat-soluble vitamin related to cholesterol, is produced in the skin from reactions of sunlight and is found in sardines, salmon, and egg yolks. In the body, vitamin D increases the absorption of calcium from the intestinal tract and regulates the amount of calcium deposited in the bones and teeth. The other fat-soluble vitamins are unsaturated hydrocarbons. Vitamin A is important in the formation of visual pigments for night vision. Vitamin K is important in blood clotting. Table 15.7 lists the fat-soluble vitamins.

Table 15.7 Fat-Soluble Vitamins

Structure	Biological Function	Deficiency	Symptoms of Excess
Retinol (vitamin A)	Formation of visual pigment; development of epithelial cells	Night blindness; dry, scaly skin; bacterial infections	Hypervitaminosis, irritability, sloughing of skin, joint pain, weight loss, liver enlargement
Calciferol (vitamin D)	Absorption of calcium and phosphate; deposition of calcium and phosphate in bone	Rickets; bone decalcification; bone deformities	Hypervitaminosis D; muscle weakness, nausea, diarrhea, anorexia; calcification in tissue, kidney damage
α-Tocopherol (vitamin E)	May prevent oxidation of unsaturated fatty acids in cell membranes	Hemolysis of red blood cells; infertility in rats	
Phylloquinone (vitamin K)	Synthesis of prothrombin for blood clotting	Bruising, hemorrhaging from minor cuts, longer clotting time	

15.6

Steroid Hormones

The word *hormone* comes from the Greek "to arouse" or "to excite." Hormones are chemical messengers that serve as a kind of communication system from one part of the body to another. The *steroid* hormones, which include the sex hormones and the adrenocortical hormones, are closely related in structure to cholesterol and depend on cholesterol for their synthesis.

Two important male sex hormones, *testosterone* and *androsterone,* promote the growth of muscle and of facial hair and the maturation of the male sex organs and of sperm. Table 15.8 gives the structures of some of the steroid hormones and their physiological effects.

Table 15.8 Examples of Steroid Hormones Derived from Cholesterol

Hormone	Biological Effects	Hormone	Biological Effects
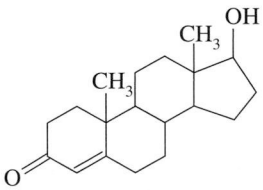 Testosterone (androgen) (produced in testes)	Development of male organs; male sexual characteristics including muscles and facial hair; sperm formation	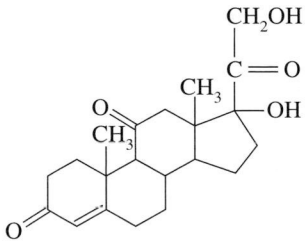 Cortisone (produced in adrenal gland)	Increases the blood glucose and glycogen levels from fatty acids and amino acids
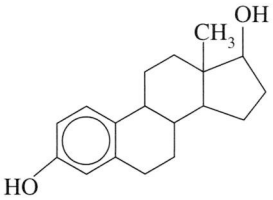 Estradiol (estrogen) (produced in ovaries)	Development of female sexual characteristics; ovulation	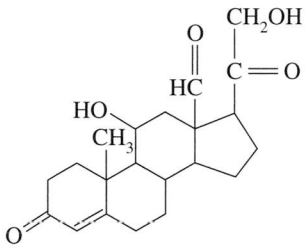 Aldosterone (mineralocorticoid) (produced in adrenal gland)	Increases the reabsorption of Na^+ in kidneys; retention of water
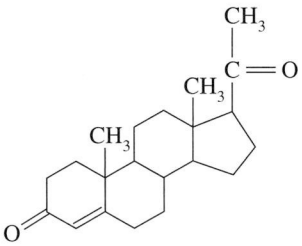 Progesterone (produced in ovaries)	Prepares uterus for fertilized egg	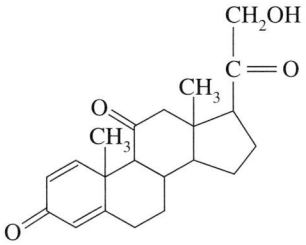 Prednisone (synthetic corticoid)	Reduces inflammation; treatment of asthma and rheumatoid arthritis
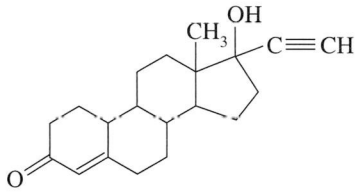 Norethindrone (synthetic progestin)	Contraceptive (birth-control) pill		

The *estrogens,* a group of female sex hormones, direct the development of female sexual characteristics: the uterus increases in size, fat is deposited in the breasts, and the pelvis broadens. *Progesterone* prepares the uterus for the implantation of a fertilized egg. If an egg is not fertilized, the levels of progesterone and estrogen drop sharply, and menstruation follows. Synthetic forms of the female sex hormones are used in birth-control pills. As with other kinds of steroids, side effects include weight gain and a greater risk of forming blood clots.

Adrenal Corticosteroids

The adrenal glands, located on the top of each kidney, produce the corticosteroids. *Aldosterone,* a mineralocorticoid, is responsible for electrolyte and water balance by the kidneys. *Cortisone,* a glucocorticoid, increases the blood glucose level and stimulates the synthesis of glycogen in the liver from amino acids. Synthetic corticoids such as *prednisone* are derived from cortisone and used medically for reducing inflammation and treating asthma and rheumatoid arthritis, although precautions are given for long-term use.

HEALTH NOTE

15.6

Anabolic Steroids

Some of the physiological effects of testosterone are to increase muscle mass and decrease body fat. Derivatives of testosterone called *anabolic steroids* that enhance these effects have been synthesized. Although they have some medical uses, anabolic steroids have been used in rather high dosages by some athletes in an effort to increase muscle mass. Such use is illegal.

Use of anabolic steroids in attempting to improve athletic strength can cause side effects including hypertension, fluid retention, increased hair growth, sleep disturbances, and acne. Over a long period of time, their use can be devastating and may cause irreversible liver damage and decreased sperm production.

Some Anabolic Steroids

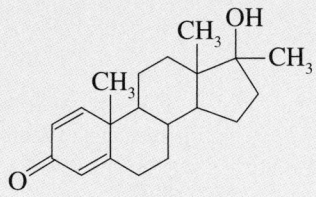

Methandienone

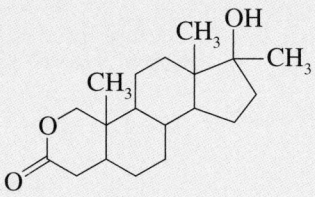

Oxandrolone

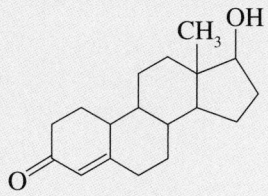

Nandrolone

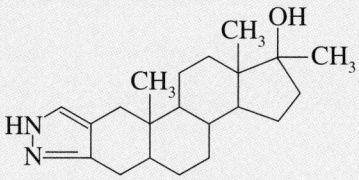

Stanozolol

S·A·M·P·L·E 15.8 P·R·O·B·L·E·M

Steroid Hormones

What are the functional groups on the steroid nucleus in the sex hormones estradiol and testosterone?

Solution

Estradiol contains a benzene ring, a methyl group, and two alcohol groups. Testosterone contains a ketone group, a double bond, two methyl groups, and an alcohol group.

Study Check

What are the similarities and differences in the structures of testosterone and the anabolic steroid nandrolone?

Q·U·E·S·T·I·O·N·S & P·R·O·B·L·E·M·S

Steroids and Cholesterol

15.41 Draw the structure for the steroid nucleus.

15.42 Which of the following compounds are derived from cholesterol?
a. glyceryl tristearate **b.** cortisone
c. bile salts **d.** testosterone
e. estradiol

15.43 What is the function of bile salts in digestion?

15.44 Why are gallstones composed of cholesterol?

15.45 What are the similarities and differences between the sex hormones estradiol and testosterone?

15.46 What are the similarities and differences between the adrenal hormone cortisone and the synthetic corticoid prednisone?

15.47 Which of the following are male sex hormones?
a. cholesterol **b.** aldosterone
c. estrogen **d.** testosterone
e. choline

15.48 Which of the following are adrenal steroids?
a. cholesterol **b.** aldosterone
c. estrogen **d.** testosterone
e. choline

Study Pack

Chem of Life Lipids: Phospholipids 6, 7

15.7 Plasma (Cell) Membranes

Learning Goal

Describe the function of the lipid bilayer in plasma membranes.

All of our cells are enclosed within a thin barrier called a **plasma,** or **cell, membrane** that keeps the cellular contents separated from the surrounding liquid environment and regulates the substances that go into and out of the cell.

The basic structure of the plasma membrane is a **lipid bilayer,** which consists of two rows of phospholipids. The nonpolar hydrocarbon tails, which are hydrophobic (water fearing), form the center of the lipid bilayer. The polar sections of the phospholipids, which are hydrophilic (water loving) are aligned along the outside of the lipid bilayer. (See Figure 15.14.) As a result the polar and nonpolar sections of the lipid bilayer create a wall separating the aqueous contents inside a cell from the surrounding fluids. Nonpolar oxygen, nitrogen, carbon dioxide, steroids, fat-soluble vitamins, and glycerol do pass through the lipid bilayer, but it is impermeable to other substances.

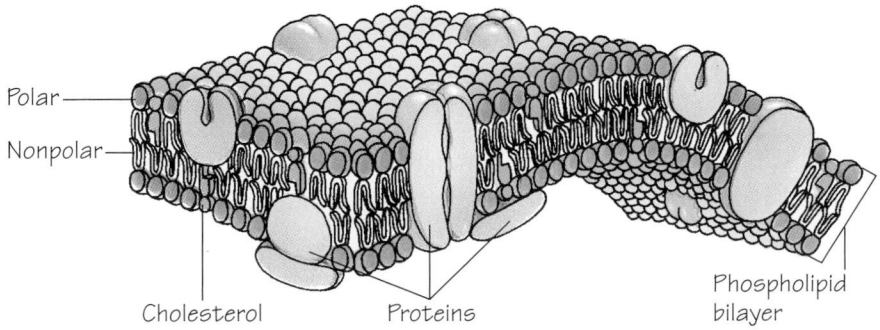

Figure 15.14 A model of a cell membrane. A double row of phospholipids forms a barrier with polar ends in contact with fluids and nonpolar ends at the center away from the water. Proteins embedded in the lipid bilayer permit passage of polar substances and act as receptors.

Most of the lipids in the lipid bilayer contain fatty acids with cis double bonds. As a result of the structure of cis double bonds, the phospholipids do not fit closely together. They can diffuse sideways, which gives a dynamic structure to the lipid bilayer.

In animal membranes, cholesterol makes up about 20% of the lipid bilayer. Embedded among the nonpolar tails of the fatty acids, cholesterol molecules help add strength and rigidity to the plasma membrane. Glycolipids make up about 5% of the membrane lipids, appearing only on the outside of the plasma membrane. Their function is not completely known, but they appear to be responsible for communication between cells and for the growth of the cell membrane.

15.7

Proteins are also found embedded in the lipid bilayer. Certain proteins extending through the lipid bilayer provide channels through which substances such as water, K^+, Cl^-, and HCO_3^- ions flow into or out of the cell. Other types of proteins emerge on the surface where they act as receptors for hormones, neurotransmitters, antibiotics, and other chemicals that are necessary for the proper functioning of the cell.

S·A·M·P·L·E 15.9 ·P·R·O·B·L·E·M

Lipid Bilayer in Plasma Membranes

Describe the role of phospholipids in the lipid bilayer.

Solution

Phospholipids consist of polar and nonpolar parts. In a plasma membrane, an alignment of the nonpolar sections toward the center with the polar sections on the outside produces a barrier that prevents the contents of a plasma membrane from mixing with the fluids on the outside of the cell.

Study Check

Why are protein channels needed in the lipid bilayer?

Q·U·E·S·T·I·O·N·S & P·R·O·B·L·E·M·S

Plasma (Cell) Membranes

15.49 What types of lipids are found in plasma membranes?

15.50 Describe the structure of a lipid bilayer.

15.51 What is the function of the lipid bilayer in a plasma membrane?

15.52 What kinds of substances can move through the lipid bilayer? Why?

15.53 How do electrolytes, water, and other polar molecules enter or leave a cell?

15.54 What are the functions of cholesterol and glycolipids found in a plasma membrane?

Chapter Review

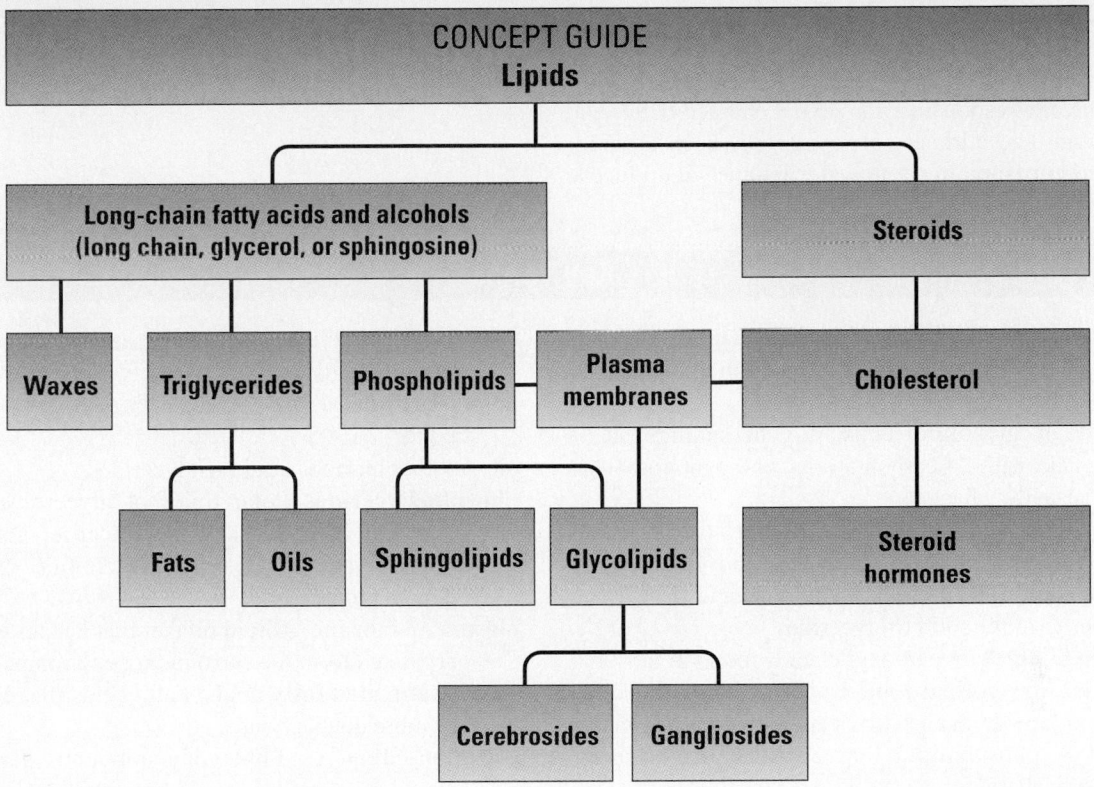

Chapter Summary

15.1 Types of Lipids

Lipids are nonpolar compounds that are not soluble in water. Classes of lipids include waxes, fats and oils, phospholipids, and steroids.

15.2 Fatty Acids

Fatty acids are unbranched carboxylic acids that typically contain an even number (12–18) of carbon atoms. Fatty acids may be saturated, monounsaturated with one double bond, or polyunsaturated with two or more double bonds. The double bonds in unsaturated fatty acids are almost always cis.

15.3 Waxes, Fats, and Oils

A wax is an ester of a long-chain fatty acid and a long-chain alcohol. The triglycerides of fats and oils are esters of glycerol with three long-chain fatty acids. Fats contain more saturated fatty acids and have higher melting points than most vegetable oils.

15.4 Chemical Properties of Triglycerides

The hydrogenation of unsaturated fatty acids converts double bonds to single bonds. The oxidation of unsaturated fatty acids produces short-chain fatty acids with disagreeable odors. The hydrolysis of the ester bonds in fats or oils produces glycerol and fatty acids. In saponification, a fat heated with a strong base produces glycerol and the salts of the fatty acids or soaps. The dual polarity of a soap permits its solubility in both water and fats.

15.5 Phospholipids and Glycolipids

Phosphoglycerides are esters of glycerol with two fatty acids and a phosphate group attached to an amino alcohol. In sphingolipids, the alcohol sphingosine forms an ester bond with one fatty acid and a phosphate–amino alcohol group. In glycolipids, glycerol or sphingosine is bonded to fatty acids and one or more monosaccharides.

15.6 Steroids and Cholesterol

Steroids are lipids containing the steroid nucleus, which is a fused structure of four rings. Steroids include cholesterol, bile salts, and vitamin D. The steroid hormones are closely related in structure to cholesterol and depend on cholesterol for their synthesis. The sex hormones, such as estrogen and testosterone, are responsible for sexual characteristics and reproduction. The adrenal corticosteroids such as aldosterone and cortisone regulate water balance and glucose levels in the cells.

15.7 Plasma (Cell) Membranes

All animal cells are surrounded by a plasma membrane that separates the cellular contents from the external fluids. The membrane is composed of two rows of phospholipids in a lipid bilayer that controls the flow of substances in and out of the cell.

Key Terms

cephalins Phospholipids found in brain and nerve tissues that incorporate the amino alcohol serine or ethanolamine.

cerebroside A glycolipid consisting of sphingosine, a fatty acid with 24 carbon atoms, and a monosaccharide (usually galactose).

cholesterol The most prevalent of the steroid compounds found in cellular membranes; needed for the synthesis of vitamin D, hormones, and bile acids.

fat Another term for solid triglycerides.

fatty acids Long-chain carboxylic acids found in fats.

ganglioside A glycolipid consisting of sphingosine, a 24-carbon fatty acid, and two or more monosaccharides.

glycolipid A phospholipid that contains one or more monosaccharides.

hydrogenation The addition of hydrogen to unsaturated fats.

lecithins Phospholipids containing choline as the amino alcohol.

lipid bilayer A model of a cell membrane in which phospholipids are arranged in two rows interspersed with proteins arranged at different depths.

lipidoses Genetic diseases in which a deficiency of an enzyme for the hydrolysis of a lipid causes the accumulation of that lipid to toxic levels.

lipids A family of compounds that is nonpolar in nature and not soluble in water; includes fats, waxes, phospholipids, and steroids.

lipoprotein A combination of nonpolar lipids with phospholipids and proteins to form a polar complex that can be transported through body fluids.

mixed triglycerides Fats or oils that contain two or three different fatty acids.

monounsaturated fatty acids Fatty acids with one double bond.

oil Another term for liquid triglycerides.

phosphoglycerides Polar lipids of glycerol attached to two fatty acids and a phosphate group connected to an amino group such as choline, serine, or ethanolamine.

plasma membrane A lipid bilayer that keeps cell contents separated from the surrounding environment.

polyunsaturated fatty acids Fatty acids that contain two or more double bonds.

prostaglandins A number of compounds derived from arachidonic acid that regulate several physiological processes.

saturated fatty acids Fatty acids that have no double bonds; they have higher melting points than unsaturated lipids and are usually solid at room temperatures.

sphingolipids Phospholipids in which sphingosine has replaced glycerol.

steroids Types of lipid composed of a multicyclic ring system.

triglycerides A family of lipids composed of three fatty acids bonded through ester bonds to glycerol, a trihydroxy alcohol.

wax The ester of a long-chain alcohol and a long-chain saturated fatty acid.

Chapter Practice

15.55 Among the ingredients of a lipstick are beeswax, carnauba wax, hydrogenated vegetable oils, capric triglyceride, and tocopherol. What types of lipids have been used? Draw the structure of capric triglyceride (capric acid is the saturated 10-carbon fatty acid).

15.56 Because peanut oil floats on the top of peanut butter, many brands of peanut butter are hydrogenated. A solid product then forms that is mixed into the peanut butter and does not separate. If a triglyceride in peanut oil that contains one palmitic acid, one oleic acid, and one linoleic acid is completely hydrogenated, what is the product?

15.57 Trans fats are produced during the hydrogenation of polyunsaturated oils.
a. What is the typical configuration of a monounsaturated fatty acid?
b. How does a trans fat compare with a cis fat?
c. Draw the structure of *trans*-oleic acid.

15.58 One mole of triolein is completely hydrogenated. What is the product? How many moles of hydrogen are required? How many grams of hydrogen? How many liters of hydrogen are needed if the reaction is run at STP?

15.59 On the list of ingredients of a cosmetic product are glyceryl stearate and lecithin. What are the structures of these compounds?

15.60 Some typical meals at fast-food restaurants are listed here. Calculate the number of kilocalories from fat and the percentage of total kilocalories due to fat (1 gram of fat = 9 kcal). Would you expect the fats to be mostly saturated or unsaturated? Why?
a. a chicken dinner, 830 kcal, 46 g of fat
b. a quarter-pound cheeseburger, 518 kcal, 29 g of fat
c. pepperoni pizza (three slices), 560 kcal, 18 g of fat
d. beef burrito, 470 kcal, 21 g of fat
e. deep-fried fish (three pieces), 480 kcal, 28 g of fat

15.61 Identify the following as fatty acids, soaps, triglycerides, wax, phosphoglyceride, sphingolipid, or steroid:
a. beeswax **b.** cholesterol
c. lecithin
d. glyceryl tripalmitate (tripalmitin)
e. sodium stearate **f.** safflower oil
g. sphingomyelin **h.** whale blubber
i. adipose tissue **j.** progesterone
k. cortisone **l.** stearic acid

Answers

Answers to Study Checks

15.1 a glycolipid

15.2 **a.** 16 **b.** unsaturated **c.** liquid

15.3

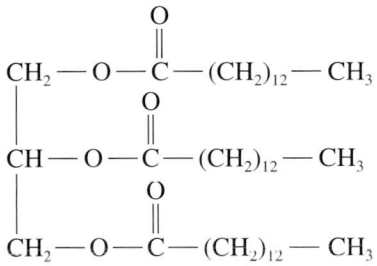

15.4 tristearin

15.5 Triglycerides contain three fatty acids connected by ester bonds to glycerol. Phospholipids contain glycerol esterified to two fatty acids and a phosphate connected to an amino alcohol.

15.6 Cerebrosides contain only one monosaccharide, and gangliosides contain two or more monosaccharide units.

15.7 Cholesterol is not soluble in water; it is classified with the lipid family.

15.8 Testosterone and nandrolone both contain a steroid nucleus with one double bond and a ketone group in the first ring, and a methyl and alcohol group on the five-carbon ring. Nandrolone does not have the second methyl group at the first and second ring fusion that is seen in the structure of testosterone.

15.9 Protein channels allow water and electrolytes to flow in and out of the cell through the lipid bilayer.

Answers to Selected Questions and Problems

15.1 Lipids provide energy and protection and insulation for the organs in the body.

15.3 Because lipids are not soluble in water, a polar solvent, they are nonpolar molecules.

15.5 All fatty acids contain a long chain of carbon atoms with a carboxylic acid group. Saturated fats contain only carbon-to-carbon single bonds; unsaturated fats contain one or more double bonds. More saturated fats are found in animal fats, whereas vegetable oils contain more unsaturated fats.

15.7

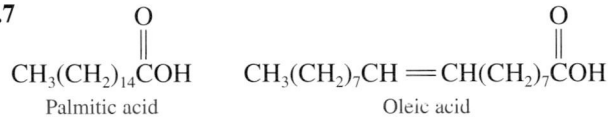

15.9 **a.** saturated **b.** unsaturated
c. unsaturated **d.** saturated

15.11 In a cis fatty acid, the hydrogen atoms are on the same side of the double bond, which produces a bend in the carbon chain. In a trans fatty acid, the hydrogen atoms are on opposite sides of the double bond, which gives a carbon chain without any bend.

15.13 In an omega-3 fatty acid, there is a double bond on carbon 3 counting from the methyl group, whereas in an omega-6 fatty acid, there is a double bond beginning at carbon 6 counting from the methyl group.

15.15

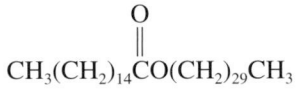

15.17

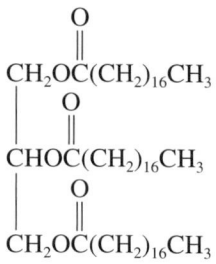

15.19

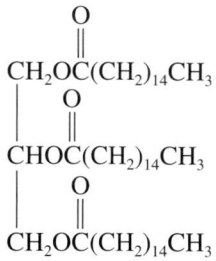

15.21 Safflower oil contains fatty acids with two or three double bonds; olive oil contains a large amount of oleic acid, which has only one (monounsaturated) double bond.

15.23 Although coconut oil comes from a vegetable, it has large amounts of saturated fatty acids and small amounts of unsaturated fatty acids.

15.25

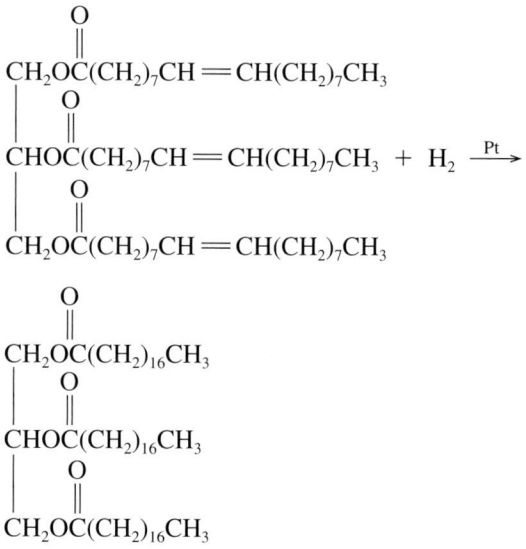

15.27 **a.** Some of the double bonds in the unsaturated fatty acids have been converted to single bonds.
b. It is mostly saturated fatty acids.

15.29 **a.**

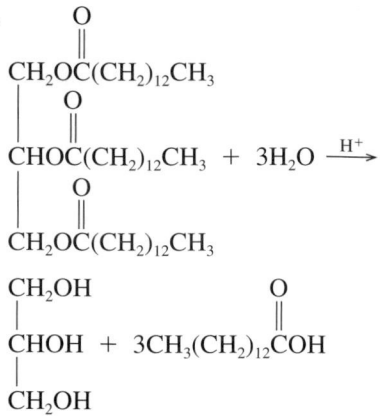

b.

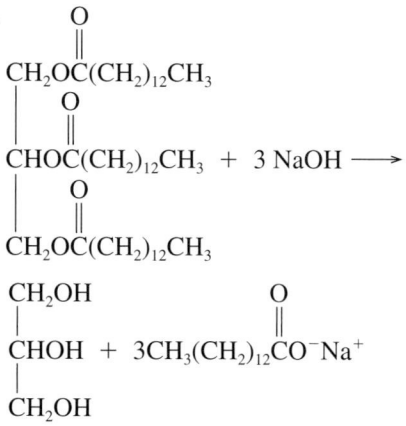

15.31 A triglyceride is composed of glycerol with three hydroxyl groups that form ester links with three long-chain fatty acids. In olestra, six to eight long-chain fatty acids form ester links with the hydroxyl groups on sucrose, a sugar. The olestra cannot be digested because our enzymes cannot break down the large olestra molecule.

15.33 **a.** A fat consists of glycerol and three fatty acids. A phosphoglyceride consists of glycerol, two fatty acids, a phosphate group, and an amino alcohol.
b. A phosphoglyceride consists of glycerol, two fatty acids, a phosphate group, and an amino alcohol. A sphingolipid contains the amino alcohol sphingosine (instead of glycerol) and only one fatty acid.

15.35

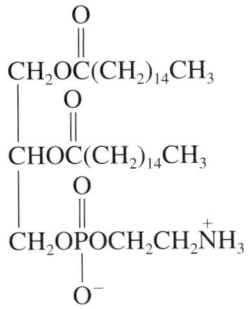

This is a cephalin

15.37

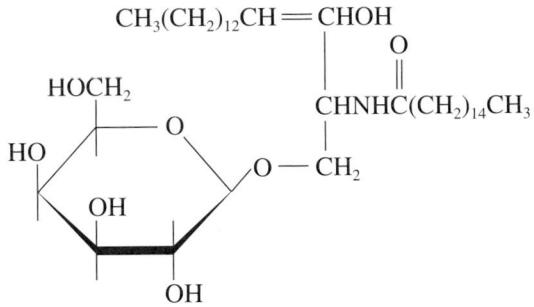

15.39 This phospholipid is a cephalin. It contains glycerol, oleic acid, stearic acid, a phosphate, and ethanolamine.

15.41

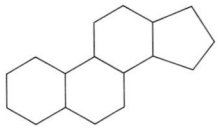

15.43 Bile salts act to emulsify fat globules, allowing the fat to be more easily digested.

15.45 Both estradiol and testosterone contain the steroid nucleus and a hydroxyl group. Testosterone has a ketone group, a double bond, and two methyl groups. Estradiol has a benzene ring, a hydroxyl group in place of the ketone, and a methyl group.

15.47 Testosterone is a male sex hormone.

15.49 Phospholipids with smaller amounts of glycolipids and cholesterol.

15.51 The lipid bilayer in a plasma membrane surrounds the cell and separates the contents of the cell from the external fluids.

15.53 Ions and polar molecules such as water enter or leave a cell through the protein channels that are embedded in the lipid bilayer and extend from one side of the cell membrane to the other.

15.55 Beeswax and carnauba are waxes. Vegetable oil and capric triglyceride are triglycerides. Tocopherol is a vitamin.

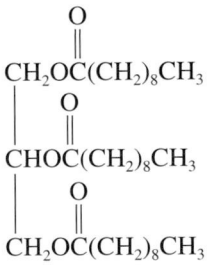

Capric triglyceride

15.57 **a.** A typical fatty acid has a cis double bond.
b. A trans fatty acid has a trans double bond.
c.

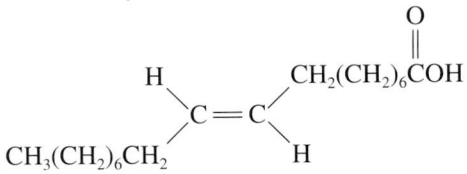

15.59

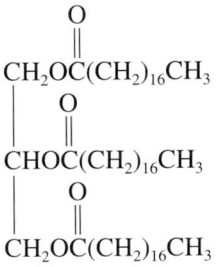

Glyceryl stearate

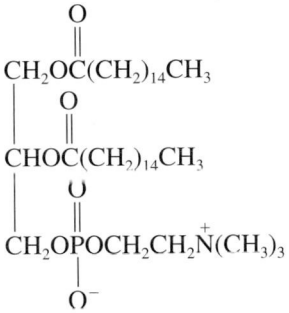

Lecithin

(Note: This lecithin structure is shown with choline, but ethanolamine or serine are also possible amino alcohols in lecithin.)

15.61 Stearic acid is a fatty acid. Sodium stearate is a soap. Glyceryl tripalmitate, safflower oil, whale blubber, and adipose tissue are triglycerides. Beeswax is a wax. Lecithin is a phosphoglyceride. Sphingomyelin is a sphingolipid. Cholesterol, progesterone, and cortisone are steroids.

16

Amino Acids, Proteins, and Enzymes

Fish in our diet provide protein to build muscles, to produce the hemoglobin that carries oxygen in the bloodstream, and to build enzymes and hormones that direct metabolic activities.

Looking Ahead

16.1	Types of Proteins
16.2	Amino Acids
16.3	The Peptide Bond
16.4	Primary and Secondary Structure
16.5	Tertiary and Quaternary Structure
16.6	Protein Hydrolysis and Denaturation
16.7	Enzymes
16.8	Enzyme Action
16.9	Factors Affecting Enzyme Action
16.10	Enzyme Inhibition

www.chemplace.com/college/timberlake

Visit the web site above to test yourself, run simulations, and review key concepts from this chapter. For even more interactive review and quizzing, use the two CD-ROMs included with the book.

The word "protein" is derived from the Greek word *proteios,* meaning "first." Made of amino acids, proteins provide structure in membranes, build cartilage and connective tissue, transport oxygen in blood and muscle, direct biological reactions as enzymes, defend the body against infection, and control metabolic processes as hormones. They can even be a source of energy.

Compared with many of the compounds we have studied, protein molecules can be gigantic. Insulin has a molar mass of 5,700, and hemoglobin has a molar mass of about 64,000. Some virus proteins are still larger, having molar masses of more than 40 million. Yet all proteins are polymers made up of 20 different amino acids. Each kind of protein is composed of amino acids arranged in a specific order that determines the characteristics of the protein and its biological action.

Enzymes are proteins that act as biological catalysts in the cells of our body. Enzymes help us digest our food, contract our muscles, and produce the biomolecules and energy we need for survival.

Proteins:
Functions of Proteins 1

Learning Goal

Classify proteins by their functions in the cells.

16.1 Types of Proteins

The many kinds of proteins perform many different functions in the body. There are proteins that form structural components such as cartilage, muscles, hair, and nails. Wool, silk, feathers, and horns are some other proteins made by animals. Proteins called enzymes regulate biological reactions such as digestion and cellular metabolism. Still other proteins, hemoglobin and myoglobin, carry oxygen in the blood and muscle. Table 16.1 gives examples of proteins that are classified by their functions in biological systems.

Table 16.1 Classification of Some Proteins and their Functions

Type of Protein	Function in the Body	Examples
Structural	Provide structural components	*Collagen* is in tendons and cartilage. *Keratin* is in hair, skin, wool, and nails.
Contractile	Movement of muscles	*Myosin* and *actin* contract muscle fibers.
Transport	Carry essential substances throughout the body	*Hemoglobin* transports oxygen. *Lipoproteins* transport lipids.
Storage	Store nutrients	*Casein* stores protein in milk. *Ferritin* stores iron in the spleen and liver.
Hormone	Regulate body metabolism and nervous system	*Insulin* regulates blood glucose level. *Growth hormone* regulates body growth.
Enzyme	Catalyze biochemical reactions in the cells	*Sucrase* catalyzes the hydrolysis of sucrose. *Trypsin* catalyzes the hydrolysis of proteins.
Protection	Recognize and destroy foreign substances	*Immunoglobulins* stimulate immune response.

S·A·M·P·L·E **16.1** ·P·R·O·B·L·E·M

Classifying Proteins by Function

Give the type of protein that would perform each of the following functions.

a. catalyzes metabolic reactions of lipids

b. carries oxygen in the bloodstream

c. stores amino acids in milk

Solution

a. Enzymes catalyze metabolic reactions.
b. Transport proteins carry substances such as oxygen through the bloodstream.
c. A storage protein stores nutrients such as amino acids in milk.

Study Check

What proteins are known as chemical messengers?

16.2

Q·U·E·S·T·I·O·N·S & P·R·O·B·L·E·M·S

Types of Proteins

16.1 Classify each of the following proteins according to its function:
a. hemoglobin, oxygen carrier in the blood
b. collagen, a major component of tendons and cartilage
c. keratin, a protein found in hair
d. amylase, an enzyme that hydrolyzes starch

16.2 Classify each of the following proteins according to its function:
a. insulin, a hormone needed for glucose utilization
b. antibodies, proteins that disable foreign proteins
c. casein, milk protein
d. lipases that hydrolyze lipids

Study Pack

Chem of Life Proteins: Levels of Protein Structure 1–3

Learning Goal

Draw the structure for an amino acid in neutral or zwitterion form.

16.2 Amino Acids

Proteins are composed of molecular building blocks called amino acids. An **amino acid** contains two functional groups, an amino group (—NH_2) and a carboxylic acid group (—COOH). In most of the amino acids found in proteins, the amino group, the carboxylic group, and a hydrogen atom are bonded to a central carbon atom. Amino acids with this structure are called *α-(alpha)-amino acids*. Different side groups (R) are attached to the center (α) carbon, giving each amino acid unique characteristics.

General Structure of an α-Amino Acid

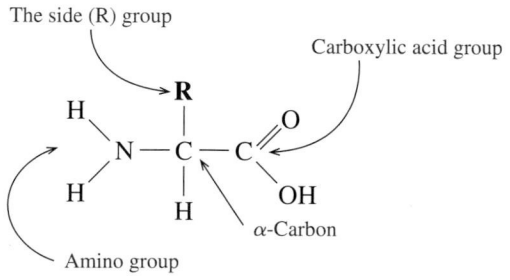

Classification of Amino Acids

Amino acids can be classified according to their side (R) groups. **Nonpolar amino acids** contain a hydrocarbon or aromatic side chain, which makes them **hydrophobic. Polar amino acids** are **hydrophilic** because their side (R) groups contain atoms such as —OH that are attracted to water. Some of the polar amino acids are *acidic* if the R group contains a carboxylic acid group, and *basic* if the R group contains an amino group. The structures of the amino acids, their R groups, common names, and their abbreviations (one or three letters) are listed in Table 16.2.

Table 16.2 The 20 Amino Acids Found in Proteins

Nonpolar Amino Acids

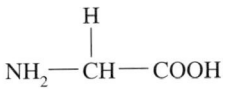
Glycine
(Gly or G)

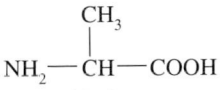
Alanine
(Ala or A)

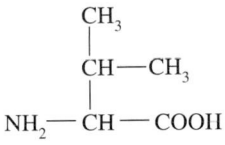
Valine
(Val or V)

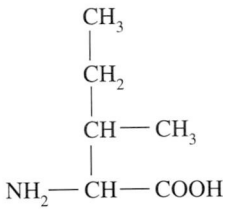
Isoleucine
(Ile or I)

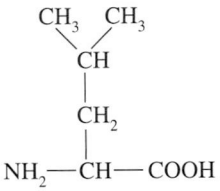
Leucine
(Leu or L)

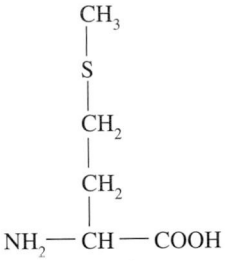
Methionine
(Met or M)

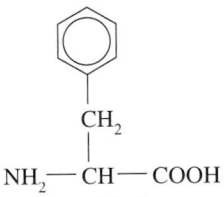
Phenylalanine
(Phe or F)

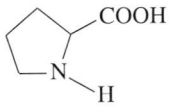
Proline
(Pro or P)

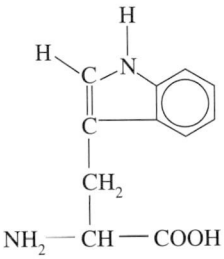
Tryptophan
(Trp or W)

Polar Amino Acids

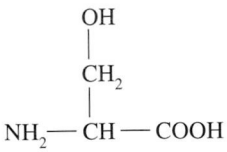
Serine
(Ser or S)

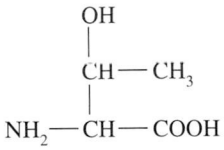
Threonine
(Thr or T)

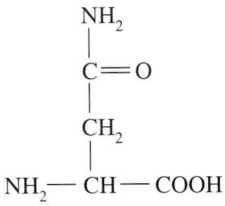
Asparagine
(Asn or N)

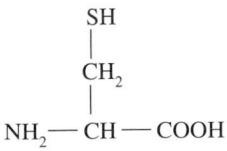
Cysteine
(Cys or C)

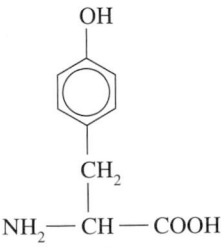
Tyrosine
(Tyr or Y)

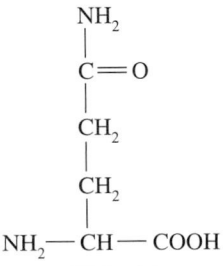
Glutamine
(Gln or Q)

Continued

16.2

16.2

Table 16.2 The 20 Amino Acids Found in Proteins—continued

Acid (Polar) Amino Acids

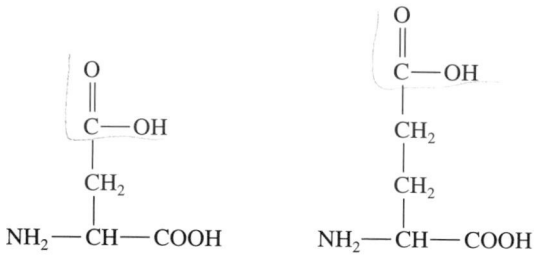

Aspartic acid (Asp or D)

Glutamic acid (Glu or E)

Basic (Polar) Amino Acids

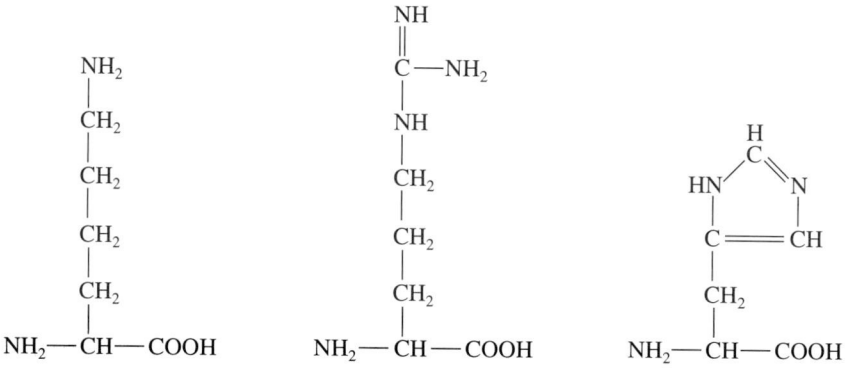

Lysine (Lys or K)

Arginine (Arg or R)

Histidine (His or H)

S·A·M·P·L·E **16.2** ·P·R·O·B·L·E·M

Structural Formulas of Amino Acids

Write the structural formulas and abbreviations for the following amino acids:

a. alanine (R = $—CH_3$) **b.** serine (R = $—CH_2OH$)

Solution

a. The structure of the amino acids is written by attaching the side group (R) to the central carbon atom of the general structure of an amino acid.

alanine (ala or A)

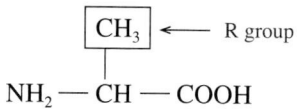

b. serine (ser or S)

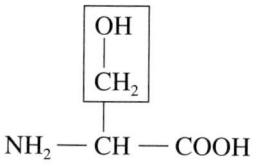

Study Check

Classify the amino acids in the sample problem as polar or nonpolar.

Figure 16.1 In our diet, amino acids are obtained from the proteins of eggs, milk, fish and poultry, corn, rice, beans, and nuts.

Essential Amino Acids

Of the 20 amino acids used to build all the proteins in our body, 10 can be synthesized in the body. However, 10 other amino acids, listed in Table 16.3, cannot be synthesized. They are **essential amino acids** and must be obtained from the proteins in the diet. All of the essential amino acids are found in the proteins of animal products such as eggs, milk, meat, fish, and poultry. Because the proteins in grains, beans, and nuts are deficient in one or more essential amino acids, different proteins must be eaten together. (See Figure 16.1.)

In a condition known as kwashiorkor, the number of calories in the diet of corn meal is sufficient, but corn is deficient in lysine and tryptophan. (See Table 16.4.) As a result, a child fails to grow, becomes lethargic, and has an enlarged abdomen from hypoprotein edema due to the inability to properly build proteins.

Table 16.3 Essential Amino Acids

Arginine	Methionine
Histidine	Phenylalanine
Isoleucine	Threonine
Leucine	Tryptophan
Lysine	Valine

Ionization of Amino Acids

An amino acid ionizes when the carboxyl group donates a proton and when the amino group attracts a proton. This ionized form of the amino acid, called a **zwitterion** or *dipolar ion,* has a net charge of zero.

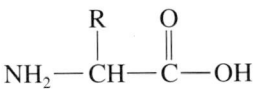

Nonionized amino acid

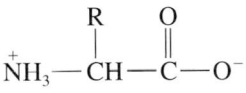

Zwitterion (a dipolar ion)

Most of the amino acids exist as zwitterions at pH values around 6 or 7, close to neutral. For example, alanine forms its zwitterion at a pH of 6.1:

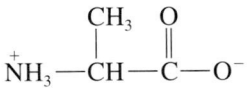

Zwitterion of alanine

Table 16.4 Amino Acids Missing in Selected Vegetables and Grains

Food Source	Amino Acid Deficiency
Eggs, milk, meat, fish, poultry	None
Wheat, rice, oats	Lysine
Corn	Lysine, tryptophan
Beans	Methionine, tryptophan
Peas	Methionine
Almonds, walnuts	Lysine, tryptophan

In an acidic solution (low pH), the ($—COO^-$) of the zwitterion accepts a proton (H^+), forming an ion with a positive charge. When placed in a basic solution (high pH), the ($—NH_3^+$) of the zwitterion donates a proton (H^+), forming an ion with a negative charge. For alanine, these changes are as follows:

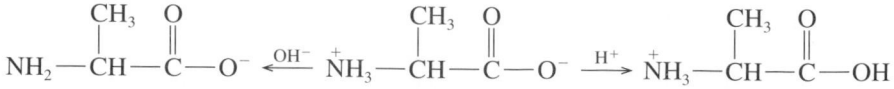

Negatively charged ion of alanine at a high pH (charge = 1−)

Zwitterion of alanine at neutral pH (charge = 0)

Positively charged ion of alanine at a low pH (charge = 1+)

16.2

S·A·M·P·L·E 16.3 ·P·R·O·B·L·E·M

Amino Acids in Acid or Base

Serine exists in its zwitterion form at a pH of 5.7. Draw the structural formula for the zwitterion of serine.

Solution

As a zwitterion, both the carboxylic acid group and the amino group are ionized.

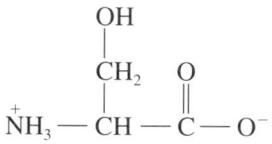

Zwitterion of serine

Study Check

At a pH of 2.0 serine is a positively charged ion. Draw its structure.

Q·U·E·S·T·I·O·N·S & P·R·O·B·L·E·M·S

Amino Acids

16.3 Describe the functional groups found in all amino acids.

16.4 What is the difference between a polar and a nonpolar side (R) group?

16.5 Draw the structural formula of each of the following amino acids:
a. alanine **b.** threonine
c. glutamic acid **d.** phenylalanine

16.6 Draw the structural formula of each of the following amino acids:
a. lysine **b.** aspartic acid **c.** leucine **d.** tyrosine

16.7 Describe the amino acids in problem 16.5 as nonpolar or polar.

16.8 Describe the amino acids in problem 16.6 as nonpolar or polar.

16.9 Give the name of the amino acid represented by each of the following three-letter abbreviations:
a. Ala **b.** Val **c.** Lys **d.** Cys

16.10 Give the name of the amino acid represented by each of the following three-letter abbreviations:
a. Trp **b.** Met **c.** Pro **d.** Gly

16.11 What is an essential amino acid? Which of the amino acids listed in problem 16.9 are essential amino acids?

16.12 How are essential amino acids obtained? Which of the amino acids listed in problem 16.10 are essential amino acids?

16.13 Write the zwitterion of each of the following amino acids:
a. glycine **b.** cysteine **c.** serine **d.** alanine

16.14 Write the zwitterion of each of the following amino acids:
a. phenylalanine **b.** methionine
c. leucine **d.** valine

16.15 Write the ionized form of the amino acids in problem 16.13 at low pH.

16.16 Write the ionized form of the amino acids in problem 16.14 at high pH.

Study Pack

Chem of Life Proteins: Levels of Protein Structure 4, 5

16.3

16.3 The Peptide Bond

Learning Goal

Describe a peptide bond; draw the structure for a peptide.

The linking of two or more amino acids forms a **peptide**. A **peptide bond** is an amide bond that forms when the carboxylic acid group (—COOH) of one amino acid reacts with the amino group (—NH_2) of the next amino acid.

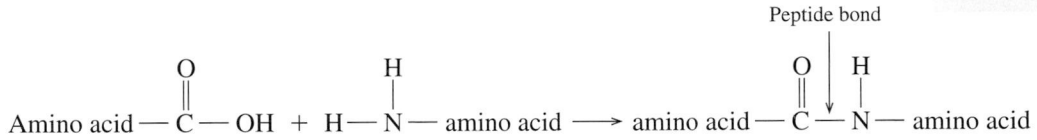

Two amino acids linked together by a peptide bond form a *dipeptide*. We can write the formation of the dipeptide between glycine and alanine as follows:

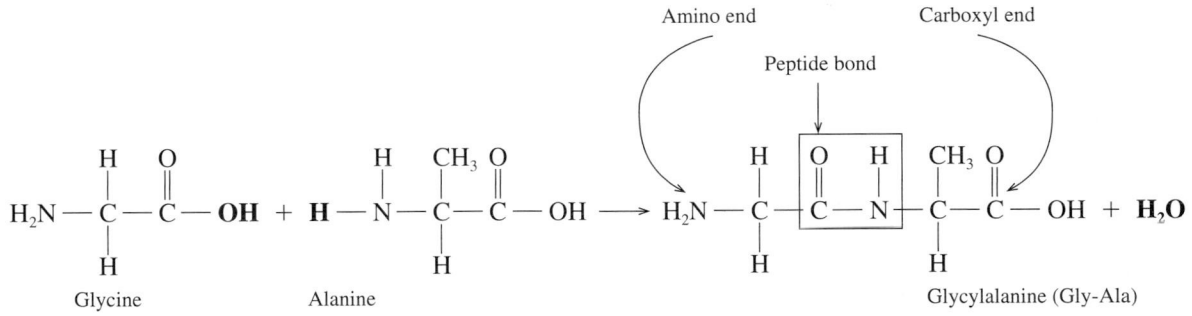

A peptide will have a free amino group at one end, and a free carboxyl group at the other end. In naming a small peptide, each amino acid beginning from the free amino group is named with a *yl* ending followed by the full name of the amino acid with the free carboxyl group. Usually the order of the amino acids in the peptide is written as the sequence of the three-letter abbreviations for the amino acids, such as Gly-Ala, rather than using the name.

In a tripeptide, three amino acids are linked by peptide bonds. For example, the tripeptide Ala-Gly-Ser is written as follows:

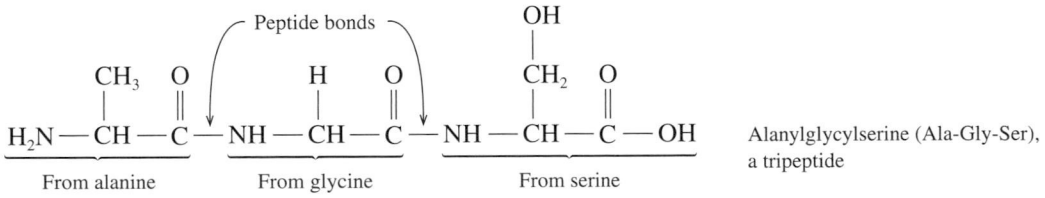

Alanylglycylserine (Ala-Gly-Ser), a tripeptide

16.3

S·A·M·P·L·E 16.4 ·P·R·O·B·L·E·M

Writing Structures of Dipeptides

Write a structural formula for the dipeptide Val-Ser.

Solution

Valine is joined to serine by a peptide bond; valine has a free amino group and serine has a free carboxyl group.

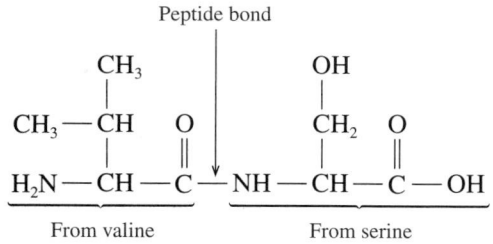

Valylserine (Val-Ser), a dipeptide

Study Check

Aspartame, an artificial sweetener 200 times sweeter than sucrose, contains the dipeptide Asp-Phe. What is the structure of the dipeptide in aspartame?

S·A·M·P·L·E 16.5 ·P·R·O·B·L·E·M

Identifying a Tripeptide

Consider the following tripeptide.

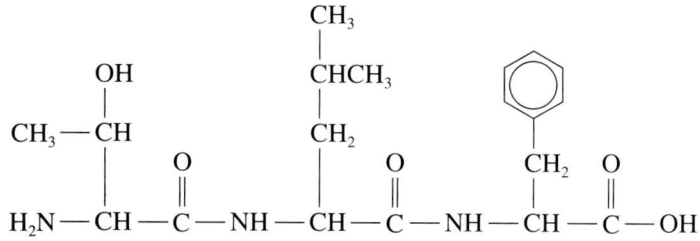

a. What amino acid has the free amino group? What amino acid has the free carboxyl group?

b. Using the three-letter abbreviations from Table 16.2, what is the order of amino acids in the peptide?

Solution

a. Threonine is the amino acid with the free amino group; phenylalanine has the free carboxyl group.

b. Thr-Leu-Phe

Study Check

What is the name of the tripeptide?

QUESTIONS & PROBLEMS

The Peptide Bond

16.17 Draw the structural formulas of the following peptides:

a. Ala-Cys **b.** Ser-Phe
c. Gly-Ala-Val **d.** Val-Ile-Trp

16.18 Draw the structural formulas of the following peptides:

a. Met-Asp **b.** Ala-Tyr
c. Met-Gln-Lys **d.** His-Gly-Glu-Ala

16.4

Proteins:
Secondary Structure 1–3

Learning Goal

Distinguish between the primary and secondary structures of a protein.

16.4 Primary and Secondary Structure

Long chains of amino acids linked by peptide bonds are called **polypeptides.** When there are more than 50 amino acids in a chain, the polypeptide is usually called a **protein.**

Primary Structure

The particular sequence of amino acids in a protein is referred to as that protein's **primary structure.** For example, the primary structure of a hormone that stimulates the thyroid to release thyroxine is the tripeptide Glu-His-Pro. Although other sequences are possible for these three amino acids, only the tripeptide with the following sequence of amino acids has hormonal activity.

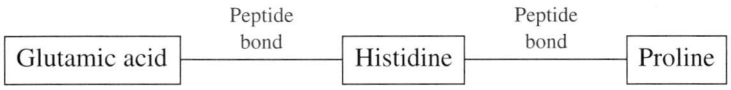

Each protein in our cells has a unique sequence of amino acids that determines the particular biological function. Figure 16.2 illustrates the primary structure of human insulin, which consists of two polypeptide chains that are held together by disulfide cross-links. There are 21 amino acids in chain A and 30 amino acids in chain B. For many years, bovine insulin was obtained from the pancreas of cows. In the past decade, human insulin produced through genetic engineering has been used to treat diabetes.

Secondary Structure

The **secondary structure** of a protein refers to the way the amino acids along the polypeptide are arranged in space. The three most common types of secondary structure are the *alpha helix*, the *beta-pleated sheet*, and the *triple helix* found in collagen. In each type of secondary structure, we will look at the hydrogen bonding between the hydrogen atom of an amino group in the polypeptide chain and the oxygen atom of the carboxyl group in another part of the chain.

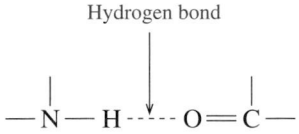

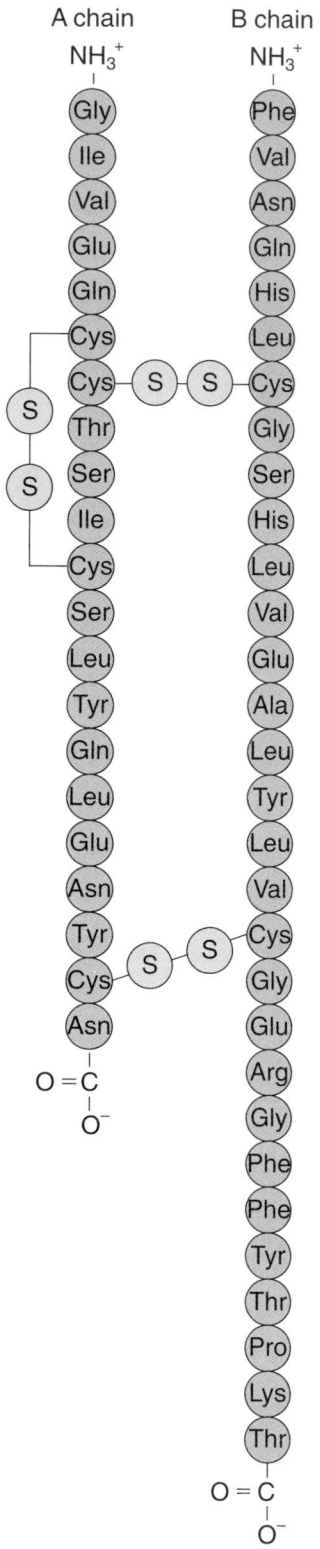

Figure 16.2 Primary structure of human insulin.

The Alpha Helix

The corkscrew shape of an **alpha helix** (α helix) is held in place by hydrogen bonds between each N—H group and the oxygen of a C=O group in the next turn of the helix, four amino acids down the chain. (See Figure 16.3.) Because many hydrogen bonds form along the peptide backbone, the protein takes the shape of a strong, tight coil that looks like a telephone cord or a Slinky toy. All the side groups (R groups) of the amino acids are located on the outside of the helix.

Beta-Pleated Sheet

Another type of secondary structure is known as the **beta-pleated sheet (β-pleated sheet)**. In a β-pleated sheet, polypeptide chains are held together side by side by hydrogen bonds between the peptide chains. In a β-pleated sheet, the small R groups of the prevalent amino acids, glycine, alanine, and serine, extend above and below the sheet. This results in a series of β-pleated sheets that can be stacked close together. The hydrogen bonds holding the β-pleated sheets tightly in place account for the strength and durability of fibrous proteins such as silk. (See Figure 16.4.)

In most proteins, only certain parts of the molecule have the α helix secondary structure. Other parts of the protein take a random shape, and sometimes a protein contains a β-pleated sheet secondary structure as well.

Triple Helix

Collagen makes up as much as one-third of all the protein in animals. It is prevalent in connective tissue, skin, tendons, ligaments, the cornea of the eye, and cartilage. The strong structure of collagen is a result of three peptide chains woven together like a braid to form a **triple helix,** seen in Figure 16.5. Collagen has a high percentage of glycine (35%) and proline (40%) and smaller amounts of hydroxyproline and hydroxylysine, forms of proline and lysine with —OH groups. Hydrogen bonding between these hydroxyl groups gives strength to the collagen triple helix. When several triple helixes wrap together, they form the fibrils of collagen that make up connective tissues.

S·A·M·P·L·E· 16.6 ·P·R·O·B·L·E·M

Identifying Secondary Structures

Indicate the type of secondary structure (α helix, β-pleated sheet, or triple helix) described in each of the following statements:

a. a coiled peptide chain held in place by hydrogen bonding between peptide bonds in the same chain

b. a structure that has hydrogen bonds between polypeptide chains arranged side by side.

Solution

a. α helix **b.** β-pleated sheet

Study Check

What is the type of secondary structure in collagen?

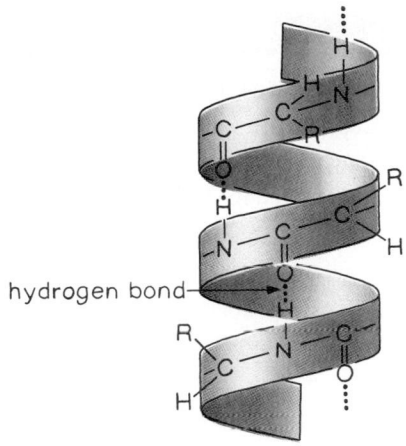

Figure 16.3 The α helix acquires a coiled shape through the hydrogen bonding between the N — H of one loop and the C = O of the next loop.

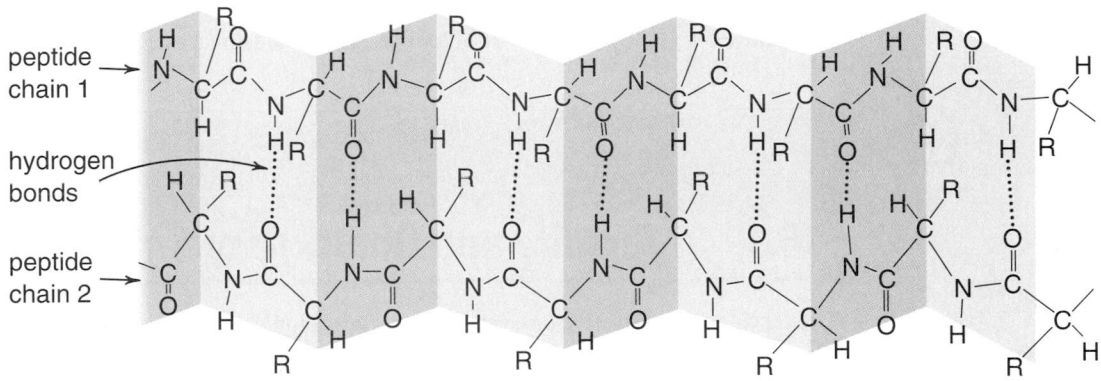

β-Pleated sheet

Figure 16.4 In a β-pleated sheet, hydrogen bonds form between the peptide bonds in adjacent rows of polypeptide chains.

HEALTH NOTE

Natural Opiates in the Body

Painkillers known as enkephalins and endorphins are produced naturally in the body. They are polypeptides that bind to receptors in the brain to give relief from pain. This effect appears to be responsible for the runner's high, for the temporary loss of pain when severe injury occurs, and for the analgesic effects of acupuncture.

The *enkephalins,* found in the thalamus and the spinal cord, are pentapeptides, the smallest molecules with opiate activity. The amino acid sequence of an enkephalin is found in the longer amino acid sequence of the endorphins.

Four groups of *endorphins* have been identified: α-endorphin contains 16 amino acids, β-endorphin contains 31 amino acids, γ-endorphin has 17 amino acids, and δ-endorphin has 27 amino acids. Endorphins may produce their sedating effects by preventing the release of substance P, a polypeptide with 11 amino acids, which has been found to transmit pain impulses to the brain.

α-Endorphin

Tyr — Gly — Gly — Phe — Met — Thr — Ser — Glu — Lys — Ser — Glu — Thr — Pro — Leu — Val — Thr

Enkephalin

Leu

Glu — Gly — Lys — Lys — Tyr — Ala — Asn — Lys — Ile — Ile — Ala — Asn — Lys — Phe

β-Endorphin

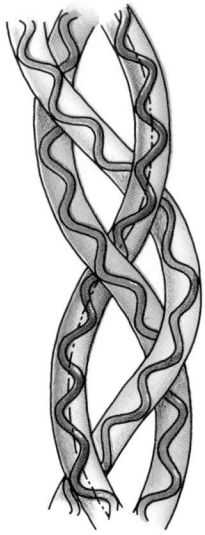

Figure 16.5 The triple helix in collagen.

16.5

Q·U·E·S·T·I·O·N·S & P·R·O·B·L·E·M·S

Primary and Secondary Structure

16.19 What type of bonding occurs in the primary structure of a protein?

16.20 How can two proteins with exactly the same number and type of amino acids have different primary structures?

16.21 Two peptides each contain one molecule of valine and two molecules of serine. What are their possible primary structures?

16.22 What are three different types of secondary protein structure?

16.23 What happens to the primary structure of a protein when a protein forms a secondary structure?

16.24 In an α helix, how does bonding occur between the amino acids in the polypeptide chain?

16.25 What is the difference in bonding between an α helix and a β-pleated sheet?

16.26 How is the secondary structure of a β-pleated sheet different from that of a triple helix?

Study Pack

Chem of Life Proteins: Tertiary Structure 1–8

Learning Goal

Distinguish between the tertiary and quaternary structures of a protein.

16.5 Tertiary and Quaternary Structure

The **tertiary structure** is the overall three-dimensional shape of a protein, which forms when attractions and repulsions between the side groups (R groups) of amino acids in the polypeptide chain twist and bend the protein into a specific shape. In **globular proteins** such as insulin, hemoglobin, enzymes, and antibodies, the tertiary structure is compact and nearly spherical in shape. Within this compact shape, the nonpolar amino acids form a hydrophobic (water-fearing) center, and the polar (water-loving) and ionized amino acids are grouped on the surface. The polar amino acids on the surface can hydrogen bond with water, which makes globular proteins soluble in water.

By contrast, the tertiary structures of **fibrous proteins** consist of long, thin, fiber-like shapes. One type of fibrous protein is alpha keratin (α-keratin), the protein that makes up hair, wool, skin, and nails. In hair, three α helixes coil together like a rope to form a fibril. The α helixes are cross-linked by disulfide (—S—S—) linkages between the many cysteine amino acids in hair, which contain sulfur in their side groups. When several of these fibrils bind together, they form one strand of hair as shown in Figure 16.6. When they are wet the fibrils in wool stretch, as hydrogen bonds are broken. They return to their original shape as the wool dries.

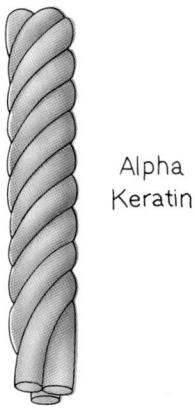

Figure 16.6 In hair and wool, which are fibrous proteins of α-keratins, many α helixes are wrapped together to form fibrils.

Cross-Links in Tertiary Structures

The tertiary structure of a protein is produced by various kinds of interactions between the R groups of amino acids in one region of the polypeptide chain with amino acids in other regions of the polypeptide chain. (See Figure 16.7.) All of these interactions play a role in stabilizing the tertiary structure of both the α helix and β-pleated sheet regions of a protein. Some of these stabilizing tertiary linkages are listed in Table 16.5.

S·A·M·P·L·E 16.7 ·P·R·O·B·L·E·M

Cross-Links in Tertiary Structures

What type of interaction would you expect between the R groups of the following amino acids?

a. cysteine and cysteine **b.** glutamic acid and lysine

Solution

a. Because cysteine has an R group containing —SH, a disulfide bond will form.
b. An ionic bond (salt bridge) can form by the interaction of the $—COO^-$ of glutamic acid and the $—NH_3^+$ of lysine.

Study Check

Would you expect to find valine and leucine in a globular protein on the outside or the inside of the tertiary structure? Why?

Table 16.5 Some Cross-Links in Tertiary Structures

Cross-Links	Nature of Bonding	Example
Disulfide bonds	Strong covalent links between sulfur atoms of two cysteine amino acids	—SH + HS— ⟶ —S—S—
Ionic bonds (salt bridges)	Ionic interactions between ionized R groups of acidic and basic amino acids	—CO⁻ ------ H—N⁺—
Hydrogen bonds	Occur between polar side group of amino acids	>C=O ------ HO— >C=O ------ H—N—
Hydrophobic interactions	Attractions between nonpolar alkyl and aromatic groups form a nonpolar center that is repelled by water	—CH_3 CH_3—

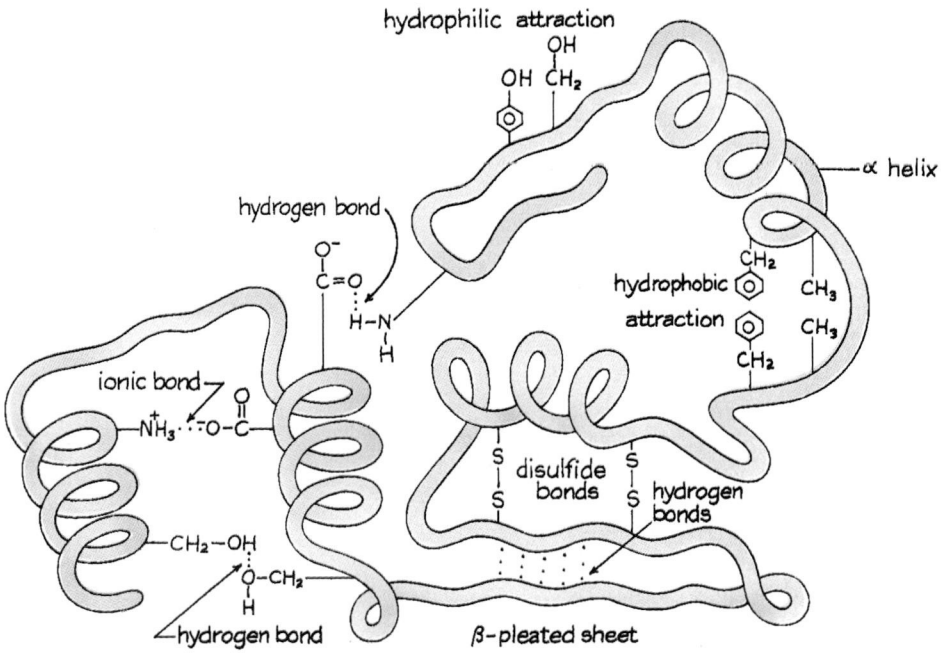

Figure 16.7 Interactions between amino acid R groups fold a protein into the tertiary structure of a globular protein.

HEALTH NOTE

Hair Permanents

Much of the protein in the hair is cross-linked by disulfide bonds between R groups of cysteine, an amino acid that is prevalent in hair.

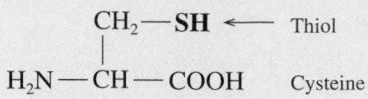

16.5 In a disulfide, two sulfur atoms are linked together to give an —S—S— group.

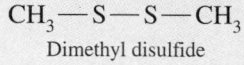

Disulfides are formed by the reaction of two thiols.

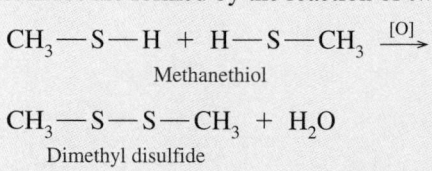

The shape of the hair, curly or straight, is determined by the way the proteins are cross-linked by disulfide bonds formed between molecules of cysteine.

When a permanent is given, a reducing substance is put on the hair to break apart the disulfide bonds. Then the hair is wrapped around curlers and an oxidizing substance is used that causes the disulfide bonds to form between protein strands that take the shape of the curlers. (See Figure 16.8.)

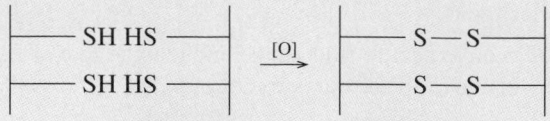

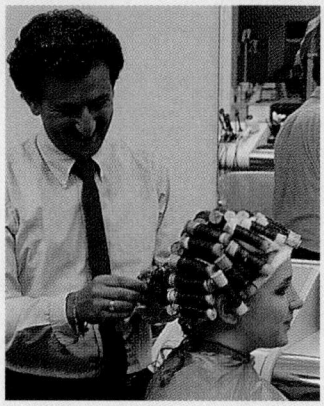

Figure 16.8 In a permanent, the disulfide bonds in hair protein are reduced so they break apart and then are oxidized so they reform disulfide bonds to give a new shape.

Myoglobin, a Tertiary Protein Structure

Myoglobin is a globular protein that stores oxygen in skeletal muscle. High concentrations of myoglobin have been found in the muscles of sea mammals, such as seals and whales, that stay under the water for long periods. Myoglobin contains 153 amino acids in a single polypeptide chain with about three-fourths of the chain in the α helix secondary structure. The polypeptide chain, including its helical regions, forms a compact tertiary structure by folding upon itself. (See Figure 16.9.) Within the tertiary structure, a nonpolar pocket of amino acids and a heme group binds and stores oxygen (O_2).

Quaternary Structure: Hemoglobin

When a biologically active protein consists of two or more polypeptide subunits, the structural level is referred to as a **quaternary structure.** Hemoglobin, a globular protein that transports oxygen in blood, consists of four polypeptide chains or subunits, two α chains and two β chains. The subunits are held together in the quaternary structure by the same interactions that stabilize the tertiary structure, such as hydrogen and ionic bonds between side groups, disulfide links, and hydrophobic attractions. (See Figure 16.10.) Each subunit of the hemoglobin contains a heme group that binds oxygen. In the adult hemoglobin molecule, all four subunits ($\alpha_2\beta_2$) must be combined for the hemoglobin to properly function as an oxygen carrier. Therefore, the complete quaternary structure of hemoglobin can bind and transport four molecules of oxygen.

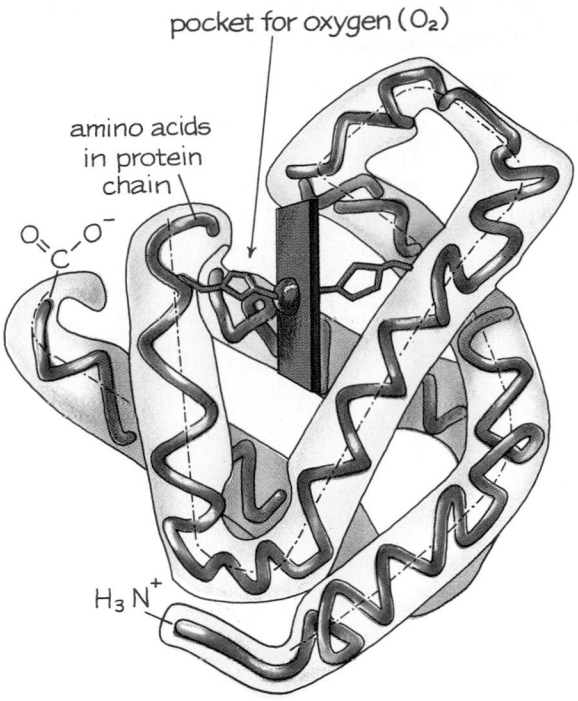

Figure 16.9 In the tertiary structure of myoglobin, a heme group and surrounding nonpolar amino acids form a nonpolar pocket for storing oxygen.

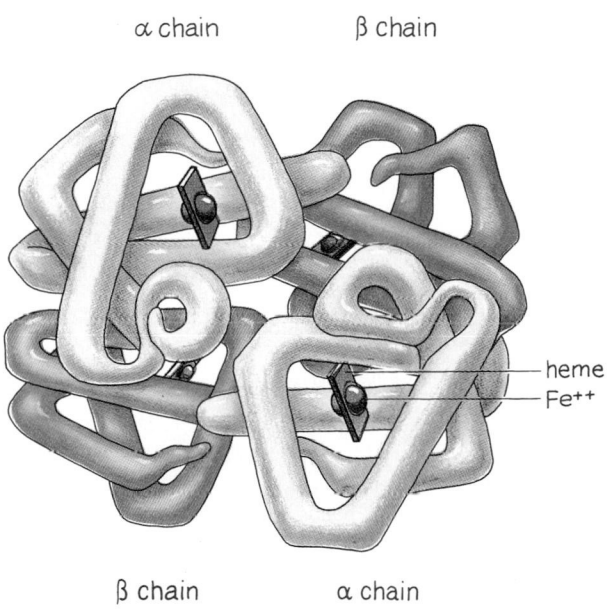

Figure 16.10 The structure of hemoglobin contains four polypeptide subunits, each containing a heme group where oxygen is stored.

Hemoglobin and myoglobin have similar biological functions. Hemoglobin carries oxygen in the blood, whereas myoglobin carries oxygen in muscle. Myoglobin, a single polypeptide chain with a molar mass of 17,000, has about one-fourth the molar mass of hemoglobin (64,000). The tertiary structure of the single polypeptide myoglobin is almost identical to the tertiary structure of each of the subunits of hemoglobin. Myoglobin stores just one molecule of oxygen, just as each subunit of hemoglobin carries one oxygen molecule. The similarity in tertiary structures allows each protein to bind and release oxygen in a similar manner. Table 16.6 and Figure 16.12 summarize the structural levels of proteins.

HEALTH NOTE

Sickle-Cell Anemia

16.5

Sickle-cell anemia is a disease caused by an abnormality in the shape of one of the subunits of the hemoglobin protein. In the β chain, the sixth amino acid, glutamic acid, which is polar, is replaced by valine, a nonpolar amino acid. An essential ionic cross-link no longer forms, causing a severe change in the shape and function of the tertiary structure of the hemoglobin. The affected red blood cells change from a rounded shape to a crescent shape, like a sickle, which interferes with their ability to transport adequate quantities of oxygen. (See Figure 16.11.)

Sickled cells are removed from circulation more rapidly than normal red blood cells, causing anemia and low oxygen concentration in the tissues. Clumps of sickled cells clog the capillaries, where they cause great pain and organ damage. Critically low oxygen levels may occur in the affected tissues.

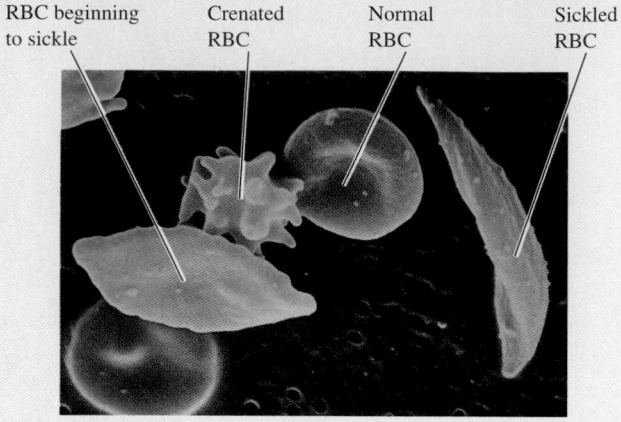

Figure 16.11 Comparison of the shape of sickled cells with a normal red blood cell (RBC).

Polar amino acid

Normal β chain: Val—His—Leu—Thr—Pro—[Glu]—Glu—Lys—

Sickled β chain: Val—His—Leu—Thr—Pro—[Val]—Glu—Lys—

Nonpolar amino acid

In sickle-cell anemia, both genes for the altered hemoglobin must be inherited. However, a few sickled cells are found in persons who carry one gene for sickle-cell hemoglobin, a condition that is also known to provide protection from malaria.

Table 16.6 Summary of Structural Levels in Proteins

Structural Level	Characteristics
Primary	The sequence of amino acids
Secondary	The coiled α helix, β-pleated sheet, or a triple helix formed by hydrogen bonding between peptide bonds along the chain
Tertiary	A folding of the protein into a compact, three-dimensional shape stabilized by interactions between side R groups of amino acids
Quaternary	A combination of two or more protein subunits to form a larger, biologically active protein

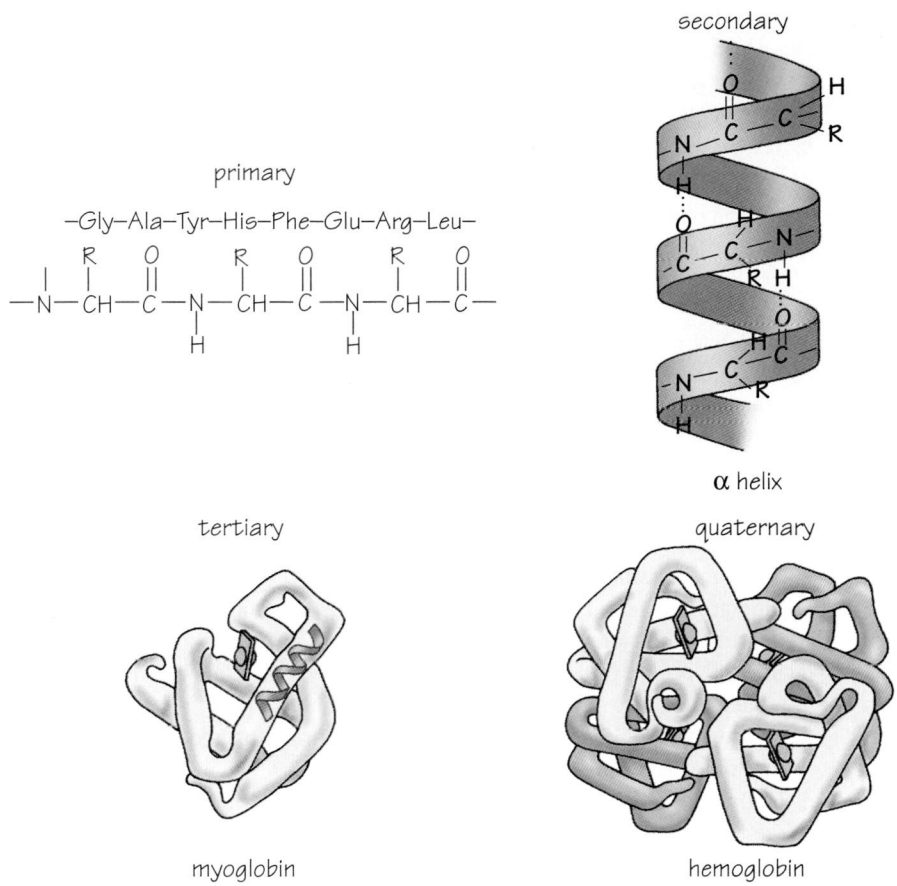

Figure 16.12 Characteristics of the four structural levels of proteins.

S·A·M·P·L·E 16.8 ·P·R·O·B·L·E·M

Identifying Protein Structure

Indicate whether the following conditions are responsible for primary, secondary, tertiary, or quaternary protein structures:

a. Disulfide bonds form between portions of a protein chain.

b. Peptide bonds form a chain of amino acids.

Solution

a. Disulfide bonds help to stabilize the tertiary structure of a protein.

b. The sequence of amino acids in a polypeptide is a primary structure.

Study Check

What structural level is represented by the grouping of two subunits in insulin?

Q·U·E·S·T·I·O·N·S & P·R·O·B·L·E·M·S

Tertiary and Quaternary Structure

16.27 What type of interaction would you expect from the following R groups in a tertiary structure?

a. two cysteine residues
b. glutamic acid and lysine
c. serine and aspartic acid
d. two leucine residues

16.28 In myoglobin, about one-half of the 153 amino acids have nonpolar side chains.

a. Where would you expect those amino acids to be located in the tertiary structure?

b. Where would you expect the polar side chains to be?

c. Why is myoglobin more soluble in water than silk or wool?

16.29 A portion of a polypeptide chain contains the following sequence of amino acid residues:

-Leu-Val-Cys-Asp-

16.6

a. Which R groups can form a disulfide cross-link?

b. Which amino acid residues are likely to be found on the inside of the protein structure? Why?

c. Which amino acid residues would be found on the outside of the protein? Why?

d. How does the primary structure of a protein affect its tertiary structure?

16.30 State whether the following statements apply to primary, secondary, tertiary, or quaternary protein structure:

a. Side groups interact to form disulfide bonds or ionic bonds.

b. Peptide bonds join amino acids in a polypeptide chain.

c. Several polypeptides are held together by hydrogen bonds between adjacent chains.

d. Hydrogen bonding between carbonyl oxygen atoms and nitrogen atoms of amide groups causes a polypeptide to coil.

e. Hydrophobic side chains seeking a nonpolar environment move toward the inside of the folded protein.

f. Protein chains of collagen form a triple helix.

g. An active protein contains four tertiary subunits.

Study Pack

16.6 Protein Hydrolysis and Denaturation

Learning Goal

Describe the hydrolysis and denaturation of proteins.

Peptides are broken apart by hydrolysis, a reaction that adds water to the peptide bonds in the presence of an acid or enzyme such as pepsin or trypsin.

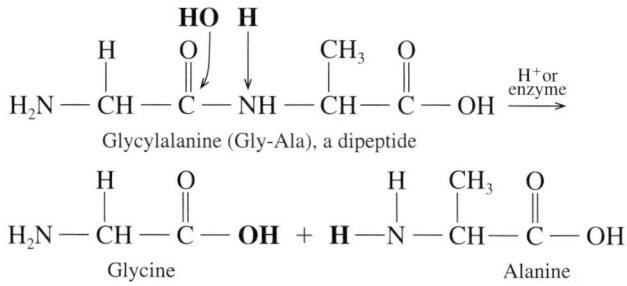

Denaturation occurs when there is a disruption of the bonds that stabilize the secondary, tertiary, or quaternary structures of proteins. When the interactions between the R groups are undone or altered, a globular protein unfolds like a loose piece of spaghetti. With the loss of its overall shape, the protein is no longer biologically active. (See Figure 16.13.)

Denaturing agents include heat, acids and bases, organic compounds, heavy metal ions, and mechanical agitation.

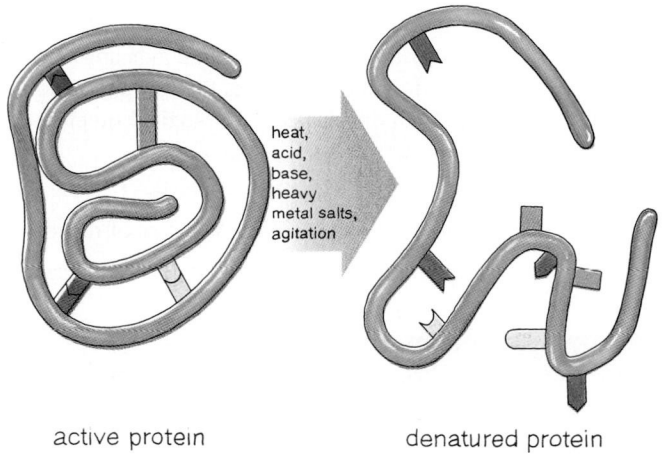

Figure 16.13 Denaturation of a protein occurs when the bonds of the tertiary structure are disrupted, destroying the shape and rendering the protein biologically inactive.

Heat. Heat denatures proteins by breaking apart hydrogen bonds and the hydrophobic attraction between nonpolar side groups. Few proteins can remain biologically active above 50°C. Whenever you cook food, you are using heat to denature protein. The nutritional value of the proteins in food is not changed, but they are made more digestible. High temperatures are also used to disinfect surgical instruments and gowns by denaturing the proteins of any bacteria present.

Acids and Bases. Placing a protein in an acid or base affects the hydrogen bonding between polar R groups and disrupts the ionic bonds (salt bridges). In the preparation of yogurt and cheese, a bacteria that produces lactic acid is added to denature the milk protein and produce solid casein. Tannic acid, a weak acid used in burn ointments, coagulates proteins at the site of the burn, forming a protective cover and preventing further loss of fluid from the burn.

Organic Compounds. Ethanol and isopropyl alcohol act as disinfectants by forming their own hydrogen bonds with a protein and disrupting the hydrophobic interactions. An alcohol swab is used to clean wounds or to prepare the skin for an injection because the alcohol passes through the cell walls and coagulates the proteins inside the bacteria.

Heavy Metal Ions. When heavy metal ions like Ag^+, Pb^{2+}, and Hg^{2+} form ionic bonds or react with the disulfide (—S—S—) bonds, the denatured protein solidifies. In hospitals, dilute (1%) solutions of $AgNO_3$ are placed in the eyes of newborn babies to destroy the bacteria that causes gonorrhea. These heavy metal ions can be toxic. If ingested, they act as poisons by severely disrupting body proteins, especially enzymes. An antidote is a high-protein food such as milk, eggs, or cheese that will tie up the heavy metal ions until the stomach can be pumped.

Agitation. The whipping of cream and the beating of egg whites are examples of using mechanical agitation to denature protein. The whipping action stretches the polypeptide chains until the bonds break apart.

S·A·M·P·L·E **16.9** ·P·R·O·B·L·E·M

Effects of Denaturation

What happens to the tertiary structure of a globular protein when it is placed in an acidic solution?

Solution

An acid causes denaturation by disrupting the hydrogen bonds and the ionic bonds between the R groups. A loss in interactions causes the tertiary structure to lose stability. As the protein unfolds, both the shape and biological function are lost.

Study Check

Why is a dilute solution of $AgNO_3$ used to disinfect the eyes of newborn infants?

16.6

EXPLORE YOUR WORLD

Denaturation of Milk Protein

Place some milk in five glasses. Add the following to the milk samples in glasses 1–4. The fifth glass of milk is a reference sample.

1. Vinegar, drop by drop. Stir.
2. One-half teaspoon of meat tenderizer. Stir.
3. One teaspoon of fresh pineapple juice. (Canned juice has been heated.)
4. One teaspoon of fresh pineapple juice that you have heated to boiling.

Questions

1. How did the appearance of the milk change in each of the samples?
2. What enzyme is listed on the package label of the tenderizer?
3. How does the effect of the heated pineapple juice compare with that of the fresh juice? Explain.
4. Why is cooked pineapple used when making gelatin (a protein) desserts?

QUESTIONS & PROBLEMS

Protein Hydrolysis and Denaturation

16.31 What products would result from the complete hydrolysis of Gly-Ala-Ser?

16.32 Would the hydrolysis products of the tripeptide Ala-Ser-Gly be the same or different from the products in problem 16.31? Explain.

16.33 What dipeptides would be produced from the partial hydrolysis of His-Met-Gly-Val?

16.34 What tripeptides would be produced from the partial hydrolysis of Ser-Leu-Gly-Gly-Ala?

16.35 What structural level of a protein is affected by hydrolysis?

16.36 What structural level of a protein is affected by denaturation?

16.37 Indicate the changes in protein structure for each of the following:

a. An egg placed in hot water at 100°C is soft boiled in about 3 min.
b. Prior to giving an injection, the skin is wiped with an alcohol swab.
c. Surgical instruments are placed in a 120°C autoclave.
d. During surgery, a wound is closed by cauterization (heat).

16.38 Indicate the changes in protein structure for each of the following:

a. Tannic acid is placed on a burn.
b. Milk is heated to 60°C to make yogurt.
c. To avoid spoilage, seeds are treated with a solution of $HgCl_2$.
d. Hamburger is cooked at high temperatures to destroy *E. coli* bacteria that may cause intestinal illness.

16.7 Enzymes

Study Pack

Chem of Life Enzyme Action: Enzymes and Pathways 1–3

Learning Goal

Describe enzymes and their functions.

16.7

In the laboratory, we can break down polysaccharides, fats, or proteins by using a strong acid or base, high temperatures, and long reaction times. In the cells of our body, the same reactions occur at incredibly fast rates under mild conditions close to pH 7 and a body temperature of 37°C. In our cells almost all the chemical reactions are catalyzed by proteins known as **enzymes.** At this moment, in each of your cells there are as many as 1500 different types of enzymes that catalyze the chemical reactions needed by your body.

In our discussion of the rates of chemical reactions in Chapter 6, we described a catalyst as a substance that lowers the activation energy of a reaction without being changed itself. Enzymes are catalysts that speed up chemical reactions in our cells by lowering the energy required for the reaction. The same reactions would eventually occur, but not at a rate fast enough for survival. For example, the hydrolysis of protein in our diet would occur without a catalyst, but not fast enough to meet the body's requirements for amino acids. However, when a digestive enzyme such as pepsin or trypsin is present, the hydrolysis occurs at a much faster rate.

In the blood, an enzyme called carbonic anhydrase converts large amounts of carbon dioxide and water to carbonic acid. Each molecule of anhydrase catalyzes about 35 million reactant molecules every minute. Typically, an enzyme can make a reaction go 10 million times faster. (See Figure 16.14.)

Classification and Names of Enzymes

There are many different enzymes because each one catalyzes one particular reaction. As enzymes were discovered, many were named by adding the suffix *ase* to the names of the reactants. The enzyme that hydrolyzes sucrose is *sucrase,* and the enzymes that hydrolyze lipids are *lipases.* However, the names of some digestion enzymes, including pepsin, renin, and trypsin, have been retained.

More recently, a systematic method of classifying and naming enzymes has been established. The name and class of each indicates the type of reaction it catalyzes. There are six main classes of enzymes, as described in Table 16.7.

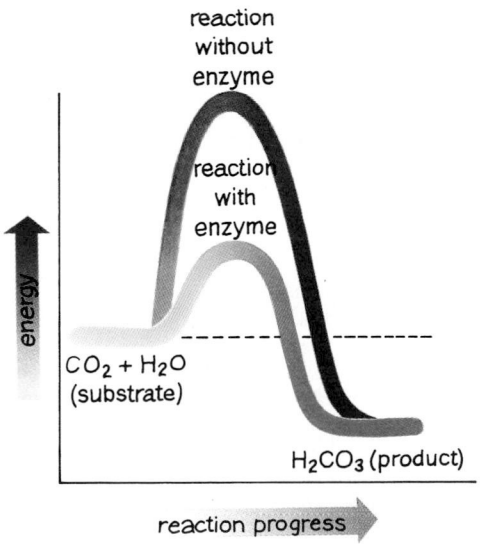

Figure 16.14 The energy needed for the reaction of CO_2 and H_2O is lowered by the enzyme carbonic anhydrase.

16.7

Table 16.7 Classes of Enzymes

Class	Reaction Catalyzed	Examples
Oxidoreductases	Oxidation–reduction reactions	*Oxidases* oxidize a substance. *Reductases* reduce a substance. *Dehydrogenases* remove 2H atoms to form a double bond.
Transferases	Transfer a group between two compounds	*Transaminases* transfer amino groups. *Kinases* transfer phosphate group.
Hydrolases	Hydrolysis reactions	*Proteases* hydrolyze peptide bonds. *Lipases* hydrolyze ester bonds in lipids. *Carbohydrases* hydrolyze glycosidic bonds in carbohydrates. *Phosphatases* hydrolyze phosphate–ester bonds. *Nucleases* hydrolyze nucleic acids.
Lyases	Add or remove groups involving a double bond without hydrolysis	*Carboxylases* add CO_2. *Deaminases* remove NH_3.
Isomerases	Rearrange atoms in a molecule to form an isomer	*Isomerases* convert cis to trans, or reverse. *Epimerases* convert D to L isomers, or reverse.
Ligases	Form bonds between molecules using ATP energy	*Synthetases* combine two molecules.

S·A·M·P·L·E 16.10 ·P·R·O·B·L·E·M

Enzyme Names

What chemical reaction would be catalyzed by the following enzymes?

a. amino transferase **b.** lactate dehydrogenase

Solution

a. catalyzes the transfer of an amino group
b. catalyzes the removal of hydrogen from lactate

Study Check

What is the class of lipase, an enzyme that catalyzes the hydrolysis of ester bonds in triglycerides?

Q·U·E·S·T·I·O·N·S & P·R·O·B·L·E·M·S

Enzymes

16.39 What is the reactant for each of the following enzymes?
a. galactase **b.** lipase **c.** aspartase

16.40 What is the reactant for each of the following enzymes?
a. peptidase **b.** cellulase **c.** lactase

16.41 What main class of enzymes catalyzes each of the following reactions?
a. hydrolysis of sucrose (sugar)
b. addition of oxygen

c. converting glucose ($C_6H_{12}O_6$) to fructose ($C_6H_{12}O_6$)
d. moving an amino group to another substance

16.42 What main class of enzymes catalyzes each of the following reactions?
a. addition of water to a double bond
b. removal of hydrogen
c. splitting ester bonds in triglycerides
d. removing CO_2 from pyruvate

16.43 Identify the main class of enzyme that would catalyze each of the following reactions:

a. $CH_3-\overset{\overset{\displaystyle O}{\|}}{C}-COOH \longrightarrow CH_3-\overset{\overset{\displaystyle O}{\|}}{C}H + CO_2$

b. $CH_3-\overset{\overset{\displaystyle NH_2}{|}}{C}H-COOH + HOOC-\overset{\overset{\displaystyle O}{\|}}{C}-CH_2CH_3 \longrightarrow$

$CH_3-\overset{\overset{\displaystyle O}{\|}}{C}-COOH + HOOC-\overset{\overset{\displaystyle NH_2}{|}}{C}H-CH_2CH_3$

16.44 Identify the main class of enzyme that catalyzes each of the following reactions:

a. $CH_3-\overset{\overset{\displaystyle O}{\|}}{C}-COOH + CO_2 + ATP \longrightarrow$

$HOOC-CH_2-\overset{\overset{\displaystyle O}{\|}}{C}-COOH + ADP + P_i$

b. $CH_3CH_2OH + NAD^+ \longrightarrow CH_3\overset{\overset{\displaystyle O}{\|}}{C}H + NADH + H^+$

16.8

16.8 Enzyme Action

Study Pack

Learning Goal

Describe the role of an enzyme in an enzyme-catalyzed reaction.

Enzymes are globular proteins, which means that they have unique three-dimensional shapes. In a reaction catalyzed by an enzyme, the reactant is called the **substrate.** One important feature of enzymes is their specificity of catalyzing reactions involving only one type of substrate. Some examples of enzyme specificity are listed in Table 16.8.

Table 16.8 Types of Enzyme Specificity

Enzyme Specificity	Reaction Type	Example
Absolute	Catalyze one type of reaction for one substrate	Urease only catalyzes the hydrolysis of urea.
Group	Catalyze one type of reaction for substrate with the same functional group	Hexokinase adds a phosphate group to sugars.
Linkage	Catalyze one type of reaction for a particular bond	Lipase catalyzes the hydrolysis of ester links of glycerol and fatty acids.

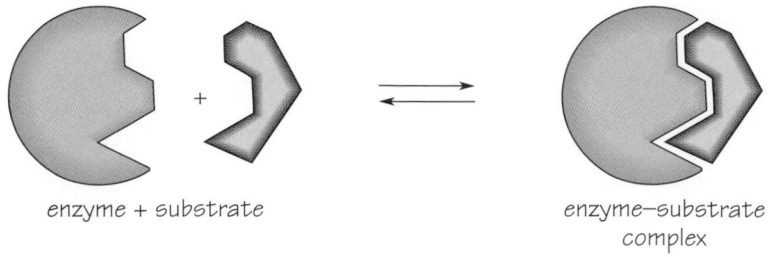

Figure 16.15 In the lock-and-key model of enzyme action, a substrate that fits the shape of the active site binds to the enzyme. [Adapted from Caret, Denniston and Topping, *Principles and Applications of Inorganic, Organic and Biological Chemistry,* 2nd ed. (Dubuque, IA: Wm. C. Brown, 1997). © 1997. Reprinted by permission of The McGraw-Hill Companies.]

16.8 Lock-and-Key Model

An enzyme recognizes a substrate when the shape of the substrate fits into a part of the three-dimensional structure of the enzyme. This part of the enzyme, called the **active site,** is a pocket that is formed by some R groups of the amino acids in the protein. The active site is that part of the enzyme where the catalyzed reaction of the substrate takes place. Enzyme action is described in a **lock-and-key model** because a substrate must have the geometry of the correct key to fit the active site, which is the lock. (See Figure 16.15.)

Enzyme-Catalyzed Reaction

In preparation for an enzyme-catalyzed reaction, a substrate (S) must first attach to the active site of an enzyme (E) to form an **enzyme–substrate complex.** Within the enzyme–substrate complex, the substrate is converted to a product, producing an **enzyme–product complex.** Because the shape of the product is different from that of the substrate, the molecules of product detach from the enzyme. The active site is available again to bind and catalyze the reaction of another substrate molecule. There has been no change in the enzyme catalyst, only in the substrate. (See Figure 16.16.) The overall enzyme-catalyzed reaction can be written as follows:

$$\underset{\text{Enzyme + substrate}}{E + S} \rightleftharpoons \underset{\text{Enzyme–substrate complex}}{ES} \rightleftharpoons \underset{\text{Enzyme–product complex}}{EP} \rightleftharpoons \underset{\text{Enzyme + product}}{E + P}$$

Figure 16.16 An enzyme-catalyzed reaction requires the formation of an enzyme–substrate complex.

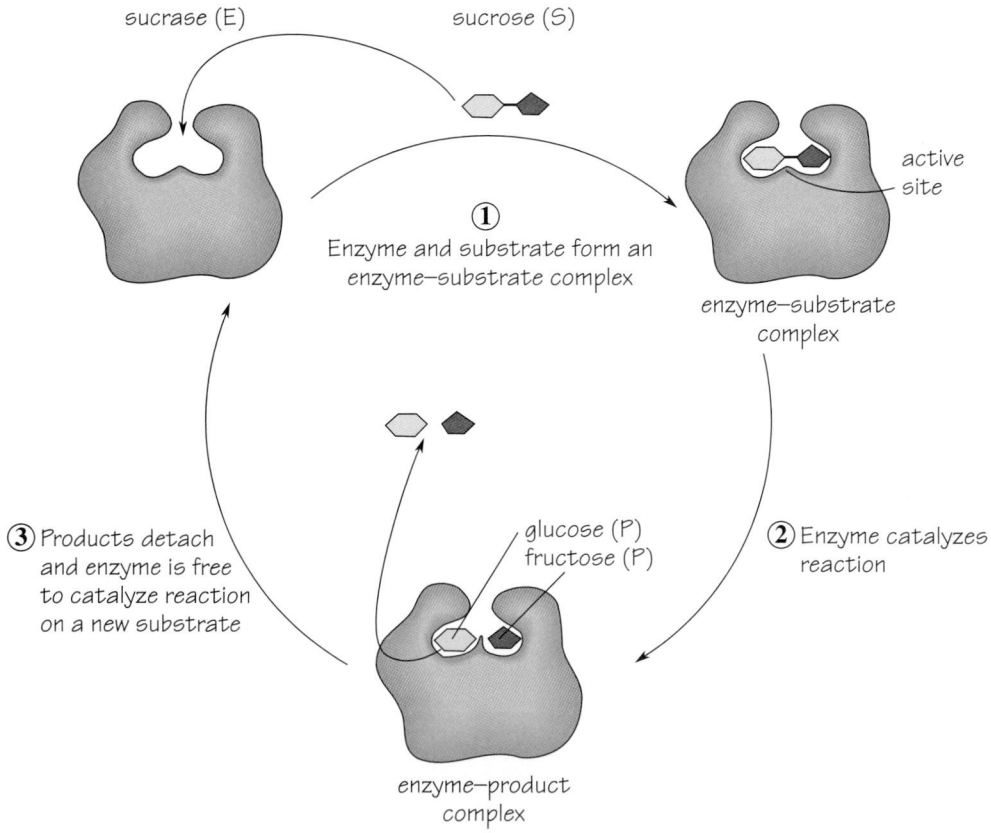

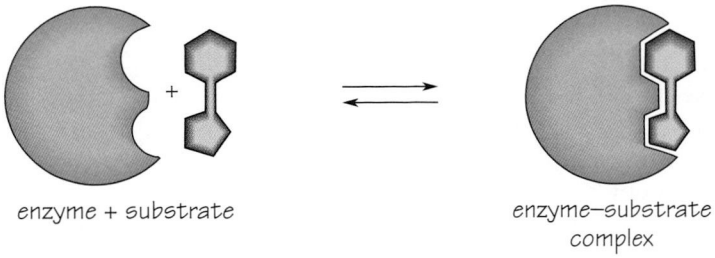

Figure 16.17 In the induced-fit model, the flexible active site adapts to the shape of the substrate. After catalysis occurs in the ES complex, the products disperse and the enzyme is free to bind another molecule of substrate. [Adapted from Caret, Denniston and Topping, *Principles and Applications of Inorganic, Organic and Biological Chemistry,* 2nd ed. (Dubuque, IA: Wm. C. Brown, 1997). © 1997. Reprinted by permission of The McGraw-Hill Companies.]

Induced-Fit Model

The lock-and-key model helps explain typical enzyme action. However, it assumes that the active site is rigid and unbending. We now know that some enzymes have a broader range of specificity and can adjust their shape to fit more than one substrate. In the **induced-fit model,** the structure of an enzyme is flexible, not rigid, which allows an enzyme to adjust to the shapes of different, but similar, substrates. (See Figure 16.17.)

S·A·M·P·L·E· **16.11** ·P·R·O·B·L·E·M

Enzyme Action

What is the function of the active site in an enzyme?

Solution

The active site in an enzyme is where the catalyzed reaction of a substrate takes place.

Study Check

How do the lock-and-key model and the induced-fit model describe the active site?

HEALTH NOTE

Enzymes as Diagnostic Tools

Enzymes that catalyze the same reactions but vary slightly in structure are called isoenzymes. Different cells in the body produce isoenzymes for the same type of reactions. For example, there are five isoenzymes for lactate dehydrogenase (LDH), an enzyme that converts pyruvic acid to lactic acid. Each isoenzyme consists of a quaternary structure containing four subunits. In heart muscle, the most prevalent subunit is designated as H. In skeletal muscle, the major subunit is designated as M. With different combinations of subunits, isoenzymes can be separated by electrophoresis.

In healthy tissues, enzymes are contained within cellular membranes. However, if a particular organ is damaged, the contents of the cells, including the enzymes, enter the blood. By identifying the isoenzyme that is elevated in the blood serum, the damaged organ can be identified. For example, liver diseases are indicated by a rise in the serum LDH_5 level. When myocardial infarction (MI), or heart attack, damages the cells in heart muscle, the serum LDH_2 level increases.

Isoenzyme	Subunits	Abundant in
LDH_1	H_4	Heart, kidneys
LDH_2	H_3M	Heart, kidneys, brain red blood cells
LDH_3	H_2M_2	Kidneys, lungs, brain
LDH_4	HM_3	Spleen, lung
LDH_5	M_4	Liver, skeletal muscle

Another isoenzyme used diagnostically is creatine kinase (CK), which consists of two types of subunits. One subunit (B) is prevalent in the brain and the other (M) predominates in skeletal muscle. Normally only the CK_3 is present in low amounts in the blood serum. However, in a patient who has suffered an MI, the levels of CK_2 will be elevated soon after the heart attack. (See Figure 16.18.)

Table 16.9 lists some enzymes used to diagnose diseases of certain organs.

16.8

Isoenzyme	Subunits	Abundant in
CK_1	BB	Brain, lung
CK_2	MB	Heart
CK_3	MM	Skeletal muscle, red blood cells

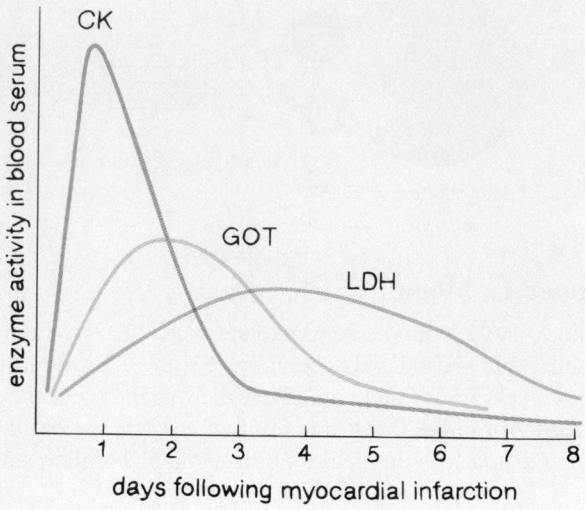

Figure 16.18 Normally, blood serum levels of LDH, CK, and GOT are low. After a heart attack, their levels rise. The serum levels of these enzymes can aid in the assessment of the severity of heart damage.

Table 16.9 Serum Enzymes Used in Diagnosis of Tissue Damage

Organ	Condition	Diagnostic Enzymes Elevated
Heart	Myocardial infarction (MI)	Lactate dehydrogenase (LDH_1) Creatinine phosphokinase (CK_2) Glutamic oxaloacetic transaminase (GOT)
Liver	Cirrhosis, carcinoma, hepatitis	Glutamic pyruvic transaminase (GPT) Lactate dehydrogenase (LDH_5) Alkaline phosphatase (ALP) Glutamic oxaloacetic transaminase (GOT)
Bone	Rickets, carcinoma	Alkaline phosphatase (ALP)
Pancreas	Pancreatic diseases	Amylase Cholinesterase Lipase (LPS)
Prostate	Carcinoma	Acid phosphatase (ACP)

Electrophoresis

In the hospital laboratory, **electrophoresis** is used to separate proteins. An electric current applied to a protein mixture, as shown in Figure 16.19, causes the positively charged proteins to move toward the negative electrode and the negatively charged proteins to move toward the positive electrode. The proteins are separated because they move at different rates toward the electrode depending on their ionic charge. A protein that is neutral does not move during electrophoresis. The proteins are identified by their direction and rate of migration in the electric field. Electrophoresis is also used to separate other charged molecules, most notably DNA in DNA fingerprinting techniques.

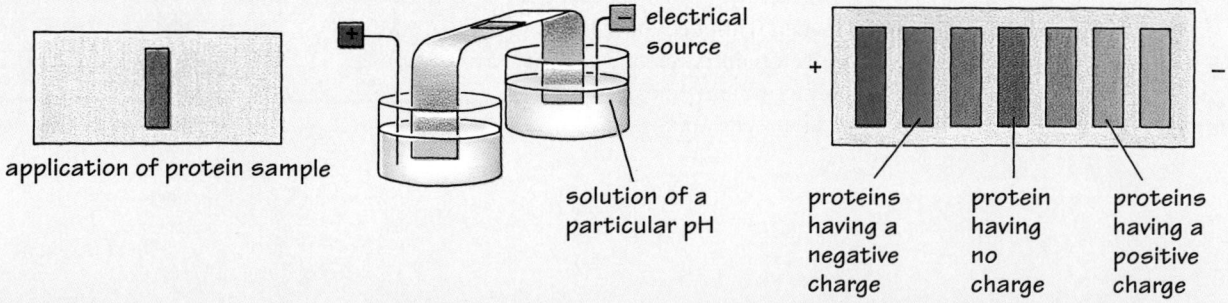

Figure 16.19 A positively charged protein moves toward the negative electrode; a negatively charged protein moves toward the positive electrode; a protein with no charge does not migrate.

Q·U·E·S·T·I·O·N·S & P·R·O·B·L·E·M·S

Enzyme Action

16.45 Match the following three terms, (1) enzyme–substrate complex, (2) enzyme, and (3) substrate, with these phrases:
a. has a tertiary structure that recognizes the substrate
b. the combination of an enzyme with the substrate
c. has a structure that complements the structure of the enzyme

16.46 Match the following three terms, (1) active site, (2) lock-and-key model, and (3) induced-fit model, with these phrases:
a. the portion of an enzyme where catalytic activity occurs
b. an active site that adapts to the shape of a substrate
c. an active site having a rigid shape

16.47 Write an equation that represents an enzyme-catalyzed reaction.
a. Identify each of the symbols used in the equation.
b. How is the active site different from the whole enzyme structure?
c. After the products have formed, what happens to the enzyme?

16.48 Lactase is an enzyme that hydrolyzes lactose to glucose and galactose.
a. What are the reactants and products of the reaction?
b. Draw an energy diagram for the reaction with and without the enzyme.
c. Why does the enzyme make the reaction go faster?

Study Pack

Chem of Life Enzyme Action: Controlling Enzymes 1–4

16.9 Factors Affecting Enzyme Action

Learning Goal

Discuss the effect of cofactors, temperature, and pH on enzyme-catalyzed reactions.

A functional enzyme that consists of only polypeptides is called a simple enzyme. However, many enzymes require the addition of a nonprotein portion to catalyze a reaction. Then the protein part of an enzyme is called the **apoenzyme** and the nonprotein is a **cofactor.** A cofactor may be a metal ion such as copper (Cu^{2+}), iron (Fe^{2+}), magnesium (Mg^{2+}), manganese (Mn^{2+}), and zinc (Zn^{2+}). For example, carbonic anhydrase requires a zinc ion for catalytic activity, and pyruvate decarboxylase needs manganese ion. Many of these metal ions are trace elements that must be obtained in our diet.

Other enzymes require cofactors called **coenzymes,** which are small organic molecules, usually vitamins. The water-soluble vitamins, B and C, supply many of the coenzymes needed for catalytic activity. Most vitamins are not synthesized in the body and must be obtained from the diet. Deficiencies of vitamins are known to cause scurvy, rickets, beriberi, and pellagra. (See Table 16.10.)

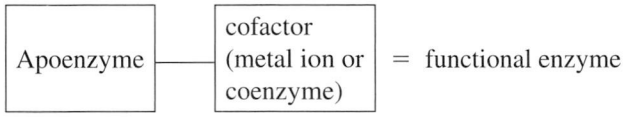

16.9

Table 16.10 The Water-Soluble Vitamins as Coenzymes

Vitamin	Coenzyme Function	Deficiency Symptoms
Thiamine (vitamin B_1)	Decarboxylation	Beriberi (fatigue, anorexia, nerve degeneration, heart failure)
Riboflavin (vitamin B_2)	Biological oxidation	Dermatitis, glossitis (tongue inflammation), cataracts
Niacin (vitaminB_3)	Biological oxidation	Pellagra (scaly skin, muscle fatigue, diarrhea, mouth sores, mental disorder)
Pyridoxine (vitamin B_6)	Transamination	Dermatitis, fatigue, anemia, irritability
Cobalamin (vitamin B_{12})	Transfer of methyl groups	Pernicious anemia, malformed red blood cells, mental disorders
Ascorbic acid (vitamin C)	Synthesis of collagen, protein metabolism, iron absorption, healing of wounds	Scurvy (bleeding gums, slow-healing wounds, muscle pain, anemia)
Biotin	Carboxylation	Dermatitis, fatigue, anemia, nausea, depression
Folic acid	Transfer of methyl groups in biosynthesis reactions	Abnormal red and white blood cells, intestinal tract disturbances
Pantothenic acid	Transfer of acetyl groups	Fatigue, anemia

Structures of the Vitamin Coenzymes

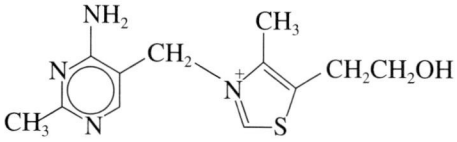

Thiamine (vitamin B_1)

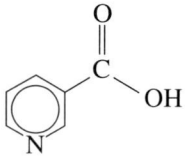

Niacin (vitamin B_3)

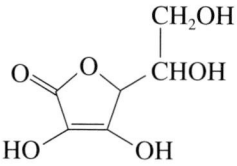

Ascorbic acid (vitamin C)

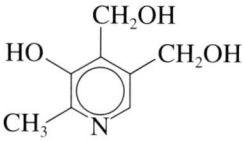

Pyridoxine (vitamin B_6)

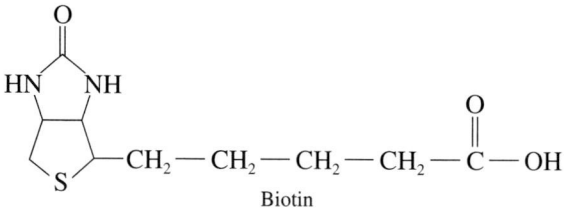

Biotin

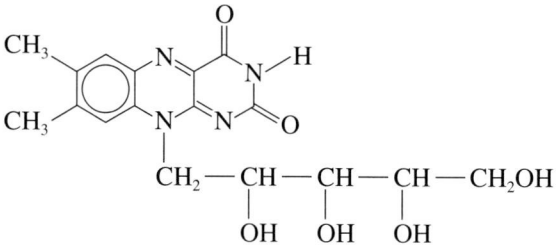

Riboflavin (vitamin B_2)

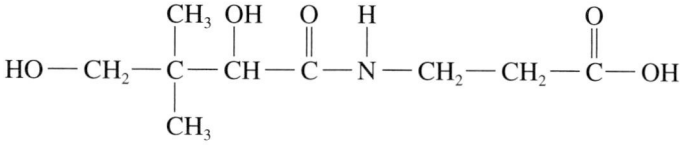

Pantothenic acid

S·A·M·P·L·E· **16.12** ·P·R·O·B·L·E·M

Enzymes and Cofactors

Indicate whether each of the following is a simple enzyme or one that requires a cofactor.

a. a polypeptide that needs Mg^{2+} for catalytic activity
b. a functional enzyme composed only of a polypeptide chain
c. an enzyme that consists of a protein portion attached to vitamin B_6

Solution

a. The metal ion Mg^{2+} is a cofactor.
b. A functional enzyme consisting only of a polypeptide chain is a simple enzyme.
c. The vitamin B_6 is a cofactor.

Study Check

Which of the nonprotein portions of the enzymes in the sample problem is a coenzyme?

Effect of Substrate Concentration. When the amount of enzyme is fixed and we add more substrate, the catalyzed reaction goes faster. However, at some point, all of the active sites on the enzymes are filled with substrate. When enzymes are saturated with substrate, there can be no further increase in the rate at which substrate is catalyzed. The reaction proceeds at its maximum rate, even if more substrate is added. (See Figure 16.20.)

Effect of Temperature. At low temperatures, most enzymes show little activity because there is not a sufficient amount of energy for the catalyzed reaction to take place. As the temperature increases, the rate of a reaction also increases. Enzymes are most active at an *optimum temperature,* usually around 37°C. At temperatures above the optimum, the protein of an enzyme begins to denature. Higher temperatures alter the structure of the enzyme further, leading to a rapid loss of catalytic activity. (See Figure 16.21.)

Effect of pH. Enzymes are most active at their **optimum pH,** the pH that maintains the proper tertiary and quaternary structure of the protein. (See Figure 16.22.) At pH values above or below the optimum, changes occurring in the ionic side groups of the enzyme disrupt the active site until it no longer binds with the substrate. For example,

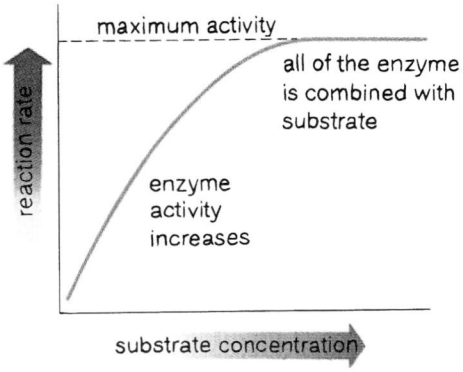

Figure 16.20 An increase in substrate concentration increases the rate of an enzyme-catalyzed reaction until all of the available enzyme has combined with substrate.

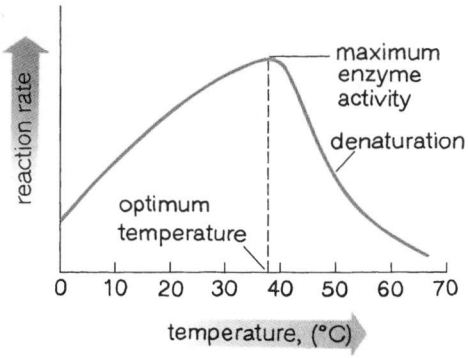

Figure 16.21 An increase in temperature increases enzyme activity to its optimum temperature. Higher temperatures cause denaturation and loss of catalytic activity.

16.9

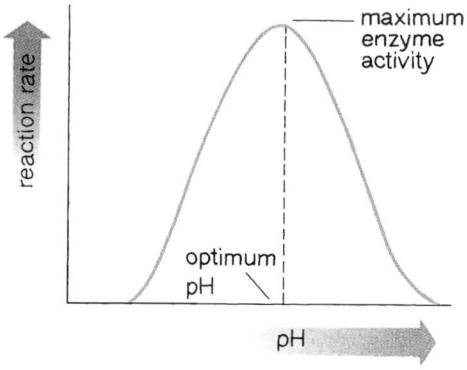

Figure 16.22 Enzymes are most active at their optimum pH. At higher or lower pH, denaturation of the enzyme causes a loss of catalytic activity.

the digestive enzymes in the stomach, such as pepsin, have an optimum pH of 2. If pH rises to pH 4 or 5, pepsin shows little or no digestive activity. At extreme pH values, the enzyme may be completely denatured. Table 16.11 lists the optimum pH values for some enzymes.

Table 16.11 Optimum pH for Selected Enzymes

Enzyme	Location	Substrate	Optimum pH
Pepsin	Stomach	Protein	1.5
Urease	Liver	Urea	5.0
Sucrase	Small intestine	Sucrose	6.2
Pancreatic amylase	Pancreas	Amylose	7.0
Trypsin	Small intestine	Polypeptides	7.7

EXPLORE YOUR WORLD

Enzyme Activity

The enzymes on the surface of a freshly cut apple, avocado, or banana react with oxygen in the air to turn the surface brown. An antioxidant, such as vitamin C in lemon juice, prevents the oxidation reaction. Cut an apple, an avocado, or a banana into several slices. Place one slice in a plastic zipper lock bag, squeeze out all the air, and close the zipper lock. Dip another slice in lemon juice and place it on a plate. Sprinkle another slice with a crushed vitamin C tablet. Leave the other slice alone as a control. Record your observations immediately, then every hour for 6 hours or longer.

Questions

1. Which slice(s) showed the most oxidation (turned brown)?
2. Which slice(s) showed little or no oxidation?
3. How was the oxidation reaction on each slice affected by treatment with an antioxidant?

S·A·M·P·L·E 16.13 ·P·R·O·B·L·E·M

Factors Affecting Enzymatic Activity

Describe what effect the changes in parts a and b would have on the rate of the reaction catalyzed by urease.

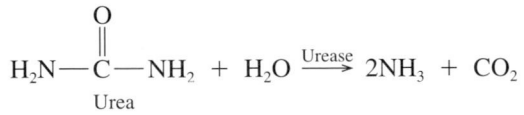

a. increasing the urea concentration
b. lowering the temperature to 10°C

Solution

a. An increase in urea concentration will increase the rate of reaction until all of the enzyme molecules are bound to the urea substrate. No further increase in rate occurs.
b. Because 10°C is lower than the optimum temperature of 37°C, the lower temperature will decrease the rate of the reaction.

Study Check

If urease has an optimum pH of 5, what is the effect of lowering the pH to 3?

Q·U·E·S·T·I·O·N·S & P·R·O·B·L·E·M·S

Factors Affecting Enzyme Action

16.49 Is the enzyme described in each of the following statements simple or one that contains a cofactor?
a. contains a sugar portion
b. requires Zn^{2+} for activity
c. contains only protein

16.50 Is the enzyme described in each of the following statements simple or one that contains a cofactor?
a. requires thiamine
b. is composed of 155 amino acids
c. produces Cu^{2+} upon hydrolysis

16.51 Trypsin, a peptidase that hydrolyzes polypeptides, functions in the small intestine at an optimum pH of 8. How is the rate of a trypsin-catalyzed reaction affected by each of the following conditions?
a. lowering the concentration of polypeptides
b. changing the pH to 3
c. running the reaction at 75°C

16.52 Pepsin, a peptidase that hydrolyzes proteins, functions in the stomach at an optimum pH of 2. How is the rate of a pepsin-catalyzed reaction affected by each of the following conditions?
a. increasing the concentration of proteins
b. changing the pH to 5
c. running the reaction at 0°C

16.53 The following graph shows the curves for pepsin, urease, and oxidase. Estimate the optimum pH for each.

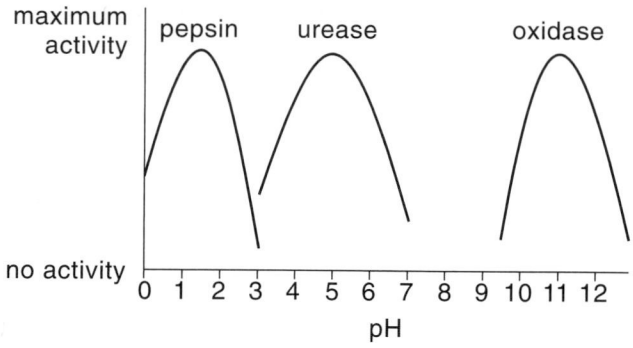

16.54 Refer to the graph in problem 16.53 to determine if the reaction rate in each condition will be at the optimum rate or not.

a. oxidase, pH 6 **b.** urease, pH 5
c. pepsin, pH 4 **d.** urease, pH 8

16.10

Study Pack

Chem of Life Enzymes and Pathways
Types of Control

Learning Goal

Describe the effect of an inhibitor upon enzyme activity.

16.10 Enzyme Inhibition

Inhibition is a loss of catalytic activity by enzymes. Chemical compounds called **inhibitors** combine with enzymes, causing a change in the protein structure. The active site is prevented from reacting with its proper substrate and the enzyme is inactive. For example, many antibiotics stop bacterial infections because they inhibit enzymes essential to the growth process of bacteria.

Competitive Inhibition

A **competitive inhibitor** has a structure similar to the substrate. Therefore, the inhibitor fits into the active site on the enzyme just like the substrate. As long as the inhibitor occupies the active site, no reaction of the substrate can take place. (See Figure 16.23.) However, the attractions for competitive inhibitors are weak, and the inhibitor is easily released from the active site. When the active site is open, a substrate molecule can bind to the enzyme. As a result, the substrate and its inhibitor are competing for the active site on the enzyme. As the number of substrate molecules increases, the active sites are more likely to be filled with substrate rather than inhibitor molecules. Increasing the substrate concentration can reverse competitive inhibition.

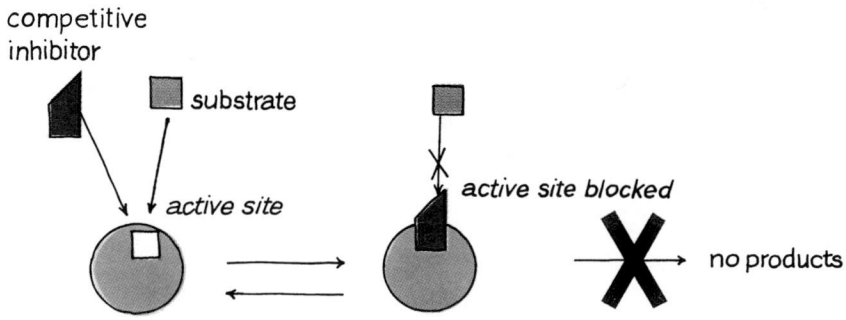

Figure 16.23 A competitive inhibitor competes for the active site because it has a structure very similar to the substrate.

Malonic acid is a competitive inhibitor of the enzyme succinic dehydrogenase. Because malonic acid has a structure similar to the substrate, succinic acid, the two substances compete for the active site on the dehydrogenase. As long as the inhibitor malonic acid occupies the active site on the enzyme, no reaction occurs. When more succinic acid is added, more active sites will fill with substrate, and there will be less inhibition.

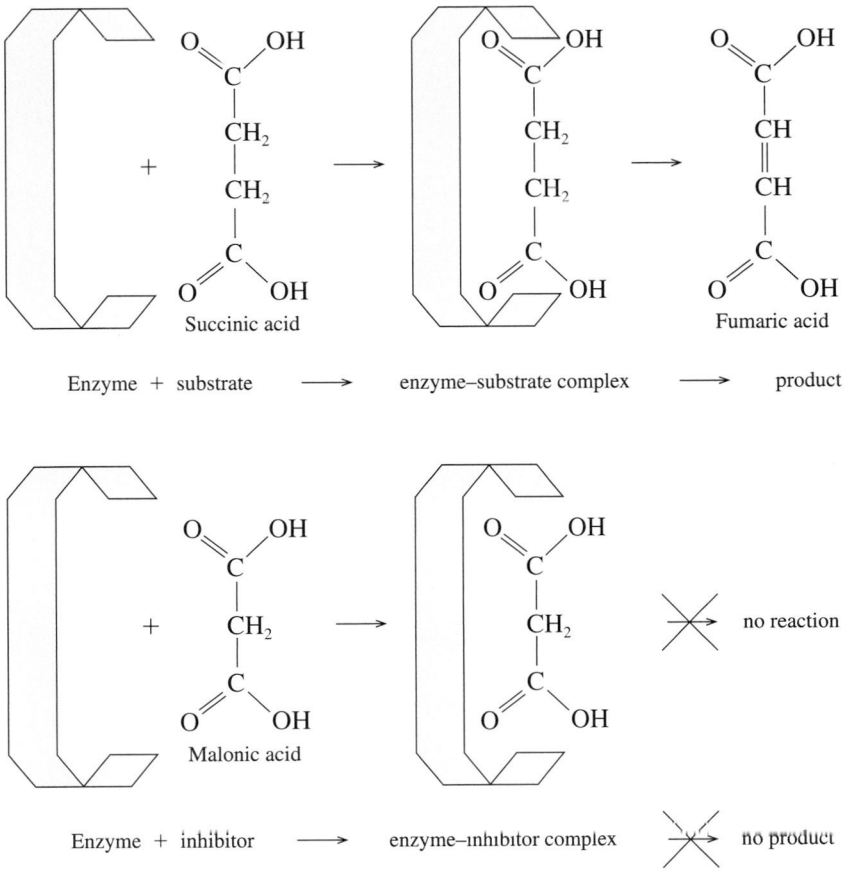

Noncompetitive Inhibition

The structure of a **noncompetitive inhibitor** does not resemble the substrate. Therefore, it does not compete with the substrate for the active site and usually binds to a part of an enzyme that is not the active site. However, when a noncompetitive inhibitor attaches to the enzyme, it changes the shape of the enzyme as well as the shape of the active site. Inhibition occurs because the substrate cannot fit into the active site and no reaction takes place. (See Figure 16.24.) Examples of noncompetitive inhibitors are the

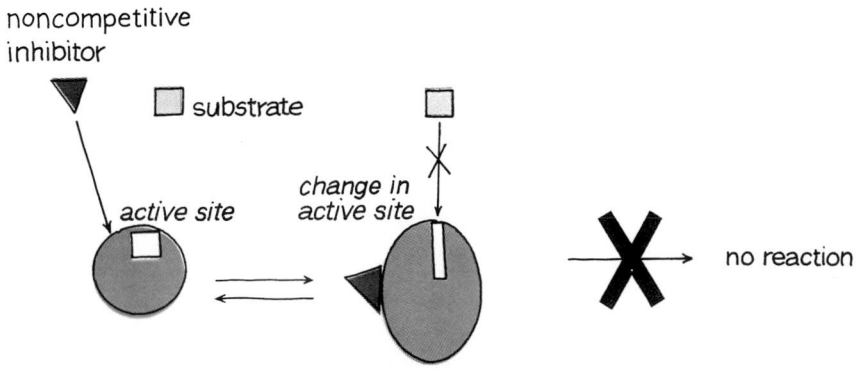

Figure 16.24 When a noncompetitive inhibitor attaches to the enzyme away from the active site, it changes the shape of the enzyme, which alters the active site and prevents the binding of the substrate.

HEALTH NOTE

Competitive Inhibitors in Medicine

Some bacterial infections are treated with competitive inhibitors called antimetabolites. Sulfanilamide, one of the first sulfa drugs, competes with PABA, which is an essential substance (metabolite) in the growth cycle of bacteria.

However, some bacteria are resistant to penicillin because they contain penicillinase, an enzyme that breaks down penicillin. New kinds of penicillin to which bacteria have not yet become resistant have been produced.

Substrate of Bacterial Growth **Inhibitor**

16.10

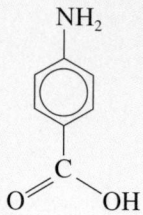

PABA
(*p*-aminobenzoic acid)

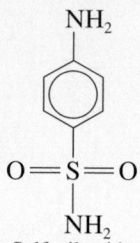

Sulfanilamide

R Groups for Penicillin Derivatives

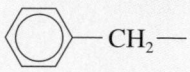

Penicillin G

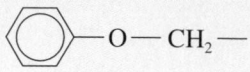

Penicillin V

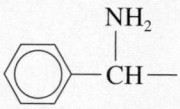

Ampicillin

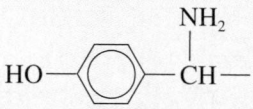

Amoxicillin

Antibiotics produced by bacteria, mold, or yeast are used to inhibit bacterial growth. For example, penicillin inhibits an enzyme needed for the formation of cell walls in bacteria, but not human cell membranes. With an incomplete cell wall, bacteria cannot survive, and the infection is stopped.

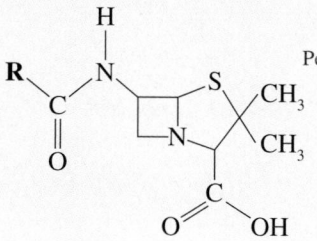

Penicillin

Another group of antibiotics, tetracycline, streptomycin, chloramphenicol, and aureomycin, stops bacterial infections by inhibiting the synthesis of protein in bacteria. They are also used to fight bacterial infections in people allergic to the penicillins.

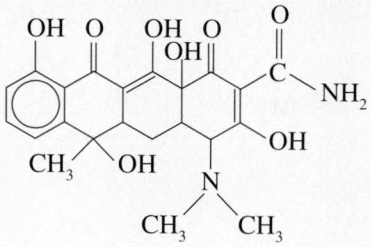

Tetracycline

heavy metal ions Pb^{2+}, Ag^{+}, and Hg^{2+} that bond with side groups such as —SH, —COO^{-}, or —OH. Catalytic activity is restored when the inhibitors leave the enzyme or are removed by chemical reagents. Because there is no competition for the active site, the addition of more substrate does not reverse this type of inhibition.

A noncompetitive inhibition can be irreversible if the inhibitor is tightly bonded at the active site and cannot be removed without destroying the protein. Nerve gases act as irreversible inhibitors by forming a covalent bond with the side group of serine (—CH_2OH), which is part of the active site in acetylcholinesterase. When this enzyme is inhibited, the transmission of nerve impulses is blocked.

S·A·M·P·L·E 16.14 P·R·O·B·L·E·M

Enzyme Inhibition

State the type of inhibition in the following:

a. The inhibitor has a structure that is similar to the substrate.

b. This inhibitor binds to the surface of the enzyme, changing its shape in such a way that it cannot bind to substrate.

Solution

a. competitive inhibition

b. noncompetitive inhibition

Study Check

Hydrogen cyanide (HCN) forms strong bonds with catalase, an enzyme that contains iron (Fe^{3+}). What type of inhibitor is HCN?

Q·U·E·S·T·I·O·N·S & P·R·O·B·L·E·M·S

Enzyme Inhibition

16.55 Indicate whether the following describe a competitive or a noncompetitive enzyme inhibitor:

a. The inhibitor has a structure similar to the substrate.

b. The effect of the inhibitor cannot be reversed by adding more substrate.

c. The inhibitor competes with the substrate for the active site.

d. The structure of the inhibitor is not similar to the substrate.

e. The addition of more substrate reverses the inhibition.

16.56 Oxaloacetic acid is an inhibitor of succinic dehydrogenase:

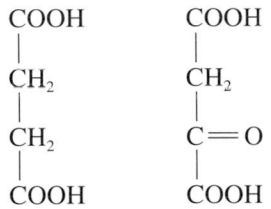

Succinic acid Oxaloacetic acid

a. Would you expect oxaloacetic acid to be a competitive or a noncompetitive inhibitor? Why?

b. Would oxaloacetic acid bind to the active site or elsewhere on the enzyme?

c. How would you reverse the effect of the inhibitor?

Chapter Review

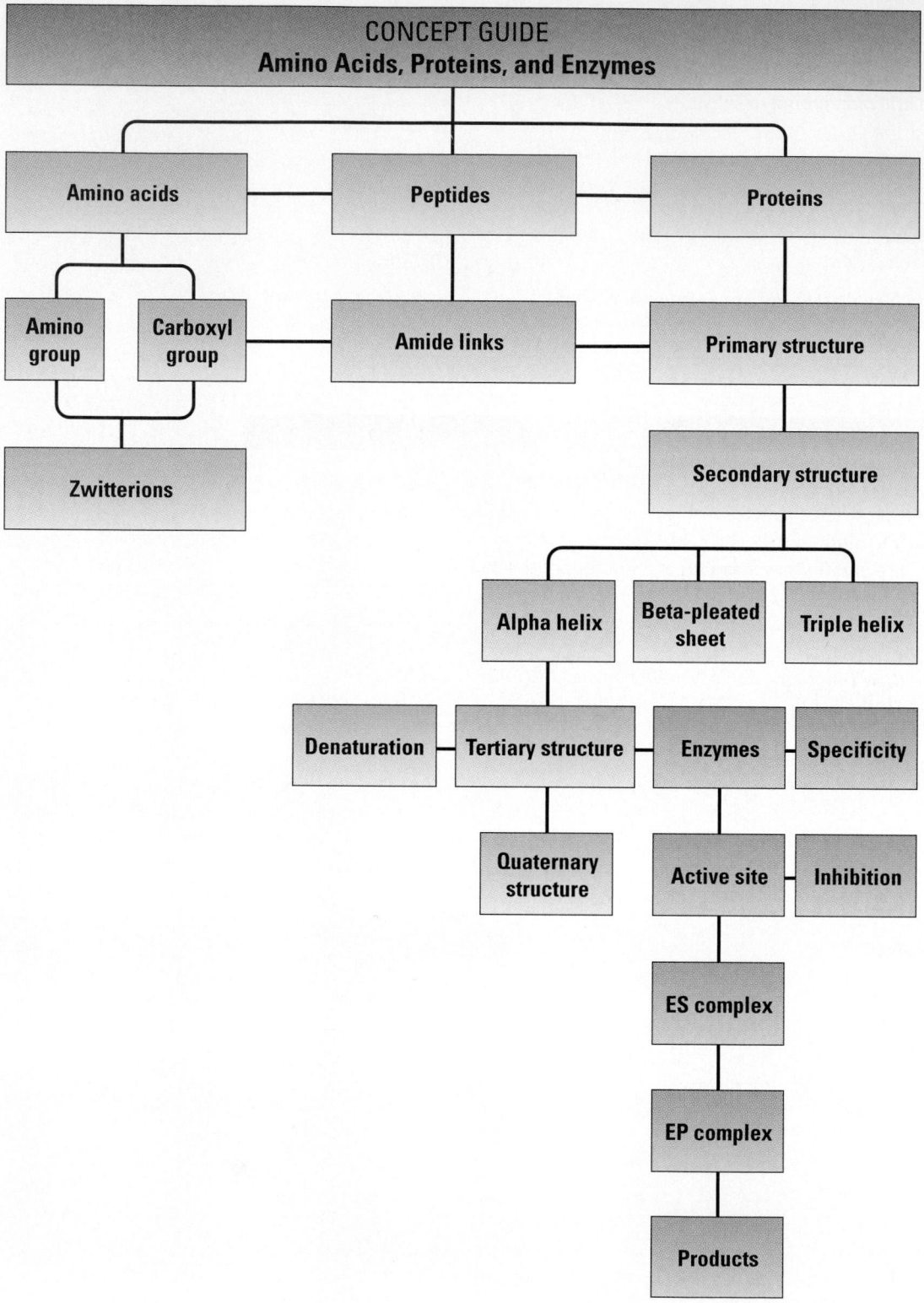

Chapter Summary

16.1 Types of Proteins

Some proteins are enzymes or hormones, whereas others are important in structure, transport, protection, storage, and muscle contraction.

16.2 Amino Acids

A group of 20 amino acids provides the molecular building blocks of proteins. Attached to the central (alpha) carbon of each amino acid is an amino group, a carboxyl group, and a unique side group (R). The R group gives an amino acid the property of being nonpolar, polar, acidic, or basic. Amino acids exist as dipolar ions called zwitterions, as positive ions at low pH, and as negative ions at high pH levels.

16.3 The Peptide Bond

Peptide bonds join amino acids to form peptides by forming an amide bond between the carboxyl group of one amino acid and the amino group of the second. Long chains of amino acids are called proteins.

16.4 Primary and Secondary Structure

The primary structure of a protein is its sequence of amino acids. In the secondary structure, hydrogen bonds between peptide bonds produce a characteristic shape such as an α helix, β-pleated sheet, or a triple helix.

16.5 Tertiary and Quaternary Structure

In globular proteins, the polypeptide chain including its α helical and β-pleated sheet regions folds upon itself to form a tertiary structure. A tertiary structure is stabilized by interactions that move hydrophobic R groups to the inside and hydrophilic R groups to the outside surface, and by attractions between R groups that form hydrogen, disulfide, and ionic bonds. In a quaternary structure, two or more tertiary subunits must combine for biological activity. They are held together by the same interactions found in tertiary structures.

16.6 Protein Hydrolysis and Denaturation

Denaturation of a protein occurs when heat or other denaturing agents destroy the structure of the protein (but not the primary structure) until biological activity is lost.

16.7 Enzymes

Enzymes are globular proteins that act as biological catalysts. The names of most enzymes are indicated by their *ase* endings. Enzymes are classified by the type of reaction they catalyze, such as oxidoreductase, hydrolase, or isomerase.

16.8 Enzyme Action

Within the structure of the enzyme, there is a small pocket called the active site, which has a specific shape that fits a specific substrate. In the lock-and-key model or the induced-fit model, an enzyme and substrate form an enzyme–substrate complex so the reaction of the substrate can be catalyzed at the active site.

16.9 Factors Affecting Enzyme Action

Simple enzymes are biologically active as a protein only, whereas other enzymes require both a protein portion (apoenzyme) and a cofactor. A cofactor may be a metal ion such as Cu^{2+} or Fe^{2+}, or an organic compound called a coenzyme, usually a vitamin.

An increase in substrate concentration increases the reaction rate of an enzyme-catalyzed reaction until all the enzyme molecules combine with substrate. Then additional substrate will not increase the rate further. Enzymes are most effective at optimum temperature and pH. The rate of an enzyme reaction decreases considerably at temperatures and pH above or below the optimum.

16.10 Enzyme Inhibition

An enzyme can be made inactive by changes in pH or temperature, or by chemical compounds called inhibitors. A competitive inhibitor has a stucture similar to the substrate and competes for the active site. When the active site is occupied, the enzyme cannot catalyze the reaction of the substrate. A noncompetitive inhibitor attaches elsewhere on the enzyme, changing the shape of both the enzyme and the active site. As long as the noncompetitive inhibitor is attached to the enzyme, the altered active site cannot bind with substrate.

Key Terms

active site The region of an enzyme that binds and catalyzes a reaction of substrate.

α (alpha) helix A secondary level of protein structure, in which hydrogen bonds connect the NH of one peptide bond with the C=O of a peptide bond later in the chain to form a coiled or corkscrew structure.

amino acid The building block of proteins, consisting of an amino group, a carboxylic acid group, and a unique side group attached to the alpha carbon.

antibiotic An inhibitor produced by a bacteria, mold, or yeast that is toxic to other bacteria.

apoenzyme The protein portion that along with a cofactor provides a functional enzyme.

β (beta)-pleated sheet A secondary level of protein structure that consists of hydrogen bonds between peptide links in parallel polypeptide chains.

coenzyme Organic molecules required as cofactors in enzyme action.

cofactor The nonprotein portion, a metal ion or an organic compound, that is necessary for a biologically functional enzyme.

competitive inhibitor A molecule that has a structure similar to the substrate and inhibits enzyme action by competing for the active site.

denaturation The loss of secondary and tertiary protein structure caused by heat, acids, bases, organic compounds, heavy metals, and/or agitation.

electrophoresis The use of electrical current to separate proteins or other charged molecules with different isoelectric points.

enzyme–product complex The intermediate in an enzyme-catalyzed reaction in which product is formed at the active site.

enzymes Proteins that catalyze biological reactions.

enzyme–substrate complex An intermediate in an enzyme-catalyzed reaction in which the substrate combines with the enzyme.

essential amino acids Amino acids that must be supplied by the diet because they are not synthesized by the body.

fibrous protein A protein that is insoluble in water; consists of polypeptide chains with α helixes or β-pleated sheets, that make up the fibers of hair, wool, skin, nails, and silk.

globular proteins Proteins that acquire a compact shape from attractions between the R group of the amino acid residues in the protein.

hydrophilic amino acid An amino acid having polar, acidic, or basic R groups that are attracted to water; "water-loving."

hydrophobic amino acid A nonpolar amino acid with hydrocarbon R groups; "water-fearing."

induced-fit model A model of enzyme action that describes a flexible active site that adapts to the structure of a substrate to form the enzyme–substrate complex.

inhibitors Substances that make an enzyme inactive by interfering with its ability to react with a substrate.

lock-and-key model A model of enzyme action that represents the substrate as a key fitting the correct lock, which is the specific shape of the active site in an enzyme.

noncompetitive inhibitor A substance that changes the shape of the enzyme and the active site, preventing the active site from binding with the substrate.

nonpolar amino acids Amino acids that are not soluble in water because they contain a nonpolar R group.

optimum pH The pH at which an enzyme is most active.

peptide The combination of two or more amino acids joined by peptide bonds; dipeptide, tripeptide, and so on.

peptide bond The amide bond that joins amino acids in polypeptides and proteins.

polar amino acids Amino acids that are soluble in water because their R group is polar: hydroxyl (OH), thiol (SH), carbonyl (C=O), amino (NH_2), or carboxyl (COOH).

polypeptides Polymers of amino acids joined by peptide bonds.

primary structure The sequence of the amino acids in a protein.

protein A term used for biologically active polypeptides that have many amino acids linked together by peptide bonds.

quaternary structure A protein structure in which two or more protein subunits form an active protein.

secondary structure The formation of an α helix, β-pleated sheet, or triple helix.

substrate The substance that reacts at the active site in an enzyme-catalyzed reaction.

tertiary structure The folding of the secondary structure of a protein into a compact structure that is stabilized by the interactions of R groups such as ionic and disulfide bonds.

triple helix The protein structure found in collagen consisting of three polypeptide chains woven together like a braid.

zwitterion The dipolar form of an amino acid consisting of two oppositely charged ionic regions, $-NH_3^+$ and $-COO^-$.

Chapter Practice

16.57 Seeds and vegetables are often deficient in one or more essential amino acids. Using the following table, state whether the following combinations would provide all the essential amino acids:

Source	Lysine	Tryptophan	Methionine
Oatmeal	No	Yes	Yes
Rice	No	Yes	Yes
Garbanzo beans	Yes	No	Yes
Lima beans	Yes	No	No
Cornmeal	No	Yes	Yes

a. rice and garbanzo beans
b. lima beans and cornmeal
c. a salad of garbanzo beans and lima beans
d. rice and lima beans
e. rice and oatmeal
f. oatmeal and lima beans

16.58 How does denaturation of a protein differ from its hydrolysis?

16.59 What are some differences between the following pairs?

a. secondary and tertiary protein structures
b. essential and nonessential amino acids
c. polar and nonpolar amino acids
d. di- and tripeptides
e. an ionic bond (salt bridge) and a disulfide bond
f. fibrous and globular proteins
g. α helix and β-pleated sheet
h. tertiary and quaternary structures of proteins

16.60 The proteins placed on a gel for electrophoresis have the following isoelectric points: albumin, 4.9, hemoglobin, pH 6.8, and lysozyme, 11.0. A buffer of pH 6.8 is placed on the gel.

a. Which protein will migrate toward the positive electrode?
b. Which protein will migrate toward the negative electrode?
c. Which protein will remain at the same place it was originally placed?

16.61 What is the class of the enzyme that would catalyze each of the following reactions?

a. $CH_3\overset{O}{\overset{\|}{C}}H \longrightarrow CH_3\overset{O}{\overset{\|}{C}}OH$

b. $NH_2-CH_2-\overset{O}{\overset{\|}{C}}-NH-\overset{CH_3}{\overset{|}{CH}}-\overset{O}{\overset{\|}{C}}OH + H_2O \longrightarrow$

$NH_2-CH_2-\overset{O}{\overset{\|}{C}}OH + NH_2-CH_2-\overset{O}{\overset{\|}{C}}OH$

c. $CH_3-CH=CH-CH_3 + H_2O \longrightarrow$

$CH_3-CH_2-\overset{OH}{\overset{|}{CH}}-CH_3$

d. $CH_3-\overset{O}{\overset{\|}{C}}-\overset{O}{\overset{\|}{C}}OH \longrightarrow$

$CH_3-\overset{O}{\overset{\|}{C}}-OH + CO_2$

e. $CH_3-\overset{O}{\overset{\|}{C}}-\overset{O}{\overset{\|}{C}}OH + CO_2 + ATP \longrightarrow$

$HO-\overset{O}{\overset{\|}{C}}-CH_2-\overset{O}{\overset{\|}{C}}-\overset{O}{\overset{\|}{C}}-OH + ADP + P_i$

f. glucose 6-phosphate $\longrightarrow$ fructose 6-phosphate

Answers

Answers to Study Checks

16.1 hormones

16.2 **a.** nonpolar **b.** polar

16.3

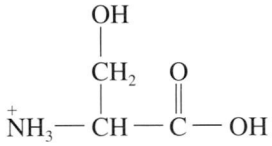

16.4

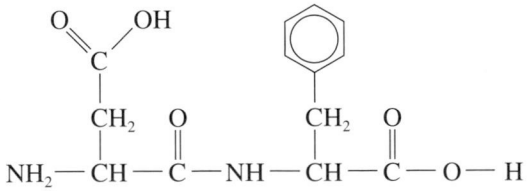

16.5 threonylleucylphenylalanine

16.6 a triple helix

16.7 Both are nonpolar and would be found on the inside of the tertiary structure.

16.8 quaternary

16.9 The heavy metal Ag^+ denatures the protein in bacteria that cause gonorrhea.

16.10 hydrolase

16.11 The active site has a rigid shape in the lock-and-key model but is flexible in the induced-fit model and adapts to the shape of the substrate.

16.12 The vitamin B_6 in part c.

16.13 Urease will lose enzymatic activity as denaturation of its protein occurs.

16.14 noncompetitive inhibitor

Answers to Selected Questions and Problems

16.1 **a.** transport **b.** structural
c. structural **d.** enzyme

16.3 All amino acids contain a carboxylic acid group and an amino group on the α carbon.

16.5

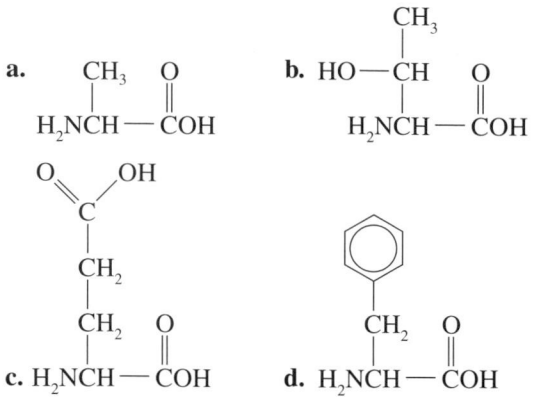

16.7 **a.** nonpolar **b.** polar
c. polar, acidic **d.** nonpolar

16.9 **a.** alanine **b.** valine
c. lysine **d.** cysteine

16.11 An essential amino acid is one that cannot be formed in the body. Essential amino acids must be supplied in the diet. The essential amino acids in 16.9 are valine and lysine.

16.13

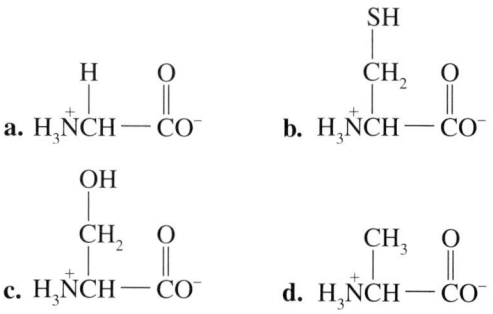

16.15

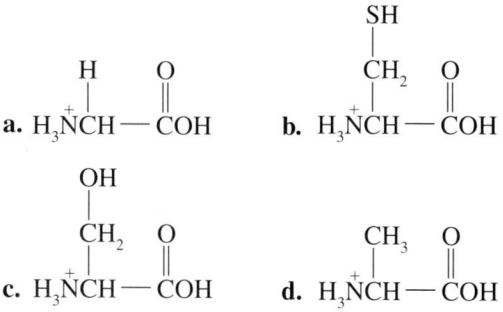

16.17

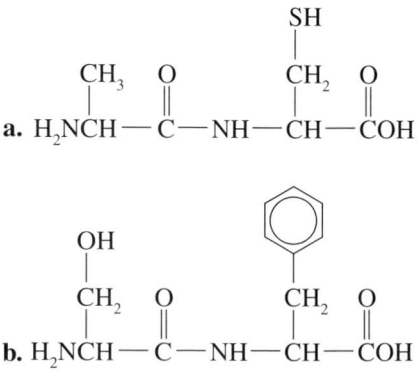

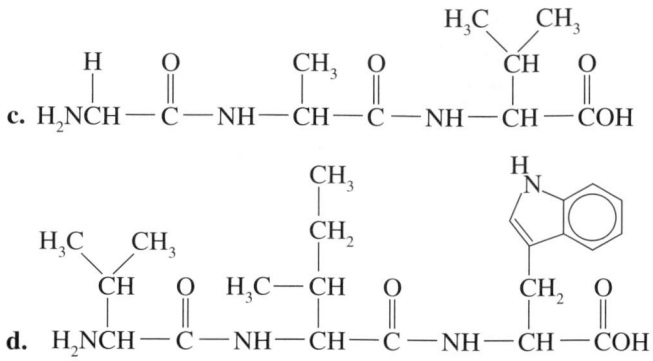

16.19 Amide bonds form to connect the amino acids that make up the protein.

16.21 Val-Ser-Ser, Ser-Val-Ser, or Ser-Ser-Val

16.23 The primary structure remains unchanged and intact as hydrogen bonds form between carbonyl oxygen atoms and amino hydrogen atoms.

16.25 In the α helix, hydrogen bonds form between the carbonyl oxygen atom and the amino hydrogen atom of the fourth amino acid in the sequence. In the β-pleated sheet, hydrogen bonds occur between parallel peptides or across sections of a long polypeptide chain.

16.27 **a.** a disulfide bond
b. an ionic bond (salt bridge)
c. hydrogen bond
d. hydrophobic interaction

16.29 **a.** cysteine
b. Leucine and valine will be found on the inside of the protein because they are hydrophobic.
c. The cysteine and aspartic acid would be on the outside of the protein because they are polar.
d. The order of the amino acids (the primary structure) provides the R groups, whose interactions determine the tertiary structure of the protein.

16.31 The products would be the amino acids glycine, alanine, and serine.

16.33 His-Met; Met-Gly; Gly-Val

16.35 Hydrolysis splits the amide linkages in the primary structure.

16.37 **a.** Placing an egg in boiling water coagulates the protein of the egg.
b. Using an alcohol swab coagulates the protein of any bacteria present.
c. The heat from an autoclave will coagulate the protein of any bacteria on the surgical instruments.
d. Heat will coagulate the surrounding protein to close the wound.

16.39 **a.** galactose **b.** lipids **c.** aspartic acid

16.41 **a.** hydrolase **b.** oxidoreductase
c. isomerase **d.** transferase

16.43 **a.** lyase **b.** transferase

16.45 **a.** enzyme **b.** enzyme–substrate complex
c. substrate

16.47 $E + S \rightleftharpoons ES \rightleftharpoons EP \rightleftharpoons E + P$

a. E = enzyme, S = substrate, ES = enzyme–substrate complex, EP = enzyme–product complex, P = products

b. The active site is the portion of the enzyme that binds to the substrate.

c. After products are formed, the enzyme can catalyze further reactions with other substrate molecules.

16.49 **a.** requires a cofactor **b.** requires a cofactor
c. simple

16.51 **a.** The rate would decrease.
b. The rate would decrease.
c. The rate would decrease.

16.53 pepsin, pH 1.5; urease, pH 5; oxidase, pH 11

16.55 **a.** competitive inhibitor
b. noncompetitive inhibitor
c. competitive inhibitor
d. noncompetitive inhibitor
e. competitive inhibitor

16.57 **a.** yes **b.** yes **c.** no
d. yes **e.** no **f.** yes

16.59 **a.** The secondary structure of a protein depends on hydrogen bonds to form a helix or a pleated sheet; the tertiary structure is determined by the interaction of R groups and determines the three-dimensional structure of the protein.

b. Nonessential amino acids can be synthesized by the body; essential amino acids must be supplied by the diet.

c. Polar amino acids have hydrophilic side groups, whereas nonpolar amino acids have hydrophobic side groups.

d. Dipeptides contain two amino acids, whereas tripeptides contain three.

e. An ionic bond is an interaction between a basic and acidic side group; a disulfide bond links the sulfides of two cysteines.

f. Fibrous proteins consist of three to seven α helixes coiled like a rope. Globular proteins form a compact spherical shape.

g. The α helix is the secondary shape like a staircase or corkscrew. The β-pleated sheet is a secondary structure that is formed by many protein chains side by side like a pleated sheet.

h. The tertiary structure of a protein is its three-dimensional structure. The quaternary structure involves the grouping of two or more peptide units for the protein to be active.

16.61 **a.** oxidoreductase **b.** hydrolase
c. lyase **d.** lyase
e. ligase **f.** isomerase

17

Metabolic Pathways and Energy Production

During physical activity, ATP energy from the metabolism of glucose stored in muscle cells is used to contract the muscles.

Looking Ahead

17.1	Metabolism and ATP Energy
17.2	Digestion of Foods
17.3	Important Coenzymes in Metabolic Pathways
17.4	Glycolysis: Oxidation of Glucose
17.5	Pathways for Pyruvate
17.6	Citric Acid Cycle
17.7	Electron Transport Chain
17.8	ATP Energy from Glucose
17.9	Oxidation of Fatty Acids
17.10	Metabolic Pathways for Amino Acids

www.chemplace.com/college/timberlake

Visit the web site above to test yourself, run simulations, and review key concepts from this chapter. For even more interactive review and quizzing, use the two CD-ROMs included with the book.

All the chemical reactions in living cells that break down or build molecules are known as *metabolism.* Many reactions, each catalyzed by a specific enzyme, link together to form a metabolic pathway. Some of the pathways produce energy for the cell; others use energy.

When we digest food, products are obtained that are small enough to be absorbed into the cells of the body. Within our cells, the oxidation of these food products releases energy that is stored in a high-energy compound called adenosine triphosphate (ATP). Our cells need ATP energy to do work such as contracting muscles, synthesizing larger molecules, sending nerve impulses, and moving substances across cell membranes.

Study Pack

Chem of Life Enzymes and Pathways: Biosynthesis 1–3

17.1

17.1 Metabolism and ATP Energy

Learning Goal

Describe the role of ATP in catabolic and anabolic reactions.

A 25-year-old college student requires a daily diet of 2000 Calories (kcal), an aerobics instructor needs 3000 Calories, and a bicyclist preparing for an international race needs 6000 Calories a day. In each example, the amount of energy expended is balanced by the amount of energy obtained from food.

The term **metabolism** refers to all of the chemical reactions that provide energy and the substances required for continued cell growth. There are two types of metabolic reactions: catabolic and anabolic. In **catabolic** reactions, complex molecules are broken down to simple ones accompanied by a release of energy. **Anabolic** reactions require energy to build larger molecules from simpler ones.

Cell Structure

To understand the relationship of metabolic reactions, we need to describe some components of cells. There are two types of cells: prokaryotic and eukaryotic. Prokaryotic ("before nucleus") cells have no nucleus and are found in single-celled organisms such as bacteria. The cells in plants and animals are eukaryotic cells, which do have a nucleus. They are more complex than prokaryotic cells and about 1000 times larger.

In eukaryotic cells, the **cytoplasm** consists of all the materials within the plasma membrane and surrounding the nucleus. (See Figure 17.1.) The nucleus contains the genetic material in an assortment of chromosomes. The important metabolic reactions

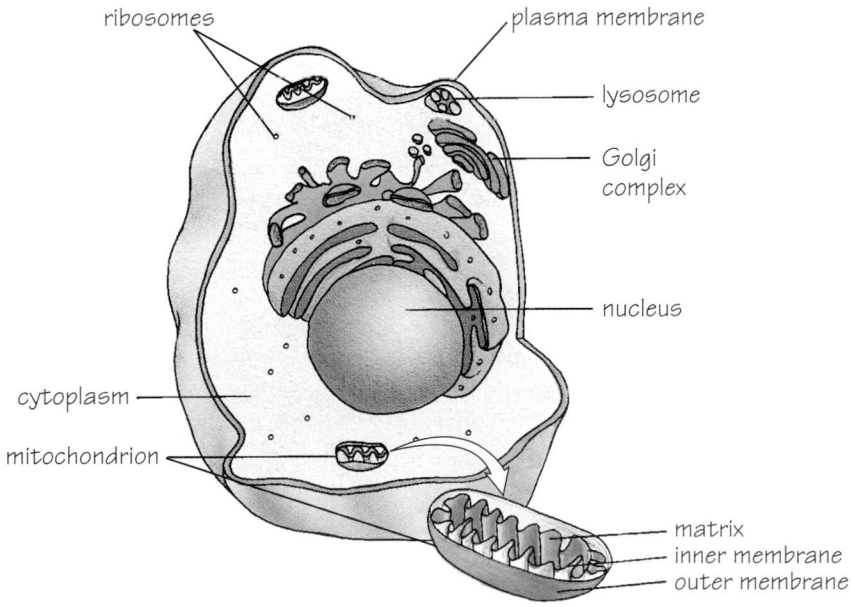

Figure 17.1 A simplified drawing of a cell. Some degradative pathways such as glycolysis occur in the cytoplasm, but most of the oxidation and energy production occurs within the mitochondria, the powerhouse of the cell.

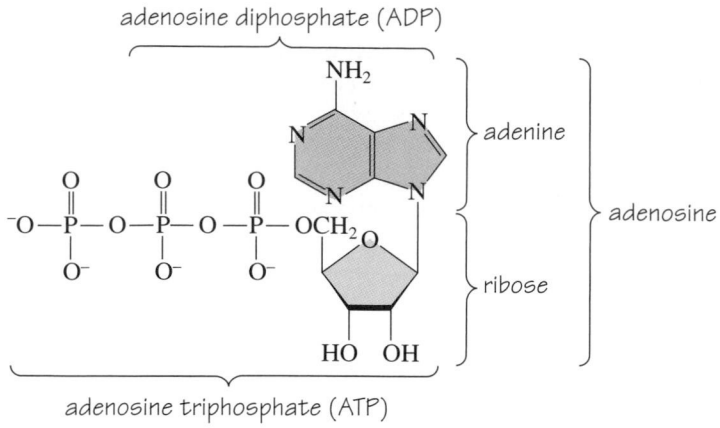

Figure 17.2 The structure of adenosine triphosphate, ATP, a compound that stores energy in the cell.

17.1

are organized among several specialized structures of the cytoplasm. The ribosomes are the sites of protein synthesis. Lysosomes contain hydrolytic enzymes that break down molecules during cellular repair and degradation. The mitochondria are the energy-producing factories of the cells. The mitochondria consist of two different types of membranes, an outer membrane and an inner membrane. The region between the inner and outer membrane is called the intermembrane space. The central compartment is the matrix. Molecules of carbohydrates, fats, and amino acids move through the outer membrane into the matrix, where enzymes can oxidize them to CO_2 and H_2O. The energy released is used to form the energy-rich compound ATP.

ATP

In our cells, the energy released from the oxidation of the food we eat is used to form a compound called *adenosine triphosphate* (**ATP**). The ATP molecule is composed of adenine, ribose, and three phosphate groups. (See Figure 17.2.) ATP is like a battery in the cell: it provides energy to run reactions, but it runs down if not recharged.

Hydrolysis of ATP Yields Energy

We can think of ATP as the storage form of energy within the cell that connects the energy-producing reactions with the energy-requiring reactions. When ATP is hydrolyzed, 7–12 kcal of energy per mole of ATP can be released. This energy is used to drive energy-requiring processes. The other products of ATP hydrolysis are **ADP** and inorganic phosphate ion (P_i).

$$\text{ATP} + H_2O \longrightarrow \text{ADP} + P_i + \text{energy}$$

Every time we contract muscles, move substances across cellular membranes, send nerve signals, or synthesize an enzyme, we use some energy from ATP hydrolysis. In a cell that is doing work, 1–2 million ATP molecules may be hydrolyzed in a second. The amount of ATP hydrolyzed in one day can be as much as our body mass, even though only about 1 gram of ATP is in all our cells at any given time.

Catabolic reactions begin with digestion, the process of breaking down carbohydrates, fats, and proteins in the intestinal tract. The products, amino acids, fatty acids, and monosaccharides, are absorbed into our cells where they are oxidized to generate energy. The energy released by catabolism is linked with the synthesis of ATP, which stores energy. (See Figure 17.3.)

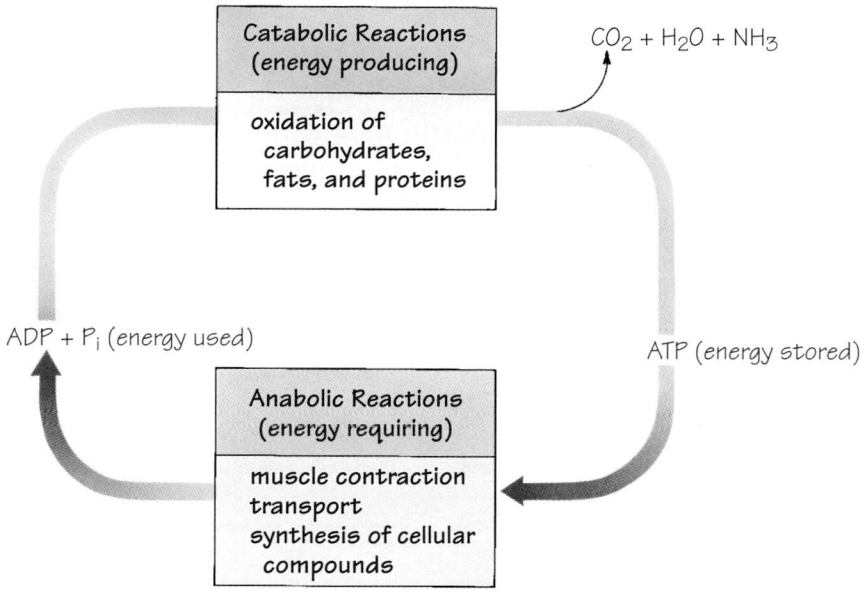

Figure 17.3 ATP, the energy-storage molecule, connects the energy-producing reactions with the energy-requiring reactions that do work in the cells.

S · A · M · P · L · E · 17.1 · P · R · O · B · L · E · M

Hydrolysis of ATP

Write an equation for the hydrolysis of ATP.

Solution

The hydrolysis of ATP produces ADP, P_i, and energy.

$$ATP + H_2O \longrightarrow ADP + P_i + \text{energy}$$

Study Check

What are the components of the ATP molecule?

HEALTH NOTE

ATP Energy and Ca^{2+} Needed to Contract Muscles

Our muscles consist of thousands of parallel fibers. Within these muscle fibers are fibrils composed of two kinds of proteins called filaments. Arranged in alternating rows, the thick filaments of myosin overlap the thin filaments containing actin. (See Figure 17.4.) During a muscle contraction, the thin filaments slide inward over the thick filaments causing a shortening of the muscle fibers.

Calcium ion, Ca^{2+}, and ATP play an important role in muscle contraction. An increase in the Ca^{2+} concentration in the muscle fibers causes the filaments to slide, while a decrease stops the process. In a relaxed muscle, the Ca^{2+} concentration is low. When a nerve impulse reaches the muscle, calcium channels in the membrane open and Ca^{2+} flows into the fluid surrounding the filaments. The Ca^{2+} combines with troponin, a protein that covers the binding site on actin. When the troponin detaches from actin, myosin can bind to actin, and start to pull the actin filaments inward. The energy used for the contraction is provided by splitting ATP that is attached to myosin to ADP + P_i and energy.

Muscle contraction continues as long as both ATP and Ca^{2+} levels are high around the filaments. Relaxation of the muscle occurs as the nerve impulse ends and the Ca^{2+} channels close. The Ca^{2+} concentration is lowered as ATP energy is used to pump the remaining calcium ions out of the area. Troponin reattaches to the actin, which causes the thin filaments to return to their relaxed state. In rigor mortis, Ca^{2+} concentration remains high within the muscle fibers causing a continued state of rigidity for approximately 24 hours. After that, cellular deterioration causes a decrease in calcium ions and muscles relax.

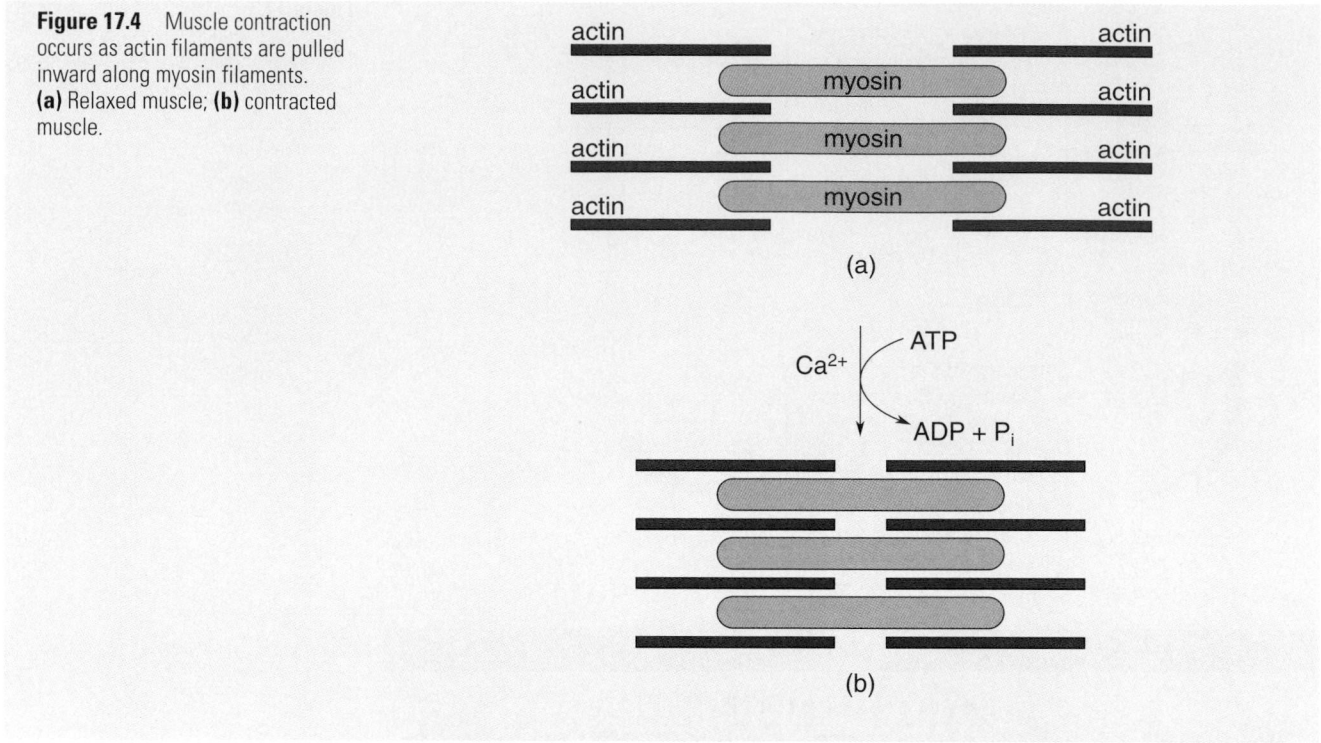

Figure 17.4 Muscle contraction occurs as actin filaments are pulled inward along myosin filaments. **(a)** Relaxed muscle; **(b)** contracted muscle.

17.2

Q·U·E·S·T·I·O·N·S & P·R·O·B·L·E·M·S

Metabolism and ATP Energy

17.1 How do prokaryotic cells differ from eukaryotic cells?

17.2 Where do the important metabolic reactions take place in eukaryotic cells?

17.3 What is a catabolic reaction in metabolism?

17.4 What is an anabolic reaction in metabolism?

17.5 How does ADP differ from ATP? Write an equation for the formation of ATP from ADP and P_i.

17.6 Why is ATP considered a high-energy compound? Write an equation to represent the hydrolysis of ATP.

Study Pack

17.2 Digestion of Foods

Learning Goal

Give the sites and products of digestion for carbohydrates, lipids, and proteins.

In the first step of catabolism, foods undergo **digestion,** a process that converts large molecules to smaller ones that can be absorbed by the body. (See Figure 17.5.)

Digestion of Carbohydrates

We begin to digest carbohydrates as soon as we chew food. An enzyme in the saliva, α-**amylase,** breaks apart some of the α-glycosidic bonds in the starches, producing smaller polysaccharides called dextrins (three to eight glucose units), maltose, and some glucose. After swallowing, the partially digested starches encounter the highly acidic environment of the stomach, where little carbohydrate digestion occurs.

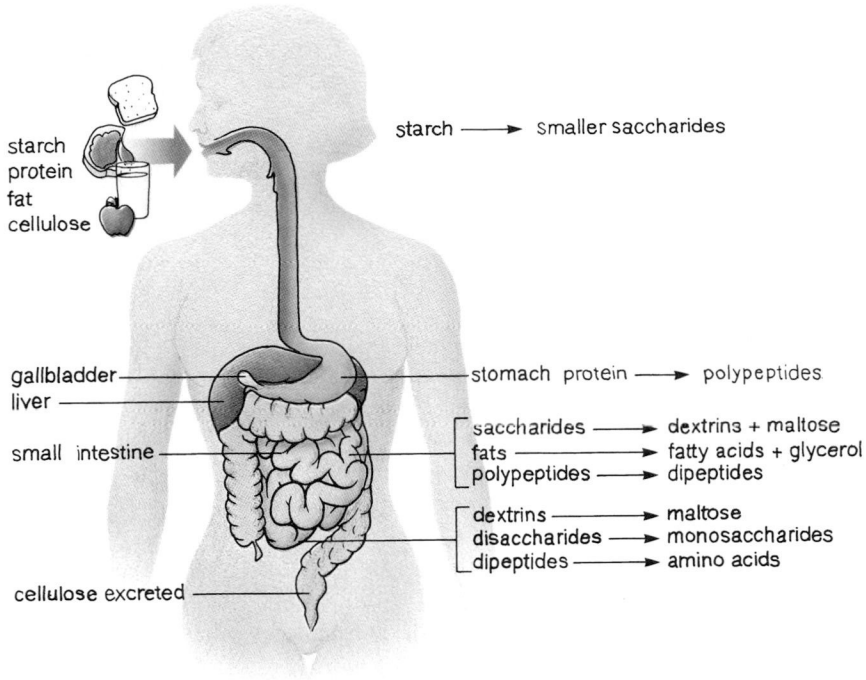

Figure 17.5 Digestion of foods along the gastrointestinal tract.

In the small intestine, which has a pH of about 8, amylase secreted by the pancreas hydrolyzes the remaining polysaccharides to maltose. Finally, the maltose and any sucrose and lactose from the diet are hydrolyzed to the monosaccharides glucose, fructose, and galactose, which are small enough to move through the intestinal wall and enter the bloodstream. (See Figure 17.6.) In the body, they may be used to synthesize ATP or stored as glycogen in the muscle and liver.

Digestion of Proteins

The digestion of proteins begins in the stomach, where hydrochloric acid (HCl) at pH 2 denatures the protein and pepsin hydrolyzes many of the peptide bonds. In the basic environment (pH 8) of the small intestine, proteases such as trypsin and chymotrypsin

Figure 17.6 Digestion of dietary saccharides.

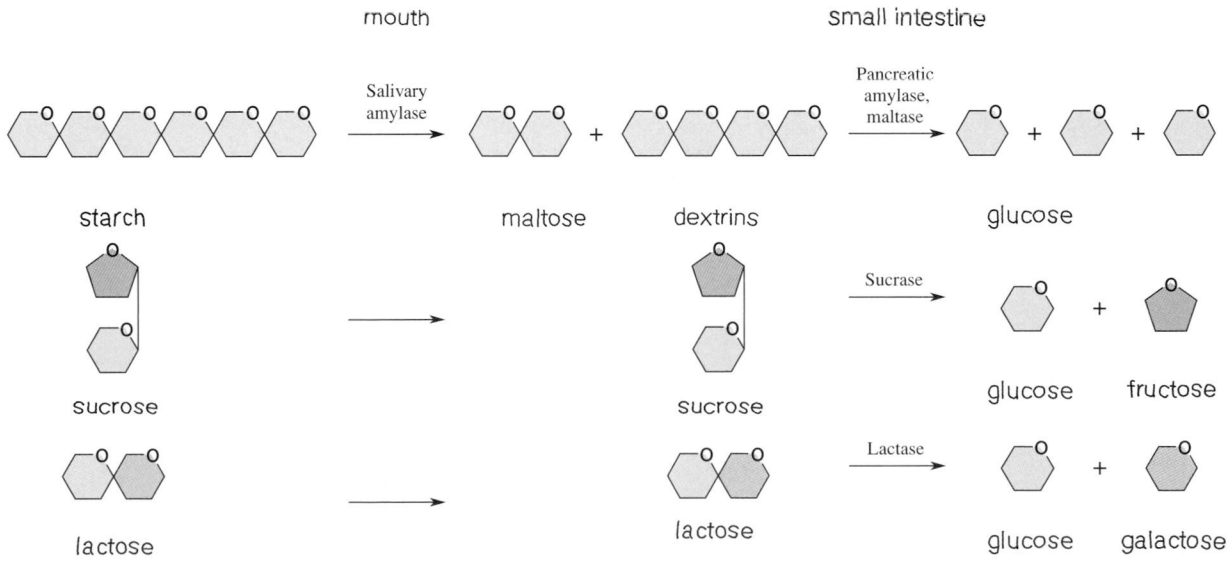

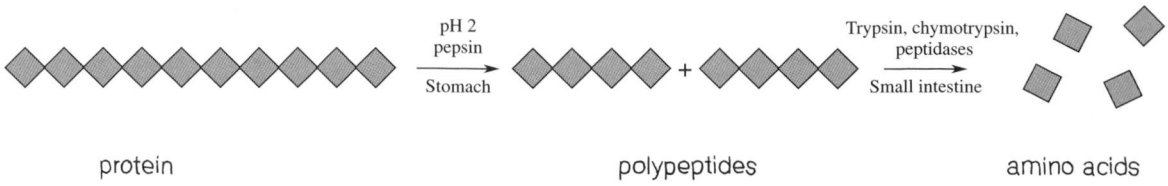

Figure 17.7 Digestion of dietary proteins.

complete the hydrolysis of the polypeptides to amino acids. The amino acids are absorbed through the intestinal walls into the bloodstream. (See Figure 17.7.)

17.2 Digestion of Lipids

The lipids in our diets begin digestion when they enter the small intestine where they mix with bile salts from the gallbladder and lipases from the pancreas. The bile salts break the fat globules into smaller droplets called micelles. In the small intestine, lipases hydrolyze the triglycerides on the surface of the micelles to yield fatty acids and glycerol. Water-soluble glycerol and short-chain fatty acids move through the intestinal lining into the bloodstream.

The insoluble fatty acids and triglycerides as well as cholesterol are absorbed into the intestinal lining where they are coated with proteins. The resulting lipoproteins carry lipids through the bloodstream to be used for energy or stored in fat cells. (See Figure 17.8.)

EXPLORE YOUR WORLD

Digestion of Foods

1. Obtain a cracker or small piece of bread. Chew the food in your mouth for 4–5 minutes. During that time observe the taste of the food.
2. Place a few drops of vegetable oil in a glass of water. Observe. Add a few drops of soap and shake well.
3. Some milk products contain Lactaid, an enzyme that digests the lactose in milk. Look for the brands of milk and ice cream that contain Lactaid.

Questions

1a. How does the taste of the carbohydrate change after it has been chewed in your mouth for 4–5 minutes?

1b. What part of carbohydrate digestion occurs in the mouth?

2a. How do the oil droplets change after soap is added to the oil layer?

2b. How is soap like the bile salts in the digestion of fats?

2c. Where does fat digestion occur in the body?

2d. People without gallbladders take lipase medication. Why is this necessary?

3a. Write the equation for the digestion of lactose.

3b. Where in the intestinal tract does the hydrolysis of proteins take place?

Figure 17.8 Digestion of dietary fats.

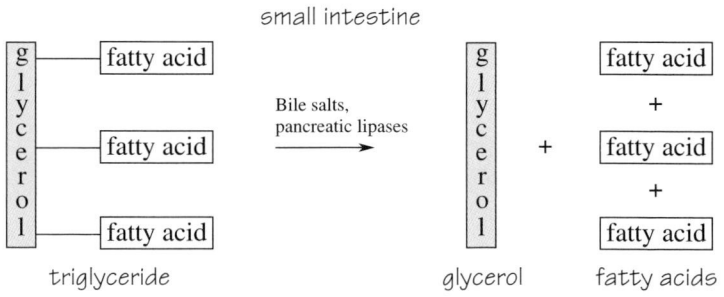

S·A·M·P·L·E· 17.2 ·P·R·O·B·L·E·M

Digestion of Foods

Describe the role of bile salts and pancreatic lipase in fat digestion.

Solution

In the small intestine, bile salts released from the gallbladder break apart fat globules to yield micelles. On the surface of the micelles, pancreatic lipase hydrolyzes triglycerides.

Study Check

Describe the digestion of amylose, a polymer of glucose molecules joined by α-glycosidic bonds.

Q·U·E·S·T·I·O·N·S & P·R·O·B·L·E·M·S

Digestion of Foods

17.7 What is the purpose of digestion?

17.8 What are the major sites of digestion?

17.9 Give the site, enzyme(s), and end products for the digestion of the following:
a. sucrose **b.** proteins **c.** amylose

17.10 Give the site, enzyme(s), and end products for the digestion of the following:
a. triglycerides **b.** peptides **c.** lactose

17.3 Important Coenzymes in Metabolic Pathways

Learning Goal

Name the major components of the coenzymes FAD, NAD⁺, and coenzyme A.

Before we look at the different metabolic reactions that extract energy from our food, we need to describe some of the substances that connect the metabolic pathways. These are the coenzymes that pick up hydrogen and electrons during oxidation or carry chemical groups from one pathway to another.

In Chapter 5, we saw that **oxidation** involves the loss of hydrogen atoms or electrons by a substance. In metabolism, oxidation is generally associated with the degradation of large molecules and the production of energy. **Reduction** is the gain of hydrogen or electrons by a substance. In metabolism, coenzymes are often reduced as they accept hydrogen from compounds in degradative processes.

FAD (flavin adenine dinucleotide) is a coenzyme that accepts hydrogen atoms. FAD is made from *riboflavin* (vitamin B_2) and adenine diphosphate. (See Figure 17.9.) FAD is the hydrogen acceptor when an oxidation produces a carbon–carbon (C═C) double bond. The hydrogen atoms from the oxidation are accepted by the flavin part of FAD, which is reduced to $FADH_2$.

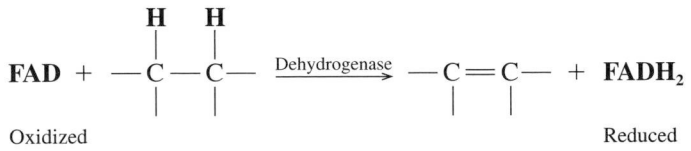

Figure 17.9 Structure of coenzyme FAD.

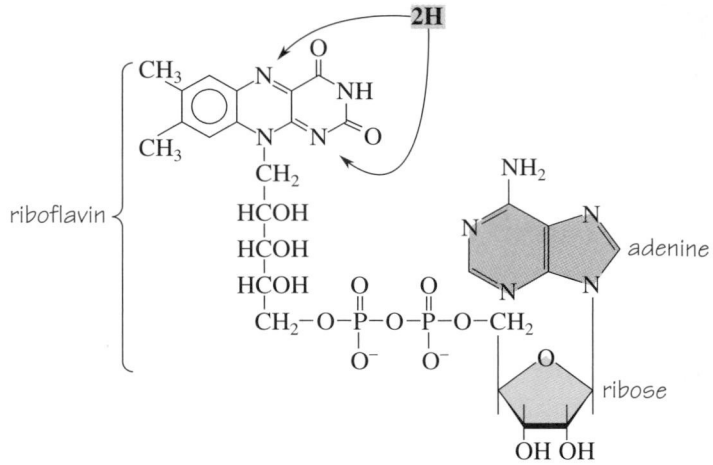

17.3

NAD^+ (nicotinamide adenine dinucleotide) is a coenzyme that also accepts electrons from oxidation reactions. NAD^+ is made from *nicotinamide* (derived from niacin) and adenine diphosphate. (See Figure 17.10.) NAD^+ is the coenzyme used when an oxidation removes hydrogen to produce a carbon–oxygen (C═O) double bond. In the process, NAD^+ is reduced to NADH + H^+.

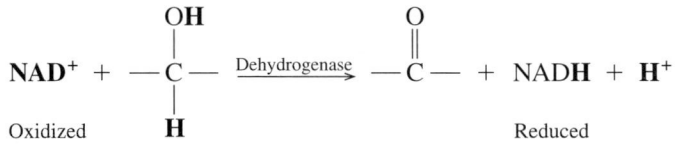

Coenzyme A (CoA) is the carrier of acetyl (two-carbon) groups, which are the degradation products of glucose, fatty acids, glycerol, and amino acids. CoA is made up of several components: pantothenic acid (vitamin B_3), adenine diphosphate (ADP) with a phosphate group, and an aminoethanethiol. Acetyl CoA forms when the —SH group of the thiol group connects the CoA structure to a two-carbon acetyl group. (See Figure 17.11.)

Figure 17.10 Structure of coenzyme NAD^+.

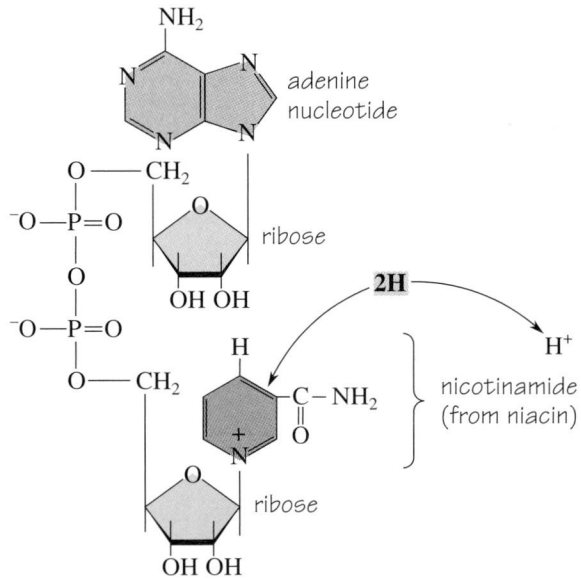

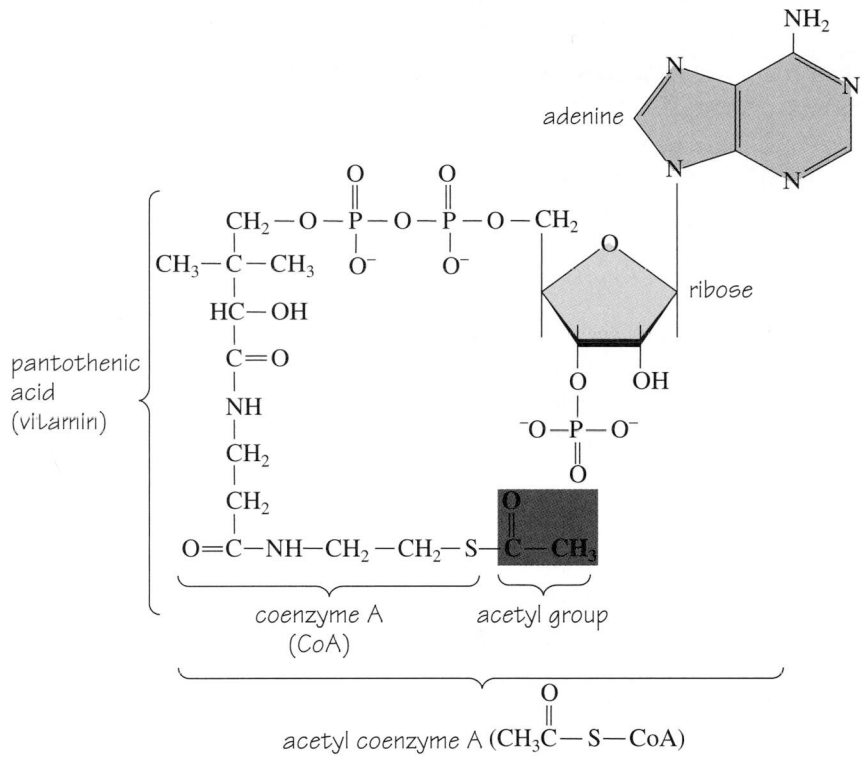

Figure 17.11 Structure of acetyl coenzyme A (CoA).

S · A · M · P · L · E · **17.3** · P · R · O · B · L · E · M

Coenzymes in Metabolic Pathways

What vitamin is part of the coenzyme FAD?

Solution

FAD, flavin adenine dinucleotide, is made from the vitamin riboflavin.

Study Check

What is the abbreviation of the reduced form of FAD?

Q · U · E · S · T · I · O · N · S & P · R · O · B · L · E · M · S

Important Coenzymes in Metabolic Pathways

17.11 What coenzyme is made from pantothenic acid?

17.12 What coenzyme is made from riboflavin?

17.13 Give the abbreviation for the following:
a. the reduced form of NAD^+ **b.** the oxidized form of $FADH_2$

17.14 Give the abbreviation for the following:
a. the reduced form of FAD **b.** the oxidized form of NADH

17.15 What coenzyme picks up hydrogen when a carbon–carbon double bond is formed?

17.16 What coenzyme picks up hydrogen when a carbon–oxygen double bond is formed?

Study Pack

17.4 Glycolysis: Oxidation of Glucose

Learning Goal

Describe the conversion of glucose to pyruvate in glycolysis.

17.4

Our major source of energy is the glucose produced during the digestion of carbohydrates. After glucose is absorbed into the bloodstream, it enters the cytoplasm of our cells for further degradation in a process called glycolysis. Glycolysis was the pathway used by early organisms to produce energy from simple nutrients, long before there was any oxygen in Earth's atmosphere. Glycolysis is an **anaerobic** process; no oxygen is required.

In the metabolic pathway of **glycolysis,** the six-carbon glucose molecule is split into two pyruvate molecules, each having three carbon atoms. (See Figure 17.12.) In steps 1–3, phosphate groups are combined with glucose, a process that uses energy from the hydrolysis of two molecules of ATP. In step 4, the six-carbon compound splits into two three-carbon molecules. The remaining steps (5–10) eventually convert the three-carbon molecules to pyruvate, the end product of glycolysis. In this part of glycolysis, ATP is synthesized, and NADH is formed in step 6.

Throughout the series of reactions, we have written the three-carbon compounds in their anion form, which is the prevalent form in physiological solutions (pH 7.4). For example, pyruvate is the ionized form of pyruvic acid.

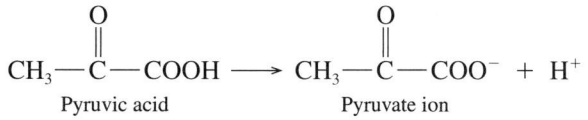

Reactions in Glycolysis

Steps 1–3 In these steps, glucose is converted into a phosphorylated compound that can be split into two smaller molecules. In step 1, a *hexokinase* enzyme adds a phosphate to glucose by converting ATP to ADP. This step is accelerated by insulin when the glucose level is high in a cell. In step 2, an *isomerase* converts the glucose to fructose. In step 3, a second phosphate group from a second hydrolysis of ATP is added to yield fructose-1,6-diphosphate. So far, there has been an energy investment of two ATP molecules.

Steps 4–5 In step 4, the enzyme *aldolase* splits the six-carbon fructose into two three-carbon molecules, dihydroxyacetone phosphate and glyceraldehyde-3-phosphate. After an *isomerase* converts dihydroxyacetone to a second molecule of glyceraldehyde-3-phosphate, there are two three-carbon compounds ready to be oxidized.

Step 6 The oxidation of the two glyceraldehyde-3-phosphate molecules are catalyzed by a *dehydrogenase* enzyme. Two molecules of NAD^+ are reduced to form two molecules of NADH as the aldehyde is oxidized. The addition of inorganic phosphate yields two molecules of 1,3-diphosphoglycerate.

Step 7 In a reaction catalyzed by *phosphokinase,* two ATP are generated when a phosphate group from each 1,3-diphosphoglycerate is transferred directly to ADP, a process known as **direct substrate phosphorylation.** At this point, glycolysis pays back the debt of two ATP used in the early steps.

Steps 8–9 In step 8, a *mutase* enzyme catalyzes an isomerization of the phosphate group to yield 2-phosphoglycerate. In step 9, *enolase* catalyzes the removal of water to yield phosphoenolpyruvate. This rearrangement provides a compound that will release more energy when it is hydrolyzed in the next step.

Step 10 The remaining phosphate groups on the three-carbon molecules are transferred by a *kinase* enzyme to ADP, which yields two ATP and two pyruvates.

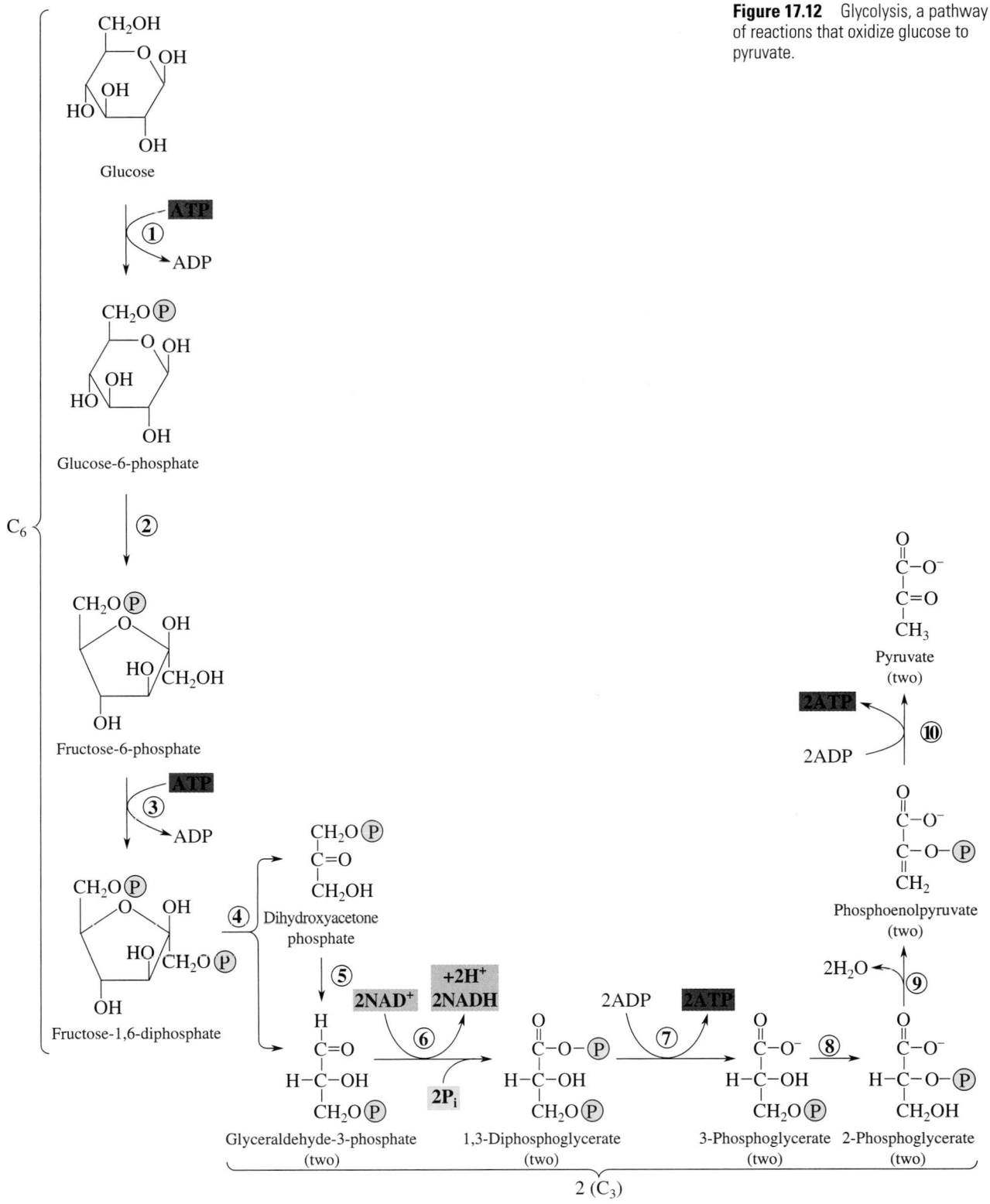

Figure 17.12 Glycolysis, a pathway of reactions that oxidize glucose to pyruvate.

Summary of Glycolysis

Overall, glycolysis yields two ATP molecules and two NADH for every glucose that is split to yield two pyruvates. Note that a total of four ATP molecules were generated in the last part of glycolysis, but there is only a net gain of two ATP molecules because two ATP molecules were required in the early part when phosphate groups were transferred to the glucose molecule.

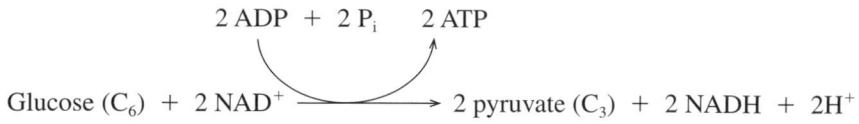

17.4

S·A·M·P·L·E 17.4 ·P·R·O·B·L·E·M

Glycolysis

What are the steps in glycolysis that generate ATP?

Solution

ATP is produced when phosphate groups are transferred directly to ADP from 1,3-diphosphoglycerate (step 7) and from phosphoenolpyruvate (step 10).

Study Check

If four ATP molecules are produced in glycolysis, why is there a net yield of two ATP?

Q·U·E·S·T·I·O·N·S & P·R·O·B·L·E·M·S

Glycolysis: Oxidation of Glucose

17.17 What is the starting product of glycolysis?

17.18 What is the end product of glycolysis?

17.19 How is ATP used in the initial steps of glycolysis?

17.20 How many ATP molecules are used in the initial steps of glycolysis?

17.21 What trioses are obtained when fructose-1,6-diphosphate splits?

17.22 Why does one of the triose products undergo isomerization?

17.23 How does direct phosphorylation account for the production of ATP in glycolysis?

17.24 Why are there two ATP molecules formed for one molecule of glucose?

17.25 Indicate the enzyme that catalyzes the following reactions in glycolysis:
a. phosphorylation
b. direct transfer of a phosphate group

17.26 Indicate the enzyme that catalyzes the following reactions in glycolysis:
a. isomerization
b. formation of a ketotriose and an aldotriose

17.27 How many ATP or NADH are produced (or required) in each of the following steps in glycolysis?
a. glucose to glucose-6-phosphate
b. glyceraldehyde-3-phosphate to 1,3-diphosphoglycerate
c. glucose to pyruvate

17.28 How many ATP or NADH are produced (or required) in each of the following steps in glycolysis?
a. 1,3-diphosphoglycerate to 3-phosphoglycerate
b. fructose-6-phosphate to fructose-1,6-diphosphate
c. phosphoenolpyruvate to pyruvate

17.29 Which step in glycolysis involves the following?
a. the first ATP molecule is hydrolyzed
b. direct substrate phosphorylation occurs
c. six-carbon sugar splits into two three-carbon molecules

17.30 Which step in glycolysis involves the following?
a. isomerization takes place
b. NAD^+ is reduced
c. a second ATP molecule is synthesized

17.5 Pathways for Pyruvate

Study Pack

Learning Goal

Give the conditions for the conversion of pyruvate to lactate, ethanol, and acetyl coenzyme A.

The pyruvate produced from glucose can now enter one of three pathways that continue to extract energy. The available pathway depends on whether there is sufficient oxygen in the cell. During aerobic conditions, oxygen is available to convert pyruvate to acetyl coenzyme A (CoA). When oxygen levels are low, pyruvate is reduced to lactate. In yeast cells, which are anaerobic, pyruvate is converted to ethanol.

Pyruvate to Acetyl CoA (Aerobic Conditions)

In glycolysis, two ATP molecules were generated when glucose was converted to pyruvate. However, much more energy is still available. The greatest amount of the energy is obtained from glucose when oxygen levels are high in the cells. Under **aerobic** conditions, pyruvate moves from the cytoplasm (where glycolysis took place) into the matrix of the mitochondria to be oxidized further. In a complex reaction, pyruvate is oxidized and a carbon atom is removed from pyruvate as CO_2. The coenzyme NAD^+ is required for the oxidation. The resulting two-carbon acetyl compound is attached to CoA, producing **acetyl CoA,** an important intermediate in many metabolic pathways. (See Figure 17.13.)

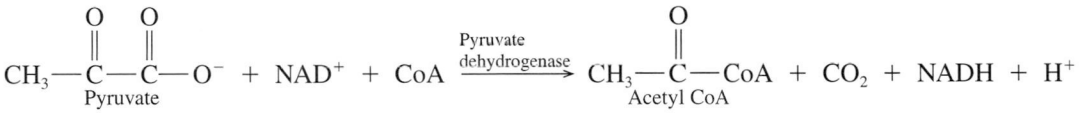

Pyruvate to Lactate (Anaerobic Conditions)

When we engage in strenuous exercise, the oxygen stored in our muscle cells is quickly depleted. Under anaerobic conditions, pyruvate remains in the cytoplasm where it is reduced to lactate. NADH is produced and used to oxidize more glyceraldehyde-3-phosphate in the glycolysis pathway, which produces a small but needed amount of ATP.

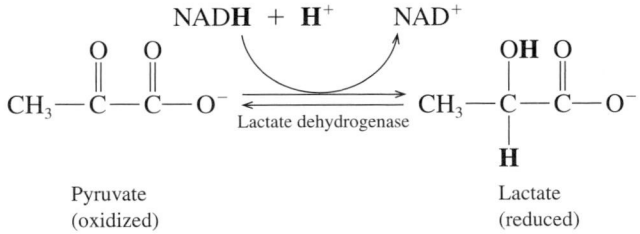

17.5

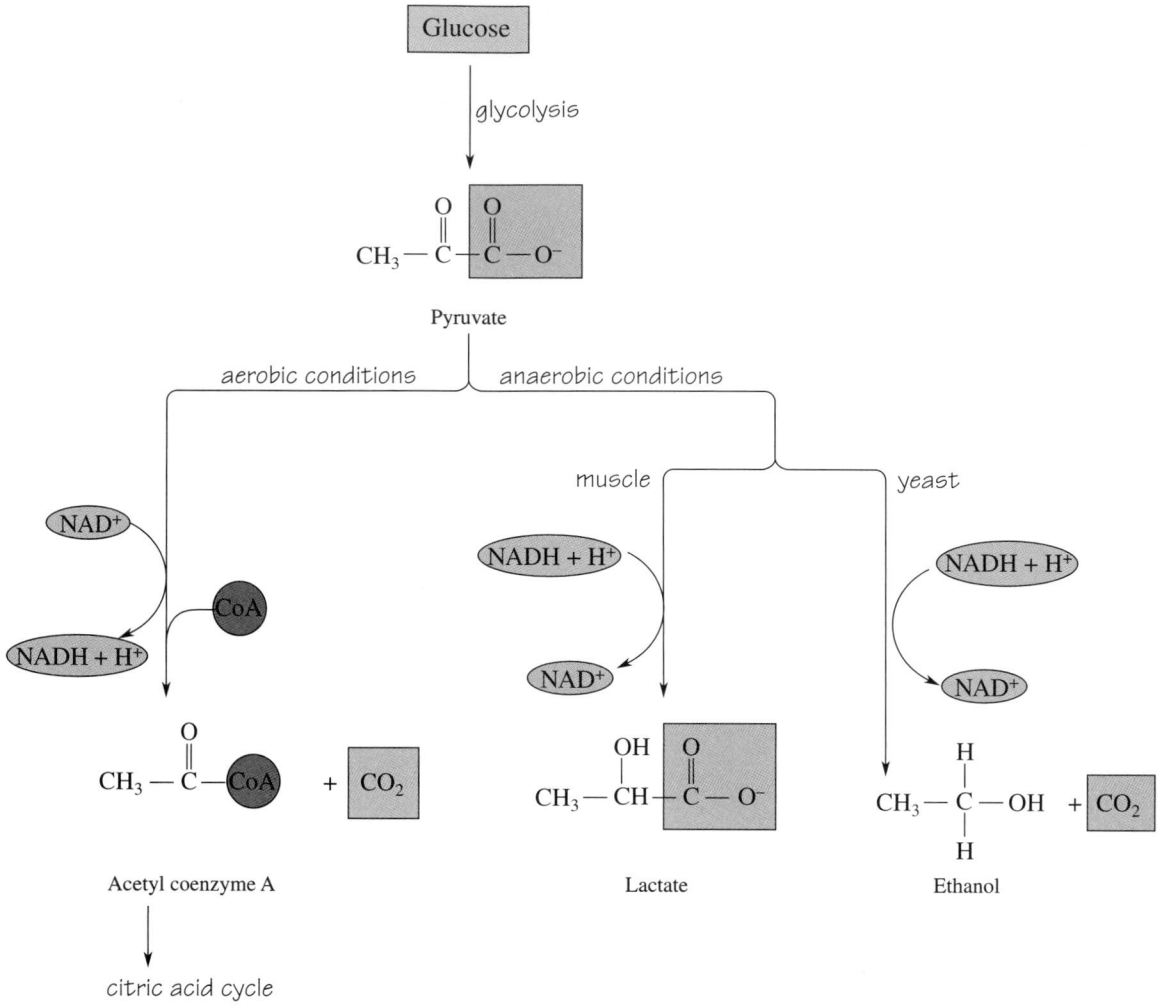

Figure 17.13 Reactions of pyruvate under aerobic and anaerobic conditions.

The accumulation of lactate causes the muscles to tire rapidly and become sore. After exercise, a person continues to breathe rapidly to repay the oxygen debt incurred during exercise. Most of the lactate is transported to the liver where it is converted back into pyruvate. Under anaerobic conditions, the only ATP production in glycolysis occurs during the steps that phosphorylate ADP directly, giving a net gain of only two ATP molecules:

$$\text{Glucose} + 2\,\text{ADP} + 2\,\text{P}_i \longrightarrow 2\,\text{lactate} + 2\,\text{ATP}$$

Bacteria also convert pyruvate to lactate under anaerobic conditions. In the preparation of kimchee and sauerkraut, cabbage is covered with a salt brine. The glucose from the starches is converted to lactate. This acid environment acts as a preservative that prevents the growth of other bacteria. The pickling of olives and cucumbers gives similar products. When cultures of bacteria that produce lactate are added to milk, the acid denatures the milk proteins to give sour cream and yogurt.

Pyruvate to Ethanol (Anaerobic)

Some microorganisms, particularly yeast, convert sugars to ethanol under anaerobic conditions by a process called **fermentation.** After pyruvate is formed in glycolysis, a carbon atom is removed in the form of CO_2 **(decarboxylation).** The NAD^+ for continued glycolysis is regenerated when the acetaldehyde is reduced to ethanol.

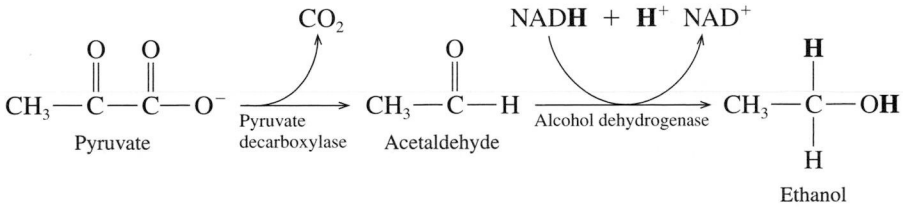

The process of fermentation by yeast is one of the oldest known chemical reactions. Enzymes in the yeast convert the sugars in a variety of carbohydrate sources to glucose and then to ethanol. The evolution of CO_2 gas produces the bubbles in beer, sparkling wines, and champagne. The type of carbohydrate used determines the taste associated with a particular alcoholic beverage. Beer is made from the fermentation of barley malt, wine from the sugars in grapes, vodka from potatoes, sake from rice, and whiskeys from corn or rye. Fermentation produces solutions up to about 15% alcohol by volume. At this concentration, the alcohol kills the yeast, and fermentation stops. Higher concentrations are obtained by distilling the alcohol.

S·A·M·P·L·E 17.5 ·P·R·O·B·L·E·M

Fates of Pyruvate

Is each of the following products from pyruvate produced under anaerobic or aerobic conditions?

a. acetyl CoA **b.** lactate

Solution

a. aerobic conditions **b.** anaerobic conditions

Study Check

After strenuous exercise, some lactate is oxidized back to pyruvate by lactate dehydrogenase using NAD^+. Write an equation to show this reaction.

Q·U·E·S·T·I·O·N·S & P·R·O·B·L·E·M·S

Pathways for Pyruvate

17.31 What condition is needed in the cell to convert pyruvate to acetyl CoA?

17.32 What coenzymes are needed for the oxidation of pyruvate to acetyl CoA?

17.33 Write the overall equation for the conversion of pyruvate to acetyl CoA.

17.34 What are the possible products of pyruvate under anaerobic conditions?

17.35 How does the formation of lactate permit glycolysis to continue under anaerobic conditions?

17.36 After running a marathon, a runner has muscle pain and cramping. What might have occurred in the muscle cells to cause this?

17.37 In fermentation, a carbon atom is removed from pyruvate. What is the compound formed by that carbon atom?

17.38 Some students decided to make some wine by placing yeast and grape juice in a container with a tight lid. A few weeks later, the container explodes. What reaction could account for the explosion?

Study Pack

17.6 Citric Acid Cycle

Learning Goal

Describe the oxidation of acetyl CoA in the citric acid cycle.

17.6

Under aerobic conditions, acetyl CoA is oxidized by enzymes in the matrix of the mitochondria to produce CO_2 in a series of reactions known as the **citric acid cycle** (or Krebs cycle). The citric acid cycle begins with acetyl CoA that usually comes from the oxidation of pyruvate, but may also be produced from the degradation of fats and proteins in our diet. (See Figure 17.14.) The end product of the citric acid cycle is oxaloacetate, a four-carbon compound that can attach to another acetyl group and go through the cycle again.

As the oxidation reactions occur in the citric acid cycle, chemical energy is stored in NADH, $FADH_2$, and ATP. Another series of reactions known as the electron transport pathway uses the energy to synthesize ATP. We will see how this is done in detail when we discuss the role of the electron transport pathway later in this chapter.

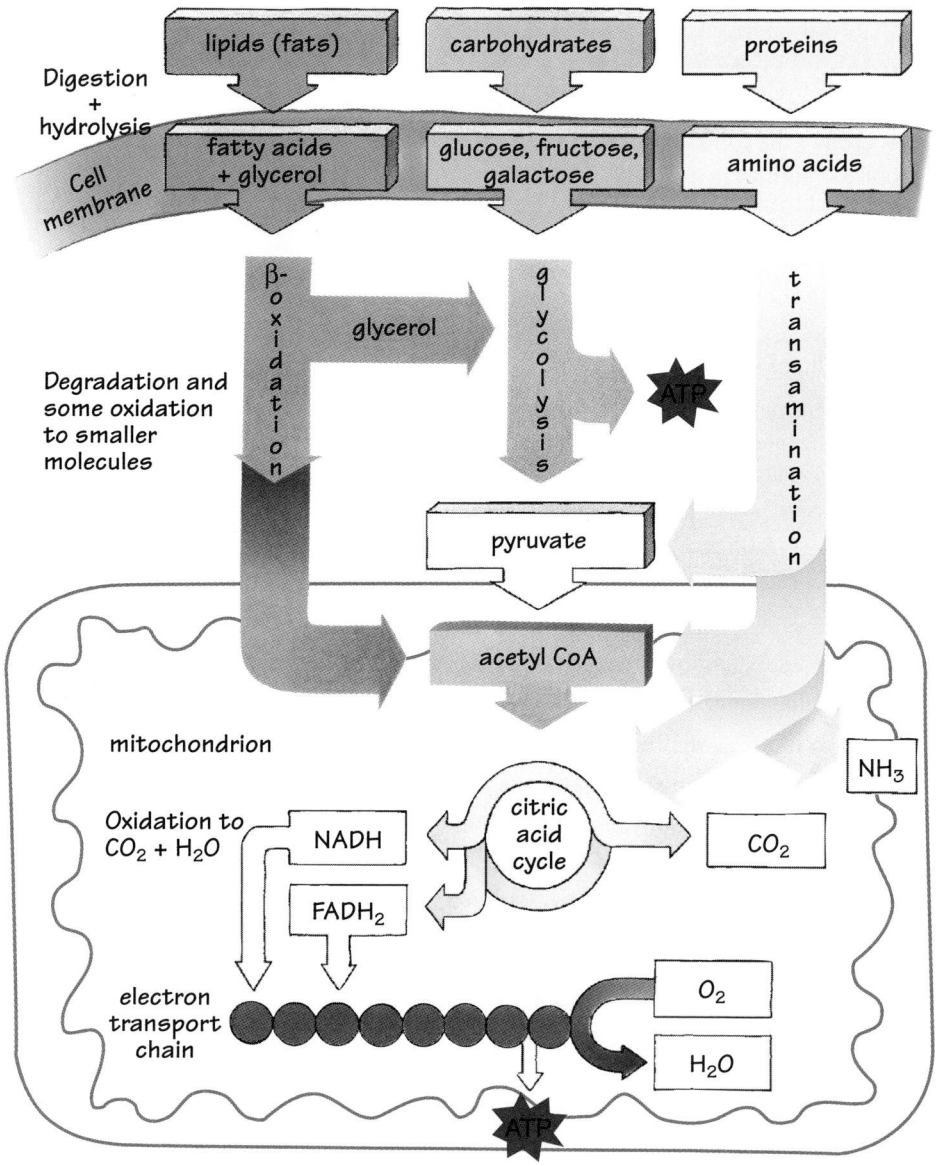

Figure 17.14 Catabolic pathways oxidize carbohydrates, lipids, and proteins and produce CO_2, H_2O, and the energy to build ATP molecules.

Reactions of the Citric Acid Cycle

We can now follow the two-carbon acetyl group as it enters the series of eight reactions that make up the citric acid cycle. (See Figure 17.15.)

Figure 17.15 The reactions of the citric acid cycle.

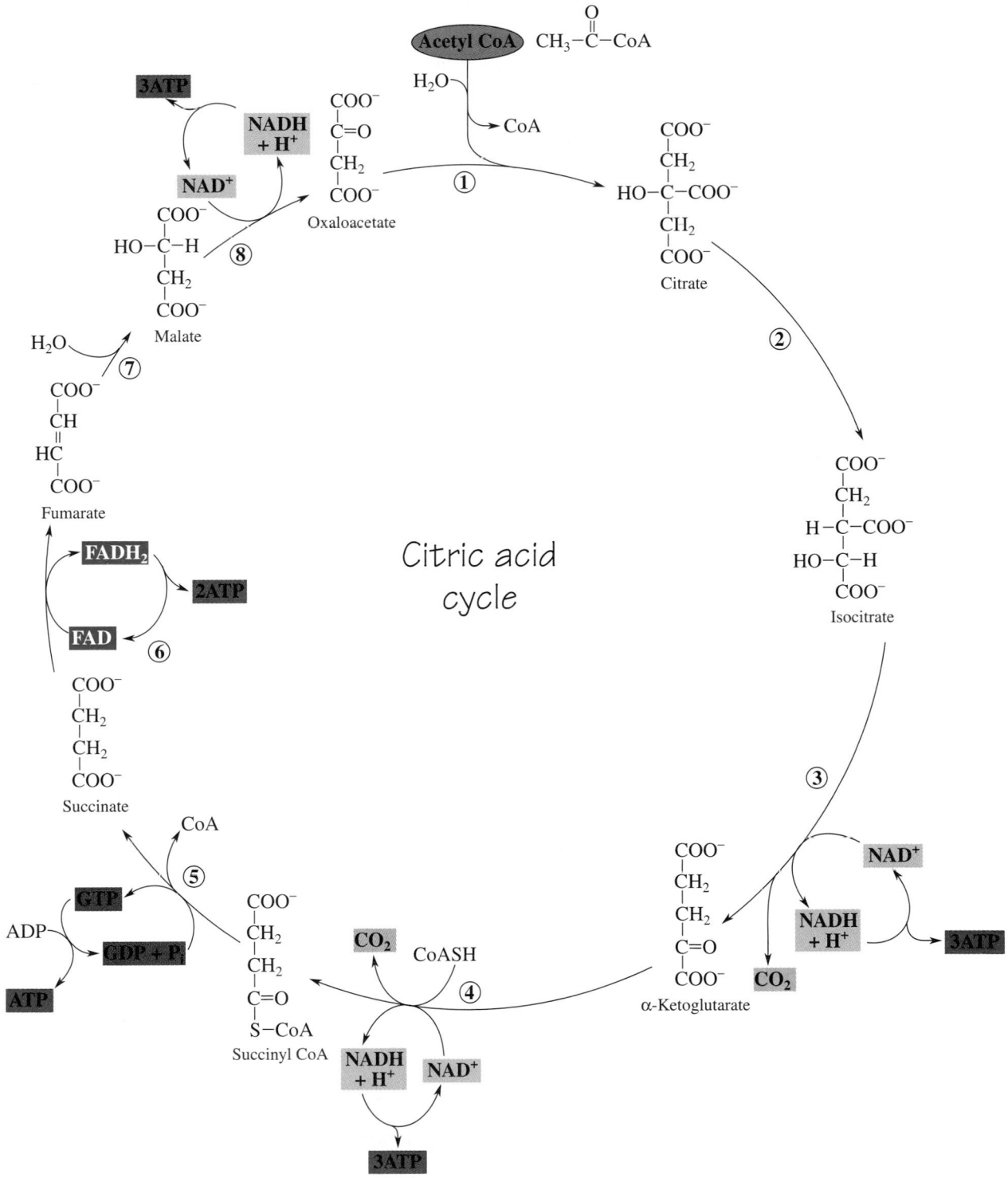

Step 1 Coenzyme A provides a two-carbon acetyl group that combines with a four-carbon molecule, oxaloacetate. The reaction, which is catalyzed by *citrate synthetase,* yields citrate, a compound with six carbon atoms. As we proceed through the citric acid cycle, two carbon atoms in the form of CO_2 will be removed from this six-carbon skeleton. In several steps, oxidation reactions will reduce NAD^+ to NADH and H^+, and FAD to $FADH_2$.

Step 2 Citrate is converted by an *aconitase* enzyme to an isomer called isocitrate. The same enzyme converts the hydroxyl group (—OH) in citrate to a secondary alcohol to be oxidized further.

17.6

Step 3 Isocitrate is now oxidized and decarboxylated in a complex step with several reactions that are catalyzed by the enzyme *isocitrate dehydrogenase*:

1. The hydroxyl group is oxidized to a ketone.
2. NAD^+ is reduced to yield NADH and H^+.
3. A carbon atom is removed as CO_2.

The formation of CO_2 shortens the carbon chain by one carbon to yield a five-carbon compound, α-ketoglutarate.

Step 4 Another series of reactions oxidize and convert α-ketoglutarate to succinate, a four-carbon compound. The enzymes and cofactors of *α-ketoglutarate dehydrogenase complex* catalyze the reactions in this step.

1. NAD^+ is reduced to yield NADH and H^+.
2. The carboxylate group is removed to yield CO_2.

An oxidation reaction combines succinate with coenzyme A to yield succinyl CoA and NADH and H^+.

Step 5 In a reaction catalyzed by *succinyl CoA synthetase,* succinyl CoA is hydrolyzed. The energy released is used to add a phosphate group directly to GDP (guanosine diphosphate) to yield GTP, which is a high-energy compound similar to ATP. This direct transfer of a phosphate group to a substrate is known as a substrate-level phosphorylation. The GDP is regenerated when *nucleotide phosphokinase* transfers the phosphate group to ADP to form ATP. This is the only place in the citric acid cycle where ATP is produced by a direct substrate phosphorylation.

$$\text{GTP} + \text{ADP} \longrightarrow \text{GDP} + \text{ATP}$$

Step 6 Succinate is oxidized by *succinate dehydrogenase* to yield fumarate. The hydrogen removed to form the carbon–carbon double bond reduces the coenzyme FAD to yield $FADH_2$. This is the only place in the citric acid cycle that reduces FAD to $FADH_2$. The enzyme for this reaction is located in the inner mitochondrial membrane.

Step 7 Water (H_2O) is added to the double bond of fumarate to yield malate. The enzyme *fumarase* catalyzes the hydration reaction.

Step 8 In the last step of the citric acid cycle, *malate dehydrogenase* catalyzes the oxidation of the hydroxyl (—OH) group to yield a ketone group in oxaloacetate. The coenzyme NAD^+ is reduced to NADH.

Summary of the Citric Acid Cycle

These eight steps complete one full turn of the citric acid cycle. The cycle starts all over when oxaloacetate combines with another acetyl CoA. In each turn of the citric acid cycle, the following are produced:

1. Two CO_2 molecules
2. Three NADH molecules
3. One $FADH_2$ molecule
4. One GTP molecule that was converted to ATP

$$CH_3-\overset{\overset{\displaystyle O}{\|}}{C}-CoA + 3\,NAD^+ + FAD + ADP + P_i \longrightarrow 2CO_2 + 3\,NADH + FADH_2 + ATP$$

SAMPLE 17.6 PROBLEM

Citric Acid Cycle

When one acetyl CoA completes the citric acid cycle, how many of each of the following are produced?

a. NADH **b.** ketone group **c.** CO_2

Solution

a. One turn of the citric acid cycle produces three molecules of NADH.
b. Two ketone groups form when the secondary alcohol groups in isocitrate and malate are oxidized by NAD^+.
c. Two molecules of CO_2 are produced by the decarboxylation of isocitric acid and α-ketoglutarate.

Study Check

What substance is a substrate in the first step of the citric acid cycle and a product in the last step?

QUESTIONS & PROBLEMS

Citric Acid Cycle

17.39 Refer to the diagram of the citric acid cycle to answer each of the following:
a. What are the six-carbon compounds?
b. How is the number of carbon atoms decreased?
c. How many NADH molecules form?
d. What are the five-carbon compounds?
e. Which steps are oxidation reactions?
f. In which steps are secondary alcohols oxidized?

17.40 Refer to the diagram of the citric acid cycle to answer each of the following:
a. What is the yield of CO_2 molecules?
b. How many $FADH_2$ molecules form?
c. What are the four-carbon compounds?
d. What is the yield of GTP molecules?
e. What are the decarboxylation steps?
f. Where does a hydration occur?

17.41 Indicate the name of the enzyme for each of the following reactions in the citric acid cycle:
a. joins acetyl CoA to oxaloacetate
b. forms a carbon–carbon double bond
c. adds water to fumarate

17.42 Indicate the name of the enzyme for each of the following reactions in the citric acid cycle:
a. isomerizes citrate
b. oxidizes and decarboxylates α-ketoglutarate
c. adds P_i to GDP

17.43 State the acceptor for hydrogen or phosphate in each of the following reactions:

a. isocitrate ⟶ α-ketoglutarate
b. succinyl CoA ⟶ succinate
c. succinate ⟶ fumarate

17.44 State the acceptor for hydrogen or phosphate in each of the following reactions:

a. malate ⟶ oxaloacetate
b. α-ketoglutarate ⟶ succinyl CoA
c. pyruvate ⟶ acetyl CoA

17.7

Study Pack

17.7 Electron Transport Chain

Learning Goal

Describe the synthesis of ATP from NADH and $FADH_2$ by the electron transport chain.

Up to now, we have seen how the energy from the oxidation of glucose has produced a collection of the reduced coenzymes NADH and $FADH_2$. In the **electron transport chain,** the hydrogen and electrons from NADH and $FADH_2$ are transferred to a series of compounds called **electron carriers.** As electrons move among the carriers in the chain, energy is made available to generate ATP. Eventually, the electrons combine with oxygen (O_2), the final acceptor in the chain. As long as oxygen is available in the cell, the electron transport chain ultimately produces almost all of the ATP energy that is possible from the oxidation of glucose.

Electron Carriers

Although there are different types of electron carriers, each one contains a group or ion that can be alternately reduced and oxidized as the carrier accepts electrons and then passes them on.

1. Flavin mononucleotide **(FMN)** is a coenzyme composed of riboflavin (vitamin B_2) similar to FAD. (See Figure 17.16.)
2. **Coenzyme Q** (Q or CoQ) is derived from quinone and contains a long carbon chain. (See Figure 17.17.)

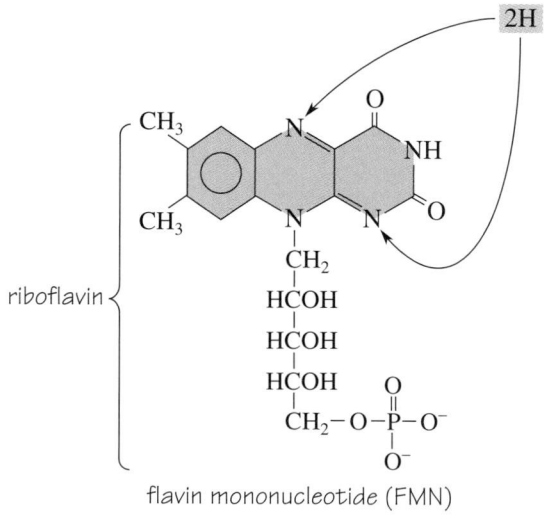

Figure 17.16 Structure of coenzyme FMN.

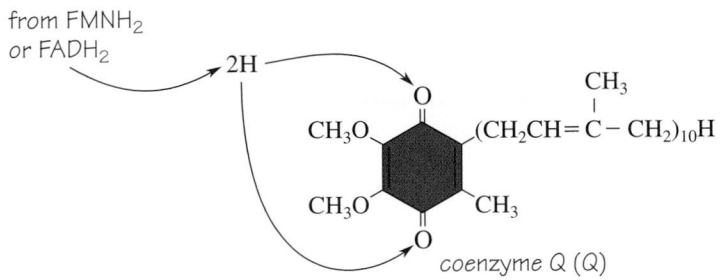

Figure 17.17 Structure of coenzyme Q.

3. Iron–sulfur proteins **(FeS)** contain iron ions attached to sulfur. The iron is reduced (Fe^{2+}) and oxidized (Fe^{3+}) as electrons pass through the carrier.
4. **Cytochromes** (cyt) are proteins that also contain iron. There are several cytochromes: cyt *b*, cyt c_1, cyt *c*, cyt *a*, and cyt a_3. As the electrons are accepted, iron is reduced to Fe^{2+}, and then oxidized back to Fe^{3+} as the electron is passed to the next cytochrome.

Chemiosmotic Model

The function of the electron transport chain is to provide energy for the synthesis of ATP and to reoxidize the coenzymes. At the beginning of the chain, energy is released when hydrogen and electrons from the reduced coenzyme NADH and H^+ are transferred to the electron carriers. The energy is used to pump protons from the matrix into the intermembrane space. (See Figure 17.18.) Because the inner membrane is impermeable to ions, H^+ ions accumulate in the intermembrane space, forming a proton gradient. In the **chemiosmotic model,** the H^+ can return to the matrix only by way of a H^+ channel in an enzyme called **ATP synthase.** As the H^+ ions flow back through ATP synthase, energy is released and used to synthesize ATP.

In the overall process called **oxidative phosphorylation,** the oxidation energy from all the steps and cycles involved in the degradation of glucose is now coupled with the formation of energy-rich ATP.

$$\text{ADP} + \text{P}_i + \text{energy} \xrightarrow{\text{ATP synthase}} \text{ATP}$$

The transfer of electrons from NADH + H^+ along the electron transport chain and the potential energy created by the proton gradient provides energy for the synthesis of three ATP molecules. However, the transfer of hydrogen and electrons by $FADH_2$ occurs at a lower energy level, which provides energy for the synthesis of only two ATP molecules, as shown by the table in the margin.

Reduced Coenzyme	ATP Produced in Electron Transport Chain
NADH	3 ATP
$FADH_2$	2 ATP

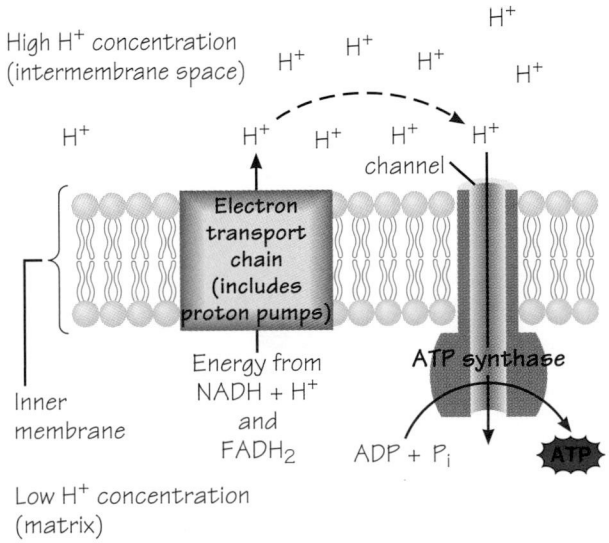

Figure 17.18 In chemiosmosis, ATP is synthesized when H^+ flows back into the mitochondrial matrix through a H^+ channel in ATP synthase.

Steps in the Electron Transport Chain

Most of the electron carriers are found in three protein complexes, which are attached to the inner membrane of the mitochondria. Each complex acts as a **proton pump,** which moves H^+ from the matrix into the intermembrane space and generates a proton gradient. Hydrogen and electrons from NADH + H^+ are shuttled between the complexes by two mobile carriers. Coenzyme Q carries electrons between the first and second complex and cytochrome *c* carries electrons between the second and third complex. Coenzyme Q also picks up hydrogen and electrons from $FADH_2$ for transport to the second complex.

17.7

Protein Clusters and Mobile Carriers	Electron Carriers
Complex 1: NADH dehydrogenase	FMN ⟶ FeS
Mobile carrier: coenzyme Q	CoQ
Complex 2: cytochrome *c* reductase	cyt *b* ⟶ cyt c_1
Mobile carrier: cytochrome *c*	cyt *c*
Complex 3: cytochrome *c* oxidase	cyt *a* ⟶ cyt a_3

Complex 1: NADH Dehydrogenase

NADH enters the electron transport chain at the first complex by transferring hydrogen and electrons to FMN. (See Figure 17.19.) FMN is reduced to $FMNH_2$ and NADH is reoxidized to NAD^+, which returns to the citric acid cycle to oxidize more substrates. The first complex does not accept electrons from $FADH_2$. Eventually, the electrons from $FMNH_2$ are passed to FeS and then coenzyme Q.

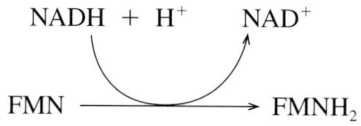

Mobile Electron Carrier Coenzyme Q

Coenzyme Q is a mobile compound not attached to a protein complex, which allows it to move freely between the complexes. Q collects the electrons from the first complex and transports them to the second complex. This is also the point in the electron chain

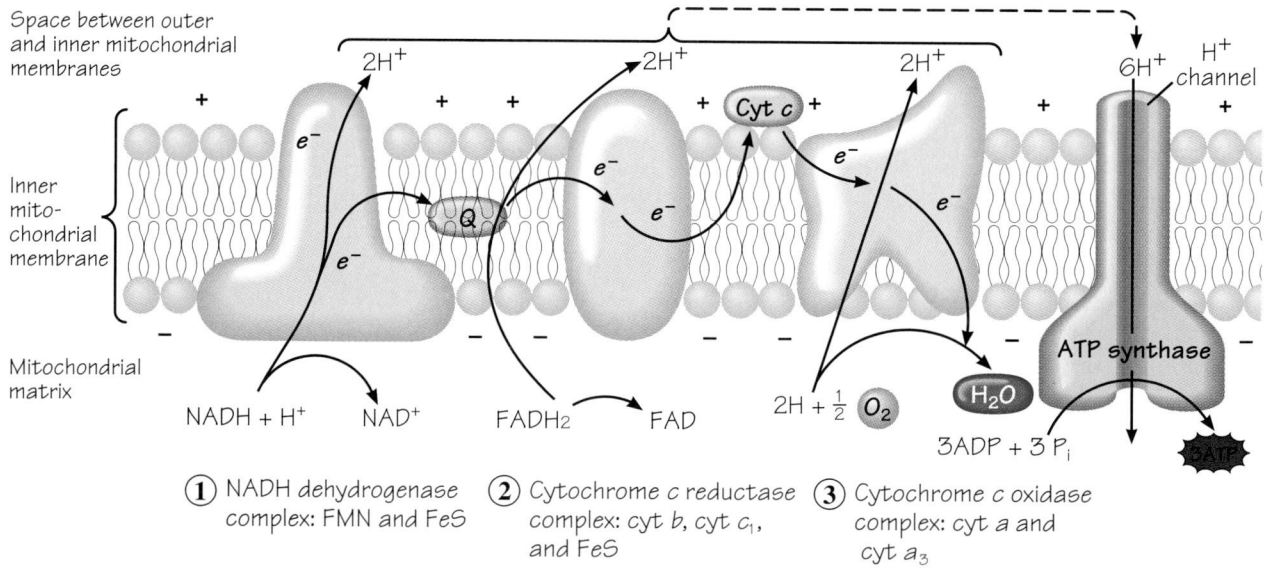

Figure 17.19 Proton pumps, which are complexes of electron carriers, move H^+ from the matrix into the intermembrane.

where hydrogen and electrons from another reduced coenzyme, $FADH_2$, are transferred to coenzyme Q. Because coenzyme Q has a lower energy level, the electrons from $FADH_2$ enter the electron transport chain at a lower energy level than electrons from $NADH_2$.

$$Q + FADH_2 \longrightarrow QH_2 + FAD$$

Complex 2: Cytochrome *c* Reductase

At the second complex, electrons from QH_2 pass to cyt *b*, and then to cyt c_1. As each cytochrome accepts an electron, the iron in the center is reduced to Fe^{2+}, and reoxidized to Fe^{3+} as the electron is passed on. Finally, the electrons are transferred to cytochrome *c*, a second mobile carrier, which carries the electrons to the third complex.

$$\underset{\text{Reduced}}{\text{Cytochrome-Fe}^{2+}} \longrightarrow \underset{\text{Oxidized}}{\text{cytochrome-Fe}^{3+}}$$

Complex 3: Cytochrome *c* Oxidase

At the third complex, electrons pass from cyt *c* to cyt *a* and then to cyt a_3. In the final step of the electron transport chain, electrons from cyt a_3 are donated to an oxygen atom, which combines with two protons to form H_2O.

$$2e^- + \tfrac{1}{2}O_2 + 2H^+ \longrightarrow H_2O$$

S·A·M·P·L·E 17.7 ·P·R·O·B·L·E·M

Electron Carriers

Give the abbreviation for each of the following carriers:
a. the oxidized form of flavin mononucleotide
b. the reduced form of coenzyme Q

Solution

a. FMN **b.** QH_2

Study Check

What is the oxidized form of iron in cyt a_3?

S·A·M·P·L·E 17.8 ·P·R·O·B·L·E·M

Oxidation and Reduction

Identify the following steps in the electron transport chain as oxidation or reduction:
a. $FMN + 2H^+ + 2e^- \longrightarrow FMNH_2$
b. Cyt *c* $(Fe^{2+}) \longrightarrow$ cyt *c* $(Fe^{3+}) + e^-$

Solution

a. The gain of hydrogen is reduction. **b.** The loss of electrons is oxidation.

Study Check

Identify each of the following as oxidation or reduction:
a. $QH_2 \longrightarrow Q$ **b.** Cyt *b* $(Fe^{3+}) \longrightarrow$ cyt *b* (Fe^{2+})

17.7

S·A·M·P·L·E 17.9 P·R·O·B·L·E·M

Electron Transport System

Why does the oxidation of NADH by the electron transport chain provide energy for the formation of three ATP whereas $FADH_2$ produces two ATP?

Solution

Electrons from the oxidation of NADH enter the electron chain at an earlier site than the $FADH_2$, providing energy to pump three pairs of protons to the intermembrane. However, $FADH_2$ transfers electrons to Q that go through two sites that pump protons and therefore provide energy for two ATP.

Study Check

Where is water (H_2O) formed in the electron chain?

Q·U·E·S·T·I·O·N·S & P·R·O·B·L·E·M·S

Electron Transport Chain

17.45 What reduced coenzymes provide the electrons for the electron transport chain?

17.46 What happens to the energy level as electrons are passed down the electron transport chain?

17.47 What are the oxidized and reduced forms of each of the following?
a. NADH **b.** FMN

17.48 What are the oxidized and reduced forms of each of the following?
a. cytochrome *c* **b.** O_2

17.49 In what order do the following appear in the electron chain: cytochrome *c*, FAD, and coenzyme Q?

17.50 In what order do the following appear in the electron chain: O_2, NAD^+, ATP synthase?

17.51 How is NADH oxidized by the electron chain? What is the ATP yield?

17.52 How is $FADH_2$ oxidized by the electron chain? What is the ATP yield?

17.53 What is meant by oxidative phosphorylation?

17.54 How is the proton gradient established?

17.55 According to the chemiosmotic theory, how does the proton gradient provide energy to synthesize ATP?

17.56 How does the phosphorylation of ADP occur?

17.57 How are glycolysis and the citric acid cycle linked to the production of ATP by the electron transport chain?

17.58 Why does $FADH_2$ have a yield of two ATP via the electron chain, but NADH yields three ATP?

17.8 ATP Energy from Glucose

Study Pack

Learning Goal

Account for the ATP produced by the complete oxidation of glucose.

Under aerobic conditions, the degradation of one glucose molecule produces two acetyl CoA that enter the citric acid cycle to yield the following molecules: two CO_2, three NADH, one $FADH_2$, and one ATP. Now that we have described the synthesis of ATP in the electron transport chain, we can calculate the total number of ATP molecules produced from the complete oxidation of one glucose molecule.

17.8

ATP and Glycolysis

In glycolysis, the oxidation of glucose stores energy in two NADH molecules as well as two ATP molecules from direct substrate phosphorylation. The electron transport chain makes it possible to convert those NADH molecules to ATP. However, glycolysis occurs in the cytoplasm, and the NADH cannot pass through the inner membrane of the mitochondria.

Therefore, in a series of reactions the hydrogen and electrons of cytoplasmic NADH are transferred to a compound that can enter the mitochondria, where they are used to reduce FAD to $FADH_2$.

$$NADH + H^+ + FAD \longrightarrow NAD^+ + FADH_2$$

In the electron transport chain, $FADH_2$ provides two ATP, rather than three, from NADH in the mitochondria. In glycolysis, four ATP from two NADH and two ATP from direct phosphorylation add up to six ATP from one glucose molecule.

$$\text{Glucose} \longrightarrow 2 \text{ pyruvate} + 6 \text{ ATP}$$

ATP and Pyruvate

Under aerobic conditions, pyruvate enters the mitochondria, where it is oxidized to give acetyl CoA, CO_2, and NADH. Because each glucose molecule yields two pyruvate, two NADH enter the electron transport chain. The oxidation of two pyruvate molecules leads to the production of six ATP molecules.

$$2 \text{ Pyruvate} \longrightarrow 2 \text{ acetyl CoA} + 2CO_2 + 6 \text{ ATP}$$

ATP and the Citric Acid Cycle

One turn of the citric acid cycle produces two CO_2, three NADH, one $FADH_2$, and one ATP by direct substrate phosphorylation. When NADH and $FADH_2$ enter the electron transport chain, three NADH produce a total of nine ATP molecules, and one $FADH_2$ produces two more ATP molecules. Thus, one turn of the citric acid cycle provides a total of 12 ATP molecules.

$$\begin{array}{lcl} 3 \text{ NADH} & \times\ 3 \text{ ATP} & = \ \ 9 \text{ ATP} \\ 1 \text{ FADH}_2 & \times\ 2 \text{ ATP} & = \ \ 2 \text{ ATP} \\ 1 \text{ GTP} & \times\ 1 \text{ ATP} & = \ \ \underline{1 \text{ ATP}} \\ & \text{Total} & = 12 \text{ ATP} \end{array}$$

From two acetyl CoA molecules, a total of 24 ATP molecules are possible.

$$2 \text{ Acetyl CoA} \longrightarrow 24 \text{ ATP} + 4CO_2$$

17.8

ATP and Complete Oxidation of Glucose

Now we can combine all the metabolic pathways that oxidize glucose to CO_2 and H_2O. (See Figure 17.20.) Adding the ATP associated with each pathway (aerobic), a total of 36 ATP molecules are possible.

Metabolic Pathway	Substrate(s) Oxidized	Products	ATP
Glycolysis	1 Glucose	2 Pyruvate, 2 ATP, 2 NADH	6 ATP
Oxidation	2 Pyruvate	2 Acetyl CoA, 2 NADH, $2CO_2$	6 ATP
Citric acid cycle	2 Acetyl CoA	6 NADH, 2 $FADH_2$, 2 ATP, $4CO_2$	24 ATP
Complete oxidation	Glucose + $6O_2$	$\longrightarrow$ $6CO_2$ + $6H_2O$ + 36 ATP	36 ATP

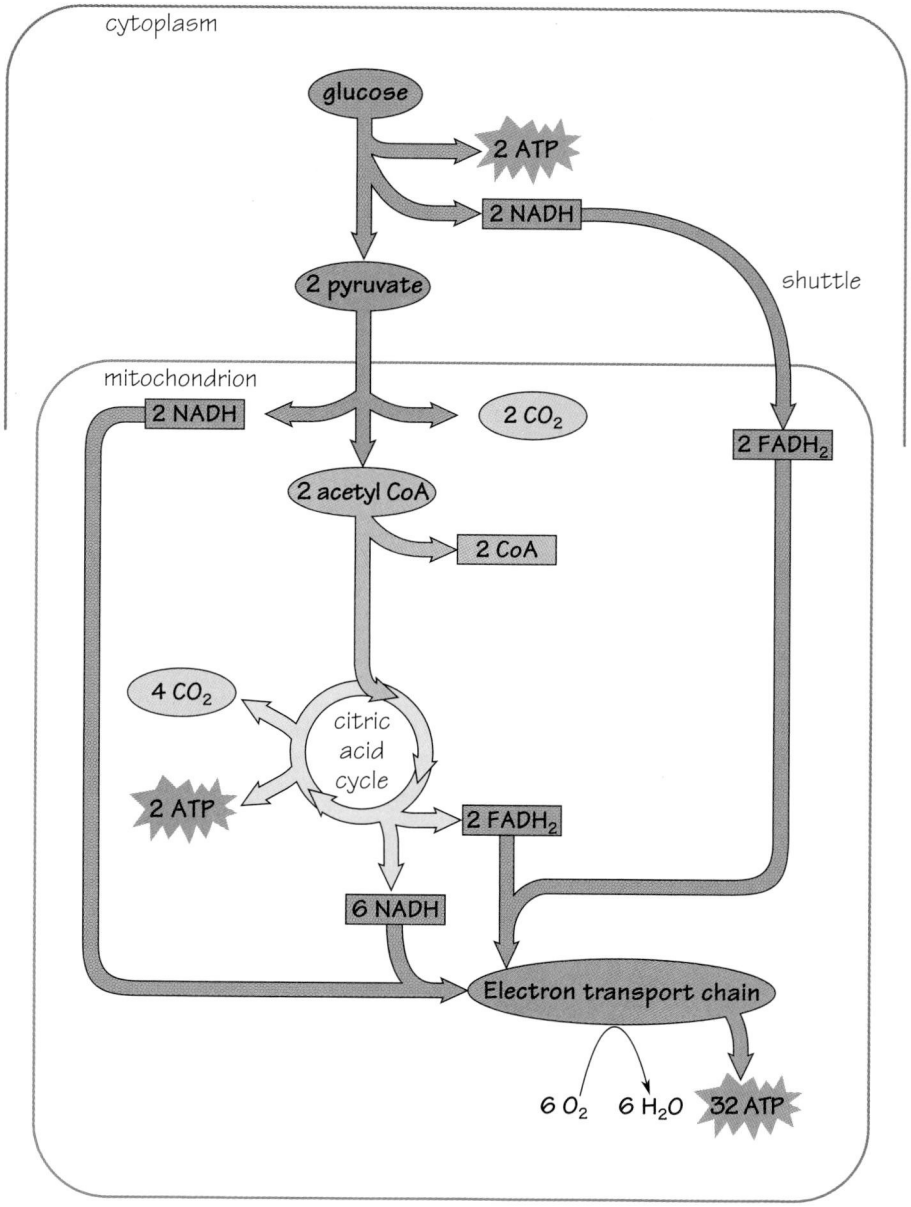

Figure 17.20 The complete oxidation of glucose yields 36 ATP.

Table 17.1 ATP Produced by the Complete Combustion of Glucose

Reaction Pathway	ATP for One Glucose
ATP from Glycolysis	
Activation of glucose	−2 ATP
Oxidation of glyceraldehyde-3-phosphate (2 NADH)	6 ATP
Transport of 2 NADH across membrane	−2 ATP
Direct ADP phosphorylation (two triose phosphate)	4 ATP
Summary: $\underset{\text{Glucose}}{C_6H_{12}O_6} \longrightarrow 2 \text{ pyruvate} + 2H_2O$	6 ATP
ATP from Acetyl CoA	
2 pyruvate $\longrightarrow$ 2 acetyl CoA (2 NADH)	6 ATP
ATP from Citric Acid Cycle	
Oxidation of 2 isocitrate (2 NADH)	6 ATP
Oxidation of 2 α-ketoglutarate (2 NADH)	6 ATP
2 direct substrate phosphorylations (2 GTP)	2 ATP
Oxidation of 2 succinate (2 $FADH_2$)	4 ATP
Oxidation of 2 malate (2 NADH)	6 ATP
Summary: 2 Acetyl CoA $\longrightarrow 4CO_2 + 2H_2O$	24 ATP
Overall ATP Production for One Glucose	
$\underset{\text{Glucose}}{C_6H_{12}O_6} + 6O_2 + 36 \text{ ADP} + 36 P_i \longrightarrow 6CO_2 + 6H_2O + 36 \text{ ATP}$	

The total amount of ATP from direct substrate phosphorylation and the oxidation of NADH or $FADH_2$ in the electron chain can be calculated as shown in Table 17.1.

S·A·M·P·L·E· **17.10** ·P·R·O·B·L·E·M

ATP Production

Indicate the amount of ATP produced by each of the following oxidation reactions:

a. pyruvate to acetyl CoA **b.** glucose to acetyl CoA

Solution

a. The oxidation of pyruvate to acetyl CoA produces one NADH, which yields three ATP.

b. Six ATP are produced from the oxidation of glucose to two pyruvate molecules. Another six ATP result from the oxidation of two pyruvate molecules to two acetyl CoA molecules.

Study Check

What are the sources of ATP in the citric acid cycle?

When glucose is not immediately used by the cells for energy, it is stored as glycogen in the liver and muscles. When the levels of glucose in the brain or blood become low, the glycogen reserves are hydrolyzed and glucose is released into the blood. If glycogen stores are depleted, some glucose can be synthesized from noncarbohydrate sources. It is the balance of all these reactions that maintains the necessary blood glucose level available to our cells and provides the necessary amount of ATP for our energy needs. (See Figure 17.21.)

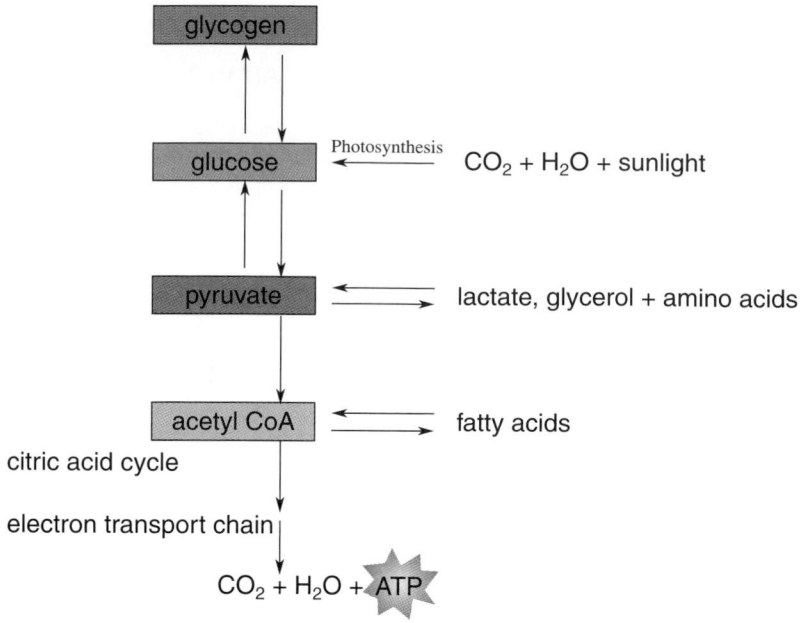

Figure 17.21 The metabolic reactions of glucose in the cell.

17.9

Q·U·E·S·T·I·O·N·S & P·R·O·B·L·E·M·S

ATP Energy from Glucose

17.59 Why does the NADH produced in glycolysis yield only two ATP?

17.60 Under anaerobic conditions, what is the maximum number of ATP molecules that can be produced from one glucose molecule?

17.61 What is the energy yield in ATP molecules associated with each of the following?

a. NADH $\longrightarrow$ NAD^+
b. glucose $\longrightarrow$ 2 pyruvate
c. 2 pyruvate $\longrightarrow$ 2 acetyl CoA + $2CO_2$
d. acetyl CoA $\longrightarrow$ $2CO_2$

17.62 What is the energy yield in ATP molecules associated with each of the following?

a. $FADH_2$ $\longrightarrow$ FAD
b. glucose + $6O_2$ $\longrightarrow$ $6CO_2 + 6H_2O$
c. glucose $\longrightarrow$ 2 lactate
d. pyruvate $\longrightarrow$ lactate

Study Pack

17.9 Oxidation of Fatty Acids

Learning Goal

Describe the role of the oxidation of fatty acids in ATP production.

We store most of our energy in the form of triglycerides (lipids) in adipose (fat) cells. (See Figure 17.22.) If our reserves of glucose and glycogen are depleted, we obtain energy by hydrolyzing the triglycerides to glycerol and free fatty acids. Glycerol, a triose, is converted to dihydroxyacetone phosphate, a compound that is oxidized by the glycolysis pathway. The fatty acids are oxidized by our tissues for energy. The oxidation of triglycerides, which produces as much as six times the ATP from glucose, is an important reserve source of ATP in the body.

Fatty Acid Activation

In the cytoplasm, a fatty acid is joined to CoA using energy from the hydrolysis of ATP to adenosine monophosphate (AMP) and inorganic phosphate (2 P_i), an energy cost that is equivalent to the hydrolysis of two ATPs. The activated fatty acyl CoA is transported across the inner membrane into the mitochondria where it can be oxidized further by the citric acid cycle and the electron transport chain.

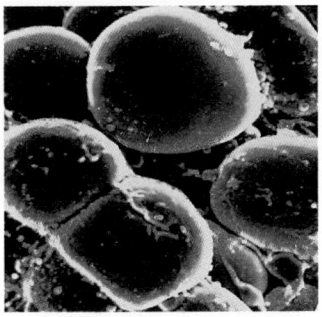

Figure 17.22 Photomicrograph of an adipose cell filled with fat.

17.9

Beta Oxidation of Fatty Acids

Beta (β) oxidation involves a cycle of four reactions that removes two-carbon segments from a fatty acid in sequential order. In each cycle, the β carbon in the fatty acid chain is oxidized, yielding an acetyl CoA unit and a shortened fatty acid chain.

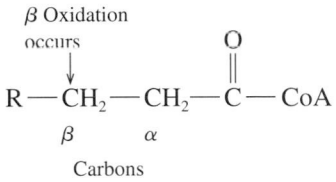

The cycle repeats until the original fatty acid is entirely degraded into two-carbon acetyl CoA units. Each acetyl CoA unit is then processed through the citric acid cycle in the same way as the acetyl CoA units derived from glucose. The number of carbon atoms in a fatty acid determines the number of times the cycle is repeated and the number of acetyl units produced. We will now take a look at the individual steps of β oxidation as shown in Figure 17.23.

Summary of Reactions in the Oxidation of Fatty Acids

Step 1 **First Oxidation** An oxidation reaction forms a double bond when a *dehydrogenase* removes two hydrogen atoms from the α and β carbons of the fatty acid. FAD picks up the hydrogen to yield $FADH_2$.

Step 2 **Hydration** Water is added across the double bond, placing a hydroxyl group (—OH) on the β carbon.

Step 3 **Second Oxidation** A *dehydrogenase* oxidizes the hydroxyl group on the β carbon to a ketone and NAD^+ reduces to NADH.

Step 4 **Release of Acetyl CoA** In the final step, a unit of acetyl CoA separates from the fatty acid chain, and another molecule of CoA bonds to the β carbon. The remaining fatty acyl CoA is two carbons shorter than the fatty acid chain that started the cycle. As long as the fatty acyl CoA contains four or more carbons, it can recycle through β oxidation again.

ATP and Fatty Acid Oxidation

We can now determine the total ATP for the complete oxidation of a particular fatty acid. Halving the number of carbon atoms in a fatty acid gives the number of acetyl CoA groups it will provide for the citric acid cycle. Because the final four-carbon chain splits into two acetyl CoA groups, the initial fatty acid must repeat the oxidation cycle one time less than the total number of acetyl CoA groups it produces. For example, caprylic acid is a fatty acid with eight carbon atoms. After caprylic acid repeats the fatty acid cycle three times, it has been split into four acetyl CoA groups.

The total ATP is the sum of the ATP produced from all the NADH and $FADH_2$ formed in the β oxidation cycle and from acetyl units entering the citric acid cycle. In each fatty acid oxidation cycle, one NADH, one $FADH_2$, and one acetyl CoA are produced. In the electron transport chain, the NADH generates three ATP, and the $FADH_2$ generates two ATP, for a total of five ATP from the coenzymes. In addition, each acetyl CoA is oxidized further by the citric acid cycle to produce 12 ATP.

17.9

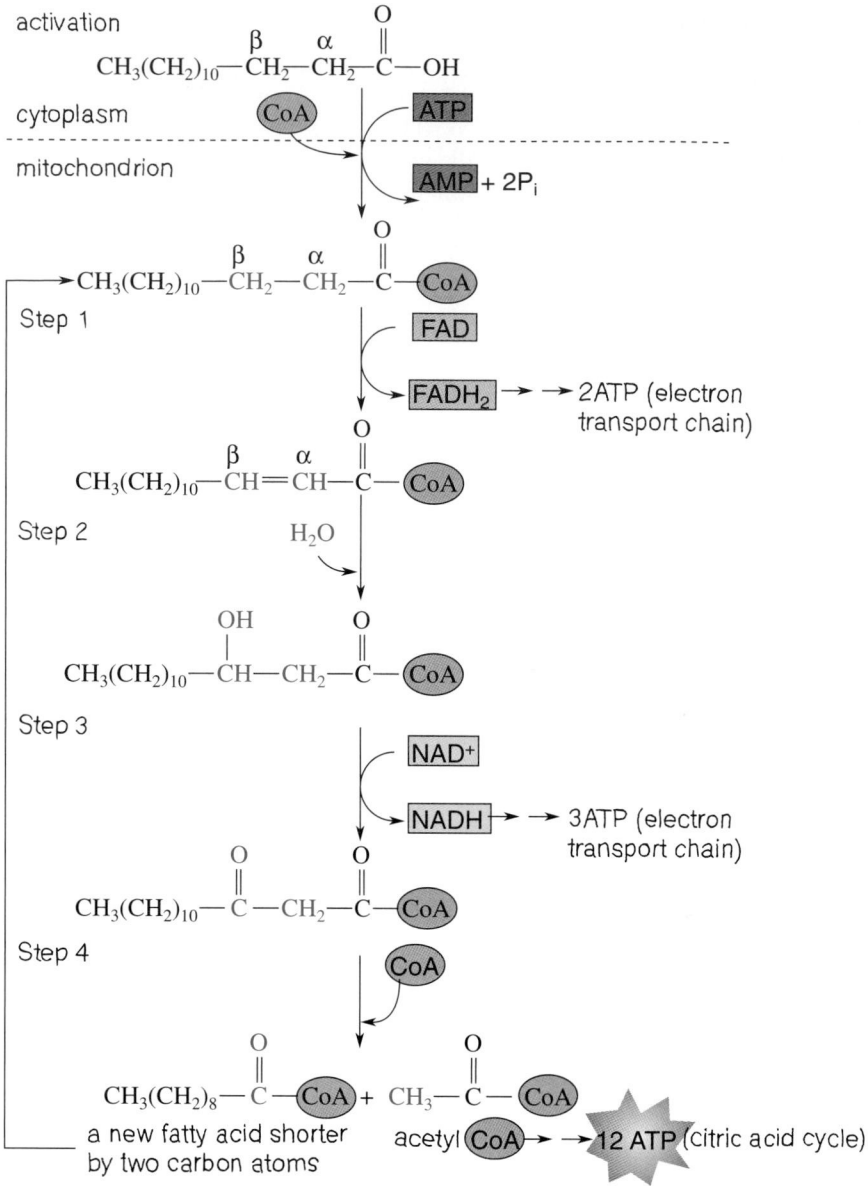

Figure 17.23 Reaction steps in the β oxidation cycle of fatty acids.

For one molecule of caprylic acid, 63 ATP molecules are generated. However, 2 ATP are required to activate β oxidation, which gives a net of 61 ATP molecules. The calculations for caprylic acid are given in Table 17.2.

S·A·M·P·L·E· **17.11** ·P·R·O·B·L·E·M

ATP Production from β Oxidation

How much ATP will be produced from the β oxidation of myristic acid, a 14-carbon fatty acid?

Solution

A 14-carbon fatty acid will produce seven acetyl CoA units and go through six β oxidation cycles. Each acetyl CoA will produce 12 ATP by the citric acid cycle for a total of 84 ATP. Each β oxidation cycle will produce one NADH and one $FADH_2$ for a total of six NADH and six $FADH_2$. In the electron transport chain, each NADH will produce energy for three ATP; each $FADH_2$ for two ATP.

ATP Production for Myristic Acid

Activation		−2 ATP
7 acetyl CoA × 12 ATP	(Citric acid cycle)	84 ATP
6 $FADH_2$ × 2 ATP	(Electron transport chain)	12 ATP
6 NADH × 3 ATP	(Electron transport chain)	18 ATP
Total ATP by β oxidation of myristic acid		112 ATP

Study Check

How much ATP is produced if stearic acid (an 18-carbon acid) goes through only two turns of the β oxidation cycle?

Table 17.2 Total ATP Produced by Caprylic Acid (Eight Carbon Atoms)

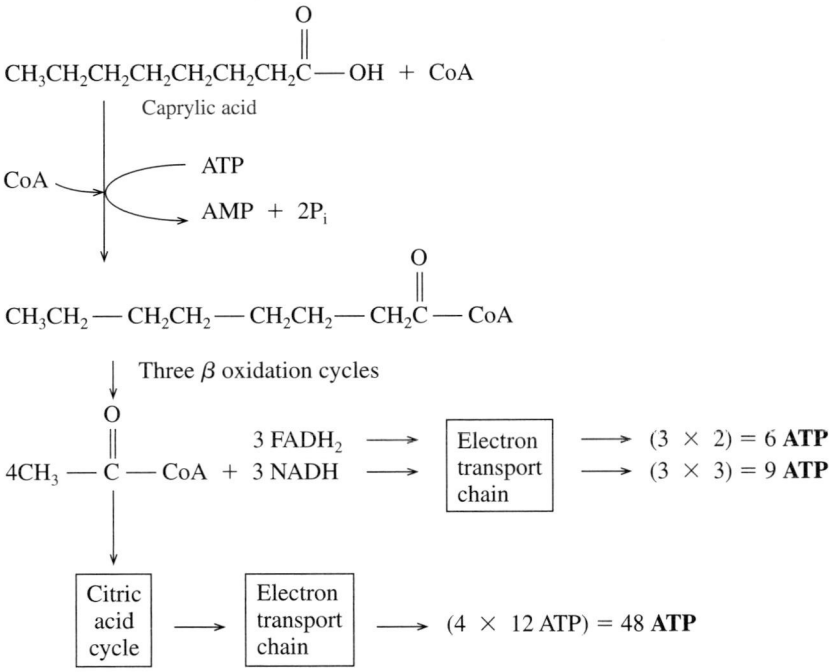

Summary of ATP in β oxidation of caprylic acid

Total ATP:	63 **ATP**
Activation:	−2 **ATP**
	61 **ATP**

HEALTH NOTE

Ketone Bodies

When carbohydrates are not available to meet energy needs, the body breaks down large quantities of body fat. The oxidation of fatty acids produces so many units of acetyl CoA that the citric acid cycle is overwhelmed. At the same time, less oxaloacetate is available for the citric acid cycle because it is used to synthesize glucose. Without oxaloacetate, acetyl CoA from oxidation cannot be used completely by the citric acid cycle. Instead, acetyl CoA enters a pathway called *ketogenesis* in which four-carbon *ketone bodies,* acetoacetate and β-hydroxybutyrate, are produced from two acetyl CoA units. Acetoacetate is then converted to acetone, another ketone body.

17.9

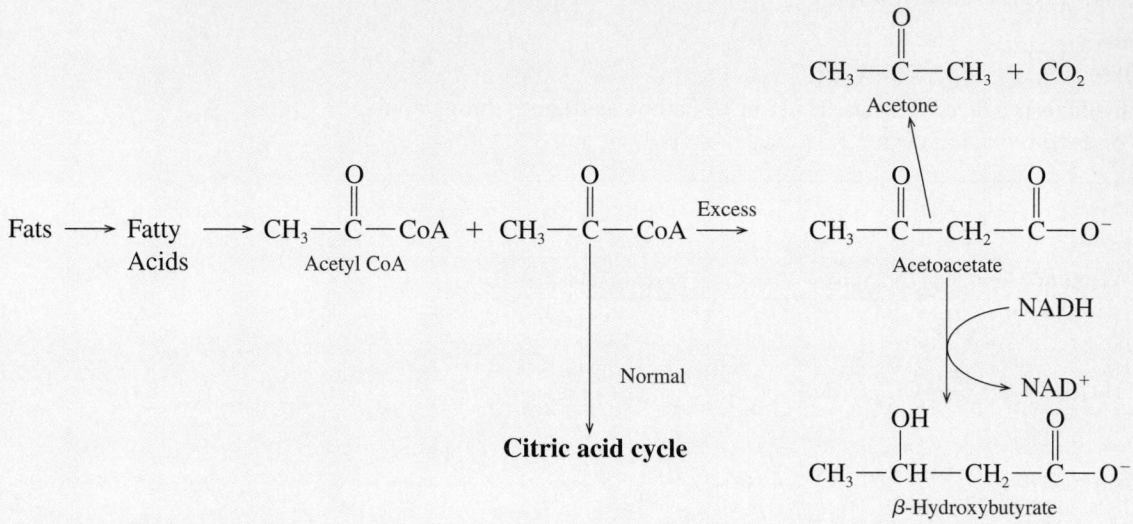

In the body, a small amount of energy is derived from ketone bodies by converting them back to acetyl CoA. When depleted of glucose, the brain uses ketone bodies for energy. However, in **ketosis,** the level of ketone bodies becomes so high that they are not completely metabolized by the body. This condition occurs in severe diabetes, diets that are high in fat and low in carbohydrate, and starvation. The fruity odor of acetone can often be detected on the breath of a person with ketosis.

Acidosis is also associated with ketosis because the acidity of the ketone bodies lowers the pH of the blood. A severe drop in blood pH can interfere with the ability of the blood to carry oxygen. As a result, breathing becomes strained. As the body tries to excrete ketone bodies in the urine, large amounts of sodium are removed from body fluids. As the sodium supply is depleted, the output of urine becomes high, accompanied by a strong thirst.

Fatty Acid Synthesis

When the body has met all its energy needs and the glycogen stores are full, the intake of additional nutrients such as carbohydrates will provide excess acetyl CoA molecules that enter pathways for the synthesis of fatty acids and fats. In a series of reactions, fatty acid synthase enzyme links two-carbon acetyl units together until there are 16 carbon atoms in the fatty acid, which is palmitic acid. Then other enzymes add more carbon atoms and double bonds. Several of the reactions are much like the reverse of the reactions we discussed in fatty acid oxidation. However, the enzyme complex is located in the cytoplasm, not in the mitochondria where oxidation takes place. In fatty acid synthesis, the coenzyme is NADPH, whereas fatty acid oxidation uses FAD and NAD^+, which are coenzymes for degradative pathways. After the new fatty acids are synthesized, they are attached to glycerol to make triglycerides, which are stored as body fat.

QUESTIONS & PROBLEMS

Oxidation of Fatty Acids

17.63 Consider the activation of a fatty acid.

a. Triglycerides provide fatty acids for β oxidation. What happens to the glycerol?

b. Where in the cells are fatty acids activated for β oxidation?

c. What is the energy cost in ATP for activation of fatty acids?

d. What is the purpose of activating fatty acids?

17.64 Consider the β oxidation pathway for fatty acids.

a. Why is the oxidation of fatty acids called β oxidation?

b. What coenzymes are required?

c. When is each coenzyme used?

d. What is the yield in ATP molecules for one cycle of β oxidation?

17.65 Lauric acid, $CH_3—(CH_2)_{10}—COOH$, is a fatty acid.

a. Write the formula of the activated lauric acid.

b. Indicate the α and β carbon atoms in lauric acid.

c. Write the equations for one cycle of β oxidation for lauric acid.

d. How many acetyl CoA units will be produced when β oxidation of lauric acid is complete?

e. How many cycles of β oxidation are needed to complete the oxidation?

f. Account for the total ATP yield from β oxidation of lauric acid by completing the following calculation:

Units Produced from Lauric Acid

____	acetyl CoA $\longrightarrow$	____	ATP
____	$FADH_2$ $\longrightarrow$	____	ATP
____	NADH $\longrightarrow$	____	ATP
	Activation of fatty acid	____	ATP
	Total	____	ATP

17.66 Stearic acid, $CH_3—(CH_2)_{16}—COOH$, is a fatty acid in butter fat and beef fat.

a. Write the formula of the activated stearic acid.

b. Indicate the α and β carbon atoms in stearic acid.

c. Write the equations for one cycle of β oxidation for stearic acid.

d. How many acetyl CoA units will be produced when β oxidation of stearic acid is complete?

e. How many cycles of β oxidation are needed to complete the oxidation?

f. Account for the total ATP yield from β oxidation of stearic acid by completing the following table:

Units Produced from Stearic Acid

____	acetyl CoA $\longrightarrow$	____	ATP
____	$FADH_2$ $\longrightarrow$	____	ATP
____	NADH $\longrightarrow$	____	ATP
	Activation of fatty acid	____	ATP
	Total	____	ATP

17.67 How does the breakdown of large amounts of fats lead to the formation of excess ketone bodies? Under what conditions does ketosis occur? What are some symptoms of high levels of ketone bodies?

17.68 Consider the lipogenesis of myristyl CoA. How many acetyl CoA units are needed?

Study Pack

Chem of Life Enzymes and Pathways Overview

Learning Goal

Describe some metabolic pathways for amino acids.

17.10

17.10 Metabolic Pathways for Amino Acids

Amino acids obtained from the digestion of proteins are absorbed by the bloodstream and carried to the cells in the body, where most are used to build proteins. Our bodies are constantly replacing old proteins with new ones. Although amino acids are not a major energy source, energy can be extracted from amino acids if glycogen and fat reserves have been depleted. If amino acids become the only source of energy as in starvation, the breakdown of the body's own proteins eventually leads to a destruction of essential body tissues.

Normally, most of our energy is provided by the oxidation of carbohydrates and fats. Only a small amount (about 10%) comes from the oxidation of the carbon atoms of amino acid. Most of the amino acids from our diet are used in the synthesis of proteins in our cells. However, when a person is fasting or starving and there is no carbohydrate intake, amino acids from the breakdown of body protein are the only available source of energy. To extract energy from amino acids, two types of reactions take place. First, the amino group is removed, and second, the resulting carboxylic acid is degraded.

Transamination

In **transamination,** amino groups are moved from amino acids to α-keto acids, usually α-ketoglutarate. The enzymes for these transfer reactions are known as amino transferases, or transaminases.

$$\text{Amino acid} + \alpha\text{-ketoglutarate} \rightleftharpoons \alpha\text{-keto acid} + \text{glutamate}$$

In the following equation, the amino group of alanine is transferred to α-ketoglutarate, an intermediate in the citric acid cycle.

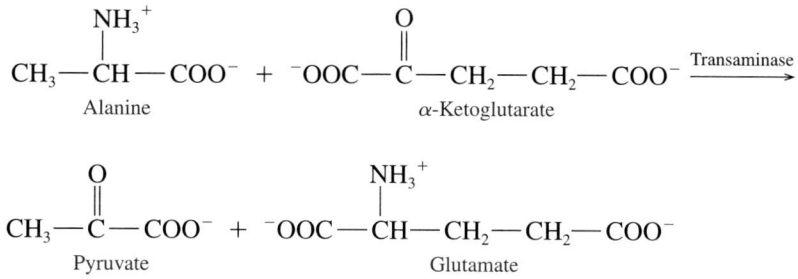

The α-keto acids that result from the loss of the amino group become an energy source as metabolic substances that can enter the citric acid cycle. (See Figure 17.24.) Reactions including transamination of alanine, serine, and cysteine give pyruvate. Glycine can be changed to serine, which leads to pyruvate. Pyruvate is oxidized to acetyl CoA, which is the fuel for the citric acid cycle. Oxaloacetate is the α-keto acid from the transamination of the four-carbon aspartate.

$$\text{Aspartate} + \alpha\text{-ketoglutarate} \longrightarrow \text{oxaloacetate} + \text{glutamate}$$

The five-carbon amino acids glutamine, proline, arginine, and histidine can be converted to α-ketoglutarate. Other amino acids are converted to succinyl CoA. Thus, in an assortment of pathways, the carbon atoms from amino acids are converted to compounds that are intermediates of the citric acid cycle where they now generate ATP.

Transamination is also used to prepare the nonessential amino acids for the body by transferring an amino group from glutamic acid to several of the α-keto acids that can be obtained from the citric acid cycle or glycolysis.

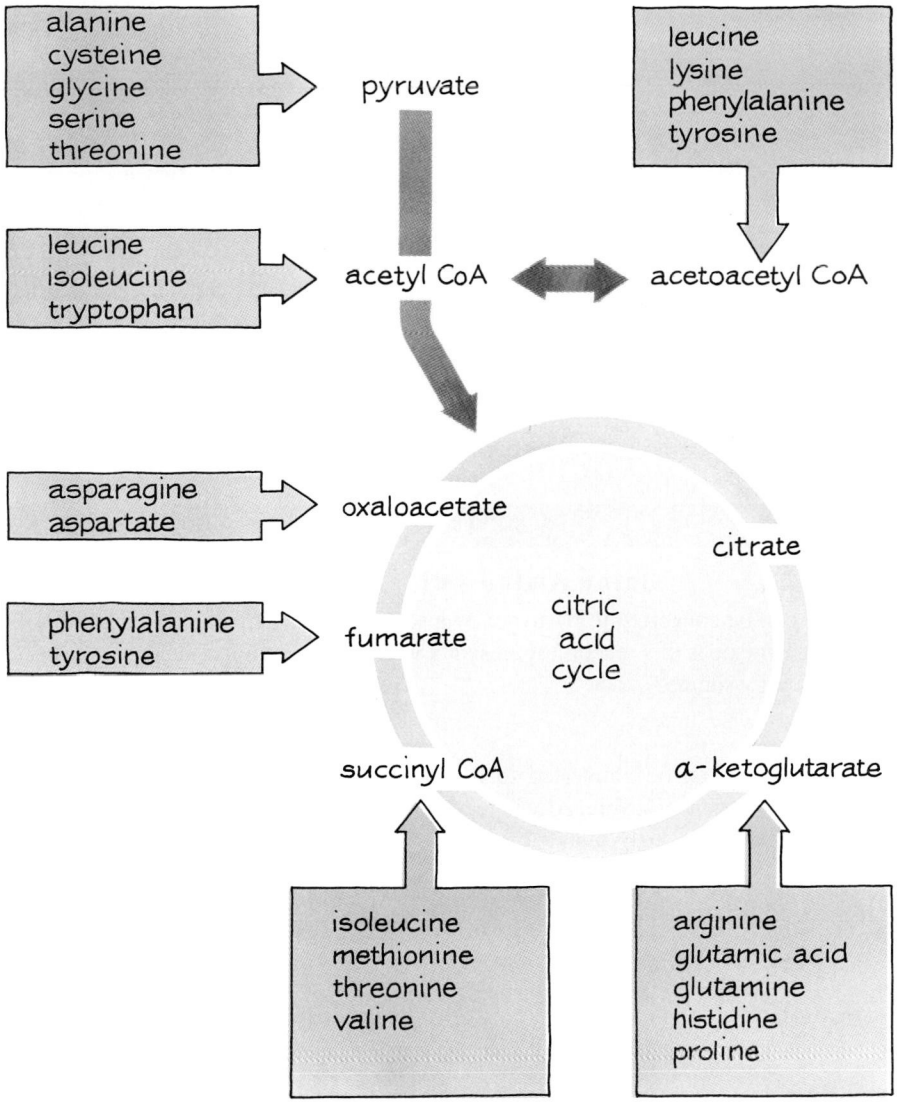

Figure 17.24 The loss of the amino group converts amino acids to intermediates of the citric acid cycle.

α-Keto Acids in Citric Acid Cycle	Nonessential Amino Acids
Pyruvate	Alanine, serine, cysteine, glycine, tyrosine
Oxaloacetate	Aspartate, asparagine
α-Ketoglutarate	Glutamate, glutamine, proline

Oxidative Deamination

In **oxidative deamination,** an amino group is removed from an amino acid as the ammonium ion, NH_4^+. This reaction is used to regenerate α-ketoglutarate from glutamate and is catalyzed by the enzyme *glutamate dehydrogenase* and the NAD^+ coenzyme.

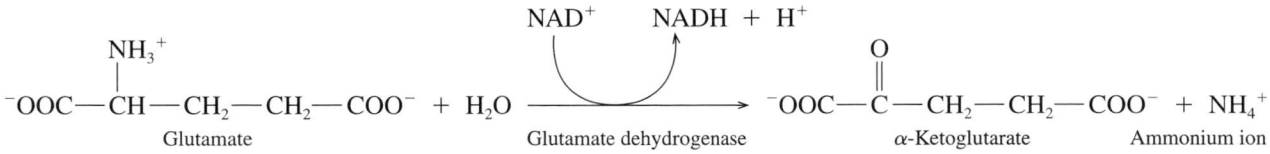

Ammonium ion is highly toxic. In another series of reactions called the urea cycle, NH_4^+ combines with CO_2 to yield urea, which is excreted in the urine. There are several steps of the urea cycle with several intermediates, but the overall reaction converts ammonium ions into urea.

$$2NH_4^+ + CO_2 \longrightarrow \underset{\text{Urea}}{H_2N-\overset{\overset{\displaystyle O}{\|}}{C}-NH_2} + 2H^+ + H_2O$$

17.10

In one day, a typical adult produces about 50 g of urea in the liver, and excretes about 25–30 g in the urine. This amount increases with increased dietary protein intake. If urea is not properly excreted, it can build up quickly to a toxic level. To detect renal disease, the blood urea nitrogen (BUN) test is used to measure the blood urea level. If the level is high, protein intake must be reduced, and hemodialysis may be needed to remove toxic nitrogen wastes from the blood.

S·A·M·P·L·E· 17.12 ·P·R·O·B·L·E·M

Using Amino Acids for Energy

Serine can be converted to pyruvate by a series of reactions including transamination. How does this reaction represent a way in which amino acids can be used as an energy source?

Solution

Amino acids can be converted to compounds that are a part of the energy-producing citric acid cycle. After the removal of the amino group of serine, it is converted to pyruvate, which is converted to acetyl CoA. The acetyl CoA enters the citric acid cycle, where it produces 12 ATP.

Study Check

What α-keto acid from the citric acid cycle would undergo reactions including transamination to yield aspartate?

HEALTH NOTE

Why Do I Get Hungry?

When you say "I'm hungry," you probably decide to find something to eat. You are responding to a stimulation of the hunger center of the brain. When you eat, the satiety center near by is stimulated, which makes you feel full, and you stop eating. The hunger center is operational all the time, but it is inhibited by the satiety center.

In the glucostatic theory, a low blood glucose level decreases the activity of the satiety center so it no longer inhibits the hunger center. We respond by eating a meal, which causes our blood glucose levels to rise until we are no longer hungry. An increase in the level of fatty acids and amino acids in the blood also depresses the hungry feeling. They act in a fashion similar to glucose and cause an inhibition of the hunger center. In general, we feel hungry and eat when the nutrient levels in the body are low.

Our body temperature also affects our food intake. In a cold environment, food intake increases, which accelerates metabolic reactions, increases body temperature, and provides more body fat for insulation. When we are hot, less body fat is needed and we tend to eat less.

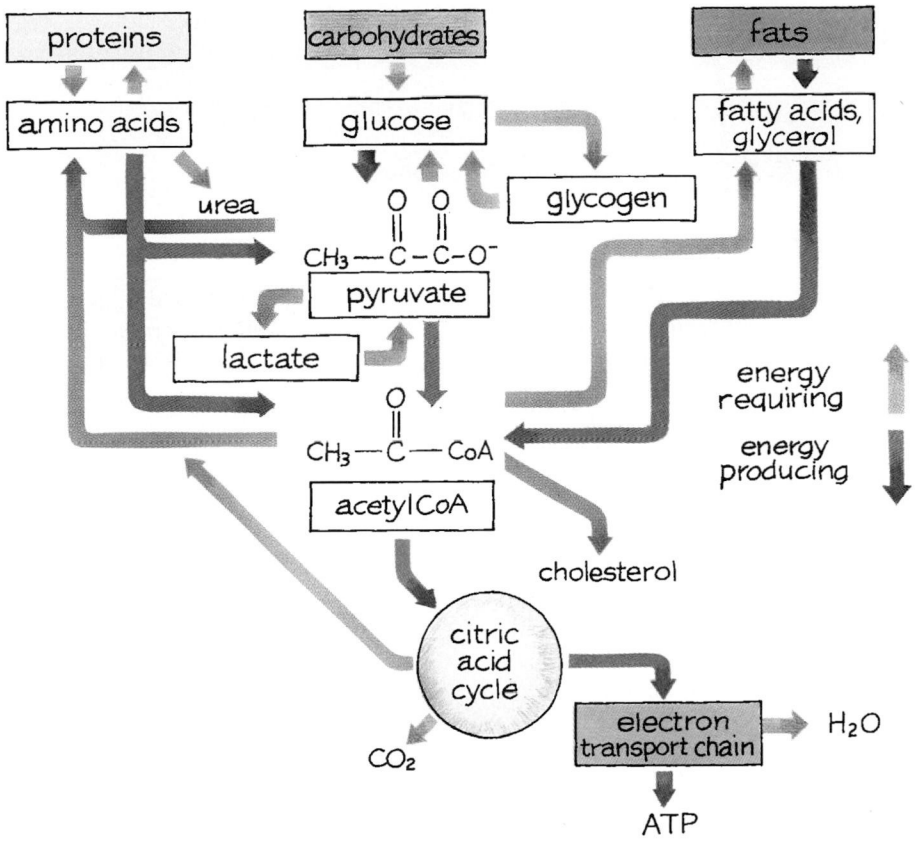

Figure 17.25 Overall relationship of energy-producing and energy-requiring pathways in metabolism.

Overview of Metabolism

The oxidative pathways degrade large molecules to small molecules that can be used for energy production via the citric acid cycle and the electron transport chain. These same molecules can also enter pathways that lead to the synthesis of larger molecules in the cell. Such synthesis is energy consuming and depends on a supply of ATP. (See Figure 17.25.)

The pathways in metabolism have several branch points from which compounds may be degraded for energy or used to synthesize larger molecules. For example, glucose can be converted to glycogen for storage, or degraded to acetyl CoA for the citric acid cycle to produce energy. When glycogen stores are depleted, fatty acids can be oxidized for energy. Amino acids are normally used to synthesize proteins for the cells, but they can be converted to intermediates of the citric acid cycle in an energy crisis. The α-keto acids of the citric acid cycle may be used in the synthesis of several nonessential amino acids by transamination reactions.

Q·U·E·S·T·I·O·N·S & P·R·O·B·L·E·M·S

Metabolic Pathways for Amino Acids

17.69 What are the products of transamination of amino acids?

17.70 What are the products of oxidative deamination of amino acids?

17.71 Write equations for the transamination of each of the following amino acids and α-ketoglutarate:
a. alanine **b.** aspartate

17.72 Write the equation for the oxidative deamination of glutamate.

17.73 If necessary, how are amino acids used as a fuel in the cells?

17.74 If a nonessential amino acid is low in a cell, how might it be prepared from intermediates of the citric acid cycle?

Chapter Review

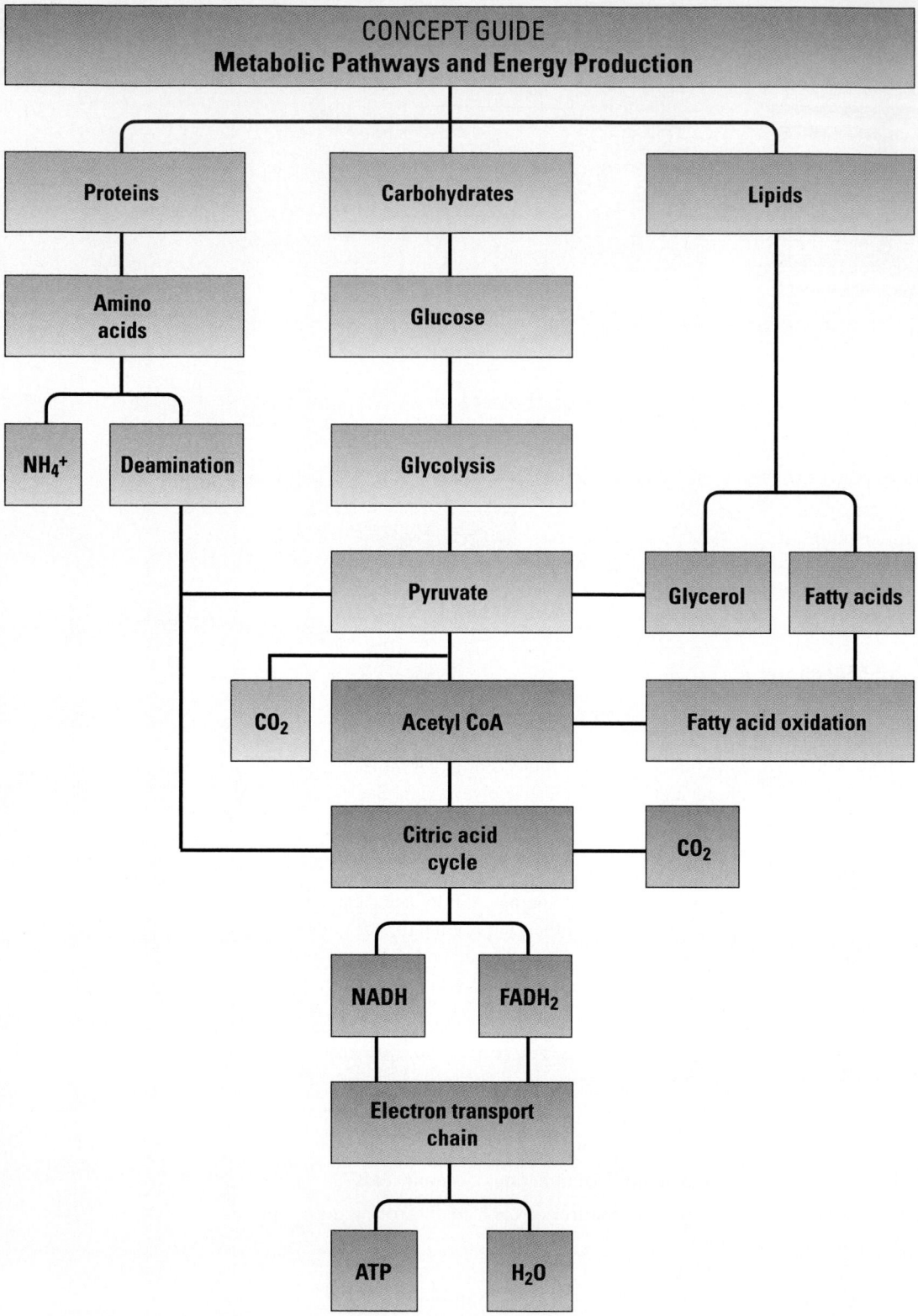

Chapter Summary

17.1 Metabolism and ATP Energy

In metabolism, catabolic reactions break down large molecules into small molecules in processes that provide energy. The energy is stored in ATP, a high-energy compound, that is hydrolyzed when energy is required by the anabolic reactions that do work in the cells.

17.2 Digestion of Foods

Digestion is a series of reactions that break down large food molecules of carbohydrates, lipids, and proteins into smaller molecules that can be absorbed and used by the cells. The end products of digestion include monosaccharides from carbohydrates, fatty acids and glycerol from triglycerides, and amino acids from proteins.

17.3 Important Coenzymes in Metabolic Pathways

FAD and NAD^+ are coenzymes that participate in oxidation reactions mostly in the citric acid cycle. They are reduced to $FADH_2$ and NADH as they pick up hydrogen and electrons to carry to the electron transport chain. Coenzyme A is attached to acetyl groups from the degradation of glucose, fatty acids, or amino acids to form acetyl CoA, which can be oxidized further in the citric acid cycle.

17.4 Glycolysis: Oxidation of Glucose

Glycolysis is the primary anaerobic pathway for the degradation of glucose to yield pyruvate. Glucose is phosphorylated and isomerized to fructose-1,6-diphosphate that is split into two triose phosphate molecules. The oxidation of the three-carbon sugars yields two molecules of the reduced coenzyme NADH and two ATP.

17.5 Pathways for Pyruvate

In the absence of oxygen, pyruvate is reduced to lactate and NAD^+ is regenerated for the continuation of glycolysis. Under aerobic conditions, pyruvate is oxidized in the mitochondria to acetyl CoA, which enters the citric acid cycle.

17.6 Citric Acid Cycle

In a sequence of reactions called the citric acid cycle, acetyl CoA combines with oxaloacetate to yield citrate. Citrate undergoes oxidation and decarboxylation to yield two CO_2, GTP, three NADH, and $FADH_2$. The phosphorylation of ADP by GTP yields ATP.

17.7 Electron Transport Chain

The reduced coenzymes from glycolysis and the citric acid cycle are oxidized to NAD^+ and FAD by transferring protons and electrons to the electron transport chain. In the electron chain, electrons are transferred to electron carriers including flavins, iron-sulfur proteins, and cytochromes with Fe^{3+}/Fe^{2+}. The flow of electrons along the electron chain pumps protons across the inner membrane, which produces a high-energy proton gradient that provides energy for the synthesis of ATP. The final acceptor, O_2, combines with protons and electrons to yield H_2O. The process of using the energy of the electron transport chain to synthesize ATP is called oxidative phosphorylation. The oxidation of NADH yields three ATP molecules, and $FADH_2$ yields two ATP.

17.8 ATP Energy from Glucose

The complete oxidation of one glucose yields a total of 36 ATP from direct phosphorylation and the oxidation of the reduced coenzymes NADH and $FADH_2$ by the electron transport chain and oxidative phosphorylation.

17.9 Oxidation of Fatty Acids

As an energy source, fatty acids are linked to coenzyme A and transported into the mitochondria where they undergo β oxidation. The fatty acetyl chain is oxidized to yield a shortened fatty acid, acetyl CoA, and the reduced coenzymes NADH and $FADH_2$. Although the energy from a particular fatty acid depends on its length, each cycle of β oxidation yields 17 ATP, and another 12 ATP come from the acetyl CoA left by the final oxidation. When there is an excess of acetyl CoA in the cell, the two-carbon units link together to synthesize fatty acids that are stored in the adipose tissue.

17.10 Metabolic Pathways for Amino Acids

Amino acids are normally used for protein synthesis. The nonessential amino acids can be prepared by transferring an amino group to an α-keto acid to yield a different amino acid and α-keto acid. Some α-keto acids that are intermediates of the citric acid cycle can be used in the synthesis of lipids or glucose or oxidized further for energy. Oxidative transamination removes an amino group as NH_3 and converts it to urea.

Summary of Key Reactions

Hydrolysis of ATP

$$ATP + H_2O \longrightarrow ADP + P_i + \text{energy}$$

Glycolysis

$$\text{Glucose} + 2\,ADP + 2\,P_i + NAD^+ \longrightarrow \text{pyruvate} + 2\,ATP + 2\,NADH + 2H^+$$

Formation of Lactate

$$\text{Pyruvate} + NADH + H^+ \xrightarrow{\text{Anaerobic}} \text{lactate} + NAD^+$$

Formation of Acetyl CoA

$$\text{Pyruvate} + NAD^+ + CoA \xrightarrow{\text{Aerobic}} \text{acetyl CoA} + NADH + CO_2$$

Citric Acid Cycle

$$\text{Acetyl CoA} + 3\,NAD^+ + FAD + ADP + P_i \longrightarrow 2CO_2 + 3\,NADH + FADH_2 + ATP$$

Electron Transport Chain

$$NADH + H^+ + 3\,ADP + 3\,P_i + \tfrac{1}{2}O_2 \longrightarrow NAD^+ + 3\,ATP + H_2O$$

$$FADH_2 + 2\,ADP + 2\,P_i + \tfrac{1}{2}O_2 \longrightarrow FAD + 2\,ATP + H_2O$$

Phosphorylation of ADP

$$ADP + P_i \longrightarrow ATP + H_2O$$

Complete Combustion of Glucose

$$\text{Glucose} + 36\,ADP + 36\,P_i \longrightarrow 6CO_2 + 36\,ATP + 6H_2O$$

Key Terms

acetyl CoA A two-carbon acetyl unit attached to coenzyme A. A product of glucose, fatty acid, and protein metabolism, acetyl CoA is oxidized to yield energy via the citric acid cycle or used as a substrate to synthesize fatty acids or glucose.

ADP Adenosine diphosphate; a compound of adenine, a ribose sugar, and two phosphate groups, it is formed by the hydrolysis of ATP.

aerobic An oxygen-containing environment in the cells.

amylase An enzyme that hydrolyzes the glycosidic bonds in polysaccharides.

anabolic reactions Metabolic reactions that are energy requiring.

anaerobic The absence of oxygen in the cells.

ATP Adenosine triphosphate; a high-energy compound that stores energy in the cells and consists of adenine, a ribose sugar, and three phosphate groups.

ATP synthase An enzyme complex that links the flow of protons along an electrical gradient with the synthesis of ATP from ADP.

beta (β) oxidation A reaction cycle that removes acetyl CoA to shorten a fatty acid and yields the reduced coenzymes NADH and $FADH_2$.

catabolic reactions Metabolic reactions that produce energy for the cell by the degradation and oxidation of glucose, fatty acids, and amino acids.

chemiosmotic model The conservation of energy from transfer of electrons in the electron transport chain by pumping protons into the intermembrane space to produce a high-energy gradient that provides the energy to synthesize ATP.

citric acid cycle A series of oxidation reactions in the mitochondria that convert acetyl CoA to CO_2 and yield NADH and $FADH_2$. It is also called the Krebs cycle.

coenzyme A A cofactor that transports acetyl groups to the citric acid cycle.

coenzyme Q A mobile carrier that accepts electrons from FeS or $FADH_2$ in the electron transport chain.

cytochromes Iron-containing proteins (Fe^{3+}, Fe^{2+}) in the electron transport chain that transfer electrons from QH_2 to oxygen.

cytoplasm The material within the plasma membrane and surrounding the nucleus in eukaroytic cells.

decarboxylation The loss of a carbon atom in the form of CO_2.

digestion The processes that break down large food molecules to small ones that can be absorbed by cells of the body.

direct substrate phosphorylation The transfer of inorganic phosphate (P_i) directly from a high-energy phosphate bond to a substrate.

electron transport chain A mitochondrial system that transfers electrons from the reduced coenzymes NADH and $FADH_2$ to clusters of electron carriers such as flavins, iron–sulfur proteins, and cytochromes. The final electron acceptor, O_2, yields H_2O. Energy changes during some of the transfers provide energy to phosphorylate ADP to yield ATP.

electron carrier In the electron transport chain, a component that is reduced when it accepts electrons, and is oxidized when electrons are passed on.

FAD A coenzyme (flavin adenine dinucleotide) for dehydrogenase enzymes that form carbon–carbon double bonds. The reduced form, $FADH_2$, transfers electrons to CoQ in the electron transport chain to regenerate FAD.

fermentation The anaerobic conversion of sugars by enzymes in yeast to yield alcohol and CO_2.

FeS An iron–sulfur protein containing an iron ion attached to sulfur that accepts hydrogen and electrons from $FMNH_2$.

FMN The initial electron acceptor (flavin mononucleotide) of the electron transport chain for NADH.

glycolysis The oxidation reactions of glucose that yield two pyruvate molecules.

ketosis A condition in which the level of ketone bodies is elevated.

metabolism All the chemical reactions in living cells that carry out molecular and energy transformations.

NAD^+ The hydrogen acceptor in oxidation reactions that forms carbon–oxygen double bonds. The reduced coenzyme NADH transfers electrons to FMN in the electron transport chain to yield three ATP molecules and regenerate NAD^+.

oxidation The loss of hydrogen atoms or electrons by a reactant. In metabolism, it is generally associated with the degradation of large molecules and the production of energy.

oxidative deamination The removal of an amino group as NH_3.

oxidative phosphorylation The synthesis of ATP in the mitochondria by combining ADP and P_i using energy from the oxidation reactions in the electron transport chain.

proton pump The movement of H^+ by the enzyme complexes from the matrix into the intermembrane space, which creates a proton gradient.

reduction The gain of hydrogen or electrons by a reactant.

transamination The transfer of an amino group from an α-amino acid to produce α-keto acids that are intermediates in the citric acid cycle.

Chapter Practice

17.75 Why is NAD^+ used for the oxidation of isocitrate and malate, whereas FAD is the coenzyme used for the oxidation of succinate?

17.76 What would be the ATP yield for the complete oxidation of maltose?

17.77 At the gym, you expend 300 kcal riding the stationary bicycle for 1 hr. Using the value of 7.3 kcal/mole for the hydrolysis of ATP, how many grams of glucose (molar mass, 180 g/mole) would you need to oxidize?

17.78 Why are the degradative pathways carried out in a series of steps, and not in one step?

17.79 If the citric acid cycle contains no reactions with oxygen, O_2, why does it only operate under aerobic conditions?

17.80 What is the ratio of ATP produced by the oxidation of glucose under aerobic versus anaerobic conditions?

17.81 When alcohol is consumed, it is oxidized to acetaldehyde and then to acetate. How does alcohol provide calories for the body?

17.82 Your friend has gone on a low-calorie diet and has developed "fruity" breath. What would you tell your friend about his diet?

Answers

Answers to Study Checks

17.1 Adenine, ribose, and three phosphate groups.

17.2 The digestion of amylose begins in the mouth when salivary amylase hydrolyzes some of the α-1,4-glycosidic bonds. In the small intestine, pancreatic amylase hydrolyzes more glycosidic bonds, and finally maltose is hydrolyzed by maltase to yield glucose.

17.3 $FADH_2$

17.4 In the initial steps of glycolysis, two ATP are used to convert glucose to fructose-1,6-diphosphate.

17.5

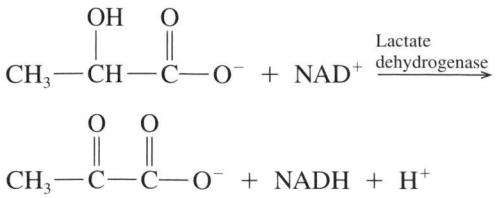

17.6 oxaloacetate

17.7 Fe^{3+}

17.8 **a.** oxidation **b.** reduction

17.9 When the final acceptor, O_2, combines with protons and electrons from cytochrome a_3.

17.10 Three NADH provide nine ATP, one $FADH_2$ provides two ATP, and one direct phosphorylation of ADP provides one ATP.

17.11 activation (−2 ATP); 2 acetyl CoA (24 ATP); 2 NADH (6 ATP); and 2 $FADH_2$ (4 ATP) = 32 ATP

17.12 oxaloacetate

Answers to Selected Questions and Problems

17.1 Prokaryotic cells, such as bacteria, do not have a nucleus. Eukaryotic cells, found in plants and animals, have a nucleus and are larger and more complex.

17.3 In metabolism, a catabolic reaction breaks apart large molecules, releasing energy for the synthesis of ATP.

17.5 ATP, adenosine triphosphate, is composed of adenine, ribose, and three phosphate groups. ADP has two phosphate groups.

$$ADP + P_i + \text{energy} \longrightarrow ATP + H_2O$$

17.7 Digestion breaks down the large molecules in food to smaller compounds that can be absorbed by the body.

17.9 **a.** small intestine—sucrase, glucose and fructose
b. stomach—pepsin; small intestine—proteases; amino acids
c. mouth—α-amylase, dextrins and maltose; small intestine—pancreatic α-amylase, maltose; maltase, glucose

17.11 Coenzyme A

17.13 **a.** NADH **b.** FAD

17.15 FAD

17.17 glucose

17.19 ATP is required in phosphorylation steps.

17.21 glyceraldehyde-3-phosphate and dihydroxyacetone phosphate

17.23 ATP is produced in glycolysis by transferring a phosphate from 1,3-diphosphoglycerate and from phosphoenolpyruvate directly to ADP.

17.25 **a.** hexokinase **b.** phosphokinase

17.27 **a.** 1 ATP required
b. 1 NADH is produced for each triose.
c. 2 ATP and 2 NADH

17.29 **a.** In step 1, a hexokinase uses ATP to phosphorylate glucose.
b. In steps 7 and 10, phosphate groups are transferred from 1,3-diphosphoglycerate and phosphoenolpyruvate directly to ADP to produce ATP.
c. In step 4, the six-carbon molecule fructose-1,6-diphosphate is split into two three-carbon molecules, glyceraldehyde-3-phosphate and dihydroxyacetone phosphate.

17.31 Aerobic (oxygen) conditions are needed.

17.33 The oxidation of pyruvate converts NAD^+ to NADH and produces acetyl CoA and CO_2.

$$\text{Pyruvate} + NAD^+ + CoA \longrightarrow \text{Acetyl CoA} + CO_2 + NADH + H^+$$

17.35 When pyruvate is reduced to lactate, NAD^+ is produced, which can be used for the oxidation of more glyceraldehyde-3-phosphate in glycolysis. This recycles NADH.

17.37 Carbon dioxide, CO_2

17.39 **a.** citrate and isocitrate
b. In a decarboxylation, a carbon atom is lost as CO_2.
c. 3 NADH
d. α-ketoglutarate
e. isocitrate $\longrightarrow$ α-ketoglutarate; α-ketoglutarate $\longrightarrow$ succinyl CoA; succinate $\longrightarrow$ fumarate; malate $\longrightarrow$ oxaloacetate
f. steps 3, 8

17.41 **a.** citrate synthetase
b. succinate dehydrogenase
c. fumarase

17.43 **a.** NAD^+ **b.** GDP **c.** FAD

17.45 NADH and $FADH_2$

17.47 **a.** reduced NADH, oxidized NAD^+
b. reduced $FMNH_2$, oxidized FMN

17.49 FAD, CoQ, cyt *c*

17.51 Electrons are transferred from NADH to FMN; 3 ATP are produced.

17.53 In oxidative phosphorylation, the energy from the oxidation reactions in the electron transport chain is used to add a phosphate to ADP for ATP production.

17.55 The release of energy from oxidation causes an accumulation of protons that creates an energy-rich proton gradient. As protons interact with ATPase, energy is released to phosphorylate ADP.

17.57 Reactions in glycolysis and the citric acid cycle produce the reduced coenzymes NADH and $FADH_2$ that are oxidized by the acceptors in the electron transport chain.

17.59 Glycolysis takes place in the cytoplasm, not in the mitochondria. Because NADH cannot cross the mitochondrial membrane, one ATP is hydrolyzed to transport the electrons from NADH to the electron transport chain. The net yield is 2 ATP for 1 NADH produced in glycolysis.

17.65 a.

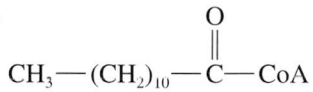

b.

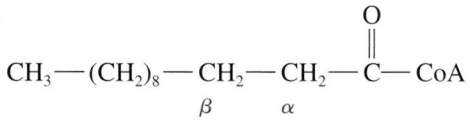

c.

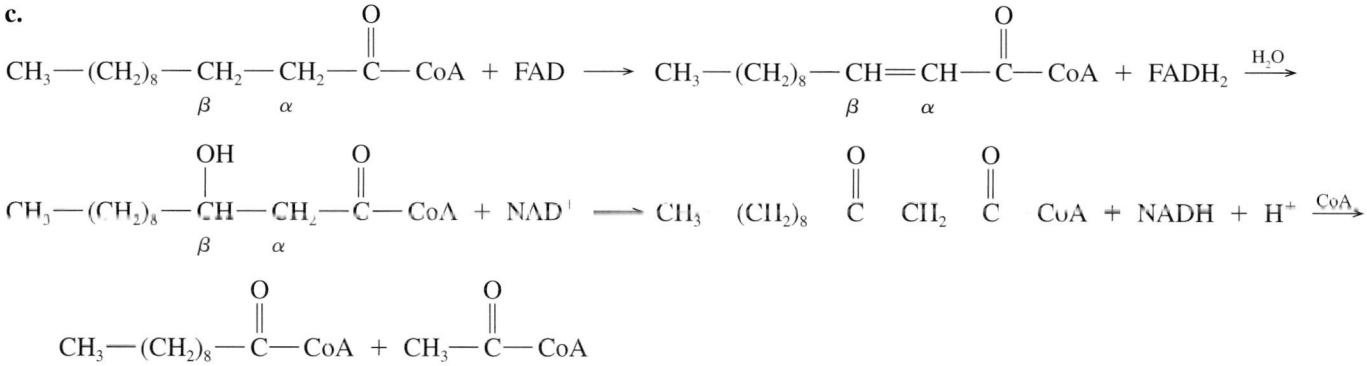

d. 6 acetyl CoA

e. five cycles

f. 6 acetyl CoA × 12 ATP/acetyl CoA ⟶ 72 ATP
5 $FADH_2$ × 2 ATP/$FADH_2$ ⟶ 10 ATP
5 NADH × 3 ATP/NADH ⟶ 15 ATP
Activation of fatty acid ⟶ −2 ATP
95 ATP

17.61 a. 3 ATP **b.** 2 ATP
c. 6 ATP **d.** 12 ATP

17.63 a. Glycerol is converted to dihydroxyacetone phosphate, a compound that can enter the glycolysis pathway for oxidation.
b. Fatty acids are activated for β oxidation in the cytoplasm.
c. 2 ATP
d. For transport to the mitochondria for β oxidation.

17.67 When large quantities of fats are oxidized, the excess acetyl CoA forms four-carbon ketones through ketogenesis. Conditions leading to ketosis include diabetes, high-fat and low-carbohydrate diets, and starvation. Because ketone bodies are acidic, blood pH is lowered and may hinder the ability of the blood to carry oxygen. A person in ketosis may experience increased urine output, sodium depletion, and a sensation of thirst.

17.69 α-keto acids and other amino acids

17.71 a.

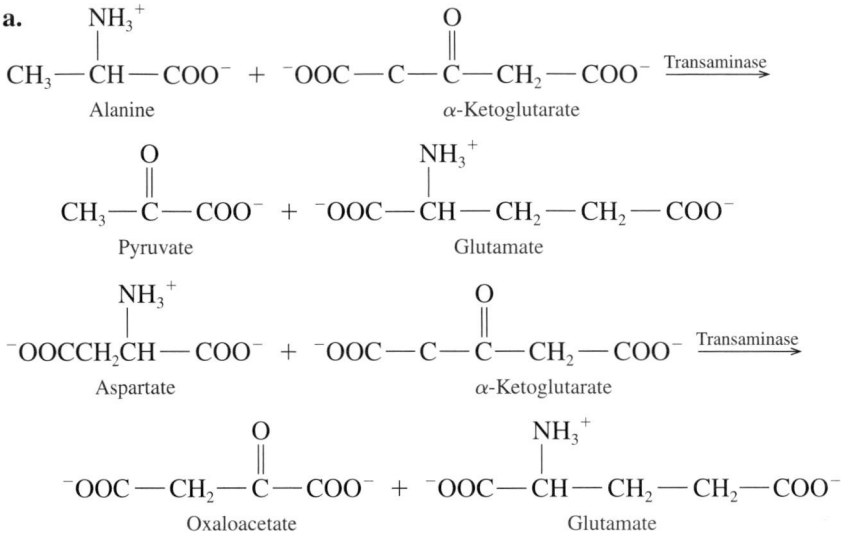

17.73 Amino acids can undergo transamination to α-keto acids such as pyruvate, oxaloacetate, and α-ketoglutarate, intermediates of the citric acid cycle that oxidize to produce ATP.

17.75 NAD^+ accepts H in oxidizations that form C═O bonds, but FAD is used when C═C bonds form.

17.77 300 kcal × 1 mole ATP/7.3 kcal = 41 moles ATP

41 moles ATP × 1 glucose/36 moles ATP = 1.1 moles glucose

1.1 moles glucose × 180 g/mole = 200 g glucose (rounded)

17.79 The oxidized coenzymes NAD^+ and FAD needed for the citric acid cycle are regenerated by the electron transport chain, which operates by transferring electrons to O_2.

17.81 Acetate can be converted to acetyl CoA and enter the citric acid cycle to be oxidized.

18

Nucleic Acids

The double helix is the characteristic shape of the DNA molecules that transmit genetic information from one generation to the next.

Looking Ahead

18.1	Nucleic Acids
18.2	Structures of Nucleic Acids
18.3	DNA Replication
18.4	RNA and Transcription
18.5	The Genetic Code
18.6	Protein Synthesis
18.7	Recombinant DNA
18.8	Genetic Mutations
18.9	Regulation of Protein Synthesis
18.10	Viruses and AIDS

www.chemplace.com/college/timberlake

Visit the web site above to test yourself, run simulations, and review key concepts from this chapter. For even more interactive review and quizzing, use the two CD-ROMs included with the book.

In the preceding chapters we discussed the digestion and metabolism of carbohydrates, lipids, and proteins essential to the survival of living organisms. One more group of biomolecules must be included in our discussion of life—the nucleic acids.

Nucleic acids are the molecules in our cells that store and direct information for cellular growth and reproduction. Deoxyribonucleic acid (DNA), the genetic material in the nucleus of the cell, contains all the information needed for the development of a complete living system. The way you grow, your hair, your eyes, your physical appearance, and the activities of the cells in your body are all determined by a set of directions contained in the DNA of each of your cells. DNA controls the functioning of the cells by coding for the amino acid sequences of enzymes. Another type of nucleic acid, RNA, carries the genetic information from the DNA to the ribosomes, the cellular factories for the synthesis of protein. When the genetic blueprint is altered, incorrect information is transmitted to the ribosomes, leading to the formation of defective proteins and malfunctioning enzymes.

18.1

Nucleic Acids: Structure 2–4

Learning Goal

Describe the nucleotides contained in DNA and RNA.

18.1 Nucleic Acids

The **nucleic acids**, large molecules first discovered in the nuclei of cells, contain all the information to direct the activities of a cell and its reproduction. There are two closely related types of nucleic acids: *deoxyribonucleic acid* (**DNA**), and *ribonucleic acid* (**RNA**). Both are linear polymers of *nucleotides.* A DNA molecule may contain several million nucleotides; smaller RNA molecules may contain up to several thousand.

Composition of Nucleotides

A DNA or RNA molecule has four types of **nucleotides.** As shown in Figure 18.1, a nucleotide has the following three parts: a nitrogen-containing base, a sugar, and a phosphate group.

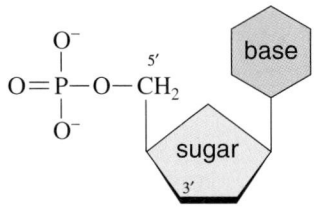

Figure 18.1 A diagram of the general structure of a nucleotide.

There are five kinds of **nitrogen-containing bases** in nucleic acids, all derived from *pyrimidine* or *purine.* The **pyrimidines** are single-ring structures of carbon and nitrogen atoms; **purines** have double-ring structures. The bases adenine (abbreviation: A), cytosine (C), and guanine (G) are found in the nucleotides of both DNA and RNA; thymine (T) is found only in DNA, and uracil (U) only in RNA. (See Figure 18.2.)

In RNA the sugar is *ribose,* a five-carbon sugar. *Deoxyribose* is found in DNA. Deoxyribose is similar to ribose except that there is no hydroxyl group (—OH) on carbon 2, thus the prefix *deoxy.*

Nucleotides in DNA and RNA

The bonding of a nitrogen-containing base with a sugar, either ribose (RNA) or deoxyribose (DNA), yields a **nucleoside**. When a phosphate group attaches to the sugar portion of the nucleoside, a *nucleotide* forms. (See Figure 18.3.) A nucleotide in DNA is named a *deoxynucleoside phosphate*; in RNA, a *nucleoside phosphate*. The names of the nucleosides and nucleotides in DNA and RNA are listed in Table 18.1. Although the letters A, G, C, U, and T represent the bases, they are often used to represent their respective nucleotides as well.

Sugar + base ⟶ nucleoside

Sugar + base + phosphate ⟶ nucleotide

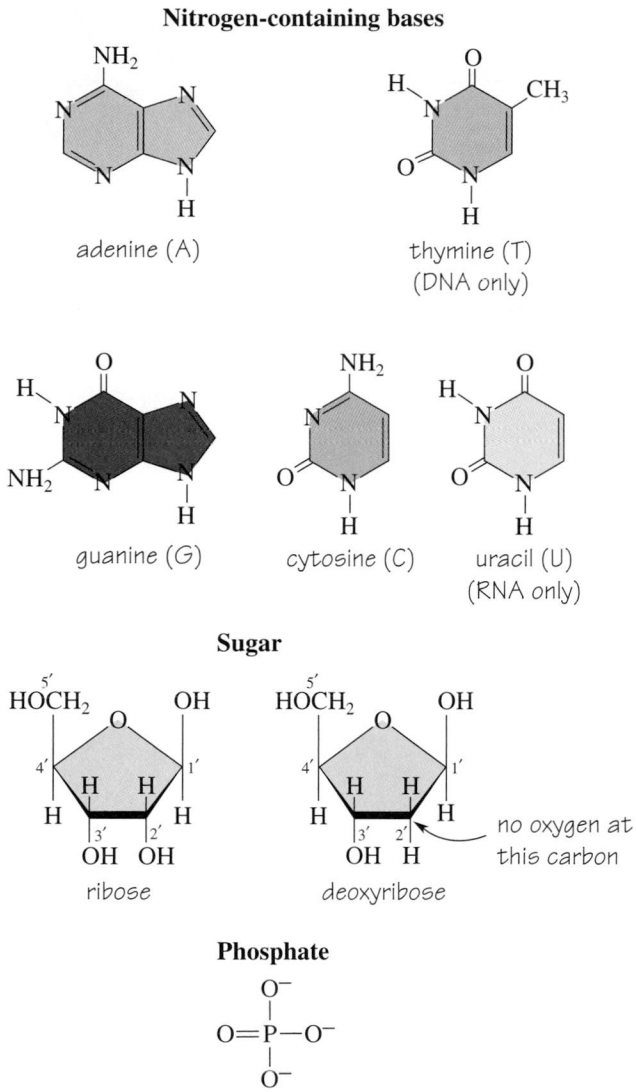

Figure 18.2 A nucleotide consists of a nitrogen-containing base, a ribose or deoxyribose sugar, and a phosphate group.

Table 18.1 Names of Nucleosides and Nucleotides in DNA and RNA

Base	Nucleosides	Nucleotides
RNA		
Adenine (A)	Adenosine (A)	Adenosine monophosphate (AMP)
Guanine (G)	Guanosine (G)	Guanosine monophosphate (GMP)
Cytosine (C)	Cytidine (C)	Cytidine monophosphate (CMP)
Uracil (U)	Uridine (U)	Uridine monophosphate (UMP)
DNA		
Adenine (A)	Deoxyadenosine (A)	Deoxyadenosine monophosphate (dAMP)
Guanine (G)	Deoxyguanosine (G)	Deoxyguanosine monophosphate (dGMP)
Cytosine (C)	Deoxycytidine (C)	Deoxycytidine monophosphate (dCMP)
Thymine (T)	Deoxythymidine (T)	Deoxythymidine monophosphate (dTMP)

Figure 18.3 The structures of the nucleotides of DNA.

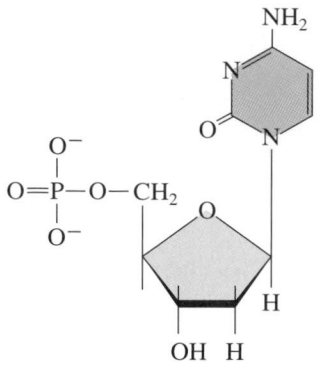

deoxycytidine monophosphate
(dCMP)

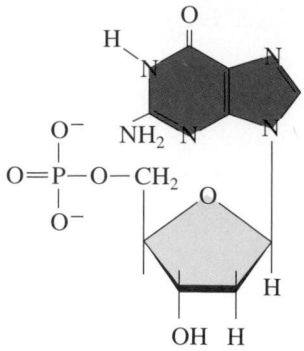

deoxyguanosine monophosphate
(dGMP)

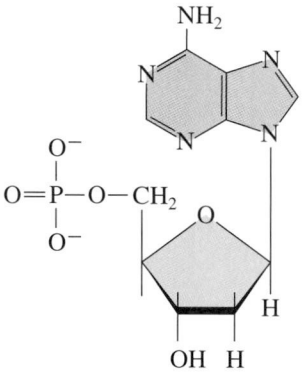

deoxyadenosine monophosphate
(dAMP)

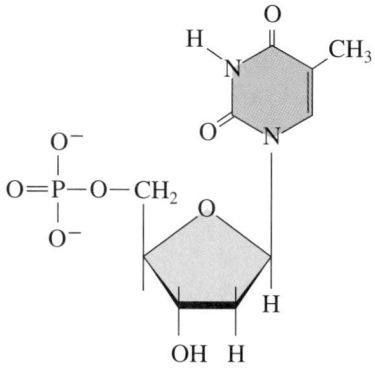

deoxythymidine monophosphate
(dTMP)

S·A·M·P·L·E· **18.1** ·P·R·O·B·L·E·M

Nucleotides

Identify the nucleic acid (DNA or RNA) that contains each of the following nucleotides; state the components of each nucleotide:

a. deoxyguanosine monophosphate (dGMP)
b. adenosine monophosphate (AMP)

Solution

a. This DNA nucleotide consists of deoxyribose, guanosine, and phosphate.
b. This RNA nucleotide contains ribose, adenine, and phosphate.

Study Check

What is the name of the DNA nucleotide of cytosine?

Q·U·E·S·T·I·O·N·S & P·R·O·B·L·E·M·S

Nucleic Acids

18.1 What bases and sugar are found in the nucleotides of DNA?

18.2 What bases and sugar are found in the nucleotides of RNA?

18.3 What are the four nucleotides in DNA?

18.4 What are the four nucleotides in RNA?

18.5 Identify each of the following as a nucleoside or nucleotide:
a. adenosine **b.** deoxycytidine
c. uridine **d.** cytidine monophosphate

18.6 Identify each of the following as a nucleoside or nucleotide in DNA or RNA:
a. deoxythymidine **b.** guanosine
c. adenosine **d.** uridine monophosphate

18.7 Draw the structure of deoxyadenosine monophosphate (dAMP).

18.8 Draw the structure of uridine monophosphate (UMP).

Study Pack

Chem of Life Nucleic Acids: The Double Helix 1, 3

18.2 Structures of Nucleic Acids

Learning Goal

Describe the structures of RNA and DNA; show the relationship between the bases in the double helix.

A nucleic acid is a polymer formed by joining many nucleotides. (See Figure 18.4.) The hydroxyl group on the 3′ carbon of the sugar in one nucleotide bonds to the phosphate group on the 5′ carbon of the sugar of the next nucleotide. As more nucleotides

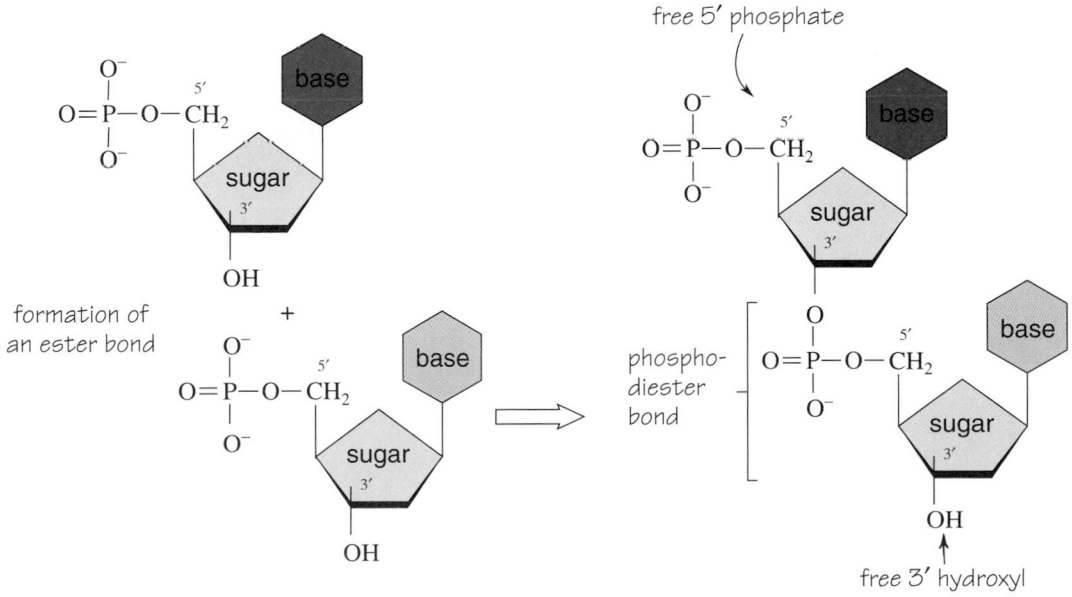

Figure 18.4 Nucleotides are joined by forming an ester bond between the 3′ hydroxyl group of one nucleotide and the phosphate group of the next nucleotide.

are added, a backbone forms with alternating sugar and phosphate groups joined by phosphodiester linkages. The bases attached to each of the sugars extend out from the backbone. In every nucleic acid chain, the sugar at one end has a free phosphate group; the sugar at the other end has a free hydroxyl group.

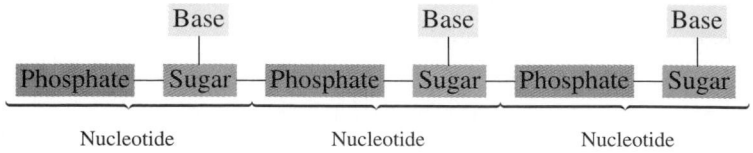

18.2

A sequence of four nucleotides in RNA is shown in Figure 18.5. In a strand of DNA, the sugar is deoxyribose; the base thymine replaces uracil.

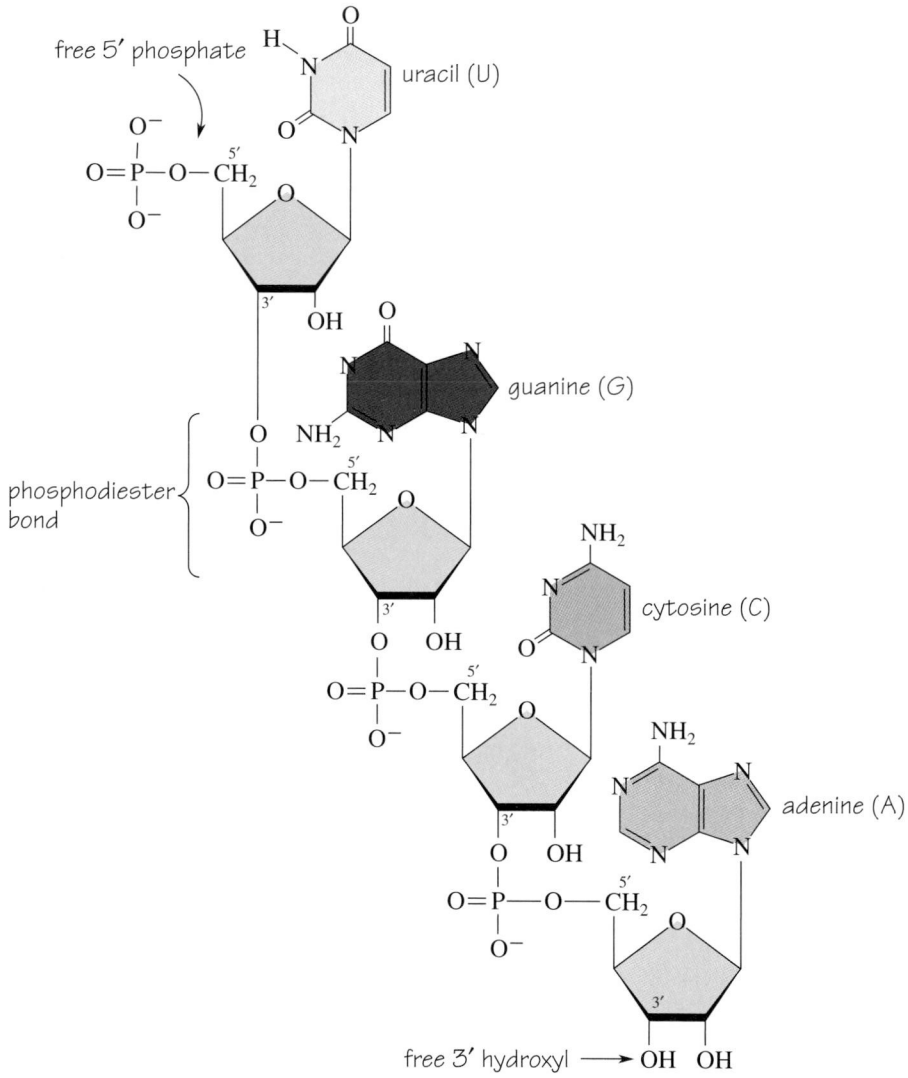

Figure 18.5 Structure of a portion of RNA with a backbone of alternating sugar and phosphate groups. In each nucleotide, the sugar is attached to one of the nitrogen bases, A, G, C, or U. There is a free phosphate group at the 5′ end and a free hydroxyl (—OH) at the 3′ end.

S·A·M·P·L·E· **18.2** ·P·R·O·B·L·E·M

Bonding of Nucleotides

Draw the structure of an RNA dinucleotide formed by two cytidine monophosphates.

Solution

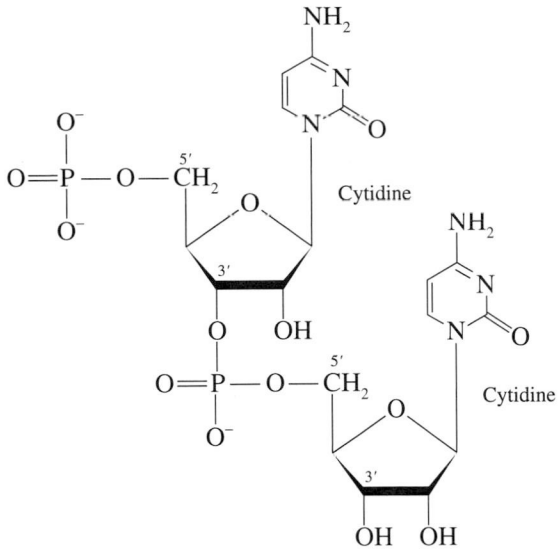

Study Check

In the dinucleotide of cytidine shown in the solution, identify the free 5′ phosphate group and the free 3′ hydroxyl (—OH) group.

Double Helix of DNA

In the early 1950s, James Watson and Francis Crick determined that DNA has a double-helical structure with two nucleotide strands winding about each other like a spiral staircase. In the DNA **double helix,** the sugar–phosphate backbones are like the railings. Holding the two strands together are hydrogen bonds that form between the bases on one strand and the bases on the other, much like stair steps in a spiral staircase. An important feature of these base pairs is they always pair the same way: adenine is always opposite thymine (A—T or T—A), and guanine is always opposite cytosine (G—C or C—G). The two strands of DNA are not identical to each other; rather they are complementary. Thus, the pairs of bases (A—T and G—C) are called **complementary base pairs.** (See Figure 18.6.) There are three hydrogen bonds between guanine and cytosine, and two hydrogen bonds are between adenine and thymine. Because other pair combinations do not fit together well, there are no C—T, C—A, G—T, or G—A pairs. As we shall see, complementary base pairing plays a crucial role in cell replication and the transfer of hereditary information.

X-ray diffraction patterns of DNA indicate that the sugar–phosphate backbone in DNA is a right-handed helix. In one complete turn, there are about 10 pairs of nucleotides. (See Figure 18.7.) In mitochondria, bacteria, and viruses, DNA molecules are compact, highly coiled molecules. In the chromosomes, these DNA strands are wrapped around proteins called *histones,* a structure that provides the most stable arrangement for the long DNA molecules.

18.2

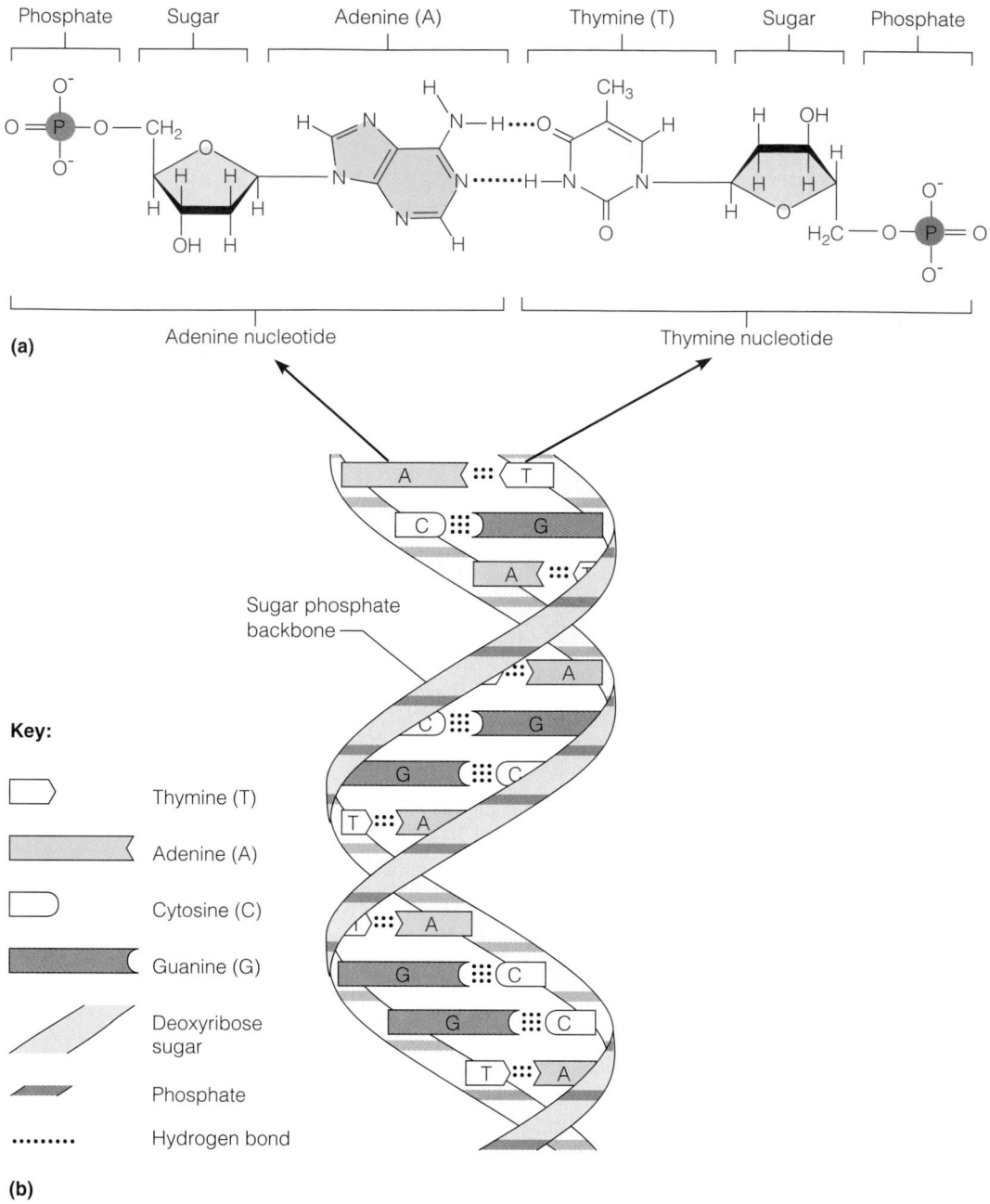

Figure 18.6 Complementary base pairs hold the DNA strands of the double helix together.

S·A·M·P·L·E **18.3** ·P·R·O·B·L·E·M

Complementary Base Pairs

Write the base sequence of the complementary segment for the following segment of a strand of DNA.

—A—C—G—A—T—C—T—

Solution

In the complementary strand of DNA, the base A pairs with T, and G pairs with C.

Given segment of DNA: —A—C—G—A—T—C—T—
: : : : : : :
Complementary segment: —T—G—C—T—A—G—A—

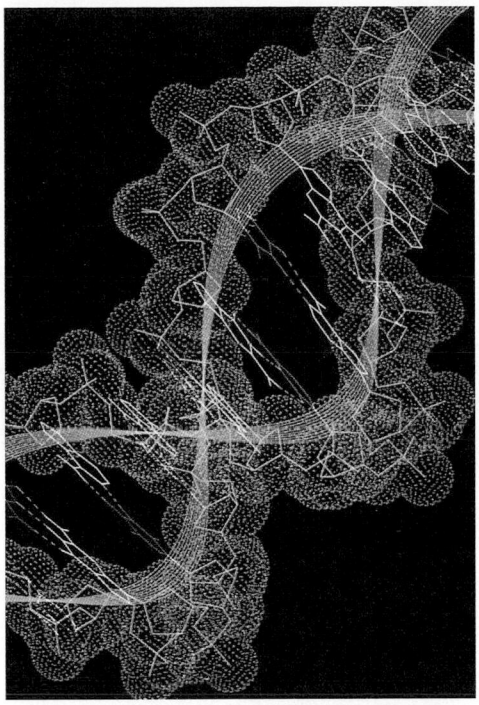

Figure 18.7 A computer-designed model of the DNA molecule.

Study Check

What is the sequence of bases that is complementary to a portion of DNA with a base sequence of —G—G—T—T—A—A—C—C—?

QUESTIONS & PROBLEMS

Structures of Nucleic Acids

18.9 How are nucleotides held together in a nucleic acid chain?

18.10 How are the two strands of nucleic acid in DNA held together?

18.11 To what part of the backbone are the bases attached?

18.12 What is meant by complementary base pairing?

18.13 Write the structure of the dinucleotide GC in RNA.

18.14 Write the structure of the dinucleotide AT in DNA.

18.15 Complete the base sequences in a second DNA strand if a portion of one strand has the following base sequence:

a. AAAAAA **b.** GGGGGG
c. AGTCCAGGT **d.** CTGTATACGTTA

18.16 Complete the base sequences in a second DNA strand if a portion of one strand has the following base sequence:

a. TTTTTT **b.** CCCCCCCCC
c. ATGGCA **d.** ATATGCGCTAAA

Study Pack

Chem of Life Nucleic Acids: DNA Replication 1

Learning Goal

Explain the process of DNA replication.

18.3

18.3 DNA Replication

All of our body cells have identical strands of DNA (egg and sperm cells contain half as much). DNA, along with some proteins (histones), makes up the material of the chromosomes, the threadlike structures observed in the nucleus during cellular division. In humans, every body cell contains 23 pairs of chromosomes, or 46 chromosomes in all (except egg and sperm cells, in which each contains only one of each pair for a total of 23). (See Figure 18.8.)

When a cell divides, DNA *replication* must provide the same genetic information for each new daughter cell. Therefore, the order of nucleotides in the parent DNA must be duplicated exactly. Central to the process of DNA **replication** is the concept of complementary base pairing. The replication process begins when hydrogen bonds between the base pairs are broken and a portion of the double helix unwinds, much like opening a zipper. (See Figure 18.9.) Then DNA polymerases begin to copy each of the open strands called the templates, by pairing each base on the DNA with its complementary base: A with T, and G with C. For example, a thymine in the parent DNA bonds to a nucleotide with adenine, not to any of the other nucleotides. As each base bonds to its complementary base, a phosphodiester bond joins the new nucleotide to the backbone of the growing DNA daughter strand.

During replication, DNA polymerase catalyzes only ester bonds between a 5′ phosphate of one nucleotide and a 3′ —OH of another. That means that DNA polymerases are moving in opposite directions along the separated strands of the DNA. On one of the templates, a new DNA strand is synthesized continuously. On the other template, new DNA is synthesized in the opposite direction in short segments of DNA. Several shorter segments are synthesized at the same time by many DNA polymerases and connected later by a DNA ligase enzyme.

Eventually the entire double helix of the original DNA is copied. In each new DNA molecule, one strand of the helix is from the original DNA and one is a newly synthesized strand. This process, called semiconservative replication, produces two new daughter DNAs that are identical to each other and exact copies of the original parent DNA. In the process of DNA replication, complementary base pairing ensures the correct placement of bases for the new identical DNA strands.

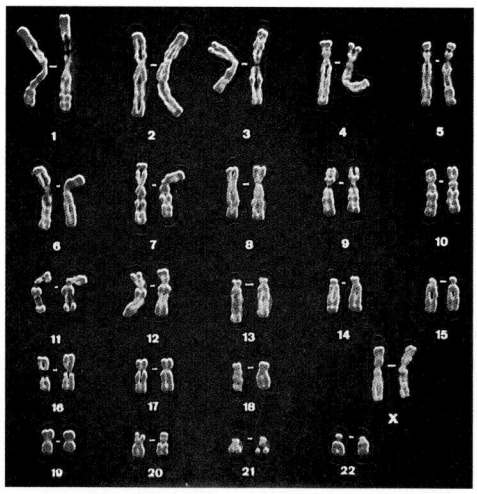

Figure 18.8 A set of 23 pairs of chromosomes stained in metaphase.

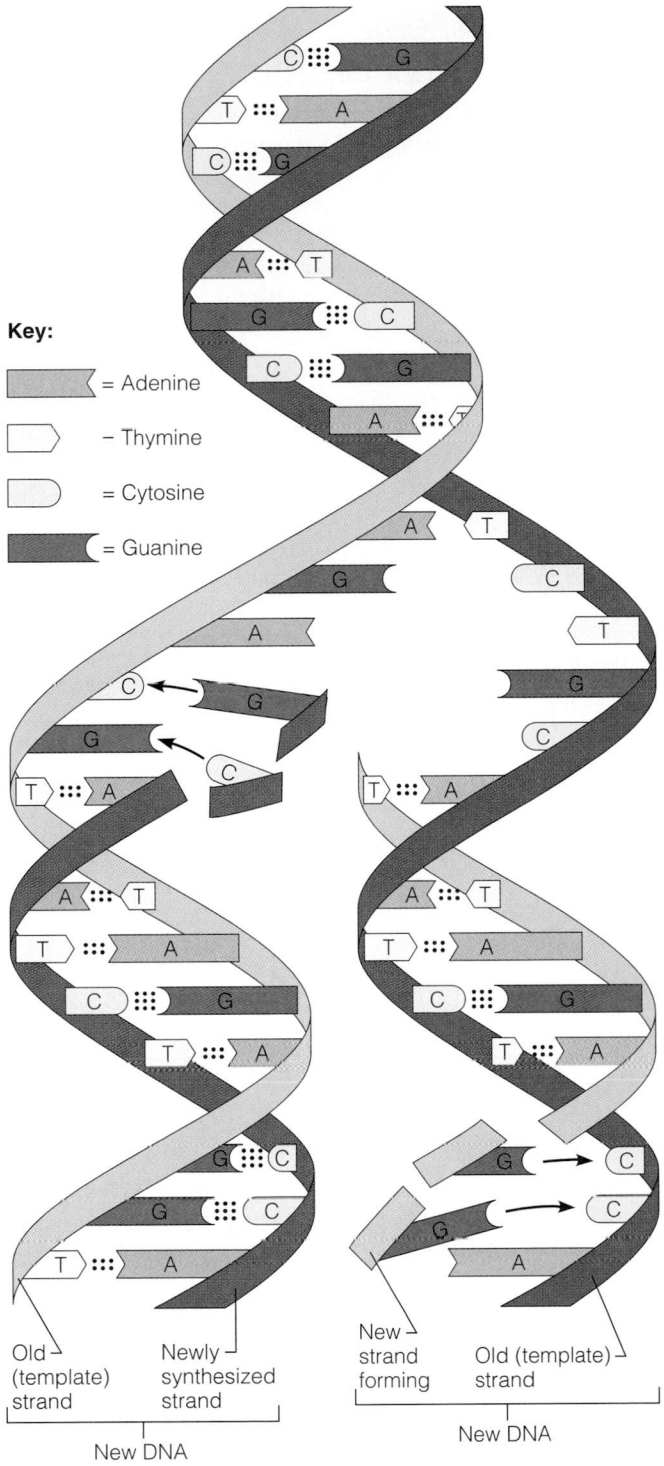

Figure 18.9 In DNA replication, two exact copies of DNA are produced. As the strands in the double helix of the original (parent) unwind and separate, new strands of daughter DNA are synthesized that contain base pairs complementary to the original DNA.

18.4

S·A·M·P·L·E 18.4 P·R·O·B·L·E·M

DNA Replication

In an original DNA strand, there is a portion with the bases A—G—T. What nucleotides are placed in a growing DNA daughter strand for this sequence?

Solution

Complementary base pairing allows only one possible nucleotide to pair with each base on the original strand. Thymine will pair only with adenine, cytosine with guanine, and adenine with thymine to give the sequence T—C—A.

Study Check

After a new base is paired with a base in an original DNA strand, how is the backbone formed?

Q·U·E·S·T·I·O·N·S & P·R·O·B·L·E·M·S

DNA Replication

18.17 Describe the process of DNA replication.

18.18 How does replication of DNA produce identical copies?

18.19 Describe the replication of the following portion of a DNA molecule:

```
—A—G—G—T—C—
 :  :  :  :  :
—T—C—C—A—G—
```

18.20 Where are the DNA strands of the original DNA found in the double helix of each of the new DNA molecules?

Study Pack

Chem of Life Nucleic Acids: Transcription 1

Learning Goal

Describe the different RNAs and the synthesis of mRNA (transcription).

18.4 RNA and Transcription

Ribonucleic acid, RNA, makes up most of the nucleic acid found in the cell. (See Figure 18.10.) Similar to DNA, RNA molecules are linear polymers of nucleotides. However, there are several important differences. The sugar in RNA is ribose rather than the deoxyribose found in DNA. The nitrogen-containing base uracil is used in place of thymine. RNA molecules are single stranded, not double, and are much smaller than DNA molecules.

Types of RNA

The three major types of RNA differ in their functions in the cells: one type of RNA carries the genetic information to synthesize protein, whereas the others are involved in the actual construction of proteins. The different types of RNA molecules found in a cell are classified according to their location and function. (See Table 18.2.)

Ribosomal RNA (**rRNA**), the most abundant RNA, makes up 60% of the structural material of the ribosomes. *Ribosomes*, the sites for protein synthesis, are small, complex structures in the cytoplasm composed of rRNA and protein.

Messenger RNA (**mRNA**) carries genetic information from the DNA in the nucleus to the ribosomes in the cytoplasm for protein synthesis. Each gene, a segment of DNA, produces a separate mRNA molecule when the cell requires the protein encoded by that gene. The size of an mRNA depends on the number of nucleotides in that particular gene.

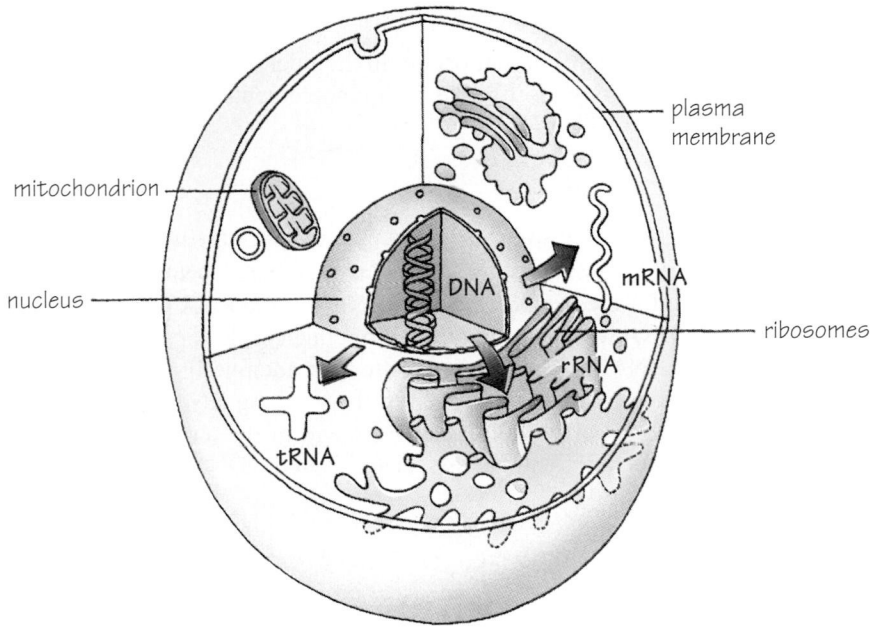

Figure 18.10 Location of the nucleic acids in a cell.

Table 18.2 Types of RNA Molecules

Type	Abbreviation	Percentage of Total RNA	Function in the Cell
Ribosomal RNA	rRNA	75	Major component of the ribosomes
Messenger RNA	mRNA	5–10	Carries information for protein synthesis from the DNA in the nucleus to the ribosomes
Transfer RNA	tRNA	10–15	Brings amino acids to the ribosomes for protein synthesis

Transfer RNA (**tRNA**), the smallest of these kinds of RNA molecules, interprets the genetic code and brings specific amino acids to the ribosome for protein synthesis. Only tRNAs link the genetic code to the amino acids in the proteins. There are different tRNAs for all the amino acids. The structures of the transfer RNAs are similar. Each tRNA consists of a single chain of about 70–90 nucleotides. Hydrogen bonds between some of the complementary nitrogen bases in the chain produce loops that give an overall cloverleaf structure.

S·A·M·P·L·E· **18.5** ·P·R·O·B·L·E·M

Types of RNA

What is the function of mRNA in a cell?

Solution

mRNA carries the instructions for the synthesis of a protein from the DNA to the ribosomes.

Study Check

What is the function of tRNA in a cell?

Transcription

In a process called **transcription** that occurs in the nucleus, the nucleotides on one strand of DNA are copied or transcribed as their complementary bases in a messenger RNA molecule.

$$\text{DNA} \xrightarrow{\text{Transcription (nucleus)}} \text{mRNA}$$

18.4

As in DNA replication, a portion of the DNA is unwound, this time by the enzyme RNA polymerase. In RNA synthesis, only one of the separated DNA strands is used as a pattern or template for the synthesis of an mRNA. As the RNA polymerase moves along the separated DNA template strand, each ribonucleotide is paired with its complementary base. In RNA, uracil is paired with each adenine in the DNA template strand. Each new ribonucleotide is attached to the growing RNA strand by a phosphoester bond. When the polymerase reaches the end of the template, transcription ends and the newly formed RNA strand dissociates from the DNA.

Bases on a DNA template:	—G—T—A—C—
	↓ ↓ ↓ ↓
Complementary base sequence in mRNA:	—C—A—U—G—

S·A·M·P·L·E· **18.6** ·P·R·O·B·L·E·M

RNA Synthesis

The sequence of bases in a part of the DNA template for mRNA is CGATCA. What is the corresponding mRNA produced?

Solution

The nucleotides in DNA pair up with the ribonucleotides as follows: G ⟶ C, C ⟶ G, T ⟶ A, and A ⟶ U.

Portion of DNA template:	—C—G—A—T—C—A—
	↓ ↓ ↓ ↓ ↓ ↓
Complementary bases in mRNA:	—G—C—U—A—G—U—

Study Check

What is the DNA template that codes for the mRNA having the ribonucleotide sequence GGGUUUAAA?

Q·U·E·S·T·I·O·N·S & P·R·O·B·L·E·M·S

RNA and Transcription

18.21 What are the three different types of RNA?

18.22 Where is DNA found in eukaryotic cells?

18.23 Where is each type of RNA found in eukaryotic cells?

18.24 What are the functions of each type of RNA?

18.25 What is meant by the term "transcription"?

18.26 What bases in mRNA are used to complement the bases in DNA?

18.27 Write the corresponding section of mRNA produced from the following section of DNA template:

—C—C—G—A—A—G—G—T—T—C—A—C—

18.28 Write the corresponding section of mRNA produced from the following section of DNA template:

—T—A—C—G—G—C—A—A—G—C—T—A—

18.5 The Genetic Code

Study Pack

Learning Goal

Describe the function of the codons in the genetic code.

The overall function of the RNAs in the cell is to facilitate the task of synthesizing protein. After the genetic information encoded in DNA is transcribed into mRNA molecules, the mRNAs move out of the nucleus to the ribosomes in the cytoplasm. At the ribosomes, the genetic information in the mRNAs is converted into a sequence of amino acids in protein. (See Figure 18.11.)

$$\text{mRNA} \xrightarrow[\text{ribosomes in the cytoplasm}]{\text{Translation at the}} \text{protein}$$

Codons

Genetic information from DNA is encoded in the mRNA as a sequence of nucleotides. In the **genetic code,** a sequence of three bases (triplet), called a **codon,** specifies each amino acid in the protein. Early work on protein synthesis showed that repeating triplets of uracil (UUU) produced a polypeptide that contained only phenylalanine. Therefore, a sequence of UUU—UUU—UUU codes is for three phenylalanines.

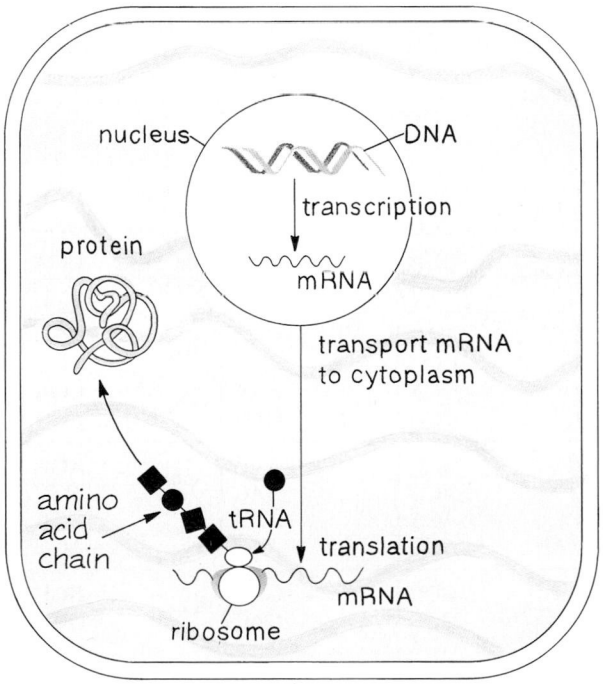

Figure 18.11 Diagram of the overall process of protein synthesis.

Codons (base sequence) in mRNA	UUU — UUU — UUU
Translation	↓ ↓ ↓
Amino acid sequence	Phe — Phe — Phe

The codons have now been determined for all 20 amino acids. A total of 64 codons are possible from the triplet combinations of A, G, C, and U. Three of these, UGA, UAA, and UAG, are stop signals that code for the termination of protein synthesis. All the other three-base codons shown in Table 18.3 specify amino acids, which means that one amino acid can have several codons. For example, glycine is the amino acid coded by the codons GGU, GGC, GGA, and GGG. The triplet AUG has two roles in protein synthesis. At the beginning of an mRNA, the codon AUG signals the start of protein synthesis. In the middle of a series of codons, the AUG codon specifies the amino acid methionine.

18.5

S·A·M·P·L·E· 18.7 ·P·R·O·B·L·E·M

Codons

What is the sequence of amino acids coded by the following codons in mRNA?

GUC — AGC — CCA

Solution

According to Table 18.3, GUC codes for valine, AGC for serine, and CCA for proline. The sequence of amino acids is Val-Ser-Pro.

Study Check

The codon UGA does not code for an amino acid. What is its function in mRNA?

Table 18.3 mRNA Codons: The Genetic Code for Amino Acids

First Letter ↓	Second Letter: U	Second Letter: C	Second Letter: A	Second Letter: G	Third Letter ↓
U	UUU, UUC Phe	UCU, UCC, UCA, UCG Ser	UAU, UAC Tyr	UGU, UGC Cys	U
	UUA, UUG Leu		UAA STOP	UGA STOP	C
			UAG STOP	UGG Trp	A
					G
C	CUU, CUC, CUA, CUG Leu	CCU, CCC, CCA, CCG Pro	CAU, CAC His	CGU, CGC, CGA, CGG Arg	U
			CAA, CAG Gln		C
					A
					G
A	AUU, AUC, AUA Ile	ACU, ACC, ACA, ACG Thr	AAU, AAC Asn	AGU, AGC Ser	U
					C
	[a]AUG Met		AAA, AAG Lys	AGA, AGG Arg	A
					G
G	GUU, GUC, GUA, GUG Val	GCU, GCC, GCA, GCG Ala	GAU, GAC Asp	GGU, GGC, GGA, GGG Gly	U
			GAA, GAG Glu		C
					A
					G

[a]Codon that signals the start of a peptide chain.
STOP codons signal the end of a peptide chain.

Q·U·E·S·T·I·O·N·S & P·R·O·B·L·E·M·S

The Genetic Code

18.29 What is a codon?

18.30 What is the genetic code?

18.31 What amino acid is coded for by each codon?
a. CUU **b.** UCA **c.** GGU **d.** AGG

18.32 What amino acid is coded for by each codon?
a. AAA **b.** GUC **c.** CGG **d.** GCA

18.33 When does the codon AUG signal the start of a protein, and when does it code for the amino acid methionine?

18.34 The codons UGA, UAA, and UAG do not code for amino acids. What is their role as codons in mRNA?

18.6 Protein Synthesis

Learning Goal

Describe the process of protein synthesis.

After an mRNA is synthesized, it migrates out of the nucleus into the cytoplasm where ribosomes, tRNA molecules, amino acids, and some enzymes use the three-base codons on the mRNA to synthesize a protein. **Translation** is the group of reactions that take place at the ribosome where the sequence of nucleotides in the mRNA directs the order of amino acids in the synthesis of a protein. These reactions include (1) the activation of tRNA, (2) initiation, (3) elongation, and (4) termination.

Activation of tRNA

Two important sections on a tRNA are one that recognizes a specific amino acid and one that recognizes a three-base codon on the mRNA. (See Figure 18.12.) A tRNA is activated when it attaches to its specific amino acid by using energy from ATP to form an ester bond with the adenine at one end of the tRNA molecule. At the loop on the opposite end of the tRNA is a series of three bases called an anticodon. This **anticodon** sequence is complementary to a codon of three bases on the mRNA.

Initiation

Protein synthesis begins when an mRNA combines with a small subunit of a ribosome. (See Figure 18.13.) The start codon AUG initiates synthesis by attaching through complementary base pairing to the anticodon of a tRNA that carries methionine. The machinery for protein synthesis is ready when the large subunit of the ribosome joins the small subunit to create two binding sites.

Elongation

The second codon on the mRNA is now ready for translation. Only a tRNA with the three-base sequence that complements the second codon can translate it. Suppose that the second codon is UCC, which codes for serine. Only the tRNAs attached to serine have the anticodon AGG that can bond with the mRNA codon. By base pairing, the three bases of the tRNA form hydrogen bonds with the complementary bases in the mRNA. This places the amino acids methionine and serine close together, which allows enzymes in the ribosome to form a peptide bond. The bond between the first amino acid, methionine, and its tRNA is broken, and the tRNA returns to the cytoplasm. When

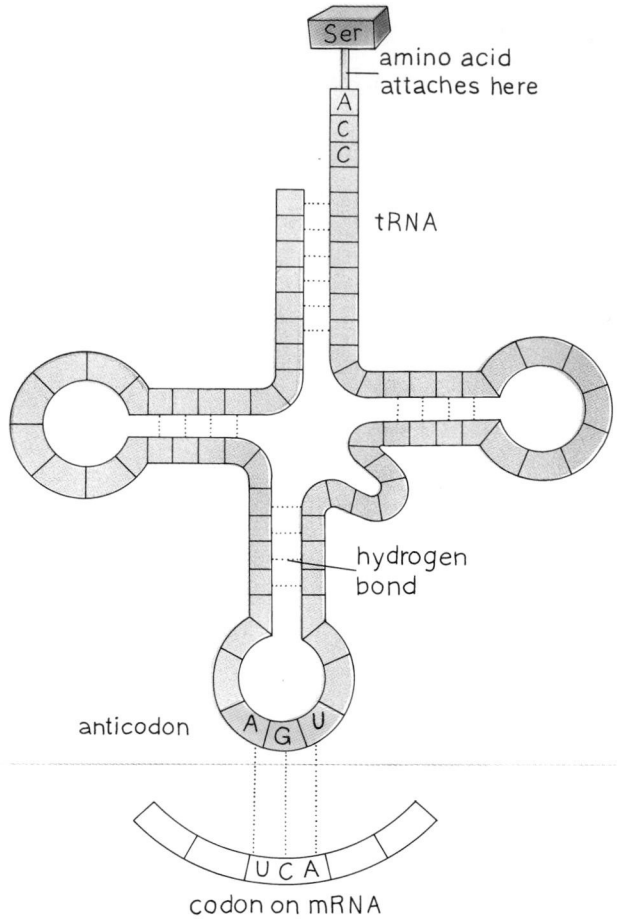

Figure 18.12 The cloverleaf structure of tRNA picks up an amino acid and carries it to an mRNA codon that is complementary to the tRNA anticodon loop.

the first tRNA detaches from the ribosome, the ribosome shifts along the mRNA until it reaches the next codon of three bases, a process called **translocation**. A new tRNA can now attach to the open binding site so that its amino acid forms a peptide bond with the previous amino acid, serine. Again the earlier tRNA detaches and the ribosome shifts down the mRNA to read the next codon.

Elongation continues as the ribosome moves along the mRNA, one codon at a time. At each new codon, a tRNA attaches to the mRNA by complementary base pairing, which places the next amino acid into the growing peptide. Several ribosomes can translate the same strand of mRNA at the same time, producing multiple copies of the peptide chain.

Termination

After all the amino acids for a particular protein have been linked together by peptide bonds, the ribosome encounters a stop codon. Because there are no tRNAs to complement the termination codon, protein synthesis ends. An enzyme releases the completed polypeptide chain from the ribosome. The initiating amino acid methionine is also removed from the beginning of the peptide chain. Now the amino acids in the chain form the three-dimensional structure that makes the polypeptide into a biologically active protein.

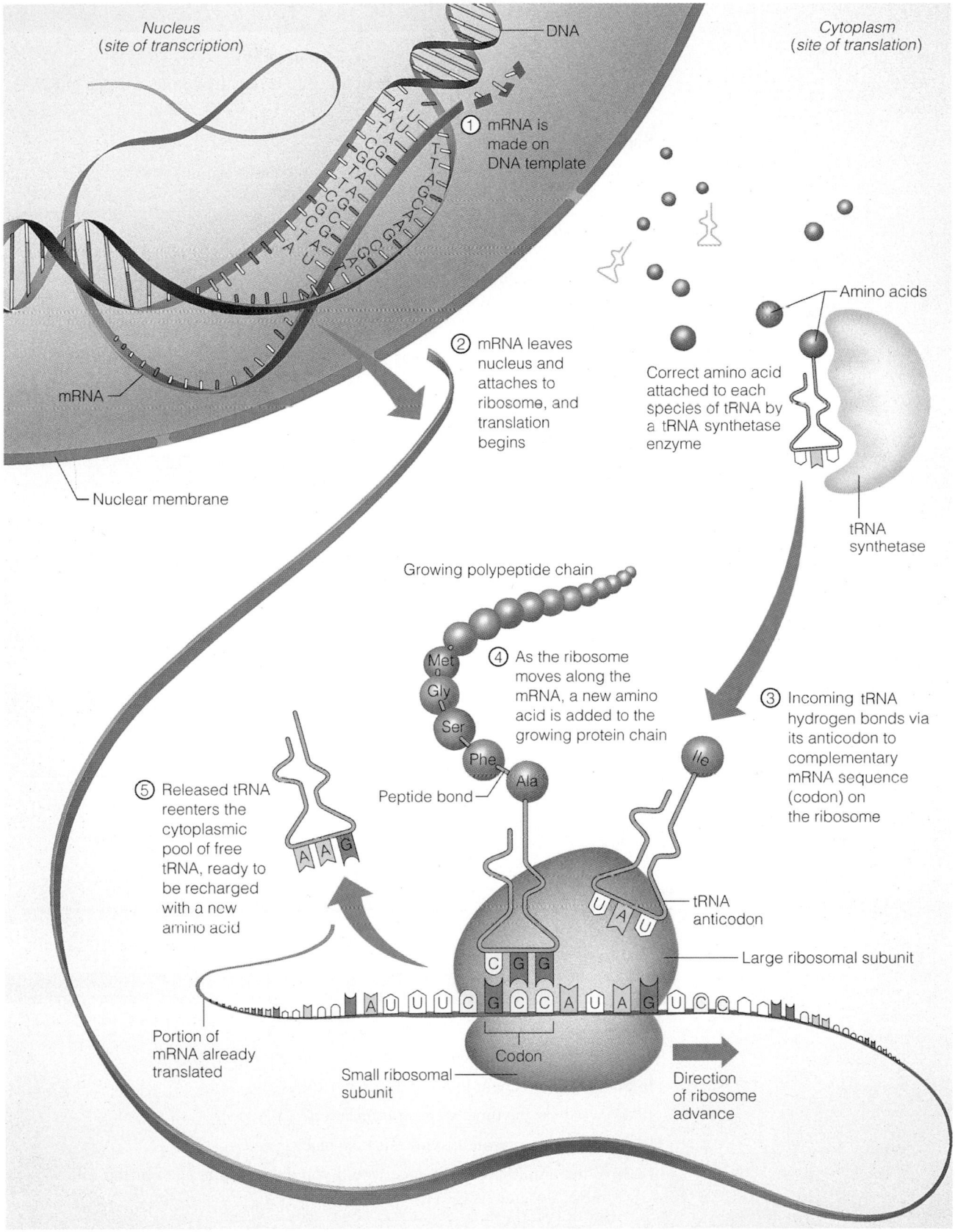

Figure 18.13 Process of translation.

SAMPLE 18.8 PROBLEM

Protein Synthesis

What order of amino acids would you expect in a peptide for the mRNA sequence of UCA—AAA—GCC—CUU?

Solution

Each of the codons specifies a particular amino acid. Using Table 18.3, we write a peptide with the following amino acid sequence:

18.6

RNA codons:	UCA—	AAA—	GCC—	CUU
	↓	↓	↓	↓
Amino acid sequence:	Ser —	Lys —	Ala —	Leu

Study Check

Where would protein synthesis stop in the following series of bases in an mRNA? GGG—AGC—AGU—UAG—GUU

HEALTH NOTE

Many Antibiotics Inhibit Protein Synthesis

Several antibiotics stop bacterial infections by interfering with the synthesis of proteins needed by the bacteria. Some antibiotics act only on bacterial cells by binding to the ribosomes in bacteria but do not act on human cells. A description of some of these antibiotics is given in Table 18.4.

Table 18.4 Antibiotics That Inhibit Protein Synthesis in Bacterial Cells

Antibiotic	Effect on Ribosomes to Inhibit Protein Synthesis
Chloramphenicol	Inhibits peptide bond formation and prevents the binding of tRNA
Erythromycin	Inhibits peptide chain growth by preventing the translocation of the ribosome along the mRNA
Puromycin	Causes release of an incomplete protein by ending the growth of the polypeptide early
Streptomycin	Prevents the proper attachment of tRNAs
Tetracycline	Prevents the binding of tRNAs

QUESTIONS & PROBLEMS

Protein Synthesis

18.35 What is the difference between a *codon* and an *anticodon*?

18.36 Why are there at least 20 different tRNAs?

18.37 What are the three steps of translation?

18.38 Where does protein synthesis take place?

18.39 What amino acid sequence would you expect from each of the following mRNA segments?

a. —A—A—A—A—A—A—A—A—A—

b. —U—U—U—C—C—C—U—U—U—C—C—C—

c. —U—A—C—G—G—G—A—G—A—U—G—U—

18.40 What amino acid sequence would you expect from each of the following mRNA segments?

a. —A—A—A—C—C—C—U—U—G—G—C—C—

b. —C—C—U—C—G—A—A—G—C—C—C—A—U—G—A—

c. —A—U—G—C—A—C—A—A—A—G—A—A—G—U—A—C—U—U—

18.41 How is a peptide chain extended?

18.42 What is meant by "translocation"?

18.43 The following portion of DNA is in the template DNA strand:

—G—C—T—T—T—T—C—A—A—A—A—A—

a. What is the corresponding mRNA section?

b. What are the anticodons of the tRNAs?

c. What amino acids will be placed in the peptide chain?

18.44 The following portion of DNA is in the template DNA strand:

—T—G—T—G—G—G—G—T—T—A—T—T—

a. What is the corresponding mRNA section?

b. What are the anticodons of the tRNAs?

c. What amino acids will be placed in the peptide chain?

18.7

18.7 Recombinant DNA

Learning Goal

Describe the steps in preparing recombinant DNA.

Over the past two decades, new techniques called genetic engineering have permitted scientists to cut, splice, and recombine DNA from genes from different kinds of cells. The new synthetic forms of DNA, which contain a DNA fragment from another organism, are known as **recombinant DNA.** Recombinant DNA technology is now used to produce human insulin for diabetics, the antiviral substance interferon, blood clotting factor VIII, and human growth hormone.

Most of the work in recombinant DNA is done with *Escherichia coli (E. coli)* bacteria. Within the bacterial cell are several small circular molecules of DNA called *plasmids.* Plasmids, which are easy to isolate, are also capable of replication. After a piece of "foreign" DNA is inserted into a plasmid, a copy is made every time the bacterium divides. In one day, one *E.coli* bacterium is capable of producing a million copies of itself including the foreign DNA, a process known as gene cloning.

Restriction Enzymes

The ability to insert foreign DNA developed after scientists discovered enzymes that can cut DNA between a specific pair of nucleotides. The resulting DNA fragments have "sticky" ends that can hydrogen bond again when they are mixed with plasmids that are also cut open using restriction enzymes. After the DNA fragment joins with the bacterial plasmids, the recombined plasmids are able to synthesize the proteins of the new DNA.

Preparing Recombinant DNA

Initially, *E. coli* cells are soaked in a detergent solution to dissolve the plasma membrane. The contents of the cells including the plasmids are released and collected. A **restriction enzyme,** which breaks sugar–phosphate bonds in DNA between specific

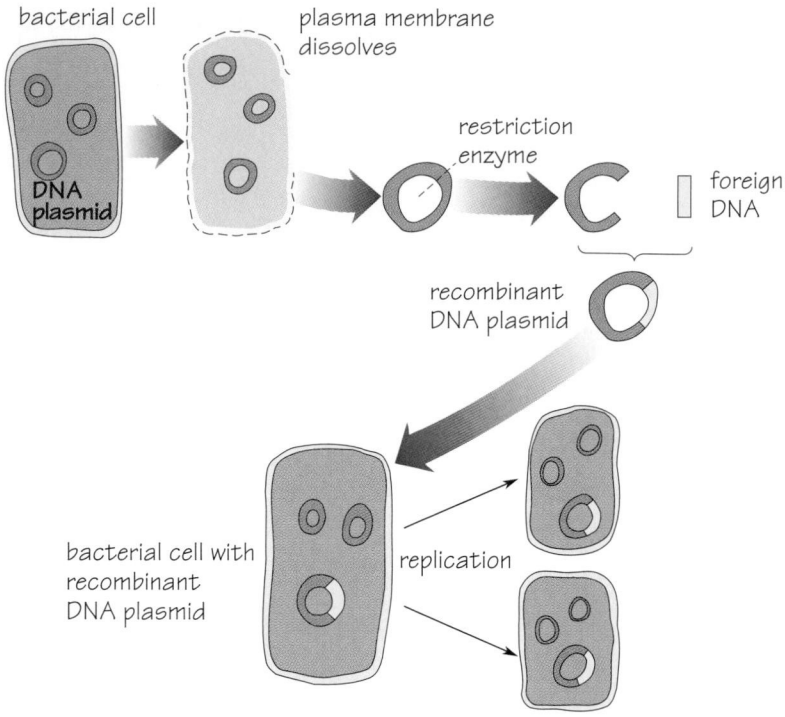

Figure 18.14 Formation of recombinant DNA by placing new DNA in plasmid DNA of bacterium. The bacterium can now produce a nonbacterial protein such as insulin or growth hormone.

18.7

nucleotides, is used to cut open the circular DNA strands in the plasmids. (See Figure 18.14.) The same enzyme is also used to cut out a foreign piece of DNA from the chromosome of a different organism, such as the gene that produces insulin or growth hormone. The cut out genes are then mixed with the plasmids that were cut open. The ends of the foreign DNA piece and the ends of the opened plasmids are joined by a DNA ligase. Then the altered plasmids containing the recombined DNA are placed in a fresh culture of *E. coli* bacteria where they can be reabsorbed into the bacterial cells. The new gene that was inserted in the plasmids is copied as the genetically engineered *E. coli* cells start to replicate. Eventually, a large number of cells with the new DNA produce the new protein. If the inserted DNA coded for the human insulin protein, the altered plasmids would begin to synthesize human insulin.

DNA Fingerprinting

Forensics uses recombinant DNA technology to determine the DNA sequence in saliva, hair, blood, skin, or semen. The DNA of an individual produces a unique set of DNA fragments that can be used like a fingerprint for comparison and identification. If only a very small sample of DNA is available, a process called **polymerase chain reaction (PCR)** makes it possible to make multiple copies of (amplify) the DNA in a short time. When the original DNA is mixed with DNA polymerase and a mixture of the deoxyribonucleotides, the polymerase synthesizes many copies of that DNA.

In a process called DNA sequencing, restriction enzymes are used to cut fragments from DNA, which are then sorted by size using gel electrophoresis. The gel with the DNA fragments is soaked in a solution containing a radioactive isotope, which adheres to specific bases in the DNA. A piece of x-ray film placed over the gel is exposed by the radioactivity emitted to give a pattern of dark and light bands known as a DNA fingerprint. (See Figure 18.15.) By matching the fragments, the order (sequence) of nucleotides for the entire DNA strand can be determined. When the sequence of nucleotides in the DNA is determined, it can be compared to the DNA sequence of a suspect. It has been estimated that the odds of two persons (who are not identical twins) producing the same DNA fingerprint is less than one in a billion.

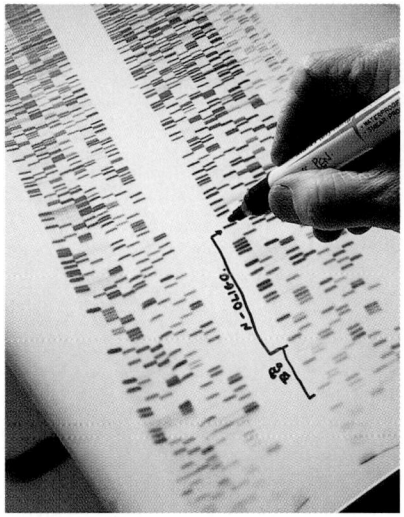

Figure 18.15 A scientist analyzes a nucleotide sequence in DNA from a human gene. Dark and light bands represent the order of nucleotides. The marked sequence is involved in the growth of melanoma cancer cells.

In another application of DNA technology, the sequence of deoxynucleotides for all the genes in a human cell is being determined (Human Genome Project) to help scientists understand more about genetic traits and diseases. The results of the genome project will help doctors diagnose and design treatments for the 3500 genetic diseases known today.

S·A·M·P·L·E 18.9 ·P·R·O·B·L·E·M

Recombinant DNA

What is the function of restriction enzymes in recombinant DNA?

Solution

Restriction enzymes are used to cut out a particular piece of DNA from a gene, and to cut open the circular plasmids in a bacterium where the foreign DNA attach.

Study Check

What is gene cloning?

Q·U·E·S·T·I·O·N·S & P·R·O·B·L·E·M·S

Recombinant DNA

18.45 Why are *E. coli* bacteria used in recombinant DNA procedures?

18.46 What is a plasmid?

18.47 How are plasmids obtained from *E. coli*?

18.48 Why are restriction enzymes mixed with the plasmids?

18.49 How is a gene for a particular protein inserted into a plasmid?

18.50 Why is DNA polymerase useful in criminal investigations?

18.51 What is a DNA fingerprint?

18.52 What beneficial proteins are produced from recombinant DNA technology?

Study Pack

18.8 Genetic Mutations

Learning Goal

Describe some ways in which DNA is altered to cause mutations.

18.8

A **mutation** is a change in the DNA nucleotide sequence that alters the sequence of amino acids, which may alter the structure and function of a protein in a cell. Some mutations are known to result from X rays, overexposure to sun (ultraviolet or UV light), chemicals called mutagens, and possibly some viruses. If a change in DNA occurs in a somatic cell, a cell other than a reproductive cell, the altered DNA will be limited to that cell and its daughter cells. If there is uncontrolled growth, the mutation could lead to cancer. If the mutation occurs in germ cell DNA (egg or sperm), then all the DNA produced in a new individual will contain the same genetic change. If the genetic change greatly affects the catalysis of metabolic reactions or the formation of important structural proteins, the new cells may not survive or the person may exhibit a genetic disease. (See Figure 18.16.)

Consider a triplet of bases such as CCC, which produces the codon GGG in the mRNA. At the ribosome, a tRNA would place the amino acid glycine in the peptide chain. Now, suppose the DNA sequence is changed to CAC. The codon in the resulting mRNA is altered to give GUG, which codes for valine, not glycine. The replacement of one base in DNA with another is called **substitution.** The change in the codon may cause a different amino acid to be inserted at that point in the polypeptide. Substitution is the most common way in which mutations occur. (See Table 18.5.)

In a **frame shift mutation,** a base is added to or deleted from the normal order of bases in the DNA. Then there is a change in the codon that lost or gained a base, and in all the codons that follow. Table 18.6 illustrates addition and deletion mutations.

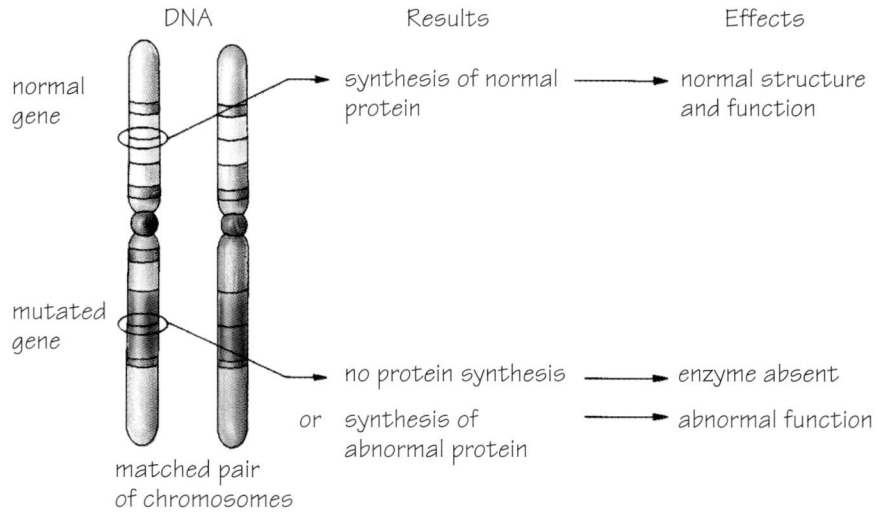

Figure 18.16 Most genetic diseases are a result of gene mutations that produce no protein or abnormal protein.

Table 18.5 Effect of Substitution Mutations on Amino Acid Sequence

Substance	Normal Sequence	Effect of Mutation
		Possible change in base ↓
DNA	ACA — **CCC** — AGG — TTT	ACA — **CAC** — AGG — TTT
↓		Change in codon
mRNA	UGU — **GGG** — UCC — AAA	UGU — **GUG** — UCC — AAA
		Change in amino acid order
Amino acid sequence	Cys — **Gly** — Ser — Lys	Cys — **Val** — Ser — Lys

Table 18.6 Addition and Deletion Mutations

	Normal Sequence	Frame Shift Effect of Mutation	
		A added	A deleted
DNA	—CCC—**AGG**—TTT—	—CCC—**AAG**—**GTT**—T	—CCC—**GGT**—TT
		Shift in codons	
mRNA	—GGG—**UCC**—AAA—	—GGG—**UUC**—**CAA**—A	—GGG—**CCA**—**AA**
		Incorrect order	
Amino acid sequence	— Gly — **Ser** — Lys —	— Gly — **Phe** — **Gln** —	— Gly — **Pro** —

18.8

Effect of Mutations

When a mutation causes a change in the amino acid sequence, the structure of the resulting protein can be altered severely and it may lose biological activity. If the protein is an enzyme, it may no longer bind to its substrate or react with the substrate at the active site. When an altered enzyme cannot catalyze a reaction, certain substances may accumulate until they act as poisons in the cell, or substances vital to survival may not be synthesized. If a defective enzyme occurs in a major metabolic pathway or in the building of a cell membrane, the mutation can be lethal. When a protein deficiency is genetic, the condition is called a **genetic disease.**

$$\text{DNA} \xrightarrow{\text{X rays, UV sunlight, mutagens, viruses}} \text{alteration of DNA} \longrightarrow \text{defective protein} \longrightarrow \text{genetic disease (germ cells) or cancer (somatic cells)}$$

EXPLORE YOUR WORLD

A Model for DNA Replication and Mutation

1. Cut out 16 rectangular pieces of paper. For strand 1 of a segment of DNA, write on two pieces this nucleotide symbol with hydrogen bonds: A═, T═, G≡, and C≡. For strand 2, write on two pieces each: ═A, ═T, ≡G, and ≡C. Mix up the pieces for strand 1 and randomly place in vertical order. From the second group, strand 2 nucleotides, select the complementary bases to complete this section of DNA.
2. Using the nucleotides from experiment 1, make a strand 1 of A—T—T—G—C—C. Arrange the corresponding bases to complete the DNA section. In the original column, change the G to an A. What does this do to the complementary strand? How could this change result in a mutation?

S·A·M·P·L·E 18.10 ·P·R·O·B·L·E·M

Mutations

An mRNA has the sequence of codons CCC—AGA—GGG. If a base substitution in the DNA changes the mRNA codon of AGA to ACA, how is the amino acid sequence affected in the resulting protein?

Solution

The initial mRNA sequence of CCC—AGA—GGG codes are for the amino acids proline, arginine, and glycine. When the mutation occurs, the new sequence of mRNA codons CCC—ACA—GGG are for proline, threonine, and glycine. The amino acid arginine is replaced by threonine.

	Normal	After Mutation
mRNA codons	CCC—[AGA]—GGG	CCC—[ACA]—GGG
Amino acids	Pro — Arg — Gly	Pro — Thr — Gly

Study Check

How might the protein made from this mRNA be affected by this mutation?

18.8 HEALTH NOTE

Genetic Diseases

A genetic disease is the result of a defective enzyme caused by a mutation in its genetic code. For example, phenylketonuria, PKU, results when DNA cannot direct the synthesis of the enzyme phenylalanine hydroxylase, required for the conversion of phenylalanine to tyrosine. In an attempt to break down the phenylalanine, other enzymes in the cells convert it to phenylpyruvic acid. The accumulation of phenylalanine and phenylpyruvic acid in the blood can lead to severe brain damage and mental retardation. If PKU is detected in a newborn baby, a diet is prescribed that eliminates all the foods that contain phenylalanine. Preventing the buildup of the phenylpyruvic acid ensures normal growth and development.

The amino acid tyrosine is needed in the formation of melanin, the pigment that gives the color to our skin and hair. If the enzyme that converts tyrosine to melanin is defective, no melanin is produced, a genetic disease known as albinism. Persons and animals with no melanin have no skin or hair pigment. (See Figure 18.17.) Table 18.7 lists some other common genetic diseases and the type of metabolism or area affected.

Figure 18.17 This peacock with albinism does not produce the melanin for the bright colors of its feathers.

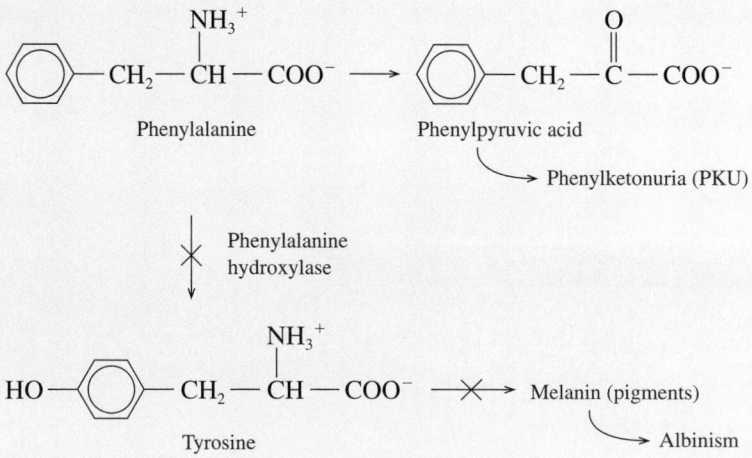

Table 18.7 Some Genetic Diseases

Genetic Disease	Result
Galactosemia	The transferase enzyme required for the metabolism of galactose-1-phosphate is absent. Accumulation of Gal-1-P leads to cataracts and mental retardation.
Cystic fibrosis	The most common inherited disease. Thick mucus secretions make breathing difficult and block pancreatic function.
Down syndrome	The leading cause of mental retardation, occurring in about 1 of every 800 live births. Mental and physical problems including heart and eye defects are the result of the formation of three chromosomes, usually chromosome 21, instead of a pair.
Familial hypercholesterolemia	A mutation of a gene on chromosome 19 results in high cholesterol levels that lead to early coronary heart disease in people 30–40 years old.
Muscular dystrophy (Duchenne)	One of 10 forms of MD. A mutation in the X chromosome results in the low or abnormal production of *dystrophin* by the X gene. This muscle-destroying disease appears at about age 5 with death by age 20 and occurs in about 1 of 10,000 males.
Huntington's disease (HD)	Appearing in middle age, HD affects the nervous system, leading to total physical impairment. It is the result of a mutation in a gene on chromosome 4, which can now be mapped to test people in families with HD.
Sickle-cell anemia	A defective hemoglobin from a mutation in a gene on chromosome 11 decreases the oxygen-carrying ability of red blood cells, which take on a sickled shape, causing anemia and plugged capillaries from red blood cell aggregation.
Hemophilia	One or more defective blood-clotting factors lead to poor coagulation, excessive bleeding, and internal hemorrhages.
Tay-Sachs disease	Hexosaminidase A is defective, causing an accumulation of gangliosides resulting in mental retardation, loss of motor control, and early death.

QUESTIONS & PROBLEMS

Genetic Mutations

18.53 What is a substitution mutation?

18.54 How does a substitution mutation in the genetic code for an enzyme affect the order of amino acids in that protein?

18.55 What is the effect of a mutation on the amino acids in the polypeptide?

18.56 How can a mutation decrease the activity of a protein?

18.57 How is protein synthesis affected if the normal base sequence —T—T—T— in the DNA template is changed to —T—T—C—?

18.58 How is protein synthesis affected if the normal base sequence —C—C—C— is changed to —A—C—C?

18.59 Consider the following portion of mRNA produced by the normal order of DNA nucleotides:

—ACA—UCA—CGG—GUA—

a. What is the amino acid order produced for normal DNA?
b. What is the amino acid order if a mutation changes UCA to ACA?
c. What is the amino acid order if a mutation changes CGG to GGG?
d. What happens to protein synthesis if a mutation changes UCA to UAA?

18.60 Consider the following portion of mRNA produced by the normal order of DNA nucleotides:

—CUU—AAA—CGA—GUU—

a. What is the amino acid order produced for normal DNA?
b. What is the amino acid order if a mutation changes CUU to CCU?
c. What is the amino acid order if a mutation changes CGA to AGA?
d. What happens to protein synthesis if a mutation changes AAA to UAA?

18.61 **a.** A base substitution changes a codon for an enzyme from GCC to GCA. Why is there no change in the amino acid order in the protein?
b. In sickle-cell anemia, a base substitution in hemoglobin replaces glutamine (a polar amino acid) with valine. Why does the replacement of one amino acid cause such a drastic change in biological function?

18.62 **a.** A base substitution for an enzyme replaces leucine (a nonpolar amino acid) with alanine. Why does this change in amino acids have little effect on the biological activity of the enzyme?
b. A base substitution replaces cytosine in the codon UCA with adenine. How would this substitution affect the amino acids in the protein?

18.9 Regulation of Protein Synthesis

Learning Goal

Explain cellular regulation of protein synthesis.

If a cell is to maximize available energy, it must operate efficiently. Therefore, materials are produced in a cell only as they are needed and degraded when they are not needed. This means that when a substance enters a cell, the enzyme required to catalyze its reactions is also produced via protein synthesis. We also need different proteins at different times in our lives. Different organs and tissues in the body have different functions, and therefore require the synthesis of different proteins. mRNA is not randomly synthesized by DNA; rather, it is synthesized in response to cellular needs for certain materials. That means that a gene can be turned on to synthesize a particular protein, and turned off again when the cell no longer requires that protein.

Operon Model

In 1961, the French biologists François Jacob and Jacques Monod proposed the *operon model* of cellular control to explain how genes in bacteria regulate protein synthesis. In this model, groups of nucleotides in DNA act to control the synthesis of mRNA and, therefore, the synthesis of protein. Each group of related proteins is regulated by an **operon,** a section of DNA containing a control site and structural genes. The **structural genes** are the DNA sections that code for proteins. The **promoter** and the **operator** make up the **control site,** which is the section of DNA ahead of the structural genes. Preceding the control site is a regulatory gene. (See Figure 18.18.)

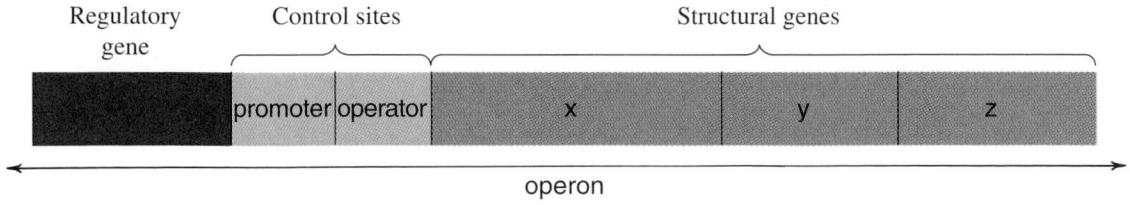

Figure 18.18 Diagram of an operon. The structural genes x, y, and z code for proteins x, y, and z.

In many cases, one operator controls several structural genes. When the operator is turned on, its structural genes produce mRNA for a specific group of proteins. Because mRNA is broken down rapidly, there must be a continuous formation (transcription) of mRNA as long as the protein is synthesized. When those proteins are no longer required by the cell, their structural genes are turned off by the operator gene and no mRNA is produced. Thus, resources are conserved by not producing enzymes when the substrates for those enzymes are not present.

Enzyme Induction

In **enzyme induction,** a substrate induces the synthesis of the enzymes for its metabolism. Suppose some bacteria are grown on glucose, and some on lactose. The bacteria grown on glucose will produce enzymes that metabolize glucose but not lactose. The bacteria grown on lactose will produce enzymes that metabolize lactose but not glucose. The enzymes synthesized were *induced* by the presence of a particular substrate.

In enzyme induction, the **regulatory gene** produces a protein called a **repressor** that binds to the operator. (See Figure 18.19.) As long as the repressor remains bound, RNA polymerase cannot read the structural genes and no mRNA is produced. For example, if there is no lactose, the structural genes that produce enzymes for the metabolism of lactose are blocked by a repressor. As long as the operator is blocked, the process of transcription is turned off. In fact, at any given time, most of the structural genes in a DNA are not operating.

Suppose now that some lactose enters the cell. The lactose binds to the repressor, altering the shape of the repressor and causing it to detach from the operator gene. With an open operator, the RNA polymerase transcribes the structural genes and protein is produced. In this case, lactose initiates (induces) the production of its own enzymes. As the enzymes metabolize the lactose, the concentration of lactose decreases. Eventually there is no longer any lactose to bind to the repressor. Then the repressor binds again to the operator gene and stops the synthesis of protein.

Enzyme Repression

In **enzyme repression,** it is the end products of an enzyme-catalyzed reaction, not the substrate, that turn off (repress) the synthesis of their enzymes. In the repression model, the operator gene is on and the structural genes are producing mRNA. (See Figure 18.20.) As long as the concentration of end product stays low, the structural genes con-

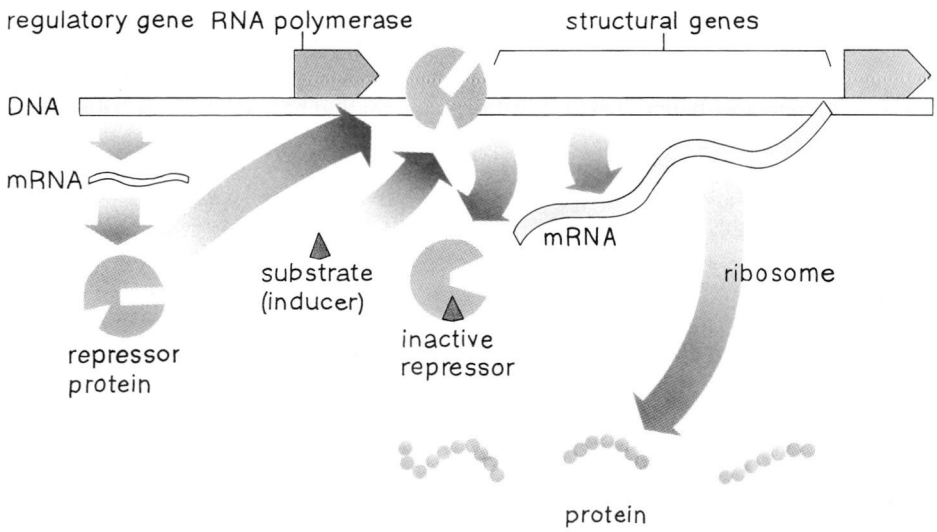

Figure 18.19 In enzyme induction, a repressor from the regulatory gene (left) blocks the operator and prevents the synthesis of mRNA by the structural genes. However, an inducer substrate (center) inactivates the repressor, the operator is turned on, and transcription occurs, producing the mRNA for protein synthesis.

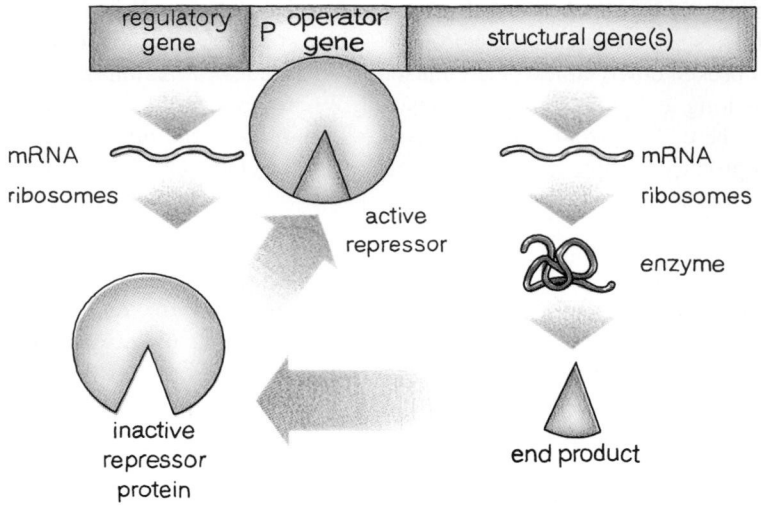

Figure 18.20 Enzyme repression. When the concentration of end products of an enzyme is elevated, the end-product molecules form an active repressor that blocks the operator and stops protein synthesis.

tinue to direct the synthesis of protein. However, when the amount of end product exceeds the needs of the cell, the end-product molecule combines with a protein produced by the regulator gene. The combination becomes an active *repressor* that binds to the operator gene and blocks the RNA polymerase. The operator is turned off and there is no further synthesis of mRNA. The end product has repressed the synthesis of the enzymes needed to produce that end product.

S·A·M·P·L·E· 18.11 ·P·R·O·B·L·E·M

Cellular Regulation

Indicate whether the following statements are *true* or *false:*

a. In enzyme induction, the operon is normally repressed and must be turned on by the substrate.

b. In enzyme repression, an active repressor forms when the concentration of the end product of an enzyme series is elevated.

Solution

a. True. A substrate binds with a repressor, causing the repressor to detach from the operator gene. With the operator gene turned on, the structural genes transcribe mRNA for the synthesis of protein.

b. True. When the concentration of the end product of an enzyme rises, there is no further need for that enzyme. An active repressor formed by the binding of end product with a protein turns off the operator and protein synthesis stops.

Study Check

When there is no sucrose present in the cell, why is there no sucrase enzyme?

HEALTH NOTE

Cancer

In an adult body, most cells do not continue to reproduce. When cells in the body begin to grow and multiply without control, they invade neighboring cells and appear as a tumor or growth (neoplasm). When tumors interfere with normal functions of the body, they are cancerous. If they are limited, they are benign. Cancer can be caused by chemical and environmental substances, by radiation, or by oncogenic viruses.

Some reports estimate that 70–80% of all human cancers are initiated by chemical and environmental substances. A carcinogen is any substance that increases the chance of inducing a tumor. Known carcinogens include dyes, cigarette smoke, and asbestos. More than 90% of all persons with lung cancer are smokers. A carcinogen causes cancer by reacting with molecules in a cell, probably DNA, and altering the growth of that cell. Some known carcinogens are listed in Table 18.8.

Radiant energy from sunlight or medical radiation is another type of environmental factor. Skin cancer has become one of the most prevalent forms of cancer. It appears that DNA damage in the exposed areas of the skin causes mutations. The cells lose their ability to control protein synthesis, and uncontrolled cell division leads to cancer. The incidence of malignant melanoma, one of the most serious skin cancers, has been rapidly increasing. Some possible factors for this increase may be the popularity of suntanning as well as the reduction of the ozone layer, which absorbs much of the harmful radiation from sunlight.

Oncogenic viruses cause cancer when cells are infected. Several viruses associated with human cancers are listed in Table 18.9. Some cancers such as retinoblastoma and breast cancer appear to occur more frequently in families. There is some indication that a missing or defective gene may be responsible. We discuss the difference between viruses containing DNA and those containing RNA in section 18.10.

Table 18.8 Some Chemical and Environmental Carcinogens

Carcinogen	Tumor Site
Asbestos	Lung, respiratory tract
Arsenic	Skin, lung
Cadmium	Prostate, kidneys
Chromium	Lung
Nickel	Lung, sinuses
Aflatoxin	Liver
Nitrites	Stomach
Aniline dyes	Bladder
Vinyl chloride	Liver

Table 18.9 Human Cancers Caused by Oncogenic Viruses

Virus	Disease
RNA viruses	
Human T-cell lymphotropic virus-type I (HTLV-I)	Leukemia
DNA viruses	
Epstein-Barr virus (EBV)	Burkitt's lymphoma (cancer of white blood B cells)
	Nasopharyngeal carcinoma
	Hodgkin's disease
Hepatitis B virus (HBV)	Liver cancer
Herpes simplex virus (type 2)	Cervical and uterine cancer
Papilloma virus	Cervical and colon cancer, genital warts

QUESTIONS & PROBLEMS

Regulation of Protein Synthesis

18.63 What is an operon?

18.64 Does the operon model control protein synthesis at transcription or translation? Why?

18.65 How is protein synthesis repressed in enzyme induction?

18.66 How is protein synthesis repressed in enzyme repression?

18.67 A bacterium grown in glucose is switched to lactose. Soon enzymes are produced that metabolize lactose. Why?

18.68 In enzyme repression, how is the active repressor formed?

Study Pack

18.10 Viruses and AIDS

Learning Goal

Describe the methods by which a virus infects a cell.

18.10

Viruses are small particles of 3 to 200 genes that cannot replicate without a host cell. A typical virus contains a nucleic acid, DNA or RNA but not both, inside a protein coat. It does not have the necessary material such as nucleotides and enzymes to make proteins and grow. The only way a virus can replicate is to invade a host cell and take over the materials necessary for protein synthesis and growth. Some infections caused by viruses invading human cells are listed in Table 18.10. There are also viruses that attack bacteria, plants, and animals.

A viral infection begins when an enzyme in the protein coat makes a hole in the host cell, allowing the nucleic acids to enter and mix with the materials in the host cell. (See Figure 18.21.) If the virus contains DNA, the host cell begins to replicate the viral DNA in the same way it would replicate its normal DNA. Viral DNA produces viral RNA, which proceeds to make the proteins for more protein coats. So many virus particles are synthesized that the cell bursts and releases new viruses to infect more cells.

A virus that contains RNA as its genetic material is a retrovirus. Once inside the host cell, it must first make viral DNA. In a reverse synthesis catalyzed by an enzyme called reverse transcriptase, the viral RNA is used as a template for the construction of viral DNA using the nucleotides and enzymes present in the host cell. (See Figure 18.22.) The viral DNA joins the DNA in the nucleus.

Table 18.10 Some Diseases Caused by Viral Infection

Disease	Virus
Common cold	Coronavirus (over 100 types)
Influenza	Orthomyxovirus
Warts	Papovavirus
Herpes	Herpesvirus
Leukemia, cancers, AIDS	Retrovirus
Hepatitis	Hepatitis A virus (HAV), hepatitis B virus (HBV)
Mumps	Paramyxovirus
Epstein-Barr	Epstein-Barr virus (EBV)

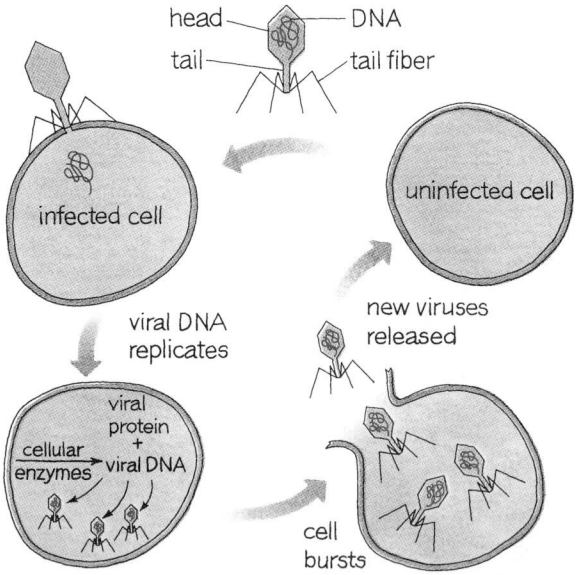

Figure 18.21 Action of a virus.

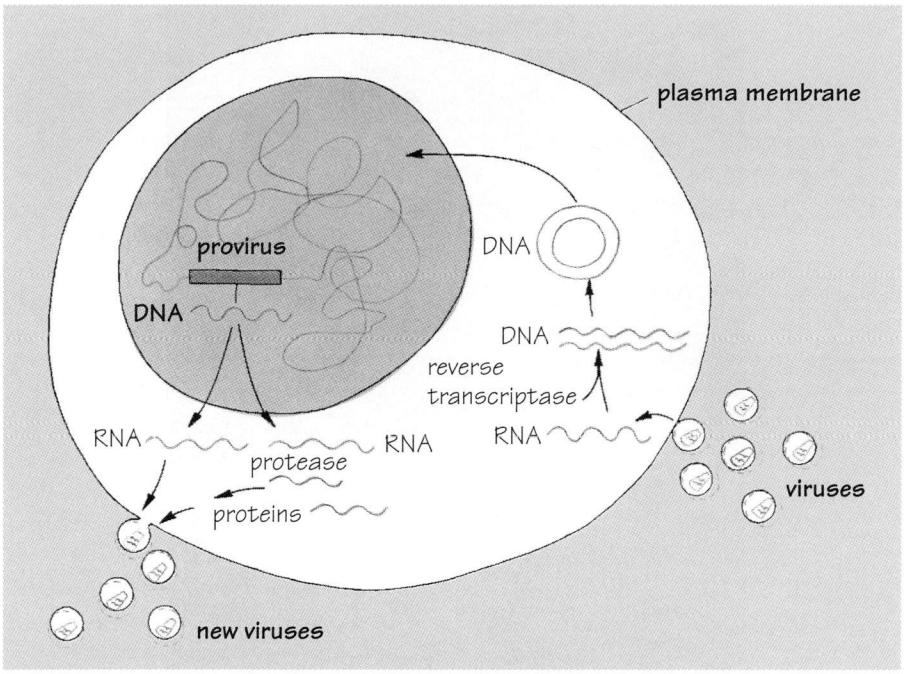

Figure 18.22 After a virus binds to the host cell, viral RNA is injected. The viral RNA is copied into a DNA provirus by reverse transcription. The provirus joins the DNA in the nucleus. When the cell reproduces, the provirus DNA duplicates the viral RNA needed to produce more virus particles.

Vaccines are inactive forms of viruses that boost the immune response by causing the body to produce antibodies to the virus. Several childhood diseases such as polio, mumps, chicken pox, and measles can be prevented through the use of vaccines.

AIDS

In the early 1980s, a disease called AIDS (acquired immune deficiency syndrome) began to claim an alarming number of lives. An HIV-1 virus (human immunodeficiency virus type 1) is now known to be the AIDS-causing agent. (See Figure 18.23.) HIV is a retrovirus that infects and destroys T4 lymphocyte cells, which are involved in the immune response. After the HIV-1 virus binds to receptors on the surface of a T4 cell, the virus injects viral RNA into the host cell. As a retrovirus, the genes of the viral RNA direct the formation of viral DNA. The gradual depletion of T4 cells reduces the ability of the immune system to destroy harmful organisms. The AIDS syndrome is characterized by opportunistic infections such as *Pneumocystis carinii,* which causes pneumonia, and *Kaposi's sarcoma*, a skin cancer.

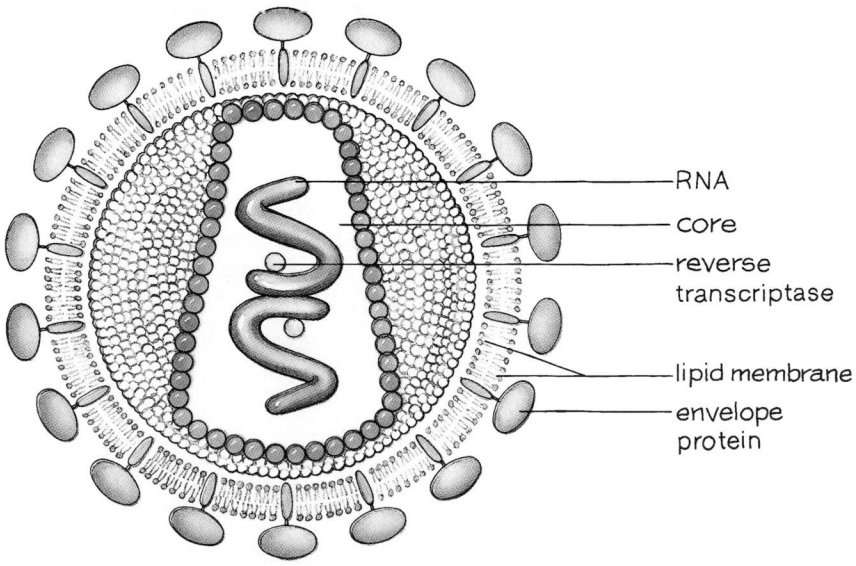

Figure 18.23 A diagram of HIV-1.

Figure 18.24 A micrograph of crystallites of the anti-AIDS drug azidothymidine (AZT). In reverse transcription, AZT causes a premature termination of the DNA chain, which prevents the DNA from combining with the host DNA.

18.10

Treatment for AIDS is based on attacking the HIV-1 at different points in its life cycle including reverse transcription and protein synthesis. Nucleoside analogs mimic the structures of the nucleosides used for DNA synthesis. For example, the drug AZT (azidothymine) is similar to thymine, and DDI (dideoxyinosine) is similar to guanosine. (See Figure 18.24.) When a nucleoside analog is incorporated into viral DNA, the lack of a hydroxyl group on the 3′ carbon in the sugar prevents the formation of the sugar–phosphate bonds, which prevents the replication of the virus.

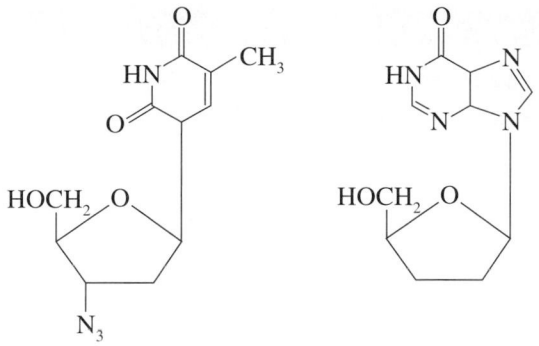

AZT (azidothymine, analog of thymine) DDI (dideoxyinosine, analog of guanosine)

The newest and most powerful anti-HIV drugs are the protease inhibitors such as saquinavir (Invirase), indinavir, and ritonavir, which modify the three-dimensional structure of the protease active site. The inhibition of protease prevents the synthesis of proteins needed to make more copies of the virus. Researchers are not yet certain how long protease inhibitors will be beneficial for a person with AIDS, but they are encouraged by the current studies.

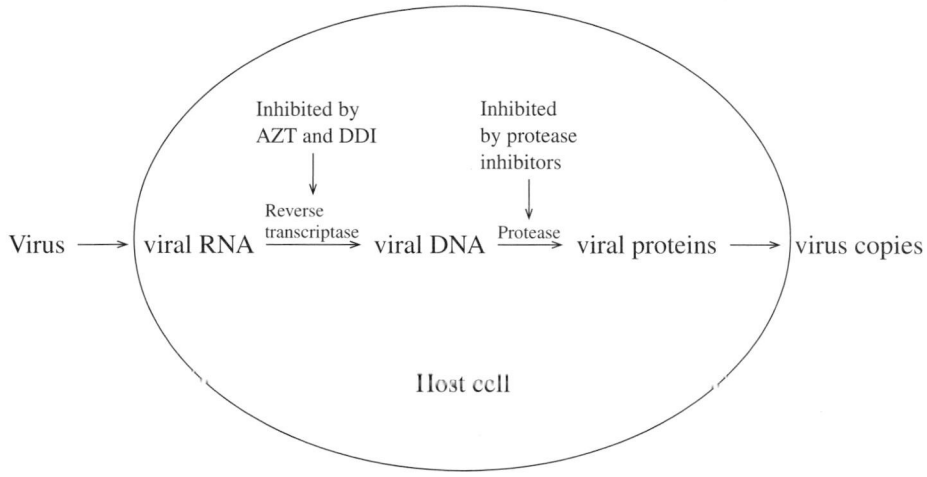

S·A·M·P·L·E· **18.12** ·P·R·O·B·L·E·M

Viruses

Why are viruses unable to replicate on their own?

Solution

Viruses only contain packets of DNA or RNA, but not the necessary replication machinery that includes enzymes and nucleosides.

Study Check

How do protease inhibitors affect the life cycle of the HIV-1 virus?

QUESTIONS & PROBLEMS

Viruses and AIDS

18.69 What type of genetic information is found in a virus?

18.70 Why do viruses need to invade a host cell?

18.71 A virus contains viral RNA.

a. Why would reverse transcription be used in the life cycle of this type of virus?

b. What is the name of this type of virus?

18.72 What is the purpose of a vaccine?

18.73 How do nucleoside analogs disrupt the life cycle of the HIV-1 virus?

18.74 How do protease inhibitors disrupt the life cycle of the HIV-1 virus?

Chapter Review

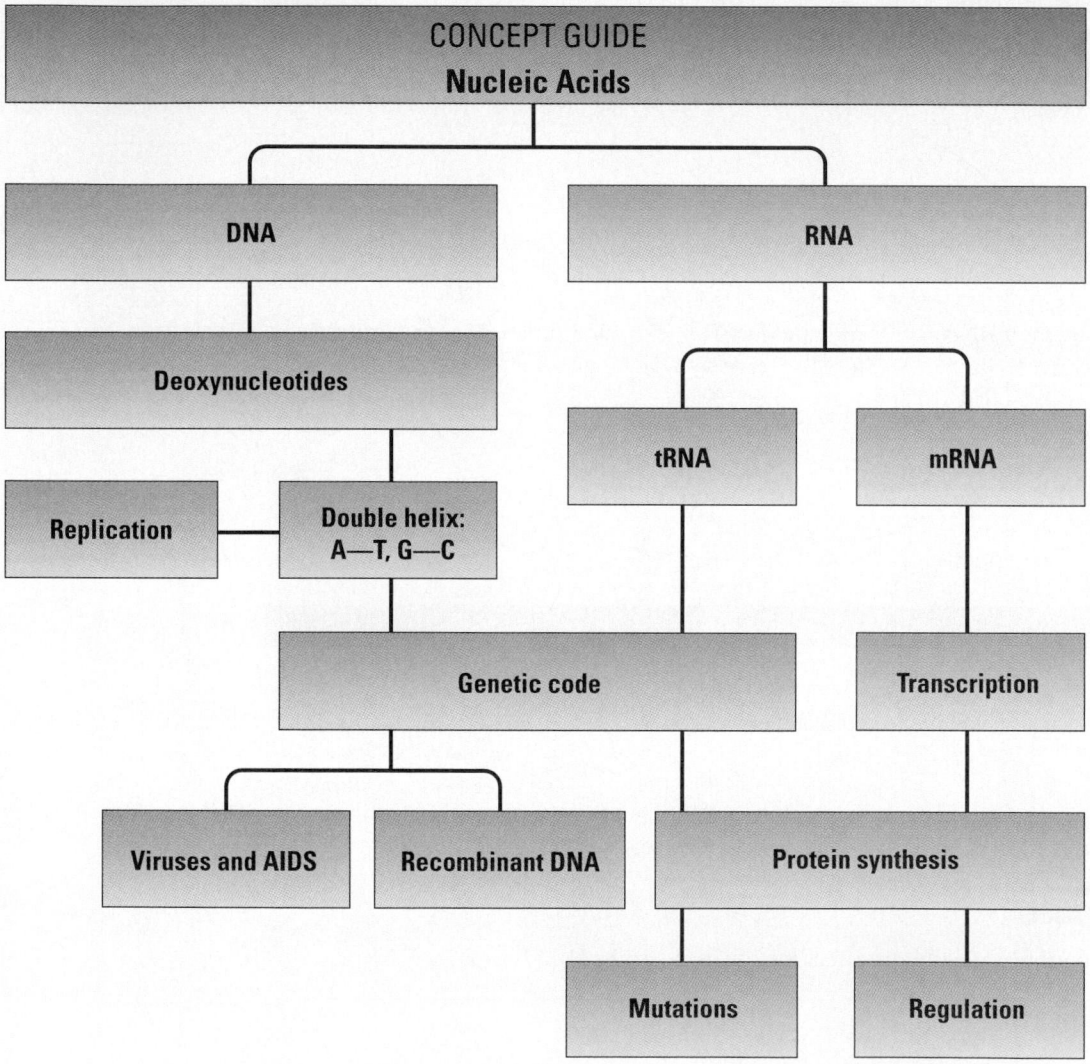

Chapter Summary

18.1 Nucleic Acids

Nucleic acids, deoxyribonucleic acid (DNA) and ribonucleic acid (RNA), are linear polymers of nucleotides. A nucleotide is composed of three parts: a nitrogenous base, a sugar, and a phosphate group. In DNA, the sugar is deoxyribose and the nitrogen-containing base can be adenine, thymine, guanosine, or cytosine. In RNA, the sugar is ribose and uracil replaces thymine.

18.2 Structures of Nucleic Acids

A DNA molecule consists of two strands of nucleotides that are wound around each other like a spiral staircase. The two strands are held together by hydrogen bonds between complementary base pairs, A with T, and G with C.

18.3 DNA Replication

During DNA replication, DNA polymerase makes new DNA strands along each of the original DNA strands that serve as templates. Complementary base pairing ensures the correct pairing of bases to give identical copies of the original DNA.

18.4 RNA and Transcription

The three types of RNA differ by function in the cell: ribosomal RNA makes up most of the structure of the ribosomes, messenger RNA carries genetic information from the DNA to the ribosomes, and transfer RNA places the correct amino acids in the protein. Transcription is the process by which RNA polymerase produces mRNA from one strand of DNA. The bases in the mRNA are complementary to the DNA, except U is paired with A in RNA.

18.5 The Genetic Code

The genetic code consists of a sequence of three bases (triplet) that specifies the order for the amino acids in a protein. There are 64 codons for the 20 amino acids, which means there are several codons for some amino acids. The codon AUG signals the start of transcription and codons UAG, UGA, and UAA signal it to stop.

18.6 Protein Synthesis

Proteins are synthesized at the ribosomes in a translation process that includes three steps: initiation, elongation, and termination. During translation, tRNAs bring the appropriate amino acids to the ribosome and peptide bonds form. When the polypeptide is released, it takes on its secondary and tertiary structures and becomes a functional protein in the cell.

18.7 Recombinant DNA

A recombinant DNA is prepared by inserting a DNA segment, a gene, into the DNA present in plasmids of *E. coli* bacteria. As the altered bacterial cells replicate, the protein expressed by the foreign DNA segment is produced. Recombinant DNA technology has been used in criminal investigations to obtain large quantities from a small amount of DNA by DNA polymerase reactions.

18.8 Genetic Mutations

A genetic mutation is a change of one or more bases in the DNA sequence that alters the structure and ability of the resulting protein to function properly. In a substitution, one base may be altered, and a frame shift mutation inserts or deletes a base and changes all the codons after the base change.

18.9 Regulation of Protein Synthesis

The production of mRNA occurs when certain proteins are needed in the cell. In enzyme induction, the appearance of a substrate in a cell removes a repressor, which allows RNA polymerase to produce mRNA at the structural genes. In enzyme repression, the end product of an enzyme-catalyzed sequence activates a repressor to inhibit the further production of mRNA.

18.10 Viruses and AIDS

Viruses containing DNA or RNA must invade host cells to use the machinery within the cell for the synthesis of more viruses. For a retrovirus containing RNA, a viral DNA is synthesized by reverse transcription using the nucleotides and enzymes in the host cell. In the treatment of AIDS, nucleoside analogs inhibit the reverse transcriptase of the HIV-1 virus, and protease inhibitors disrupt the catalytic activity of protease needed to produce proteins for the synthesis of more viruses.

Key Terms

anticodon The triplet of bases in the center loop of tRNA that is complementary to a codon on mRNA.

codon A sequence of three bases in mRNA that specifies a certain amino acid to be placed in a protein. A few codons signal the start or stop of transcription.

complementary base pairs In DNA, adenine is always paired with thymine (A—T or T—A), and guanine is always paired with cytosine (G—C or C—G). In forming RNA, adenine is paired with uracil (A—U).

control site A section of DNA composed of a promoter and operator that regulates protein synthesis.

DNA Deoxyribonucleic acid; the genetic material of all cells containing nucleotides with deoxyribose sugar, phosphate, and the four nitrogenous bases adenine, thymine, guanine, and cytosine.

double helix The helical shape of the double chain of DNA that is like a spiral staircase with a sugar–phosphate backbone on the outside and base pairs like stair steps on the inside.

enzyme induction A model of cellular regulation in which protein synthesis is induced by a substrate.

enzyme repression A model of cellular regulation in which protein synthesis is repressed by elevated levels of end product.

frame shift mutation A mutation that inserts or deletes a base in a DNA sequence.

genetic code The sequence of codons in mRNA that specifies the amino acid order for the synthesis of protein.

genetic disease A physical malformation or metabolic dysfunction caused by a mutation in the genetic code.

mRNA Messenger RNA; produced in the nucleus by DNA to carry the genetic information to the ribosomes for the construction of a protein.

mutation A change in the DNA base sequence that alters the formation of a protein in the cell.

nitrogen-containing base Nitrogen-containing compounds found in DNA and RNA: adenine (A), thymine (T), cytosine (C), guanine (G), and uracil (U).

nucleic acids Large molecules composed of nucleotides, found as a double helix in DNA, and as the single strands of RNA.

nucleoside The combination of a pentose sugar and a nitrogen-containing base.

nucleotides Building blocks of a nucleic acid consisting of a nitrogen-containing base, a pentose sugar (ribose or deoxyribose), and a phosphate group.

operator The section of DNA that precedes the structural genes and controls the synthesis of an mRNA.

operon A group of genes, including a control site and structural genes, whose transcription is controlled by the same regulatory gene.

polymerase chain reaction (PCR) A strand of DNA is copied many times by mixing it with DNA polymerase and a mixture of deoxyribonucleotides.

promoter A component of the control site on an operon that regulates DNA synthesis.

purines The double-ringed nitrogen-containing compounds of adenine and guanine found in nucleic acids.

pyrimidines The single-ring nitrogen-containing compounds of cytosine, uracil, and thymine found in nucleic acids.

recombinant DNA The splicing of DNA from different organisms to form new, synthetic DNA.

regulatory gene A gene in front of the control site that produces a repressor.

replication The process of duplicating DNA by pairing the bases on each parent strand with their complementary base.

repressor A protein that interacts with the operator gene in an operon to prevent the transcription of mRNA.

restriction enzyme An enzyme that cuts a large DNA strand into smaller fragments to isolate a gene or to remove a portion of the DNA in the plasmids of *E. coli.*

RNA Ribonucleic acid, a type of nucleic acid that is a single strand of nucleotides containing adenine, cytosine, guanine, and uracil.

rRNA Ribosomal RNA; the most prevalent type of RNA; a major component of the ribosomes.

structural genes The sections of DNA that code for the synthesis of proteins.

substitution A mutation that replaces one base in a DNA with a different base.

transcription The transfer of genetic information from DNA by the formation of mRNA.

translation The interpretation of the codons in mRNA as amino acids in a peptide.

translocation The shift of a ribosome along mRNA from one codon (three bases) to the next codon during translation.

tRNA Transfer RNA; an RNA that places a specific amino acid into a peptide chain at the ribosome. There is one or more tRNA for each of the 20 different amino acids.

virus Small particles containing DNA or RNA in a protein coat that require a host cell for replication.

Chapter Practice

18.75 Which amino acids have the fewest number of codons? Which have the highest number of codons?

18.76 A protein contains 35 amino acids. How many nucleotides would be found in the mRNA for this protein?

18.77 Leu-enkephalin is a brain pentapeptide with the amino acid sequence of Tyr-Gly-Gly-Phe-Leu. What is a possible mRNA for it?

18.78 Why does the cell regulate protein synthesis from DNA?

Answers

Answers to Study Checks

18.1 deoxycytidine monophosphate

18.2

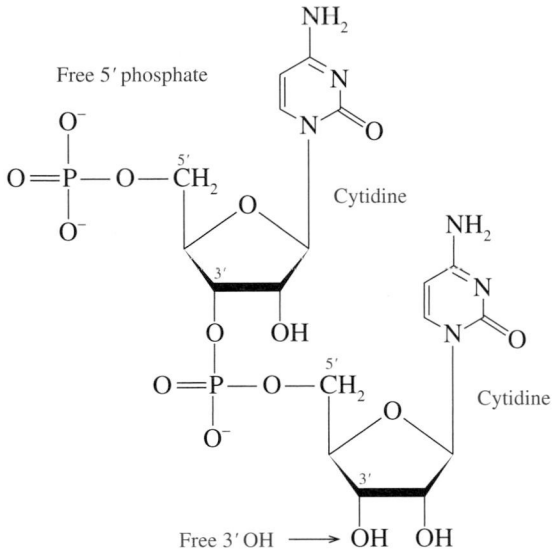

18.3 —C—C—A—A—T—T—G—G—

18.4 DNA polymerase forms a phosphodiester bond between the newly aligned nucleotides.

18.5 Each type of tRNA matches a specific codon to a specific amino acid.

18.6 CCCAAATTT

18.7 UGA is a stop codon that signals the termination of translation.

18.8 at UAG

18.9 Gene cloning is the process by which recombinant DNA technology inserts the DNA of a gene into the plasmid of *E. coli* bacteria that multiply rapidly to make many copies of the gene.

18.10 If the substitution of an amino acid in the polypeptide affects an interaction essential to functional structure on the binding of a substrate, the resulting protein could be less effective or nonfunctional.

18.11 With no sucrose present, the operator for sucrase is turned off. Without the mRNA, the enzyme is not synthesized.

18.12 A protease inhibitor modifies the three-dimensional structure of protease, which prevents the enzymes from synthesizing the proteins needed to produce more viruses.

Answers to Selected Questions and Problems

18.1 adenine (A), thymine (T), cytosine (C), guanine (G), and deoxyribose

18.3 deoxyadenosine monophosphate, deoxythymidine monophosphate, deoxycytidine monophosphate, and deoxyguanosine monophosphate

18.5 **a.** nucleoside **b.** nucleoside **c.** nucleoside **d.** nucleotide

18.7

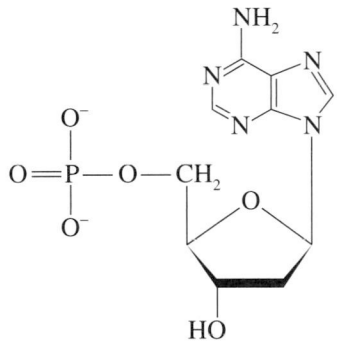

18.9 The phosphate group of each nucleotide connects to the hydroxyl group on carbon 3′ of the next nucleotide.

18.11 The sugar.

18.13

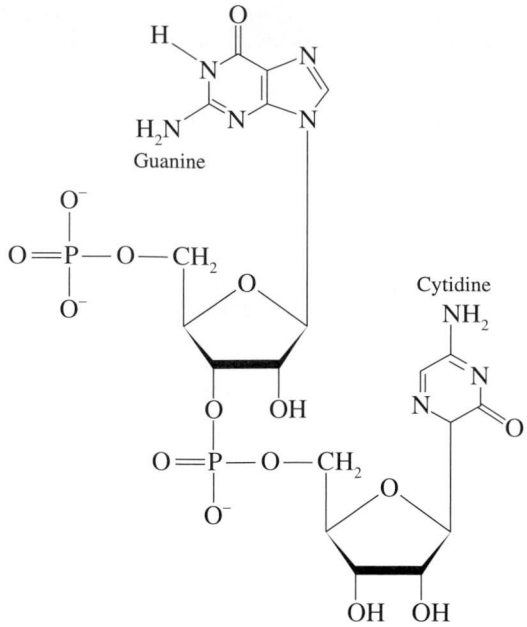

18.15 **a.** TTTTTT **b.** CCCCCC
c. TCAGGTCCA **d** GACATATGCAAT

18.17 The DNA strands separate and the DNA polymerase pairs each of the bases with its complementary base and produces two exact copies of the original DNA.

18.19 The pairs of bases along the DNA strands are separated and paired with new bases.

DNA copies

—A—G—G—T—C— ⟶ —A—G—G—T—C—
: : : : :
—T—C—C—A—G—

Separated strands
pick up complementary DNA

—T—C—C—A—G— ⟶ —A—G—G—T—C—
: : : : :
—T—C—C—A—G—

18.21 Ribosomal RNA, messenger RNA, and transfer RNA

18.23 Ribosomal RNA is found in the ribosomes, mRNA made in the nucleus goes to the ribosomes, and tRNA is found in the cytoplasm picking up amino acids to bring to the ribosomes.

18.25 In transcription, the sequence of nucleotides on a DNA template (one strand) is used to produce the base sequences of a messenger RNA.

18.27 —G—G—C—U—U—C—C—A—A—G—U—G—

18.29 A three-base sequence in mRNA that codes for a specific amino acid in a protein.

18.31 **a.** leucine **b.** serine
c. glycine **d.** arginine

18.33 When AUG is the first codon, it signals the start of protein synthesis. Thereafter, AUG codes for methionine.

18.35 A codon is a base triplet in the mRNA template. An anticodon is the complementary triplet on a tRNA for a specific amino acid.

18.37 Initiation, elongation, and termination.

18.39 **a.** —Lys—Lys—Lys—
b. —Phe—Pro—Phe—Pro—
c. —Tyr—Gly—Arg—Cys—

18.41 The new amino acid is joined by a peptide bond to the peptide chain. The ribosome moves to the next codon, which attaches to a tRNA carrying the next amino acid.

18.43 **a.** —C—G—A—A—A—A—G—U—U—U—U—U—
b. GCU, UUU, CAA, AAA
c. Using codons in mRNA: Arg—Lys—Val—Phe

18.45 *E. coli* bacterial cells contain several small circular plasmids of DNA that can be isolated easily. After the recombinant DNA is formed, *E. coli* multiply rapidly, producing many copies of the recombinant DNA in a relatively short time.

18.47 *E. coli* are soaked in a detergent solution that dissolves the cell membrane and releases the cell contents including the plasmids, which are collected.

18.49 When a gene has been obtained using restriction enzymes, it is mixed with the plasmids that have been opened by the same enzymes. When mixed together in a fresh *E. coli* culture, the sticky ends of the DNA fragments bond with the sticky ends of the plasmid DNA to form a recombinant DNA.

18.51 In DNA fingerprinting, restriction enzymes cut a sample DNA into fragments, which are sorted by size by gel electrophoresis. After tagging the DNA fragments with a radioactive isotope, a piece of x-ray film placed over the gel is exposed by the radioactivity to give a pattern of dark and light bands known as a DNA fingerprint.

18.53 A base in DNA is replaced by a different base.

18.55 If the resulting codon still codes for the same amino acid, there is no effect. If the new codon codes for a different amino acid, there is a change in the order of amino acids in the polypeptide.

18.57 The normal triplet TTT forms a codon AAA, which codes for lysine. The mutation TTC forms a codon AAG, which also codes for lysine. There is no effect on the amino acid sequence.

18.59 **a.** —Thr—Ser—Arg—Val—
b. —Thr—Thr—Arg—Val—
c. —Thr—Ser—Gly—Val—
d. —Thr—STOP Protein synthesis would terminate early. If this occurs early in the formation of the polypeptide, the resulting protein will probably be nonfunctional.

18.61 **a.** GCC and GCA both code for alanine.
b. A vital ionic cross-link in the tertiary structure of hemoglobin cannot be formed when the polar glutamine is replaced by valine, which is nonpolar. The resulting hemoglobin is malformed and less capable of carrying oxygen.

18.63 An operon is a section of DNA that regulates the synthesis of one or more proteins.

18.65 In enzyme induction, protein synthesis is repressed when the regulatory gene produces a repressor protein that binds to the operator and inhibits the ability of the RNA polymerase to produce mRNA.

18.67 Lactose induced the synthesis of the enzymes for its metabolism.

18.69 DNA or RNA, but not both.

18.71 **a.** A viral RNA is used to synthesize a viral DNA to produce the proteins for the protein coat, which allows the virus to replicate and leave the cell.
b. retrovirus

18.73 Nucleoside analogs such as AZT and DDI are similar to the nucleosides required to make viral DNA in reverse transcription. However, they interfere with the ability of the DNA to form and thereby disrupt the life cycle of the HIV-1 virus.

18.75 Tryptophan (Trp) and methionine (Met) each has only a single codon. Leucine (Leu), serine (Ser), and arginine (Arg) each have six codons.

18.77 An mRNA could be:

UAU—GGU—GGU—UUU—CUU

Credits

ILLUSTRATIONS

2.10, 9.4: Adapted from Michael Matta, Dennis Staley, and Anthony Wilbraham, *Addison-Wesley Chemistry* (Menlo Park, CA: Addison-Wesley, 1990) © 1990 Addison-Wesley. Reprinted by permission.

2.17, 6.1: Adapted from *Scott Foresman Science: Discover the Wonder*, Grade 5 (Glenview, IL: Scott Foresman, 1993) © 1993, 1994 Scott Foresman and Company. Reprinted by permission of Addison Wesley Longman, Inc.

15.10, 16.16, 17.2, 17.18, 17.19, 18.6, 18.9, 18.11, 18.14: Adapted from Gerard Tortora and Sandra Grabowski, *Principles of Anatomy and Physiology,* Eighth Edition, Figs. 2.10, 2.15, 2.16, 2.19, 3.22, 3.23, 3.25, 25.7, 25.9 (New York: HarperCollins, 1996) © 1996 by Biological Sciences Textbooks, Inc. and Sandra Reynolds Grabowski. Reprinted by permission of Addison Wesley Longman, Inc.

15.14, 18.12, 18.13, 18.17, 18.23, 18.24: Adapted from Michael C. Mix, Paul Farber, and Keith I. King, *Biology The Network of Life* (New York: HarperCollins, 1992) © 1992 by Michael C. Mix, Paul Farber, and Keith I. King. Reprinted by permission of Addison Wesley Longman, Inc.

16.09, 16.10, 16.12: © Irving Geis.

PHOTOGRAPHS

Chapter 1 Opener: © 1997 PhotoDisc/Janis Christie. **1.4:** © SIU/Visuals Unlimited. **1.5:** © John Wilson White/Addison Wesley Longman. **1.9:** © 1997 PhotoDisc. **1.16:** © Tom Pantages. **1.17:** David R. Frazier Photolibrary. **Page 20:** © John Wilson White/Addison Wesley Longman.

Chapter 2 Opener: © 1997 PhotoDisc/Hisham Ibrahim. **2.1 (top left):** © 1997 PhotoDisc/David Buffington. **2.1 (top right):** © 1997 PhotoDisc/Siede Preis. **2.1 (bottom left):** © 1997 PhotoDisc/Arthur S. Aubry. **2.1 (bottom right):** © 1997 PhotoDisc/John A. Rizzo. **2.2:** © Daniel Aubry/The Stock Market. **2.4 (all):** © Andy Washnik/Addison Wesley Longman. **2.5 (all):** © Andy Washnik/Addison Wesley Longman. **2.6:** © 1997 Photo Disc. **2.8:** Courtesy of Lawrence Berkeley Laboratory, University of California. **2.16:** Courtesy of the National Museum of American History/Smithsonian Institution. **2.18:** Courtesy of the Anderson Cancer Center Univeristy of Texas, Austin.

Chapter 3 Opener: © 1997 PhotoDisc/National Cancer Institute. **3.2:** © John Coletti/Stock Boston. **3.8:** © David Parker/SPL/Photo Researchers, Inc. **3.10a:** © John Wilson White/Addison Wesley Longman. **3.11:** © SIU/Visuals Unlimited. **3.12:** © 1997 PhotoDisc/Lawrence Berkeley Laboratory. **3.13:** © 1997 PhotoDisc/Lawrence Berkeley National Library. **3.14:** © GJLP/CNRI/Phototake, NYC. **3.19:** © Jessica Ehler/Bruce Coleman.

Chapter 4 Opener: © 1997 PhotoDisc/Siede Preis. **4.2:** © Robert E. Lyons/Visuals Unlimited. **4.14 (all):** © Professor P. Motta, Department of Anatomy, University "La Sapienza," Rome/SPL/Photo Researchers, Inc.

Chapter 5 Opener: © 1997 PhotoDisc/Arthur S. Aubry. **5.4, 5.6, 5.7:** © Andy Washnik/Addison Wesley Longman. **5.5:** © Tom Pantages. **5.9:** © 1997 PhotoDisc/K. Knudson/Photolink.

Chapter 6 Opener: © 1997 PhotoDisc/L. Hobbs/Photolink. **6.4:** © Custom Medical Stock Photo. **6.5:** Courtesy of United States Department of Agriculture. **6.6 (top left):** William McCoy/Rainbow. **6.6 (top right):** Runk/Schoenberger/Grant Heilman Photography. **6.6 (bottom):** © 1997 PhotoDisc. **6.8:** © 1997 PhotoDisc/Spike Mafford. **6.14:** © John Wilson White/Addison Wesley Longman. **6.16 (top):** © 1997 PhotoDisc/Jack Star/Photolink. **6.16 (bottom):** © 1997 PhotoDisc/Sean Thompson.

Chapter 7 Opener: © 1997 PhotoDisc/T. O'Keefe/Photolink. **7.2:** © 1997 PhotoDisc/Don Farrall. **7.6:** © 1997 PhotoDisc/Larry Brownstein. **7.15:** © SIU/Visuals Unlimited.

Chapter 8 Opener: © 1997 PhotoDisc/Mitch Hrdlicka. **8.3:** © George I. Bernard/Animals Animals. **8.4:** © Tom Pantages. **8.5, 8.13 (all):** © Andy Washnik/Addison Wesley Longman. **8.14:** © CNRI/SPL/Photo Researchers, Inc. **8.17:** © Matt Meadows/Peter Arnold, Inc. **8.23:** © SIU/Bruce Coleman, Inc.

Chapter 9 Opener: © 1997 PhotoDisc/S. Pearce/Photolink. **9.8a:** © NYC Parks Photo Archive/Fundamental Photographs. **9.8b:** © Kristen Brochmann/Fundamental Photographs. **9.15:** © Andy Washnik/Addison Wesley Longman.

Chapter 10 Opener: © 1997 PhotoDisc/Doug Menuez. **10.5:** © 1997 PhotoDisc/K. Knudson/Photolink. **10.6:** © Michael Newman/Photo Edit. **10.11:** © Dianora Niccolini/Medical Images, Inc.

Chapter 11 Opener: © 1997 PhotoDisc/Lawrence Lawry. **11.1:** © Stephen Kline/Bruce Coleman, Inc. **11.2:** © 1997 Photo Disc/F. Schussler/Photolink. **11.3:** Courtesy of NASA. **11.6:** © 1997 PhotoDisc/Tim Hall. **11.7:** © 1997 PhotoDisc/Don Tremain. **11.10:** © Runk/Schoenberger/Grant Heilman Photography. **11.12 (all):** © Tom Pantages. **11.13 (left):** © SIU/Visuals Unlimited. **11.13 (right):** © Ellen Harmon/Medical Images, Inc. **11.14, 11.15:** © John Wilson White/Addison Wesley Longman.

Chapter 12 Opener: © 1997 PhotoDisc. **12.7:** © Tom Pantages. **12.8:** © Al Assid/The Stock Market. **12.11:** © Karl Weidmann/Photo Researchers, Inc. **12.12:** © Jim Pickerell/Tony Stone Images.

Chapter 13 Opener: © 1997 PhotoDisc. **13.5a:** © John Wilson White/Addison Wesley Longman. **13.7:** © Custom Medical Stock. **13.8:** © John D. Cunningham/Visuals Unlimited. **13.10:** © 1997 PhotoDisc. **13.13:** © 1998 Courtesy of Karen C. Timberlake. **13.14:** © 1997 PhotoDisc/Jules Frazier.

Chapter 14 Opener: © 1997 PhotoDisc/Greg Kuchik. **14.1:** © 1997 PhotoDisc/Steve Cole. **14.3, 14.5, 14.8:** © John Wilson White/Addison Wesley Longman. **14.10:** © Ed Drews/Photo Researchers, Inc.

Chapter 15 Opener: © 1997 PhotoDisc/Sami Sarkis. **15.3:** © 1997 PhotoDisc. **15.4:** © 1997 PhotoDisc/Don Farrall. **15.7:** © 1997 PhotoDisc/Steve Cole. **15.8:** © Matt Meadows/Peter Arnold, Inc. **15.11:** © C. Raines/Visuals Unlimited. **15.12 (all):** Courtesy of the National Heart, Lung, and Blood Institute/National Institutes of Health.

Chapter 16 Opener: © 1997 PhotoDisc. **16.11:** © Lewin/Royal Free Hospital/Photo Researchers, Inc.

Chapter 17 Opener: © 1997 PhotoDisc/Sean Thompson. **17.22:** © Fred Hossler/Visuals Unlimited.

Chapter 18 Opener: © 1997 PhotoDisc/M. Freeman/Photolink. **18.8:** © CNRI/SPL/Photo Researchers, Inc. **18.16:** © James King-Holmes/Photo Researchers, Inc. **18.18:** © Jessica Ehlers/Bruce Coleman, Inc. **18.25:** © 1992 Michael W. Davidson, Institute of Molecular Biophysics, The Florida State University, Tallahassee, Florida. All rights reserved.

Glossary/Index

A

ABO blood group, 503
Absolute zero The lowest possible temperature that matter can attain. 207, 239
Accuracy How close a measurement is to the true value. 9, 41
Accutane, 71
Acetaldehyde, 452, 461, 519
Acetals The product of the addition of a second molecule of alcohol to a hemiacetal. 467–68, 472, 499
Acetaminophen, 391, 545, 563
Acetic acid, 341, 461, 516–17, 519
Acetoacetate, 668
Acetoacetyl CoA, 671
Acetone, 437, 452, 668
Acetylcholinesterase, 628
Acetyl coenzyme (CoA) A two-carbon acetyl unit attached to coenzyme A. A product of glucose, fatty acid, and protein metabolism, acetyl CoA is oxidized to yield energy via the citric acid cycle or used as a substrate to synthesize fatty acids or glucose. 645, 652–55, 663, 665, 668, 671, 673, 676
  from fatty acid oxidation, 665
  from pyruvate, 649–50
Acetylene, 409, 411
N-Acetylgalactosamine, 503
N-Acetylglucosamine, 503, 509
Acetylsalicylic acid, 528
Achiral object, 484
Acidic amino acids, 596–98
Acidic solution, 332–35, 337
Acidosis A physiological condition in which the blood pH is lower than 7.35. 352, 358, 668
  metabolic, 352–53
  respiratory, 352–53
Acid phosphatase (ACP), 620
Acid rain, 172, 245, 339
Acids A substance that dissolves in water and produces hydrogen ions (H^+), according to the Arrhenius theory. All acids are proton donors, according to the Brønsted–Lowry theory. 327–30, 358
  acid-base neutralization, 347–48
  acid-base titration, 354
  amino. *See* Amino acids
  Arrhenius, 327
  Brønsted, 329–30
  carboxylic. *See* Carboxylic acids
  characteristics of, 328
  denaturation of proteins, 613
  fatty. *See* Fatty acids
  naming of, 327–28
  pH scale, 334–38
  strong, 341, 358
  weak, 341, 358, 521–22
Aconitase, 654
Acquired immune deficiency syndrome (AIDS), 712–15
Actin, 595, 639–40
Activation energy The energy needed upon collision to break apart the bonds of the reacting molecules. 228–31, 239, 615
Active site The region of an enzyme that binds and catalyzes a reaction of substrate. 618–19, 626–27, 631
Acupuncture, 605
Addition (mathematics), significant figures in, 14–15
Addition mutation, 705
Addition reactions A reaction in which atoms or groups of atoms bond to a double bond. Addition reactions include the addition of hydrogen (hydrogenation), halogens (halogenation), hydrogen halides (hydrohalogenation), or water (hydration). 418–23, 430
Adenine, 682–83, 686–88
Adenosine diphosphate (ADP) A compound of adenine, a ribose sugar, and two phosphate groups, it is formed by the hydrolysis of ATP. 638–39, 673
Adenosine triphosphate (ATP) A high-energy compound that stores energy in the cells and consists of adenine, a ribose sugar, and three phosphate groups. 637, 676
  energy from glucose, 661–64
  energy from hydrolysis of, 638–39
  metabolism and, 637–39
  production of
    from acetyl CoA, 663
    in citric acid cycle, 652–63
    in complete oxidation of glucose, 662–63
    in electron transport chain, 656–59
    in fatty acid oxidation, 665–66
    in glycolysis, 646–48, 661
    in pyruvate metabolism, 661
  structure of, 638
  use in muscle contraction, 639–40
Adipose tissue, 558
Adrenaline. *See* Epinephrine
Adrenocortical hormones, 584–86
Aerobic conditions An oxygen-containing environment in the cells. 649, 676
Aerosols, 307
Aflatoxin, 711
Agitation, denaturation of proteins by, 613
Air, composition of, 270–71
Air pollution, 245
Alanine, 597, 670–71
Alar, 547
Albinism, 706
Alcoholic beverages, 437, 461–62, 499, 651
Alcohols Organic compounds that contain the hydroxyl (—OH) functional group, attached to a carbon chain, R—OH. 403, 437–43, 472
  classification of, 459
  combustion of, 448–49
  dehydration of, 449–51, 472
  esterification, 527–28
  formation of ethers, 450–51
  in health and medicine, 441
  IUPAC system for naming, 437–38
  oxidation of, 458, 460–62, 519
  reactions of, 448–51, 466–67
  solubility in water, 438–39
Aldehydes An organic compound with a carbonyl functional group and a hydrogen on the end carbon of the chain. 403, 437, 452–55, 472
  in health and medicine, 455–56
  IUPAC system for naming, 453–54
  oxidation of, 458, 460, 462–63, 519
  oxidation of primary alcohols to, 460
  reaction with alcohols, 466–68
  reduction of, 464
  solubility in water, 453
Aldolase, 646
Aldose Monosaccharides that contain an aldehyde group. 481, 512
Aldosterone, 586
Aldrin, 405
Alkali metals Elements of Group 1 except hydrogen; these are soft, shiny metals with one outer shell electron. 53, 76
Alkaline earth metals Group 2 elements, which have 2 electrons in their outer shells. 53, 76
Alkaline phosphatase (ALP), 620
Alkaloids Amines having physiological activity that are produced in plants. 540–42, 552
Alkalosis A physiological condition in which the blood pH is higher than 7.45. 352, 358
  metabolic, 353
  respiratory, 352–53
Alkanes A hydrocarbon that has only single carbon–carbon bonds. 364, 396, 403
  branched, 376–79, 396
  combustion of, 373–74
  IUPAC system for naming, 370–72, 377–79
  properties of, 373–74
  structural formulas for, 382–85
  structural isomers of, 383–85
  unbranched, 371–72
Alkenes An unsaturated hydrocarbon containing a carbon–carbon double bond. 364, 403, 409–11, 430
  addition reactions of, 418–23
  from dehydration of alcohol, 449
  fragrant, 412
  geometric isomers of, 413–14
  halogenation of, 418, 420
  hydration of, 418, 423
  hydrogenation of, 418
  hydrohalogenation of, 418, 421

(Alkenes, continued)
IUPAC system for naming, 410
polymers of, 425–27
structural isomers of, 413–14
Alkoxy group, 446
Alkyl groups A group derived from an alkane by removing one hydrogen atom. The ending of the alkane name is changed to *yl* to name it as an alkyl branch. 396
in branched alkanes, 376–77
cycloalkanes with, 387–88
Alkynes An unsaturated hydrocarbon containing a carbon–carbon triple bond. 403, 409–11, 430
α-amino acids, 596
Alpha emitters, 87–88, 91
Alpha helix A secondary level of protein structure, in which hydrogen bonds connect the NH of one peptide bond with the C═O of a peptide bond later in the chain to form a coiled or corkscrew structure. 603–6, 631
Alpha hydroxy acids, 518–19
Alpha particles A nuclear particle identical to a helium ($^4_2$He, or α) nucleus (2 protons and 2 neutrons). 59, 81–85, 87–88, 113
Altitude, atmospheric pressure and, 249–50
Aluminum, 48, 69, 73, 119, 124, 131
Aluminum carbonate, 150
Aluminum hydroxide, 349
Aluminum oxide, 131
Aluminum phosphate, 151
Aluminum sulfate, 151
Amidation The formation of an amide from a carboxylic acid and ammonia or an amine. 538–39, 552
Amides Organic compounds containing the carbonyl group attached to an amino group or a substituted nitrogen atom. 403, 538–39, 543–44, 552
acid hydrolysis of, 546
base hydrolysis of, 546–47
in health and medicine, 544–45
hydrolysis in nature, 547
IUPAC system for naming, 543–44
reactions of, 546–47
Aminase, 616
Amines Organic compounds containing a nitrogen atom attached to one, two, or three hydrocarbon groups. 403, 533–35, 552
amidation, 538–39
aromatic, 535
classification of, 533
in health and medicine, 536
ionization of, 538
IUPAC system for naming, 534–35
neutralization of, 538
in plants, 540–42
reactions of, 537–42
solubility of, 533–34
Amino acids The building block of proteins, consisting of an amino group, a carboxylic acid group, and a unique side group attached to the alpha carbon. 487, 596–600, 631
acidic, 596–98
basic, 597–98
classification of, 596–98
digestion of proteins, 641–42
essential, 599, 632
genetic code for, 695–96
ionization of, 599–600
metabolic pathways for, 670–72
nonpolar, 596–97, 632
oxidative deamination of, 671–72
polar, 596–97, 606, 632
structure of, 596–97
transamination, 670–71
Amino group, 534, 596, 601
naming substituents on, 535
Ammonia, 138, 329–30, 342, 537
Ammonium ion, 148, 537–38, 672
Ammonium nitrate, 229
Amoxicillin, 628
Amphetamines, 536
Ampicillin, 628
Amylase An enzyme that hydrolyzes the glycosidic bonds in polysaccharides. 494, 505, 620, 624, 640–41, 676
Amylopectin A branched-chain polymer of starch composed of glucose units joined by α-1,4- and α-1,6-glycosidic bonds. 505–7, 512
Amylose An unbranched polymer of starch composed of glucose units joined by α-1,4-glycosidic bonds. 505–6, 512
Anabolic reactions Metabolic reactions that are energy requiring. 637, 639, 676
Anabolic steroids, 586
Anaerobic conditions The absence of oxygen in the cells. 646, 649–51, 676
Androsterone, 584
Anemia
pernicious, 622
sickle-cell, 610, 707
Anesthetics, 404, 541
ethers, 447–48
haloalkanes, 405
Aniline, 535
Aniline dyes, 711
Animal starch, 505
Anions A negatively charged ion such as Cl^-, O^{2-}, or SO_4^{2-}. 123–24, 156
Anomers The isomers that occur when the formation of a hemiacetal or hemiketal of monosaccharides produces a new hydroxyl group on carbon 1 (or carbon 2). In the a anomer, the OH is drawn downward; in the β isomer the OH is up. 494, 512
Antabuse, 461
Antacids, 161, 308, 342, 349
Anthracene, 392
Antibiotics An inhibitor produced by a bacteria, mold, or yeast that is toxic to other bacteria. 628, 631, 700
Antibodies, 606
Anticoagulant, 509
Anticodon The triplet of bases in the center loop of tRNA that is complementary to a codon on mRNA. 697–98, 718
Antidote, 613
Antifreeze, 441, 462
Antihistamine, 536
Antimetabolites, 628
Antioxidants, 442
Antiseptics, 441–42
Apoenzyme The protein portion that along with a cofactor provides a functional enzyme. 621, 631
Apple juice, 547
Arachidonic acid, 561–63
Arginine, 598–99, 670–71
Argon, 48, 54, 68–69, 73, 121
Aroma. *See* Fragrance
Aromatic amines, 535
Aromatic carboxylic acid, 518
Aromatic compounds A compound that contains the ring structure of benzene. 364, 396, 403
in health and medicine, 391
Aromatic hydrocarbons, 391–92
naming of, 392
polycyclic, 392
Arrhenius concept, 327–28
Arsenic, 711
Arteriosclerosis, 558
Artificial kidney, 314–15
Artificial sweetener, 502
Asbestos, 711
Ascorbic acid, 621–22
Asparagine, 597, 671
Aspartame, 502
Aspartate, 598, 670–71
Asphalt, 371
Aspirin, 391, 516, 528, 563
Aspirin substitutes, 544–45
Astringent, 441
Atherosclerosis, 562, 582–83
Athletes, use of anabolic steroids, 586
Atmosphere, 245
Atmosphere (atm) The pressure exerted by a column of mercury 760 mm high. 249–50, 277
Atmospheric pressure The pressure exerted by the atmosphere. 248, 251–52, 277
altitude and, 249–50
Atomic energy, 108
Atomic mass The average mass of all the naturally occurring isotopes of an element. 64–65, 76
average, 65
Atomic mass unit (amu) A small mass unit used to describe the mass of very small particles such as atoms and subatomic particles; 1 amu is equal to one-twelfth the mass of a carbon-12 atom. 58, 76
Atomic number A number that is equal to the number of protons in an atom. 60–61, 76, 81
Atomic symbol An abbreviation used to indicate the mass number and atomic number of an isotope. 64, 76
Atomic theory, 57
Atoms The smallest particle of an element that retains the characteristics of the element. 57–59, 76
electrical neutrality of, 61
electron arrangement in, 67–72
isotopes of, 64–65
mass number for, 61
nucleus of, 59, 76
parts of, 58
structure of, 59
ATP. *See* Adenosine triphosphate (ATP)
ATP synthase An enzyme complex that links the flow of protons along an electrical gradient with the synthesis of ATP from ADP. 657–58, 676
Atropine, 540–41
Aureomycin, 628
Autoclave, 252
Average atomic mass, 65
Avogadro, Amedeo, 179, 266
Avogadro's law A gas law that states that the volume of gas is directly related to the number of moles of gas in the sample when

pressure and temperature do not change. 266–68, 277
Avogadro's number The number of items in a mole, equal to 6.02×10^{23}. 179, 196
AZT (azidothymine), 714–15

B

Background radiation, 97
Balance, 4
Balanced equations The final form of a chemical reaction that shows the same number of atoms of each element in the reactants and products. 164, 196
Balloon, helium-filled, 122
Barbiturates, 545
Barium, 48, 53, 131
Barium hydroxide, 342
Barium sulfate, 151, 299
Barometer An instrument used to measure atmospheric pressure. 248–50, 277
Bases A substance that dissolves in water and produces hydroxide ions (OH^-) according to the Arrhenius theory. All bases are proton acceptors, according to the Brønsted–Lowry theory. 327–30, 358
acid-base neutralization, 347–48
acid-base titration, 354
Arrhenius, 328
Brønsted, 329–30
characteristics of, 328
denaturation of proteins, 613
pH scale, 334–38
strong, 342, 358
weak, 342, 358
Basic amino acids, 597–98
Basic solution, 332–33, 335, 337
Becquerel (Bq) A unit of the activity of a radioactive sample equal to one disintegration per second. 95, 113
Beeswax, 565
Bends, 274
Benedict's solution, 496–97
Benzaldehyde, 455
Benzamide, 543
Benzedrine, 536
Benzene, 391, 396
Benzene ring A ring of six carbon atoms each of which is attached to a hydrogen atom, C_6H_6. 364
Benzoic acid, 518
Benzpyrene, 392
Beriberi, 622
Beryllium, 53, 69, 73
Beta emitters, 89–91
in medicine, 91
Beta oxidation, of fatty acids A reaction cycle that removes acetyl CoA to shorten a fatty acid and yields the reduced coenzymes NADH and $FADH_2$. 665–66, 676
Beta particles A particle identical to an electron ($-^0_1e$, or β) that forms in the nucleus when a neutron changes to a proton and an electron. 81–85, 89–90, 113
Beta-pleated sheet A secondary level of protein structure that consists of hydrogen bonds between peptide links in parallel polypeptide chains. 603–5, 631
Bicarbonate, 148, 294, 351
Bicarbonate buffer, 352
Bile salts, 581, 583, 642
Biogenic amines, 536
Biotin, 622
Birth-control pills, 585
Birth defects, 487
Bisulfite, 148
Bladder cancer, 711
Blimp, 122
Blood
buffers in, 351–52
pH of, 350, 668
Blood alcohol level, 461–62
Blood clots, 562
Blood gases, 272–74
Blood group, 503
Blood pressure, 563
Blood type, 503
Blood urea nitrogen (BUN), 672
Body
cholesterol in, 581–82
colloids and solutions in, 309
ions in, 125
natural opiates in, 605
oxidation of alcohol in, 460–62
trace elements in, 55
water in, 283
Body fat, determination of percentage of, 37
Body fluids, 282
electrolytes in, 294
osmolarity of, 319
saturation in, 297
Body odor, 570
Body temperature, 672
regulation of, 236
variation in, 205
Body weight, 507
losing and gaining weight, 215
Boiling The formation of bubbles of gas throughout a liquid. 162, 221–23, 239
Boiling point (bp) The temperature at which a substance exists as a liquid and gas; liquid changes to gas (boils), and gas changes to liquid (condenses). 222, 239
vapor pressure and, 251–52
Bonds
covalent. *See* Covalent bonds
disulfide, 606–8
double. *See* Double bonds
glycosidic. *See* Glycosidic bonds
in hydrocarbons, 365–66
hydrogen. *See* Hydrogen bonds
ionic, 118, 127, 145–46, 156, 607
peptide, 601, 632, 697–98
phosphodiester, 685 86, 690
polarity of, 143–46
review of bonding, 145–46
rules for predicting bond type, 146
triple, 138, 156, 364, 409, 420
Bone, polyatomic ions in, 149
Boron, 48, 69, 73
Boyle's law A gas law stating that the pressure of a gas is inversely related to the volume, when temperature and moles of the gas do not change; that is, if volume decreases, pressure increases. 253–54, 277
Brain hemorrhage, 102
Brain tumor, 101–2
Branch An alkyl group attached to a parent carbon chain or ring. 376, 396
Branched alkanes A hydrocarbon containing at least one alkyl group attached to the parent chain. 376–79, 396
IUPAC system for naming, 377–79
Breast cancer, 558, 571, 711
Breathalyzer test, 461–62
Breathing
partial pressure of gases during, 273
pressure–volume relationship in, 254
Bromide, 131
Bromine, 48, 53–54, 124, 137, 420
Brønsted–Lowry concept An acid is a proton donor, and a base is a proton acceptor. 329–30, 358
Buffers A mixture of a weak acid or a weak base and its salt that resists changes in pH when small amounts of an acid or base are added. 327, 350–53, 358
Buoyant force, 37
Burkitt's lymphoma, 711
Burning, 162
Burns
steam, 223
treatment in hyperbaric chamber, 274
Butane, 370, 372–73, 383
Butanedione, 456
Butanone, 453
2-Butene, 414
Butylated hydroxyanisole (BHA), 442
Butylated hydroxytoluene (BHT), 442
Butyric acid, 516–17

C

Cadmium, 48, 711
Caffeine, 540
Calamine lotion, 308
Calciferol. *See* Vitamin D
Calcium, 48–49, 53, 68–69, 73, 119, 124, 131
in body, 125
in muscle contraction, 639–40
in plasma, 294
strontium and, 54
Calcium carbonate, 148, 151, 349
Calcium chloride, 131
Calcium hydroxyapatite, 149
Calcium oxalate, 300
Calcium phosphate, 151, 300
Calcium sulfate, 151
Calcium sulfite, 151
Californium, 47
Caloric value The kilocalories obtained per gram of the food types: carbohydrate, fat, and protein. 213–14, 239
Calorie (cal) The amount of heat energy that raises the temperature of exactly 1 g of water exactly 1°C. 209, 212, 239
Calorie (Cal) The dietary unit of energy, equal to 1000 cal, or 1 kcal. 209, 212, 239
Calorimeter An instrument used to measure the amount of heat released by burning a sample. 212, 239
Cancer, 620, 704, 711
bladder, 711
breast, 558, 571, 711
cervical, 711
colon, 507, 558, 711
liver, 711
lung, 88, 101, 711
nasopharyngeal, 711
pancreatic, 558
prostate, 711
radiation therapy in, 83, 91, 101
renal, 711
skin, 101, 407, 711

(Cancer, continued)
stomach, 711
uterine, 711
Canola oil, 567, 571
Caprylic acid, 665–66
Carbohydrase, 616
Carbohydrates A simple or complex sugar composed of carbon, hydrogen, and oxygen. 512
blood types and, 503
caloric value of, 213–14
classification of, 479–80
digestion of, 640–41
disaccharides, 479–80, 499–501, 512
monosaccharides, 479–81, 490–97
polysaccharides, 479–80, 505–7
types of, 479–80
Carbolic acid, 439
Carbon, 48–49, 54, 57, 64, 68–69, 73, 119, 137, 363
chiral carbon atoms, 483–86
Carbon-11, 101, 104
Carbon-14, 90, 97, 104, 106
Carbonated beverages, 271–72, 282, 298, 341
Carbon cycle, 479
Carbon dating A technique used to date ancient specimens that contain carbon. The age is determined by the amount of active C-14 that remains in the sample. 106, 113
Carbon dioxide, 141
in atmosphere, 203
bicarbonate buffer system, 352
composition of air, 271
from fermentation, 501
partial pressure of
during breathing, 273
in blood and tissues, 272–74
Carbon disulfide, 141
Carbonic acid, 341, 351
Carbonic anhydrase, 615, 621
Carbon monoxide poisoning, 235, 274, 375
Carbon tetrachloride, 405
Carbonyl group A functional group that contains a carbon–oxygen double bond. 452, 472, 521
Carboxylase, 616
Carboxylate ion, 521–22
Carboxyl group A functional group found in carboxylic acids composed of carbonyl and hydroxyl groups. 516, 524, 552
Carboxylic acid group, 596, 601
Carboxylic acids A family of organic compounds containing the carboxyl group. 403, 460, 496–97, 516–19, 552
amidation of, 538–39
aromatic, 518
esterification, 527–28
IUPAC system for naming, 516–18
neutralization of, 522–23
preparation of, 519
properties of, 521–23
solubility of, 521
as weak acids, 521–22
Carcinogens, 392, 507, 547, 711
Car exhaust, 172, 392
Carnauba wax, 565
Carotenoids, 568
Carvone, 456, 487
Casein, 595, 613
Catabolic reactions Metabolic reactions that produce energy for the cell by the degradation and oxidation of glucose, fatty acids, and amino acids. 637–39, 676
Catalyst A substance that increases the rate of reaction by lowering the activation energy. 230–31, 239, 418. *See also* Enzymes
Cations A positively charged ion such as Na^+, Mg^{2+}, or Al^{3+}. 123, 156
Cell membrane, 575–76, 587–88
Cell structure, 637–38
Cell surface receptor, 487
Cellulose An unbranched polysaccharide composed of glucose units linked by β-1,4-glycosidic bonds that cannot be hydrolyzed by the human digestive system. 479, 490, 494, 505–7, 512
Celsius (°C) temperature scale A temperature scale on which water has a freezing point of 0°C and a boiling point of 100°C. 4, 41, 204–5, 208, 239
Centimeter (cm) A unit of length in the metric system; there are 2.54 cm in 1 in. 2, 16, 41
Cephalins Phospholids found in brain and nerve tissues that incorporate the amino alcohol serine or ethanolamine. 575, 590
Cerebrosides, 578–79, 590
Cervical cancer, 711
Cesium, 53
Cesium-137, 96
Cesium-141, 100
Chain reaction A fission reaction that will continue once it has been initiated with a high-energy neutron bombarding a heavy nucleus such as U-235. 108–9, 113
Changes of state The transformation of one state of matter to another; for example, solid to liquid, liquid to solid, liquid to gas. 219–20, 239
diagrams for, 224–26
Charge balance, in ionic compounds, 127
Charles, Jacques, 259
Charles' law A gas law stating that the volume of a gas changes directly with a change in Kelvin temperature, when pressure and moles of the gas do not change. 259–60, 277
Cheese making, 613
Chemical changes The formation of a new substance with a different composition and properties than the initial substance. 161–62, 196
Chemical energy, 202
Chemical equations A shorthand way to represent a chemical reaction using chemical formulas to indicate the reactants and products. 163–64, 196
balanced, 164, 196
conservation of mass in, 187
mole relationships in, 186–89
steps for balancing, 166–68
symbols used in, 163
writing an equation, 163–64
Chemical equilibrium The point at which the forward and reverse reactions take place at the same rate so that there is no further change in amounts of reactants and products. 201, 233–35, 239
concentration effects on, 234–35
temperature effects on, 235
Chemical quantities, 178–81
Chemical reactions The process by which a chemical change takes place. 161, 196
additions reactions, 418–23, 430
calculations using molar mass, 183
combination reactions, 170, 173, 196
combustion reactions, 171–73, 196, 396
decomposition reactions, 170, 173, 196
endothermic, 201, 228–29, 239
energy in, 228–31
exothermic, 201, 228–29, 239
ionization reactions, 327
mass calculations for, 191–93
mole relationships in chemical equations, 186–89
oxidation-reduction reactions, 174–77
rate of, 201, 230–31, 240, 615
replacement reactions, 170–71, 173
reversible, 233, 240
types of, 170–73
Chemical symbols An abbreviation that represents the name of an element. 47–48, 76
Chemiosmotic model The conservation of energy from transfer of electrons in the electron transport chain by pumping protons into the intermembrane space to produce a high-energy gradient that provides the energy to synthesize ATP. 657, 676
Chiral carbon atoms A carbon atom that is attached to four different atoms or groups of atoms; the mirror images of a chiral molecule are not superimposable. 483–86, 512
chiral compounds in nature, 487
D and L notation for enantiomers, 486
Fischer projections, 485–86
stereoisomers, 484
Chiral object, 483–84
Chiral technology, 487
Chitin, 509
Chloramphenicol, 628, 700
Chlorate, 148
Chlordane, 405
Chloride, 131, 294
Chlorinated cyclic compounds, as pesticides, 405–6
Chlorine, 47–49, 53–54, 62, 64–65, 69, 73, 119, 123–25, 137, 420
Chlorine dioxide, 141
Chlorine trifluoride, 141
Chlorite, 148
Chloroethane, 405
Chloroethene, 426–27
Chlorofluorocarbons (CFC), 404, 407
Chlorpromazine, 542
Cholesterol The most prevalent of the steroid compounds found in cellular membranes; needed for the synthesis of vitamin D, hormones, and bile acids. 441, 580–86, 590, 642
blood, 562, 571, 582–83
in body, 581–82
content of foods, 582
dietary, 558, 582
in plasma membranes, 587
Choline, 575
Cholinesterase, 620
Chromium, 711
Chromium-51, 104
Chromosomes, 687, 690
Chylomicrons, 582–83
Chymotrypsin, 641
Cinnamaldehyde, 455
Cis double bond, 588
Cis isomer A geometric isomer in which similar groups are connected on the same side of the double bond. 414, 416–17, 431
Citrate synthetase, 654
Citric acid, 341, 516, 518, 538, 671
Citric acid cycle A series of oxidation reactions in the mitochondria that convert acetyl

CoA to CO_2 and yield NADH and $FADH_2$. It is also called the Krebs cycle. 652–55, 670–71, 673, 676
ATP and, 661–63
reactions of, 653–54
Coagulant, 308
Coal tar, 392
Cobalamin, 622
Cobalt, 48
Cobalt-60, 96, 100
Cocaine, 540–41
Coconut oil, 566–67, 571, 573
Codeine, 541
Codon A sequence of three bases in mRNA that specifies a certain amino acid to be placed in a protein. A few codons signal the start or stop of transcription. 695–96, 718
Coefficients Whole numbers placed in front of the formulas in an equation to balance the number of atoms or moles of atoms of each element in an equation. 164, 196
Coenzyme A (CoA) A cofactor that transports acetyl groups to the citric acid cycle. 644, 676
Coenzyme Q A mobile carrier that accepts electrons from FeS or $FADH_2$ in the electron transport chain. 656–59, 676
Coenzymes Organic molecules required as cofactors in enzyme action. 621–22, 631
in metabolic pathways, 643–45
Cofactor The nonprotein portion, a metal ion or an organic compound, that is necessary for a biologically functional enzyme. 621, 631
Colchicine, 297
Cold packs, 229
Collagen, 595, 604, 606
Colloids A mixture having particles that are moderately large. Colloids scatter light (the Tyndall effect) and pass through filters but cannot pass through semipermeable membranes. 282, 306–9, 322
Colon cancer, 507, 558, 711
Combination reactions A reaction in which reactants combine to form a single product. 170, 173, 196
Combined gas law A relationship that combines several gas laws relating pressure, volume, and temperature: $P_iV_i/T_i = P_fV_f/T_f$. 265, 277
Combustion A chemical reaction in which an alkane reacts with oxygen to produce CO_2, H_2O, and heat. 171–73, 196, 396
of alcohols, 448–49
of alkanes, 373–74
incomplete, 375
Common cold, 712
Competitive inhibitors A molecule that has a structure similar to the substrate and inhibits enzyme action by competing for the active site. 626–28, 632
in medicine, 628
Complementary base pairs In DNA, adenine is always paired with thymine (A—T or T—A), and guanine is always paired with cytosine (G—C or C—G). In forming RNA, adenine is paired with uracil (A—U). 687–88, 690–91, 694, 697–98, 718
Complete structural formulas A type of formula in which each bond in the compound is shown as a line, C—H or C—C, to give the arrangement of all the atoms in the compound. 365–66, 396
Complex carbohydrates, 479
Compounds A combination of atoms in which noble gas arrangements are attained through electron transfer or electron sharing. 118, 121, 156
containing polyatomic ions
formulas for, 148–49
naming of, 150–51
covalent, 141
inorganic, 363–64
ionic, 118, 127–29, 135, 156, 299
moles of elements in, 180
organic, 363–64, 396
Computed tomography (CT), 100, 102
Concentration A measure of the amount of solute that is dissolved in a specified amount of solution. 302, 322
dilution, 343–44, 358
effect on chemical equilibrium, 234–35
percent, 302–4
Condensation The change of state of a gas to a liquid. 221–23, 239
heat of, 223
Condensed structural formulas A type of formula that illustrates the arrangement of the carbon atoms in an organic compound by grouping each carbon atom and its hydrogen atoms. 366, 382, 396, 409
Cones, 416
Coniine, 540
Conservation of mass, 186–87
Contractile proteins, 595
Control site A section of DNA composed of a promoter and operator that regulates protein synthesis. 708–9, 718
Conversion factors A ratio in which the numerator and denominator are quantities from an equality or given relationship. For example, the conversion factors for the relationship 1 kg = 2.20 lb are written as 2.20 lb/1 kg and 1 kg/2.20 lb. 20–22, 25–31, 41
clinical calculations using, 29–30
metric–metric, 21–22
metric–U.S. system, 21–22
molarity, 317–18
molar volume, 268
mole–mole, 187–89, 196
from percent concentration, 304
using percents as, 30
use of, 25–31
using two or more, 28–29
Cooling curve A diagram that illustrates temperature changes and changes of states for a substance as heat is removed. 224–26, 239
Copper, 48, 50, 54–55, 64, 132–33
Copper chloride, 133–34
Copper sulfate, 285
Corn oil, 566–67
Coronavirus, 712
Corticosteroids, 586
Cortisone, 456, 586
Cosmic radiation, 97
Cottonseed oil, 566
Covalent bonds A sharing of valence electrons by atoms. 118, 136–39, 156
double, 138, 156
multiple, 138–39
nonpolar, 143–46, 156
polar, 144–46, 156
sharing electrons between atoms of different elements, 137–38
triple, 138, 156
Covalent compounds, 141
electron-dot structure of, 137–39
naming of, 141, 151–53
"Crank," 536
Creatine kinase (CK), 620
Crenation The shriveling of a cell due to water leaving the cell when the cell is placed in a hypertonic solution. 311–12, 322
Crick, Francis, 687
Cross-links, in tertiary structure, 606–8
Crude oil, 371
Cubic centimeter (cm^3, cc) The volume of a cube that has 1-cm sides, equal to 1 mL. 18–19, 41
Curie, Marie, 95
Curie (Ci) A unit of radiation equal to 3.7×10^{10} disintegrations/s. 95, 113
Curium, 47
Cyanide ion, 148
Cyclic hemiacetal, 467, 493–97
Cycloalkanes An alkane that exists as a ring or cyclic structure. 387–88, 396
with alkyl groups, 387–88
naming of, 387–88
Cycloalkanol, 438
Cycloalkenes A cyclic hydrocarbon that contains a double bond in the ring. 409, 431
condensed structural formulas of, 409
naming of, 411
Cyclobutane, 388
Cyclohexane, 388
Cyclopentane, 388
Cyclopropane, 387–88
Cysteine, 597, 606–8, 670–71
Cystic fibrosis, 707
Cytochrome *c*, 658–59
Cytochrome *c* oxidase, 659
Cytochrome *c* reductase, 659
Cytochromes Iron-containing proteins (Fe^{3+}, Fe^{2+}) in the electron transport chain that transfer electrons from QH_2 to oxygen. 657, 659, 676
Cytoplasm The material within the plasma membrane and surrounding the nucleus in eukaryotic cells. 637–38, 676
Cytosine, 682–83, 686–88

D

Dacron, 441, 529
Dalton, John, 58
Dalton, 58
Dalton's law A gas law stating that the total pressure exerted by a mixture of gases in a container is the sum of the partial pressures that each gas would exert alone. 269–73, 277
DDI (dideoxyinosine), 714–15
DDT, 405–6
Deamination, oxidative, 671–72, 677
Decane, 370
Decarboxylation The loss of a carbon atom in the form of CO_2. 650–51, 676
Decay curve, 104
Deciliter (dL), 17
Decomposition reactions A reaction in which a single reactant splits into two or more simpler substances. 170, 173, 196
Decompression, 274
Deep-sea diver. *See* Diver
Deforestation, 203

Dehydration A reaction that removes water from an alcohol in the presence of an acid to form alkenes at high temperature, or ethers at lower temperatures. 283
of alcohols, 449–51, 472
Dehydrogenase, 616, 646, 665
Deletion mutation, 705
Demerol, 541
Denaturation of proteins The loss of secondary and tertiary protein structure caused by heat, acids, bases, organic compounds, heavy metals, and/or agitation. 612–14, 632
Density The relationship of the mass of an object to its volume expressed as grams per cubic centimeter (g/cm^3), grams per milliliter (g/mL), or grams per liter (g/L). 33–38, 41
of gases, 33
of liquids, 33
problem solving using, 36
of solids, 33–34
Deoxynucleoside phosphate, 682
Deoxyribonucleic acid (DNA) The genetic material of all cells containing nucleotides with deoxyribose sugar, phosphate, and the four nitrogenous bases adenine, thymine, guanine, and cytosine. 682, 718
double helix of, 687–88, 718
electrophoresis of, 620, 702–3
genetic code, 695–96
mutations in, 704–7
nucleosides in, 683
nucleotides in, 682–84
recombinant, 701–3, 718
replication of, 690–91, 718
structure of, 685–89
transcription, 694
Deoxyribose, 682–83
Deposition The change of a gas directly to a solid; the reverse of sublimation. 220, 239
Desalinization, 310
Detergent, 574
Dextrins, 505, 640–41
Diabetes mellitus, 490–91, 497, 558, 668
Dialysis A process in which water and small solution particles pass through a semipermeable membrane. 282, 313, 322
by kidney and artificial kidney, 314–15
Dialyzing membrane, 313
Diastolic pressure, 249
Diatomic molecules, 137
Diazepam, 545
Dieldrin, 405
Dienes, 411
Diesel fuel, 371
Dietary fiber, 309, 479, 507, 558
Diethyl ether, 447
Dieting, 215
Diffusion The movement of gas particles from a more concentrated area (high pressure) to a less concentrated area (low pressure). 254, 273, 277
Digestion The processes that break down large food molecules to small ones that can be absorbed by cells of the body. 309, 479, 677
of carbohydrates, 494, 505, 640–41
of fats, 530, 583, 642
of foods, 640–42
of proteins, 547, 641–42, 670
Dihydrogen phosphate, 148
Dilution of solutions A process by which water (solvent) is added to a solution to increase the volume and decrease (dilute) the concentration of the solute. 343–44, 358
Dimethylamine, 533
Dimethyl ether, 446
2,2-Dimethylpropane, 385
Diols, 438
Dipeptide, 601
Diphenhydramine, 536
Dipolar ions, 599–600
Direct relationship A relationship in which two properties increase or decrease together. 260, 277
Direct substrate phosphorylation The transfer of inorganic phosphate (P_i) directly from a high-energy phosphate bond to a substrate. 646, 677
Disaccharides Carbohydrates composed of two monosaccharides joined by a glycosidic bond. 479–80, 499–501, 512
Disinfectants, 442, 613
Dissociation, 290
Disulfide bonds, 606–8
Disulfiram, 461
Diver, 122, 249, 274
Diverticulitis, 507
Division, significant figures in, 13
DNA. *See* Deoxyribonucleic acid (DNA)
DNA fingerprinting, 620, 702–3
DNA ligase, 690, 702
DNA polymerase, 690, 702
DNA sequencing, 702–3
DNA virus, 711–12
D notation for enantiomers, 486
Docosahexaenoic acid (DHA), 562
Dopamine, 536
Double bonds A sharing of two pairs of electrons by two atoms. 138, 156, 364, 409, 418, 420, 560
cis, 588
trans, 571
Double helix The helical shape of the double chain of DNA that is like a spiral staircase with a sugar–phosphate backbone on the outside and base pairs like stair steps on the inside. 687–88, 718
Double replacement reaction A reaction in which parts of two different reactants exchange places. 171, 173, 196
Down syndrome, 707
Dry ice, 220
Dubnium, 93
Dynamite, 441
Dystrophin, 707

E

Eicosapentaenoic acid (EPA), 562
Einstein, Albert, 108
Electrical energy, 202
Electric current, 70
Electrolytes A substance that produces ions when dissolved in water; its solution conducts electricity. 290–93, 322
in body fluids, 294
equivalents of, 292–93
in intravenous replacement solutions, 294
strong, 290–91, 322
weak, 290–91, 322
Electromagnetic spectrum, 70
Electron arrangements An organization of electrons within the atom by increasing energy shells. 67–69, 76, 118–19
Electron carriers In the electron transport chain, a component that is reduced when it accepts electrons, and is oxidized when electrons are passed on. 656–57, 677
mobile, 658–59
Electron-dot structures The representation of an atom that shows each valence electron as a dot above, below, or beside the symbol of the element. 118–20, 137–39, 156
Electronegativity value A number that indicates the relative ability of an element to attract electrons. 143–44, 156
Electrons A negatively charged subatomic particle having a very small mass that is usually ignored in calculations; its symbol is e^-. 58–69, 76
energy-level changes, 70
loss and gain in oxidation-reduction reactions, 174–75
sharing between atoms of different elements, 137–48
valence, 118–20, 137, 156
Electron shell. *See* Shell
Electron transport chain A mitochondrial system that transfers electrons from the reduced coenzymes NADH and $FADH_2$ to clusters of electron carriers such as flavins, iron–sulfur proteins, and cytochromes. The final electron acceptor, O_2, yields H_2O. Energy changes during some of the transfers provide energy to phosphorylate ADP to yield ATP. 656–59, 661, 673, 677
chemiosmotic model of, 657, 676
steps in, 658–59
Electrophoresis The use of electrical current to separate proteins or other charged molecules with different isoelectric points. 632
of DNA, 620, 702–3
of proteins, 620
Elements A primary substance that cannot be separated into any simpler substances. 76
atomic number for, 60–61, 76
chemical symbols for, 47–48
essential to health, 49
isotopes of, 64–65
Latin names in clinical usage, 49
mass number for, 61
moles of elements in a compound, 180
periodic law for, 67–72
in periodic table, 51–55
physical properties of, 50, 76
rules for naming, 153
trace elements in body, 55
Emphysema, 352
Emulsion, 307
Enantiomers Stereoisomers that are mirror images of each other. 484, 487, 512
D and L notation for, 486
Endorphins, 605
Endothermic reactions A reaction that requires heat; the energy of the products is higher than the energy of the reactants. 201, 228–29, 239
Energy The ability to do work. 201–2, 239
activation, 228–31, 239, 615
chemical, 202
in chemical reactions, 228–31
electrical, 202
expenditure for physical activities, 215
form of, 202
heat, 201–2, 209–10
kinetic, 201, 240

mechanical, 202
nutrition and, 212–14
potential, 201, 240
radiant, 202
Energy-level changes, 70
Enflurane, 447–48
Enkephalins, 605
Enolase, 646
Environmental issues
acid rain, 339
biological reactions to UV light, 71
carbon dating, 106
CFCs and ozone depletion, 407
chlorinated cyclic compounds as pesticides, 405–6
fragrant alkenes, 412
global warming, 203
petroleum products, 370–71
pheromones in insect communication, 416
plastics, 529
polysaccharides, 509
radon in homes, 88
soaps, 573–74
vanilla, 456
Enzyme-catalyzed reaction, 618–19
Enzyme induction A model of cellular regulation in which protein synthesis is induced by a substrate. 709, 718
Enzyme inhibition, 626–28
competitive inhibitors, 626–28, 632
noncompetitive inhibitors, 627–28, 632
Enzyme–product complex The intermediate in an enzyme-catalyzed reaction in which product is formed at the active site. 618, 632
Enzyme repression A model of cellular regulation in which protein synthesis is repressed by elevated levels of end product. 709–10, 718
Enzymes Proteins that catalyze biological reactions. 230, 487, 595, 606, 615–16, 632
action of, 617–19
alterations due to mutations, 705
classification of, 615–16
as diagnostic tools, 619–20
factors affecting enzyme action, 621–24
induced-fit model of, 619, 632
lock-and-key model of, 618, 632
optimum pH, 623–24, 632
optimum temperature, 623–24
specificity of, 617
Enzyme substrate complex An intermediate in an enzyme-catalyzed reaction in which the substrate combines with the enzyme. 618, 632
Epilepsy, 101
Epimerase, 616
Epinephrine, 236, 536
Epstein-Barr virus (EBV), 711–12
Equalities A relationship between two units that measure the same quantity. 16–17, 21, 41
Equanil, 545
Equations, chemical, 163–64, 196
Equilibrium, chemical, 201, 233–35
Equivalent dose The measure of biological damage from an absorbed dose that has been adjusted for the type of radiation. 95–96, 113
Equivalents (Eq) The amount of a positive or negative ion that supplies 1 mole of electrical charge. 292–93, 322
Erythromycin, 700
Escherichia coli, 96, 701–2
Essential amino acids Amino acids that must be supplied by the diet because they are not synthesized by the body. 599, 632
Essential fatty acids, 561
Essential oils, 442
Esterification The formation of an ester from a carboxylic acid and an alcohol with the elimination of a molecule of water in the presence of an acid catalyst. 527–28, 552
Esters A family of organic compounds in which an alkyl group replaces the hydrogen atom in a carboxylic acid. 403, 516, 524–26, 552
acid hydrolysis of, 530
base hydrolysis of, 531
esterification, 527–28
IUPAC system for naming, 524–25
in plants, 525–26
waxes, 564
Estradiol, 585
Estrogens, 585
Ethanamide, 543
Ethane, 366, 370, 373
Ethanethiol, 442
Ethanol, 423, 437, 441, 448–49, 461, 519, 613
from fermentation, 501
metabolism of, 461
from pyruvate, 650–51
Ethanolamine, 575
Ethene, 409–10
Ethers An organic compound, R—O—R, in which an oxygen atom is bonded to two alkyl or two aromatic groups, or a mix of the two. 403, 437, 445–47, 472
as anesthetics, 447–48
formation of, 450–51
IUPAC system for naming, 446
solubility of, 447
uses of, 447
Ethrane, 447–48
Ethyl alcohol. *See* Ethanol
Ethyl butanoate, 526
Ethylene, 427
Ethylene glycol, 441
Ethyl ether, 437
Ethyl group, 376–77
Ethyl methanoate, 526
Ethyne, 411
Eugenol, 442
Eukaryotic cells, 637
Evaporation The formation of a gas (vapor) by the escape of high-energy molecules from the surface of a liquid. 221–23, 239
Exact numbers A number obtained by counting or definition. 8–9, 41
Exercise
energy expenditure for, 215
lactic acid accumulation in, 463
Exhalation, 272–73
Exothermic reactions A reaction that releases heat; the energy of the products is lower than the energy of the reactants. 201, 228–29, 239
Expiration, 254
Extracellular fluids, 283

F

Fabry's disease, 579
FAD (flavin adenine dinucleotide) A coenzyme for dehydrogenase enzymes that form carbon–carbon double bonds. The reduced form, $FADH_2$, transfers electrons to CoQ in the electron transport chain to regenerate FAD. 643–44, 652–53, 656–59, 661, 665, 677
Fahrenheit (°F) temperature scale A temperature scale on which ice melts at 32°F and water boils at 212°F. 4, 204–5, 208, 239
Familial hypercholesterolemia, 707
Family, of organic compounds, 364
Faraday, Michael, 391
Fats Another term for solid triglycerides. 558–59, 564–67, 590
caloric value of, 213–14
dietary, 571
digestion of, 530, 583
hydrogenation of, 419
melting points of, 566–67
unsaturated, 419
Fat-soluble vitamins, 558, 568, 583–84
Fat substitute, 568
Fatty acids Long-chain carboxylic acids found in fats. 560–62, 590
activation of, 665
beta oxidation of, 665–66, 676
essential, 561
melting points of, 561–62
monounsaturated, 560–61, 566, 590
omega-3, 562
oxidation of, 664–68
physical properties of, 561–62
polyunsaturated, 560–62, 566, 590
saturated, 560–62, 566–67, 571, 582, 590
synthesis of, 668
trans, 571
in triglycerides, 565
unsaturated, 560–62, 566–67
Fatty acid synthase, 668
Fatty acyl CoA, 665
Fermentation The anaerobic conversion of sugars by enzymes in yeast to yield alcohol and CO_2. 162, 437, 460, 499, 501, 512, 650–51, 677
Ferric ion, 132
Ferritin, 595
Ferrous ion, 132
Fiber, dietary, 309, 479, 507, 558
Fibrous proteins A protein that is insoluble in water; consists of polypeptide chains with a helixes or β-pleated sheets, that make up the fibers of hair, wool, skin, nails, and silk. 606, 632
Fight-or-flight situation, 536
Film badge, 98
Fingernail polish remover, 453
Fischer projections A two-dimensional straight-chain representation of a molecule. 485–86, 512
Fish oils, 562
Fishy odor, 535, 538
Fission, nuclear A process in which large nuclei are split into smaller pieces, releasing large amounts of energy. 108, 113
Flavorings, 455–56, 487, 516, 525–26
Floc, 308
Fluoride, 131
Fluorine, 48, 53, 55, 69, 73, 124, 137
Fluothane, 405
FMN (flavin mononucleotide) The initial electron acceptor of the electron transport chain for NADH. 656, 658, 677
Foams, 307
Folic acid, 622

Food
caloric value of, 212–14
carbohydrates in, 479
cholesterol content of, 582
digestion of, 640–42
freeze-dried, 220
hydrogenation of unsaturated fats, 419
preservation of, 442, 522–23
radiation treatment of, 96
water content of, 283
Food chain, 406
Food intake, 215
Forane, 447–48
Formaldehyde, 437, 441, 452, 462
Formic acid, 462, 516–17
Formulas The group of symbols that represent the atoms or ions in a compound. 127, 156
for compounds containing polyatomic ions, 148–49
ionic, 127, 129–35
molecular, 366
structural. *See* Structural formulas
subscripts in, 128
Fossil fuels, 172, 203, 339
Fragrances, 412, 456, 516, 525–26
Frame shift mutation A mutation that inserts or deletes a base in a DNA sequence. 704, 718
Francium, 53
Free radicals, 83
Freeze-dried foods, 220
Freezing A change of state from liquid to solid. 219–20, 239
Freezing point (fp) The temperature at which the solid and liquid forms of a substance are in equilibrium; a liquid changes to a solid (freezes), a solid changes to a liquid (melts). 219, 240
Freon, 407
Fructose A monosaccharide found in honey and fruit juices; it is combined with glucose in sucrose. Also called levulose and fruit sugar. 481, 487, 490, 492, 495, 512, 641
Fruit, ripening of, 410
Fucose, 503
Fumarase, 654
Fumarate, 671
Functional groups An atom or group of atoms that has a strong influence on the overall chemical behavior of an organic compound. 402–3, 431, 437
Fusion A reaction in which large amounts of energy are released when small nuclei combine to form larger nuclei.
heat of, 219, 240
nuclear, 109–11, 113

G

Galactose A monosaccharide that occurs combined with glucose in lactose. 487, 490–91, 494–95, 503, 512, 577–78, 641
Galactosemia, 491, 707
α-Galactosidase, 579
Gallium, 100
Gallstones, 441, 582
Gamma emitters, 91
Gamma rays High-energy radiation (γ) emitted to make a nucleus more stable. 70, 81–85, 91, 113
Gangliosides A glycolipid consisting of sphingosine, a 24-carbon fatty acid, and two more monosaccharides. 578–79, 590
Garlic, 443
Gases A state of matter characterized by no definite shape or volume. Particles in a gas move rapidly. 216–17, 240
amount of, 246–47
Avogadro's law of, 266–68, 277
blood, 272–74
Boyle's law of, 253–54, 277
Charles' law of, 259–60, 277
combined gas law, 265, 277
Dalton's law of, 269–73, 277
density of, 33
dissolved in liquid, 271–72
Gay-Lussac's law of, 262, 277
Henry's law of, 271–72, 277
kinetic theory of, 245, 277
liquid-gas equilibrium, 222
molar volume of, 267–68
moles of, 266–68
partial pressure of, 269–73, 277
pressure of, 246–54, 262
properties of, 245–47
solubility of, 298
STP conditions, 267–68
temperature of, 246–47, 259–60, 262
volume of, 246–47, 253–54, 259–60, 266–68
Gasoline, 371, 373
Gas solutions, 286
Gaucher's disease, 578–79
Gay-Lussac's law A gas law stating that the pressure of a gas changes directly with a change in temperature when the number of moles of a gas and its volume are held constant. 262, 277
Geiger counter, 95
General anesthetics, 405, 447
Genes
regulatory, 709, 718
structural, 708–10, 718
Genetic code The sequence of codons in mRNA that specifies the amino acid order for the synthesis of protein. 695–96, 718
Genetic disease A physical malformation or metabolic dysfunction caused by a mutation in the genetic code. 704–6, 718
Genetic engineering, 701
Genetic material, 682
Genetic mutations, 704–7
Genital warts, 711
Geometric isomers Two distinct arrangements of unsymmetrical alkenes made possible by the lack of rotation around the double bond. In a cis isomer, the groups are attached to the same side of the double bond; in the trans isomer, they appear on opposite sides. 413–14, 431
Geraniol, 412
Global warming, 203, 245
Globular proteins Proteins that acquire a compact shape from attractions between the R group of the amino acid residues in the protein. 606–7, 632
Glomerulus, 314
Glucocorticoids, 586
Glucosamine, 509
Glucose The most prevalent monosaccharide in the diet. An aldohexose that is found in fruits, vegetables, corn syrup, and honey. Also known as blood sugar and dextrose. Combines in glycosidic bonds to form most of the polysaccharides. 467, 487, 490, 492, 512, 578, 641
ATP energy from, 661–64
blood, 490–91, 497, 663, 672
cyclic structures of, 493–94
oxidation of, 646–48
urine, 497
Glucose oxidase, 497
Glucose tolerance test, 490–91
β-Glucosidase, 579
Glucosuria, 497
Glucuronic acid, 509
Glutamate, 598, 670–71
Glutamate dehydrogenase, 671
Glutamic oxaloacetic transaminase (GOT), 620
Glutamic pyruvic transaminase (GPT), 620
Glutamine, 597, 670–71
Glyceraldehyde, 481
Glycerol, 441, 565, 664
Glycine, 597, 604, 670–71
Glycogen A polysaccharide formed in the liver and muscles for the storage of glucose as an energy reserve. It is composed of glucose in a highly branched polymer joined by α-1,4- and α-1,6-glycosidic bonds. 490, 505, 507, 512, 663
Glycolic acid, 518
Glycolipids A phospholipid that contains one or more monosaccharides. 558–59, 574–78, 588, 590
Glycolysis The oxidation reactions of glucose that yield two pyruvate molecules. 646–48, 677
ATP and, 661–64
reactions in, 646–48
Glycosides Acetal products of a monosaccharide reacting with an alcohol or another sugar. 499, 512
Glycosidic bonds The acetal bond that forms when an alcohol or a hydroxyl group of a monosaccharide adds to a hemiacetal. It is the type of bond that links monosaccharide units in di- or polysaccharides. 499–501, 505–6, 512
Gold, 48, 54, 61, 133
Gold-foil experiment, 59
Gout, 297
Gram (g) The metric unit used in measurements of mass. 3, 19, 41
Gray (Gy) A unit of absorbed dose equal to 100 rads. 95, 113
Group number A number that appears at the top of each vertical column (group) in the periodic table and indicates the number of electrons in the outermost shell. 72–73, 76, 118–19, 124
Group of elements A vertical column in the periodic table that contains elements having similar physical and chemical properties. 52, 76
classification of groups, 53
Growth hormone, 595
Guanine, 682–83, 686–88

H

Hair, 606
Hair permanents, 608
Half-life, of radioactive isotope The length of time it takes for one-half of a radioactive sample to decay. 103–4, 113
Hallucinogens, 541

Haloalkanes A type of alkane in which one or more hydrogen atoms has been replaced by a halogen atom. 403–4, 431
as anesthetics, 405
IUPAC system for naming, 404
Halogen Group 7 elements of fluorine, chlorine, bromine, and iodine. 53, 76, 404
Halogenation The addition of Cl_2 or Br_2 to an alkene to form halogen-containing compounds. 418, 420, 431
Halothane, 405, 447
Haworth structure The cyclic structure that represents the closed chain of a monosaccharide. 493, 513
HDL (high-density lipoproteins), 571, 582–83
HDPE (high-density polyethylene), 426
Health issues
alcohols in health and medicine, 441
alpha hydroxy acids, 518–19
amide hydrolysis in nature, 547
amides in health and medicine, 544–45
amines in health and medicine, 536
anabolic steroids, 586
antacids, 349
antibiotics inhibit protein synthesis, 700
aromatic compounds, 391
ATP and calcium in muscle contraction, 639–40
beta emitters in medicine, 91
biological effects of radiation, 83
blood pressure, 249
blood types and carbohydrates, 503
calcium and strontium, 54
cancer, 711
carbon dioxide and bicarbonate buffer, 352
chiral compounds in nature, 487
cholesterol in body, 581–82
cis-trans isomers for night vision, 416–17
colloids and solutions in body, 309
competitive inhibitors, 628
dialysis by kidney and artificial kidney, 314–15
electrolytes in body fluids, 294
elements essential to health, 49
enzymes as diagnostic tools, 619–20
ethers as anesthetics, 447–48
feeling hungry, 672
fiber in the diet, 507
genetic diseases, 706
hair permanents, 608
haloalkanes as anesthetics, 405
hot packs and cold packs, 229
hydrogenation of unsaturated fats, 419
hyperbaric chambers, 274
hyperglycemia and hypoglycemia, 490–91
imaging methods, 102
incomplete combustion, 375
ions in body, 125
ketone bodies, 668
ketones in health and medicine, 455–56
kidney stones, 300
Latin names for elements in clinical usage, 49
lipid diseases, 578–79
lipoproteins, 582–83
losing and gaining weight, 215
natural opiates in body, 605
olestra, 568
omega-3 fatty acids in fish oils, 562
osmolarity of physiological solutions and body fluids, 319
oxygen, hemoglobin, and carbon monoxide poisoning, 235
percentage of body fat, 37
phenol derivatives in antiseptics, essential oils, and antioxidants, 442
polyatomic ions in bone and teeth, 149
polycyclic aromatic hydrocarbons, 392
pressure-volume relationship in breathing, 254
prostaglandins, 563
radiation and food, 96
radiation dose in diagnostic and therapeutic procedures, 101
regulation of body temperature, 236
salicylic acid and aspirin, 528
saturation in body fluids, 297
sickle-cell anemia, 610
smog, 172
specific gravity of urine, 39
steam burns, 223
stomach acid, 336
sweeteners, 502
testing for glucose in urine, 497
trace elements in body, 55
trans fatty acids, 571
uses for noble gases, 122
variation in body temperature, 205
water in body, 283
Heart attack. *See* Myocardial infarction
Heart disease, 507, 562, 571, 582–83
Heat The energy associated with the motion of particles in a substance. 240
of condensation, 223
of fusion The energy required to melt exactly 1 g of a substance. For water, 80. cal are needed to melt 1 g of ice; 80. cal are released when 1 g of water freezes. 219
of reaction The energy released or absorbed during a chemical reaction equal to the energy difference between the reactants and products. 228, 240
specific, 209–10, 240
of vaporization The energy required to vaporize 1 g of a substance. For water, 540 calories are needed to vaporize exactly 1 g of liquid: 1 g of steam gives off 540 cal when it condenses. 223, 240
Heat energy, 201–2
measurement of, 209–10
Heating curve A diagram that shows the temperature changes and changes of state of a substance as it is heated. 224–26, 240
Heating oil, 371
Heatstroke, 205
Helium, 48, 54, 61, 69, 73, 121–22
Hemiacetal The product of the addition of an alcohol to the double bond of the carbonyl group in an aldehyde. 466–67, 472
cyclic, 467, 493–97
Hemiketal The product of the addition of an alcohol to the double bond of the carbonyl group in a ketone. 466–67, 472
Hemodialysis A mechanical cleansing of the blood by an artificial kidney using the principle of dialysis. 314–15, 322, 672
Hemoglobin, 235, 273, 375, 595, 606, 707
functions of, 609
sickle-cell anemia, 610, 707
structure of, 608–9, 611
Hemolysis A swelling and bursting of red blood cells in a hypotonic solution due to an increase in fluid volume. 311–12, 322
Hemophilia, 707
Hemorrhoids, 507
Henry's law A gas law stating that the amount of gas dissolved in a liquid at a certain temperature is directly related to the pressure of the gas above the liquid. 271–72, 277
Heparin, 509
Hepatitis A virus (HAV), 712
Hepatitis B virus (HBV), 711–12
Heptane, 370
Heroin, 541
Herpesvirus, 712
Hexane, 370
Hexokinase, 617, 646
Hexosaminidase A, 579, 707
Hexose, 481
4-Hexylresorcinol, 442
Histamine, 536
Histidine, 598–99, 670–71
Histones, 687
HIV-1 (human immunodeficiency virus type 1), 713–15
Homeostasis, 236
Homes, radon in, 88
Honeycomb, 565
Hormones
protein, 595
steroid, 581, 584–85
Hot packs, 229
Human Genome Project, 703
Human T-cell lymphotrophic virus-type 1 (HTLV-1), 711
Hunger, 672
Huntington's disease (HD), 707
Hyaluronic acid, 509
Hydrated ions, 287, 290
Hydration An addition reaction in which the components of water, H— and —OH, bond to the carbon–carbon double bond to form an alcohol. 287, 431
of alkenes, 418, 423
Hydrocarbons An organic compound consisting of only carbon and hydrogen atoms. 364, 396
aromatic, 391–92
bonds in, 365–66
chlorinated, 405–6
mixtures of, 370
saturated, 364, 396, 409, 431
unsaturated, 364, 409, 431
Hydrochloric acid, 327, 329–30, 336, 341, 641
Hydrogen, 48–49, 54, 61–62, 69, 73, 136–37
loss and gain in biological systems, 176–77
Hydrogenation The addition of hydrogen (H_2) to the double bond of alkenes to yield alkanes. 431, 590
of alkenes, 418
trans fatty acids and, 571
of triglycerides, 569–71
of unsaturated fats, 419
Hydrogen bonds The attraction between a partially positive hydrogen atom in one water molecule and the partially negative oxygen atom of another. 322
between amines and water, 533–34
between carboxylic acids and water, 521
between water and ether, 447
in DNA, 687–88
in proteins, 604–5, 607
in water, 283
Hydrogen ions, 327
Hydrogen phosphate, 148

Hydrohalogenation, of alkenes The addition of a hydrogen halide such as HCl or HBr to a double bond. 418, 421, 431
Hydrolase, 530, 616
Hydrolysis The splitting of a molecule by the addition of water. Esters hydrolyze to produce a carboxylic acid and an alcohol; amides to yield the corresponding carboxylic acid and amine or their salts. 552
of amides, 546–47
of ATP, 638–39
of esters, 530–31
of proteins, 612, 615
of triglycerides, 571
Hydrometer, 37
Hydronium ions The H_3O^+ ion formed by the attraction of a proton (H^+) to an H_2O molecule. 329–33, 337–38, 358, 521–22
Hydrophilic, 573
Hydrophilic amino acids An amino acid having polar, acidic, or basic R groups that are attracted to water; "water-loving." 596, 632
Hydrophobic, 573
Hydrophobic amino acids A nonpolar amino acid with hydrocarbon R groups; "water-fearing." 596, 606, 632
Hydroxide ions, 148, 327–28, 331–33
Hydroxyapatite, 149
β-Hydroxybutyrate, 668
Hydroxyl group The —OH functional group. 437–39, 472, 521
Hydroxylysine, 604
Hydroxyproline, 604
Hyperbaric chamber, 235, 274
Hypercalcemia, 125
Hypercholesterolemia, familial, 707
Hyperglycemia, 490–91
Hyperkalemia, 49, 125
Hypernatremia, 49, 125
Hyperthyroidism, 100
Hypertonic solutions A solution that has a higher osmotic pressure than the red blood cells of the body. 311–12, 322
Hyperventilation, 352
Hypervitaminosis, 584
Hypocalcemia, 125
Hypoglycemia, 490–91
Hypokalemia, 49, 125
Hyponatremia, 49, 125
Hypothermia, 205
Hypotonic solutions A solution that has a lower osmotic pressure than the red blood cells of the body. 311–12, 322
Hypoventilation, 352

I

Ibuprofen, 391, 487
Imaging methods, 101–2
Immunoglobulins, 595
Inches of mercury, 250
Incomplete combustion, 375
Indicator, pH A substance added to a solution that changes color when acids or bases are added. 335–36, 354, 358
Indinavir, 715
Induced-fit model, of enzyme action A model that describes a flexible active site that adapts to the structure of a substrate to form the enzyme–substrate complex. 619, 632
Inducer, 709
Industrial smog, 172
Influenza, 712
Infrared rays, 70
Inhalation, 272–73
Inhibitors, enzyme Substances that make an enzyme inactive by interfering with its ability to react with a substrate. 626–28, 632
Inorganic compounds, 363–64
Insecticides, 405
Insects
pheromones in communication, 416
stings, 517
Insoluble salts An ionic compound that does not dissolve in water. 299, 322
kidney stones, 300
Inspiration, 254
Insulin, 595, 603–4, 606, 702
International System of Units (SI units), 1
International Union of Pure and Applied Chemistry (IUPAC), system for naming
alcohols, 437–38
aldehydes, 453–54
alkanes, 370–72
alkenes, 410
alkynes, 410
amines, 534–35, 543–44
branched-chain alkanes, 377–79
esters, 524–25
ethers, 446
haloalkanes, 404
ketones, 455
thiols, 442–43
Interstitial fluid, 283
Intracellular fluids, 283
Intravenous fluids, 294, 311
Inverse relationship A relationship in which two properties change in opposite directions. 253, 277
Invirase, 715
Iodide, 131
Iodine, 47–48, 53–55, 124, 137
tincture of, 282
Iodine-123, 100
Iodine-125, 81, 100, 104
Iodine-131, 81, 97, 99–100, 103–4
Iodine test A test for amylose that forms a blue-black color after iodine is added to the sample. 507, 513
Ion(s) An atom or group of atoms having an electrical charge because of a loss or gain of electrons. 156
in body, 125
hydrated, 287, 290
negative, 123–24
octet rule and, 121–24
polyatomic, 147–53, 156
positive, 122–23, 132–33
solutions containing, 290–92
Ionic bonds The attraction between oppositely charged ions that results from the transfer of electrons. 118, 127, 145–46, 156, 607
Ionic charge The difference between the number of protons (positive) and the number of electrons (negative), written in the upper right corner of the symbol for the element. 122, 156
from group numbers, 124
writing ionic formulas from, 129
Ionic compounds A compound of positive and negative ions held together by ionic bonds. 118, 127–29, 156
charge balance in, 127
containing two elements, 131
naming and writing, 130–35, 151–53
solubility rules for, 299
writing formulas from name of, 135
Ionic formulas, 127
naming and writing, 130–35
writing formulas from ionic charges, 128
Ionization, 327
of amines, 538
of amino acids, 599–600
of water, 331–33
Ionizing radiation, 83
Ion-product constant, for water The product of $[H_3O^+]$ and $[OH^-]$ in solution; K_{w5} $[H_3O^+][OH^-]$. 332–33, 358
Iron, 48, 54–55, 61–62, 132–33
Iron-59, 91, 104
Iron phosphate, 150–51
Iron-sulfur proteins (FeS), 657–58, 677
Isobutyl methanoate, 526
Isocitrate dehydrogenase, 654
Isoenzymes, 619
Isoeugenol, 442
Isoflurane, 447–48
Isoleucine, 597, 599, 671
Isomerase, 616, 646
Isomers
cis, 414, 416–17, 431
geometric. *See* Geometric isomers
stereoisomer, 484
structural. *See* Structural isomers
trans, 414, 416–17, 431
Isopropyl alcohol, 441, 613
Isopropyl group, 376–77
Isotonic solutions A solution that has the same osmotic pressure as that of the red blood cells of the body. 311–12, 322
Isotopes An atom that differs only in mass number from another atom of the same element. Isotopes have the same atomic number (number of protons) but different numbers of neutrons. 64–65, 76
radioactive. *See* Radioactive isotopes

J

Jacob, François, 708
Jefferson, Thomas, 456
Jojoba wax, 565
Joule (J) The SI unit of heat energy; 4.18 J = 1 cal. 209, 240

K

Kaopectate, 308
Kaposi's sarcoma, 713
Kekulé, August, 391
Kelvin (K) temperature scale An SI temperature scale that assigns zero to the lowest possible temperature (absolute zero) which makes the freezing point of water 273 K and the boiling point of water 373 K. A kelvin (K) represents a unit of temperature on the Kelvin scale: K = °C + 273. 4, 41, 207–8, 240
Keratin, 595, 606
Kerosene, 371
Ketals The product of the addition of a second molecule of alcohol to a hemiketal. 467–68, 472
Ketogenesis, 668
α-Ketoglutarate, 654, 670–71
α-Ketoglutarate dehydrogenase complex, 654

Ketone bodies, 668
Ketones An organic compound in which the carbonyl functional group is bonded to two alkyl groups. 403, 437, 452–55, 472
in health and medicine, 455–56
IUPAC system for naming, 455
oxidation of secondary alcohols to, 462–63
reaction with alcohols, 466–67
reduction of, 464
solubility in water, 452
Ketose A monosaccharide that contains a ketone group. 481, 513
Ketosis A condition in which the level of ketone bodies is elevated. 668, 677
Kidney
artificial, 314–15
dialysis by, 314–15
Kidney failure, 314–15
Kidney stones, 300
Kilocalorie (kcal) An amount of heat energy equal to 1000 calories. 209, 240
Kilogram (kg) A metric mass of 1000 g, equal to 2.20 lb. The kilogram is the SI standard unit of mass. 3, 41
Kilojoule (kJ), 212
Kimchee, 650
Kinase, 616, 646
Kinetic energy A type of energy that is required for actively doing work; energy of motion. 201, 240
Kinetic theory A model used to explain the behavior of gases. 245, 277
Krebs cycle. *See* Citric acid cycle
Krypton, 54, 121
Kwashiorkor, 599

L

Lachrymator, 443
Lactase, 500
Lactate dehydrogenase (LDH), 619–20, 649–50
Lactic acid, 465, 516, 518, 613, 673
accumulation in muscle, 463
from pyruvate, 649–50
Lactose 479, 490–92, 499–500, 513, 641, 709
Lanolin, 565
Lauric acid, 560
LDL (low-density lipoproteins), 571, 582–83
L-dopamine, 487
LDPE (low-density polyethylene), 426
Lead, 48, 133, 628
LeChâtelier's principle When a stress is placed on a system at equilibrium, the reaction shifts to relieve that stress. 234, 240
Lecithins Phospholipids containing chlorine as the amino alcohol. 575, 590
Length, 2
measurement of, 16–17
Lethal dose, 98
Leucine, 597, 599, 671
Leukemia, 91, 711–12
Libby, Willard, 106
Lidocaine, 541
Ligase, 616
Lighting tube, 122
"Like dissolves like," 285–86
Limonene, 412
Lindane, 405
Linoleic acid, 561–62
Linolenic acid, 561–62
Lipase, 571, 583, 615–17, 642
Lipid bilayer A model of a cell membrane in which phospholipids are arranged in two rows interspersed with proteins at different depths. 587–88, 590
Lipidoses Genetic diseases in which a deficiency of an enzyme for the hydrolysis of a lipid causes the accumulation of that lipid to toxic levels. 578–79, 590
Lipids A family of compounds that is nonpolar in nature and not soluble in water; includes fats, waxes, phospholipids, and steroids. 590
cholesterol, 580–86
digestion of, 642
fats, 558–59, 564–67
fatty acids, 560–62
glycolipids, 574–78
oils, 558–59, 564–67
phospholipids, 558–59, 574–78
in plasma membranes, 587–88
steroids, 558–59, 580–86
triglycerides, 569–72
types of, 558–59
waxes, 558, 564–67
Lipoproteins A combination of nonpolar lipids with phospholipids and proteins to form a polar complex that can be transported through body fluids. 582–83, 590, 595, 642
Liquids A state of matter that takes the shape of its container but has a definite volume. 216–17, 240
density of, 33
gas dissolved in, 271–72
liquid-gas equilibrium, 222
Liquid solutions, 286
Liter (L) The metric unit for volume that is slightly larger than a quart. 3, 17–18, 41
Lithium, 48, 53, 64, 68–69, 72–73, 119, 124, 131
Liver cancer, 711
Liver disease, 619–20
L notation for enantiomers, 486
Local anesthetics, 405, 541
Lock-and-key model, of enzyme action A model of enzyme action that represents the substrate as a key fitting the correct lock, which is the specific shape of the active site in an enzyme. 618, 632
Logarithm, 337
LSD (lysergic acid diethylamide), 487, 541
Lubricating oil, 371
Luminal, 545
Lung cancer, 88, 101, 711
Lyase, 616
Lye, 342
Lymphoma, 71, 91, 101, 711
Lysine, 598–99, 671
Lysol, 442
Lysosome, 638

M

Magnesium, 47–49, 53–54, 64, 69, 73, 119, 123–25, 131, 294
Magnesium-23, 81
Magnesium-27, 81
Magnesium bromide, 131
Magnesium chloride, 128, 290
Magnesium hydroxide, 342, 349
Magnesium nitrate, 149
Magnesium sulfate, 151
Magnesium sulfide, 131
Magnetic resonance imaging (MRI), 102
Malate dehydrogenase, 654
Malic acid, 518
Malonic acid, 627
Maltase, 505
Maltose A disaccharide consisting of two glucose units; it is obtained from the hydrolysis of starch and in germinating grains. 490, 499–500, 505, 513, 640–41
Manganese, 55
Margarine, 570–71
Markovnikov's rule When adding HX or HOH to alkenes with different numbers of groups attached to the double bonds, the H adds to the carbon that has the greater number of hydrogen atoms. 421–23, 431
Mass A measure of the quantity of material in an object. 3–4, 41
calculations for reactions, 191–93
conservation of, 186–87
finding mass of substances in chemical reaction, 193
measurement of, 19
molar, 180–81, 183, 196
of subatomic particles, 58
Mass/mass percent The grams of solute in exactly 100 grams of solution. 302, 322
Mass number The total number of neutrons and protons in the nucleus of an atom. 61, 76, 81
Mass/volume percent The grams of solute in exactly 100 ml of solution. 303–4, 322
Matter Anything that has mass and occupies space. 201, 240
states of, 201, 216–17
Maximum permissible dose (MPD), 98
Measured numbers A number obtained when a quantity is determined by using a measuring device. 7–8, 10, 41
Measurement
conversion factors, 20–22, 25–31, 41
of density, 33–38
of heat energy, 209–10
of length, 16–17
of mass, 19
measured and exact numbers, 7–8, 41
of pressure, 250
of radiation, 95–98
scientific notation, 5–6, 41
SI and metric prefixes, 16–19
significant figures, 9–10, 12–14, 41
of specific gravity, 37–38, 41
of temperature, 204–8
units of, 1–3
of volume, 17–19
Mechanical energy, 202
Melanin, 706
Melanoma, 711
Melting The conversion of a solid to a liquid. 162, 219–20, 240
Melting point (mp) The temperature at which a solid becomes a liquid (melts). It is the same temperature as the freezing point. 219, 240
of fats and oils, 566–67
of fatty acids, 561–62
Membrane, semipermeable, 306, 309–10, 313, 322
Mendeleev, Dmitri, 51
Menthol, 437, 441
Meperidine, 541
Meprobamate, 545
Mercury, 48, 628
Mescaline, 541
Messenger RNA (mRNA) Produced in the nucleus by DNA to carry the genetic

(Messenger RNA, continued)
information to the ribosomes for the construction of a protein. 692–700, 709–10, 718
Metabolic acidosis, 352–53
Metabolic alkalosis, 353
Metabolic pathways
for amino acids, 670–72
citric acid cycle, 652–55
coenzymes in, 643–45
digestion of foods, 640–42
electron transport chain, 656–59
glycolysis, 646–48
oxidation of fatty acids, 664–68
for pyruvate, 649–51
Metabolic rate, 236
Metabolism All the chemical reactions in living cells that carry out molecular and energy transformations. 677
ATP energy and, 637–39
definition of, 637
overview of, 673
Metal ions, denaturation of proteins, 613
Metals An element that is shiny, malleable, and a good conductor of heat and electricity. The metals are located to the left of the zigzag line in the periodic table. 54, 76, 124
alkali, 53, 76
alkaline earth, 53, 76
forming two positive ions, 132–33
Meter (m) The metric unit for length that is slightly longer than a yard. The meter is the SI standard unit of length. 2, 16, 41
Methandienone, 586
Methane, 138, 365, 370, 373, 375
Methanol, 176–77, 437, 441, 462
Methedrine, 536
Methionine, 597, 599, 671
Methoxyflurane, 447–48
Methyl acetate, 525
Methylamine, 533, 538
2-Methylbutane, 385
Methyl group, 376–77
Methyl orange, 391
Methylpropane, 383
Methyl salicylate, 528
Methyl thiobutanoate, 526
Metric-metric conversion factors, 21–22
Metric system A system of measurement used by scientists and in most countries of the world. 1, 41
prefixes, 16–19
Metric-U.S. system conversion factors, 21–22
Micelles, 573, 642
Microgram (μg), 19
Microliter (μL), 18
Microwaves, 70
Milk of magnesia, 342
Milk sugar. *See* Lactose
Milligram (mg), 19
Milliliter (mL) A metric unit of volume equal to one-thousandth of a L (0.001 L). 3, 18, 41
Millimeter, 16
Millirem (mrem), 96
Mineralocorticoids, 586
Mineral oil, 370
Mitochondria, 638, 656–59
Mixed triglycerides Fats or oils thatcontain two or three fatty acids. 566, 590
Mixture, 282
Mobile electron carriers, 658–59
Molarity (M) The number of moles of solute in exactly 1 L of solution. 316–18, 322
as conversion factor, 317–18
Molar mass The mass in grams of 1 mole of an element equal numerically to its atomic mass. The molar mass of a compound is equal to the sum of the masses of the elements in the formula. 180–81, 196
calculations using, 183
Molar volume A volume of 22.4 L occupied by 1 mole of a gas at STP conditions of 0°C (273 K) and 1 atm. 267–68, 277
conversion factors, 268
Molecular formula, 366
Molecule The smallest unit of two or more atoms held together by covalent bonds. 136, 156
diatomic, 137
Mole–mole conversion factors A conversion factor that relates the number of moles of two compounds in an equation derived from their coefficients. 187–89, 196
Moles A group of atoms, molecules, or formula units that contains 6.02×10^{23} of these items. 178–81, 196
of elements in a compound, 180
finding moles of substances in chemical reaction, 193
of gases, 266–68
relationships in chemical equations, 186–89
Molybdenum-99, 93
Monod, Jacques, 708
Monomers The small organic molecule that is repeated many times in a polymer. 425, 431
Monosaccharides A polyhydroxy compound that contains an aldehyde or ketone group. 479–81, 513
α and β anomers of, 494
cyclic structures of, 493–97
fermentation of, 501
mutarotation, 494
oxidation of, 496–97
structures of, 490–92
Monosodium glutamate (MSG), 522–23
Monounsaturated fatty acids Fatty acids with one double bond. 560–61, 566, 590
Morphine, 541
Mothballs, 392
Motor oil, 370
Mucopolysaccharides, 509
Multiple covalent bonds, 138–39
Multiple sclerosis, 102, 578
Multiplication, significant figures in, 13
Mumps, 712
Muscles
contraction of, 639–40
lactic acid accumulation in, 650
Muscone, 456
Muscular dystrophy (Duchenne), 707
Mutagens, 704
Mutarotation The conversion between α and β anomers. 494, 513
Mutase, 646
Mutations A change in the DNA base sequence that alters the formation of a protein in the cell. 704–7, 718
addition, 705
deletion, 705
effect of, 705
frame shift, 704, 718
substitution, 704, 718
Myelin sheath, 575, 577–79
Mylar, 529
Myocardial infarction, 558, 583, 619–20
Myoglobin, 595, 608–9, 611
Myosin, 595, 639–40
Myrcene, 412
Myristic acid, 560

N

NAD^+ (nicotinamide adenine dinucleotide) The hydrogen acceptor in oxidation reactions that forms carbon–oxygen double bonds. The reduced coenzyme NADH transfers electrons to FMN in the electron transport chain to yield three ATP molecules and regenerate NAD^+. 644, 649–51, 671, 677
NADH (nicotinamide adenine dinucleotide), 464, 646, 648, 652–53, 656–59, 661, 665
NADH dehydrogenase, 658–59
NADPH (nicotinamide adenine dinucleotide phosphate), 668
Nails, 606
Nandrolone, 586
Naphthalene, 392
Nasopharyngeal carcinoma, 711
Natural gas, 371, 375
Negative ions, 123–24
Nembutal, 545
Neon, 48, 54, 68–69, 73, 121
Neo-Synephrine, 536
Nephron, 314
Nerve gas, 628
Net ionic equation, 300
Neutralization A reaction between an acid and a base to form a salt and water.
acid-base, 347–48, 358
of amines, 538
of carboxylic acids, 522–23
Neutralization endpoint, 354
Neutral solution A solution with equal concentrations of $[H_3O^+]$ and $[OH^-]$. 332–33, 335, 337, 358
Neutron A neutral subatomic particle having a mass of 1 amu and found in the nucleus of an atom; its symbol is n or n°. 58–59, 76
Niacin, 622, 644
Nickel, 48, 711
Nicotinamide, 644
Nicotine, 540
Niemann-Pick disease, 579
Night blindness, 584
Night vision, 416–17
Nitrate, 148
Nitric acid, 327, 341
Nitride, 131
Nitrite, 148, 711
Nitrogen, 48–49, 54, 62, 69, 73, 119, 124, 137
composition of air, 271
covalent compounds formed by, 141
Nitrogen-13, 101
Nitrogen-containing bases Nitrogen-containing compounds found in DNA and RNA: adenine (A), thymine (T), cytosine (C), guanine (G), and uracil (U). 682–83, 718. *See also* Complementary base pairs
Nitrogen oxides, 172
Nitrogen trichloride, 141
Noble gases An element in Group 8 of the periodic table, generally unreactive and sel-

dom found in combination with other elements. 54, 76, 119, 121–22
Nonane, 370
Noncompetitive inhibitor A substance that changes the shape of the enzyme and the active site, preventing the active site from binding with the substrate. 627–28, 632
Nonelectrolytes A substance that dissolves in water as molecules; its solution will not conduct an electrical current. 290–93, 322
Nonmetals An element with little or no luster that is a poor conductor of heat and electricity. The nonmetals are located to the right of the zigzag line in the periodic table. 54, 76, 124
Nonpolar amino acids Amino acids that are not soluble in water because they contain a nonpolar R group. 596–97, 632
Nonpolar covalent bond A covalent bond in which the electrons are shared equally. 143–46, 156
Noradrenaline. *See* Norepinephrine
Norepinephrine, 536
Norethindrone, 585
Novocaine, 541
Nuclear equations, 87–91
Nuclear fission, 108, 113
Nuclear fusion, 109–11, 113
Nuclear medicine, 99–102
  radiation dose in diagnostic and therapeutic procedures, 101
Nuclear power plant, 98, 109–10
Nuclear radiation. *See* Radiation
Nuclear testing, 97–98
Nuclease, 616
Nucleic acids Large molecules composed of nucleotides, found as a double helix in DNA, and as the single strands of RNA. 682–84, 718
Nucleoside The combination of a pentose sugar and a nitrogen-containing base. 682–83, 718
Nucleoside analogs, 714
Nucleoside phosphate, 682
Nucleotide phosphokinase, 654
Nucleotides Building blocks of a nucleic acid consisting of a nitrogen-containing base, a pentose sugar (ribose or deoxyribose), and a phosphate group. 682, 718
  composition of, 682–83
  in DNA and RNA, 682–84
Nucleus, atomic The compact, very dense center of an atom, containing the protons and neutrons of the atom. 59, 76
Nucleus, cellular, 637, 693
Numbers
  exact, 8–9, 41
  measured, 7–8, 10, 41
Nutrition, energy and, 212–14

O

Octane, 370
Octet rule Elements in Groups 1–7 react with other elements by forming ionic or covalent bonds to produce a noble gas arrangement, usually 8 electrons in the outer shell. 121–24, 156
Octyl ethanoate, 526
Oil of wintergreen, 528
Oil refinery, 371
Oils Another term for liquid tryglycerides. 558–59, 564–67, 590
  melting points of, 566–67
Oil spill, 373
Oleic acid, 419, 561
Olestra, 568
Olive oil, 566–67, 571
Omega-3 fatty acids, 562
Oncogenic virus, 711
Onions, 443
Operator The section of DNA that precedes the structural genes and controls the synthesis of an mRNA. 708–10, 718
Operon A group of genes, including a control site and structural genes, whose transcription is controlled by the same regulatory gene. 708–9, 718
Operon model, 708–9
Opiates, natural, 605
Opium, 541
Opportunistic infection, 713
Organic chemistry, 363
Organic compounds Compounds that contain carbon in covalent bonds. Most are nonpolar with low melting and boiling points, insoluble in water, soluble in nonpolar solvents, and flammable. 363–64, 396
  denaturation of proteins by, 613
  families of, 364
  functional groups, 402–3, 431
Orthomyxovirus, 712
Osmolarity
  of body fluids, 319
  of physiological solutions, 319
Osmosis The flow of a solvent, usually water, through a semipermeable membrane into a solution of higher solute concentration. 282, 310–13, 322
  reverse, 310
Osmotic pressure The pressure that prevents the flow of water into the more concentrated solution. 310–11, 319, 322
Osteoporosis, 149
Oxaloacetate, 652, 668, 671
Oxandrolone, 586
Oxidase, 616
Oxidation The loss of hydrogen atoms or electrons by a reactant. In metabolism, it is generally associated with the degradation of large molecules and the production of energy. 175–77, 196, 458–65, 472, 643, 677
  of alcohols, 458, 460–62
  of aldehydes, 458, 460, 462–63
  of fatty acids, 664–68
  of glucose, 646–48
  of methanol, 462
  of monosaccharides, 496–97
  of triglycerides, 570
Oxidation-reduction reactions A reaction in which the oxidation of one reactant is always accompanied by the reduction of another reactant. 174–77, 196
  gain and loss of oxygen in, 176
  loss and gain of electrons in, 174–75
  loss and gain of hydrogen in, 176–77
Oxidative deamination The removal of an amino group as NH_3. 671–72, 677
Oxidative phosphorylation The synthesis of ATP in the mitochondria by combining ADP and P_i using energy from the oxidation reactions in the electron transport chain. 657, 677
Oxide, 131
Oxidoreductase, 616
Oxygen, 48–49, 54, 61, 64, 69, 73, 124, 137
  composition of air, 271
  covalent compounds formed by, 141
  in electron transport chain, 659
  gain and loss in oxidation-reduction reactions, 176
  partial pressure of
    during breathing, 273
    in blood and tissues, 272–73
  transport by hemoglobin, 235
Oxygen-15, 101, 104
Oxygen debt, 650
Ozone, 172, 245, 404, 407, 711

P

PABA, 628
Pain, 605
Pain control, 528, 541, 545
Paint remover, 452, 462
Palmitic acid, 560
Palmitoleic acid, 561
Palm oil, 566, 571
Pancreatic cancer, 558
Pancreatic disease, 620
Pantothenic acid, 622, 644
Papilloma virus, 711
Papovavirus, 712
Paramyxovirus, 712
Parent chain The longest unbranched chain of carbon atoms in a formula. 377, 396
Parkinson's disease, 101, 487, 536
Partially hydrogenated vegetable oil, 570–71
Partial pressure, of gas The pressure exerted by a single gas in a gas mixture. 269–73, 277
Pascal (Pa), 250
Peanut oil, 566
Pellagra, 622
Penicillin, 308, 628
Penicillinase, 628
Penicillin derivatives, 628
Pentane, 370, 385
Penthrane, 447–48
Pentobarbital, 545
Pentose, 481
Pentyl butanoate, 526
Pepsin, 336, 615, 624, 641
Peptic ulcer, 349
Peptide The combination of two or more amino acids joined by peptide bonds; dipeptide, tripeptide, and so on. 601, 632
Peptide bonds The amide bond that joins amino acids in polypeptides and proteins. 601, 632, 697–98
Percent, as conversion factor, 30
Percent concentration, 302–4
Period A horizontal row of elements in the periodic table. 52, 76
Periodic law The repetition of similar chemical and physical properties with increasing atomic number due to the reappearance of the same number of electrons in the outermost shells of atoms of those elements. 67–72, 76
Periodic table An arrangement of elements by increasing atomic number such that elements having similar chemical behavior are grouped in vertical columns. 51–55, 76
  classification of groups, 53
  groups and periods in, 52
  metals and nonmetals, 54
Permanents, hair, 608
Pernicious anemia, 622
Perspiration, 236, 283

Pesticides, 404, 416
chlorinated cyclic compounds, 405–6
PETE (polyethyleneterphthalate), 426, 529
Petrolatum, 370
Petroleum products, 370–71
pH A measure of the $[H_3O^+]$ in a solution; pH = $-\log[H_3O^+]$. 327, 334–38, 358
blood, 350
buffers, 350–53
effect on enzyme action, 623–24
pH meter, 335
pH values of common substances, 335
significant figures in pH values, 337
urine, 352
Phenacetin, 545
Phenanthrene, 392
Phenobarbital, 545
Phenolphthalein, 328, 354
Phenols An organic compound that has an —OH group attached to a benzene ring. 437–43, 472
derivatives of, 442
naming of, 439
properties of, 439
Phenylalanine, 391, 597, 599, 671
Phenylalanine hydroxylase, 706
Phenylephrine, 536
Phenylethene, 427
Phenylketonuria (PKU), 706
Phenylphenol, 442
Phenylpyruvic acid, 706
Pheromones, insect, 416
Phosphatase, 616
Phosphate, 148, 294
Phosphide, 131
Phosphite, 148
Phosphodiester bonds, 685–86, 690
Phosphoglycerides Polar lipids of glycerol attached to two fatty acids and a phosphate group connected to an amino group such as choline, serine, or ethanolamine. 575–76, 590
Phosphokinase, 646
Phospholipids, 558–59, 574–78
in plasma membranes, 587–88
Phosphorus, 48–49, 62, 69, 73, 124, 137
Phosphorus-32, 91, 100, 104
Phosphorylation
oxidative, 657, 677
substrate, 646, 677
Photochemical smog, 172
Photosensitivity, 71
Photosynthesis, 479
Phototherapy, 71
Phylloquinone. *See* Vitamin K
Physical change A change in which the physical properties change but the chemical composition does not change. 161, 196
Physical properties A characteristic that can be observed or measured without affecting the identity of an element, including shape, color, odor, taste, density, hardness, melting point, and boiling point. 50, 76
Physical state, 216–17
Physiological solution A solution that exerts the same osmotic pressure as normal body fluids. 311, 319, 322
Pickling, 650
Plants
amines in, 540–42
esters in, 525–26
Plant starch, 505
Plaque, 582
Plasma, 283
Plasma membrane A lipid biplayer that keeps cell contents separated from the surrounding environment. 587–88, 590
Plasmids, 701–2
Plastics, 425–27, 529
recycling of, 426, 529
Pneumocystis carinii pneumonia, 713
Polar amino acids Amino acids that are soluble in water because their R group is polar: hydroxyl (OH), thiol (SH), carbonyl (C—O), amino (NH_2), or carboxyl (COOH). 596–97, 606, 632
Polar covalent bonds A covalent bond in which the electrons are shared unequally. 144–46, 156
Polarity A measure of the unequal sharing of electrons, indicated by the difference in electronegativity values. 143–46, 156
of bonds, 143–46
of soap, 573
Polar substance, 282
Polonium-218, 88
Polyatomic ions A group of covalently bonded nonmetal atoms that has an overall electrical charge. 147–53, 156
in bone and teeth, 149
formulas for compounds containing, 148–49
naming compounds containing, 150–51
naming of, 148
Polycyclic aromatic hydrocarbons, 392
Polycythemia vera, 91
Polyester, 529
Polyethylene, 426–27
Polymerase chain reaction (PCR) A strand of DNA is copied many times by mixing it with DNA polymerase and a mixture of deoxyribonucleotides. 702, 718
Polymers A very large molecule that is composed of many small, repeating structural units that are identical. 425, 431
of alkenes, 425–27
Polypeptide Polymers of amino acids joined by peptide bonds. 603, 632
Polypropylene (PP), 426–27
Polysaccharides Polymers of many monosaccharide units, usually glucose. Polysaccharides differ in the types of glycosidic bonds and the amount of branching in the polymer. 479–80, 505–7, 509, 513
structural, 505–7
Polystyrene (PS), 426–27
Polyunsaturated fatty acids Fatty acids that contain two or more double bonds. 560–62, 566, 590
Polyvinyl chloride (PVC), 426–27
Positive ions, 122–23, 132–33
Positron emission tomography (PET), 101
Potassium, 48–49, 53, 68–69, 72–73, 118, 124–25, 131, 294
Potassium-40, 97, 104
Potassium-42, 104
Potassium carbonate, 151
Potassium hydroxide, 342
Potassium oxide, 131
Potential energy An inactive type of energy that is stored for use in the future. 201, 240
Pounds per square inch, 250
Powers of ten, 5–6
Precipitate, 300
Precision Ability to repeat a measurement and obtain values close together. 9, 41
Prednisone, 586
Prefix The part of the name of a metric unit that precedes the base unit and specifies the size of the measurement. All prefixes are related on a decimal scale. 16–19, 41
Preservatives, 522–23
Pressure The force exerted by gas particles that hit the walls of a container. 277
atmospheric, 248–52, 277
blood, 249
definition of, 248
of gas, 246–54, 262, 269–73
measurement of, 250
osmotic, 310–11, 319, 322
partial, 269–73, 277
vapor, 250–52, 277
Pressure cooker, 252
Pressure gradient A difference in pressure causing gas particles to move from the area of high pressure to the area of low pressure. 254, 277
Pressure-volume relationship, in breathing, 254
Primary (1°) alcohols An alcohol that has one alkyl group attached to the alcohol carbon atom, R—CH_2OH. 459, 472
oxidation of, 460
reduction of aldehydes to, 464
Primary amines, 533
Primary structure, of proteins The sequence of the amino acids in a protein. 603–4, 610, 632
Procaine, 541
Products The substances formed as a result of a chemical reaction. 163–64, 196
Progesterone, 456, 585
Prokaryotic cells, 637
Proline, 597, 604, 670–71
Promoter A component of the control site on an operon that regulates DNA synthesis. 708–9, 718
Propane, 366, 370, 373
Propanone, 452
Propellants, 407
Propionic acid, 516–17
Propyl ethanoate, 526
Propyl group, 376–77
Prostaglandins A number of compounds derived from arachidonic acid that regulate several physiological processes. 563, 590
Prostanoic acid, 563
Prostate cancer, 711
Protease, 547, 616, 641
Protease inhibitors, 715
Protective proteins, 595
Protein hormones, 595
Proteins A term used for biologically active polypeptides that have many amino acids linked together by peptide bonds. 603, 632. *See also* Enzymes
alterations due to mutations, 705
amino acids in, 596–600
caloric value of, 213–14
denaturation of, 612–14, 632
digestion of, 547, 641–42, 670
electrophoresis of, 620
fibrous, 606, 632
functions in body, 595
globular, 606–7, 632
hydrolysis of, 612, 615
peptide bonds in, 601
in plasma, 294
in plasma membranes, 587–88
primary structure of, 603–4, 610, 632
quaternary structure of, 606–11, 632

secondary structure of, 603–5, 610, 632
synthesis of, 695–700
elongation, 697–99
inhibition by antibiotics, 700
initiation, 697–99
regulation of, 708–10
termination, 698–99
tertiary structure of, 606–11, 632
types of, 595
Proton A positively charged subatomic particle having a mass of 1 amu and found in the nucleus of an atom; its symbol is p or p^+. 58–59, 76
Proton pump The movement of H^+ by the enzyme complexes from the matrix into the intermembrane space, which creates a proton gradient. 657–58, 677
Psoriasis, 71
Purines The double-ringed nitrogen-containing compounds of adenine and guanine found in nucleic acids. 682–83, 718
Puromycin, 700
Pyridoxine, 622
Pyrimidines The single-ring nitrogen-containing compounds of cytosine, uracil, and thymine found in nucleic acids. 682–83, 718
Pyruvate, 646, 648, 670–71, 673
to acetyl CoA, 649–50
to ethanol, 650–51
to lactate, 649–50
metabolism of, 649–51, 661
Pyruvate decarboxylase, 621
Pyruvate dehydrogenase, 649
Pyruvic acid, 463, 465, 516

Q

Quaternary structure, of proteins A protein structure in which two or more protein subunits form an active protein. 606–11, 632
Quinine, 540

R

Radiant energy, 202
Radiation, 81
background, 97
biological effects of, 83, 95–96
cosmic, 97
detection and measurement of, 95–98
dose in diagnostic and therapeutic procedures, 101
food treatment, 96
ionizing, 83
lethal dose of, 98
maximum permissible dose of, 98
medical applications of, 99–102
natural, 81–85
sources of, 97–98
types of, 82
Radiation protection, 83–85
Radiation sickness, 98
Radiation therapy The use of high doses of radiation to destroy harmful tissues in the body. 83, 91, 101, 113
Radioactive decay The process by which a radioactive nucleus breaks down with the release of high-energy radiation. 87, 113
Radioactive hazardous waste, 109
Radioactive iodine uptake (RAIU), 99–100
Radioactive isotopes A radioactive atom of an element. 81, 113
activity of, 95
half-life of, 103–4, 113
production of, 92–93
Radiological dating, 106
Radiowaves, 70
Radium, 53
Radon, 54
in homes, 88
Radon-222, 97
Radon-226, 104
Rad (radiation absorbed dose) A measure of an amount of radiation absorbed by the body. 95, 113
Rancidity, 517, 570
Rate of reaction The speed at which reactants are used to form product(s). 201, 230–31, 240, 615
Reactants The initial substances that undergo change in a chemical reaction. 163–64, 196
Recipe, 163
Recombinant DNA The splicing of DNA from different organisms to form new, synthetic DNA. 701–3, 718
Recycling, of plastics, 426, 529
Red blood cells, 311–12, 503, 610
Reducing sugar A carbohydrate with a free aldehyde group capable of reducing the Cu^{2+} in Benedict's reagent. 496–97, 513
Reductase, 616
Reduction The addition of hydrogen to a carbonyl bond. Aldehydes are reduced to primary alcohols; ketones to secondary alcohols. Reduction can also be the loss of an oxygen atom. 175–77, 196, 458–65, 472, 643, 677
of aldehydes, 464
of ketones, 464
Refrigeration, 230
Regulatory gene A gene in front of the control site that produces a repressor. 709–10, 718
Rem (radiation equivalent in humans) A measure of the biological damage caused by the various kinds of radiation (rad × radiation biological factor). 95–96, 113
Renal cancer, 711
Replacement reactions, 170–71, 173
double, 171, 173, 196
single, 170–71, 173, 196
Replication, of DNA The process of duplicating DNA by pairing the bases on each parent strand with their complementary base. 690–91, 718
Representative elements An element found in Groups 1 through 8 of the periodic table. 52, 76
Representative metals, 132
Repressor A protein that interacts with the operator gene in an operon to prevent the transcription of mRNA. 709–10, 718
Reserpine, 542
Resorcinol, 442
Respiration, 479
Respiratory acidosis, 352–53
Respiratory alkalosis, 352–53
Respiratory tract cancer, 711
Restriction enzymes An enzyme that cuts a large DNA strand into smaller fragments to isolate a gene or to remove a portion of the DNA in the plasmids of *E. coli.* 701–3, 718
Retin-A, 71
Retinal, 416–17
Retinoblastoma, 711
Retinol. *See* Vitamin A
Retrovirus, 712–13
Reverse osmosis, 310
Reverse transcriptase, 712–15
Reversible reaction A reaction in which a forward reaction occurs from reactants to products, and a reverse reaction occurs from products back to reactants. 233, 240
R groups, 596–98
Rhodopsin, 416
Riboflavin, 622, 643
Ribonucleic acid (RNA) A type of nucleic acid that is a single strand of nucleotides containing adenine, cytosine, guanine, and uracil. 682, 692–94, 718
messenger, 692–700, 709–10, 718
nucleosides in, 683
nucleotides in, 682–84
ribosomal, 692–93, 718
structure of, 685–89
transcription, 694
transfer, 693, 695, 697–98, 719
types of, 692–93
Ribose, 481, 682–83
Ribosomal RNA (rRNA) The most prevalent type of RNA; a major component of the ribosomes. 692–93
Ribosomes, 638, 692–93, 697–98
Rickets, 584, 620
Rigor mortis, 639
Ringer's solution, 294
Ritonavir, 715
RNA. *See* Ribonucleic acid (RNA)
RNA polymerase, 694
RNA virus, 711–12
Rods, 416
Rounding off, 12
Rubbing alcohol, 441
Rubidium, 53
Runner's high, 605
Rust, 161–62, 176
Rutherford, Ernst, 59

S

Saccharide A term from the Latin word *saccharum,* meaning "sugar"; it is used to describe the carbohydrate family. 479, 513
Saccharin, 502, 544–45
Safflower oil, 566–67
Salicylic acid, 528
Saliva, 640
Salts An ionic compound that contains a metal ion and a nonmetal or polyatomic ion. 347, 358, 522
insoluble, 299, 322
soluble, 299, 322
Saponification The hydrolysis of an ester with a strong base to produce a salt of the carboxylic acid and an alcohol. 531, 552, 572
Saquinavir, 715
Saran, 427
Satiety, 672
Saturated fatty acids Fatty acids that have no double bonds; they have higher melting points than unsaturated lipids and are usually solid at room temperature. 558, 560–62, 566–67, 571, 582, 590

Saturated hydrocarbon A compound of carbon and hydrogen that has only single carbon–carbon bonds (C—C). 364, 396, 409, 431
Saturated solution A solution containing the maximum amount of solute that can dissolve at a given temperature. Any additional solute will remain undissolved in the container. 296, 322
body fluids, 297
Sauerkraut, 650
Scan The image of a site in the body created by the detection of radiation from radioactive isotopes that have accumulated in that site. 99, 113
Scanning tunneling microscope, 57
Scientific notation A form of writing large and small numbers using a coefficient between 1 and 10, followed by a power of 10. 5–6, 41
Scuba diver. *See* Diver
Scurvy, 622
Secobarbital, 545
Seconal, 545
Secondary (2°) alcohols An alcohol that has two alkyl groups attached to the carbon atom bonded to the —OH group. 459, 472
oxidation of, 462–63
reduction of ketones to, 464
Secondary amines, 533
Secondary structure, of proteins The formation of an α helix, β-pleated sheet, or triple helix. 603–5, 610, 632
Semiconservative replication, 690–91
Semipermeable membrane A membrane that permits the passage of certain substances while blocking or retaining others. 306, 309–10, 313, 322
Serine, 575, 597, 670–71
Serotonin, 542
Serum, osmolarity of, 319
Sex attractant, 416
Sex hormones, 584–85
Shell An energy level containing electrons of similar energies. 67–68, 76
Shielding Materials used to provide protection from radioactive sources. 83–85, 113
Sickle-cell anemia, 610, 707
Sievert (Sv) A unit of biological damage (equivalent dose) equal to 100 rems. 96, 113
Significant figures The numbers recorded in a measurement. 9–10, 12–14, 41
adding significant zeros, 13
in addition, 14–15
in division, 13
in multiplication, 13
in pH values, 337
rounding off, 12
rules for counting, 10–11
in subtraction, 14–15
Silicon, 48, 69, 73, 137
Silicon dioxide, 141
Silk, 604
Silver, 48, 54, 132, 628
Silver nitrate, 300, 613
Simple enzyme, 621
Single replacement reaction An element replaces a different element in a compound. 170–71, 173, 196
SI units An International System of units that modifies the metric system. 1, 41
prefixes, 16–19
Skin, 606
reactions to UV light, 71
regulation of body temperature, 236
Skin cancer, 101, 407, 711
Skin care products, 441, 518–19
Skunk spray, 443
Smog, 172
Soaps, 572–74
cleaning action of, 573–74
Sodium, 48–49, 53–54, 61, 68–69, 72–73, 122–25, 131, 294
Sodium-24, 104
Sodium benzoate, 522–23
Sodium bicarbonate, 151, 349, 351
Sodium chloride, 127–28, 286–87, 291, 294
Sodium fluoride, 131
Sodium hydroxide, 342
Sodium lauryl sulfate, 574
Sodium nitride, 131
Sodium propionate, 522–23
Sodium sulfate, 148–50
Sodium sulfide, 129
Sol, 307
Solid A state of matter that has its own shape and volume. 216–17, 240
density of, 33–34
formation of, 300
solubility of, 297
Solid solutions, 286
Solubility The maximum amount of solute that can dissolve in exactly 100 g of solvent, usually water, at a given temperature. 296–300, 322
of alcohols, 438–49
of aldehydes and ketones, 452
of amines, 533–34
of carboxylic acids, 521
of ethers, 447
of gases, 298
of ionic compounds, 299
of solids, 297
temperature effects on, 297–98
Soluble salts An ionic compound that dissolves in water. 299, 322
Solute The component in a solution that changes state upon dissolving; if no change in state occurs, it is the component present in the smaller quantity. 285, 322
dissolving a solute, 287
polarity of, 286
types of, 285
Solutions A homogeneous mixture in which the solute is made up of small particles (ions or molecules) that can pass through filters and semipermeable membranes. 285–88, 322
acidic, 332–35, 337
basic, 332–33, 335, 337
in body, 309
buffers, 350–53
colloids, 306–9, 322
containing ions, 290–92
dialysis, 313
dilutions, 343–44
formation of, 287–88
gas, 286
hypertonic, 311–12, 322
hypotonic, 311–12, 322
isotonic, 311–12, 322
liquid, 286
molarity, 316–18
neutral, 332–33, 335, 337, 358
osmosis, 310–13, 322
percent concentration, 302–4
physiological, 311, 319, 322
properties of water, 282–84
saturated, 296, 322
solid, 286
suspensions, 306–9, 322
unsaturated, 296, 322
Solvent The substance in which the solute dissolves; usually the component present in greatest amount. 285, 322, 453
polarity of, 286
types of, 285
Sour cream, 650
Soybean oil, 567
Specific gravity (sp gr) A relationship between the density of a substance and the density of water: Sp gr = density of sample/density of water. 27–28, 41
of urine, 39
Specific heat A quantity of heat that changes the temperature of exactly 1 g of a substance by exactly 1°C. 209–10, 240
calculations using, 210
"Speed," 536
Sphingolipids Phospholipids in which sphingosine has replaced glycerol. 558–59, 577, 579, 590
Sphingomyelin, 577–78
Sphingomyelinase, 579
Sphingosine, 558, 577
Sphygmomanometer, 249
Standard temperature and pressure (STP) conditions Standard conditions of 0°C (273 K) temperature and 1 atm pressure used for the comparison of gases. 267–68, 277
Stanozolol, 586
Starch, 490
animal, 505
digestion of, 494, 505
iodine test for, 507, 513
plant, 505
Starvation, 670
States of matter, 201, 216–17
changes of, 219–20, 239
diagrams for, 224–26
Steam burns, 223
Stearic acid, 419, 560
Stereoisomers Molecules with the same structural formulas and attached atoms but having different three-dimensional arrangements of the atoms. 484, 487, 524
Sterno, 462
Steroids Types of lipid composed of a multicyclic ring system. 558–59, 580–86, 590
anabolic, 586
hormones, 456, 581, 584–85
Sterol, 581
Stirring, 288
Stomach acid, 336, 342, 349
Stomach cancer, 711
Stop codon, 698
Storage proteins, 595
STP conditions, 267–68, 277
Streptomycin, 628, 700
Stroke, 101
Strong acids An acid that completely ionizes in water. 341, 358
Strong bases A base that completely ionizes in water. 342, 358
Strong electrolytes An ionic compound that ionizes completely when it dissolves in water. Its solution is a good conductor of electricity. 290–91, 322

Strontium, 53–54
Strontium-85, 100, 104
Strontium-89, 100
Strontium-90, 97
Structural formulas
for alkanes, 382–85
complete, 365–66, 396
condensed, 366, 382, 396
Structural gene The sections of DNA that code for the synthesis of proteins. 708–10, 718
Structural isomers An organic compound that has the same molecular formula as another organic compound but a different structure. 306
of alkanes, 383–85
of alkenes, 413–14
drawing of, 384–85
Structural polysaccharides, 505–7
Structural proteins, 595
Styrene, 427
Subatomic particles A particle within an atom; protons, neutrons, and electrons are subatomic particles. 58–59, 76
Sublimation The change of state in which a solid is transformed directly to a gas without forming a liquid. 220, 240
Subscripts in formulas, 128, 149
Substance P, 605
Substituent, 376
Substitution mutations A mutation that replaces one base in a DNA with a different base. 704, 718
Substrate The substance that reacts at the active site in an enzyme-catalyzed reaction. 617, 632
concentration effects on enzyme action, 623
Substrate phosphorylation, 646, 677
Subtraction, significant figures in, 14–15
Succinate dehydrogenase, 654
Succinic acid, 627
Succinic dehydrogenase, 627
Succinyl CoA, 670–71
Succinyl CoA synthetase, 654
Sucralose, 502
Sucrase, 595, 615, 624
Sucrose A disaccharide composed of glucose and fructose; a nonreducing sugar, commonly called table sugar or "sugar." 291–92, 479, 490, 492, 499, 501, 513, 641
Sugar, reducing, 496–97, 513
Sulfanilamide, 628
Sulfide, 131
Sulfite, 148
Sulfur, 48–49, 54, 64, 68–69, 73, 119, 124, 137
Sulfur dioxide, 141, 172, 339
Sulfur group, 437, 442–43
Sulfur hexafluoride, 141
Sulfuric acid, 172, 339, 341
Sulfur trioxide, 141
Sun exposure, 71, 407, 519, 711
Sunflower oil, 566–67
Sunscreen, 519
Sunset, 307
Surface tension A characteristic of water in which the hydrogen bonding between water molecules pulls its surface molecules together so tightly that they form a "skin." 284, 322
Suspensions A mixture in which the solute particles are large enough and heavy enough to settle out and be retained by both filters and semipermeable membranes. 282, 306–9, 322
Sweetener, artificial, 502
Synthetase, 616
Systolic pressure, 249

T

Table sugar. *See* Sucrose
Tannic acid, 613
Tar, 371
Tarnish, 161–62, 187
Tartaric acid, 518
Tay-Sachs disease, 579, 707
T4 cells, 713
Technetium-99m, 91, 93, 100, 104
Teeth, polyatomic ions in, 149
Teflon, 427
Temperature An indication of the direction of heat flow, high to low. 4, 41, 204, 240
body. *See* Body temperature
effect on chemical equilibrium, 235
effect on enzyme action, 623–24
effect on formation of solutions, 288
effect on solubility, 297–98
of gas, 246–47, 259–60, 262
heat denaturation of proteins, 613
measurement of, 204–8
rate of reaction and, 230
Template strand, 690–91
Terephthalic acid, 529
Terpenes, 412
Tertiary (3°) alcohols An alcohol that has three alkyl groups attached to the carbon atom attached to the —OH. 459, 463, 473
Tertiary amines, 533
Tertiary structure, of proteins The folding of the secondary structure of a protein into a compact structure that is stabilized by the interactions of R groups such as ionic and disulfide bonds. 606–11, 632
Testosterone, 456, 584–86
Tetracycline, 628, 700
Tetrafluoroethene, 427
Tetrahedral structure, 365–66
Tetrose, 481
Thalidomide, 487
Thallium-201, 100
Thermal pollution, 298
Thermometer, 4
Thiamine, 622
Thick filaments, 639–40
Thin filaments, 639–40
Thiols Organic compounds that contain a thiol group (—SH) in place of the —OH group of an alcohol. 403, 437–43, 473
IUPAC system for naming, 442–43
Threonine, 597, 599, 671
Threose, 481
Thymine, 682–83, 686–88
Thymol, 442
Thyroid function, 99–100
Tin, 48, 54, 133
Titanium, 47
Titration, acid-base The addition of base to an acid sample to determine the concentration of the acid. 354, 358
TNT, 391
Tobacco smoke, 392, 711
α-Tocopherol. *See* Vitamin E
Toluene, 391
Torr A unit of pressure equal to 1 mm Hg; 760 torr = 1 atm. 249–50, 277
Torricelli, Evangelista, 249
Trace elements, in body, 55
Tranquilizers, 542
Transaminase, 616, 670
Transamination The transfer of an amino group from an α-amino acid to produce α-keto acids that are intermediates in the citric acid cycle. 670–71, 677
Transcription The transfer of genetic information from DNA by the formation of mRNA. 694, 718
Trans fatty acids, 571
Transferase, 616, 707
Transfer RNA (tRNA) An RNA that places a specific amino acid into a peptide chain at the ribosome. There is one or more tRNA for each of the 20 different amino acids. 693, 719
activation of, 695
anticodon of, 697–98, 718
Transfusion, 503
Trans isomer A geometric isomer in which similar groups are connected to opposite sides of the double bond in an alkene. 414, 416–17, 431
Transition elements, 52, 76
Transition metals, 132
naming compounds that include, 133–34
Translation The interpretation of the codons in mRNA as amino acids in a peptide. 695–700, 718
Translocation The shift of a ribosome along mRNA from one codon (three bases) to the next codon during translation. 698, 718
Transmutation The formation of a radioactive nucleus by bombarding a stable nucleus with fast-moving particles. 93, 113
Transport proteins, 595
Triacylglycerols, 565
Triglycerides A family of lipids composed of three fatty acids bonded through ester bonds to glycerol, a trihydroxy alcohol. 565–67, 590
chemical properties of, 569–72
hydrogenation of, 569–71
hydrolysis of, 571
mixed, 566, 590
oxidation of, 570, 664–68
saponification, 572
Triols, 438
Triose, 481
Tripeptide, 601
Triple bonds A sharing of three pairs of electrons by two atoms. 138, 156, 364, 409, 420
Triple helix The protein structure found in collagen consisting of three polypeptide chains woven together like a braid. 603–4, 606, 632
Troponin, 639
Trypsin, 595, 615, 624, 641
Tryptophan, 597, 599, 671
Tylenol, 545
Tyndall effect The opaque path of a light beam through a colloid as light is scattered by the colloidal particles. 307, 322
Tyrosine, 597, 671, 706

U

UDMH, 547
Ultraviolet (UV) radiation, 71, 407, 519, 704
Unbranched alkanes, 371–72
Unsaturated fats, hydrogenation of, 419
Unsaturated fatty acids, 560–62, 566–67

Unsaturated hydrocarbon A compound of carbon and hydrogen in which the carbon chain contains at least one double (alkene) or triple carbon–carbon bond (alkyne). An unsaturated compound is capable of an addition reaction with hydrogen, which converts the double or triple bonds to single carbon–carbon bonds. 364, 409, 431
Unsaturated solution A solution that contains less solute than can be dissolved. 296, 322
"Uppers," 536
Uracil, 682–83, 686, 694
Uranium, 47
Uranium-235, 81, 108–9
Uranium-238, 81, 87–88, 104, 106
Urea, 314, 363, 544, 672
Urea cycle, 672
Urease, 617, 624
Uremia, 544
Uric acid, 297
Urine, 283, 314
    glucose levels in, testing for, 497
    osmolarity of, 319
    pH of, 352
    specific gravity of, 39
Uterine cancer, 711

V

Vaccine, 713
Valence electrons The electrons present in the outermost shell of an atom, which are largely responsible for the chemical behavior of the element. 118–20, 137, 156
Valine, 597, 599, 671
Valium, 545
Vanilla, 456
Vanillin, 391, 442, 455–56
Vaporization, heat of, 223, 240
Vapor pressure The pressure exerted by the particles of vapor above a liquid. 250–52, 277
    boiling point and, 251–52
    of water, 251
Vaseline, 370
Vegetable oils, 558, 562
    hydrogenation of, 570–71
Vinegar, 341, 516–17, 519
Vinyl chloride, 426–27, 711
Virus Small particles containing DNA or RNA in a protein coat that require a host cell for replication. 704, 712–15, 718
    DNA, 711–12
    oncogenic, 711
    RNA, 711–12
Visible spectrum, 70
Vision, cis-trans isomers for night vision, 416–17
Vitamin A, 416, 568, 583–84
Vitamin B_1. *See* Thiamine
Vitamin B_2. *See* Riboflavin
Vitamin B_3. *See* Niacin
Vitamin B_6. *See* Pyridoxine
Vitamin B_{12}. *See* Cobalamin
Vitamin C. *See* Ascorbic acid
Vitamin D, 441, 568, 581, 583–84
Vitamin E, 568, 583–84
Vitamin K, 568, 583–84
Vitamins
    fat-soluble, 558, 568, 583–84
    water-soluble, 622
VLDL (very-low-density lipoproteins), 582–83
Volatile oils, 412, 442
Volume The amount of space occupied by a substance. 3, 41
    of gas, 246–47, 253–54, 259–60, 266–68
    measurement of, 17–19
    molar, 267–68, 277
Volume/volume percent, 303

W

Warts, 711–12
Water
    in body, 283
    composition of air, 271
    in foods, 283
    heat of fusion of, 219, 240
    heat of vaporization of, 223
    hydrogen bonds in, 283
    ionization of, 331–33
    ion-product constant for, 332–33, 358
    osmosis, 310–13
    properties of, 282–84
    solubility of alcohols in, 438–39
    solubility of aldehydes and ketones in, 452
    structure of, 138
    surface tension of, 284, 322
    vapor pressure of, 251
Water-soluble vitamins, 622
Water-treatment plant, 308
Watson, James, 687
Waxes The ester of a long-chain alcohol and a long-chain saturated fatty acid. 558–59, 564–67, 590
Weak acids An acid that ionizes only slightly in solution. 341, 358, 521–22
Weak bases A base that ionizes only slightly in solution. 342, 358
Weak electrolytes A substance that produces only a few ions along with many molecules when it dissolves in water. Its solution is a weak conductor of electricity. 290–91, 322
Weight, 3
    body. *See* Body weight
Windshield washer fluid, 462
Wood alcohol, 441, 462
Work An activity that requires energy. 201, 240

X

Xenon, 54
X rays, 70, 98, 704
Xylocaine, 541

Y

Yogurt, 613, 650

Z

Zero, absolute (temperature), 207, 239
Zeros, adding significant zeros, 13
Zinc, 48, 55, 132
Zinc phosphate, 151
Zwitterion The dipolar form of an amino acid consisting of two oppositely charged ionic regions, $—NH_3^+$ and $—COO^-$. 599–600, 632

Functional Groups in Organic Compounds

Type	Functional Group	Example: Structural Formula	Example: IUPAC Name	Example: Common Name
alkane	$-\overset{\vert}{\underset{\vert}{C}}-\overset{\vert}{\underset{\vert}{C}}-$	CH_3CH_3	ethane	ethane
haloalkane	F, Br, Cl, or I	CH_3CH_2Cl	chloroethane	ethyl chloride
alkene	$-CH{=}CH-$	$CH_2{=}CH_2$	ethene	ethylene
alkyne	$-C{\equiv}C-$	$CH{\equiv}CH$	ethyne	acetylene
aromatic	benzene ring		benzene	benzene
alcohol	$-OH$	CH_3CH_2OH	ethanol	ethyl alcohol
thiol	$-SH$	CH_3SH	methanethiol	methylmercaptan
ether	$-O-$	CH_3-O-CH_3	methoxymethane	dimethyl ether
aldehyde	$-\overset{O}{\overset{\Vert}{C}}-H$	$CH_3-\overset{O}{\overset{\Vert}{C}}-H$	ethanal	acetaldehyde
ketone	$-\overset{O}{\overset{\Vert}{C}}-$	$CH_3-\overset{O}{\overset{\Vert}{C}}-CH_3$	propanone	dimethyl ketone (acetone)
carboxylic acid	$-\overset{O}{\overset{\Vert}{C}}-OH$	$CH_3-\overset{O}{\overset{\Vert}{C}}-OH$	ethanoic acid	acetic acid
ester	$-\overset{O}{\overset{\Vert}{C}}-O-$	$CH_3-\overset{O}{\overset{\Vert}{C}}-OCH_3$	methyl ethanoate	methyl acetate
amine	$-NH_2$	$CH_3CH_2NH_2$	aminoethane	ethylamine
amide	$-\overset{O}{\overset{\Vert}{C}}-NH_2$	$CH_3-\overset{O}{\overset{\Vert}{C}}-NH_2$	ethanamide	acetamide

Important Common Ions

1+		2+		3+		3−		2−		1−	
H^+	hydrogen ion	Mg^{2+}	magnesium ion	Al^{3+}	aluminum ion	N^{3-}	nitride ion	O^{2-}	oxide ion	F^-	fluoride ion
Li^+	lithium ion	Ca^{2+}	calcium ion			P^{3-}	phosphide ion	S^{2-}	sulfide ion	Cl^-	Chloride ion
Na^+	sodium ion	Sr^{2+}	strontium ion							Br^-	bromide ion
K^+	potassium ion	Ba^{2+}	barium ion							I^-	iodide ion
Ag^+	silver ion	Zn^{2+}	zinc ion								

Important Ions with Variable Valence

1+ or 2+		2+ or 3+		1+ or 3+		2+ or 4+	
Cu^+	copper (I) ion	Fe^{2+}	iron (II) ion	Au^+	gold (I) ion	Sn^{2+}	tin (II) ion
Cu^{2+}	copper (II) ion	Fe^{3+}	iron (III) ion	Au^{3+}	gold (III) ion	Sn^{4+}	tin (IV) ion
						Pb^{2+}	lead (II) ion
						Pb^{4+}	lead (IV) ion